MHCC WITHDRAWN

TH 420 .T46 2013
Thornburg, Douglas W.
2012 International building
  code handbook

2012 INTERNATIONAL BUILDING CODE® HANDBOOK

About the Authors

Douglas W. Thornburg, AIA, is the Vice President and Technical Director of Product Development and Education, where he provides leadership in the technical development and positioning of support products, educational activities, and certification programs for the Code Council (ICC). Prior to joining the ICC in 2004, Mr. Thornburg served as a code consultant and educator for building codes. He has been involved extensively in building code activities since 1980.

John R. Henry, P.E., is the Principal Staff Engineer with the International Code Council (ICC) and has been with the association since 1997. During his tenure with ICC, he has worked for the code development department, technical services, government relations, training and education, and product development. Mr. Henry is currently responsible for the research, development, and authoring of technical resources related to the structural engineering provisions of the IBC and IRC and is an instructor of structural and framing seminars for the Code Council.

About the International Code Council

The International Code Council® (ICC®) is a member-focused association dedicated to helping the building safety community and construction industry provide safe, sustainable and affordable construction through the development of codes and standards used in the design, build and compliance process. Most U.S. communities and many global markets choose the International Codes. ICC Evaluation Service (ICC-ES), a subsidiary of the International Code Council, has been the industry leader in performing technical evaluations for code compliance fostering safe and sustainable design and construction.

Headquarters: 500 New Jersey Avenue, NW, 6th Floor, Washington, DC 20001-2070
District Offices: Birmingham, AL; Chicago. IL; Los Angeles, CA
1-888-422-7233
www.iccsafe.org

2012 INTERNATIONAL BUILDING CODE® HANDBOOK

Douglas W. Thornburg, AIA | John R. Henry, P.E.

New York Chicago San Francisco Lisbon London Madrid
Mexico City Milan New Delhi San Juan Seoul
Singapore Sydney Toronto

McGraw-Hill Education and ICC books are available at special quantity discounts to use as premiums and sales promotions, or for use in corporate training programs. To contact a representative please e-mail us at bulksales@mcgraw-hill.com.

2012 International Building Code® Handbook

Copyright © 2013 by the International Code Council. All rights reserved. Printed in the United States of America. Except as permitted under the United States Copyright Act of 1976, no part of this publication may be reproduced or distributed in any form or by any means, or stored in a data base or retrieval system, without the prior written permission of the publisher.

1 2 3 4 5 6 7 8 9 0   DOW/DOW   1 9 8 7 6 5 4 3

ISBN 978-0-07-180131-7
MHID 0-07-180131-6

This book is printed on acid-free paper.

Sponsoring Editor
Michael McCabe,
McGraw-Hill Education
(MHE)

Acquisitions Coordinator
Bridget Thoreson, MHE

Editorial Supervisor
David E. Fogarty, MHE

Project Manager
Harleen Chopra,
Cenveo® Publisher
Services

Copy Editor
Cenveo Publisher Services

Proofreader
Paula Garber

Production Supervisor
Richard C. Ruzycka, MHE

Composition
Cenveo Publisher Services

Art Director, Cover
Jeff Weeks, MHE

ICC Staff

Executive VP and Director of Business Development
Mark Johnson

VP and Technical Director of Product Development and Education
Doug Thornburg

Director of Products and Special sales
Suzane Nunes

Manager of Product Development
Cindy Rodriguez

Information contained in this work has been obtained by McGraw-Hill Education, LLC from sources believed to be reliable. However, neither McGraw-Hill Education, the International Code Council, nor its authors guarantee the accuracy or completeness of any information published herein, and neither McGraw-Hill Education nor its authors shall be responsible for any errors, omissions, or damages arising out of use of this information. This work is published with the understanding that McGraw-Hill Education and its authors are supplying information but are not attempting to render engineering or other professional services. If such services are required, the assistance of an appropriate professional should be sought.

Dedication

This book is dedicated to the memory of the late James E. Bihr, P.E., past chief executive officer of the International Conference of Building Officials (one of the three legacy founding members of the International Code Council) and past chairman of the board of directors of ASTM International (formerly the American Society of Testing and Materials). Mr. Bihr provided tremendous vision and leadership in various roles during his long and distinguished career. His integrity, professionalism, and compassion provided support and inspiration to the authors as well as so many others.

Online Bonus Resources

Enhance Your 2012 IBC Handbook Reading Experience. The *2012 International Building Code® Handbook* includes a variety of helpful bonus resources just right for you. These online bonus resources are designed to enhance your expertise and knowledge of various building code provisions, including:

- FEMA/NEHRP/NIST Publications related to earthquake safety and seismic design
- Articles by expert engineers on changes to the latest structural material standards
- ATC Wind Speed site provides site specific wind speeds for use with ASCE 7-10
- ICC Guidelines for acoustics and commissioning
- Helpful Articles from ICC's Building Safety Journal
- Helpful ICC YouTube Videos on various code related topics
- Resources related to accessible means of egress and fire protection

To access your bonus features, visit www.iccsafe.org/2012IBChandbook.

Contents

Foreword xv

Preface xvii

Acknowledgments xix

Chapter 1
Scope and Administration 1

- Section 101 General 2
- Section 102 Applicability 4
- Section 103 Department of Building Safety 5
- Section 104 Duties and Powers of Building Official 6
- Section 105 Permits 9
- Section 107 Submittal Documents 11
- Section 108 Temporary Structures and Uses 12
- Section 109 Fees 12
- Section 110 Inspections 13
- Section 111 Certificate of Occupancy . . .14
- Section 112 Service Utilities 15
- Section 113 Board of Appeals 15
- Section 114 Violations 15
- Section 115 Stop Work Order 16
- Section 116 Unsafe Structures and Equipment 17
- KEY POINTS 17

Chapter 2
Definitions 19

- Section 201 General 20
- Section 202 Definitions 20
- KEY POINTS 45

Chapter 3
Use and Occupancy Classification 47

- Section 302 Classification 48
- Section 303 Assembly Group A 50
- Section 304 Business Group B 54
- Section 305 Educational Group E 55
- Section 306 Factory Group F 56
- Section 307 High-Hazard Group H 57
- Section 308 Institutional Group I 63
- Section 309 Mercantile Group M 65
- Section 310 Residential Group R 66
- Section 311 Storage Group S 68
- Section 312 Utility and Miscellaneous Group U 69
- KEY POINTS 69

Chapter 4
Special Detailed Requirements Based on Use and Occupancy 71

- Section 402 Covered Mall and Open Mall Buildings 72
- Section 403 High-Rise Buildings 81
- Section 404 Atriums 88
- Section 405 Underground Buildings 90
- Section 406 Motor-Vehicle-Related Occupancies 92
- Section 407 Group I-2 96
- Section 408 Group I-3 100
- Section 409 Motion-Picture Projection Rooms 101
- Section 410 Stages, Platforms, and Technical Production Areas 101
- Section 411 Special Amusement Buildings 104
- Section 412 Aircraft-Related Occupancies 105
- Section 413 Combustible Storage 106
- Section 414 Hazardous Materials 107
- Section 415 Groups H-1, H-2, H-3, H-4, and H-5 110
- Section 416 Application of Flammable Finishes 114
- Section 417 Drying Rooms 114
- Section 418 Organic Coatings 114
- Section 419 Live/Work Units 115
- Section 420 Groups I-1, R-1, R-2, and R-3 116
- Section 422 Ambulatory Care Facilities 117
- Section 423 Storm Shelters 119
- Section 424 Children's Play Structures 119
- KEY POINTS 120

Chapter 5
General Building Heights and Areas.... 123

- Section 501 — General.................124
- Section 503 — General Building Height and Area Limitations.....124
- Section 504 — Building Height.........128
- Section 505 — Mezzanines and Equipment Platforms.....130
- Section 506 — Building Area Modifications..........135
- Section 507 — Unlimited-Area Buildings 148
- Section 508 — Mixed Use and Occupancy.........155
- Section 509 — Incidental Uses.........165
- Section 510 — Special Provisions......168
- KEY POINTS 172

Chapter 6
Types of Construction 175

- Section 602 — Construction Classification 176
- Section 603 — Combustible Material in Type I and II Construction...........186
- KEY POINTS 188

Chapter 7
Fire and Smoke Protection Features.... 189

- Section 703 — Fire-Resistance Ratings and Fire Tests..........190
- Section 704 — Fire-Resistance Rating of Structural Members ...197
- Section 705 — Exterior Walls..........201
- Section 706 — Fire Walls 215
- Section 707 — Fire Barriers 226
- Section 708 — Fire Partitions 228
- Section 709 — Smoke Barriers.........231
- Section 710 — Smoke Partitions 232
- Section 711 — Horizontal Assemblies....232
- Section 712 — Vertical Openings 235
- Section 713 — Shaft Enclosures........237
- Section 714 — Penetrations 241
- Section 715 — Joint Systems 249
- Section 716 — Opening Protectives......250
- Section 717 — Ducts and Air Transfer Openings..................255
- Section 718 — Concealed Spaces 260
- Section 719 — Fire-Resistance Requirements for Plaster............266
- Section 720 — Thermal- and Sound-Insulating Materials...............266
- Section 721 — Prescriptive Fire Resistance..............266
- Section 722 — Calculated Fire Resistance..............270
- KEY POINTS 272

Chapter 8
Interior Finishes.................. 275

- Section 801 — General................276
- Section 803 — Wall and Ceiling Finishes 276
- Section 804 — Interior Floor Finish.................279
- Section 805 — Combustible Materials in Types I and II Construction............280
- Section 806 — Decorative Materials and Trim..............281
- KEY POINTS 281

Chapter 9
Fire Protection Systems 283

- Section 901 — General................284
- Section 903 — Automatic Sprinkler Systems 286
- Section 904 — Alternative Automatic Fire-Extinguishing Systems 304
- Section 905 — Standpipe Systems 305
- Section 907 — Fire Alarm and Detection Systems........310
- Section 908 — Emergency Alarm Systems 317
- Section 909 — Smoke-Control Systems 317
- Section 910 — Smoke and Heat Vents ...320
- Section 911 — Fire Command Center.................322
- Section 914 — Emergency Responder Safety Features.........322
- KEY POINTS 323

Chapter 10
Means of Egress 325

- Section 1001 — Administration 327
- Section 1002 — Definitions 327
- Section 1003 — General Means of Egress 327
- Section 1004 — Occupant Load 332
- Section 1005 — Means of Egress Sizing . . . 344
- Section 1006 — Means of Egress Illumination 351
- Section 1007 — Accessible Means of Egress 353
- Section 1008 — Doors, Gates, and Turnstiles 358
- Section 1009 — Stairways 376
- Section 1010 — Ramps 387
- Section 1011 — Exit Signs 390
- Section 1012 — Handrails 393
- Section 1013 — Guards 400
- Section 1014 — Exit Access 405
- Section 1015 — Exit and Exit Access Doorways 408
- Section 1016 — Exit Access Travel Distance 414
- Section 1017 — Aisles 416
- Section 1018 — Corridors 417
- Section 1019 — Egress Balconies 423
- Section 1020 — Exits 423
- Section 1021 — Number of Exits and Exit Configuration 424
- Section 1022 — Interior Exit Stairways and Ramps 428
- Section 1023 — Exit Passageways 432
- Section 1024 — Luminous Egress Path Markings 434
- Section 1025 — Horizontal Exits 435
- Section 1026 — Exterior Stairways and Ramps 439
- Section 1027 — Exit Discharge 441
- Section 1028 — Assembly 444
- Section 1029 — Emergency Escape and Rescue 457
- KEY POINTS 460

Chapter 11
Accessibility 463

- Section 1101 — General 466
- Section 1102 — Definitions 466
- Section 1103 — Scoping Requirements . . . 467
- Section 1104 — Accessible Route 469
- Section 1105 — Accessible Entrances 470
- Section 1106 — Parking and Passenger Loading Facilities 472
- Section 1107 — Dwelling Units and Sleeping Units 473
- Section 1108 — Special Occupancies 476
- Section 1109 — Other Features and Facilities 478
- Section 1110 — Signage 481
- KEY POINTS 482

Chapter 12
Interior Environment 483

- Section 1203 — Ventilation 484
- Section 1204 — Temperature Control 488
- Section 1205 — Lighting 488
- Section 1206 — Yards or Courts 489
- Section 1207 — Sound Transmission 490
- Section 1208 — Interior Space Dimensions 490
- Section 1209 — Access to Unoccupied Spaces 491
- Section 1210 — Toilet and Bathroom Requirements 492
- KEY POINTS 493

Chapter 13
Energy Efficiency 495

Chapter 14
Exterior Walls 497

- Section 1402 — Definitions 498
- Section 1403 — Performance Requirements 498
- Section 1404 — Materials 499
- Section 1405 — Installation of Wall Coverings 499
- Section 1406 — Combustible Materials on the Exterior Side of Exterior Walls 505
- Section 1407 — Metal Composite Materials 505
- Section 1408 — Exterior Insulation and Finish Systems (EIFS) 506
- KEY POINTS 506

Chapter 15
Roof Assemblies and Rooftop Structures 507

- Section 1502 — Definitions 508
- Section 1503 — Weather Protection 508
- Section 1504 — Performance Requirements 509
- Section 1505 — Fire Classification 511
- Section 1506 — Materials 512
- Section 1507 — Requirements for Roof Coverings 513
- Section 1508 — Roof Insulation 519
- Section 1509 — Rooftop Structures 519
- Section 1510 — Reroofing 521
- KEY POINTS 522

Introduction to the Structural Provisions 523

Structural Design (Chapter 16): Nonseismic 524
Structural Design (Chapter 16): Seismic ... 525
Structural Design (Chapter 16): Load Combinations 527

Chapter 16
Structural Design 529

- Introduction 530
- Section 1601 — General 530
- Section 1602 — Definitions and Notations 531
- Section 1603 — Construction Documents 531
- Section 1604 — General Design Requirements 532
- Section 1605 — Load Combinations 539
- Section 1606 — Dead Loads 541
- Section 1607 — Live Loads 541
- Section 1608 — Snow Loads 552
- Section 1609 — Wind Loads 555
- Section 1610 — Soil Lateral Loads 562
- Section 1611 — Rain Loads 563
- Section 1612 — Flood Loads 563
- Section 1613 — Earthquake Loads 565
- Section 1614 — Atmospheric Ice Loads 570
- Section 1615 — Structural Integrity 570
- KEY POINTS 571

- Example 16-1 — Design Axial Force, Shear Force, and Bending Moment for Shear Wall Due to Lateral and Gravity Loads (Strength Design) 573
- Example 16-2 — Design Axial Force, Shear Force, and Bending Moment for Shear Wall Due to Lateral and Gravity Loads (Allowable Stress Design Using Basic Load Combinations) 574
- Example 16-3 — Design Axial Force, Shear Force, and Bending Moment for Shear Wall Due to Lateral and Gravity Loads (Allowable Stress Design Using Alternate Basic Load Combinations) 575
- Example 16-4 — Calculations of Live Load Reduction 576

Chapter 17
Structural Tests and Special Inspections ... 579

- Introduction 580
- Section 1701 — General 581
- Section 1702 — Definitions 582
- Section 1703 — Approvals 584
- Section 1704 — Special Inspections, Contractor Responsibility, and Structural Observations 590
- Section 1705 — Required Verification and Inspection 596
- Section 1706 — Design Strengths of Materials 612
- Section 1707 — Alternate Test Procedures 612
- Section 1708 — Test Safe Load 612
- Section 1709 — In Situ Load Tests 613
- Section 1710 — Preconstruction Load Tests 613
- Section 1711 — Material and Test Standards 614
- KEY POINTS 618

Chapter 18
Soils and Foundations 621

- Introduction 622
- Section 1801 — General 622

☐ Section 1802	Definitions	623
☐ Section 1803	Geotechnical Investigations	623
☐ Section 1804	Excavation, Grading, and Fill	628
☐ Section 1805	Dampproofing and Waterproofing	628
☐ Section 1806	Presumptive Load-Bearing Values of Soils	637
☐ Section 1807	Foundation Walls, Retaining Walls, and Embedded Posts and Poles	639
☐ Section 1808	Foundations	644
☐ Section 1809	Shallow Foundations	649
☐ Section 1810	Deep Foundations	654
☐ KEY POINTS		680

Chapter 19
Concrete 683

☐ Introduction		684
☐ Section 1901	General	684
☐ Section 1902	Definitions	685
☐ Section 1903	Specifications for Tests and Materials	685
☐ Section 1904	Durability Requirements	691
☐ Section 1905	Modifications to ACI 318	692
☐ Section 1906	Structural Plain Concrete	696
☐ Section 1907	Minimum Slab Provisions	696
☐ Section 1908	Anchorage to Concrete—Allowable Stress Design	697
☐ Section 1909	Anchorage to Concrete—Strength Design	698
☐ Section 1910	Shotcrete	699
☐ Section 1911	Reinforced Gypsum Concrete	700
☐ Section 1912	Concrete-Filled Pipe Columns	700
☐ KEY POINTS		701

Chapter 20
Aluminum 705

☐ Introduction		706
☐ Section 2002	Materials	706
☐ KEY POINTS		707

Chapter 21
Masonry 709

☐ Introduction		710
☐ Section 2101	General	711
☐ Section 2102	Definitions and Notations	714
☐ Section 2103	Masonry Construction Materials	714
☐ Section 2104	Construction	717
☐ Section 2105	Quality Assurance	718
☐ Section 2106	Seismic Design	718
☐ Section 2107	Allowable Stress Design	723
☐ Section 2108	Strength Design of Masonry	725
☐ Section 2109	Empirical Design of Masonry	727
☐ Section 2110	Glass Unit Masonry	728
☐ Section 2111	Masonry Fireplaces	728
☐ Section 2112	Masonry Heaters	729
☐ Section 2113	Masonry Chimneys	729
☐ KEY POINTS		729

Chapter 22
Steel 733

☐ Introduction		734
☐ Section 2201	General	734
☐ Section 2202	Definitions	734
☐ Section 2203	Identification and Protection of Steel for Structural Purposes	734
☐ Section 2204	Connections	735
☐ Section 2205	Structural Steel	737
☐ Section 2206	Composite Structural Steel and Concrete Structures	739
☐ Section 2207	Steel Joists	739
☐ Section 2208	Steel Cable Structures	741
☐ Section 2209	Steel Storage Racks	742
☐ Section 2210	Cold-Formed Steel	742
☐ Section 2211	Cold-Formed Steel Light-Framed Construction	745
☐ KEY POINTS		747

Chapter 23
Wood 749

☐ Introduction		750
☐ Section 2301	General	750

Section 2302	Definitions 752
Section 2303	Minimum Standards and Quality............755
Section 2304	General Construction Requirements 767
Section 2305	General Design Requirements for Lateral-Force-Resisting Systems 778
Section 2306	Allowable Stress Design................780
Section 2307	Load and Resistance Factor Design 786
Section 2308	Conventional Light-Frame Construction..........786
KEY POINTS	829

Chapter 24
Glass and Glazing 833

Section 2403	General Requirements for Glass 834
Section 2404	Wind, Snow, Seismic and Dead Loads on Glass....834
Section 2405	Sloped Glazing and Skylights 835
Section 2406	Safety Glazing 838
Section 2407	Glass in Handrails and Guards...........847
Section 2408	Glazing in Athletic Facilities.............847
KEY POINTS	848

Chapter 25
Gypsum Board and Plaster........... 849

Section 2501	Scope 850
Section 2502	Definitions 850
Section 2504	Vertical and Horizontal Assemblies 851
Section 2506	Gypsum Board Materials.............852
Section 2508	Gypsum Construction...853
Section 2509	Gypsum Board in Showers and Water Closets......854
Section 2510	Lathing and Furring for Cement Plaster (Stucco)........855
Section 2511	Interior Plaster 856
Section 2512	Exterior Plaster 857
KEY POINTS	859

Chapter 26
Plastic 861

Section 2603	Foam Plastic Insulation 862
Section 2604	Interior Finish and Trim..............868
Section 2605	Plastic Veneer 869
Section 2606	Light-Transmitting Plastics 869
Section 2607	Light-Transmitting Plastic Wall Panels......870
Section 2608	Light-Transmitting Plastic Glazing..........870
Section 2609	Light-Transmitting Plastic Roof Panels 870
Section 2610	Light-Transmitting Plastic Skylight Glazing 871
KEY POINTS	871

Chapter 27
Electrical 873

| Section 2702 | Emergency and Standby Power Systems ....874 |

Chapter 28
Mechanical 877

Chapter 29
Plumbing 879

| Section 2902 | Minimum Plumbing Facilities ....880 |
| KEY POINTS | ....883 |

Chapter 30
Elevators and Conveying Systems...... 885

Section 3002	Hoistway Enclosures....886
Section 3003	Emergency Operations 888
Section 3004	Hoistway Venting 889
Section 3006	Machine Rooms........889
Section 3007	Fire Service Access Elevator..............889
Section 3008	Occupant Evacuation Elevators..............891
KEY POINTS	892

Chapter 31
Special Construction 893

☐ Section 3102	Membrane Structures	...894
☐ Section 3104	Pedestrian Walkways and Tunnels	895
☐ Section 3105	Awnings and Canopies	896
☐ Section 3106	Marquees	896
☐ Section 3109	Swimming Pool Enclosures and Safety Devices	896
☐ KEY POINTS		898

Chapter 32
Encroachments in the Public Right-of-Way 899

☐ Section 3201	General	900
☐ Section 3202	Encroachments	900

Chapter 33
Safeguards During Construction 905

☐ Section 3302	Construction Safeguards	906
☐ Section 3303	Demolition	906
☐ Section 3304	Site Work	906
☐ Section 3306	Protection of Pedestrians	908
☐ Section 3307	Protection of Adjoining Property	910
☐ Section 3308	Temporary Use of Streets, Alleys and Public Property	911
☐ Section 3309	Fire Extinguishers	911
☐ Section 3310	Means of Egress	911
☐ Section 3311	Standpipes	911
☐ KEY POINTS		912

Chapter 34
Existing Structures 913

☐ Section 3404	Alterations	914
☐ Section 3408	Change of Occupancy	...915
☐ Section 3411	Accessibility for Existing Buildings	915
☐ Section 3412	Compliance Alternatives	916
☐ KEY POINTS		917

Chapter 35
Referenced Standards................ 919

Appendixes 921

☐ Appendix A	Employee Qualifications	...922
☐ Appendix B	Board of Appeals	922
☐ Appendix C	Group U Agricultural Buildings	923
☐ Appendix D	Fire Districts	923
☐ Appendix E	Supplementary Accessibility Requirements	925
☐ Appendix F	Rodentproofing	925
☐ Appendix G	Flood-Resistant Construction	926
☐ Appendix H	Signs	926
☐ Appendix I	Patio Covers	926
☐ Appendix J	Grading	927
☐ Appendix K	Administrative Provisions	927
☐ Appendix L	Earthquake Recording Instrumentation	927
☐ Appendix M	Tsunami-Generated Flood Hazard	927

Metric Conversion Table.............. 929

☐ Metric Units, System International (SI) 930
☐ Soft Metrication 930
☐ Hard Metrication 930

Index 937

Foreword

How often have you heard these questions when discussing building codes: "What is the intent of this section?" or, "How do I apply this provision?" This publication offers the code user a resource that addresses much of the intent and application principles of the major provisions of the *2012 International Building Code®* (IBC®).

It is impossible for building codes and similar regulatory documents to contain enough information, both prescriptive and explanatory narrative, to remove all doubt as to the intent of the various provisions. If such a document were possible, it would be so voluminous that it would be virtually useless.

Because the IBC must be reasonably brief and concise in its provisions, the user must have knowledge of the intent and background of these provisions to apply their intent appropriately. The IBC places great reliance on the judgment of the building official and design professional for the specific application of its provisions. Where the designer and building official have knowledge of the rationale behind the provisions, the design and enforcement of the code will be based on informed judgment rather than arbitrariness or rote procedure.

The information that this handbook provides, coupled with the design professional's and building official's experience and education, will result in better use of the IBC and more uniformity in its application. As lengthy as this document may seem, it still cannot provide all of the answers to questions of code intent, that is why the background, training, and experience of the reader must also be called on to properly apply, interpret, and enforce the code provisions.

The preparation of a document of this nature requires consulting a large number of publications, organizations, and individuals. Even so, the intent of many code provisions is not completely documented. Sometimes the discussion is subjective; therefore, individuals may disagree with the conclusions presented. It is, however, important to note that the explanatory narratives are based on many decades of experience by the authors and the other contributors to the manuscript.

Preface

Internationally, code officials and design professionals recognize the need for a modern, up-to-date building code addressing the design and installation of building systems through requirements emphasizing performance. The *International Building Code®* (IBC®) meets those needs by providing model code regulations that safeguard the public health and safety in all communities, large and small. The *IBC Handbook* is a valuable resource for those who design, plan, review, inspect, or construct buildings or other structures regulated by the 2012 IBC.

The IBC is one of a family of codes published by the International Code Council® (ICC®) that establishes comprehensive minimum regulations for building systems using prescriptive and performance-related provisions. It is founded on broad-based principles that use new materials and new building designs. Additionally, the IBC is compatible with the entire family of International Codes® published by the ICC.

There are three major subdivisions to the IBC:

1. The text of the IBC
2. The referenced standards listed in Chapter 35
3. The appendices

The first 34 chapters of the IBC contain both prescriptive and performance provisions that are to be applied. Chapter 35 contains those referenced standards that, although promulgated and published by separate organizations, are considered as a part of the IBC as applicable. The provisions of the appendix do not apply unless specifically included in the adoption ordinance of the jurisdiction enforcing the code.

The *2012 IBC Handbook* is designed to present commentary only for those portions of the code for which commentary is helpful in furthering the understanding of the provision and its intent. This handbook uses many drawings and figures to help clarify the application and intent of many code provisions.

This handbook examines the intent and application of many code provisions for both the nonstructural- and structural-related aspects of the IBC. It addresses in detail many requirements that are considered as "fire- and life-safety" provisions of the code. Found in IBC Chapters 3 through 10, these provisions focus on the important considerations of occupancy and type of construction classification, allowable building size, fire and smoke protection features, fire protection systems, interior finishes, and means of egress.

The discussion of the structural provisions in this handbook is intended to help code users understand and properly apply the requirements in Chapters 16 through 23 of the 2012 IBC. Although the discussion is useful to a broad range of individuals, the discussion of the structural provisions was written primarily so that building officials, plans reviewers, architects, and engineers can get a general understanding of the IBC's structural requirements and gain some insight into their underlying basis and intent. To that end, the numerous figures, tables, and examples are intended to illustrate and help clarify the proper application of many structural provisions of the IBC.

As the IBC adopts many national standards by reference rather than transcribing the structural provisions of the standards into the code itself, in some cases the discussion in this handbook pertains to the provisions found in the referenced standard such as

ASCE 7 and ACI 318 rather than the IBC. The structural provisions addressed focus on the general design requirements related to structural load effects; special inspection and verification, structural testing, and structural observation; foundations and soils; and specific structural materials design requirements for concrete, masonry, steel, and wood.

Questions or comments concerning this handbook are encouraged. Please direct any correspondence to *handbook@iccsafe.org*.

Acknowledgments

The publication of this handbook is based on many decades of experience by the authors and other contributors. Since its initial publication, the handbook has become a living document subject to changes and refinements as newer code editions are released. This latest edition reflects extensive modifications based on the requirements found in the *2012 International Building Code.*

The initial handbook, on which the nonstructural portions of this document are based, was published in 1988. It was authored by Vincent R. Bush. In developing the discussions of intent, Mr. Bush drew heavily on his 25 years of experience in building safety regulation. Mr. Bush, a structural engineer, was intimately involved in code development work for many years.

In addition to the expertise of Mr. Bush, major contributions were made by John F. Behrens. Mr. Behrens' qualifications were as impressive as the original author's. He had vast experience as a building official, code consultant, and seminar instructor. Mr. Behrens provided the original manuscript of the means of egress chapter and assisted in the preparation of many other chapters.

Revisions to the handbook occurred regularly over the years, with content based on the provisions of the *International Building Code* authored by Doug Thornburg, AIA, C.B.O. Mr. Thornburg, a certified building official and registered architect, has over 32 years of experience in the building regulatory profession. Previously a building inspector, plans reviewer, building code administrator, seminar instructor, and code consultant, he is currently vice-president and technical director/Product Development and Education for the International Code Council (ICC). In his present role, Mr. Thornburg develops and reviews technical publications, reference books, resource materials, and educational programs relating to the International Codes. He continues to present building code seminars nationally and has developed numerous educational texts, including *Significant Changes to the IBC, 2012 Edition.* Mr. Thornburg was presented with ICC's inaugural Educator of the Year Award in 2008, recognizing his outstanding contributions in education and training.

The basis of the discussion on the 2012 structural requirements is the *2000 IBC Handbook—Structural Provisions,* authored by S. K. Ghosh, Ph.D., and Robert Chittenden, S.E. Both authors have extensive knowledge, expertise, and experience in the development of many of the structural provisions of the IBC. Dr. Ghosh initially authored Chapters 16 and 19, and Mr. Chittenden authored Chapters 17, 18, 20, 21, 22, and 23. John Henry, ICC principal staff engineer, is the author of the current commentary addressing IBC Chapters 16 through 23. Mr. Henry, a registered civil engineer and certified plans examiner, has over 25 years of experience in structural-related aspects of building code safety including responsibilities as a design engineer in private practice, plans check engineer, and code consultant. A member of the Product Development group with the ICC, he provides technical support for the *International Building Code* and has developed and presented many seminars on the structural provisions of the IBC. Mr. Henry was presented with ICC's John Nosse Award for Technical Excellence in 2011, recognizing his outstanding contributions and technical expertise.

Special acknowledgement goes to Alan Carr, S.E., ICC senior staff engineer/Codes and Standards, who updated and contributed a significant portion of the discussion in Chapters 16 and 19, and to Sandra Hyde, P.E., ICC staff engineer/Product Development, who updated the discussion in Chapter 18. Such recognition also goes to Scott Stookey,

Acknowledgments

engineering associate for the Austin, Texas, Fire Department, and ICC product Development group member Jay Woodward, senior staff architect, for their contributions to discussions of Chapters 9 and 10, respectively.

The information and opinions expressed in this handbook are those of the present and past authors, as well as the many contributors, and do not necessarily represent the official position of the International Code Council. Additionally, the opinions may not represent the viewpoint of any enforcing agency. Opinions expressed in this handbook are only intended to be a resource in the application of the IBC, and the building official is not obligated to accept such opinions. The building official is the final authority in rendering interpretations of the code.

CHAPTER 1

SCOPE AND ADMINISTRATION

Section 101 General
Section 102 Applicability
Section 103 Department of Building Safety
Section 104 Duties and Powers of Building Official
Section 105 Permits
Section 107 Submittal Documents
Section 108 Temporary Structures and Uses
Section 109 Fees
Section 110 Inspections
Section 111 Certificate of Occupancy
Section 112 Service Utilities
Section 113 Board of Appeals
Section 114 Violations
Section 115 Stop Work Order
Section 116 Unsafe Structures and Equipment
Key Points

1 Scope and Administration

Section 101 *General*

In addition to the code's scope, Chapter 1 covers general subjects such as the purpose of the code, the duties and powers of the building official, performance provisions relating to alternative methods and materials of construction, applicability of the provisions, and creation of the department of building safety. This chapter also contains requirements for the issuance of permits, subsequent inspections, and certificates of occupancy. The provisions in Chapter 1 are of such a general nature as to apply to the entire *International Building Code*® (IBC®).

101.2 Scope. The intent of the code as outlined in this section is that the IBC applies to virtually anything that is built or constructed. The definitions of "Building" and "Structure" in Chapter 2 are so inclusive that the code intends that any work of any kind that is accomplished on any building or structure comes within the scope of the code. Thus, the code would apply to a major high-rise office building as well as to a retaining wall creating an elevation change on a building site. However, certain types of work are exempt from the permit process as indicated in the discussion of required permits in this chapter.

Whereas initially the IBC appears to address all construction-related activities, the design and construction of detached one- and two-family dwellings and townhouses, as well as their accompanying accessory structures, are intended to be regulated under the *International Residential Code*® (IRC®). However, in order for such structures to fall under the authority of the IRC, two limiting factors have been established. First, each such building is limited to a maximum height of three stories above grade plane as established by the definition of "Story above grade plane" in Section 202. In broad terms, where a floor level is located predominantly above the adjoining exterior ground level, it would be considered in the total number of stories above grade plane for evaluation of its regulation by the IRC. It is quite possible that a residential unit with four floor levels will be regulated by the IRC, provided that the bottom floor level is established far enough below the exterior grade that it would not qualify as a story above grade plane, but rather as a basement. A fifth occupiable floor level is also permitted under the allowances in the IRC for habitable attics. For further discussion on the determination of a story above grade plane as similarly regulated in the IBC, see the commentary on Section 202. Secondly, each dwelling unit of a two-family dwelling or townhouse must be provided with a separate means of egress. Although the definition of an IBC means of egress would require travel extending to the public way, for the purpose of this requirement it is acceptable to provide individual and isolated egress only until reaching the exterior of the dwelling at the required egress door. Once reaching the exterior, the building occupants could conceivably share a stairway, sidewalk, or similar pathway to the public way. The IRC does not regulate egress beyond the structure itself; thus, any exit discharge conditions would only be applicable to IBC structures.

Townhouse design and construction is also regulated by the IRC. Section 202 defines a townhouse as a grouping of three or more single-family dwelling units in the same structure. The units must each extend individually from the ground to the sky, with open space provided on at least two sides of each dwelling unit. The effect of such limitations maintains the concept of "multiple single-family dwellings."

The requirement for open space on a minimum of two sides of each townhouse unit allows for interpretation regarding the degree of openness. Although not specific in language, the provision intends that each townhouse be provided with a moderate degree of exterior wall, thus allowing for adequate fire department access to each individual unit.

Structures such as garages, carports, and storage sheds are also regulated by the IRC where they are considered accessory to the residential buildings previously mentioned. Such accessory buildings are limited to 3,000 square feet (279 m^2) in floor area and two stories in height.

Even though the IRC may use the IBC as a reference for certain design procedures, the intent is to use only the IRC for the design and construction of one- and two-family dwellings, multiple single-family dwellings (townhouses), and their accessory structures. This does not preclude the use of the IBC by a design professional for the design of the types of residential buildings specified. However, unless specifically directed to the IBC by provisions of the IRC, it is not the intent of the IRC to utilize the IBC for provisions not specifically addressed. For example, the maximum allowable floor area of a residence based on the building's type of construction is not addressed in the IRC. Therefore, there is no limit to the floor area permitted in the dwelling unit. It would not be appropriate to use the IBC to limit the residence's floor area based on construction type.

Appendices. A number of subjects are addressed in Appendices A through M. The topics range from detailed information on the creation of a board of appeals to more general provisions for grading, excavation, and earthwork construction. Although the code clearly indicates that the appendices are not considered a part of the IBC unless they are specifically adopted by the jurisdiction, this does not mean they are of any less worth than those set forth in the body of the code. Although there are several reasons why a set of code requirements is positioned in the appendix, the most common reason is that the provisions are limited to a small geographic location or are of interest to only a small number of jurisdictions. **101.2.1**

Jurisdictions have the ability to adopt any or all of the appendices based on their own needs. However, just because an appendix has not been adopted does not lessen its value as a resource. In making decisions of interpretation of the code, as well as in evaluating alternate materials and methods, the provisions of an appendix may serve as a valuable tool in making an appropriate decision. Even in those cases where a specific appendix is not in force, the information it contains may help in administering the IBC.

Intent. Various factors are regulated that contribute to the performance of a building in regard to the health, safety, and welfare of the public. The IBC identifies several of these major factors as those addressing structural strength, egress capabilities, sanitation and other environmental issues, fire- and life-safety concerns, and energy conservation. In addition, the safety of fire fighters and emergency personnel responding to an emergency situation is an important consideration. The primary goal of the IBC is to address any and all hazards that are attributed to the presence and use of a jurisdiction's buildings and structures, and to safeguard the public from such hazards. **101.3**

The intent of the code is more inclusive than most people realize. A careful reading will note that in addition to providing for life safety and safeguarding property, the code also intends that its provisions consider the general welfare of the public. This latter item, *general welfare*, is not so often thought of as being part of the purpose of a building code. However, in the case of the IBC, safeguarding the public's general welfare is a part of its intent, which is accomplished, for example, by provisions that ameliorate the conditions found in substandard or dangerous buildings. Moreover, upon the adoption of a modern building code such as the IBC, the general level of building safety and quality is raised. This in turn contributes to the public welfare by increasing the tax base and livability. Additionally, substandard conditions are reduced, and the subsequent reduction of unsanitary conditions contributes to safeguarding the public welfare. For example, the maintenance requirements of Section 3401.2 apply to all buildings, and as a result, the continued enforcement of the IBC slows the development of substandard conditions. A rigorous enforcement of Section 3401.2 will actually reduce the conditions that contribute to the deterioration of the existing building stock. Thus, public welfare is enhanced by the increased benefits that inure to the general public of the jurisdiction as a result of the code provisions.

The concept of "minimum" requirements is the established basis for the technical provisions set forth in the IBC. The requirements are intended to identify the appropriate level of regulation to achieve a balanced approach to the design and construction of buildings.

On one hand, it is critical that an appropriate degree of safety be established in order to protect the general public. Conversely, it is also important that the economic impact of the regulations be considered. It is this balance of concerns that provides for the necessary degree of public health, safety, and welfare within appropriate economic limits. The establishment of multiple occupancy classifications with varying requirements for each is a basic example of this philosophy.

101.4 Referenced codes. A number of other codes are promulgated by the *International Code Council*® (ICC®) in order to provide a full set of coordinated construction codes. Seven of those companion codes are identified in this section, as they are specifically referenced in one or more provisions of the IBC. The adoption of the IBC does not automatically include the full adoption of the referenced codes, but rather only those portions specifically referenced by the IBC.

For example, Section 903.3.5 requires that water supplies for automatic sprinkler systems be protected against backflow in accordance with the *International Plumbing Code*® (IPC®). As a result, when the IBC is adopted, so are the backflow provisions of the IPC. The extent of the reference is backflow protection; therefore, that is the only portion of the IPC that is applicable. Broader references are also provided, such as many of the references to the *International Fire Code*® (IFC®). Section 307.1.1 requires that hazardous materials in any quantity conform to the requirements of the IFC. Although the entire IFC may not be adopted by the jurisdiction, the provisions applicable to hazardous materials are in force with the adoption of the IBC.

Section 102 *Applicability*

102.1 General. Where there is a conflict between two or more provisions found in the code as they relate to differences of materials, methods of construction, or other requirements, the most restrictive provision will govern. Typically, the code will identify how the varying requirements should be applied. For example, the occupant load along with the appropriate factor from Section 1005.3.1 is used to calculate the total width required for egress stairways—often referred to as the *calculated* width. Section 1009.4 also addresses the minimum required width for a stairway based on the absolute width necessary for use of a stairway under any condition, deemed to be the *component* width. When determining the proper width required by the code, the more restrictive, or wider, stairway width would be used. See Application Example 102-1.

In addition, where a conflict occurs between a specific requirement and a general requirement, the more specific provision shall apply. Again, the IBC provisions typically clarify the appropriate requirement that is to be applied. As an example, Section 1009.7.2 limits the height of stair risers to 7 inches (178 mm) as a general requirement for stairways. However, Section 1028.11.2 allows for a maximum riser height of 8 inches (203 mm) for aisle stairs serving assembly seating areas. Because the greater riser height is only permitted for a specific stair condition, rather than for all stairways in general, it is intended to apply where those special means of egress provisions established in Section 1028 are applicable.

Occasionally it is difficult, during the comparison of two different code provisions, to determine which is the general requirement and which is the specific requirement. In some cases, both requirements are specific, but one is more specific than the other. It is important that the intent of this section be applied in reviewing the proper application of the code. Where it can be determined that one provision is more specific in its scope than the other provision, the more specific requirement shall apply, regardless of whether it is more or less restrictive in application.

> **Application Example 102-1**
>
> **GIVEN:** An occupant load of 130 assigned to each of two stairways in a nonsprinklered office building.
> **DETERMINE:** The required minimum width of each stairway.
> 1. Based on Section 1005.3.1, the minimum calculated width would be:
>   0.3 inches/occupant × 130 occupants = 39 inches
> 2. Based on Section 1009.4, the minimum required width would be 44 inches.
>
> **SOLUTION:** Therefore, the more restrictive condition, 44 inches, would apply.
> For SI: 1 inch = 25.4 mm.
>
> **CONFLICTING REQUIREMENTS**

102.4

Referenced codes and standards. Differences between the code and the various standards it references are to be expected. Unlike the companion International Codes®, there is not necessarily a conscious effort to see that the publications are completely compatible with each other. As a result, it is critical that the code indicate that its provisions are to be applied over those of a referenced standard where such differences exist. For example, the provisions of NFPA 13R addressing sprinkler systems in residential occupancies allow for the omission of sprinklers at specified exterior locations, including porches and balconies. However, the provisions of IBC Section 903.3.1.2.1 mandate sprinkler protection for such areas where specific conditions exist. In this case, the provisions of the IBC for sprinkler protection would apply regardless of the allowances contained in NFPA 13R.

There are also times when the standard being referenced includes subject matter that falls within the scope of the IBC or the other International Codes. It is intended that the requirements of a referenced standard supplement the IBC provisions in those areas not already addressed by the code. In those areas where parallel or conflicting requirements occur, the IBC provisions are always to be applied. For example, IBC Section 415.8.4 mandates that "the construction and installation of dry cleaning plants shall be in accordance with the requirements of the IBC, the *International Mechanical Code*, the *International Plumbing Code,* and NFPA 32. Although NFPA 32 addresses construction and installation criteria for dry cleaning plants, only those portions of the standard that are not addressed within the IBC, IMC, and IPC are applicable.

Section 103 *Department of Building Safety*

This section establishes the department of building safety as the jurisdictional enforcement agency charged with administering the IBC. The term *building official*, as used in the IBC, represents the individual appointed by the jurisdiction to head the department of building safety. Although many jurisdictions utilize the title *building official* to recognize the individual in charge of the building safety department, there are many other titles that are used. These include *Chief Building Inspector, Superintendent of Central Inspection, Director of Code Enforcement*, and various other designations. Regardless of the title selected for use by the individual jurisdiction, the IBC views all of these as equivalent to the term *building official.*

The building official, in turn, appoints personnel as necessary to carry out the duties and responsibilities of the department. Such staff members (deputies), including inspectors, plan examiners, and other employees, are empowered by the building official to carry out

1 Scope and Administration

those functions set forth by the jurisdiction. Where the IBC references the building official in any capacity, the code reference also includes any deputies who have been granted enforcement authority by the building official. Where an inspector or plan reviewer makes a decision of interpretation, they are assuming the role of building official in arriving at that decision. There is an expectation on behalf of the jurisdiction that such employees possess the knowledge and experience to take on this responsibility. The failure to grant appropriate authority will often result in both ineffective and inefficient results.

For those jurisdictions desiring guidelines within the text of the code for the selection of department personnel, Appendix A addresses minimum employee qualifications for various positions. Experience and certification criteria for building officials, chief inspectors, inspectors, and plans examiners are set forth in this appendix chapter.

Section 104 *Duties and Powers of Building Official*

104.1 General. The IBC is designed to regulate in both a prescriptive and performance manner. An extensive number of provisions have been intentionally established to allow for jurisdictional interpretation based on the specifics of the situation. This section establishes the building official's authority to render such interpretations of the IBC. In addition, the building official may adopt policies and procedures that will help clarify the application of the code. Although having no authoring to provide variances or waivers to the code requirements, the building official is charged with interpreting and clarifying the provisions found in the IBC, provided that such decisions are in conformance with the intent and purpose of the code.

The authority to interpret the intended application of the IBC is a powerful tool available to the building official. With such authority comes a great degree of responsibility. Such interpretations must be consistent with the intent and purpose of the code. It is therefore necessary that all reasonable efforts be made to determine the code's intent in order to develop an appropriate interpretation. Various sources should be consulted to provide a broad background from which to make a decision. These could include discussions with peers, as well as information found in various educational texts and technical guides. However, it must be stressed that the ultimate responsibility for determining the appropriateness of an interpretation lies with the jurisdictional building official, and all other opinions, both verbal and written, are just that, opinions. The building official must never relinquish his or her authority to others in the administration of these very important interpretive powers. See also the discussion on alternative materials, design, and methods of construction in Section 104.11.

104.4 Inspections. Those inspections required under the provisions found in Section 110 are to be performed by the building official or by authorized representatives of the building official. It is also acceptable that outside firms or individuals be utilized for inspections, provided such firms or individuals have been approved by the building official. This option may allow for better use of available resources. Written reports shall be provided for each inspection that is made.

104.5 Identification. For the benefit of all individuals involved, inspection personnel of the department of building safety are mandated to carry proper identification. The display of an identification card or badge, an example of which is shown in Figure 104-1, signifies the function and authority of the individual performing the inspection.

Duties and Powers of Building Official

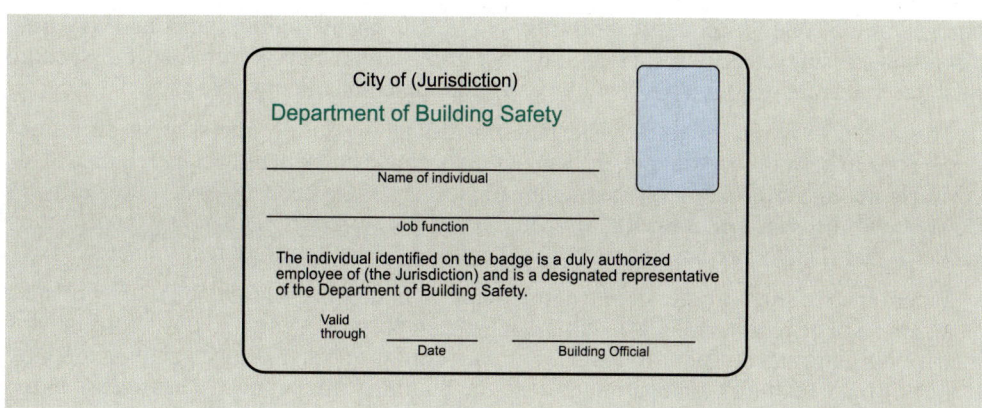

Figure 104-1
Personnel identification badge.

Right of entry. This section is compatible with Supreme Court decisions since the 1960s regarding acts of inspection personnel seeking entry into buildings for the purpose of making inspections. Under present case law, an inspection may not be made of a property, whether it be a private residence or a business establishment, without first having secured permission from the owner or person in charge of the premises. If entry is refused by the person having control of the property, the building official must obtain an inspection warrant from a court having jurisdiction in order to secure entry. The important feature of the law regarding right of entry is that entry must be made only by permission of the person having control of the property. Lacking this permission, entry may be gained only through the use of an inspection warrant. — **104.6**

If entry is again refused after an inspection warrant has been obtained, the jurisdiction now has recourse through the courts to remedy this situation. One avenue is to obtain a civil injunction in which the court directs the person having control of the property to allow inspection. Alternatively, the jurisdiction can initiate proceedings in criminal court for punishment of the person having control of the property. It cannot be repeated too strongly that criminal court proceedings should never be initiated against an owner or other person having control of the property if an inspection warrant has not been obtained. Because the consequences of not following proper procedures can be so devastating to a jurisdiction if a suit is brought against it, the jurisdiction's legal officer should always be consulted in these matters.

Liability. It is the intent of the IBC that the building official not become personally liable for any damage that occurs to persons or property as a result of the building official's acts so long as he or she acts in good faith and without malice or fraud. This protection is also extended to any member of the Board of Appeals, as well as any jurisdictional employee charged with enforcement of the IBC. Nevertheless, there continues to be a trend in the courts to find civil officers personally liable for their acts. This section requires that the jurisdiction defend the building official or other protected party if a suit is brought against him or her. Furthermore, the code requires any judgment resulting from a suit to be assumed by the jurisdiction. However, regardless of this language in the code, the jurisdiction may elect to not defend the building official on the basis, for example, that he or she acted carelessly. — **104.8**

Case law regarding tort liability of building officials is constantly in a state of flux, and old doctrines may not now be applicable. Therefore, the legal officer of the jurisdiction should always be consulted when there is any question about liability.

Modifications. The provisions of this section allow the building official to make modifications to the requirements of the code under certain specified circumstances. The building official may modify requirements if it is determined that strict application of the code is impractical and, furthermore, that the modification is in conformity with the intent and — **104.10**

purpose of the code. Without this provision in the IBC, the building official has very little discretionary enforcement authority and, therefore, would have to enforce the specific wording in the code, no matter how unreasonable the application would be.

The code does not intend to allow the building official to issue a variance to the provisions of the code to permit, for example, the use of only a single exit where two are required. This is clearly not in conformity with the intent and purpose of the code, no matter how difficult it may be to meet the requirements. In fact, the code is very specific that any modification cannot reduce health, accessibility, structural, and fire- and life-safety requirements.

Where the building official grants a modification under this section, the details of such an action shall be recorded. This document must then be entered into the files of the department of building safety. By providing a written record of the action taken and maintaining a copy of that action in the department files, the building official always has access to the decision-making process and final determination of his or her action should there be a need to review the decision.

Although it is expected that a permanent record be available for future reference when a modification is accepted, there is perhaps an even more important reason for the recording and filing of details of the approving action. The willingness to document and archive the modification action indicates the confidence of the building official in the decision that was made. A reluctance to maintain a record of the action taken typically indicates a lack of commitment to the action taken.

104.11 Alternative materials, design, and methods of construction and equipment. This section of the IBC may be one of the most important. It allows for the adoption of new technologies in materials and building construction that currently are not covered by the code. Furthermore, it gives the code even more of a performance character. The IBC thus encourages state-of-the-art concepts in design, construction, and materials as long as they meet the performance intended by the code. When evaluating the alternative methods under consideration, the building official must review for equivalency in quality, strength, effectiveness, fire resistance, durability, and safety. It is expected that all alternatives, once presented to the building official for review and approval, be thoroughly evaluated by the building department for compliance with this section. If such compliance can be established, the alternatives are deemed to be acceptable.

The provisions of this section, similar to those of Sections 104.1 and 104.10, reference the *intent* of the code. It is mandated that the building official, when evaluating a proposed alternative to the code, only approve its use where it can be determined that it complies with the intent of the specific code requirements. Thus, it is the responsibility of the building official to utilize those resources necessary to understand the intended result of the code provisions. Only then can the code be properly applied and enforced.

Similar to the approach taken where modifications are requested under the criteria of Section 104.10, the request for acceptance under this section should be made in writing to the building official. At a minimum, the submittal should include: (1) the specific code section and requirement, (2) an analysis of the perceived intent of the provision under review, (3) the special reasons as to why strict compliance with the code provision is not possible, (4) the proposed alternative, (5) an explanation of how the alternative meets or exceeds the intended level of compliance, and (6) a request for acceptance of the alternative material, design, or method of construction.

104.11.1 Research reports. Whereas the provisions of Section 104.11 grant the building official broad authority in accepting alternative materials, designs, and methods of construction, the process of evaluating such alternatives is often a difficult and complicated task. Valid research reports, including those termed *evaluation reports*, can address and delineate a review of the appropriate testing procedures to support the alternative as code compliant. The use of a research report may be helpful in reducing additional testing or documentation that is necessary to indicate compliance. It is important that the building official evaluate not only the information contained within the research report, but also the technical expertise of the individual or firm issuing the report. It must be noted, however, that a research report is simply a resource to the building official to assist in the decision-making process. The research report itself does not grant approval, as acceptance is still under the sole authority of the building official.

One of the most commonly utilized research reports is the ICC Evaluation Service (ICC-ES) Evaluation Report. ICC-ES is a nonprofit, limited liability corporation that does technical evaluations of building products, components, methods, and materials. If it is found that the subject of an evaluation complies with code requirements, then ICC-ES publishes a report to that effect and makes the report available to the public. However, ICC-ES Evaluation Reports are only advisory. The authority having jurisdiction is always the final decision maker with respect to acceptance of the product, material, or method in question.

Tests. The provisions of this section provide the building official with discretionary authority to require tests to substantiate proof of compliance with code requirements. The application of these provisions should be restricted to those cases where evidence of compliance is either nonexistent or involves actions considered to be impractical. Certainly, when the use of an alternative material, design, or method of construction is requested under the provision of Section 104.11, test information can be quite beneficial to the building official. There may also be insufficient evidence of compliance that can be substantiated through alternative tests. **104.11.2**

An example would be the placement of concrete that the quality-control measures (i.e., cylinder tests) did not prove to be complying with minimum strength requirements. Testing of core samples or perhaps use of nondestructive test methods might be appropriate to demonstrate compliance.

The provisions also specify that the tests be those that are specifically enumerated within the adopted construction regulations or, as an alternative, be those of other recognized national test standards. Where test standards do not exist, the building official has the authority to determine the test procedures necessary to demonstrate compliance. In addition to determining appropriate test methods or procedures, the building official is mandated to maintain records of such tests in accordance with local or state statutes.

Section 105 *Permits*

This section covers those requirements related to the activities of the building department with respect to the issuance of permits. The issuance of permits, plan review, and inspection of construction for which permits have been issued constitute the bulk of the duties of the typical department of building safety. It is for this reason that the code goes into detail regarding the permit-issuance process. Additionally, the code provides detailed requirements for the inspection process in order to help ensure that the construction for which the inspections are made complies with the code in all respects.

Required. Prior to obtaining a permit, the owner of the property under consideration, or the owner's authorized agent, must apply to the building official for any necessary permits that are required by the jurisdiction. One or more permits may be required to cover the various types of work being accomplished. In addition to building permits, which address new construction, alterations, additions, repairs, moving of structures, demolition, or change in occupancy, trade permits are required to erect, install, enlarge, alter, repair, remove, convert, or replace any electrical, gas, mechanical, or plumbing system. It is evident that almost any work, other than cosmetic changes, must be done under the authority of a permit. **105.1**

Typically, a permit is required each time a distinct activity occurs that is regulated under the code. However, certain alterations to previously approved systems can be performed

Scope and Administration

under an annual permit authorized by the building official. Electrical, gas, mechanical, or plumbing installations are eligible for such consideration when one or more qualified trade persons are employed by the person, firm, or corporation who owns or operates the building, structure, or premises where the work is to take place. In addition, the qualified individuals must regularly be present at the building or site.

105.2 Work exempt from permit. It would seem that the IBC should require permits for any type of work that is covered by the scope of the code. However, this section provides limited applications for exempted work. This section not only exempts certain types of building construction from permits, but also addresses electrical, gas, mechanical, and plumbing work that is of such a minor nature that permits are not necessary.

It is further the intent of the IBC that even though work may be exempted from a permit, such work done on a building or structure must still comply with the provisions of the code. As indicated in Section 101.2, the scope of the IBC is virtually all-inclusive. This may seem to be a superfluous requirement where a permit is not required. However, this type of provision is necessary to provide that the owner, as well as any design professional or contractor involved, be responsible for the proper and safe construction of all work being done.

A common example is a small, one-story detached accessory structure such as a storage shed. Although the code does not require a permit for an accessory building not exceeding 120 square feet (11 m^2) in floor area, all provisions in the code related to a Group U occupancy must still be followed.

105.3 Application for permit. In this section, the IBC directs that a permit must be applied for, and describes the information required on the permit application. The permit-issuance process, as envisioned by the IBC, is intended to provide records within the code enforcement agency of all construction activities that take place within the jurisdiction and to provide orderly controls of the construction process. Thus, the application for permit is intended to describe in detail the work to be done. In this section, the building official is directed to review the application for permit. This review is not a discretionary procedure, but is mandated by the code.

The code also charges the building official with the issuance of the permit when it has been determined that the information filed with the application shows compliance with the IBC and other laws and ordinances applicable to the building at its location in the jurisdiction. The building official may not withhold the issuance of a permit if these conditions are met. As an example, the building official would be in violation in withholding the issuance of a building permit for a swimming pool because an adjacent cabana was previously constructed without a permit.

105.4 Validity of permit. The code intends that the issuance of a permit should not be construed as permitting a violation of the code or any other law or ordinance applicable to the building. In fact, the IBC authorizes the building official to require corrections if there were errors in the approved plans or permit application at the time the permit was issued. The building official is further authorized to require corrections of the actual construction if it is in violation of the code, although in accordance with the plans. Moreover, the building official is further authorized to invalidate the permit if it is found that the permit was issued in error or in violation of any regulation or provision of the code.

Although it may be poor public relations to invalidate a permit or to require corrections of the plans after they have been approved, it is clearly the intent of the code that the approval of plans or the issuance of a permit may not be done in violation of the code or of other pertinent laws or ordinances. As the old saying goes, "Two wrongs do not make a right."

105.5 Expiration. The IBC anticipates that once a permit has been issued, construction will soon follow and proceed expeditiously until completion. However, this ideal procedure is not always the case and, therefore, the code makes provisions for those cases where work has not started, or alternatively where the work, after being started, has been suspended for a period of time. In these cases, the IBC allows a period of 180 days to transpire before the permit becomes void. The code then requires that a new permit be obtained. It is assumed by the code that the department of building safety will have expended some effort in follow-up inspections of the work, etc., and, therefore, the original permit fee must be retained in order to compensate the agency for the work. The building official has the authority to grant one or more extensions of time, provided the permit holder can demonstrate a justifiable cause as to why the permit should not be invalidated. The time period for such extensions cannot exceed 180 days; however, additional extensions may be granted if approved by the building official.

There are several reasons why it is important to establish a limitation on the validity of a permit, such as purging the department files of inactive permits. Additionally, it keeps the project on track with the code edition in effect at the time of permit issuance.

Section 107 *Submittal Documents*

Plans, specifications, and other construction documents, along with other applicable data, must be filed with the permit application. Such submittal documents are intended to graphically depict the construction work to be done. The IBC sets forth the necessary information that must be provided to the building official at the time of plans submittal, as well as procedures for deferred submittals.

General. In this section, the IBC directs that at the time of application for a permit, construction documents and other essential data on the project be submitted. The code requires that plans, engineering calculations, and any other information necessary to describe the work to be done be filed along with the application for a permit. A statement of special inspections that may be required by Section 1704.2.3 is also to be submitted with the application for a permit. Based on the statutes of the jurisdiction, the plans, specifications, and other documents may need to be prepared by a registered design professional. Under special circumstances, the building official may also require that the submittal documents be prepared by a registered design professional even when not mandated by jurisdictional statute. The building official is permitted to waive the requirement for the filing of plans and other data, where not required by statute to be prepared by a registered design professional, provided the building official ensures that the work for which the permit is applied is of such a nature that plans or any other data are not necessary in order to indicate and obtain compliance with the code. **107.1**

Examination of documents. In this section, the building official is directed to review the plans, specifications, and other submittal documents filed with the permit. The building official is not at liberty to check only a portion of the plans. On this basis, the structural drawings, as well as the engineering calculations, must be checked in order for the building official to provide a full examination. **107.3**

Design professional in responsible charge. The building official has the authority to require the designation of a design professional to act in responsible charge of the work being performed. The function of such an individual is to review and coordinate submittal documents prepared by others and, if necessary, coordinate any deferred or phased submittal items. In some cases, the design of portions of the building has been completed and may be dependent on the manufacturer of proposed prefabricated elements, such as for truss drawings. Therefore, the code specifically allows deferring the submittal of portions of plans and specifications. **107.3.4**

1 Scope and Administration

Section 108 *Temporary Structures and Uses*

This section authorizes the erection of temporary buildings and structures such as those erected at construction sites. The regulation of temporary viewing stands and other miscellaneous, temporary structures would also fall under this section. The following are key provisions for temporary buildings and structures:

1. They are erected by special permit.
2. They are erected for a limited period of time.
3. They must conform to the requirements of the IBC for structural strength, fire safety, means of egress, accessibility, light, ventilation, and sanitation.

Section 109 *Fees*

Permits required by the jurisdiction are not valid until the appropriate fees have been paid. The IBC provides for each individual jurisdiction to establish its own schedules for permit fees. The fees collected by the department of building safety are typically set at a level to adequately cover the cost to the department for services rendered.

109.3 Building permit valuations. The code uses the concept of valuation to establish the permit fee. This concept is based on the proposition that the valuation of a project is related to the amount of work to be expended in the various aspects of administering the permit. Also, there should be some excess in the permit fee to cover departmental overhead. Essentially, the valuation is considered as the cost of replacing the building. The valuation also includes any electrical, plumbing, and mechanical work, even though separate permits may be required for the mechanical, plumbing, and electrical trades.

To provide some uniformity in the determination of valuation so that there is a consistent base for the assignment of fees, the IBC directs the building official to determine the value of the building. To assist in obtaining uniformity, the ICC periodically publishes "Building Valuation Data (BVD)" in its *Building Safety Journal*®. In addition, the valuation data can be accessed on ICC's website. Thus, building officials may utilize a common base in their determination of the value of buildings. However, ICC strongly recommends that all jurisdictions and other interested parties actively evaluate and assess the impact of the BVD table before utilizing it in their current code enforcement activities. As an option, any other appropriate method may also be used by the jurisdiction as the basis for determining the proper valuation of the work.

109.4 Work commencing before permit issuance. When work requiring a permit is started without such a permit, the IBC allows the building official to assess an additional fee above and beyond the required permit fees. This fee is intended to compensate the department of building safety for any additional time and effort necessary in evaluating the work initiated without a permit. It may often be necessary for the building official to cause an investigation to be made of the work already done. The intent of the investigation is to determine to what extent the work completed complies with the code, and to describe with as much detail as possible the work that has been completed.

109.6 Refunds. This section authorizes the building official to establish a policy for partial or complete refunds of fees paid to the department. Although not specified in the code, there are a variety of reasons why some level of a fee refund would be appropriate. One instance would be where the permit fee is collected in error. Another reason for authorizing a refund of the fees paid would be that circumstances beyond the control of the applicant caused delays and the eventual expiration of the building permit. Typically, the building official will withhold from the refund any monies expended by the department for related administrative activities that have taken place. It should be noted that the building official, when approving a fee refund, is authorizing the disbursement of public funds. Therefore, the building official must be sure that there is good cause for the refund to the applicant.

Section 110 *Inspections*

The inspection function is arguably the most critical aspect of building department operations. An important concept views that, as with the issuance of permits, inspections that presume to give authority to violations of the code are invalid. In general, the IBC charges the permit applicant with the responsibility of ensuring that the work to be inspected remain accessible and exposed until it is approved. Any expense incurred in the process of providing for an accessible and exposed inspection, such as the removal of gypsum board or insulation, is not to be borne by the building official or jurisdiction. Therefore, it is critical that work should not proceed beyond the point where an inspection is required.

Required inspections. The code mandates seven typical inspections during the progress of construction of a building, as follows: **110.3**

1. Footing and foundation inspection
2. Concrete slab or under-floor inspection
3. Frame inspection
4. Lath or gypsum board inspection
5. Fire-resistant penetrations inspection
6. Energy efficiency inspection
7. Final inspection

The IBC also gives the building official authority to require and make other inspections where necessary to determine compliance with the code and other laws and regulations enforced by the department. This includes those special inspections addressed in Section 1705. The need for either periodic or continuous inspection is established for construction that requires special expertise to ensure compliance. Another type of inspection is required in flood hazard areas. After the lowest floor level is established, but prior to additional vertical construction, the lowest floor elevation must be determined.

In each case for the required inspections identified in Section 110.3, the IBC is very specific as to how far the construction must have progressed prior to a request for an inspection. If it is necessary to make a re-inspection because work has not progressed to the point where it is ready for inspection, the building official may charge an additional fee under the general provisions of Section 109. A re-inspection fee is not specifically mandated, however, and normally would not be assessed unless the person doing the work continually calls for inspections before the work is ready for inspection. Such actions would typically cause increased costs to the jurisdiction not covered by the original building permit.

Inspection requests. In general, the IBC charges the individual holding the building permit with notifying the department of building safety when it is time to make an inspection. The permit holder may also authorize one or more agents for this responsibility. The code places a duty upon the permit holder or their authorized agent to have the work to be inspected accessible and exposed, so it can be evaluated as to code compliance. **110.5**

Approval required. The code intends that no work shall be done beyond the point where an inspection is required until the work requiring inspection has been approved. Moreover, it is intended that work requiring inspection not be covered until it has been inspected and approved. **110.6**

1 Scope and Administration

Section 111 Certificate of Occupancy

The tool that the building official utilizes to control the uses and occupancies of the various buildings and structures within the jurisdiction is the certificate of occupancy. The IBC makes it unlawful to use or occupy a building or structure unless a certificate of occupancy has been issued for that use. Furthermore, the code imposes the duty of issuing a certificate of occupancy upon the building official when he or she is satisfied that the building, or portion thereof, complies with the code for the intended use and occupancy.

Prior to use or occupancy of the building, the building official shall perform a final inspection as addressed in Section 110.3.10. If no violations of the code and other laws enforced by the department of building safety are found, the building official is required to issue a certificate of occupancy. Figure 111-1 illustrates the information that must be provided on the certificate of occupancy.

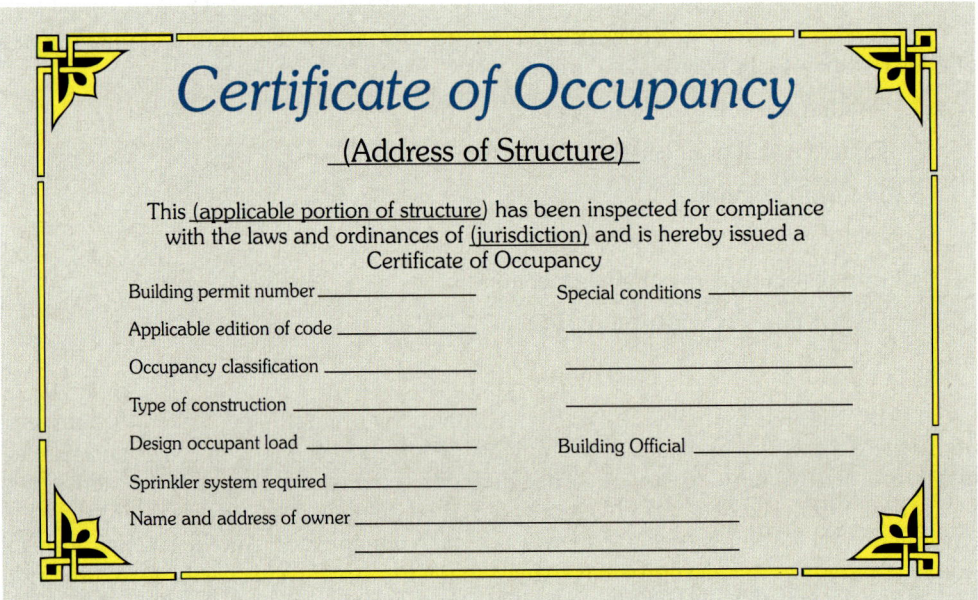

Figure 111-1
Certificate of occupancy.

Where a portion of a building is intended to be occupied prior to occupancy of the entire structure, the building official may issue a temporary certificate of occupancy. This situation would occur where partial occupancy is requested prior to completion of all work authorized by the building permit. Prior to issuance of a temporary certificate of occupancy, it is critical that the building official ascertain that those portions to be occupied provide the minimum levels of safety required by the code. In addition, the building official shall establish a definitive length of time for the temporary certificate of occupancy to be valid.

In essence, the certificate of occupancy certifies that the described building, or portion thereof, complies with the requirements of the code for the intended use and occupancy, except as provided for existing buildings undergoing a change of occupancy as regulated by Chapter 34. However, any certificate of occupancy may be suspended or revoked by the building official under one of three conditions: (1) where the certificate is issued in error, (2) where incorrect information is supplied to the building official, or (3) where it is determined that the building or a portion of the building is in violation of the code or any other ordinance or regulation of the jurisdiction. The most common reason for suspending or revoking a certificate of occupancy is that the building is being used for a purpose other than that intended when approval for occupancy was granted. When a permit is issued, plans reviewed, inspections made, and a certificate of occupancy given, there is an expectation that the building will be used for specific activities. As such, the hazards associated with such activities can be addressed. However, where an unanticipated use occurs, there is the potential that the necessary safeguards are not in place to address the related hazards. An unauthorized use can be intentional or unintentional; however, it is no less a concern until the use is discontinued or the necessary remedies are in place.

Section 112 *Service Utilities*

The building official is authorized to control the connection release for any service utility where the connection occurs to a building or system regulated by the IBC. Such authority also includes the temporary connection of the service utility. Perhaps even more important, the building official is also granted authority for the disconnection of service utilities where it has been determined that an immediate hazard to life or property exists. As in all administrative functions, it is important that due process be followed.

Section 113 *Board of Appeals*

The IBC intends that the board of appeals have the authority to hear and decide appeals of orders and decisions of the building official relative to the application and interpretations of the code. However, the code specifically denies the authority of the board relative to waivers of code requirements. Based on the qualification level set forth by the code for board membership, it can be assumed that the board is intended as fundamentally a technical body. Any broader authority may place the board outside of its area of expertise, such as addressing internal administrative issues of the department. It should be noted that the board's role is to hear and decide on appeals of decisions made by the building official. Until the building official makes his or her determination, the issue is not subject to board review.

The importance of a board of appeals should not be taken lightly. Its role is equivalent to that of the building official when it comes to technical questions placed before it, thus the need for highly qualified board members. The board is not merely advisory in its actions, but rather is granted the authority to overturn the decisions of the building official where the determination is within the scope as described by the code. An example would be an appeal requesting the use of an alternative method of construction previously denied by the building official. The alternative method would be permitted if found by the board to be equivalent to or better than that set forth in the code. Where the board of appeals is desired to take on a more expansive role, such as code adoption functions or contractor licensing oversight, those duties should be specifically granted by the jurisdiction. More detailed information on the qualifications and duties of a board of appeals is found in IBC Appendix B.

Section 114 *Violations*

The provisions of this section establish that violations of the code are considered unlawful and such violations shall be abated. Where necessary, a notice of violation may be served by the building official to the individual responsible for the work that is in violation of the code. Further action may be taken where there is a lack of compliance with the notice. The jurisdiction shall establish penalties based on the various specified violations.

1 Scope and Administration

Section 115 *Stop Work Order*

The stop work order is a tool authorized by the IBC to enable the building official to demand that work on a building or structure be temporarily suspended. Intended to be utilized only under rare circumstances, this order may be issued where the work being performed is dangerous, unsafe, or significantly contrary to the provisions of the code. A stop work order is often a building official's final method in obtaining compliance on issues of extreme importance. All other reasonable avenues should be considered prior to the issuance of such an order.

The stop work order shall be a written document indicating the reason or reasons for the work to be suspended. It shall also identify those conditions where compliance is necessary before the cited work is allowed to resume. All work addressed by the order shall immediately cease upon issuance. It is important that the stop work order be presented to the owner of the subject property, the agent of the owner, or the individual doing the work. Only after all issues have been satisfied may work continue. Because of the potential consequences involved with *shutting down* a construction project, it is critical that the appropriate procedures be followed. Although a stop work order can be issued at any time during the construction process, the most common application of the provision is where work has commenced before or without the issuance of a building permit. The order is given to notify those affected parties that all work be stopped until a valid permit is obtained. An example of a stop work order is shown in Figure 115-1.

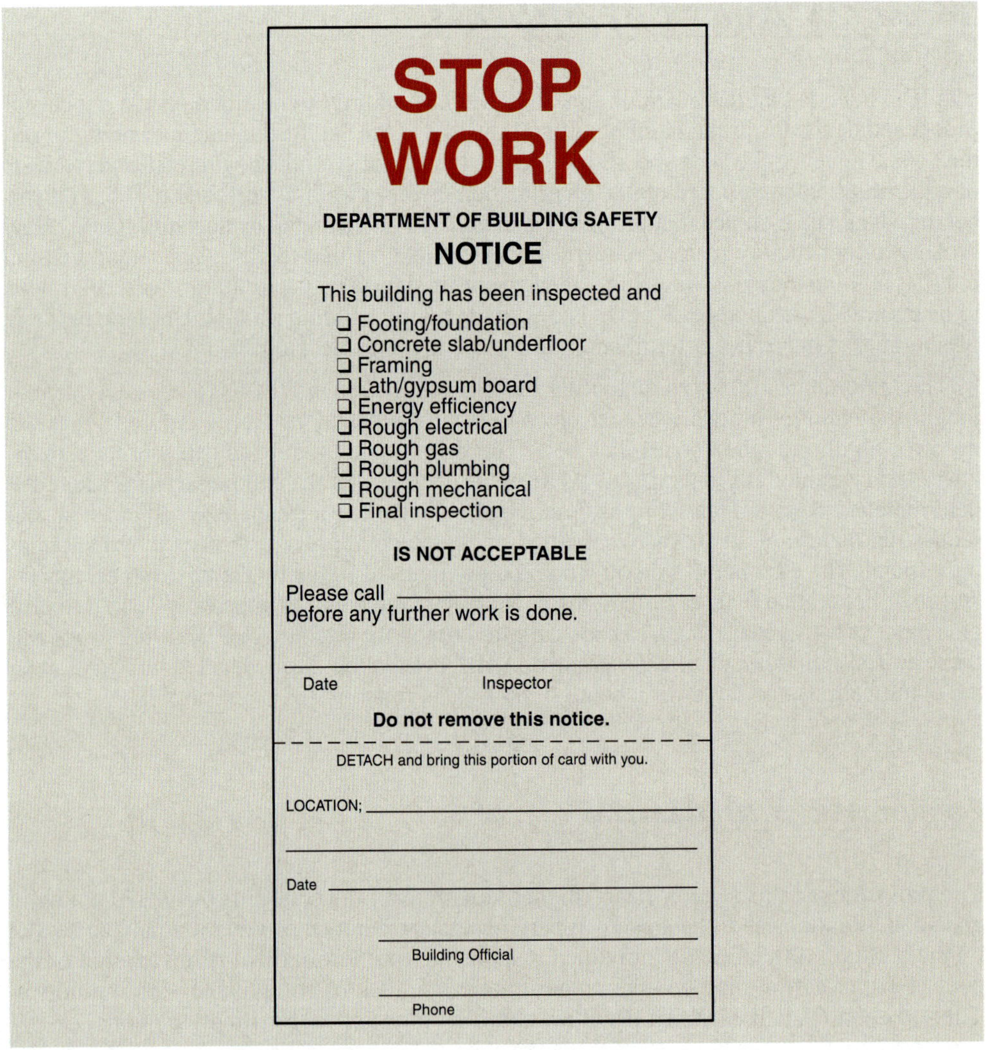

Figure 115-1 Stop work order.

Section 116 *Unsafe Structures and Equipment*

The provisions of this section are intended to define what constitutes an unsafe building and unsafe use of a building. Unsafe buildings and structures are considered public nuisances and require repair or abatement. The abatement procedures are indicated in this section, including the creation of a report on the nature of the unsafe condition. Written notice shall be provided and the method of service of such notice is specified in Section 116.4. Where restoration or repair of the building is desired, the provisions of Chapter 34 for existing buildings are appropriate.

KEY POINTS

- The IBC provides minimum standards to safeguard the health, safety, and welfare of the public.
- Rendering interpretations of specific code provisions is the responsibility of the building official, to be based on the purpose and intent of the code.
- Modifications to specific IBC provisions may be acceptable where strict application of the code is impractical.
- Alternative designs, materials, and methods of construction to those detailed in the code are to be evaluated by the building official based on an equivalency to the prescribed regulations.
- Tests may be mandated by the building official in order to verify compliance with the code.
- The issuance of permits, plan review, and inspection of construction for which permits have been issued constitutes the bulk of the duties of the typical department of building safety.
- The code intends that the issuance of a permit should not be construed as allowing a violation of the code or any other local ordinance applicable to the building.
- Submittal documents must be appropriately submitted to the building official for review, and approved again prior to receiving a building permit.
- The building official has the authority to require the designation of a registered design professional to review and coordinate submittal documents prepared by others.
- A certificate of occupancy, granted by the building official indicates that the structure is lawfully permitted to be occupied.
- The board of appeals provides a mechanism for individuals to challenge the interpretation of the building official on technical issues in the IBC.

CHAPTER 2

DEFINITIONS

Section 201 General
Section 202 Definitions
Key Points

2 Definitions

Section 201 *General*

A number of definitions are applicable specifically to the *International Building Code®* (IBC®) and may not have an appropriate definition for code purposes in the dictionary. Therefore, over 700 words and terms are defined in Chapter 2 to assist the user in the proper application of the requirements. Those words and terms specifically defined in the code are *italicized* where they occur throughout the text in order to identify to the user that a specific definition can be found in Chapter 2. Although infrequent, the definitions of some terms are contained within the text of the requirement. For example, the definition of *day care* is implied in the description of Group E occupancies. Other frequently used and significant terms are undefined (i.e., 1-hour fire-resistance-rated construction), and their meaning can be discerned only from their context. There are numerous definitions in Chapter 2, but only selected definitions are included in this commentary.

An important feature of this section is the requirement that ordinarily accepted meanings be used for definitions that are not provided in the code. Such meanings are based on the context in which the term or terms appear. The code defines terms that have specific intents and meanings insofar as the code is concerned, and leaves it up to the user to apply all undefined terms in the manner in which they are ordinarily used. Where there is any question as to the meaning or application of a particular definition, the building official shall make the determination under the interpretive authority granted in Section 104.1 based on the context and intent of the term's usage in the code.

Section 202 *Definitions*

ACCESSIBLE MEANS OF EGRESS. In concert with efforts to make buildings accessible and usable for persons with disabilities, it is necessary that safe egress for physically disabled persons also be provided. Therefore, accessible exit paths of continuous and unobstructed travel are to be provided. Generally consistent with the provisions of other exiting systems, an accessible means of egress shall begin at any accessible point within the building and continue until reaching the public way. The primary provisions regulating accessible means of egress are located in Section 1007.

AISLE. Where furniture, fixtures, and equipment limit the potential travel paths within the means of egress system, aisles and aisle accessways are created. Typically, aisles accept the contribution of occupant travel from adjoining aisle accessways. At times, multiple aisles may converge into a main aisle, which then may lead to exit access doorways or exits. Aisles are common throughout most buildings and are considered portions of the exit access.

AISLE ACCESSWAY. The path of travel from an occupiable point in a building to an aisle is considered to be an aisle accessway. Often called a "row" in everyday language when adjacent to seats and fixtures, an aisle accessway is typically used by small numbers of people prior to converging with other persons at an aisle. An example of aisle accessways is shown in Figure 202-1.

ALTERNATING TREAD DEVICE. By appearance more of a ladder than a stairway, an alternating tread device is shown in Figure 202-2. Typically, steps are supported by a center rail placed at a severe angle from the floor. The key difference between this device and other forms of stairways addressed in the IBC is that the user, by nature of the design of the device, can never have both feet on the same level at the same time.

AMBULATORY CARE FACILITY. Facilities where individuals are provided with medical care on less than a 24-hour basis are classified as Group B occupancies. However, two separate definitions highlight the fact that there are two unique types of persons that

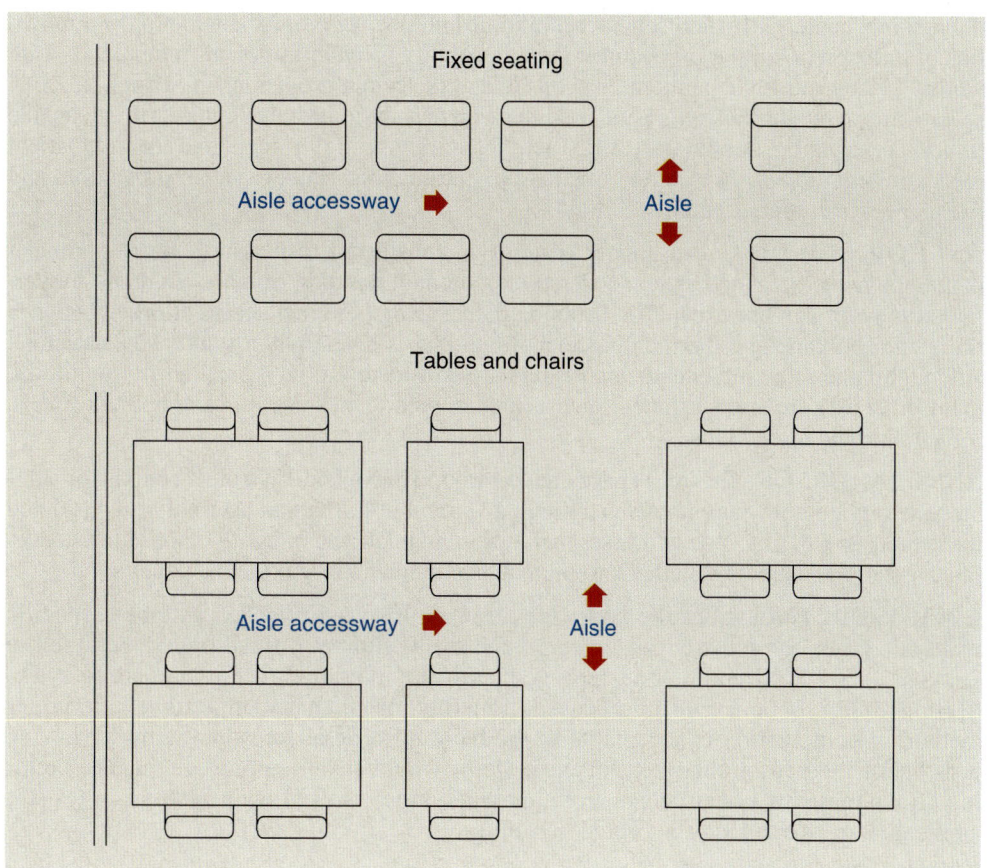

Figure 202-1
Aisle accessways.

occupy such facilities. The important difference involves the self-preservation capabilities of the individuals. Where the occupants are capable of self-preservation (the ability to respond to emergency situations without physical assistance from others) throughout the period of time in which they are at the facility, the building is considered an "outpatient clinic" as

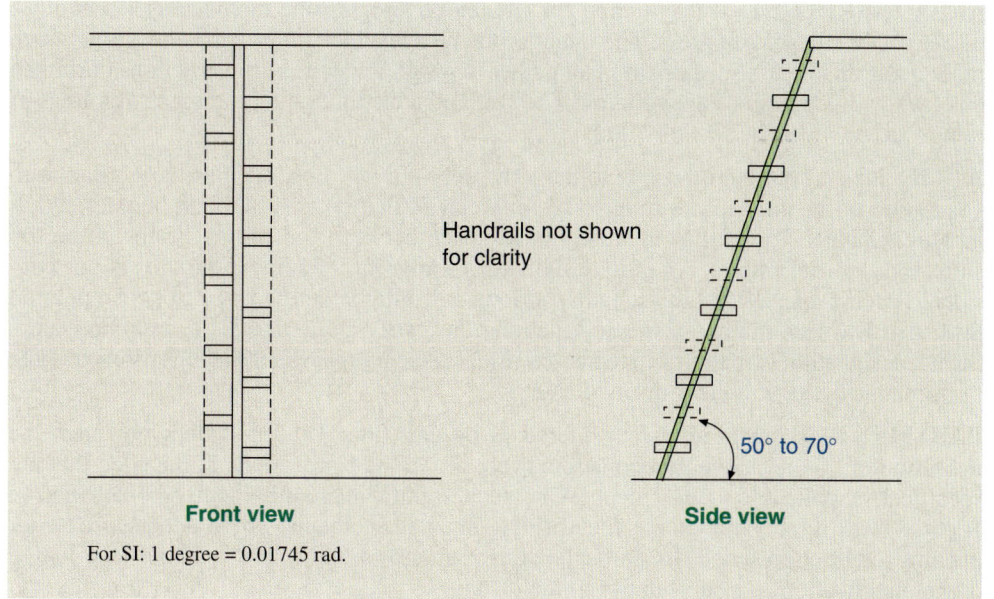

For SI: 1 degree = 0.01745 rad.

Figure 202-2
Alternating tread device.

2 Definitions

defined in Chapter 2. If such self-preservation cannot be accomplished due to the application of sedation or similar procedures, then the facility is by definition an "ambulatory care facility." The need for separate definitions is due to special provisions in Section 422 that specifically regulate those types of facilities where individuals are temporarily incapable of self-preservation. Ambulatory care facilities are more highly regulated than outpatient clinics in regard to smoke compartmentation, automatic sprinkler system protection, and fire alarm system requirements.

ANCHOR BUILDING. An anchor building is considered to be any exterior perimeter building having direct access to a covered or open mall building. Anchor buildings are permitted to be of any use except for Group H. Other than the direct access afforded between the anchor building and the mall, an anchor building is essentially a stand-alone structure. Although large retail merchandizing centers are considered to be typical anchor buildings, many other uses are also permitted, such as motion picture theaters, ice-skating rinks, hotels, restaurants, and night clubs.

ANNULAR SPACE. The open space created around the outside of a pipe, conduit, or similar penetrating item where it passes through a vertical or horizontal assembly is considered the annular space. The code addresses methods to maintain the integrity of a fire-resistance-rated assembly, including methods to protect any annular space around a penetration.

APPROVED. Throughout the code, the term *approved* is used to describe a specific material or type of construction, such as approved automatic flush bolts mentioned in Section 1008.1.9.3, Item 3, or an approved barrier in interior exit stairways addressed in Section 1022.8. Where *approved* is used, it merely means that such design, material, or method of construction is acceptable to the building official (or other authority having jurisdiction), based on the intent of the code. It would seem appropriate that the building official base his or her decision of approval on the result of investigations or tests, if applicable, or by reason of accepted principles.

APPROVED SOURCE. One provision mandating the use of an approved source is found in Section 104.11.1, which specifically identifies the use of valid research reports as an acceptable method the building official can use to evaluate alternative methods and materials of construction. It is expected that the authors of such reports be technically competent and appropriately experienced in the subject under consideration. The building official is designated as the individual solely charged with ascertaining that the person, firm, or corporation providing the technical evaluation meets the necessary qualifications.

AREA OF REFUGE. It is common for an area of refuge to be included as a portion of the accessible means of egress. The intent of the refuge area is to provide a location where individuals unable to use stairways can gather to await assistance or instructions during an emergency evacuation of the building. The size and construction requirements for areas of refuge are provided in Section 1007.6.

ATTIC. Several provisions apply to the attic area of a building, such as those relating to ventilation of the attic space. In order to fully clarify that portion of a building defined as an attic, Chapter 2 identifies an attic as that space between the ceiling beams at the top story and the roof rafters. An attic designation is appropriate only if the area is not considered occupiable. Where this area has a floor, it would be defined as a story. A common misuse of IBC terminology is the designation of a space as a *habitable* or *occupiable* attic. Such a designation is inappropriate insofar as once such a space is utilized for some degree of occupancy, it is no longer deemed an attic.

BASEMENT. A basement is considered to be any floor level that does not meet the definition of "Story above grade plane." There are limited provisions in the code that are specifically applicable to basements. One such significant requirement is established in Section 903.2.11.1.3 mandating the sprinklering of basements where adequate exterior openings are not provided. In short, the code regulates a below-ground floor level based on its qualification as a story above grade plane.

BLEACHERS. Structures designed for seating purposes containing tiered or stepped seating two or more rows high are considered bleachers. The definition of "Grandstand" is identical to that of bleachers, recognizing that the terms are interchangeable. Bleachers may, or may not, be provided with backrests. In addition, bleachers may be located either inside or outside a structure. The specific provisions for bleachers are found in ICC 300, *Bleachers, Folding and Telescopic Seating, and Grandstands*. Similar seating areas that are considered as building elements would not be defined as bleachers and are not regulated by ICC 300. A building element, as defined in Section 702.1, is deemed to be a fundamental component of the building's construction as listed in Table 601.

BUILDING AREA. The term *building area* describes that portion of the building's floor area to be utilized in the determination of whether or not a structure complies with the provisions of Chapter 5 for allowable building size. It is not to be confused with the term *floor area*, which is the basis for occupant-load determination in Chapter 10 for means of egress evaluation, nor the term *fire area* as used in the application of automatic sprinkler requirements in Chapter 9.

The definition of building area is the area included within the surrounding exterior walls of the building, and the definition further states that the floor area of a building or portion thereof not provided with surrounding exterior walls shall be the usable area under the horizontal projection of the roof or floor above. The intent of this latter provision is to address where a structure may not have exterior walls or may have one or more sides open without an enclosing exterior wall. Examples would include a canopy covering pump islands at a service station, or the drive-through area of a fast food restaurant. Where a column line establishes the outer perimeter of the usable space under the roof, it is also typically the extent of building area. Beyond the column line, the overhead cover is simply viewed as a projection. See Figure 202-3. If all of the area beneath the roof above can be considered usable space, then the building area is measured to the leading edge of the roof above. See Figure 202-4.

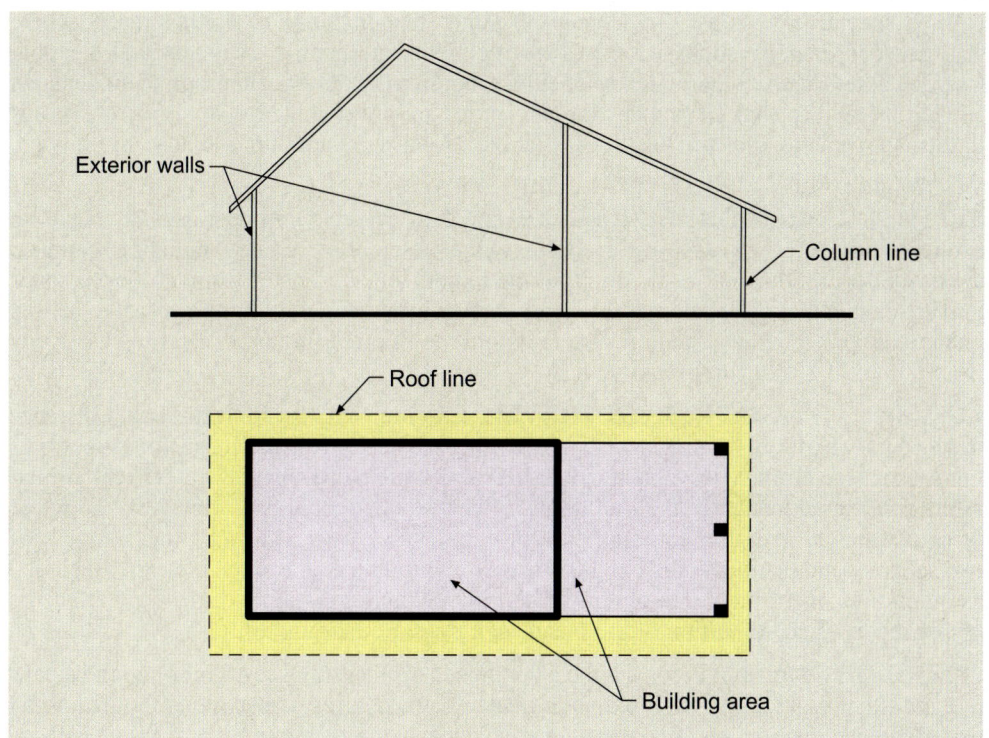

Figure 202-3
Building area.

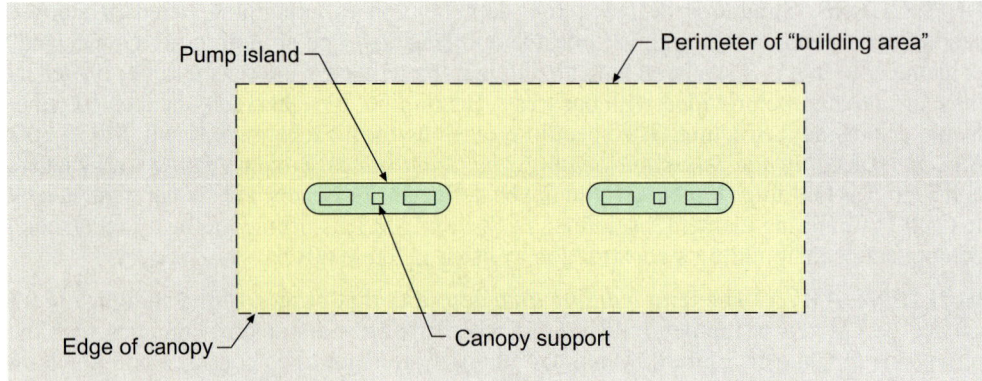

Figure 202-4
Building area.

BUILDING ELEMENT. Primary structural frame members, bearing walls, nonbearing walls and partitions, floor construction including secondary members, and roof construction including secondary members are considered to be building elements for the purposes of the IBC. Such elements are primarily regulated based on two criteria: fire resistance and combustibility. In determining a building's type of construction, the building elements are evaluated based on the criteria previously mentioned.

BUILDING OFFICIAL. Regardless of title, the individual who is designated by the jurisdiction as the person who administers and enforces the IBC is considered by the code to be the building official. In addition, all other individuals who have been given similar enforcement authority, such as plans examiners and inspectors, are also considered building officials to a limited degree under the IBC. A further discussion of the duties and responsibilities of the building official is found in the commentary on Section 104.

CARE SUITE. Special means of egress provisions are provided in Section 407.4.3 for care suites in Group I-2 occupancies. The definition of "care suite" establishes the scope of such special provisions. The concept of suites recognizes those arrangements where staff must have more supervision of care recipients in specific treatment and sleeping rooms. Therefore, the general means of egress requirements are not appropriate under such conditions. The special allowances for care suites are not intended to apply to day rooms or business functions of the health care facility. Intensive-care and critical-care units are typically configured as care suites.

CEILING RADIATION DAMPER. Designed to protect air openings that occur in fire-resistance-rated roof/ceiling or floor/ceiling assemblies, ceiling radiation dampers are listed devices that automatically limit the radiative heat transfer from a room or space into the cavity above the ceiling. The damper, in conjunction with the fire-resistive ceiling membrane, protects the structural system within the floor/ceiling or roof/ceiling assembly from failure that is due to excessive heat.

COMMON PATH OF EGRESS TRAVEL. It is important to limit the travel distance that occurs where only a single path is available to the user of the means of egress. For this reason, such limited travel is regulated as a common path of egress travel. Very similar to the concept addressed in the limitations of dead-end conditions in corridors, the intent of regulations regarding a common path of egress travel is based on the lack of at least two separate and distinct paths of egress travel toward two or more remote exits. Included as the initial part of the permitted travel distance, a common path of egress travel occurs within the exit access portion of the means of egress. See Figure 202-5.

CORRIDOR. A corridor is considered a component of the exit access portion of the means of egress that provides an enclosed and directed path of egress travel along the path to an exit. Regulated by Section 1018, corridors may or may not be required to be fire-resistance

Definitions 2

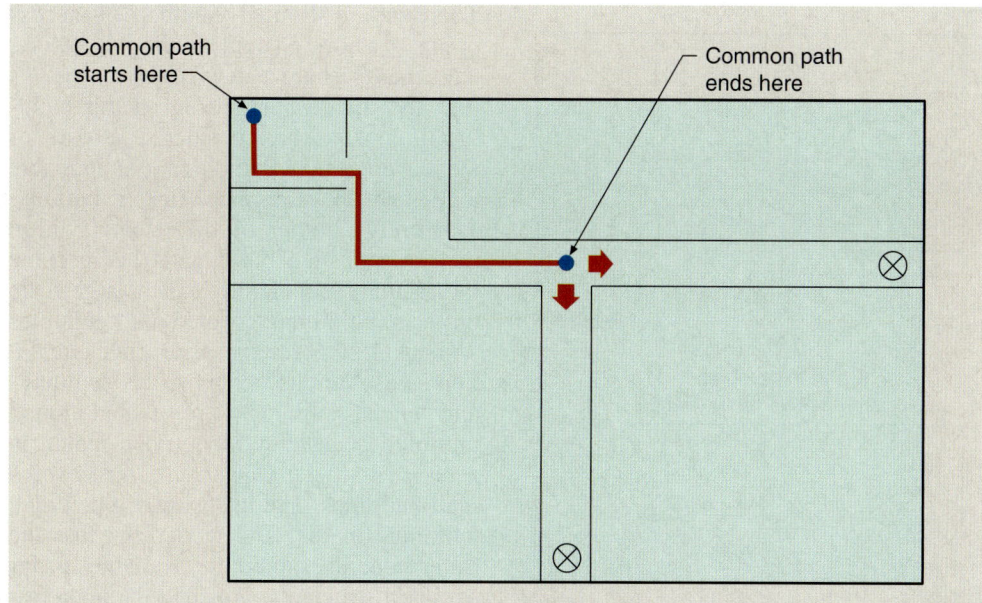

Figure 202-5
Common path of egress travel.

rated, but in all cases, an enclosure of some sort is anticipated. The determination as to whether or not a design element is to be regulated as a corridor is to be made by the building official. Many factors may be considered prior to making this determination, which could possibly include length, degree of enclosure, length-to-width ratio, adjacent spaces, and other considerations. However, because of the reliance upon corridors as an important egress element, it is critical that they be appropriately regulated. A corridor's primary purpose is for the movement of occupants, both as a part of the building's circulation and its use as a means of egress. Although some spaces may have one or more of the characteristics of a corridor as previously mentioned (length, degree of enclosure, length-to-width ratio, etc.), their primary function is that of rooms, and they should not be considered corridors. A classic example can be found in many observation buildings in zoos, where a very long, narrow element is used as a means for occupants to view displays of wildlife. Although in plan view it may appear to be a corridor, it actually functions as a room and should be regulated as such. It should be noted that the placement of a few pieces of furniture or equipment within a corridor in an effort to consider the space a room rather than a corridor is not appropriate. It should be noted that corridors are never mandated by the IBC, but rather are utilized as design elements. Where provided, however, such components must comply with the code.

COURT. Open and unobstructed to the sky above, an exterior area is considered a court where it is enclosed on at least three sides by exterior walls of the building or other enclosing elements, such as a screen wall. Regulations for courts, including those used for egress purposes, are found throughout the code. Examples of courts are shown in Figure 202-6. Although the IBC does not mandate a minimum depth for consideration as a court, it is expected that certain design and structural features of the building that create minor exterior wall offsets would not require designation as a court. The determination of the presence of a court under such conditions is subject to the building official's discretion.

COVERED MALL BUILDING. A covered mall building consists of various tenants and occupancies, as well as the common pedestrian areas that provide access to the tenant spaces. Although the most common tenants of covered mall buildings are retail stores, various other uses are also commonly found in such a structure. Restaurants, drinking establishments, entertainment and amusement facilities, passenger transportation terminals such as airports and bus stations, offices, and other similar uses are often located in a covered mall building.

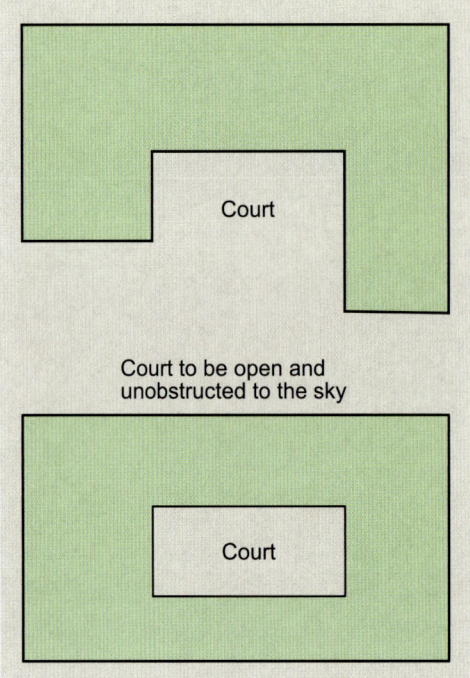

Figure 202-6
Definition of courts.

There are a limited number of access points to a covered mall building; therefore, the tenants share a few major entrances into the mall. Although an anchor building is not to be considered a part of the covered mall building, it is possible to design an exterior perimeter building as merely another of the tenants within the covered mall building. It has become increasingly popular to create large-scale projects resembling covered mall buildings without roofs over the pedestrian circulation areas. Various "tenant space" buildings and "anchor buildings" are situated around unroofed pedestrian ways (open malls) in a manner very similar to that for covered mall buildings. The inclusion of open mall buildings in the IBC recognizes that the same benefits should be available as for enclosed structures, provided the appropriate measures are taken. Where the mall area is open to the sky, equivalent or better life safety and property protection is provided.

When one thinks of a covered mall building, its use is typically associated with retail sales and related activities. However, the provisions of Section 402 may also lend themselves to other occupancies such as office uses and transportation facilities. Recently, more attention has been given to the application of the covered mall building concept to large educational buildings. Many of the characteristics of an exciting and efficient school environment are consistent with those of a covered mall building, including spacious areas open to each other, vertical interaction between floor levels, ease of supervision, and limited points of entry. Educational facilities even have their own *anchor buildings* such as the auditorium, gymnasium, media center, and other spaces. However, because the criteria of Section 402 are more traditional in their approach, some of the provisions cannot be directly applied to an educational occupancy. It will be necessary to utilize the alternate methods and design allowance in Section 104.11 to more specifically address those issues.

DOOR, BALANCED. Balanced doors are a special type of double-pivoted door in which the pivot point is located some distance in from the door edge, thus creating a counterbalancing effect.

DRAFTSTOP. Required by the code only in concealed areas of combustible construction, draftstops are utilized in large concealed spaces to limit air movement, which is accomplished through the subdivision of such spaces. Draftstops are to be constructed of those materials or construction identified by the IBC that effectively create smaller compartments within attics and similar areas.

DWELLING UNIT AND DWELLING. A *dwelling unit* is considered a single unit that provides living facilities for one or more persons. Dwelling units include permanent provisions for living, sleeping, eating, cooking, and sanitation, thus providing a complete independent living arrangement. A dwelling unit, while typically addressed in the IBC as a portion of a Group R-2 occupancy, may also be classified as Group I-1, R-1, or R-3. A *dwelling* is a building that contains either one or two dwelling units. Dwellings are typically regulated under the provisions of the *International Residential Code*® (IRC®), as noted in the exception to Section 101.2.

EGRESS COURT. That portion of the exit discharge at ground level that extends from a required exit to the public way is considered an egress court. An egress court may be a yard, a court, or a combination of a yard and a court that extends from the end point of an exit, typically an exterior door at grade level, until it reaches a public way. An example is illustrated in Figure 202-7.

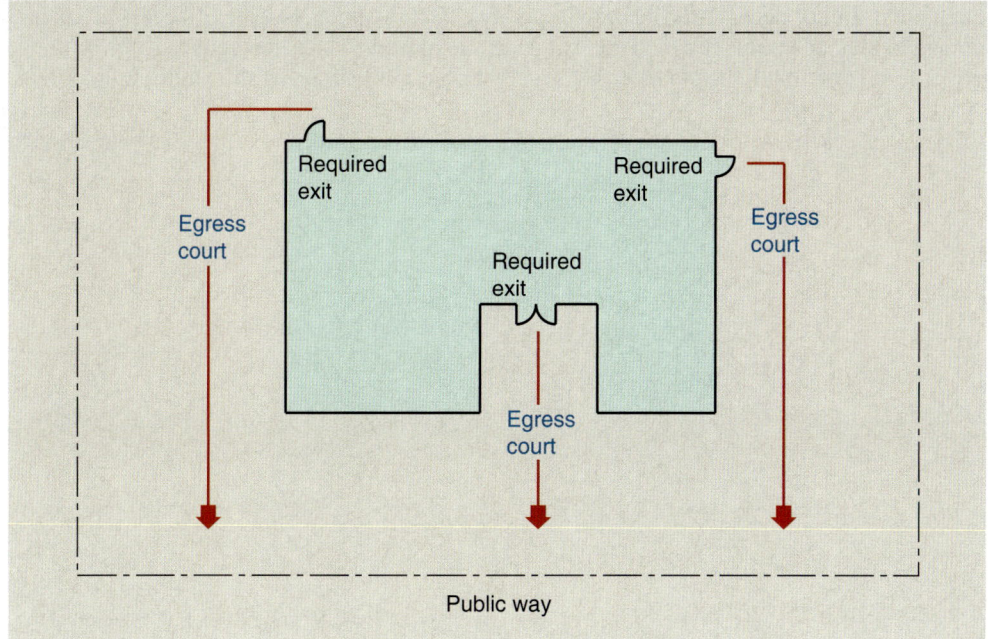

Figure 202-7
Egress court.

EMERGENCY ESCAPE AND RESCUE OPENING. Required in sleeping rooms and basements in limited residential occupancies, emergency escape and rescue openings are intended to allow for a secondary means of escape or rescue in the event of an emergency. Typically an operable window or door, such an opening is regulated by Section 1029 for minimum size, maximum height from the floor, and operational constraints. Although the provisions are found in Chapter 10 of the code, the intent of the opening is for emergency escape or rescue access, and it is not intended to be considered an element of a complying means of egress.

EXIT. An exit is the first portion of egress travel where the code assumes that the occupant has obtained an adequate level of safety so that travel distance measurements are no longer a concern. In addition, an exit typically provides only single-direction egress travel. The adequate level of safety is provided by building elements that completely separate the means of egress from other interior spaces in the building. Inside the building, fire-resistance-rated construction and opening protectives are utilized to provide a protective path of egress travel between the exit access and exit discharge portions of the means of egress. Building elements that are considered exits include exterior exit doors at the level of exit discharge, interior exit stairways and ramps, exit passageways, exterior exit stairs, exterior exit ramps, and horizontal exits.

EXIT ACCESS. The exit access is identified as the initial component of the means of egress system, the portion between any occupied point in a building and the exit. Leading to one or more of the defined exit components, the exit access makes up the vast majority of any building's floor area. Because the exit access begins at any point that may potentially be occupied, it is probable that only those concealed areas, such as penthouses, attics, and under-floor spaces that are typically unoccupied, fall outside of the definition of the means of egress.

2 Definitions

EXIT ACCESS DOORWAY. The term "exit access doorway" is commonly used in IBC Chapter 10 to establish a reference point within the exit access for applying various means of egress provisions, including those addressing arrangement, number, separation, opening protection, and exit sign placement. Although one would expect the term "doorway" to be limited to those situations where an actual door opening, either with or without a door, is present, the IBC definition expands this traditional meaning by including certain access points that do not necessarily include doorways, such as unenclosed exit access stairs and ramps. In fact, any point at which the exit access is narrowed so as to create a single point of travel could potentially be considered as an exit access doorway, as shown in Figure 202-8.

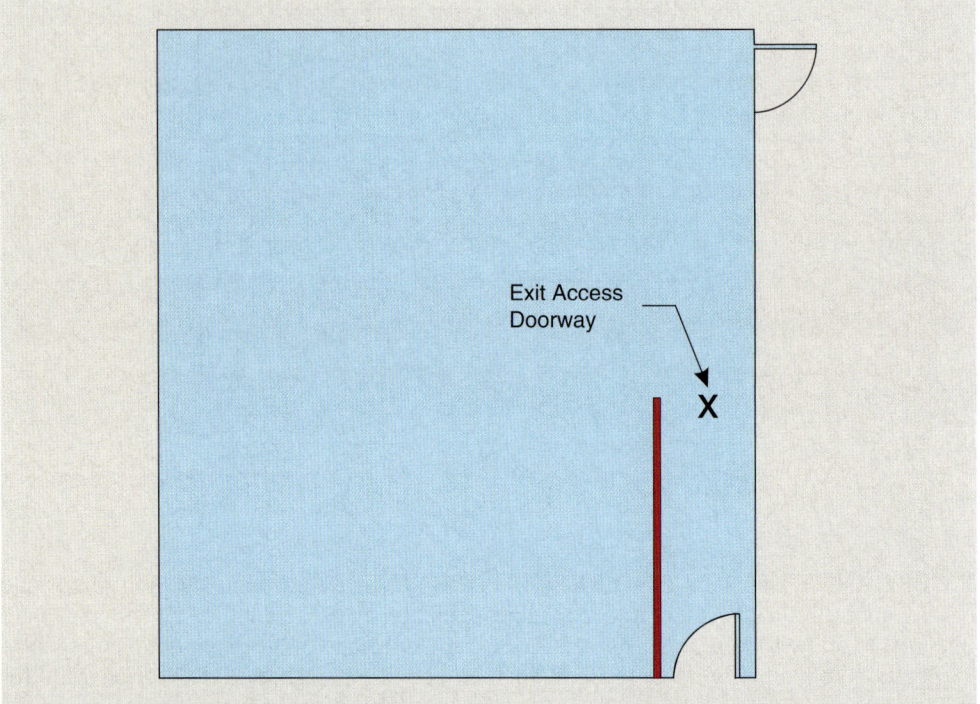

Figure 202-8
Exit access doorway.

EXIT ACCESS STAIRWAY. Stairways intended to serve as required exit components within the means of egress are considered as interior exit stairways. All other stairways are regulated as exit access stairways. As expected, an exit access stairway is considered an exit access component of the egress system. Exit access stairways, like most other exit access components, are not intended to provide any type of fire-resistive protection for the building occupants as they travel through the means of egress. Thus, most exit access stairways are permitted to be unenclosed under the allowances of Section 1009. However, since stairways create vertical openings between stories, enclosure may be required to restrict the vertical spread of fire, smoke, and gases.

EXIT DISCHARGE. Exit discharge is the last portion of the three-part means of egress system and is that portion between the point where occupants leave an exit and the point where they reach a public way. For conceptual ease, all exterior travel at ground level is considered a part of the exit discharge. An egress court is the primary component of exit discharge.

EXIT DISCHARGE, LEVEL OF. The story within a building where an exit terminates and the exit discharge begins creates a condition defined as the level of exit discharge. Because exit discharge occurs substantially at ground level, the level of exit discharge typically provides a horizontal path of travel toward the public way. At times, the code uses this term to define another floor level's relationship to egress at grade level. For example, in most buildings, the first story above the level of exit discharge is typically the building's second story above grade plane, as shown in Figure 202-9.

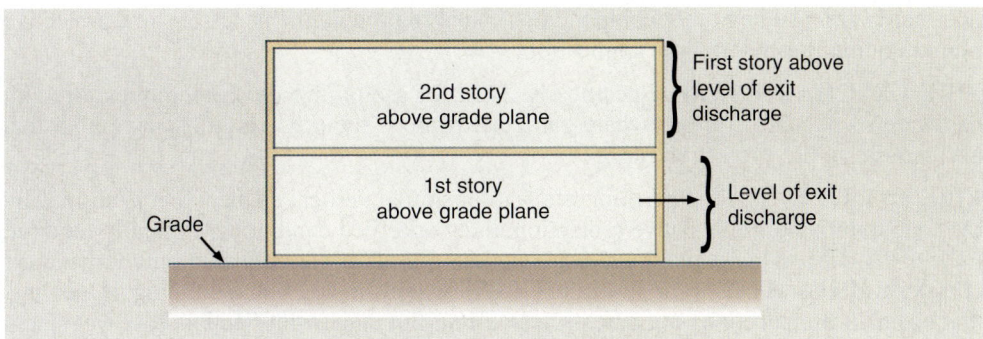

Figure 202-9
Level of exit discharge.

EXIT, HORIZONTAL. The horizontal exit is considered a component of the exit portion of the means of egress. The concept of a horizontal exit is to provide a refuge compartment adequately separated from fire, smoke, and gases generated in the area of a fire incident. The separation may occur from one building to another at approximately the same level, or more typically, through a fire-resistance-rated wall within a single building. As the occupants pass through the horizontal exit, they enter an area intended to afford safety from the fire and smoke of the area from which they departed.

EXIT PASSAGEWAY. Much like an interior exit stairway or ramp in providing smoke and fire protection, an exit passageway is an exit component that is separated from the remainder of the building by fire-resistance-rated construction and opening protectives. The horizontal path of protected travel afforded by an exit passageway shall extend to the exit discharge or the public way. An exit passageway is often utilized as the horizontal extension of egress travel connecting an interior exit stairway or ramp to the exterior of the building. Its function is to maintain a level of occupant protection equivalent to that of the interior exit stairway or ramp to which it is connected. An exit passageway is also to be used for protected horizontal travel where the continuity of an interior exit stairway or ramp enclosure must extend to another enclosure.

F RATING. Penetration firestop systems are provided with F ratings to indicate the time periods in which they resist the spread of fire through the penetration. Tested to the requirements of ASTM E 814 or UL 1479, penetration firestop systems for fire-resistance-rated assemblies will typically have an F rating of at least 1 hour.

FIRE AREA. Many of the provisions of Section 903 requiring the installation of automatic sprinkler systems utilize the fire area concept. A fire area is considered a compartment that will contain a fire such that the maximum fire size will be limited to the size of the compartment. An in-depth review of fire areas is found in the discussion of Section 901.7.

FIRE BARRIER. Fire-resistance-rated walls are considered fire barriers if constructed under the provisions of Section 707. The purpose of such assemblies is to create a barrier that will restrict fire spread to and from other portions of the building. All openings within a fire barrier must be protected with a fire-protective assembly. Fire barriers are often utilized to separate incompatible uses within the building, to create fire areas, or to provide for egress through a protected exit system.

FIRE DAMPER. Regulated by test standard UL 555, fire dampers are devices located in ducts and air-transfer openings to restrict the passage of flames. Fire dampers close automatically upon the detection of heat, maintaining the integrity of the fire-resistance-rated assembly that is penetrated. The actuation of fire dampers creates some restriction to airflow from migrating throughout the duct system or through transfer openings, although not to the level of that required for a smoke damper.

Definitions

FIRE DOOR ASSEMBLY. Where openings occur in fire-resistance-rated assemblies, fire door assemblies are permitted as a method to protect the openings. Addressed in Section 716, fire door assemblies include not only the door but also the door frame, the door hardware, and any other components needed to provide the necessary fire-protective rating required for the specific application.

FIRE EXIT HARDWARE. Specifically listed for use on fire door assemblies, fire exit hardware is mandated for use where panic hardware is required on a door assembly that also requires a fire-protection rating.

FIRE PARTITION. A fire partition is a wall or similar vertical element that is utilized by the code to provide fire-resistive protection under specified conditions. Typically required to be of 1-hour fire-resistance-rated construction, fire partitions are commonly utilized for corridor walls, as well as walls separating dwelling units in apartment buildings. Openings that occur in fire partitions must be protected in accordance with Section 716. Regulated in Section 708, a fire partition is considered a lower type of fire-resistance-rated assembly than a fire barrier; thus, it is not permitted as an enclosure element for defining a fire area.

FIRE-PROTECTION RATING. Opening protectives, such as fire doors, fire windows, and fire dampers, are assigned a fire-protection rating in order to identify the time period in which the protective is expected to confine fire spread. The specific rating, in either hours or minutes, varies based on the details of the fire-resistance-rated assembly in which the opening protective is located.

FIRE-RATED GLAZING. Fire separation elements such as fire barriers and fire walls will often include glazing in some form, such as glazed wall assemblies, fire windows, and/or fire doors with vision panels. The definition of "fire-rated glazing" encompasses both types of such glazing addressed by the code, fire-resistance-rated glazing, and fire-protection-rated glazing. Fire-resistance-rated glazing, addressed in Sections 703.5 and 716.2, must be tested in accordance with ASTM E119 or UL 263 as a wall assembly. Fire-protection-rated glazing, established for use by Sections 716.5.8 and 716.6, is to be tested in accordance with NFPA 257 or UL 9 as an opening protective. Both types of glazing are collectively referred to as fire-rated glazing.

FIRE-RESISTANCE RATING. Identified by a specific time period, a fire-resistance rating is assigned to a tested component or assembly based on its ability to perform under fire conditions. A fire-resistance-rated component or assembly is intended to restrict the spread of fire from a specified area or provide the necessary protection for the continued performance of a structural member. This performance is based on fire resistance, defined as the property of materials or assemblies that prevents or retards the passage of excessive heat, hot gases, or flames.

FIRE-RESISTANT JOINT SYSTEM. Where a linear opening is placed in or between adjacent fire-resistance-rated assemblies to allow independent movement of the assemblies, it is considered by the IBC as a joint. One or more joints may be provided to address movement caused by thermal, seismic, wind, or any other similar loading method. Designed to protect a potential breach in the integrity of a fire-resistance-rated horizontal or vertical assembly, a fire-resistant joint system is a tested assembly of specific materials designed to restrict the passage of fire through joints. The fire-resistance ratings required for the joint systems, as well as other requirements, are addressed in Section 715.

FIRE SEPARATION DISTANCE. The fire separation distance describes that distance between the exterior surface of a building and one of three locations—the nearest interior lot line; the centerline of a street, alley, or other public way; or an imaginary line placed between two buildings on the same lot. The method of measurement is based on the distance as measured perpendicular to the face of the building. See Figure 202-10. The fire separation distance is important in the determination of exterior wall and opening protection based on the proximity to the lot lines. See the discussion in Table 602 for a further analysis of this subject.

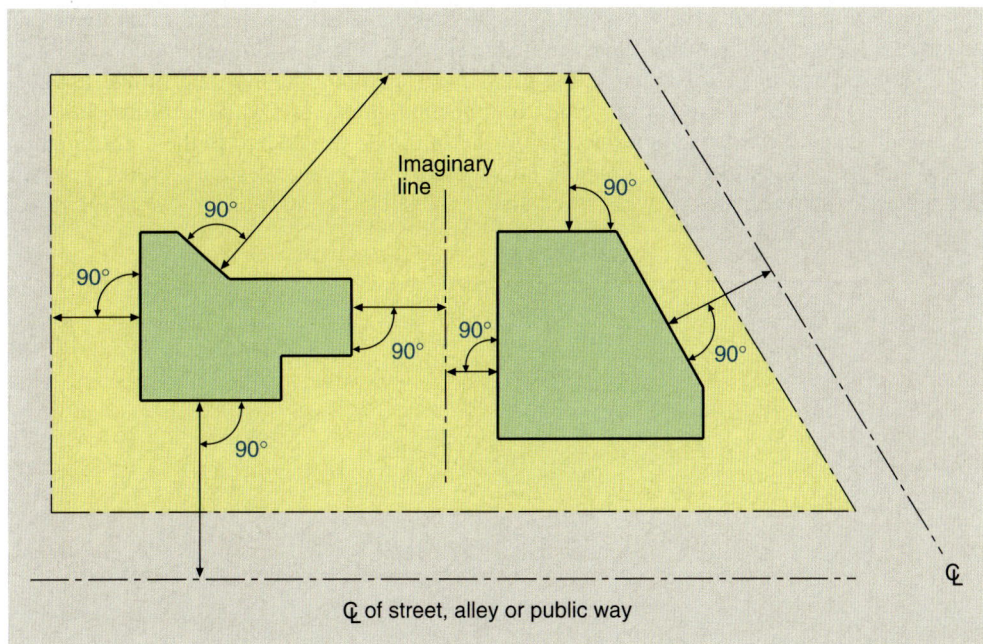

Figure 202-10
Fire separation distance.

FIRE WALL. Fully addressed in Section 706, one or more fire walls are building elements used to divide a single building into two or more buildings for the purpose of applying the IBC. Starting at the foundation and continuing vertically to or through the roof, a fire wall is intended to fully restrict the spread of fire from one side of the wall to the other. Fire walls are higher-level fire-resistance-rated elements than both fire barriers and fire partitions. Because the concept of fire walls is to create smaller buildings within one larger structure, with the code regulating each small building individually rather than collectively, it is critical that a fire wall be capable of maintaining structural stability under fire conditions. If construction on either side of a fire wall should collapse, such a failure should not cause the fire wall to collapse for the prescribed time period of the rating of the wall.

FIRE WINDOW ASSEMBLY. Consistent with the purpose of a fire door assembly, a fire window assembly provides protection against the spread of fire through a glazed opening.

FIREBLOCKING. Fireblocking is mandated by the code to address the spread of fire through concealed spaces of combustible construction. Experience has shown that the greatest damage occurs to conventional wood-frame buildings during a fire when the fire travels unimpeded through concealed draft openings. Materials identified by the IBC as effective in resisting fire spread through concealed spaces include 2-inch (51-mm) nominal lumber, gypsum board, and glass-fiber batts.

FLIGHT. The use of the term "flight" is specifically defined in order to establish its use within the code. The definition addresses two separate issues. First, a flight is made up of the treads and risers that occur between landings. As an example, a stairway connecting two stories that includes an intermediate landing consists of two flights. Second, the inclusion of winders within a stairway does not create multiple flights. Winders are simply treads within a flight and are often combined with rectangular treads within the same flight.

FLOOR AREA, GROSS. As evidenced in Table 1004.1.2, the determination of the occupant load in the design of the means of egress system for most building uses is typically based on the gross floor area. This term describes the total floor area included within the surrounding exterior walls of a building, and the definition further states that the floor area of the building or portion thereof not provided with surrounding exterior walls shall be the usable area under the horizontal projection of the roof or floor above. The intent of this latter provision is to cover where a structure may not have exterior walls or may have one or more sides open without an enclosing exterior wall. Where buildings are composed of both enclosed and unenclosed areas, the gross floor area is typically determined as illustrated in Figure 202-11. Projections extending beyond an exterior wall or column line that are not intended to create usable space below are not to be considered in the determination of gross floor area. Areas often considered accessory-type spaces, such as closets, corridors, elevator shafts, and stairways, must also be considered a part of the gross floor area.

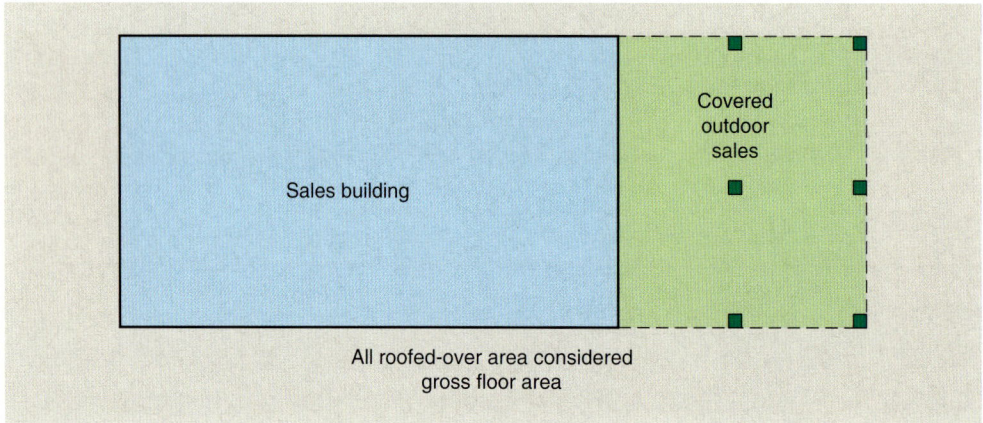

Figure 202-11
Gross floor area.

FLOOR AREA, NET. The net floor area is considered that portion of the gross floor area that is typically occupied. Normally unoccupied accessory areas such as corridors, stairways, closets, toilet rooms, equipment rooms, and similar spaces are not to be included in the calculation of net floor area. In addition, the measurements are based on clear floor space, allowing for the deduction of building construction features such as interior walls and columns, as well as elevator shafts and plumbing chases. The use of net floor area in the calculation of design occupant load is typically permitted only in assembly and educational uses as set forth in Table 1004.1.2. It is important to note that in calculating net floor area, as well as gross floor area, the floor space occupied by furniture, fixtures, and equipment is not to be excluded in the calculation. The floor-area-per-occupant factor established in Table 1004.1.2 includes any such anticipated furnishings in the establishment of an appropriate density estimate.

FLOOR FIRE DOOR ASSEMBLY. Although a fire door is typically viewed as an element protecting an opening in a vertical building element such as a wall, it is possible that such doors can be effective if installed horizontally for the protection of an opening in a fire-resistance-rated floor. The floor fire door assembly, like other fire door assemblies, includes the door, frame, hardware, and other accessories that make up the assembly, and provides a specified level of fire protection for the opening.

FOAM PLASTIC INSULATION. Considered to be an expanded plastic produced through use of a foaming agent, this reduced-density plastic contains voids consisting of open or closed cells distributed throughout the plastic, providing for thermal insulation or acoustic control. The density of the material is to be less than 20 pounds per cubic foot (320 kg/m^3).

FOLDING AND TELESCOPING SEATING. Folding and telescoping seats are structures that provide tiered seating, which can be reduced in size and moved without dismantling. Utilized quite often in school gymnasiums, such seating presents the same concerns and risks as permanently installed bleacher seating when occupied. Such seating is regulated by ICC 300, *Standard for Bleachers, Folding and Telescopic Seating, and Grandstands*.

GRADE PLANE. The code indicates that the grade plane is a reference plane representing the average of the finished ground level adjoining the building at its exterior walls. Under conditions where the finished ground level slopes significantly away from the exterior walls, that reference plane is established by the lowest points of elevation of the finished surface of the ground within an area between the building and lot line, or where the lot line is more than 6 feet (1829 mm) from the building, between the building and a line 6 feet (1829 mm) from the building. Where the slope away from the building is minimal (typically provided only to drain water away from the exterior wall), the elevation at the exterior wall provides an adequate reference point.

The method for calculating grade plane can vary based on the site conditions. Where the slope is generally consistent as it passes across the building site, it may only require the averaging of a few points along the exterior wall of a rectangular-shaped building, as illustrated in Figure 202-12. Where the slope is inconsistent or retaining walls are utilized, or where the building footprint is complex, the determination of grade plane can be more complicated. In such cases, a more exacting method for calculating the grade plane must be utilized. In addition, where fire walls are present, the elevation points should be taken at the intersections of the fire wall and the exterior walls.

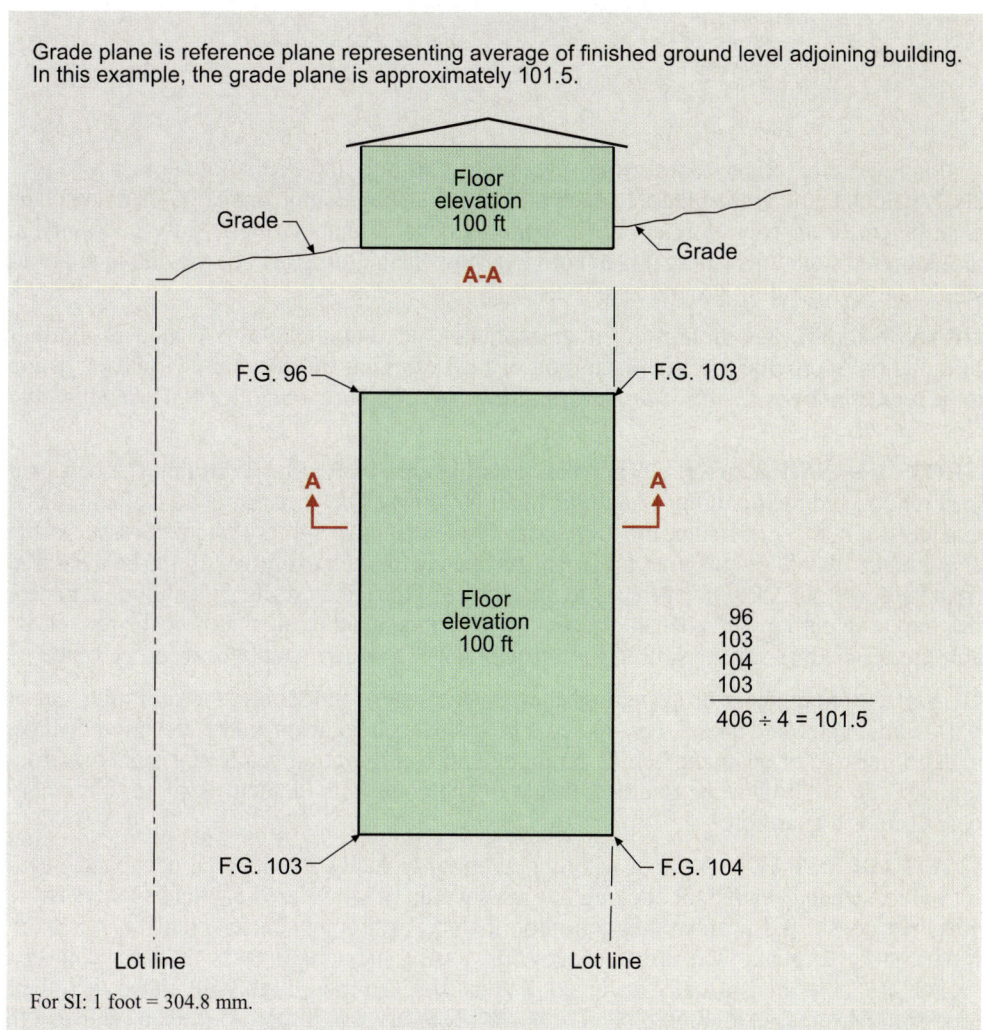

Figure 202-12
Grade plane calculation.

This definition is important in determining the number of stories above grade plane within a building as well as its height in feet. In some cases, the finished surface of the ground may be artificially raised with imported fill to create a higher grade plane around a building so as to decrease the number of stories or height in feet. The code does not prohibit this practice, and as long as a building meets the code definition and restriction for height or number of stories, the intent of the code is met. See Figure 202-13.

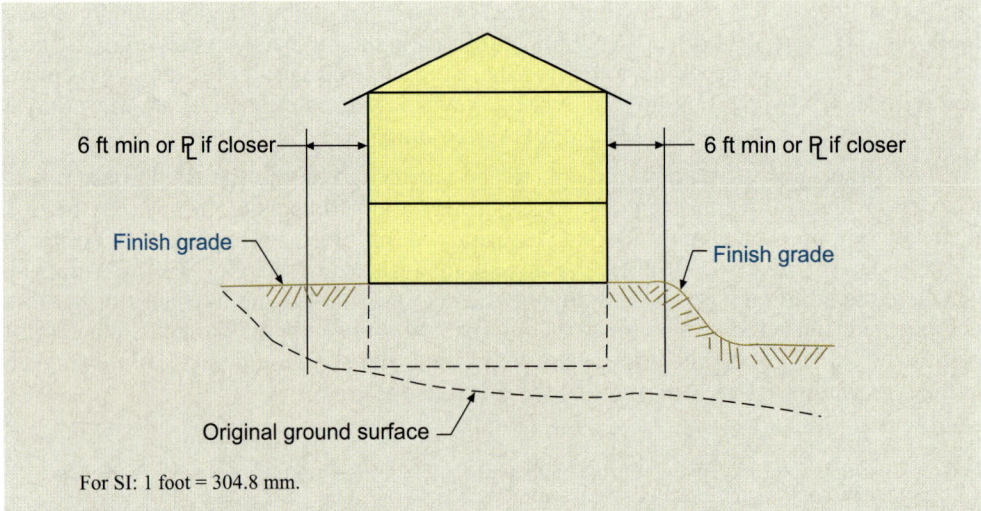

Figure 202-13
Use of built-up soil to raise finished grade.

It is important to note that for the vast majority of buildings, it is not necessary to precisely calculate the grade plane. In such buildings, a general approximation of grade plane is sufficient to appropriately apply the code. A detailed calculation is only necessary in those limited situations where it is not obvious how the building is to be viewed in relationship to the surrounding ground level.

GRANDSTAND. The definition of grandstand is also applicable to bleachers. Further information is provided in the discussion of the definition of bleachers. Grandstands are to be regulated by ICC 300, *Standard for Bleachers, Folding and Telescopic Seating, and Grandstands*.

GROSS LEASABLE AREA. The gross leasable area is used to determine the occupant load for covered mall buildings and open mall buildings. The occupant load is then utilized to address a number of issues, most importantly the means of egress system. Gross leasable area includes that floor area used for tenant occupancy for their exclusive use, to be measured from the center line of joint partitions to the outside of the tenant walls. It is important to note that the entire portion of the tenant space is to be included in the calculation of the gross leasable area, including any areas used for tenant storage, restroom facilities, or office spaces.

GUARD. A component or system of components whose function is the minimization of falls from an elevated area is considered a guard. Placed adjacent to the elevation change, a guard must be of adequate height, strength, and configuration to prevent someone from falling over or through the guard. Outside of the code, this element is more commonly described as a guardrail.

HABITABLE SPACE. An area within a building, typically a residential occupancy, used for living, sleeping, eating, or cooking purposes would be considered habitable space. Those areas not considered to meet this definition include bathrooms, closets, hallways, laundry rooms, storage rooms, and utility spaces. Obviously, habitable spaces as defined in this section are those areas usually occupied, and as such are more highly regulated than their accessory use areas. Although typical, it is not necessary that a room or area be finished in order to be considered habitable space. It is not uncommon for a dwelling unit to have a large basement that is not completely finished-out. Nevertheless, the basement may be used as living space, particularly for children who use it as a playroom. Such a basement would be considered habitable space, as the definition is simply based on the use of the room or area.

HANDRAIL. Typically used in conjunction with a ramp or stairway, a handrail is intended to provide support for the user along the travel path. A handrail may also be used as a guide to direct the user in a specified direction.

Definitions 2

HEIGHT, BUILDING. Once the elevation of the grade plane has been calculated, it is possible to determine the building's height. This height is measured vertically from the grade plane to the average height of the highest roof surface. Examples of this measurement are shown in Figure 202-14.

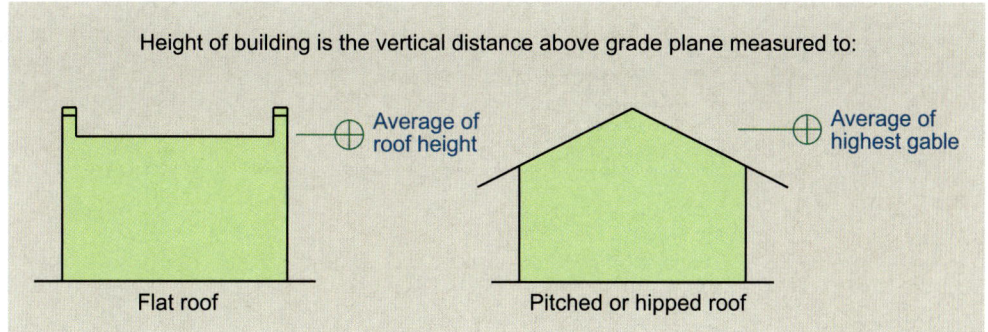

Figure 202-14
Height of building.

Where the building is stepped or terraced, it is logical that the height is the maximum height of any segment of the building. It may be appropriate under certain circumstances that the number of stories in a building be determined in the same manner. Because of the varying requirements of the code that are related to the number of stories, such as means of egress, type of construction, fire resistance of shaft enclosures, and so on, each case should be judged individually based on the characteristics of the site and construction. In addition to those factors better related to the number of stories, other items to consider are fire department access, location of exterior exit doors, routes of exit travel, and types of separation between segments.

Figure 202-15 illustrates one example in which the height of the building and number of stories are determined for a stepped or terraced building. In the case of a stepped or terraced building, the language *total perimeter* is used to define the situation separating

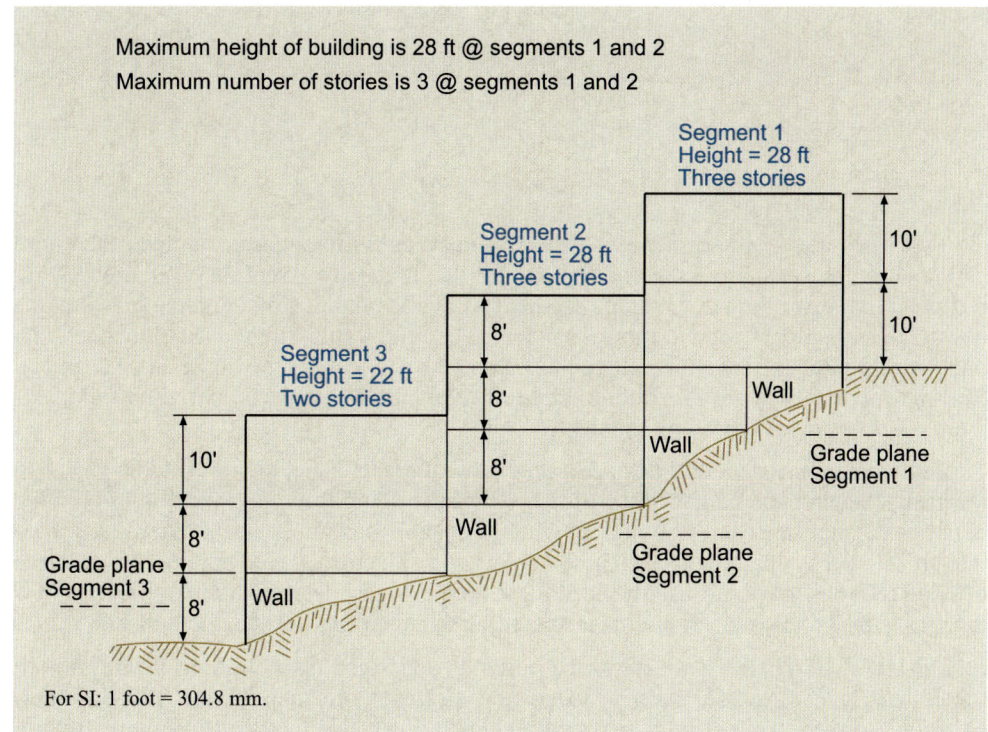

Figure 202-15
Terraced building.

2012 International Building Code Handbook 35

2 Definitions

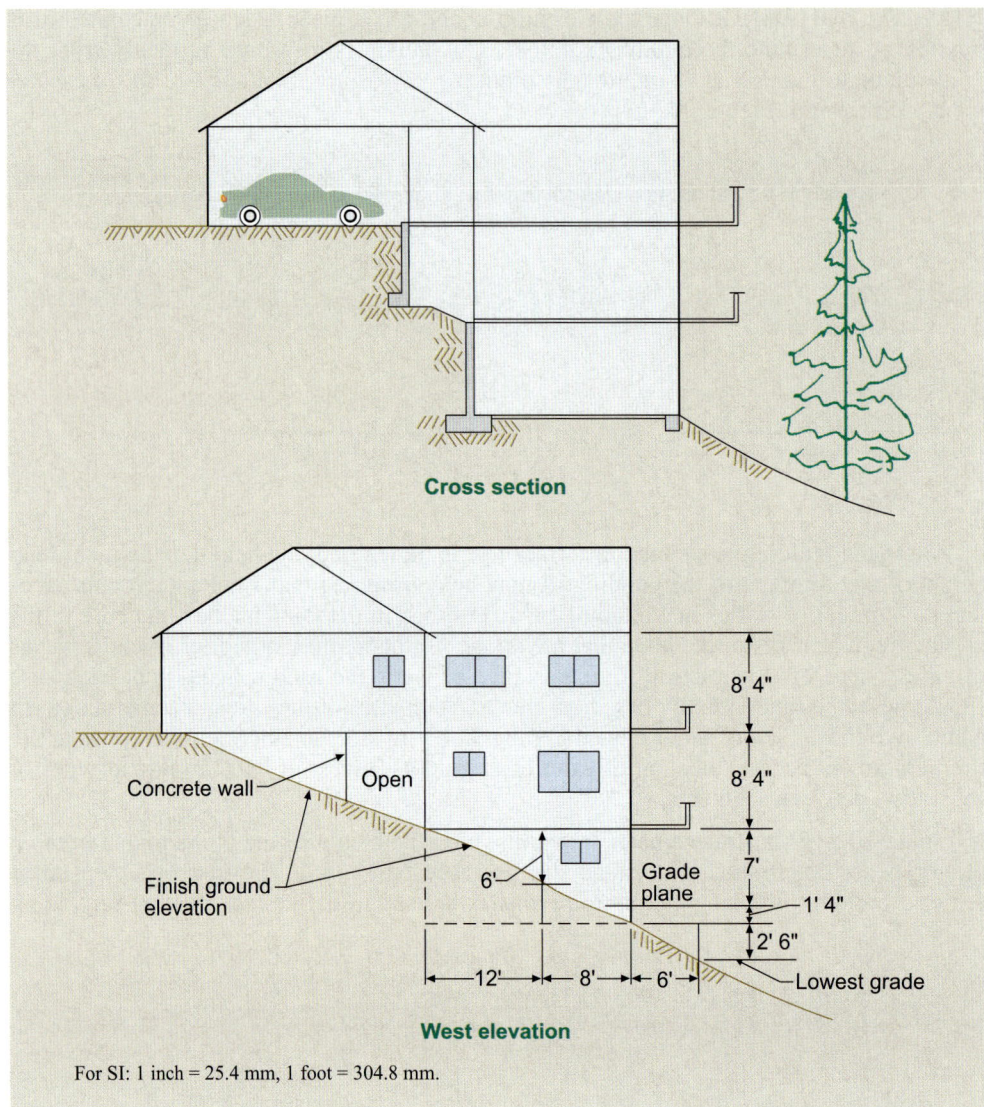

**Figure 202-16
Three-story building.**

the first story above grade plane from a basement and is intended to include the entire perimeter of the segment of the building. Therefore, in the cross section of Figure 202-16, the total perimeter of the down-slope segment would be bounded by the retaining wall, the down-slope exterior wall, and the east and west exterior walls. In the case illustrated, the building has three stories above grade plane and no basement for the down-slope segment. The measurement for the maximum height of the building would be based on the maximum height of the down-slope segment.

Similar to an unnecessarily detailed calculation of grade plane, there is seldom a need to precisely calculate the height of a building. Typically, a general determination of building height is adequate to ensure compliance with the code. For example, it is not necessary to go into great detail evaluating the average roof elevation of a built-up roof that has a low degree of slope for drainage purposes. The need for a more exacting determination of roof height is directly related to any uncertainty that may occur in reviewing for code compliance.

HIGH-RISE BUILDING. A high-rise building is defined as a building having one or more floor levels used for human occupancy located more than 75 feet (22 860 mm) above the lowest level of fire department vehicle access, as illustrated in Figure 202-17.

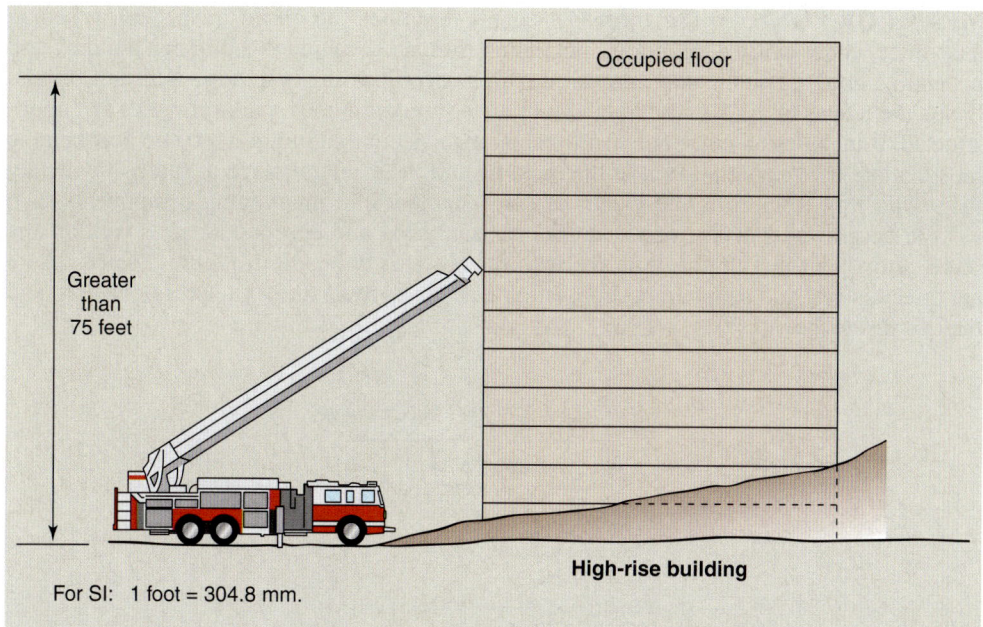

Figure 202-17
Definition of high-rise building.

Most moderately large and larger cities have apparatus that can fight fires up to about 75 feet (22 860 mm); thus, the fire can be fought from the exterior. Any fires above this height will require that they be fought internally. Also, in some circles, 75 feet (22 860 mm) is considered to be about the maximum height for a building that could be completely evacuated within a reasonable period of time. Thus, the fire department's capability plus the time for evacuation of the occupants constitute the criteria used by the IBC for defining a high-rise building.

The determining measurement is to be taken from the lowest point at which the fire department will locate their fire apparatus to the highest floor level that is viewed as occupiable. This would include a mezzanine floor level that may occur within the highest story, as well as an occupied roof (such as a tennis court, swimming pool, or sun deck on the roof of a condominium building).

HORIZONTAL ASSEMBLY. A horizontal assembly is the horizontal equivalent of a fire barrier. It is utilized to restrict vertical fire spread through an established degree of fire resistance as mandated through various provisions in the IBC. The specifics for horizontal assemblies, which include both floor and roof assemblies, are found in Section 711.

INTERIOR EXIT STAIRWAY. One of several exit components established in the code, an interior exit stairway provides a fire-resistance-rated enclosure for vertical egress in buildings. Section 1022 identifies the requirements for an interior exit stairway, including its construction and termination; permissible openings, penetrations, and ventilation; and necessary signage. One or more interior exit stairways are typically required in multi-story buildings in order to satisfy the various means of egress requirements in Chapter 10.

L RATING. An L rating is used in the evaluation of air leakage at penetrations and joints in smoke barriers. The IBC requires that penetrations and joints in those walls and floors that are intended to provide smoke separations be provided with a complying L rating as indicated in Sections 714.5 and 715.6.

LIGHT-TRANSMITTING PLASTIC WALL AND ROOF PANELS. This definition addresses materials that are fastened to structural members, or to structural panels or sheathing, and that are used to transmit light at the exterior walls or the roof.

Definitions

MEANS OF EGRESS. The means of egress describes the entire travel path a person encounters under exiting conditions, beginning from any occupiable point in a building and not ending until the public way is reached. Often encompassing both horizontal and vertical travel, the means of egress should be direct, obvious, continuous, undiminished, and unobstructed. It includes all components of the exiting system that might intervene between the most remote occupiable portion of the building and the eventual place of safety—typically the public way. Therefore, the means of egress includes all intervening components such as aisle accessways, aisles, doors, corridors, stairways, and egress courts, as well as any other component that might be in the path of travel, as depicted in Figure 202-18. There are three distinct and separate portions of a means of egress—the exit access, the exit, and the exit discharge.

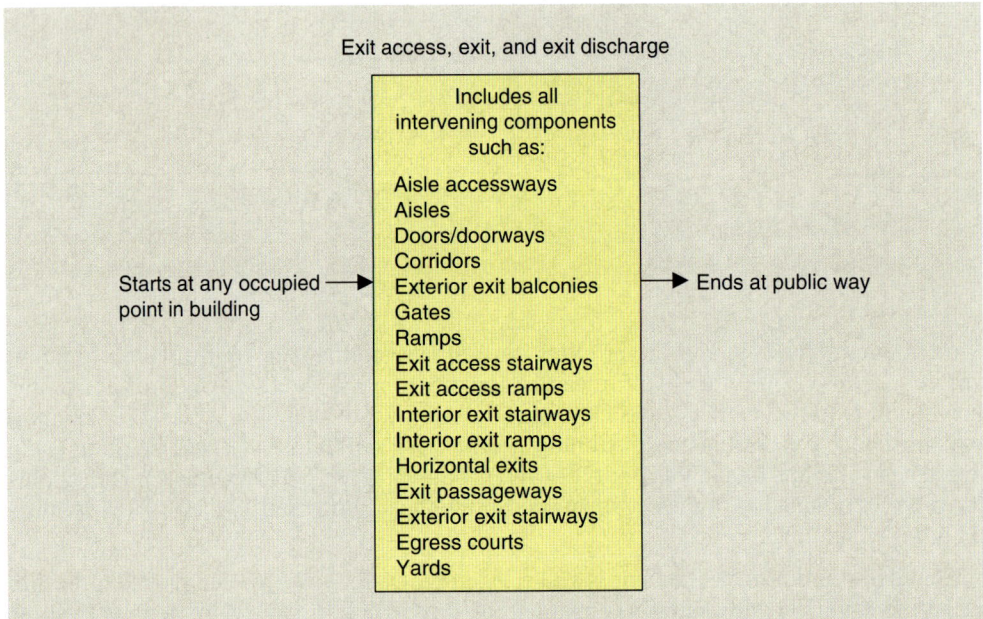

**Figure 202-18
Means of egress.**

MEMBRANE PENETRATION. Similar to a through penetration in its performance requirements, a membrane penetration is an opening through only one membrane of a wall, ceiling, or floor. An example of a very common membrane penetration is an electrical box.

MEMBRANE-PENETRATION FIRESTOP. Where a penetrating item such as a pipe or conduit passes through a single membrane of a fire-resistance-rated wall, floor/ceiling, or roof/ceiling assembly, a membrane-penetration firestop may be required in order to adequately protect the penetration. Such a firestop consists of a device or construction that would effectively resist the passage of flame and heat through the opening in the membrane created by the penetrating item. A fire-resistance rating is assigned to a membrane-penetration firestop to indicate the time period for which the firestop is listed.

MERCHANDISE PAD. A merchandise pad is created where racks, displays, shelving units, and similar fixtures are grouped into a specific area. Aisle accessways are provided within the merchandise pad to allow for customer circulation and are utilized as exit access elements. Such aisle accessways connect to the aisles that partially or entirely surround the merchandise pad. An example is shown in Figure 202-19.

MEZZANINE. A mezzanine is merely a code term for an intermediate floor level placed within a room between the floor and ceiling of a story. Typically limited in floor area to $33^1/_3$ percent of the area of the room or space into which it is located, a mezzanine is regulated under the provisions of Section 505. A floor level fully complying with the provisions

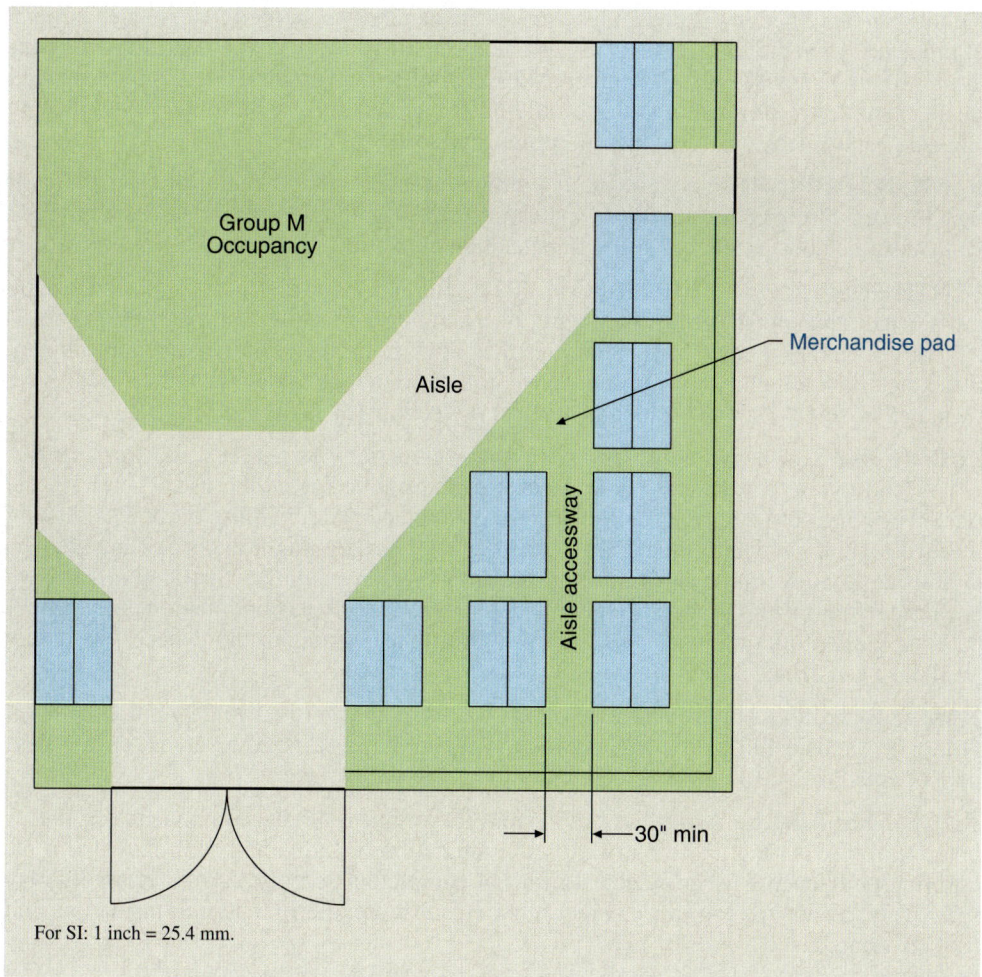

Figure 202-19
Merchandise pad.

of Section 505 is permitted to be considered a mezzanine in order to utilize those special allowances applicable to such a condition. The use of the mezzanine provisions is a design option, as an elevated floor level complying with the scoping provisions of Section 505 is always permitted to be considered as an additional story rather than a mezzanine.

NOSING. The leading edge portions of stair treads are considered nosings. Nosings may also be found where a landing is the top step of a stairway run.

OCCUPANT LOAD. Viewed as the basis for the design of the means of egress system, the occupant load is that number of persons considered for means of egress purposes both within and from any space, area, room, or building. Variations in code requirements are often caused based on the anticipated occupant load served by the specific building egress element. Although the establishment of an occupant load is critical in the application of many means of egress provisions located in Chapter 10, it is also often necessary when addressing minimum requirements for fire alarm systems, sprinkler systems in Group A occupancies, and plumbing fixture counts.

OCCUPIABLE SPACE. A number of provisions in the code apply only to those spaces, rooms, or areas typically occupied during the course of a building's use. This definition clarifies that an occupiable space is intended for human occupancy, and as such is provided with means of egress, as well as light and ventilation facilities. Occasionally the code refers to the term *normally occupied* space. For example, Section 1022.4 limits openings in interior exit stairways to those necessary for exit access from *normally occupied spaces*.

Although not defined in the code, these spaces are generally occupied for extended periods of time during the building's use. There is an expectation that a fire or other hazardous condition would be quickly identified and addressed, rather than go unnoticed for an extended time. Examples of those spaces that would not be considered *normally occupied* include storage rooms, mechanical equipment rooms, and toilet rooms.

PANIC HARDWARE. The term *panic hardware* describes an unlatching device that will operate even during panic situations, so that the force of individuals against the egress door will cause the door to unlatch without manual manipulation of the device.

PENTHOUSE. Regulated by Section 1509.2, penthouses are structures placed on the roofs of buildings to shelter various types of machinery and equipment, as wells as vertical shaft openings. The definition clarifies that a penthouse is intended to be unoccupied. Structures such as tanks, towers, spires, and domes are not considered penthouses, but are regulated to some degree by other provisions in Section 1509.

PUBLIC WAY. The code defines a public way essentially as a street, alley, or any parcel of land that is permanently appropriated to the public for public use. Therefore, the public's right to use such a parcel of land is guaranteed. The building occupants, having reached a public way, are literally free to go wherever they might choose. They are certainly free to go so far as to escape any fire threats in any building that they might have been occupying. There is an expectation that the public way be a continuation of the egress path, providing for continued exit travel away from the building until the necessary safety level has been achieved.

RAMP. Where the slope of a travel path exceeds 1 unit vertical to 20 units horizontal (1:20), it is considered a ramp. Where the travel path has a slope of 5 percent or less (less than or equal to 1:20), it is considered merely a walking surface.

ROOF ASSEMBLY. A roof assembly may be either a single component serving as both the roof covering and the deck, or a combination of individual roof deck and roof-covering components used together to form a complete assembly. A roof assembly may selectively include the roof deck, vapor retarder, substrate, thermal barrier, insulation, and roof covering.

ROOF COVERING. The roof covering is considered the covering placed upon a roof to provide the building with weather protection, fire retardancy, or decoration.

SCISSOR STAIR. A unique design element, the scissor stair allows for two independent paths of travel within a single interior exit stairway. A scissor stair is considered a single means of egress because the exit paths are not isolated from each other with the required level of fire resistance.

SELF-CLOSING. In order to eliminate a portion of the human element in maintaining the integrity of fire-resistance-rated wall assemblies, doors in the assemblies are typically required to be provided with self-closing devices that will ensure closing of the doors after having been opened. Occasionally, automatic-closing fire doors are installed in specific locations on account of the nature of the situation. Under such conditions, the automatic-closing fire assemblies are to be regulated by NFPA 80.

SHAFT. A shaft is considered the enclosed space that extends through one or more stories of a building. Its function is to connect vertical openings in successive floors that have been created to accommodate elevators, dumbwaiters, mechanical equipment, or similar devices, as well as for the transmission of natural light or ventilation air.

SHAFT ENCLOSURE. A shaft enclosure is the building element defined by the boundaries of a shaft, which typically includes its surrounding walls and other forms of construction. Regulated by Section 713, it is required to be of fire-resistance-rated construction. A shaft enclosure is the most common application listed in Section 712.1 for addressing vertical openings in buildings.

SLEEPING UNIT. The single required characteristic of a sleeping unit is that it is used as the primary location for sleeping purposes. The room or space that has sleeping facilities may also provide for eating and living activities. It could have a bathroom or a kitchen but not both, as this would qualify it as a dwelling unit. Guestrooms of Group R-1 hotels and motels would typically be considered sleeping units. Sleeping units are also commonly found in congregate living facilities, such as dormitories, sorority houses, and fraternity houses, and are regulated as Group R-2 occupancies.

Group R occupancies are not the only types of uses where sleeping units are located. Several of the varied uses classified as Group I occupancies also contain resident or patient sleeping units. The proper designation of these spaces as sleeping units is important in the application of Section 420 mandating the separation of sleeping units in Group R and I-1 occupancies, as well as addressing the appropriate accessibility provisions of Chapter 11.

SMOKE BARRIER. Required under various circumstances identified by the code, smoke barriers are either vertical or horizontal membranes, or a combination of both, intended to restrict the movement of smoke. Walls, floors, and ceiling assemblies may be considered smoke barriers where they are designed and constructed in accordance with the provisions of Section 709.

SMOKE COMPARTMENT. Where smoke barriers totally enclose a portion of a building, the enclosed area is considered a smoke compartment. By completely isolating the compartment from the remainder of the building by walls, floors, and similar elements, smoke can be either contained within the originating area or prevented from entering other areas of the building. The use of smoke compartments is predominant in Group I-2 and I-3 occupancies.

SMOKE DAMPER. Test standard UL 555S states that leakage-rated dampers (smoke dampers in the terminology of the IBC) are intended to restrict the spread of smoke in heating, ventilating, and air-conditioning (HVAC) systems that are designed to automatically shut down in the event of a fire, or to control the movement of smoke within a building when the HVAC system is operational in engineered smoke-control systems. The IBC simply identifies smoke dampers as listed devices designed to resist the passage of air and smoke through ducts and air-transfer openings. Smoke dampers must operate automatically unless manual control is desired from a remote command station.

SMOKE-PROTECTED ASSEMBLY SEATING. Where the means of egress for assembly seating areas is designed to be relatively free of the accumulation of smoke, the seating is considered to be smoke protected. In order to qualify as smoke protected, the seating area and its exiting system must comply with the provisions of Section 1028.6.2, which addresses the methods of smoke control, the minimum roof height, and the possible installation of sprinklers in adjacent enclosed spaces. Exterior seating facilities such as stadiums or amphitheaters are commonly considered to have smoke-protected assembly seating due to the natural ventilation that is available.

STAIR. Where one or more risers are provided to address a change in elevation, a stair is created. A stair may simply be a slight change in height from one floor level to another, commonly referred to as a step, or may be a series of treads and risers connecting one floor or landing to another. Also described in the code as a flight of stairs, a stair does not include the landings and floor levels that interrupt stairway travel.

STAIRWAY. Where one or more flights of stairs occur, including any intermediate landings that connect the stair flights, a stairway is created. The term *stairway* describes the entire vertical travel element that is made up of stairs, landings, and platforms.

STAIRWAY, EXTERIOR. To be classified as an exterior stairway, it must be open on at least one side. The open side must then adjoin an open area such as a yard, egress court, or public way. By limiting the number of enclosed sides, an exterior stairway will be sufficiently open to the exterior to prevent the accumulation of smoke and toxic gases. Additional criteria for defining an exterior stairway used as a means of egress are found in Section 1026.3.

2 Definitions

STAIRWAY, INTERIOR. By definition, a stairway that does not comply with the definition for an exterior stairway is considered an interior stairway. In other words, if all sides of a stairway are enclosed by the building's construction, it is considered interior. Stairways that fail to meet the openness criteria of Section 1026.3 are, by default, considered interior stairways.

STAIRWAY, SPIRAL. A spiral stairway is a stairway configuration where the treads radiate from a central pole. The treads are uniform in shape, with a tread length that varies significantly from the inside of the tread to the outside. The dimensional characteristics of a spiral stairway cause it to be limited in its application.

STORY. Although seemingly quite obvious, the definition of a story is that portion of a building from a floor surface to the floor surface or roof above. In the case of the topmost story, the height of a story is measured from the floor surface to the top of the ceiling joists, or to the top of the roof rafters where a ceiling is not present. The critical part of the definition of a story involves the definition of *story above grade plane* as described in the following discussion.

It is not uncommon for a roof level to be utilized for purposes other than weather protection or mechanical equipment. A roof patio, garden, or sports area is sometimes provided in order to utilize as much of the building as possible. Although an occupied roof does not meet the definition of "story," there are certain provisions in the IBC that would be applicable due to the fact occupants can be present. For example, an occupied roof must be provided with a complying means of egress designed for the anticipated occupant load of the roof level. Required fire alarm system protection should also be extended to such occupied roofs. However, the roof level would not be considered as an additional story above grade plane for the purpose of allowable building height, nor would it be considered part of the building area for allowable area purposes. A careful analysis should be made when determining which provisions are applicable to an occupied roof.

STORY ABOVE GRADE PLANE. Throughout the code, the number of qualifying stories in a building is a contributing factor to the proper application of the provisions. As an example, a building's allowable types of construction are based partly on the limits in story height placed on various occupancy groups. In this case, the code is limiting construction type based on the number of stories above grade plane. The code defines a story above grade plane as any story having its finished floor surface entirely above grade plane. However, floor levels partially below the grade at the building's exterior may also fall under this terminology. The critical part of the definition involves whether or not a floor level located partially below grade is to be considered a story above grade plane. There are two criteria that are important to the determination if a given floor level is to be considered a story above grade plane:

1. If the finished floor level above the level under consideration is more than 6 feet (1829 mm) above the grade plane as defined in Section 502.1, the level under consideration is a story above grade plane, or
2. If the finished floor level above the level under consideration is more than 12 feet (3658 mm) above the finished ground level at any point, the floor level under consideration shall be considered a story above grade plane.

Where either one of these two conditions exists, the level under consideration is to be considered a story above grade plane.

Conversely, if the finished floor level above the level under consideration is 6 feet (1829 mm) or less above the grade plane, and does not exceed 12 feet (3658 mm) at any point, the floor level under consideration is not considered a story above grade plane. By definition, it is regulated as a basement. Figures 202-20 and 202-21 illustrate the definitions of "Story," "Basement," and "Story above grade plane."

Definitions 2

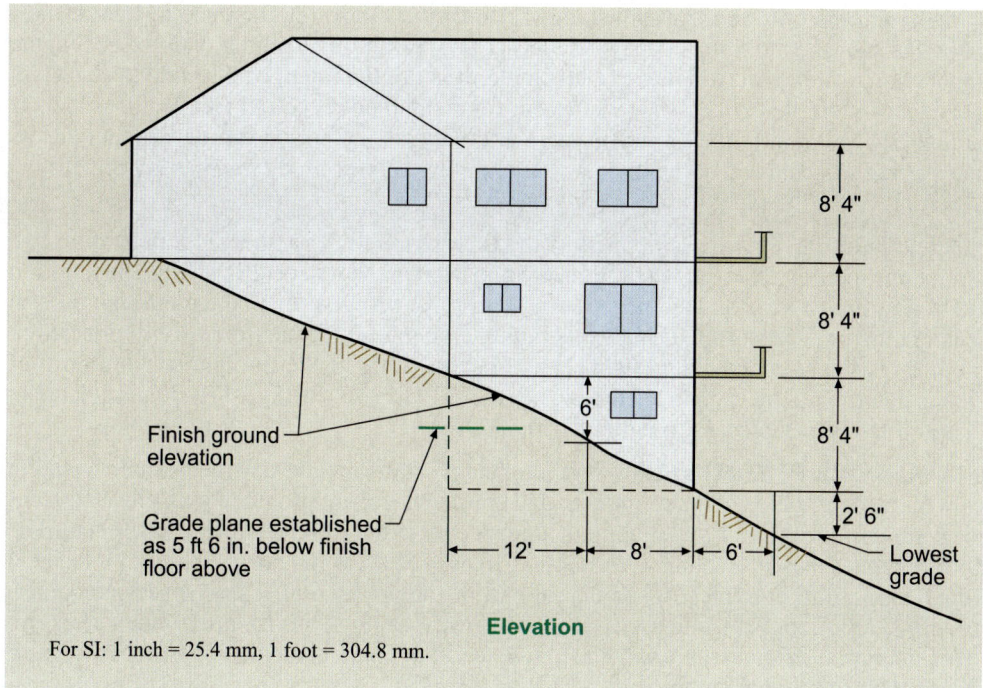

Figure 202-20
Building with two stories above grade plane and one basement level.

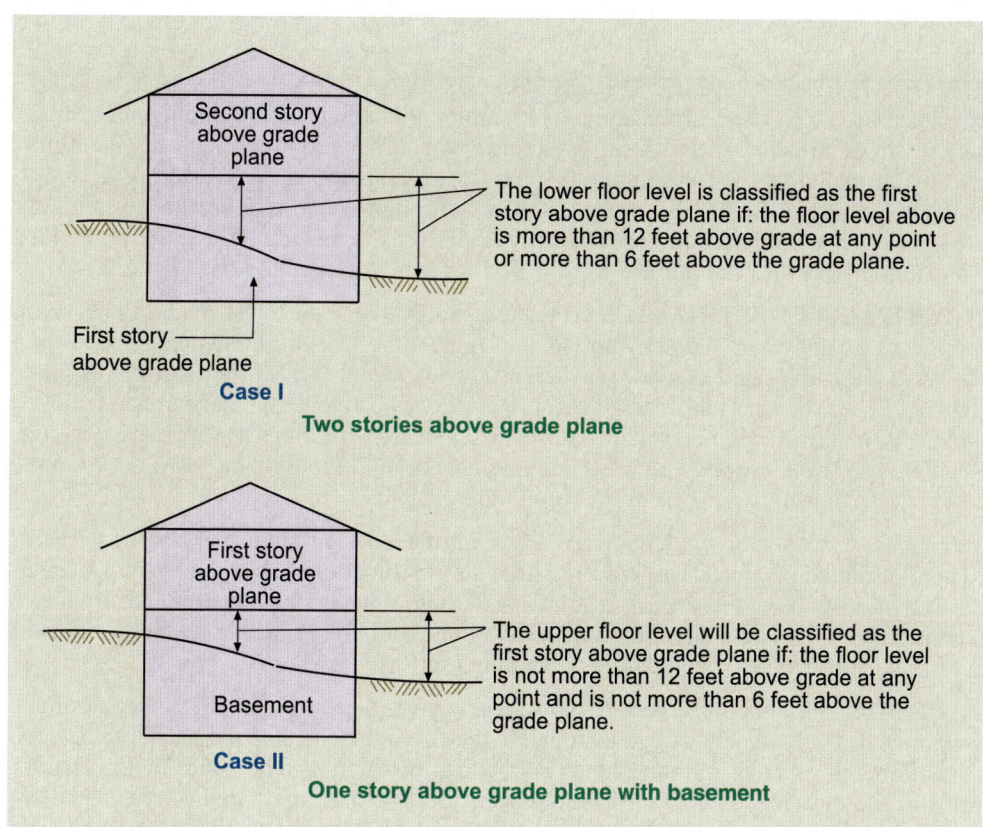

Figure 202-21
Multistory building.

Although the criteria for establishing the first story above grade plane in Item 2 indicates that such a condition occurs where the 12-foot (3658-mm) limitation is exceeded, the application of this provision is not that simple. It is not the intent of the code to classify a story that is completely below grade except for a small entrance ramp or loading dock as a *story above grade plane*, provided there is no adverse effect on fire department access and staging. An analysis of the impact of such limited elevation differences is necessary to more appropriately apply the code's intended result.

T RATING. The T rating is defined as the time required for a specific temperature rise on the unexposed side of a penetration firestop system. More specifically, the penetration firestop system as well as the penetrating item must provide for a maximum increase in temperature of 325°F (163°C) above its initial temperature for the time period reflected in the fire-resistance rating. The establishment of the rating of the through-penetration firestop is determined by tests in accordance with ASTM E 814 or UL 1479. A more detailed discussion of the subject is found in Chapter 7.

TECHNICAL PRODUCTION AREA. Many auditoriums and performance halls, as well as other types of entertainment and sport venues, are provided with elevated technical support areas used for lighting, sound, scenery, and other performance effects. The code regulates these as technical production areas. The areas may or may not be associated with a stage, but are typically an integral part of the production. These spaces are generally limited in floor area, and access is always restricted to authorized personnel. The term "technical production areas" is intended to encompass all technical support areas, regardless of their traditional name.

THROUGH PENETRATION. A through penetration is considered an opening that passes through an entire assembly, accommodating various penetrating items such as cables, conduit, and piping. Where membrane construction is provided, such as gypsum board applied to both sides of a stud wall, a through penetration would pass entirely through both membranes and the cavity of the wall.

THROUGH-PENETRATION FIRESTOP SYSTEM. In order to adequately protect the penetration of a fire-resistance-rated assembly by conduit, tubing, piping, and similar items, a through-penetration firestop system is sometimes required. Such a system may selectively include various materials or products that have been designed and tested to resist the spread of fire through the penetration. Through-penetration firestop systems are fire-resistance rated based on the criteria of ASTM E 814 or UL 1479, and are provided with an hourly rating for both fire spread (F rating) and temperature rise (T rating).

WALKWAY, PEDESTRIAN. Described as a walkway used exclusively as a pedestrian trafficway, a pedestrian walkway provides a connection between buildings. A pedestrian walkway may be located at grade, as well as above ground level (bridge) or below grade (tunnel). The provisions addressing pedestrian walkways are optional in nature and utilized primarily to allow for the consideration of the connected buildings as separate structures. Regulations for pedestrian walkways and tunnels are found in Section 3104. An example of a pedestrian walkway is shown in Figure 202-22.

WINDER. A winder, or winder tread, is a type of tread that is used to provide for a gradual change in direction of stairway travel. Although the directional change created by winders is typically 90 degrees (1.57 rad), other configurations are also acceptable. Owing to a reduced level of safe stairway travel, winders may only be used in a required means of egress stairway when located within a dwelling unit.

YARD. Used throughout the code to describe an open space at the exterior of a building, a yard must be unobstructed from the ground to the sky and located on the same lot on which the building is situated. A court, which is bounded on three or more sides by the exterior walls of the building, is not considered a yard. Both a yard and a court are expected to provide adequate openness and natural ventilation so that the accumulation of smoke and toxic gases will not occur.

Figure 202-22
Pedestrian walkway.

It is not intended that exterior areas devoted to parking, landscaping, or signage be prohibited to qualify as a yard, provided access to and from the building is available and maintained for both the occupants and fire department personnel. It is also important to recognize that the code provisions sometimes require a *yard* and at other times an *open space*, as well as references to *fire separation distance*. Although the differences may appear to be subtle, each term is applied somewhat differently.

KEY POINTS

- The IBC defines terms that have specific intents and meanings insofar as the code is concerned.
- All terms specifically defined in the IBC are listed in Section 202 along with their definitions.
- Words and terms specifically defined in Chapter 2 are italicized wherever they appear within the code.
- Ordinarily accepted meanings are to be used for definitions that are not provided in the code, based on the context in which the term or terms appear.

CHAPTER 3

USE AND OCCUPANCY CLASSIFICATION

Section 302 Classification
Section 303 Assembly Group A
Section 304 Business Group B
Section 305 Educational Group E
Section 306 Factory Group F
Section 307 High-Hazard Group H
Section 308 Institutional Group I
Section 309 Mercantile Group M
Section 310 Residential Group R
Section 311 Storage Group S
Section 312 Utility and Miscellaneous Group U
Key Points

3 Use and Occupancy Classification

> Chapter 3 of the code is simply an extension of Chapter 2 in that there are no technical requirements provided but rather descriptions of various uses and their assignment into specific classifications. These "occupancy" classifications are used as the basis for many of the provisions established elsewhere in the code. Potential uses are assigned an occupancy classification based on the specific occupant and content hazards that are anticipated to exist. Uses having similar hazard characteristics are all grouped into the same occupancy classification. Such characteristics may include the number, density, mobility, and awareness of the probable building occupants, as well as the combustibility, quantity, and environment of the building's contents. Proper occupancy classification is a major factor in achieving the code's intended purpose of establishing "minimum requirements to safeguard the public health, safety, and general welfare" as set forth in Section 101.3.

Section 302 *Classification*

302.1 General. Every structure, or portion of a structure, must be classified with respect to its use by placing it into one of the 26 specific occupancy groups identified in the code. These groups are used throughout the *International Building Code®* (IBC®) to address everything from building size to fire-protection features. The occupancy groups are organized into 10 categories of a more general nature, representing the following types of uses: assembly, business, educational, factory/industrial, high-hazard, institutional, mercantile, residential, storage, and utility/miscellaneous.

The provisions of Section 302.1 provide direction to:

1. Classify all buildings into one or more of the 26 groups identified in the IBC.

The occupancy classification is typically established by the design professional during the code analysis phase. Most of the time, the designer's determination is consistent with that of the building department. However, where there is disagreement as to the proper classification of the various uses within the building, it is the building official's responsibility to make the final decision. This authority is granted in Section 104.1 dealing with the interpretive powers of the building official. Although the IBC lists in some detail the uses allowed within a specific occupancy classification, the building official will at times also be called upon to judge whether or not a selected classification is appropriate under specific conditions. Assigning occupancy classification often not only depends on the use, but also on the extent and intensity of that use. A use may be so incidental to the overall occupancy that its effect on fire and life safety is negligible. As an example, the administrative office area in a high school performs a business-type function, but such a use is so incidental to the general operation and activities of the school that assigning it to a separate occupancy group would quite probably be unproductive. Therefore, the building official's judgment will often be relied upon to classify occupancies that could potentially fall into more than one group.

2. Address any room or space within the building that will be occupied at different times for different purposes.

Although an uncommon occurrence, a building space may at times be used for an activity that is considerably different from its typical use. It is important that the hazards associated with that different use also be addressed. The code is basically asking that such a space be assigned multiple occupancy classifications, with the requirements of each assigned occupancy group to be applied. For practical purposes, the classification of the space should probably be based on the more restrictive of the occupancies involved. This would account for most of the requirements that would be in place. Any additional requirements that would be applicable because of the other occupancy classification would then be layered on top of the other provisions. Another method would be to establish the occupancy classification of the major use, that which has the greatest occurrence, with a layering of the other occupancy requirements on top. Whatever the procedure, it is important that all anticipated uses, and the hazards these uses pose to the building's occupants, be taken into account.

A common example of a space used for various functions is a high school gymnasium. The space takes on various occupancy classifications based on the varied activities that occur, including Group E (physical education classroom), Group A-3 (community activities), Group A-4 (spectator gymnasium), and Group M (weekend craft shows). The classification of the space, which will result in the necessary safeguards being put in place, requires a comprehensive review of the anticipated activities and the hazards involved. A seasonal change in occupancy is another occurrence that must be considered. The creation of a *haunted house* for Halloween activities in a space typically used for other purposes is not uncommon. Regardless of the occupancy classification assigned, it is important that all of the anticipated uses be identified in order to apply the necessary code requirements.

3. Regulate buildings having two or more distinct occupancy classifications under the provisions of Section 508.

Many buildings cannot simply be classified under a single designation. A hotel, considered a Group R-1 occupancy, typically has assembly spaces classified as Group A. In addition, Group M and B occupancies may be present. Each distinct occupancy will be regulated based on the specific hazards that the individual uses create. The relationship between one occupancy and another is also very important. Where multiple occupancy conditions exist, the provisions of Section 508 are applicable.

There are two basic approaches to assigning occupancy classifications where buildings have multiple uses. One approach is to evaluate the building as individual areas, assigning classifications specific to the use that is under evaluation. Once this process is complete, a re-evaluation should occur to determine which classifications can be revised to reflect that of the major use. The other option is to initially classify the building as a single occupancy. Then each anticipated use that cannot be adequately addressed under the major classification will be assigned its own classification. Whatever approach is used, the goal is to make sure that the code provisions that are intended to address the anticipated uses of the space, and their potential hazards, are put in place.

There is an expectation that small support and circulation elements be included within the occupancy classification of the area in which they are located. This would include toilet rooms, storage closets, mechanical equipment rooms, and corridors. These spaces do not take on a unique classification unless they pose a unique hazard that can only be addressed with a different classification. A small closet of 20 square feet (1.86 m^2) would certainly not be considered a Group S occupancy where it is a portion of a Group B office space. On the other hand, a 3,600-square-foot (334.5-m^2) storage room could hardly be considered as merely an extension of the Group B classification. At some point, the use of space and its relationship to other spaces in the building provide for a need to assign a separate classification.

4. Classify a use into the group that the occupancy most nearly resembles, based on life and fire hazard, when the use is not described specifically in the code.

The code intends to divide the many uses possible in buildings and structures into 10 separate groupings where each group by itself represents a broadly similar hazard. The perils contemplated by the occupancy groupings are of the fire- and life-safety types and are broadly divided into two general categories: those related to people and those related to property. The people-related hazards are divided further by activity, by number of occupants, their ages, their capability of self-preservation, and the individual's control over the conditions to which he or she is subjected. The property-related hazards are divided further by the quantity of combustible, flammable, or explosive materials and by whether such materials are in use or in storage.

3 Use and Occupancy Classification

The uses to which a building may be put are obviously manifold, and as a result the building official will, on more than one occasion, either find or be presented with a use that will not conveniently fit into one of the occupancy classifications outlined in the code. As indicated previously in this commentary, under these circumstances, the IBC directs the building official to place the use in that classification delineated in the code that it most nearly resembles based on its life and fire risk. This requirement gives the building official broad authority to use judgment in the determination of the hazard of the affected group and, as a result of this evaluation, determine the occupancy classification that the hazards of the use most nearly resemble.

Occasionally, there may be a question as to which classification is to be assigned to a specific use. The owner of a building and the building official may have a difference of opinion as to the proper occupancy classification, or the building official may face a use that appears to fit into one of the code-described groups, but after further analysis it is determined that the hazards representative of the code-defined group are not present in the use proposed. In such situations, the building official should use his or her authority to place the use in the occupancy classification that it most nearly resembles based on its life- and fire-hazard characteristics. It must be remembered that the purpose of occupancy classification is solely to have the ability to regulate for the hazards associated with the building's expected use.

Section 303 *Assembly Group A*

An important item to be considered in this section is the description of an assembly occupancy as "the use of a building or structure, or portion thereof, for the gathering of persons for purposes such as civic, social or religious functions, recreation, food or drink consumption, or awaiting transportation." The description of an assembly occupancy is further defined by the numerous examples of Group A uses listed in Section 303.

The concerns unique to assembly uses are based primarily on two factors: the large occupant loads and the concentration of those occupant loads into very small areas. Both conditions must exist to warrant a Group A classification. For example, a Group M big-box store also has the potential for a sizable occupant load; however, the anticipated density of occupants within the store is not as concentrated as those densities regulated under the Group A provisions. Assembly uses are further divided based on other factors unique to the activities that take place.

A review of the uses specifically listed as Group A occupancies indicates that although some are general in nature, such as libraries and arenas, others are specific to a room or area within a building (courtroom, lecture hall, and waiting areas in transportation terminals). This concept of identifying uses is not limited to Group A occupancies, but rather is consistent throughout all of the occupancy group descriptions. It is critical that the building official thoroughly evaluate the uses that are anticipated and assign occupancy classifications based on the hazards that have been identified.

303.1.1 Small buildings and tenant spaces. Classification of a small assembly building or tenant space as a Group A occupancy is considered unnecessary due to the limited hazards that are expected to be present. An assembly use with an occupant load of fewer than 50 persons and not accessory to another use, such as a small free-standing chapel or a café located in a strip shopping center, is to be classified as Group B. As a result, a Group A occupancy classification is typically only assigned to an assembly use containing at least 50 occupants.

Where a Group B café or similar establishment has an associated kitchen, the kitchen is also to be classified as a portion of the Group B occupancy. This approach is consistent with the classification of commercial kitchens associated with Group A-2 dining facilities as indicated in Section 303.3.

303.1.2 Small assembly spaces. Where the area used for assembly purposes is accessory to another occupancy in the building and contains an occupant load of fewer than 50 persons, the room or space can merely be considered an extension of the other occupancy. As an example, a break room with an occupant load of 30 in a large manufacturing facility could simply be considered a portion of the Group F occupancy. See Figure 303-1. As an option, a Group B classification can be assigned to the break room.

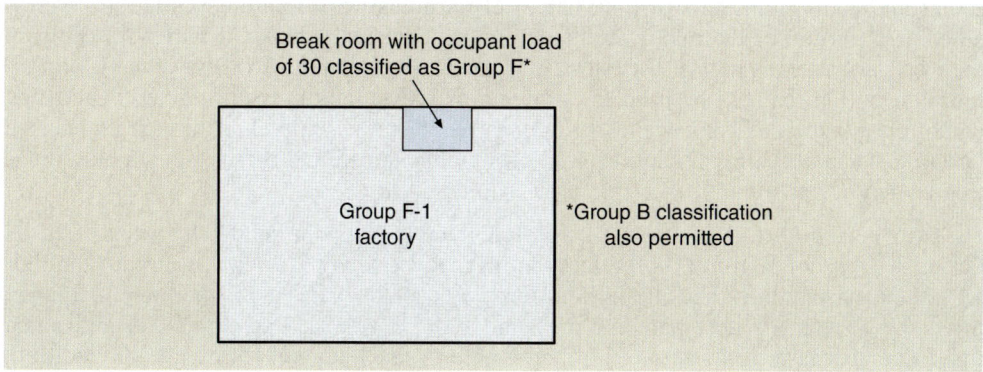

Figure 303-1
Group F classification.

The same options for classification are available where the accessory assembly space has less than 750 square feet (70 m²) in floor area, regardless of occupant load. Figure 303-2 illustrates how a small accessory assembly space with an occupant load of 50 or more need not be classified as a Group A occupancy. In the example, a small 480-square-foot (44 m²) chapel, having an occupant load of 68 and located within an assisted living facility classified as a Group I-1 occupancy, can simply be classified as an extension of the Group I-1 classification. As an alternative, a classification of Group B is also permitted to be assigned to the chapel.

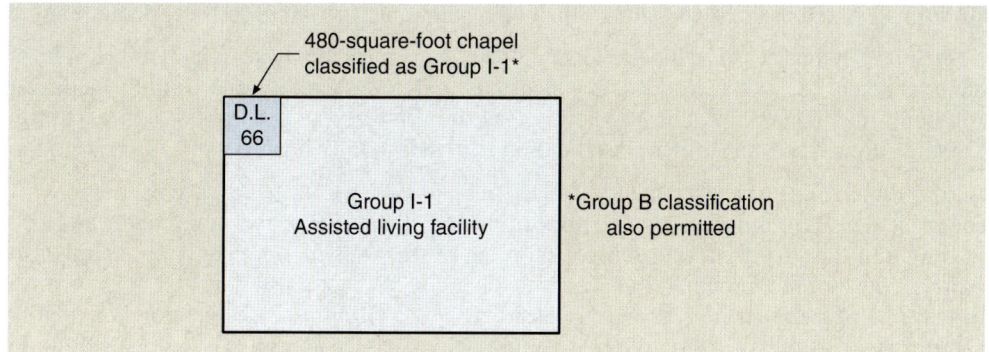

Figure 303-2
Group I-1 classification.

303.1.3 Associated with Group E occupancies. The classification of those assembly areas associated with a Group E occupancy as part of the Group E is permitted; however, the application of the Group E classification for the entire educational facility is based on the assembly areas being subsidiary to the school function. This would seem to indicate that the users of the associated assembly spaces are limited to students, teachers, relatives of students, administrators, and others directly involved in educational activities. A typical example of an associated assembly space is a library or media center that is used almost exclusively by students of the school. On the other hand, gymnasiums and auditoriums located in

3 Use and Occupancy Classification

high school buildings are often used for community functions and other outside activities such as sports tournaments, craft shows, and community theater productions that have no relationship to normal educational uses. In such cases, a classification of Group A is often viewed as more appropriate based on these unrelated uses. Even in those situations where a Group E classification is appropriate for associated assembly uses, it is important that the assembly accessibility provisions of Chapter 11 be applied. In addition, the assembly means of egress provisions set forth in Section 1028 should also be used in designing the means of egress system for the assembly areas, regardless of occupancy classification.

303.1.4 Accessory to places of religious worship. Small and moderately sized educational rooms and auditoriums accessory to places of religious worship are permitted to be classified as a portion of the major occupancy rather than individually. This would result in their classification as part of the overall Group A-3 occupancy. The allowance is simply a design option that can be used to eliminate or reduce any potential mixed occupancy conditions. If the designer wishes to classify such spaces individually, such as using a Group E occupancy classification for any religious educational rooms, such a classification is also permissible.

303.2 Assembly Group A-1. A factor involving human behavior in theaters classified as Group A-1 assembly rooms is the fact that in many cases the occupants are not familiar with their surroundings and the lighting level is usually low. Thus, when an emergency arises, the occupants may perceive the danger to be greater than presented, and panic may occur because of the fear of not being able to reach an exit for escape. In addition, the concentration of occupants in such uses is quite dense. The presence of a stage and its distinctive hazards that occur in some Group A-1 occupancies cause unique concerns, addressed by the special provisions of Section 410.

303.3 Assembly Group A-2. Group A-2 occupancies include uses primarily intended for the consumption of food or drink, and include dining rooms, cafeterias, restaurants, cafes, nightclubs, taverns, and bars. The fire record in occupancies of this type is not very good, based in part on the delay in responding to a fire or other emergency incident. Because of the common presence of loose tables and chairs, aisles are often difficult to maintain, resulting in obstructions to egress travel. Overcrowding conditions, low-lighting levels, and the consumption of alcoholic beverages also increase the risks associated with many of these types of occupancies. The gaming floor areas of casinos are also classified as Group A-2 based in great part due to the congestion and distractions often encountered.

Included in the uses classified as Group A-2 occupancies are those commercial kitchens directly associated with Group A-2 restaurants, cafeterias, and similar dining facilities. Although a commercial kitchen does not pose the same types of hazards as an assembly use, the allowance for a similar classification has traditionally been viewed as appropriate.

303.4 Assembly Group A-3. Occupancies classified as Group A-3 have varying degrees of occupant density, numerous types and numbers of furnishings and equipment, and fire loading that can vary from low to high. The hazards for uses in this category are similar to most of those of the Group A-1 and A-2 occupancies. Where a use does not conveniently fit into one of the other four Group A classifications, a Group A-3 designation is typically appropriate. The classification of an assembly occupancy as a Group A-3 is also common where varying assembly uses are likely to occur at different times within the same space. For example, a meeting room at a hotel is typically used at differing times for various functions, including seminar presentations, dining activities, trade shows, and wedding receptions. Although these functions may have different Group A designations when viewed individually, as a group they pose a hazard level that can be appropriately addressed with the Group A-3 classification. Therefore, most multipurpose rooms are simply classified as Group A-3 occupancies.

The mere presence of recreational activities does not necessarily warrant the classification of a building or space as a Group A-3 occupancy. Rather, it is the concentration of occupants participating or viewing the activity that should be used to determine the appropriate classification. The example illustrated in Figure 303-3 is based on a golf instruction academy with various "stations" at which students work on their skills individually or with an instructor. Support facilities such as restrooms, locker rooms, merchandise sales, and a snack bar are also provided. Although the occupant load of the building exceeds 49, there is little or no concentration of the occupants in a single area. Therefore, classification as Group A occupancy would not be warranted. A more appropriate classification of Group B would better describe the hazards and conditions found within a use of this type.

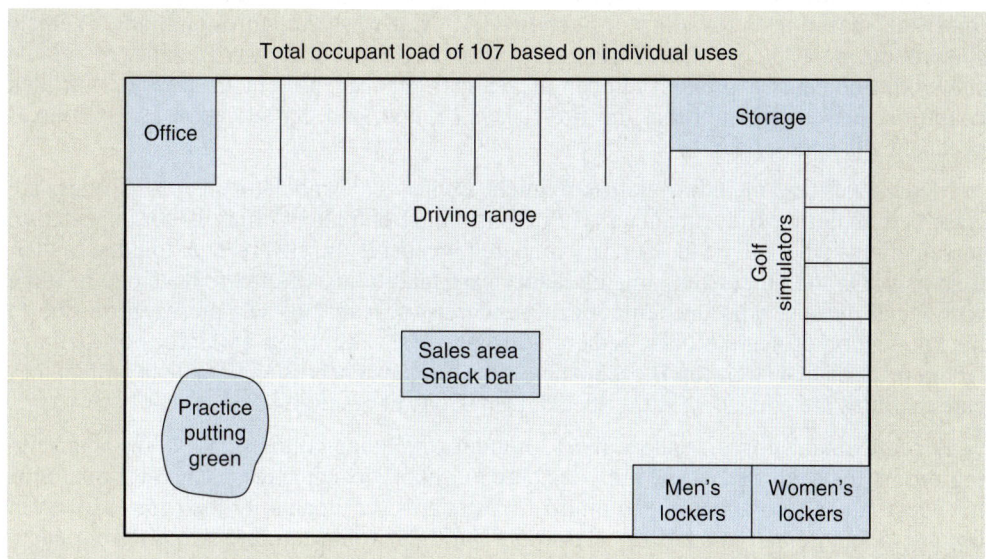

Figure 303-3
Group B Golf Institute Academy.

Assembly Group A-4. The combination of spectator seating and sporting events creates a condition within a building that warrants a specific occupancy classification within the Group A classification. A Group A-4 facility contains those occupant-related hazards found in other assembly occupancies, namely high occupant loads in concentrated areas, along with large areas having limited occupants and little, if any, fire loading conditions. The focus of the Group A-4 designation is that there is a significant number of spectators present in a relatively concentrated environment.

303.5

Assembly Group A-5. Uses classified as Group A-5 are similar in nature to Group A-4 occupancies, with the controlling difference being that Group A-5 occupancies are outdoor structures. Therefore, the fire hazard for Group A-5 occupancies is less than that for those classified as Group A-4 occupancies, and significantly lower than the other assembly Group A occupancies. It is also expected that there would be little to no smoke accumulation under fire conditions in the assembly areas of a Group A-5 occupancy. However, there still exist the hazards of crowding a large number of occupants within a relatively small space. The hazard of panic is assumed to be a large portion of the overall concern for Group A-5 structures. Generally, associated spaces such as concession stands, locker rooms, storage areas, press boxes, and toilet rooms are included as a portion of the Group A-5 classification. However, where uses within the building create conditions more hazardous than anticipated by the Group A-5 designation, such uses must be classified according to their individual characteristics. For example, an enclosed 400-seat restaurant within a sports stadium should be appropriately classified as a Group A-2 occupancy.

303.6

3 Use and Occupancy Classification

Section 304 *Business Group B*

The most common use classified as a Group B occupancy is an office building, or a portion of a building containing office tenants or office suites. The portions of such business occupancies where records and accounts are stored are also considered part of the Group B use. Examples of uses involving office, professional, or service-type transactions are listed in the code.

Airport traffic control towers are considered Group B occupancies and are regulated under the special-use provisions of Section 412.3. Car wash structures and motor vehicle showrooms are also considered business occupancies. Since a car wash facility or vehicle showroom contains a limited number of vehicles that are present in a very controlled condition, it is anticipated that the fire risk is limited, and classification as a Group B occupancy is appropriate.

Medical offices, including both outpatient clinics and ambulatory care facilities, are classified as Group B occupancies. By definition, an outpatient clinic is not expected to serve patients who will be temporarily rendered incapable of self-preservation due to their treatment. On the other hand, an ambulatory care facility is expected to have one or more individuals present who are temporarily rendered incapable of self-preservation due to the application of nerve blocks, sedation, or anesthesia. Although both types of facilities would be classified as Group B occupancies, the unique concerns applicable to ambulatory care facilities are addressed in the special provisions of Section 422.

Educational occupancies above the 12th grade, including college classrooms and training rooms for adult education, are considered Group B occupancies. This designation is not specified as incumbent on the number of students (occupants) in the room. However, because a lecture hall in a college classroom building would fall into two different occupancy classifications, Groups A-3 and B, it is important that the concerns of both classifications be considered. As a part of the overall building analysis, a Group B classification would typically be appropriate. It can be viewed as simply an extension of the Group B uses that occur throughout the building. This might include using the Group B criteria for construction type, mixed-occupancy provisions, and other more general categories. On the other hand, a Group A classification would seem to be necessary in evaluating the means of egress and fire protection requirements. The code is silent to this type of analysis; however, the intent should be to regulate the building in a manner consistent with the hazards that are anticipated.

Where not exceeding the maximum quantities of hazardous materials allowed by the code in Section 307, testing and research laboratories may be considered Group B occupancies. Where the allowable quantities are exceeded, such laboratories would be classified as Group H.

The Group B classification of training and skill development programs outside of a school or academic program includes uses such as tutoring programs and instrumental music training. Although there are no occupant load or occupant age criteria associated with classification as a Group B occupancy, it may be necessary to apply a higher level classification where the hazards anticipated based on the specific use fall outside of those typically encountered within a Group B occupancy. For example, a training program conducted in assembly-type conditions where the occupant load is 50 or more would typically be classified as a Group A occupancy rather than a Group B. Skill development activities conducted for very young children could more appropriately be considered as a Group E or I-4 occupancy if the potential hazard level is consistent with that of a day-care facility. As with any occupancy designation, the intent of the classification is to be able to correctly apply the appropriate minimum standards of health, safety, and public welfare.

Section 305 *Educational Group E*

The Group E classification is assigned to schools, including primary, middle, and high schools, as well as day-care facilities. All Group E occupancies have three features in common: they are limited to the education, supervision, or personal care of persons at an educational level no higher than the 12th grade; the occupants are only in the facility for a limited time each day; and there are at least six persons being educated, supervised, or cared for at the same time.

Educational Group E. The classification of school classrooms as Group E occupancies is typically a straightforward decision. It is also common to classify the administrative offices within a school building as an extension of the Group E function. Even the media center and lunchroom are generally viewed as just additional areas of the Group E occupancy, but what about other assembly spaces such as the gymnasium and auditorium? There is a unique feature involved in educational occupancies—the use of school buildings for assembly purposes outside the scope of the educational use. For example, many school auditoriums are used for community theater and other productions to which the public at large is invited. Also, the school gymnasium in many cases is used for neighborhood recreation activities or sporting events where all ages of occupants are present. It is not uncommon for school auditoriums and gymnasiums to be rented out to groups for special functions. When these additional uses are anticipated, there is adequate reason that they be classified as Group A occupancies. Therefore, on account of these multipurpose uses in many school buildings, it is necessary that the code requirements applicable to all expected uses be enforced in order to satisfy the safety requirements for each use. See the additional discussion in Section 303 on multi-purpose areas within educational facilities.

305.1

Accessory to places of religious worship. It is common in buildings of religious worship that support spaces are provided in addition to the main worship hall. Such spaces typically include rooms for educational activities for persons of all ages, including children. Such educational areas are permitted to be classified as an extension of the Group A-3 worship hall rather than Group E, thus eliminating a mixed-occupancy condition. The limitation of the provision to rooms and auditoriums with occupant loads of less than 100 has limited, if any, application, as larger assembly spaces would typically be considered Group A-3 occupancies as well. This allowance is also addressed in Section 303.1.4.

305.1.1

Group E day-care facilities. Day-care facilities considered as Group E occupancies are limited to those facilities where children are provided with educational, supervision, or personal care services for periods of less than 24 hours per day. Facilities that provide such services for adults more appropriately are typically assigned a Group I-4 classification. In addition, full-time care facilities cannot be considered Group E occupancies. The number of children housed in a day-care operation classified as Group E is not limited; however, where the number of children is five or fewer, the use is to be classified as part of the primary occupancy.

305.2

The provisions of Section 305.2 only address care facilities for children over the age of 2½ years. However, the Group E classification is also applicable to facilities that provide infant/toddler care (2½ years of age or younger) where the conditions established in Section 308.6.1 are met. In such cases, the rooms housing the infants and/or toddlers must be located on the level of exit discharge, and each of such rooms must have an exit door directly to the exterior of the building. See Figure 305-1.

3 Use and Occupancy Classification

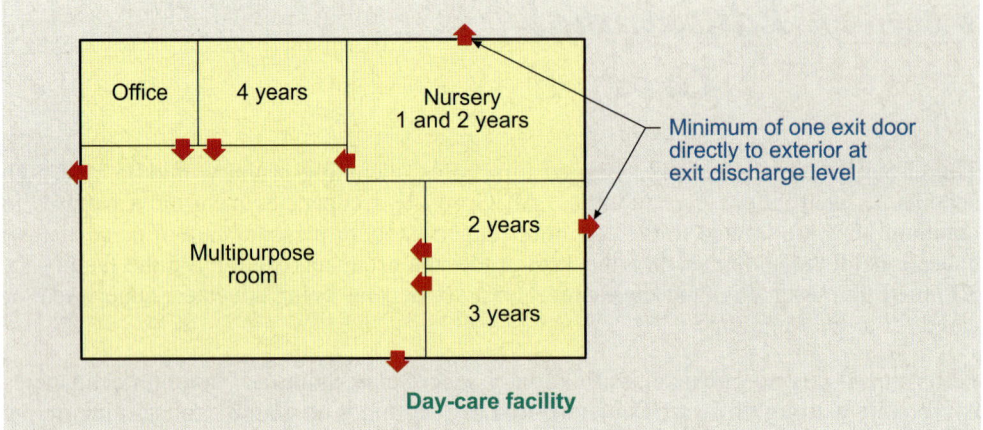

Figure 305-1
Group E classification.

305.2.1 Within places of religious worship. It is common for child care to be available at places of religious worship during worship services and other activities. Nurseries, "cry rooms," and other child care spaces are occupied during the same limited time period as the other activities within the worship facility. The occupancy classification of such spaces is permitted to be consistent with that of the main occupancy so that a mixed-occupancy condition need not be created. Although this allowance is limited to only those child care facilities where the children are older than 2½ years of age, a similar allowance for facilities caring for younger children and adults is found in Section 308.6.2.

305.2.2 Five or fewer children. Where five or fewer children are receiving day care within a building housing another use that can be considered the primary occupancy, the day-care operation is simply to be considered a portion of that occupancy. For example, a small day-care activity within an office environment would be classified as a portion of the Group B office use. Where there is no use in the building other than a day-care facility serving five or fewer children, it is assumed that a Group R-3 classification is to be applied. One of the listed Group R-3 uses in Section 310.5 is "care facilities that provide accommodations for five or fewer persons receiving care."

305.2.3 Five or fewer children in a dwelling unit. Where care is being provided to no more than five children and such care occurs within a dwelling unit, the occupancy classification of the dwelling unit is not to be modified due to the presence of the child care activities. For example, child care for five or fewer children in a dwelling unit complying with the scope of the *International Residential Code®* (IRC®) is permitted with no additional requirements applicable to the day-care operation. An occupancy classification of Group R-2 would continue to be appropriate for such child care activities in a dwelling unit classified as a Group R-2 occupancy. It is intended that the presence of a child care use in a dwelling unit where the number of children receiving care does not exceed five will have no effect on the occupancy classification determination or the applicable code that is to be enforced.

Section 306 *Factory* Group F

Although the potential hazard and fire severity of the multiple uses in the Group F occupancy classification is quite varied, these uses share common elements. The occupants are adults who are awake and generally have enough familiarity with the premises to be able to exit the building with reasonable efficiency. Public occupancy is usually quite limited, and most occupants are aware of the potential hazards the use creates. Group F occupancies are generally regarded as factory and industrial uses. The degree of hazard between the uses is very broad, and therefore the occupancy is divided into two categories.

Many of the Group F-1 uses contain some degree of hazardous material as a necessary part of the manufacturing process. However, where the amount of hazardous material does not exceed the maximum allowable quantities set forth in Table 307.1(1) or 307.1(2), the lower classification of Group F is appropriate. Because of the similarity between the names of the uses in Group F-1 occupancies and those in Group H occupancies, care must be exercised when determining the appropriate classification, and operators of Group F-1 occupancies should be apprised of the limitations on the quantities of hazardous materials that are allowed.

Some of the activities specifically listed as Group F occupancies also occur in a limited sense as accessory functions, and as such are not to be classified as Group F. For example, *food processing* is identified as a Group F-1 occupancy, but this is not to say that a kitchen serving a restaurant or cafe should be classified as such. Kitchens are considered to be classified as a portion of the major occupancy that they serve, typically either a Group A or a Group B occupancy. The food processing operations designated as Group F-1 occupancies primarily include large factories that produce canned or packaged items in bulk, as well as commercial kitchens that support catering operations or similar activities.

The hazard from uses in Group F-2 occupancies is very low; in fact, the activities are deemed as among the lowest hazard groups in the code. It is assumed that the fabrication or manufacturing of noncombustible materials will pose little if any fire risk to the building or its occupants. Foundries would be considered Group F-2 occupancies, as would facilities used for steel fabrication or assembly. Manufacturing operations producing ceramic, glass, or gypsum products are also included in this classification. Although very high temperatures are critical to the processes and operations of these types of uses, such heat is controlled and not a concern due to any combustibles or other fire loading that may be present.

Section 307 *High-Hazard Group H*

High-hazard Group H occupancies are characterized by an unusually high degree of explosion, fire, or health hazard as compared to typical commercial and industrial uses. The identification of hazardous occupancies is provided in this section.

There is one common feature about all Group H occupancies—they are designated as Group H based on excessive quantities of hazardous materials contained therein. Where the quantities of hazardous material stored or used in a building exceed those set forth in Section 307, a Group H classification is warranted. On the other hand, where such quantities are not exceeded, a Group H classification is not appropriate.

Because of the technical nature of the operations and materials found in Group H occupancies, a number of specific terms are defined by the IBC in Chapter 2. The definitions are intended to assist the code user in applying the provisions of this chapter, as well as other portions of the code relating to high-hazard uses.

Group H-1 occupancies are those buildings containing high-explosion hazard materials. Materials that have the potential for detonation must be housed in buildings regulated in a very special manner, and designed and constructed unlike any other occupancies described in the code. Examples of detonable materials include explosives and Class 4 oxidizers.

Group H-2 generally includes those occupancies that contain materials with hazards of accelerated burning or moderate explosion potential, including materials with deflagration hazards. Common occupancies included in this category are those operations where flammable or combustible liquids are being used, mixed, or dispensed. The potential for a hazardous incident is increased because of the materials' exposure to the surrounding area. Occupancies containing combustible dusts are also considered Group H-2, as dusts in suspension, or capable of being put into suspension, in the atmosphere are a deflagration hazard. In the determination of occupancy classification for a facility where combustible dusts are anticipated, a technical report and opinion must be provided to the building official that provides all necessary information for a qualified decision as to the potential combustible dusts hazard.

Buildings containing materials that present high-fire or heat-release hazards are classified as Group H-3 occupancies. Where flammable or combustible liquids are present in such occupancies, they must be stored in normally closed containers or used in low-pressure systems. Because of the enclosed nature of these liquids, the hazard level is not nearly as severe as it is for Group H-2 occupancies. Other hazardous materials such as organic peroxides and oxidizers, based on their hazard classification, may also be used or stored in Group H-3 occupancies.

Group H-4 occupancies are those containing health-hazard materials such as corrosives and toxics. Section 202 defines "Health hazards" as those "chemicals for which there is statistical significant evidence that acute or chronic health effects are capable of occurring in exposed persons." Quite often, a material considered a health hazard also possesses the characteristics of a physical hazard. It is important that all hazards of materials be addressed.

Occupancies classified as Group H-5 are those uses containing semiconductor fabrication facilities, including the ancillary research and development areas. The Group H-5 category was created in order to address the explosive and highly toxic materials used in semiconductor fabrication by providing specific requirements for the particular operations conducted, while at the same time providing a level that allows reasonable transaction of the fabrication process.

A more complete commentary on hazardous materials is provided under the discussion of Section 414.

307.1 High-hazard Group H. The concept of maximum allowable quantities of hazardous materials as the basis for occupancy classification is further extended through the use of control areas as regulated by Section 414.2.

Maximum allowable quantities. Occupancy classifications of buildings containing hazardous materials are based on the *maximum allowable quantities* concept. Tables 307.1(1) and 307.1(2), together with their appropriate footnotes, identify the maximum amounts of hazardous materials that may be stored or used in a control area before the area must be designated as a Group H occupancy. The maximum quantities of hazardous materials permitted in non-Group H occupancies vary for different states of materials (solid, liquid, or gas) and for different situations (storage or use). The allowable quantities are also varied based on protection that is provided, such as fire-extinguishing systems and storage cabinets.

Control areas. Areas in a building that are designated to contain less than or equal to the maximum allowable quantities of hazardous materials and that are properly separated from other areas containing hazardous materials are called *control areas*. Any combination of hazardous materials, up to the maximum allowable quantities, is permitted in a control area. A control area may be an entire building or only a portion of the building. It can be part of a story, an entire story, or even include multiple stories.

The control-area method is based on the concept of fire-resistance-rated compartmentation. It regulates quantities of hazardous materials per compartment (control area), rather than per building. The limit for the entire building, using control areas, is then established by limiting the total number of control areas allowed per building and the quantities of hazardous materials that are located in each control area. The control-area concept was introduced in an effort to regulate buildings of different sizes in a consistent manner. It is based on a premise that the storage and use of limited quantities of hazardous materials (not exceeding maximum allowable quantities) in areas that are separated from each other by fire-resistance-rated separations do not substantially increase the risk to the occupants or change the character of the building to that of a hazardous occupancy, subject to a limitation on the number of control areas. The fire-resistive separations are relied on to minimize the risk of having multiple control areas involved simultaneously during an emergency.

The occupancy classification of a control area is the same as the occupancy classification of the portion of the building in which the control area is located. There is no special occupancy designation for a control area. For example, a control area in a manufacturing occupancy is merely part of the Group F occupancy. Further discussion of control areas is addressed in the commentary of Section 414.2.

Increased quantities. Given this basic understanding of maximum allowable quantities and control areas, the various options in the code for increasing the quantities of hazardous materials within a building are as follows:

1. Buildings are generally allowed to have up to the basic maximum allowable quantities of hazardous materials without restriction with respect to separations or protection. In this case, the entire building is designated as a control area. The boundaries of the control area are the boundaries of the building (i.e., exterior walls, roof, and foundation). See Figure 307-1.

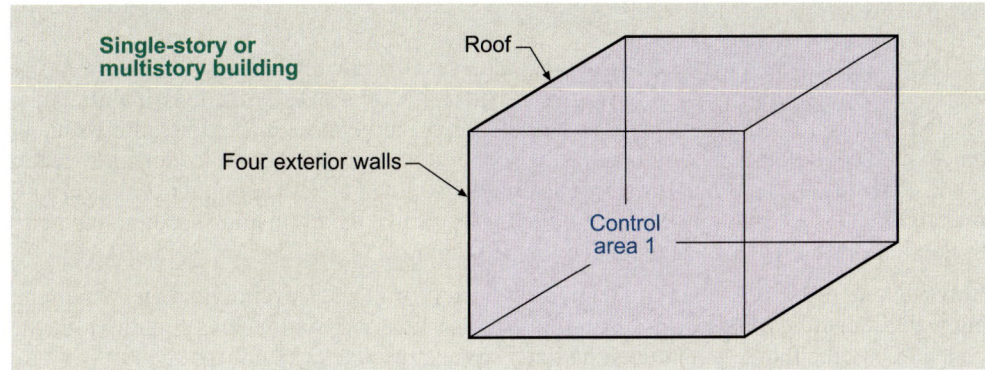

Figure 307-1
Control area boundaries for one control area.

2. Using the footnotes to Tables 307.1(1) and 307.1(2), the maximum allowable quantities can often be increased by adding sprinklers throughout the building or by using approved storage cabinets, safety cans, or other code-approved enclosures to protect the hazardous materials. It is important that the increases identified in the footnotes only be used where applicable.

3. Two other options are available to further increase the quantities of hazardous materials in any building:

   3.1. Provide additional control areas, or

   3.2. Construct the building as required for a Group H occupancy.

4. Assuming additional control areas are used, each additional control area must be separated from all other control areas by minimum 1-hour fire barriers, or 2-hour fire barriers if required by Section 414.2.4. Vertical isolation of control areas must be accomplished by floors having a minimum 2-hour fire-resistance rating. Under limited conditions, the floor construction of the control area separation may be reduced to 1 hour. Its application is limited to fully sprinklered two- or three-story buildings of Type IIA, IIIA, or VA construction. In all cases, construction supporting such floors shall have an equivalent fire-resistance rating. A designated percentage of the maximum allowable quantities of hazardous materials is allowed in each control area per Table 414.2.2. See Figure 307-2.

3 Use and Occupancy Classification

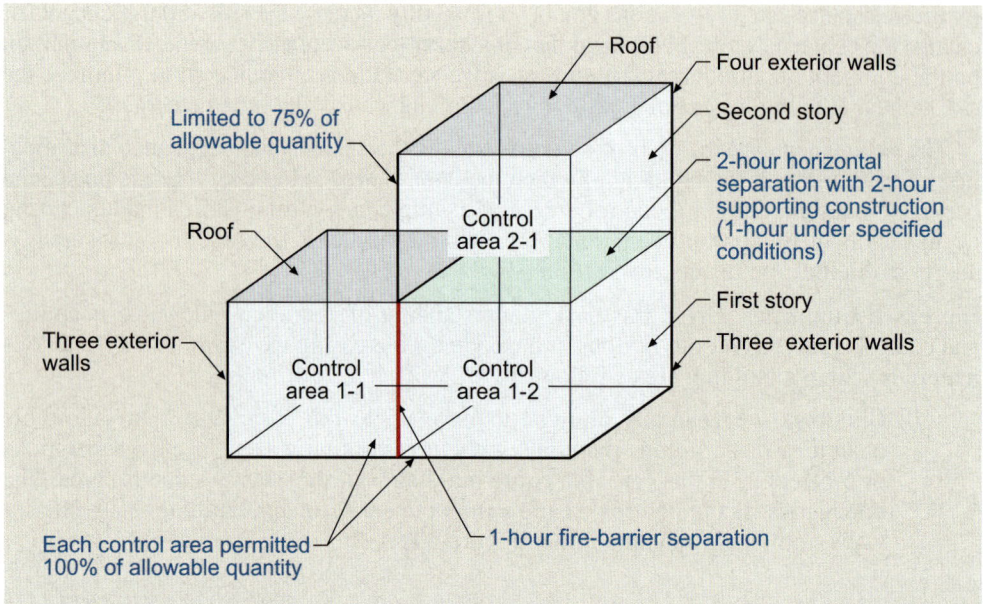

Figure 307-2
Multistory areas.

The permitted number of control areas decreases vertically through the building, as does the quantities of hazardous materials per control area. As hazardous materials are located higher and higher above ground level, they become more difficult to address under emergency conditions. As with many other conditions regulated by the code, a key factor is the ability of the fire department to access the incident area. The higher the hazardous materials are located in the building, the more restrictive the provisions become, owing to the limitations on fire department access and operations.

Storage and use. One other fundamental concept involved in applying the maximum allowable quantities is *situation of material*. The maximum allowable quantities in the code are based on three potential situations: storage, use-closed, and use-open.

Though not defined by the code, the term *storage* is generally considered to include materials that are idle and not immediately available for entering a process. The term *not immediately available* can be thought of as requiring direct human intervention to allow a material to enter a process or, alternatively, as using approved supervised valving systems that separate stored material from a process. In the case of liquids and gases, storage is generally considered to be limited to materials in closed vessels (not open to the atmosphere). For example, materials kept in closed containers such as drums or cans are in storage because deliberate action (opening the drum or can) would be required to use the material. However, when a container or tank is connected to a process, the question arises of whether the material in the container or tank is in storage or in use.

In general, the quantity of material that would be considered to be in use is the quantity that could normally be expected to be involved in a process, or that could reasonably be expected to be released or involved in an incident as a result of a process-related emergency. Consider, for example, a process having hazardous materials that are piped from an underground storage tank outside of a building to a dispensing outlet within a building. Because the tank is connected to a process within the building, it could be argued that the contents of the tank are available for use in the building (see definition of "Use" in Section 202) and that the amount should be counted toward the maximum allowable quantities. However, if an approved, reliable arrangement of valving is provided between the supply and the point where the material is dispensed, it would be reasonable to conclude that the quantity on the supply side of such valving that is outside of the building would be unlikely to impact incidents occurring within the building and, therefore, need not be counted toward the maximum allowable quantities. This reliable arrangement of valving can be considered an interruption of the connection between the confined material (storage) and the point where material is placed into action or made available for service, as discussed in the definition of "Use." See Figure 307-3.

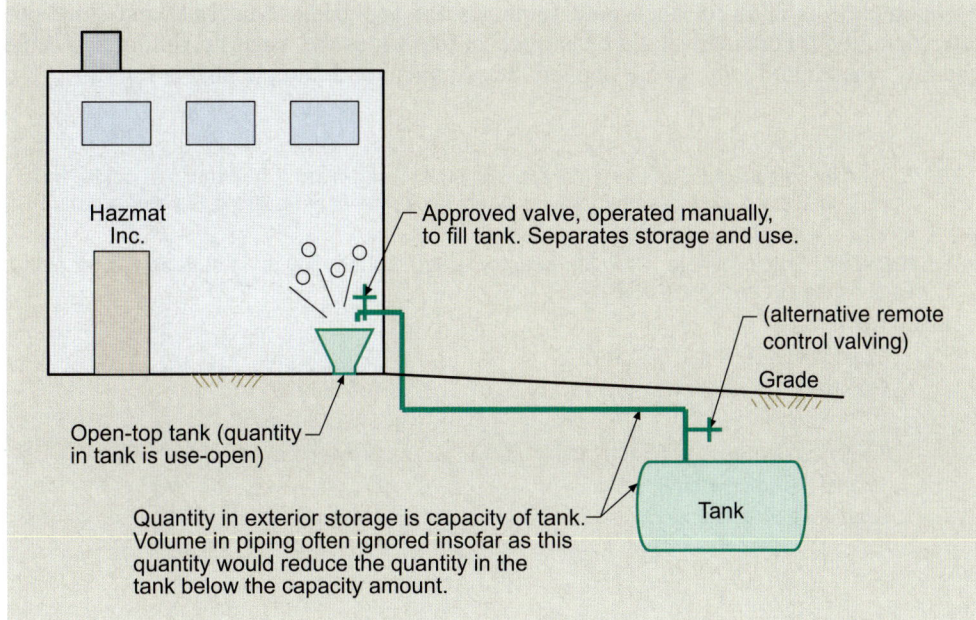

Figure 307-3
Example of storage versus use.

The difference between use-closed and use-open is basically whether the hazardous material in question is exposed to the atmosphere during a process, with the exception that gases are defined as always being in closed systems when used insofar as they would be immediately dispersed (unless immediately consumed) if exposed to the atmosphere without some means of containment.

Table 307.1(1)—Maximum Allowable Quantity per Control Area of Hazardous Materials Posing a Physical Hazard. This table sets forth maximum allowable quantities for physical hazard materials. All three situations (storage, use-closed, and use-open) are considered. For the specific case of gases, maximum allowable amounts are all listed under storage and use-closed because the definition of "Use" includes all gases.

With two exceptions, any combination of materials or situations listed in this table is allowed in each control area. These two exceptions are (1) as provided by Footnote h, the aggregate of IA, IB, and IC flammable liquids, and (2) as provided by Footnote b, which requires that aggregate quantities of materials in both use and storage must not exceed the allowable quantity for storage.

Specific footnotes to the table provide the following information:

Footnote a. This footnote references Section 414.2 for the use of control areas. For additional information, see the discussion of control areas in Sections 307.1 and 414.2.

Footnote b. This footnote requires quantities of materials that are in use to be counted as both storage and use when comparing quantities to those permitted. For example, a single control area in a manufacturing facility would be permitted up to 30 gallons (116.25 L) of a Class II combustible liquid in use and a maximum of 120 gallons (454.2 L) in storage without being considered a Group H occupancy. However, a total of 150 gallons (567.75 L) would be prohibited. The total of both use and storage is limited to 120 gallons (454.2 L), with no more than 30 gallons (116.25 L) permitted in use.

Footnote c. This footnote exempts small size containers of certain consumer products that are considered to present minimal hazards based on the types of materials and the container sizes.

Footnote d. This footnote allows certain materials to have exempt amounts doubled when stored or used in sprinklered buildings. Compounding with the increases provided by Footnote e is allowed when both footnotes are applicable. Materials and situations referencing both Footnotes d and e can receive four times the listed maximum allowable quantity when both footnotes are applied. See Application Example 307-1.

Application Example 307-1

GIVEN: A fully-sprinklered Group F-1 storage building housing Class II combustible liquids. The Class II liquids are all stored in approved safety cans. The entire building is a single control area.

DETERMINE: The maximum allowable quantity of the Class II liquids in storage in order to maintain the Group F-1 classification.

SOLUTION:
Basic MAQs per Table 307.1(1)	120 gallons
Sprinkler increase per Footnote d (100%)	+ 120 gallons
	240 gallons
Safety can increase per Footnote e (100%)	+ 240 gallons
Total of maximum permitted for Group F-1 classification	480 gallons

Footnote e. This footnote allows exempt amounts for certain materials in storage to be doubled when approved storage cabinets, gas cabinets, safety cans, etc., as applicable, are employed. Also, see Footnote d.

Footnote f. This footnote allows certain materials to be stored or used in unlimited quantities in sprinklered buildings. When the building is fully sprinklered, it has the effect of classifying the building as other than a Group H occupancy. Application of this footnote is limited to the storage or use of Class IIIB combustible liquids or Class I oxidizers. An unlimited quantity of each of these materials is permitted in a fully sprinklered building without requiring a designation of Group H.

Footnote g. This footnote limits storage and use of certain materials to sprinklered buildings. Where a non–Group H classification is desired and the quantity does not exceed the maximum allowable, the building is required to be sprinklered throughout.

Footnote h. Where flammable liquids are concerned, the maximum allowable quantities are regulated both individually and cumulatively. To be considered an occupancy other than Group H, the control area must not contain more than the maximum allowable quantities for each type of flammable liquid (Class IA, IB, and IC), as well as the combination of such limits established by Table 307.1(1).

Footnote i. The threshold of 660 gallons (2,498 L) is commonly used in the *International Fire Code®* (*IFC®*) where regulating fuel oil storage and piping systems. This footnote allows for additional quantities above those set forth in the table.

Footnote k. Substantial increases in the maximum allowable quantities are permitted for Class 3 oxidizers used for maintenance and sanitation purposes. A common application of the increased quantities occurs in health-care facilities.

Footnote l. If the net weight of the pyrotechnic composition of fireworks is known, that weight is used in applying the limitations of the table. Otherwise, 25 percent of the gross weight of the fireworks is to be used, including the packaging materials.

Footnote o. Cotton is almost exclusively pressed and stored as densely packed baled cotton meeting the weight and dimension requirements of ISO 8115. In this form, the fibers are not easily ignitable, and the regulation by this table as a hazardous material is deemed unnecessary.

Footnote p. This note further clarifies the application of Exception 3 to Section 307.1. Where liquid or gaseous fuel is used in the operation of machinery or equipment, including vehicles, the quantities are not to be included in the determination of maximum allowable quantities.

Footnote q. The listing of combustible dusts in Table 307.1(1) is unique in that it is the only material listed where a maximum allowable quantity has not been established in order to determine occupancy classification. The classification of Group H-2 is to be based solely on a determination by a qualified person, firm, or corporation, along with concurrence of the building official. Reference is made to Section 414.1.3 where the general information that is required to make such a determination is set forth.

Table 307.1(2)—Maximum Quantity per Control Area of Hazardous Materials Posing a Health Hazard. This table is similar in nature to Table 307.1(1), except that the maximum allowable quantities listed in Table 307.1(2) are for health hazard materials. For discussions of specific footnotes, see the above discussion of Table 307.1(1).

Multiple hazards. As previously noted, most hazardous materials possess the characteristics of more than one hazard. This section requires that all hazards of materials must be addressed. For example, if a material possesses the hazard characteristics of a Class 2 oxidizer and a corrosive, the material would be regulated under the provisions for both Group H-3 and H-4 occupancies. The more restrictive provisions of each occupancy must be satisfied.

307.8

Section 308 *Institutional Group I*

Group I occupancies are institutional uses and in the IBC are considered to be basically of two broad types. The first are those facilities where individuals are under supervision and care because of physical limitations of health or age. The second category includes those facilities in which the personal liberties of the occupants are restricted.

In both types, the occupants are either restricted in their movements or require supervision in an emergency, such as a fire, to escape the hazard by proceeding along an exit route to safety. There is actually a third category in which the occupants enjoy mobility and are reasonably free of constraints but do require a measure of professional care and are asleep for a portion of the day.

As Group I occupancies are people-related occupancies, the primary hazard is from the occupants' lack of free mobility needed to extricate themselves from a hazardous situation. On the other hand, the hazard from combustible contents is typically very low and, as a result, the occupancy requirements for Group I occupancies are essentially based on the limited free mobility of the occupants. Also, the occupants of most Group I occupancies are usually institutionalized for 24 hours or longer and therefore are asleep at some point during their stay. Thus, the protection requirements of the code are more comprehensive than in almost any other people-related occupancy. It should be noted that institutional occupancies described as Groups I-1, I-2, and I-4 may be classified as Group R-3 where care is provided for five or fewer persons. As an option for such small institutional uses, the structure need only comply with the IRC. The code typically recognizes that such small occupant loads in educational or institutional environments can be adequately addressed for fire and life safety through the provisions for dwelling units.

Although the Group I-3 classification is only applicable where six or more individuals are restrained or secured, there is no indication as to the proper classification where five or fewer persons are involved. Where the intended detention or restraint occurs as an accessory use within some other occupancy, classification would be based on that of the major occupancy. For example, up to five individuals could be restricted in areas such as interrogation rooms for alleged shoplifters in a covered mall building, jewelry viewing rooms for customers of a retail store, and time-out rooms in a school, without classifying that portion of the building a Group I-3 occupancy. Further discussion on allowances for locking devices for such spaces can be found in the commentary on Section 1008.1.9.5, Exception 1.

308.3 Institutional Group I-1. The occupants housed in buildings classified as Group I-1 occupancies are ambulatory but live in a supervised environment where custodial care services are provided, such as assistance with cooking, bathing, and other daily tasks. There is no need for the occupants to receive any physical staff assistance should a fire or other emergency exist that would require the occupants to evacuate the building or relocate within the building. Types of uses included in this category are halfway houses, alcohol and drug rehabilitation facilities, assisted living facilities, convalescent facilities, and group homes. Several of the listed uses may rather be considered Group I-2 or I-3 occupancies if the residents are incapable of self-preservation because of injury, illness, or incarceration. For example, an alcohol treatment center may provide lockdown for a number of persons under care. Where this number exceeds five, a Group I-3 classification would be more appropriate than a Group I-1. The same concern occurs where a convalescent facility provides care for individuals who are not able to respond to emergency conditions without physical assistance. Again, once the threshold of five persons incapable of self-preservation is exceeded, the proper occupancy classification would not be Group I-1.

The specific classification is based on the number of residents. In this case, a Group I-1 is the proper classification where more than 16 occupants are receiving custodial care. The threshold of 17 or more is based on the number of supervised individuals who reside in the facility and does not include associated staff members. Where the number of residents is between six and 16 inclusive, a Group R-4 classification is appropriate. Those supervised residential facilities that provide custodial care services for five or fewer occupants can be either considered a Group R-3 occupancy or designed and constructed under the provisions of the IRC.

308.4 Institutional Group I-2. The primary feature that distinguishes the Group I-2 occupancy from the others is that it is a medical care facility in which the patients are, in general, nonambulatory. This classification includes hospitals, detoxification facilities, psychiatric hospitals, and nursing homes. The nursing homes included in this category are deemed to provide intermediate care or skilled nursing care. Foster care facilities for the full-time care of infants (under the age of 2½ years) are included in this classification, as the code assumes that the very young require the same protection as is provided for those individuals whose capability of self-preservation is severely restricted. Where care is provided for a limited time period, such as at an outpatient health-care clinic, a classification of Group I-2 is not appropriate. In such cases, a Group B classification is warranted, even in those cases where some of the patients are incapable of self-preservation.

308.5 Institutional Group I-3. The uses of Group I-3 occupancies encompass jails, prisons, reformatories, detention centers, and other buildings where the personal liberties of the inmates are similarly restricted. For guidance on the classification of detention facilities having occupant loads of five or less, see the general discussion of Section 308. The classification of Group I-3 buildings shall also include one of five occupancy conditions. Several provisions specific to Group I-3 occupancies vary based on which condition is anticipated, such as the manner of subdividing resident housing areas. The conditions are described as follows:

1. The highest level of freedom assigned to a Group I-3 occupancy is considered Condition 1. Free movement is permitted throughout the sleeping areas and the common areas, including access to the exterior for egress purposes. A facility classified as Condition 1 is permitted to be constructed as a Group R occupancy, most likely a Group R-2. There may also be cases where it is more appropriate to classify the use as some occupancy other than Group R. For example, an industrial building included within a Condition 1 facility would most probably be classified as a Group F-1 occupancy.

2. Condition 2 buildings permit free movement between smoke compartments; however, access to the exterior for egress purposes is restricted because of locked exits.

3. Access between smoke compartments is not permitted in Condition 3 occupancies, except for the remote-controlled release of locked doors for necessary egress travel. Movement within each individual smoke compartment is permitted, including access to individual sleeping rooms and group activity spaces.

4. Condition 4 buildings restrict the movement of occupants to their own space, with no freedom to travel to other sleeping areas or common areas. Movement to other sleeping rooms, activity spaces, and other compartments is controlled through a remote release system.

5. The lack of freedom provided in Condition 5 facilities is consistent with that of Condition 4. However, staff-controlled manual release is necessary to permit movement throughout other portions of the building.

Institutional Group I-4, day-care facilities. Where custodial care is provided for persons for periods less than 24 hours at a time, an occupancy classification of Group I-4 is appropriate. The code further restricts this category by limiting the care to individuals other than parents, guardians, or relatives. This occupancy classification is appropriate for day-care facilities with children no older than 2½ years of age, as well as older persons who are deemed to be incapable of self-preservation. The need for supervision and custodial care services is the primary factor that contributes to this type of use being classified as an institutional occupancy.

308.5

Day-care facilities for children above the age of 30 months are considered educational occupancies (Group E). Where certain conditions are met, such a facility caring for infants/toddlers (30 months or less in age) may also be classified as Group E. See the discussion on Section 305.2. Where the care activities are for adults who can physically respond to an emergency situation without physical assistance, an institutional classification is not appropriate.

Section 309 *Mercantile Group M*

The mercantile uses listed in this section are mostly self-explanatory. For the most part, occupants of this type of use are ambulatory adults, with younger children supervised by parents or other adults. Although the occupancies may contain a variety of combustible goods, the possibility of ignition is limited. High-hazard materials may be present in small quantities, but not enough of the hazard material is present to be considered a Group H occupancy. For the limitations on hazardous materials in a Group M occupancy, see the discussion on Section 414.

3 Use and Occupancy Classification

As a sales operation, a service station whose primary function is the fueling of motor vehicles is considered a Group M use. The Group M designation applies to any building or kiosk used to support the function of vehicle fueling. This classification would also apply to a canopy constructed over the pump islands. By assigning the structures associated with the fueling of motor vehicles a classification the same as that for other sales operations, such as convenience stores, there is no question as to the proper application of the code. Through the design and construction of a motor-vehicle service station in conformance with Section 406.7 and the IFC, there is no distinct uncontrolled hazard that would cause separate and unique occupancies to be assigned. Conversely, where service or repair activities are involved, a Group S-1 classification is warranted. This would include operations limited to the exchange of parts, such as tire and muffler shops, as well as service-oriented activities (oil change and lubrication work).

Section 310 *Residential Group R*

Group R occupancies are residential occupancies and are characterized by:

1. Use by people for living and sleeping purposes.
2. Relatively low potential fire severity.
3. The worst fire record of all structure fires.

The basic premise of the provisions in this section is that the occupants of residential buildings will be spending about one-third of the day asleep and that the potential for a fire getting out of control before the occupants are awake is quite probable. Furthermore, once awakened, the occupants will be somewhat confused and disoriented, particularly in hotels.

310.1 Residential Group R. The four unique residential classifications are based on occupant density as well as the permanency of the occupants. Therefore, hotels, motels, and similar uses in which the occupants are essentially transient in nature are distinct in classification from apartment houses. The reason for this is the occupants' lack of familiarity with their surroundings. This in turn leads to confusion and disorientation when a fire occurs while the occupants are asleep. Because of this key difference, hotels and motels are considered Group R-1 occupancies, whereas apartment houses are designated as Group R-2. The uses classified as Group R-3 and R-4 include both transient- and nontransient-type facilities housing limited numbers of occupants.

310.3 Residential Group R-1. In addition to hotels and motels for transient guests, the Group R-1 classification also includes boarding houses and congregate living facilities, such as bed-and-breakfast establishments, where the number of transient occupants exceeds 10 persons. Such facilities are permitted to be classified as Group R-3 rather than Group R-1 where the occupant load does not exceed 10; however, they are not permitted to be constructed in accordance with the IRC. Consider a transient lodging operation consisting of a large number of single-family cabins containing living, cooking, sleeping, and sanitation facilities. The cabins could be classified in whole as a Group R-1 occupancy, or more probably, each individual cabin would be considered as a Group R-3 structure, provided the occupant load of the cabin does not exceed 10 persons. However, it would not be appropriate to apply the provisions of the IRC to this type of transient use.

310.4 Residential Group R-2. Included in the Group R-2 occupancy classification along with apartment buildings are dormitories, fraternity and sorority houses, convents, and monasteries. These types of uses are considered congregate living facilities. By definition, they contain one or more sleeping units where the residents share bathroom and/or kitchen facilities. An occupancy classification of Group R-2 is only appropriate for congregate living facilities where the occupant load of the facility is 17 or more persons. A lesser number will result in a Group R-3 classification. On the other hand, apartment buildings containing three or more dwelling units are always considered Group R-2 occupancies regardless of the building's occupant load. The Group R-2 classification is also appropriate for live/work units as regulated by Section 419. Live/work units include those dwelling units or sleeping units where a significant portion of the space includes a nonresidential use operated by the tenant.

Residential Group R-3. Group R-3 occupancies are generally limited to mixed-occupancy buildings containing only one or two dwelling units, as well as those small facilities used for adult or child care. It is expected that the occupant load of a Group R-3 occupancy will be quite low. Typically, dwellings would not be classified as Group R-3 occupancies, as they will be regulated by the IRC. Only where the dwelling falls outside the scope of the IRC will the Group R-3 classification for such structures be appropriate. For example, a four-story dwelling would be regulated as an R-3 occupancy, as would a single dwelling unit located above a small retail store. As previously indicated, boarding houses and congregate living facilities, both transient and nontransient in nature, may be classified as Group R-3 where the occupant load is relatively low.

310.5

Where the use of a single-family dwelling classified as a Group R-3 occupancy consists of adult or child care activities, the IRC may be used, provided the building falls under the scoping provisions of the IRC. It should be noted that Group R-3 uses constructed under the provisions of the IRC are required either to be protected by an NFPA 13D automatic sprinkler system or to comply with the IRC residential sprinkler system provisions of Section P2904.

Residential Group R-4. Assisted living facilities and other 24-hour custodial care facilities are to be classified as Group R-4 occupancies, provided the number of residents under custodial care does not exceed 16. Where the number of residents is five or less, the use is considered a Group R-3 occupancy. The list of Group R-4 uses is fully consistent with those designated as Group I-1 (more than 16 persons receiving care), as the only difference between the two classifications is the number of persons who receive custodial care.

310.6

The Group R-4 classification applied to supervised custodial care facilities differentiates them from Group R-3 care facilities based solely on the number of persons receiving care. For the most part, there is no difference between the hazards that are anticipated at such facilities. Therefore, the code mandates that those uses classified as Group R-4 comply with the requirements for Group R-3 occupancies except for those requirements specific to Group R-4 occupancies.

As an example, assume an assisted living facility has accommodations for up to 12 persons who receive custodial care. The facility would be classified as a Group R-4 occupancy since the number of care recipients is more than five but less than 17. However, most of the applicable code requirements would be based on a Group R-3 classification. Stair riser heights would be limited to 7¾ inches with minimum tread runs of 10 inches required, as set forth in Exception 5 to Section 1009.7.2. A reduction in the required 42-inch height of guards would be permitted per Section 1013.3, Exceptions 1 through 3. In addition, the emergency escape and rescue opening provisions of Section 1029 would be applicable. Unless the code specifically identifies a requirement as applicable to a Group R-4 occupancy, the provisions for Group R-3 shall be met.

There are, however, a number of provisions specifically established for Group R-4 occupancies. Using the example in the preceding paragraph, Table 503 is to be used as the basis for allowable building heights and areas based on the Group R-4 classification, with a tabular allowable area of 7,000 square feet if the building is of Type VB construction. A fire alarm system and smoke alarms are generally required under the provisions of Section 907.2.10, and an Accessible unit would be mandated per Section 1107.6.4.

Section 311 *Storage Group S*

In general, the Group S designation includes storage occupancies that are not highly hazardous and uses related to the storage, servicing, or repair of motor vehicles. Such storage uses are classified into two divisions based on the hazard level involved. Group S-1 describes those buildings used for moderate-hazard storage purposes, whereas low-hazard uses make up the Group S-2 classification.

Before discussing the two different types of storage uses, consideration should be given to the classification of borderline uses. An exclusionary rule is used to assist in determining those moderate-hazard storage uses that are to be classified as Group S-1. For example, a Group S-1 occupancy is used for storage of materials not classified as a Group S-2 or H occupancy. The building official will often be called upon to decide which classification is most appropriate when a use can fall within the two Group S occupancy classifications. As guidance in making this decision, it is usually more appropriate to choose the most restrictive occupancy, which is the Group S-1. Classifying the use into the more restrictive category would allow the building to be protected at a higher level and address the worst-case situations that might occur. By classifying the use into the least restrictive category, it would typically reduce the required controls, causing a potential problem where the building operator chooses to store combustible materials within the building.

311.2 Moderate-hazard storage, Group S-1. Group S-1 occupancies are typically used for the storage of combustible commodities. A complete list of all products allowed in this use would be very lengthy; however, many of the more common storage items are identified by the code. In addition, repair garages for motor vehicles are considered Group S-1 occupancies. In general, buildings classified in this manner would be used for the storage of commodities that are manufactured within buildings classified as Group F-1 occupancies. Commodities that constitute a high physical or health hazard, and exceed the maximum allowable quantities set forth in Section 307, would be stored in the appropriate Group H occupancy.

311.3 Low-hazard storage Group S-2. Group S-2 occupancies include the storage of noncombustible commodities, as well as open or enclosed parking garages. Buildings in which noncombustible goods are packaged in film or paper wrappings or cardboard cartons, or stored on wooden pallets are still considered Group S-2 occupancies. This also includes any products that have minor amounts of plastics, such as knobs, handles, or similar trim items. It is important, however, that the commodities being stored are essentially noncombustible, insofar as the provisions that regulate Group S-2 occupancies are based on an anticipated minimal fire load.

The classification of a warehouse or similar storage building as a Group S-2 occupancy is incumbent upon the expectation that combustible materials will not be stored within the building. This expectation is almost always difficult to achieve due to the transient nature of the materials being stored. It is not uncommon for the types of materials being stored to vary significantly from one month to the next, particularly for those storage activities not directly related to a specific manufacturing function. Therefore, the most common type of Group S-2 storage condition would be an extension of a Group S-2 manufacturing operation, where the types of materials being stored are consistent with those being manufactured or produced.

Section 312 *Utility and Miscellaneous Group U*

This section covers those utility occupancies that are not normally occupied by people, such as sheds and other accessory buildings, carports, small garages, fences, tanks and towers, and agricultural buildings. The fire load in these structures and uses varies considerably but is usually not excessive. Because they are normally not occupied, the concern for fire load is not very great, and as a group these uses constitute a low hazard. It is also important to note that a Group U occupancy is not expected to have any public use.

Group U occupancies can generally be divided into two areas. The first includes those buildings that are accessory to other major-use structures. Although these accessory-use buildings will at times be occupied, the time period for occupancy is typically limited to short intervals. The second type of Group U occupancy is those miscellaneous structures that cannot be properly classified into any other listed occupancy. The structures are not intended to be occupied, but must be classified in order to regulate any hazards they may pose to property or adjoining structures and persons.

The uses classified as Group U have been deemed to pose little, if any, risk to persons who may be present. Where hazards exist that are not typical of those represented by a Group U classification, it is important that another occupancy classification be assigned. For example, an agricultural building used as an arena for horse shows or livestock auctions should typically be regulated as a Group A assembly occupancy rather than a Group U agricultural structure.

If the jurisdiction has adopted Appendix C, then it will govern the design and construction of agricultural buildings that come under its purview; however, many urban jurisdictions do not adopt this appendix chapter. In this case, should an occasional agricultural building be constructed, it would be regulated by Section 312.

KEY POINTS

- Proper occupancy classification is a critical decision in determining code compliance.
- Uses are classified by the code into categories of like hazards, based on the risk to occupants of the building as well as the probability of property loss.
- Group A occupancies include rooms and buildings with an occupant load of 50 or more, used for the gathering together of persons for civic, social, or religious functions; recreation, food, or drink consumption; or similar activities.
- The hazards unique to assembly uses are based primarily on the large occupant loads and the concentration of occupants into very small areas.
- Business uses, such as offices, are classified as Group B occupancies and are considered moderate-hazard occupancies.
- Group E occupancies are limited to schools for students through 12th grade and most day-care operations.
- Manufacturing occupancies, classified as Group F, are classified based on whether or not the materials being produced are combustible or noncombustible.
- Group H occupancies are heavily regulated because of the quantities of hazardous materials present in use or storage.

- Where amounts of hazardous materials are limited in control areas to below the maximum allowable quantities, the occupancy need not be considered a Group H.
- Both physical hazards and health hazards are addressed under the requirements for Group H occupancies.
- Institutional occupancies, classified as Group I, are facilities where individuals are under supervision and care because of physical limitations of health or age, or that house individuals whose personal liberties are restricted.
- Group M occupancies include both sales rooms and motor fuel–dispensing facilities.
- Residential Group R occupancies are partially regulated based on occupant load or number of units, as well as the occupants' familiarity with their surroundings.
- Group S occupancies for storage are viewed in a manner consistent with Group F manufacturing uses.
- Group U occupancies are utilitarian in nature and are seldom, if ever, occupied.

CHAPTER 4

SPECIAL DETAILED REQUIREMENTS BASED ON USE AND OCCUPANCY

Section 402 Covered Mall and Open Mall Buildings
Section 403 High-Rise Buildings
Section 404 Atriums
Section 405 Underground Buildings
Section 406 Motor-Vehicle-Related Occupancies
Section 407 Group I-2
Section 408 Group I-3
Section 409 Motion-Picture Projection Rooms
Section 410 Stages, Platforms, and Technical Production Areas
Section 411 Special Amusement Buildings
Section 412 Aircraft-Related Occupancies
Section 413 Combustible Storage
Section 414 Hazardous Materials
Section 415 Groups H-1, H-2, H-3, H-4, and H-5
Section 416 Application of Flammable Finishes
Section 417 Drying Rooms
Section 418 Organic Coatings
Section 419 Live/Work Units
Section 420 Groups I-1, R-1, R-2, and R-3
Section 422 Ambulatory Care Facilities
Section 423 Storm Shelters
Section 424 Children's Play Structures
Key Points

4 Special Detailed Requirements Based on Use and Occupancy

> This chapter provides specific detailed regulations for those types of buildings and uses that have very unique characteristics. The uses in Chapter 4, though encompassing only a very small fraction of the uses commonly encountered, require special consideration. Some of the provisions address conditions that could occur in various occupancy classifications such as covered and open mall buildings, high-rise buildings, and underground buildings. Concerns associated with motor-vehicle-related uses, hazardous occupancies, and institutional uses are specifically addressed. Special elements within a building, such as stages, platforms, motion picture projection rooms, and atriums, are also regulated by Chapter 4.
>
> In all cases it should be remembered that the provisions found in Chapter 4 deal in a more detailed manner with uses and occupancies also addressed elsewhere in the code. Some of the provisions in this chapter may be more restrictive than the general requirements of the code, whereas others may be less restrictive. The general rules found in other areas of the *International Building Code®* (IBC®) will govern unless modifications from this chapter are utilized.

Section 402 Covered Mall and Open Mall Buildings

Provisions for covered mall and open mall buildings included in the IBC set forth specific code requirements for a specific building type. Provisions in this section only apply to covered and open mall buildings having a height of not more than three levels at any one point and not more than three stories above grade plane. Furthermore, the provisions are only those that are considered to be unique to covered and open mall buildings. For those features that are not unique, the general provisions of the code apply. Covered and open mall buildings that comply in all respects with other provisions of the code are not required to comply with these provisions. It should be noted that foyers and lobbies of office buildings, hotels, and apartment buildings should not be considered covered or open mall buildings. An example of an open mall and associated buildings is shown in Figure 402-1.

402.1.1 Open space. The special provisions of Section 402 are applicable only where the entire building, including the anchor buildings and attached parking structures, are surrounded by permanent open space at least 60 feet (18 288 mm) in width and the anchor buildings are no more than three stories in height above grade plane. See Figure 402-2. The allowance provided in the exception to Section 402.1.1 for reduced open space surrounding covered mall buildings, including their associated parking garages and anchor stores, is consistent with that for other unlimited area buildings as permitted by Section 507.5 since a covered mall building contains similar characteristics of those buildings addressed in Section 507. They are limited in height, contain similar occupancies, are fully sprinklered, and require significant open space surrounding the building. A limit is placed on the occupancies contained within the covered mall building to maintain consistency with the occupancy groups addressed in Section 507.5. The reduction in required open space is not permitted where the covered mall building or anchor stores include Group B, H, I, or R occupancies. Where a building is considered an open mall building, the required permanent open space is regulated based on the open mall building perimeter line as described in Section 402.1.2.

Covered Mall and Open Mall Buildings

Figure 402-1
Open mall.

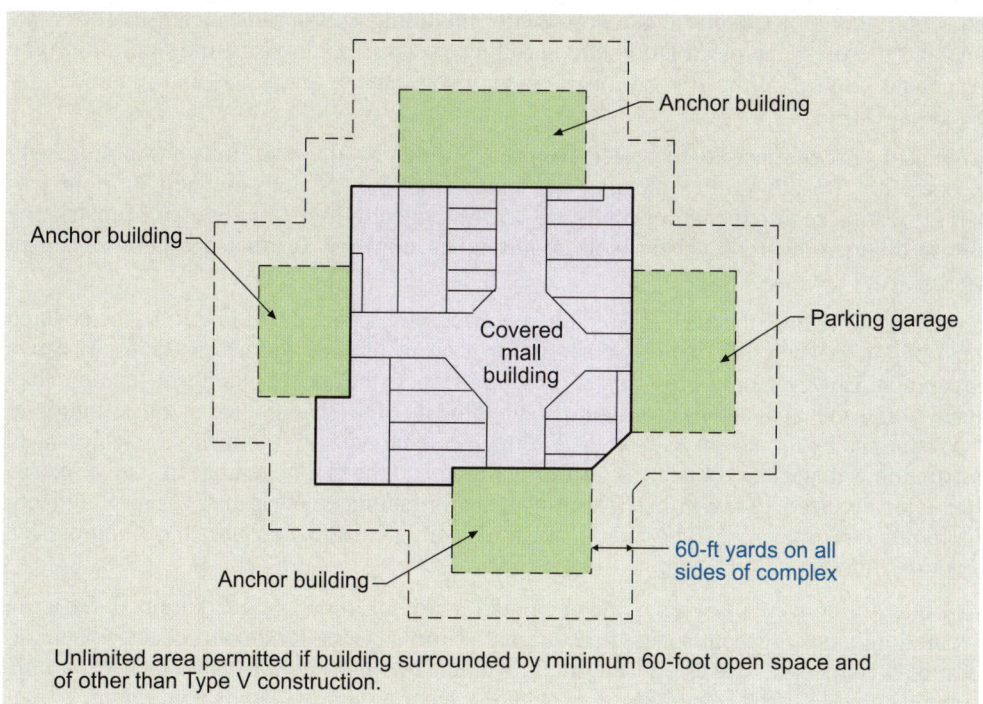

Unlimited area permitted if building surrounded by minimum 60-foot open space and of other than Type V construction.

Figure 402-2
Covered mall building.

Open mall building perimeter line. It is necessary that a perimeter line for an open mall building be established in order to apply various requirements in Section 402. The line creates the boundary of the tenant spaces, service areas, pedestrian paths, and similar spaces that make up the open mall building, but does not include any anchor buildings or parking garages. An example is shown in Figure 402-3.

402.1.2

2012 International Building Code Handbook   73

4 Special Detailed Requirements Based on Use and Occupancy

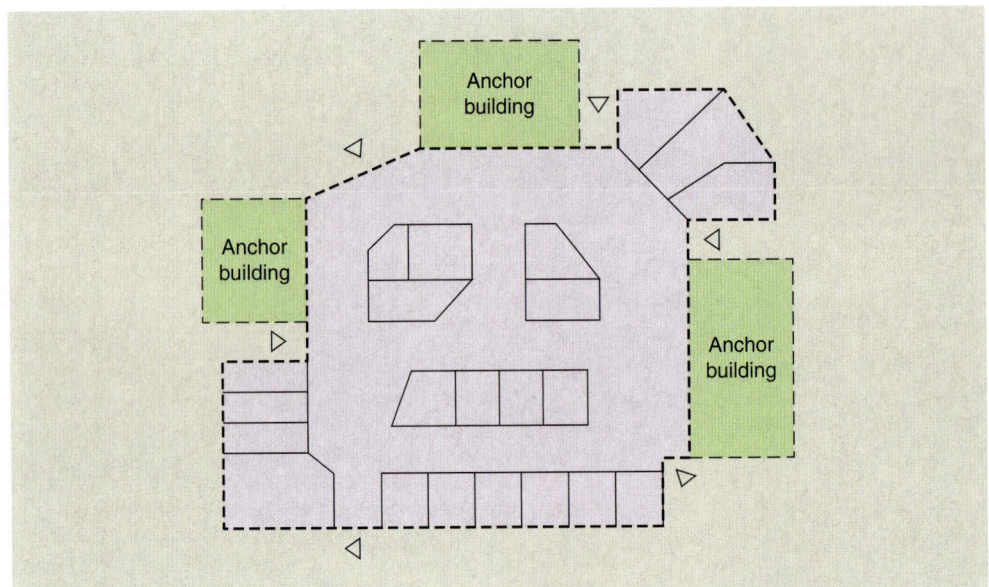

Figure 402-3
Open mall building perimeter line.

402.4 Construction. Where covered and open mall buildings are constructed of other than Type V construction, they may be of unlimited area. Associated anchor buildings may also have unlimited floor area provided they are no more than three stories in height above grade plane. For those anchor buildings exceeding three stories, the general height and area limitations of Chapter 5 are applicable, including appropriate increases for frontage and the presence of an automatic sprinkler system. In all cases, the minimum type of construction required for any open or enclosed parking garage is that mandated by Section 406.

Tenant spaces must be separated from each other by fire partitions complying with Section 708. In addition to protecting one tenant from the activities of a neighbor, the tenant separation requirements for malls are also intended to assist in the goal of restricting fire to the area of origin. There is no requirement, however, for the separation of tenant spaces from the mall itself.

As a general rule, the anchor building is viewed as a separate building from the covered mall building. Therefore, a fire wall must be used to provide the necessary fire-resistive separation. However, only a fire barrier is required where the anchor building is no more than three stories in height and its use is consistent with one of those identified in the definition of "Covered mall building" in Section 202. Although some type of 2-hour fire-resistance-rated separation is mandated between an anchor building and the mall, openings in such a separation typically need no fire-protection rating. Anchor buildings, other than Group R-1 sleeping units, constructed of Type I or II noncombustible construction may have unprotected openings into the mall.

Under the concept of covered and open mall buildings, there is no requirement for a fire separation between tenant spaces and the mall. Similarly, the food court needs no separation between tenant spaces and the mall. The hazards presented by an attached parking garage, however, must be addressed through the separation provided by a minimum 2-hour fire-resistance-rated fire barrier.

402.5 Automatic sprinkler system. An automatic fire-sprinkler system is the primary means of fire protection for a covered mall building. The system is required throughout all portions of the covered mall building other than open parking garages. Additionally, the code requires a standpipe system in accordance with Section 905. Because of the reliance placed on the sprinkler system, this section requires the following additional safeguards:

Covered Mall and Open Mall Buildings

1. The code requires that the sprinkler system be complete and operative throughout all of the covered mall building before occupancy of any of the tenant spaces. In those areas that are unoccupied, an alternative protection method may be approved by the building official.

2. The mall and the tenant spaces shall be protected by separate sprinkler systems, except that the code will permit spaces to be supplied by the same system as the mall, provided they can be independently controlled.

Sprinkler protection is also required for open mall buildings. This protection must extend to beneath any exterior circulation balconies located adjacent to the open mall.

Interior finish. The interior finish requirements for tenant spaces and anchor buildings are regulated for interior finishes based on their specific occupancy classification in accordance with Section 803.9. The common areas, including the mall and exits, are to have wall and ceiling finishes that have a minimum Class B flame spread rating.   **402.6.1**

Kiosks. Kiosks and similar structures, both temporary and permanent in nature, are regulated for construction materials and fire protection owing to their presence in an established egress path. Such structures shall be noncombustible or constructed of fire-retardant-treated wood, complying foam plastics, or complying aluminum composite materials. Active fire protection is provided by required fire suppression and detection devices. Kiosks are also limited in size, and their relationship to other kiosks is regulated. Multiple kiosks can be grouped together, provided their total area does not exceed 300 square feet (28 m^2). At that point, a separation of at least 20 feet (6,096 mm) is required from another kiosk or grouping of kiosks.   **402.6.2**

Plastic signs. In this section, the IBC limits plastic panels and plastic signs because they are within an exitway (the mall) and they are combustible (even though of approved plastic). It is important to note that the percentage of wall covered is based on the area common to each single tenant space. Thus, for a tenant space whose common wall with the mall is 60-feet (18,288-mm) wide and 11-feet (3,353-mm) high, the total area is 660 square feet (61.3 m^2). As the code permits 20 percent of that area to be of plastic panels or signs, the sum of all of the plastic signs and panels on the common wall is limited to 132 square feet (12.3 m^2). Figure 402-4 illustrates the code limitations for plastic signs and panels.   **402.6.4**

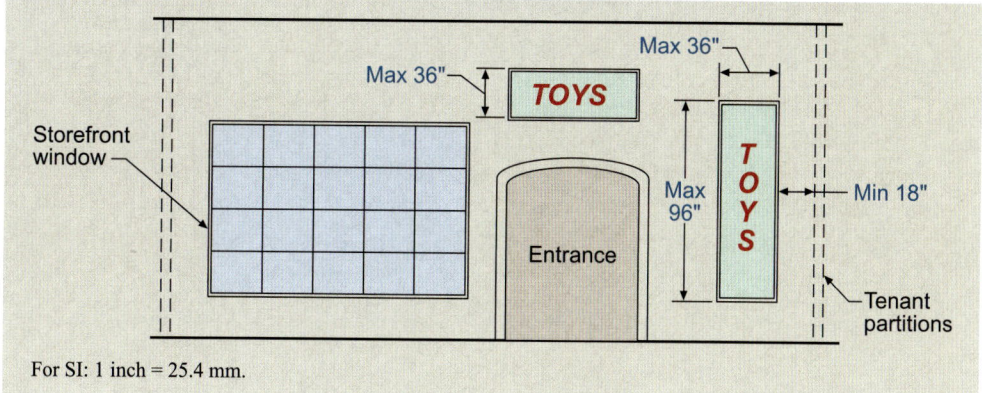

Figure 402-4 **Limitations for plastic signs and panels in malls.**

This section also requires that the use of foam plastic in signs be based on testing in accordance with UL 1975, *Fire Tests for Foamed Plastics Used for Decorative Purposes*, or NFPA 289, *Standard Method of Fire Test for Individual Fuel Packages*.

Smoke control. A smoke-control system need not be provided for a covered mall building unless an atrium is provided that connects at least three floor levels of the building. In covered mall buildings of one or two stories, no smoke-control system is required. In addition, a smoke-control system is not required for an open mall building due to the lack of   **402.7.2**

4 Special Detailed Requirements Based on Use and Occupancy

a roof over the mall area. The minimum 20-foot (6,096-mm) width mandated for an open mall, extending from the floor to the roof, provides an equivalent level of protection as a smoke-control system in the maintenance of a tenable environment in the mall area.

402.8 Means of egress. One of the significant areas in which the provisions for covered mall and open mall buildings differ from the general provisions applied to the majority of buildings is the means of egress. Issues such as occupant load determination and travel distance are modified specifically in this section owing to the unique features of a covered or open mall building. It is important to remember that where this section conflicts with the general requirements of the code in Chapter 10, the provisions of this section are applicable.

402.8.1 Mall width. With its added life-safety systems, the mall may be considered a corridor without meeting the width requirements of Section 1005.1 when the mall complies with the conditions of this section as depicted in Figure 402-5. In this case, the code requires that the minimum mall width be 20 feet (6,096 mm), and this typical cross section shows that the minimum required width may be divided so that a clear width of 10 feet (3,048 mm) is provided separately on each side of any kiosks, vending machines, benches, displays, etc., contained in the mall. In addition, food court seating in the mall would have to be located so as not to encroach upon any required mall width. Understandably, the mall width shall also accommodate the occupant load immediately tributary thereto.

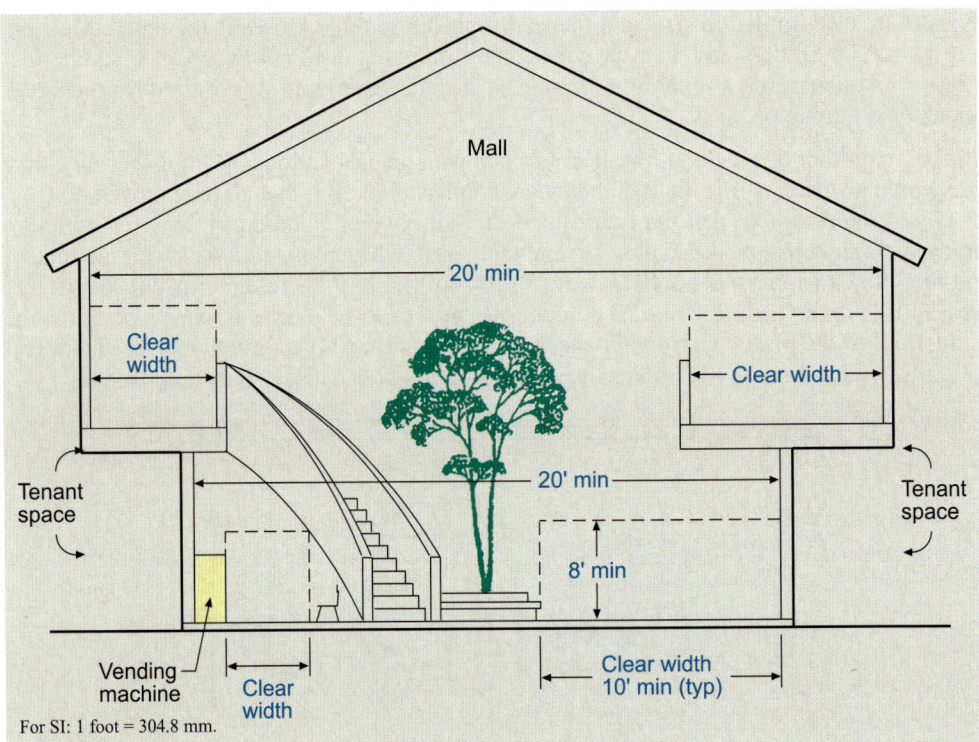

Figure 402-5
Mall width requirements.

402.8.2 Determination of occupant load. The determination of the occupant load and minimum required number of exits can be divided into two areas:

1. Tenant spaces.
2. The covered or open mall building.

The maximum occupant load of any individual tenant space is determined in a manner consistent with its use as regulated by Chapter 10. Means of egress requirements for individual tenant spaces are to be based on the occupant load as typically determined. Figure 402-6 depicts the method for determination of the occupant load in a tenant space

Covered Mall and Open Mall Buildings

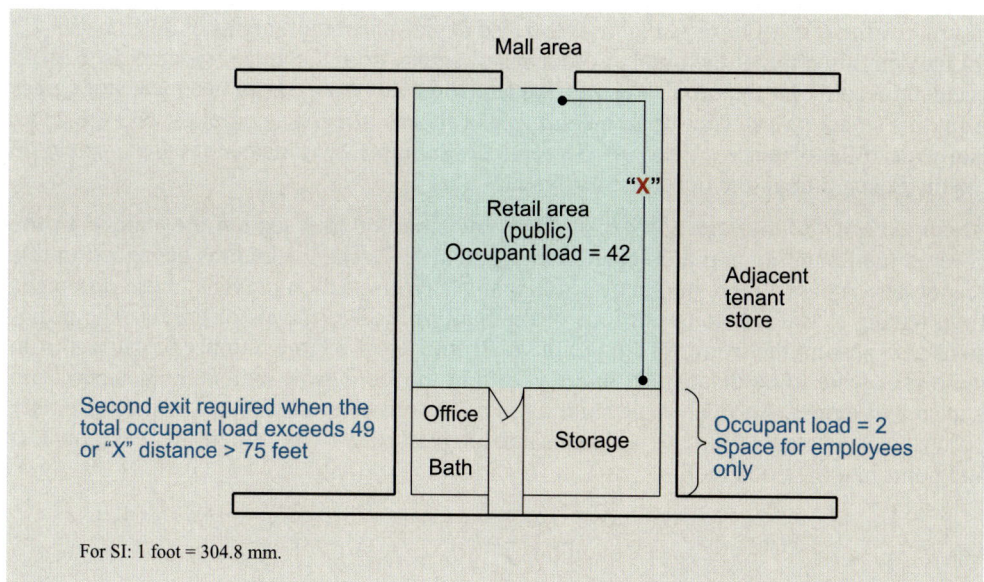

Figure 402-6
Occupant load and means of egress from tenant space.

having a retail area, a storage room, an office, and a bathroom. Although the total tenant space contains 1,500 square feet (139.3 m^2) of floor area, each individual use has a designated occupant load based on the appropriate factor from Table 1004.1.2. In this example, the occupant load would be calculated at 44.

Not only must the occupant load be determined for each individual tenant space, but the occupant load for the entire covered or open mall building must also be determined. It is highly unlikely that all tenant spaces will be fully occupied at the same time in a covered or open mall building. Therefore, a different method is used to determine the number of occupants from which to base the means of egress from the mall itself. This occupant load is to be determined based on the gross leasable area of the covered or open mall building, excluding any anchor buildings and those tenant spaces having an independent means of egress, with the occupant-load factor determined by the following formula:

Occupant-load factor = (0.00007) (gross leasable area) + 25

As a result, the net effect is that the total occupant load computed for the covered or open mall building will be something less than the summation of the occupant loads previously determined for each individual tenant space.

The occupant-load factor used for egress purposes shall not exceed 50, nor is it ever required to be less than 30. Where there is a food court provided within the covered or open mall building, the occupant load of the food court is to be added to the occupant load of the covered or open mall building as previously calculated in order to determine the total occupant load.

In utilizing several examples, assume a building contains 600,000 square feet (55,740 m^2) of gross leasable area. The occupant-load factor, when calculated, would be 67. However, a factor of 50 would be used in determining an occupant load of 12,000. Should a food court be present that seats 600 occupants, the occupant load of 12,000 would be increased accordingly. Where a covered mall building contains 100,000 square feet (9,290 m^2) of gross leasable area, the occupant-load factor would be 32. A factor of 32 would then be used to calculate the occupant load of the covered mall building, which would be 3,125 occupants. As the provision is applied to a smaller covered mall building, an occupant-load factor of 30 will be used when the gross leasable area of the covered mall building is less than 71,500 (6,642.3 m^2) square feet. As a final note, because anchor buildings are not considered a part of a covered mall building, their occupant load shall not be included in computing the total number of occupants for the mall.

4 Special Detailed Requirements Based on Use and Occupancy

402.8.3 Number of means of egress. Figure 402-6 also depicts the requirements of Section 402.4.2 for the determination of the number of means of egress from the tenant space. Based on an occupant load of 44, the provisions of Chapter 10 would require only one means of egress from this tenant space. Therefore, the number of means of egress complies with the code. However, if the distance, x, exceeds 75 feet (22,860 mm), two means of egress would be required even though the occupant load is less than 50.

402.8.4 Arrangements of means of egress. The provisions of this section are unique to the covered mall building and are depicted in Figure 402-7. The limitations described in this section encompass a large number of occupants, and this section prevents those occupants from having to traverse long portions of the mall to reach a means of egress. The provisions also prevent the overcrowding of the mall, such as if a large number of patrons from these uses were to be discharged into the mall at the same time and some distance from a means of egress. An open mall building is addressed a bit differently in that assembly occupancies located within the perimeter line are permitted to have their main exit open at any point into the open mall.

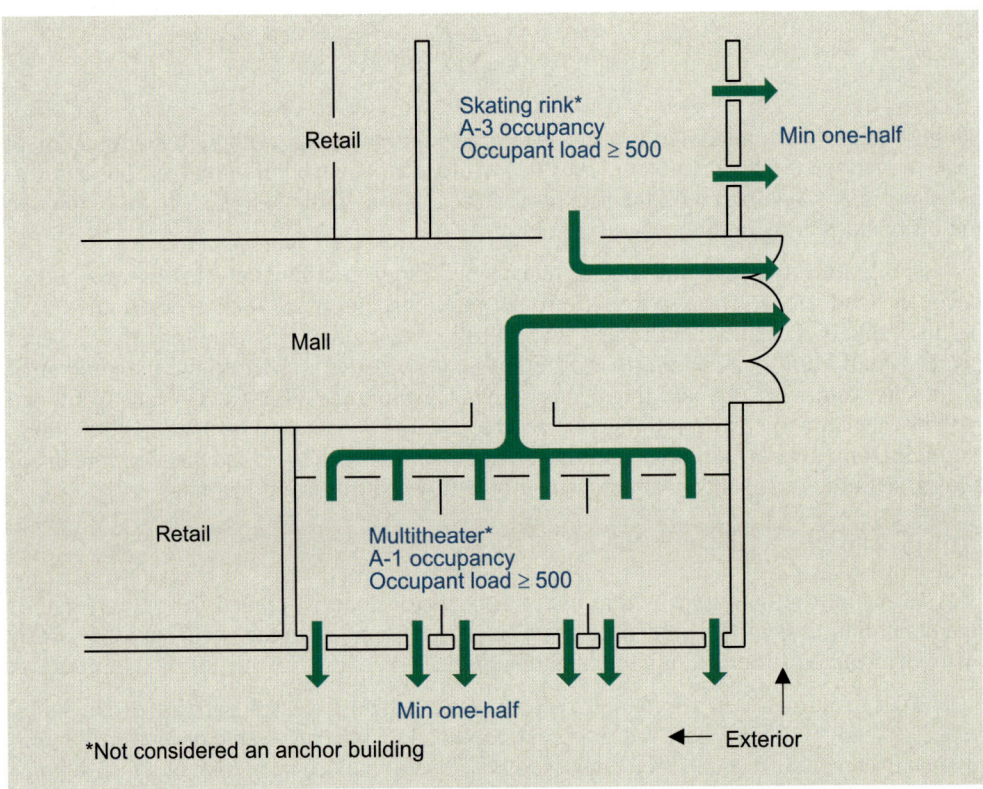

Figure 402-7 Arrangement of exits.

In securing the intent of Section 402, the provisions of Section 402.8.4 establish the requirement that means of egress for anchor buildings shall be provided independently from the mall exit system. Furthermore, the mall shall not egress through the anchor buildings. Moreover, the termination of a mall at an anchor building where no other means of egress has been provided except through the anchor buildings shall be considered to be a dead end, which is limited in accordance with Section 402.8.6.

402.8.5 Distance to exits. Figure 402-8 depicts the multifaceted provisions of this section as it relates to a covered mall building as follows:

1. The first case is illustrated for travel within the tenant space and includes the provisions applicable to tenant spaces A and B. For tenant space A, the diagram depicts the application of the code for a tenant space with a closed front with only

Covered Mall and Open Mall Buildings

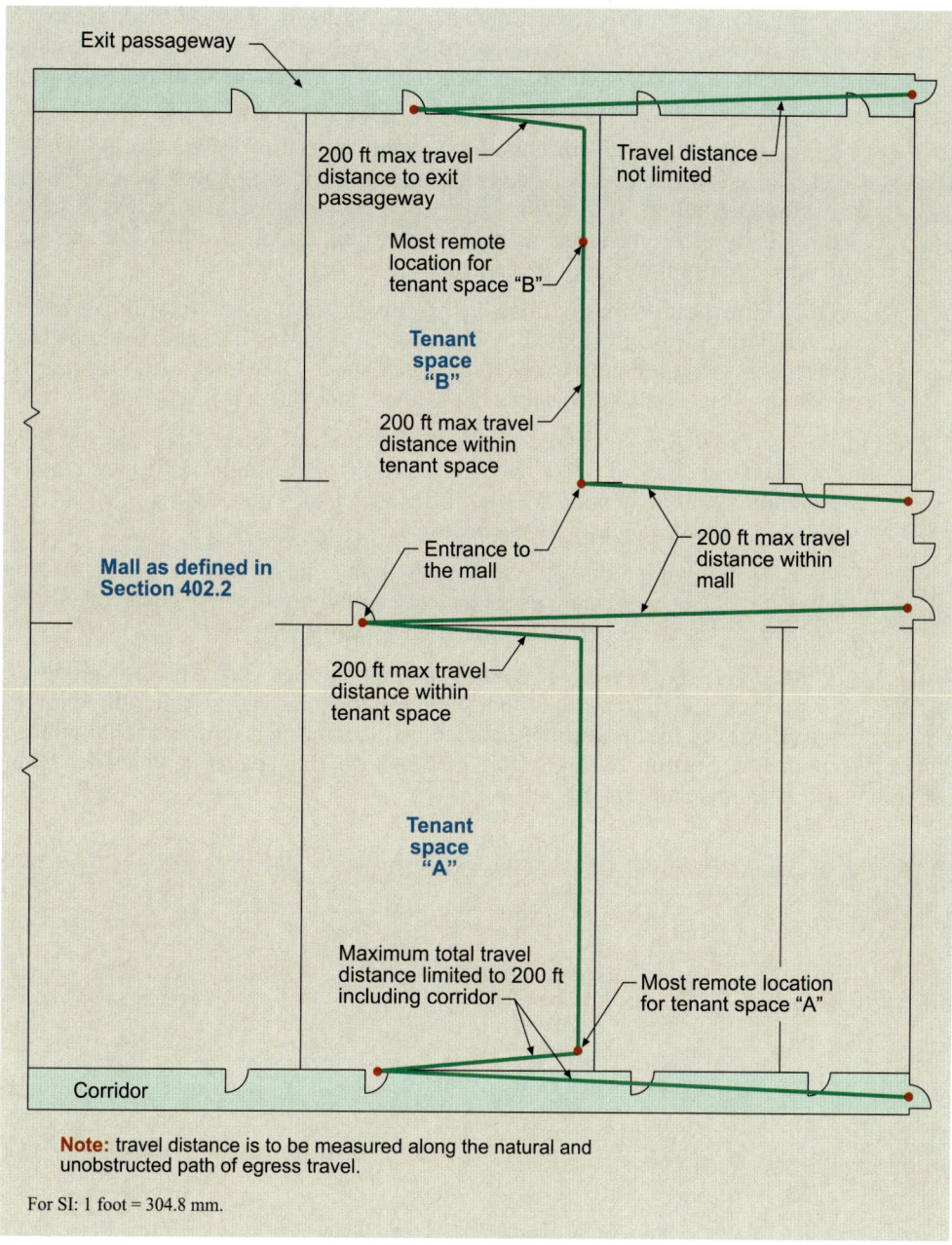

Figure 402-8
Travel distance.

a swinging exit door to the mall. The entrance to the mall would be the point at which occupants from the tenant space pass through the exit door from the tenant space to the mall.

Tenant space B represents the condition for an open storefront using a security grille instead of a standard exit door. The entrance to the mall in this case is the point at which occupants of the tenant space pass by an imaginary plane that is common to both the tenant space and the pedestrian mall. The location of the assumed required clear exit width along the open front of tenant space B may be placed at any point along the front, and its location would depend only on that which would render the least-restrictive application of the provisions.

For either tenant space A or B, the code permits the travel distance within the tenant space to the entrance to the mall to be a maximum of 200 feet (60,960 mm).

2. After the occupants exit from a tenant space into the mall, the code permits another 200 feet (60,960 mm) of exit travel distance to one of the exit elements described in Section 202. This travel limitation also applies to all other locations in the mall where occupants may be located when an exiting condition occurs.

It can be seen from this discussion, plus the perusal of Figure 402-8, that the travel distances permitted for a covered mall building are generally more liberal than those permitted by Section 1016.1. This liberalized and increased travel distance is based on the rationale that travel within a mall will be within an area where special fire protection features are provided.

3. Another limitation of these provisions regarding travel distance within covered mall buildings is also illustrated by Figure 402-8. In this instance, if the path of travel is through a secondary exit from tenant space B to an exit (in this case an exit passageway), travel distance is not limited once the exit is reached.

4. In the case of exiting via the corridor as depicted for tenant space A, the total travel distance is limited to 200 feet (60,960 mm). This limitation is based on the consideration that a corridor does not offer as much protection as either an exit (such as an exit passageway) or the mall.

For an open mall building, travel from an individual tenant space to the mall is addressed in the same manner as for a covered mall building. Once the open mall is reached, travel to the perimeter line is then regulated.

402.8.6 Access to exits. This section uses the same approach as does Chapter 10 of requiring that the means of egress be arranged so that the occupants may go in either direction to a separate exit. However, in this section the dead end is measured in a manner similar to that of Exception 3 of Section 1018.4. Figure 402-9 shows the manner in which dead-end conditions are measured and limited.

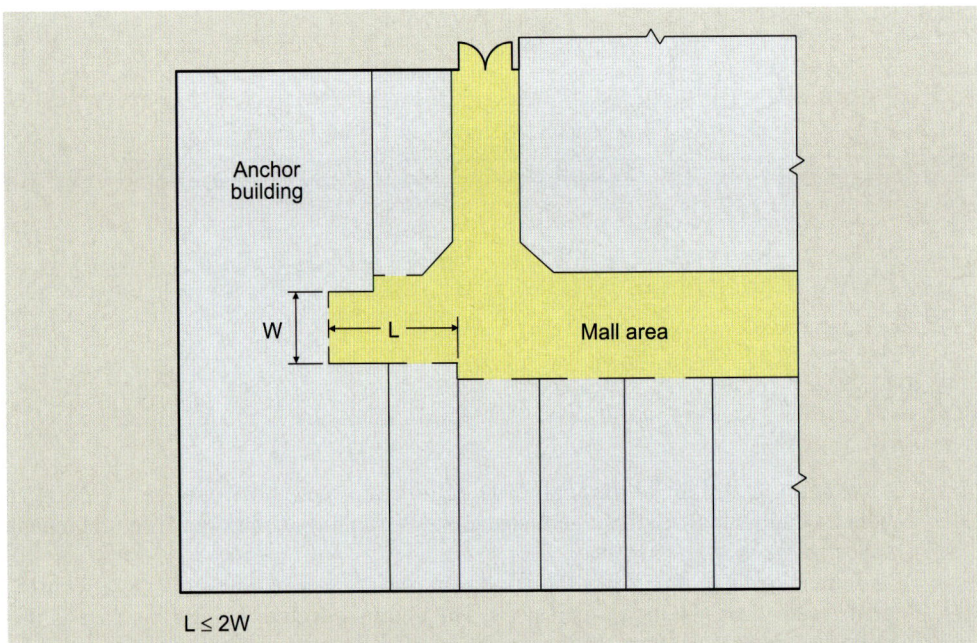

Figure 402-9 Dead-end mall criteria.

Regardless of the occupant load served, the minimum width of an exit passageway or corridor from a mall is to be 66 inches (1,676 mm). The exit passageway shown in Figure 402-10 must be at least 66-inches (1,676-mm) wide. The main entrances shown in the same figure are not subject to this requirement insofar as the exit width limitations of Section 1005.1 will most often require a greater exit width.

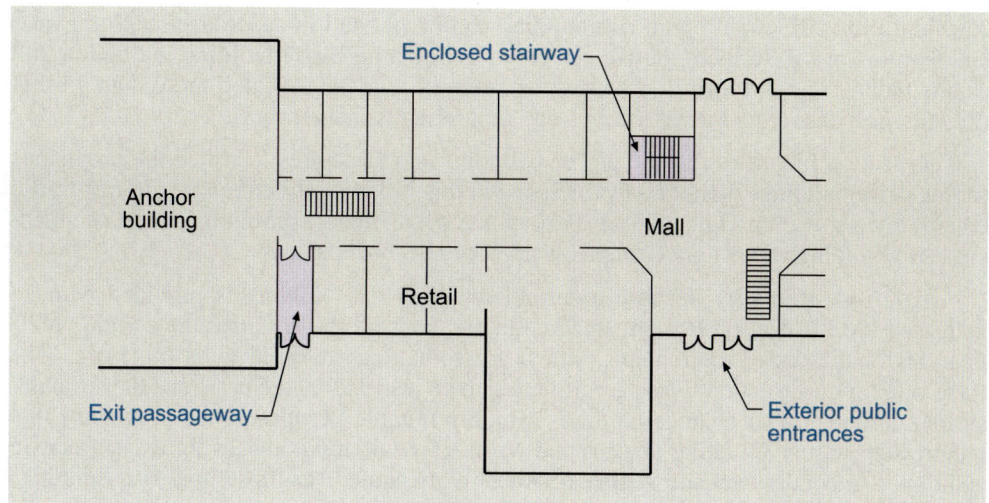

Figure 402-10
Exit passageway width.

Another exiting provision that is unique to a covered or open mall building regards exit passageway enclosures. Section 402.8.6.1 allows mechanical and electrical equipment rooms, building service areas, and service elevators to open directly into exit passageways, provided the minimum 1-hour fire-resistance-rated separation is maintained.

Security grilles and doors. Quite often, mall tenants wish to have the dividing plane between the mall and the tenant space completely open during business hours. Horizontal sliding or vertical security grilles or doors are usually placed across this opening. This section permits their use, provided they do not detract from safe exiting from the tenant space into the mall. To secure that intent, the code requires four limitations outlined in this section.

402.8.8

Section 403 High-Rise Buildings

This section encompasses special life-safety requirements for high-rise buildings. The comparatively good fire record notwithstanding, particularly in office buildings, fires in high-rise buildings have prompted government at all levels to develop special regulations concerning life safety in high-rise buildings. The potential for disaster that is due to the large number of occupants in high-rise buildings has resulted in the provisions included in this section.

The high-rise building is characterized by several features:

1. It is impractical, if not impossible, to completely evacuate the building within a reasonable period of time.
2. Prompt rescue will be difficult, and the probability of fighting a fire in upper stories from the exterior will be low.
3. High-rise buildings are occupied by large numbers of people, and in certain occupancies the occupants may be asleep during an emergency.
4. A potential exists for stack effect. The stack effect can result in the distribution of smoke and other products of combustion throughout the height of a high-rise building during a fire.

The provisions in this section are designed to account for the features described above.

4 Special Detailed Requirements Based on Use and Occupancy

403.1 Applicability. Although a high-rise building can be defined in accordance with the special features just described, the IBC elects to define a high-rise building in Section 202 as one having one or more floors used for human occupancy located more than 75 feet (22,860 mm) above the lowest level of fire department vehicle access.

This section identifies those types of buildings and structures to which the provisions for high-rise buildings do not apply. Included in this group of structures are aircraft-traffic control towers, open parking garages, Group A-5 occupancies, special industrial occupancies, and buildings housing a Group H-1, H-2, or H-3 occupancy.

403.2 Construction. Primarily because a sprinklered high-rise building is provided with an increased level of fire protection supervision and control, the IBC permits certain modifications of the code requirements, which are sometimes referred to as trade-offs. The trade-offs for construction type are considered to be justified on the basis that the sprinkler system, although a mechanical system, is highly reliable because of the provisions that require supervisory initiating devices and water-flow initiating devices for every floor. In addition, a secondary on-site supply of water is mandated for those high-rise buildings subject to a moderate to high level of seismic risk.

This section permits some degree of reduction in the required fire-resistance ratings of building elements required to be protected on account of type of construction. In the evaluation of the maximum allowable height and area permitted for the building, the original construction type would remain applicable.

In addition to the reductions permitted for building elements identified in Table 601 based on the building's type of construction, the fire-resistance rating of shaft enclosures may be reduced to 1 hour where sprinklers are installed within the shafts at the top and at alternative floor levels.

The reduction in fire-resistance ratings is not applicable in all cases. In all high-rise buildings of Type IA construction, the 3-hour rating for structural columns supporting floors must be maintained. The critical role of columns in the structural integrity of a high-rise structure during fire conditions mandates that their fire resistance not be lessened. For buildings that exceed 420 feet (128 m) in height, no reduction to the required fire resistance of any building elements is permitted. In addition, the required vertical shaft protection cannot be reduced from the general 2-hour requirement. The increased risk of catastrophic damage associated with these very tall buildings requires an increased level of fire resistance.

403.3 Automatic sprinkler system. The automatic fire-sprinkler system required by Section 403.3 must be completely reliable, as must the other life-safety systems. As part of that reliability effort, a high-rise building must be provided with a secondary water supply where required by Section 903.3.5.2. In Seismic Design Category C, D, E, or F, an on-site automatic secondary water supply shall be provided, with a supply of water equal to the hydraulically calculated sprinkler design demand, including the hose stream requirement, for a duration of at least 30 minutes. As fires can (and do) break out as a result of earthquake damage to the various mechanical systems within a building, it is imperative that the reliability of the sprinkler system be such that any resulting fires can be automatically extinguished.

403.4 Emergency systems. Among the more important life-safety features required by the code are the alarm and communications systems required by this section. Where it is expected that people will be unable to evacuate the building, it is imperative that they be informed as to the nature of any emergency that may break out, as well as the proper action to take to exit to a safe place of refuge. Furthermore, a system is necessary in most cases to provide for communication between the fire officer in charge at the scene and the fire fighters throughout the building. Section 911 provides details of the required fire command center utilized by the fire department to coordinate fire suppression and rescue operations.

In order to provide for efficient and reliable communications among fire fighters, police officers, medical personnel, and other emergency responders, an emergency responder radio communications system must be installed in all high-rise buildings. The details for such systems are established in Section 510 of the *International Fire Code*® (IFC®). IFC Section 510.4.1 specifies that acceptable radio coverage is satisfied when 95 percent of all areas on each floor of the building meet the signal strength requirements in Sections 510.4.1.1 and 510.4.1.2 for signals transmitted into and out of a building.

In addition, one of the fire department's duties during a fire event is to expel the smoke after the fire has occurred. Three methods for smoke exhaust are available: through natural means, through the use of mechanical air-handling equipment, or through an equivalency approach. Where the method of natural ventilation is used for the removal of smoke, openable windows or panels are required to be distributed around the perimeter of each floor level. The fire departments can open the appropriate windows as necessary and provide pressurization through the use of fans. The use of fixed tempered glass panels is also acceptable if they are not coated in a manner that will modify the natural breaking characteristics of the glass. Where mechanical air-handling equipment is used for smoke removal purposes, the building's HVAC system is equipped with appropriate dampers at each floor that are arranged in a manner that will stop the recirculation of air through the use of 100 percent fresh air intake and outside exhaust. The panel for controlling this system is to be located in the building's fire command center. In addition, the building official has the authority to approve any other means of smoke removal provided it accomplishes the intended goal of the prescriptive mechanical or natural ventilation approaches described by the code. See Figure 403-1. It must be noted that this smoke exhaust system is for fire department use only and is not intended to be a part of the occupant-related life-safety systems placed in high-rise buildings.

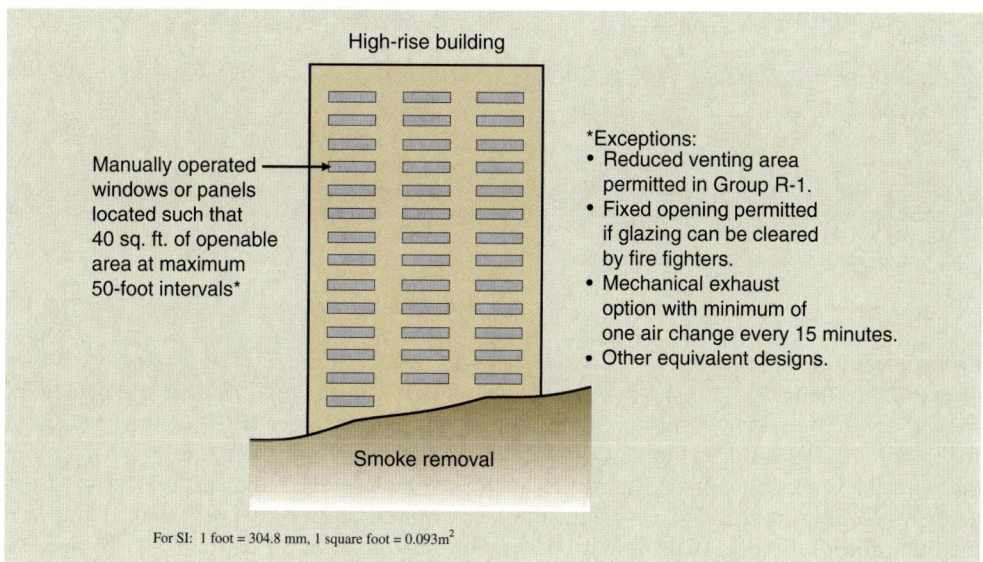

Figure 403-1
Smoke removal.

Furthering the intent of the IBC that life-safety systems in high-rise buildings be highly reliable, the code requires that the power supply to the life-safety systems be regulated by the appropriate provisions of NFPA 70, more specifically, Articles 700 and 701. The basis of the reliability is that the building's power be automatically transferable to a standby or emergency power system in the event of the failure of the normal power supply. Those standby power loads required by the code include power and lighting for the fire command center, ventilation and automatic fire detection equipment for smokeproof enclosures, and elevators. Some of the requirements and details for the standby power system are depicted in Figure 403-2.

4 Special Detailed Requirements Based on Use and Occupancy

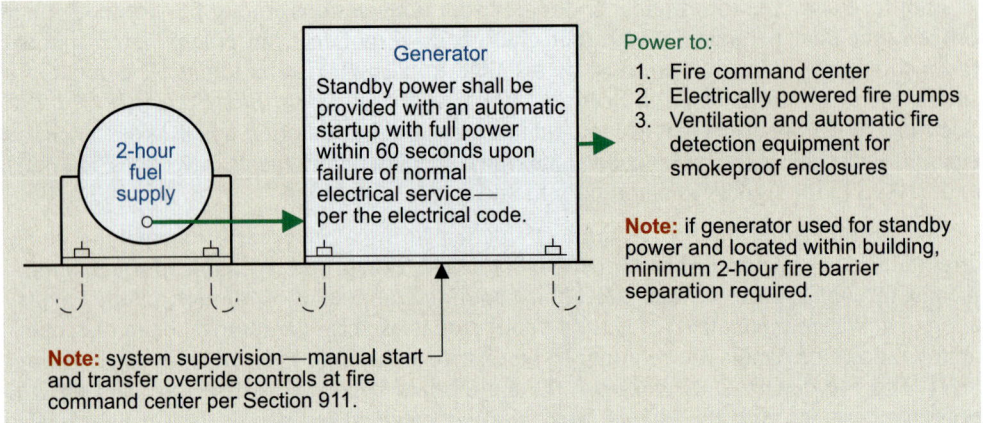

Figure 403-2
Standby power.

The code further requires that lighting for exit signs, means of egress illumination, and elevator car lighting be automatically transferable to an emergency power system capable of operation within 10 seconds of the failure of the normal power supply. Additionally, all emergency voice/alarm communications systems, automatic fire detection systems, fire alarm systems, and electrically powered fire pumps are to be provided with emergency power. Figure 403-3 illustrates one means of transfer to emergency power.

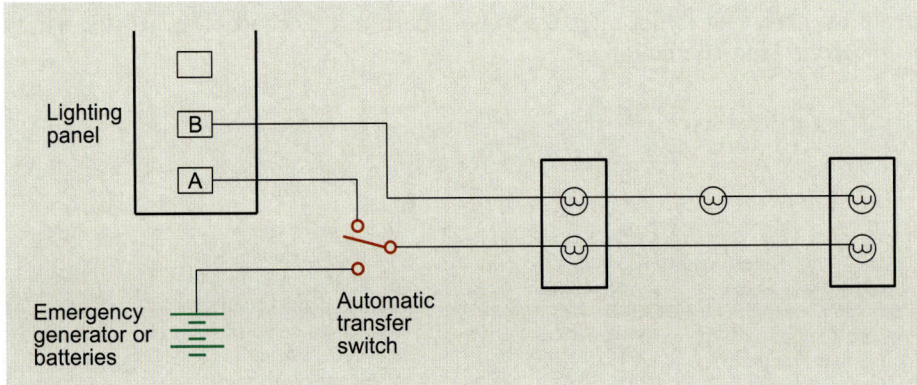

Figure 403-3
Emergency lighting system.

403.5.1 Remoteness of interior exit stairways. The general requirements for separation of exit or exit access doorways as established in Sections 1015.2.1 and 1015.2.2 are supplemented in this section by adding a minimum required separation distance between the enclosures for interior exit stairways. In addition to maintaining a minimum separation between the doors to the enclosures of one-third the length of the overall diagonal dimension of the area served, the interior exit stairways must be located at least 30 feet apart or not less than one-fourth the diagonal dimension, whichever is less. See Figure 403-4. If three or more interior exit stairway enclosures are mandated, at least two of the enclosures must be separated as indicated.

403.5.2 Additional exit stairway. During a fire that requires a full evacuation of a building of extensive height, the fire-fighting operations will reduce the capacity of the egress system. The extended period of time needed to fully evacuate a very tall building means that people will still be evacuating while full fire-fighting operations are taking place. Sound high-rise fire-fighting doctrine provides that the fire department take control of one stair, the one most appropriate to the circumstances of the given fire condition. This can result in a significant reduction in egress capacity of the stairway system. For example, in order to conduct suppression activities in a building with two required stairs of the same width,

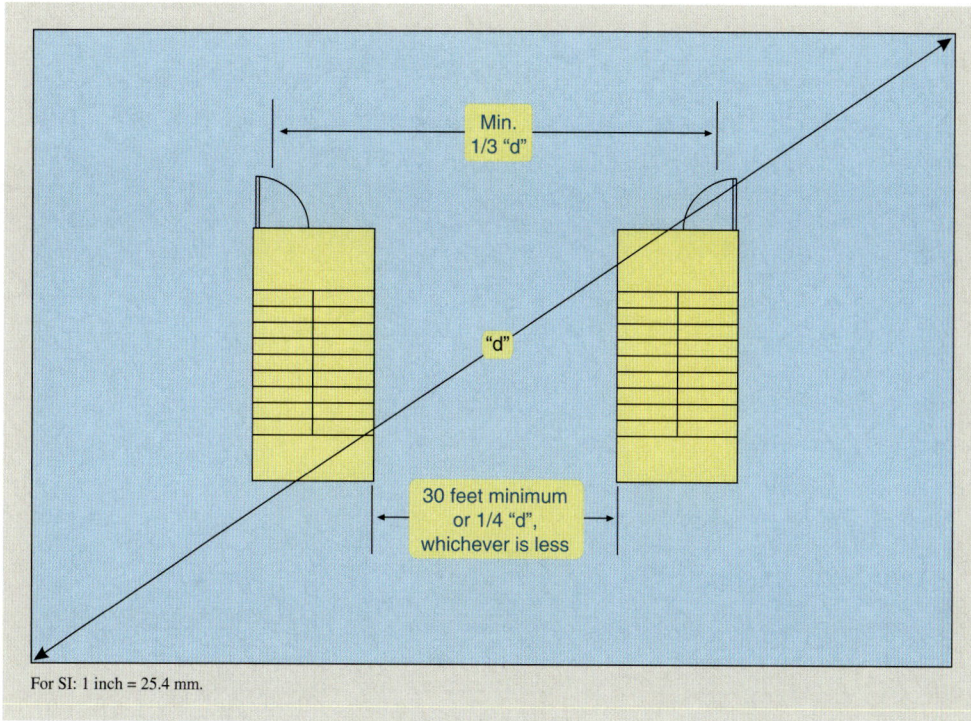

Figure 403-4
Remoteness of exit enclosure.

one-half of the exit capacity is unavailable while the building is still being evacuated. An additional stair is required so that egress capacity will be maintained through the time that full evacuation is complete.

It is important to note that this additional stair is not required to be a dedicated fire department stair. The fire department should be able to choose the stair that is most appropriate for the actual fire event. As a result, it will be necessary for emergency responders to manage evacuation flow to the available stairs.

The application of this requirement is limited to only those buildings over 420 feet (128 m) in height. In addition, it does not apply to Group R-2 occupancies due to the limited occupant load of such uses. In determining if the required egress width is provided by the stairway system, it must be assumed that the widest of the stairways is the one that is unavailable for means of egress travel. The remaining stairways must be sized to accommodate the total required egress width. The additional stairway's sole purpose is to provide additional egress capacity. Therefore, other means of egress design issues, such as travel distance and exit separation, are not regulated. See Application Example 403-1.

The additional exit stairway is not required where occupant evacuation elevators are provided in accordance with Section 3008. The availability of elevators for evacuation purposes provides for a reasonable alternative to an additional stairway.

Stairway door operation. In those cases where it is impractical to totally evacuate the occupants from the building through the stairway system, it must be possible to move the occupants to different floors of the building that are safe by way of the stairway system. For this to happen, the doors to the interior exit stairway enclosures must either be unlocked or be designed for automatic unlocking from the fire command station. **403.5.3**

The IBC further requires that a telephone or other two-way communication system (such as a two-way system with speaker and microphone) be located at every fifth floor in each required stair enclosure for those cases where the stair enclosure doors are to be locked. Moreover, the code requires that this communication system be connected to an approved station that is constantly attended. Thus, anyone trapped in the stairway during a

4 Special Detailed Requirements Based on Use and Occupancy

Application Example 403-1

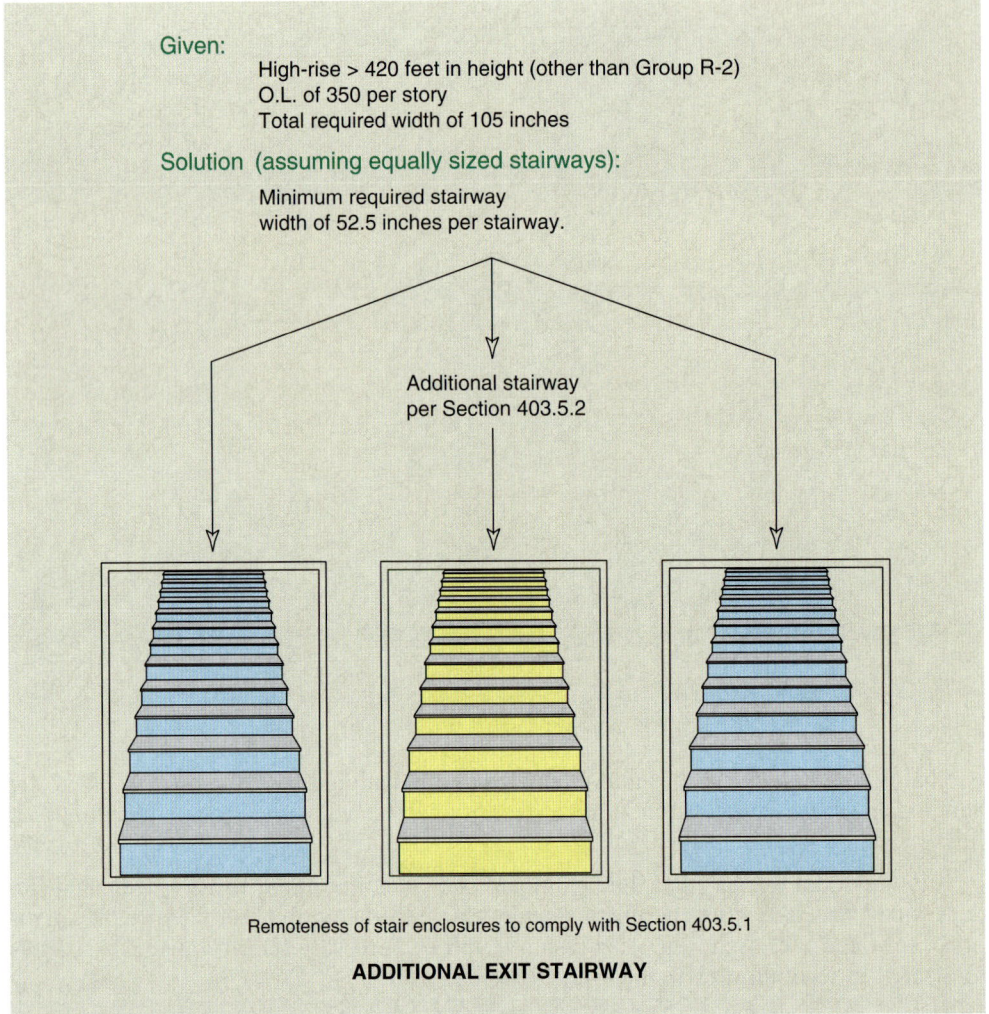

ADDITIONAL EXIT STAIRWAY

nonfire emergency may call for help without traversing more than two levels. In the case of office buildings and apartment houses, the attended station may be considered the office of the building as long as the office has continuous attendance by responsible individuals who are familiar with the life-safety systems. For hotel buildings, the most likely choice for the attendance station will probably be the hotel telephone operators, and, again, they must be trained to assist the persons trapped within the stair enclosure.

403.5.4 Smokeproof enclosures. Those interior exit stairway enclosures in a high-rise building that serve floor levels located more than 75 feet (22,860 mm) above the lowest level of fire department vehicle access must be designed as smokeproof enclosures. Exit stairways that do not serve floors above the height indicated are not regulated by this section. Section 1022.10 regulates the access, extension, and termination relating to the utilization of smokeproof enclosures and pressurized stairways as a part of the means of egress system. Section 909.20 provides the construction and ventilation criteria for smokeproof enclosures, as well as establishing stair pressurization as an acceptable alternative.

403.5.5 Luminous egress path markings. In high-rise buildings, increased visibility for travel on stairways and through exit passageways is important due to the extreme conditions that may be encountered under emergency conditions. The use of photoluminescent or self-illuminating materials to delineate the exit path is required in high-rise buildings housing Group A, B, E, I, M, and R-1 occupancies. Specific requirements related to these egress path markings are set forth in Section 1024 and include the regulation of striping on steps, landings, and handrails; perimeter demarcation lines on the floor and walls, including their transition; acceptable materials; and illumination periods. An example of such markings is shown in Figure 403-5.

Figure 403-5
Luminous egress path markings.

Elevators. In order to facilitate the rapid deployment of fire fighters, at least two fire service access elevators are required in high-rise buildings that have an occupied floor more than 120 feet (36,576 mm) above the lowest level of fire department vehicle access. Usable by fire fighters and other emergency responders, the specific requirements for the elevators are set forth in Section 3007. There are a number of key features that allow fire fighters to use the elevator for safely accessing an area of a building that may be involved in a fire. A complying lobby is required adjacent to the elevator hoistway opening, creating a protected area from which to stage operations. Access to standpipe hose valves is required, as are two-way communication features. A single fire service access elevator is permitted in the unlikely situation where a building regulated by this section has only one elevator. **403.6**

The mandate for multiple fire service access elevators is based on information that indicates at least two elevators are necessary for firefighting activities in high-rise buildings. In addition, past experience has shown that on many occasions elevators are not available due to shut downs for various reasons, including problems in operation, routine maintenance, modernization programs, and EMS operations in the building prior to firefighter arrival. A minimum of two fire service elevators provided with all of the benefits afforded to such elevators better ensures that there will be a fire service access elevator available for the fire fighters' use in the performance of their duties.

A more comprehensive discussion of the requirements for fire service access elevators is found in the analysis of Section 3007.

The use of elevators as an evacuation element for occupants of a high-rise building is possible provided the elevators are in compliance with the requirements established in

4 Special Detailed Requirements Based on Use and Occupancy

Section 3008. The controls and safeguards provided in Section 3008 create a suitable environment to allow complying elevators to be used for occupant self-evacuation purposes. The presence of such elevators does not reduce the general means of egress requirements established in Chapter 10; however, the additional exit stairway mandated for very tall high-rise buildings by Section 403.5.2 is no longer required. It is important to note that the installation of occupant evacuation elevators in high-rise buildings is not mandated by the code; however, such elevators are permitted for use for occupant self-evacuation and may be utilized as an alternative to the additional stairway requirement.

Section 404 *Atriums*

This section was developed to fill a need for code provisions applicable to the trends in the architectural design of buildings where the designer makes use of an atrium. Prior to the early 1980s, building codes did not provide for atriums, and, moreover, atriums were prohibited because of the requirements for protection of vertical openings. They were, however, permitted on an individual basis, usually under the provisions in the administrative sections of the code permitting alternative designs and alternative methods of construction. The general concept of alternative protection is to provide for both the equivalence of an open court and at the same time provide protection somewhat equivalent to shaft protection to prevent products of combustion from being spread throughout the building via the atrium.

An atrium is considered "an opening connecting two or more stories other than enclosed stairways, elevators, hoistways, escalators, plumbing, electrical, air-conditioning or other equipment, which is closed at the top and not defined as a mall." This section permits large unprotected vertical openings through floors without the need for a shaft enclosure or other means of vertical opening protection. The use of atriums is permitted in all buildings other than those classified as Group H occupancies.

Note that most cases where two floors are open to each other do not create atrium conditions. That is because Section 712.1.8 permits two stories to be open to each other where seven conditions are met. The atrium provisions are typically only utilized for open two-story spaces where they cannot fully comply with the conditions of Section 712.1.8 and are too large to qualify as a mezzanine as permitted by Section 712.1.10. Addressed in Section 712.1.6, the atrium provisions are provided only as one of many applications addressed in Section 712.1 for the protection of vertical openings.

404.3 Automatic sprinkler protection. One of the basic requirements for atriums is that the building be provided with an automatic sprinkler system throughout. See Figure 404-1. Two exceptions modify this general requirement. Those areas of the building adjacent to or above the atrium are not required to be sprinklered if appropriately separated from the atrium. This separation must consist of minimum 2-hour fire barriers, horizontal assemblies, or both. In addition, sprinkler protection is not required at an atrium ceiling located more than 55 feet (16,764 mm) above the atrium floor.

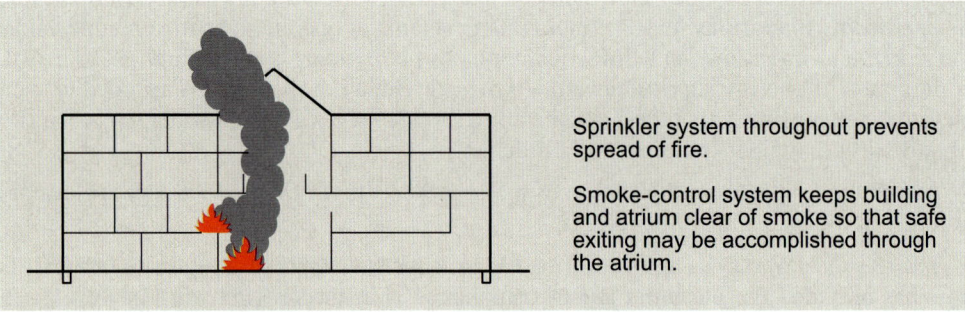

Figure 404-1 Atrium concept.

Atriums

404.5 **Smoke control.** Another major component of the life-safety system for a building containing an atrium is the required smoke-control system. The design of the smoke-control system is to be in accordance with Section 909. Although the exhaust method is typically used as the means of accomplishing smoke control, the code would not prohibit the use of the airflow or pressurization methods where shown to be suitable. One of these methods is often used where the ceiling height makes it difficult to maintain the smoke layer at least 6 feet (1,829 mm) above the floor of the means of egress.

An exception eliminates the requirement for smoke control in those atriums that connect only two stories. However, as previously addressed, most situations where two floors are open to each other are not regulated under the provisions of Section 404. Typically, Section 712.1.8 is utilized to permit an opening between two floor levels without requiring compliance with any of the atrium provisions.

404.6 **Enclosure of atriums.** With some exceptions, an enclosure separation is required between the atrium and the remainder of the building. See Figure 404-2. The basic requirement is for a 1-hour fire-resistance-rated fire barrier with openings protected in accordance with Tables 716.5 and 716.6. This degree of enclosure, in addition to the other special conditions of Section 404, is intended to provide protection somewhat equivalent to the otherwise mandated shaft protection. Two alternative methods of atrium separation are described in the exception. The special sprinkler-wetted glass enclosure as depicted in Figure 404-3 provides a prescriptive method of achieving equivalency. In addition, the separation may consist of a ¾-hour-rated glass-block wall assembly.

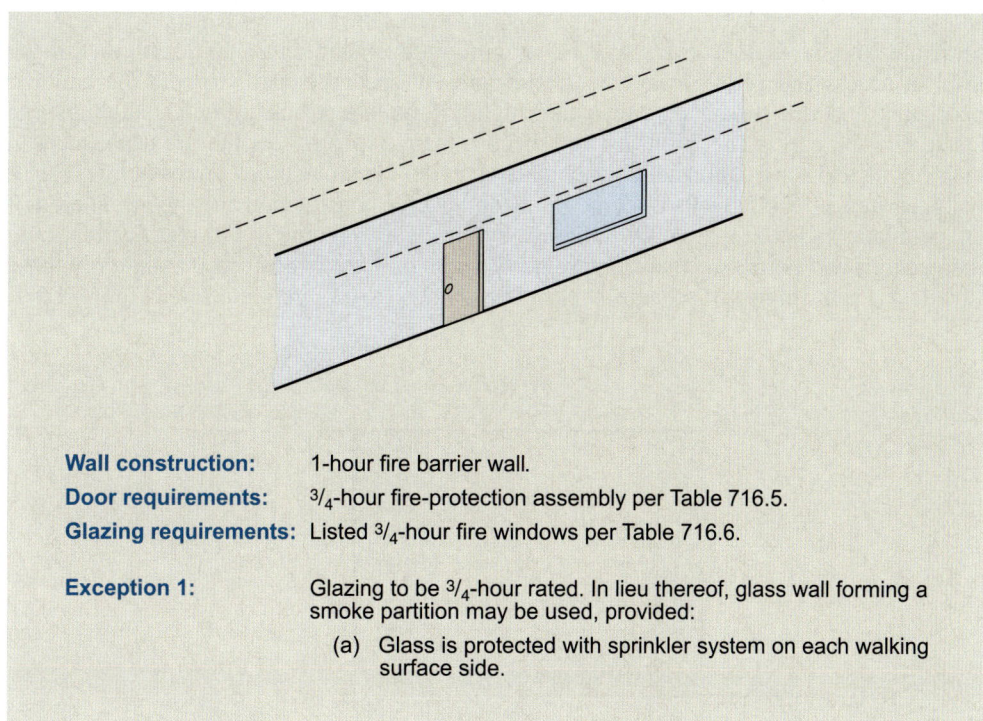

Wall construction:	1-hour fire barrier wall.
Door requirements:	¾-hour fire-protection assembly per Table 716.5.
Glazing requirements:	Listed ¾-hour fire windows per Table 716.6.
Exception 1:	Glazing to be ¾-hour rated. In lieu thereof, glass wall forming a smoke partition may be used, provided:
	(a) Glass is protected with sprinkler system on each walking surface side.

Figure 404-2
Atrium enclosure.

The separation between adjacent spaces and the atrium may be omitted on a maximum of any three floor levels, provided the remaining floor levels are separated as provided in this section. In computing the atrium volume for the design of the smoke-control system, the volume of such open spaces shall be included.

4 Special Detailed Requirements Based on Use and Occupancy

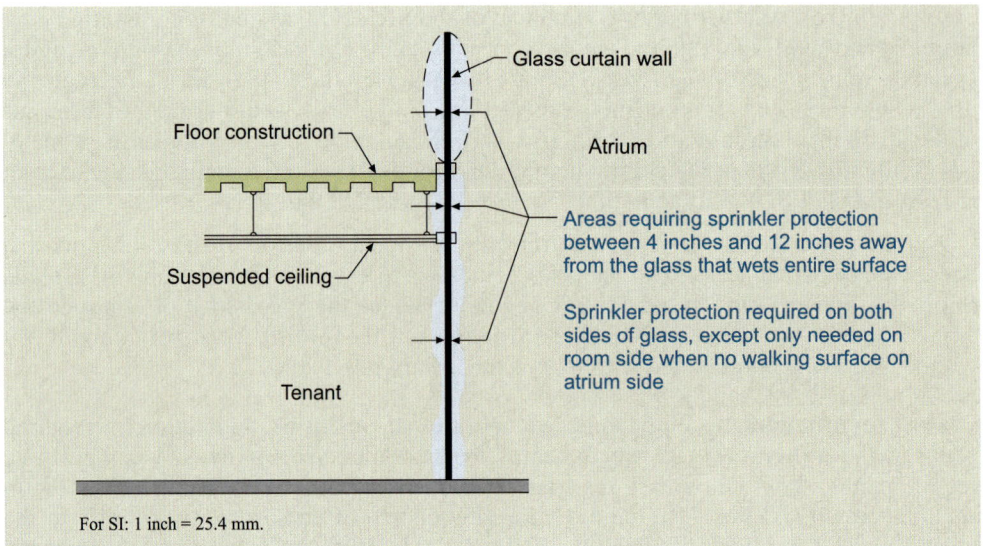

Figure 404-3
Glass protection.

Section 405 Underground Buildings

Structures that have floor levels well below ground level, and thus significantly below the level of access and egress from the exterior, present special hazards to both the building occupants and fire personnel. Much like high-rise buildings, underground buildings can create difficult egress conditions as well as pose many problems for the fire department in their rescue and suppression activities. Fundamental to the protection features of this type of building are the requirements for Type I noncombustible construction and the installation of an automatic sprinkler system. A standpipe system is also required. For clarification, only the underground portion of the structure needs to be of Type I construction, and only those floor levels at the highest discharge level and below need to be sprinklered. See Figure 405-1.

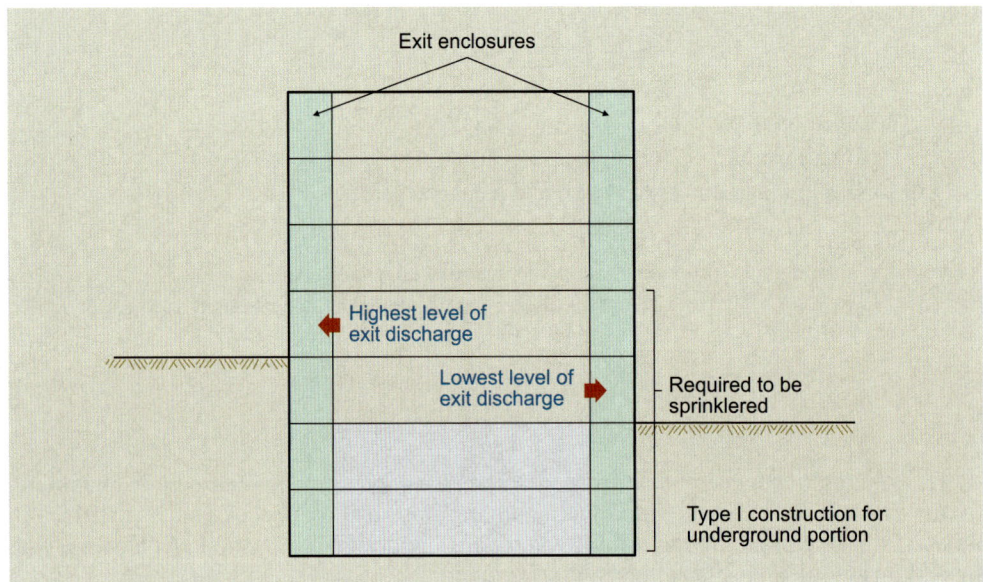

Figure 405-1
Underground buildings.

The basic criteria for consideration as an underground building is that a floor level used for human occupancy be located more than 30 feet (9,144 mm) below the finished floor of the lowest level of exit discharge. Exempted from the requirements of Section 405 are sprinklered dwellings; parking garages having fire suppression systems; fixed guideway transit systems such as subways; stadiums, arenas, and similar assembly uses; those buildings where only a very limited amount of floor area would qualify by the definition; and mechanical spaces that are typically unoccupied.

A valuable concept in fire protection is utilized in the provisions for underground buildings that extend even deeper into the ground. Where an occupied floor level is located more than 60 feet (18,288 mm) below the finished floor of the lowest level of exit discharge, at least two compartments of approximately equal size must be created. The compartmentation must extend throughout the underground portion of the structure, up to and including the highest level of exit discharge. The separation between the two areas is intended to allow for horizontal egress travel to a refuge area if necessary, while also permitting the use of the compartment as a staging area for fire-suppression activities. A smoke barrier is required as the separation element, with door openings also protected in a manner to restrict smoke leakage. Other openings and penetrations are strictly limited. Air supply and exhaust systems, where provided, must be independent of the other compartments. Where the underground portion of the building is served by elevators, each compartment must have access to at least one elevator. An elevator lobby, enclosed by a smoke barrier, may be used to allow a single elevator to serve more than one compartment. Doors into the elevator lobby shall be gasketed, have a drop sill, and be automatic closing by smoke detection. See Figure 405-2.

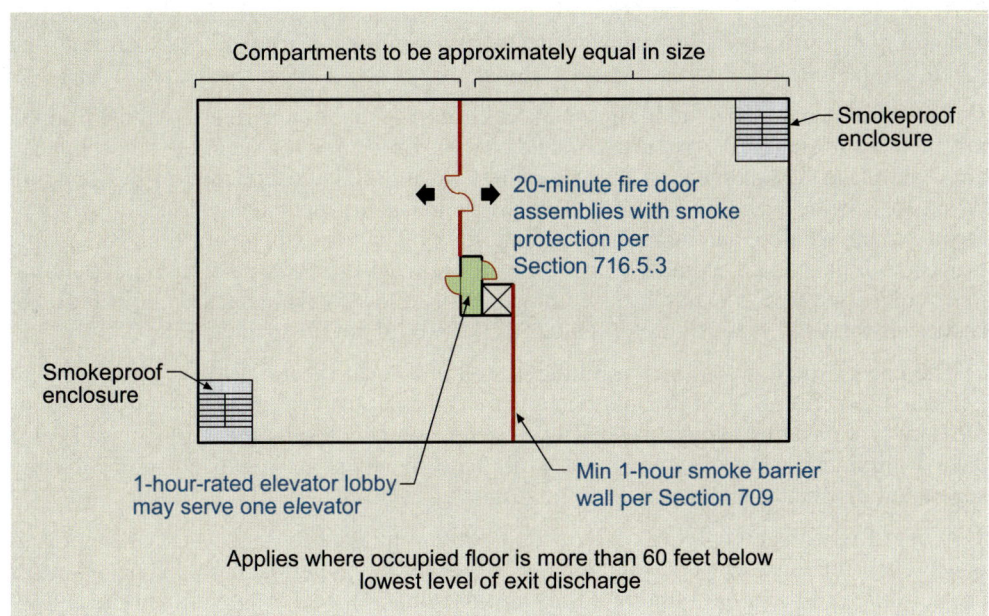

Figure 405-2
Compartmentation of underground buildings.

Smoke control is also an important part of the overall fire-protection package. By limiting the spread of smoke to only the originating area of the fire, the remainder of the underground building should be provided with acceptable egress paths. Each compartment shall be provided with its own smoke-control system. A manual fire-alarm system and complying communications system is also an integral part of the fire- and life-safety concept for underground buildings.

Stairways serving the floor levels of an underground building that are more than 30 feet (9,144 mm) below the discharge level are to be smokeproof enclosures, with at least two means of egress from each floor level. If multiple compartments are formed, each compartment must have at least one exit, with a second egress path available into an adjoining compartment.

4 Special Detailed Requirements Based on Use and Occupancy

It is mandatory that multiple exits be provided within enclosures designed to resist the penetration of smoke.

Both standby and emergency power shall be provided to specific loads identified by this section. See the discussion of Section 403.4.

Section 406 *Motor-Vehicle-Related Occupancies*

Although uncommon, fire hazards related to motor vehicles are a concern, particularly where associated with other occupancies. The code regulates occupancies containing motor vehicles, whether they be parked, under repair, or being fueled. The hazards are primarily related to the fuel in the vehicles, as the overall fire loading related to vehicle occupancies is typically quite low.

406.3 Private garages and carports. In order to secure the low fire hazard intended by the Group U occupancy classification, the code initially limits buildings in this category to 1,000 square feet (93 m^2) in floor area and one story in height. However, the code goes on to develop exceptions to the 1,000-square-foot (93-m^2) limitation for those private garages that are not used for repair or fuel-dispensing operations. In such cases, a limit of 3,000 square feet (279 m^2) is established for the Group U occupancy. Because it is doubtful that a garage containing a repair or fuel-dispensing use would receive a Group U classification, the maximum permissible floor area would almost always be 3,000 square feet (279 m^2).

For a mixed-occupancy building such as a Group R-2 apartment house with a Group U private parking garage, the exterior wall and opening protection for the Group U occupancy are required to be the same as those required for the major occupancy in the building (Group R-2 in the example). Thus, the increase in area is justified on the basis of restricting the use of the building and increasing the exterior wall and opening protection. Where this provision of the code is utilized, it also limits the allowable floor area of the total building to that of the major occupancy. This would generally allow for an increased allowable area, as the *unity formula* provision of Section 508.4.2 would not be applicable.

The code also allows an increase in area to 3,000 square feet (279 m^2) if the building is limited to a Group U as its only occupancy and is used for the parking of private- or pleasure-type vehicles. Under such single-occupancy conditions, the exterior wall must have a minimum 1-hour fire-resistance rating and openings are regulated where the fire separation distance is less than 5 feet (1,524 mm).

Where fire walls complying with Section 706 are utilized to divide the parking occupancy into floor areas of 3,000 square feet (279 m^2) or less, multiple Group U occupancies are permitted to be in the same building. In this situation, each compartment created by the fire walls would be restricted in size in order to classify the entire structure a Group U occupancy.

406.4 Public parking garages. Public parking garages, which are simply those garages that are not considered as private garages based on the limitations of Section 406.3, will fall into one of two categories, either open or enclosed. The special characteristics of each type of parking structure are addressed in Sections 406.5 and 406.6. This section addresses the general requirements for public parking garages, whether they be open or enclosed.

There are fundamentally two types of parking garages regulated by the IBC—private garages and public garages. Although there is no specific definition for either type of garage, the basis for both classifications is Section 406.3 addressing private garages and carports. Those parking structures that fall outside of the scope of Section 406.3 are considered as public parking garages. The primary difference between private and public garages is the size of the facility, rather than the use. Strictly limited in permissible height and area, private parking garages are typically not commercial in nature. They generally serve only a specific tenant or building and are not open for public use. It is important to note that there is no implication that public parking garages must be open to the public, as they are only considered public in comparison to private garages. A public parking garage is then further characterized as one of two types, either an enclosed parking garage or an open parking garage, and regulated accordingly.

To allow access to the garage to other than high-profile vehicles, the clear height of each level is to be at least 7 feet (2,134 mm). Note that the minimum height of the means of egress system is 7 feet, 6 inches (2,286 mm), based on the general provisions of Section 1003.2. However, Exception 7 to Section 1003.2 allows the clear height to be reduced to 7 feet (2,134 mm) in those areas of parking garages used for vehicular and pedestrian traffic.

Guards must also be provided in accordance with the general provisions of Section 1013. In addition, all parking areas more than 12 inches (305 mm) above adjacent levels shall be provided with vehicle barriers at the ends of parking spaces and drive lanes. The height of the vehicle barriers cannot be less than 2 feet, 9 inches (835 mm). Where the parking garage is connected directly to a room containing a fuel-fired appliance, a vestibule is mandated to separate the two spaces. This provides at least two doorways to isolate the equipment from the vehicle area. The vestibule is not necessary where the appliance ignition sources are placed at least 18 inches (457 mm) above the floor. The *International Mechanical Code*® (IMC®) and/or *International Fuel Gas Code*® (IFGC®) should also be consulted for those requirements regulating the installation of mechanical equipment within parking garages.

Open parking garages. Studies and tests of fires in open parking garages have shown that, in addition to a low fire loading, the potential for a large fire is exceedingly remote. Based on this data, the IBC establishes special provisions for open parking garages in this section, which in general are less restrictive than those for enclosed parking structures addressed in Section 406.6. The key is that the open parking garage is well ventilated naturally, and as a result, the products of combustion dissipate rapidly and do not contribute to the spread of fire.

406.5

To secure the proper amount of openness, the code, as illustrated in Figure 406-1, specifies the following:

1. The building must have openings on at least two sides.
2. The openings must be uniformly distributed along each side.
3. The area of openings in the exterior walls on any given tier must be at least equal to 20 percent of the wall area of the total perimeter of each tier.
4. Unless the required openings are uniformly distributed over two *opposing* sides of the building, the aggregate length of openings considered to provide natural ventilation shall constitute a minimum of 40 percent of the wall length of the perimeter of that tier.
5. The area of openings in the interior walls must be at least 20 percent of the area of the interior walls with openings uniformly distributed.

There are situations where the required openings of open parking garages are located below the surrounding grade. Section 406.5.2.1 mandates that a clear horizontal space be provided adjacent to the garage's exterior openings that allows for adequate air movement through the opening. The dimensional requirements are based on the provisions of Section 1203.4.1.2, which addresses openings below grade when such openings are used for the required natural ventilation of a building's occupied spaces. Where openings in the exterior wall of an open parking garage are located below grade level, some degree of clear space must be provided at the exterior of the openings. As the distance of the openings below

4 Special Detailed Requirements Based on Use and Occupancy

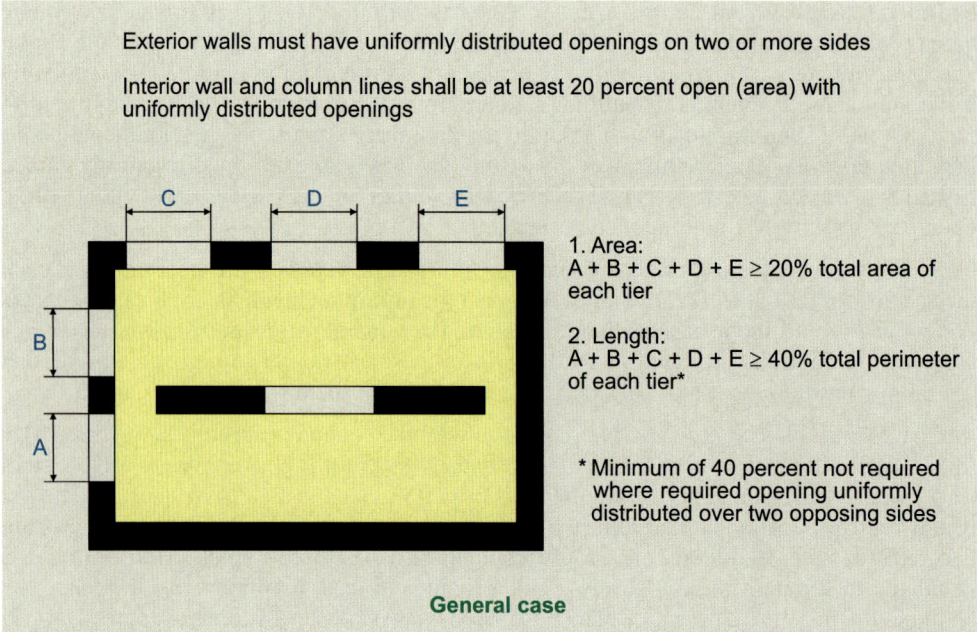

**Figure 406-1
Open parking garages.**

the adjoining ground increases, the minimum required exterior clear space also increases proportionately. The horizontal clear space dimension, measured perpendicular to the exterior wall opening, must be at least one and one-half times the distance between the bottom of the opening and the average adjoining ground level above. The extent of the required clear space allows for adequate exterior open space to meet the intent and dynamics of natural ventilation requirements for open parking garages. See Figure 406-2.

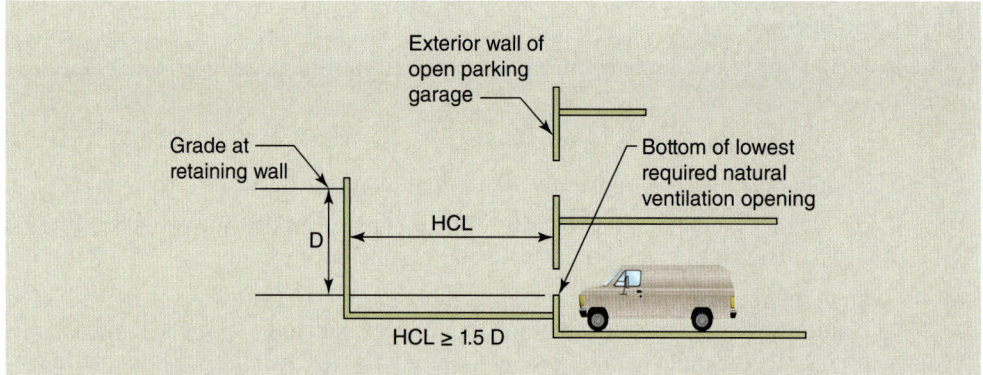

**Figure 406-2
Parking garage openings below grade.**

As a general rule, the maximum allowable height and area of open parking garages is calculated in the same manner as for other buildings. Classified as a Group S-2 occupancy, the provisions of Chapter 5 would apply. However, where the open parking garage contains no uses other than the parking of private motor vehicles, the specific size limitations of Table 406.5.4 and Section 406.5.5 take effect. Because the potential fire severity of an open parking structure used solely for vehicle parking is extremely low, the code permits area and height limitations in excess of those for other Group S-2 occupancies. For example, a stand-alone open parking garage of Type IIB construction would be permitted a floor area of 50,000 square feet (4,645 m^2) per tier based on Table 406.5.4, with a height limit of eight tiers for a ramp-access garage. For such an open parking garage exceeding three stories in height, the total area of the multistory building is not limited to three times that for a one-story building, as is required by Section 506.4.1, Item 3, but rather can be computed as the permitted area per tier times the number of tiers. Therefore, in the example just given, the total area permitted would be 400,000 square feet (37,160 m²) for a stand-alone Type IIB open parking garage. The maximum height in tiers has been limited somewhat arbitrarily by the code, based on the length of time it would take for fire-department personnel to reach the top of the structure for fire-suppression purposes.

The area and height increases above the tabular limits listed in Table 406.5.4 for single-use open parking garages are those outlined in Section 406.5.5, and are basically keyed to the provision of more natural ventilation area than the minimum required by the code. For unlimited-area buildings permitted by this section, see Figure 406-3.

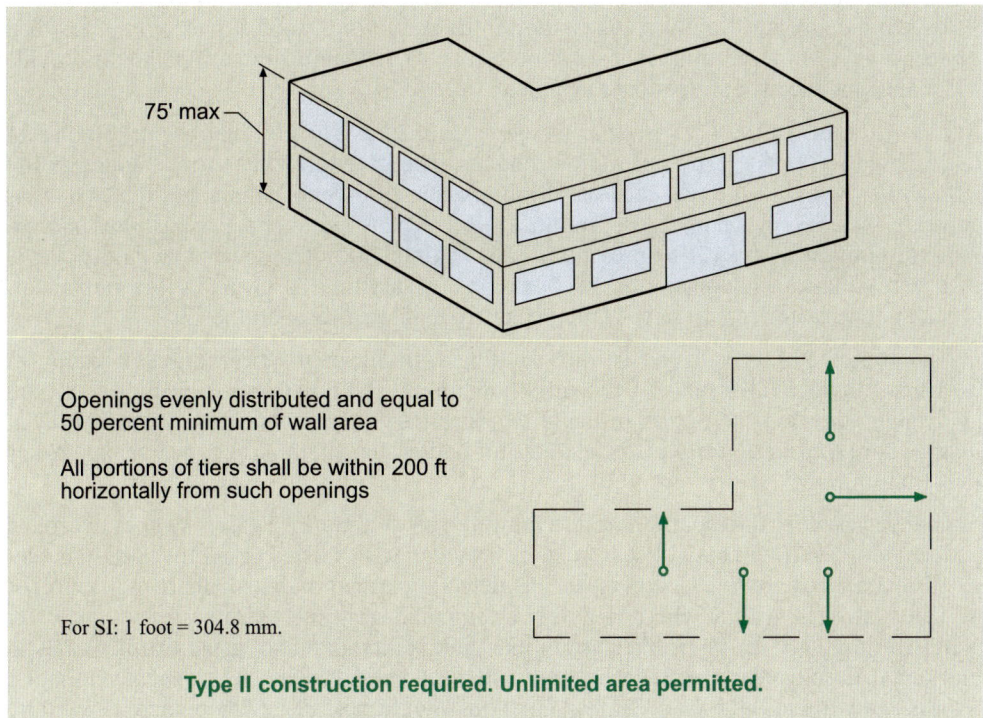

Figure 406-3
Unlimited area open parking garages.

In the classification of a Group S-2 parking structure as an open parking garage, the code identifies the following prohibitions:

1. There shall be no automobile repair work performed in the building.

2. There shall be no parking of buses, trucks, or similar vehicles.

3. There shall be no partial or complete closing of the required exterior wall openings by tarpaulins or by any other means.

4. There shall be no dispensing of fuel.

The intent of these limitations is to further ensure low fire loading, low possibility of fire spread, and natural cross ventilation.

Enclosed parking garages. Any vehicle parking garage that does not meet the criteria of an open parking garage or a Group U private garage is to be regulated under the general provisions for a Group S-2 occupancy. Table 503, along with any applicable height and area increases, will limit the height and floor area of an enclosed parking garage, with an allowance for use of the roof for parking purposes. Ventilation of an enclosed parking garage must be provided in accordance with the IMC. **406.6**

4 Special Detailed Requirements Based on Use and Occupancy

406.7 Motor-fuel-dispensing facilities. Because most of the hazards involved with a fuel-dispensing operation are due to the storage and dispensing of flammable liquids, the majority of regulations are addressed by the IFC. The primary provisions of this section apply to canopies that are placed over the fueling areas for the purpose of customer convenience. Because of the potential exposure of gasoline and vehicle fires during fuel-dispensing operations, the canopies and supports over pumps are required by this section to be of noncombustible construction or, alternatively, constructed of fire-retardant-treated wood, complying heavy-timber members, or be of 1-hour fire-resistance-rated combustible construction. Occasionally, combustible materials may be used in or on a canopy under limited conditions. The allowance for approved plastic panels installed in canopies over motor-vehicle pumps is intended to isolate the combustible plastic materials from other buildings so that if the materials become ignited, they will not present an exposure problem to other buildings.

To avoid damage to vehicles and canopies, the height of canopies must not be less than 13 feet, 6 inches (4,115 mm). The 13-foot, 6-inch (4,115-mm) dimension should provide adequate clearance for recreational vehicles.

406.8 Repair garages. In the IFC, a repair garage is defined as any building or portion thereof that is used to service or repair motor vehicles. The potential exists for a moderate fire hazard that is due to the presence of various combustible and flammable liquids such as solvents, cleaning products, and gasoline. During repair operations, it is also not uncommon for ignition sources to be present. It is this combination that creates the highest level of hazards that are addressed by Section 406.8. Classified as Group S-1 occupancies, special concerns for repair garages are primarily regulated through the IFC.

The presence of a repair garage in a building with different types of uses is addressed no differently than other mixed-occupancy conditions. The provisions of Section 508.1 are applicable, allowing the option of using the *accessory occupancies*, *nonseparated occupancies*, or *separated occupancies* method for addressing the multiple occupancy groups in the building.

Garages used for the repair of vehicles powered by nonodorized gases, such as hydrogen and nonodorized liquid natural gas, are to be provided with a listed or approved flammable-gas-detection system. The design of the gas-detection system will provide for activation of the safeguards at the point where the level of flammable gas exceeds 25 percent of the lower explosive limit. At this point, the system must initiate audible and visual alarm signals in the garage, deactivate the garage's heating systems, and activate any mechanical ventilation interlocked with gas detection. Similar functions should take place where there is a failure of the detection system.

Section 407 *Group I-2*

In institutional occupancies, particularly those classified as Group I-2, it is important to balance the fire-safety concerns with the functional concerns of the health-care operations. This section modifies the general code provisions in an effort to achieve such a balance.

407.2 Corridors continuity and separation. Corridors are intended to provide a direct egress path adequately separated from hazards in adjoining spaces. However, in hospitals, nursing homes, and other Group I-2 occupancies, a number of necessary modifications are provided to facilitate the primary functioning of these types of health-care facilities. These modifications recognize the special needs of these occupancies to provide the most efficient and effective health-care services. See Figure 407-1.

In order to provide appropriate waiting spaces for visitors, Section 407.2.1 allows such waiting spaces to be unseparated from the corridors. One reason for this is to permit the waiting areas and similar spaces to be so located as to permit direct visual supervision by

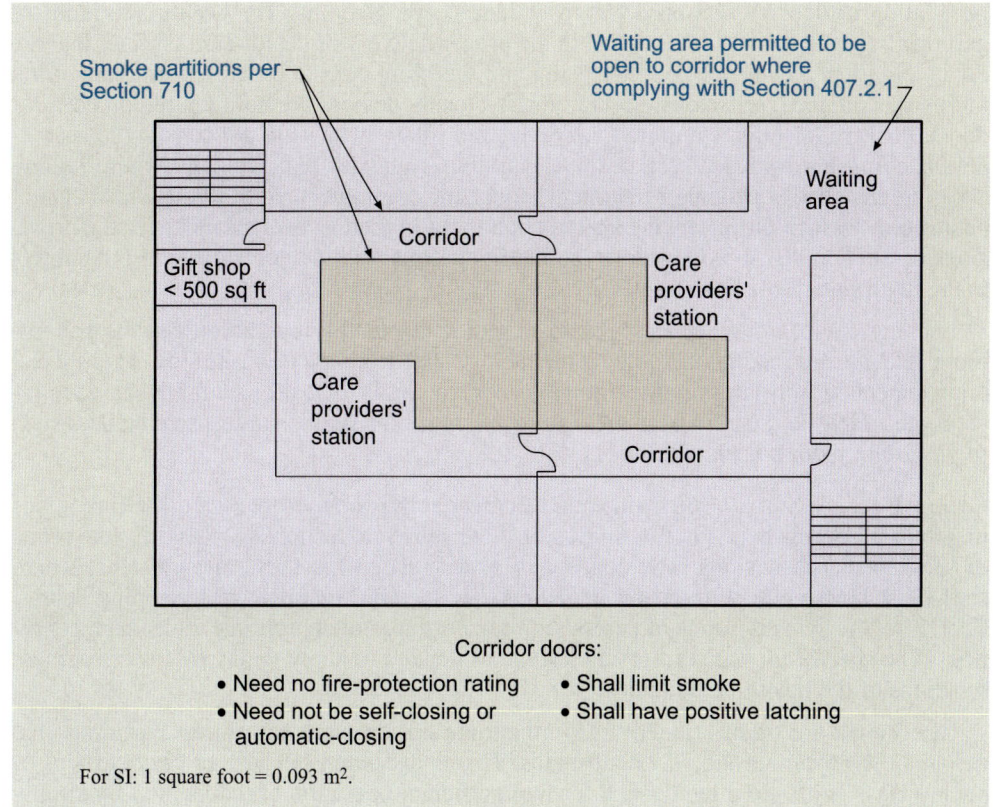

Figure 407-1
Corridors in Group I-2 occupancies.

health-care facility staff. In exchange for the elimination of the corridor separation, certain conditions are imposed on the location of such waiting spaces. Although the scoping language only includes waiting areas and similar spaces, the primary criteria limiting those spaces that can be open to a corridor are identified in Item 1. Health-care facilities will often create alcoves adjacent to the corridor for the temporary storage of medical supplies, linen carts, food carts, etc., that are necessary to the daily functions of the facility. Without the alcoves, the corridors would be obstructed by these uses. Therefore, the code makes an allowance for such spaces. Allowances are also made for areas associated with the treatment of mental-health patients. Provided the areas are under continuous supervision by facility staff, they may be open to the corridor where six conditions are met.

Similarly, Section 407.2.2 makes provisions for the location of clerical stations and similar spaces necessary for doctors' and nurses' charting and communications in positions that need not be separated from the corridors. Essentially, these special-use areas are permitted to be located in the corridor. When this arrangement occurs, however, it is necessary that the construction surrounding the clerical station be as that required for corridors.

Corridor wall construction. Walls enclosing corridors and other spaces permitted by Section 407.2 to be open into corridors are intended to provide a relatively smoke-free environment during the relocation of patients during a fire emergency. Therefore, such walls must be constructed in accordance with the provisions of Section 710 as smoke partitions. The walls may extend either tight to the floor or roof deck above, or extend tight to the ceiling, provided the ceiling is also constructed to limit smoke transfer.

407.3

Corridor doors protecting those spaces adjacent to the corridor are not required to have a fire-protection rating, nor are they required to be self-closing assemblies. They must, however, be able to limit the transfer of smoke through the opening but need not be tested for air leakage under UL 1784. One of the most controversial issues relative to the arrangement of health-care facilities such as hospitals and nursing homes is the matter of the installation of door closers on doors to patient sleeping rooms. The health-care industry has long believed it is more important to the proper delivery of health-care services that the doors to patient rooms not be self-closing and therefore constantly closed. In recognition of this special need, self-closing or automatic-closing devices are not required on corridor doors. Positive latching is required, however, and roller latches are not considered acceptable latching hardware. Where positive latching is not desired, typically where sliding doors are installed at patient or treatment rooms, the common corridor arrangement cannot be utilized. In such instances, the spaces could be designed as care suites under the provisions of Section 407.4.3. Corridor-type configurations within such suites are not subject to the requirements of Section 407.3.

Locking devices may be arranged so that they are readily operable from the patient-room side and are readily operable by the facility staff from the opposite side. This special arrangement permits keys or other limited access methods to be utilized for the care recipient rooms. However, egress from the care recipient rooms shall be unrestricted unless such rooms are in mental-health facilities.

407.4.3 Group I-2 care suites. Special means of egress provisions are provided for care suites in Group I-2 occupancies. The definition of "care suite" in Section 202 identifies the scope of such special provisions. The concept of suites recognizes those arrangements where staff must have more supervision of patients in specific treatment and sleeping rooms. Therefore, the general means of egress requirements are not appropriate under such conditions. The special allowances for suites are not intended to apply to day rooms or business functions of the health-care facility.

Care suites are often utilized to eliminate the requirement for smoke partitions that define a corridor in Group I-2 occupancies. This is necessary because privacy curtains or sliding glass doors are often desired for more efficient operation of the facility. By simply considering the connected space an intervening room rather than a corridor, the limitations imposed by the corridor provisions do not apply.

As a general rule, exiting from habitable rooms in Group I-2 occupancies must be directly to a corridor. The potential for obstructions is high where egress in such occupancies must pass through other use areas. There is an exception for those rooms arranged as care suites in accordance with this section.

Where the area of the care suite does not contain care recipient sleeping areas, travel through intervening rooms is limited to one of the following conditions:

1. Occupants of rooms, other than those used for care recipient sleeping purposes, that are located within a suite may travel through a single intervening room if the travel distance does not exceed 100 feet (30,480 mm).
2. Occupants of rooms within suites designed for other than sleeping purposes may travel through up to two intervening rooms, provided the travel distance is restricted to 50 feet (15,240 mm).

Additional requirements are illustrated in Figure 407-2.

407.4.3.5 Care suites containing sleeping room areas. Travel through one intervening room is permitted in suites of patient sleeping room areas provided one of two specified conditions is met:

1. Care recipient sleeping rooms are permitted to egress through one intervening space, provided no more than eight care recipient beds are served.
2. For care suites with more than eight beds, the code allows for egress travel through one intervening room where direct and continuous supervision is provided.

Limitations on floor area, single means of egress, and travel distance are shown in Figure 407-2.

Group I-2

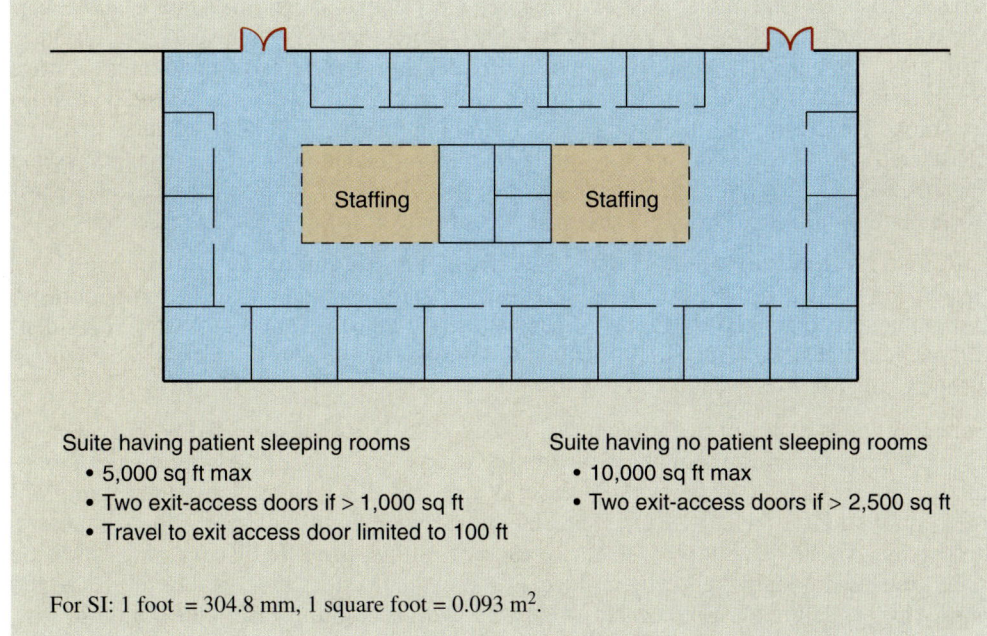

Figure 407-2
Care suites.

It should be noted that for care suites, the threshold for the minimum number of means of egress is based on floor area as opposed to occupant load. Typically, surgery recovery areas (post-op) and intensive care units (ICUs) are not considered patient sleeping rooms as they are under constant monitoring and supervision.

Smoke barriers. Evacuation of a building such as a hospital or nursing home is a virtual impossibility in the event of a fire, particularly in multistory structures. Horizontal evacuation, on the other hand, is possible with a properly trained staff. As a result, the code makes provisions for horizontal compartmentation as illustrated in Figure 407-3, so that if necessary, care recipients can be moved from one compartment to another. This intent is secured by this section wherein, under most conditions, each story of a Group I-2 occupancy is

407.5

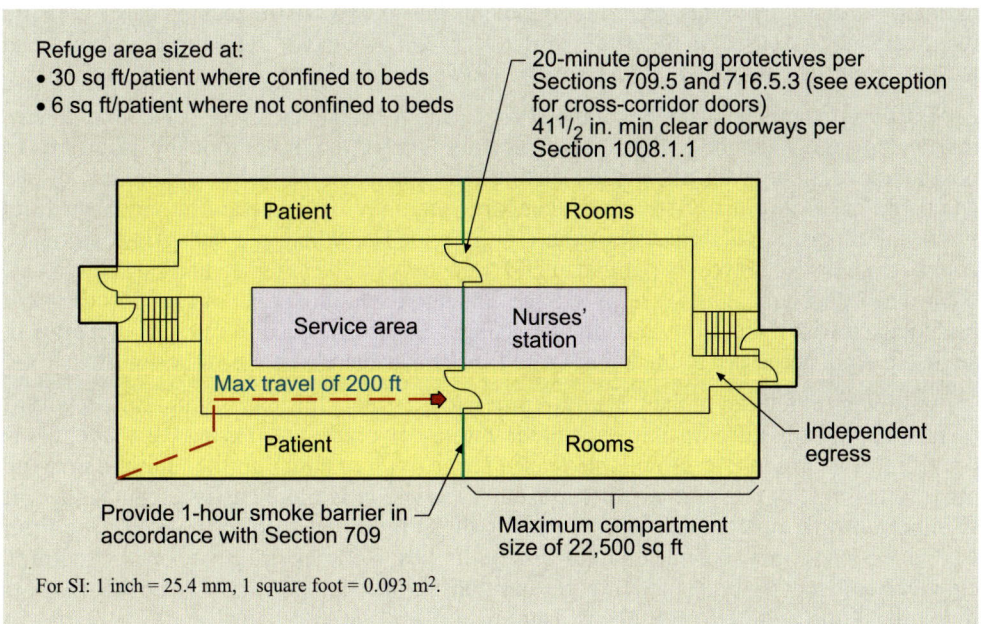

Figure 407-3
Hospital compartmentation.

2012 International Building Code Handbook 99

required to be divided into at least two approximately equal compartments by a smoke barrier constructed in accordance with Section 709. Limited by floor area and travel distance, each compartment shall be sized to permit the housing of patients from adjoining smoke compartments. It is expected that in multistory buildings, the floor construction also provides for smoke compartmentation vertically. As such, the concept of smoke resistance must be considered relative to vertical openings and penetrations, including interior exit stairways and shaft enclosures. A more detailed analysis of the requirements for vertical smoke compartmentation in multistory Group I-2 occupancies is found in the discussion of Section 711.9.

407.6 Automatic sprinkler system. Owing primarily to the very limited mobility of patients in this type of occupancy, the code requires every compartment containing sleeping rooms to be equipped with an automatic fire-sprinkler system. Approved quick-response or residential sprinklers shall be provided throughout the smoke compartments, not just in the patient sleeping units.

407.9 Secured yards. It is not uncommon that a secured exterior area or yard be provided for Group I-2 occupancies, particularly where the facility specializes in the treatment of mental disabilities such as Alzheimer's disease. Where such fencing and locked gates prohibit the continuation of the exit discharge to the public way, the use of safe dispersal areas is acceptable. To adequately provide for temporary refuge, the safe dispersal area must be sized to accommodate the occupant load of the egress system it serves. In all cases, the entire dispersal area must be located at least 50 feet (15,240 mm) from the building.

Section 408 *Group I-3*

The concerns for both security and fire safety must be balanced when it comes to Group I-3 detention facilities. Special consideration must be given to the secured areas without sacrificing an unreasonable degree of fire and life safety for the occupants. This section addresses the unique conditions that occur in these types of buildings.

Section 408.3 modifies the general requirements for the means of egress found in Chapter 10. A major difference is the allowance for glazing in the doors and walls of required exit stairways, provided a number of conditions are met. As would be expected, the most dramatic variation from the general requirements has to do with the locking hardware. The requirements vary based on the nature of the detention occupancy. Reference must be made to the occupancy conditions of Section 308.5 to determine the appropriate egress criteria.

Similar to the provisions of Section 407 for Group I-2 occupancies, smoke compartments must be created where the occupant load per story is 50 or more. Additionally, regardless of occupant load, floor levels utilized as sleeping areas must be divided into a minimum of two compartments. More than two smoke compartments may be necessary on any floor level where the dictated travel distances cannot be provided or where the occupant load of the compartment is excessive. No more than 200 occupants can be assigned to a single compartment. The refuge area must be sized to accommodate the total number of residents that may be contained within the compartment. Independent egress is needed from each compartment so that it is not necessary to travel back into the compartment where travel originated.

An important feature of the Group I-3 provisions is the allowance for multiple levels of residential housing to be open to each other without an enclosure. Through the safeguards provided, it is possible to provide increased security by opening up the multiple housing areas to a single common area where visual supervision is more easily accomplished. It is important that independent egress to an exit be provided from each level. The limit of 23 feet (7,010 mm) between the lowest and highest floor levels, as well as the required egress directly out of each story, provide additional qualifications that must be met in order to eliminate the required vertical enclosure protection.

As an additional allowance for security purposes, the fire-protection rating is not required for security glazing installed in 1-hour fire barriers, fire partitions, and smoke barriers that may be present. Rather, equivalent protection is provided through compliance with four specific conditions addressing the glazing and its frame. The use of security glazing is necessary in such facilities to track and contain inmate movement for the protection of other inmates and administrative personnel. Three of the most common types of fire separations are addressed: fire barriers, fire partitions, and smoke barriers. The allowance is not applicable to fire walls, nor is it permissible where the fire separation wall has a required fire-resistance rating of more than 1 hour. The conditions imposed on the security glazing limit the area of each individual glazed panel, mandate sprinkler protection that will wet the entire glazing surface on both sides, regulate the gasketed frame for deflection, and prohibit the installation of obstructions between the sprinklers and the glazing.

Section 409 *Motion-Picture Projection Rooms*

Prior to the 1970s, building codes addressed the subject of motion picture projection rooms based on the hazard of the cellulose nitrate film being used at that time. Actually, production of cellulose nitrate film ceased around 1950, although its use continued thereafter. In fact, even today, some cellulose nitrate film is used at film festivals and special occasions requiring the projection of historically significant films that are still imprinted on cellulose nitrate film. Where this type of film continues to be utilized or stored, it will be regulated under the provisions of NFPA 40. Although the provisions in the codes since 1970 are based on the use of safety film, some of the protection requirements for cellulose nitrate film have been retained in the present requirements, such as ventilation requirements for the projection room.

The intent of the current provisions regulating motion picture projection rooms is to provide safety to the occupants of a theater from the hazards consequent on the light source where electric arc, xenon, or other light-source projection equipment is used. Although not used to any extent today, electric-arc projection lamps emit hazardous radiation. Xenon lamps, which have been highly prevalent as projection lamps, emit ozone. As a result, the provisions of Section 409 are based on the lamps used for projection of the film rather than the type of film to be used, as long as the film is not nitrate based.

The provisions intend to isolate the projection room so that it does not present a danger to the theater audience. As the room is designed for the projection of safety film, there is no intent to provide a special fire-resistive enclosure, and fire protection of openings between the projection room and the auditorium is not required. However, due to the projection lamps, it is the intent of the code to provide an emission-tight separation so that any opening should be sealed with glass or other approved material such that emissions from the projection lamps will not contaminate the auditorium.

Section 410 *Stages, Platforms, and Technical Production Areas*

The provisions in Section 410 are continuously reviewed in an attempt to bring the code requirements in line with the present methods and technologies regarding the use of stages and platforms, as well as related accessory and support areas. Although the basic provisions for life safety have remained essentially unchanged over the years, occasional modifications have been made that are due to the need to accommodate state-of-the-art performances.

410.2 Definitions. Although the definitions listed in this section are complete and reasonably understandable, there are terms unique to the performing arts that are not generally understood, such as fly gallery, gridiron, and pinrail, which fall under the general term "technical production area." The distinctions between the definitions of a stage and a platform are also very important because of the specific requirements for each. The primary difference between a stage and a platform is the presence of overhead hanging curtains, drops, scenery, and other stage effects. The amount of combustible materials associated with a stage is typically greater than that for a platform. Thus, the fire-severity potential is much higher.

410.3 Stages. An assembly occupancy considered among the most hazardous is a Group A-1 containing a large occupant load and a performance stage. The hazard created by the stage is the presence of combustibles in the form of hanging curtains, drops, leg drops, scenery, etc., which in the past have been the source of ignition for disastrous fires in theaters. Modern stages also have an increased hazard from special effects such as pyrotechnics, utilized in so-called *spectaculars*.

Where the stage height exceeds 50 feet (15,240 mm), the fire hazard is even greater because the fly area that is usually above the stage is a large blind space containing combustible materials that have a fuel load considerably greater than that normally associated with an assembly occupancy. Many of the construction requirements for stages are depicted in Figure 410-1.

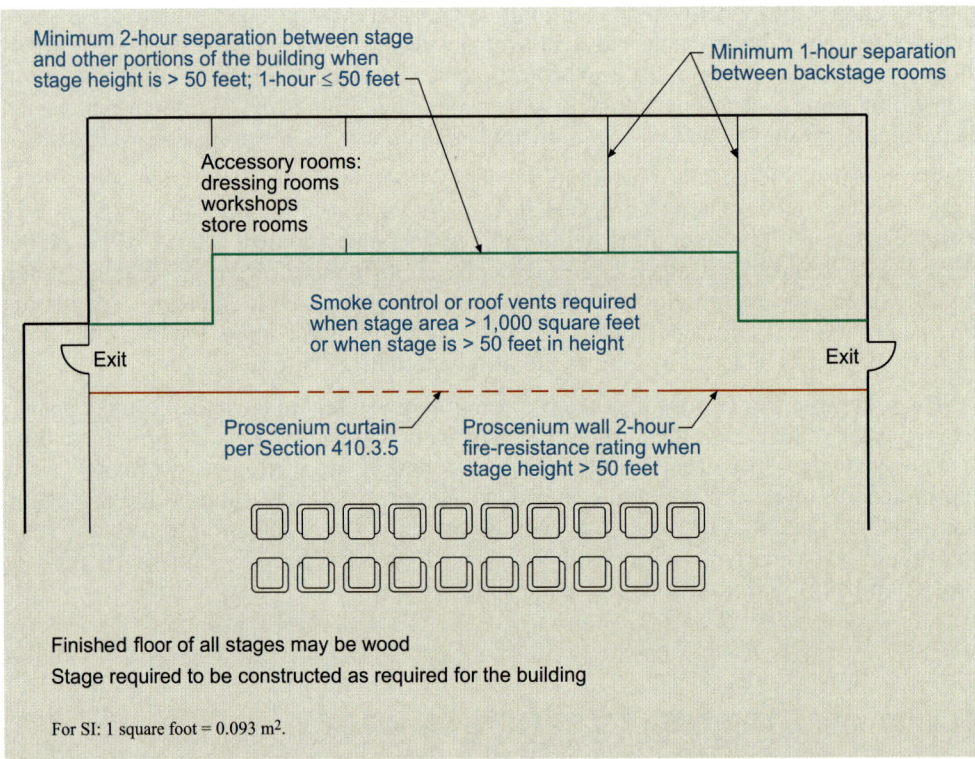

Figure 410-1 Stages.

410.3.1 Stage construction. In addition to the features shown in Figure 410-1, any stage may have a finished floor of wood, provided construction of the stage floor or deck is in compliance with this section. As the area above and at the sides of stages can be filled with combustible materials that can be moved both vertically and horizontally, such as curtains,

Stages, Platforms, and Technical Production Areas

drops, leg drops, scenery, and other stage effects, the code requires that such stages be constructed of the same materials as required for floors for the type of construction of the building and separated from the balance of the building.

Proscenium wall. Where the stage height exceeds 50 feet (15,240 mm), measured from the lowest point on the stage floor to the highest point of the roof or floor deck above, a proscenium wall must be provided. The proscenium wall is intended by the IBC to provide a complete fire separation between the stage and the auditorium. Extending from the foundation continuously to the roof, the wall is to have a minimum fire-resistance rating of 2 hours. **410.3.4**

Proscenium curtain. Because the opening in the proscenium wall described in Section 410.3.4 is too large to protect with any usual type of fire assembly, the code requires that it be protected with a fire-resistive fire curtain or water curtain. Where a fire curtain is installed, it must comply with the provisions for fire-safety curtains set forth in NFPA 80 *Fire Doors and Other Opening Protectives*. A fire curtain or water curtain is not required where a complying smoke-control system or natural ventilation is provided. The purpose of the proscenium curtain protection is to provide occupants with additional time to exit the assembly seating area if there is a fire in the stage area. With the benefits afforded by an engineered smoke-control system or natural ventilation, the occupants should be equally or better protected from the hazards of fire than with a proscenium curtain or water curtain. By providing a performance-based alternative to a proscenium curtain, more design options are available where the use of fire-safety curtains is considered impractical or causes obstructions of the production. It is important to note that the elimination of the proscenium curtain is not permitted if the smoke-protected seating provisions of Section 1028.6.2 are being utilized, for example, a decrease in the required egress widths of the assembly seating area. **410.3.5**

The requirement for a complying fire curtain is triggered solely by the proscenium wall provisions of Section 410.3.4. Where a proscenium wall is fire-resistance rated for a different purpose, such as a bearing wall in a Type IB building, the fire curtain is not required.

Stage ventilation. The Iroquois Theater fire in 1903 was directly responsible for the requirement for automatic vents in the roofs of theater stages. Because of the presence of large amounts of combustible materials, excessive quantities of smoke will accumulate in and above the stage area unless it is automatically vented or removed by a smoke-control system. The removal of smoke is necessary for fire fighting as well as the prevention of panic by drawing off the smoke so that it will not infiltrate the theater auditorium. **410.3.7**

The maximum floor area of stages that is permitted without the installation of venting is 1,000 square feet (93 m^2). The stage area to be considered includes the performance area and adjacent backstage and support areas not separated from the performance area by fire-resistance-rated construction. In addition, stages must be equipped with smoke-removal equipment or roof vents where they are greater than 50 feet (15,240 mm) in height. If either of these two conditions exist, stage ventilation is required. The detailed requirements for smoke vents in the IBC are intended to provide reliability and a reasonable assurance that after many years of operation the vents will operate when needed.

Platform construction. Materials used in the construction of permanent platforms must be consistent with those materials permitted based on the building's type of construction. Therefore, in noncombustible buildings, the platforms must be of noncombustible construction. However, in buildings of Type I, II, and IV construction, the use of fire-retardant-treated wood is permitted where all of the following conditions are met: **410.4**

1. The platform is limited in height to 30 inches (762 mm) above the floor.
2. The floor area of the platform does not exceed one-third the floor area of the room in which it is located.
3. The platform does not exceed 3,000 square feet (279 m^2) in floor area.

4 Special Detailed Requirements Based on Use and Occupancy

In those situations where the concealed area below the platform is to be used for storage or any purpose other than equipment, wiring, or plumbing, the floor construction of the platform is to be fire-resistance rated for a minimum of 1 hour. Otherwise, no protection of the platform floor is necessary.

As it is often impractical to construct temporary platforms of fire-resistive materials, the code permits temporary platforms to be constructed of any materials, but restricts the use below the platform to that of electrical wiring or plumbing to operate platform equipment. Therefore, no storage of any kind is permitted beneath temporary platforms, because of the potential for a fire to start and spread undetected.

410.5 Dressing and appurtenant rooms. Not only must a stage exceeding 50 feet (15,240 mm) in height be separated from the adjoining seating area by a minimum 2-hour fire-resistance-rated proscenium wall, but such a separation is also required between the stage and all other portions of the building, including all related backstage areas. Dressing rooms, property rooms, workshops, storage rooms, and all other areas must be separated from the stage with minimum 2-hour fire-resistance-rated fire barriers and/or horizontal assemblies, and all openings must be appropriately protected. A minimum 1-hour fire-resistance-rated separation is required where the stage height does not exceed 50 feet (15,240 mm).

In addition to their required fire separation from the stage, dressing rooms and all other related backstage areas must be separated from each other. One-hour fire-resistance-rated fire barriers and/or horizontal assemblies, along with opening protectives, satisfy the minimum requirements. The hazards caused by the significant fire loading that occurs in conjunction with stages are greatly reduced through the use of compartments.

410.7 Automatic sprinkler system. One of the special areas mentioned in Table 903.2.11.6 that requires a suppression system is stages. The general requirement mandates the sprinklering of not only the stage area but also all support and backstage areas serving the stage. An automatic sprinkler system is an effective tool in limiting the exposure of a fire to the area of origin. Sprinklers are not required for a stage having both a small floor area and a low roof height. Under such conditions, the amount of combustibles in the stage area is typically very limited.

Section 411 Special Amusement Buildings

Amusement buildings are usually classified as Group A occupancies but should be classified as Group B where the occupant load is less than 50. The major factors contributing to the loss of life in fires within amusement buildings has been the failure to detect and extinguish the fire in its incipient stage, the ignition of synthetic foam materials and subsequent fire and smoke spread, and the difficulty of escape. Provisions for the detection of fires, the illumination of the exit path, and the sprinklering of the structures are required to protect the occupants in such structures. However, amusement buildings or portions thereof without walls or a roof are not required to comply with this section, provided they are designed to prevent smoke from accumulating in the assembly areas. Approved smoke-detection and alarm systems are also required in amusement buildings. A provision of Section 411.7 is that on the activation of the system as described, an approved directional exit-marking system shall activate in those areas where the configuration of the space is such as to disguise the path and make the egress route not readily apparent.

Section 412 *Aircraft-Related Occupancies*

Because of the unique nature of occupancies related to aircraft manufacture, repair, storage, and even flight control, provisions have been developed to address the special conditions that may exist. Although the various uses fall into different occupancy classifications, they all have one thing in common—they are related to aircraft. Additional requirements related directly to aviation facilities are found in Chapter 20 of the IFC.

Airport-traffic control towers. These provisions are intended to reconcile the differences between the life-safety needs of air-traffic control towers and the life-safety requirements in the body of the code. The life and property loss in these towers has been very small even though they have not complied completely with all of the code requirements in the past. In developing these provisions, consideration was given to the inherent qualities of the use, which makes the general requirements of the IBC inappropriate. For example, air-traffic control personnel are required to undergo medical examinations to ensure they are of sound body and mind. Recognition was also given to the life-safety record of these uses and specific limitations, which are imposed on the allowable size, type of construction, etc. The provisions also require automatic fire-detection systems. Because of the critical nature of the facility, a standby power system is required for towers over 65 feet (19,812 mm) in height. — **412.3**

Aircraft hangar. Aircraft hangars are intended to be classified as Group S-1 occupancies. All aircraft hangars are to be located at least 30 feet (9,144 mm) from any public way or lot line, providing adequate spatial separation for neighboring areas. Otherwise, their exterior walls must have a minimum 2-hour fire-resistance rating. Because of the concerns about below-grade spaces under any facility where flammable and combustible liquids are commonly present, the code requires the hangar floor over a basement to be liquid and air tight with absolutely no openings. Floor surfaces must also be sloped to allow for drainage of any liquid spills. — **412.4**

Fire suppression. In order to minimize the fire hazards associated with aircraft hangars, fire suppression is required based on the criteria of Table 412.4.6. The table determines the hangar classification (Group I, II, or III) to which the fire suppression must be designed in accordance with NFPA 409, *Aircraft Hangars*. The classification is based on the hangar's type of construction and fire area size. Fire area size is based on the aggregate floor area bounded by specified fire separation elements that have a fire-resistance rating in accordance with Section 707.3.10. For the purposes of hangar classification, ancillary uses located within the fire area are not required to be included in the fire area size provided they are separated from the aircraft serving area by minimum 1-hour fire barriers. See Figure 412-1. — **412.4.6**

Residential aircraft hangars. Where a private aircraft hangar is accessory to a dwelling, it is classified as a residential aircraft hangar, provided it meets the criteria of this section. Where the hangar is less than 20 feet (6,096 mm) in height and less than 2,000 square feet (186 m²) in floor area, it is considered to be no greater a hazard than any private garage housing several motor vehicles. — **412.5**

The fire separation between a dwelling and an attached hangar is to be at least a 1-hour fire-resistance-rated fire barrier. Self-closing doors between the dwelling and the hangar are the only permitted openings, and each door must have a minimum 4-inch-high (102-mm) noncombustible sill. Two means of egress from the hangar are required, only one of which may pass through the dwelling. Smoke alarms shall be installed within the hangar, and the mechanical and DWV (drainage, waste, and vent) systems installed for the hangar shall be independent of the dwelling's systems. Every reasonable effort is being made to isolate the residential aircraft hangar from the dwelling to reduce the likelihood that a fire in the hangar will be a life-safety threat to occupants of the dwelling.

4 Special Detailed Requirements Based on Use and Occupancy

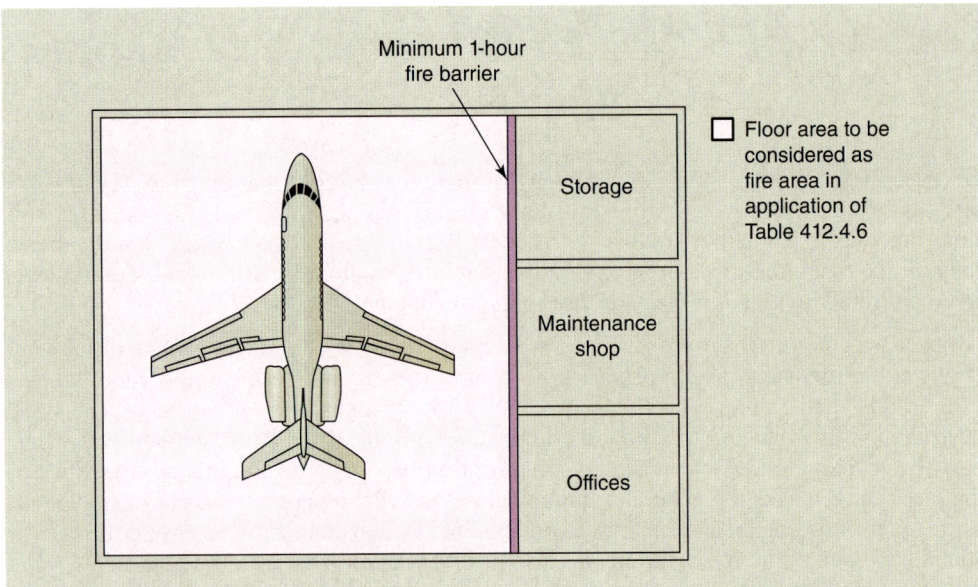

**Figure 412-1
Aircraft hanger fire area.**

412.6 **Aircraft paint hangars.** The hazards involved with the application of flammable paint or other liquids cause aircraft painting operations to be highly regulated. Where the quantities of flammable liquids exceed the exempt quantities listed in Table 307.1(1), such hangars are classified as Group H-2 occupancies. They must be built of noncombustible construction, provided with fire suppression per NFPA 409, and ventilated in the manner prescribed by the IMC. Where the amount of flammable liquids within the hangar does not exceed the maximum allowable quantities set forth in Table 307.1(1), the classification is most appropriately a Group S-1 occupancy, and the provisions of this section do not apply.

412.7 **Heliports and helistops.** Helistops are differentiated from heliports by the presence of refueling facilities, maintenance operations, and repair and storage of the helicopters; thus, helistops pose similar hazards to those posed by aircraft repair hangars. The minimum size of a helicopter landing area is addressed, as are requirements for construction features and egress. Where heliports and helistops are constructed in compliance with the provisions of this section, they may be erected on buildings regulated by this code.

Section 413 *Combustible Storage*

Any occupancy group containing high-piled stock or rack storage is subject to the provisions of the IFC as well as the IBC. Chapter 32 of the IFC regulates combustible storage based on a variety of conditions, including the type of commodities stored, as well as the height and method of storage and the size of the storage area.

This section also specifically addresses any concealed spaces within buildings, including attics and under-floor spaces, that are used for the storage of combustible material. Where combustible storage occurs in areas typically considered unoccupiable, the storage areas are to be separated from the remainder of the building by 1-hour fire-resistance-rated construction on the storage side. The protective membrane need only be applied on the storage side insofar as the location of the hazard has been identified as the storage area only. Openings are to be protected with self-closing door assemblies that are either of non-combustible construction or are a minimum 1¾-inch (45-mm) solid wood. This separation is not necessary in Group R-3 and U occupancies. In addition, those combustible storage areas protected with sprinkler systems need not be separated. The provisions are not intended to apply to those storage rooms that are constructed and regulated as usable spaces within the building.

Section 414 *Hazardous Materials*

Figure 414-1 outlines the process for determining the code requirements that are a function of the quantities of hazardous materials stored or used. The outline is useful for both design and review. To begin, one must determine the hazardous processes and materials involved in a given occupancy and gain a thorough understanding of the operations taking place. Once the hazardous processes and materials have been identified, it is necessary to classify the materials based on the categories used by the code.

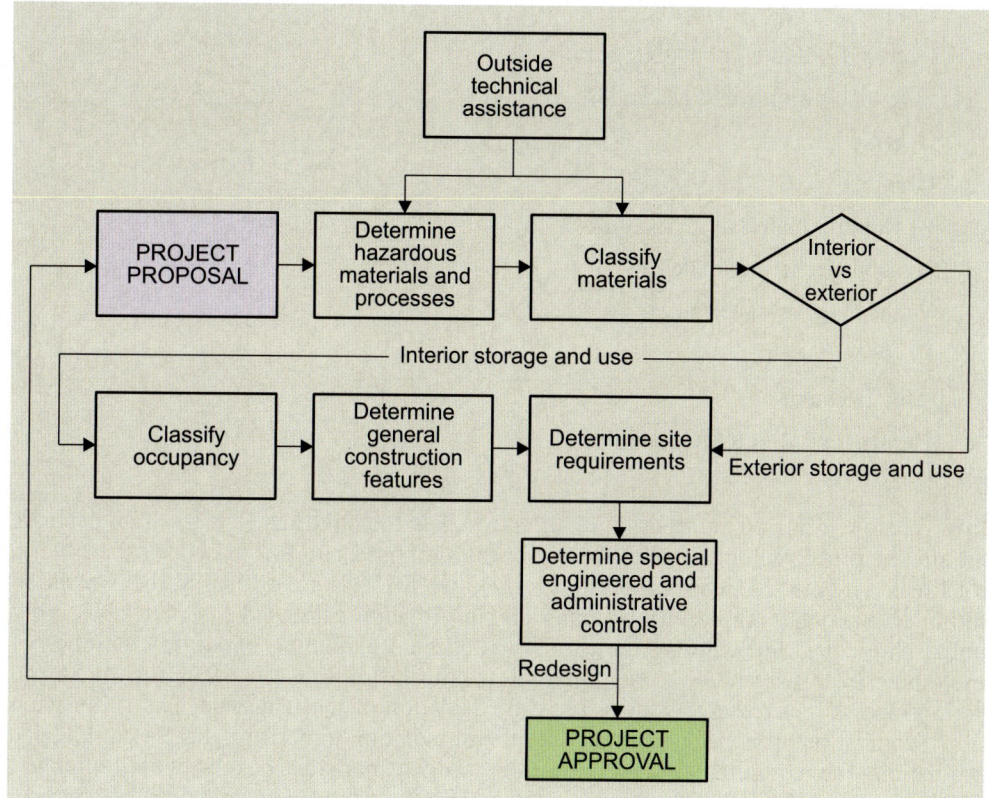

Figure 414-1
Code approach to hazardous materials.

Section 414.1.3 provides the means for the building official to acquire outside technical assistance to assist in the review of a project. Such assistance is often critical in assuring that appropriate decisions are made.

Classifying materials is a subjective science, requiring judgment decisions by an expert familiar with the characteristics of a particular material to categorize it within the categories used by the IBC and IFC. Accordingly, material classifications must be determined by qualified individuals, such as industrial hygienists, chemists, or fire-protection engineers. Though some jurisdictions employ individuals qualified to make these determinations, most jurisdictions rely on outside experts acceptable to the jurisdiction to submit a report detailing classifications compatible with the system used by the code.

4 Special Detailed Requirements Based on Use and Occupancy

Often, a permit applicant will attempt to submit a cadre of Material Safety Data Sheets (MSDSs) as a means of identifying material classifications. Though these may contain the information necessary to determine the proper classification, they do not normally contain a complete designation of classifications that is compatible with the system used by the IBC. Therefore, MSDSs are not normally acceptable as a sole means of providing material classifications to a jurisdiction. The building official should understand that it is not the responsibility of the jurisdiction to provide classifications for hazardous materials. Rather, it is the responsibility of the permit applicant to provide material classification information. In this way, potential liability of the jurisdiction for improper classification of materials is avoided.

In the classification system used by the *International Codes*®, hazardous materials are generally divided into two major categories, physical and health hazards, and 12 subcategories, as follows:

Physical Hazards

- Explosives and fireworks
- Combustible dusts and fibers
- Flammable and combustible liquids
- Flammable solids and gases
- Organic peroxides
- Oxidizers
- Pyrophoric materials
- Unstable (reactive) materials
- Water-reactive materials
- Cryogenic liquids

Health Hazards

- Highly toxic and toxic materials
- Corrosives

414.1.3 Information required. A report is required to allow the building department to evaluate the presence of hazardous materials within the proposed building based on the criteria established by the IBC. Since Tables 307.1(1) and 307.1(2) are critical in the evaluation of buildings containing hazardous materials, information is needed in order to properly utilize the tables. Such information must include the maximum expected quantities of each material in use and/or storage conditions, those fire-protection features that are to be in place, and any use of control areas for isolation of the materials. The submission of a technical report is necessary to allow the jurisdiction to perform a code compliance evaluation. The requirement for a technical report gives jurisdictions the benefit of expert opinions provided by knowledgeable persons in the particular hazard field of concern. Technical reports are required to be prepared by an individual, firm, or corporation acceptable to the jurisdiction, and must be provided without charge to the jurisdiction. Where the quantities of hazardous materials are such that a Group H occupancy is warranted, floor plans must be submitted to the building official identifying the locations of hazardous contents and processes.

414.2 Control areas. As addressed previously in the discussion of Section 307, areas in a building that are designated to contain the maximum allowable quantities of hazardous materials in use, storage, dispensing, or handling are considered control areas. At a minimum, 1-hour fire barriers shall be used to separate control areas from each other. Where required by Table 414.2.2 for the fourth story above grade plane and all stories above, a minimum 2-hour fire-resistance rating is required for such fire barriers. Openings in fire barriers are to be protected in accordance with Section 716. As a general rule, all floor construction that forms the boundaries of control areas is to have a minimum 2-hour fire-resistance rating. Building elements structurally supporting the 2-hour floor construction shall have an equivalent fire-resistance rating. There is an allowance for those two-story and three-story sprinklered buildings that are primarily of 1-hour fire-resistive construction (Types IIA, IIIA, and VA), which permits 1-hour floor construction of the control area and the supporting construction. It is apparent that a considerable level of fire separation must be achieved in order to increase the quantity of hazardous materials in non-Group H buildings. An example of this provision is illustrated in Figure 414-2.

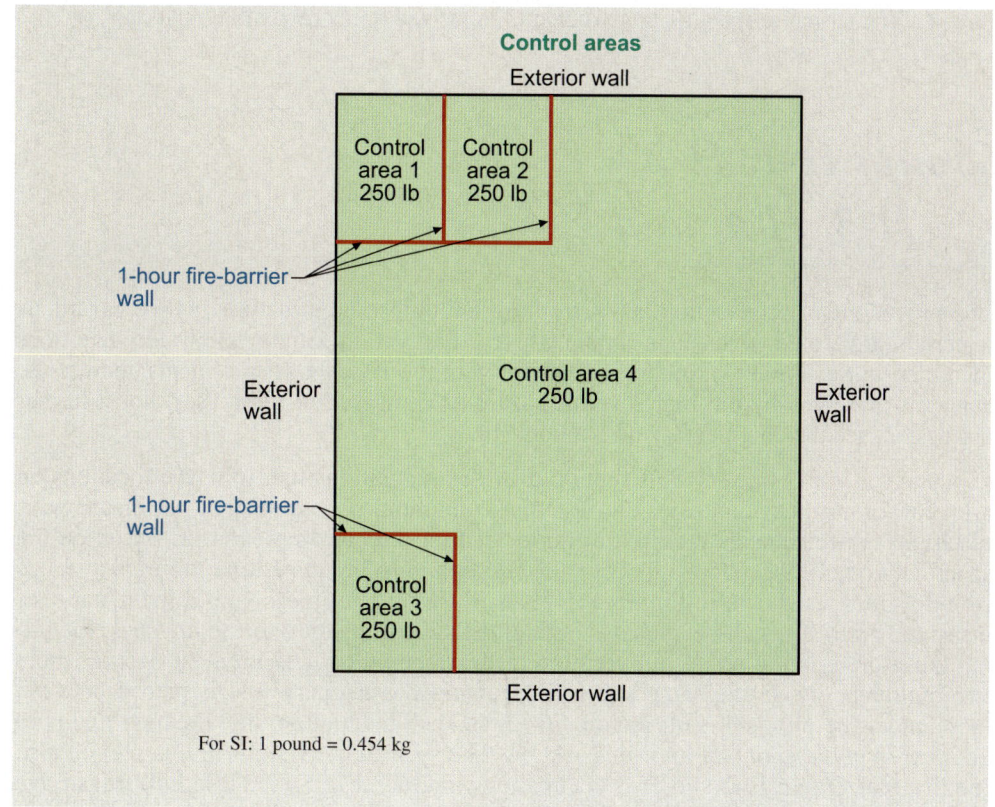

Figure 414-2
Control areas.

414.5.1 **Explosion control.** Table 414.5.1 indicates, based on material, the explosion control methods that must be provided where hazardous materials exceed the allowable quantities specified in Table 307.1(1). Explosion control is also required in any structure, room, or space occupied for purposes involving explosion hazards. Once some type of explosion control is required, Section 911 of the IFC must be referenced to identify the details for controlling explosion hazards.

414.5.4 **Spill control, drainage, and containment.** The intent of this section is the prevention of the accidental spread of hazardous material releases to locations outside of containment areas. Applicable to rooms, buildings, or areas used for the storage of both solid and liquid hazardous materials, the specifics for spill control, drainage, and containment are contained in the IFC.

414.6.1 **Weather protection.** In order to be considered outside storage or use in the application of the IFC, hazardous material storage or use areas must be primarily open to the exterior.

4 Special Detailed Requirements Based on Use and Occupancy

If it is necessary to shelter such areas for weather protection purposes, the enclosure and its location are limited by the following requirements:

1. No more than one side of the perimeter of the area may be obstructed by enclosing walls and structural supports unless the total obstructed perimeter is limited to 25 percent of the structure's total perimeter.

2. The minimum clearance between the structure and neighboring buildings, lot lines, or public ways shall be equivalent to that required for outside storage or use areas without weather protection.

3. Unless increased by the provisions of Section 506, the maximum area of the overhead structure shall be 1,500 square feet (140 m^2).

4. The structure must be constructed of approved noncombustible materials.

Section 415 *Groups H-1, H-2, H-3, H-4, and H-5*

The provisions of this section apply to those buildings and structures where hazardous materials are stored or used in amounts exceeding the maximum allowable quantities identified in Section 307. Applied in concert with the IFC, the requirements address the concerns presented by the high level of hazard as compared to other uses. For a further discussion, see the commentary on Section 414.

415.5 Fire separation distance. This section provides regulations that limit the locations on a lot for Group H occupancies and establish minimum percentages of perimeter walls of Group H occupancies required to be located on the building exterior. Based on the specific Group H occupancy involved, the building must be set back a minimum distance from lot lines, as shown in Figure 415-1. As illustrated in Figure 415-2, the distance is measured from the walls enclosing the high-hazard occupancy to the lot lines, including those on a public way. An exception to this method of measurement occurs where two buildings are on the same site and an assumed imaginary line is placed between them under the provisions of Section 705.3. In such a situation, the assumed line is to be ignored in the application of this section. The specific provisions in this section also require that Group H-2 and H-3 occupancies included in mixed-use buildings have 25 percent of the perimeter wall of the Group H occupancy on the exterior of the building. The access capability for fire personnel is greatly enhanced where the hazardous conditions are located in such a manner that allows for exterior fire-fighting operations. Exceptions are provided for smaller, liquid use, dispensing, and mixing rooms; liquid storage rooms; and spray booths. See Figure 415-3. It should be noted that where a detached building, required by Table 415.5.2, is located on the lot in accordance with this section, wall construction and opening protection is not regulated based on location on the lot. A minimum fire separation distance of 50 feet (15,240 mm) is required for such buildings. Therefore, the exterior wall and opening requirements of Table 602 have no application.

415.6 Special provisions for Group H-1 occupancies. Because of the extreme hazard presented by Group H-1 occupancies, this section requires that such occupancies be used for no other purpose, and it prohibits basements, crawl spaces, and under-floor spaces where flammable or explosive material might gather. Roofs are required to be of lightweight construction so that, in case of an explosion, they will rapidly vent with minimum destruction to the building. In addition, thermal insulation is sometimes required to prevent heat-sensitive materials from reaching decomposition temperatures.

Groups H-1, H-2, H-3, H-4, and H-5

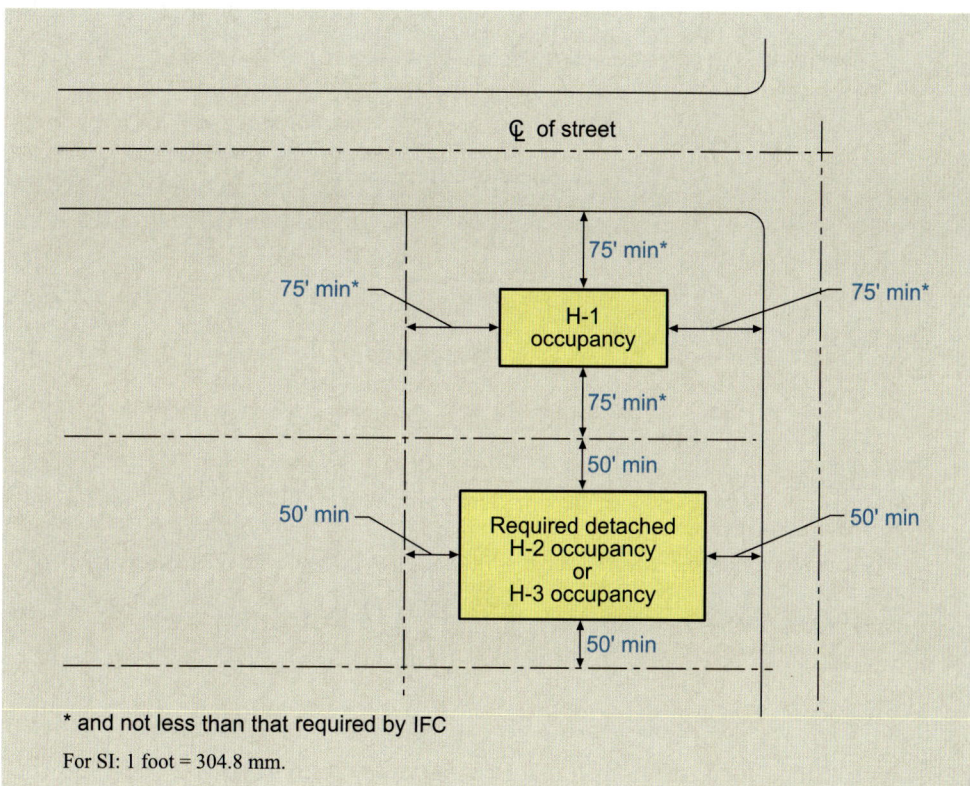

Figure 415-1
Location on property for the detached buildings.

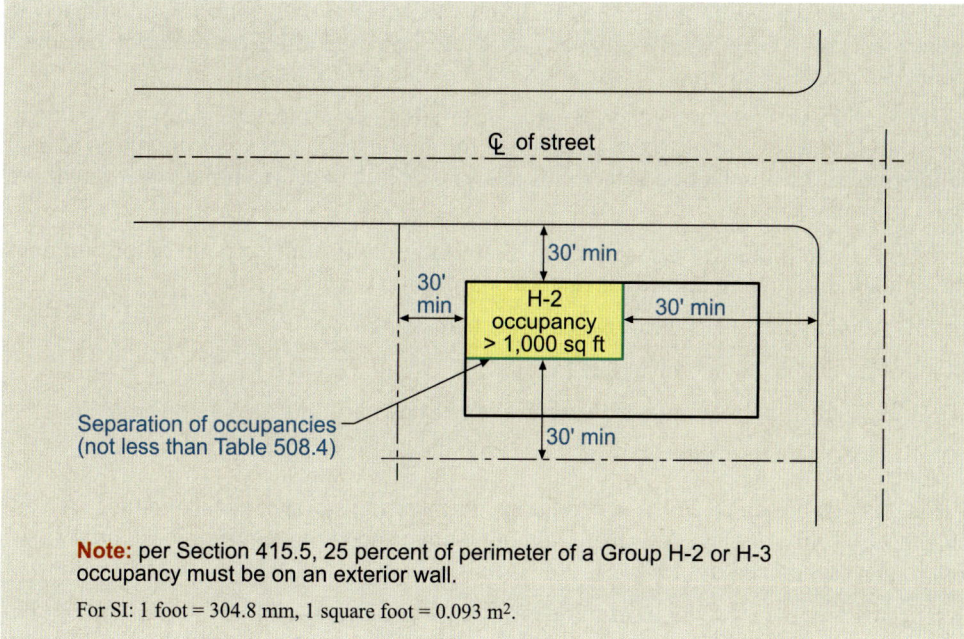

Figure 415-2
Location on property for mixed occupancies that include a Group H-2 occupancy.

This section also requires that Group H-1 occupancies that contain materials possessing health hazards in amounts exceeding the maximum allowable quantities for health-hazard materials in Table 307.1(2) also meet the requirements for Group H-4 occupancies. This provision is parallel to Section 307.8, which requires multiple hazards classified in more than one Group H occupancy to conform to the code for each of the occupancies classified.

2012 International Building Code Handbook   111

4 Special Detailed Requirements Based on Use and Occupancy

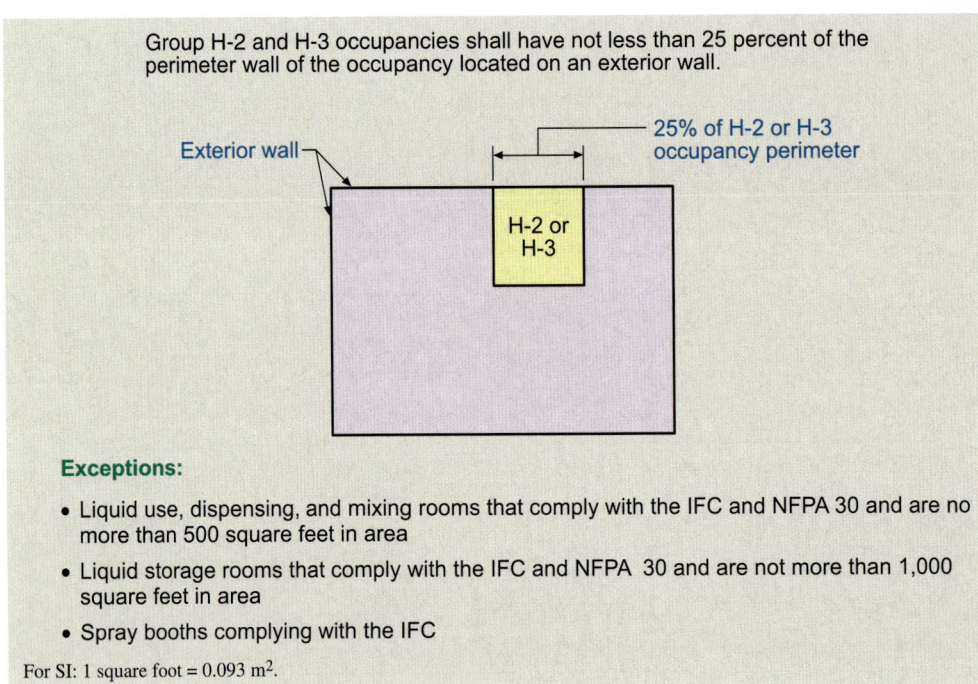

Figure 415-3
Perimeter wall on exterior.

415.7 Special provisions for Group H-2 and H-3 occupancies. Group H-2 and H-3 occupancies containing large quantities of the more dangerous types of physical hazard materials are considered to present unusual fire or explosion hazards that warrant a separate and distinct occupancy in a detached building used for no other purpose, similar to the requirements for a Group H-1 occupancy. The threshold quantities for requiring detached Group H-2 and H-3 occupancies are set forth in Table 415.5.2.

This section also requires water-reactive materials to be protected from water penetration or liquid leakage. Fire-protection piping is allowed in such areas in recognition of both the integrity of fire-protection system installations and the need to protect water-reactive materials from exposure fires.

415.8 Group H-2. Both this section and the IFC are to be used in the regulation of buildings containing the following hazardous materials operations:

1. Combustible dusts, grain processing, and storage
2. Flammable and combustible liquids
3. Liquefied petroleum gas distribution facilities
4. Dry cleaning plants

The hazards presented in these operations, through the storage, use, handling, processing, or transporting of hazardous materials, are unique enough to require special provisions, both in this section and in the IFC.

415.9 Groups H-3 and H-4. This section identifies several specific issues in Group H-3 and H-4 occupancies. Group H gas rooms shall be isolated from other areas of the building by minimum 1-hour fire barriers and/or horizontal assemblies. Highly toxic solids and liquids must also be separated from other hazardous material storage by fire barriers and/or horizontal assemblies having a minimum 1-hour fire-resistance rating, unless the highly toxic materials are stored in approved hazardous material storage cabinets. A related provision requires liquid-tight noncombustible floor construction in areas used for the storage of corrosive liquids and highly toxic or toxic materials.

Groups H-1, H-2, H-3, H-4, and H-5

415.10

Group H-5. The Group H-5 occupancy category was created to standardize regulations for semiconductor manufacturing facilities. This section provides the specific regulations for these occupancies. The Group H-5 category requires engineering and fire-safety controls that reduce the overall hazard of the occupancy to a level thought to be equivalent to a moderate-hazard Group B occupancy. Accordingly, the areas permitted for Group H-5 occupancies are the same as those for Group B occupancies.

The code requires that special ventilation systems be installed in fabrication areas that will prevent explosive fuel-to-air mixtures from developing. The ventilation system must be connected to an emergency power system. Furthermore, buildings containing Group H-5 occupancies are required to be protected throughout by an automatic fire-sprinkler system and fire and emergency alarm systems. Fire and emergency alarm systems are intended to be separate and distinct systems, with the emergency alarm system providing a signal for emergencies other than fire. This section also provides requirements for piping and tubing that transport hazardous materials that allow piping to be located in exit corridors and above other occupancies subject to numerous, stringent protection criteria. The provisions for Group H-5 occupancies are correlated with companion provisions in Chapter 27 of the IFC.

Table 415.5.2—Detached Building Required. This table establishes the threshold quantities of hazardous materials requiring detached buildings. Once the quantities listed in the table are exceeded, a detached building is required. The limitations placed on detached buildings required by Table 415.5.2 and classified as Group H-2 or H-3 are essentially the same as those applied to H-1 occupancies. The detached building must contain no other occupancy classification, be limited to one story in height, and have no basement, crawl space, or similar under-floor area. The applicability of the table is based solely on the total quantity of material present. The material can be in use, in storage, or both. See Application Example 415-1.

Application Example 415-1

GIVEN: A manufacturing operation requires up to 300 pounds of a Class 3 water reactive (in use).

DETERMINE: The maximum amount of the Class 3 material that can be stored in the building without detached storage being required.

Group H-2 | Group F-1 Manufacturing

Group H-2 Detached storage — Required where aggregate quantity in use and storage exceeds 2,000 pounds

Limited to 2,000 pounds of Class 3 water reactives

```
2,000 lb   total permitted in mixed-occupancy building
  300 lb   in use
-----------
1,700 lb   maximum quantity allowed in storage
```

For SI: 1 pound = 0.454 kg.

Otherwise, a single-occupancy Group H-2 detached building is required.

4 Special Detailed Requirements Based on Use and Occupancy

Section 416 Application of Flammable Finishes

This section applies to those buildings used for the spraying of flammable finishes such as paints, varnishes, and lacquers. In addition, Chapter 24 of the IFC contains extensive requirements for these types of operations. The IFC addresses a variety of spraying arrangements, each of which is specifically defined and regulated. These include spray rooms, spray booths, spraying space, and limited spraying space. The IBC provides limited construction provisions only for those arrangements determined to be spray rooms, as well as ventilation and surfacing requirements for all spraying spaces. An automatic fire-extinguishing system is mandated for all areas where the application of flammable finishes occurs, including all spray, dip, and immersing spaces, and storage rooms.

The occupancy classification of buildings, rooms, and spaces utilized for flammable finish application is not specifically addressed. Certainly, where the quantity of hazardous materials exceeds the maximum amounts established by Table 307.1(1) or 307.1(2), a Group H classification is warranted. However, where the maximum amounts are not exceeded, the analysis would be no different than that for other types of uses. In a manufacturing building, a Group F-1 occupancy classification would be appropriate. Spraying operations within a vehicle repair garage would most likely be considered part of the Group S-1 classification. A spray room, designed and constructed to house the spraying of flammable finishes, must be adequately separated from the remainder of the building. The enclosure must consist of fire barriers, horizontal assemblies, or both, each having a minimum fire-resistance rating of 1 hour. Spraying rooms must be frequently cleaned; thus, all of the interior surfaces must be smooth and easily maintained. The smooth surfaces also allow for the free passage of air in order to maintain efficient ventilation. The room construction must also be tight in order to eliminate the passage of residues from the room, which should be easily accomplished because of the fire separation required. Spraying spaces not separately enclosed shall be provided with noncombustible spray curtains to restrict the spread of vapors.

Section 417 Drying Rooms

Where the manufacturing process requires the use of a drying room or dry kiln, the room or kiln containing the drying operations must be of noncombustible construction. It must also be constructed in conformance with the specific and general provisions of the code as they relate to the special type of operations, processes, and materials that are involved. Clearance between combustible contents that are placed in the dryer and any overhead heating pipes must be at least 2 inches (51 mm). In addition, methods are addressed to insulate high-temperature dryers from adjacent combustible materials.

Section 418 Organic Coatings

Defined in Section 202 of the IFC, organic coatings are those compounds that are applied for the purpose of obtaining a finish that is protective, durable, and decorative. Used to protect structures, equipment, and similar items, organic coatings provide a surface finish that resists the effects of harsh weather. The concern for occupancies where organic coatings are manufactured or stored is based primarily on the presence of flammable vapors. As such, this type of use is highly controlled, both by the IBC and the IFC.

The manufacturing of organic coatings creates a high probability that flammable vapors will be present. Therefore, buildings where such materials are manufactured shall be without basements or pits because of the heavier-than-air nature of the vapors. In addition, no other occupancies are permitted in buildings used for the manufacture of organic coatings. The processing of flammable or heat-sensitive material must be done in a noncombustible or detached structure. Tank storage of flammable and combustible liquids inside a building must also be located above grade. In order to isolate the various hazard areas, the storage tank area must be separated from the remainder of the processing areas by minimum 2-hour fire barriers and/or horizontal assemblies. Because of the extreme hazards involved with nitrocellulose storage, it must also be separated by 2-hour fire-resistance-rated fire barriers and/or horizontal assemblies, or preferably located on a detached pad or in a separate structure.

Section 419 *Live/Work Units*

An increasingly popular concept of building use combines a residential unit with a small business activity. Residential live/work units typically include a dwelling unit along with some public service business, such as an artist's studio, coffee shop, or chiropractor's office. There may be a small number of employees working within the residence and the public is able to enter the work area of the unit to acquire service. Live/work units are a throwback to 1900-era community planning where residents could walk to all of the needed services within their neighborhood. These types of units began to re-emerge in the 1990s through a development style known as "Traditional Neighborhood Design." More recently, adaptive reuse of many older urban structures in city centers incorporated the same live/work tools to provide a variety of residential unit types. Provisions specifically addressing live/work units recognize the uniqueness of this type of use.

By definition, a "live/work unit" is primarily residential in nature but has a sizable portion of the space devoted to nonresidential activities. Often service-related in nature, the nonresidential portion is limited in several respects. The unit itself, including both the residential and nonresidential portions, is limited to 3,000 square feet (279 m^2) in total floor area. In addition, the nonresidential activities cannot take up more than 50 percent of the unit's total floor area. The portion dedicated to nonresidential use must be located on the first floor of the unit, or where applicable, on the unit's main floor level. In addition to the unit's residents, a limit of 5 workers or employees is permitted at any one time. An overview of the limitations is shown in Figure 419-1.

The occupancy classification of a live/work unit is Group R-2 based upon the primary use of the unit. Although differing uses are typically classified based on the characteristics of the varying uses involved and considered as mixed-occupancy conditions, in this case a single classification is considered acceptable. The potential hazards created due to the nonresidential uses are addressed through the special requirements of Section 419 that are to be applied in addition to those required due to the Group R-2 classification. Since live/work units are regulated as single-occupancy conditions, the provisions of Section 508 for mixed-occupancy buildings do not apply. In addition to the other limitations on use of a live/work unit, significant storage uses and those activities involving hazardous materials are prohibited. The increased fire load found in many storage uses is not considered in the live/work provisions, nor is the potential physical or health

4 Special Detailed Requirements Based on Use and Occupancy

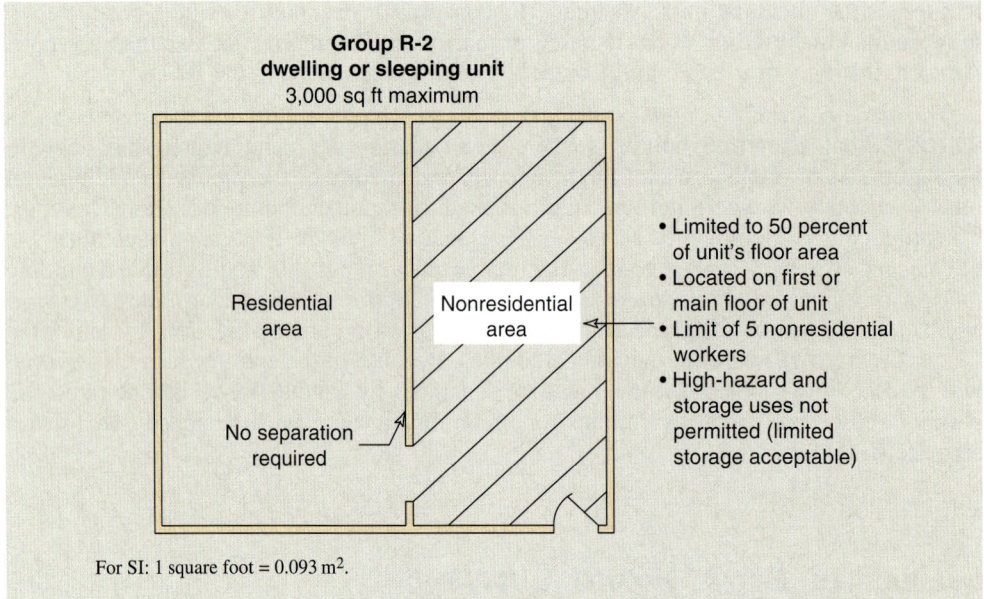

**Figure 419-1
Live/work unit.**

hazard that is due to the use or storage of hazardous materials. A very small amount of storage is permitted if it is deemed to be accessory to the nonresidential use.

Even though a live/work unit is classified as a Group R-2 occupancy, there are several issues where the residential and nonresidential portions are regulated independently. Structural loading conditions, accessibility features, and ventilation rates are all to be based on the individual function of each space within the unit, as are the design of the means of egress and the determination of required plumbing facility. In all other cases, the provisions applicable to a Group R-2 occupancy are to be applied to the entire live/work unit.

Section 420 *Groups I-1, R-1, R-2, and R-3*

In residential-type uses, it is important that any fire conditions created in one of the dwelling units or sleeping units does not spread quickly to any of the other units. As residential fires are the most common of fire incidents, it is critical that neighboring units be isolated from the unit of fire origin. The need for an adequate level of fire resistance is enhanced because of the lack of immediate awareness of fire conditions when the building's occupants are sleeping. The provisions are applicable to hotels and other Group R-1 occupancies, apartment buildings, dormitories, fraternity and sorority houses, and other types of Group R-2 occupancies, and between dwelling units of a Group R-3 two-family dwelling. Sleeping units and dwelling units of a supervised residential care facility classified as Group I-1 must also be provided with such fire-resistive separations.

Where dwelling units or sleeping units are adjacent to each other horizontally, the minimum required separation is a fire partition. The wall serving as a fire partition is regulated by Section 708 and typically must have a minimum 1-hour fire-resistance rating. A reduction to a ½-hour fire partition is permitted under the special conditions set forth in Exception 2 of Section 708.3. Where dwelling or sleeping units are located on multiple floors of a building, they must be separated from each other with minimum 1-hour fire-resistance-rated horizontal assemblies as described in Section 711. An allowance for a ½-hour reduction, similar to that permitted for fire partitions, is also available under specified conditions.

In addition to the required separation between adjoining units, dwelling units and sleeping units must also be separated by complying fire partitions and/or horizontal assemblies from other adjacent occupancies. Applicable in mixed-use buildings, this requirement takes precedence over the allowances in Sections 508.2 and 508.3 for accessory occupancies and nonseparated occupancies, respectively. Even in those cases where the mixed-occupancy provisions of Section 508.2 or 508.3 are applied, the separation requirements of Section 420 must be followed. In those cases where the separated occupancy provisions of Section 508.4 are utilized, the more restrictive fire-resistive rating is applied.

It should be noted that the separation of dwelling units and sleeping units from other types of spaces in the building appears to only apply if those spaces are of a different occupancy than that of the residential units. For example, a separation is not required between a Group R-1 sleeping unit in a hotel and the adjacent hotel lobby if the lobby is classified as a portion of the Group R-1 occupancy. However, it would seem that the intent of this section suggests that a separation be provided in order to isolate each individual dwelling unit or sleeping unit, regardless of the classification of the adjacent space. See Figure 420-1.

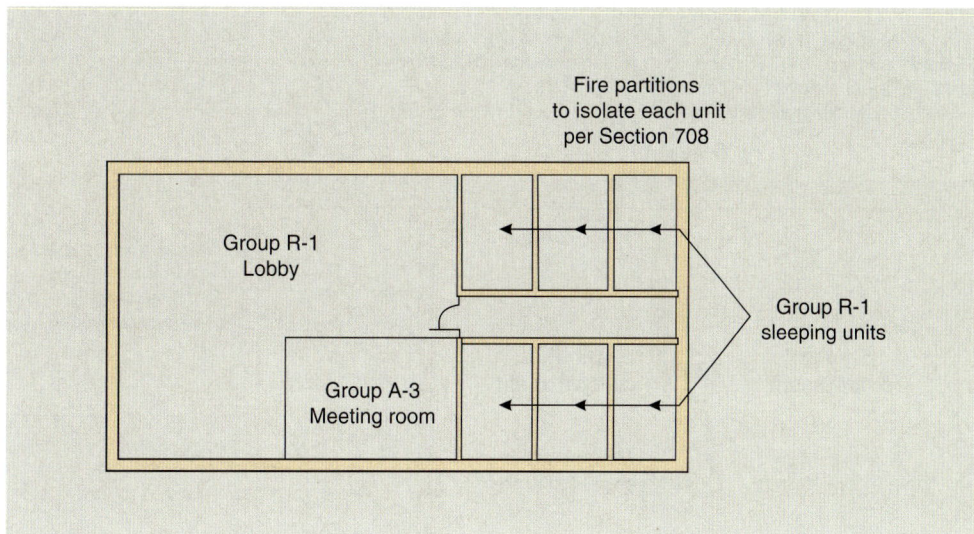

Figure 420-1
Separation of dwelling units and sleeping units.

Section 422 *Ambulatory Care Facilities*

Ambulatory care facilities, often referred to as ambulatory surgery centers or day surgery centers, are defined in Chapter 2 as a building or portion of a building "used to provide medical, surgical, psychiatric, nursing or similar care on a less than 24-hour basis to individuals who are rendered incapable of self-preservation by the services provided." Classified as Group B occupancies, such facilities are generally regarded as moderate in hazard level due to their office-like conditions. However, additional hazards are typically present due to the presence of individuals who are temporarily rendered

4 Special Detailed Requirements Based on Use and Occupancy

incapable of self-preservation due to the application of nerve blocks, sedation, or anesthesia. While the occupants may walk in and walk out the same day with a quick recovery time after surgery, there is a period of time where a potentially large number of people could require physical assistance in case of an emergency that would require evacuation or relocation.

Although classified as a Group B occupancy in the same manner as an outpatient clinic or other health-care office, an ambulatory care facility poses distinctly different hazards to life and fire safety, such as:

- Patients incapable of self-preservation require rescue by other occupants or emergency responders.
- Medical staff must stabilize the patient prior to evacuation, possibly resulting in delayed staff evacuation.
- Use of oxidizing medical gases such as oxygen and nitrous oxide.
- Potential for surgical fires.

As a result of the increased hazard level, additional safeguards have been put in place. Smoke compartments must be provided in larger facilities, and the installation of fire-protection systems is typically mandated. See Figure 422-1.

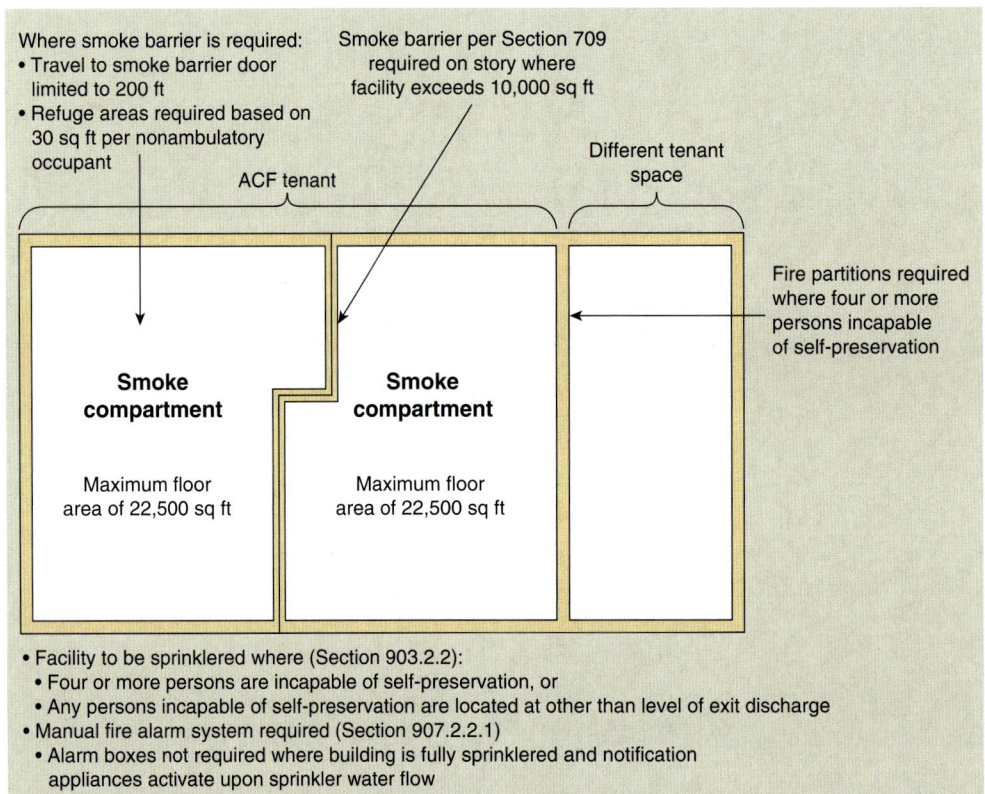

Figure 422-1
Ambulatory care facility.

Any story containing ambulatory care facilities having more than 10,000 square feet (929 m²) of floor area must be subdivided into at least two smoke compartments by smoke barriers in accordance with Section 709. The limit on compartment size of

22,500 square feet (2,092 m^2) may require that three or more smoke compartments be provided. Additional compartments may also be required due to travel distance limitations. Any point within a smoke compartment must be no more than 200 feet (60,960 mm) in travel distance from a smoke barrier door. Each smoke compartment must be large enough to allow for 30 square feet (2.8 m^2) of refuge area for each nonambulatory patient. In addition, at least one means of egress must be available from each smoke compartment without the need to return back through the original compartment.

As a general rule, Group B occupancies do not require a sprinkler system based solely on their occupancy classification. However, Section 903.2.2 mandates that a Group B ambulatory care facility be provided with an automatic sprinkler system when either of the following conditions exist at any time:

- Four or more care recipients are incapable of self-preservation, or
- One or more care recipients who are incapable of self-preservation are located at other than the level of exit discharge.

The extent of the sprinkler protection is detailed in Section 903.2.2. In addition, the fire alarm requirements are more stringent than those of other Group B occupancies. Section 907.2.2 requires the installation of a manual fire alarm system in all Group B fire areas containing an ambulatory health-care facility. The manual fire alarm boxes are not required if the building is fully sprinklered and the occupant notification appliances activate upon sprinkler water flow.

Section 423 *Storm Shelters*

ICC-500, *ICC/NSSA Standard on the Design and Construction of Storm Shelters*, establishes minimum requirements for structures and spaces designated as hurricane, tornado, or combination shelters. The standard addresses the design of such shelters from the perspective of the structural requirements for high wind conditions, and addresses minimum requirements for the interior environment during a storm event. Although the IBC does not mandate that storm shelters be provided, it does regulate their design if they are constructed.

Section 424 *Children's Play Structures*

Play structures for children's activities were regulated for some time by the IBC only where such structures were located within covered mall buildings. The primary concern, consistent with that of other structures located within a covered mall building, was the combustibility of such play structures. Due to the potential fire hazards associated with children's play structures, the regulations are now applicable where such structures are located within any building regulated by the IBC, regardless of occupancy classification.

Children's play structures must be constructed of noncombustible materials or, as an option if combustible, must comply with the appropriate criteria established in Section 424. Such alternative methods include the use of fire-retardant-treated wood, textiles complying with the designated flame propagation performance criteria, and plastics exhibiting an established maximum peak rate of heat release.

4 Special Detailed Requirements Based on Use and Occupancy

KEY POINTS

- Special uses such as covered and open mall buildings, atriums, high-rise buildings, underground buildings, and parking garages are so unique in the type of hazards presented that specialized regulations are provided in the IBC.
- A covered mall building or open mall building consists of various tenants and occupants, as well as the common pedestrian area that provides access to the tenant spaces.
- For those features that are not unique to a covered or open mall building, the general provisions of the code apply.
- The means of egress provisions for a covered or open mall building are typically more liberal than those for other buildings.
- High-rise buildings are characterized by the difficulty of evacuation or rescue of the building occupants, the difficulty of fire-fighting operations from the exterior, high occupant loads, and potential for stack effect.
- There are a number of provisions for high-rise buildings that are less restrictive than the general requirements, including the reduction in fire resistance for certain building elements.
- The special allowance for a reduction in construction type is not applicable to any high-rise building exceeding 420 feet (128 m) in height.
- Smoke detection, alarm systems, and communications systems are important characteristics of a high-rise building.
- Occupant egress and evacuation, as well as fire department access, are addressed in high-rise buildings through provisions for stairway enclosure remoteness, an additional stairway, luminous egress path markings, fire service access elevators, and occupant evacuation elevators.
- The use of the atrium provisions is typically limited to those multistory applications where compliance with the other vertical opening applications established in Section 712.1 is not possible.
- Buildings containing atriums, high-rise buildings, covered mall buildings, and open mall buildings must be provided with automatic sprinkler systems throughout.
- Another component of the life-safety system for a building containing an atrium is a required smoke-control system.
- An underground building is regulated in a manner similar to that for a high-rise building, as the means of egress and fire department access concerns are similarly extensive.
- Fundamental to the protection features for an underground building are the requirements for Type I construction for the underground portion and the installation of an automatic sprinkler system.
- Private garages and carports are regulated to a limited degree based on the hazards associated with the parking of motor vehicles.
- Special provisions for open parking garages are typically less restrictive than those for enclosed parking structures because of the natural ventilation that is available.
- In Group I-2 and I-3 occupancies, the functional concerns of the health-care operations must be balanced with the fire-safety concerns.

Key Points

- Stages exceeding 50 feet (15,240 mm) in height present additional risks that are due to the expected presence of high combustible loading such as curtains, scenery, and other stage effects.
- Under-floor areas and attic spaces used for the storage of combustible materials must be isolated from other portions of the building with fire-resistance-rated construction.
- Hazardous materials that are used or stored in any quantity are subject to regulation by Sections 307 and 414, and the IFC.
- An increase in the maximum allowable quantities of hazardous materials in a building not classified as Group H is permitted through the proper use of control areas.
- Group H occupancies are highly regulated because of the hazardous processes and materials involved in such occupancies.
- Special provisions are applicable to the construction, installation, and use of buildings for the spraying of flammable finishes in painting, varnishing, and staining operations.
- Special allowances and conditions are applicable to live/work units where a dwelling unit or sleeping unit includes a significant amount of nonresidential use operated by the tenant.
- Dwelling units and sleeping units must be separated from each other and from other portions of the building through the use of fire partitions and/or horizontal assemblies.
- Health-care offices where care is provided to individuals who are rendered incapable of self-preservation are considered to be ambulatory care facilities.
- Storm shelters, where provided for safe refuge from hurricanes, tornados, and other high-wind events, must be constructed in accordance with ICC-500.

CHAPTER 5

GENERAL BUILDING HEIGHTS AND AREAS

Section 501 General
Section 503 General Building Height and
 Area Limitations
Section 504 Building Height
Section 505 Mezzanines and Equipment Platforms
Section 506 Building Area Modifications
Section 507 Unlimited-Area Buildings
Section 508 Mixed Use and Occupancy
Section 509 Incidental Uses
Section 510 Special Provisions
Key Points

5 General Building Heights and Areas

Chapter 5 provides general provisions that are applicable to all buildings. These include requirements for allowable floor area, including permitted increases for open spaces and for the use of automatic sprinkler systems; unlimited-area buildings; and allowable height of buildings with acceptable increases. Buildings containing multiple uses and occupancies are regulated through the provisions for incidental uses, accessory occupancies, nonseparated occupancies, and separated occupancies. Miscellaneous topics addressed in Chapter 5 include premises identification and mezzanines.

In addition to the general provisions set forth in Chapter 5, there are several special conditions under which the specific requirements of Chapter 5 can be modified or exempted, including the horizontal building separation allowance and unique provisions for buildings containing a parking garage.

Section 501 *General*

501.2 Address identification. In this section, the *International Building Code*® (IBC®) intends that buildings be provided with plainly visible and legible address numbers posted on the building or in such a place on the property that the building may be identified by emergency services such as fire, medical, and police. The primary concern is that responding emergency forces may locate the building without going through a lengthy search procedure. In furthering the concept, the code intends that the approved street numbers be placed in a location readily visible from the street or roadway fronting the property if a sign on the building would not be visible from the street. Address numbers may be required in multiple locations to help eliminate any confusion or delay in identifying the location of the emergency. The fire code official can require, and must approve, additional address identification locations. Regardless of the sign's location, the minimum height of letters or numbers used in the address is to be at least 4 inches (102 mm) and have a color in contrast to the color of the background itself.

Section 503 *General Building Height and Area Limitations*

The IBC regulates the size of buildings in order to limit to a reasonable level the magnitude of a fire that potentially may develop. The size of a building is controlled by its floor area and height, and both are limited by the IBC. Whereas floor-area limitations are concerned primarily with property damage, life safety is enhanced as well by the fact that in the larger building there are typically more people at risk during a fire. Height restrictions are imposed to address egress concerns and fire department access limitations.

The essential ingredients in the determination of allowable areas are:

1. The amount of combustibles attributable to the use that determines the potential fire severity.

2. The amount of combustibles in the construction of the building, which contributes to the potential fire severity.

General Building Height and Area Limitations

In addition to the two factors just itemized, there may be other features of the building that have an effect on area limitations. These include the presence of built-in fire protection (an automatic fire-sprinkler system), which tends to prevent the spread of fire, and open space (frontage) adjoining a sizable portion of the building's perimeter, which decreases exposure from adjoining properties and provides better fire department access.

A desirable goal of floor-area limitations in a building code is to provide a relatively uniform level of hazard for all occupancies and types of construction. A glance through Table 503 of the IBC will reveal that, in general, the higher hazard occupancies have lower permissible areas for equivalent types of construction and, in addition, the less fire-resistant and more combustible types of construction have more restrictive area limitations.

The IBC also limits the maximum height and number of stories based on similar reasons discussed for area limitations. In addition, the higher the building becomes, the more difficult access for firefighting becomes. Furthermore, the time required for the evacuation of the occupants increases; therefore, the fire resistance of the building should also be increased.

The code presumes that when the height of the highest floor used for human occupancy exceeds 75 feet (22,860 mm), the life-safety hazard becomes even greater because most fire departments are unable to adequately fight a fire above this elevation from the outside. Furthermore, the evacuation of occupants from the building is often not feasible. Thus, Section 403 prescribes special provisions for these high-rise buildings. Similar concerns for buildings with occupied floors well below the level of exit discharge are addressed in Section 405 for underground buildings.

Coming back to this section, the code specifies in Table 503 both the maximum allowable height in feet (mm) and the maximum number of stories. The maximum height in feet is regulated solely by the building's construction type, with no regard for the occupancy or multiple occupancies located in the building. However, the maximum height in stories varies based on the occupancy group involved. Where multiple occupancies are located in the same building, and the provisions of Section 508.4 for separated occupancies are utilized, each individual occupancy can be located no higher than set forth in the table. See Figure 503-1. Where the nonseparated-occupancies provisions of Section 508.3 are applied, the most restrictive height limitations of the nonseparated occupancies involved will limit the number of stories in the entire building. See Figure 503-2. In general, the greater the potential fire- and life-safety hazard, the lower the permitted overall height in feet (mm), as well as the fewer the number of permitted stories.

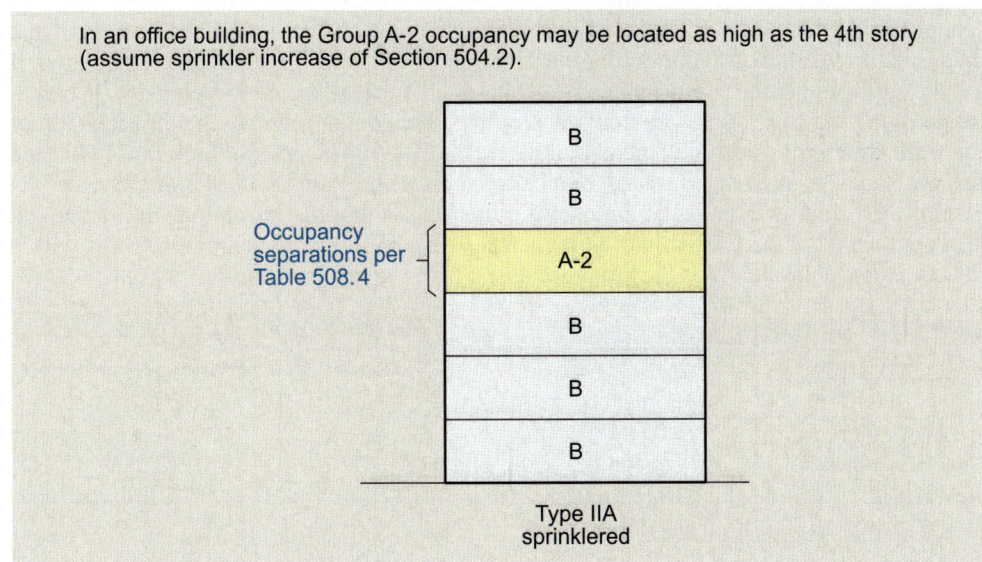

Figure 503-1
Height limitations—separated occupancies.

5 General Building Heights and Areas

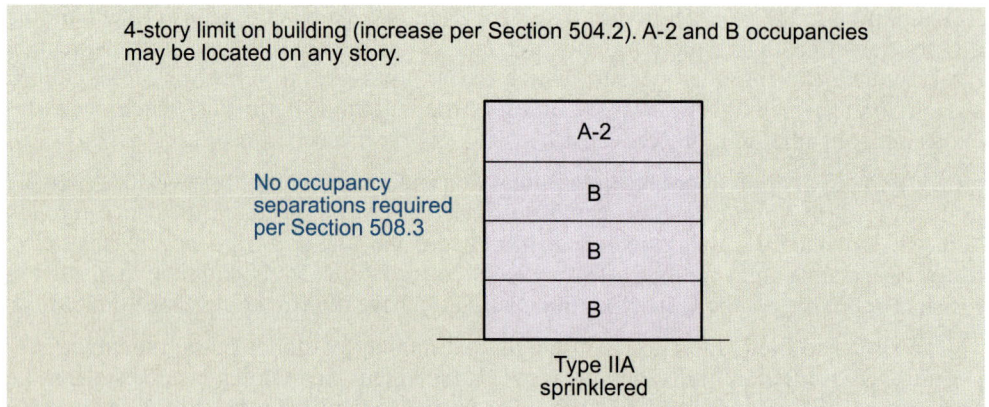

Figure 503-2
Height limitations—nonseparated occupancies.

503.1 General. Height and area limitations for buildings are set forth in Table 503. As indicated in the table, the height limits are expressed as both stories and feet above the grade plane. For allowable area purposes, the numbers in the table refer to the building area (in square feet), as defined in Section 202, per story. Therefore, based on occupancy group and type of construction, Table 503 identifies the limitations on building size permitted by the IBC. Examples of the use of this table, without any permitted increases, are shown in Figure 503-3.

The IBC establishes in Table 503 what is commonly referred to as tabular values for height and area. By use of the term *tabular*, the code recognizes that the building in question has no features that might be considered to improve the overall fire hazard (such as a fire-sprinkler system or adjacent open areas); alternatively, where such features exist, there are also conditions under which such features provide no benefit for the situation under consideration. An example is found in Section 508.2.1 for accessory occupancies, where the floor area of the accessory occupancy cannot exceed the tabular values in Table 503. Although the building being analyzed may qualify for height and area increases established elsewhere in Chapter 5, they are not considered in the evaluation of accessory occupancy compliance. For the more typical application of the provision, this section states that the height of the building and the area of any floor within the building shall not be greater than the tabular (basic) area specified in Table 503, unless the building is entitled to height or area, which are described in other sections of Chapter 5.

In this section, the IBC indicates that fire walls, in addition to exterior walls, create separate buildings when evaluating for allowable height and area. Defined and regulated under the provisions of Section 706, the function of a fire wall is to separate one area of a building from another with a fire-resistance-rated vertical separation element. Where a fully complying fire wall is provided, it provides two compartments, one on each side of the wall, which may each be considered under the IBC to be separate buildings. Multiple fire walls may be utilized to create a number of separate buildings within a single structure. One of the resulting benefits of the use of a fire wall is that the limitations on height and area are then addressed individually for each separate building created by fire walls within the structure, rather than for the structure as a whole. See Figure 503-4.

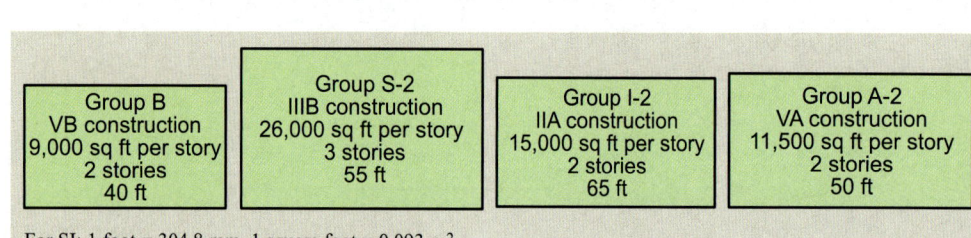

Figure 503-3
Allowable area and height.

For SI: 1 foot = 304.8 mm, 1 square foot = 0.093 m².

General Building Height and Area Limitations

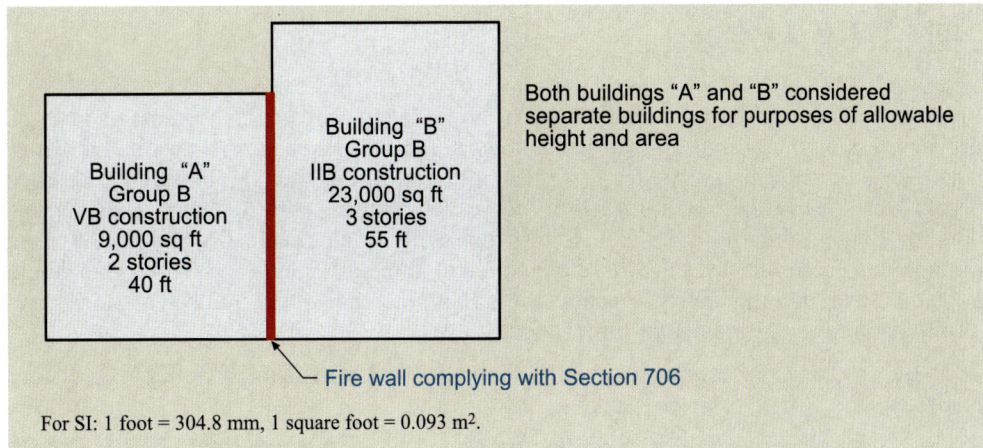

Figure 503-4
Fire walls.

Special industrial occupancies. This special provision exempts certain types of buildings from both the height limitations and the area limitations found in Table 503. Thus, the type of construction is not limited, regardless of building height or area. It is also not necessary to comply with the provisions of Section 507 for unlimited-area buildings to utilize this provision. Applicable to structures housing low-hazard and moderate-hazard industrial processes that often require quite large areas and heights, the relaxation of the general provisions recognizes the limited fire severity, as well as the need for expansive buildings to house operations such as rolling mills, structural metal-fabrication shops, foundries, and power distribution. It is not the intent that buildings classified as Group H occupancies be addressed under the allowances of Section 503.1.1.

503.1.1

Buildings on the same lot. Where two or more buildings are located on the same lot, they may be regulated as separate buildings in a manner consistent with buildings situated on separate parcels of land. See Figure 503-5A.

503.1.2

As an option, multiple buildings on a single site may be considered one building, provided the limitations of height and area based on Table 503 are met. The height of each building and the aggregate area of all buildings are to be considered in the determination.

Under this method, the provisions of the code applicable to the aggregate building shall also apply to each building individually. See Figure 503-5B. Further regulations for buildings on the same lot are discussed in the commentary for Section 705.3.

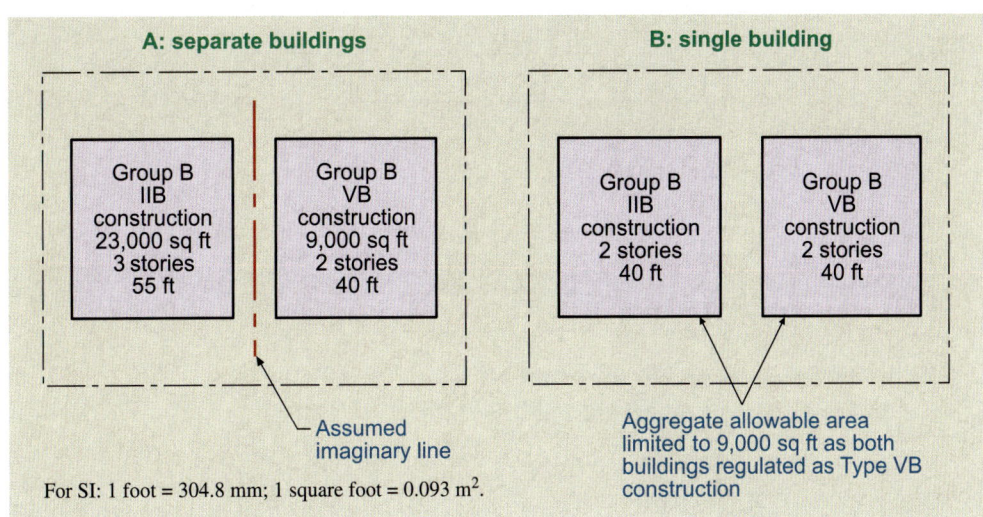

Figure 503-5
Buildings on the same lot.

2012 International Building Code Handbook

5 General Building Heights and Areas

Section 504 *Building Height*

Because automatic fire-sprinkler systems have exhibited an excellent record of in-place fire suppression over the years, the IBC allows height increases as well as area increases, where an automatic fire-sprinkler system is installed throughout the building. The code permits an increase of one story in the number of stories, and 20 feet (6,096 mm) in building height, where the building is provided with an automatic fire-sprinkler system throughout. These increases are directly applied to Table 503. See Figure 504-1. It should be emphasized that this increase applies both to an increase in the number of stories and also to an increase of the height limit in feet (mm).

There are basically four variations to the general requirements for height and story increases:

1. Such increases are not permitted for Group I-2 occupancies of Type IIB, III, IV, or V construction, or for Groups H-1, H-2, H-3, and H-5 occupancies of any construction type. These occupancies present unusual hazards that limit their heights even where a sprinkler system is present. The increases may also not be taken where the provisions of Table 601, Footnote d, for 1-hour fire-resistance rating substitution are utilized.

2. One-story aircraft manufacturing buildings and hangars may be of unlimited height when sprinklered and surrounded by adequate open space. Such uses require very large structures and through the safeguards provided, should be adequately protected.

3. For Group R buildings provided with an NFPA 13R sprinkler system, the increases in height and number of stories apply only up to a maximum of 60 feet

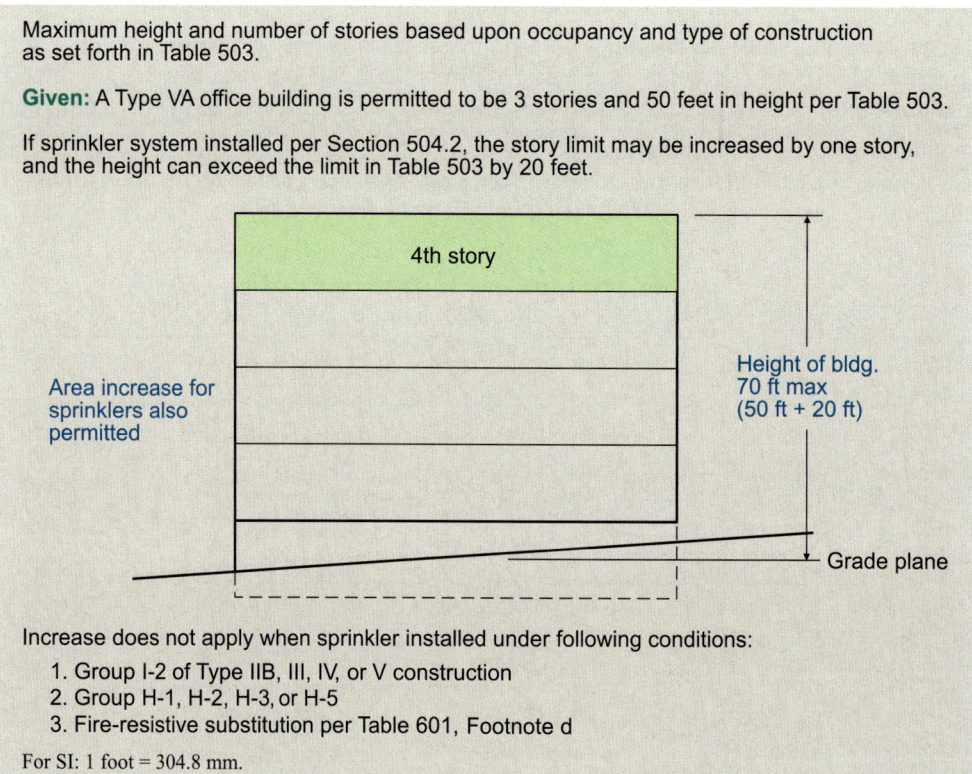

Figure 504-1
Allowable height increase.

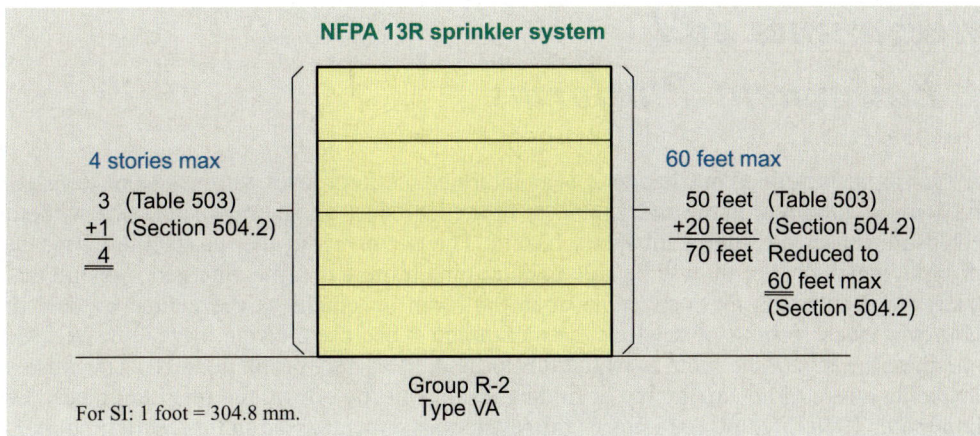

Figure 504-2
Height increase—Group R.

(18,288 mm) and four stories, respectively. The limitation of four stories and 60 feet for buildings sprinklered with a 13R system cannot be exceeded under any circumstances. See Figure 504-2. In those residential buildings where an NFPA 13, rather than an NFPA 13R system, is installed, the limitations of 60 feet (18,288 mm) and four stories do not apply.

4. Roof structures such as towers and steeples may be of unlimited height when constructed of noncombustible materials, whereas combustible roof structures are limited in height to 20 feet (6,096 mm) above that permitted by Table 503. See Figure 504-3. In all cases, such roof structures are to be constructed of materials based on the building's type of construction. These requirements are not based on the presence of the sprinkler system. Additional requirements for roof structures can be found in Section 1509.

As an important note, except for those buildings provided with an NFPA 13R system, the increases in building height and number of stories permitted by this section for a sprinklered building may be taken in addition to those floor-area increases permitted by Sections 506.2 and 506.3.

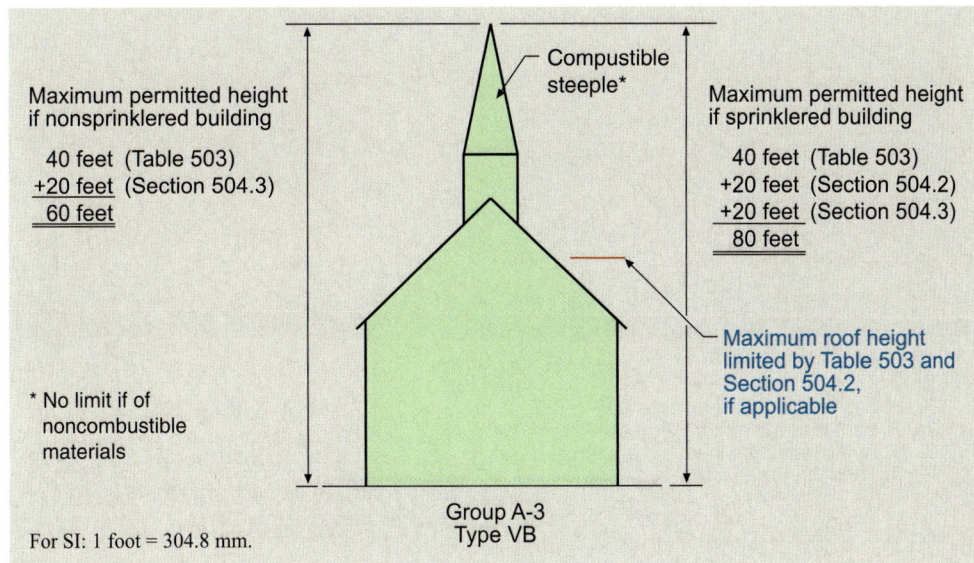

Figure 504-3
Roof structures.

2012 International Building Code Handbook   129

5 General Building Heights and Areas

Section 505 *Mezzanines and Equipment Platforms*

A mezzanine is defined in Chapter 2 as an intermediate floor level within a room or space. As long as the area of the mezzanine is limited in size, an intermediate floor without enclosure causes no significant safety hazard. The occupants of the mezzanine by means of sight, smell, or hearing will be able to determine if there is some emergency or fire that takes place either on the mezzanine or in the room in which the mezzanine is located. However, once portions of or all of the mezzanine is enclosed, or the mezzanine exceeds one-third the area of the room in which it is located, life-safety problems such as occupants not being aware of an emergency or finding a safe exit route from the mezzanine become important. Therefore, the code places the restrictions encompassed in this section on mezzanines to ameliorate the life-safety that can be created.

505.2 Mezzanines. By virtue of the conditions placed on mezzanines in Section 505, a complying mezzanine is not considered to create additional building area or an additional story for the purpose of limiting building size. The floor area of a complying mezzanine need not be added to the area of the floor below for the purpose of limiting building area by Section 503. This allowance essentially provides for free floor area in the comparison of the total actual area to the total allowable area. As previously mentioned, complying mezzanines also do not contribute to the actual number of stories in relationship to the allowable number of stories permitted by Sections 503 and 504. The limitations imposed on mezzanines are deemed sufficient to permit such benefits.

In contrast to the above allowances, the floor area of mezzanines must be included as a part of the aggregate floor area in determining the fire area. Because the size of a fire area is based on a perceived level of fire loading present within the building, the contribution of a mezzanine's fire load to the fire loading in the room in which the mezzanine is located cannot be overlooked. Figure 505-1 depicts the proper use of these provisions. The clear height above and below the floor of the mezzanine is also regulated at a minimum height of 7 feet (2,134 mm).

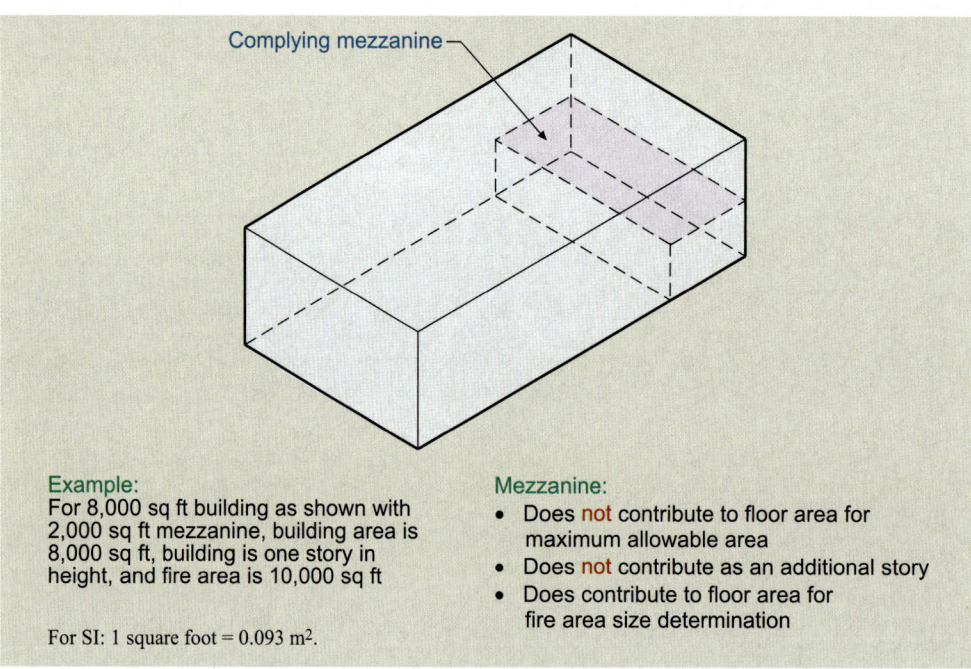

Figure 505-1 Mezzanine height and area.

Example:
For 8,000 sq ft building as shown with 2,000 sq ft mezzanine, building area is 8,000 sq ft, building is one story in height, and fire area is 10,000 sq ft

For SI: 1 square foot = 0.093 m².

Mezzanine:
- Does not contribute to floor area for maximum allowable area
- Does not contribute as an additional story
- Does contribute to floor area for fire area size determination

Mezzanines and Equipment Platforms 5

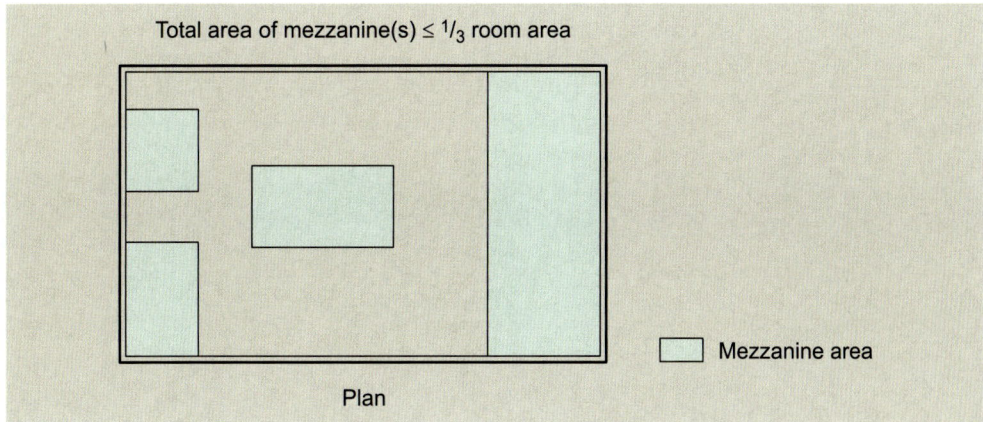

Figure 505-2
Mezzanine area.

Area limitation. There is no limit on the number of mezzanines that may be placed within a room; however, the total floor area of all mezzanines must typically not exceed one-third the floor area of the room in which they are located. See Figure 505-2. As illustrated in Figure 505-3, any enclosed areas of the room in which the mezzanine is located are not to be utilized in the calculations for determining compliance with the one-third rule. **505.2.1**

Where two specific conditions exist, the aggregate floor area of mezzanines may be increased up to two-thirds of the floor area of the room below. First, the building must contain special industrial processes as identified in Section 503.1.1, and second, the building shall be of Type I or Type II construction. Intermediate floor levels are very common in buildings of this kind because of the nature of their operations. By limiting the increased mezzanine size to noncombustible buildings housing primarily noncombustible processes, fire safety is not compromised.

A second exception also permits an increase in allowable mezzanine size, up to a maximum of one-half of the area or room in which the mezzanine is located. See Figure 505-4. The increased size takes into consideration the enhancements of noncombustible construction, automatic sprinkler system protection, and occupant notification. By limiting construction to Type I or II, there is no contribution to the fire hazard that is due to the materials of construction. The automatic sprinkler protection will increase the potential for limiting fire spread. In addition, the occupant notification system increases occupant awareness of a fire condition and allows for evacuation during the early fire stages.

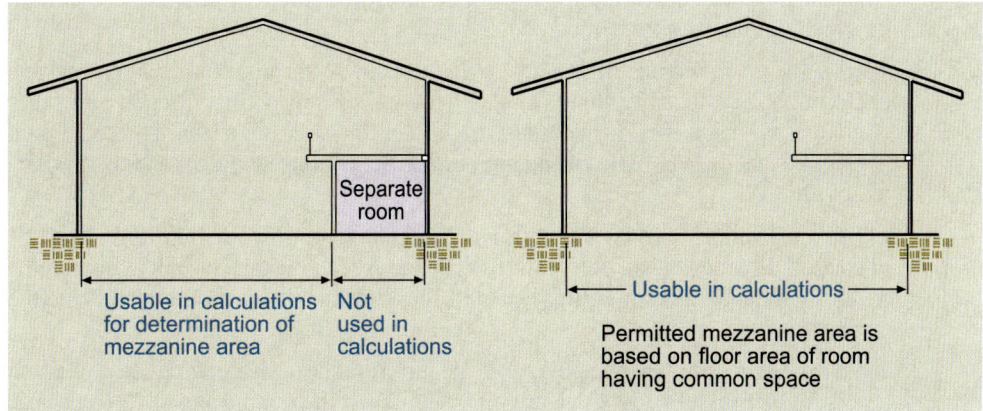

Figure 505-3
Mezzanine area.

2012 International Building Code Handbook 131

5 General Building Heights and Areas

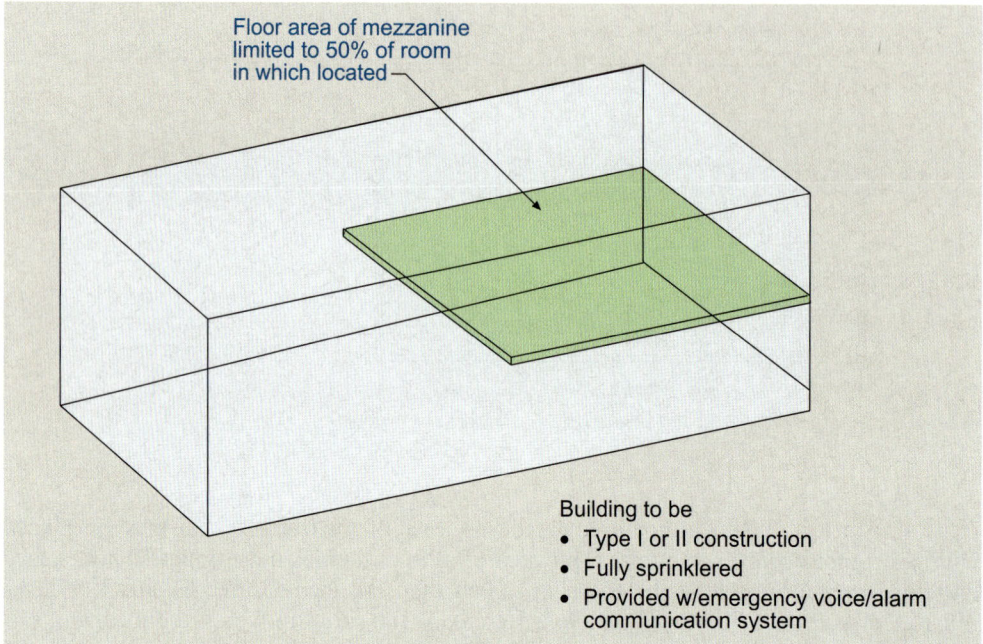

Figure 505-4.
Maximum floor area of mezzanine.

505.2.3 Openness. As a general rule, a mezzanine must be open to the room in which it is located. Any side that adjoins the room will be considered open if it is unobstructed, other than by walls or railings not more than 42 inches (1,067 mm) in height, or columns and posts. There are, however, five exceptions to the requirement for openness that result in most mezzanines being permitted to be partially or completely enclosed. If in compliance with any one of the five exceptions, the mezzanine need not be enclosed.

1. Illustrated in Figure 505-5, this criterion is that the enclosed area contains a maximum occupant load of 10 persons. The limitation on occupant load is based on the aggregate area of the enclosed space. This exception is consistent with other provisions of the code that relax the requirements where the occupant load is expected to be relatively small. This exception may result in the enclosure of an entire mezzanine or just a portion of it.

2. As shown in Figure 505-6, a mezzanine may also be fully or partially enclosed if it has two means of egress, one of which gives direct access to an exit component. While one means of egress must be provided directly to an exit, such as an interior exit stairway or exterior exit stairway, the second means of egress may enter an adjacent room as shown in the drawing or may be a stairway down into the room in which the mezzanine is located.

3. Also depicted in Figure 505-5, up to 10 percent of the mezzanine area may be enclosed. This is usually done for toilet rooms, closets, utility rooms, and other similar uses that must, of necessity, be enclosed. As long as the aggregate area does not exceed 10 percent of the area of the mezzanine, the enclosure is permitted by the code.

4. Those mezzanines used for control equipment in industrial buildings may be glazed on all sides. This exception is necessary because of the delicate nature of much of today's control equipment and the fact that it may require a dust-free environment.

5. This exception allows enclosure of the entire mezzanine, provided specified egress and sprinkler provisions are met. Similar to the conditions of Exception 2,

Mezzanines and Equipment Platforms 5

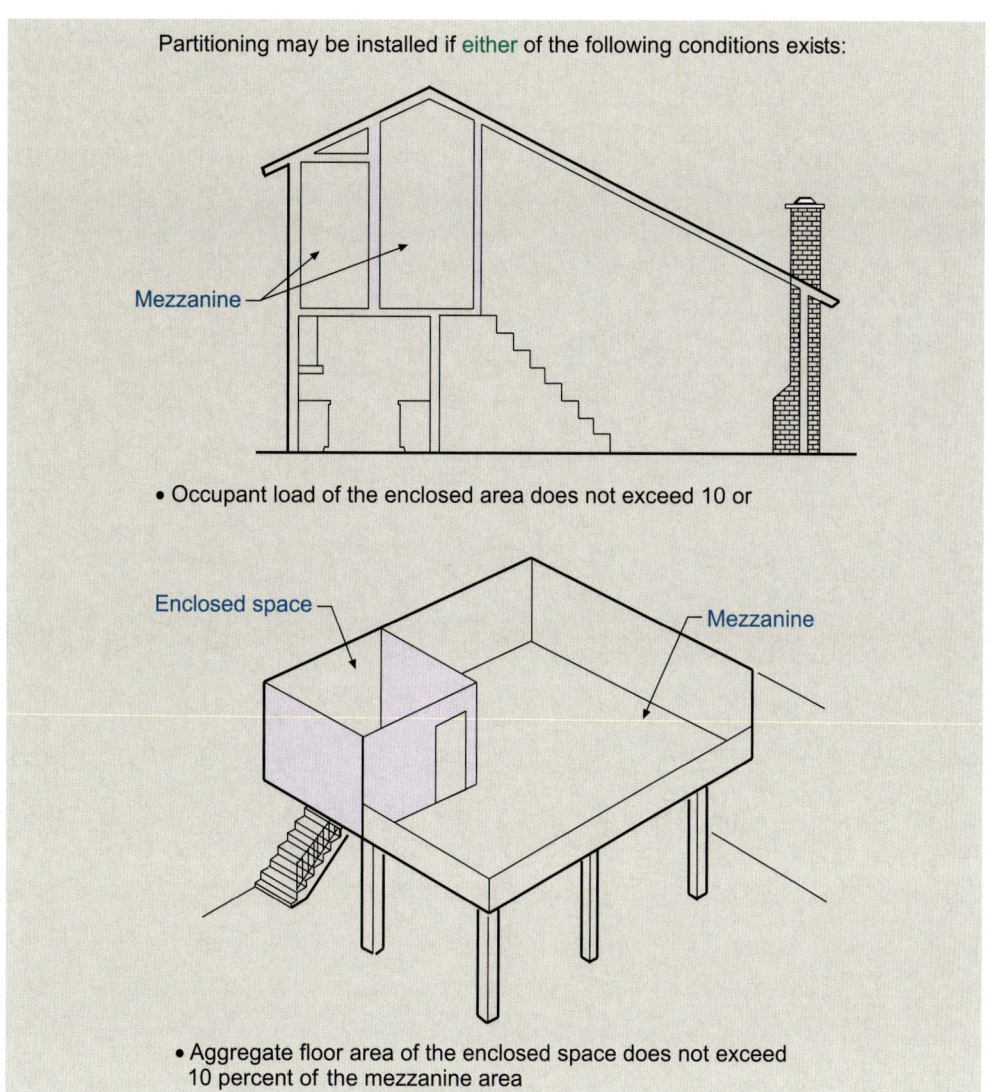

Figure 505-5
Enclosed mezzanines.

this exception requires a minimum of two means of egress from the mezzanine. However, it differs in that none of the egress travel is required to reach an exit component on the mezzanine level. Applicable only to buildings that are sprinklered throughout, the exception allows for all required egress from an enclosed mezzanine to occur on unenclosed stairways leading to the floor level below. The exception is not intended to apply to Group H and I occupancies or buildings more than two stories in height above grade plane. See Figure 505-7.

Equipment platforms. In buildings containing platforms that house equipment, such platforms need not be considered stories or mezzanines, provided they conform with the provisions of this section regulating platform size, extent of automatic sprinkler protection, and guards. In addition, the equipment platforms cannot serve as any portion of the exiting system from the building. Complying platforms are not deemed to be additional stories, do not contribute to the building floor area, and furthermore, need not be included in determining the size of the fire area.

505.3

5 General Building Heights and Areas

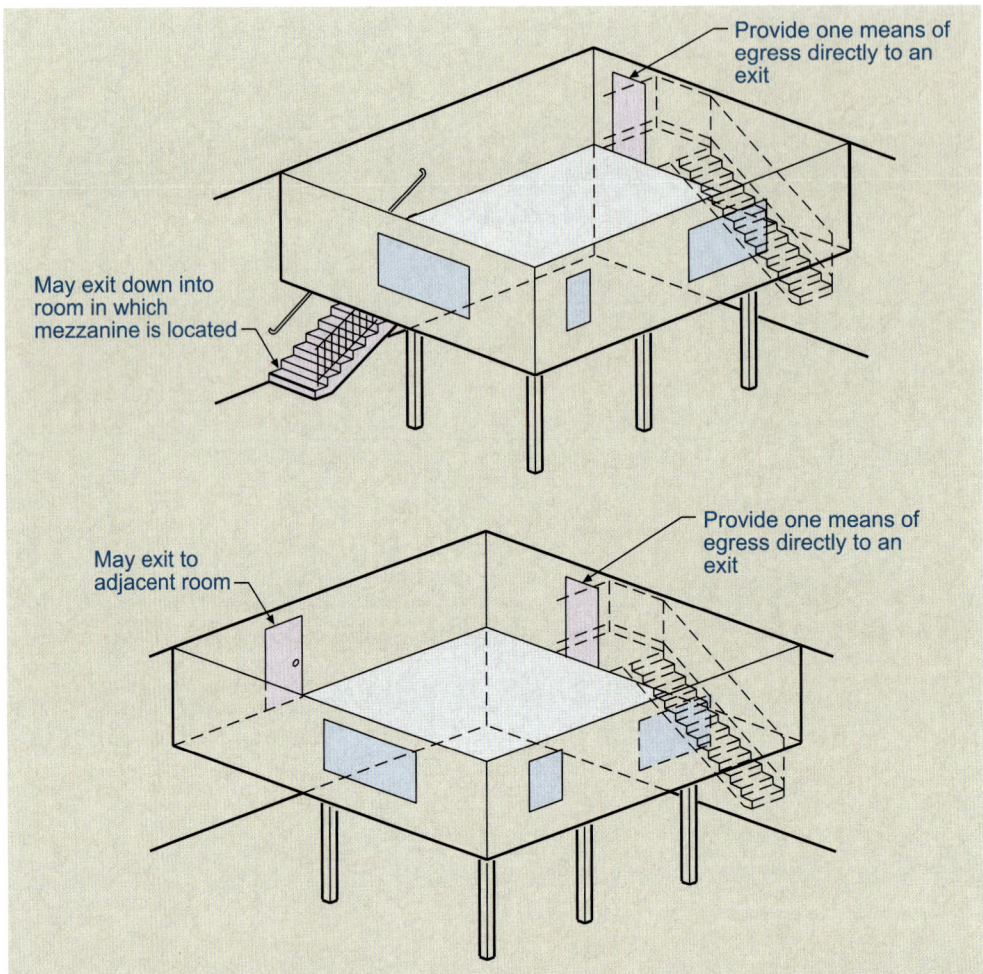

Figure 505-6 Enclosed mezzanines.

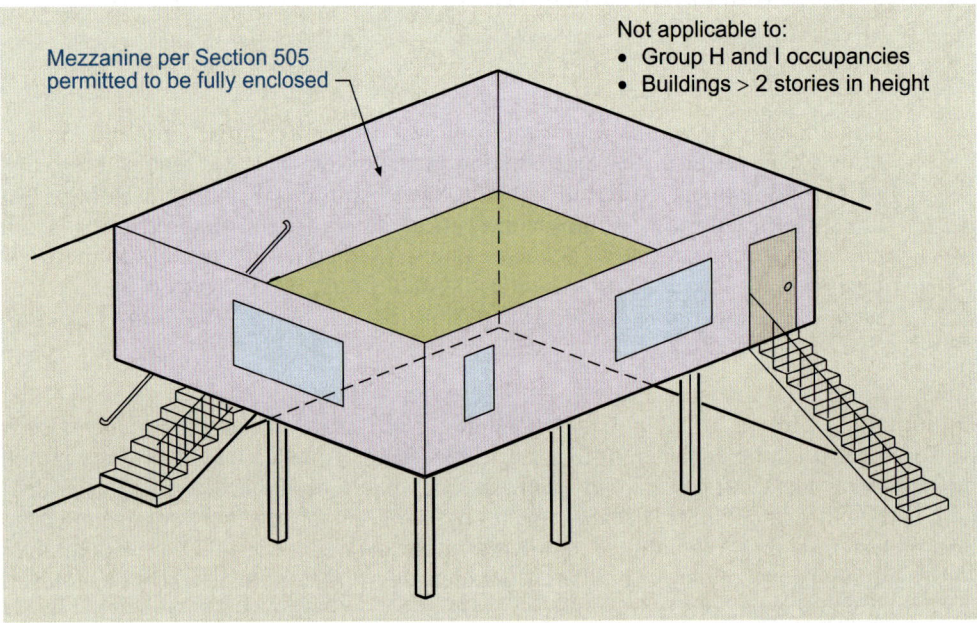

Figure 505-7 Enclosure of mezzanines.

Section 506 *Building Area Modifications*

Whereas the basic allowable building area per floor is regulated by Table 503, increases to those areas are permitted based on the presence of adequate open space on one or more sides of the building, as well as the protection of the structure with an automatic sprinkler system. In addition, the overall allowable building area is permitted to be increased in multistory buildings.

General. The formula for the calculation of the maximum allowable area per floor (in square feet) is additive, determined from the sums of the tabular area based on Table 503, any increase that is due to building frontage per Section 506.2, and any increase that is due to automatic fire-sprinkler protection as established in Section 506.3. A simple example is shown in Application Example 506-1. **506.1**

Application Example 506-1

GIVEN: One-story office building
Type IIB construction
Fully sprinklered
Yards and streets as shown

DETERMINE: Maximum allowable area per floor (A_a)

$A_a = A_t + [A_t I_f] + [A_t I_s]$
$A_t = 23{,}000$ sq ft (Table 503)
$I_f = \left[\dfrac{220}{320} - 0.25\right]\dfrac{30}{30} = 100\,[0.69 - 0.25]\,1.00 = 44\% = 0.44$
$I_s = 300\%$ (single-story building)
$A_a = 23{,}000 + [23{,}000(0.44)] + [23{,}000\,(3.0)]$
$= 23{,}000 + 10{,}120 + 69{,}000$
$= 102{,}120$ sq ft per floor

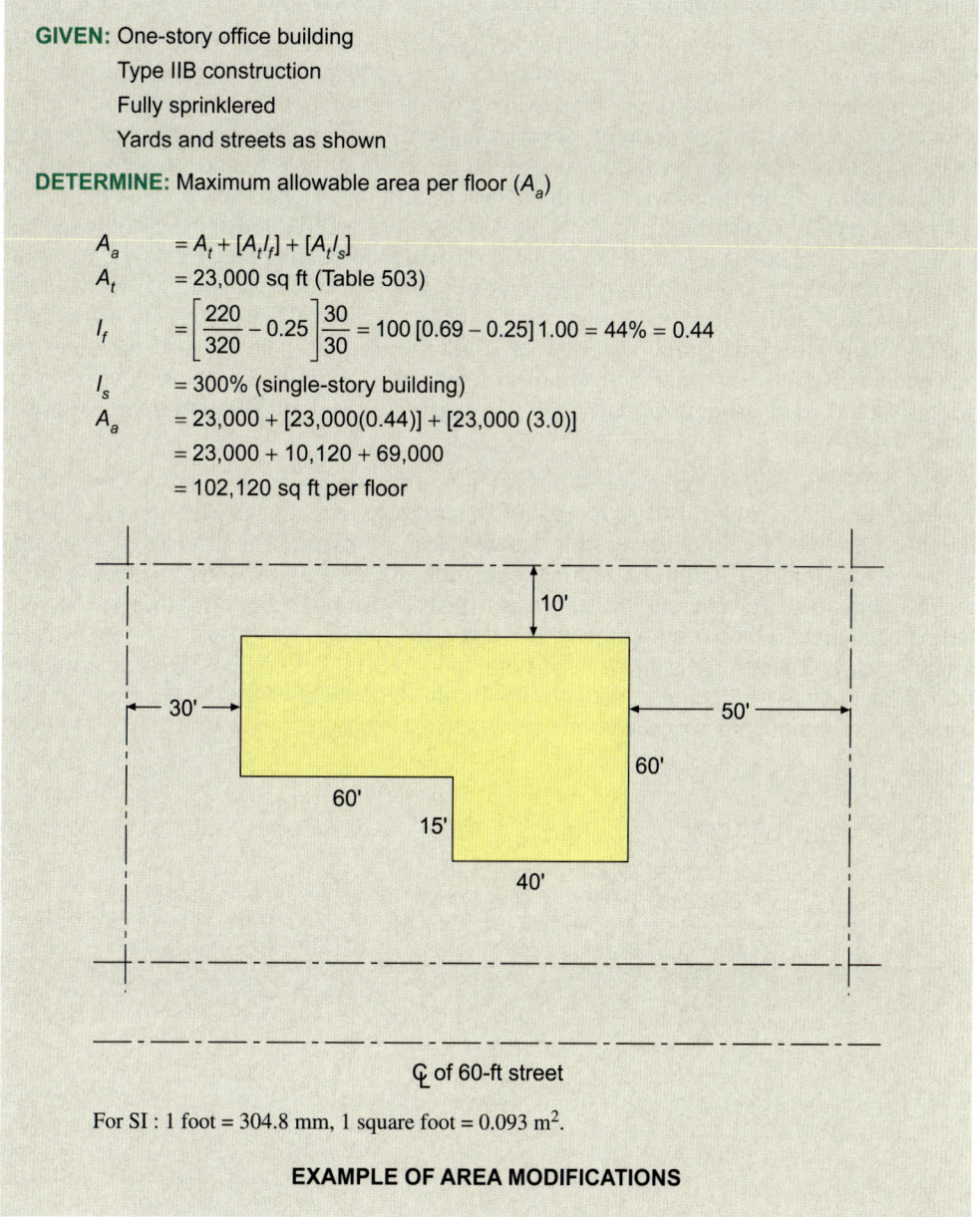

For SI : 1 foot = 304.8 mm, 1 square foot = 0.093 m².

EXAMPLE OF AREA MODIFICATIONS

5 General Building Heights and Areas

506.2 Frontage increase. The initial requirement of the code, insofar as a frontage increase is concerned, is that it adjoin or have access to a public way. Thus, the structure could extend completely between side lot lines and to the rear lot line, and be provided with access from only the front of the building, and still potentially be eligible for a small frontage increase. Therefore, it follows that if a building is provided with frontage consisting of public ways and/or open space for an increased portion of the perimeter of the building, some benefit should accrue based on better access for the fire department. Also, if the yards or public ways are wide enough, there will be a benefit that is due to the decreased exposure from adjoining properties.

Because of the beneficial aspects of open space adjacent to a building, the IBC permits increases in the tabular areas established from Table 503 based on the amount of open perimeter and width of the open space and public ways surrounding the building. For any open space to be effective for use by the fire department, it is mandated that it be accessed from a public way or a fire lane so that the fire department will have access to that portion of the perimeter of the building that is adjacent to open space. See Figure 506-1.

Open space and public ways—what can and can't be used. In addition to allowances for public ways, the IBC uses the term *open space* where related to frontage increases in the determination of allowable floor areas. Although the term *open space* is not specifically defined in the IBC, the definition of a *yard* is an open space unobstructed from the ground to the sky that is located on the lot on which the building is situated. It is logical that this definition is consistent with the intended description of open space. This definition seems to preclude the storage of pallets, lumber, manufactured goods, home improvement materials, or any other objects that similarly obstruct the open space. However, it would seem reasonable to permit automobile parking, low-profile landscaping, fire hydrants, light standards, and similar features to occupy the open space. These types of obstructions can be found within the public way, so their allowance within the open space provides for consistency. Because a yard must be unobstructed from the ground to the sky, open space widths should be measured from the edge of roof overhangs or other projections, as shown in Figure 506-2.

Regarding the use of public ways for providing frontage increases, the width of public way that should be used for determining area increases seems to cause confusion. Should the full width of the public way or only the distance to the centerline be used? The confusion evolves from the definition of fire separation distance as established in Chapter 2, which states that fire separation distance is measured from the building face to the centerline of a street, alley, or public way. However, the requirement to use the centerline is limited to fire separation distance and is not applicable to Section 506.2. For determining frontage increases for open space, the full width of the public way may be used by buildings located on both sides of the public way.

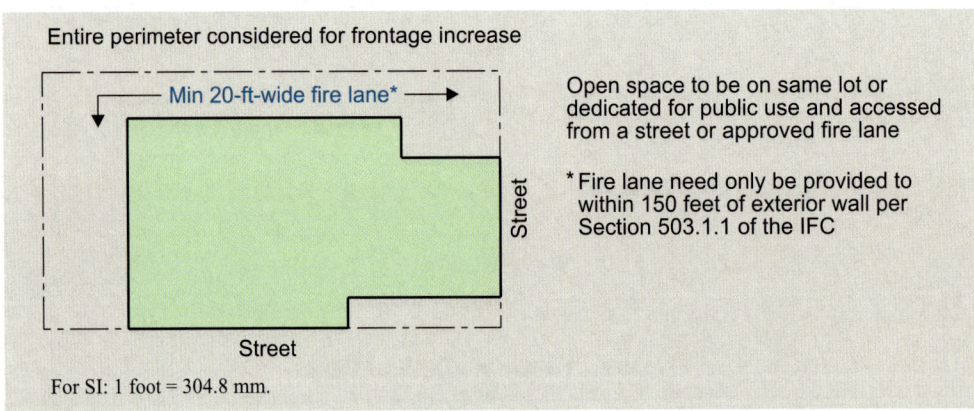

Figure 506-1 Open space access.

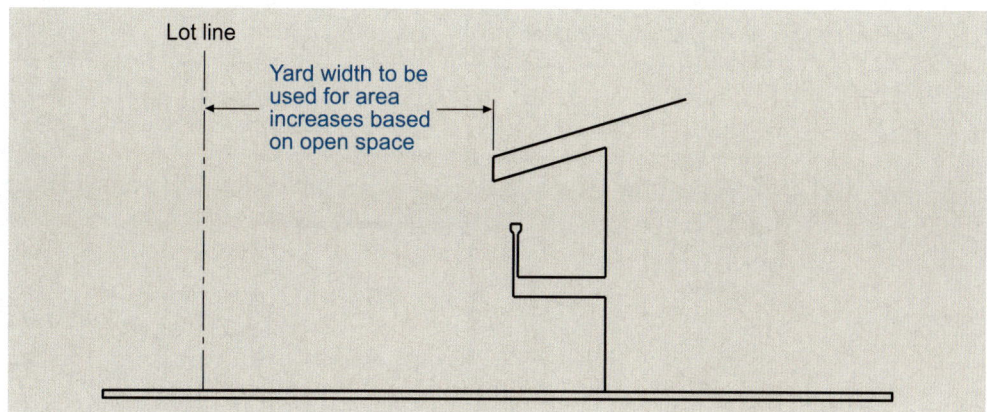

Figure 506-2
Measurement of open space.

The following type of question is also sometimes asked: "Why can't I use the big open field next door for area increases?" Section 506.2.2 specifically mandates that open space used for a frontage increase must be on the same lot as the building under consideration, or alternatively, dedicated for public use. There is a good reason for this limitation, insofar as the owner of one parcel lacks control over a parcel owned by another and, thus, the open space can disappear when the owner of "the big open field" decides to build on it. One method by which some jurisdictions have allowed such large open spaces to be used is by accepting joint use of shared yards. It is typically necessary that a recorded restrictive covenant be executed to ensure that the shared space will remain open and unoccupied as long as it is required by the code. The creation of a no-build zone does not seem unreasonable insofar as the aim is to maintain open spaces between buildings. Any covenant should be reviewed by legal counsel to be sure it will accomplish what is intended. In addition, it should clearly describe the reason and applicable code section so that any future revisions or deletions may be considered if the owners wish to terminate such an agreement. In such an event, each building should be brought into current code compliance, or the agreement would be required to remain in effect.

Whereas use of a public way as open space is permitted by the IBC, other publicly owned property is generally not, because the building official usually has no control over the long-range use of publicly owned property, and there is little assurance that such property will be available as open space for the life of the building. Remember that what is today's publicly owned open parking lot could become tomorrow's new city hall, and the open space used to justify area increases would no longer exist. Whereas Section 506.2.2 allows publicly owned property to be considered open space, the intent is such that the property be permanently dedicated for public use and maintained as unobstructed. The term *public way* was used in place of streets because its definition allows the use of a broader range of publicly owned open space while still allowing the building official some discretion as to the acceptability of a particular parcel. *Public way* usually conjures up visions of streets and alleys, but how about other open spaces such as power line right-of-ways, flood-control channels, or railroad rights-of-way? Many such open spaces are generally acceptable, provided there is a good probability that they will remain as open space during the life of the building for which they will serve. Power lines and flood-control channels are usually good bets for longevity, but railroad routes are often abandoned and, therefore, may not be as good a bet. There is also an expectation that the public way is maintained in an unobstructed condition to allow for fire department access, which potentially would disallow the use of waterways and similar features. If the public way does not provide for fire department access, its use for a frontage increase is prohibited. It should be noted that the definition for public way requires any such public parcel of land, other than a street or alley, to lead to a street. Figure 506-3 provides a visual summary of open space and public ways that could be used for open-space area increases.

5 General Building Heights and Areas

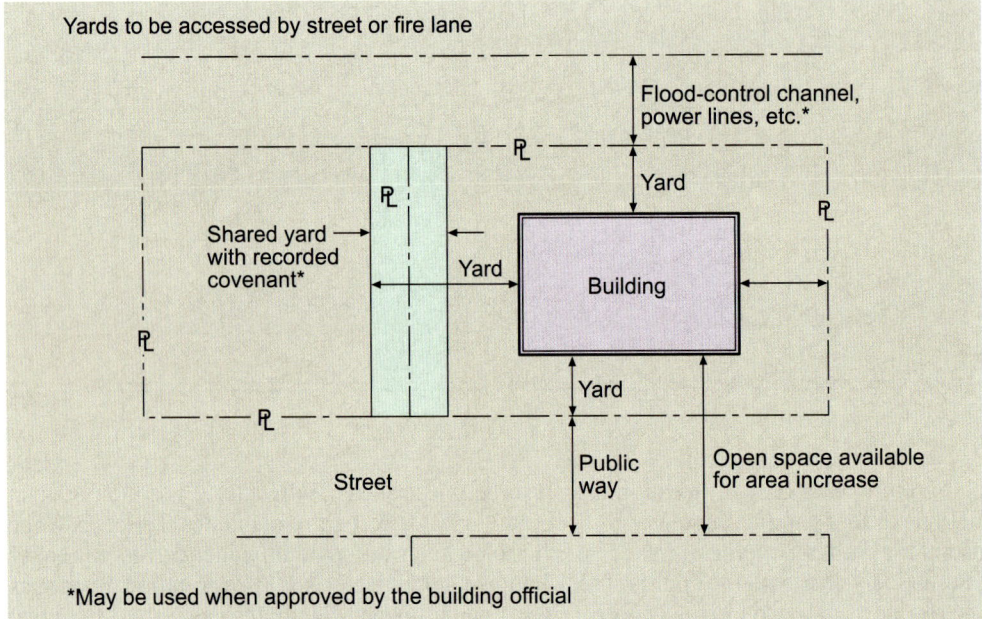

**Figure 506-3
Yards and public ways available for area increases.**

How much increase? In the case where public ways or open space adjoin more than 25 percent of the building's perimeter, the code permits an increase in the building area per story as shown in Table 503. The amount of the increase is based on the percentage of open perimeter having a width of at least 20 feet (6,096 mm). By utilizing the formula shown below, the area increase that is due to frontage (I_f) can be determined by Equation 5-2:

$$I_f = [F/P - 0.25]W/30$$

WHERE:

I_f = Area increase due to frontage (percent)
F = Building perimeter that fronts on a public way or open space having 20-foot-minimum (6,096 mm) open width
P = Perimeter of entire building
W = Width of public way or open space in accordance with Section 506.2.1

Based on this method of calculation, the maximum area increase permitted will typically be 75 percent, as shown in Application Example 506-2. This is based on the general requirement of Section 506.2.1 that requires a value of 30 feet (9,144 mm) to be used for the value W in those cases where W exceeds 30 feet (9,144 mm). As this figure illustrates, the entire perimeter of the building must adjoin a public way or open space having a width of at least 30 feet (9,144 mm). Where less than the entire perimeter has adequate open area, the area increase for frontage will be reduced as illustrated in Application Example 506-3.

Where the open space at the building's perimeter is between 20 feet (6,096 mm) and 30 feet (9,144 mm) in width, the code permits the use of the weighted average of such width in relation to the entire perimeter. This approach allows for the width W in Equation 5-2 to be more representative of the availability of open space around the building, rather than basing the frontage increase on simply the smallest open space of 20 feet (6,096 mm) or more.

Building Area Modifications

Application Example 506-2

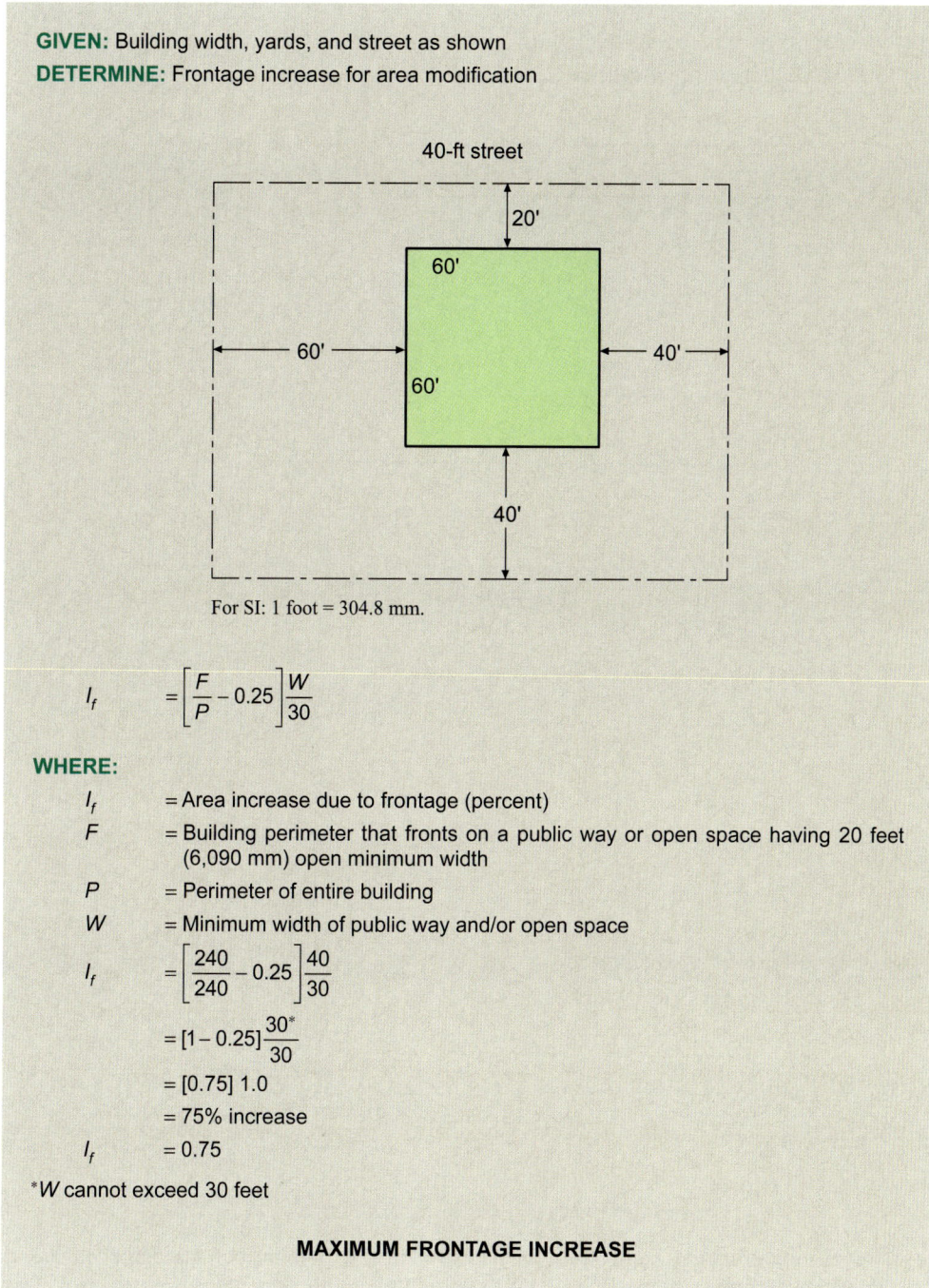

GIVEN: Building width, yards, and street as shown
DETERMINE: Frontage increase for area modification

For SI: 1 foot = 304.8 mm.

$$I_f = \left[\frac{F}{P} - 0.25\right]\frac{W}{30}$$

WHERE:

- I_f = Area increase due to frontage (percent)
- F = Building perimeter that fronts on a public way or open space having 20 feet (6,090 mm) open minimum width
- P = Perimeter of entire building
- W = Minimum width of public way and/or open space

$$I_f = \left[\frac{240}{240} - 0.25\right]\frac{40}{30}$$

$$= [1 - 0.25]\frac{30^*}{30}$$

$$= [0.75]\,1.0$$

$$= 75\% \text{ increase}$$

$$I_f = 0.75$$

*W cannot exceed 30 feet

MAXIMUM FRONTAGE INCREASE

Whereas in most cases, the increase available based on the weighted average method is minimal, it does provide for some degree of allowable area adjustment. An example of calculating W by weighted average is shown in Application Example 506-4.

Whereas 75 percent is generally the largest allowable frontage increase, a greater area increase is permitted for those buildings that comply with all of the requirements for unlimited-area buildings as described in Section 507, other than compliance with the 60-foot (18,288-mm) open space or public way requirement.

5 General Building Heights and Areas

Application Example 506-3

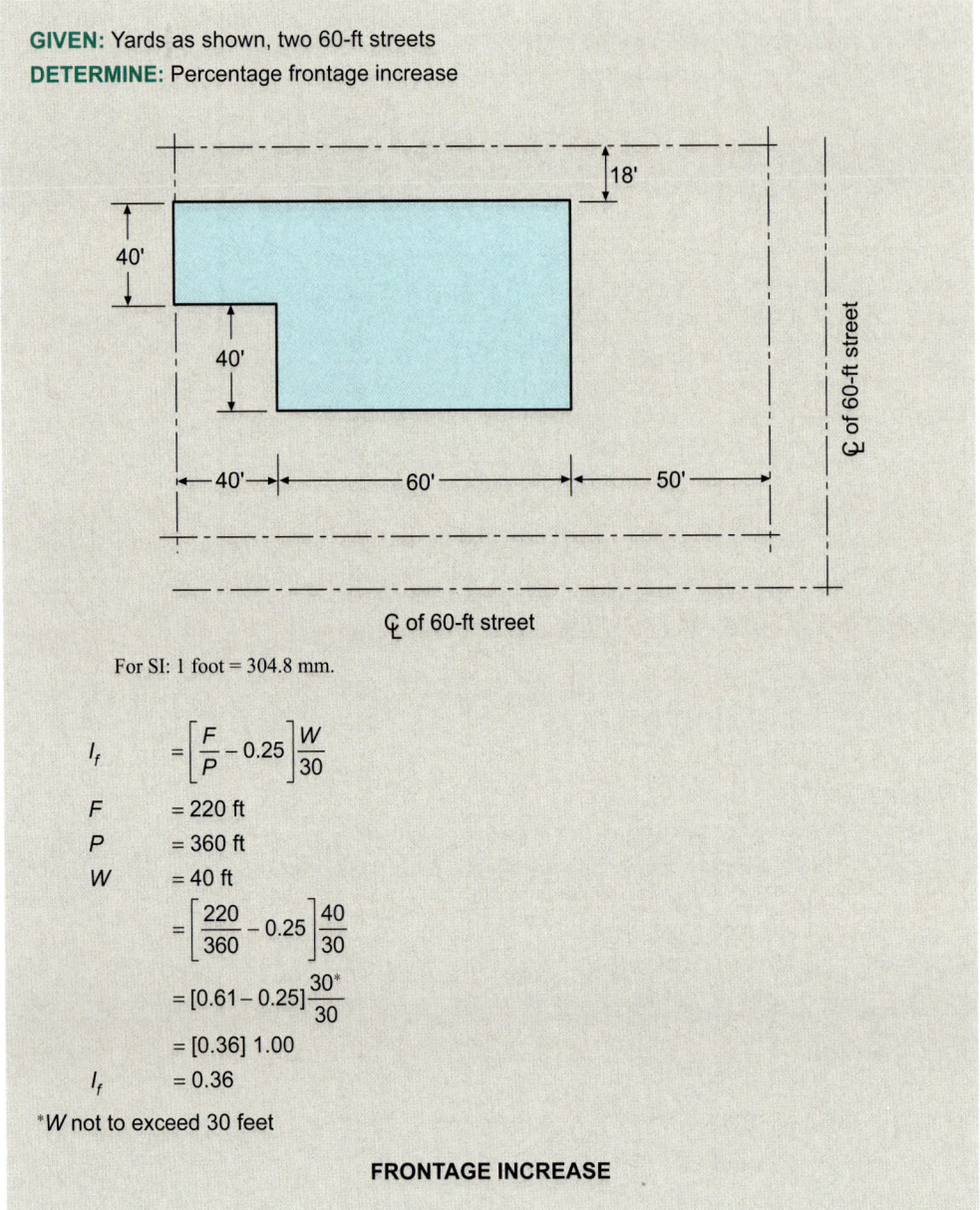

GIVEN: Yards as shown, two 60-ft streets
DETERMINE: Percentage frontage increase

For SI: 1 foot = 304.8 mm.

$$I_f = \left[\frac{F}{P} - 0.25\right]\frac{W}{30}$$

F = 220 ft
P = 360 ft
W = 40 ft

$$= \left[\frac{220}{360} - 0.25\right]\frac{40}{30}$$

$$= [0.61 - 0.25]\frac{30^*}{30}$$

$$= [0.36]\,1.00$$

I_f = 0.36

*W not to exceed 30 feet

FRONTAGE INCREASE

A maximum frontage increase of just less than 150 percent can be achieved based on the entire perimeter being open with a minimum width of slightly less than 60 feet (18,288 mm). Once 60 feet (18,288 mm) of accessible open space and public ways is obtained for 100 percent of the building's perimeter, the provisions of Section 507 are applicable and the frontage increase formula is not to be used. An example of the increased frontage increase is shown in Application Example 506-5, which also includes the calculated increase based on weighted average.

How must access be provided to an open space? The IBC provides no details as to the degree of fire department accessibility required in order to consider open space for an allowable area increase; it only mandates that access be provided from a street or approved fire line. It is clearly not the intent of the provisions to mandate a street or fire line completely around a building in order to acquire the maximum frontage increase.

Application Example 506-4

GIVEN: A building fronted by 60-foot street and three yards as shown

DETERMINE: The quantity W to be used in the calculation of I_f (area increased due to frontage)

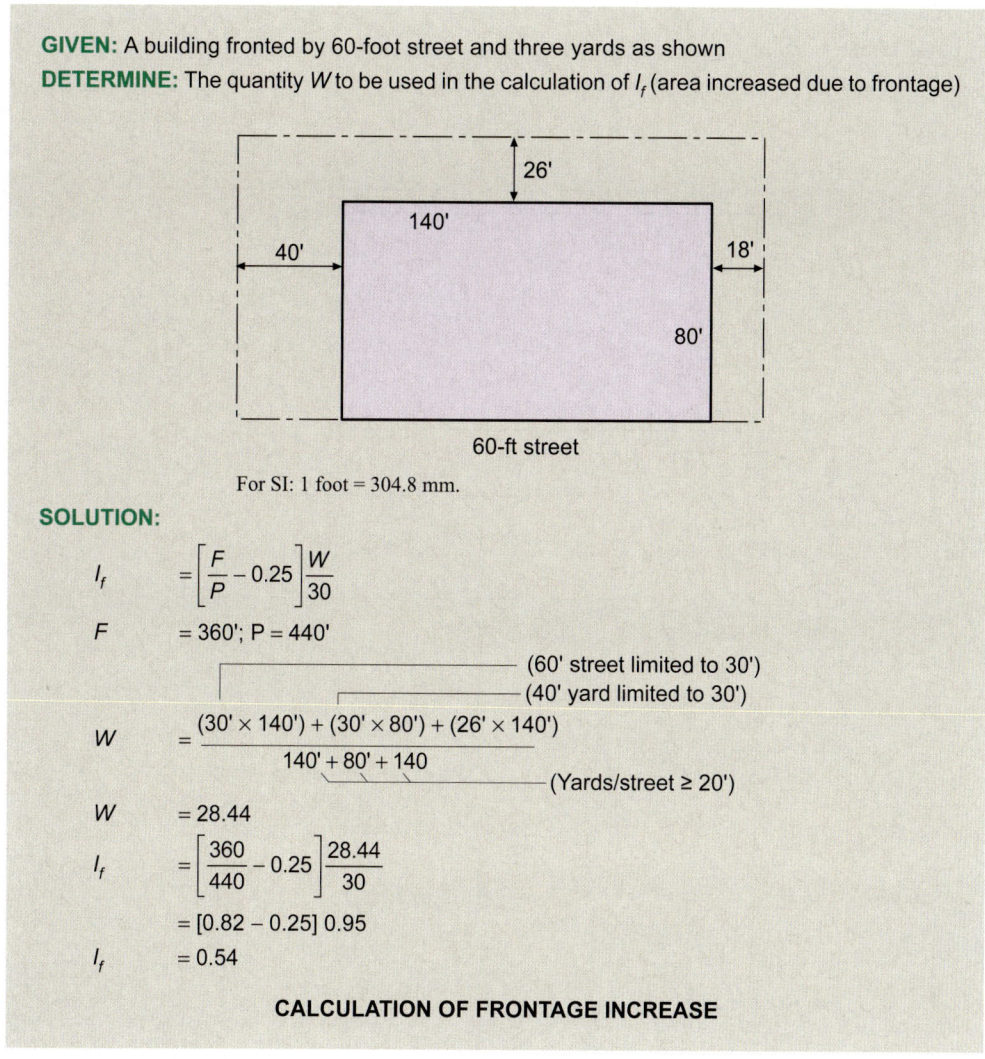

For SI: 1 foot = 304.8 mm.

SOLUTION:

$$I_f = \left[\frac{F}{P} - 0.25\right]\frac{W}{30}$$

$F = 360'; P = 440'$

$$W = \frac{(30' \times 140') + (30' \times 80') + (26' \times 140')}{140' + 80' + 140}$$

- (60' street limited to 30')
- (40' yard limited to 30')
- (Yards/street ≥ 20')

$W = 28.44$

$$I_f = \left[\frac{360}{440} - 0.25\right]\frac{28.44}{30}$$

$= [0.82 - 0.25]\,0.95$

$I_f = 0.54$

CALCULATION OF FRONTAGE INCREASE

However, fire personnel access from such streets or fire lanes is necessary. Although it is not a requirement to provide access around a building for fire department apparatus, other than that required by IFC Section 503.1.1, the frontage increase is based on the ability of fire personnel to physically approach the building's exterior under reasonable conditions. For example, where the space adjacent to the building is heavily forested or steeply sloped, the frontage increase addressed in Section 506.2 is not permitted. The presence of a lake or similar water feature next to a building would also prohibit an area increase. The evaluation of each individual building and its site conditions is necessary to properly apply the code for fire department access.

506.3 Automatic sprinkler system increase. Because of the excellent record of automatic sprinkler systems for the early detection and suppression of fires, the IBC allows quite large floor-area increases where an automatic fire sprinkler system is installed throughout the building. In this case, the maximum allowable area of a one-story building based on Table 503 may be increased by an additional 300 percent, and for a building of two or more stories in height, the tabular allowable area may be increased by an additional 200 percent. This restriction of permitting a smaller increase in area for multistory buildings protected by an automatic fire sprinkler system is based on the assumption by the code that fire department suppression activities are still going to be required even where an automatic

Application Example 506-5

GIVEN: Fully sprinklered, one-story retail sales building, yards as shown, 40-ft street
DETERMINE: Percentage increase for area purpose (I_f)

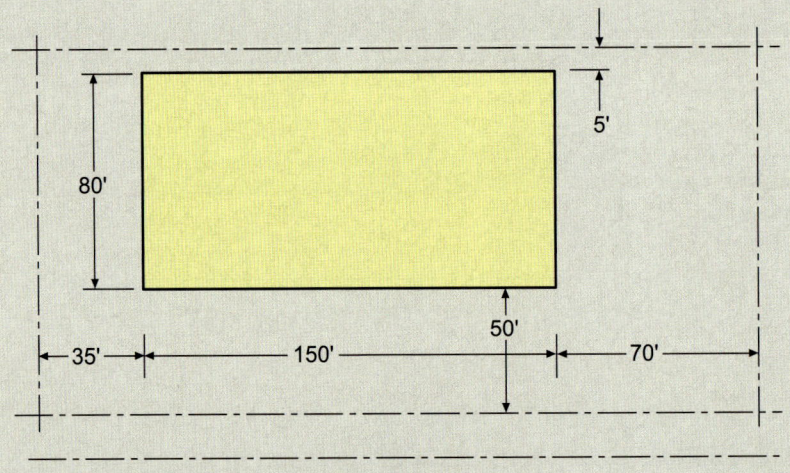

For SI: 1 foot = 304.8 mm.

$$I_f = 100\left[\frac{F}{P} - 0.25\right]\frac{W}{30}$$

$F = 310$ ft
$P = 460$ ft
$W = 35$ ft

$$I_f = \left[\frac{310}{460} - 0.25\right]\frac{35}{30} I_f$$
$$= [0.67 - 0.25]\, 1.17^*$$
$$= [0.42]\, 1.17$$
$$I_f = 0.49$$

Using the weighted average concept:

$$W = \frac{150(60') + 80(35') + 80(60')}{310}$$

$W = 53.22$

$I_f = [0.42]\, 1.77^*$

$I_f = 0.74$

*Cannot exceed 2.0 per Section 506.2.1

ALLOWABLE AREA DETERMINATION

fire-sprinkler system is installed. Therefore, a multistory building presents more problems to the fire department than a one-story building, and a smaller increase in area is permitted. These increases for sprinkler protection, like any for increased frontage, are to be added to the tabular area found in Table 503. Examples of the automatic sprinkler system increase are illustrated in Application Example 506-6.

The IBC permits area increases for buildings protected throughout with an NFPA 13 sprinkler system, but not those buildings protected with an NFPA 13R system. In addition, the increase is not applicable to Group H-1, H-2, or H-3 occupancies. The code considers that the conditions requiring the installation of sprinkler systems in these three high-hazard occupancies are such that the sprinkler system should not also be used to increase the allowable area. In a mixed-occupancy building, this restriction is only applicable to the Group H-2 and H-3 portions. Other occupancies within the building are permitted to utilize the automatic sprinkler system increase. Similar to the limitations on allowable height

Application Example 506-6

- Area increase for sprinklered single-story building to be 300% of area in Table 503 ($I_s = 3.0$)
- Area increase for sprinklered multistory building to be 200% of area in Table 503 ($I_s = 2.0$)
- Area increase permitted with height increase

EXAMPLE

GIVEN: Group B occupancy single-story
Type VB construction
No open yards available

Find: Total allowable area
Basic allowable area = 9,000 sq ft (Table 503)
Sprinkler increase (I_s) = 27,000 sq ft (3.0 × 9,000)
Total allowable area = 36,000 sq ft

GIVEN: Same situation, however two stories in height

Find: Total allowable area
Basic allowable area = 9,000 sq ft (Table 503)
Sprinkler increase (I_s) = 18,000 sq ft (2.0 × 9,000)
Total allowable area = 27,000 sq ft/story

AREA INCREASE FOR SPRINKLERS

For SI: 1 square foot = 0.093 m².

increases for sprinklered buildings, an area increase is not permitted where the sprinkler system is used for the substitution of 1-hour construction as permitted by Footnote d of Table 601.

What other conditions affect the determination of allowable floor areas? As expected, there are myriad situations that can arise involving the determination of the allowable area for a building. It should be noted that, as illustrated in Application Example 506-7, the introduction of a fire wall in a large-area building will result in the loss of a portion of the open space at the building perimeter. Application Example 506-8 illustrates the permitted increase for an open space shared by two buildings on the same lot.

Single occupancy buildings with more than one story. Where a single-occupancy building includes more than one story above grade plane, each story must be analyzed for allowable area purposes. In addition, calculations may need to be performed in order to evaluate if the building is within the total allowable building area. The aggregate area of all stories in the building, other than the basement, must not exceed the total allowable building area.

506.4

The code intends that a basement that does not exceed the allowable floor area permitted for a one-story building need not be included in determining the total allowable area of the building. This provision is a holdover from many years back when basements were most commonly used for service of the building. Today, it is not uncommon to find basements occupied for the same uses as the upper floors; consequently, the office building with three stories above grade plane and a basement, as illustrated in Figure 506-4, is permitted by the IBC to have an area of up to four times that allowed for a one-story building of like occupancy and construction type. There apparently has been no adverse experience for cases of this type (most likely because of the fire-sprinkler protection required in many basements), and therefore the code provision appears to be satisfactory. The code does not address how to handle a basement that exceeds the area permitted for a one-story building. However, as the code does not permit any story to exceed that permitted for a one-story building, it

5 General Building Heights and Areas

Application Example 506-7

AREA INCREASES WITH FIRE WALL

GIVEN: BUILDING A: Group S-2 Occupancy
Type IIB Construction
One Story

BUILDING B: Group B Occupancy
Type IIB Construction
One Story

DETERMINE: Maximum allowable floor area for Building A and Building B

SOLUTION: When the separation wall is a fire wall, each building is to be evaluated individually, and Building A would not be eligible for an area increase for open space. This is because Section 503.1 considers that each portion of a building separated by a fire wall is a separate building. As separate buildings, Buildings A and B do not reflect similar conditions. Building B is bounded on the right side by a 50-foot-wide yard and at the front by a 90-foot-wide open space consisting of a 30-foot yard and 60-foot street (public way). However, Building A is bounded by open space only at the front, which provides only a 25 percent frontage. The right side is occupied by another building (Building B).

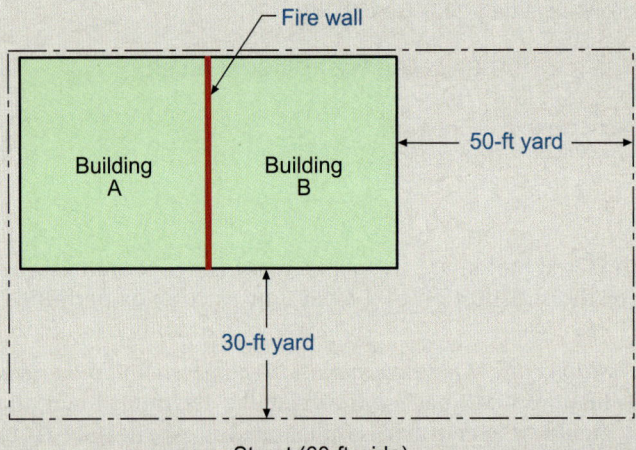

For SI: 1 foot = 304.8 mm.

Building A
  No increases for open space
  Allowable Area = 26,000 square feet
Building B
  25 percent increase (assuming exterior walls of equal length)
  Allowable Area = 23,000 square feet + 5,750 square feet = 28,750 square feet

would seem logical that a basement should also be limited to the same area. If anything, a fire in a basement is more difficult to fight than one in an above-ground story. Thus, it does not seem reasonable or appropriate to permit a single basement with an area exceeding that allowed for a one-story building. This section also does not detail the method of regulating allowable areas where multiple basements are present. However, the provisions of the exception to Section 506.4 indicate that only a single level of basement is exempt from inclusion in the building area calculations. Additional requirements for basements extending well below grade level may be found in Section 405 for underground buildings.

Building Area Modifications   5

Application Example 506-8

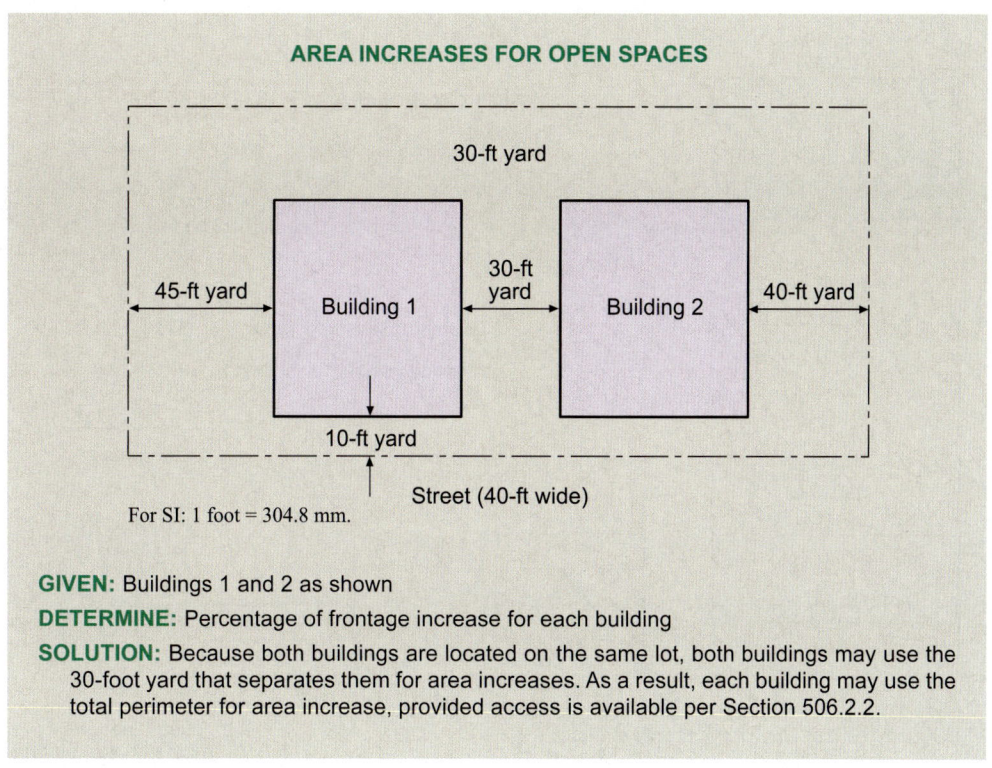

AREA INCREASES FOR OPEN SPACES

GIVEN: Buildings 1 and 2 as shown

DETERMINE: Percentage of frontage increase for each building

SOLUTION: Because both buildings are located on the same lot, both buildings may use the 30-foot yard that separates them for area increases. As a result, each building may use the total perimeter for area increase, provided access is available per Section 506.2.2.

506.4.1

Area determination. In addition to the floor-area limits placed on each story of a building, the entire building is also limited in size. For multistory buildings other than those consisting of only two stories, the IBC permits the total combined floor area to be three times the total allowable area permitted per floor. Two-story structures are limited in size to a total combined floor area of twice that permitted per floor. Examples of these limitations are shown in Figure 506-5.

The general limitation on allowable building area is not applicable to unlimited-area buildings in compliance with Section 507. In addition, four-story residential occupancies provided with an NFPA 13R automatic sprinkler system are permitted additional allowable area beyond that permitted under the general provisions. As the installation of a 13R sprinkler system in a residential occupancy does not provide for an increase in allowable area for sprinklered buildings, and the use of an NFPA 13R system is limited to buildings no more than four stories in height, it is considered appropriate to permit the maximum allowable area per story for each of the stories in the residential building. This would include four-story structures as depicted in Figure 506-6.

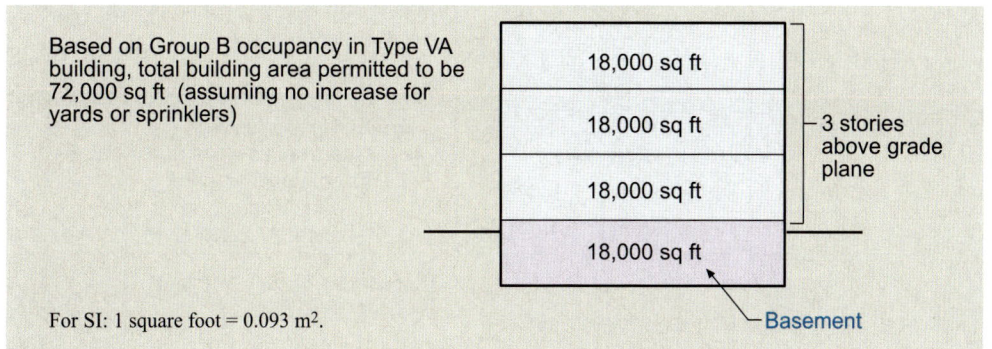

Figure 506-4 Allowable area basement.

2012 International Building Code Handbook   145

5 General Building Heights and Areas

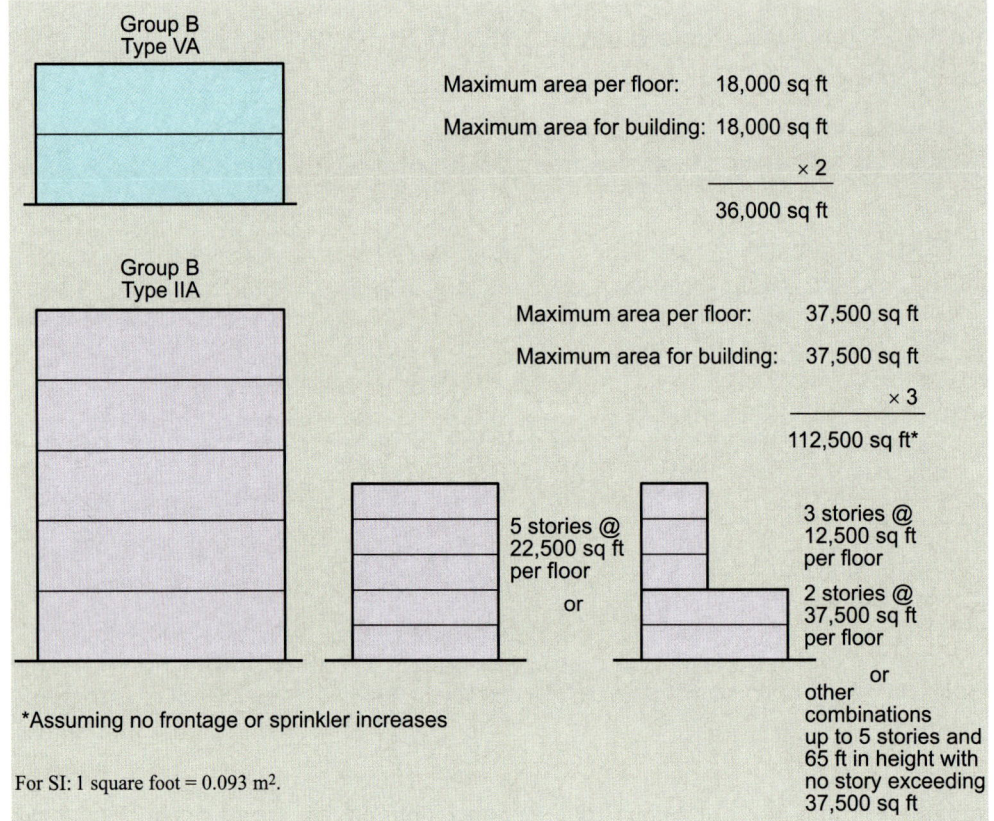

Figure 506-5
Building area limitations.

506.5 Mixed-occupancy area determination. For allowable area purposes, the introduction of multiple occupancies into the building requires an additional level of scrutiny. The methods for determining allowable area compliance in mixed-occupancy buildings differ based on the manner in which the various occupancies are addressed. Where the accessory occupancy method is utilized, the allowable area of the main occupancy will govern. Where occupancies are regulated under the nonseparated-occupancy method, the allowable area is limited to that of the most restrictive occupancy involved. In the case of separated occupancies, the unity formula must be used to determine if the allowable area is exceeded. Further discussion can be found in the narratives of Section 508 for each of the three mixed-occupancy methodologies.

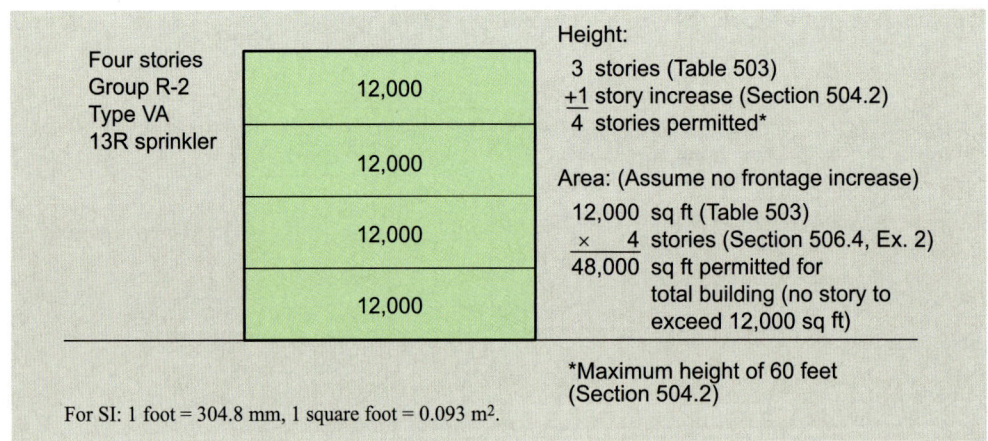

Figure 506-6
Group R area determination.

Building Area Modifications

Where a single-story mixed-occupancy building is being evaluated for allowable area compliance, the process is addressed in Section 508.1. In similar fashion, a mixed-occupancy building with more than one story above grade plane must be analyzed on a story-by-story basis and each individual story must be shown to be in compliance with the applicable provisions of Section 508.1. In all cases, regardless of the building's number of stories, each individual story must comply for allowable area purposes. For those buildings that are two and three stories in height, if each such story complies with the allowable area limitations, then the entire building complies. However, the approach to buildings four or more stories in height above grade plane differs from that used for two-story and three-story conditions. Again, each story in the building must comply individually. In addition, the aggregate sum of the actual areas of each story divided by the allowable areas of such stories cannot exceed three. The procedure is illustrated in Application Example 506-9.

Application Example 506-9

GIVEN: A fully sprinklered, four-story, Type IIA Hotel containing a Group A-2 restaurant, Group A-3 meeting rooms, and Group M retail stores. The floor areas of each occupancy are as shown. Inadequate frontage provides for no area increase.

DETERMINE: If the building complies with the allowable area provisions of Chapter 5 if the occupancies are separated under the provisions of Section 508.4 (separated occupancies).

A-2 8,000 sq ft	R-1 38,000 sq ft	
	R-1 46,000 sq ft	
	R-1 46,000 sq ft	
A-3 24,000 sq ft	R-1 8,000 sq ft	M 14,000 sq ft

SOLUTION FOR TOTAL BUILDING AREA:

Allowable area per occupancy

- A-2: 15,500 + 15,500 (200%) = 46,500
- A-3: 15,500 + 15,500 (200%) = 46,500
- M: 21,500 + 21,500 (200%) = 64,500
- R-1: 24,000 + 24,000 (200%) = 72,000

Sum of ratios calculation per story

1st story $\dfrac{24,000}{46,500} + \dfrac{8,000}{72,000} + \dfrac{14,000}{64,500} = 0.52 + 0.11 + 0.22 = 0.85 \leq 1.0$

2nd story $\dfrac{46,000}{72,000} = 0.64 \leq 1.0$

3rd story $\dfrac{46,000}{72,000} = 0.64 \leq 1.0$

4th story $\dfrac{8,000}{46,500} + \dfrac{38,000}{72,000} = 0.17 + 0.53 = 0.70 \leq 1.0$

Each story complies

Sum of ratios calculation for building

$0.85 + 0.64 + 0.64 + 0.70 = 2.83 \leq 3.0$

Entire building also complies

ALLOWABLE AREA OF MULTISTORY MIXED-OCCUPANCY BUILDING

For SI: 1 square foot = 0.093 m².

5 General Building Heights and Areas

In all cases for mixed-occupancy conditions, reference is made to the provisions of Section 508.1, which identify the three options for mixed-occupancy determination (accessory occupancies, nonseparated occupancies, and separated occupancies). Since each story is to be evaluated independently in a multistory condition, it is possible for the designer to utilize different mixed-occupancy options on various stories within the building.

Section 507 *Unlimited-Area Buildings*

There are many cases where very large undivided floor areas are required for efficient operation in such facilities as warehouses and industrial plants. Through the use of adequate safeguards, the IBC recognizes this necessity and allows unlimited areas for these uses under various circumstances. Large open floor space is also desirable for other applications; therefore, such allowances are also permitted for business and mercantile occupancies, as well as specific assembly and educational uses. The use of this section is typically intended to eliminate fire-resistive construction of the building that would be mandated based on the area limitations of Section 503. Contrary to the general philosophy that as a building increases in floor area the allowable types of construction become more restrictive, many of the unlimited-area uses permit the use of any construction type.

507.1 General. Historically, structures constructed under the provisions for unlimited area have performed quite well in regard to fire and life safety. A number of occupancy groups, particularly those relating to institutional, residential, and high-hazard occupancies, are excluded from the benefits derived from the provisions for unlimited-area buildings. Such occupancies pose unacceptable risks that are due to their unique characteristics. As a general rule, only those occupancy classifications specifically identified in this section are permitted to be housed in buildings allowed to be unlimited in area by Section 507. For example, Group I occupancies are not specifically permitted by any of the provisions addressing unlimited-area buildings. Therefore, it would appear no amount of Group I is permitted in such structures. However, the exception identifies one method that allows for a limited degree of such prohibited occupancies. Section 508.2.3 indicates that the allowable area for an accessory occupancy is to be based on the allowable area of the main occupancy. If the main occupancy is permitted by Section 507 to be in an unlimited-area building, a complying accessory occupancy also enjoys the same benefit. Figure 507-1 illustrates this condition.

It is also not uncommon for two or more occupancies regulated under the provisions of Section 507 to be located within the same building. For example, assume a one-story building contains a Group M furniture store and its associated S-1 warehouse. Because both Group M and S-1 occupancies are permitted in an unlimited-area building complying with Section 507.3, both occupancies are permitted to be located in the same unlimited-area building. See Figure 507-2. Any fire-resistive separation requirement would be based on the applicable mixed occupancy method of Section 508 applied to the building.

All buildings regulated under the unlimited-area building provisions of Section 507 are required to be surrounded by public ways and yards of substantial width. This continuous open space provides a means for the fire service to access the building as necessary from the exterior, while at the same time maintains a sizable separation from any other structures on the site. The required open space is to be measured from all points along the building's exterior wall in all directions, ensuring that the full perimeter of the building is provided with continuous open space. This differs somewhat from the right-angle method established for an allowable area frontage increase as set forth in Section 506.2 where such continuity of open space is not required at the building corners. See Figure 507-3. Consistent with the measurement method of Section 506.2 to gain a frontage increase, the

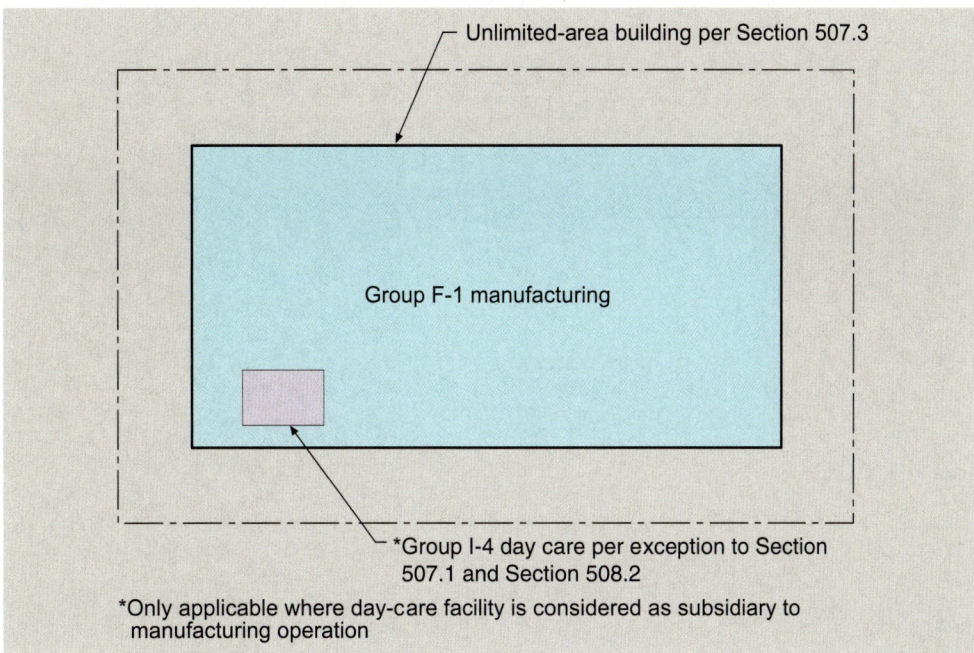

Figure 507-1
Accessory occupancy in an unlimited-area building.

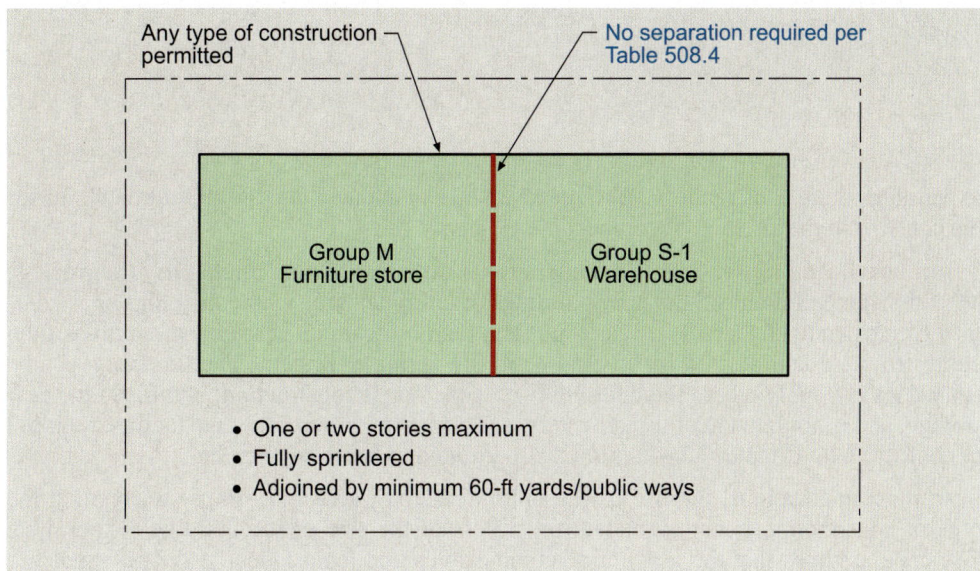

Figure 507-2
Mixed-occupancy unlimited-area buildings.

open space width adjacent to those exterior walls fronting on a public way is permitted to include the entire width of the public way.

Nonsprinklered, one-story. This section addresses a Group F-2 or S-2 occupancy in a one-story building of any type of construction. Both Group F-2 and S-2 occupancies by definition are low-hazard manufacturing or storage uses, which the code considers to be low fire risks. Fire risk is further reduced by requiring that the building be surrounded by yards or streets with a minimum width of 60 feet (18,288 mm). The relatively low fire loading expected in such occupancies is why the code does not require the installation of an automatic fire-sprinkler system for this application. The use of this provision is applicable

507.2

5 General Building Heights and Areas

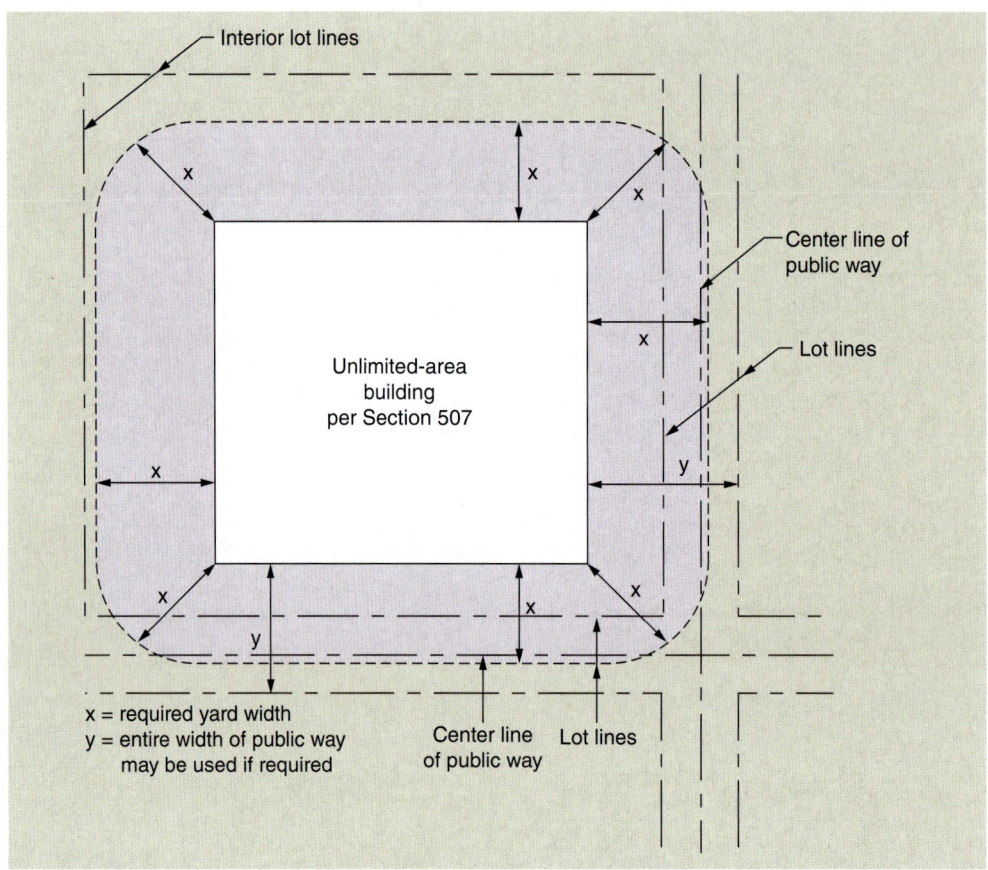

**Figure 507-3
Open space measurement for unlimited-area building.**

to buildings of all construction types; however, it is anticipated that the structure will mirror the contents in the absence of combustible elements.

507.3 Sprinklered, one story. Specific moderate-hazard occupancies, limited to Groups B, F, M, and S, are permitted single-story buildings of unlimited area where the building is completely surrounded by streets or yards not less than 60 feet (18,288 mm) in width and the entire structure is protected by an automatic fire-sprinkler system. The limitation of one story does not apply where the building is of Type I or II construction, is utilized for rack storage, and is not intended for public access. This unlimited-area storage facility, required to conform with Chapter 32 of the IFC, is permitted to be of any height.

In most applications, the use of the unlimited-area provisions simply means that the type of construction is not regulated, regardless of the size of the building's floor area. The code assumes that the amount of combustibles and, consequently, the potential fire severity are relatively moderate. In addition, the protection provided by the automatic fire-sprinkler system plus the fire-department access furnished by the 60-foot (18,288-mm) yards or streets surrounding the building reduce the potential fire severity to such a level that unlimited area is reasonable.

One-story Group A-4 occupancies are also permitted to be of unlimited area where a sprinkler system is provided throughout and a minimum 60-foot (18,288-mm) open space surrounds the building. However, because of the increased risk posed by the anticipated high number and concentration of occupants in such a structure, the construction type of the building is limited to Type I, II, III, or IV. The automatic sprinkler system required in an unlimited-area building housing a Group A-4 occupancy may be omitted in those specific areas occupied by indoor participant sports, including tennis, skating, swimming, and equestrian activities. Such an omission mandates that exit doors from the participant sports areas lead directly to the outside, and the installation of a fire-alarm system with manual fire-alarm boxes is required. It is anticipated that such sports areas will have little, if any, combustible loading if the uses are limited to those described in the code. If there is a reasonable expectation that other types of uses could occur, it would be inappropriate to omit the sprinkler system in such areas.

Group A-1 and A-2 occupancies are permitted in a mixed-occupancy building when in compliance with the general limitations of Section 507.3 plus four additional criteria as established in Section 507.3.1. This allowance does not grant the designated Group A occupancies unlimited area, but rather allows such assembly occupancies to be located within a Group B, F, M, S, or A-4 unlimited-area building under specified conditions. The building must be classified as Type I, II, III, or IV construction. In addition, the Group A assembly occupancies must be separated with fire barriers from other occupancies within the building, in accordance with the separated occupancy provisions of Section 508.4.4. For example, in an unlimited-area retail sales building, a Group A-2 restaurant would be required to be separated from the Group M sales area by a minimum 2-hour fire-resistance-rated fire barrier. No reduction in the minimum required fire-resistance rating of the fire barrier is permitted for the presence of an automatic sprinkler system. Each individual Group A tenant would also be limited in area by the provisions of Section 503.1, which would include any applicable increases to Table 503 permitted by Sections 506.2 and 506.3. No additional height increases would apply to the Group A occupancies because of the specific limitation of one story for the entire building. A fourth requirement mandates that all required means of egress from the assembly spaces exit directly to the exterior of the building. See Figure 507-4. Application of this provision does not require the assembly occupancy to be accessory to the major use of the building, nor is the assembly floor area limited to 10 percent of the floor area. As an additional note, it would seem appropriate that the provision also extend to Group A-3 occupancies, as such uses are typically considered equal or lesser in hazard level to the Group A-1 and A-2 classifications.

Two-story. In Groups B, F, M, and S, the unlimited-area provisions also apply to structures that are two stories in height. Minimum 60-foot (18,288-mm) open space or public ways must surround the building, and an automatic fire-sprinkler system is required throughout the structure.

507.4

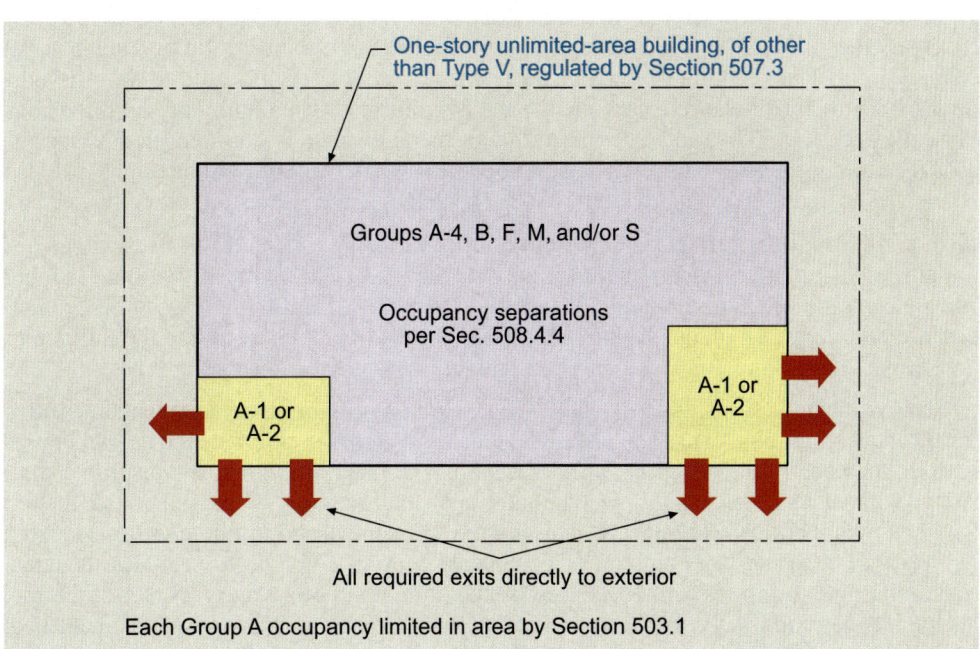

Figure 507-4
Group A occupancies in unlimited-area buildings.

5 General Building Heights and Areas

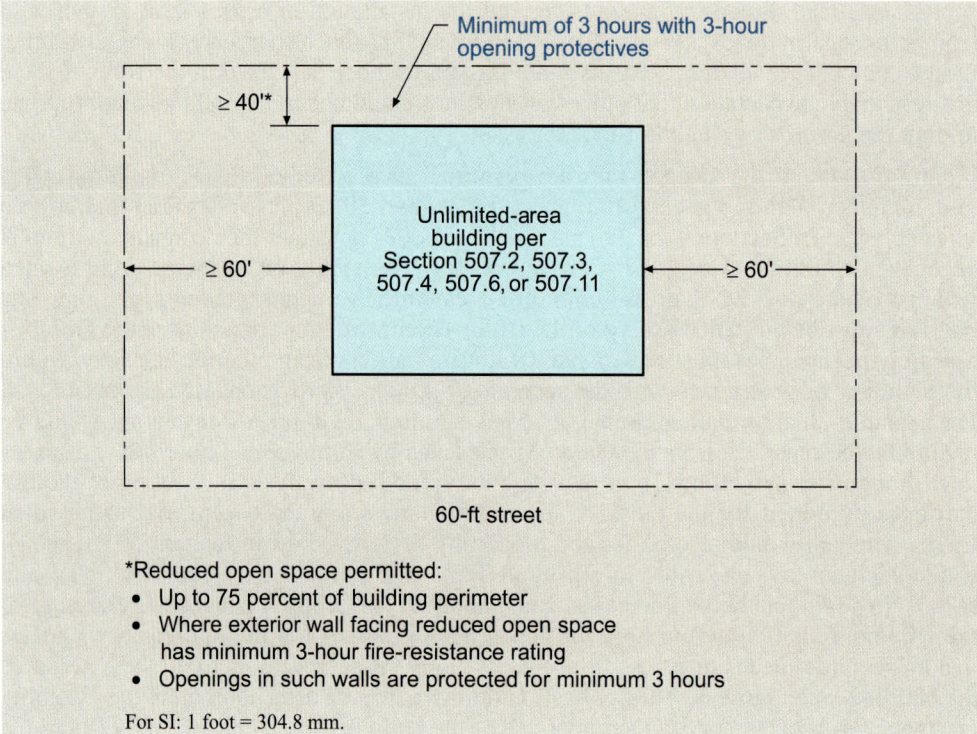

Figure 507-5
Reduced open space.

507.5 Reduced open space.
There may be situations where the full 60 feet (18,288 mm) of open space or public ways surrounding an unlimited-area building cannot be obtained or is undesirable. The IBC permits a reduction in the required open space under very specific conditions, as illustrated in Figure 507-5. In no case may the permanent open space be reduced to less than 40 feet (12,192 mm) in width. By limiting the amount of reduced open space, requiring a high degree of exterior wall fire resistance, and mandating opening protectives in all openings in the exterior wall facing the reduced open space, the code provides protection equivalent to full 60-foot (18,288-mm) yards or public ways. A final point: The permitted reduction in open space is only applicable to specific portions of Section 507. The reduced open space is prohibited when applying the provisions of Sections 507.8, 507.9, and 507.10. Although the allowance is not identified as acceptable for buildings complying with Section 507.7, it would seem the reduction could also be viewed as appropriate since Section 507.7 is a companion provision to Section 507.6.

507.6 Group A-3 buildings of Type II construction.
Although most assembly occupancies are viewed as relatively high hazard because of the concerns associated with the number and concentration of the occupants, certain types of uses classified as Group A-3 occupancies are considered as moderate hazard. Therefore, it is possible to utilize the unlimited-area provisions for specific Group A-3 uses where the specified criteria are met.

Such buildings allowed to be of unlimited area are limited to one story in height, must be of Type II noncombustible construction, and contain only those specific types of assembly uses listed in the code. By limiting the types of buildings to places of religious worship, gymnasiums (without spectator seating), lecture halls, and similar uses, it is anticipated that the fire loading is relatively low. Buildings such as libraries, museums, and similar uses pose a higher risk that is due to the large amount of combustibles expected to be present. The potential for combustible loading is further reduced by the prohibition of a stage as a part of the use, although a platform is acceptable. Installation of an automatic sprinkler system is mandated, as is the presence of a minimum 60-foot (18,288-mm) open space around the building.

Group A-3 buildings of Types III and IV construction. Buildings of Type III or IV construction are also granted unlimited-area status when housing specified Group A-4 occupancies, as also identified in Section 507.6. The requirements for sprinkler protection and adequate open space are also applicable. In addition, the assembly use must be located relatively close to the exterior ground level to expedite the exiting process. As a part of this requirement, any elevation change from the building to the grade level must be accomplished by ramps rather than stairs. This further provides for an efficient means of egress from the assembly building. The additional limitation regarding floor-level location is mandated due to the combustible nature of the building's construction. **507.7**

Group H occupancies. Because many large industrial operations, both manufacturing and warehousing, have a need to utilize a limited quantity of high-hazard materials in some manner, it is necessary that Group H-2, H-3, and H-4 occupancies be permitted to a small degree in Group F and S unlimited-area buildings. Because of the allowances given to buildings of unlimited area, it is critical that the high-hazard occupancies be strictly limited in floor area and adequately separated from the remainder of the building. **507.8**

There are three factors that limit the allowable floor area of the permitted Group H occupancies: the building's type of construction, its floor area, and the location of the high-hazard uses in relationship to the building. Where high-hazard occupancies are located on the perimeter of the building, fire-department access is enhanced, and exposure from interior areas is reduced. Accordingly, the permitted floor area of the Group H occupancies located on the building's perimeter is considerably greater than that allowed for such high-hazard uses completely surrounded by the unlimited-area building.

Where Group H-2, H-3, and H-4 occupancies are located at the perimeter of the unlimited-area Group F or S building, their size is restricted by the area limitations of Table 503 as modified by Section 506.2. In a condition where the high-hazard occupancy is totally enclosed by the unlimited-area building, the size of the occupancy is limited to only 25 percent of the area limitations specified in Table 503. Both of these conditions are shown in Application Example 507-1. The example also illustrates that multiple Group H occupancies that are not located at the perimeter of the building are limited in size based on the aggregate floor area of such occupancies. Similarly, where multiple Group H occupancies do occur on the building's perimeter, the maximum permitted size is also based on the total of all such occupancies. In all situations, the appropriate fire-barrier assemblies mandated by Table 508.4 must be provided. Regardless of the location of the applicable Group H occupancies, in no case may such aggregate floor areas exceed 10 percent of the area of the entire building. The provisions addressing liquid use, mixing and dispensing rooms, liquid storage rooms, and spray-paint booths simply replicate those of Section 415.5, Exceptions 1 through 3, in order to remind code users of the limitations and allowances placed on such rooms and areas.

Aircraft paint hangar. The provisions of Section 412.6 address aircraft painting operations where the amount of flammable liquids in use exceeds those maximum allowable quantities listed in Table 307.1(1). Classified as Group H-2, such aircraft paint hangars must be fully suppressed in accordance with NFPA 409 and of noncombustible construction. One-story hangars may be unlimited in floor area where complying with Section 412.6, provided they are surrounded by public ways or yards having a width of at least one and one-half times the height of the building. **507.9**

Group E buildings. Because of the various fire- and life-safety concerns associated with educational occupancies, buildings housing uses classified as Group E are typically not eligible for consideration as unlimited-area buildings. Only when the following six criteria are met does the IBC permit the area of a Group E educational building to be unlimited: **507.10**

1. The building is limited to one story in height.
2. The building is of Type II, IIIA, or IV construction.
3. Two or more means of egress are provided from each classroom.

5 General Building Heights and Areas

Application Example 507-1

GIVEN: A 130,000 square-foot Group F-1 of Type IIB construction having unlimited area under the provisions of Section 507.3. One H-3 storage room is located on the building's perimeter. Multiple H-3 storage rooms are located such that they are not located along an exterior wall.

DETERMINE: The maximum allowable floor areas for the H-3 storage rooms.

Group H occupancies in unlimited-area building

Aggregate of Group H: (rooms A, B, and C)	10 percent of 13,000 13,000 sf max, *nor*
	Table 503 with frontage 14,000 + 3,500 = 17,500 sf
	∴ Aggregate limit of 13,000 sf
Located within building: (rooms A and B)	25 percent of 14,000 ∴ Aggregate limit of 3,500 sf
Located on perimeter: (room C)	13,000 − (area of rooms A and B) ∴ 9,500 sf to 13,000 sf (depending on actual area of rooms A and B)

UNLIMITED AREA GROUP F OR S BUILDING WITH GROUP H OCCUPANCIES

For SI: 1 square foot = 0.093 m².

4. At least one means of egress from each classroom is a direct exit to the exterior of the building.

5. An automatic sprinkler system is provided throughout the building.

6. The building is surrounded by open space at least 60 feet (18,288 mm) in width.

507.11 Motion-picture theaters. Because of their limited combustible loading, motion-picture theaters are granted unlimited floor areas in a manner relatively consistent with other moderate-hazard uses. This specific allowance is not extended to the other uses classified as Group A-1, such as performance theaters, because of their higher fire-severity potential.

In order to address the concerns related to the high-density, high-volume occupant loads often encountered in motion-picture theaters, unlimited area is only permitted where the building is of Type II noncombustible construction. This restriction further limits the fire load contained within the building construction. In concert with Section 507.3, a fire-sprinkler system must be installed throughout, and minimum 60-foot (18,288-mm) open areas must completely surround the building.

The application of this provision differs from the allowance granted in Section 507.3.1. That provision permits any Group A-1 occupancy, including motion-picture theaters, to be located in an unlimited-area building complying with Section 507.3, provided the limitations of the exception are met. However, it does not allow Group A-1 occupancies themselves to be unlimited in area. On the other hand, this section permits a Group A-1 theater complex to be unlimited in area, provided it is fully sprinklered, of Type II construction, and surrounded by adequate open space.

Covered and open mall buildings and anchor buildings. The provisions of Section 402.4.1 for unlimited-area covered and open mall buildings are referenced for convenience purposes. Note that although the reduction in open space permitted by Section 507.5 is not applicable to covered or open mall buildings, a similar reduction is permitted by the exception to Section 402.1.1. **507.12**

Section 508 *Mixed Use and Occupancy*

Multiple uses commonly occur within a single building. Each use creates its own distinct hazards, many of which are addressed by the code. However, many of the hazards are similar in nature, which allows the varied uses to be grouped into categories that recognize the common concerns. These categories are identified in Chapter 3 as occupancy groups. Where two or more occupancy groups share a single building, it is necessary to evaluate their relationship to each other as a mixed-occupancy condition. This section provides various methods to address such relationships in regard to occupancy classification, allowable height and area, and fire-resistance-rated separation.

General. A mixed-occupancy condition exists where two or more distinct occupancy groups are determined to exist within the same building. In fact, it is quite common for a building to contain more than one occupancy group. For example, hotel buildings of various sizes not only house the residential sleeping areas, but may contain administrative offices, retail and service-oriented spaces, parking garages, and, in many cases, restaurants, conference rooms, and other assembly areas. Each of these uses typically constitutes a distinct and separate occupancy as far as Chapter 3 of the IBC is concerned. Because this situation is not uncommon, the code specifies requirements for buildings of mixed occupancies. Under such circumstances, the designer has available several methodologies (accessory occupancies, nonseparated occupancies, and separated occupancies) to address the mixed-occupancy concerns. The methods that have been established represent a hierarchy of design prerogatives that may be utilized at the discretion of the design professional. Although compliance is required with only one of the three mixed-occupancy methods, it is acceptable to utilize two or even all three methods in the same building, as shown in Figure 508-1. The common format utilized in presenting the requirements for each of the methods allows for a comparison of the provisions. This should assist in determining **508.1**

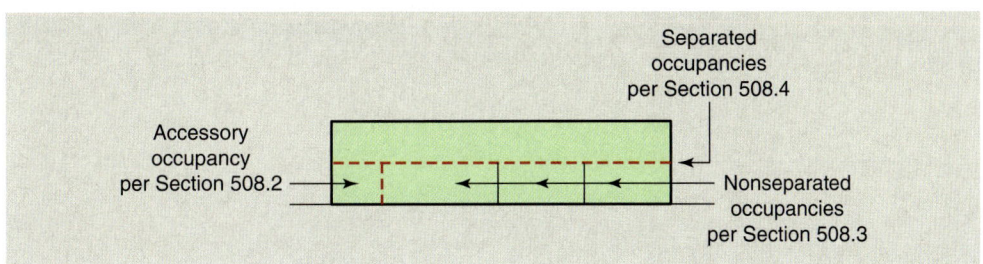

Figure 508-1 Combination of methods.

5 General Building Heights and Areas

	Accessory Occupancies Section 508.2	Nonseparated Occupancies Section 508.3	Separated Occupancies Section 508.4
Occupancy Classification	Individually classified	Individually classified	Individually classified
Allowable Area	Based on allowable area of main occupancy	Based on most restrictive of occupancies within building	Determined such that sum of the ratios cannot exceed 1.0
Allowable Height	Based on tabular values of Table 503	Based on most restrictive of occupancies within building	Based on general provisions of Section 503.1
Separation	No separation required	No separation required	Separation as required by Table 508.4
Special Conditions	1. Subsidiary to main occupancy 2. Aggregate area ≤ 10 percent of story 3. Aggregate area ≤ value in Table 503 4. Not applicable to Groups H-2, H-3, H-4, and H-5	1. Most restrictive provisions of Ch. 9 apply to entire building 2. Not applicable to Groups H-2, H-3, H-4, and H-5	

Figure 508-2 Summary of mixed-occupancy methods.

the most appropriate method, or methods, for the building under consideration. A simple comparison of the three mixed-occupancies methods is shown in Figure 508-2.

It is important to recognize that there is no relationship between the mixed-occupancy provisions of Section 508.3 and the fire area concept utilized in Section 903.2 for automatic sprinkler systems. Compliance with any of the three mixed-occupancy methods does not relieve the responsibility to comply with Section 901.7 and Table 707.3.10 regarding the proper separation of fire areas. An example is shown in Figure 508-3.

508.2 Accessory occupancies. Those minor uses in a building that are not considered consistent with the major occupancy designation can potentially be considered accessory occupancies. They often are necessary or complimentary to the function of the building's major use, but have few characteristics of the major occupancy in regard to fire hazards and other concerns. Therefore, accessory occupancies must each be assigned to an occupancy group established in Chapter 3 based on their own unique characteristics. While maintaining the philosophy of a mixed-occupancy building, this section permits such relatively small accessory uses to be considered merely a portion of the major occupancy for fire separation purposes. A good example would be a lunchroom seating 120 persons and located in a large manufacturing facility. Whereas the individuals using the lunchroom are generally the same individuals who work elsewhere in the factory, the hazards encountered while they are occupying the lunchroom are quite different from those created in the Group F-1 manufacturing environment. Therefore, the lunchroom must be appropriately classified as a Group A-2 occupancy, creating a mixed-occupancy condition. However, through compliance

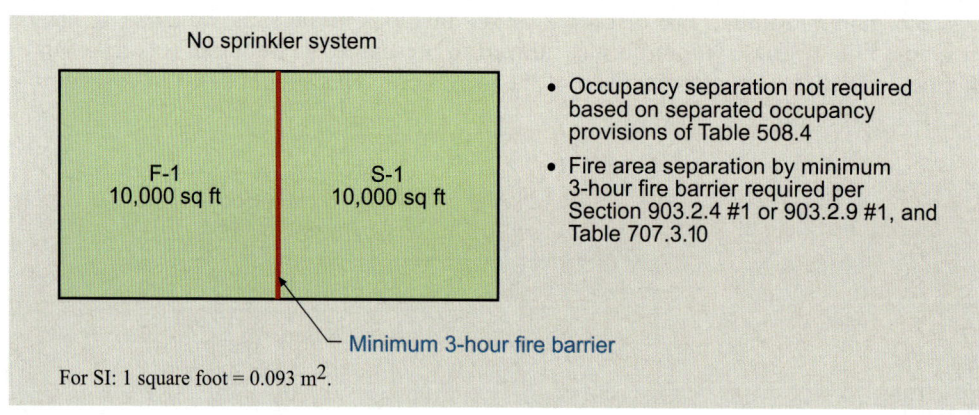

Figure 508-3 Occupancy separation versus fire area separation.

156 2012 International Building Code Handbook

with the accessory occupancy provisions of this section, the need for a fire-resistance-rated separation between the lunchroom and manufacturing area is eliminated. In fact, no physical separation of any kind would be mandated. It is important to note, however, that in spite of the absence of a fire separation, the two areas would maintain their unique occupancy classifications. They would continue to be classified as a Group A-2 and F-1, respectively, and the building would be considered a mixed occupancy.

As previously indicated, consideration as an accessory occupancy is only possible where the use is subsidiary to the main occupancy of the building. There are several additional criteria that must also be met in order to utilize the accessory occupancy method. The occupancy under consideration cannot exceed 10 percent of the floor area of the story on which the accessory occupancy is located, nor more than that permitted by Table 503 without area increases for frontage and sprinkler protection. See Figure 508-4. It is specified that the 10 percent limitation is based on the aggregate floor areas of all accessory use areas, not individually. Multiple minor uses that cumulatively make up more than 10 percent of the total floor area could pose a hazard that the code does not anticipate. There are unique situations—such as where minor uses are adequately separated spatially or are of such different types of uses that their aggregate area is not relevant—where the regulation of minor uses as individual areas could potentially be considered. See Figure 508-5. The application of these limits is subject to the interpretation of the building official, based on conditions unique to each building under consideration. The general limitations applied to accessory occupancies are shown in Application Example 508-1. As an additional limitation, the height of any accessory occupancy cannot be located higher than that specified in Table 503 without adjustment for the height increase for sprinklers typically permitted by Section 504.2. See Figure 508-6.

Although the primary allowance provided by the accessory occupancy provisions is the lack of a required fire separation between the accessory occupancy and the remainder of the building, there is also a potential benefit that is due to the manner in which allowable area is regulated. The code calls for the allowable area of the accessory occupancy to be based on the main occupancy of the building. This approach typically allows for a greater allowable area than would be permitted under the conditions for both nonseparated

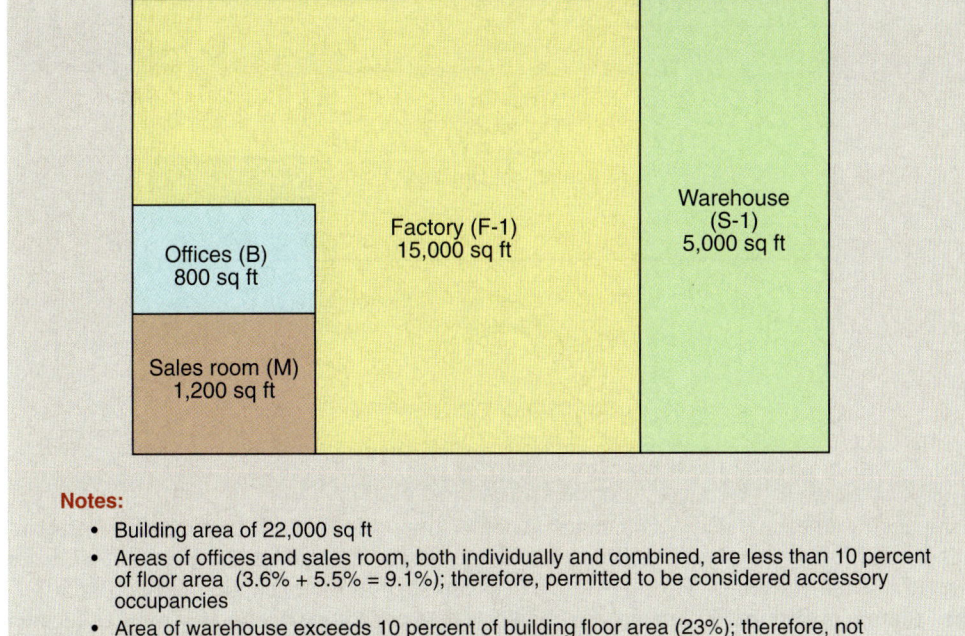

Figure 508-4 Aggregate accessory occupancies.

5 General Building Heights and Areas

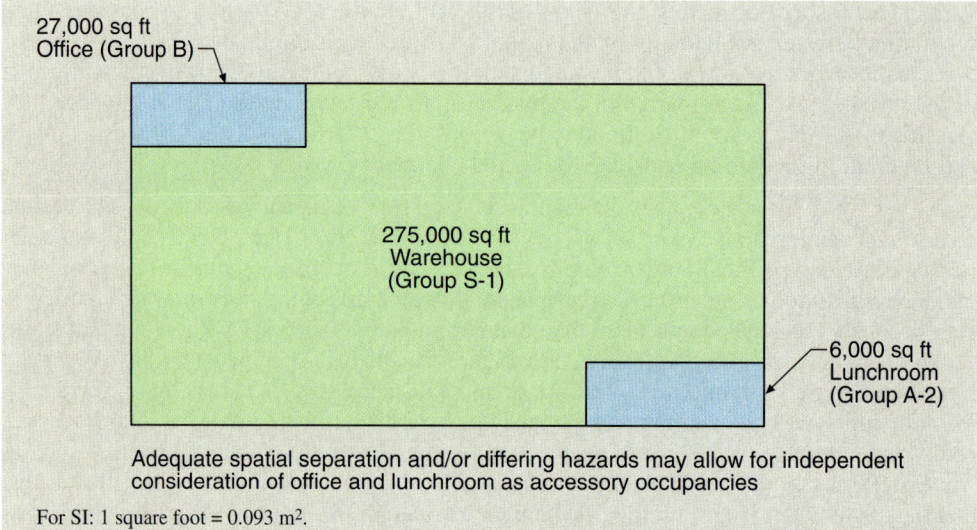

Figure 508-5
Individual accessory occupancies.

Adequate spatial separation and/or differing hazards may allow for independent consideration of office and lunchroom as accessory occupancies

For SI: 1 square foot = 0.093 m².

Application Example 508-1

General criteria for consideration as accessory occupancy:

√ 1. **Accessory to major use.** Conference room's primary function limited to use by employees of factory.

√ 2. **Does not exceed 10 percent of floor area.** Conference room does not exceed 7,225 square feet, 10 percent of total floor area.

√ 3. **Does not exceed tabular allowable area.** Conference room does not exceed 9,500 square feet allowed by Table 503 for Group A-3 occupancy of Type IIB construction.

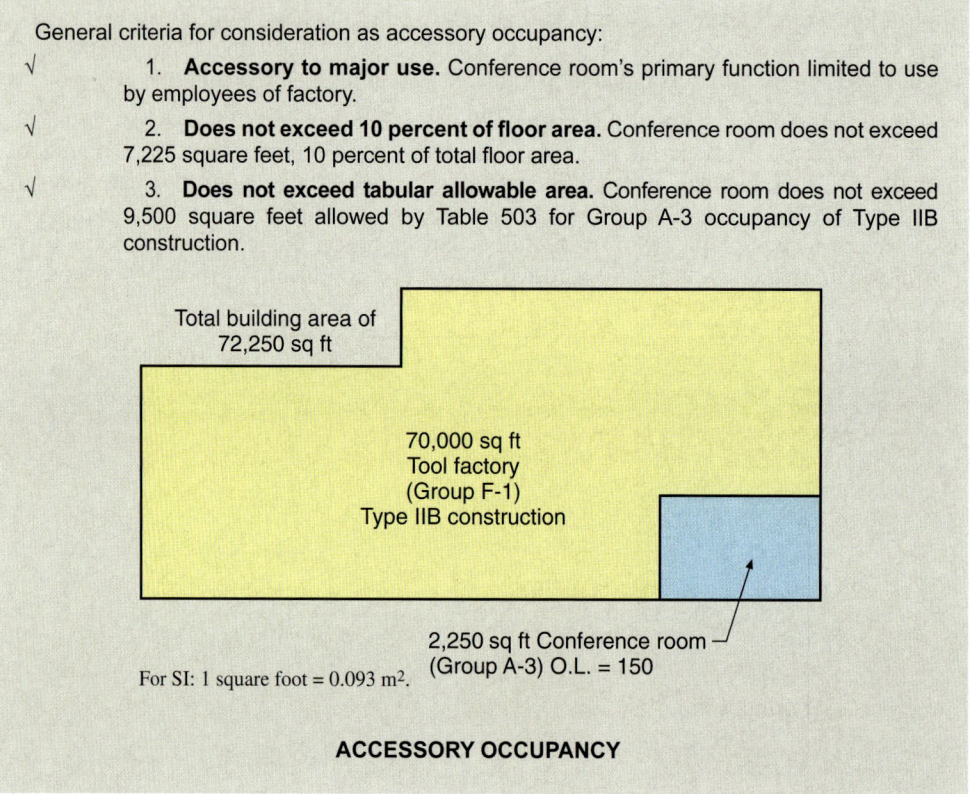

For SI: 1 square foot = 0.093 m².

ACCESSORY OCCUPANCY

occupancies and separated occupancies. Another important benefit provides for occupancies not normally permitted in unlimited-area buildings, as regulated by Section 507, to be located in such buildings. This allowance is established in the exception to Section 507.1. For example, a Group A-2 lunchroom considered an accessory occupancy may be located in a two-story Group B unlimited-area building complying with Section 507.4 with no separation required between the two occupancies. This allowance, along with the limitations for accessory occupancies, is shown in Application Example 508-2.

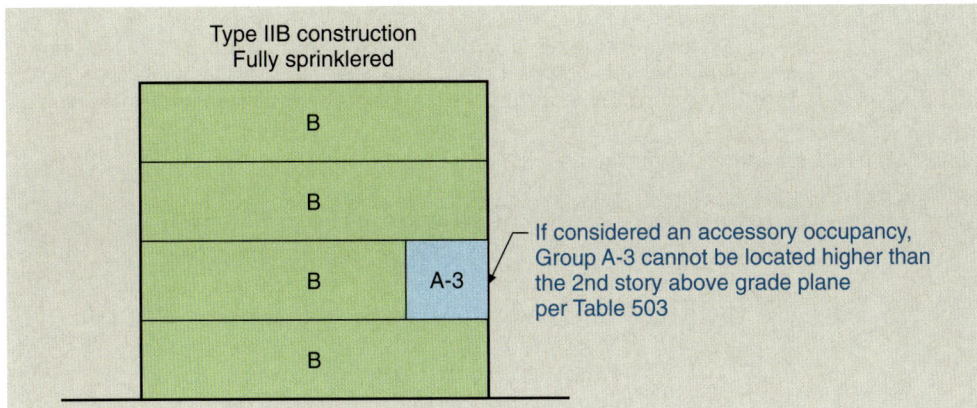

Figure 508-6 **Maximum height of accessory occupancy.**

Two exceptions indicate those conditions related to accessory occupancies under which some degree of fire-resistance-rated separation is mandated. The first exception indicates that the fire separation cannot be eliminated where the accessory occupancy is classified as Group H-2, H-3, H-4, or H-5. Where such occupancies occur in a mixed-occupancy building, it is necessary to apply the separated occupancy provisions of Section 508.4. The second exception mandates that the separation elements addressed in Section 420 for buildings containing dwelling units and sleeping units be provided.

508.3 Nonseparated occupancies. This section presents another of the available methods addressing the relationship between different occupancies in a mixed-occupancy building. Under the specific conditions of this methodology, fire-resistance-rated separations are not mandated between adjacent occupancies. The fundamental concept behind this provision assumes that if the building is designed in part to address the most restrictive and most hazardous conditions that are expected to occur based on the occupancies contained in the building, a fire-resistance-rated separation is not needed. In fact, no physical separation of any type is required.

Utilizing the nonseparated-occupancy method, the building must be individually classified for each unique occupancy that exists. The height and area limitations for those occupancies will be used to determine the required type of construction for each occupancy, with the most restrictive type of construction required for the entire building. In addition, the most restrictive fire-protection system requirements (automatic sprinkler systems and fire alarm systems) that apply to an occupancy in the building shall apply to the entire structure. For the application of other code provisions, each individual occupancy will be regulated by only the specific requirements related to that occupancy. A special condition mandates that if a high-rise building is regulated under the nonseparated-occupancy provisions, even those portions of the building that are not considered high rise must comply with the high-rise requirements. See Figure 508-7. Another important limitation is that the use of the nonseparated-occupancy method is not permitted for Group H-2, H-3, H-4, and H-5 occupancies. In addition, the nonseparated-occupancy provisions cannot be utilized to eliminate any required dwelling unit or sleeping unit separation required under the special provisions of Section 420.

The approach to understanding the rationale for the nonseparated-occupancy method may be better understood by viewing the structure as multiple single-use facilities. In evaluating the building described in Application Example 508-3, assume that the building is entirely a Group B occupancy. Based on that assumption, determine its maximum allowable height and area, as well as any required fire protection features. Now evaluate the building as if it were entirely a Group E occupancy, again addressing the maximum allowable height and area, along with the requirements for fire protection systems. By applying the most restrictive height, area, and fire protection

5 General Building Heights and Areas

Application Example 508-2

GIVEN: A 150,000-square-foot two-story office building (75,000 sf/story) housing a 3,750-square-foot employee lunchroom with an occupant load of 250 persons. The building is fully sprinklered, is of Type VB construction, and qualifies for unlimited area under the provisions of Section 507.4.

DETERMINE: Application of the accessory occupancy provisions of Section 508.2 for the lunchroom.

1. Is the accessory occupancy subsidiary to the building's major occupancy?
   Yes, the lunchroom is intended to serve the employees of the office space.
2. Is the accessory occupancy no more than 10 percent of the floor area of the story?
   Yes, 10 percent of 75,000 sq ft = 7,500 sq ft maximum; the lunchroom is 3,750 sq ft.

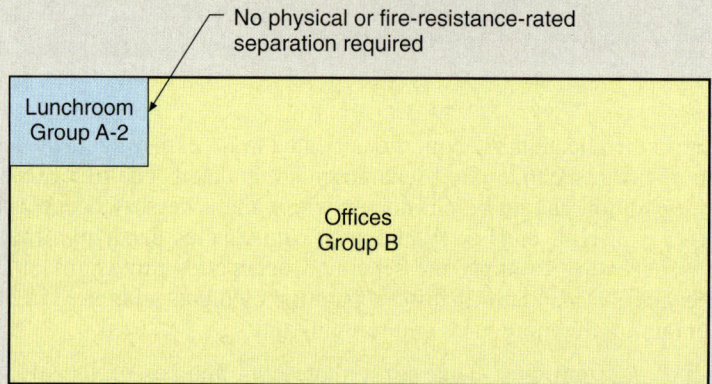

For SI: 1 square foot = 0.093m².

3. Is the accessory occupancy no larger than the tabular values in Table 503?
   Yes, the tabular value for a Group A-2 of VB construction is 6,000 sq ft; the lunchroom is 3,750 sq ft.
4. What is the occupancy classification of the lunchroom?
   Group A-2, based on the individual classification of the use.
5. How are the other requirements of the IBC applied?
   The provisions for each occupancy are applied only to that specific occupancy.
6. What is the allowable height and area of the building?
   The building's allowable height and area are based on the major occupancy involved; in this case it is Group B. Based on the criteria of Section 507.4, the building is permitted to be unlimited in area and limited to two stories in height.
7. What is the allowable height of the lunchroom?
   One story, based on Table 503 for Group A-2 in a Type VB building. The lunchroom must be located on the first story.
8. What is the minimum required separation between the lunchroom and the office?
   There is no fire-restrictive or physical separation required, owing to compliance with the provisions of Section 508.2 for accessory occupancies.

provisions of both Group B and E occupancies to the entire building, the conditions for nonseparated occupancies can be determined.

508.4 Separated occupancies. Using this method of addressing mixed-occupancy buildings, the code directs that each portion of the building housing a separate occupancy be individually classified and comply with the requirements for that specific occupancy. Furthermore, the code intends that each pair of occupancies be evaluated through Table 508.4 as to the relationship of the hazards involved, often mandating a fire-resistance-rated separation

Mixed Use and Occupancy

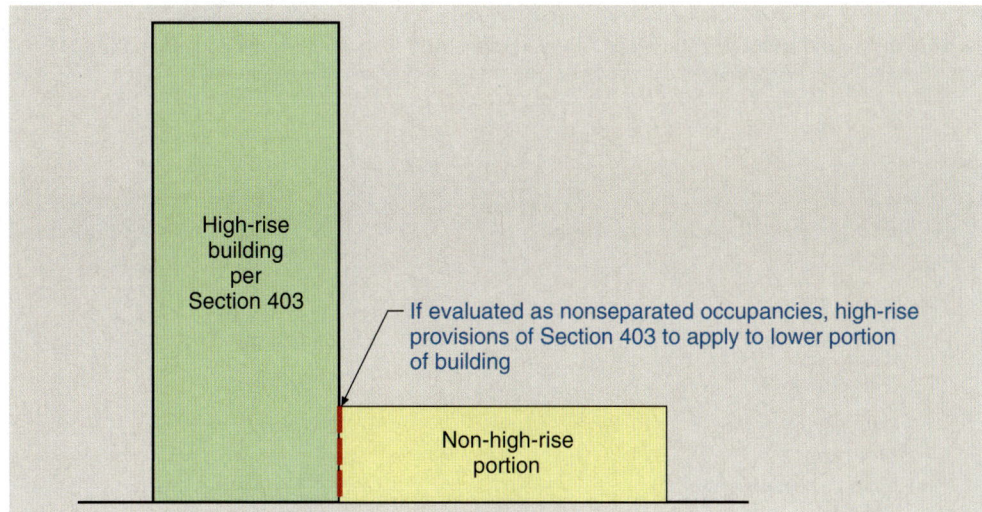

Figure 508-7
Nonseparated occupancies in a high-rise building.

between them. Measured from the grade plane, the allowable height is limited based on the type of construction of the building as shown in Table 503. See Figure 508-8.

For allowable floor-area considerations in a building having multiple occupancies, the code uses a formula that is very similar to the interaction formula used in structural engineering where two different types of stress are imposed on a member at the same time.

In the case of a mixed-occupancy building, the code uses this type of formula for the calculation of the allowable building area for each floor. For example, if there are three different occupancies in a building, the formula is as follows:

$$a1/A1 + a2/A2 + a3/A3 < 1.0$$

WHERE: $a1$, $a2$, and $a3$ represent the actual areas for the three separate occupancies, and $A1$, $A2$, and $A3$ represent the allowable areas for the three separate occupancies.

See Application Examples 508-4 and 508-5 for examples of this computation.

This formula essentially prorates the areas of the various occupancies so that the sum of percentages must not exceed 100 percent.

It is also appropriate to utilize a variation of this formula for the determination of the allowable area for the total building as evidenced by the provisions of Section 506.5.2. See Application Example 506-9. For all practical purposes, the need to evaluate the building as a whole is only necessary in buildings of four or more stories above grade plane. For two- and three-story buildings, if each story is compliant for allowable area purposes, the entire building will always comply.

Table 508.4 Required Separation of Occupancies. The code has established several alternative methods for addressing mixed-occupancy buildings regarding fire separations between the various occupancies involved. Both Sections 508.2.4 and 508.3.3 for accessory occupancies and nonseparated occupancies, respectively, do not require any fire-resistance-rated separation between occupancies. However, Table 508.4, establishing fire separations under the separated occupancy method of Section 508.4, indicates varying degrees of separation. Fire-resistive separations of 1, 2, 3, and 4 hours are selectively required based on the occupancies involved. In some cases, however, no fire separation is mandated. The intent of the table is to provide for relative separation requirements based primarily on dissimilar risk. The fire-resistance ratings, including the lack of such required ratings in many circumstances, appropriately recognize the degree of dissimilarity between the various occupancies.

5 General Building Heights and Areas

Application Example 508-3

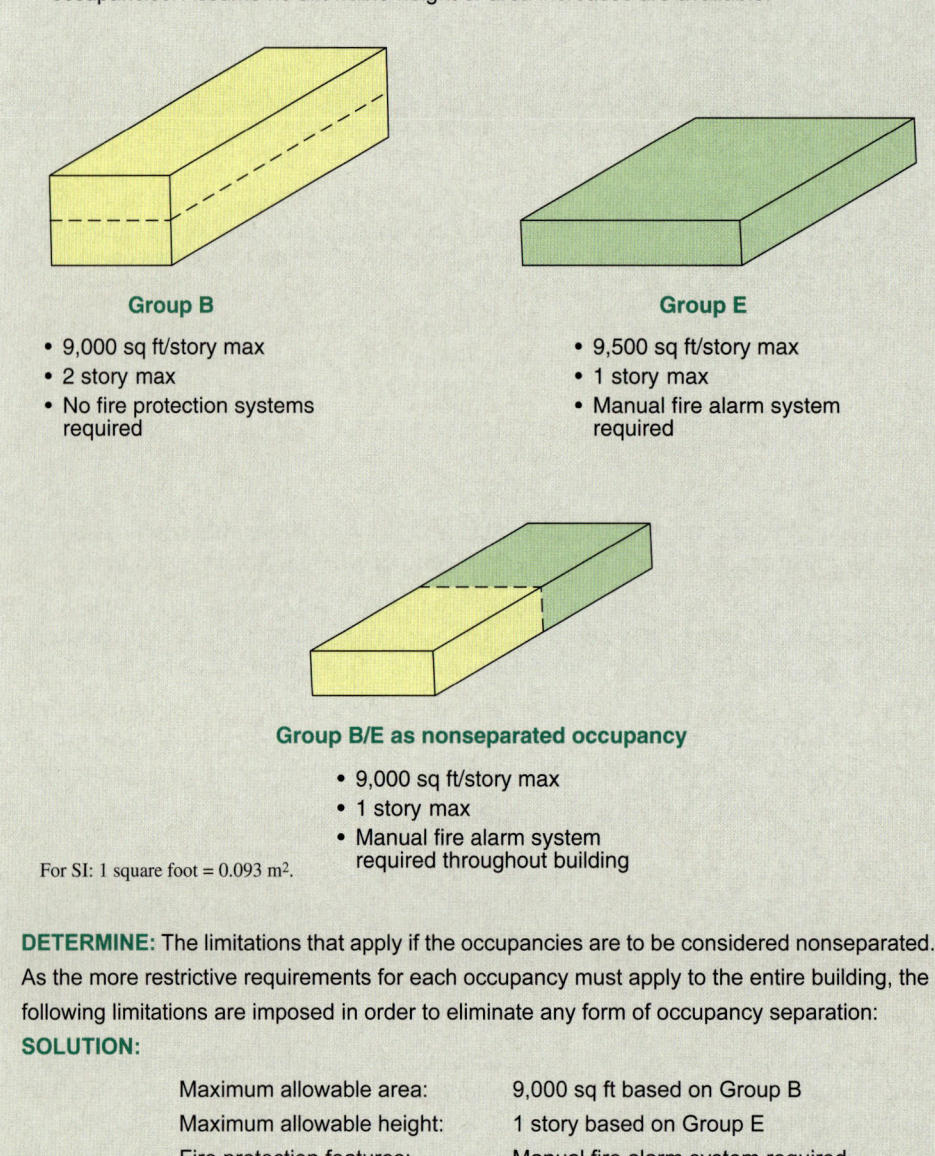

GIVEN: A mixed-occupancy building of Type VB construction, housing both Group B and Group E occupancies. Assume no allowable height or area increases are available.

Group B
- 9,000 sq ft/story max
- 2 story max
- No fire protection systems required

Group E
- 9,500 sq ft/story max
- 1 story max
- Manual fire alarm system required

Group B/E as nonseparated occupancy
- 9,000 sq ft/story max
- 1 story max
- Manual fire alarm system required throughout building

For SI: 1 square foot = 0.093 m².

DETERMINE: The limitations that apply if the occupancies are to be considered nonseparated. As the more restrictive requirements for each occupancy must apply to the entire building, the following limitations are imposed in order to eliminate any form of occupancy separation:

SOLUTION:

Maximum allowable area:	9,000 sq ft based on Group B
Maximum allowable height:	1 story based on Group E
Fire protection features:	Manual fire alarm system required throughout building, based on Group E

NONSEPARATED OCCUPANCIES

In a general sense, the following logic was utilized in formulating the table. High-hazard (Group H) occupancies are required to be separated from each other and from all other occupancies. Ordinary or moderate-hazard commercial/industrial (Groups B, F-1, M, and S-1) occupancies require no separation from each other; however, they are required to be separated from all other occupancies. People-intensive (Groups A and E) occupancies also require no occupancy separation between each other but must be separated from all other occupancies except for fully sprinklered Group F-2 and S-2 occupancies.

Mixed Use and Occupancy 5

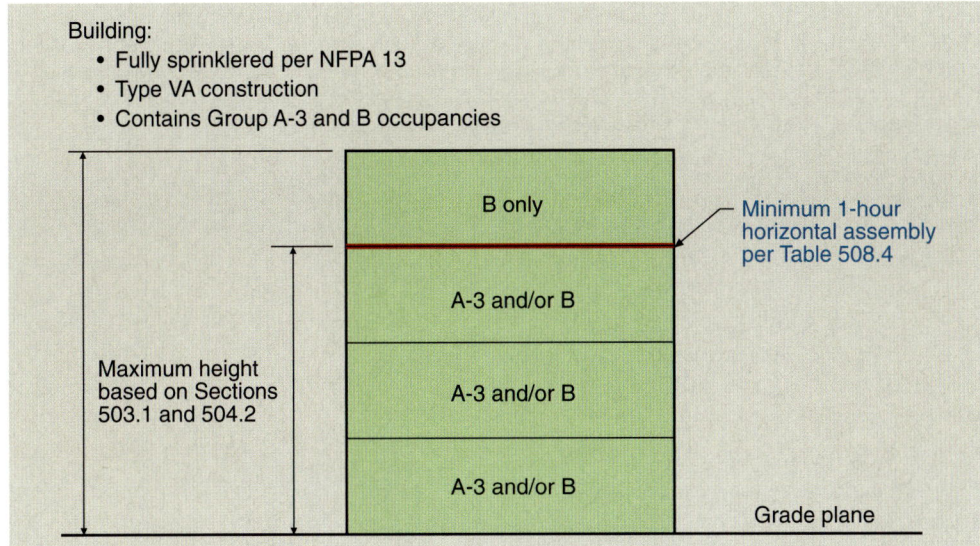

Figure 508-8
Maximum height of separated occupancies.

Application Example 508-4

GIVEN: A one-story building housing day care classified as Group E, offices, and a conference center. The building is of Type V-A construction. No yards or sprinklers are available for area increase purposes. Floor areas are as follows:

Office (B)	4,500 square feet
Assembly (A-3)	3,000 square feet
E	6,000 square feet

DETERMINE: If the building area is within the allowable area.

SOLUTION: In accordance with Section 508.4.2:

$$\frac{\text{Actual area of office}}{\text{Allowable area of office}} + \frac{\text{Actual area of assembly}}{\text{Allowable area of assembly}} + \frac{\text{Actual area of E}}{\text{Allowable area of E}} \leq 1$$

$$\frac{4{,}500}{18{,}000} + \frac{3{,}000}{11{,}500} + \frac{6{,}000}{18{,}500} \stackrel{?}{\leq} 1$$

$$0.25 + 0.26 + 0.32 \stackrel{?}{\leq} 1$$

$$0.83 < 1, \text{ therefore OK}$$

For SI: 1 square foot = 0.093 m². Building is within the allowable area.

DETERMINING ALLOWABLE AREAS FOR A MIXED-OCCUPANCY BUILDING

Group R occupancies require no separation from other Group R occupancies, but such separations are mandated between all other occupancy classifications. Similar criteria apply to the Group I occupancies with a modification for Group I-2. The philosophy of the provisions set forth in the table dictates increased fire-resistance ratings on the basis of greater inherent dissimilar risk. Where Table 508.4 mandates some degree of occupancy separation, an occupancy shall be physically separated from the other occupancy through the use of fire barriers, horizontal assemblies, or a combination of both vertical and horizontal fire-resistance-rated assemblies.

5 General Building Heights and Areas

Application Example 508-5

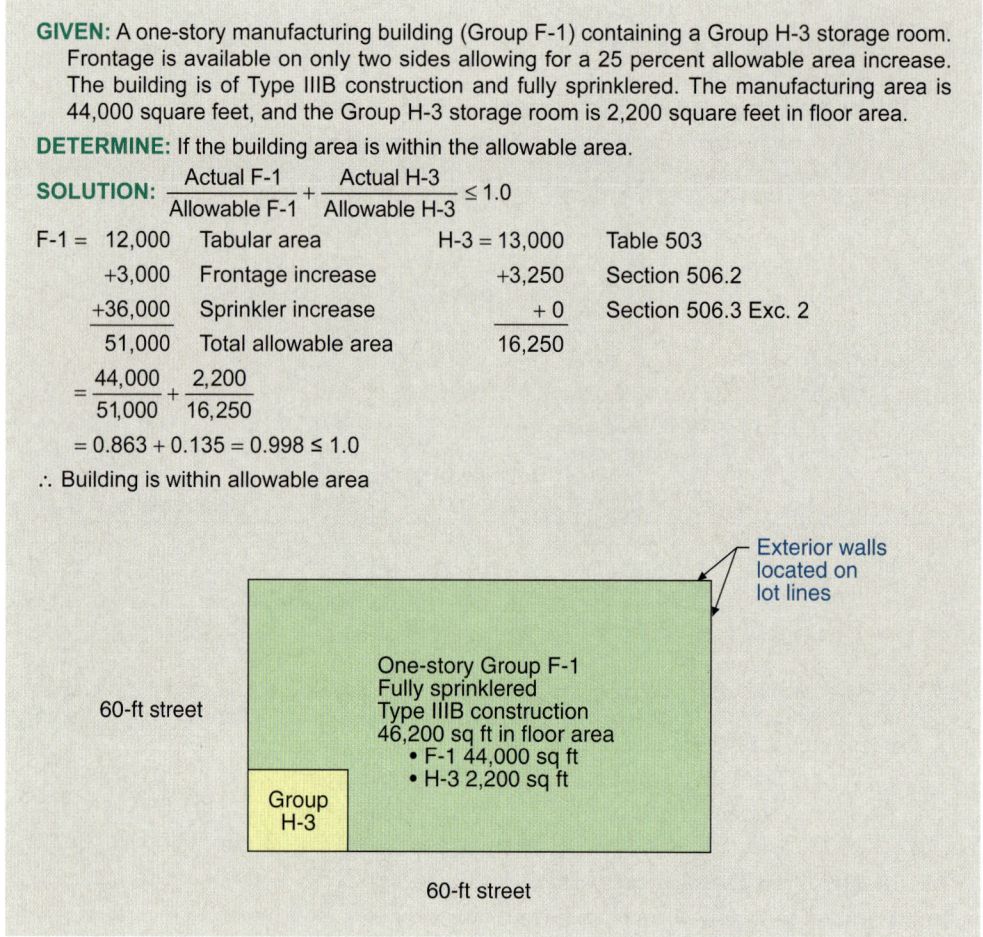

GIVEN: A one-story manufacturing building (Group F-1) containing a Group H-3 storage room. Frontage is available on only two sides allowing for a 25 percent allowable area increase. The building is of Type IIIB construction and fully sprinklered. The manufacturing area is 44,000 square feet, and the Group H-3 storage room is 2,200 square feet in floor area.

DETERMINE: If the building area is within the allowable area.

SOLUTION: $\dfrac{\text{Actual F-1}}{\text{Allowable F-1}} + \dfrac{\text{Actual H-3}}{\text{Allowable H-3}} \leq 1.0$

F-1 =	12,000	Tabular area	H-3 =	13,000	Table 503
	+3,000	Frontage increase		+3,250	Section 506.2
	+36,000	Sprinkler increase		+0	Section 506.3 Exc. 2
	51,000	Total allowable area		16,250	

$= \dfrac{44,000}{51,000} + \dfrac{2,200}{16,250}$

$= 0.863 + 0.135 = 0.998 \leq 1.0$

∴ Building is within allowable area

In some cases, the rationale behind the use of fire-resistance-rated separations between incompatible occupancies concerns itself with the amount of combustibles encompassed in the adjoining occupancies and is termed in fire-protection circles as *fire loading*. Thus, if the amount of combustibles or fire loading in one occupancy is quite high while there is a limited fire load anticipated in the other occupancy, some degree of fire separation between the two distinct occupancies is necessary. However, the relationship of fire loads is not the only factor in determining an appropriate occupancy separation. In some cases, the separation is specified to be of 1 hour or more in duration mainly because of what the code implies to be incompatibility between the activities that occur within the two occupancies. For example, the code requires a separation of 1 hour between a Group I-2 occupancy and a Group S-2 occupancy in a fully sprinklered building. The limited amount of combustibles in either occupancy does not justify a fire-resistive separation; however, because of the presumed incompatibility between the two occupancies, the 1-hour separation is considered to be justified.

As a general rule, a reduction of the fire-resistance ratings in Table 508.4 by 1 hour is permitted in buildings equipped throughout with an automatic sprinkler system. The potential of a sizable fire spreading throughout a building is minimal under fully sprinklered conditions.

The fire-resistance ratings set forth by the table are further modified by Footnote b. The required separation for storage occupancies may be reduced by 1 hour where the storage is limited to the parking of private or pleasure vehicles, but may never be less than 1 hour.

It is necessary to again emphasize that there is no relationship between the separated occupancy provisions of Section 508.4 and the fire area concept utilized in Section 903.2

for automatic sprinkler systems. Compliance with Table 508.4 does not relieve the responsibility to comply with Section 901.7 and Table 707.3.10 regarding the proper separation of fire areas. An example is shown in Figure 508-3.

Section 509 *Incidental Uses*

There are times where the hazards associated with a particular use do not rise to the level of requiring a different occupancy classification; however, such hazards must still be addressed because of their impact on the remainder of the building. These functions are identified as incidental uses, which are regulated independently of the mixed-occupancy provisions. Incidental uses are uniquely addressed through the use of fire-resistance-rated separations or automatic sprinkler system protection.

What is an incidental use? There are occasionally one or more rooms or areas in a building that pose risks not typically addressed by the provisions for the general occupancy group under which the building is classified. However, such rooms or areas may functionally be an extension of the primary use. These types of spaces are considered in the IBC to be incidental uses and are regulated according to their hazard level. These areas are not ever intended to be considered different occupancies, creating a mixed-use condition, but rather are classified in accordance with the main occupancy of the portion of the building where the incidental use is located. There is no specific definition for an incidental use, as it is simply described as any of those rooms or areas listed in Table 509. If it is not listed, it is not considered an incidental use for code purposes.

The designation and regulation of incidental uses do not apply to those areas within and serving a dwelling unit. Otherwise, the special hazards that may be found in buildings of various uses and occupancies are addressed through the construction of a fire barrier and/or horizontal assembly separating the incidental use from the remainder of the building, the installation of an automatic sprinkler system in the incidental use space, or, in special cases, both the fire separation and automatic sprinkler system.

Incidental uses are listed in Table 509. Most of the rooms or areas identified in the table are regulated where located in any of the occupancy groups established by the code, other than dwelling units as previously noted. A few of the incidental uses are to be regulated only where located within a specific occupancy or a limited number of occupancies.

It is common for many of the listed incidental uses to be unoccupied for extended periods of time, creating the potential for a fire to grow unnoticed. Oftentimes, combustible or hazardous materials are present in such areas. Because of the potentially high fuel load and lack of constant supervision, spaces such as furnace rooms, machinery rooms, laundry rooms, and waste collection rooms are selectively considered incidental uses. Other uses, such as paint shops, laboratories, and vocational shops, may cause concern to the point where they too must be protected or separated from other areas of the building.

How are incidental uses regulated? The fire-resistance-rated separations required by Table 509 are to be fire barriers and/or horizontal assemblies, typically having a minimum fire-resistance rating of 1 hour. For incinerator rooms and paint shops, the minimum required rating is greater. Where an automatic sprinkler system provides the necessary protection, it need only be installed within the incidental use under consideration. Examples of the requirements are illustrated in Figure 509-1.

As shown in Figure 509-2, there is a variety of combinations regarding the separation and protection methods identified in Table 509. It should be noted that where an automatic sprinkler system is utilized without a fire barrier as the protective element, the incidental use must still be separated from the remainder of the building. However, this separation need only consist of construction capable of resisting the passage of smoke. Although not

General Building Heights and Areas

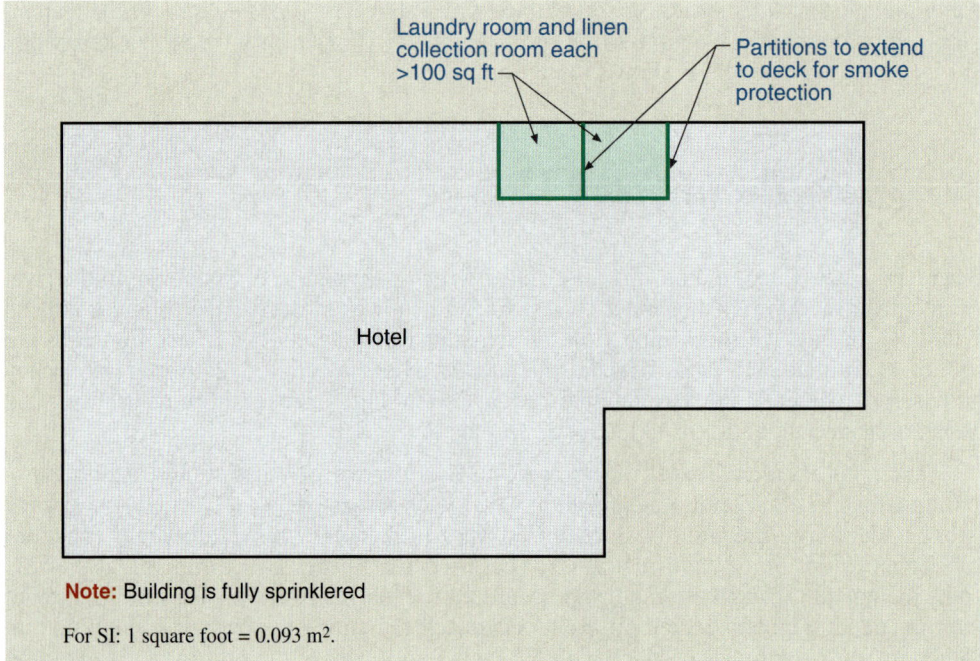

Figure 509-1 Incidental use.

Room or Area[1]	1-Hour Separation or Fire-Sprinker System	1-Hour Separation	2-Hour Separation	1-Hour Separation and Fire-Sprinker System	2-Hour Separation and Fire-Sprinker System
Furnace rooms	x				
Boiler rooms	x				
Refrigerant machinery rooms	x				
Laboratories	x				
Vocational shops	x				
Laundry rooms	x				
Waste and linen collection rooms	x				
Group I-3 cells		x			
Group I-2 and ACF[2] waste and linen collection rooms		x			
Hydrogen cut-off rooms[3]		x	x		
Battery system storage areas[3]		x	x		
Paint shop[4]			x	x	
Incinerator rooms					x

[1] See Table 509 for specifics of each use or area.
[2] Ambulatory care facilities.
[3] Varies based on type of occupancy.
[4] Option of two methods.

Figure 509-2 Incidental use requirements.

required to have a fire-resistance rating, partitions must either extend to the underside of the floor or roof deck above, or to the underside of a fire-resistance-rated floor/ceiling or roof/ceiling assembly. Doors are to be self-closing or automatic-closing upon detection of smoke, with no air-transfer openings or excessive undercuts. See Figure 509-3. The room must be tightly enclosed, providing for the containment of smoke while assisting in the heat increase necessary to activate the automatic sprinkler system.

Incidental Uses 5

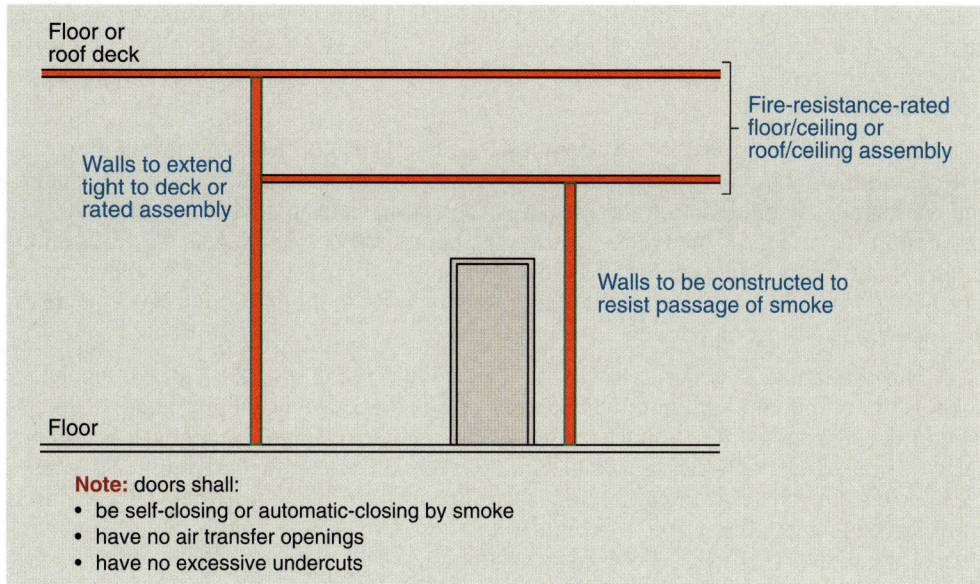

Figure 509-3
Incidental use smoke separations.

Table 509 Incidental Uses. Incidental uses are limited to those rooms or spaces listed in this table. The listed rooms have been selected for inclusion because of the increased hazard they present to the other areas of the building. However, it is recognized that the degree of hazard is such that a separate occupancy classification is not warranted. In fact, such a classification may be overly restrictive. The intent of the fire separation and fire protection requirements is to provide safeguards because of the increased hazard level presented by the incidental use.

Furnace rooms and boiler rooms. The hazard potential for fuel-fired heating equipment is addressed once the thresholds established in the table have been exceeded. It should be noted that the limitations are based on individual pieces of equipment, rather than the aggregate amounts from all equipment within the space. For example, the requirements of Table 509 are not applicable where there are two furnaces within the furnace room, each with an input rating of 300,000 Btu per hour (87,900 watts). Because no furnace exceeds the 400,000 Btu/hr (117,200 watts) threshold, there is no fire barrier separation or automatic sprinkler system protection required.

The regulated furnace or boiler is anticipated to be located within a room isolated from the remainder of the building. More specifically, the furnace room or boiler room must either be separated from other portions of the building with a complying fire barrier and/or horizontal assembly, or isolated by construction capable of resisting the passage of smoke and provided with a fire-extinguishing system. In either case, it is not permissible to eliminate the equipment enclosure. Because the intent of the requirement is to address the hazards associated with specified boilers and furnaces, the prescribed degree of separation and/or protection must always be provided.

Hydrogen cut-off rooms. Hydrogen cut-off rooms are defined in Chapter 2 as rooms or spaces that are intended exclusively to house a gaseous hydrogen system. Special requirements applicable to such rooms are set forth in Section 421. Where the quantities of materials would cause a hydrogen cut-off room to be classified as a Group H occupancy, the separation requirements of Section 508.4 for separated occupancies will apply rather than those of Table 509.

The reference in the table exempting those rooms classified as Group H is not strictly limited to hydrogen cut-off rooms. In fact, where the quantities of hazardous materials in any of the rooms or areas designated by Table 509 exceed those permitted by Section 307.1

5 General Building Heights and Areas

causing classification as a Group H occupancy, the use of the table is not appropriate. In a mixed-occupancy building, all spaces with a Group H classification must be separated from the remainder of the building in accordance with Section 508.4 for separated occupancies.

Paint shops. The provisions of Section 416 control the construction, installation, and use of rooms for spraying paints, varnishes, or other flammable materials used for painting, varnishing, staining, or similar purposes. The paint shops regulated by Table 509 are those rooms where the same types of spraying operations occur. As a general rule, a minimum 1-hour fire-resistive enclosure is mandated by Section 416.2 to isolate the spraying operations from the remainder of the building. However, in all but Group F occupancies, Table 509 mandates a higher degree of protection.

Laboratories and vocational shops. In educational buildings, particularly secondary schools, it is common to find multiple laboratories and vocational shops that are an extension of the educational function. Because of the presence of some quantities of hazardous materials in such laboratories, as well as in those labs associated with Group I-2 occupancies, the code mandates some degree of separation and/or protection. Where the quantities of hazardous materials warrant a Group H classification, the use of this table is inappropriate. Vocational shops in schools also pose a hazard to the remainder of the building that is due to the hazardous processes and combustible materials involved. Where no such hazards exist, such as in a computer lab or design lab, the table is not intended to apply.

Section 510 Special Provisions

The provisions of this section allow for modifications or exceptions to the general provisions for building heights and areas as regulated by Chapter 5 of the IBC. These special provisions are viewed as specific in nature and, based on Section 102.1, take precedence over any general provisions that may apply. Because this section permits, rather than requires, the use of the special conditions established in Section 510, the provisions are optional. Much like the application of the mezzanine provisions of Section 505, only where the designer elects to utilize the special allowances does this section apply.

It is evident that several of the provisions overlap in their scope. For example, Sections 510.2, 510.4, and 510.7 all address a potential condition where an open parking garage is located below a Group R occupancy. It is the choice of the designer which of the three methods to use where such a condition exists, based on the benefits and consequences of each method. Or, as stated above, none of the methods need to be applied. The requirements could simply be based on the general provisions of Chapter 5 for building height and area.

510.2 Horizontal building separation allowance. This section is one of several that contain provisions that might be considered the only exceptions to the principle that a fire wall is strictly a vertical element without horizontal offsets. In Item 1, the code makes provisions for the use of a minimum 3-hour fire-resistance-rated horizontal assembly as an equivalent construction feature, in certain aspects, to a fire wall. This methodology is often referred to as "podium" or "pedestal" buildings.

The provisions create, in effect, an exception that allows a Group S-2 parking garage, limited-size Group A, B, M, and/or R occupancies, along with operational areas, in the basement(s) and/or first story above grade plane of a building to be considered a separate building for specific purposes. The stories above the horizontal separation must only house Group A, B, M, R, and/or S occupancies. As these occupancies are quite common in Type V construction, the typical application is to grant the maximum number of stories for Type V construction without the penalty of sacrificing one story for the parking garage or other permitted uses located on the grade level. Distinct buildings are also created for area limitations and fire wall continuity.

It is fairly common in terrain that has a rolling or hillside character to erect apartment houses and small office/retail buildings with a garage in the basement or first story. Because of the slope of the ground surface, the lowest level is usually partially within the ground; therefore, the walls are normally designed as reinforced concrete or reinforced masonry-retaining walls. The construction of the lowest level is thereby easily able to conform to the code construction requirements for a Type IA building.

If this lowest level is classified as the first story above grade plane, it would typically be included in the number of total stories permitted by the code. However, as depicted in Figure 510-1, the first story would not be included where the level below the 3-hour horizontal separation is of Type IA construction.

The code lists seven conditions that must be met in order to take advantage of these provisions. The first condition regulates the fire-resistance rating of the horizontal assembly. The second condition identifies the maximum height at which the horizontal separation can be located. Third, the minimum type of construction for portions below the horizontal assembly is established. The fourth condition addresses the methods for protection of openings that will occur in the fire-resistance-rated horizontal assembly. Fifth, only Group A occupancies having an occupant load of less than 300, and Groups B, M, R, and S are permitted to be located above the horizontal assembly. The sixth condition limits the use of the building below the horizontal assembly to a Group S-2 parking garage; multiple Group A occupancies, each with an occupant load under 300; Group B, M, and R occupancies; and/or entry lobbies, mechanical rooms, storage rooms, and similar areas that are necessary for operation of the building. An often-forgotten provision is the seventh condition, requiring that the overall height in feet (mm) of both buildings not exceed the height limits set forth in Section 503 for the least type of construction that occurs in the building.

510.3 Group S-2 enclosed parking garage with Group S-2 open parking garage above. The provisions of this section are similar in nature to those of Section 510.2, insofar as the two different parking uses in a single structure, one located above the other, may be considered two separate and distinct buildings for the purpose of determining the type of construction. Five specific conditions must be met in order for an open parking garage, located above an enclosed parking garage, to be regulated on its own for construction type. Details of this special situation are shown in Figure 510-2.

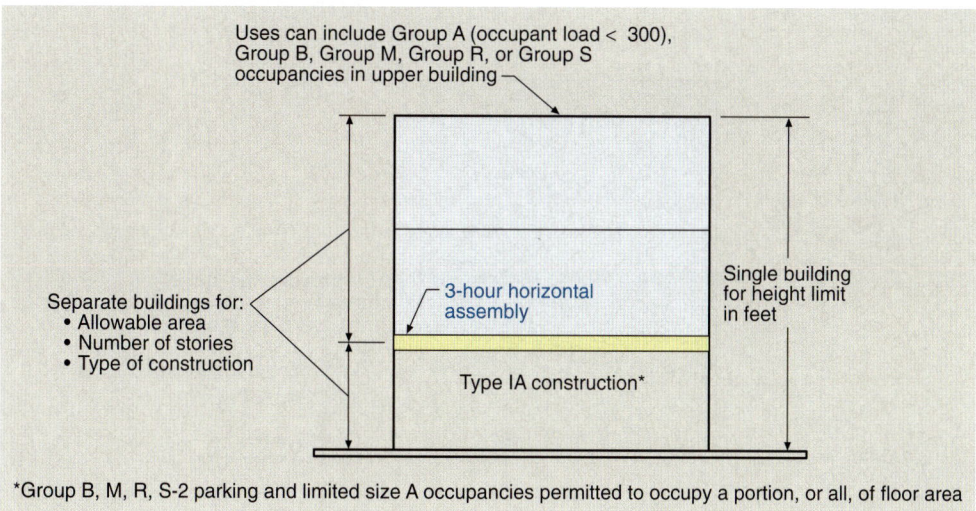

Figure 510-1
Horizontal building separation.

5 General Building Heights and Areas

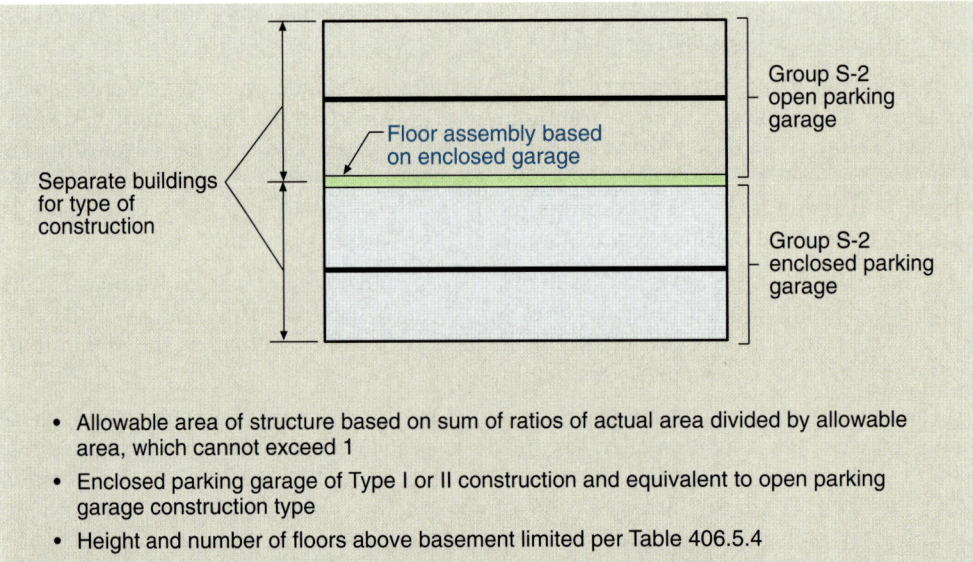

Figure 510-2
Open/enclosed parking.

510.4 Parking beneath Group R. Consistent with the concepts expressed by Sections 510.2 and 510.3, the provisions of this section address residential uses located above a first story used as a parking garage. Where parking is limited to the first story, the number of stories used in the determination of the minimum type of construction may be measured from the floor above the garage. The construction type of the parking garage and the floor assembly between the residential area and the garage are further regulated. See Figure 510-3.

510.5 Group R-1 and R-2 buildings of Type IIIA construction. Typically applicable to hotels and apartment houses, this section permits an increase in height, both in stories and feet (mm), where fire walls are utilized to create compartments having a maximum floor area of 3,000 square feet (279 m^2). The fire walls must have a minimum fire-resistance rating of 2 hours. Where there is a basement under the first floor, the first-floor construction must have a fire-resistance rating of at least 3 hours.

Group R-1 and R-2 occupancies are limited by Table 503 to four stories and 65 feet (19,825 mm) in Type IIIA construction. This provision permits an increase to six stories

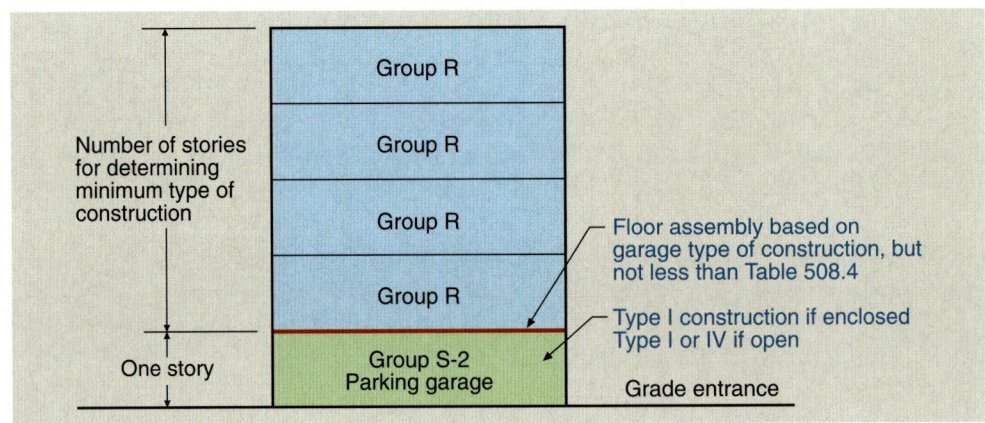

Figure 510-3
Parking beneath Group R.

170 2012 International Building Code Handbook

and 75 feet (22,860 mm), based primarily on the benefits derived from dividing the structure into small compartments separated by fire-resistance-rated construction.

510.6 Group R-1 and R-2 buildings of Type IIA construction. Because the Group R-1 and R-2 buildings addressed in this section are of noncombustible, rather than combustible, construction, it is not necessary to divide the structure into small compartments to receive a height increase, as was the case in Section 508.5. Rather, the substantial increase from four stories to nine stories, and from 65 feet (19,825 mm) to 100 feet (30,480 mm), is based on three other conditions that must occur:

1. An open area of at least 50 feet (15,240 mm) must be maintained from the Group R-1 and R-2 buildings to any other buildings on the same lot and from all lot lines.

2. A minimum 2-hour fire-resistance-rated fire wall must segregate the means of egress.

3. The floor construction of the first floor requires a minimum fire-resistance rating of $1^1/_2$ hours.

510.7 Open parking garage beneath Groups A, I, B, M, and R. The excellent fire-safety record for open parking garages is the basis for this modification in the general provisions for allowable floor area and allowable height. Where located below assembly, institutional, business, mercantile, or residential occupancies, an open parking garage is regulated for height and area by Section 406.5. Those permitted occupancies located above the parking garage are independently regulated by Section 503 for height and area. The only exception requires that the height of the portion of the building above the open parking garage, both in feet (mm) and stories, be measured from the grade plane.

The details of construction type are applicable to each of the occupancies involved; however, the structural-frame members shall be of fire-resistance-rated construction according to the most restrictive fire-resistive assemblies of the occupancy groups involved. Egress from the areas above the parking garage shall be isolated from the garage, with the level of protection at least 2 hours.

Because the provisions of Section 510.2 can also address open parking garages below similar occupancies, the application of this section may be limited. There are several minor differences between the two provisions; however, the general concept remains consistent.

510.8 Group B or M with Group S-2 open parking garage. A desirable feature in high-density areas is to have offices and/or retail stores on the first floor of open parking structures. This provision allows for the type of construction for the ground floor, as well as any basement used for Group B or M uses, to be evaluated separately from that of the open parking garage above, provided the uses are properly separated and the egress from the garage is independent from that of the first floor (and basement if applicable). This provision reverses the conditions addressed by other provisions of Section 510 where the parking garage is located below other occupancies. The resulting benefit, shown in Figure 510-4, provides for a potential reduction in the type of construction by permitting the evaluation of allowable floor areas independently for the open parking garage and the Group B and/or M occupancies.

510.9 Multiple buildings above a horizontal assembly. Where the varying provisions of Section 510 are utilized to create separate buildings above and below a complying horizontal separation, it is acceptable for two or more buildings to be located above the separation while only one building (a parking garage) is located below. For example, a condominium building is permitted to be regulated as a separate building from an adjacent office building even though both are located above a single parking facility designed under the special provisions of Sections 510.2, 510.3, or 510.8, as applicable. An example is shown in Figure 510-5.

5 General Building Heights and Areas

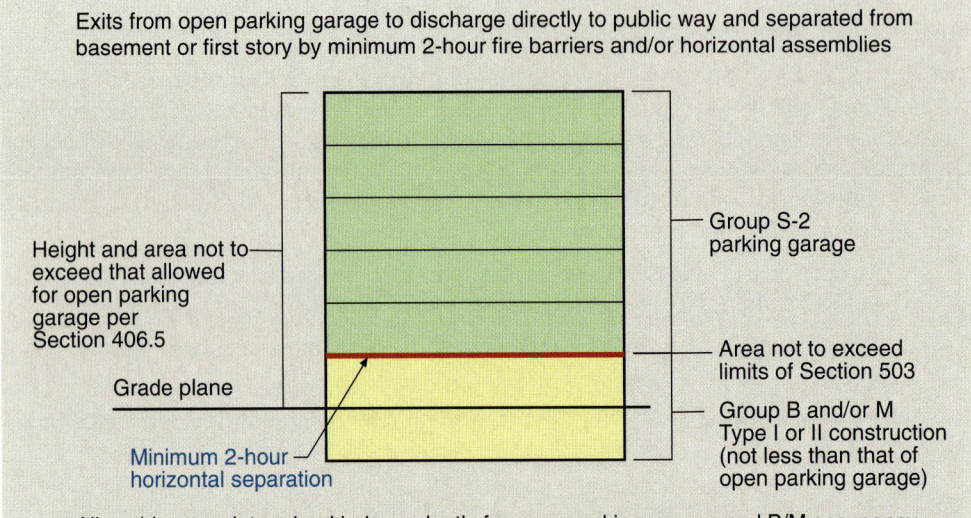

Figure 510-4 Group B and M uses below an open parking garage.

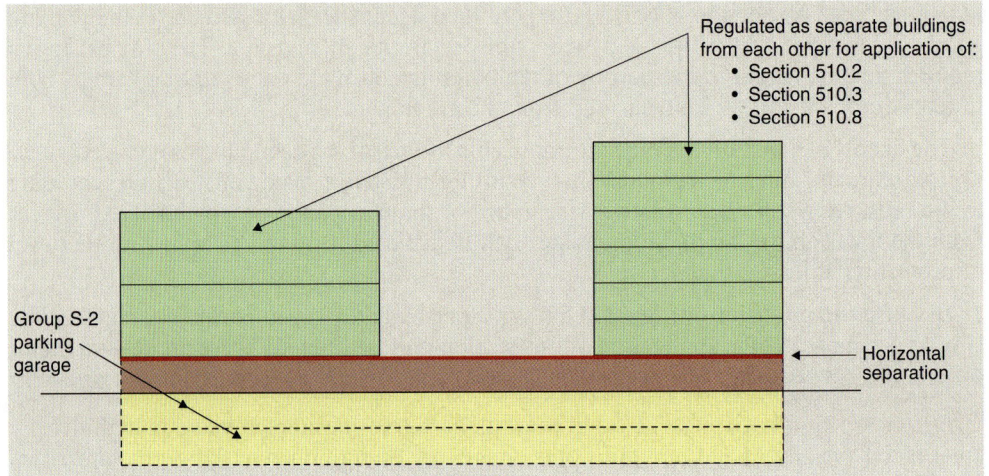

Figure 510-5 Separate buildings above horizontal separation.

KEY POINTS

- Table 503 provides the basic allowable area per floor for all occupancies and types of construction.
- The maximum height of a building is regulated by Table 503 for both the total height in feet and the maximum height in stories.
- Buildings of three or more stories in height are permitted to be three times the allowable area permitted for a single floor.
- Stepped or terraced buildings should be reviewed carefully to determine the height, as often each segment can be viewed independently of the others.
- Mezzanines are defined by the code in a very specific manner.
- The area of the basement typically does not need to be included in the total allowable area of the building.

Key Points

- The installation of automatic sprinkler systems throughout a building typically provides for a sizable increase in the allowable area and building height.
- Sufficient open yards and public ways may be used to increase the allowable area.
- Through the use of adequate safeguards, the IBC allows certain types of buildings and occupancies to have unlimited floor areas.
- In mixed-occupancy buildings, three different methods (accessory occupancies, nonseparated occupancies, and separated occupancies) are available for addressing occupancy classification, allowable height and area, and separation.
- Incidental uses, classified as a part of the building where the use is located, shall be separated, protected, or both.
- The special provisions of Section 510 provide for alternative approaches to the specific height and area requirements of Chapter 5.

CHAPTER 6

TYPES OF CONSTRUCTION

Section 602 Construction Classification
Section 603 Combustible Material in
             Type I and II Construction
Key Points

6 Types of Construction

> As its title implies, this chapter develops requirements for the classification of buildings by type of construction. In addition to identifying fire-resistance rating requirements for the major building elements, the *International Building Code*® (IBC®) regulates exterior walls for fire resistance based on their fire separation distance. The use of combustible materials in otherwise noncombustible buildings is also addressed.

Section 602 *Construction Classification*

Since early in the last century, the fire protection required for the various types of construction has been based on hourly fire-endurance ratings as established by the American Society for Testing and Materials. Prior to this time, fire-resistance requirements were developed by specifying the type and thickness of materials used.

Many of the concepts in previous building codes that have carried over to today were developed from the reports issued by the committee known as the Department of Commerce Building Code Committee, which was appointed by Herbert Hoover, then Secretary of Commerce. The committee was also dubbed the *Little Hoover Commission* and was appointed to investigate building codes. This was an outgrowth of the findings of the Senate Committee on Reconstruction and Production, which was appointed in 1920 to study the various factors entering into the recovery of our economy from the depression of the early 1920s. Although the committee studied a wide-ranging set of those institutions and groups affected by the economy, it was especially interested in construction. During its tenure, the committee held numerous hearings and expressed the following sentiment at their conclusion: "The building codes of this country have not been developed on scientific data, but rather on compromise; they are not uniform in practice and in many instances involve an additional cost of construction without assuring more useful or more durable buildings." Thus, the stage was set for improvement in building regulations, and the timing was especially favorable for the model codes to take advantage of the reports of the Department of Commerce Building Code Committee.

The IBC classifies construction into five basic categories, listed in a somewhat descending order from the most fire resistant to the least fire resistant. These five types are based on two main groupings, noncombustible (required) construction (Types I and II) and combustible (permitted) construction (Types III, IV, and V). The various types of construction within the five categories are further subdivided based on fire protection and are represented as follows:

1. Noncombustible, protected—Types IA, IB, and IIA
2. Noncombustible, unprotected—Type IIB
3. Combustible and/or noncombustible, protected—Types IIIA, IV, and VA
4. Combustible and/or noncombustible, unprotected—Types IIIB and VB

Although Types III, IV, and V are commonly considered combustible construction, the use of noncombustible materials, either in part or throughout the building, is certainly acceptable. The reference to combustible construction more simply indicates that such construction is acceptable in Types III, IV, and V but not mandated. A perusal of Table 503 will show the reader that the IBC considers Type II, III, and IV buildings to be of comparable protection. For example, Types IIA, IV, and, to some degree, IIIA are permitted the same approximate areas and heights for most occupancy classifications. The same is also true for Types IIB and IIIB.

Differing from the concept of mixed-occupancy buildings, the code does not permit a building to be considered to have more than one type of construction. In simple terms, classification of a building for construction type is based on the *weakest link* concept. If a building does not fully conform to the provisions of Chapter 6 for type of construction classification, it must be classified into a lower type into which it does conform. Unless specifically permitted elsewhere by the code, the presence of any combustible elements regulated by Table 601 prohibits its classification as Type I or II construction. Similarly, the lack of required fire resistance in any element required by Table 601 to be protected will result in a fully nonrated building.

Table 601 identifies the required fire-resistance ratings of building elements based on the specified type of construction. Exterior walls are further regulated by Table 602 based on the building's location in relation to adjoining lot lines and public ways. Reference is made to Section 703.2 for those building elements required to have a fire-resistance rating by Table 601. Section 703.2 establishes the appropriate test procedures for building elements, components, and assemblies that are required to have a fire-resistance rating.

The provisions of Chapter 6 in regard to fire resistance are intended to address the structural integrity of the building elements under fire conditions. Unlike those fire-resistance-rated assemblies, such as fire walls and fire barriers, whose intent is to safeguard against the spread of fire, the protection afforded by the provisions of Chapter 6 is solely that of structural integrity. As such, the protection of door and window openings, ducts, and air transfer openings is not required for building elements required to be fire-resistance rated by Table 601 unless mandated by other provisions of the IBC.

The IBC intends that the provisions of the code are minimum standards. Thus, Section 602.1.1 directs that buildings not be required to conform to the requirements for a type of construction higher than the type that meets the minimum requirements of the IBC based on occupancy. A fairly common case in this regard is where a developer may construct an industrial building that complies in most respects to the requirements of the code for a Type IIIB building, but the occupancy provisions are such that a Type VB building would meet the requirements of the code. In this latter case, it would be clearly inappropriate and, in fact, a violation of the code for the building official to require full compliance with requirements for a Type IIIB building. However, where the building does comply in all respects to Type IIIB, the building official may so classify it.

Types I and II. Buildings classified as Type I and II are to be constructed of noncombustible materials unless otherwise modified by the code. The various building elements in these noncombustible buildings are regulated by Table 601. Although Type I and II buildings are defined as noncombustible, it is evidenced by Section 603 that combustible materials are permitted in limited quantities. Wood doors and frames, trim, and wall finish are permitted, as well as combustible flooring, insulation, and roofing materials. Where these combustibles are properly controlled, they have proven, over the years, to not add significantly to the fire hazard.

602.2

Furthermore, Type I buildings are to be of the highest levels of fire-resistance-rated construction. The fire-resistance ratings required for Type I buildings historically have provided about the same protection over the years and, thus, have proved to be satisfactory for occupancies with low to moderate fire loadings, such as office buildings, hotels, and retail stores. Type IB construction is very similar to Type IA construction except for a reduction of 1 hour in the required ratings for interior and exterior bearing walls, and the structural frame, while providing a $1/2$-hour reduction for roof construction. Thus, and particularly because of the reduction in the fire-resistance rating required for the structural frame, the Type IB building does not enjoy all of the unlimited height and areas that accrue to the Type IA building. It will be noted from Table 503 that Type IB construction typically has height and, to some degree, area limits placed on it.

Buildings of Type II construction, although noncombustible, may be of either protected (Type IIA) or unprotected (Type IIB) construction. The building elements of a Type IIA building are typically required to be protected to a minimum fire-resistance rating of 1 hour. Such elements in a Type IIB structure may be nonrated.

602.3 Type III. The Type III building grew out of the necessity to prevent conflagrations in heavily built-up areas where buildings were erected side by side in congested downtown business districts. After the severe conflagrations of years past in Chicago and Baltimore, it became apparent that some control must be made to prevent the spread of fire from one building to another. As a result, the Type III building was defined. The Type III building is, in essence, a wood-frame building (Type V) with fire-resistance-rated noncombustible exterior walls.

Around the turn of the 20th century, and prior to the promulgation of modern building codes, Type III buildings were known as ordinary construction. They later became known in some circles as ordinary masonry construction. However, as stated previously, the intent behind the creation of this type of construction was to prevent the spread of fire from one combustible building to another. Thus, the early requirements for these buildings were for a certain thickness of masonry walls, such as 13 inches (330 mm) of brick for one-story and 17 inches (432 mm) for two-story buildings of bearing-wall construction. Later, the required fire endurance was specified in hours. Thus, any approved noncombustible construction that would successfully pass the standard fire test for the prescribed number of hours was permitted.

In spite of the requirement for noncombustible exterior walls, Type III buildings are considered combustible structures and are either protected (Type IIIA) or unprotected (Type IIIB). Interior building elements are permitted to be either combustible or noncombustible. There is an allowance for the use of fire-retardant-treated wood as a portion of the exterior wall assembly, provided such wall assemblies have a fire-resistance rating of 2 hours or less.

602.4 Type IV. Type IV buildings are designated as heavy-timber buildings. In the eastern United States during the 1800s, a type of construction evolved that was known as mill construction. Mill construction was developed by insurance companies to reduce the heavy losses they were facing in the heavy industrialized areas of the Northeast.

This type of construction has also been known as slow burning. Wood under the action of fire loses its surface moisture, and when the surface temperature reaches about 400°F (204°C), flaming and charring begin. Under a continued application of the heat, charring continues, but at an increasingly slower rate, as the charred wood insulates the inner portion of the wood member. There is quite often enough sound wood remaining during and after a fire to prevent sudden structural collapse. In recognition of these characteristics, the insurance interests reasoned that replacement of light-wood framing on the interior of factory buildings with heavy-timber construction would substantially decrease their fire losses.

The Type IV building is essentially a Type III building with a heavy-timber interior. It is of interest to see how the 1943 edition of the *National Building Code*, developed by the National Board of Fire Underwriters, defined heavy-timber construction:

> "Heavy-timber construction," as applied to buildings, means that in which walls are of approved masonry or reinforced concrete; and in which the interior structural elements, including columns, floors and roof construction, consist of heavy timbers with smooth, flat surfaces assembled to avoid thin sections, sharp projections and concealed or inaccessible spaces; and in which all structural members which support masonry walls shall have a fire-resistance rating of not less than 3 hours; and other structural members of steel or reinforced concrete, if used in lieu of timber construction, shall have a fire-resistance rating of not less than 1 hour.

From this definition, it can be seen that in the early development of heavy-timber construction, not only did the heavy-timber members have large cross sections to achieve the slow-burning characteristic, but, furthermore, surfaces were required to be smooth and flat. Sharp projections were to be avoided, as well as concealed and inaccessible spaces. Thus, the intent of the concept is to provide open structural framing without concealed spaces and without sharp projections or rough surfaces, which are more easily ignitable. In this case, flame spread along the surface of heavy-timber members is reduced, and without concealed blind spaces, there is no opportunity for fire to smolder and spread undetected.

In accordance with Table 601 and Section 602.4, modern-day heavy-timber construction can be a mixture of heavy-timber floor and roof construction and 1-hour fire-resistance-rated bearing walls and partitions. Although heavy-timber construction is not generally recognized as equivalent to 1-hour fire-resistance-rated construction, the code considers heavy timber to provide equivalent protection in Type IV buildings.

In keeping with the concept of slow-burning construction by means of wood members with large cross sections, the IBC specifies minimum nominal dimensions for wood members used in heavy-timber construction. As the code specifies the size of members as nominal sizes, the actual net surfaced sizes may be used. For example, an 8-inch by 8-inch (203-mm by 203-mm) member nominally will actually be a net size of $7^1/_4$ inches by $7^1/_4$ inches (185 mm by 185 mm). Therefore, even though the code calls for a nominal 8-inch by 8-inch (203-mm by 203-mm) member, the net $7^1/_4$-inch by $7^1/_4$-inch (185-mm by 185-mm) member meets the intent of the code. As indicated earlier, the minimum sizes for heavy-timber construction as listed in this section are based on experience and the good behavior in fire of heavy-timber construction.

Wherever framing lumber or sawn timber is specified, structural glued-laminated timber may also be used, as all have the same inherent fire-resistive capability. However, because solid sawn wood members and glued-laminated timbers are manufactured with different methods and procedures, they do not have the same dimensions. Table 602.4 compares the solid sawn sizes with those of glued-laminated members to indicate equivalency in regard to compliance with the Type IV construction criteria.

Section 602.4.6 specifies that partitions shall be of either solid-wood construction or 1-hour fire-resistance-rated construction. However, various provisions of the code address the use of fire partitions and fire barriers. In these cases, the fire-resistant-rated fire partitions or fire barriers in heavy-timber buildings should be constructed as required by the code for the required rating. For example, where there is a requirement for fire-resistance-rated corridors in heavy-timber buildings, 1-hour fire-resistance-rated construction must be used rather than solid-wood construction for the partitions.

It is highly unusual for any building designed and constructed today to be considered compliant as a Type IV structure. As previously addressed, in order for a building to be properly classified, all portions must be in conformance with the established criteria. Many buildings may have some heavy-timber elements that qualify as Type IV; however, the floor construction and/or roof construction does not fully comply with the prescriptive requirements of Sections 602.4.4 and 602.4.5, respectively. In such cases, the building cannot be classified as Type IV. Such buildings are most likely Type III or V construction. However, even if the building as a whole is not considered a Type IV structure, the recognition of individual heavy-timber elements is very important. For example, the provisions of Section 705.2.3 recognize Type IV heavy-timber projections for use in locations where unprotected combustible construction is not permitted. For this and other reasons, the requirements for Type IV heavy timber must be fully understood.

Type V. Type V buildings are essentially construction systems that will not fit into any of the other higher types of construction and may be constructed of any materials permitted by the code. The usual example of Type V construction is the light wood-frame building consisting of walls and partitions of 2-inch by 4-inch (51-mm by 102-mm) or 2-inch by 6-inch (51-mm by 152-mm) wood studs. The floor and ceiling framing are usually of light wood joists of 2-inch by 6-inch (51-mm by 152-mm) size or deeper. Roofs may also be framed with light wood rafters of 2-inch by 4-inch (51-mm by 102-mm) size or deeper

602.5

cross sections or, as is now quite prevalent, framed with pre-engineered wood trusses of light-frame construction. Wood-frame Type V buildings may be constructed with larger framing members than just described, and these members may actually conform to heavy-timber sizes. Such structures sometimes have a limited number of noncombustible building elements. However, unless the building complies in all respects to one of the other four basic types of construction, it is still a Type V building.

Type V construction is divided into two subtypes:

1. Type VA. This is protected construction and required to be of 1-hour fire-resistance-rated construction throughout.

2. Type VB. This type of construction has no general requirements for fire resistance and may be of unprotected construction, except where Section 602.1 and Table 602 require exterior wall protection because of proximity to a lot line.

Table 601—Fire-Resistance Rating Requirements for Building Elements. This table provides the basic fire-resistance rating requirements for the various types of construction. It also delineates those fire-resistance ratings required to qualify for a particular type of construction. As previously discussed, even though a building may have some features that conform to a higher type of construction, the building shall not be required to conform to that higher type of construction as long as a lower type will meet the minimum requirements of the code based on occupancy. Nevertheless, any building must comply with all the basic fire-resistance requirements in this table if it is indeed the intent to classify it for that particular type of construction. For example, in order for a building to be classified as Type IIA noncombustible construction, a minimum 1-hour fire-resistance rating is required for any and all structural frame members, bearing walls, floor construction, and roof construction. In addition, and with limited exceptions, all of these building elements must be recognized as noncombustible.

Footnote a. Limited to buildings of Type I construction, the fire-resistance ratings of primary structural frame elements and interior bearing walls supporting only a roof may be reduced by 1 hour. In other words, primary structural-frame members or interior bearing walls providing only roof support shall have a minimum fire-resistance rating of 2 hours in Type IA buildings and 1 hour in Type IB construction. Additional provisions addressing the protection of certain primary structural frame members are found in Section 704.

Footnote b. This footnote, an exception to the general rule for roof construction, addresses those situations where the roof and its components are 20 feet (6,096 mm) or more above any floor immediately below. Under these circumstances, the roof and its components, including roof framing and decking, may be of unprotected construction. The reduction of the fire-resistance rating would apply to buildings of Type IA, IB, IIA, IIIA, and VA construction. The footnote mandates that all portions of the roof construction must be located at or above the 20-foot (6,096-mm) height requirement. For example, in a sloped roof condition, it is not acceptable to merely protect those portions below the 20-foot (6,096-mm) point and leave the remainder unprotected. See Figure 601-1. It is important to note that the elimination of any required fire-resistance rating is not applicable to elements of the roof construction considered as primary structural frame members.

The reduction in rating applies to all occupancies other than Groups F-1, H, M, and S-1, where fire loading is typically higher. In all occupancies other than those just listed, the relaxation of the requirements is based on the fact that where the roof is at least 20 feet (6,096 mm) above the nearest floor, the temperatures at this elevation during most fire incidents are quite low. As a result, fire protection of the roof and its members, including the structural frame, is not necessary. For those occupancies where the fire loading and the consequent potential fire severity is relatively high, such as factory-industrial, hazardous, mercantile, or storage uses, the code does not permit a reduction in roof protection. It is also quite common in these occupancies for combustible or hazardous materials to be located in close proximity to the roof structure, as in the case of high-piled storage.

Construction Classification 6

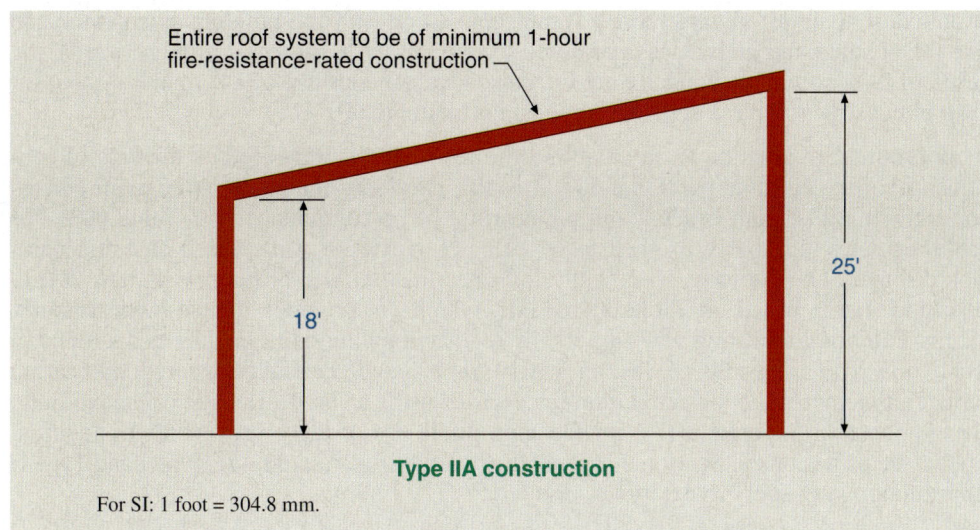

Figure 601-1
Fire-resistive ratings for roof construction.

Footnote c. Applicable to Types IB, IIA, IIB, IIIA, and VA construction, the code permits heavy-timber members complying with Section 602.4 to be used in the roof construction without any fire-resistance rating as required by the table. It is assumed that roof members sized and constructed in compliance with the details of heavy-timber construction are equivalent to roof construction having a 1-hour fire-resistance rating. In addition, heavy-timber members are permitted to be utilized in the roof construction of an otherwise noncombustible Type IB, IIA, or IIB building.

Footnote d. In this footnote, the code permits, for Type IIA, IIIA, and VA buildings, the substitution of an automatic fire-sprinkler system for the 1-hour fire-resistance-rated construction, provided the sprinkler system is not otherwise required by other provisions of the IBC. The footnote cannot be applied where the sprinkler system has been utilized for allowable height or area increases under Sections 504.2 and 506.3.

It should be noted that the substitution of an automatic fire-sprinkler system under this provision does not waive or reduce the required fire-resistance rating of exterior walls under the provisions of this footnote. Such a component represents a specific fire-resistive requirement to counter specific hazards, whereas, on the other hand, the 1-hour fire-resistance-rated construction required for Type IIA, IIIA, and VA buildings applies generally to the entire building.

In the final analysis, the IBC intends through this footnote that a building utilizing the automatic fire-sprinkler system as a substitution for 1-hour fire-resistance-rated construction may still be classified as a Type IIA, IIIA, or VA building.

It is not entirely clear what the phrase "not otherwise required by other provisions of the code" is intended to address. The most conservative approach would be based on Section 901.2, which states that "any fire protection system for which an exception or reduction to the provisions of this code has been granted shall be considered to be a required system." Therefore, the fire-resistive-substitution provisions of Footnote d would not be applicable where the sprinkler is used to modify or eliminate a code requirement. With the wide variety of allowances in the IBC for a sprinklered building, the use of this footnote would seem to be quite limited.

It is also clarified that when fire sprinklers are used for 1-hour substitution, any increase in area or height that is due to the presence of an automatic sprinkler system is not permitted. An automatic sprinkler system may not be used to increase both the area and the height, and still be used to eliminate the 1-hour fire-resistance-rated construction—but one or the other use may be selected, at the designer's option. Where fire sprinklers are installed, it is almost always more advantageous to classify the building as nonrated and use the benefits and allowances permitted by the IBC in Chapter 5. In other words, the result of this footnote is typically not the most advantageous use of a sprinkler system for type of construction purposes. See Application Example 601-1.

Footnote f. In addition to any required fire-resistance rating based on the type of construction of the building per Table 601, it is also necessary that such rating requirements for exterior walls, both bearing and nonbearing, be in compliance with Table 602. The table regulates the hourly fire-resistance ratings for exterior walls based on fire separation distance. This footnote specifically indicates that exterior bearing walls have a fire-resistance rating based on Table 601 or 602, whichever provides for the highest hourly rating. Exterior nonbearing walls are totally regulated by the rating requirements found in Table 602. The provisions of Section 704.10 must also be consulted where load-bearing structural members are located within the exterior walls or on the outside of the building. See Application Example 601-2 for the appropriate use of these provisions. In addition, applicable provisions in Section 603 regulating combustible material in Type I and Type II construction may apply to exterior walls.

Table 602—Fire-Resistance Rating Requirements for Exterior Walls Based on Fire Separation Distances.

The IBC, as far as exterior wall protection is concerned, operates on the philosophy that an owner can have no control over what occurs on an adjacent lot and, therefore, the location of buildings on the owner's lot must be regulated relative to the lot line. In fact, the location of all buildings and structures on a given piece of property is addressed in relation to the real lot lines as well as any assumed or imaginary lines between buildings on the same lot. The assumption of imaginary lines is discussed with other exterior wall provisions in Section 705.

The lot-line concept provides a convenient means of protecting one building from another insofar as exposure is concerned. Exposure is the potential for heat to be transmitted from one building to another under conditions in the exposing building. Radiation is the primary means of heat transfer.

Application Example 601-1

GIVEN: Fully sprinklered Group B office building of nonrated combustible construction.

DETERMINE: Maximum allowable height and area.

- Constructed as VB, using permitted sprinkler increases of Chapter 5
- Classified as Type VB
- Determination based on:
  - Table 503
  - Section 504.2 (height increase)
  - Section 506.3 (area increase)
- 9,000 sq ft, 2 stories (T503)
  - 18,000 sq ft increase (200 percent)
  - 1 story height increase
- ∴ Maximum height – 3 stories
  Maximum area – 27,000 sq ft/story

- Constructed as VB, using sprinkler system to classify as Type VA
- Classified as Type VA
- Determination based on:
  - Table 503
  - Table 601, Footnote d (increases of Sections 504.2 and 506.3 not applicable)
- 18,000 sq ft, 3 stories (T503)
  - No further increases permitted
- ∴ Maximum height – 3 stories
  Maximum area – 18,000 sq ft/story

For SI: 1 square foot = 0.093 m².

Thus, it is more advantageous to ignore use of Footnote d.

TYPE OF CONSTRUCTION DETERMINATION

Construction Classification

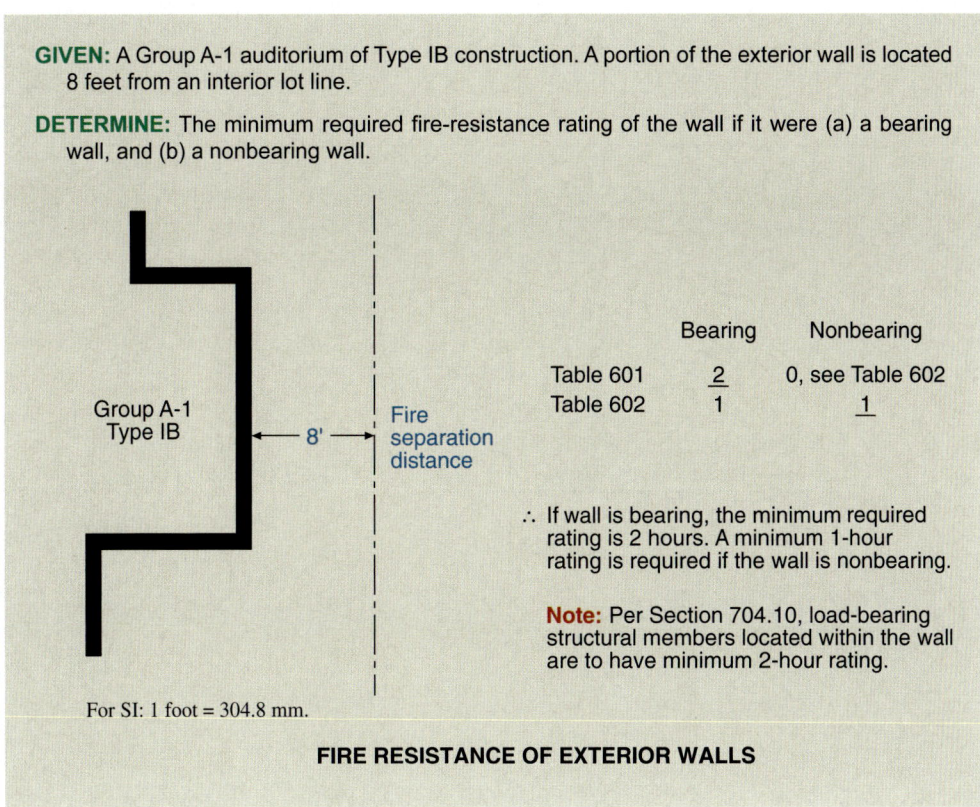

Application Example 601-2

FIRE RESISTANCE OF EXTERIOR WALLS

The code specifically provides that the fire separation distance be measured to the center line of a street, alley, or public way. As the code refers to public way, this would also be applicable to appropriate open spaces other than streets or alleys that the building official may determine are reasonably likely to remain unobstructed through the years.

The regulations for exterior wall protection based on proximity to the lot line are contained in Table 602. The IBC indicates that the distances are measured at right angles to the face of the exterior wall (see definition of "Fire separation distance" in Section 202), which would result in the fire-resistive requirements for exterior walls not applying to walls that are at right angles to the lot line. See Figure 602-1.

In order to properly utilize Table 602, it is necessary to identify the fire separation distance, the occupancies involved, and the building's type of construction. As the fire separation distance increases, the fire-resistance rating requirements are reduced, based on the occupancy group under consideration. Figure 602-2 illustrates the application of exterior wall protection where the exterior walls of the building are parallel and perpendicular to the lot line. In this case, the illustration assumes that the building is one story of Type VB construction and used for offices (Group B). Referring to Table 602, it is noted that exterior walls less than 10 feet (3,048 mm) from the lot line must be of minimum 1-hour fire-resistance-rated construction. Figure 602-3 depicts a similar building located such that the exterior walls are not parallel and perpendicular to the lot line, but are at some angle other than 90 degrees (1.57 rad). The regulation of openings in exterior walls is set forth in Section 704.8. Several footnotes to the table address modifications to the general requirements. Footnote a repeats a previous requirement that load-bearing exterior walls must comply with both Tables 601 and 602. Footnote b refers the user to Section 406.1.2 for exterior wall requirements for Group U occupancies. Where used solely for the parking of private motor vehicles, a Group U garage or carport is regulated by either Item 1

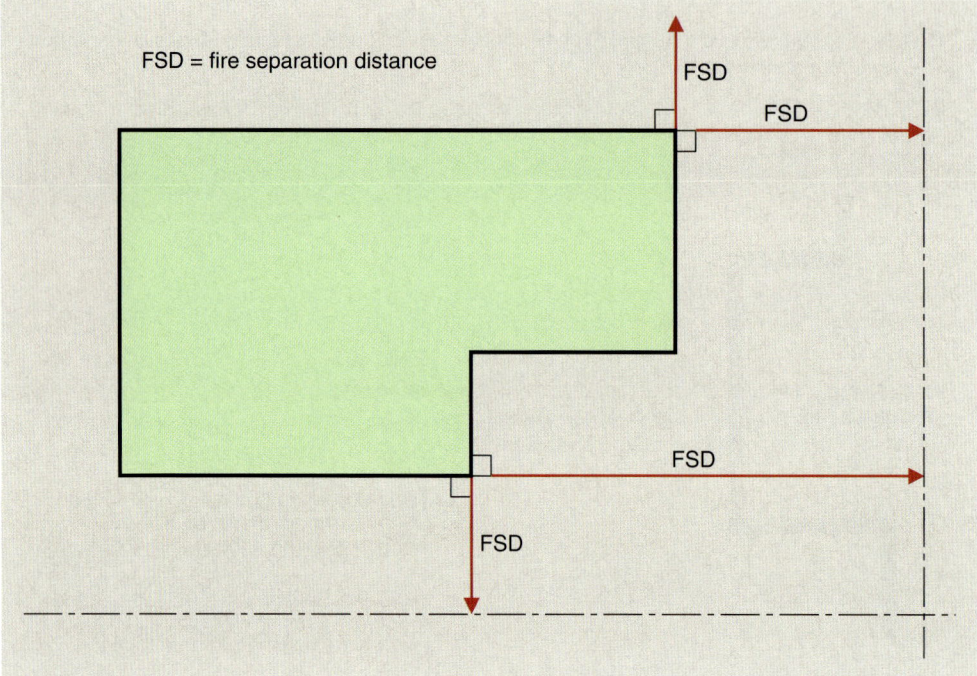

Figure 602-1 Fire separation distance.

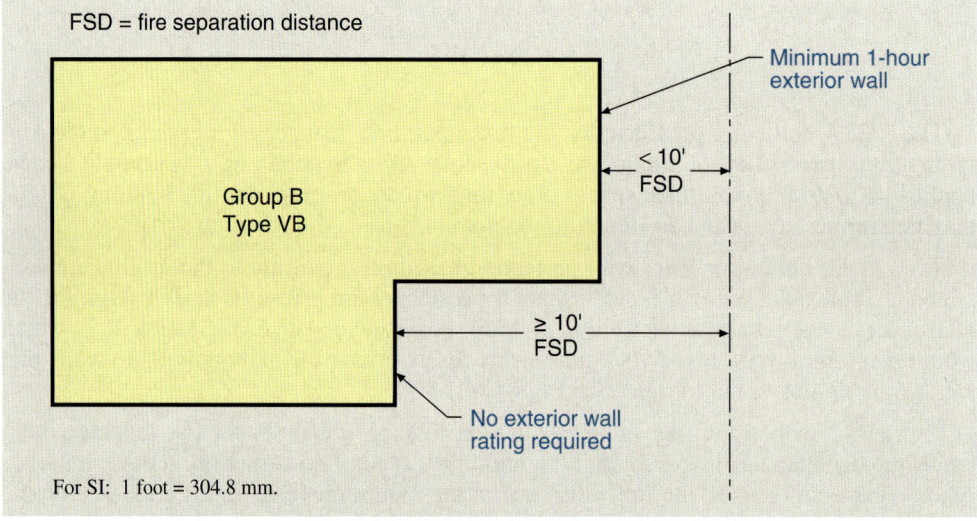

Figure 602-2 Exterior wall rating.

(mixed-occupancy building) or Item 2 (building only contains the Group U) of Section 406.3.2 rather than by Table 602.

Although Table 602 requires a Group S-2 occupancy of Type I, II, or IV construction to have a minimum 1-hour exterior wall where the fire separation distance is less than 30 feet (9,144 mm), Footnote d reduces that distance significantly where it is a complying open parking garage. Under such conditions, a minimum 1-hour fire-resistance-rated exterior wall is required only where the fire separation distance is less than 10 feet. Footnote e indicates that each story of the building is regulated independently for the fire separation distance provisions, as shown in Figure 602-4.

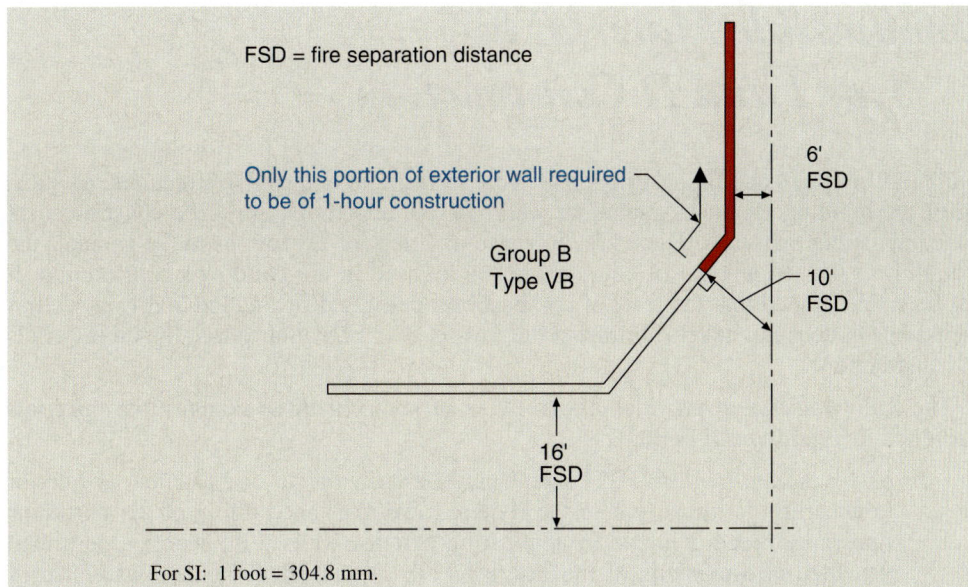

Figure 602-3
Exterior wall rating.

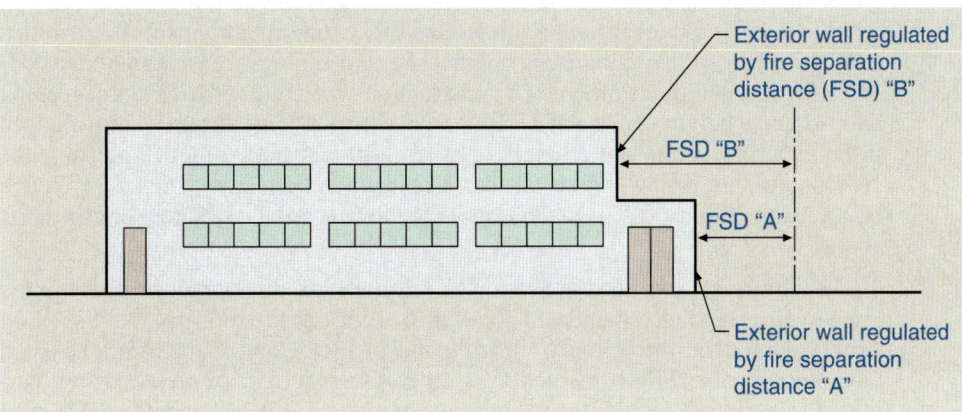

Figure 602-4
Fire separation distance measurement.

Table 705.8 allows for an unlimited amount of unprotected openings in exterior walls of a sprinklered building that has a fire separation distance of at least 20 feet (6,096 mm). However, in certain buildings, Table 602 requires those same exterior walls to be fire-resistance rated for a minimum of 1 hour. Footnote g recognizes that any nonbearing exterior wall permitted to be entirely open due to the unlimited unprotected opening allowances of Table 705.8 need not be required to have a fire-resistance rating due to fire separation distance.

There are only a small percentage of buildings where this footnote is applicable. It has no effect on:

- Exterior bearing walls
- Group H-1, H-2, and H-3 occupancies
- Nonsprinklered buildings
- Buildings of Type IIB and VB construction, other than Groups H-4 and H-5
- Exterior walls with a fire separation distance of less than 20 feet (6,096 mm)
- Exterior walls with a fire separation distance of 30 feet or more (9,144 mm)

6 Types of Construction

Section 603 *Combustible Material in Type I and II Construction*

Buildings of Type I and II construction are considered noncombustible structures. As such, all of the building elements, including walls, floors, and roofs, are to be constructed of noncombustible materials. There are, however, a variety of exceptions to the general rule that allow a limited amount of combustibles to be used in the building's construction. It has been determined that the level of combustibles permitted by Section 603.1, as well as their control, does not adversely impact the fire-severity potential caused by the materials of construction.

The following listing provides an overview of some of those combustible materials permitted in Type I and II buildings:

1. Fire-retardant-treated (FRT) wood may be used in the construction of interior nonbearing partitions where the required fire-resistance rating of the partitions does not exceed 2 hours. In nonbearing exterior walls, FRT wood is permitted provided no fire rating of the exterior walls is mandated. Roofs constructed of FRT wood are also acceptable in most buildings. This would include roof girders, trusses, beams, joists, or decking, as well as blocking, nailers, or similar components that may be a part of the roof system. Where the building is classified as other than Type IA construction, the use of FRT wood roof elements is permitted in all cases, regardless of building height. The same allowance is permitted in one- and two-story buildings of Type IA construction. For Type IA buildings exceeding two stories in height, the use of FRT wood in the roof construction is only allowed if the uppermost story has a height of at least 20 feet (6,096 mm). Logically, the 20-foot measurement would be taken in a manner consistent with that described in Footnote b of Table 601, from the floor to the lowest point of the roof construction above.

    The allowances provided in Section 603.1, Item 1.3, do not reduce any required level of fire resistance mandated for wall or roof construction as established by Table 601. Rather, they simply allow the use of FRT wood in the locations listed where noncombustible construction is otherwise required. In reviewing the permitted use of FRT wood, there are two obvious building elements where such materials are not permitted in Type I or II construction. In Type I or II buildings, FRT wood is not permitted to be used in the floor construction and any bearing wall assemblies.

2. Combustible insulation used for thermal or acoustical purposes is acceptable, provided the flame-spread index is limited. Additional regulations addressing the use of thermal- and sound-insulating materials within buildings are found in Section 720.

3. Foam plastics installed under the limitations of Chapter 26 are permitted, as are roof coverings having an A, B, or C classification as specified in Section 1505.

4. Wood doors, door frames, window sashes and frames, trim, and other combustible millwork and interior surface finishes are acceptable, as is blocking for handrails, grab bars, cabinets, window and door frames, wall-mounted fixtures, and similar items. Combustible stages and platforms are also permitted when complying with Section 410, and wood-finish flooring may be used when applied directly to the floor slab or installed over wood sleepers and fireblocked in accordance with Section 805.1.2.

Combustible Material in Type I and II Construction

Another allowable use of combustible elements in noncombustible buildings, detailed in Item 11, addresses the situation where nonbearing partitions divide portions of stores, offices, or similar spaces occupied by one tenant only. The key words in this item are "occupied by one tenant only." It is the intent of the IBC that this expression applies to an area or building that is under the complete control of one person, organization, or other occupant. This would be contrasted to multitenant occupancies, where the various tenant spaces in the building would be under the control of two or more individuals, companies, or occupants. In such a multitenant space, the walls common to the public areas and to other tenants would not be regulated under this allowance. However, within each of the tenant spaces, those nonbearing walls and partitions not common with other tenants or public areas could utilize the optional construction methods of Item 11.

A multistory building owned by a large company would also qualify as being occupied by one tenant only if the large company that owned the building also completely occupied the building. A government office building owned by a city or county occupied by several departments of the government would also be considered occupied by one tenant only. If the government office building contains an assembly room, the assembly room itself would not qualify for the special provisions of Section 603.1 unless accessory to the office use. These exceptions typically apply only to stores, offices, and similar uses. The intent of these provisions is to provide exceptions to the construction requirements of Chapter 6. Thus, one of the three types of partitions addressed in Item 11 of Section 603 could be constructed regardless of other requirements of Chapter 6 regulating the construction of partitions. For example, a 1-hour fire-resistance-rated combustible partition constructed of ordinary wood studs would be permitted to be installed in a Type IA building in accordance with the provisions of this section.

These provisions are based on the common practice in offices and stores to create large, open areas. When subdivided, low-height partitions are often utilized. Sometimes a few areas are completely partitioned off with full ceiling-height partitions in order to provide privacy or security for storage, etc. As these partitions are nonbearing and are subject to being moved to create various space configurations, the code permits the modification. Except for the partial-height panels of light construction, the other permitted partitions do provide some type of barrier to the spread of fire. They are to be constructed of FRT wood or be of a minimum 1-hour fire-resistance-rated construction. In the case of the partial-height partitions, the concept is that being only a portion of the room height, persons in one portion of the area are aware of what is going on in the other portions, and if a fire develops, the occupants would be aware of that fact and take appropriate action.

It should also be noted that combustible partitions permitted by the code under this provision are not to be used in the construction of corridor walls where the corridor serves an occupant load of 30 or more. The allowance for combustible materials is intended to address types of construction concerns; therefore, construction elements of a corridor must comply with the general requirements.

Reference is also made under Item 24 to Section 718.5 for the allowance of specific combustible elements within concealed spaces. The allowance for combustible items in concealed spaces is limited because of the increased potential for fire spread. Therefore, the flame spread index and smoke-developed index of the permitted items are often highly regulated. Combustible piping is permitted to be installed within partitions, shaft enclosures, and concealed ceiling spaces of noncombustible buildings. Various combustible materials are also permitted in plenums of Type I and II buildings, including wiring, fire-sprinkler piping, pneumatic tubing, and foam plastic insulation under the limitations imposed by Section 602 of the *International Mechanical Code®* (IMC®).

6 Types of Construction

KEY POINTS

- Buildings are classified in general terms as combustible or noncombustible, as well as protected or unprotected.
- Table 601 identifies the required fire-resistance ratings of building elements based on the specified type of construction.
- Unless a fire wall is utilized, structures can be classified into only one type of construction.
- The structural frame is regulated in a manner apart from that of walls, floors, and roofs.
- Type I and II buildings are considered noncombustible (required), whereas Type III, IV, and V buildings are viewed as combustible (permitted) construction.
- Very few structures fully comply with the provisions for heavy-timber construction; however, many buildings contain some Type IV elements.
- Type V buildings are by far the most common type of construction.
- Various reductions in fire resistance are permitted for nonbearing partitions.
- Table 602 regulates the protection of exterior walls insofar as exposure to an adjacent building is concerned.
- Combustible materials identified in Section 603 are permitted in otherwise noncombustible construction (Types I and II).

CHAPTER 7

FIRE AND SMOKE PROTECTION FEATURES

Section 703 Fire-Resistance Ratings and Fire Tests
Section 704 Fire-Resistance Rating of Structural Members
Section 705 Exterior Walls
Section 706 Fire Walls
Section 707 Fire Barriers
Section 708 Fire Partitions
Section 709 Smoke Barriers
Section 710 Smoke Partitions
Section 711 Horizontal Assemblies
Section 712 Vertical Openings
Section 713 Shaft Enclosures
Section 714 Penetrations
Section 715 Joint Systems
Section 716 Opening Protectives
Section 717 Ducts and Air Transfer Openings
Section 718 Concealed Spaces
Section 719 Fire-Resistance Requirements for Plaster
Section 720 Thermal- and Sound-Insulating Materials
Section 721 Prescriptive Fire Resistance
Section 722 Calculated Fire Resistance
Key Points

7 Fire and Smoke Protection Features

> The types of construction and the fire-resistance requirements of the *International Building Code®* (IBC®) are based on the concept of fire endurance. Fire endurance is the length of time during which a fire-resistive construction assembly will confine a fire to a given area, or continue to perform structurally once exposed to fire, or both. In the IBC, the fire endurance of an assembly is usually expressed as a "___-hour fire-resistance-rated assembly." Chapter 7 prescribes test criteria for the determination of the fire-resistance rating of construction assemblies and components, details of construction of many assemblies and components that have already been tested, and other information necessary to secure the intent of the code as far as the fire resistance and the fire endurance of construction assemblies and components are concerned. Additionally, Chapter 7 addresses other construction items that must be incorporated into a building's design in order to safeguard against the spread of fire and smoke.

Section 703 *Fire-Resistance Ratings and Fire Tests*

It is the intent of the IBC that materials and methods used for fire-resistance purposes are limited to those specified in this chapter. Materials and assemblies tested in accordance with ASTM E 119 or UL 263 are considered to be in full compliance with the code, as are building components whose fire-resistance rating has been achieved by one of the alternative methods specified in Section 703.3.

703.2 Fire-resistance ratings. This section indicates that building elements are considered to have a fire-resistance rating when tested in accordance with the procedures of ASTM E 119 or UL 263. Figures 703-1 through 703-5 depict the fundamental testing requirements of the two standards. The intent of the IBC is that any material or assembly that successfully passes the end-point criteria depicted for the specified time period shall have its fire-endurance rating accepted and the assembly classified in accordance with the time during which the assembly successfully withstood the test.

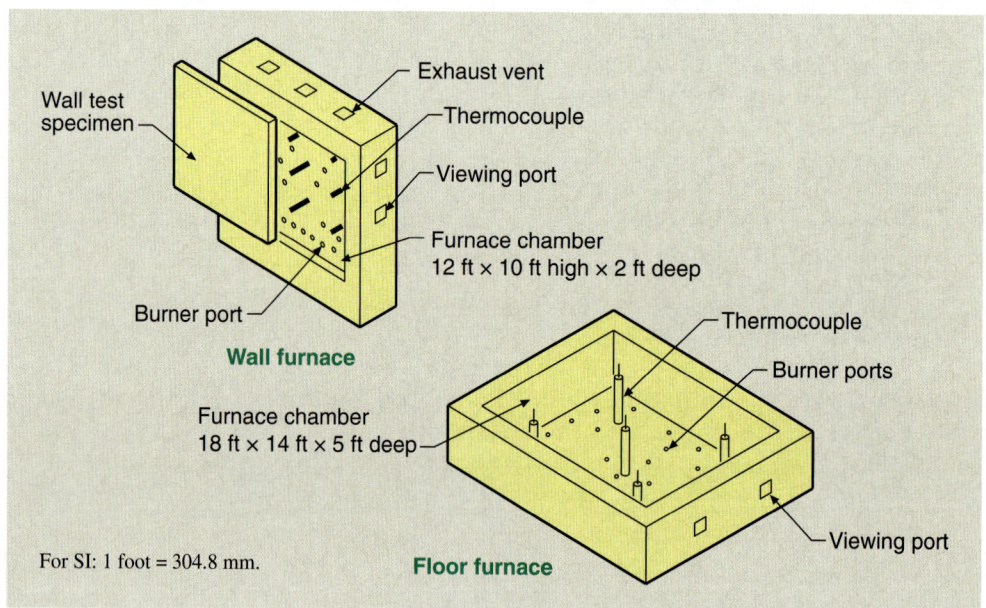

Figure 703-1
Test furnaces.

Fire-Resistance Ratings and Fire Tests

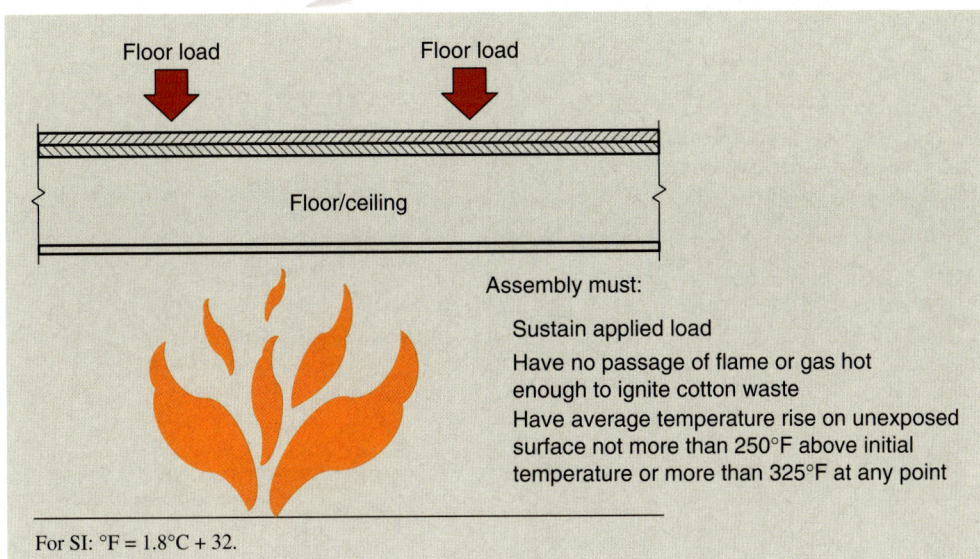

Figure 703-2
Floor assembly fire test.

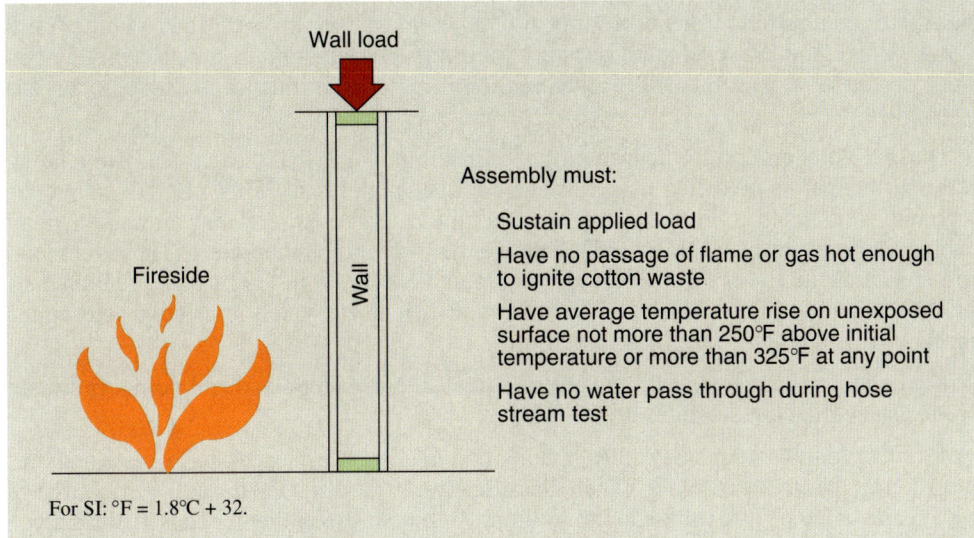

Figure 703-3
Conditions of acceptance—wall fire test.

Although early fire testing in the United States began as long ago as the 1890s, the standard fire-endurance test procedure using a standard time-temperature curve and specifying fire-endurance ratings in hours was developed in 1918. The significance of 1918 and later standards is the fact that they were and are intended to be reproducible so that the test conducted at Underwriters Laboratories (UL) can be compared with the test of the same assembly conducted at the University of California, Ohio State University, or other testing facility. An often-expressed criticism of a standard such as ASTM E 119 or UL 263 is that "it does not represent the real world." This is true in many cases, and for that reason it should not be thought of as representing the absolute behavior of a fire-resistance-rated assembly under most actual fires in buildings. There are too many variables that affect the fire endurance of an assembly during an actual fire, such as fuel load, room size, rate of oxygen supply, and restraint, to consider that the test establishes absolute values of the real-world fire endurance of an assembly. However, it is a severe test of the fire-resistive qualities of a material or an assembly, and because of its reproducibility, it provides a means of comparing assemblies.

7 Fire and Smoke Protection Features

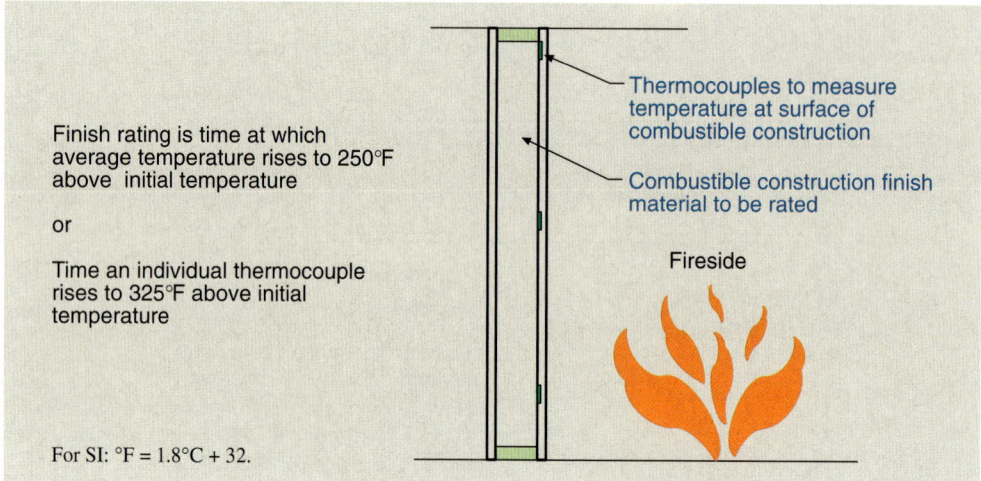

Figure 703-4
Combustible assembly for determining finish rating.

In addition to the fire-endurance fire ratings obtained from the standard fire tests of ASTM E 119 and UL 263, it is also possible to obtain, as expressed in the standard, the protective membrane performance for walls, partitions, and floor or roof assemblies. In the case of combustible walls or floor or roof assemblies, it is also referred to as the finish rating. Although the test standard does not limit the determination of the protective membrane performance to combustible assemblies, its greatest significance is with combustible assemblies.

The end-point criteria for determining the finish rating are that the average temperature at the surface of the protected materials shall not be greater than 250°F (121°C) above the beginning temperature. Furthermore, the maximum temperature at any measured point shall not be greater than 325°F (163°C) above the beginning temperature. These temperatures relate to the lower limit of ignition temperatures for wood. Figure 703-4 illustrates the determination of the finish rating for a wall assembly, which is usually determined during a fire-endurance test of the assembly.

The condition of acceptance, also referred to as failure criteria and end-point criteria, of fire-resistance-rated assemblies are as follows:

1. For load-bearing assemblies, the applied load must be successfully sustained during the time period for which classification is desired. There shall be no passage of flame or gases hot enough to ignite cotton waste on the unexposed surfaces.

2. The average temperature rise on the unexposed surface shall not be more than 250°F (121°C) above the initial temperature during the time period of the test.

3. The maximum temperature on the unexposed surface shall not be more than 325°F (163°C) above the initial temperature during the time period of the test.

4. Walls or partitions shall withstand the hose-stream test without passage of flame or gases hot enough to ignite cotton waste on the unexposed side or the projection of water from the hose stream beyond the unexposed surface.

In addition to the conditions of acceptance just described, load-carrying structural members in roof and floor assemblies are subject to special end-point temperatures for:

1. Structural steel beams and girders—1,100°F (593°C) average at any cross section and 1,300°F (704°C) for any individual thermocouple, for unrestrained assemblies.

2. Reinforcing steel in cast-in-place reinforced concrete beams and girders—1,100°F (593°C) average at any section.

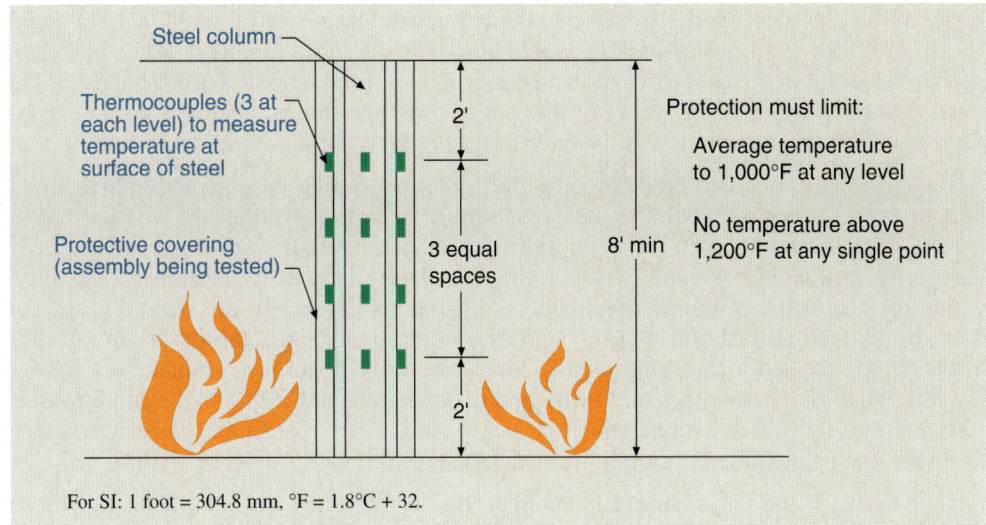

Figure 703-5
Alternative fire test of steel column protection.

3. Prestressing steel in prestressed concrete beams and girders—800°F (427°C) average at any section.

4. Steel deck floor and roof units—1,100°F (593°C) average on any one span.

As columns are exposed to fire on all surfaces, the standard has special temperature and testing criteria for these members:

1. The column is loaded so as to develop (as nearly as practicable) the working stresses contemplated by the structural design. The condition of acceptance is simply that the column sustain the load for the duration of the test period for which a classification is desired.

2. Alternatively, a steel column may be tested without load, and the column will be tested in the furnace to determine the adequacy of the protection on the steel column. The test and end points are depicted in Figure 703-5.

The exception to this section is intended to modify the acceptance criteria for exterior bearing walls so that the walls will receive a rating based on which of the two following sets of criteria occurs first during the test:

1. Heat transmission or flame and hot gases transmission for nonbearing walls.

2. Structural failure or hose-stream application failure.

The first set of end points measures the wall's ability to prevent the spread of fire from one side to the opposite side. It is considered overly restrictive to require that exterior bearing walls comply with this first set of end points for a longer time than would be required for a nonbearing wall located at the same distance from the lot line if it is still structurally capable of carrying the superimposed loads.

703.2.1

Nonsymmetrical wall construction. At times, an interior wall or partition is constructed nonsymmetrically as far as its fire protection is concerned, with the membrane on one side of the wall differing from that on the opposing side. Where the wall is to be fire-resistance rated, it must be tested from both sides in order to determine the fire-resistance rating to be assigned to the assembly. Based on the two tests, the shortest time period is determined to be the wall's rating. An assembly tested from only one side may be approved by the building official, provided there is adequate evidence furnished to show that the wall was tested with the least fire-resistive side exposed to the furnace. The provisions for exterior walls of nonsymmetrical construction differ somewhat from those addressing interior walls and are regulated by Section 705.5.

7 Fire and Smoke Protection Features

703.2.3 Restrained classification. A dual classification system is used in ASTM E 119 and UL 263 for roof and floor assemblies, including their structural members. This dual classification system involves the use of the terms *restrained* and *unrestrained*. The use of the word *restrained* entails the concept of thermal restraint (restrained against thermal expansion as well as against rotation at the ends of an assembly or structural member).

For example, if a structural beam of a uniform cross section is subjected to heat on its bottom surface, such as would be the case in the standard test furnace, it will attempt to expand in all directions with the longitudinal expansion being the primary component. If the beam is restrained at the ends so that it cannot expand, compressive stresses will build up within the beam, and it will in effect behave in similar fashion as a prestressed beam. As a result, the thermal restraint will be beneficial in terms of improving the beam's ability to sustain the applied load during the fire test. If the same beam is restrained only for the lower one-half of its cross section, it will tend to deflect upward owing to the conditions of restraint. This upward deflection tendency is also considered to enhance the beam's ability to sustain the applied load during a fire-endurance test.

Conversely, if the end restraint is applied only to the upper half of the beam's cross section, the beam will tend to deflect downward and, in this case, the restraint will be detrimental to the beam's ability to sustain the applied load during the fire-endurance test. As the heat is applied to the bottom surface during a fire, it creates a downward deflection, and the two downward deflections are additive. In an actual building, this could lead to premature failure. It can be seen, then, that thermal restraint may be either beneficial or detrimental to the fire-resistant assembly, depending on its means of application in the building.

General guidance for the building official is provided in ASTM E 119 and UL 263 as to what conditions in the constructed building provide restraint. It is generally agreed that an interior panel of a monolithically cast-in-place reinforced-concrete floor slab would be considered to have thermal restraint. Also, Footnote k to Table 721.1(1) provides that "interior spans of continuous slabs, beams and girders may be considered restrained." Conversely, because the restraint present in many construction systems cannot be determined so neatly, the IBC requires that these assemblies be considered unrestrained unless the registered design professional shows by the requisite analysis and details that the system qualifies for a restrained classification. Furthermore, the code requires that any construction assembly that is to be considered restrained be identified as such on the drawings.

703.3 Alternative methods for determining fire resistance. In addition to those assemblies and materials considered fire-resistance-rated construction based on compliance with ASTM E 119 or UL 263, a number of alternative methods for determining fire resistance are set forth in this section. Where it can be determined that the fire-resistance rating of a building element is in conformance with one of the five listed methods or procedures, such a rating is considered acceptable.

703.4 Automatic sprinklers. As a general rule, the fire-resistance ratings of building elements, components, and assemblies established through the code are to be determined in accordance with the test procedures set forth in ASTM E 119 or UL 263. In addition, alternative methods for determining fire resistance listed in Section 703.3 are acceptable where such methods are based on the fire exposure and acceptance criteria specified in ASTM E 119 or UL 263. A fire suppression system is not permitted to be included as part of the tested element, component, or assembly in order to establish the fire-resistance rating. It has been generally accepted that the various fire-resistance ratings mandated throughout the code have been established based on an assumption that the fire assembly would pass the standardized tests without the assistance of water cooling during fire exposure. This provision clarifies the assumption.

It is important to note that these provisions are not intended to limit the use of Section 104 by building officials for the approval of alternative methods on a case-by-case basis. While the

Fire-Resistance Ratings and Fire Tests   7

prescriptive provisions of the code are based on fire-resistance ratings established without the benefit of any automatic fire-suppression system, the building official has the authority to evaluate and approve alternative materials, designs, and methods of construction that meet the intent and purpose of the code.

Noncombustibility tests. Throughout the IBC, particularly in Chapter 6, the terms *combustible* and *noncombustible* are used. Under many different conditions, limits are placed on the use of combustible building materials, particularly in buildings of Type I or II construction. This section sets forth the two methods for determining if a material is noncombustible.   **703.5**

For most materials, ASTM E 136 is the test standard used to determine if a material is noncombustible. Composite materials such as gypsum board are also considered noncombustible if they comply with the criteria of Section 703.5.2. Such materials must have a structural base of noncombustible materials with a surfacing limited in thickness and flame spread.

Note that the term *noncombustible* does not apply to surface finish materials.

Fire-resistance-rated glazing. The use of fire-resistance-rated glazing typically only occurs where the limitations placed on fire-protection-rated glazing make it undesirable or impractical. Fire-resistance-rated glazing is subjected to the ASTM E 119 or UL 263 testing criteria, which includes stringent limitations on temperature rise through the assembly. Because the glazing is regulated as a wall assembly rather than an opening protective, its use is not limited by any of the provisions of Section 716. It is only regulated under the appropriate code requirements for a fire-resistance-rated wall assembly.   **703.6**

The labeling requirements specific to fire-resistance-rated glazing are set forth in Table 716.3. The table indicates that glazing intended to meet the wall assembly criteria be identified with the marking "W-XXX." The "W" indicates that the glazing meets the requirements of ASTM E 119 or UL 263, thus qualifying the glazing to be used as part of a wall assembly. It also indicates that the glazing meets the fire-resistance, hose-stream, and temperature-rise requirements of the test standard. The fire-resistance rating of the glazing will then follow the "W" designation. See Figure 703-6.

Marking and identification. The integrity of fire and/or smoke separation walls is subject to compromise during the life of a building. During maintenance and remodel activities, it is not uncommon for new openings and penetrations to be installed in a fire or smoke   **703.7**

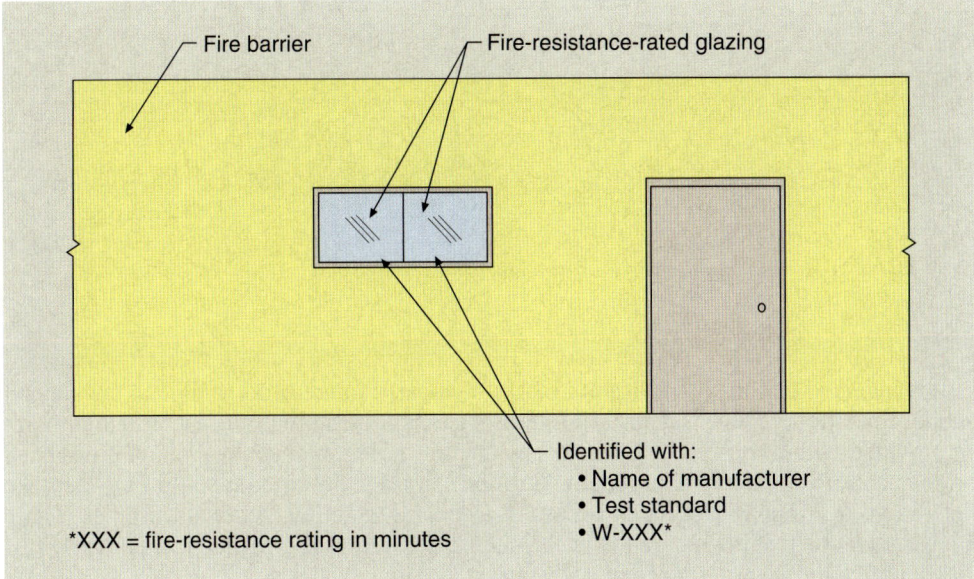

Figure 703-6
Fire-resistance-rated glazing in fire barriers.

2012 International Building Code Handbook   195

7 Fire and Smoke Protection Features

separation without the recognition that the integrity of the construction must be maintained or that some type of fire or smoke protective is required. The reduction or elimination of protection that occurs is typically not malicious. Rather, the installation of an inappropriate air opening, or the penetration of the separation without the proper firestopping, is often done due to the lack of information regarding the wall assembly's function and required fire rating.

Through the identification of fire and smoke separation elements, it is possible for tradespeople, maintenance workers, and inspectors to recognize the required level of protection that must be maintained. The requirements apply to all wall assemblies where openings or penetrations are required to be protected. This would include exterior fire-resistance-rated walls as well as fire walls, fire barriers, fire partitions, smoke barriers, and smoke partitions. The identifying markings must be located within 15 feet (4,572 mm) of the ends of the wall and at maximum 30-foot (9,144-mm) intervals to increase the possibility that they would be visible during any work on the wall assemblies. A minimum letter height of 3 inches (76 mm) is also prescribed along with sample language for the marking. See Figure 703-7.

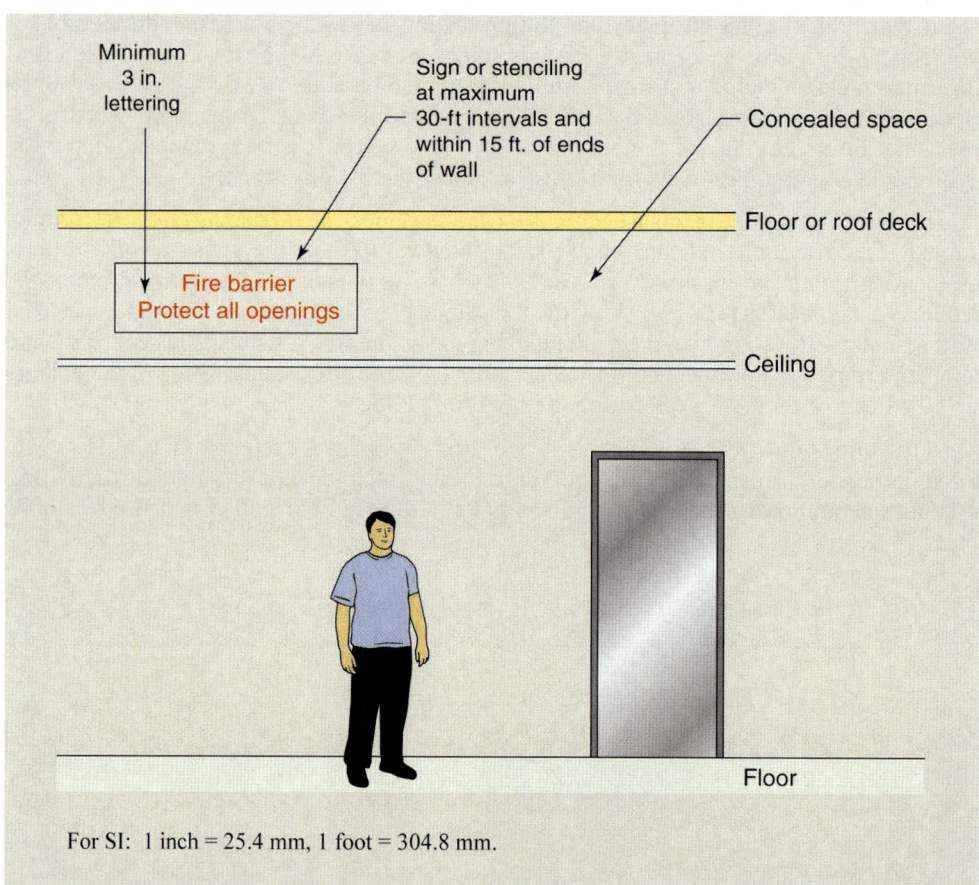

Figure 703-7
Identification sign for fire barrier.

It is intended that the identification marks be located in areas not visible to the general public. Specific locations set forth in the provisions indicate that the identification is to be provided within those concealed spaces that are accessible, such as above suspended ceilings and in attic areas. The requirement for markings does not apply to fire or smoke separations in Group R-2 occupancies unless a decorative ceiling system is installed. There is an expectation that the separation walls in these types of uses are seldom altered and thus the identification markings are not necessary.

Section 704 Fire-Resistance Rating of Structural Members

Structural frame members such as columns, beams, and girders are regulated for fire resistance based on a building's type of construction. Some types of constructions mandate a higher level of fire endurance for structural members and assemblies on account of the critical nature of their function. Type of construction considerations is based primarily on the potential for building collapse when subjected to fire. Therefore, the structural frame is specifically addressed in Table 601 as to the required fire-resistance ratings. This section provides further details for the protection of structural members.

Figure 704-1 provides simple details of fire protection of structural members that indicate the principle of *mass effect*. Mass effect is beneficial to the protection requirements for structural members of a heavy cross section. In the case of steel members, the amount of protection depends on the weight of the structural steel member. A heavy, massive structural steel cross section behaves such that the heat applied to the surface during a fire is absorbed away from the surface, resulting in lower steel surface temperatures. Thus, the insulating thicknesses indicated by tests or in Table 721.1(1) should not be used for members with a smaller weight than that specified in the test or table.

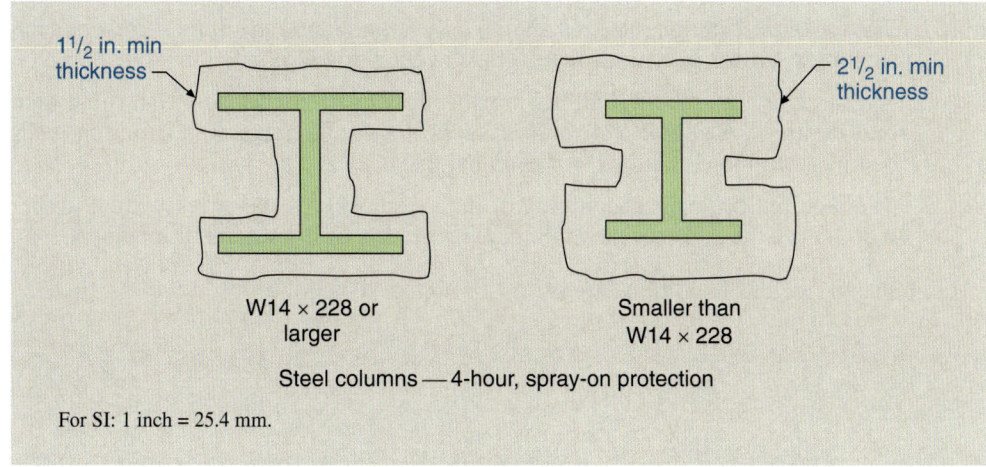

Figure 704-1
Mass effect.

704.2 Column protection. Primary structural frame members require fire-resistive protection in buildings of Type I, IIA, IIIA, and VA construction. Under all conditions, columns considered as a part of the primary structural frame system must be protected by individual encasement. This protection must occur on all sides of the column and extend for the column's full height. Where a ceiling is provided, the fire resistance of the column is to be continuous from the top of the foundation or floor/ceiling assembly below through the ceiling space to the top of the column. The fire protection required for the column shall also be provided at the connections between the column and any beams or girders. Where the column is located within a fire-resistance-rated wall assembly as shown in Figure 704-2, the mandated column protection must still be provided through individual encasement. It is not acceptable to simply place an unprotected column within a fire-resistance-rated wall assembly and consider the column as fire-resistant rated.

704.3 Protection of the primary structural frame other than columns. The code intends that the fire-resistive protection for primary structural frame members be applied to the individual structural member. This is based on the differences in both the testing procedure and the conditions of acceptance that were discussed in Section 703. In other words, the code

7 Fire and Smoke Protection Features

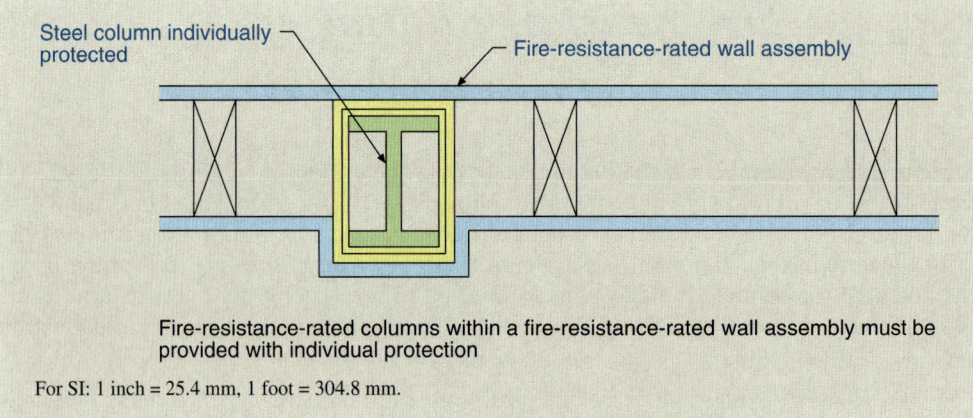

Figure 704-2
Individual protection of structural columns.

Fire-resistance-rated columns within a fire-resistance-rated wall assembly must be provided with individual protection

For SI: 1 inch = 25.4 mm, 1 foot = 304.8 mm.

does not intend that a primary structural frame member be protected by a wall assembly or fire-resistance-rated horizontal assembly, except as permitted by this section.

Under certain restrictions, the code allows the use of a floor/ceiling or roof/ceiling assembly to provide protection for structural members, rather than requiring that they be individually protected. The criteria for use of alternative membrane protection in lieu of individual encasement are depicted as follows:

1. The use of the ceiling protection applies only to horizontal structural members, such as girders, trusses, beams, or lintels. (See Section 704.2 for column protection.)

2. The structural members shall not support directly applied loads from more than two floors or one floor and roof, or support a load-bearing wall or a non-load-bearing wall more than two stories in height.

3. The required fire-resistance rating of the assembly shall be at least equal to that required by the code for the individual protection of the structural members.

Examples of various conditions are shown in Figure 704-3.

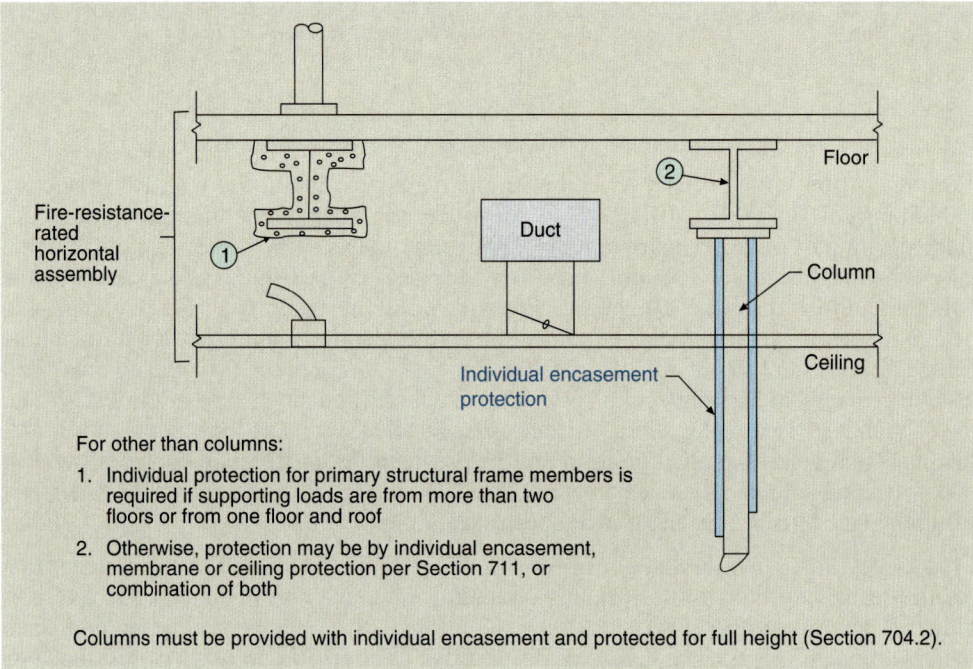

Figure 704-3
Protection of primary structural frame members.

For other than columns:
1. Individual protection for primary structural frame members is required if supporting loads are from more than two floors or from one floor and roof
2. Otherwise, protection may be by individual encasement, membrane or ceiling protection per Section 711, or combination of both

Columns must be provided with individual encasement and protected for full height (Section 704.2).

Fire-Resistance Rating of Structural Members

704.4
Protection of secondary members. Secondary members, as defined in Section 202, may be protected in the same manner as primary structural frame members where a fire-resistance rating is required. Such elements can be individually encased or protected by a membrane or ceiling of a horizontal assembly. Floor joists and roof joists are examples of secondary members that are permitted to be protected by the horizontal assembly in which they are located. In light-frame construction, membrane protection is also permitted for king studs and similar elements that are integral elements in load-bearing walls.

704.5
Truss protection. It is the intent of the IBC that this provision be applied to trusses that are a part of the primary structural frame as defined in Section 202. In this case, the code permits the encapsulation with fire-resistive materials of the entire truss assembly. It is the intent of the code that the thickness and details of construction of the fire-resistive protection be based on the results of full-scale tests or of tests on truss components. Approved calculations based on such tests that show that the truss components provide the fire endurance required by the code are also acceptable. One application of this concept is in the use of the encapsulated trusses as dividing partitions between hotel rooms in multistory steel-frame buildings. Because the truss becomes part of the primary structural frame where it used to span between exterior wall columns, it provides a column-free interior. The fire-resistive design of the encapsulated protection can be based either on tests or on analogies derived from fire tests.

Additional criteria for the protection of primary structural members are illustrated in Figures 704-4 and 704-5, which depict details for attached metal members and reinforcing discussed in Sections 704.6 and 704.7. The provisions of Section 704.9 for impact protection are also illustrated in Figure 704-6.

704.10
Exterior structural members. The code provides that structural frame elements in the exterior wall or along the outer lines of a building must be protected based on the higher rating of three criteria. The minimum fire-resistance rating is determined by evaluating the

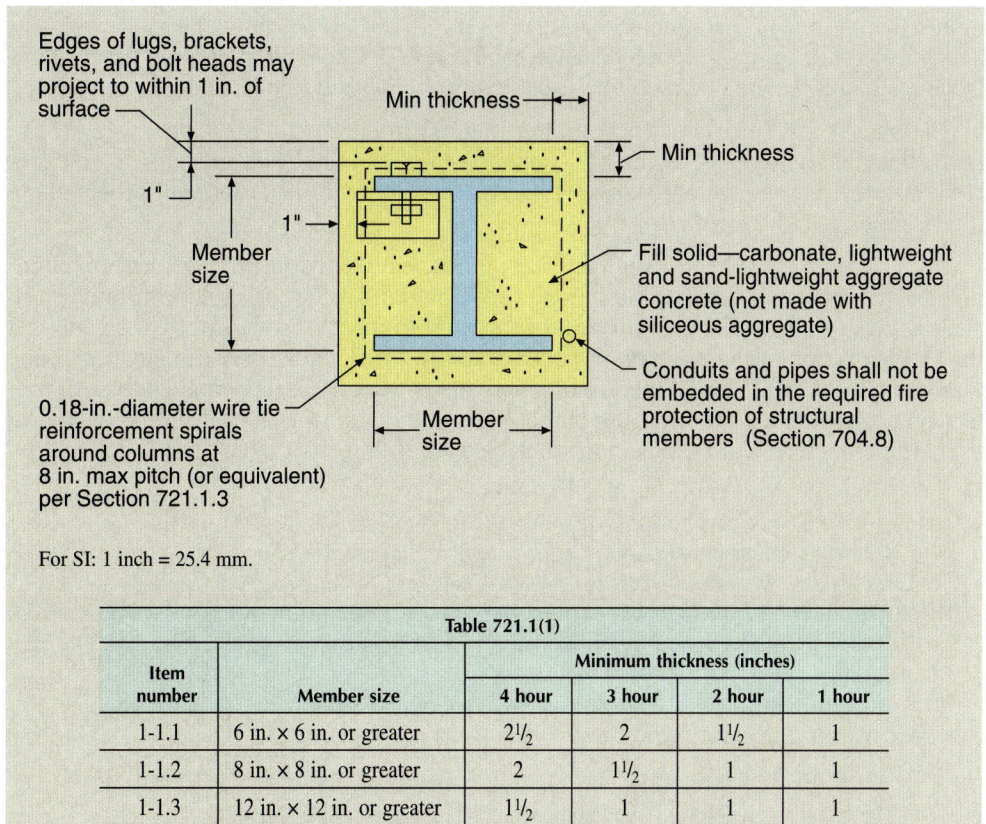

Figure 704-4 Protection of structural steel column.

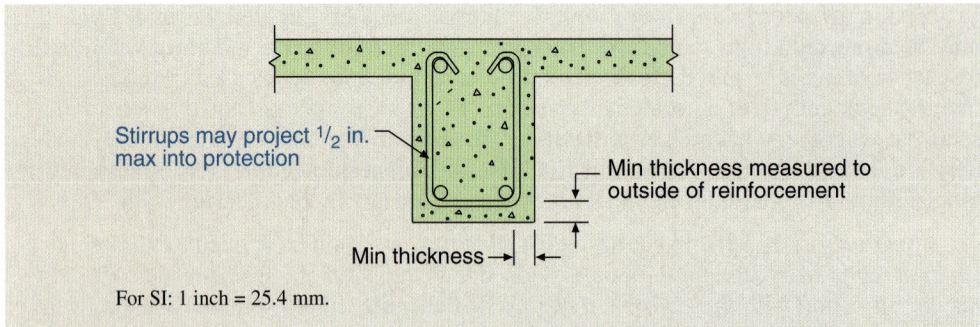

Figure 704-5
Reinforcing steel in concrete joists.

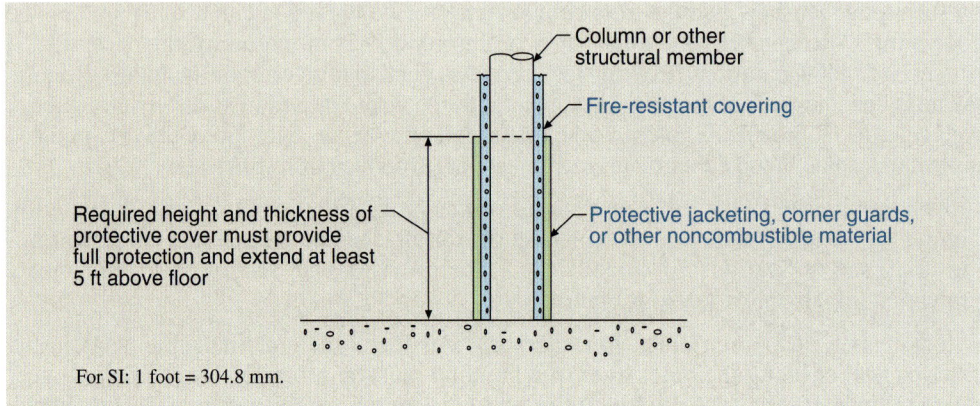

Figure 704-6
Impact protection.

requirements for (1) the structural frame per Table 601, (2) exterior bearing walls per Table 601, and (3) fire separation distance per Table 602. The highest of these three ratings is the minimum required rating of the structural members. See Application Example 704-1.

The intent of the provisions is that the structural frame should never have a lower fire rating than that required to protect the frame from internal fires. Nevertheless, if the exposure hazard from an external source is so great as to require exterior wall protection, a higher rating may be required.

704.11 Bottom flange protection. Exempted from the requirements for fire protection in buildings of fire-resistance-rated construction are the bottom flanges of short-span lintels, and shelf angles or plates that are part of the structural frame. It is assumed by the code that the arching action of the masonry or concrete above the lintel will prevent anything more than just a localized failure. Furthermore, only the bottom flange is permitted to be unprotected and, as a result, the wall supported by the lintel will act as a heat sink to draw heat away from the lintel and thereby increase the length of time until failure that is due to heat.

Application Example 704-1

GIVEN: An exterior nonbearing wall in a Type IIIB building housing a Group M occupancy. The wall has a fire separation distance of 15 feet to an interior lot line.

DETERMINE: The minimum required fire-resistance rating for structural columns located within the exterior wall.

SOLUTION:

Per Table 601 for structural frame members, a minimum of 0 hours

Per Table 601 for exterior bearing walls, a minimum of 2 hours

Per Table 602 for a fire separation distance (FSD) of 15 feet, a minimum of 1 hour

∴ The columns shall have a minimum fire-resistance rating of 2 hours.

This latter rationale also applies to shelf angles and plates that are not considered a part of the structural frame. The limitation to spans no greater than 6 feet 4 inches (1,931 mm) is intended to allow such unprotected lintels and angles where a pair of 36-inch (914-mm) doors is installed in the opening.

Section 705 *Exterior Walls*

Because of the potential for radiant heat exposure from one building to another, either on adjoining sites or on the same site, the IBC regulates the construction of exterior walls for fire resistance. Opening protection in such walls may also be required based on the fire separation distances involved. In addition to the regulation of exterior walls and openings in such walls, the code addresses associated projections, parapets, and joints.

Projections. Architectural considerations quite often call for projections from exterior walls such as cornices, eave overhangs, and balconies. Where these projections are from walls that are in close proximity to a lot line, they create problems that are due to trapping the convected heat from a fire in an adjacent building. As this trapped heat increases the hazard for the building under consideration, the code mandates a minimum distance the leading edge of the projecting element must be separated from the line used to determine fire separation distance. The permitted extent of projections is established by Table 705.2 and based solely on the clear distance between the building's exterior wall and an interior lot line, centerline of a public way, or assumed imaginary line between two buildings on the same lot. Where the distance is less than 2 feet (609 mm), all types of projections are prohibited. As the clear distance increases to 2 feet (609 mm) and beyond, projections are permitted; however, the extent of such projections is regulated. Three examples showing the application of Table 705.2 are shown in Figures 705-1(a), (b), and (c).

705.2

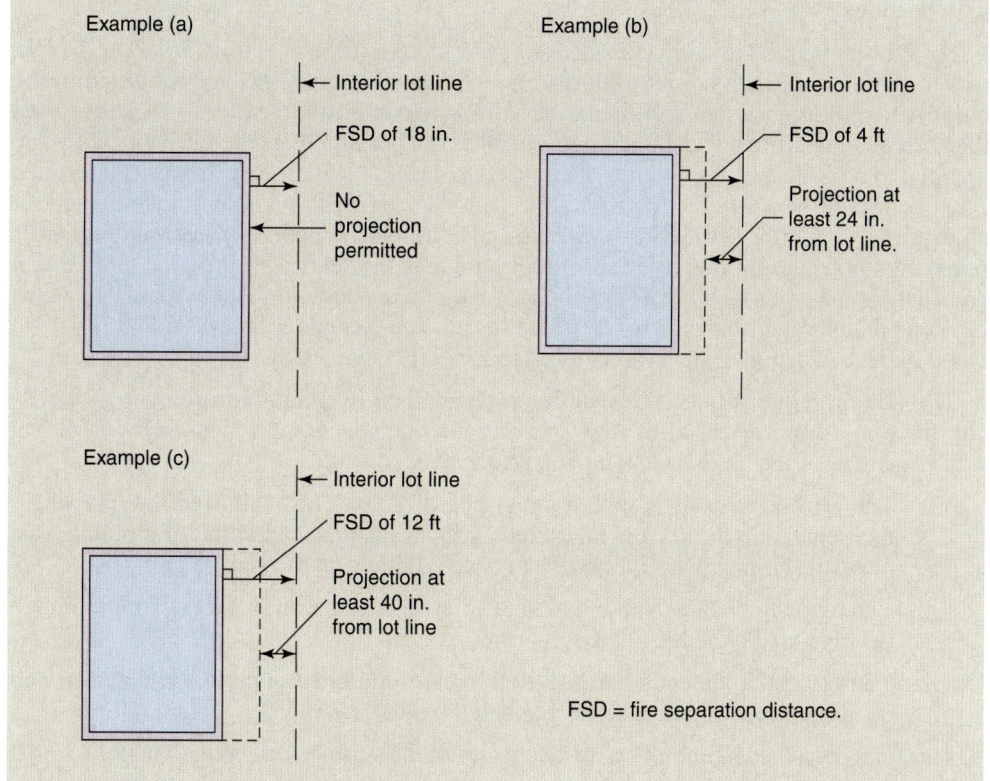

Figure 705-1
Maximum permitted projections.

The reference to multiple buildings on the same lot is intended to address only those projections that extend beyond the opposing exterior walls of the adjacent buildings. For those exterior walls that directly oppose each other, the limits on projecting elements are not applicable where the two buildings are being considered as a single building under the exception to Section 705.3. However, those projections that occur at exterior walls not located in opposition to those exterior walls of an adjacent building are to be regulated by the provisions of Section 705.2. The application of the exception to Section 705.2 is shown in Figure 705-2.

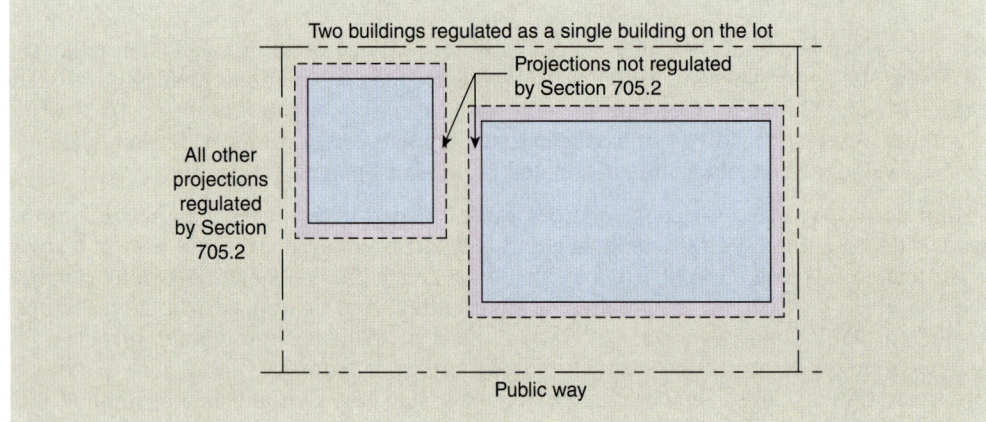

Figure 705-2
Two buildings regulated as a single building on the lot.

Projections from buildings are further regulated in order to prevent a fire hazard from inappropriate use of combustible materials attached to exterior walls. Thus, the IBC requires that projections from walls of Type I or II buildings be of noncombustible materials. However, it should be noted that certain combustible materials are permitted for balconies and similar projections as well as bay windows and oriel windows in accordance with Sections 1406.3 and 1406.4.

For buildings that the code considers to be of combustible construction (Type III, IV, or V construction), both combustible and noncombustible materials are permitted in the construction of projections. See Figure 705-3. Where combustible projections are used and they extend into an area where openings are either not permitted or where any openings are required to be protected, the code requires that they be of at least 1-hour fire-resistance-rated construction, of heavy-timber construction, constructed of fire-retardant-treated wood, or as required in Section 1406.3 for balconies and similar projections. This requirement is based on a potential for a severe exposure hazard and, consequently, the code intends that combustible materials be protected or, alternatively, be of heavy-timber construction, which has comparable performance when exposed to fire. An example is shown in Figure 705-4, based on an evaluation of the criteria as established in Table 705.8.

The criteria requires that combustible projections be of 1-hour construction, Type IV construction, or fire-retardant-treated wood, or alternatively comply with Section 1406.3, where any one of the three following conditions exists:

1. The projection extends within a distance of 5 feet (1,524 mm) to the line where fire separation distance is measured (interior lot line, centerline of a public way, or assumed imaginary line between two buildings on the same lot).

2. The projection extends into the zone where exterior wall openings are prohibited (as regulated by Table 705.8).

3. The projection extends into the zone where unlimited unprotected exterior wall openings are not permitted (as regulated by Table 705.8).

The zone described in item #3 is always inclusive of those described in items #1 and #2; therefore, it becomes the regulating provision. In order to determine the point at which a

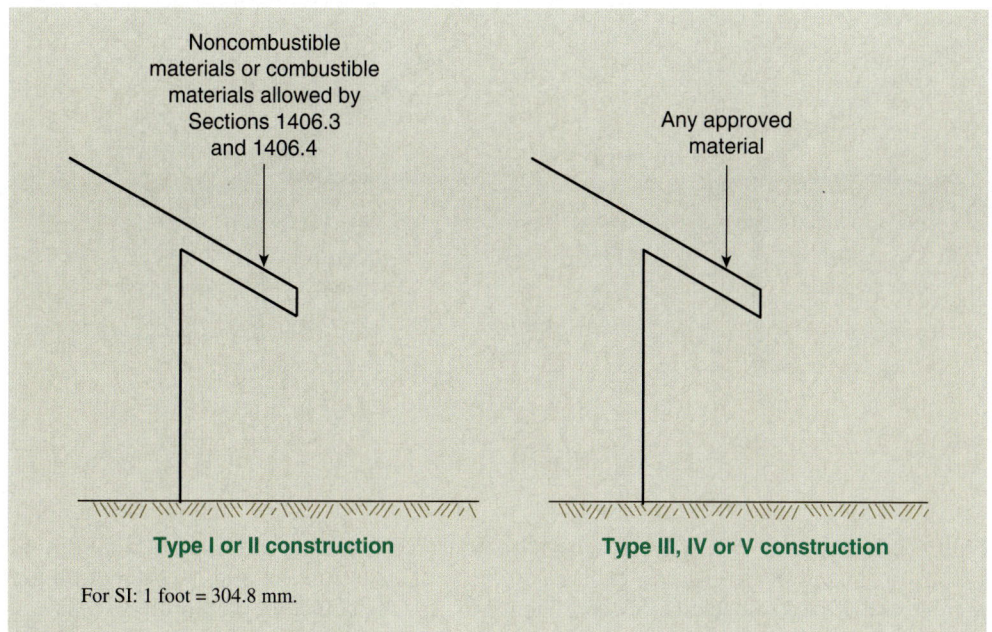

Figure 705-3
Combustibility of projections.

combustible projection is no longer regulated by Section 705.2.3, it is necessary to apply the information established in Tables 602 and 705.8. Only where the tables provide for no limit on the allowable area of unprotected exterior wall openings is a combustible projection considered outside the scope of Section 705.2.3.

Because projections are typically regulated independent of the roof construction, it is entirely possible that their construction types may be inconsistent. For example, Figure 705-5 shows two situations where the roof construction and resulting projections may differ in their required protection. Figure A relates a Type VA building with a 1-hour fire-resistance-rated roof system but a nonrated projection. On the other hand,

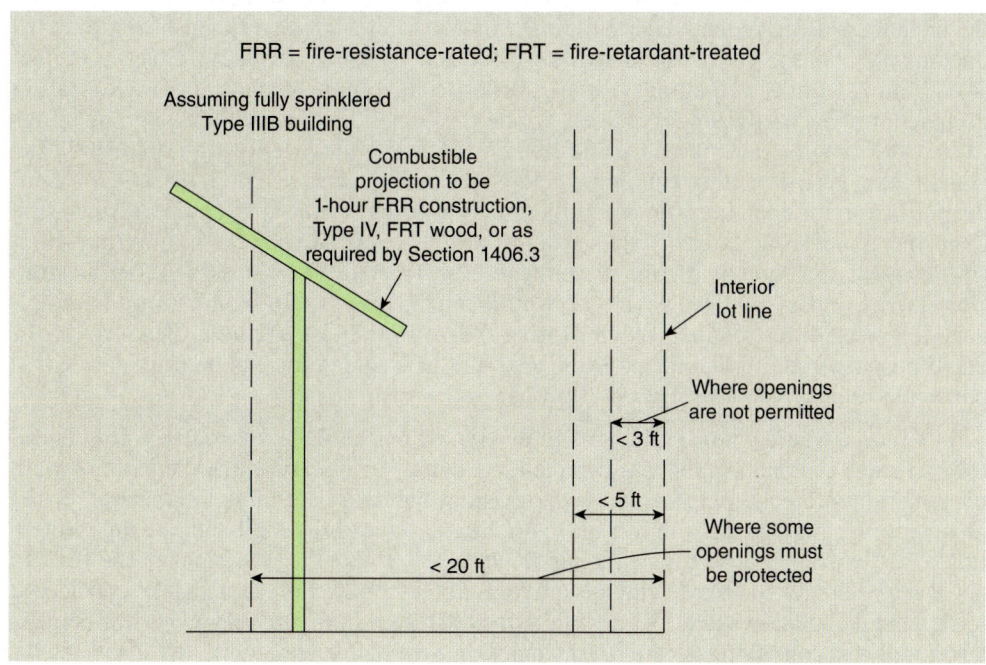

Figure 705-4
Protection of combustible projections.

7 Fire and Smoke Protection Features

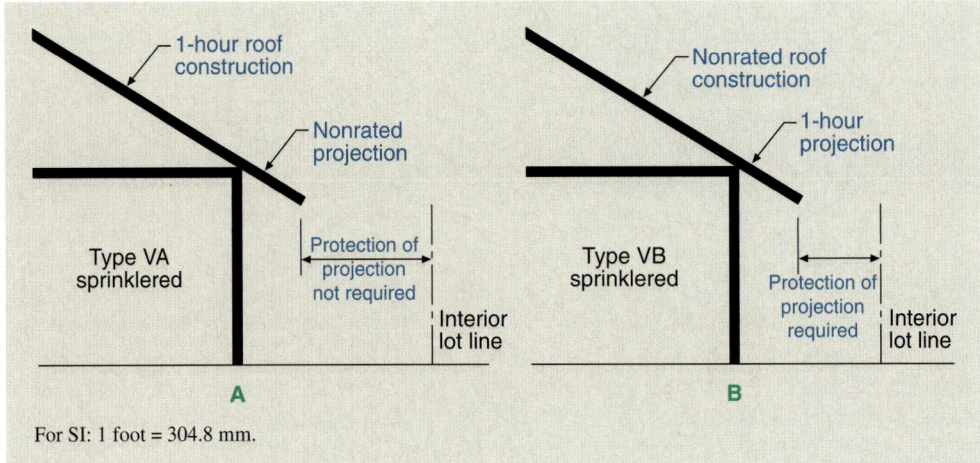

Figure 705-5 Projection versus roof protection.

Figure B indicates a Type VB building with nonrated roof construction but a minimum 1-hour-protected projection. In each case, the roof construction and its projection are regulated differently because of the concept of fire resistance being applied.

705.3 Buildings on the same lot. The IBC regulates exterior wall construction, opening protection, and projection extent and protection based on the proximity of the exterior walls to lot lines, either real or assumed. This section provides the code requirements for the establishment of imaginary lines between buildings on the same lot. Where two or more buildings are to be erected on the same site, the determination of the code requirements for protection of the exterior walls is based on placing an assumed imaginary line between buildings. Figure 705-6 illustrates an example of two nonsprinklered Type IIIB buildings housing Group S-2 occupancies sharing a 32-foot-wide (9,760-mm) yard, and it is noted that the imaginary line can be located anywhere between the two buildings so that the best advantage can be taken of wall and opening protection, depending on the use and architectural considerations for the exterior walls of the buildings. For example, if unprotected openings amounting to 25 percent of the area of the exterior walls of each nonsprinklered building were desired, the imaginary line would be located so that the distance between it and each building would permit such an amount of unprotected openings. Thus, the code would require that each building be placed at least 15 feet (4,572 mm) from the imaginary line in order to have unprotected openings totaling 25 percent of each opposing wall area. If one of the buildings were to have no openings in the exterior wall, the imaginary line could be placed at the exterior wall of the building without openings. The other building would be located at a distance of 32 feet (9,760 mm) or more from the imaginary line and the other building. In the first case described, the opposing nonbearing exterior walls would both be required to be of minimum 1-hour fire-resistance-rated construction as they are each located less than 30 feet (9,144 mm) from the imaginary line. However, the wall located 32 feet (9,760 mm) from the imaginary line would not require any fire rating. Also, in the last example, Section 705.11 could possibly require that the exterior wall on the assumed lot line be provided with a parapet. See discussion of Section 705.11.

In the case where a new building is to be erected on the same lot as an existing building, the same rationale applies, as depicted in Figure 705-6, except that the exterior wall, opening, and projection protection of the existing building determine the location of the assumed imaginary line. As shown in Figure 705-7, the exterior wall and opening protection of the existing building must remain in compliance with the provisions of the IBC. In any case, where two or more buildings are located on the same lot, they may be considered to be a single building subject to the limitations depicted in Figure 705-8. For further discussion of this condition, see the commentary on Section 503.1.2.

Exterior Walls 7

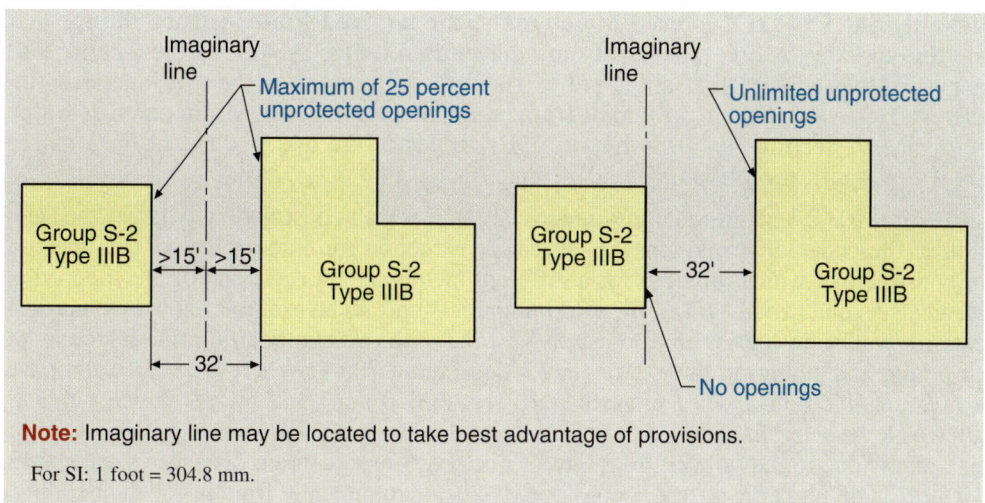

Figure 705-6
Buildings on the same lot.

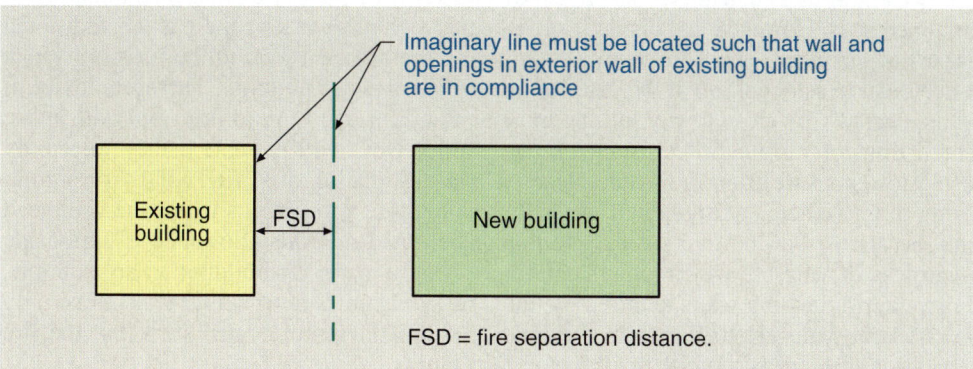

Figure 705-7
Buildings on the same lot.

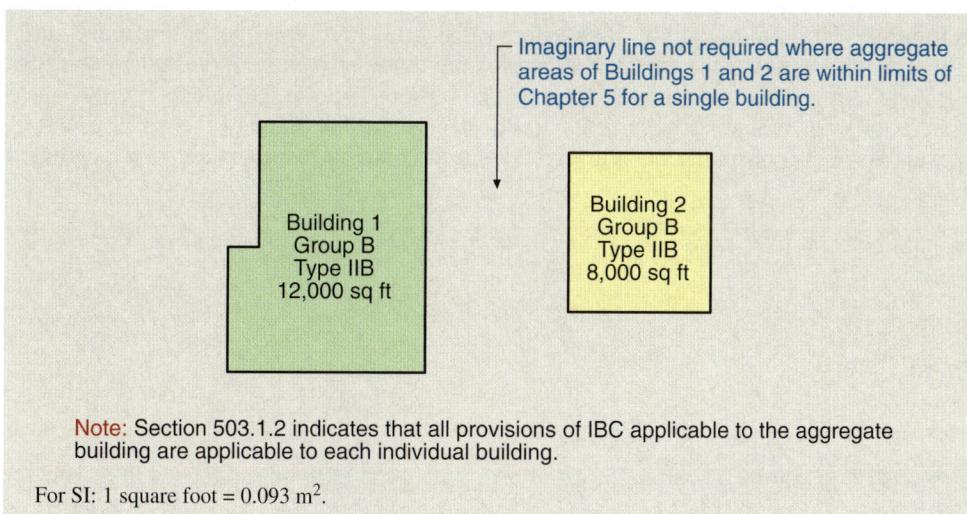

Figure 705-8
Multiple buildings considered as a single building.

705.5 **Fire-resistance ratings.** The IBC requires that exterior walls conform to the required fire-resistance ratings of Tables 601 and 602. Bearing walls must comply with the more restrictive requirements of both tables, whereas nonbearing exterior walls need only comply with Table 602. Table 601 is intended to address the fire endurance of bearing walls necessary to prevent building collapse that is due to fire for a designated

7 Fire and Smoke Protection Features

time period. Table 602 is used to determine the required fire-resistance ratings that are due to exterior fire exposure from adjacent buildings, as well as the interior fire exposure that adjacent buildings are exposed to on account of the uses sheltered by the exterior walls. Where structural frame members are located within exterior walls, reference to Section 704.10 is required. Examples of the use of these provisions are shown in Application Examples 601-2 and 704-1.

Section 703.2.1 addresses nonsymmetrical interior wall construction, whereas this section of the code addresses nonsymmetrical construction for exterior walls. This method of construction, which provides for a different membrane on each side of the supporting elements, is much more typical for exterior applications. As an example, a nonsymmetrical exterior wall may consist of wood studs covered with gypsum board on the inside, with sheathing and siding on the exterior side. See Figure 705-9. Where exterior walls have a fire separation distance of more than 10 feet (3,048 mm), the fire-resistance rating is allowed to be determined based only on interior fire exposure. This recognizes the reduced risk that is due to the setback from the lot line. For fire separation distances greater than 10 feet (3,048 mm), the hazard is considered to be predominantly from inside the building. See Figure 705-10. Thus, fire-resistance-rated construction whose tests are limited to interior fire exposure is considered sufficient evidence of adequate fire resistance under these circumstances. However, at a distance of 10 feet (3,048 mm) or less, there is the additional hazard of direct fire exposure from a building on the adjacent lot and the possibility that it may lead to self-ignition at the exterior face of the exposed building. Therefore, exterior walls located very close to any lot line must be rated for exposure to fire from both sides. The listings of various fire-resistance-rated exterior walls will indicate if they were only tested for exposure from the inside, usually by a designation of "FIRE SIDE" or similar terminology. Where so listed, their use is limited to those applications where the wall need only be rated from the interior side. It should be noted that this allowance is applicable regardless of why the wall requires a rating. Using a Type VA building as an example, those exterior bearing walls required by Table 601 to be minimum 1-hour walls need only be rated for exposure to fire from the interior side if they are located such that the fire separation distance is more than 10 feet (3,048 mm).

705.6 Structural stability. This section refers the code user to Section 705.11 for parapets in determining the required height of exterior walls. It also requires the wall to have sufficient structural stability to remain in place for the duration of time indicated by the fire-resistance rating. The provisions address two conditions regarding elements used to brace the exterior wall. Where such elements are located within the plane of the wall (as part of the wall assembly) or on the outside of the wall, the bracing members are to be regulated

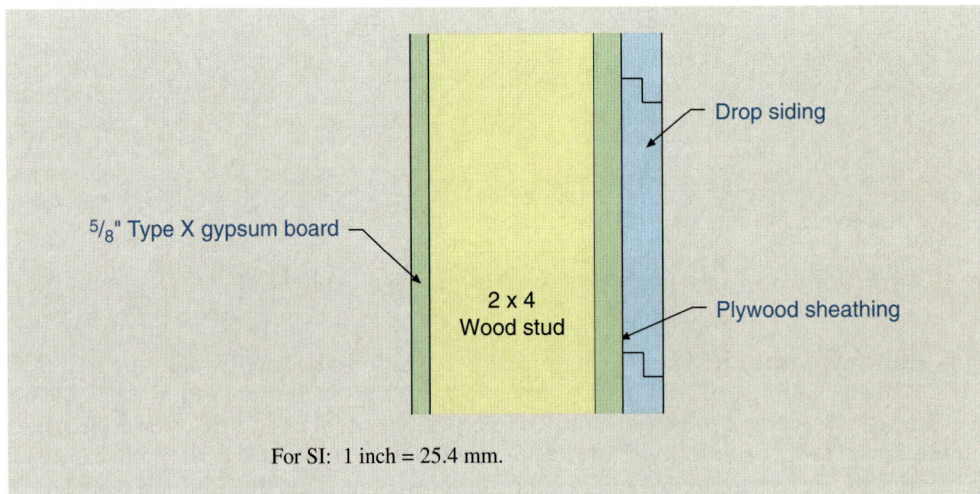

Figure 705-9
Nonsymmetrical exterior wall construction.

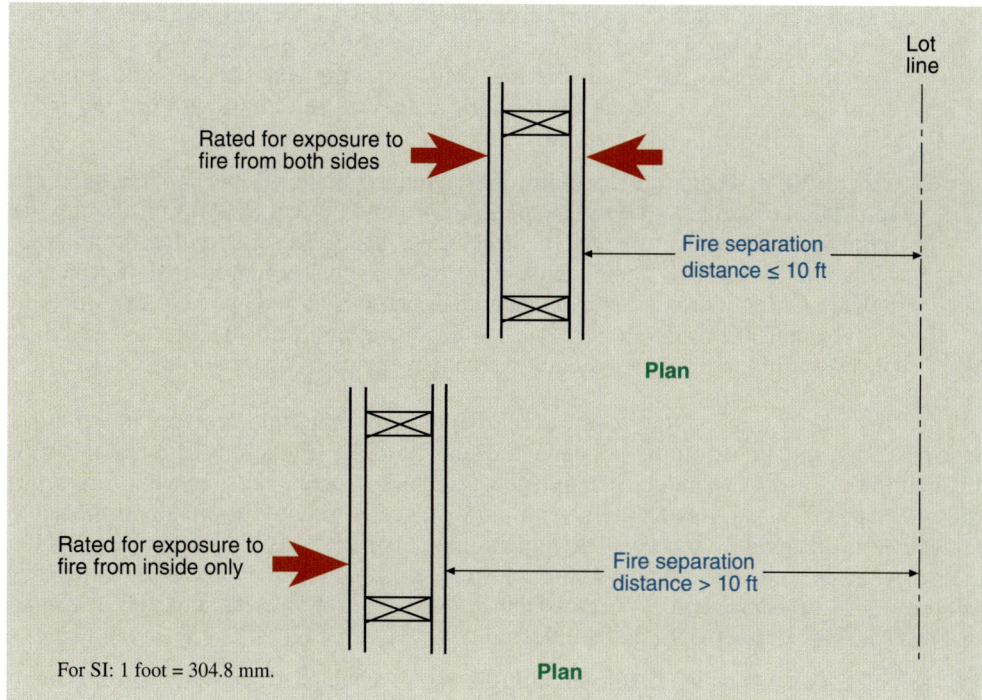

Figure 705-10
Nonsymmetrical exterior wall ratings.

for both external fire exposure and internal fire exposure based on the references to Tables 601 and 602. Those bracing elements that occur within the building, such as floor joists and/or roof joists that frame into the exterior wall, are also required to be protected from internal fire exposure if mandated by Table 601. However, external fire exposure concerns addressed in Table 602 are not specifically addressed. The reference to "exterior walls having a minimum fire separation distance of not less than 30 feet (9144 mm)" creates some confusion because such walls are not regulated under the provisions of Table 602. It appears that the intended application of the provisions be such that floor systems, roof systems, and other elements that provide lateral bracing from the interior side of an exterior wall are only required to be fire-resistance rated when required by Table 601.

Unexposed surface temperature. The provisions of this section provide a reduction of the prescriptive fire-resistance requirements for exterior walls under certain conditions. A fire-resistance-rated wall is generally required to meet the conditions of acceptance of ASTM E 119 or UL 263 for fire endurance and hose-stream tests on the surface exposed to the test fire, and heat-transmission limits on the unexposed surface. At fire separation distances beyond the point where no openings are allowed, typically 3 feet (914 mm), two more options are available:

705.7

1. Where opening protection is required, but the percentage of opening protection is not limited [typically a fire-separation distance of more than 20 feet (6,096 mm)], compliance with the heat-transmission limits of ASTM E 119 or UL 263 is not required. This recognizes that, although heat transmission is an important consideration for interior walls, the fire hazard that the limit addresses is substantially reduced once the exterior wall of a building is set back far enough that the fire hazard it presents to (and receives from) a building on an adjacent lot does not warrant a limit on the percentage of opening protection to limit the hazard. It has the effect of compliance with the conditions of acceptance for fire assemblies of the same hourly rating.

   According to NFPA 252 for fire door assemblies and NFPA 257 for fire window assemblies, nearly identical conditions of acceptance for fire endurance and hose-stream tests are required, but not limits on heat transmission. Because an unlimited percentage of opening protection is allowed, the lack of a heat transmission limit for exterior walls is consistent with that for fire door and fire window assemblies. An exterior wall that does not meet the heat transmission limits is considered equivalent to an opening protective of the same hourly rating in its reduced ability to limit heat transmission.

2. Where the percentage of opening protection is limited [typically having a fire separation distance between 3 feet (914 mm) and 20 feet (6,096 mm)], a similar reduction is possible, provided a correction is made according to the formula presented in this section. The formula converts the actual proposed area of protected openings to an increased equivalent area in proportion to the area of exterior wall surface under consideration that lacks adequate control of heat transmission. It places additional limits on the allowable percentage of opening protection.

The formula increases the required percentage of opening protection, whereas Section 705.8 sets limits on the percentage. Relative to the limitations of Section 705.8, this method allows for a smaller percentage of opening protection at the same fire separation distance. Thus, a greater fire separation distance is required to maintain the same percentage of opening protection. The reduction of the heat transmission capacity of the exterior walls is compensated by a reduction in the allowable percentage of opening protection. Without this provision, a fire-resistance-rated exterior wall that does not meet the heat transmission limits would not be allowed.

If actual test results or other substantiating data are available, they may be used in the computations. In their absence, the standard time-temperature curve of ASTM E 119 and UL 263 would be used, which results in an equivalent area of protected openings equal to the actual area of protected openings plus the exterior wall area without adequate control of heat transmission per ASTM E 119 and UL 263. This is converted into a percentage of opening protection and compared to the limits of Section 705.8. The use of actual test results may reduce this effect. See Section 705.8 for the basic limits prior to modification by this provision.

705.8.1 Allowable area of openings. Openings in an exterior wall typically consist of windows and doors. Occasionally, air openings such as vents are also present. The maximum area of either protected or unprotected openings permitted in each story of an exterior wall is regulated by this section. In addition, both unprotected and protected openings are permitted in the same exterior wall based on a unity formula. The term *protected* in this section refers to those elements such as fire doors, fire windows, and fire shutters regulated in Section 716. Protected openings have the mandated fire-protection rating necessary to perform their function. *Unprotected* openings are simply those exterior openings that do not qualify as protected openings. Opening protection presents a higher fire risk than fire-resistance-rated construction insofar as it does not meet the heat transmission limits of ASTM E 119 or UL 263, as previously discussed. At increasing distances from where openings are no longer prohibited, the hazard from heat radiation decreases, allowing the percentage of openings, both protected and unprotected, to increase. The high hazard of heat exposure at small fire separation distances justifies the prohibition of openings in order to limit the percentage of wall area without adequate heat transmission limits. As the fire separation distance increases, the percentage of openings is allowed to increase in compensation. At greater distances, the limit on the percentage of opening protection is eliminated. This recognizes that, at greater distances, the lack of adequate control of heat transmission does not pose a significant hazard to adjacent buildings, but containment of the fire to its origin inside the exposed building is still important.

There is a distance from lot lines where the hazard is reduced to such a degree that all opening limitations are no longer warranted. At this point, exposure to and from adjacent buildings is not significant and the need for fire resistance at exterior walls is reduced to fire protection of bearing walls and structural members in order to delay building collapse in the

Exterior Walls 7

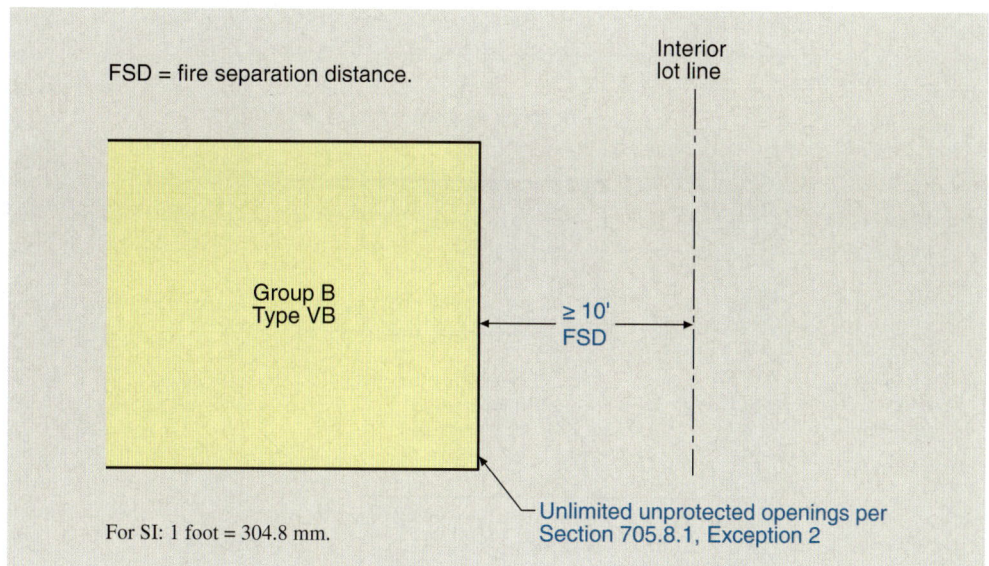

**Figure 705-11
Unlimited unprotected openings.**

event of fire. Arguably the most important provision is Exception 2 to Section 705.8.1. It indicates that if the exterior wall of the building and its primary exterior structural frame are not required by the code to have a fire-resistance rating, then unlimited unprotected openings are permitted. In other words, if the wall does not require a rating, any openings in the wall are unregulated for area and fire protection. An example is shown in Figure 705-11.

Although not stated in the exception, only when Table 602 requires a fire-resistance rating does Table 705.8 limit the maximum area of exterior openings. Where some other provision of the code mandates a fire-resistance-rated exterior wall, such as an exterior bearing wall supporting a fire-resistance-rated horizontal assembly, the limitations of Table 705.8 do not apply. It is also not necessary to review Table 601 to apply the exception as the conditions are established in such a manner that Table 602 provides all of the necessary information. Directly stated, if Table 602 does not mandate a fire-resistance-rated exterior wall, an unlimited amount of unprotected openings are permitted.

The limitation on exterior openings is also not applicable for first-story openings in buildings of other than Group H as indicated in Exception 1 to Section 705.8. Limited in application, this exception allows an unlimited amount of unprotected openings at the first story under specified circumstances. Applicable only to those buildings that require a fire-resistance-rated exterior wall where the fire separation distance equals or exceeds 10 feet (3,048 mm), the provisions are often used for opposing buildings having storefront systems. See Figure 705-12.

How are exterior openings regulated in fully sprinklered buildings? Table 705.8 also recognizes an increase in the allowable area of unprotected exterior openings for those buildings that are provided with an automatic sprinkler system throughout. For example, the fire-resistance-rated exterior wall of a fully sprinklered building having a fire separation distance of 15 feet (4,572 mm) may have 75 percent of its surface area consisting of unprotected openings. If the building is not sprinklered, the limit on unprotected openings is only 25 percent. In other than higher-level Group H occupancies, the maximum permitted area of unprotected openings in an exterior wall is allowed to be the same as the tabulated limitations for protected openings, provided the building is protected throughout with an NFPA 13 automatic sprinkler system. The increased areas permitted due to sprinkler protection are all incorporated directly into Table 705.8. It is important to note that the presence of an automatic sprinkler system does not increase the maximum allowable opening area for protected openings. Whereas the benefits of such an increase would seem

2012 International Building Code Handbook 209

7 Fire and Smoke Protection Features

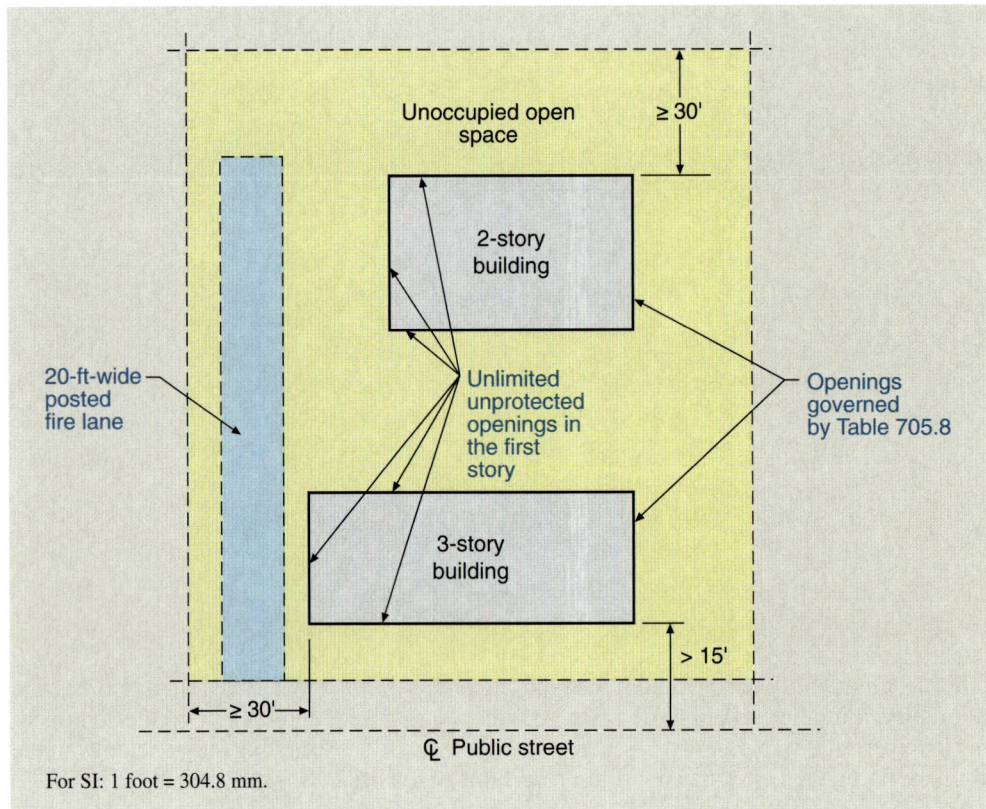

Figure 705-12
Unlimited openings in the first story exterior wall.

justifiable because of the increased level of protection, such an allowance is not addressed in the code. In addition, the unity formula (Equation 7-2) is not applicable to fully sprinklered buildings insofar as the code provides an increased allowance for unprotected openings to the amount permitted for protected openings.

705.8.2 Protected openings. Section 716.6 is referenced for identifying the level of protection required for windows needing opening protection. It further references Section 716.5 for fire doors and fire shutters. The use of sprinklers and water curtains to eliminate the required opening protection is addressed in the exception. It indicates that where the building is sprinklered throughout, those openings protected by an approved water curtain do not need to be fire-protective assemblies. However, the exception has virtually no application when the provisions of Table 705.8 are implemented, as the table allows for the elimination of protected openings in sprinklered buildings without the need for water curtains. There are a number of provisions throughout the IBC where the exception could be used. For example, Section 1027.4.2 typically mandates $3/4$-hour fire-protected openings in walls of egress courts less than 10 feet (3,048 mm) in width. In a fully sprinklered building, the use of a complying water curtain would eliminate the need for such openings to have a fire-protection rating. Other provisions where the exception might be applied include Section 1007.7 for exterior areas for assisted rescue and Section 1022.7 for interior exit stairways and ramps.

705.8.4 Mixed openings. Table 705.8 specifies the maximum allowable percentage of protected and unprotected openings, considered separately and based on fire separation distance alone. The unity formula (Equation 7-2) as set forth in Section 705.8.4 determines the maximum allowable area of protected and unprotected openings where they are proposed together in an exterior wall at an individual story of a nonsprinklered building. It offers a traditional interaction relationship, namely, the sum of the actual divided by the sum of the allowable cannot exceed one. An example of the determination of the maximum area of exterior wall openings, where both protected and unprotected openings are used, is

Application Example 705-1

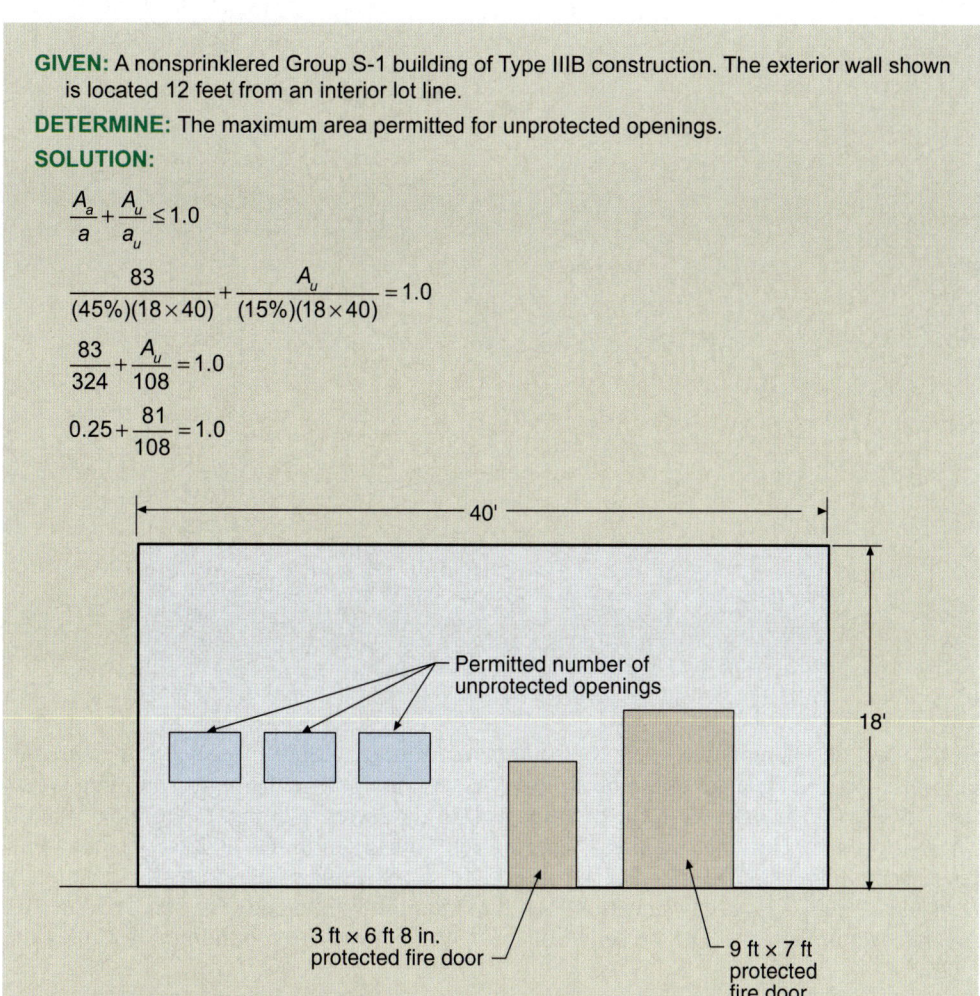

GIVEN: A nonsprinklered Group S-1 building of Type IIIB construction. The exterior wall shown is located 12 feet from an interior lot line.

DETERMINE: The maximum area permitted for unprotected openings.

SOLUTION:

$$\frac{A_a}{a} + \frac{A_u}{a_u} \leq 1.0$$

$$\frac{83}{(45\%)(18 \times 40)} + \frac{A_u}{(15\%)(18 \times 40)} = 1.0$$

$$\frac{83}{324} + \frac{A_u}{108} = 1.0$$

$$0.25 + \frac{81}{108} = 1.0$$

For SI: 1 inch = 25.4 mm, 1 foot = 304.8 mm, 1 square foot = 0.0929 m².

81 square feet of unprotected openings are permitted

provided in Application Example 705-1. The use of this table is limited to those buildings that are not provided with an NFPA sprinkler system throughout. Where the building is fully sprinklered, the code provides no advantage where protected openings are provided.

705.8.5 Vertical separation of openings. The intent of this section is to limit the vertical spread of fire from floor to floor at the exterior wall of the building. The code requires exterior flame barriers projecting out either from the wall or in line with the wall. These flame barriers are intended to prevent the leap-frogging effect of a fire at the outside of a building. See Figure 705-13. However, there are three exceptions that eliminate the required barriers. The first is for buildings three stories or less in height. The second is for fully sprinklered buildings. The third exception is for open parking garages. It is probable that this provision will have very limited application, as it is doubtful there will be much new construction of four stories or more without sprinkler protection. Provisions addressing the spread of fire from floor to floor on the interior side of an exterior wall, such as at the intersection of a floor and curtain wall system, are found in Section 715.4.

705.8.6 Vertical exposure. The scope of this section is limited to buildings located on the same lot and to the issue of the protection of openings in the exterior wall of a higher building above the roof of a lower building. It requires each opening in the exterior wall that is less

7 Fire and Smoke Protection Features

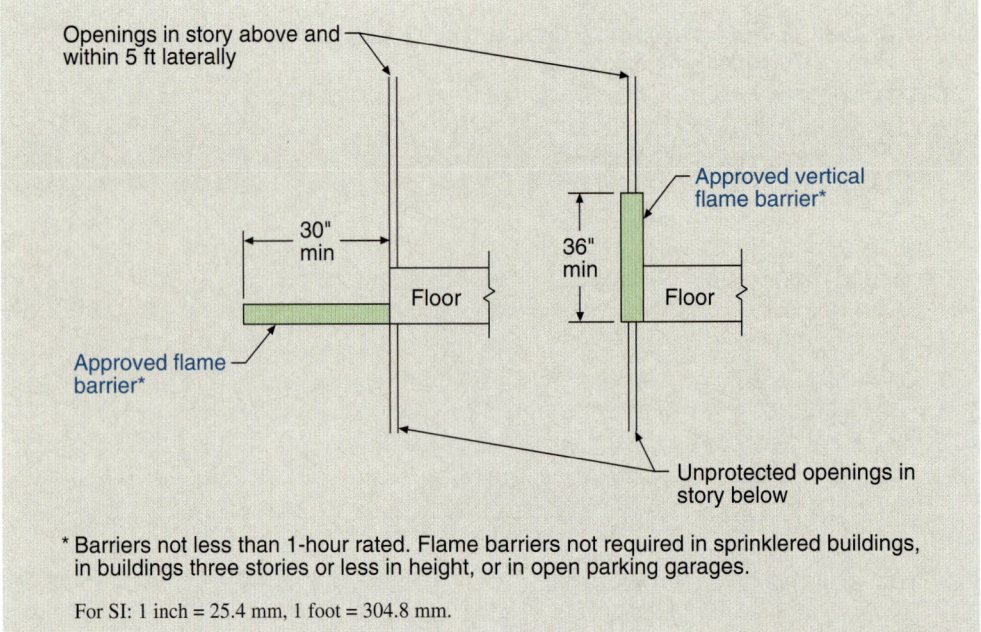

Figure 705-13
Flame barriers.

than 15 feet (4,572 mm) above the roof of the lower building to be protected if the horizontal fire separation distance of the opening in the exterior wall of the higher building is less than 15 feet (4,572 mm) from the exterior wall of the lower building. See Figure 705-14. There is an exception that applies where the roof construction has at least a 1-hour fire-resistance rating, also illustrated in Figure 705-14. Application of this provision potentially mandates a higher level of protection than that required by the code for two buildings on separate adjoining lots. The presence of a lot line between two buildings institutes the

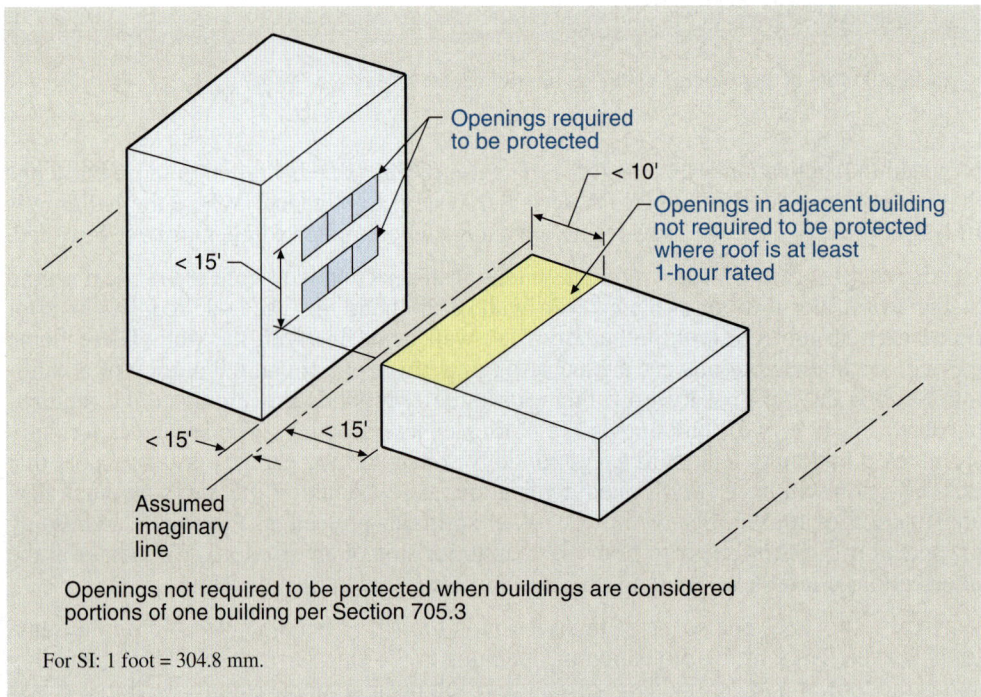

Figure 705-14
Vertical exposure.

concept of fire separation distance in the regulation of the opposing exterior walls and any openings in such walls. On a single lot with two buildings, the same concept is applied owing to the requirement for the placement of an assumed imaginary line between the buildings. This line is also the basis for regulating exterior wall and opening protection that is due to fire separation distance. The provisions of Section 705.8.6 introduce additional requirements that may not be mandated on account of the fire separation distance concept. In addition, where two buildings are located on the same lot, the provisions of Sections 705.3 and 503.1.2 permit them to be considered a single building if the aggregate area of the buildings is within the limits of Chapter 5 for a single building. For consistent application of the fire separation distance concept, it would appear that the methodology for buildings on the same lot could be permitted to be used rather than the vertical exposure provisions of Section 705.8.6.

It appears that the provision is intended to mirror the termination requirements and allowances for fire walls where located in stepped buildings as established in Section 706.6.1. It would be logical to assume that if a fire wall is not necessary to obtain code compliance, resulting in no required application of the fire wall termination requirements, the same concept should be considered when applying this provision to buildings on the same lot that can be regulated as a single building.

705.11 Parapets. This section intends that the exterior walls of buildings shall extend a minimum of 30 inches (762 mm) above the roof to form a parapet. There are two reasons for the parapet:

1. To prevent the spread of fire from the roof of the subject building to a nearby adjacent building.

2. To protect the roof of a building from exposure that is due to a fire in an adjacent nearby building.

Most buildings do not have complying parapets, and those that do typically use them to hide the roof slope or roof-top equipment. Therefore, the exceptions to this section tend to become the general rule. Three of the six exceptions listed in the code—1, 3, and 6—involve cases where the parapet would serve no useful purpose. In Exception 2, a concession is made to the small-floor-area building, and in Exceptions 4 and 5, an alternative method for providing equivalent protection is delineated. It is not necessary that all of the exceptions listed apply. Compliance with only one of the exceptions is all that is necessary for the elimination of a complying parapet.

Certainly, walls not required to be of fire-resistance-rated construction would not benefit from a parapet. In the case of walls that terminate at 2-hour fire-resistance-rated roofs or roofs constructed entirely of noncombustible materials, the parapet would be of little benefit, as the construction of the roof would prevent the spread of fire from or into the building. The exception for noncombustible roof construction is not intended to preclude the use of a classified roof covering.

In the case of walls permitted to have unprotected openings in conformance with Exception 6, the code assumes that the exterior wall will be far enough away from either an exposing building or an exposed building so that the protection provided by the parapet will not be necessary. This distance will vary based on the presence of a sprinkler system in the building, as shown in Figure 705-15.

The fourth exception makes a provision for 1-hour fire-resistance-rated exterior walls that are constructed similar to 2-hour fire walls that terminate at the underside of the roof sheathing, deck, or slab. This provides designers with an alternative to the use of parapets while recognizing that these walls provide adequate protection of the structure and its occupants as well as consistency with Section 706.6 for fire walls. See Figure 705-16. Exception 5 applies only to Group R-2 and R-3 occupancies and is intended to protect at the roof line through the use of a noncombustible roof deck, fire-retardant-wood sheathing, or a gypsum-board underlayment.

7 Fire and Smoke Protection Features

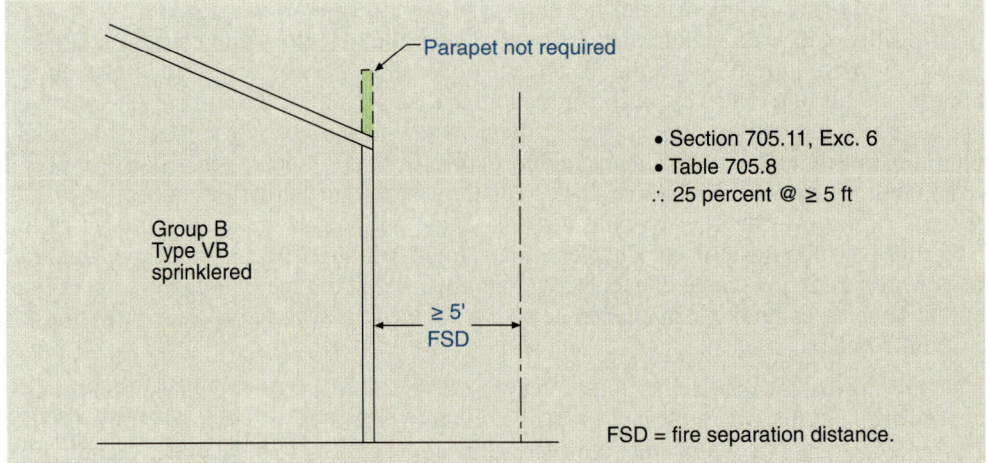

Figure 705-15
Parapet exception.

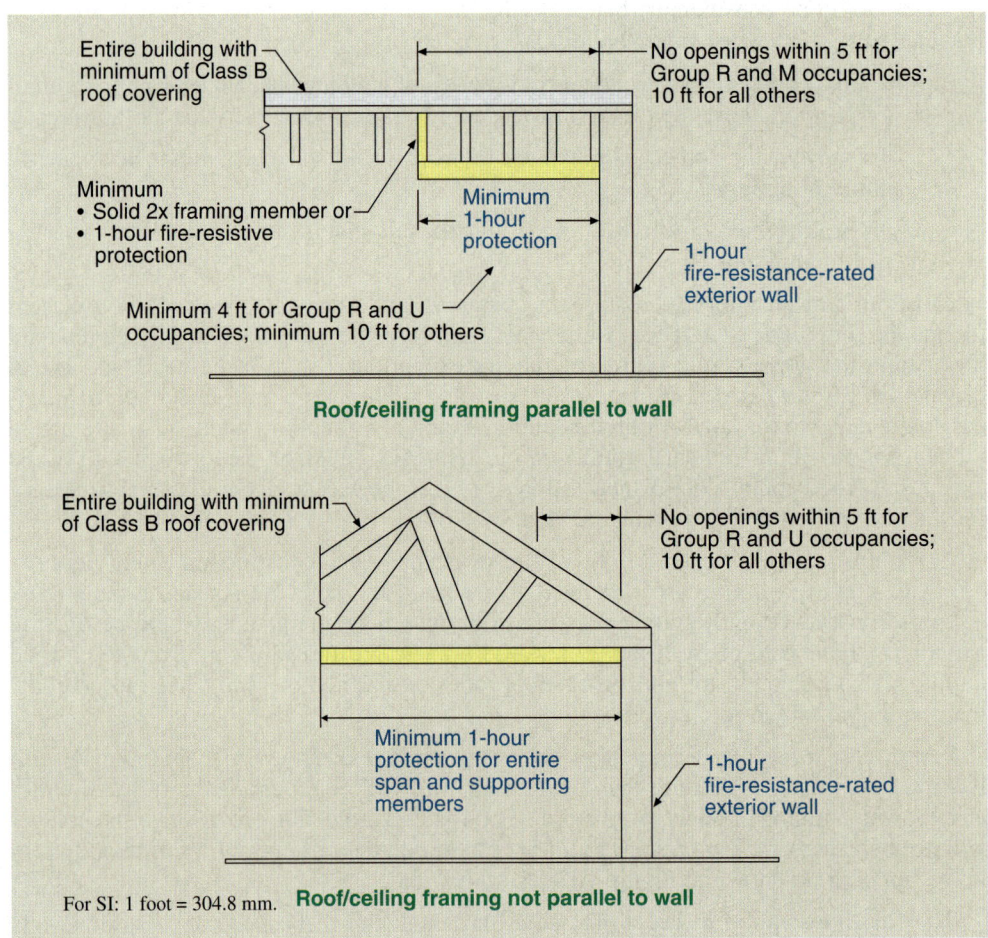

Figure 705-16
Parapet alternative.

705.11.1 Parapet construction. In addition to having the same degree of fire resistance as required for the wall, the code also requires that the surface of the parapet that faces the roof be of noncombustible materials for the upper 18 inches (457 mm). Thus, a fire that might be traveling along the roof and reaching the parapet will not be able to continue upward along the face of the parapet and over the top and expose a nearby adjacent building. The requirement only applies to the upper 18 inches (457 mm) of the parapet to allow for extending the roof covering up the base of the parapet so that it can be effectively flashed. The 18-inch (457-mm) figure is based on a parapet height of at least 30 inches (762 mm).

As stated in the code, the 30-inch (762-mm) requirement is measured from the point where the roof surface and wall intersect. Therefore, when a cricket is installed adjacent to the parapet, the 30-inch (762-mm) dimension would be taken from the top of the cricket.

In those cases where the roof slopes upward away from the parapet and slopes greater than 2 units vertical in 12 units horizontal (16.7-percent slope), the parapet is required to extend to the same height as any portion of the roof that is within the distance where protection of openings in the exterior wall would be required. However, in no case shall the height of the parapet be less than 30 inches (762 mm). See Figure 705-17 for an illustration of this requirement.

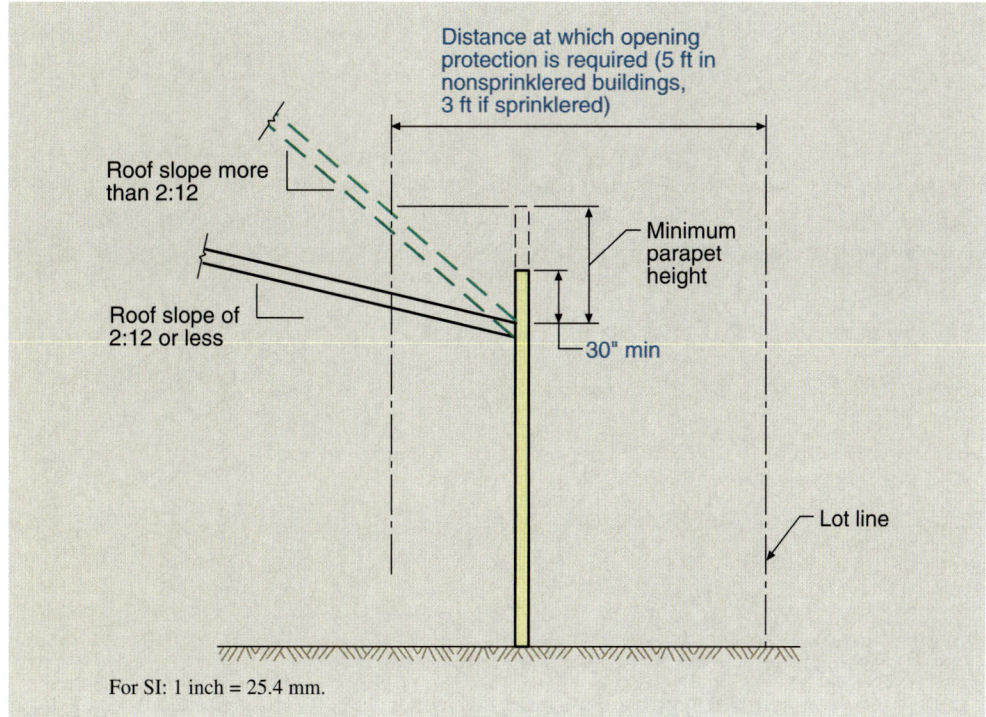

Figure 705-17
Parapet requirements.

Section 706 *Fire Walls*

The IBC permits fire walls to be installed within a building, thereby creating one or more smaller-area buildings. It further intends that each portion of the structure so separated may be considered a separate building for all purposes of the code. The concept is based on buildings on adjoining lots having a common party wall or two separate fire-resistance-rated walls located on the lot line. The high level of fire-resistance-rated construction between the two buildings, along with other controls, is deemed adequate for the protection of one building from its neighboring building. The use of one or more fire walls within a building is optional, based on a decision by the designer. The code never mandates a fire wall be used, but rather offers it as an alternative to other mandated provisions. There are various reasons for using fire walls within buildings; however, there are three such reasons that are quite common:

1. Allowable building area. The installation of one or more fire walls reduces the floor area in each of the separated buildings. Smaller floor areas can result in a reduction in the type of construction for one or more of the smaller buildings.

2. Multiple construction types. By separating a structure into separate buildings, they each are regulated independently for type of construction. Thus, not all of the structure would need to be classified based on the lowest construction type involved.

3. Automatic sprinkler systems. Fire walls can be used to reduce building size for the purpose of eliminating a requirement for the installation of an automatic sprinkler system.

Examples of these various uses of fire walls are shown in Application Examples 706-1 through 706-3. A fourth application of the fire wall concept is found in Appendix B of the *International Fire Code*® (IFC®) relating to fire-flow requirements for buildings. Where structures are separated by fire walls without openings, the divided portions may be considered separate fire-flow calculation areas.

706.1 General. As previously mentioned, one or more fire walls may be constructed in a manner such that the code considers the portions separated by the fire walls to be separate buildings.

Application Example 706-1

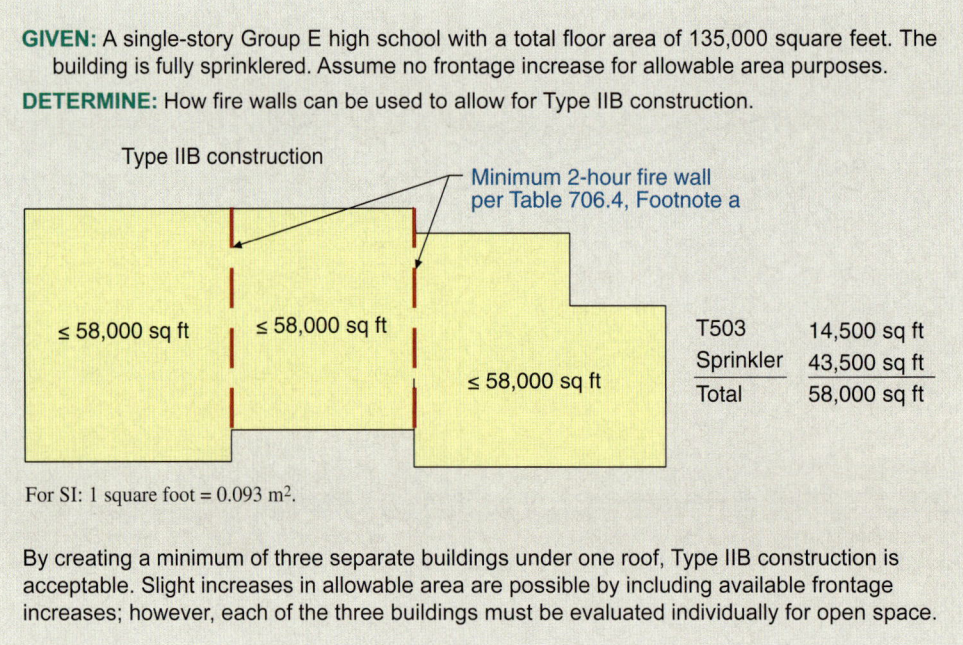

GIVEN: A single-story Group E high school with a total floor area of 135,000 square feet. The building is fully sprinklered. Assume no frontage increase for allowable area purposes.
DETERMINE: How fire walls can be used to allow for Type IIB construction.

For SI: 1 square foot = 0.093 m².

By creating a minimum of three separate buildings under one roof, Type IIB construction is acceptable. Slight increases in allowable area are possible by including available frontage increases; however, each of the three buildings must be evaluated individually for open space.

Application Example 706-2

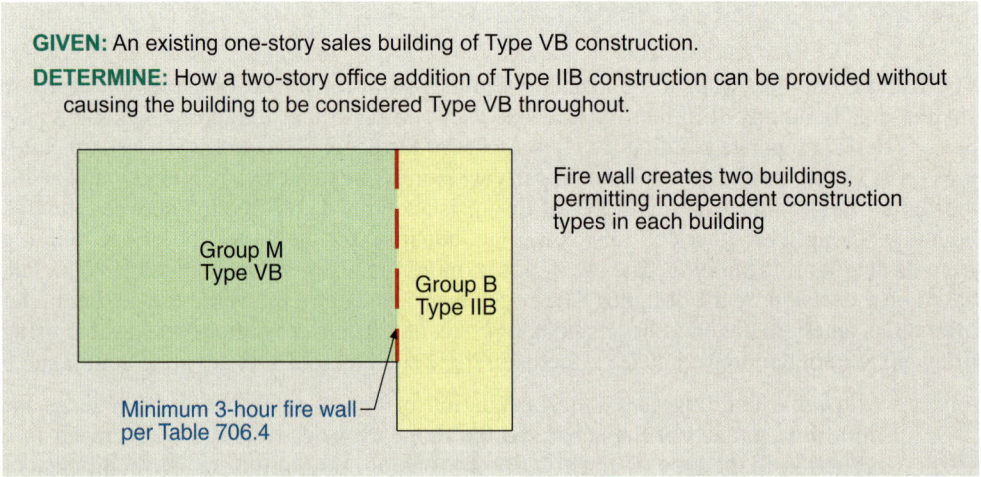

GIVEN: An existing one-story sales building of Type VB construction.
DETERMINE: How a two-story office addition of Type IIB construction can be provided without causing the building to be considered Type VB throughout.

Fire Walls 7

Application Example 706-3

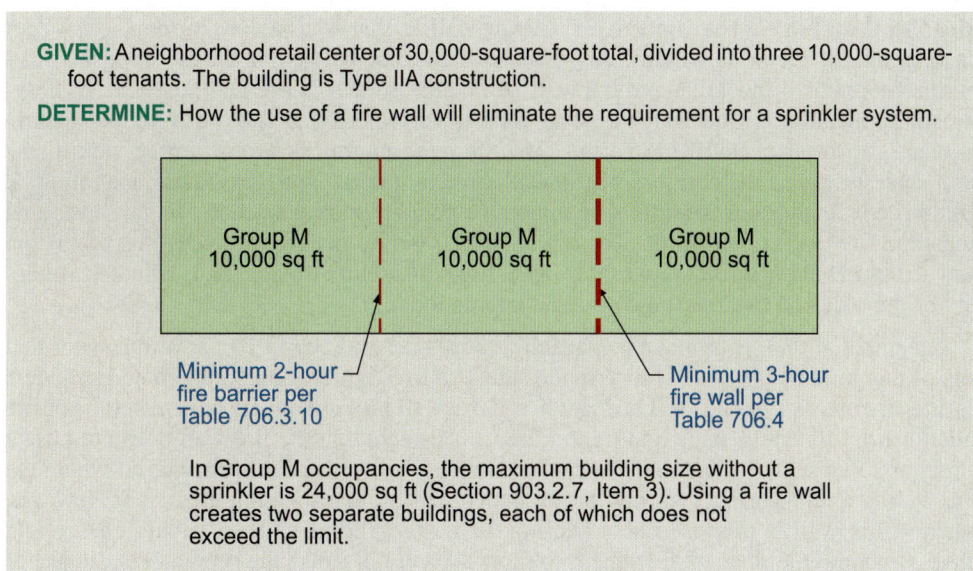

GIVEN: A neighborhood retail center of 30,000-square-foot total, divided into three 10,000-square-foot tenants. The building is Type IIA construction.

DETERMINE: How the use of a fire wall will eliminate the requirement for a sprinkler system.

In Group M occupancies, the maximum building size without a sprinkler is 24,000 sq ft (Section 903.2.7, Item 3). Using a fire wall creates two separate buildings, each of which does not exceed the limit.

Because a fire wall is such a critical element in the prevention of the spread of fire from one separated building to another, it is of great importance that the wall be situated and constructed properly. It must provide a complete separation. It should be noted that when a wall serves both as a fire barrier separating occupancies and a fire wall, the most restrictive requirements of each separation shall apply. The code also prohibits any openings in fire walls that are constructed on lot lines (defined as party walls).

Party walls. A common wall located on the lot line between two adjacent buildings is considered a party wall under this provision of the code. Regulated as a fire wall in accordance with the provisions of Section 706, a party wall can be used in lieu of separate and distinct exterior walls adjacent to the lot line, as depicted in Figure 706-1. The hazard created by neighboring buildings adjacent to each other is further addressed through the requirement that no openings be permitted in a party wall. For purposes of this section, and consistent with the general provisions of Section 503.1 for structures containing fire walls, separate buildings are created.

706.1.1

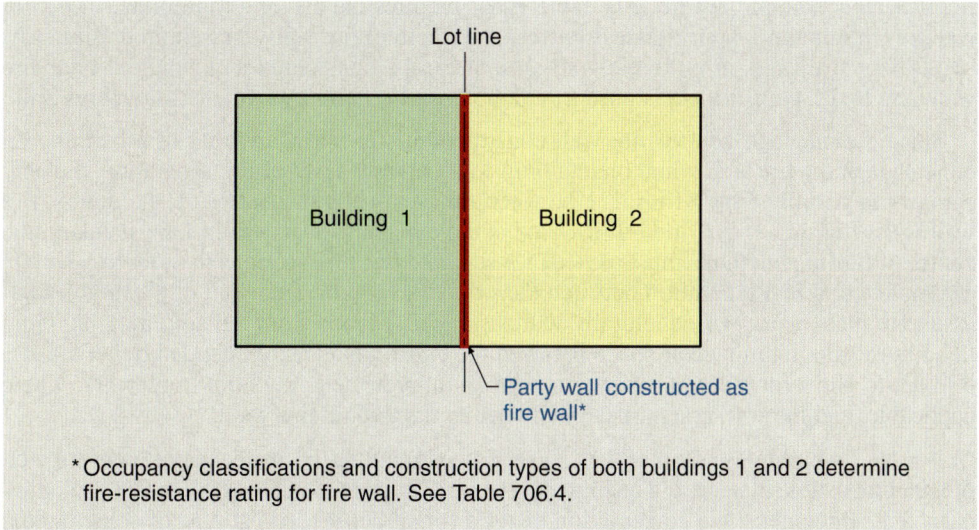

*Occupancy classifications and construction types of both buildings 1 and 2 determine fire-resistance rating for fire wall. See Table 706.4.

**Figure 706-1
Party wall.**

2012 International Building Code Handbook   217

706.2 Structural stability. The objective of a fire wall is that a complete burnout can occur on one side of the wall without any effects of the fire being felt on the opposite side. Furthermore, the only damage to the wall will be the effects of fire and the shock of hose-stream application on the fire side. The code is very clear that fire walls should remain in place for the expected time period. Therefore, structural failure on either side of the wall shall not cause the collapse of the wall, nor can the required fire-resistance rating be diminished. In addition, structural members (especially members that conduct heat) that penetrate fire walls could limit their effectiveness and do not comply with this provision. Any structural member that passes through a fire wall could also adversely affect the integrity of the required fire-resistance-rated construction.

The intent of this section can be partially traced back to Section 101.3, which states that one of the goals of the code is to provide safety to fire fighters and emergency responders during emergency operations. During a fire, a fire wall provides a safe haven on the nonfire side for fire fighters to stage and fight a fire. It is critical that the fire wall does not pose a threat of collapse to the fire department personnel. This is more easily achieved where the fire wall is a nonbearing wall and is not penetrated by load-bearing elements. However, where a fire wall is proposed as a bearing wall, the building official should ensure that those structural members that frame into the wall will not cause the premature collapse of the fire wall prior to the hourly rating established for the wall. The structural engineer of record should provide evidence to this fact. If all structural elements framing into the fire wall, as well as their supporting members, have the same fire-resistance rating as the fire wall, it is reasonable to assume that the intent of the provision has been met.

As an option to a single fire wall, the code permits the use of a double fire wall if designed and constructed in accordance with NFPA 221. Double fire walls are simply two back-to-back walls, each having an established fire-resistance rating. While acceptable for use in a new structure, double fire walls are most advantageous where an addition is being constructed adjacent to an existing building and the intent is to regulate the addition as a separate building under the fire wall provisions. The exterior wall of the existing building, if compliant, can be used as one wall of the double wall system, with the new wall of the addition providing the second wall.

Double fire wall assemblies are to comply with the applicable provisions of NFPA 221, *Standard for High Challenge Fire Walls, Fire Walls,* and *Fire Barrier Walls*. This standard addresses a number of criteria for double fire walls, including fire-resistance rating, connections, and structural support. In order to meet the minimum fire-resistance rating for a fire wall as set forth in IBC Table 706.4, each individual wall of a double fire wall assembly is permitted to be reduced to 1 hour less than the minimum required rating for a single fire wall. For example, where IBC Table 706.4 requires the use of a minimum 3-hour fire wall, two minimum 2-hour fire-resistance-rated (double) fire walls are required. Similarly, two 3-hour fire walls in a double wall system can be considered as a single 4-hour fire wall, and two 1-hour fire walls used as a double wall qualify as a single 2-hour fire wall.

Since the intended goal of fire wall construction is to allow collapse of a building on either side of the fire wall while maintaining an acceptable level of fire separation, the only connection permitted by NFPA 221 between the two walls that make up the double fire wall is the flashing, if provided. Illustrated in the explanatory material to the standard, the choice of flashing methods must provide for separate flashing sections in order to maintain a complete physical separation between the walls. Each individual wall of the double wall assembly must be supported laterally without any assistance from the adjoining building. In addition, a minimum clear space between the two walls is recommended by NFPA 221 in order to allow for thermal expansion between unprotected structural framework, where applicable, and the wall assemblies that make up the double fire wall.

706.3 Materials. In buildings of other than Type V construction, fire walls shall be constructed of noncombustible materials. The high degree of protection expected from a fire wall mandates that noncombustible construction be used for all but the lowest type of construction.

Fire-resistance rating. It is obvious that a fire wall performs the very important function of acting as a barrier to fire spread so that a fire on one side of the wall will not be transmitted to the other. On this basis, the fire wall must have a fire-resistance rating commensurate with the occupancy and type of construction of which it is constructed. The IBC provides that fire walls be of either 2-hour, 3-hour, or 4-hour fire-resistance-rated construction as specified in Table 706.4. Where the type of construction and/or occupancy group that occurs on one side of a fire wall is inconsistent with that on the other side, the more restrictive fire-resistance rating set forth in Table 706.4 shall apply. See Figure 706-2. Permitted openings in fire walls are addressed in Section 706.8.

706.4

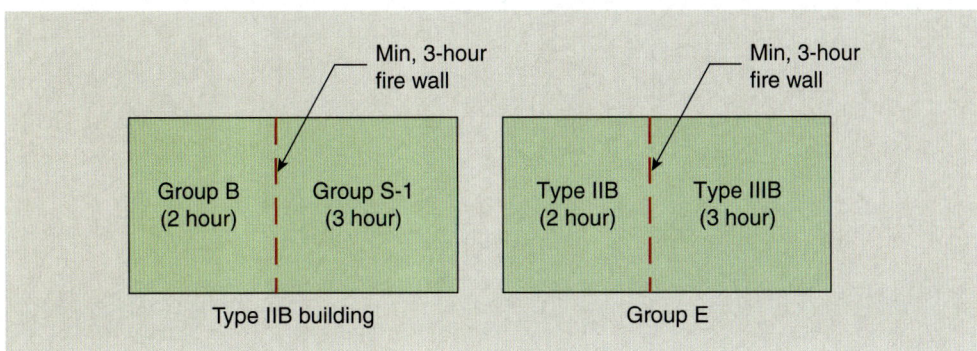

Figure 706-2
Fire-resistance rating.

Horizontal continuity. A fire wall must not only separate the interior portions of the building but must also extend at least 18 inches (457 mm) beyond the exterior surfaces of exterior walls. See Figure 706-3. A number of exceptions permit the fire wall to terminate at the interior surface of the exterior finish material, with Exception 1 illustrated in Figure 706-4. Where combustible sheathing or siding materials are used, the wall must be protected for at least 4 feet (1,220 mm) on both sides of the fire wall by minimum 1-hour construction with any openings protected at least 45 minutes. If the sheathing, siding, or other finish material is noncombustible, such noncombustible materials shall extend at least 4 feet (1,220 mm) on both sides of the fire wall; however, unlike the previous exception, no opening protection is required. As an option, where the separate buildings created by the fire wall are sprinklered, the fire wall may simply terminate at the interior surface of noncombustible exterior sheathing.

706.5

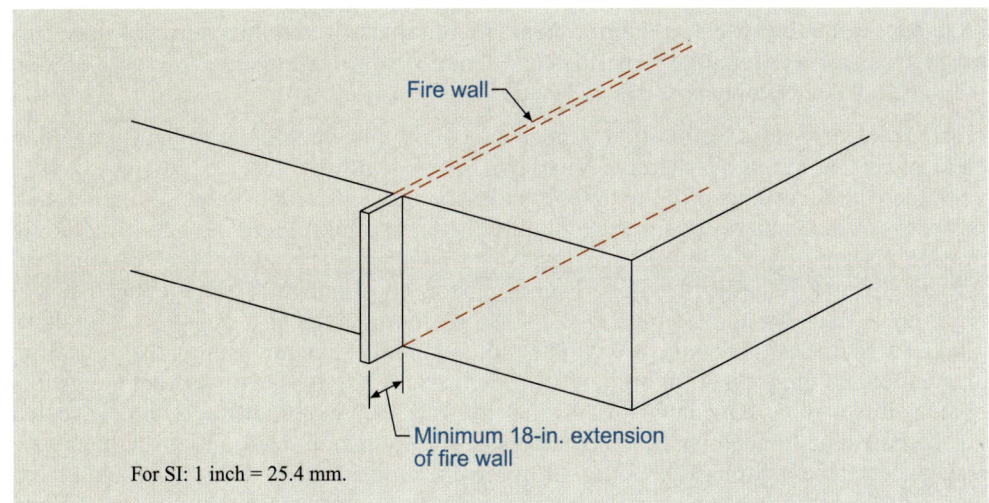

Figure 706-3
Horizontal continuity.

2012 International Building Code Handbook

7 Fire and Smoke Protection Features

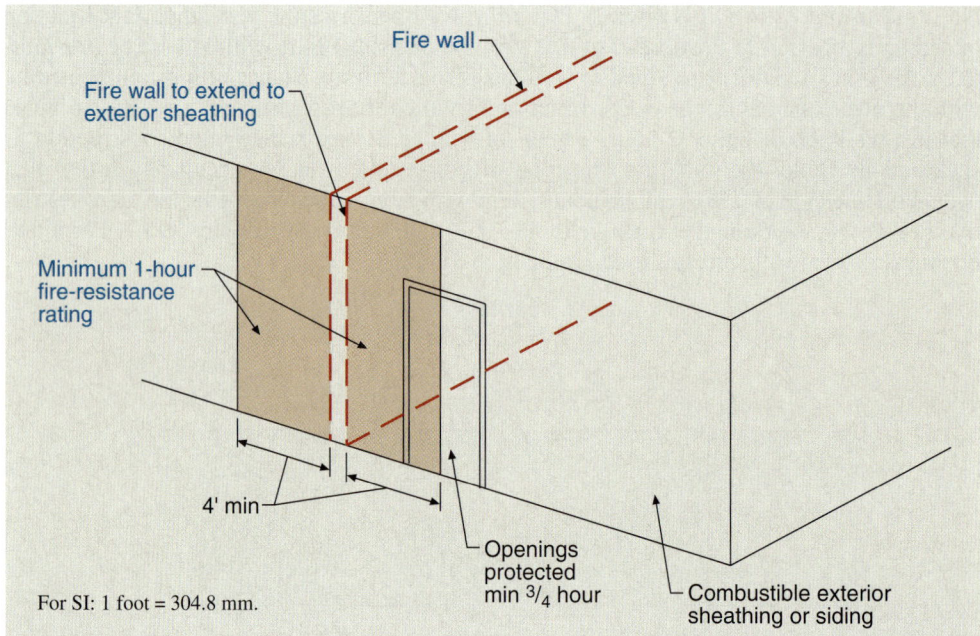

Figure 706-4
Horizontal continuity.

706.5.1 Exterior walls. Where a fire wall creating separate buildings intersects with the exterior wall, there is the potential for direct fire exposure between the buildings at the exterior. Unless the intersection of the exterior wall and the fire wall forms an angle of at least 180 degrees (3.14 rad), such as a straight exterior wall with no offsets, a condition occurs similar to that of two buildings located on the same site. The proximity of the two buildings may be such that the distance between them would allow for direct fire or substantial radiant heat to be transferred from one building to the other. This condition is also possible where the two buildings on the lot are portions of a larger structure with fire wall separations.

Where the fire wall intersects the exterior wall to form an angle of less than 180 degrees (3.14 rad), the exterior wall for at least 4 feet (1,220 mm) on both sides of the fire wall shall be of minimum 1-hour fire-resistance-rated construction, and all openings within the 4-foot (1,220-mm) portions of the exterior wall are to be protected with 45-minute fire assemblies. See Figure 706-5. As an option, an imaginary lot line may be assumed between the two buildings created by the fire wall and the exterior wall, and opening protection would be based on the fire separation distances to the imaginary lot line. This method is consistent with the provisions of Section 705.3 for addressing two buildings on the same lot. An example is shown in Figure 706-6.

706.5.2 Horizontal projecting elements. Under the conditions where a horizontal projecting element such as a roof overhang or balcony is located within 4 feet (1,220 mm) of a fire wall, the wall must extend to the outer edge of the projection. This general requirement provides for a complete separation by totally isolating all building elements, including projections, on either side of a fire-resistance-rated wall. However, such a condition is typically not visually pleasing. Therefore, the code indicates the fire wall is not required to extend to the leading edge of the projecting element if constructed in compliance with one of three exceptions. The protection must extend through the projecting element unless the projection has no concealed spaces. Where the projecting element is combustible and has concealed spaces, the fire wall shall extend through the concealed area, whereas in noncombustible construction, the extension need only be 1-hour fire-resistance-rated construction. Under all of the exceptions, the exterior wall behind and below the projecting element is to be of 1-hour fire-resistance-rated construction for

Fire Walls 7

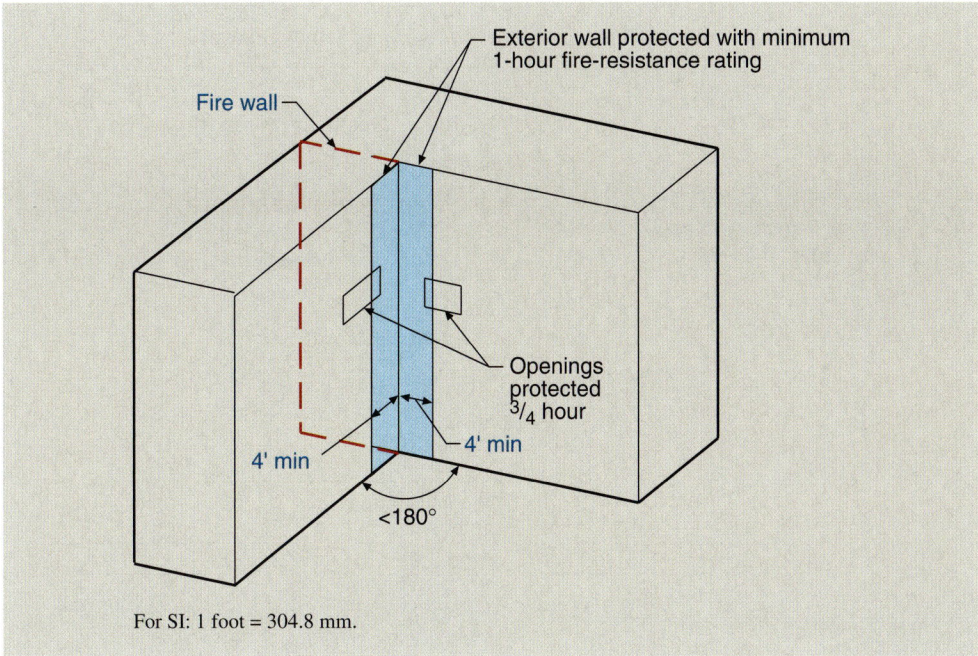

Figure 706-5
Fire wall intersection with exterior walls.

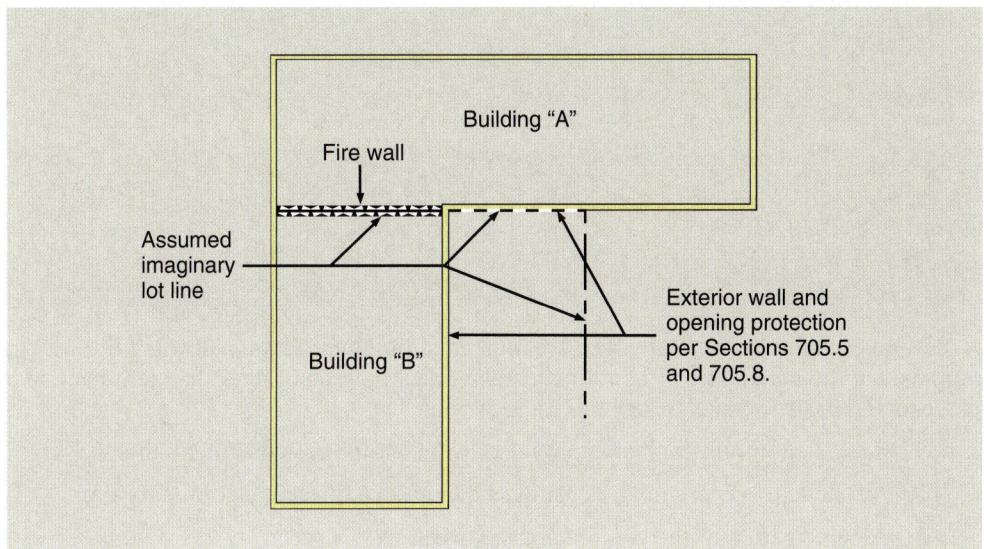

Figure 706-6
Imaginary lot line at extension of fire wall.

a distance not less than the depth of the projecting element on both sides of the wall. All openings within the rated exterior wall are to be protected by fire assemblies having a minimum fire-protection rating of 45 minutes. Figure 706-7 depicts these various conditions addressed in the exceptions.

Vertical continuity. Having established the intent of the IBC that fire walls prevent the spread of fire around or through the wall to the other side, the IBC further ensures the separate building concept by specifying that the wall shall extend continuously from the foundation to (and through) the roof to a point 30 inches (762 mm) or more above the roof. The 30-inch (762-mm) parapet prevents the spread of fire along the roof surface from the fire side to the other side of the wall.

706.6

7 Fire and Smoke Protection Features

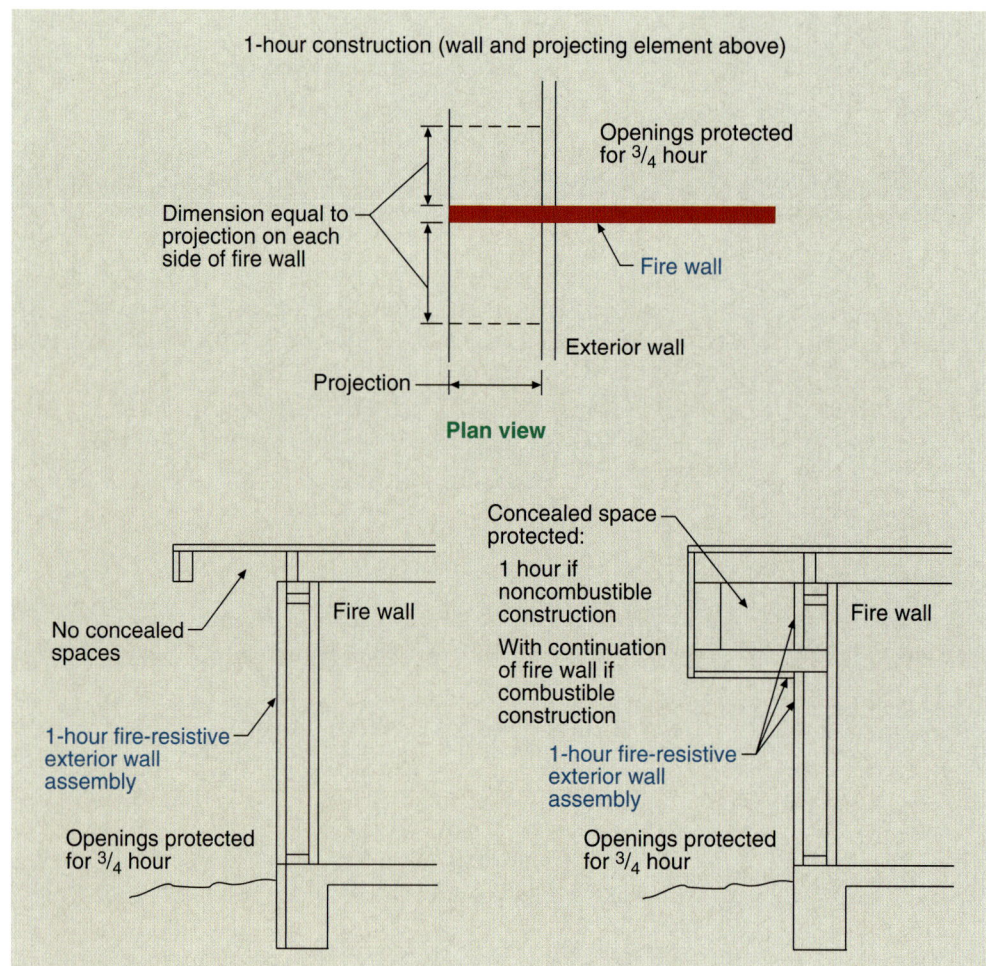

Figure 706-7
Horizontal projecting elements.

Several exceptions, some of which are illustrated in Figure 706-8, allow the fire wall to terminate at the underside of the roof sheathing, deck, or slab, rather than terminate in a parapet. The basis for such exceptions includes:

1. Equivalent protection being provided by an alternative construction method;
2. Aesthetic considerations, as parapets disrupt the appearance of the roof; or
3. A combination of the two previous reasons.

It is emphasized that the term *fire wall* also limits its use to vertical walls. Therefore, there can be no horizontal offsets nor can the plan view of the wall change from level to level. See Figure 706-9.

706.6.1 Stepped buildings. Quite often, a fire wall is provided at a point in the building where the roof changes height. Under such conditions, the fire wall must extend above the lower roof for a minimum height of 30 inches (762 mm). In addition, the exterior wall shall be of at least 1-hour fire-resistance-rated construction for a total height of 15 feet (4,572 mm) above the lower roof. The exterior wall shall be of fire-resistance-rated construction from both sides. Any opening that is located in the lower 30 inches (762 mm) of the wall shall be regulated based on the rating of the fire wall. Openings above the 30-inch (762-mm) height, but not located above a height of 15 feet (4,572 mm), shall have a minimum fire-protection rating of 45 minutes. An illustration of this provision is depicted in Figure 706-10.

Fire Walls 7

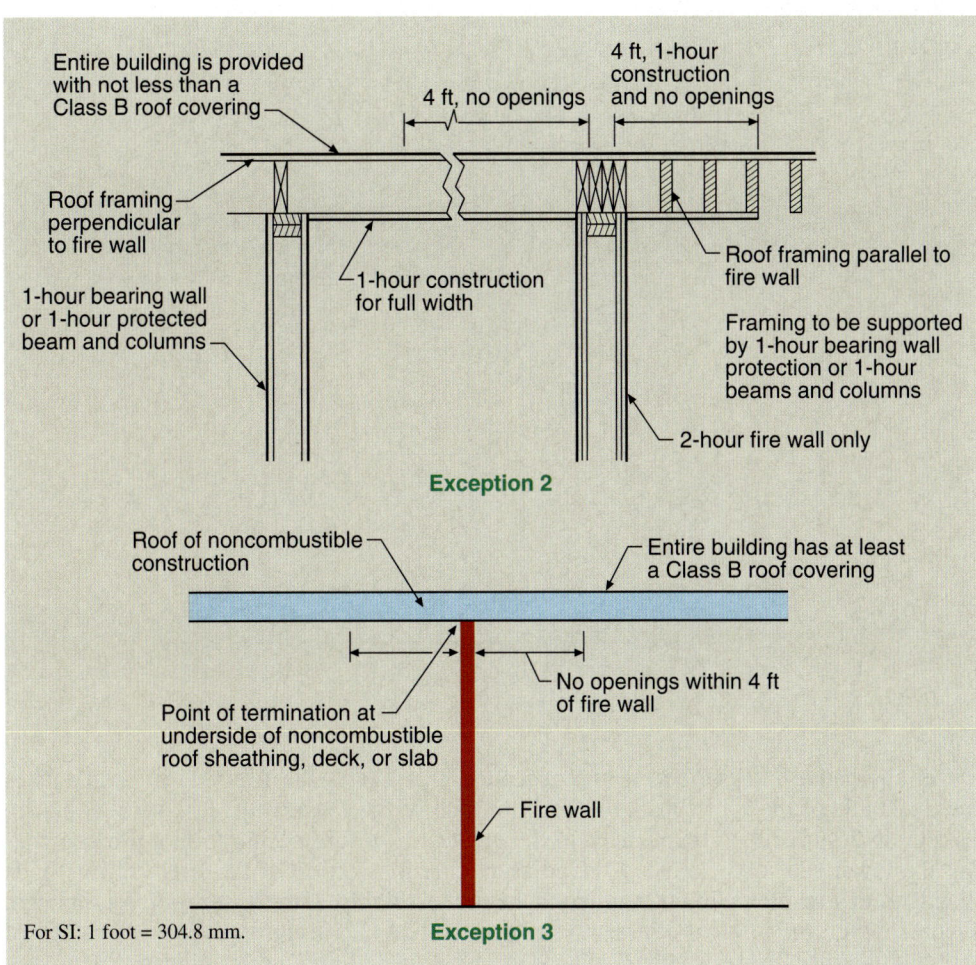

Figure 706-8
Termination of fire walls.

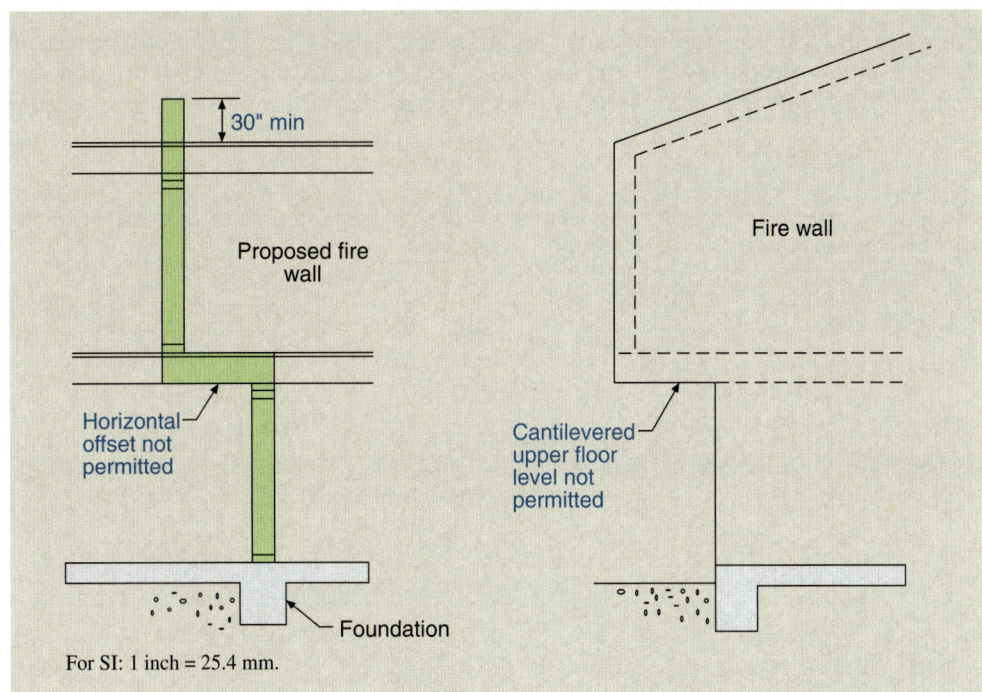

Figure 706-9
Fire wall vertical community.

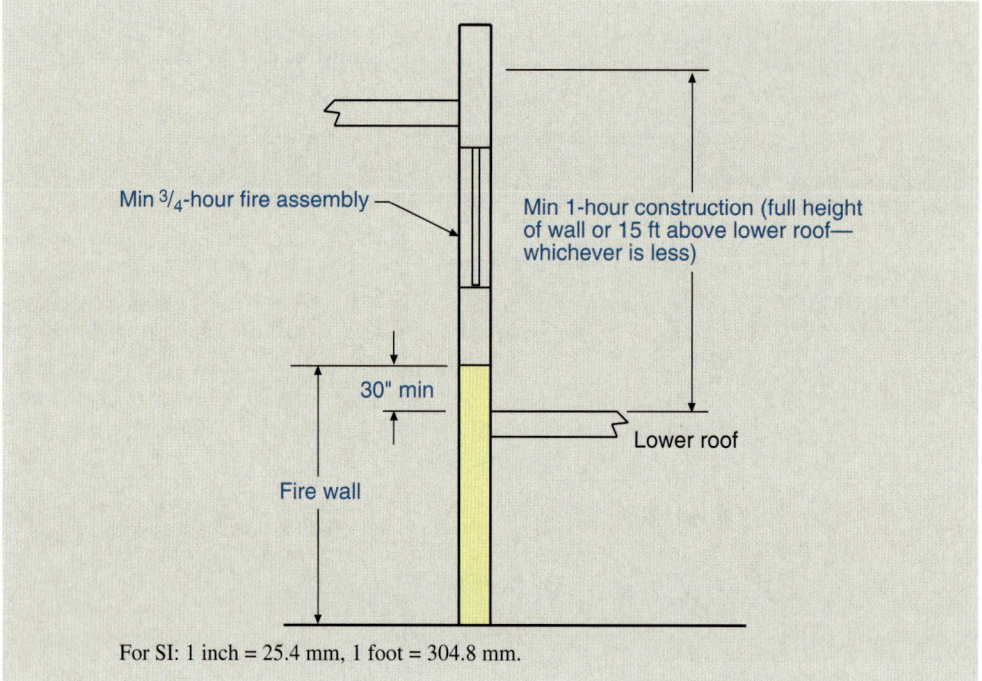

Figure 706-10 Stepped buildings.

An alternative is described in the exception that allows the fire wall to terminate at the underside of the roof sheathing, deck, or slab of the lower roof. It is very similar to Exception 2 in Section 706.6. Because the greatest exposure occurs from a fire penetrating the lower roof and exposing the adjacent exterior portion of the fire wall, the code mandates that all protection be applied to the roof assembly of the lower roof. As shown in Figure 706-11, the lower roof assembly within 10 feet (3,048 mm) of the wall shall be of minimum 1-hour fire-resistance-rated construction. In addition, no openings are permitted in the lower roof within 10 feet (3,048 mm) of the fire wall.

706.7 Combustible framing in fire walls. This section defines the limitations placed on the penetration of combustible framing into concrete or masonry fire walls. Adjacent combustible members framing into a concrete or masonry fire wall from opposite sides

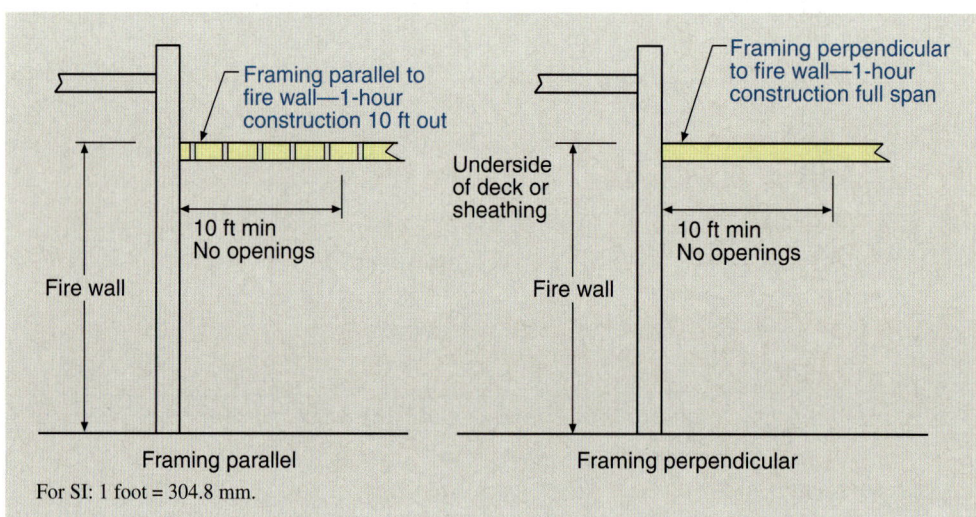

Figure 706-11 Stepped buildings.

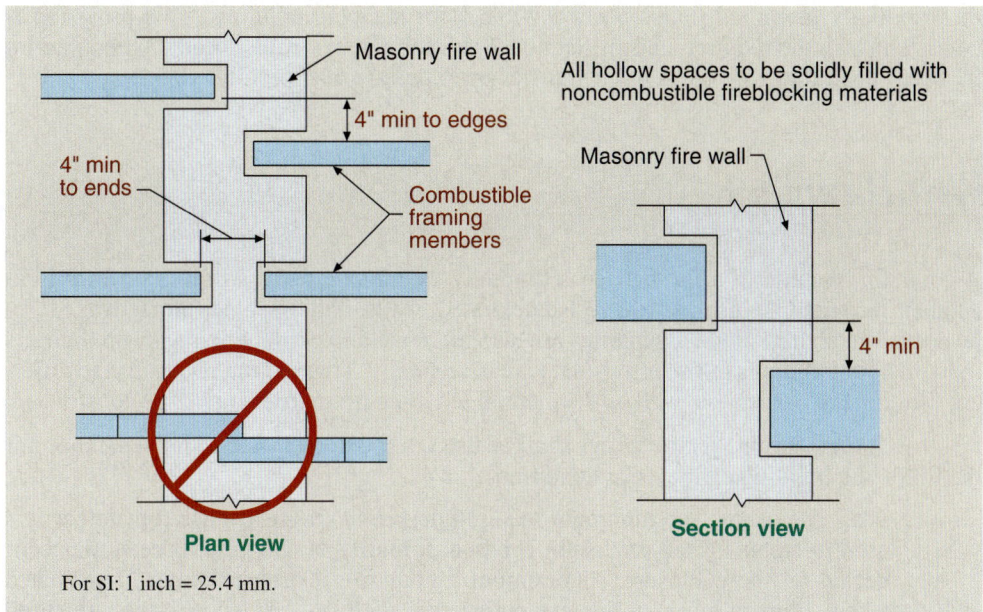

Figure 706-12
Combustible framing in fire walls.

require a minimum distance of 4 inches (102 mm) between embedded ends, as shown in Figure 706-12. All hollow spaces at this location shall be solidly filled for the full thickness of the wall and for a distance not less than 4 inches (102 mm) above, below, and between the structural members with noncombustible materials approved for fireblocking.

Openings. In order to provide for the efficient use of a structure containing one or more fire walls, the code permits their penetration with protected openings. The fire-protection rating required for openings is addressed in Table 716.5. Each opening is limited to 156 square feet (15 m²), with multiple openings permitted. However, the aggregate width of all openings at any floor level is limited to 25 percent of the length of the fire wall. See Figure 706-13. The limit of 156 square feet (15 m²) per opening does not apply in fully sprinklered buildings.

706.8

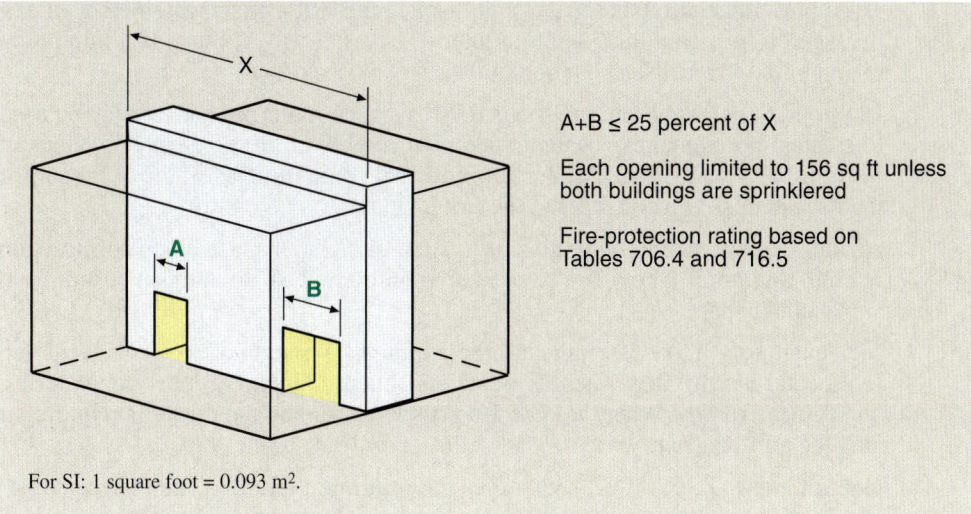

Figure 706-13
Openings in fire walls.

A party wall, defined as a fire wall located on a lot line between two adjacent buildings and used for joint service of the buildings, is prohibited by Section 706.1.1 from having any doors, windows, or other openings. This would include any openings above the lower roof normally permitted under Section 706.6.1 for stepped buildings. This provision is consistent with the requirements of Section 705.8 for exterior walls located on an interior lot line. Penetrations by ducts or air-transfer openings are also prohibited.

Section 707 *Fire Barriers*

A common function of a fire barrier is to totally isolate one portion of a floor level from another through the use of fire-resistance-rated walls and opening protectives. Fire-resistance-rated horizontal assemblies are also often used in conjunction with fire barriers in multistory buildings in order to isolate areas vertically. This section identifies the different uses for fire barriers, as well as the method in which fire barriers are to be constructed.

707.3 Fire-resistance rating. A fire barrier shall be used to provide the necessary separation for the following building elements or conditions:

1. Shaft enclosure. The minimum required degree of fire-resistance for fire barriers used to create a shaft enclosure is based primarily on the number of stories connected by the enclosure. A minimum 2-hour fire-resistance rating is mandated where four or more stories are connected, with only a 1-hour rating required where connecting only two or three stories. In all cases, the rating of the fire barriers creating a shaft enclosure must equal or exceed that of the floor assembly that is penetrated by the enclosure.

2. Interior exit stairway construction. The separation between an interior exit stairway (stair enclosure) and the remainder of the building shall be accomplished with fire barriers having either a 1- or 2-hour fire-resistance rating, as required by Section 1022.2. Similar enclosures are required for interior exit ramps.

3. Exit access stairway enclosures. Where exit access stairways are required to be enclosed by Section 1009.3, the enclosure shall include the use of fire barriers.

4. Exit passageway. An exit passageway must be isolated from the remainder of the building by minimum 1-hour fire-resistance-rated fire-barrier walls. Where horizontal enclosure is also required, minimum 1-hour fire-resistance-rated horizontal assemblies must also be used to totally isolate the exit passageway. Where an exit passageway is a continuation of an interior exit stairway, it must, at a minimum, maintain the fire-resistance rating of the stairway enclosure.

5. Horizontal exit. A minimum 2-hour fire-resistance-rated fire barrier may be used to create a horizontal exit when in compliance with all of the other provisions of Section 1025. The fire barrier creates protected compartments where occupants of the building can travel to escape the fire incident.

6. Atrium. Unless a complying glazing system or $^3/_4$-hour glass block construction is used, minimum 1-hour fire barriers are required when isolating an atrium from surrounding spaces.

7. Incidental uses. Table 509 indicates the required separation or protection required for special hazard areas such as waste and linen collection rooms, laboratories, and furnace rooms. Where a 1- or 2-hour fire-resistance-rated wall is required, it shall be a fire barrier.

8. Control areas. Table 414.2.4 identifies the minimum required fire-resistance rating for fire barriers used to create control areas in buildings housing hazardous materials. A minimum rating of 1 hour is mandated for separating control areas located on the first three floor levels above grade plane, whereas minimum 2-hour fire barriers are required for control area separations on all floor levels above the third level.

9. Separated occupancies. The separation of dissimilar occupancies in the same building is accomplished by fire barriers. Table 508.4 is used to determine the required fire-resistance rating of the required fire barriers, ranging from 1 hour through 4 hours.

10. Fire areas. Where a building is divided into fire areas by fire barriers in order to not exceed the limitations of Section 903.2 for requiring an automatic sprinkler system, the minimum required fire-resistance ratings of the fire barriers are set forth in Table 707.3.10. Ranging from a minimum of 1 hour to a maximum of 4 hours, the fire-resistive requirements are based solely on the occupancy classification of the fire areas. The provisions are applicable to both single-occupancy and mixed-occupancy conditions. See the discussion on Section 901.7 for further information.

Note also that fire barriers are required as separation elements in other miscellaneous locations identified by the code, such as stage accessory areas (Section 410.5) and flammable finish spray rooms (Section 416.2). Throughout the code, references are made to fire barriers as the method of providing the appropriate fire-resistance-rated separation intended. In addition, many of the other *International Codes* also address the use of fire barriers to create protected areas.

Continuity. Fire barriers must begin at the floor and extend uninterrupted to the floor or roof deck above. Where there is a concealed space above a ceiling, the fire barrier must continue through the above-ceiling space. See Figure 707-1. Fireblocking, required only in combustible construction, must be installed at every floor level if the fire barrier contains hollow vertical spaces. The intent of a fire barrier is to provide a continuous separation so as to completely isolate one area from another. As with many other fire-resistance-rated elements, the supporting construction must be of an equivalent rating to the fire barrier supported. A reduction relates to 1-hour incidental use separations in nonrated construction.

707.5

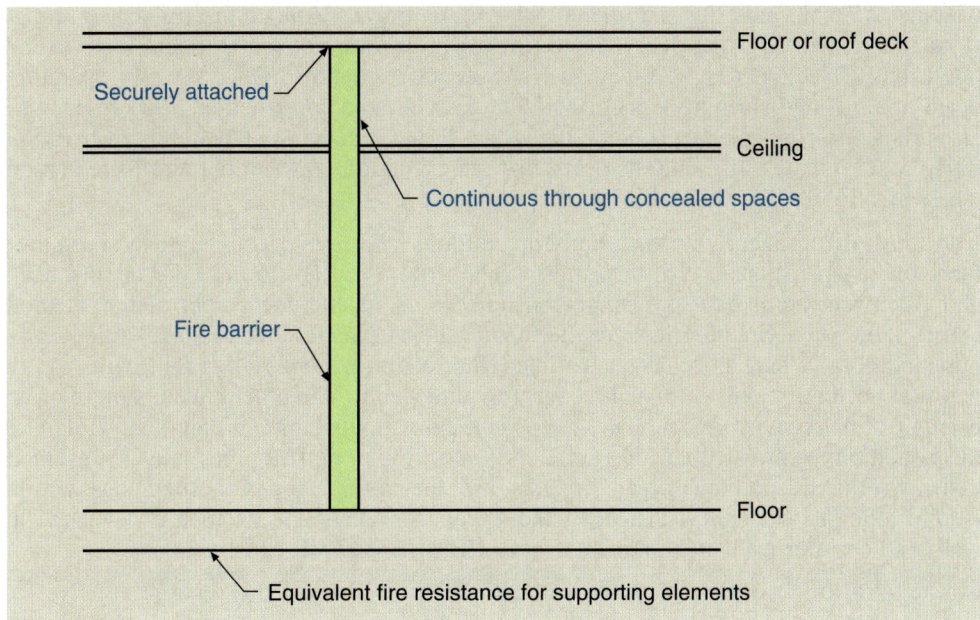

Figure 707-1
Fire barrier continuity.

Openings. The provisions of Section 716 regulate the protection of openings in fire barriers. The fire-protection ratings mandated for fire-barrier openings in Tables 716.5 and 716.6 vary depending on the fire-resistance rating of the fire barrier as well as its purpose. The required rating may be as little as $3/4$ hour to as much as 3 hours.

707.6

707.9 Voids at intersections. It is not uncommon for a void to be created at the joint between a fire barrier and the floor or roof deck above. Where the joint occurs at a fire-resistance-rated floor or roof deck, Section 715.1 mandates that the joint be protected by an approved fire-resistant joint system. Section 715.1 is also applicable where the joint occurs between a fire barrier and a nonfire-rated floor. Section 707.9 is only intended to address those situations where the roof assembly is not fire-resistance rated. The void need only be protected with an approved material that is securely installed and capable of retarding the passage of fire and hot gases.

Section 708 *Fire Partitions*

This section regulates the design and construction of fire partitions installed in the listed locations. The IBC identifies five locations where fire partitions are required:

1. Walls separating dwelling units per Section 420.2
2. Walls separating sleeping units per Section 420.2
3. Walls separating tenant spaces in covered and open mall buildings as required by Section 402.4.2.1
4. Walls of fire-resistance-rated corridors per Section 1018.1
5. Elevator lobby separation as required by Section 713.14.1

708.3 Fire-resistance rating. The minimum fire-resistance rating of fire partitions is to be 1 hour, unless a reduction is permitted by one of two exceptions. Exception 1 refers to Table 1018.1, which identifies the required fire-resistance rating of a corridor based on three factors—the occupancy classification of the area served by the corridor, the occupant load the corridor serves, and whether or not the building is sprinklered. Where conditions warrant, the table indicates that the corridor needs only a $1/2$-hour fire-resistance rating. If no fire-resistance rating is mandated, fire partitions are not required and none of the provisions of Section 708 are applicable. The second exception applies to walls separating dwelling units and sleeping units in buildings of nonrated construction. The presence of an automatic sprinkler system complying with NFPA 13 reduces the required fire-partition rating to 30 minutes. It should be noted that the exception does not permit this reduction where an NFPA 13R system is installed.

708.4 Continuity. Consistent with the required continuity of fire barriers, the general requirement for fire partitions is that they must extend from the floor to the floor or roof deck above. However, unlike the provisions for fire barriers, an alternative construction method is permitted where fire partitions may terminate short of the floor or roof deck under various conditions. Where a fire-resistance-rated floor/ceiling or roof/ceiling is provided, a fire partition need only extend to, and be securely attached to, the ceiling membrane. For an example of this provision as it relates to corridor construction, refer to Figure 708-1. Under this condition in combustible construction, fireblocking or draftstopping must be installed at the partition line in the concealed space above the ceiling. Any supporting construction is to be at least 1-hour fire-resistance rated, except for tenant and sleeping unit separation walls and corridor walls in buildings of Type IIB, IIIB, and VB construction.

The following exceptions modify the continuity provisions of this section:

1. Where a crawl space exists below a floor assembly of at least 1-hour fire-resistance-rated construction, the fire partition does not need to extend into the underfloor space. See Figure 708-2.
2. The arrangement shown in Figure 708-3 would meet the code requirement for adequately enclosing a corridor. The corridor walls are protected on the side of the occupied use spaces by a fire-resistance-rated membrane extending from the

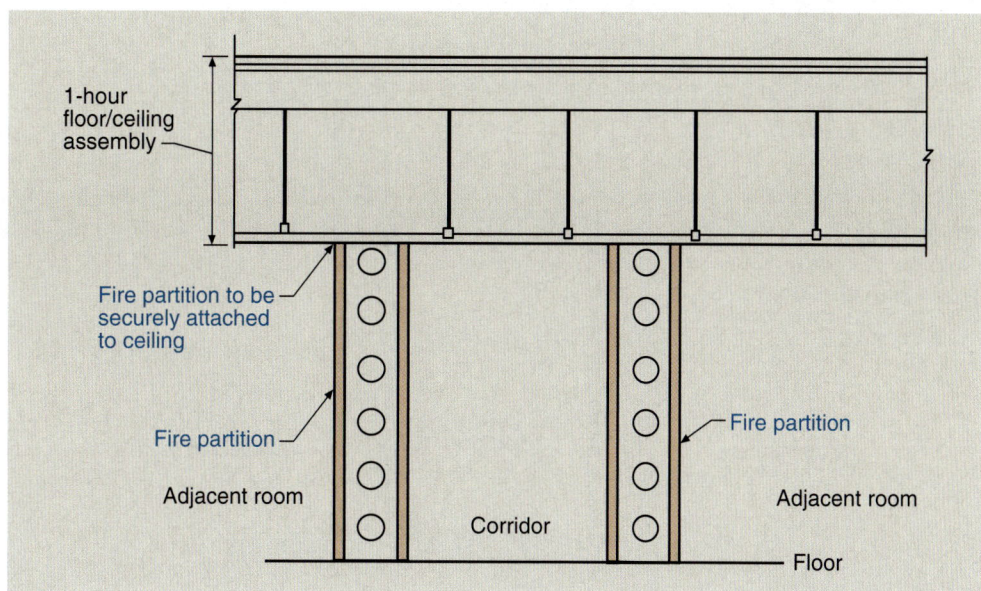

Figure 708-1
Corridor fire partitions.

floor to the floor or roof above. In this case, the ceiling over the corridor may be considered part of a fire-resistance-rated floor or roof assembly, and the corridor side of the ceiling protected by appropriate ceiling materials would satisfy the fire-resistance rating for the assembly.

3. The code provides that the corridor ceiling may be of the same construction as permitted for corridor walls as shown in Figure 708-4. In all probability, typical wall construction might not pass the 1-hour test when tested in a horizontal position. However, this arrangement, generally referred to as tunnel construction, is considered to be adequate protection for the corridor separating it from the spaces above.

By establishing various methods for the enclosure of fire-resistance-rated corridors, the code is essentially attempting to get a minimal separation between the exit corridor and the occupied-use spaces. Any arrangement of the 1-hour fire-resistance-rated construction that effectively intervenes between these use spaces and the corridor would satisfy this requirement.

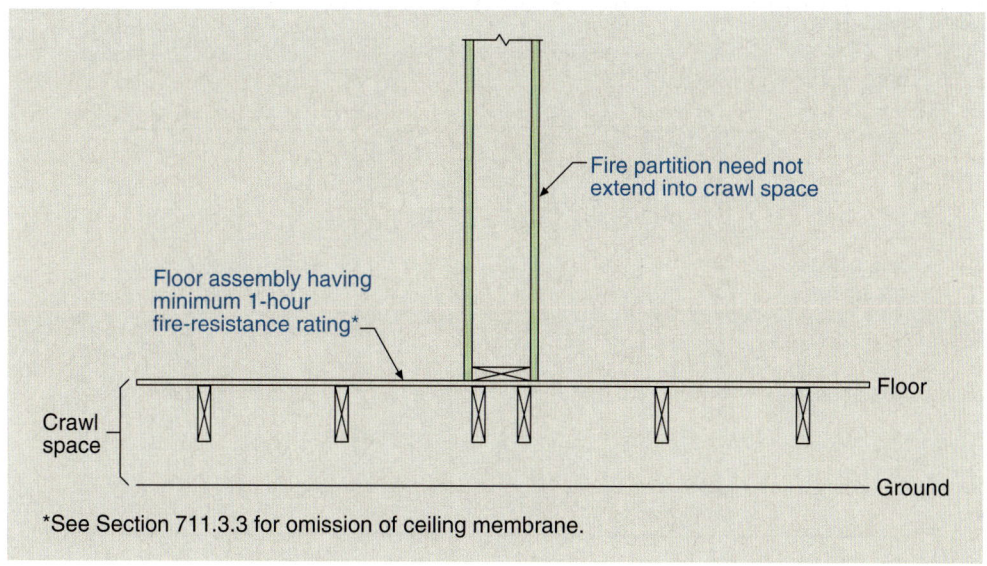

Figure 708-2
Fire partitions above a crawl space.

7 Fire and Smoke Protection Features

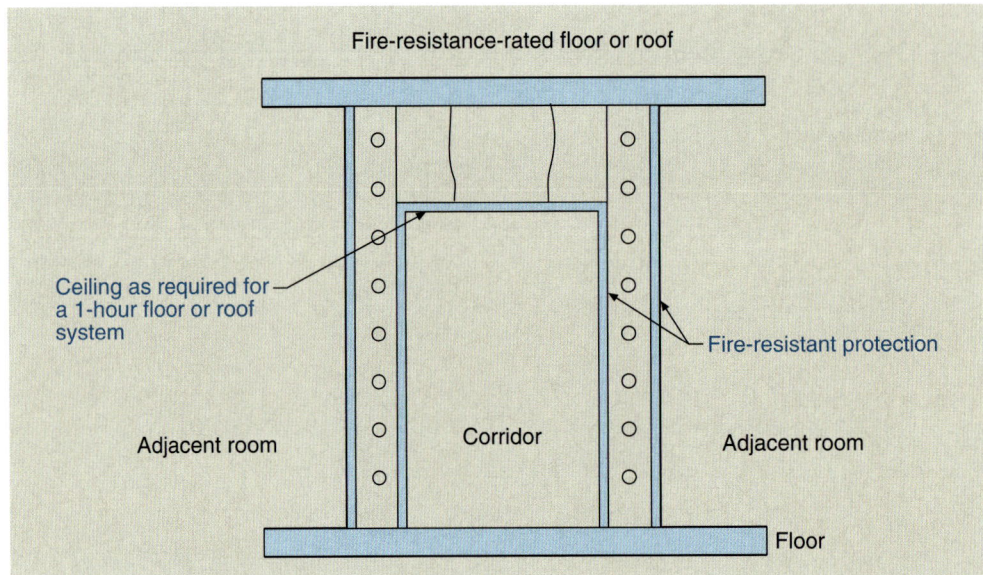

Figure 708-3
Corridor construction.

4. In covered mall buildings, fire partitions separating tenant spaces may terminate at the underside of a ceiling, even if the ceiling is not part of a fire-resistance-rated assembly. No type of extension of the fire partition is required by this section for attics and similar spaces above the ceiling.

5. In attic areas of Group R-2 occupancies less than five stories in height above grade plane, the draftstopping or fireblocking required by this section may be omitted where the attic area is subdivided by draftstopping into areas not exceeding two dwelling units or 3,000 square feet (279 m^2), whichever is less. This exception is also found in Section 718.4.2.

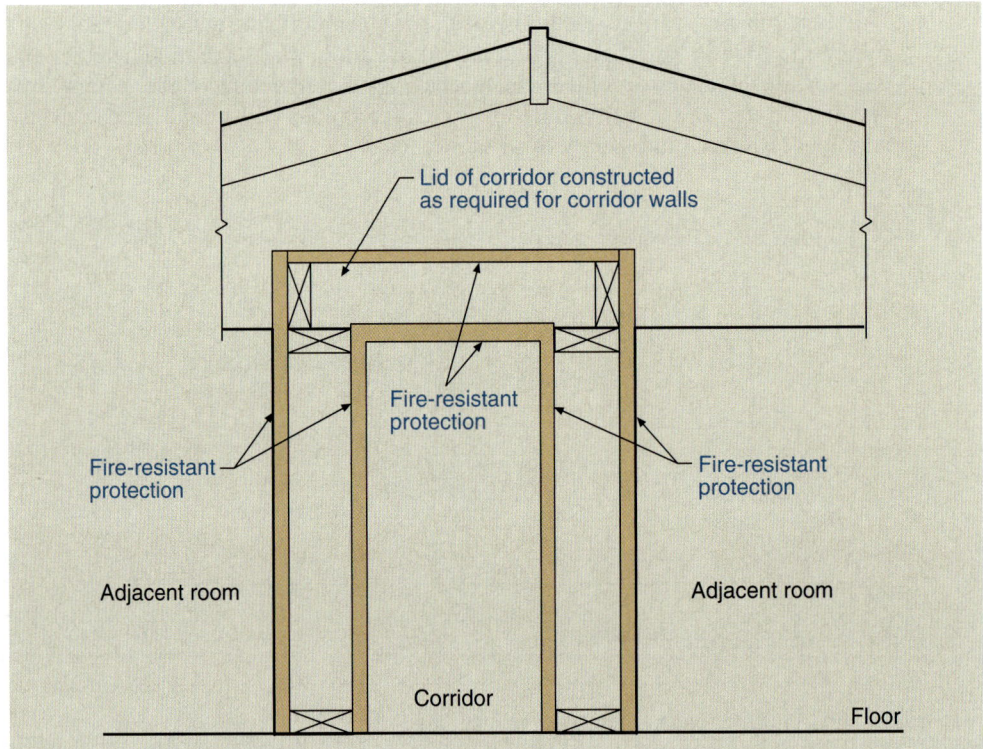

Figure 708-4
Tunnel corridor.

6. In combustible buildings where the fire partitions stop at the fire-resistance-rated ceiling membrane, fireblocking or draftstopping is not required at the partition line if the building is fully sprinklered. Under this exception, sprinklers must be installed in the combustible floor/ceiling and roof/ceiling spaces. Exceptions in Sections 718.3.2 and 718.4.2 provide the same criteria.

Exterior walls. This section clarifies that where a fire-resistance-rated separation is bounded by one or more exterior walls, the exterior wall portions of the enclosure need only comply with the provisions of Section 705. See Figure 708-5 for an example of a fire-resistance-rated corridor located along an exterior wall. Additional provisions are established in Chapter 10 for exterior egress balconies, interior exit stairways and ramps, and exterior exit stairways and ramps.

708.5

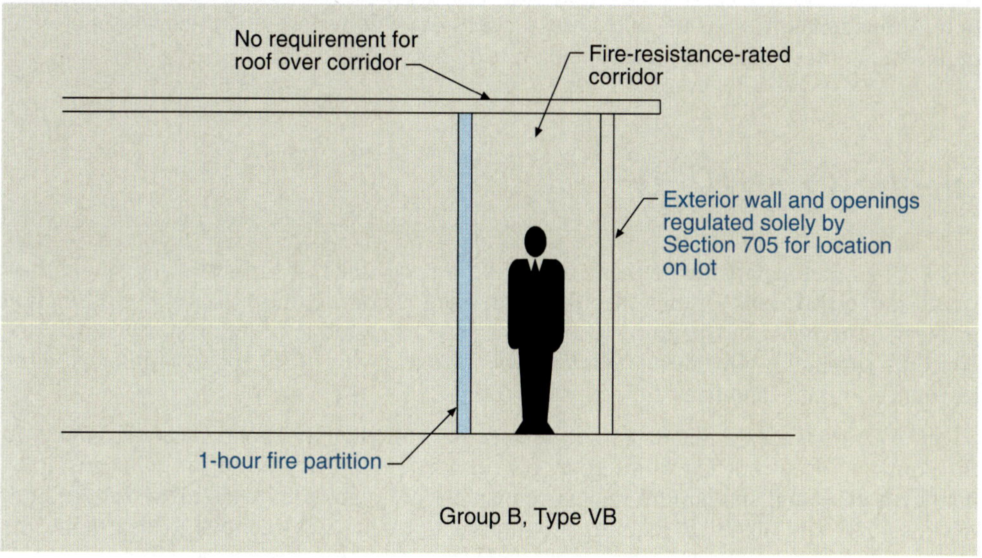

Figure 708-5
Corridor protection at exterior walls.

Section 709 *Smoke Barriers*

Smoke barriers are occasionally mandated by the code to resist the passage of smoke from one area to another. The use of smoke barriers is assigned to those portions of buildings intended to provide refuge to occupants who may not be able to exit the building in a timely manner. In such cases, relocation, rather than evacuation, is the initial approach to an emergency condition. For example, smoke barriers are used in areas of refuge (Section 1007.6.2), in smoke-control systems (Section 909.5), in Group I-3 occupancies (Section 408.6), and in various other building areas where smoke transmission is a concern. By far the most common use of smoke barriers is in Group I-2 occupancies, where they are used to create smoke compartments (Section 407.5). Smoke barriers must not only resist the passage of smoke, they must also be of minimum 1-hour fire-resistance-rated construction. In Group I-3 occupancies, an exception permits the use of 0.10-inch-thick (2.5 mm) steel in lieu of 1-hour construction.

The key to the construction of a smoke barrier is that all avenues for smoke to travel outside of the compartment created by the smoke barrier are eliminated. This requires the membrane to be continuous from outside wall to outside wall where creating smoke compartments in Group I-2 and I-3 occupancies, and from the floor slab to the floor or roof deck above. The smoke barriers must continue through all concealed spaces, such as those above ceilings, unless the ceilings provide the necessary resistance against fire and smoke passage. In buildings of rated construction, all smoke barriers shall be supported by construction consistent with the fire-resistance rating of the wall or floor supported.

All door openings in smoke barriers are to be protected with assemblies having a minimum fire-protection rating of 20 minutes, per Table 716.5. In cross-corridor situations in Group I-2 occupancies, the code mandates a pair of opposite-swinging doors installed without a center mullion. Such doors shall be provided with an approved vision panel; be close fitting; have no louvers or grilles; and undercuts are limited to $^3/_4$ inch. Although positive latching is not required, the doors are to have head and jamb stops, astragals, or rabbets at meeting edges, and automatic-closing devices.

Smoke barriers, like smoke partitions regulated by Section 710, are only mandated where specifically identified by the code. As an example, Section 509.4.2 requires the use of "construction capable of resisting the passage of smoke" as a potential physical separation for incidental uses. Thus, only construction that will perform the intended function is required, and not necessarily a smoke barrier or smoke partition.

Section 710 Smoke Partitions

Unlike the other separation elements used in the code, such as fire walls and smoke barriers, smoke partitions are not specifically defined. Their definition is simply a function of the requirements of Section 710. The purpose of a smoke partition is limited to the concerns of smoke movement under fire conditions, with no intent to regulate for the resistance to flame and heat.

Smoke partitions are mandated by the code in limited applications, most commonly in the construction of corridors in Group I-2 occupancies. As such, corridors in hospitals, nursing homes, and similar Group I-2 occupancies are regulated by the provisions of both Sections 710 and 407.3. It requires a comparison of the two sections to determine the requirements for corridor systems, particularly corridor doors and air openings.

710.5 Openings. This section prohibits the installation of louvers in doors in smoke partitions. This is consistent with the provisions of Section 407.3.1 mandating an effective barrier against the transfer of smoke. The provisions of this section also require that doors in smoke partitions be tested in accordance with UL 1784 and be self-closing or automatic closing, but only where required elsewhere in the code. A review of Section 407.3.1 regulating corridor doors in Group I-2 occupancies does not require the UL test, and it specifically states that self-closing or automatic-closing devices are not required. In this case, the provisions in Section 407.3.1 take precedence.

710.8 Ducts and air transfer openings. The provisions of both Sections 717.5.7 and 710.8 mandate the need for smoke dampers in air transfer openings that occur in smoke partitions. Smoke dampers are not required at duct penetrations of smoke partitions, but only at unducted air openings.

Section 711 Horizontal Assemblies

This section is applicable where floor and roof assemblies are required to have a fire-resistance rating. This will occur where the type of construction mandates protected floor and roof assemblies, such as in Type I, IIA, IIIA, and VA construction, and where the floor assembly is used to separate occupancies or create separate fire areas. For example, in a building of Type IIA construction, Table 601 requires minimum 1-hour fire-resistance-rated

Horizontal Assemblies

floor construction. As another example, where the floor separates a Group A-2 occupancy from a Group B, Table 508.4 addressing separated occupancies mandates a 1- or 2-hour separation.

As referenced in Section 420.3, complementary to the provisions of Section 708 for fire partitions, floor assemblies separating dwelling units or sleeping units are required to be of at least 1-hour fire-resistance-rated construction. An exception reduces the required level of protection for the floor assembly to $1/2$ hour in buildings of Type IIB, IIIB, or VB construction, provided the building is protected by an automatic sprinkler system.

711.3.1 Ceiling panels. The protection of a ceiling membrane also includes the adequacy of the panelized ceiling system to withstand forces generated by a fire and other forces that may try to displace the panels. These forces can generate positive pressures in a fire compartment that need to be counteracted. As a result, lay-in ceiling panels that provide a portion of the fire resistance of the floor/ceiling or roof/ceiling assembly should be capable of resisting this upward or positive pressure so that the panels stay in position and continue to maintain the integrity of the system. The code defines the pressure to be resisted as 1 pound per square foot (48 Pa). Section 711.3.2 permits the installation of access doors in ceilings where they are tested in accordance with ASTM E 119 or UL 263 as horizontal assemblies.

711.3.3 Unusable space. Figure 711-1 illustrates how this provision is applied in regard to unusable spaces such as crawl spaces and attics. For 1-hour fire-resistance-rated floor construction over a crawl space, the ceiling membrane is not necessary in the crawl-space area. Similarly, in 1-hour fire-resistance-rated roof construction, the floor membrane is not required in the attic. Note that the elimination of the membranes in the attic and crawl space is only applicable where the required rating of the floor or roof assembly is a maximum of 1 hour.

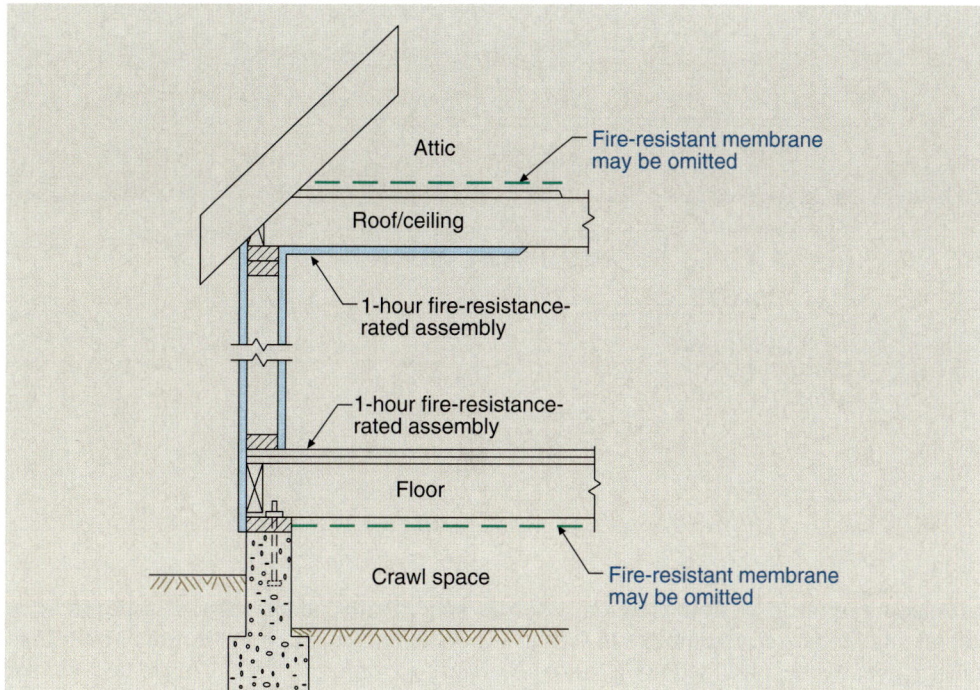

Figure 711-1
Omission of ceiling or floor membrane.

711.4 Continuity. Unless otherwise permitted by the provisions for vertical openings, penetrations, fire-resistive joint systems, or stairway enclosures, horizontal fire-resistance-rated assemblies are to be continuous without any openings. The installation of unprotected skylights and other penetrations through the roof deck of a fire-resistance-rated roof assembly is permitted, provided the structural integrity of the roof construction is maintained.

This section also mandates that the horizontal assembly be supported by structural members or walls having at least the equivalent fire rating as that for the horizontal assembly. For example, in a Type IIA school building of two stories where the floor construction is required to be a 2-hour fire-resistance-rated assembly in order to separate fire areas, any walls or structural members in the first story supporting the second floor would be required to also be of 2-hour fire-resistance-rated construction. This would be the case even though the building generally is required to be only of 1-hour fire-resistance-rated construction. Obviously, if the horizontal assembly is not supported by equivalent fire-resistance-rated construction, the intent and function of the separation are negated if its supports fail prematurely.

711.9 Smoke barrier. Horizontal assemblies used for smoke barrier purposes must have penetrations and joints protected in a manner similar to smoke barrier walls, as established in Sections 714.5 and 715.6. Both of these code sections mandate compliance with the appropriate test standards for air leakage purposes with the maximum air leakage rate established in the code. Penetrations shall meet the requirements of UL 1479 and joint systems shall be tested in accordance with UL 2079. As a general rule, no unprotected vertical openings are permitted.

Where a shaft enclosure housing an elevator passes through a horizontal smoke barrier assembly, the hoistway opening must be protected in accordance with the provisions of Section 713.14.1 addressing elevator lobbies. Since the purpose of the provisions in 713.14.1 is to limit the spread of smoke from floor to floor, the protection afforded by an elevator lobby is necessary to protect the opening created by the elevator shaft. It is important to note that the mandate for an elevator lobby applies to all multistory buildings regardless of the number of stories, not just those four or more stories in height. See Figure 711-2.

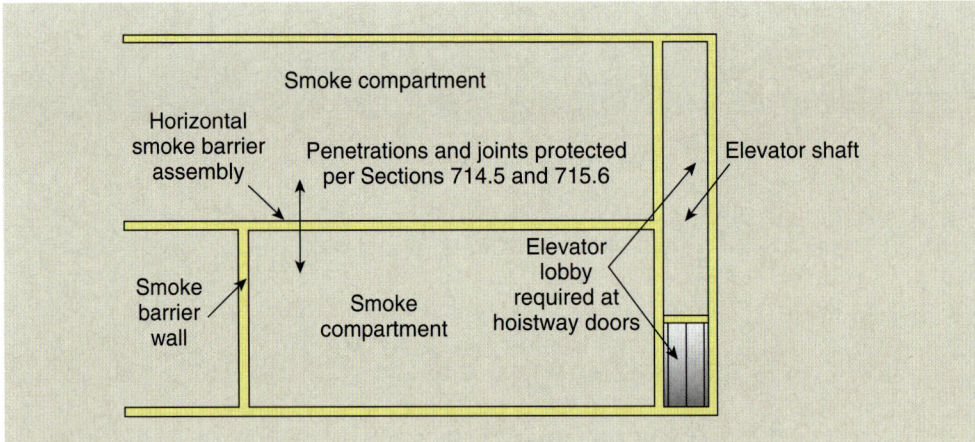

Figure 711-2
Horizontal smoke barrier.

A companion provision, Section 407.5.3, specifically addresses the issue of horizontal smoke barrier assemblies in Group I-2 occupancies. Such institutional buildings must typically be provided with smoke barriers in order to create mandated smoke compartments. Where the building is multistory, it is important that both smoke barrier walls and horizontal smoke barrier assemblies be used in order to create the necessary smoke compartmentation. The floor/ceiling assemblies between stories provide the horizontal limits of each smoke compartment when constructed in accordance with Section 711.9.

Section 712 *Vertical Openings*

It is well known that one of the primary means for the spread of fire in multistory buildings, particularly older buildings, has been the transmission of hot gases and fire upward through unprotected or improperly protected vertical openings. The primary cause of death in the hotel portion of the MGM Grand Hotel in Las Vegas, Nevada, as a result of the fire in November 1980, was the upward transmission of smoke through inadequately protected elevator shafts, stair shafts, and heating and ventilating shafts. It is because of this potential for fire spread vertically through buildings that this section requires that vertical openings be appropriately regulated. This section identifies the following 18 applications for addressing vertical openings:

1. The use of a shaft enclosure to protect vertical openings is a time-honored method recognizing that the fire-resistance-rated enclosure of a floor opening provides a near equivalency to a floor with no openings. The shaft enclosure is intended to replace the floor construction with an equal or better degree of fire resistance. A shaft enclosure is often used for vertical opening protection simply because none of the other applications listed in this section are applicable. The specific requirements for shaft enclosures are set forth in Section 713.

2. Within individual dwelling units, unconcealed vertical openings are permitted provided they connect no more than four stories. The allowance for openings is intended to apply to units that are open vertically with unenclosed stairways and floor spaces, such as lofts. Concealed spaces used for the installation of ducts, piping, and conduit between stories do not fall under this allowance.

3. In fully sprinklered buildings, the vertical openings created for an escalator may be protected by one of the following two methods:

    3.1. By limiting the size of the openings, along with the installation of draft curtains and closely spaced sprinklers, the code assumes that the vertical openings in sprinklered buildings do not present an untenable condition. The purpose of the required draft curtain is to trap heat so that the sprinklers will operate and cool the gases that are rising. The curtain is not a fabric; it is constructed of materials consistent with the type of construction of the building. In other than mercantile and business occupancies, the use of this method is limited to openings connecting four or fewer stories.

    3.2. Approved power-operated automatic shutters may also be used to cut off the openings between floors. Required to have a minimum fire-protection rating of at least $1^1/_2$ hours, the shutters shall close immediately upon activation of a smoke detector. Obviously, operation of an escalator must stop once the shutter begins to close.

4. Penetrations of cables, conduit, tubing, piping, vents, and similar penetrating items are permitted provided the penetrations are protected in conformance with the provisions of Section 714.4. The allowance for vent penetrations only applies to those vents that convey products of combustion as defined in the *International Mechanical Code*® (IMC®) and is not intended to apply to exhaust ducts.

5. As an alternative to the use of shaft enclosures for ducts penetrating floor systems, the provisions of Section 717.6 also regulate the penetration of ducts through horizontal assemblies. If the provisions of Section 717.6 do not mandate the installation of a damper, the provisions of Sections 713.1.1 and Sections 714.2 through 714.3.3 are applicable. Where Section 717.6 mandates a damper in the duct or air transfer opening, then Section 717 applies.

6. Atriums are intended to be open vertically. Where such special building features are designed and constructed in compliance with their own unique provisions,

other means of vertical opening protection are not required. The use of an atrium is a design option, voluntarily applied by the designer, and thus typically provided as an alternative to a shaft enclosure. For further information, see the discussion of Section 404.

It is a widely held belief that the allowance for atriums is also applicable to covered mall buildings that meet the special requirements of Section 402. However, there is no specific provision in Section 712 that specifically addresses multistory covered mall buildings. Vertical openings in the mall portion of the building could be regulated under the provisions of Sections 712.1.6, 712.1.8, and 712.1.12, but a general allowance for the entire building is not available.

The use of Section 712.1.6 is typically limited to those buildings where the floor openings connect three or more stories. Where a floor opening connects only two stories, Sections 712.1.8 and 712.1.12 are commonly applied.

7. A masonry chimney extending through one or more floor levels is permitted where the annular space around the chimney is protected in the manner specified by Section 718.2.5.

8. This provision permits two adjacent stories to intercommunicate with each other without protection of the openings between the two stories, except in the case of Group I-2 and I-3 occupancies. As long as these intercommunicating openings serve only the one adjacent floor, no protection is required. This provision is commonly used where office buildings have a lobby that extends up through the second story so that individuals on the second floor may look down over a guard into the lobby below. It is important that the unprotected floor opening be appropriately separated from floor openings serving other floors.

In addition, the opening between floor levels cannot be concealed within the construction of a wall or floor-ceiling assembly. The limitation on concealment is intended to prevent unprotected openings that are completely enclosed by walls, partitions, chases, or floor/ceiling assemblies. Where the openings are concealed in this manner, they permit a fire within the concealed space to burn undetected and distribute products of combustion to the upper floor.

Other conditions indicate that the floor opening cannot be open to a fire-resistance-rated corridor in Group I-1 and R occupancies, cannot be open to a fire-resistance-rated corridor on nonsprinklered floors in any occupancy, and cannot be a part of the required means of egress. However, Section 712.1.12 permits unenclosed stairways that are in compliance with Section 1009.3, and fire-resistance-rated corridors are permitted to be connected by unenclosed exit access stairways per Section 1018.6. Under the conditions established by these two sections, the limitations of Items 2, 5, and 6 are not applicable.

9. Automobile ramps in both enclosed and open parking garages, when in compliance with the provisions of Sections 406.5 and 406.6, respectively, are permitted. Because the nature of these uses makes it impractical for enclosures, other safeguards are provided by the code in Section 406.

10. By definition, a complying mezzanine is intended to be open into the room below. As such, unenclosed floor openings between the mezzanine and the lower floor are permitted.

11. Joints protected by a fire-resistant joint system, like other building elements protected by approved methods, do not create any additional hazard that needs to be addressed.

12. Exit access stairways in compliance with an exception to Section 1009.3 do not require enclosure. For example, under Exception 1, stairways may be unenclosed provided they connect no more than two stories and are not open to any other stories in the building.

13. A floor fire door assembly complying with Section 711.8 is permitted to protect vertical openings.

14. In a Group I-3 occupancy, openings in floors within a housing unit are permitted without a shaft enclosure provided four specific conditions established in Section 408.5 are met.

15. The enclosure of elevator hoistways in open parking garages and enclosed parking garages is not required for those hoistways that only serve the parking garage.

16. Vertical opening protection is not mandated for the enclosure of mechanical exhaust and supply duct systems in both types of garage facilities. The protection of vertical openings provided to accommodate elevators, as addressed in Section 712.1.15, as well as exhaust ducts and supply ducts, is unnecessary since the vehicle ramps of open and enclosed parking garages are permitted to be open at all levels.

17. Joints in or between nonfire-resistance-rated floor assemblies require no further protection other than that required by Section 711.4.1.

18. Throughout the code, there may be other allowances for floor penetrations or openings that are adequately regulated as vertical openings. Where permitted, these openings must comply with the specifics of their use.

Section 713 *Shaft Enclosures*

The use of a shaft enclosure has long been an acceptable means to protect vertical openings between stories. The enclosure construction is considered to be equivalent to the floor system and thus is permitted to protect any opening that occurs. Although using a shaft enclosure is just one of many applications listed in Section 712.1 for addressing vertical openings, its use is very common, particularly in buildings with a substantial number of stories. In some buildings, the use of a shaft enclosure is the only viable application available.

Fire-resistance rating. To provide an acceptable level of protection for vertical openings between floors, this section mandates that all shaft enclosures have a fire-resistance rating at least equivalent to the rating of the floor being penetrated, but never less than 1 hour. Therefore, in Type I construction, or where the shaft enclosure connects four or more stories, a minimum 2-hour enclosure is mandated. A shaft enclosure is never required to have a higher fire-resistance rating than 2 hours. **713.4**

Continuity. Shaft enclosures are required to be constructed as fire barriers, extending from the top of the floor/ceiling assembly below to the underside of the floor or roof deck above, except as permitted by Sections 713.11 and 713.12. It is important that the walls continue through any concealed spaces such as the area above a ceiling, and that any hollow vertical spaces within the shaft wall be fireblocked at each floor level in buildings of combustible construction. In addition, the supporting elements of any shaft enclosure construction must be of fire-resistance-rated construction equivalent to that of the shaft construction. The enclosure, fire-resistance-wise, should be continuous from the lowest floor opening through to its termination. **713.5**

Exterior walls. Unless required to be fire-resistance rated because of the proximity to an exterior exit balcony, interior exit stairway or ramp, or an exterior exit stairway or ramp, the exterior walls of a shaft enclosure need only be protected because of their location on the lot as regulated by Section 705. **713.6**

Enclosure at the bottom. Many shafts do not extend to the bottom of the building or structure. Therefore, it is necessary to provide an approved method for maintaining the integrity of the shaft enclosure at its lowest point. This section identifies three methods **713.11**

for enclosing the bottom of a shaft enclosure. First, the shaft can be enclosed with fire-resistance-rated construction equivalent to that for the lowest floor penetrated, with a minimum rating consistent with that of the shaft enclosure. Second, a termination room related to the purpose of the shaft can be considered to be the enclosure at the bottom, provided the room is separated from the remainder of the building with fire-resistance-rated construction and opening protectives equivalent to those of the shaft enclosure. Third, approved horizontal fire dampers can be used to protect openings at the lowest floor level in lieu of the enclosure at the bottom of the shaft enclosure. See Figures 713-1a, b, and c.

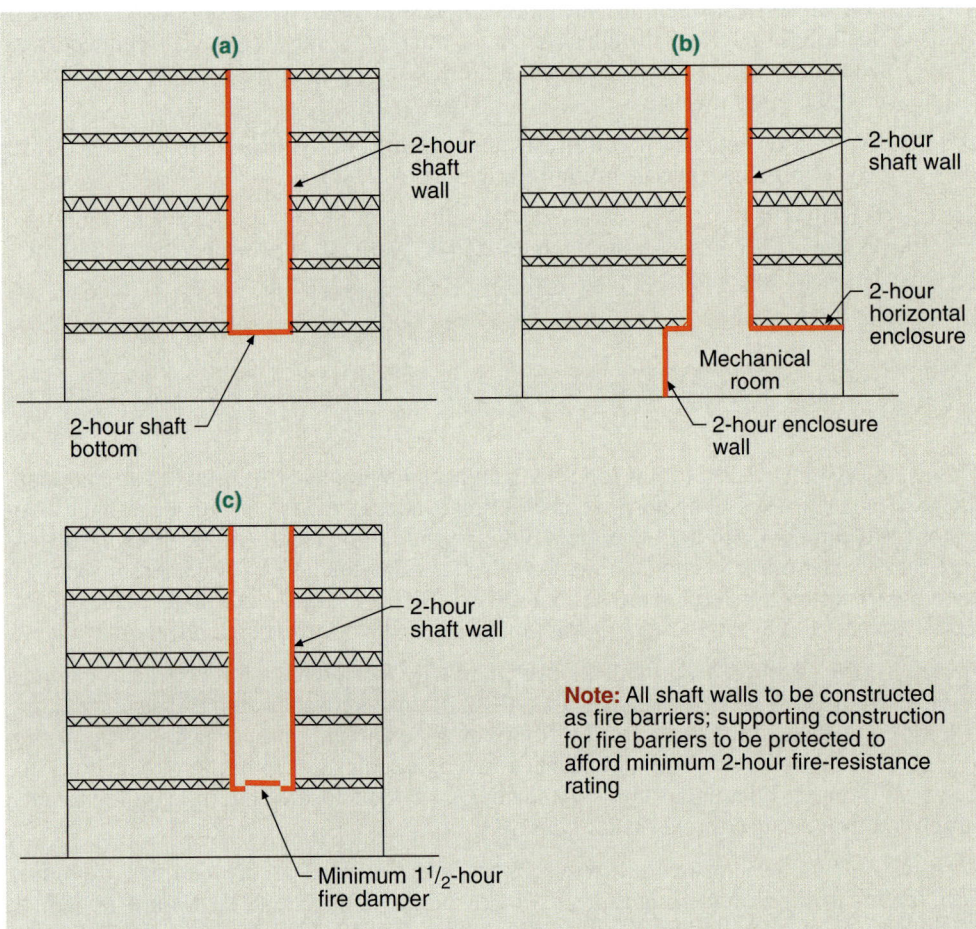

Figure 713-1 Enclosure at shaft bottom.

The first of the three exceptions eliminates the fire-resistance-rated room separation where there are no other openings into the shaft enclosure other than at the bottom. All portions of the enclosure bottom must be closed off except for the penetrating items, unless the room is provided with an automatic fire-suppression system. An example of this concept would be a vent enclosure. The second exception requires that a shaft enclosure containing a refuse or laundry chute be used for no other purpose and shall end in a termination room per Section 713.13.4. Exception 3 applies where the shaft enclosure contains no combustible materials. In this situation, there is no need for either the fire-resistance-rated room separation or protection at the bottom of the enclosure. An example would be a light well that extends through several floor levels to the roof, as illustrated in Figure 713-2. It would be considered an extension of the floor below the level of the floor opening.

Shaft Enclosures 7

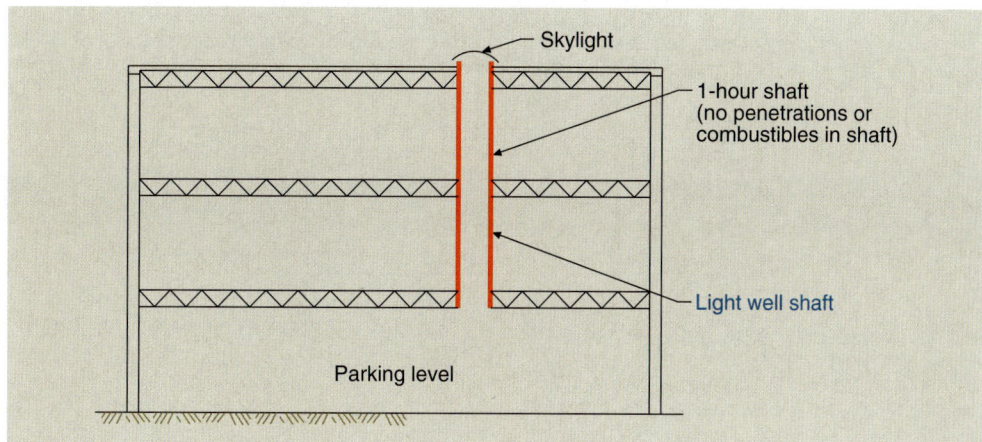

Figure 713-2 Vertical shafts—bottom enclosure.

Enclosure at the top. Most shafts extend to or through the roof deck at the exterior, where there is no requirement to maintain the fire-resistance rating of the shaft enclosure construction. However, where the enclosure does not extend to the roof, the top of the shaft must be enclosed. The required fire-resistance rating of the shaft lid shall be equivalent to the rating of the topmost floor penetrated by the shaft, but in no case less than the fire-resistance rating required for the shaft. See Figure 713-3.

713.12

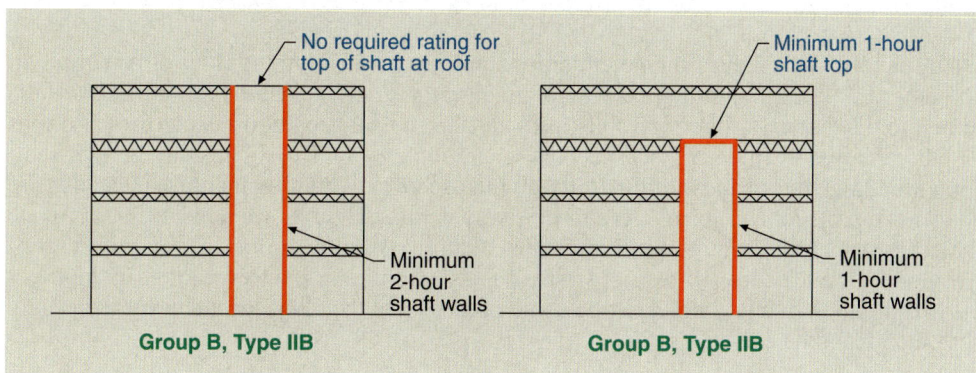

Figure 713-3 Top enclosure of shaft.

Refuse and laundry chutes. The requirements of this section are intended to further strengthen the shaft-enclosure provisions where chutes and termination rooms for refuse or laundry are constructed. Refuse and laundry areas are often poorly maintained, with a greater potential for a fire incident than most other areas of a building. Coupled with the shaft conditions that are created by the chutes, these types of areas pose hazards that exceed those typically encountered. See Figure 713-4. To further secure the intent, Section 903.2.11.2 requires sprinkler protection for the chutes and termination rooms.

713.13

These requirements addressing refuse and laundry chutes are not applicable in Group I-2 occupancies, as chutes in such institutional occupancies are regulated by Chapter 5 of NFPA 82, *Standard on Incinerators and Waste and Linen Handling Systems and Equipment*. The construction of hospitals accredited by the Joint Commission must comply not only with the IBC but also with NFPA 101, *Life Safety Code*. NFPA 101 further references NFPA 82 for provisions applicable to linen and rubbish chutes. In order to eliminate any inconsistencies between the IBC and NFPA 82 regarding such chutes, the standard is directly referenced in lieu of the IBC for the applicable construction requirements in regard to hospitals, nursing homes, and other Group I-2 occupancies. By requiring refuse and linen chutes in Group I-2 occupancies to meet the requirements of NFPA 82, the provisions are consistent with the current Joint Commission rules.

7 Fire and Smoke Protection Features

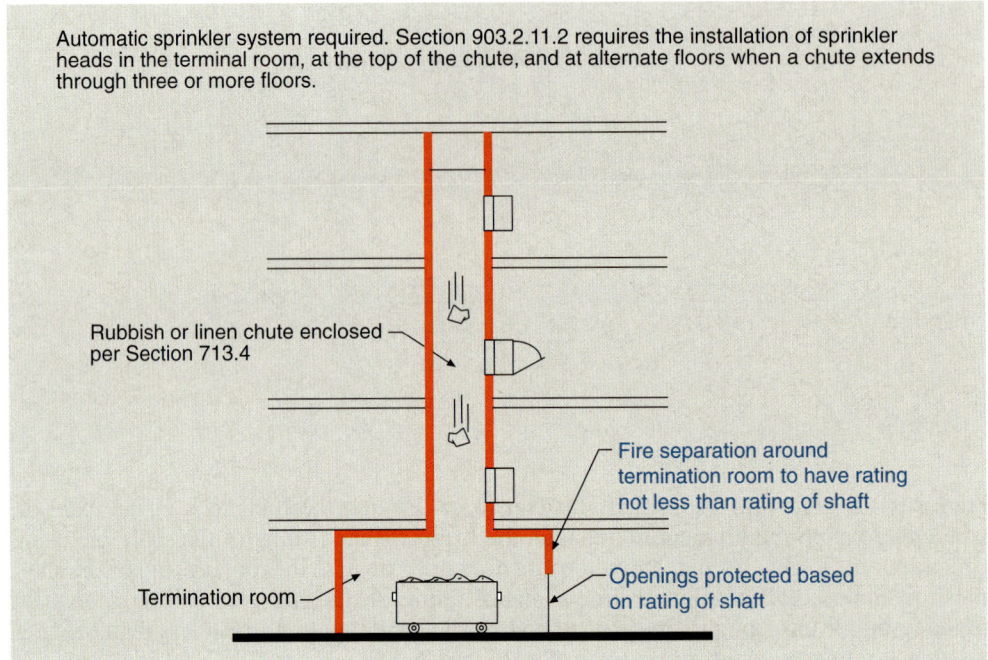

Figure 713-4
Rubbish and linen chutes.

713.14.1 Elevator lobby. To reduce the potential for smoke to travel from the floor of fire origin to any other floor of the building by way of an elevator shaft enclosure, elevator lobbies are to be provided at each floor and must provide a complete separation using fire partitions and the required opening protection. This separation essentially isolates the elevator shaft from the remainder of the building through the use of construction that provides both fire and smoke protection. Typical elevator hoistway doors, although fire rated, cannot provide the necessary barrier required to keep smoke from passing from floor to floor through an elevator shaft enclosure. To restrict smoke passage from one floor level to other floor levels within the building, there must be some point where the floor level can be adequately separated from the elevator shaft.

Elevator lobbies are not required in all multistory buildings, as the requirement for elevator lobbies is not applicable where the elevator shaft connects only two or three stories. The threshold of four stories is consistent with a number of other code provisions that increase the level of protection where four or more stories are involved. In application, the scope of four or more stories is just one of many exceptions to the elevator lobby requirement. It should be noted that the four-story threshold is not applicable where the conditions of Section 711.9 for smoke barriers are applicable.

There are also seven specific exceptions to the requirement for elevator lobbies. The first exempts the level(s) of exit discharge, provided the discharge level(s) is protected with an automatic sprinkler system. The second exception applies where elevator-shaft enclosures are not required to protect vertical openings because some other application of Section 712.1, such as Section 712.1.8, is used. Third, additional doors may be used in lieu of an elevator lobby, provided they are installed in accordance with Section 3002.6. In this scenario, the lobby is not necessary, because the smoke-infiltration problem is addressed by the additional door. The fourth exception exempts fully sprinklered buildings, provided the building is not a Group I-2 or I-3 occupancy. In addition, where the building is considered a high rise, elevator lobbies are mandated for all elevators that serve floor levels more than 75 feet (22,860 mm) above the lowest level of fire department vehicle access. The sixth exception permits pressurization of the elevator shaft enclosure as an alternative to creating an elevator lobby. The criteria for the pressurization method are established in Section 909.21.

The seventh exception eliminates the lobby requirement for elevators that serve open parking garages. This exception should also extend to enclosed parking garages since Section 712.1.15 indicates that vertical opening protection is not required in both open and enclosed parking garages for elevator hoistways serving only the parking garage. Since enclosed elevator lobbies are simply an extension of the shaft enclosure protection, such lobbies are not necessary if a shaft enclosure is not required.

Exception 5 differs from the other exceptions in that it does not provide an alternative to an elevator lobby, but rather modifies the means for constructing the lobby. Where the building is protected with an automatic sprinkler system, the exception allows for the use of smoke partitions in lieu of fire partitions. Described in Section 710, smoke partitions are not required to have a fire-resistance rating but must be constructed to limit the transfer of smoke. It is important to note that the exceptions to Section 713.14.1 do not apply where an elevator lobby is required by some other provision of the code, such as the area of refuge requirement in Section 1007.4.

Section 714 *Penetrations*

The integrity of fire-resistance-rated horizontal and vertical assemblies is jeopardized where penetrations of such assemblies are not properly addressed. Cables, cable trays, conduit, tubing, vents, pipes, and similar items are those types of penetrating items regulated by the code. This section of the IBC identifies the appropriate materials and methods of construction used to protect both membrane penetrations and through penetrations.

Ducts and air transfer openings. Section 717.5 identifies the various conditions under which fire-resistance-rated wall assemblies penetrated by ducts or air transfer openings must be provided with fire and/or smoke dampers. There are a limited number of locations where a damper is not required, such as that permitted by Exception 3 of Section 717.5.2 for fire barrier penetrations. See Figure 714-1. In such situations, it is necessary that the penetrations be protected in accordance with the appropriate provisions of Section 714 in order to maintain the integrity of the fire-resistive assembly.

714.1.1

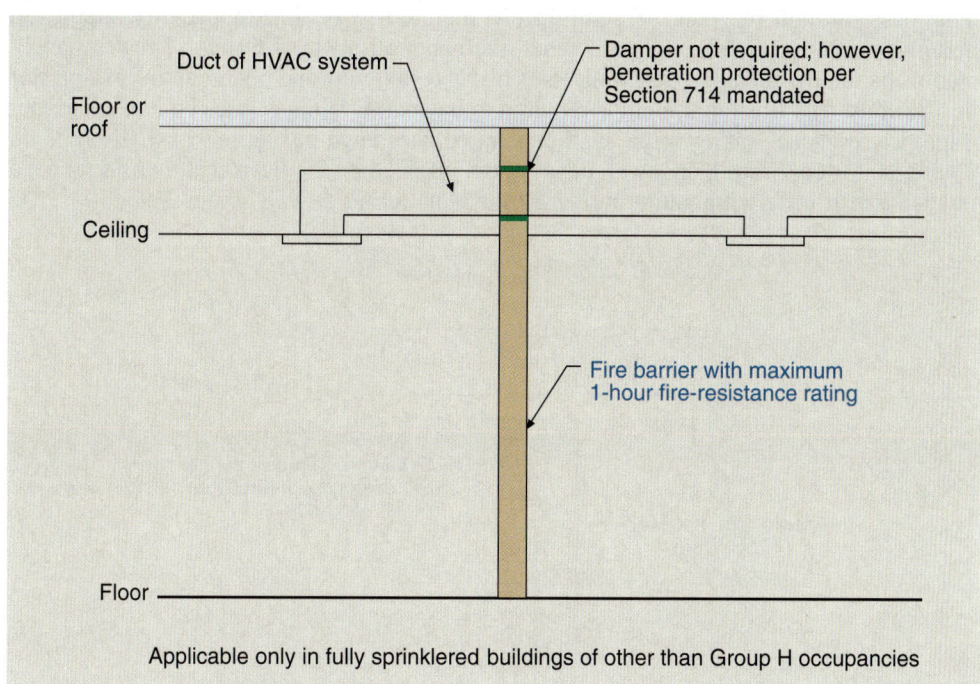

Figure 714-1
Duct penetration of 1-hour fire barrier.

7 Fire and Smoke Protection Features

714.2 Installation details. As illustrated in Figure 714-2, sleeves used in the process of creating a through-penetration of a fire-resistance-rated building element must be properly installed. They must be securely fastened to the assembly that is being penetrated. In addition, both the space between the sleeve and the assembly and the space between the sleeve and the penetrating item must be appropriately protected.

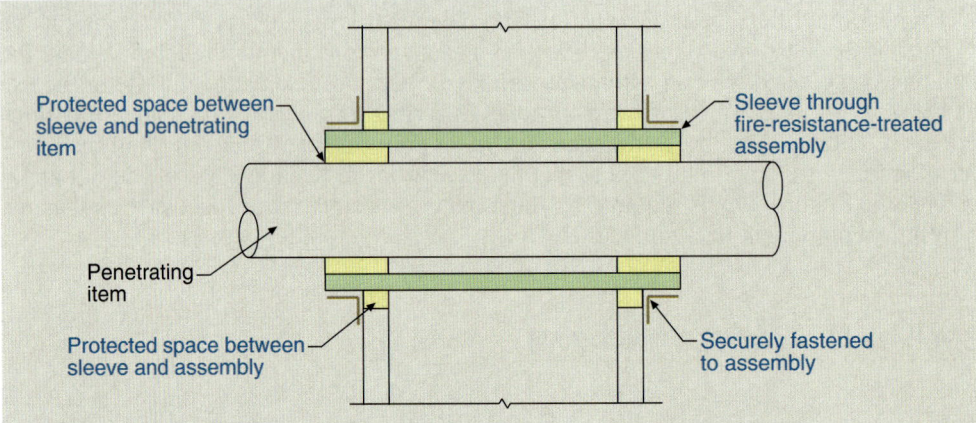

Figure 714-2
Penetration sleeve.

714.3 Fire-resistance-rated walls. This section regulates the penetration into or through fire walls, fire barriers, fire partitions, and smoke barrier walls. The protection of penetrations in fire-resistance-rated exterior walls is not addressed; however, where such exterior walls are bearing walls it is necessary to consider penetration protection in order to maintain the structural integrity of the walls during fire conditions. Fire-resistance-rated interior bearing walls are also not specifically identified as elements regulated for penetrating items; however, any penetrations of such walls should be addressed in order to maintain the necessary structural fire resistance. For the most part, membrane penetrations are addressed in the same manner as through penetrations.

714.3.1 Through-penetrations. As a general rule, through-penetrations (where the penetrating items pass through the entire assembly) are required to be firestopped with approved through-penetration firestop systems when the penetrations pass through fire-resistance-rated walls, unless the approved wall assembly is tested with the penetrations as a part of the assembly. The firestop system is required to have an F rating at least equivalent to that of the fire-resistance rating of the wall penetrated, as shown in Figure 714-3. There is no requirement for a T rating on a wall penetration, justified on the basis that there is no need for such a restrictive temperature rating for the penetration of wall assemblies.

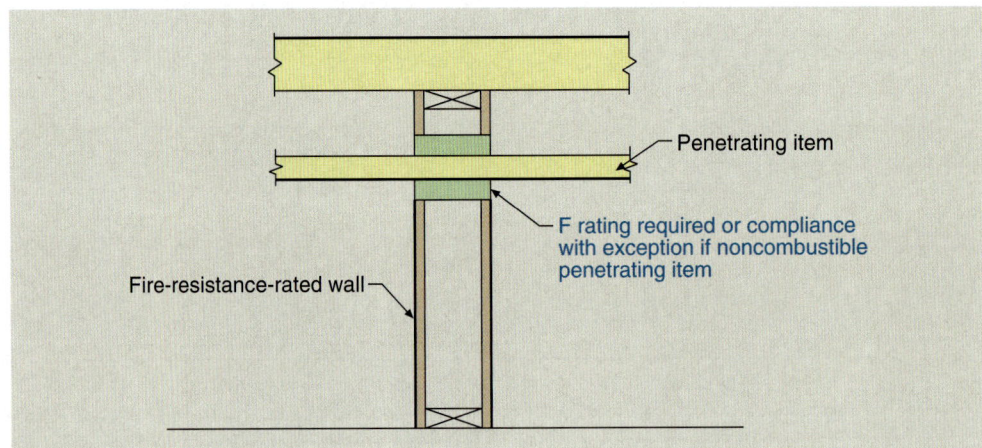

Figure 714-3
F rating required.

242 2012 International Building Code Handbook

The IBC contains an exception to the general rule for firestopped wall penetrations allowing small noncombustible penetrating items no larger than 6-inch (152-mm) nominal diameter to penetrate concrete or masonry walls, provided the full thickness of the wall, or the thickness required to maintain the fire resistance, is filled with concrete, grout, or mortar. The size of the opening is limited to 144 square inches (0.0929 m^2). A second exception that is used extensively will allow the annular space around the same type of noncombustible penetrating item to be filled with a material that prevents the passage of flame or hot gases sufficient to ignite cotton waste when tested under the time-temperature fire conditions of ASTM E 119 or UL 263, and under a positive pressure differential of 0.01-inch (0.25-mm) water column. When properly installed around the penetrations of noncombustible items, these materials provide adequate firestopping between the penetrating item and the fire-resistive membrane of the wall. See Figure 714-4.

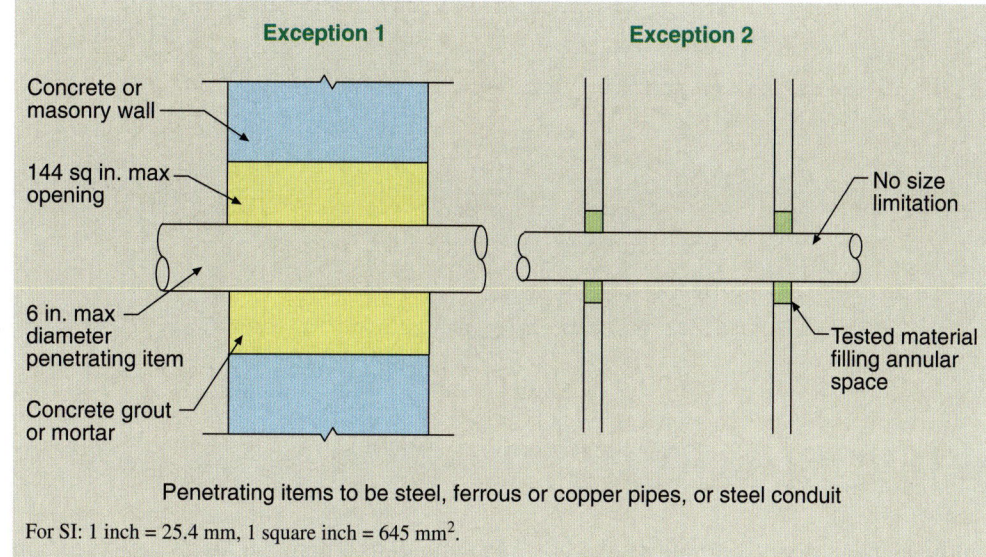

Figure 714-4
Through-penetrations of wall.

Membrane penetrations. This section addresses penetrations through a single membrane of fire-resistance-rated walls. For the most part, a membrane penetration is to be protected by one of the methods established for through-penetrations as previously described. However, there are some membrane penetrations that are allowed without a specific firestopping material in the annular space around such penetrations. Openings for steel electrical boxes are specifically addressed where located in walls with a maximum 2-hour rating, provided that they are no more than 16 square inches (0.0103 m^2) in area and the aggregate area of the boxes does not exceed 100 square inches (0.0645 m^2) for any 100 square feet (9.29 m^2) of wall area. The annular space between the wall membrane and any edge of the electrical box is limited to $^1/_8$ inch (3.1 mm). Also, to prevent an indirect through-penetration, electrical boxes on opposing sides of a fire-resistance-rated wall shall be horizontally separated by no less than 24 inches (610 mm). As an alternative, boxes may be separated horizontally by the depth of the cavity if the cavity is filled with cellulose loose-fill, rockwool, or slag mineral wool insulation; by solid fireblocking in accordance with Section 718.2.1; by protection of both outlet boxes with listed putty pads; or by any other listed methods and materials. Examples of several of these methods are illustrated in Figure 714-5.

714.3.2

A second exception for membrane penetrations of electrical-outlet boxes allows outlet boxes of any material, provided they are tested for use in fire-resistance-rated assemblies and installed in accordance with the instructions for the listing. Limitations are also placed

7 Fire and Smoke Protection Features

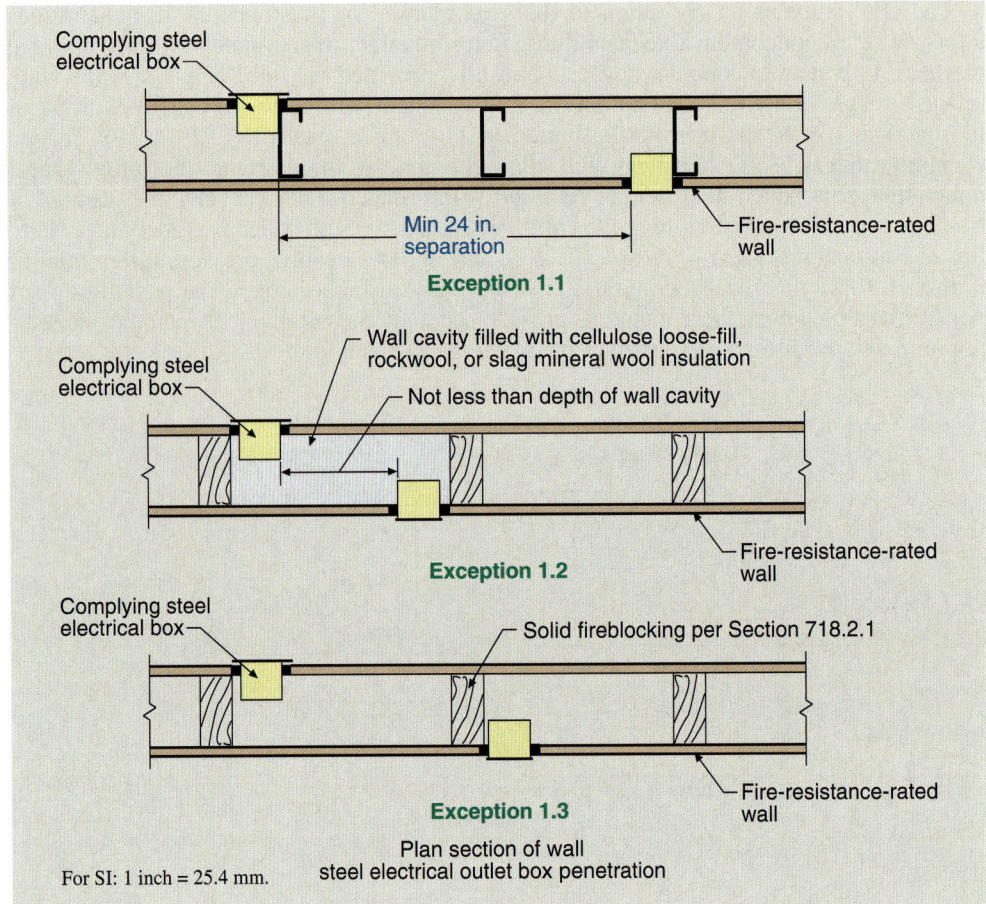

Figure 714-5
Penetration of fire-resistance-rated walls.

on the annular space surrounding the box and conditions where the boxes are placed on opposite sides of the wall. Exception 3 allows for penetrations by electrical boxes of any size or type provided they are listed as a part of a wall opening protective material system, while Exception 4 addresses boxes, other than electrical boxes, that have annular space protection provided by an approved membrane penetration firestop system. The fifth exception permits the annular space created by the penetration of a fire sprinkler to be unprotected, provided that such a space is covered by a metal escutcheon plate. Because the escutcheon is a part of the listed sprinkler, it is inappropriate to require firestopping at this location. It should be noted that this exception applies to the penetration of sprinklers, not simply sprinkler piping or cross mains that might be penetrating fire-resistance-rated construction. See Figure 714-6.

714.3.3 Dissimilar materials. This provision is intended to limit the occasional practice of using a noncombustible penetrating item (such as a short metal coupling) to penetrate a fire-resistance-rated wall, then connect to a combustible item (such as plastic piping or conduit) on the room side of the wall. The building official can accept such a condition where it is demonstrated that the fire-resistive integrity of the wall will be maintained. See Figure 714-7.

714.4 Horizontal assemblies. The shaft enclosure provisions of Section 713 intend to maintain a level of protection that is compromised when one or more openings occur in a floor or floor/ceiling assembly. However, penetrations by pipes, tubes, conduit, wire, cable, and vents are permitted without shaft enclosure protection where in compliance with this section. In addition, this section addresses penetrations that occur in the ceiling of a roof/ceiling assembly. Penetrations occurring in both fire-resistance-rated horizontal assemblies and nonfire-resistance-rated assemblies are addressed.

Penetrations 7

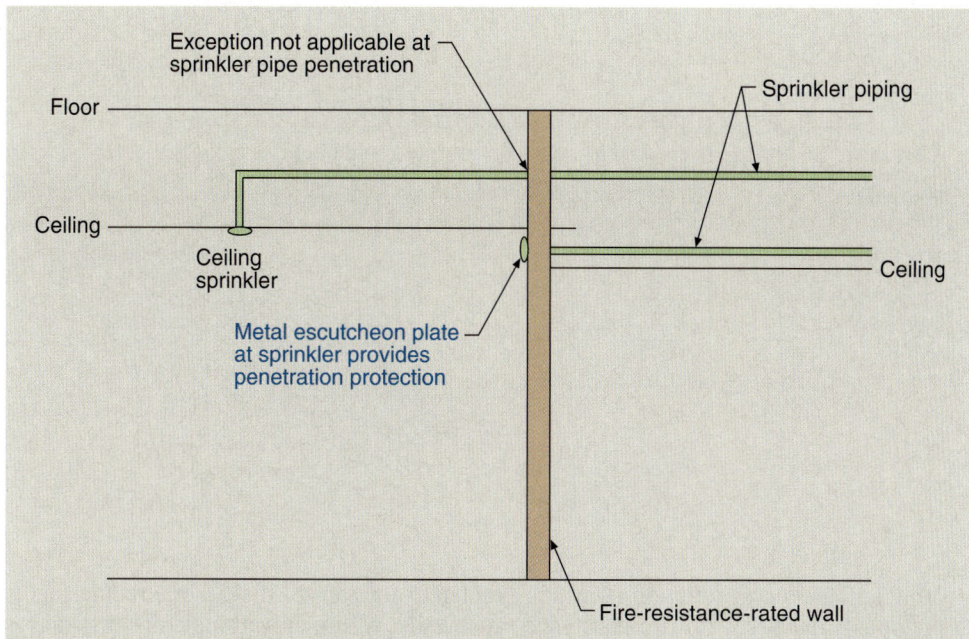

Figure 714-6
Membrane penetration protection.

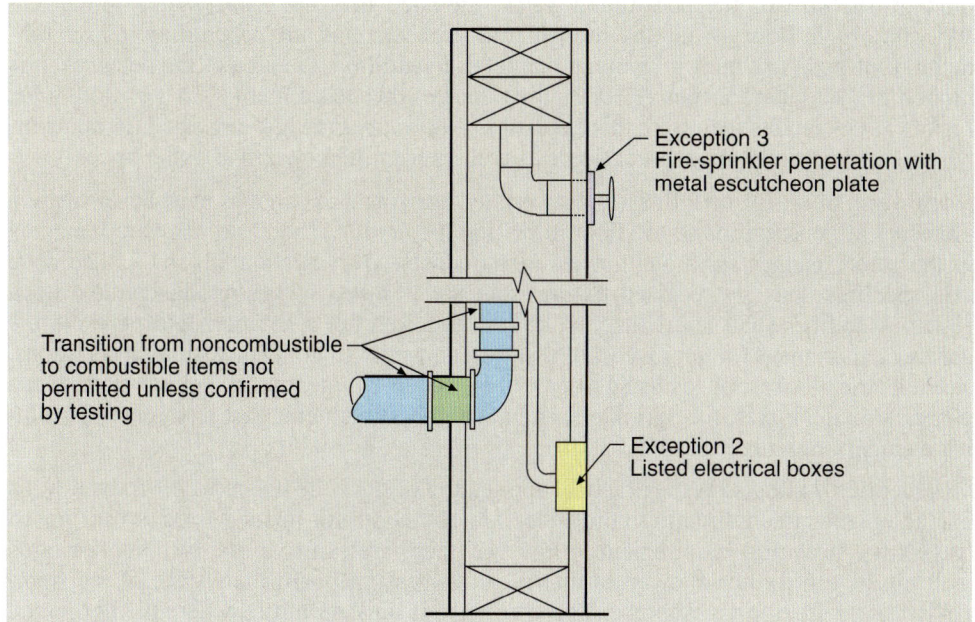

Figure 714-7
Membrane penetrations of walls.

Through-penetrations. The protection requirements for the through-penetration of fire-resistance-rated horizontal assemblies are very similar to those required for vertical elements. The general provisions state that the penetrations are to be installed as tested in an approved fire-resistance-rated assembly or protected by an approved through-penetration firestop system. Where a firestop system is used, it must have both an F rating and a T rating equivalent to the floor penetrated, but in no case less than 1 hour. Only an F rating is needed if the penetrating item, as it passes through the floor, is contained within a wall cavity above or below the floor. See Figure 714-8. A T rating is also not required where the floor penetration is a floor drain, tub drain, or shower drain.

714.4.1.1

2012 International Building Code Handbook   **245**

7 Fire and Smoke Protection Features

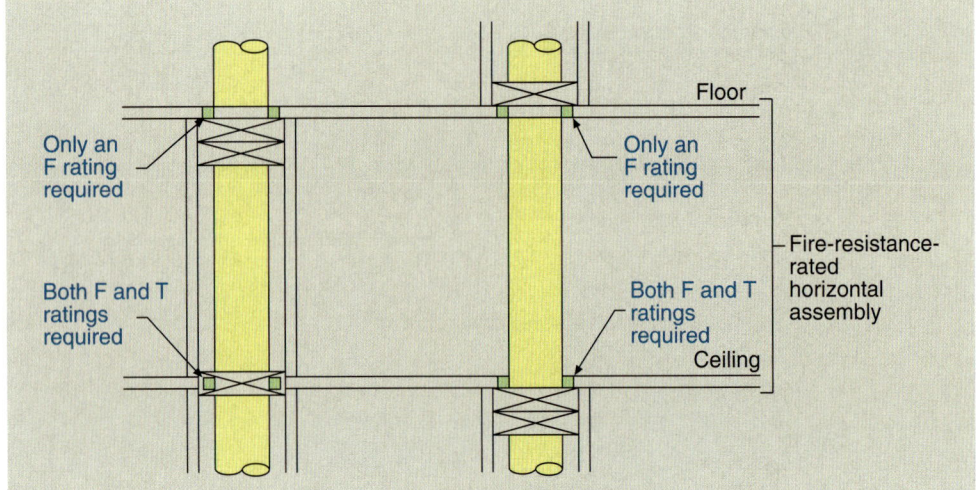

Figure 714-8
Through-penetrations of horizontal assemblies.

Noncombustible penetration items are granted exceptions to the general requirements as previously discussed for fire-resistance-rated walls. Where only a single fire-resistance-rated floor is penetrated, the annular space around the noncombustible penetration item need only to be protected with an approved material that fills the opening. There is no limit on the size of the penetrating items, provided they are appropriately protected. Where multiple floor assemblies are penetrated, the size of any penetrating item is limited to 6 inches (152 mm) in nominal diameter. In addition, the area of the penetration is limited to 144 square inches (92,900 mm^2) in any 100 square feet (93 m^2) of floor area. Figure 714-9 depicts the use of this exception. Allowances are also provided for noncombustible penetrations of concrete floors as well as for tested electrical outlet boxes.

714.4.1.2 Membrane penetrations. Fire-resistance-rated horizontal assemblies must be adequately protected at penetrations of the floor or ceiling membrane. Therefore, they are regulated in the same manner as through-penetrations addressed in Section 714.4.1.1. The code also specifies that any recessed fixtures that are installed in fire-resistance-rated horizontal assemblies shall not reduce the level of required fire resistance. Exceptions to the general requirement for approved firestop systems apply to noncombustible penetrations, steel electrical boxes, boxes listed as part of an opening protective material system, listed electrical-outlet boxes, fire sprinklers, and noncombustible items cast into concrete building elements. See Figure 714-10.

Exception 7 allows for the practical application of the code where wood-framed walls extend up and attach directly to the underside of wood floor joists/trusses or roof joists/trusses for structural requirements. However, there are limits to its use. Nonfire-rated wall top plates are not allowed to interrupt the gypsum-board membrane of the floor/ceiling or roof/ceiling membrane. The allowance is only permitted where the horizontal assembly has a required fire-resistance rating of 2 hours or less, and the intersecting wall must have a fire-resistance rating equal to or greater than that of the horizontal assembly. Piping, conduit, and similar items within the fire-resistance-rated wall must be adequately protected where they penetrate the double-wood top plate. Compliance with the established criteria is deemed to provide for an equivalent degree of fire resistance at the discontinuous portion of the ceiling membrane at the intersection of a horizontal assembly and a fire-resistance-rated wall with a double-wood top plate. See Figure 714-11.

714.4.2 Nonfire-resistance-rated assemblies. Figure 714-12 illustrates the provisions for penetrations of those horizontal assemblies not required to have a fire-resistance rating. Section 713 for shaft enclosures will regulate such penetrations where the allowances set forth in this

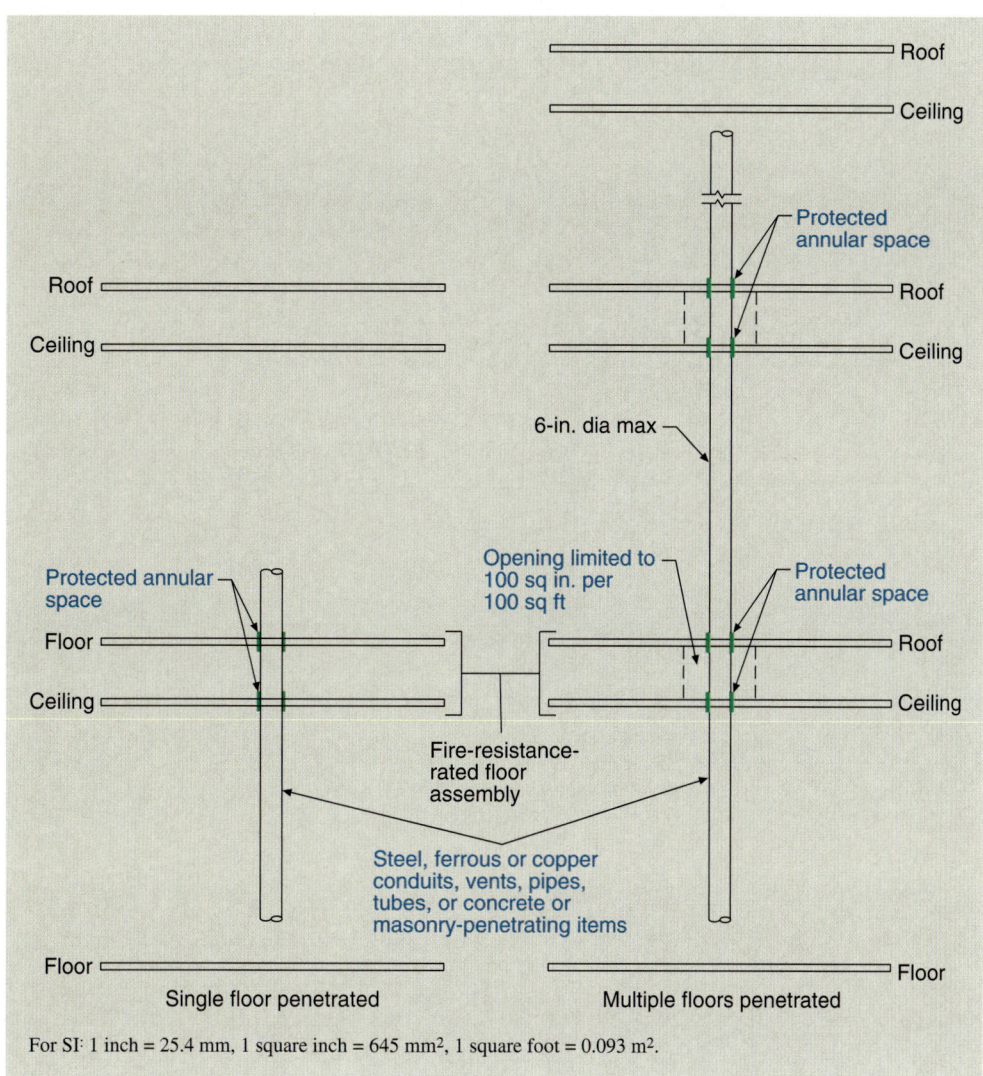

Figure 714-9
Penetrations of horizontal assemblies.

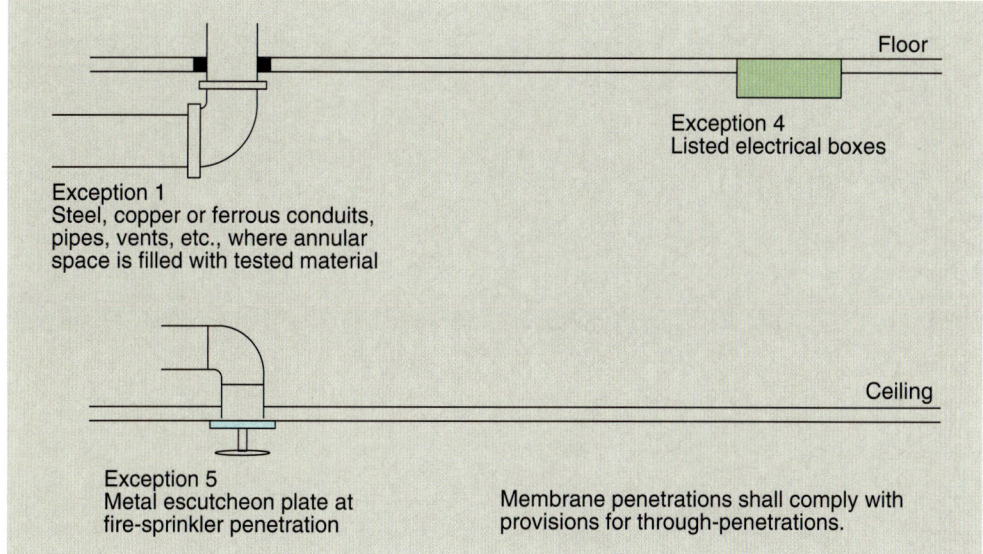

Figure 714-10
Membrane penetrations of horizontal assemblies.

7 Fire and Smoke Protection Features

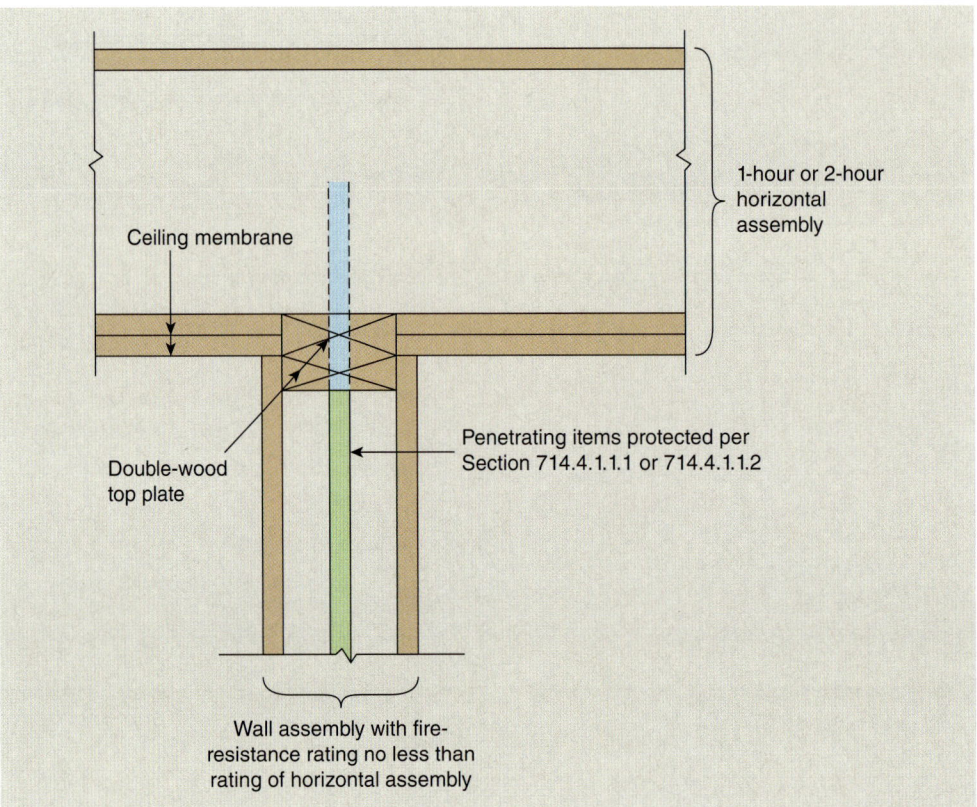

Figure 714-11 Horizontal assembly continuity at fire-rated wall.

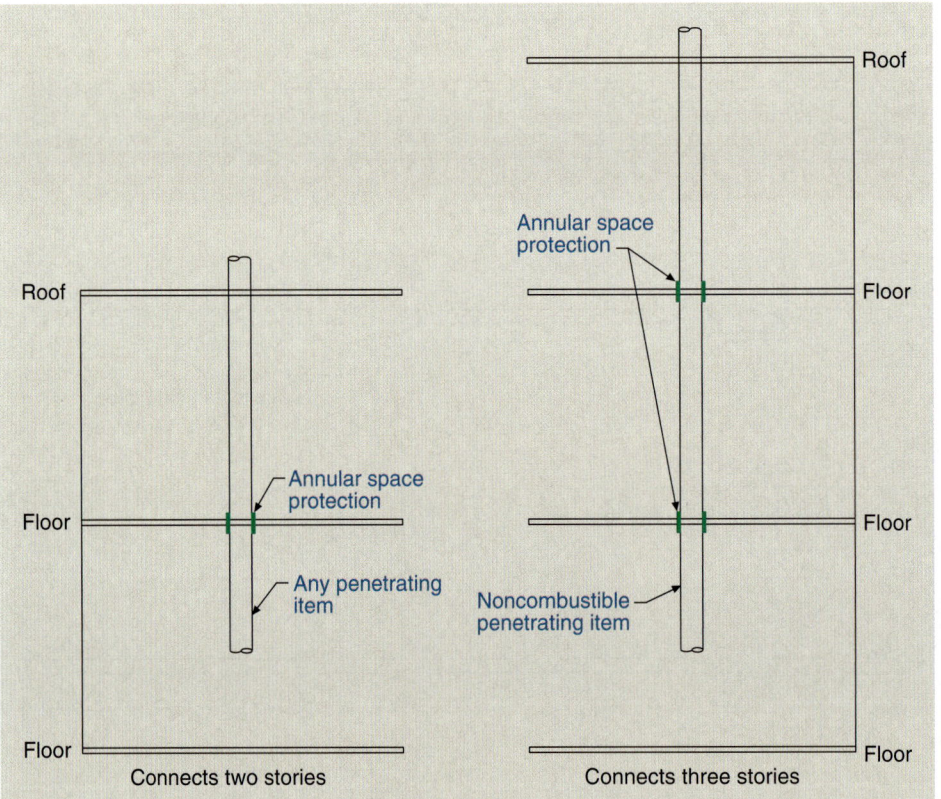

Figure 714-12 Nonfire-resistance-rated assemblies.

248 2012 International Building Code Handbook

section are not applicable. Where penetrations connect only two stories, the annular space around the penetrating items must simply be protected with a material that resists the free passage of fire and smoke. If the penetrating items are noncombustible, up to three stories may be connected, provided the annular space is filled appropriately.

Section 715 *Joint Systems*

A fire-resistant joint system is defined in Section 202 as "an assemblage of specific materials or products that are designed, tested, and fire-resistance rated in accordance with either ASTM E 1966 or UL 2079 to resist, for a prescribed period of time, the passage of fire through joints made in or between fire-resistance-rated assemblies." The term *joint* is also defined as "the linear opening in or between adjacent fire-resistance-rated assemblies that is designed to allow independent movement of the building, in any plane, caused by thermal, seismic, wind or any other loading." The approved joint system should be designed to resist the passage of fire for a time period not less than the required fire-resistance rating of the floor, roof, or wall in or between which it is installed. See Figure 715-1.

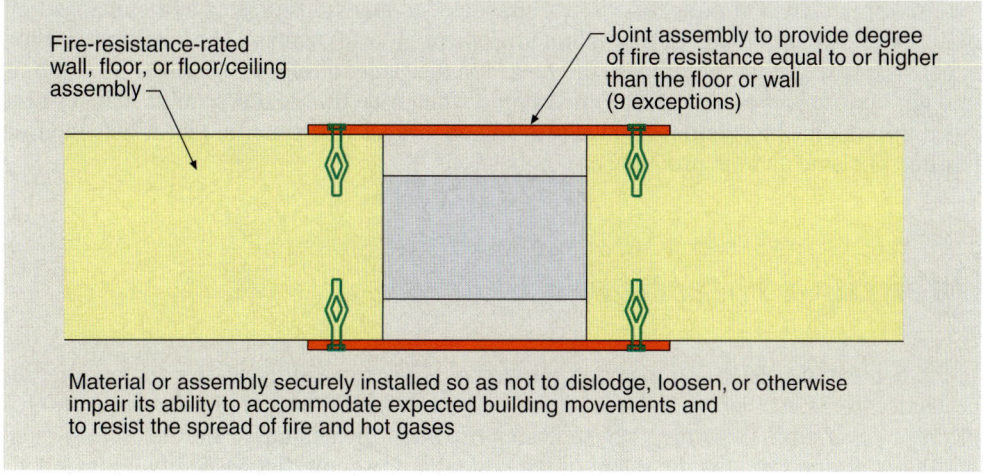

Figure 715-1
Fire-resistant joint systems.

The code lists nine locations where it is not necessary to provide fire-resistant joint systems. For most of the applications listed, they are also locations where fire assemblies are not required to protect openings in the horizontal or vertical assemblies. Item 9 references maximum $5/8$-inch (15.9-mm) control joints when tested in accordance with ASTM E 119 or UL 263.

Exterior curtain wall/floor intersection. Vertical passages without barriers allow fire and hot gases to circumvent the protection for occupants in the floors above. When floors or floor/ceiling assemblies do not extend to the exterior face of a building, this section requires an approved barrier at the intersection at least equal to the fire resistance of the floor or floor/ceiling assembly. See Figure 715-2. **715.4**

ASTM E 2307 is identified for the specification and testing methods used to determine the necessary fire resistance. This test method measures the performance of a perimeter fire-barrier system and its ability to maintain a seal to prevent fire spread during the deflection and deformation of the exterior wall assembly and floor assembly expected during a fire condition, while resisting fire exposure from both an interior compartment and the flame plume emitted from a window burner below.

7 Fire and Smoke Protection Features

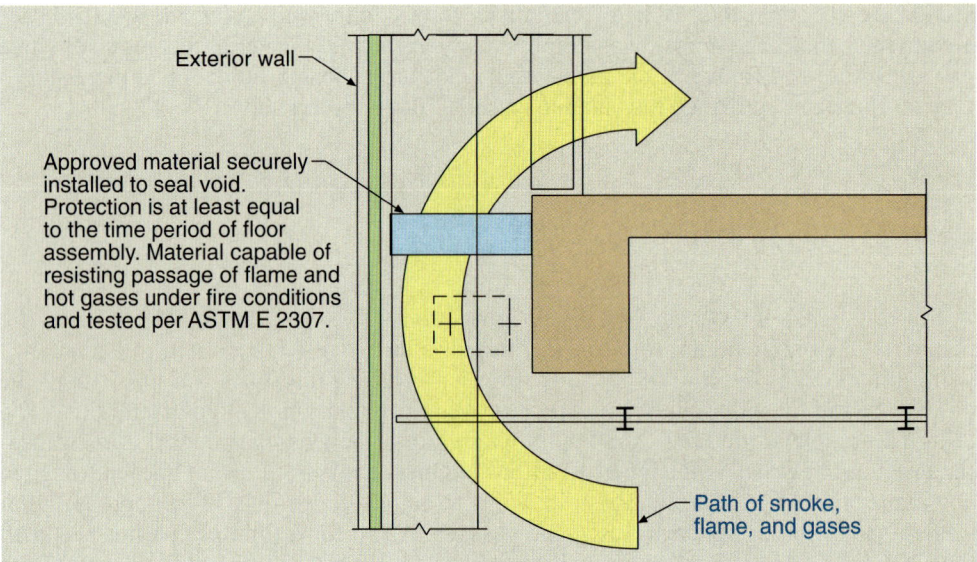

Figure 715-2 Exterior curtain wall/ floor intersection.

A minimum level of protection is also mandated at any voids created at the intersection of an exterior curtain wall and a nonrated floor or floor assembly. The required method is consistent with that required in the code for the penetration of ducts and other items through nonfire-resistance-rated floor systems. The protection of the annular space is provided through the installation of an approved noncombustible material that resists the free passage of flame and the products of combustion.

Section 716 *Opening Protectives*

In the context of the IBC, an opening protective refers to a fire door, fire shutter, or fire-protection-rated glazing, including the required frames, sills, anchorage, and hardware for its proper operation. Generally, whenever any fire door, fire shutter, or fire-protection-rated glazing is referred to, it is the intent of the code that the entire fire assembly be included.

716.2 Fire-resistance-rated glazing. This section is an extension of the provisions in Section 703.6 regarding the use of fire-resistance-rated glazing. Where such glazing is appropriately tested and labeled, its use is permitted in fire doors and fire window assemblies as specified in Table 716.5. Vision panels of fire-resistance-rated glazing are only prohibited in 3-hour fire door assemblies.

716.3.1 Fire-rated glazing that exceeds the code requirements. Both fire-resistance-rated glazing and fire-protection-rated glazing must be appropriately identified for verification of their appropriate application. These markings establish compliance with hose-stream and temperature rise requirements, while also identifying the minimum assembly rating in minutes. It is not unusual for such glazing to be marked indicating a higher degree of protection than mandated by the code. This provision clarifies that the use of glazing marked to indicate a higher level of compliance is permitted for use where such compliance is not required.

Table 716.3 defines and relates the various test standards for fire-rated glazing to the designations used to mark such glazing. The table reflects the use of the designations "W," "OH," "D," "T," "H," and "XXX" as markings for fire-rated glazing. Tables 716.5 and 716.6 set forth the markings required for acceptance in specified applications. The marking of fire-rated glazing does not include the "NH" (not hose-stream tested) and "NT" (not temperature-rise tested) designations, as these designations correspond with test standards, not end uses.

716.5 Fire door and shutter assemblies. This section sets forth the test standards and additional criteria necessary for the acceptance of fire door and fire shutter assemblies. In addition, Table 716.5 identifies the minimum fire-protection rating for an opening protective based on the type of assembly in which it is installed. For example, a door assembly in a 1-hour fire barrier wall separating hazardous material control areas would need to have a minimum $^3/_4$-hour fire-protective rating, whereas a 1-hour fire-resistance-rated interior exit stairway enclosure would require a minimum 1-hour door assembly.

Side-hinged or swinging doors are to be tested for conformance with NFPA 252 or UL 10C. It is important that the NFPA 252 test provides for positive pressure in the furnace as established by this section. See Figure 716-1. For other types of doors, the pressure level need only be maintained as nearly equal to the atmosphere's pressure as possible.

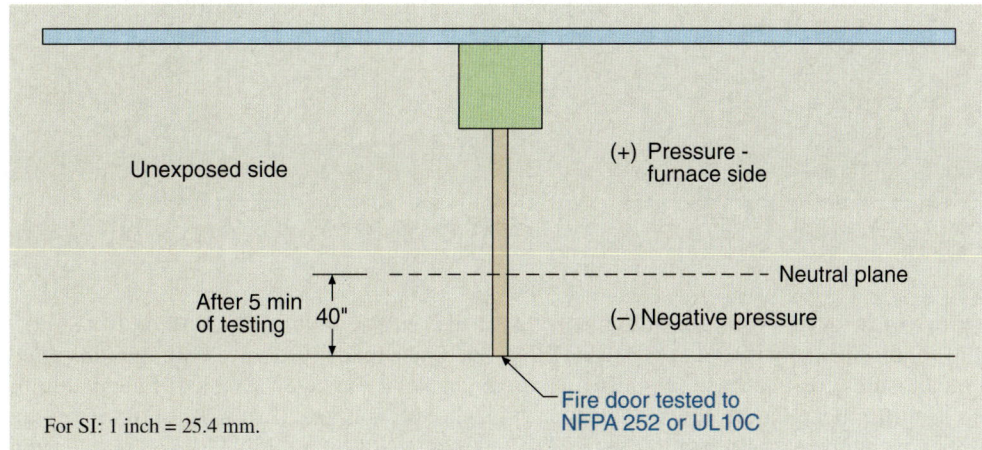

Figure 716-1 Fire test of door assemblies.

716.5.3 Door assemblies in corridors and smoke barriers. Fire door assemblies located in fire-resistance-rated corridor walls or smoke barrier walls are further regulated where required by Table 716.5 to have a 20-minute fire-protection rating. They are commonly referred to as smoke- and draft-control assemblies. Their primary purpose is to minimize smoke leakage around the door and through the opening. For this reason, these doors shall not contain louvers and must be installed in accordance with NFPA 105.

The protection of fire-rated corridors is intended to be a two-way protection. Although the general intent is to protect the corridor from smoke that might be generated by a fire occurring within the adjacent use spaces, there are occasions where it is just as important to protect the occupied use spaces from smoke in the corridor. The fire test for corridor and smoke-barrier doors is essentially the same test of the door as for other fire-door assemblies, except that the fire test for the 20-minute assembly does not include the hose-stream test. In addition, Section 716.5.3.1 requires the door assembly to be tested for smoke infiltration through the UL 1784 air leakage test and identifies the criteria for acceptance. Note that glazing other than in the door itself, such as in sidelites or transoms, must be tested with the hose-stream test as set forth in NFPA 257 or UL 9.

An exception permits the installation of a viewport through the door for purposes of observation. These viewports must be installed under the limitations of, and in accordance with, the conditions specified in the exception. Corridor door provisions are modified in Section 407.3.1 for Group I-2 occupancies and in multitheater complexes as shown in Figure 716-2. In addition, where horizontal sliding doors are used in smoke barriers of Group I-3 occupancies as specified, the 20-minute fire-protection rating is not required.

7 Fire and Smoke Protection Features

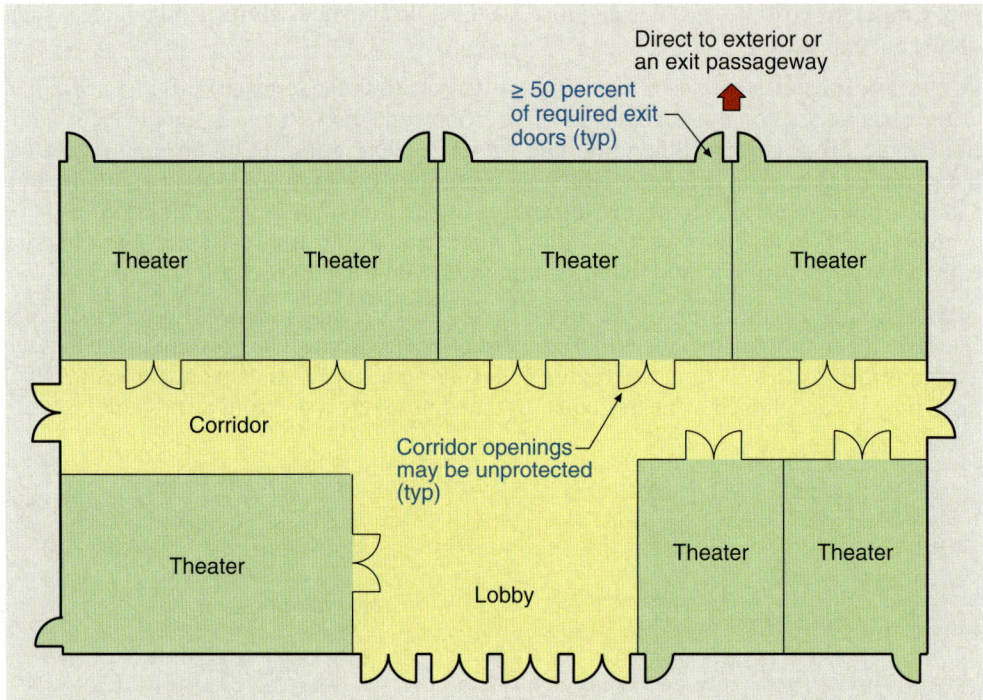

**Figure 716-2
Corridor doors in a multi-theater complex.**

716.5.5 Doors in interior exit stairways/ramps and exit passageways. In addition to the normal requirement for fire doors, the IBC is concerned that fire door assemblies installed in enclosures for interior exit stairways/ramps and exit passageways shall be capable of limiting the temperature transmission through the door. It specifies that the temperature rise above ambient temperature shall be limited to a maximum of 450°F (232°C) at the end of 30 minutes of the normal fire test. However, in buildings equipped with an automatic sprinkler system, the temperature limitation is not applicable.

The purpose of these highly protected exit elements and their openings is to protect the building occupants while they are exiting the building. It is intended that in a properly enclosed and protected interior exit stairway or ramp enclosure, building occupants from the floors above the fire floor will be able to pass through the fire floor inside the enclosure and eventually pass down and out of the building. The end-point limitation on temperature transmission through the fire door, then, is literally to protect the person inside the enclosure from excessive heat radiation from the fire door as he or she passes through the fire floor. In sprinklered buildings, the maximum transmitted temperature end point is not required. It is expected that a sprinkler system will limit the fire growth to the point where such extra care is unnecessary.

716.5.7 Labeled protective assemblies. Fire doors are required to have an approved label or listing mark permanently affixed at the factory. The label must contain information that identifies the manufacturer, the third-party inspection agency, and the fire-protection rating. Where applicable, the maximum transmitted temperature end point or the smoke- and draft-control designation must be identified.

Listing agencies will typically only label door assemblies that have been tested. However, some door assemblies are too large to be tested in available furnaces. As a result, the code permits the installation of oversized fire doors under the conditions of this section. As oversized fire doors are not subjected to the standard fire test, an approved testing agency must provide a certificate of inspection from them certifying that, except for the fact that the doors are oversized, they comply with the requirements for materials, design, and construction for a fire door of the specified fire-endurance rating. An approved agency may also provide a label on the door indicating it is oversized. Where the certificate or label of an approved agency has been provided, there is assurance that the fire door will protect the opening as required by the code.

The letter "S" on a fire door indicates that it is in compliance with UL 1784, the air leakage testing. Through this identifying mark, it is possible to quickly identify the door as appropriate where smoke and draft control doors are mandated.

Individual components, such as vision panels, may be installed in labeled fire doors provided such components are listed and the installation is done through a listing program of a third-party agency.

Size limitations. Fire-protection-rated glazing is permitted in wall assemblies rated at 1 hour or less when in compliance with the size limitations of NFPA 80. Where the wall assembly requires a rating greater than 1 hour, such glazing is prohibited except for two conditions. First, fire-protection-rated glazing may be used as vision panels in swinging fire door assemblies serving as horizontal exits when limited in size. Owing to the use of a horizontal exit as a required means of egress, it is often beneficial to provide a glass light of limited size so that occupants may view the egress path ahead of them. Second, the maximum size of all types of fire-protection-rated glazing in $1^1/_2$-hour fire doors is limited to 100 square inches (0.065 m^2) when installed in a fire barrier.

716.5.8.1

Identification. Glazing used in fire door assemblies must be identified for verification of its appropriate application. The "D" designation indicates the glazing can be used in a fire door assembly, with the remaining identifiers providing specific information as to the glazing's capability to meet the hose-stream test and temperature limits. See Figure 716-3.

716.5.8.3.1

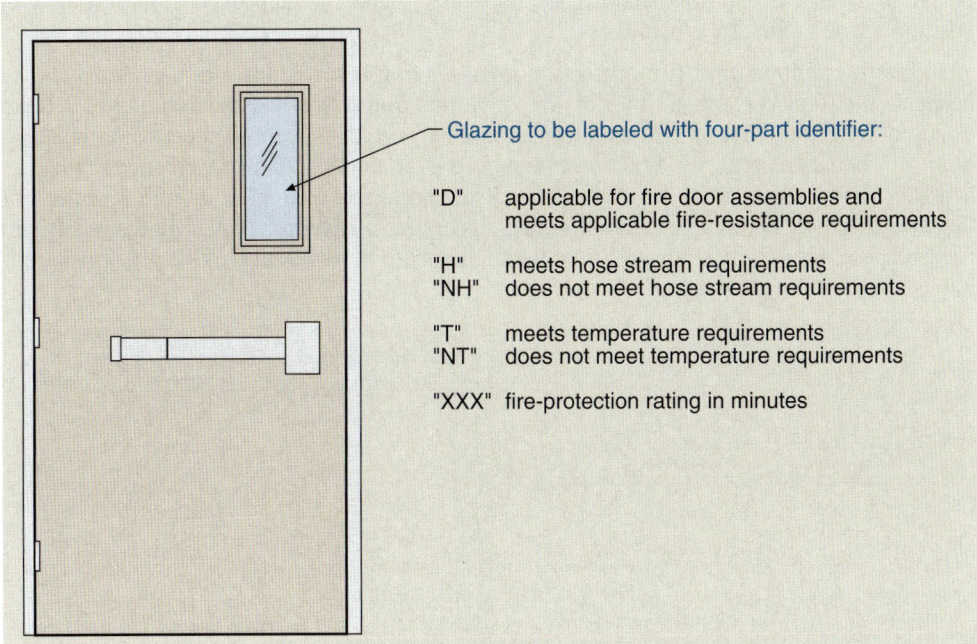

Figure 716-3
Identification of glazing in fire doors.

Door closing. The code mandates that fire doors be provided with closers to allow them to shut and protect the opening without manual operation. One exception to this broad-based requirement applies to those fire doors located in the common walls between sleeping units of hotels and motels. These doors are so seldom open that it is unreasonable to require door-closing hardware.

716.5.9

716.5.9.2 Automatic-closing fire door assemblies. Where automatic-closing devices are used instead of self-closing devices on fire doors, they must also comply with the provisions of NFPA 80 for self-closing action. Although they are generally held in an open position, doors equipped with automatic-closing devices become self-closing when actuated. The use of automatic-closing devices is typically a design decision; however, the code does mandate such devices in two applications. Automatic-closing devices are required by Exception 1 to Section 709.5 on cross-corridor doors located in smoke barriers of Group I-2 occupancies. They must also be installed on cross-corridor doors located in a horizontal exit as set forth in Section 1025.3.

716.5.9.3 Smoke-activated doors. This section identifies 11 locations where a smoke detector is to be used to actuate the closing operation for an automatic-closing fire door where such a closing device is provided. The detectors must be installed in accordance with the provisions of Section 907.3 and, furthermore, they must be of an approved type that will release the door in the event of a power failure. Automatic-closing fire door assemblies are often used to increase the reliability of the opening protection. Swinging fire doors with self-closers are all too often propped open with wood blocks or wedges. Although this section regulates the method for activating automatic-closing fire doors, it does not identify where automatic-closing doors are mandated.

716.6 Fire-protection-rated glazing. In many situations, it is necessary to provide glazed openings in fire-resistance-rated walls. The provisions of this section address fire window assemblies installed as opening protectives in fire partitions and exterior walls, as well as in some 1-hour fire barriers.

Fire-protection-rated glazing in fire window assemblies must be tested in accordance with NFPA 257 or UL 9. In addition, they must be installed and sized in accordance with NFPA 80. In all cases, a fire-window assembly must include an approved frame, be fixed in position, or be automatic closing.

In interior applications, fire-protection-rated glazing is limited to fire partitions, smoke barriers, and two types of fire barriers having a maximum fire-resistance rating of 1 hour. The total aggregate area of fire windows cannot exceed 25 percent of the area of the common wall between areas, as shown in Figure 716-4. In making this 25-percent calculation, it is permissible to assume the entire area of the common wall even though a portion of that area might be taken up by doors. This gross area is usable in calculating the maximum percentage of area for windows. Where the ceilings are of different heights, the lower ceiling establishes the gross area.

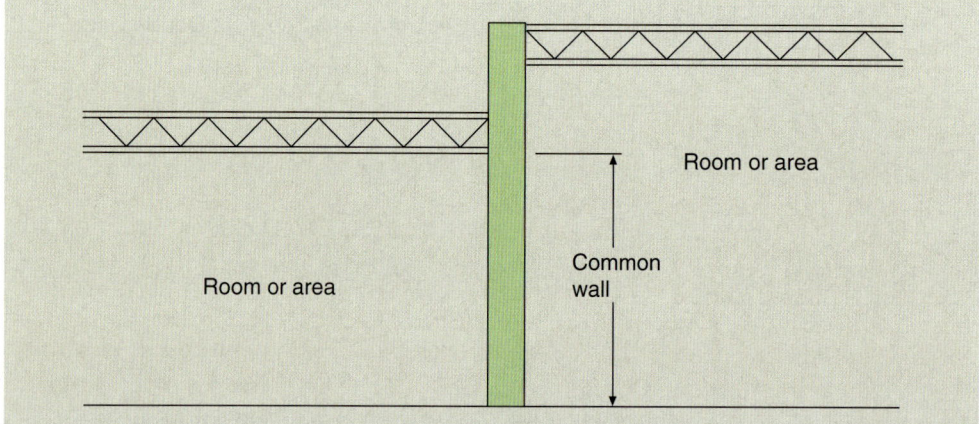

Figure 716-4 Glazing limitations.

Fire-protection-rated glazing is not permitted to be located in any fire wall or fire barrier (with two exceptions); however, Section 716.2 recognizes that glazing tested as a part of a wall assembly is permitted in all applications and, therefore, not regulated by this section. This type of glazing is referred to as fire-resistance-rated glazing, rather than fire-protection-rated glazing, and is further addressed in Section 703.6. In addition, the 25-percent area limitation is not applicable. Glazing that has a fire-resistance rating under ASTM B 119 or UL 263 that meets the fire-resistance rating of the wall may be used in more applications than fire-protection-rated glazing complying with NFPA 257 or UL 9. Also see the discussion of Section 703.6.

Section 717 *Ducts and Air Transfer Openings*

Where a duct or air transfer opening penetrates a fire-resistance-rated assembly, it is often necessary that a method of protection be provided to maintain the integrity of the assembly. Many times, dampers are used to protect the opening created by the duct or transfer opening. If dampers are not required to be provided under the provisions of this section, it is still necessary to protect the penetration under the provisions of Section 714.4.1.

Installation. This section states that fire and smoke dampers shall be installed in accordance with their listing. The test standards for each of the types of damper carry specific requirements that manufacturers provide installation and operating instructions, and that a reference to these instructions shall be a part of the required marking information on the damper. **717.2**

Damper testing. Dampers must not only be listed but also bear a label indicating that the damper is in compliance with the appropriate standard as identified by this section. For example, for fire dampers the required information on the damper includes the hourly rating; the words "Fire Damper"; whether or not the damper is to be in a dynamic or static system (or both); maximum rated airflow and pressure differential across the closed damper for dampers intended for use in dynamic systems; an arrow showing direction of airflow for dampers intended for use in dynamic systems; the intended mounting position (vertical, horizontal, or both); top of damper; and, of course, the manufacturer's name and model number. UL 555 (which applies to fire dampers) requires that all of this information shall be available on the damper label, which is installed at the factory, and that all labels shall be located on the internal surface of the damper and be readily visible after the damper is installed. UL 555 indicates that fire dampers tested under that standard are intended for use in HVAC duct systems passing through fire-resistive walls, partitions, or floor assemblies. **717.3.1**

Just as fire dampers are tested for different hourly ratings, they are also tested for different installation positions. A damper listed for vertical installation cannot arbitrarily be installed in the horizontal position.

This section also states that only fire dampers labeled for use in dynamic systems shall be installed in systems intended to operate with fans on during a fire. The test standard for fire dampers states that fire dampers are intended for use in either static systems that are automatically shut down in the event of a fire, or in dynamic systems that are operational in the event of a fire. If the HVAC system has not been designed and constructed to shut down in case of a fire, then dynamically listed fire dampers are necessary. Special attention should be paid to damper listings when smoke-control systems are installed under the provisions of Section 909.

Test standard UL 555S is to be used to determine the compliance of smoke dampers. This standard states that leakage-rated dampers (smoke dampers in the IBC) are intended to restrict the spread of smoke in HVAC systems that are designed to automatically shut down in the event of a fire, or to control the movement of smoke within a building when the HVAC system is operational in engineered smoke-control systems.

In addition to fire dampers and smoke dampers, two other types of dampers are referenced in the IBC. Where combination fire/smoke dampers are provided, they must comply with the requirements of both UL 555 for fire dampers and UL 555S for smoke dampers. Figure 717-1 illustrates the installation of an automatic-closing combination fire and smoke damper. Ceiling radiation dampers, intended for installation in air-handling openings penetrating the ceiling membranes of fire-resistance-rated floor/ceiling and roof/ceiling assemblies, are to meet the conditions of UL 555C.

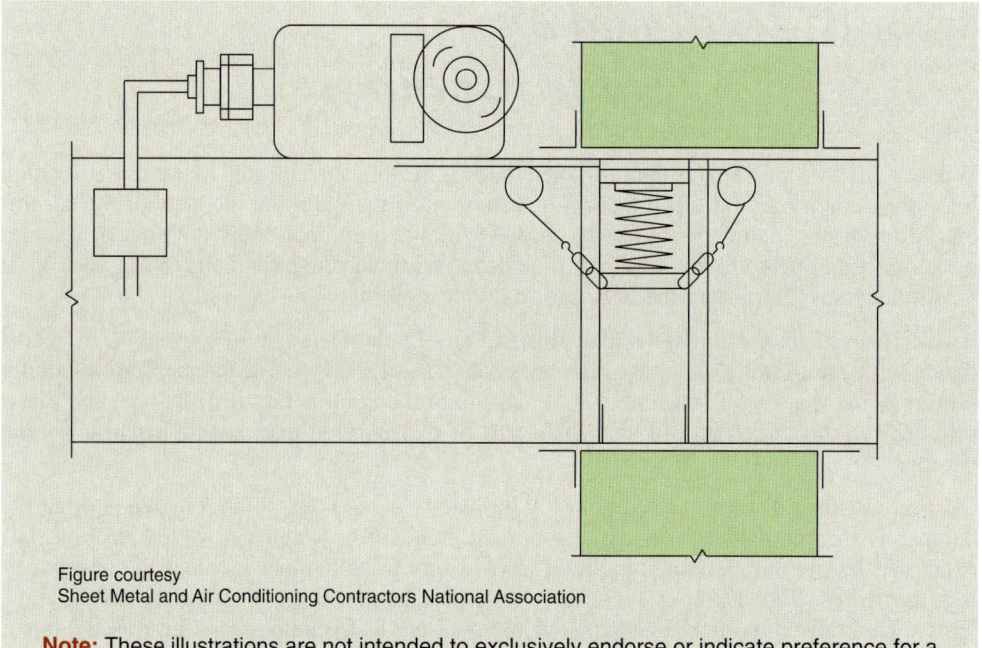

Figure 717-1
Combination fire and smoke dampers.

Figure courtesy
Sheet Metal and Air Conditioning Contractors National Association

Note: These illustrations are not intended to exclusively endorse or indicate preference for a combination fire and smoke damper. Two separate dampers that satisfy the requirements for the respective functions may also be used for fire and smoke control.

717.3.2.1 Fire damper ratings. Test standard UL 555 covers fire dampers ranging from $1/2$ hour to 3 hours. Because fire dampers carry an hourly rating, plans should reflect the rating required at a particular location if more than one rating is required within a building. Table 717.3.2.1 indicates whether a $1^1/_2$-hour or 3-hour rating is required for a fire damper, based on the fire-resistance rating of the assembly in which it is installed. In all applications rated at 3 hours or greater, a 3-hour-rated damper is mandated; otherwise, a $1^1/_2$-hour damper is acceptable.

717.3.2.2 Smoke damper ratings. A Class I or Class II leakage rating is required for smoke dampers, which also must have an elevated temperature rating of at least 250°F (121°C). The class designation indicates the maximum leakage permitted in cubic feet per minute per square foot (cubic mm per minute per mm^2) for the particular class. The four classes progress from Class I (least leakage or best performance) through Class IV (greatest leakage or poorest performance). The IBC requires conformance with Class I or II, so a damper rated as Class III or IV would not be acceptable. These leakage ratings are determined at ambient temperature after exposing the damper to temperature degradation at an elevated temperature, with 250°F (121°C) being the lowest elevated temperature allowed by the code. Dampers can be tested using higher degradation temperatures [one as high as 850°F (454°C)], but most listed dampers seem to have been tested at either 250°F (121°C) or 350°F (177°C).

The provisions of Section 717.3.3.2 specifically instruct the designer or installer on how to control smoke dampers. Smoke dampers are required to be closed by activation of smoke detectors installed in accordance with Section 907.3 for fire-detection systems and any of the five specified methods of control listed in this provision. These methods of control, each having benefits and drawbacks, were proposed by those individuals involved with damper installation and should provide consistent and logical control methods for the dampers.

Access and identification. Both fire dampers and smoke dampers shall be installed so that they are accessible for inspection and servicing. It is important that any access openings in a fire-resistance-rated assembly be adequately protected in order to maintain the integrity of the assembly. This will typically involve the use of an access door having the required fire-protection rating. Permanent identification of the access points to fire-damper and smoke-damper locations is also mandated. **717.4**

Where required. This section lists those specific locations where the various dampers are required. Dampers need only be installed in ducts and transfer openings where specifically identified by this section. In some locations, both a fire damper and a smoke damper are required. This means that either two dampers must be installed or a damper listed for both heat and smoke control must be used. See Figure 717-2 for an overview of the required locations for smoke and fire dampers. **717.5**

Fire walls. Because of the importance of maintaining the separation provided by fire walls used to divide a structure into two or more separate buildings, the code requires the use of approved fire dampers under all conditions. Such dampers are to be installed at all permitted duct penetrations and air-transfer openings of fire walls. Where the fire wall serves as a party wall as addressed in Section 706.1.1, ducts and air transfer openings are prohibited. There is no requirement for smoke dampers at duct penetrations and air openings through fire walls except for those fire walls serving as horizontal exits. **717.5.1**

Fire barriers. Much like fire walls, fire barriers are designed to totally isolate one area of a building from another. Therefore, the general requirement is that all duct penetrations and air transfer openings of fire barriers be protected by complying fire dampers. There are, however, several exceptions that may eliminate the need for dampers. Of special note is the elimination of fire dampers in certain sprinklered buildings. Fire dampers are not required for duct penetrations and air openings in fire barriers where all of the following conditions exist: **717.5.2**

1. The penetration consists of a duct that is a portion of a ducted HVAC system.
2. The fire-resistance rating of the fire barrier is 1 hour or less.
3. The area is not a Group H occupancy.
4. The building is fully protected by an automatic fire-sprinkler system.

A fire barrier that serves as a horizontal exit must also be provided with a listed smoke damper at each point a duct or air transfer opening penetrates the fire barrier.

Shaft enclosures. Both a fire damper and a smoke damper, or a combination fire/smoke damper, must be installed where a duct or air transfer opening penetrates a shaft enclosure. Five exceptions identify conditions under which fire dampers are not required. As shown in Figure 717-3, the first exception permits fire dampers to be eliminated where steel exhaust subducts enter an exhaust shaft. The subducts must extend vertically at least 22 inches (559 mm), and there must be continuous air flow upward to the outside through the shaft. Smoke dampers are also not required under similar conditions, but the exception is limited to fully sprinklered Group B and R occupancies where the fan providing continuous airflow is on standby power. Exception 5 exempts fire dampers and fire/smoke dampers in kitchen and clothes dryer exhaust systems due to potential obstruction hazards. Smoke dampers are not addressed in this exception, but may be omitted when Exception 2 is applied. **717.5.3**

7 Fire and Smoke Protection Features

Fire and Smoke Damper Location			
Location		Fire damper	Smoke damper
Fire walls		Required	Not Required[26]
Fire barriers		Required[1,2,3]	Not Required[26]
Shaft enclosure[7]		Required[1,2,4,5,29]	Required[2,5,6,29]
Fire partitions		Required[3,8,9,22]	Not Required[2,27]
Fire-resistance-rated corridors[23]		Not Required[28]	Required[10,18]
Smoke partitions		Not Required	Required[2,21]
Smoke barriers		Not Required	Required[11]
Horizontal assemblies[12]	Through-penetrations	Required[13,19]	Not Required
	Membrane penetrations[24]	Required[14,20,25]	Not Required
	Nonfire-resistance-rated assemblies	Required[15,16,17]	Not Required

[1] Not required for penetrations tested in accordance with ASTM E 119 or UL 263 as part of the rated assembly.
[2] Not required for ducts used as a part of an approved smoke control system in accordance with Section 909.
[3] Not required in sprinklered buildings of other than Group H for maximum 1-hour walls penetrated by ducted HVAC systems.
[4] Not required for steel exhaust subducts extending at least 22 inches vertically in exhaust shafts having continuous airflow upward to the outside.
[5] Not required in parking garage supply or exhaust shafts that are separated from other building shafts by a minimum of 2-hour fire-resistance-rated construction.
[6] Not required in fully sprinklered Group B and R occupancies for kitchen, clothes dryer, bathroom, and toilet room exhaust openings with steel exhaust subducts that extend at least 22 inches vertically and an exhaust fan installed at the upper terminus of the shaft is powered continuously per Section 909.11 with a continuous upward airflow to the outside.
[7] See Section 1022.6 for permitted ventilation of interior exit stairways.
[8] Not required in sprinklered buildings of other than Group H for corridor walls, provided duct protected per Section 714 as a penetration.
[9] Not required in buildings of other than Group H where duct penetration is limited to 100 square inches; is of minimum 0.0217-inch steel; does not have communicating openings between a corridor and adjacent spaces; is installed above a ceiling; does not terminate at a wall register of the fire-resistance-rated wall; and minimum 12-inch-long steel sleeve centered and secured in opening.
[10] Not required for corridor penetrations of minimum 0.019-inch steel ducts with no openings into corridor.
[11] Not required where openings in steel ducts are limited to a single smoke compartment.
[12] Section 712 addresses permissible openings in floor and roof systems.
[13] In other than Group I-2 and I-3, fire dampers are permitted in lieu of shaft enclosures for penetration of fire-resistance-rated horizontal assembly that connects two floors.
[14] Where shaft enclosure is not provided, an approved ceiling damper is required at the ceiling line of a fire-resistance-rated floor/ceiling or roof/ceiling assembly where duct penetrates ceiling or diffuser is installed without a duct.
[15] Not required where duct does not connect more than two stories and the annular space around the duct is filled with noncombustible material.
[16] Limited to three connected stories without shaft enclosure, provided fire dampers are installed at each floor line and annular space is filled.
[17] Not required in ducts within individual dwelling units.
[18] Not required in building with a smoke control system if not necessary for operation and control of system.
[19] Not required at each floor where (1) penetrating three floors or less, (2) duct of steel construction and within wall cavity, (3) duct opens into only one dwelling unit or sleeping unit and is continuous from unit to building exterior, (4) duct limited to 4 inches in diameter and total area limited to 100 square inches per 100 square feet, (5) annular space around duct protected with materials that prevent passage of flame and hot gases, and (6) grille openings in ceiling of fire-resistance-rated floor/ceiling or roof/ceiling assembly protected with ceiling radiation damper.
[20] Ceiling radiation dampers not required for exhaust duct penetrations protected per Section 714.4.1.2, where ducts are located within wall cavity and do not pass through another dwelling unit or tenant space.
[21] Only required to protect air transfer openings.
[22] Not required for tenant partitions in covered mall buildings where walls not required to extend to floor or roof deck above.
[23] Fire partition provisions also applicable to fire-resistance-rated corridors.
[24] Applicable to penetrations of ceiling membrane of fire-resistance-rated floor/ceiling or roof/ceiling assembly.
[25] Ceiling radiation dampers not required where tests per ASTM E 119 or UL 263 have shown that dampers are not necessary in order to maintain the fire-resistive rating of the assembly.
[26] Only required where duct or air transfer opening penetrates a horizontal exit wall.
[27] Only required where penetration is an air transfer opening.
[28] Only required for duct or air transfer openings in fire-resistance-rated corridor walls in a nonsprinklered building.
[29] Not required in kitchen and clothes dryer exhaust systems when installed per IMC.

Figure 717-2
Fire and smoke damper location.

717.5.4 Fire partitions. Fire partitions are not regulated by the code as highly as fire barriers or fire walls, so it is consistent that the requirements for dampers through such partitions are not as restrictive. The general rule is that a fire damper is required in any duct or air transfer opening that penetrates a fire partition. However, where the building is fully sprinklered, ducts penetrating tenant separation walls in covered mall buildings or fire-resistance-rated corridor walls need not be fire dampered. In addition, fire dampers are not necessary for

Ducts and Air Transfer Openings 7

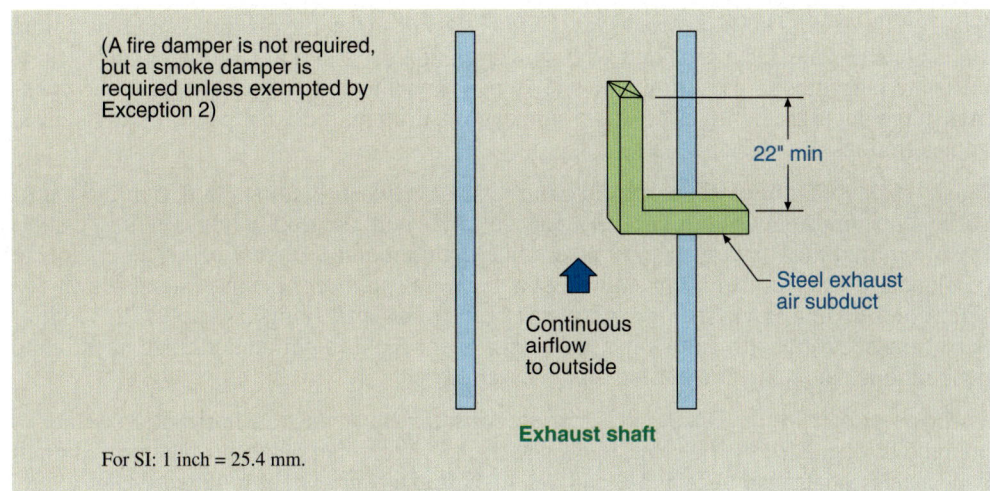

Figure 717-3
Exhaust subducts penetrating shafts.

small steel ducts installed above a ceiling, provided the duct does not communicate between a corridor and adjacent rooms and does not terminate at a register in a fire-resistance-rated wall. The fourth exception is consistent with Exception 3 of Section 717.5.2 addressing fire barriers.

Corridors. Because a fire-resistance-rated corridor is intended to be an exit access component providing a limited degree of occupant protection during egress activities, it is logical that air openings into the corridor be addressed. This section mandates, with exceptions, that all corridors required to be protected with smoke- and draft-control doors shall also be provided with smoke dampers where ducts or air transfer openings penetrate the corridor enclosures. As illustrated in Figure 717-4, an important exception eliminates the need for smoke dampers where steel ducts pass through, but do not serve, the corridor.

717.5.4.1

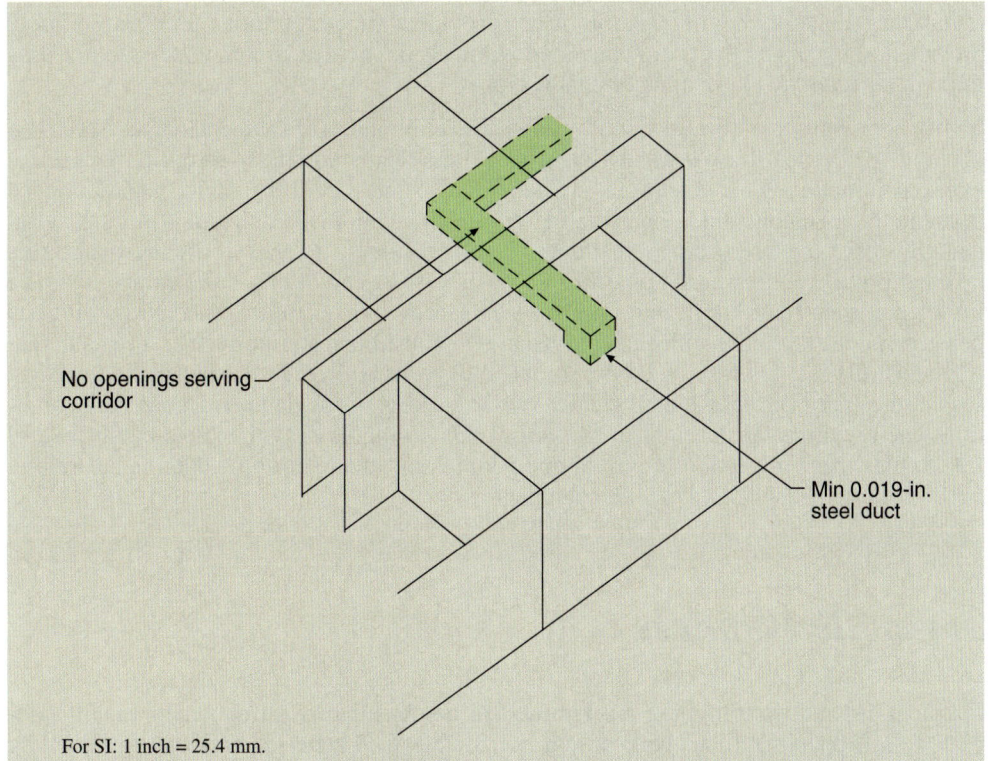

Figure 717-4
Ducts crossing corridor.

717.5.5 Smoke barriers. Those air openings, both ducts and transfer openings, that penetrate smoke barriers are to be provided with smoke dampers at the points of penetration. Steel ducts are permitted to pass through a smoke barrier without a damper, provided the openings in the ducts are limited to a single smoke compartment. Fire dampers are not required at penetrations of smoke barriers.

717.6 Horizontal assemblies. The code is quite restrictive when it comes to the protection of vertical openings between floor levels, particularly where the floor or floor/ceiling assembly is required to be fire-resistance rated. This section requires the use of a shaft enclosure to address the hazard that is created where a duct or air transfer opening extends through a floor, floor/ceiling assembly, or ceiling membrane of a roof/ceiling assembly. The remainder of the section modifies this general requirement for through-penetrations, membrane penetrations, and nonfire-resistance-rated assemblies.

717.6.1 Through-penetrations. Ducts and air transfer openings that penetrate horizontal assemblies are initially regulated by the provisions of Section 713 for shaft enclosures. Permitted in all occupancies other than Groups I-2 and I-3, a shaft enclosure is not required where a duct that connects only two stories is provided with a fire damper installed at the floor line of the fire-resistance-rated floor/ceiling assembly that is penetrated. As an option, the duct may be protected in a manner prescribed in Section 714.4 for the penetration of horizontal assemblies. The code's intent to limit fire and smoke migration between smoke compartments vertically in Group I-2 and I-3 occupancies is maintained through the limitation imposed in this section. The exception goes on to eliminate the fire damper requirement as it relates to dwelling units and sleeping units, provided the duct penetrates no more than three floors.

717.6.2 Membrane penetrations. A shaft enclosure need not be provided where an approved ceiling radiation damper is installed at the ceiling line of a fire-resistance-rated floor/ceiling or roof/ceiling assembly penetrated by a duct or air-transfer opening. Designed to protect the construction elements of the floor or roof assembly, the ceiling damper is not required where fire tests have shown that ceiling radiation dampers are not necessary to maintain the fire-resistance rating of the assembly. Additionally, ceiling radiation dampers are not required at penetrations of exhaust ducts, provided the penetrations are appropriately protected, the exhaust ducts are contained within wall cavities, and the ducts do not pass through adjacent dwelling units or tenant spaces.

717.6.3 Nonfire-resistance-rated floor assemblies. The elimination of shaft enclosures at vertical openings is also possible where the floor assemblies are not required to be of fire-resistance-rated construction. Two conditions are identified using the filling of the annular space between the assembly and the penetrating duct with an approved noncombustible material that will resist the free passage of fire and smoke. Where only two stories are connected, no other protective measures are necessary. In three-story conditions, a fire damper must be installed at each floor line. The code also mandates that the annular space surrounding the penetrating duct be filled with an approved noncombustible material, such as a sealant, that will resist the free passage of flame, smoke, and gases. However, the installation of such sealant or other material would typically void the listing of the damper. Under such conditions, the use of the damper's steel mounting angles would satisfy the intended purpose of the annular space protection. An exception permits this method of protection in a dwelling unit without the installation of a fire damper.

Section 718 *Concealed Spaces*

Fireblocking and draftstopping are required in combustible construction to cut off concealed draft openings (both vertical and horizontal). The code requires that fireblocking form an effective barrier between floors and between the top story and attic space. The code

Concealed Spaces 7

also requires that attic spaces be subdivided, as will be discussed later, along with concealed spaces within roof/ceiling and floor/ceiling assemblies. Figures 718-1 through 718-4 depict IBC requirements for fireblocking.

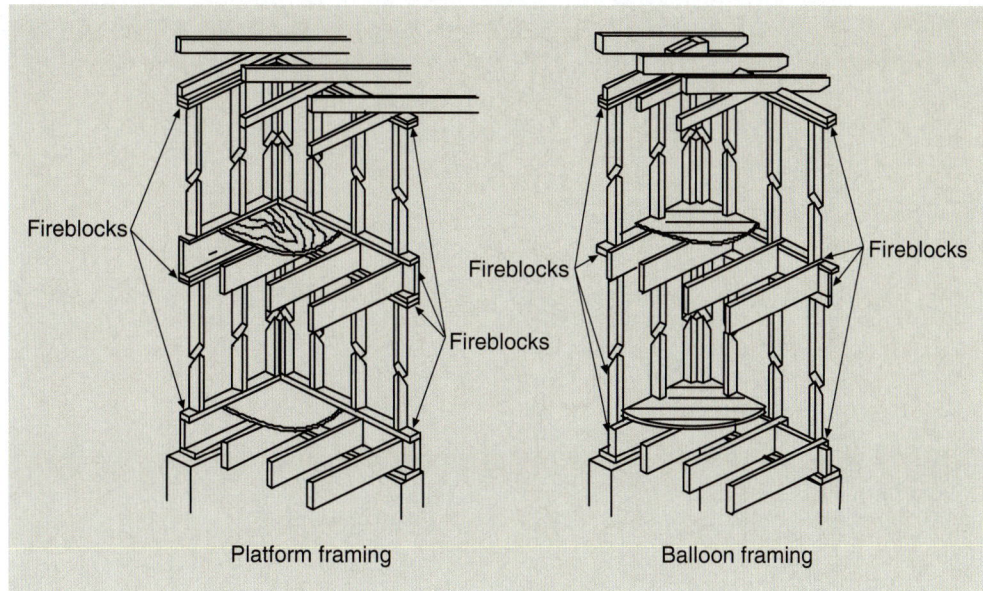

Figure 718-1
Fireblocking.

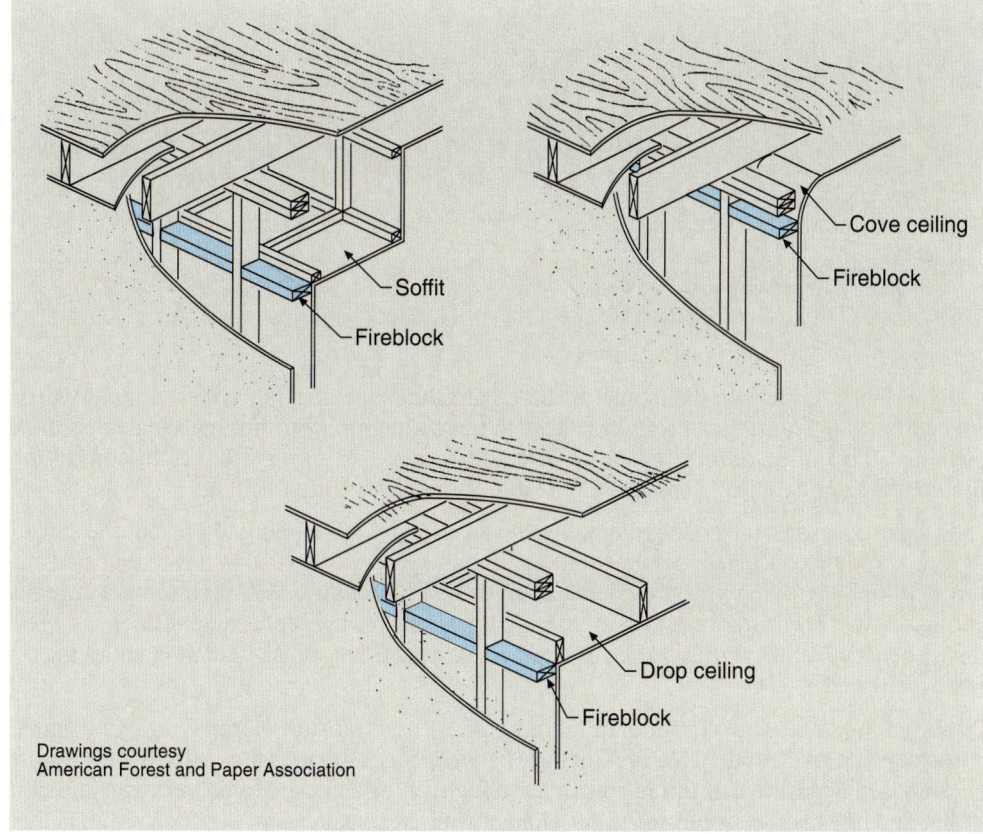

Figure 718-2
Fireblocks—vertical and horizontal space connections.

7 Fire and Smoke Protection Features

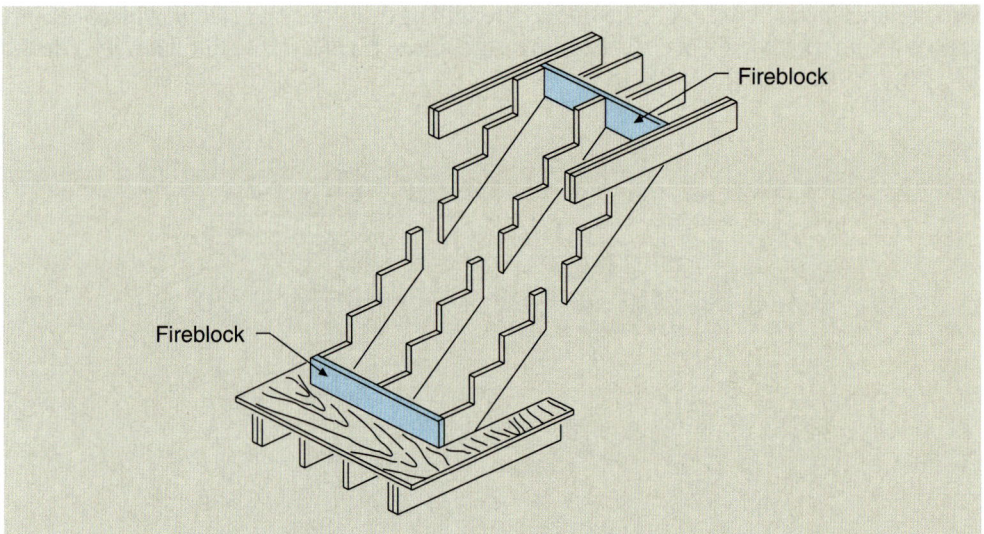

**Figure 718-3
Fireblocks—
stairs.**

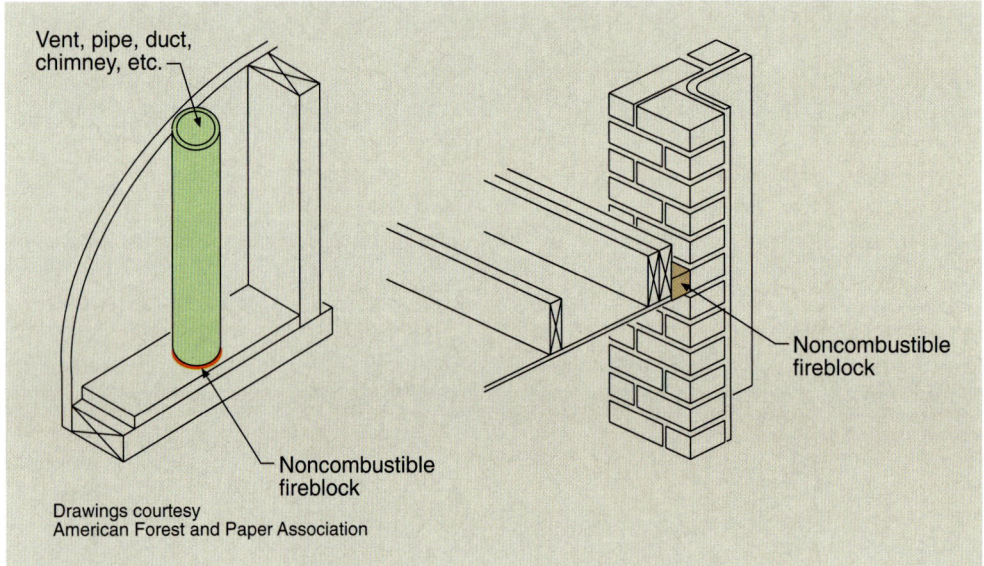

**Figure 718-4
Fireblocks—
pipes,
chimneys, etc.**

Experience has shown that some of the greatest damage occurs to conventional wood-framed buildings during a fire when the fire travels unimpeded through concealed draft openings. This often occurs before the fire department has an opportunity to control the fire, and greater damage is created as a result of the lack of fireblocking.

For these reasons, the code requires fireblocking and draftstopping to prevent the spread of fire through concealed combustible draft passageways. Virtually any concealed air space within a building will provide an open channel through which high-temperature air and gases will spread. Fire and hot gases will spread through concealed spaces between joists, between studs, within furred spaces, and through any other hidden channel that is not fireblocked.

718.2 Fireblocking. The platform framing method that is used most often today in wood-frame construction provides adequate fireblocking between stories in the stud walls, but care must be exercised to ensure that furred spaces are effectively fireblocked to prevent transmission of fire and hot gases between stories or along a wall. For this reason, the code requires that fireblocking be provided at 10-foot (3,048-mm) intervals horizontally along walls that are either furred out, of double-wall construction, or of staggered-stud construction.

Fireblocking provisions for wood flooring used typically in gymnasiums, bowling alleys, dance floors, and similar uses containing concealed sleeper spaces are found in Section 718.2.7. As long as the wood flooring described in this section is in direct contact with a concrete or masonry fire-resistance-rated floor, there is no significant hazard. However, if there is a void between the wood flooring and the fire-resistance-rated floor, a blind space is created that is enclosed with combustible materials and provides a route for the undetected spread of fire. Therefore, the code requires that where the wood flooring is not in contact with the fire-resistance-rated floor, the space shall be filled with noncombustible material or shall be fireblocked. Two exceptions to these fireblocking requirements are:

1. The first exception exempts slab-on-grade floors of gymnasiums. In this case, the code presumes a low hazard, as gymnasiums are usually only one story in height. If the floor is at or below grade, it is unlikely that any ignition sources would be present to start a fire that would spread through the blind space under the wood flooring.

2. Bowling lanes are exempted from fireblocking except as described in the code, which provides for areas larger than 100 square feet (93 m^2) between fire blocks. Fireblocking intermittently down a bowling lane would create problems for a consistent lane surface.

Fireblocking materials are required to consist of lumber or wood structural panels of the thicknesses specified, gypsum board, cement fiber board, mineral wool, glass fiber, or any other approved materials securely fastened in place.

Batts or blankets of mineral wool and glass fiber materials are allowed to be used as fireblocking and work especially well where parallel or staggered-stud walls are used. Loose-fill insulation should not be used as a fireblocking material unless specifically tested for such use. It must also be shown that it will remain in place under fire conditions. Even in the case where it fills an entire cavity, a hole knocked into the membrane enclosure for the cavity could allow the loose-fill insulation material to fall out, negating its function. Therefore, loose-fill insulation material shall not be used as a fireblock unless it has been properly tested to show that it can perform the intended function. The main concern is that the loose-fill material, even though it may perform adequately in a fire test to show sufficient fire-retardant characteristics to meet the intent of this section, would not be adequately evaluated for various applications because of the physical instability of the material in certain orientations.

Draftstopping in floors. Draftstops are often used to subdivide large concealed spaces within floor/ceiling assemblies of combustible construction. Figure 718-5 shows IBC requirements for draftstopping in these locations. Gypsum board, wood structural panels, particleboard, mineral wool, or glass fiber batts and blankets, and other approved materials are considered satisfactory for the purpose of subdividing floor/ceiling areas, provided the materials are of adequate thickness, are adequately supported, and their integrity is maintained. **718.3**

Draftstops are to be installed in floor/ceiling assemblies as follows:

1. Residential occupancies. The code requires that draftstops be installed in line with the wall separating tenants or dwelling units from each other and the remainder of the building, consistent with the provisions of Section 718.3.2 for dwelling unit and sleeping unit separations. In this case, a fire originating in a dwelling unit or hotel room will find draftstops in the concealed space blocking the transmission of fire and hot gases into another hotel room or apartment. Where the residential occupancy is fully sprinklered, draftstopping is not required. Where a residential sprinkler system is used, automatic sprinklers must also be installed in the combustible concealed floor areas.

7 Fire and Smoke Protection Features

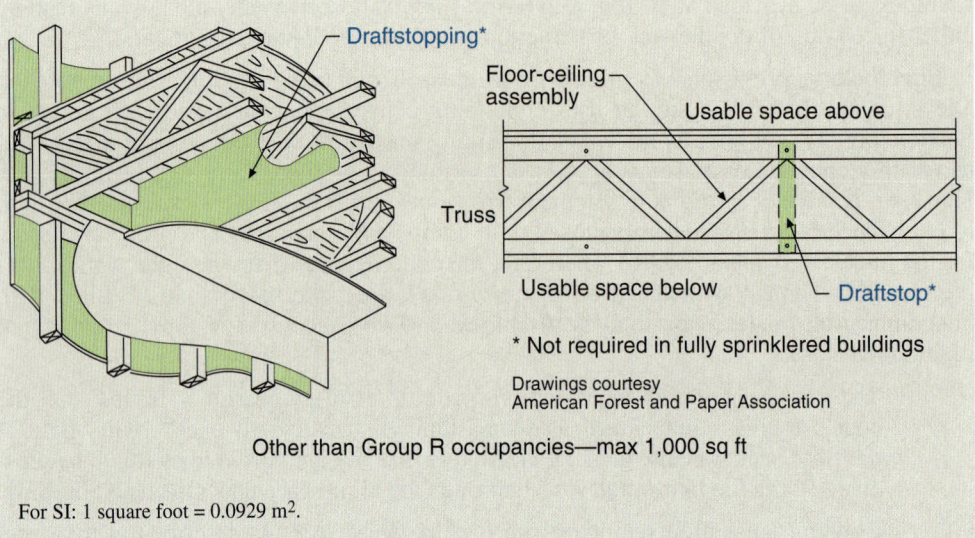

Figure 718-5
Draftstops—floors.

2. All other occupancies. For uses other than residential occupancies, the code intends that the concealed space within the floor/ceiling assembly be separated by draftstopping so that the area of any concealed space does not exceed 1,000 square feet (93 m²). An exception permits the elimination of draftstopping where automatic fire sprinklers are installed throughout the building.

718.4 Draftstopping in attics. In attics and concealed roof spaces of combustible construction, the code requires draftstopping under certain circumstances. Consistent with the requirements for fireblocking, draftstopping is not required for spaces constructed entirely of noncombustible materials. Materials used for draftstopping purposes, such as gypsum board, plywood, or particleboard, are to be installed consistent with the provisions of Section 718.3.1 for the draftstopping of floors. The following locations are identified as those requiring draftstopping:

1. Groups R-1 and R-2. Draftstops are to be installed above and in line with the walls separating dwelling units or between walls separating sleeping units. Figure 718-6 explains the intent of Exception 1. Exception 3 applies to Group R-2 occupancies less than five stories in height. In this case, attic areas may be increased by installing draftstops to subdivide the attic into a maximum of 3,000-square-foot (279-m²) spaces, with no area to exceed the inclusion of two dwelling units. Exception 2 eliminates the need for draftstopping in fully sprinklered buildings, while Exception 4 considers the installation of a residential sprinkler system with sprinklers in the attic as an acceptable alternative to draftstopping.

2. Other uses. Draftstops are required by the code to be installed in attics and similar concealed roof spaces of buildings other than Groups R-1 and R-2 so that the area between draftstops does not exceed 3,000 square feet (279 m²). As permitted by the exception, draftstopping of the attic space is not required in any building equipped throughout with an automatic sprinkler system. See Figure 718-7.

718.5 Combustibles in concealed spaces in Type I or II construction. Where buildings are intended to be classified as noncombustible, it is intended that combustibles not be permitted, particularly in concealed spaces. The six exceptions to this limitation identify conditions under which a limited amount of combustible materials is acceptable. Exception 1 references Section 603, which identifies 25 applications where combustible materials are permitted in buildings of Type I or Type II construction. It is felt that the low level of combustibles permitted, as well as their control, does not adversely impact the fire-severity

Concealed Spaces 7

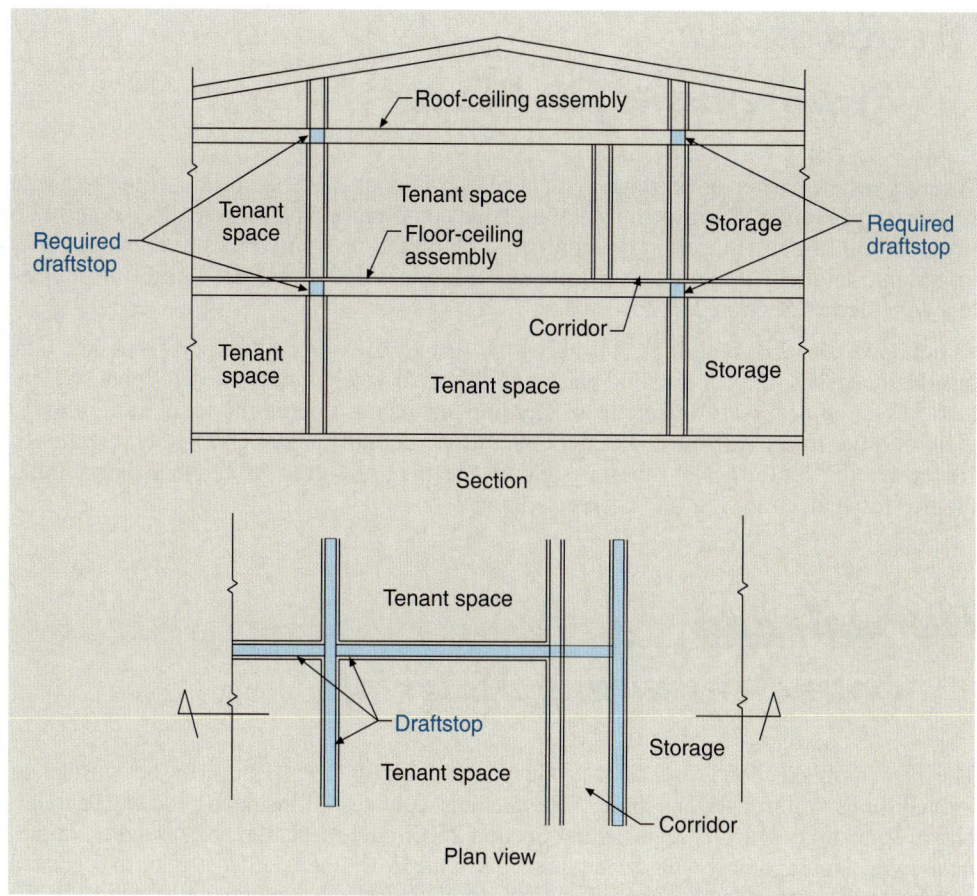

Figure 718-6
Attic draftstop—Groups R-1 and R-2.

potential caused by the combustible materials. Exception 2 permits the use of combustible materials in plenums under the limitations and conditions imposed by IMC Section 602. The third exception allows the concealment of interior finish materials having a flame-spread index of Class A. Exception 4 addresses combustible piping, provided it is located within partitions or enclosed shafts in a complying manner. As an example, the presence of plastic pipe within the wall construction of a Type I or II building does not cause the building to be considered combustible construction. Exception 5 allows for the installation of combustible piping within concealed ceiling areas of Type I and II buildings, while Exception 6 permits combustible insulation on pipe and tubing in all concealed spaces other than plenums.

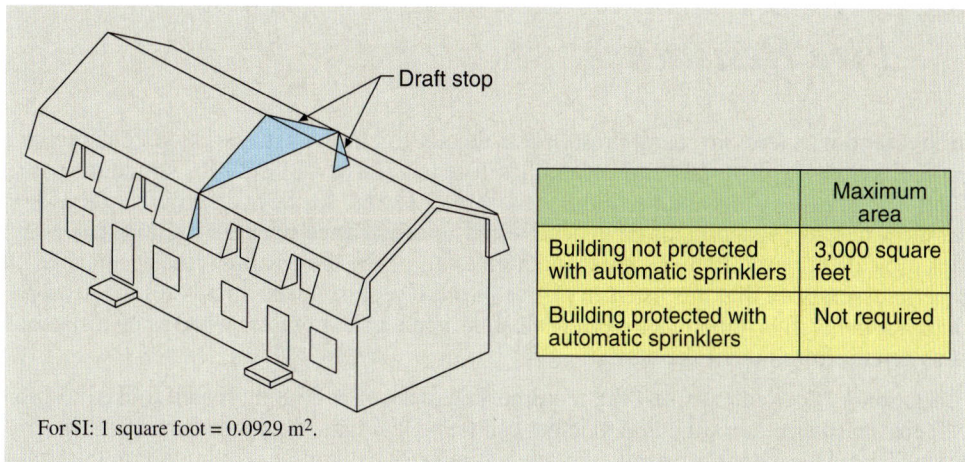

Figure 718-7
Attic draftstop—other than Group R occupancies.

2012 International Building Code Handbook   265

7 Fire and Smoke Protection Features

Section 719 Fire-Resistance Requirements for Plaster

Where gypsum plaster or portland cement plaster is considered a portion of the required fire-resistance rating of an assembly, it must be in compliance with this section. Appropriate fire tests shall be referenced in determining the minimum required plaster thickness. It is important that the material under consideration is addressed in the test, unless the equivalency method of Section 719.2 is used.

In noncombustible buildings, it is necessary that all backing and support be of noncombustible materials. Except for solid plaster partitions or where otherwise determined by fire tests, it is also necessary in certain plaster applications to double the required reinforcement in order to provide for additional bonding, particularly under elevated temperatures. Under specific conditions, it is permissible to substitute plaster for concrete in determining the fire-resistance rating of the concrete element.

Section 720 Thermal- and Sound-Insulating Materials

The intent of this section is to establish code requirements for thermal and acoustical insulation located on or within building spaces. This section regulates all insulation except for foam-plastic insulation, which is regulated by Section 2603: duct insulation and insulation in plenums, which must comply with the requirements of the IMC; fiberboard insulation as regulated by Chapter 23; and reflective plastic core insulation, which must comply with Section 2613.

As a general requirement, insulation, including facings used as vapor retarders or as vapor permeable membranes, must have a flame spread index not in excess of 25 and a smoke-developed index not to exceed 450. Section 720.2.1 waives the flame-spread and smoke-developed limitations for facings on insulation installed in buildings of Type III, IV, and V construction, provided that the facing is installed behind and in substantial contact with the unexposed surface of the ceiling, floor, or wall finish.

Section 721 Prescriptive Fire Resistance

In this section, there are many prescriptive details for fire-resistance-rated construction, particularly those materials and assemblies listed in Table 721.1(1) for structural parts, Table 721.1(2) for walls and partitions, and Table 721.1(3) for floor and roof systems. For the most part, the listed items have been tested in accordance with the fire-resistance ratings indicated. In addition, a similar footnote to all of the tables allows the acceptance of generic assemblies that are listed in GA 600, the Gypsum Association's *Fire-Resistance Design Manual*. It is important to review all of the applicable footnotes when using a material or assembly from one of the tables.

Section 721.1.1 intends that the required thickness of insulating material used to provide fire resistance to a structural member cannot be less than the dimension established by Table 721.1(1), except for permitted modifications. An example of the minimum thickness of concrete required for a structural-steel column is shown in Figure 721-1. Note that Figure 704-4 illustrates that the edges of such members are to be adequately reinforced in compliance with the provisions of Section 721.1.3. Figure 721-2 illustrates the minimum concrete-thickness requirements for protecting reinforcing steel in concrete columns, beams, girders, and trusses. Refer to Section 704 for additional provisions regarding structural members.

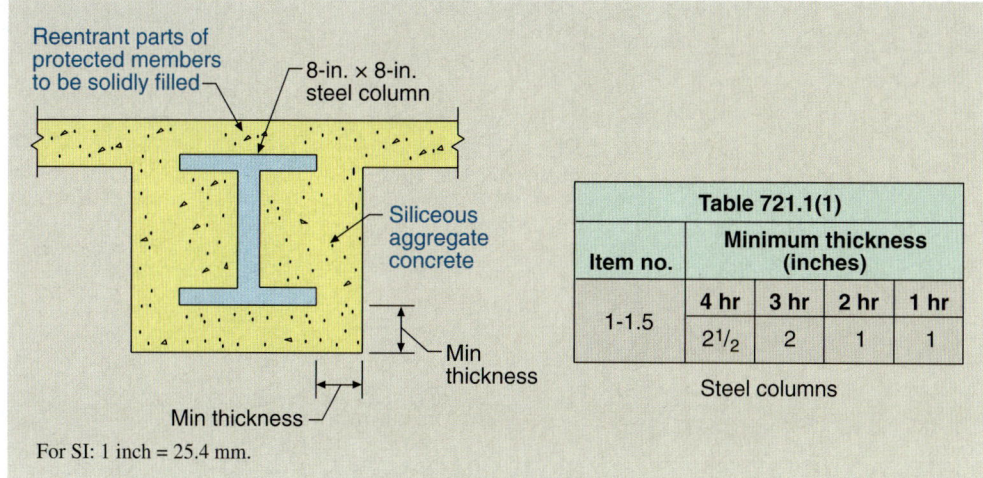

Figure 721-1
Prescriptive fire resistance.

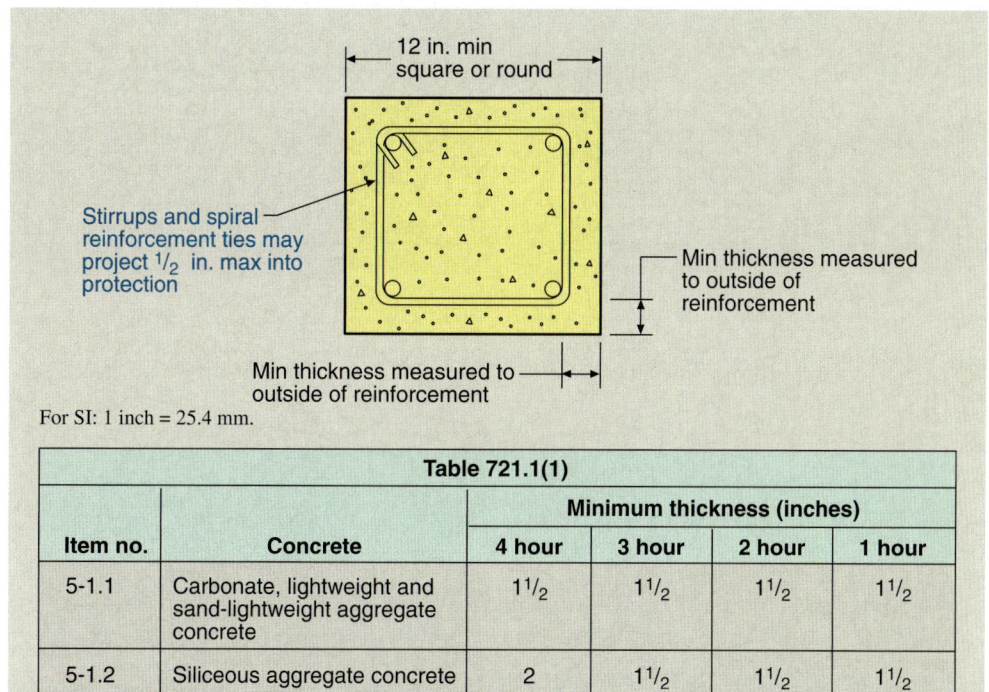

Figure 721-2
Reinforcing steel in concrete columns, beams, girders, and trusses.

As previously mentioned, the fire-resistance ratings for the fire-resistance-rated walls and partitions outlined in Table 721.1(2) are based on actual tests. Figure 721-3 shows two samples from the table. For reinforced concrete walls, it is important to note the type of aggregate as discussed earlier in this chapter. The difference in aggregates is quite significant for a 4-hour fire-resistance-rated wall, as it amounts to a difference in thickness of almost

2 inches (51 mm). For hollow-unit masonry walls, the thickness required for a particular fire-endurance rating is the equivalent thickness as defined in Section 722.3.1 for concrete masonry and Section 722.4.1.1 for clay masonry. Figure 721-4 outlines the manner in which the equivalent thickness is determined.

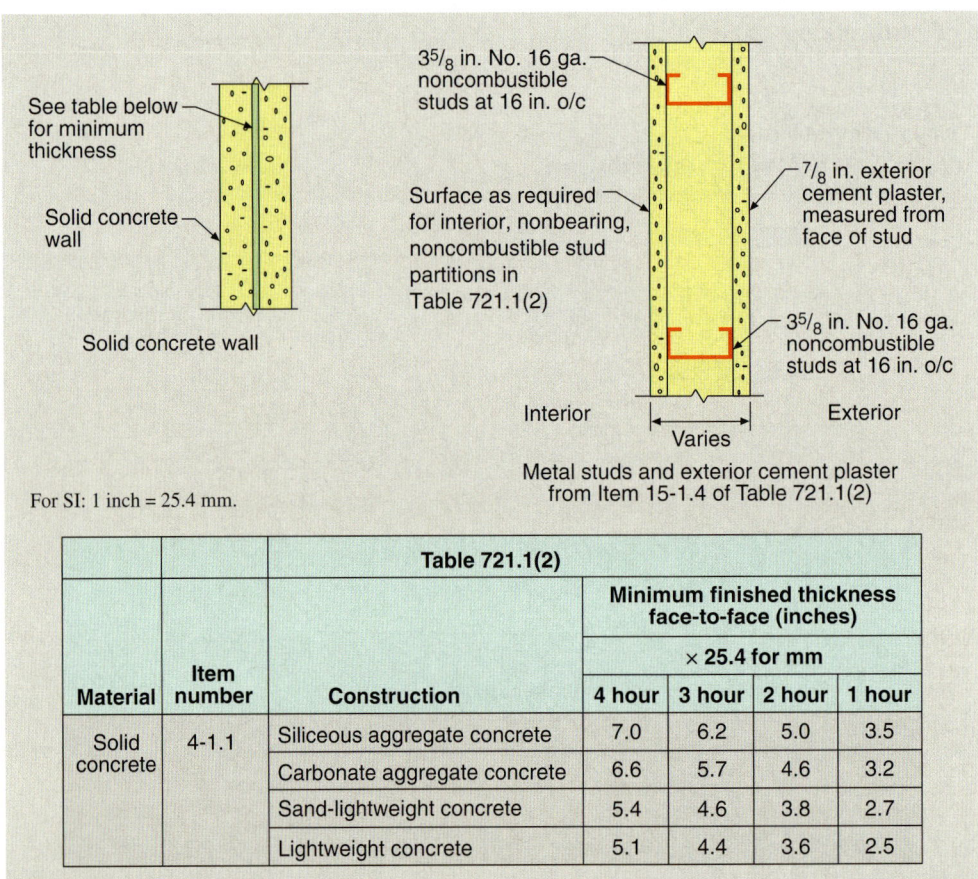

Figure 721-3
Walls and partitions.

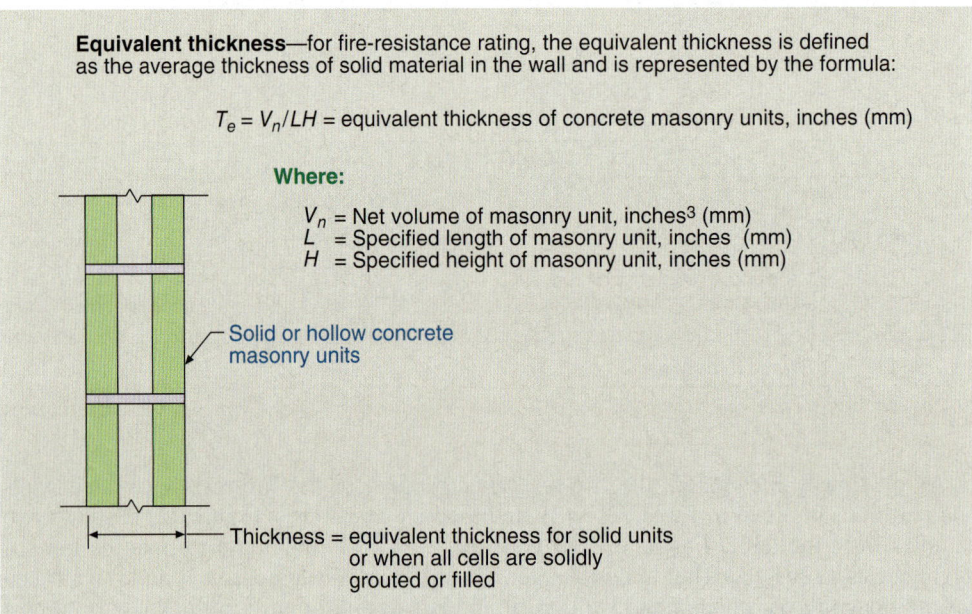

Figure 721-4
Equivalent thickness of masonry walls.

Table 721.1(3) of the IBC provides fire-resistance ratings for floor/ceiling and roof/ceiling assemblies, and Figure 721-5 depicts the construction of a 1-hour fire-resistance-rated wood floor or roof assembly. Of special note is Footnote n, which exempts unusable space from the flooring and ceiling requirements. See Figure 711-1.

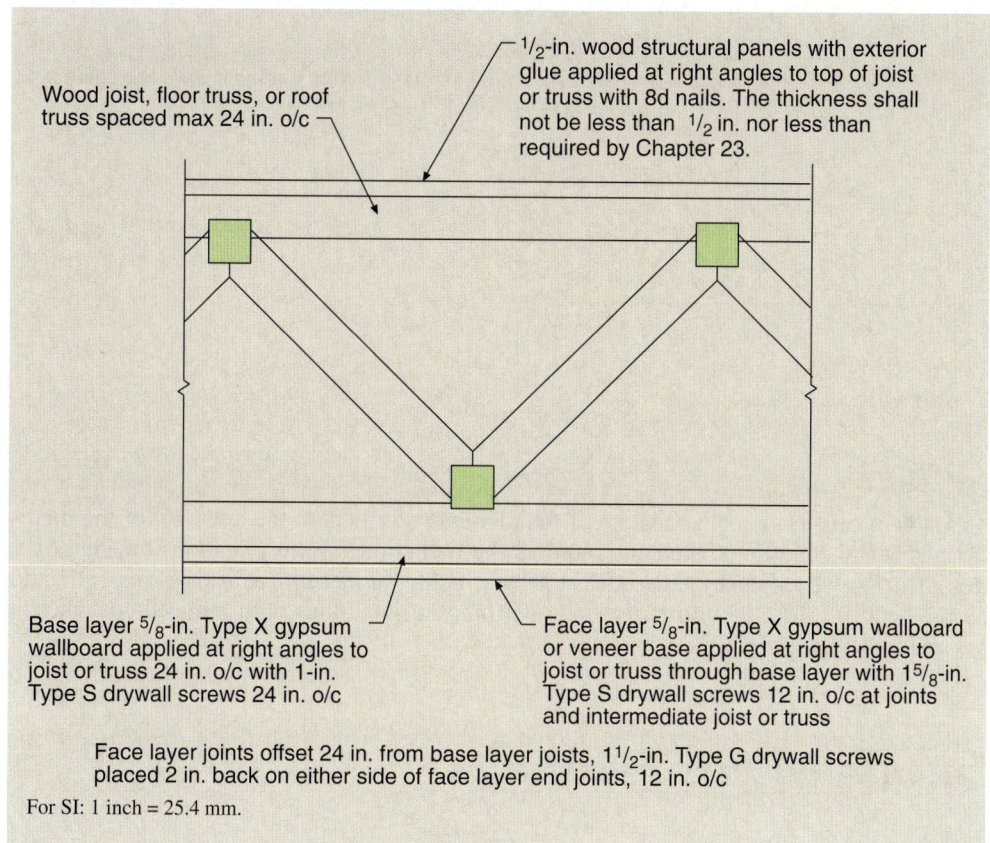

Figure 721-5
One-hour wood floor or roof assembly item number 21-1.1.

Often, materials such as insulation are added to fire-resistance-rated assemblies. It is the intent of the IBC to require substantiating fire test data to show that when the materials are added, they do not reduce the required fire-endurance time period. As an example, adding insulation to a floor/ceiling assembly may change its capacity to dissipate heat and, particularly for noncombustible assemblies, the fire-resistance rating may be changed. Although the primary intent of the provision is to cover those cases where thermal insulation is added, the language is intentionally broad so that it applies to any material that might be added to the assembly.

Bonded prestressed concrete tendons. Figure 721-6 depicts the requirements specified in Items 1 and 2 for variable concrete cover for tendons. It must be noted that for all cases of variable concrete cover, the average concrete cover for the tendons must not be less than the cover specified in IBC Table 721.1(1).

721.1.5

As prestressed concrete members are designed in accordance with their ultimate-moment capacity, as well as with their performance at service loads, Item 3 provides two sets of criteria for variable concrete cover for the multiple tendons:

1. Those tendons having less concrete cover than specified in Table 721.1(1) shall be considered to be furnishing only a reduced portion of the ultimate-moment capacity of the member, depending on the cross-sectional area of the member.

2. No reduction is necessary for those tendons having reduced cover for the design of the member at service loads.

7 Fire and Smoke Protection Features

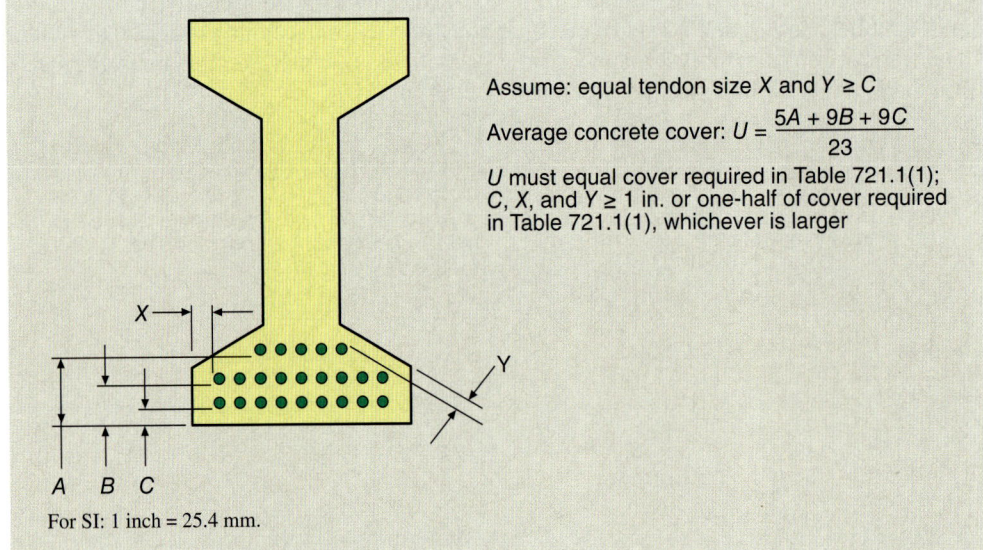

Figure 721-6 Variable protection of bonded prestressed tendons, multiple tendons.

As the ultimate-moment capacity of the member is critical to the behavior of the member under fire conditions, the code requires the reduction for those tendons having cover less than that specified by the code. However, behavior at service loads is less affected by the heat of a fire; therefore, the code permits those tendons with reduced cover to be assumed as fully effective.

Section 722 *Calculated Fire Resistance*

Fire research and the theory of heat transmission have combined to make it possible with the present state-of-the-art technology to calculate the fire endurance for certain materials and assemblies. As a result of this testing and research, this section permits the calculation of the fire-resistance rating for assemblies of structural steel, reinforced concrete, wood, concrete masonry, and clay masonry.

At the present time, it is doubtful that the fire resistance of many buildings will be based on calculations. Even so, the code users should be aware of the useful information presented in this section, including:

Reference	Subject
Section 722.5.1.2	Attachment of gypsum wallboard around structural-steel columns
Section 722.2.1.3.1	Thickness of ceramic blanket joint material for precast concrete wall panels
Section 722.6.2	Wood wall, floor, and roof assemblies
Section 722.6.3	Design of exposed wood members for 1-hour fire-resistance rating

The procedure set forth in Section 722.6.3 should be used when someone wishes to consider exposed heavy timber as a 1-hour construction in something other than a Type IV building. One of the factors affecting a wood member's fire-resistance rating is the load on the member as a percentage of its allowable structural capacity. See Application Example 722-1.

Calculated Fire Resistance 7

Application Example 722-1

DETERMINE: 8-inch by 8-inch timber column's fire-resistance rating at 50 percent, 75 percent, and 100 percent of its structural capacity.

SOLUTION: According to Section 722.6.3, the column fire-resistance rating is calculated from the formula, $2.54\, Z_d\, [3 - (d/b)]$ for columns that may be exposed to fire on four sides.

WHERE:

b = larger side of column [inches (mm)]

d = smaller side of column [inches (mm)]

$Z = 0.9 + 30/r$ [load factor based on Figure 722.6.3(1)]

r = ratio of applied load to allowable load expressed as a percent of allowable

When $r = 50\%$ or less, $Z = 1.5$

Calculated Z =
- 50% capacity $Z = 1.5$
- 75% $Z = 0.9 + 30/75 = 1.3$
- 100% $Z = 0.9 + 30/100 = 1.2$

Calculating rating:

$2.54Z\, (7.5)\, [3 - (7.5/7.5)] = 38.1Z$

50% capacity $38.1\, (1.5) = 57.15$ minutes

75% capacity $38.1\, (1.3) = 49.53$ minutes

100% capacity $38.1\, (1.2) = 45.72$ minutes

The same column loaded to 50 percent of capacity, but exposed on only three sides, would have a fire-resistance rating of 100 minutes.

For SI: 1 inch = 25.4 mm.

HEAVY-TIMBER CALCULATION

722.1.1 Definitions. Several definitions that pertain to Section 722 are presented in this section. Of these, four definitions for concrete made from different aggregate types are important not only for structural considerations but also from a fire-endurance standpoint. Because the aggregate type bears on concrete performance, minimum concrete thickness listings in this section, as well as in Tables 721.1(1), 721.1(2), and 721.1(3), provide listings for different aggregates. Generally, the use of siliceous aggregate results in lower fire-resistance ratings, whereas structural lightweight concretes have better fire resistance than normal-weight concrete. This is illustrated by the equivalent thicknesses for concrete walls as shown in Table 722.2.1.1. For a 4-hour wall, the following minimum thicknesses are required:

Concrete type	Thickness (inches)
Siliceous aggregate	7.0
Carbonate aggregate	6.6
Sand-lightweight	5.4
Lightweight	5.1

For SI: 1 inch = 25.4 mm.

7 Fire and Smoke Protection Features

KEY POINTS

- Fire endurance is the basis for the fire-resistance requirements in the IBC.
- Materials and assemblies tested in accordance with ASTM E 119 or UL 263 are considered to be in full compliance with the code for fire-resistance purposes.
- Elements required to be fire-resistance rated include structural frame members, walls and partitions, and floor/ceiling and roof/ceiling assemblies.
- The method for protecting fire-resistance-rated elements must be in exact compliance with the desired listing.
- In many cases, fire-resistance protection for structural members must be applied directly to each individual structural member.
- Exterior walls of buildings located on the same lot are regulated by the placement of an assumed line between the two buildings.
- Where an exterior wall is located an acceptable fire separation distance from the lot line, the wall's fire-resistance rating is allowed to be determined based only on interior fire exposure.
- Opening protection presents a higher fire risk than fire-resistance-rated construction insofar as it does not need to meet the heat-transmission limits of ASTM E 119 or UL 263.
- The maximum area of both protected and unprotected openings permitted in each story of an exterior wall is regulated by the fire separation distance.
- The code intends that each portion of a structure separated by a fire wall be considered a separate building.
- The objective of fire walls is that a complete burnout can occur on one side of the wall without any effects of the fire being felt on the opposite side.
- The purpose of a fire barrier is to totally isolate one portion of a floor from another through the use of fire-resistance-rated walls and opening protectives as well as fire-resistance-rated horizontal assemblies.
- Fire barriers are used as the separating elements for interior exit stairways, exit access stairways, exit passageways, horizontal exits, incidental use separations, occupancy separations, and other areas where a complete separation is required.
- Fire barriers must begin at the floor and extend uninterrupted to the floor or roof deck above.
- The potential for fire spread vertically through buildings mandates that openings through floors be protected with fire-resistance-rated shaft enclosures.
- Various modifications and exemptions for the enclosure of horizontal openings are found in the IBC.
- Fire partitions are used to separate dwelling units, sleeping units, tenant spaces in covered mall buildings, and fire-resistance-rated corridors from adjacent spaces.
- Fire partitions are permitted to extend to the membrane of a fire-resistance-rated floor/ceiling or roof/ceiling assembly.
- Smoke barriers are required in building areas where smoke transmission is a concern.
- The membrane of smoke barriers must be continuous from outside wall to outside wall and from floor slab to the floor roof deck above, to eliminate all avenues for smoke to travel outside of the compartment created by the smoke barriers.

Key Points

- Smoke partitions are intended to solely restrict the passage of smoke.
- Horizontal assemblies are required to have a fire-resistance rating where the type of construction mandates protected floor and roof assemblies, and where the floor assembly is used to separate occupancies or create separate fire areas.
- Penetration firestop systems are approved methods of protecting openings created through fire-resistance-rated walls and floors for piping and conduits.
- A limited level of protection is permitted for penetrations of noncombustible items.
- Both through-penetrations and membrane penetrations are regulated, typically in similar fashion.
- Joints, such as the division of the building designed for movement during a seismic event, must often be protected if they occur in a fire-resistance-rated vertical or horizontal element.
- An opening protective refers to a fire door, fire shutter, or fire-protection-rated glazing, including the required frames, sills, anchorage, and hardware for its proper operation.
- Table 716.5 identifies the minimum fire-protection rating for a fire door assembly based on the type of wall assembly in which it is installed.
- In interior applications, fire-protection-rated glazing is limited to fire partitions and fire barriers having a maximum fire-resistance rating of 1 hour.
- In addition to fire dampers and smoke dampers, ceiling radiation dampers and combination fire/smoke dampers are referenced in the IBC.
- Fireblocking and draftstopping are required in combustible construction to cut off concealed draft openings.
- Prescriptive methods for fire-resistance-rated construction are detailed for structural parts, walls and partitions, and floor and roof systems.
- The calculation of fire resistance is permitted for structural steel, reinforced concrete, wood, concrete masonry, and clay masonry.

CHAPTER 8

INTERIOR FINISHES

Section 801 General
Section 803 Wall and Ceiling Finishes
Section 804 Interior Floor Finish
Section 805 Combustible Materials in
            Types I and II Construction
Section 806 Decorative Materials and Trim
Key Points

8 Interior Finishes

> Unfortunately, a number of building code provisions are enacted only after a disaster (usually with a large loss of life) indicates the need to regulate in a specific area. This is true of the interior wall and ceiling finish requirements of Chapter 8. In this case, the 1942 Cocoanut Grove nightclub fire in Boston, with a loss of almost 500 lives, provided the impetus to develop code requirements for the regulation of interior finish. Based on fire statistics, lack of proper control over interior finish (and the consequent rapid spread of fire) is second only to vertical spread of fire through openings in floors as a cause of loss of life during fire in buildings.
>
> The dangers of unregulated interior finish are as follows:
>
> *The rapid spread of fire.* Rapid spread of fire presents a threat to the occupants of a building by either limiting or denying their use of exitways within and out of the building. This limitation on the use of exits can be created by:
>
> 1. The rapid spread of the fire itself so that it blocks the use of exitways.
> 2. The production of large quantities of dense, black smoke (such as smoke created by certain plastic materials), which obscures the exit path and exit signs.
>
> *The contribution of additional fuel to the fire.* Unregulated finish materials have the potential for adding fuel to the fire, thereby increasing its intensity and shortening the time available to the occupants to exit safely. However, because ASTM E 84 and UL 723 do not require the determination of the amount of fuel contributed, the *International Building Code®* (IBC®) does not regulate interior finish materials on this basis.

Section 801 *General*

It is the intent of the IBC to regulate the interior finish materials on walls and ceilings, as well as coverings applied to the floor. In addition, limitations on the use of trim and decorative materials are found in Section 806, with the exception of foam plastics used as trim or finish material. These are addressed in Chapter 26. Combustible materials are permitted as finish materials in buildings of any type of construction, provided the wall, ceiling, or floor finishes are in compliance with this chapter.

As established in Section 803.2, it is not the intent of the IBC to regulate thin materials such as wallpaper that are less than 0.036-inch (0.9-mm) thick. These thin materials behave essentially as the backing to which they are applied and, as a result, are not regulated. In some cases, however, repeated applications of wallpaper where the original materials are not removed can accumulate to a thickness of such magnitude that they must be regulated. The IBC, as stated in Section 803.3, also does not regulate the finish of exposed heavy-timber members complying with Section 602.4 insofar as this type of construction is not subject to rapid flame spread.

Section 803 *Wall and Ceiling Finishes*

803.1.1 Interior wall and ceiling finish materials. The standard test for the determination of flame spread and smoke-development characteristics is set forth in both ASTM E 84 and UL 723. This test is commonly known as the Steiner Tunnel Test. Based on the results of this test, interior wall and ceiling materials are classified. These classifications are divided into three groups as Class A, Class B, and Class C.

Room corner test for interior wall or ceiling finish materials. As an option to testing interior wall and ceiling finish materials under the criteria of ASTM E 84 or UL 723, such materials, other than textiles, can be tested per NFPA 286. Described as the "room corner" test, NFPA 286 describes the conditions of the test, whereas the code sets forth the minimum acceptance criteria.

803.1.2

Textile wall coverings. This section regulates carpet—as well as other textiles that are napped, tufted, looped, woven, or nonwoven—where applied as wall finish materials. Textile wall coverings present a unique hazard because of the potential for extremely rapid fire spread. The code provides three options for the acceptance of textiles used as interior wall finish. One method of testing includes the surface burning characteristics test of ASTM E 84 or UL 723. The textile must have a flame-spread index of Class A and be protected by automatic sprinklers. A second option is based on the room corner test for textiles as established in NFPA 265, where the testing must be done in accordance with the Method B protocol. It is important that the testing be done in the same manner as the intended use of the textile materials. A third approach is the use of the ceiling and wall finish room corner test as set forth in NFPA 286. This test is also based on the intended application of the textile material and must include the product-mounting system. Where textiles are intended to be applied as ceiling finish materials, only the methods using ASTM E 84, UL 723, and NFPA 286 are to be used.

803.5

Interior finish requirements based on group. Table 803.9 is divided based on the presence, or lack of, an automatic sprinkler system. Note that an extensive number of notes modify the general provisions of Table 803.9.

803.9

As a general rule, interior exit stairways, interior exit ramps, and exit passageways are regulated at the highest level because of their importance as exit components in the means of egress. These types of exits permit unlimited travel distance and are typically single-directional. Corridors and exit access stairways are also highly regulated, but not to the extent of the higher-level exit components. Interior finish requirements for rooms and other enclosed areas are not as restrictive as for exitways; however, the wall and ceiling finishes are still regulated to some degree.

When it comes to occupancy groups, the high-hazard, institutional, and assembly occupancies typically have the most restrictive flame-spread classifications. On the other hand, utility occupancies have no restrictions. Using a nonsprinklered office building as an example, the maximum flame spread classification of finish materials, based on occupancy group and location within the building, is shown in Figure 803-1.

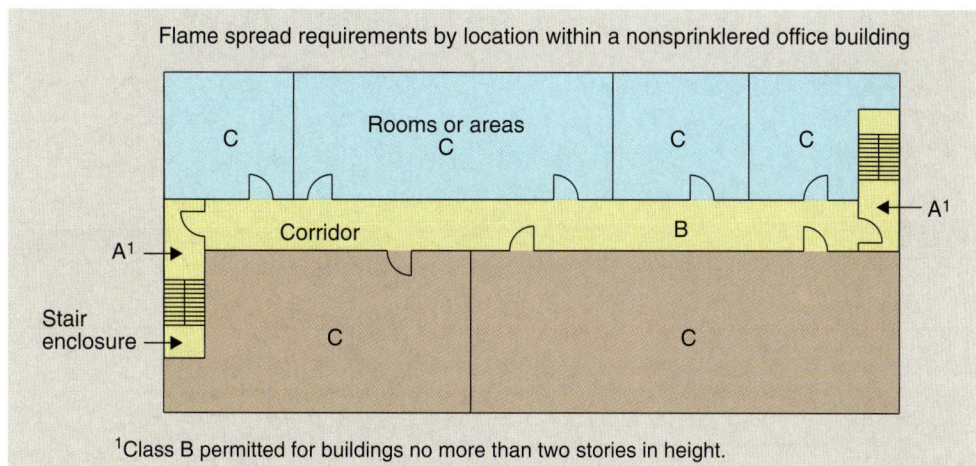

Figure 803-1
Flame-spread requirements by location within a nonsprinklered office building.

Interior Finishes

Based on the cooling that an automatic sprinkler system provides, the code permits a reduction of one classification for many of the occupancies and locations. However, this is not a standard reduction. The table must be referenced to identify those reductions available for sprinklered applications. There is also no allowance for reducing a Class C requirement to a lower classification based on sprinkler protection. It should be noted that the required sprinkler system need only be provided in those exitways or rooms where the classification reduction is taken, and not throughout the entire building.

Figure 803-2 provides flame-spread classifications of woods commonly used in construction and finish work. A glance at the chart will show that most species of wood qualify for a Class C rating. There are few wood species that warrant a Class B rating, and no species is shown that qualifies for a Class A rating. However, there are many paints and varnishes on the market that manufacturers refer to as fire-retardant coatings. Because of intumescence, these paints or coatings bubble up or swell up under the action of flame and heat to provide an insulating coating on the surface of the material treated. Certain intumescent paints and varnishes can reduce the flame spread of combustible finishes to as low as Class A and the smoke density to considerably below 450. These flame-spread-reducing intumescent coatings are particularly useful when correcting an existing nonconforming combustible interior finish.

Species of wood	Flame spread	Source*
Birch, yellow	105–110	UL
Cedar, eastern red	110	HUD/FHA
Cedar, Pacific Coast yellow	78	CWC
Cedar, western red	70	HPMA
Cottonwood	115	UL
Cypress	145–150	UL
Fir, Douglas	70–100	UL
Gum, red	140–155	UL
Hemlock, West Coast	60–75	CWC
Lodgepole	93	CWC
Maple flooring	104	CWC
Oak, red or white	100	UL
Pine, eastern white	85	UL
Pine, Idaho white	72	CWC
Pine, northern white	120–215	HPMA
Pine, ponderosa	105–200	UL
Pine, red	142	HUD/FHA
Pine, southern yellow	130–190	CWC
Pine, western white	75	HUD/FHA
Poplar	170–185	UL
Redwood	70	UL
Spruce, northern	65	UL
Spruce, western	100	UL
Spruce, white	65	UL
Walnut	130–140	CWC
Plywoods		UL
Douglas fir, 1/4-inch	120	HUD/FHA
Lauan, three-ply urea glue, 1/4-inch	110	HUD/FHA
Particleboard, 1/2-inch	135	HPMA
Redwood, 3/8-inch	95	CRA
Redwood, 5/8-inch	75	CRA
Walnut, 3/4-inch	130	HUD/FHA

*Source:
CRA: California Redwood Association
  Data Sheet-2D2-7L (Lumber)
  Data Sheet-2D2-7P (Plywood)
CWC: Canadian Wood Council
  Data File FP-6
HPMA: Hardwood Plywood Manufactures Association
  Test No. 337, Test No. 592, Test No. 596
HUD/FHA: Flame-spread Rating for Various Material
UL: Underwriters Laboratories
  UL 527, May 1971

Figure 803-2 Flame-spread classification of woods.

803.10 Stability. The IBC requires that the method of fastening the finished materials to the interior surfaces be capable of holding the material in place for 30 minutes under a room temperature of 200°F (93°C). If there is any question as to the adequacy of the fastening, appropriate tests should be required to determine compliance with this provision of the code.

803.11 Application of interior finish materials to fire-resistance-rated or noncombustible building elements. This section is applicable only where finish materials are applied on walls, ceilings, or structural elements required to have a fire-resistance rating, or where such building elements must be of noncombustible construction (typically Type I or II construction). The greatest concern is where interior finish is not applied directly to a backing surface, creating concealed spaces that provide the opportunity for fire to originate and spread without detection until the interior finish material has burned through. The installation of furring strips is permitted, provided they are installed directly against the surface of the wall, ceiling, or structural member. In addition, the concealed space created by the furring strips must be either fireblocked at maximum 8-foot (2,438-mm) intervals, or filled completely with a Class A, noncombustible, or organic material. The maximum depth of the concealed space is limited to $1^3/_4$ inches (44 mm). This section is also referenced by Section 803.11.3 for fireblocking in heavy-timber construction.

Where interior finish materials are set out or suspended more than the $1^3/_4$ inches (44 mm) specified in Section 803.11.1, the potential exists for the fire to gain access to the space through joints or imperfections and to spread along the back surface as well. In this case, the flame spreads at a much faster rate than on one surface, as the flame front will be able to feed on the material from two sides. Therefore, the provisions of Section 803.11.2 are intended to protect against this type of hazard. In the case where a wall is set out, the wall, including the portion that is set out, is required by the code to be of fire-resistance-rated construction as would be required by the code for the occupancy and type of construction.

It should be noted that the provisions of Sections 803.11.1 and 803.11.2 are applicable only where the walls and ceiling assemblies are required to be of either fire-resistance-rated or noncombustible construction. Where the walls and ceiling assemblies are of unprotected combustible construction, only the fireblocking provisions of Section 718.2 are applicable.

Section 803.11.4 requires that thin materials—no more than $1/_4$-inch (6.4-mm) thick—other than noncombustible materials, be applied directly against a noncombustible backing unless they are qualified by tests where the material is furred out or suspended from the noncombustible backing. The reason for this requirement is similar to that in Section 803.11.2. There are some buildings where thin paneling, such as luan plywood, is installed on walls and ceilings. When not installed against a noncombustible backing, these materials readily burn through and permit an almost uncontrolled rapid flame spread because the flame proceeds on both surfaces of the material.

Section 804 *Interior Floor Finish*

Floor finishes such as wood, terrazzo, marble, vinyl, and linoleum present little, if any, hazard that is due to the spread of fire along the floor surface. However, other flooring materials such as carpeting are highly regulated by the IBC because of their potential for helping increase the growth of a fire.

804.2 Classification. For the purpose of regulating floor finishes based on the occupancy designation of the area where the finishes are installed, the code identifies two classes: Class I and Class II. Determined by test standard NFPA 253, the classifications are

8 Interior Finishes

based on the critical radiant flux. The critical radiant flux is determined as that point where the heat flux level will no longer support the spread of fire. Class I is considered to have a critical radiant flux of 0.45 watts per square centimeter or greater, where Class II need only exceed 0.22 watts per square centimeter. Therefore, the Class I material is more resistant to flame spread because of the higher heat-flux-level characteristics.

804.4 Interior floor finish requirements. Interior floor finishes are regulated differently by the code based on two factors: the occupancies in the building and where the finish materials are located in relationship to the means of egress. The IBC selectively requires that interior exit stairways and ramps, exit passageways, and corridors be provided with a floor finish exceeding the critical radiant flux level established by the "pill test" used in DOC (U.S. Department of Commerce) FF-1 and test standard ASTM D 2859. In addition, the floor finish materials of all rooms or spaces unseparated from a corridor by full-height partitions are regulated in a like manner, as there is evidence that corridor floor coverings can propagate flame when exposed to a fully developed fire in a room that opens into a corridor. Those rooms that have no direct connection with a corridor are simply regulated for the DOC FF-1 criteria, as are rooms that are separated from a corridor by full-height partitions.

The DOC FF-1 and ASTM D 2859 "pill tests" require a minimum radiant flux of 0.04 watts per square centimeter and are used to regulate all carpeting sold in the United States. Fire tests have demonstrated that carpet on the floor that passes the pill test is not likely to become involved in a room fire until the fire has reached or approached flashover. Only those finish materials having a Class I or II classification may be installed in corridors, exit passageways, and exit enclosures of fully sprinklered Group I-1, I-2, and I-3 occupancies. The same limitation holds true for floors of exitways in areas of nonsprinklered buildings housing Group A, B, E, M, or S occupancies. In other occupancies, the interior floor finish in the listed exitways need only comply with DOC FF-1 or ASTM D 2859 listing. The commentary above assumes the reduction in the classifications of floor finishes that is permitted where the building is fully sprinklered. Class II floor finish materials are permitted in lieu of Class I materials, whereas materials complying with the DOC FF-1 or ASTM D 2859 "pill test" may be used instead of Class II materials. The entire building, and not just the area where the floor finish is located, must be provided with an automatic sprinkler system.

It is important to note that all fibrous floor coverings in those occupancies and locations not required by the IBC to have a Class I or II classification must comply with the "pill test" requirements of DOC FF-1 or ASTM D 2859.

Section 805 *Combustible Materials in Types I and II Construction*

805.1 Application. Where combustible flooring materials are installed in or on floors in noncombustible buildings, they are regulated by this section. Combustible sleepers may only be installed where the space between the fire-resistance-rated floor deck and the sleepers is completely filled with approved noncombustible materials or is fireblocked in the manner described in Section 718. Finish flooring of wood shall be attached to sleepers, which, if not imbedded, shall be appropriately fireblocked. As long as the wood flooring is in direct contact with a fire-resistance-rated floor, there is no significant hazard. However, if there is a space between the wood flooring and the fire-resistance-rated floor, a concealed space is created that is enclosed with combustible materials and provides a route for the undetected spread of fire. Therefore, the code requires that where wood flooring is not in direct contact with the fire-resistance-rated floor, the space be filled with noncombustible material or be fireblocked. Based on the controls placed on wood sleepers and finish flooring, it is also reasonable that combustible insulating boards be permitted where installed in a similar manner.

Section 806 *Decorative Materials and Trim*

Curtains, hangings, and other decorative materials can potentially assist the spread of flame through a room or an area. Therefore, for certain occupancies, the code regulates such materials as to their flame resistance and limits their use. In Groups A, E, I, R-1, and dormitories of Group R-2 occupancies, decorative materials hanging from the walls or ceilings must be flame resistant or noncombustible. In Group I-3 occupancies, only noncombustible materials are permitted. Test standard NFPA 701 is to be used in order to determine the effectiveness of the materials for flame resistance. In other than auditoriums of Group A, the total amount of flame-resistant decorative materials is limited to 10 percent of the total area of the walls and ceiling within the space. The amount of decorative material in auditoriums may be up to 75 percent of the aggregate wall and ceiling area, provided the building is fully sprinklered and the material is applied in a manner consistent with Section 803.11. As in all cases involving interior finish materials, documentation for such materials must be available for review and analysis by the building official.

Although typically not an issue, the amount of combustible trim in a room is limited to 10 percent of the aggregate wall or ceiling area of the room. By excluding handrails and guardrails, the code is only concerned with those rare situations where an extensive amount of decorative trim could present a problem. All trim must have a minimum flame-spread index and smoke-developed index of Class C.

KEY POINTS

- Regulation of finish materials by the IBC includes those on walls and ceilings, as well as floor coverings.
- Unregulated interior finish materials contribute to the rapid spread of fire, presenting a threat to the occupants by limiting or denying their use of exitways.
- Interior exit stairways and exit passageways are the most highly regulated building elements for the application of interior finish materials, with corridors and exit access stairways being moderately controlled, and rooms or areas being the least-regulated portions of the building.
- Installation of an automatic sprinkler system often allows a one-class reduction in the requirement for flame-spread classification.
- Textile wall and ceiling coverings are more highly regulated than other finish materials because of the potential for extremely rapid fire spread.
- Carpeting and similar floor covering materials are highly regulated in specific locations of specific occupancies.
- In all occupancies, fibrous floor coverings must, at a minimum, comply with the requirements of the DOC FF-1 "pill test" or with ASTM D 2859.
- In certain occupancies, the code regulates curtains, hangings, and other decorative materials as to their flame resistance and limits their use.

CHAPTER 9

FIRE PROTECTION SYSTEMS

Section 901 General
Section 903 Automatic Sprinkler Systems
Section 904 Alternative Automatic
               Fire-Extinguishing Systems
Section 905 Standpipe Systems
Section 907 Fire Alarm and Detection Systems
Section 908 Emergency Alarm Systems
Section 909 Smoke-Control Systems
Section 910 Smoke and Heat Vents
Section 911 Fire Command Center
Section 914 Emergency Responder
               Safety Features
Key Points

9 Fire Protection Systems

> Chapter 9 provides requirements for three distinct systems considered vital for the creation of a safe building environment. The first of these systems is intended to control and limit fires and to provide building occupants and fire fighters with the means for fighting fire. Included are fire-extinguishing and standpipe systems. The detection and notification of a fire condition is addressed by the second system. Manual fire alarms, automatic fire detection, and emergency alarm systems are included in this grouping. The third system is intended to control smoke migration. Included are design installation standards for smoke-control systems required by other chapters of the *International Building Code®* (IBC®) as well as smoke- and heat-venting systems. In addition to the provisions for fire protection systems, criteria are provided to increase the efficiency and safety of fire department personnel during emergency operations. Topics addressed include emergency responder safety features and radio coverage, the fire department command center, fire department connections, and fire pump rooms.

Section 901 General

It is the intent of this chapter to require fire protection systems in those buildings and with those uses that through experience have been shown to present hazards requiring the additional protection provided by fire protection systems. The installation, repair, operation, and maintenance of such systems are based on the provisions of the IBC and the *International Fire Code®* (IFC®). Furthermore, it is the intent of the code to prescribe standards for those systems that are required. However, there are times when the installation of a fire-protection system is not based on a code mandate. In such situations, the nonrequired system must still meet the provisions of the code. Once fire protection is provided to some degree, it is expected that the system is properly installed.

An exception to this section permits a fire protection system or any portion of that system that is not required by the code to be installed for partial or complete protection, provided that the installation meets the code requirements. As an example, fire-sprinkler protection may be provided only in a specific area of a building, based on a request by the owner rather than on a requirement of the code. Although the sprinkler system must be installed in accordance with the proper design standard (in most cases NFPA 13), it is not necessary that the sprinkler system extend into other areas of the building.

More than likely, however, a fire protection system is used to gain exceptions to, or reductions in, other code requirements. Under these conditions, the fire protection system is considered a required system and is subject to all of the requirements imposed by the IBC and IFC.

901.7 Fire areas. The fire area concept is based on a time-tested approach to limiting the spread of fire in a building. Through the use of fire-resistive elements, compartments can be created that are intended to contain a fire for a prescribed period of time. The floor area that occurs within each such compartment is considered to be the fire area. By definition, a fire area is the aggregate floor area enclosed and bounded by fire walls, fire barriers, exterior walls, and/or fire-resistance horizontal assemblies of a building. See Figure 901-1. In addition, any areas beyond the exterior wall that are covered with a floor or roof above, such as a canopy extending from the building, are considered part of the building for fire area purposes. This approach is consistent with the determination of building area in Chapter 2. An example is shown in Application Example 901-1. By isolating a fire condition to a single fire area through the use of fire separation elements, only a portion of the building is considered at risk because of a single fire incident.

General

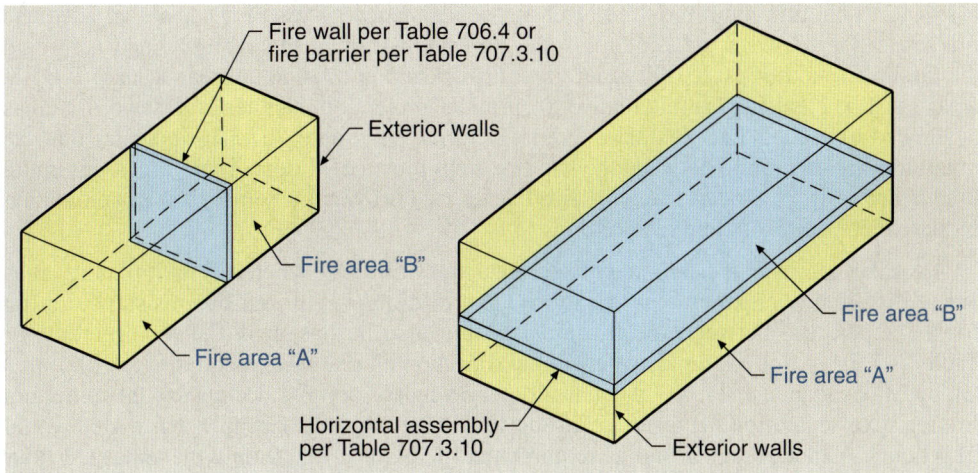

Figure 901-1
Fire areas.

The use of fire areas as a fire protection tool is limited almost exclusively to the requirements for automatic sprinkler systems. Other fire protection systems, such as fire alarm systems, are for the most part regulated by methods that are not based on the fire area concept. Even within the automatic sprinkler provisions of Section 903.2, only a portion of the requirements use the fire area approach as an alternative means of protection. The fire area methodology set forth in the IBC, applicable only in limited occupancy groups under limited conditions, allows for the omission of automatic sprinkler protection.

As an example, the provisions of Section 903.2.3, Item 1 require that a fire area containing a Group E occupancy that exceeds 12,000 square feet (1,115 m²) in floor area be provided with an automatic sprinkler system. Conversely, where the fire area size does not exceed the established threshold of 12,000 square feet (1,115 m²), a sprinkler system is not required unless mandated by another code provision. Where the building under consideration is limited to a maximum of 12,000 square feet (1,115 m²), it can be viewed as a single fire area, and no sprinkler system is mandated. However, where the building exceeds 12,000 square feet (1,115 m²) in floor area, two or more fire areas must be established to eliminate the sprinkler requirement. Table 707.3.10 is referenced because

Application Example 901-1

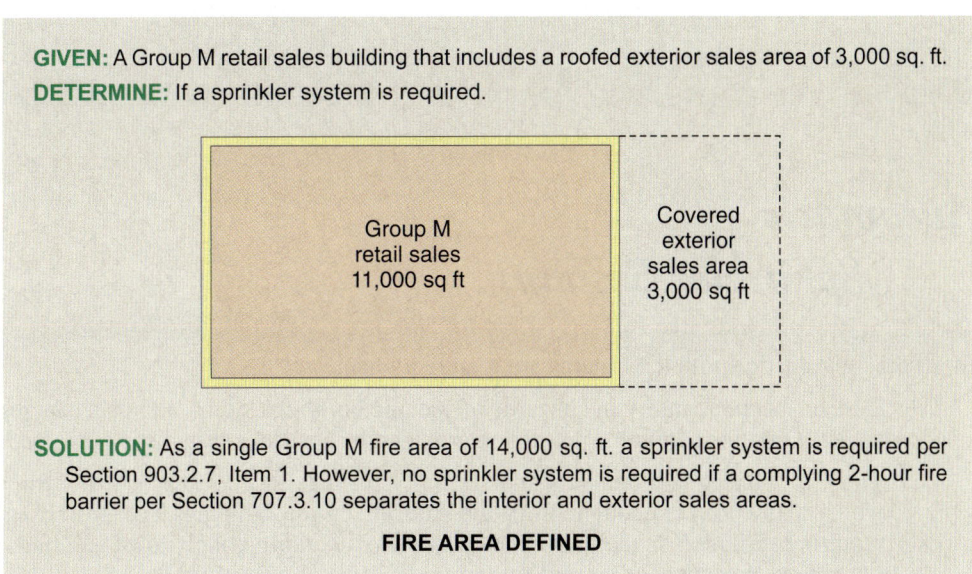

2012 International Building Code Handbook 285

Fire Protection Systems

it sets forth the minimum required level of fire resistance necessary to create an adequate fire separation between the fire areas that are established. In the example, and assuming the Group E building is 20,000 square feet (1,858 m^2) in total floor area, at least two fire areas must be created as an alternative to sprinkler protection. Neither of the two fire areas is allowed to exceed 12,000 square feet (1,115 m^2), and Table 707.3.10 indicates that the minimum fire separation between the two fire areas must be 2 hours. Therefore, a minimum 2-hour fire-resistance-rated fire wall, fire barrier, or horizontal assembly, or a combination of these elements, would be required.

A similar approach is taken in a mixed-occupancy building where the multiple fire areas are of different occupancy classifications. The minimum required fire-resistance rating for the separation between the fire areas would also be based on the requirements of Table 707.3.10. For example, where a building contains a 10,000-square-foot (929-m^2) Group M occupancy and a 6,000-square-foot (558-m^2) Group F-1 occupancy, the minimum fire-resistive separation between the Group M fire area and the Group F-1 fire area would be 3 hours. Although the Group M requirement in Table 707.3.10 only mandates a 2-hour separation, a minimum 3-hour fire separation is required for a Group F-1 occupancy. For further information, see Application Example 901-2 and the discussion of Table 707.3.10.

Application Example 901-2

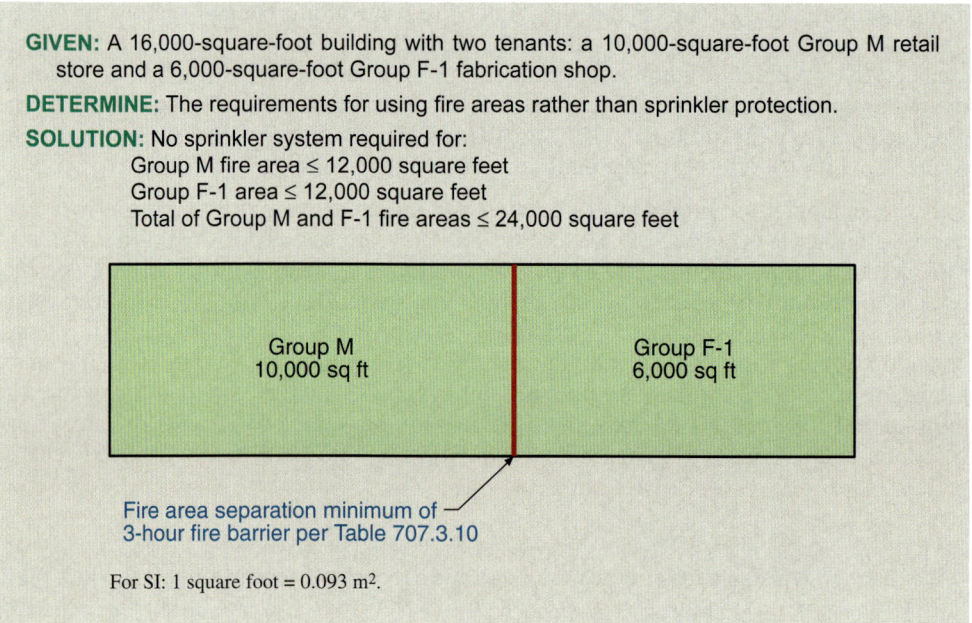

Section 903 *Automatic Sprinkler Systems*

In general, automatic sprinkler systems are required when:

1. Certain special features and hazards of specific buildings, areas, and occupancies are such that the additional protection provided by sprinkler systems is warranted.

2. There are inadequate numbers and sizes of openings in the exterior walls from which a fire may be fought from the exterior of the building. The provisions requiring sprinklers in these so-called *windowless buildings* apply to all buildings, regardless of occupancy, except for Group R-3 and U occupancies.

There are three general situations in which sprinkler or other fire suppression systems are to be provided within a building. An automatic sprinkler may be required throughout the building, throughout a fire area, or only in the specific room or space where sprinkler protection is necessary. Examples are depicted in Figure 903-1.

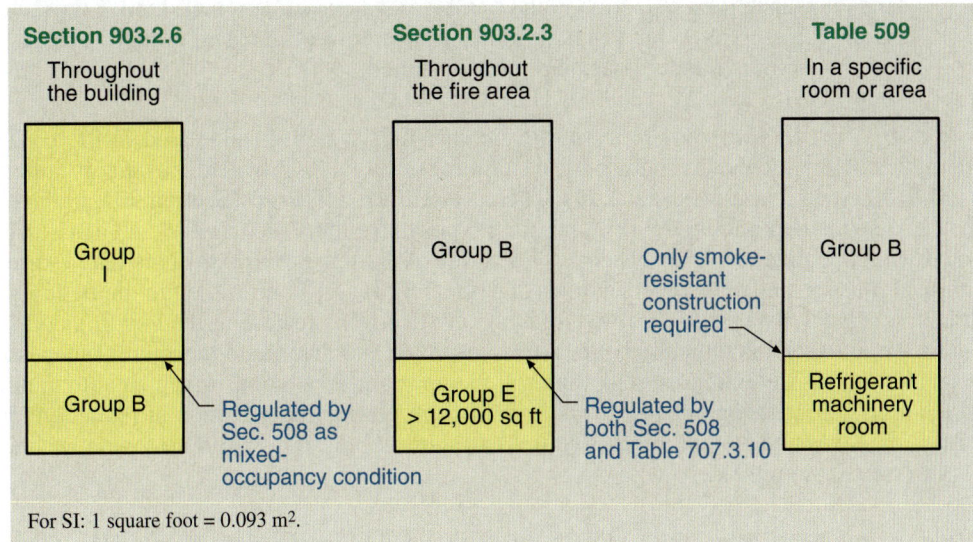

Figure 903-1
Sprinkler requirements.

1. *Throughout the building.* There are numerous applications of the code that require the entire building to be sprinklered, either mandated because of a code requirement or used as a substitute for other fire- and life-safety features. Examples include the requirement for sprinklers throughout all buildings containing a Group I fire area per Section 903.2.6 and the elimination of corridor fire protection in some occupancies based on Table 1018.1. The extent of required sprinkler protection in Group A occupancies is described in a different manner, but quite often results in a fully sprinklered condition. Initially, only the specific Group A occupancy must be provided with sprinklered protection where exceeding the limits established by Sections 903.2.1.1 through 903.2.1.4. However, as mandated by Section 903.2.1, the sprinkler system must also be provided throughout *the floor area* where the Group A occupancy is located. In application, it is appropriate that the sprinkler system be installed throughout the entire story or floor level containing the Group A occupancy. In addition, all floor levels between the Group A occupancy and the level of exit discharge must be sprinklered. This will commonly result in a requirement that the entire building be provided with an automatic sprinkler system. Additional commentary is provided in the discussion of Section 903.2.1.

2. *Throughout a fire area.* In Section 903.2, a variety of provisions require only those fire areas that exceed a certain size or occupant load, or are located in a specific portion of the structure to be sprinklered. The sprinkler requirements based on fire area include the provisions of Section 903.2.3 for Group E occupancies and Section 903.2.4.1 for woodworking operations.

   A variation of this requirement occurs in those mixed-occupancy buildings containing a Group H-2, H-3, or H-4 occupancy. The code only mandates that the sprinkler system be provided in the Group H portion, not the entire building. However, since all other occupancies in a mixed-occupancy building must be appropriately isolated from the Group H occupancy because of the *separated occupancies* provisions of Section 508.4, the result is basically a requirement to sprinkler the Group H compartment.

3. Specific rooms or areas. Occasionally, only a specific portion of the building requires the protection provided by a sprinkler system. The sprinkler addresses the particular hazard that occurs at possibly only a single location. For example, the allowance for a reduction in the flame-spread classification of interior finishes from Table 803.9 is based on sprinkler protection in the room, area, or exitway where the finish under consideration is installed.

903.1.1 Alternative protection. Where an automatic sprinkler system is addressed in the IBC, alternative automatic fire-extinguishing systems are acceptable, provided they are installed in accordance with approved standards. These systems, regulated by Section 904, include special systems required by the IFC and other types of systems such as dry chemical, carbon dioxide, or aqueous-foam systems. This is one of the few provisions in the IBC where approval must come from someone other than the building official. Although the building official is almost always charged with making any decisions regarding the building code, the fire code official is typically better able to evaluate and determine the appropriateness of an alternative fire-extinguishing system. It is important to note that where an automatic sprinkler system is recognized by the code for the purpose of an exception or reduction to a requirement, the use of an alternative fire-extinguishing system will not provide such a benefit. See Section 904.2.

903.2 Where required. It is the intent of this section to specify those occupancies and locations where automatic sprinkler systems are required. A fire-extinguishing system is a system that discharges an approved fire-extinguishing agent such as water, dry chemicals, aqueous foams, or carbon dioxide onto or in the area of a fire. A fire sprinkler system discharges water only. The code specifies a fire sprinkler system in this section, as it is the intent of the code that water be applied and not one of the other extinguishing agents. Generally, water is the most effective extinguishing agent for fires. Only where water creates problems, such as in magnesium or calcium carbide storage areas, would some other type of extinguishing agent be required. The allowance for the installation of a system other than an automatic sprinkler system is subject to approval by the fire code official. Section 904.2 states that where an automatic fire-extinguishing system is installed as an alternative to the required automatic sprinkler system of Section 903, it cannot be used for the purposes of exceptions or reductions allowed by other code provisions. Alternative automatic fire-extinguishing systems are addressed in Section 904.

Fire areas. Most of the requirements of this section are based on the concept of fire areas. Where a fire area exceeds a specified size, is located in a certain portion of the building, or exceeds a specified occupant load, the code often requires the installation of an automatic sprinkler system to address the increased hazards and concerns that exist. The provisions for fire areas can be found in various sections of the IBC.

The definition of "Fire area" is located in Chapter 2. A fire area is "the aggregate floor area enclosed and bounded by fire walls, fire barriers, exterior walls or fire-resistance-rated horizontal assemblies of a building." Complete isolation and separation of a portion of a building from all other interior areas is provided for a fire area through the use of fire-resistance-rated construction and opening protectives. The total floor area within the enclosed area, including the floor area of any mezzanines or basements, is considered the size of the fire area. See Application Example 903-1. It is also important to note that "areas of the building not provided with surrounding walls shall be included in the fire area if such areas are included within the horizontal projections of the roof or floor next above."

Where fire walls are used, Section 503.1 indicates that each portion of the structure included within the fire walls is considered a separate building. This concept would be

Automatic Sprinkler Systems

Application Example 903-1

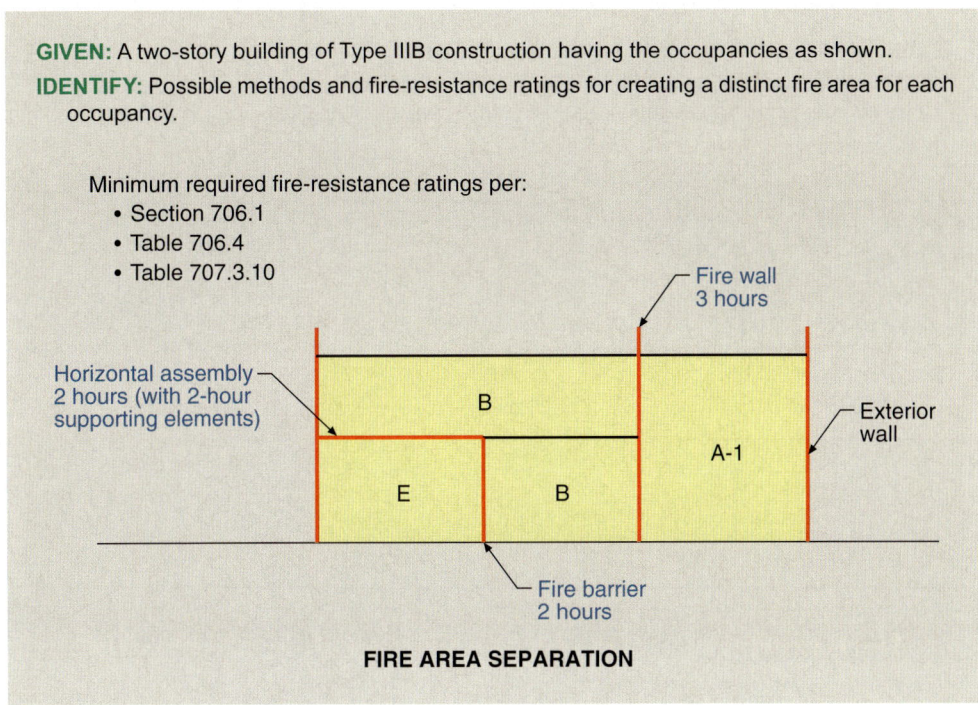

GIVEN: A two-story building of Type IIIB construction having the occupancies as shown.
IDENTIFY: Possible methods and fire-resistance ratings for creating a distinct fire area for each occupancy.

Minimum required fire-resistance ratings per:
- Section 706.1
- Table 706.4
- Table 707.3.10

FIRE AREA SEPARATION

consistent with that of separate and distinct fire areas being created through the use of fire walls. The fire-resistance rating of the wall and the fire-protective ratings of any openings in the fire wall are identified in Chapter 7. Also see the discussion of Section 706.1 where the fire wall separates different occupancies.

Fire barriers and fire-resistance-rated horizontal assemblies may also be used to create fire areas, provided the fire-resistance-rated construction totally separates one interior area from another. In order to determine the minimum fire-resistance rating of the vertical and horizontal elements, the occupancy classifications of the areas being separated must be identified. Table 707.3.10 is then referenced to determine the minimum fire-resistance rating of the separation. The use of this table is applicable to both single-occupancy and mixed-occupancy buildings, as illustrated in Application Example 903-2. Where the fire area separation occurs between two fire areas of the same occupancy, the hourly rating established by Table 707.3.10 for that single occupancy classification is applied. If the fire areas are of different occupancy classifications, the controlling fire-resistance rating of the fire barrier or horizontal assembly separating the occupancies is based on the higher of the ratings as established by Table 707.3.10 for the occupancies involved. For further information, see the discussions of Section 901.7 and Table 707.3.10.

Because the majority of sprinkler provisions are based on the size of the fire area, it is sometimes possible for the designer to eliminate the requirement for sprinklers by reducing the floor area within the surrounding fire-resistance-rated construction. The use of fire walls or fire barriers and fire-resistance-rated horizontal assemblies can subdivide a structure into smaller, less hazardous areas that are of such a size that sprinklers are not necessary. See Application Example 903-3. This concept of compartmentation has been used in building codes for decades as an effective method of reducing the loss of life and property in fires.

The exception to Section 903.2 eliminates the sprinkler requirement in telecommunications occupancies in those rooms or areas dedicated solely for essential telecommunications and power equipment. The alternative protection is provided through the required installation of an automatic fire alarm system, as well as fire-resistance-rated separation from other areas of the building.

Fire Protection Systems

Application Example 903-2

GIVEN: The various occupancies housed in a building as shown below.

DETERMINE: The required fire-resistance ratings of the assemblies separating the occupancies in order to create different fire areas for the purpose of applying Section 903.2.

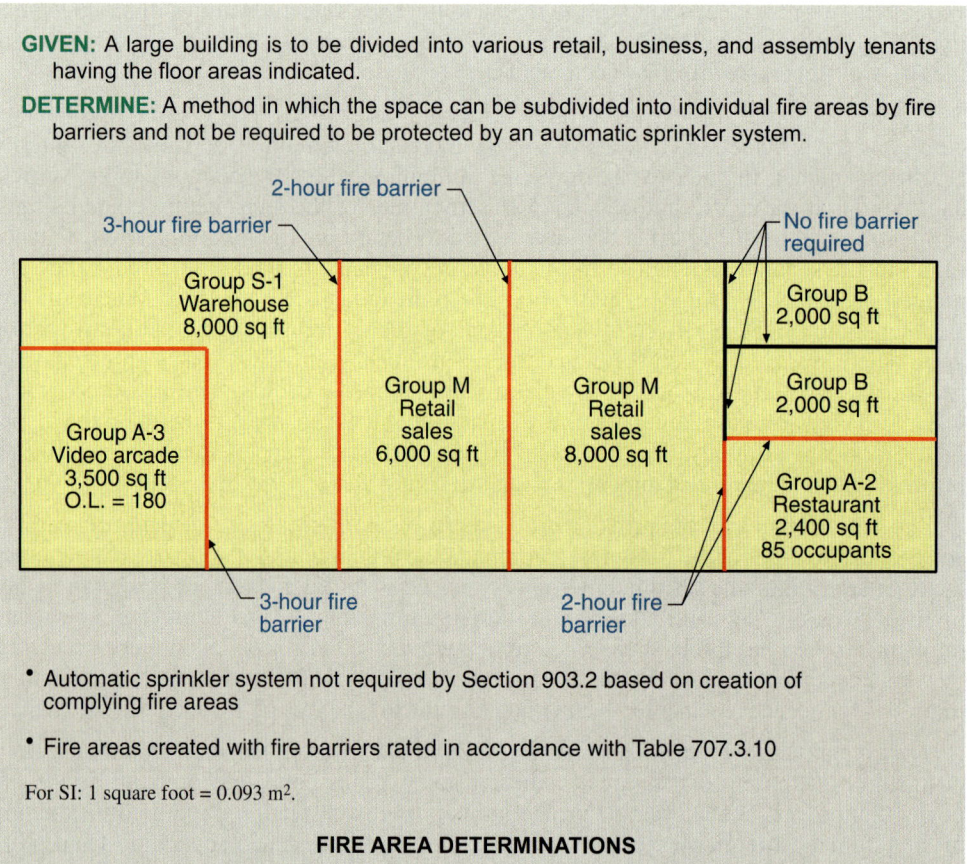

*Required minimum fire-resistance rating for fire barrier based on higher of ratings as established by Table 707.3.10

Application Example 903-3

GIVEN: A large building is to be divided into various retail, business, and assembly tenants having the floor areas indicated.

DETERMINE: A method in which the space can be subdivided into individual fire areas by fire barriers and not be required to be protected by an automatic sprinkler system.

- Automatic sprinkler system not required by Section 903.2 based on creation of complying fire areas
- Fire areas created with fire barriers rated in accordance with Table 707.3.10

For SI: 1 square foot = 0.093 m².

FIRE AREA DETERMINATIONS

Group A. Because of the potentially high occupant load and density anticipated in Group A occupancies, coupled with the occupants' probable lack of familiarity with the means of egress system, various assembly uses must be protected by an automatic sprinkler system. Where an automatic sprinkler system is required for a Group A occupancy, the system must be installed throughout the entire floor level or story where the Group A occupancy is located. In addition, where the Group A occupancy requiring a sprinkler system is located on a floor level other than the level of exit discharge, all floor levels between the Group A occupancy and the nearest level of exit discharge must be sprinklered as well. By expanding the areas of the building required to be protected by an automatic sprinkler system beyond just the assembly areas, the code provides for protection adjacent to the Group A areas as well as throughout the means of egress. Figures 903-2 and 903-3 illustrate these fundamental provisions.

903.2.1

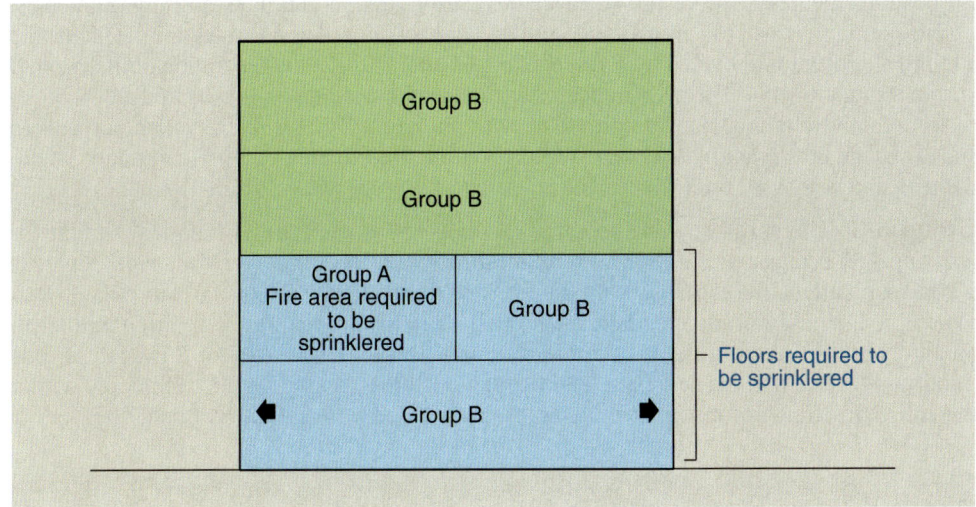

Figure 903-2
Group A sprinkler.

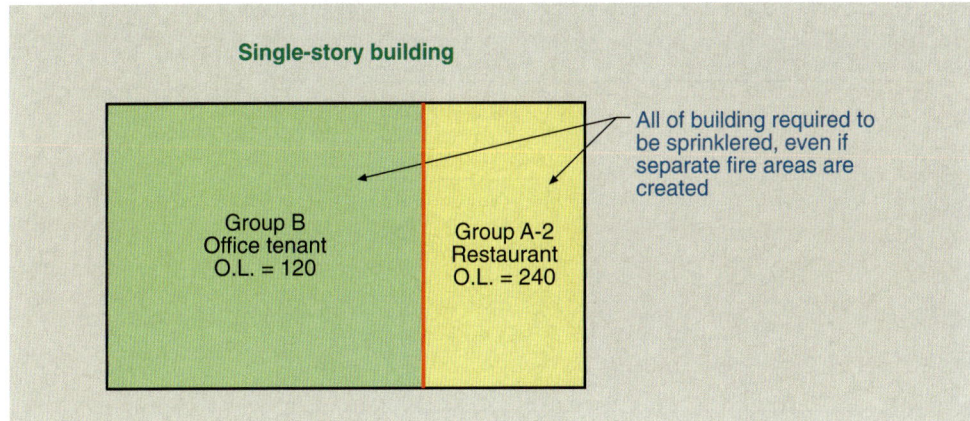

Figure 903-3
Group A fire area.

Group A-1. The combination of highly concentrated occupant loads, high numbers of occupants, reduced lighting levels, and potentially high fuel loads create a level of hazard that justifies the need for sprinkler protection. Therefore, fire areas containing theaters and similar assembly uses intended for the viewing of motion pictures or the performing arts shall be provided with an automatic sprinkler system where any one of the following conditions exists:

903.2.1.1

1. The fire area containing the Group A-1 occupancy exceeds 12,000 square feet (1,115 m²).

2. The occupant load of the fire area exceeds 299.

3. The fire area is located on any floor level other than that of the exit discharge.

It should also be noted that any fire area containing a multitheater complex, defined as two or more theaters served by a common lobby, shall be provided with a sprinkler system throughout the fire area.

903.2.1.2 Group A-2. Fire areas housing uses intended for food or drink consumption are regulated for sprinkler protection at a higher level than other enclosed assembly occupancies. Even where the occupant load is not excessive, the hazards associated with such uses warrant the protection provided by a sprinkler system. Oftentimes, the consumption of alcohol beverages by the building's occupants creates an environment more likely to be unsafe. The reduced lighting levels in some uses, along with the probability of loose chairs and tables, also increase the risk for obstructed egress. The record of casualties during fires in buildings housing nightclubs, casino gaming areas, restaurants, and similar types of uses demonstrates the need for the additional protection provided by fire sprinklers or, alternatively, the separation of the use into smaller compartments. The code intends that fire areas exceeding 5,000 square feet (465 m^2) that contain Group A-2 uses be provided with an automatic sprinkler system, as well as such uses having an aggregate occupant load within the fire area of 100 or more, or where the fire area is located on a floor level other than the level of exit discharge.

903.2.1.3 Group A-3. The sprinkler threshold for a Group A-3 occupancy is identical to that for a Group A-1 occupancy. As such, where any fire area in a Group A-3 occupancy exceeds 12,000 square feet (1,115 m^2), where the fire area has an occupant load greater than 299 or where the assembly occupancy is located on any floor other than the exit discharge level, an automatic sprinkler system is required. In applying the provisions of this section, it is important to note that the occupant load threshold is based on the number of people within the entire fire area, not just in each assembly room. See Application Example 903-4.

Application Example 903-4

GIVEN: A mixed-occupancy building containing a Group B office area and four Group A-3 conference rooms (each with an occupant load of 88)

DETERMINE: An appropriate method of fire area separation as an alternative to installation of a sprinkler system

- Conference room O.L. = 88
- Group B Office area O.L. = 80
- Separate fire areas must be created to eliminate sprinkler requirement*
- Occupant load of 88 per conference room Total of 264 occupants

*Each Group A fire area to be less than 12,000 sq ft with no more than 299 occupants

Note: Mixed occupancy conditions must also comply with Section 508.

For SI: 1 square foot = 0.093 m^2.

SOLUTION: A minimum of two fire areas must be created so as not to exceed the 299 occupant load and 12,000 square foot limitations.

AGGREGATE GROUP A OCCUPANT LOADS

The code requirements for these types of uses, specifically for exhibition and display rooms, can be strongly attributed to the McCormick Place fire in Chicago on January 16, 1967. McCormick Place was not sprinklered and consisted of three levels, including a main exhibit area of 320,000 square feet (29,728 m^2) on the upper level. Both the upper and lower levels were in the final stages of readiness for a housewares exhibition and were heavily laden with combustibles when the fire broke out. The fire was reported to have originated in the storage area behind an exhibit booth on the upper level. The upper level was almost totally destroyed, and considerable damage occurred to the lower level.

Ordinarily, assembly occupancies are considered to have a very low fire loading; however, the need for built-in fire suppression for an assembly use that is used for exhibition or display purposes was clearly demonstrated by the McCormick Place fire. Display booths are most often constructed with combustible materials, and the storage area behind the booths is a receptacle for combustible materials and packing boxes. Thus, without built-in fire suppression, the large quantities of combustible materials and large areas combine to create an excessive hazard. Many other assembly occupancies classified as Group A-3 also present significant fire loading such as art galleries, libraries, and museums. Therefore, sprinkler protection is beneficial for all large Group A-3 occupancies.

903.2.1.4

Group A-4. The fire-sprinkler requirements for Group A-4 occupancies (those assembly uses provided with spectator seating for the viewing of indoor activities and sporting events) are identical to the provisions for Group A-1 and A-3 uses. See Section 903.2.1.3 for a discussion of the sprinkler requirements.

903.2.1.5

Group A-5. The fire loading in stadiums and grandstands is typically quite low except for specific accessory areas such as concession stands, storage and equipment rooms, press boxes, and ticket offices. Therefore, assembly occupancies classified as Group A-5 do not require the installation of an automatic sprinkler system except for those support areas exceeding 1,000 square feet (92.9 m^2) in floor area. The limitation of 1,000 square feet (92.9 m^2) is based on the floor area of each individual area and not on the aggregate area of all such spaces. Where such accessory spaces are of a considerable size, the hazards posed by the potentially large quantities of combustible materials can be reduced where such areas are sprinklered.

903.2.2

Ambulatory care facilities. As a general rule, Group B occupancies do not require a sprinkler system based solely on their occupancy classification. However, Section 903.2.2 mandates that a Group B ambulatory care facility be provided with an automatic sprinkler system when either of the following conditions exists at any time:

- Four or more care recipients are incapable of self-preservation, or
- One or more care recipients who are incapable of self-preservation are located at other than the level of exit discharge.

Although such facilities are generally regarded as moderate in hazard level due to their office-like conditions, additional hazards are typically created due to the presence of individuals who are temporarily rendered incapable of self-preservation due to the application of nerve blocks, sedation, or anesthesia. While the occupants may walk in and walk out the same day with a quick recovery time after surgery, there is a period of time where a potentially large number of people could require physical assistance in case of an emergency that would require evacuation or relocation. The installation of an automatic sprinkler provides an important safeguard that enables the moderate-hazard classification of Group B.

The sprinkler system, when required, must extend throughout the entire story on which the ambulatory care facility is located. In addition, in multistory buildings where ambulatory care is provided above and/or below the exit discharge level, the sprinkler system must be installed on those stories between the level of ambulatory care and the level of exit discharge, inclusive.

903.2.3 Group E. History has shown that educational occupancies perform quite well when it comes to fire- and life-safety concerns. Much of this can be attributed to the continuous control and supervision that takes place within schools, as well as the students' knowledge of egress responsibilities in case of a fire or other emergency. However, because of the potential for moderate to high combustible loading, fire areas in Group E occupancies that exceed 12,000 square feet (1,115 m^2) in floor area must be provided with an automatic sprinkler system. In addition, fire sprinklers are required for those portions of educational buildings located below the lowest level of exit discharge regardless of floor area.

An exception to the sprinkler provision for Group E uses located below the lowest level of exit discharge is based on increased criteria for exiting. Where every classroom within an educational building has at least one exit door to the exterior and such exit doors are located at ground level, an automatic sprinkler system need not be provided. By providing direct egress for the students from their classrooms to the exit discharge, a very high level of occupant protection has been attained. It should be noted that although this exception is only applicable to Item 2, its application to the general requirement for sprinklers based on fire area size also seems appropriate. The benefits of providing direct egress from each classroom to the exterior at grade level should also be extended to areas at or above the level of exit discharge.

903.2.4 Group F-1. Without an automatic sprinkler system to limit the size of a fire, the fire can spread very quickly to other portions of the structure. This is particularly true for large floor-area buildings containing combustible materials such as manufacturing facilities, warehouses, and retail sales buildings. The IBC requires an automatic sprinkler system to be installed throughout any building containing a Group F-1 occupancy where the fire area containing the Group F-1 exceeds 12,000 square feet (1,115 m^2). A fire sprinkler is also required where the building housing the Group F-1 occupancy is four stories or more in height or has an aggregate of Group F-1 fire areas in the building of more than 24,000 square feet (2,230 m^2). The aggregate Group F-1 fire area would also include the floor area of any mezzanines involved. See Figure 903-4. The 24,000-square-foot (2,230-m^2) limitation would also be applicable in single-story structures, as shown in Figure 903-5.

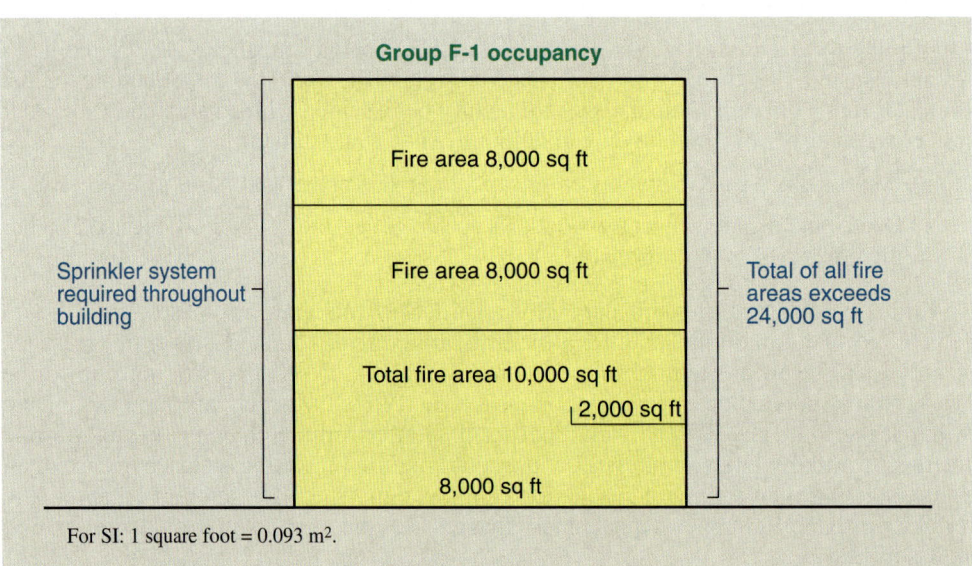

Figure 903-4
Multistory Group F-1 occupancies.

The potential for increased fire loading in a Group F-1 occupancy where upholstered furniture or mattresses are manufactured causes the threshold for automatic sprinkler protection to be reduced well below the general requirement. The sprinkler mandate is based on only that floor area of the facility devoted to the manufacture of upholstered furniture

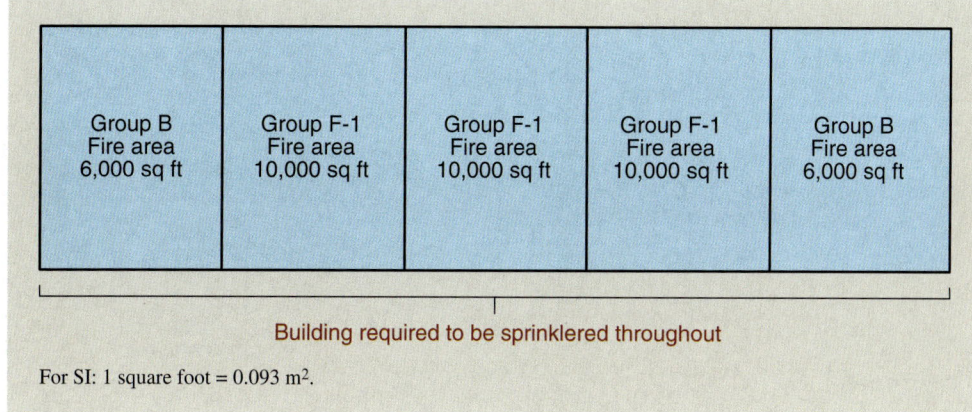

Figure 903-5
Sprinklered Group F-1 occupancies.

or mattresses, rather than the size of the entire fire area in which the manufacturing process occurs. The 2,500-square-foot (232-m^2) threshold, although arbitrary in nature, represents a reasonable top-end limit where sprinkler protection is not required. It is also consistent with the sprinkler requirements of IFC Table 3206.2 addressing high-piled combustible storage of high-hazard commodities in buildings not typically accessible to the public. The code is unclear as to the appropriate method for determining the boundaries of the area used for the manufacture of upholstered furniture or mattresses. It would seem appropriate that all such manufacturing areas that could be quickly involved in a fire event originating at a single location be used in applying the code provision. Those manufacturing areas with adequate spatial or fire-resistive separation could possibly be excluded in the aggregate floor area determination. It is incumbent upon the authority having jurisdiction to apply the sprinkler requirement in a manner that reflects its intent based on the potential hazard that could be created in the manufacturing process.

Woodworking operations. Because of the special hazards involving dusts created during woodworking operations such as sanding and sawing, this section requires that an automatic fire-sprinkler system be installed in fire areas of Group F-1 woodworking occupancies where the floor area of such operations exceeds 2,500 square feet (232 m^2). Where equipment, machinery, or appliances that generate finely divided combustible waste or that use finely divided combustible materials are a portion of a woodworking operation, the size of the operation is strictly limited unless sprinklers are installed. An example of this provision is shown in Figure 903-6. The provision is based on the size of the area where only the sanding, sawing, and similar operations occur, not necessarily the floor area of the entire woodworking operation. However, because these types of operations occur quite often as an integral part of the overall woodworking activities, rather than isolated in their own room or area, some means of regulating and controlling the hazard should be provided.

903.2.4.1

Group H occupancies. Group H occupancies are high-hazard uses, and one special feature is that, in addition to presenting a local hazard within the building, it has a potential for presenting a high level of hazard to the surrounding properties. Therefore, the code requires sprinkler protection for all Group H occupancies. Note that the sprinkler system is not necessarily required throughout the entire building that contains a Group H-2, H-3, or H-4 occupancy. Only such Group H areas must be provided with a sprinkler system. In the case of a Group H-1 occupancy, no other occupancies are permitted in the same building. Therefore, a building containing a Group H-1 occupancy must be sprinklered throughout. In addition, buildings containing Group H-5 occupancies require sprinklers in other portions of the building in addition to the high-hazard area proper. This requirement is based on the original premise that the primary protection feature of this highly protected use is the automatic sprinkler system. For the purpose of sprinkler-system design, all areas

903.2.5

9 Fire Protection Systems

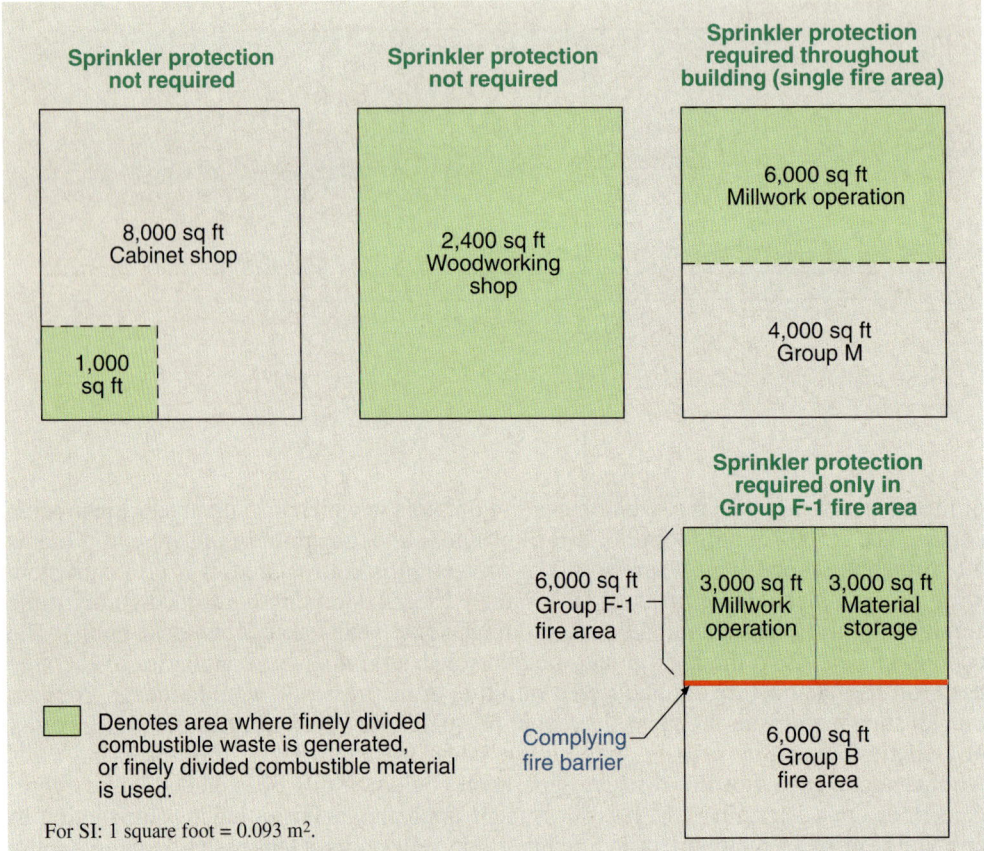

Figure 903-6 Woodworking operations.

of a Group H-5 are considered Ordinary Hazard Group 2, except for storage rooms with dispensing operations, which are considered Extra Hazard Group 2.

903.2.6 Group I. Because the mobility of the occupants of Group I occupancies is greatly diminished (in the case of hospitals and detention facilities, the self-mobility is essentially nonexistent), the code requires an NFPA 13 automatic sprinkler system throughout any building where a Group I fire area exists. For supervised residential facilities classified as Group I-1 occupancies, allowances are made for the use of an approved NFPA 13D or 13R system. The similarities between this Group I use and those uses classified as Group R justify the reduction in sprinkler protection.

Exception 3 provides the only instance in which a building containing a Group I occupancy is not required to be provided with some type of automatic sprinkler system. Day-care facilities classified as Group I-4 occupancies are not required to be sprinklered where the day-care operations only occur at the level of exit discharge and every room where care is provided has an exit door that leads directly to the exterior. This allowance is similar in some respects to the allowance in Section 308.6.1 where day-care operations consistent with a Group I-4 classification may be classified as a Group E occupancy.

903.2.7 Group M. The typical American supermarket evolved during the construction boom that followed World War II. At that time, the typical supermarket consisted of a one-story building of moderately large area, e.g., 15,000 to 25,000 square feet (1,394 m² to 2,323 m²). During the 1950s, fire statistics indicated that large-area supermarkets without sprinkler protection were subject to a larger proportion of fires than were usually attributable to this use in the past. As a result, building codes began requiring sprinklers in larger retail sales occupancies. The present requirements, detailed in the discussion in Section 903.2.4, are based on any of three factors: the size of the fire area, the number of stories, or the combined fire area on all floors. In addition, reference to the IFC is made for sprinkler protection in mercantile buildings where merchandise is placed in high-piled or rack storage. The installation of an automatic sprinkler system is also mandated in any Group M occupancy that is used for the display and sale of upholstered furniture or mattresses where the floor area devoted to such goods exceeds 5,000 square feet (464 m^2). This provision is not based on the size of the fire area of the Group M occupancy—instead, it is based strictly on the amount of floor area devoted to the specific contents of the mercantile occupancy. The requirement does not apply to the display and sale of furniture that is not upholstered, such as furniture constructed entirely of wood, plastic, or metal. Similar provisions are established for Group F-1 and S-1 occupancies where upholstered furniture or mattresses are manufactured or stored, respectively. See the discussion of Section 903.2.4 for further commentary. The increased threshold for sprinkler protection in Group M occupancies is based in part on an anticipated lower density of goods within a display area.

Group R. In hotels, apartment buildings, dormitories, and other Group R occupancies, occupants may be asleep at the time of a fire, and may experience delay and disorientation in trying to reach safety. In addition, fire hazards in residential uses are often unknown to most occupants of the building, as they are created within an individual dwelling unit or guestroom. This helps to explain why these occupancies have a poor fire record when it comes to injury and loss of life. Therefore, an automatic sprinkler system is required throughout any building containing a Group R occupancy. The sprinkler requirement applies to the entire building and not just the fire area containing the Group R occupancy. **903.2.8**

Group S-1. In a manner consistent with that for Group F-1 and M occupancies, buildings containing combustible storage and warehousing uses must be provided with an automatic sprinkler system where the floor area or height exceeds the specified threshold. The sprinkler requirement is based on the probable presence of large amounts of combustible materials, typically arranged in a highly concentrated manner. **903.2.9**

Although the storage of commercial trucks and buses is typically regulated under the provisions of Section 903.2.10.1, there are situations where the parking of such vehicles occurs in the same area with other storage uses. These multipurpose spaces, such as fire station bays, are more appropriately classified as Group S-1 occupancies. In such cases, a more restrictive threshold of 5,000 square feet (464 m^2) is used to require sprinkler protection.

The storage of upholstered furniture and mattresses poses much the same hazard as in buildings where such goods are manufactured or displayed. Therefore, the sprinkler requirements are to a great degree consistent with those for Group F-1 and M occupancies where upholstered furniture or mattresses are present. Additional information is provided in the commentary on Sections 903.2.4 and 903.2.7.

Repair garages. The unique hazards associated with vehicle repair garages may be addressed in part through the installation of an automatic sprinkler system. However, the requirement for sprinklers is limited only to those repair garages that present a high level of concern based on size or location. By locating the repair garage above grade in a building of one or two stories, the size of the fire area containing the garage becomes the controlling factor in the determination of whether or not a sprinkler system is required. Where there is vehicle parking in the basement of a building used for vehicle repair, the building must be sprinklered regardless of fire area size. The sprinkler requirement is applicable even where the repair activity occurs only above the basement level. In buildings where commercial trucks or buses are repaired, the threshold for sprinkler protection is consistent with that established in Section 903.2.10.1 for commercial parking garages. **903.2.9.1**

Group S-2 enclosed parking garages. Because the bulk of the uses designated as Group S-2 occupancies present very low fire-load potential, there is generally no requirement for these low-hazard occupancies to be sprinklered. However, where the Group S-2 **903.2.10**

portion of a building is an enclosed parking garage, the hazard level is increased. There is a need to protect other uses housed above an enclosed parking garage; thus, a Group S-2 enclosed parking garage is required to be sprinklered where the garage is located below another occupancy. In fact, in such a situation the entire building must be sprinklered, regardless of the size of the garage itself. There is an exception to the sprinkler requirement where an enclosed parking garage is located beneath a Group R-3 occupancy. Where the enclosed parking garage has no uses above, the required point at which an automatic sprinkler system is required is consistent with the threshold established for other moderate-hazard occupancies. The installation of an automatic fire sprinkler system for enclosed parking garages is required where the fire area containing the garage exceeds 12,000 square feet (1,115 m^2) in floor area. The fire behavior in an enclosed parking garage, although similar to that in an open parking garage, is of greater concern since smoke ventilation will be more difficult due to the lack of sufficient exterior openings. This concern is addressed by the required installation of an automatic sprinkler system once the 12,000-square-foot (1,115-m^2) fire area threshold is exceeded. The sprinkler requirement is not applicable to open parking garages.

903.2.10.1 Commercial parking garages. Where the vehicles stored within a building consist of commercial trucks and buses, the code mandates stringent floor areas when it comes to the requirement for an automatic sprinkler system. Where a fire area containing commercial parking exceeds 5,000 square feet (464 m^2) in floor area, the building housing the vehicles must be sprinklered throughout. The provision is intended to address those facilities housing larger vehicles. It is generally not applicable where pick-up trucks and similar-sized vehicles are being used for business activities.

903.2.11.1 Stories without openings. The provisions of this section make specific the intent of the code to require automatic sprinkler protection in *windowless buildings*. A structure having inadequate openings on the exterior wall as determined by this section such that fire department access is insufficient is considered a *windowless building*. The requirements of this section apply to all occupancies except Groups R-3 and U. The provisions are applicable on a floor-by-floor basis and do not apply to any story above grade plane or basement having a floor area of 1,500 square feet (139.4 m^2) or less:

- **On the basis of each individual story above ground.** Each individual story is analyzed for the size and the number of exterior wall openings. Thus, in a multistory building, it is possible to have a requirement that a sprinkler system be installed in one story and not in another.

The code requires that the openings be:

1. **Installed entirely above the adjoining ground level.** This provision is necessary so that effective fire suppression and rescue can be accomplished from the exterior of the building.

    Where the openings cannot be located entirely above the adjoining ground level, the code permits the use of exterior stairways or ramps that lead directly to grade.

2. **Of adequate size and spacing.** Although it may be argued that the openings required by the code are not the equivalent of automatic fire-sprinkler protection, the access for fire fighting provided by the openings has proven satisfactory.

    Although not expressly stated in the code, there is an expectation that a below-grade opening used to satisfy this provision be simply a typical 3-foot by 6-foot, 8-inch (914-mm by 2032-mm) door leading directly to the exterior stairway or ramp. However, above-grade openings are more specifically addressed. A total of 20 square feet (1.86 m^2) of openings is mandated in each 50 lineal feet (15,240 mm) of exterior wall. It is not necessary to obtain all 20 square feet (1.86 m^2) from a single opening, as long as the minimum dimension requirement of 30 inches (762 mm) is met. Multiple 30-inch by 30-inch (762-mm by 762-mm) openings would comply; however, they may not be as effective as a larger single opening.

The intent of the code is that there shall be at least one opening in each 50 linear feet (15,240 mm) of exterior wall. It may be better stated that *any* wall section of 50 feet in length be provided with complying openings. Thus, an exterior wall 100-feet (30,480-mm) long with 20-square-foot (1.86-m^2) openings located at third points along the wall would comply, as shown in Figure 903-7. There is no portion of the wall that is 50 feet (15,240 mm) in length that does not contain the necessary openings. However, the same wall with such openings located at each end, as depicted in Figure 903-8, will not comply with the intent of the code insofar as there is a length of wall that exceeds the 50-foot (15,240-mm) dimension without an opening. Certainly, the same wall with only one 40-square-foot (3.72-m^2) opening at one end also would not comply.

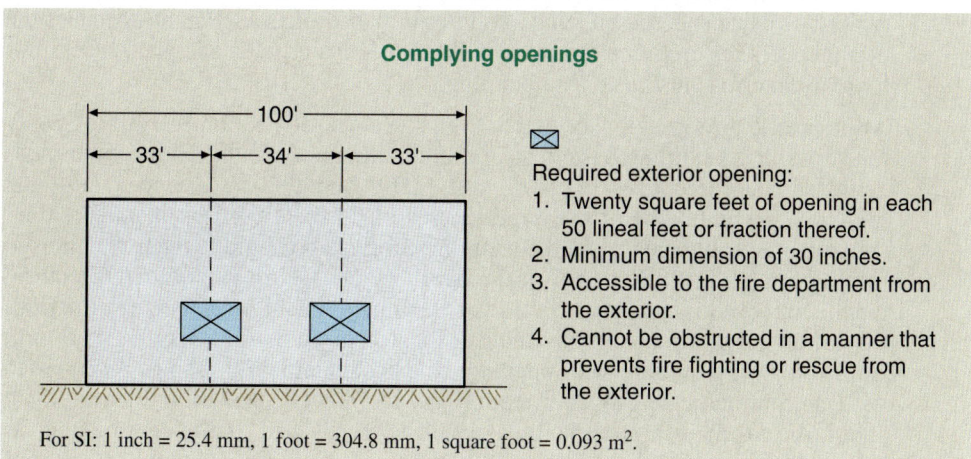

Figure 903-7
Required exterior openings.

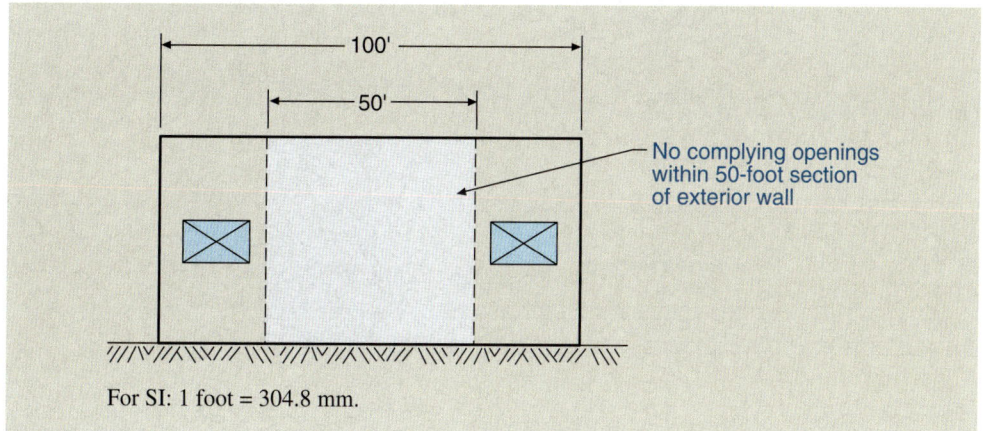

Figure 903-8
Required exterior openings.

3. **Accessible to the fire department from the exterior.** Surely, the openings would be of no value for fire fighting if the fire-fighting forces could not gain access. The mere fact that the openings may be 30 or 40 feet (9,144 or 12,192 mm) above grade does not mean the openings are inaccessible. However, if, with the resources available to the fire department, access cannot be obtained to the openings, they would be considered inaccessible. The determination of accessibility rests with the building official. However, personnel in a fire department should be consulted for their professional opinions and also for their knowledge of the capabilities of their equipment.

Fire Protection Systems

4. **Adequate to allow access for fire fighting to all portions of the interior of the building.** For this reason, the code requires that where openings are provided on only one side and the opposite exterior wall is more than 75 feet (22,860 mm) away, sprinklers shall be provided, or, as an alternative, openings shall be provided on at least two sides. The 75-foot (22,860-mm) distance is a straight-line measurement taken between the two opposing walls. Where complying openings are required in two exterior walls because of the 75-foot (22,860-mm) limitation, the openings are permitted on either two adjacent sides or opposite sides on the assumption that, with two exterior sides having openings, adequate access may be gained to effectively fight the fire.

In other than basements, the provisions requiring openings in exterior walls do not extend beyond the exterior wall line into the building. Thus, the code does not typically dictate specific openings for interior partition arrangements, because the normal openings provided through interior partitions provide adequate accessibility to all interior portions of the building.

5. **Applicability.** As previously noted, the provisions of Section 903.2.11.1 apply to every story, including basements, of all buildings where the floor area exceeds 1,500 square feet (139.4 m^2). Figures 903-9 and 903-10 provide additional graphic representations of the requirements of this section. Basements are considered to be somewhat more difficult than stories above grade when it comes to fighting fires from the exterior of the building. Therefore, an additional requirement is imposed in addition to those of Section 903.2.11.1.2. The code provides that when any portion of a basement is located more than 75 feet (22,860 mm) from complying exterior wall openings, the basement is required to be provided with an automatic sprinkler system. The 75-foot (22,860-mm) measurement should be taken in a straight line, resulting in the use of the arc method, as shown in Figure 903-11. The two methods of providing complying exterior

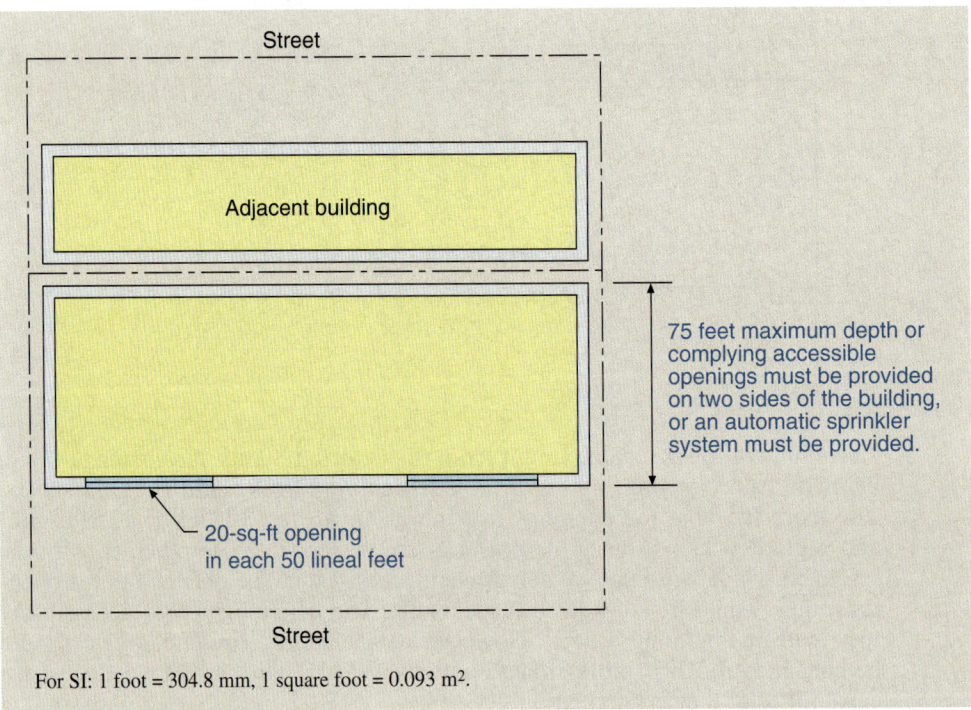

Figure 903-9 Maximum distance between walls.

For SI: 1 foot = 304.8 mm, 1 square foot = 0.093 m^2.

Automatic Sprinkler Systems 9

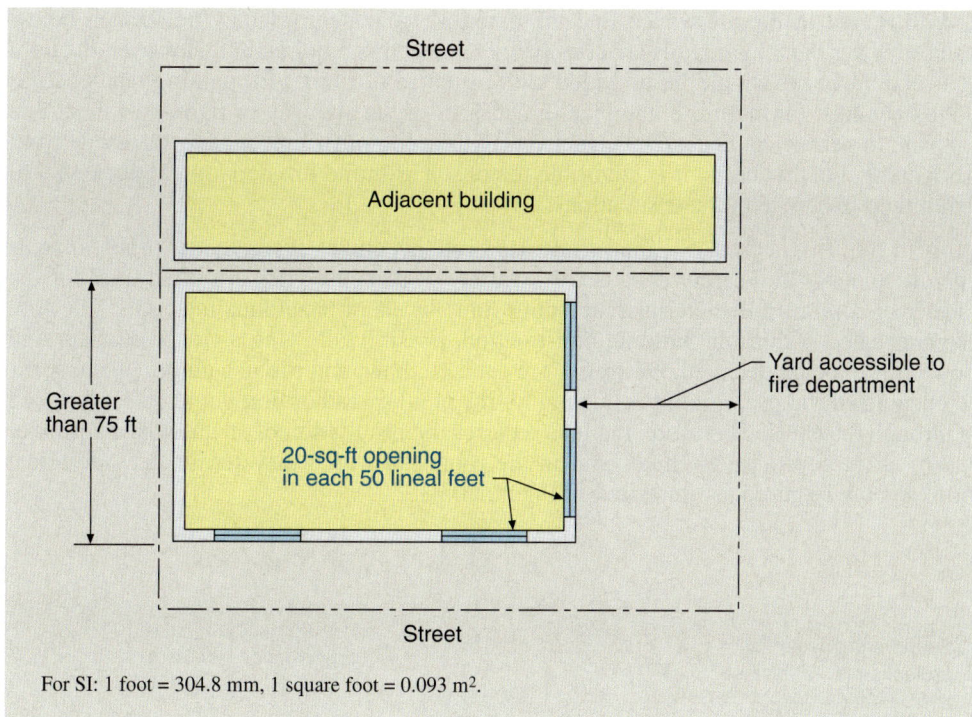

Figure 903-10
Access to required openings.

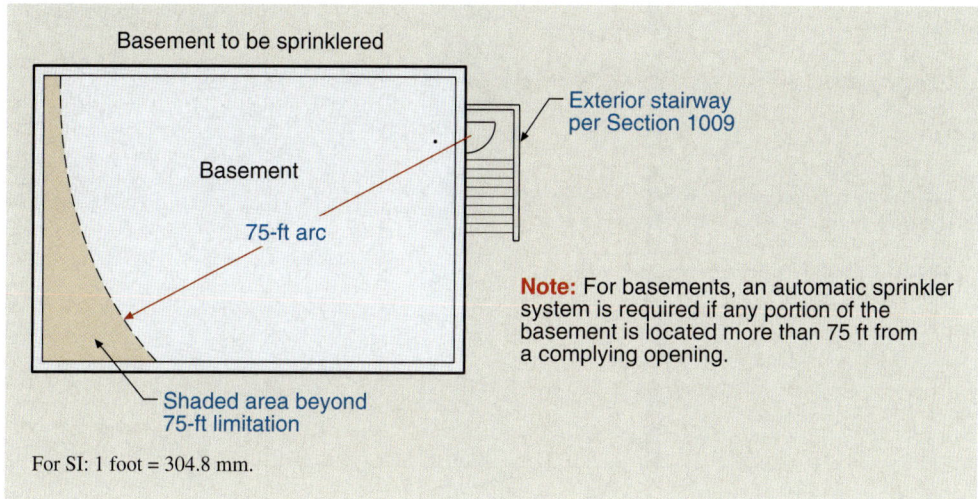

Figure 903-11
Openings in basement.

openings set forth in Section 903.2.11.1 are both available for a basement condition. If the openings are available entirely above the adjoining ground level, they are regulated in the same manner as for floor levels above grade. Otherwise, the openings must lead directly to a complying exterior stairway or ramp. The 75-foot (22,860-mm) criterion is only applicable where the basement is a wide-open space with no interior walls or partitions that could obstruct a fire-hose water stream. Where the basement contains such walls or partitions, a sprinkler system is always required in the basement if it exceeds 1,500 square feet (139.4 m²) in floor area.

2012 International Building Code Handbook 301

With regard to the allowance that the exterior wall openings may be located below grade, areaways and light wells are considered to meet this requirement. However, the light wells and areaways should be provided with a stairway or ramp for gaining ready access to the openings. Furthermore, the plan dimensions of the areaway or light well should be adequate to permit the necessary maneuvering to accomplish fire fighting or rescue from the opening. On this basis, it is advisable to consult with the fire department personnel to obtain their expertise in these situations.

903.2.11.2 Rubbish and linen chutes. Linen chutes and rubbish chutes are potential problem areas when it comes to fire safety because of a variety of reasons. They are often used for the transfer of combustible materials, including some levels of hazardous materials. They are also concealed within the building construction, possibly allowing a fire to smolder and grow prior to being detected. Of even more concern, linen and rubbish chutes create vertical openings through a building, allowing for the rapid spread of fire, hot gases, and smoke up through the chute. Therefore, the IBC requires the installation of an automatic sprinkler system at the top of such chutes and in the rooms in which they terminate. Additional sprinklers are required as illustrated in Figure 903-12.

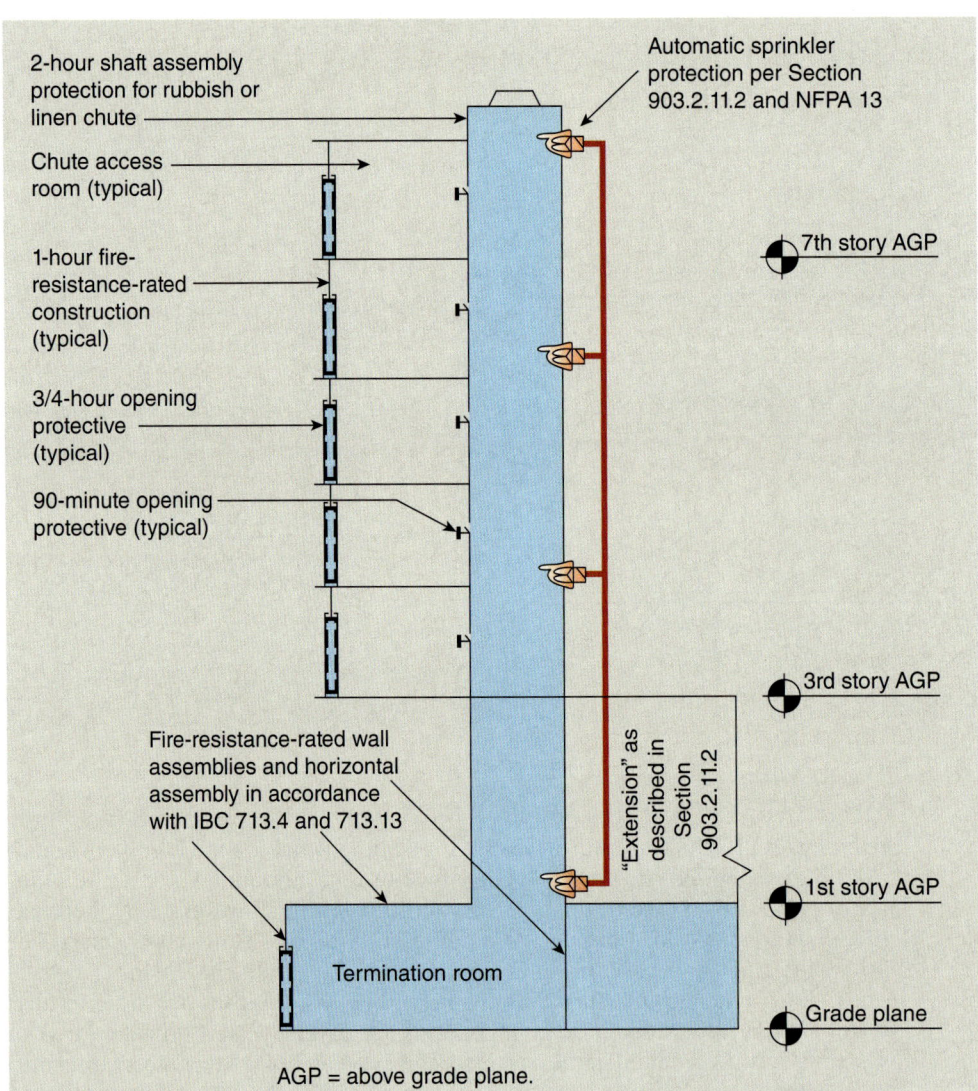

Figure 903-12
Rubbish and linen chute construction and protection.

903.2.11.3
Buildings 55 feet or more in height. Fire fighting in buildings that are over 55 feet (16,764 mm) in height is difficult, and many jurisdictions do not have the personnel or equipment to rescue occupants and control fires on upper floors. Therefore, this section was developed in recognition of this problem. The provision applies to all buildings other than airport control towers, open parking garages, and Group F-2 occupancies. This provision lowers the threshold at which automatic fire sprinklers are required for high-rise buildings from 75 feet (22,860 mm) to 55 feet (16,764 mm), but it does not alter the applicability of any of the other special requirements set forth in Section 403. These still do not apply until the structure has occupiable floors more than 75 feet (22,860 mm) above the lowest level of fire-department vehicular access.

903.2.11.6
Other required suppression systems. A number of additional locations and uses require suppression systems based on requirements located throughout other portions of the IBC. Table 903.2.11.6 provides a cross-reference to the sprinkler requirements for such subjects as atriums, stages, application of flammable finishes, and unlimited area buildings. In addition, a reference is made to Section 903.2.11.6 of the IFC for many other specific buildings and areas where the installation of a fire-extinguishing system is mandated.

903.3.1.1
NFPA 13 sprinkler systems. Where the code requires the installation of an automatic sprinkler system in a building, it typically is referring to a system designed and installed in accordance with the criteria of NFPA 13. This standard is also applicable for those provisions that use a sprinkler system as an alternative to other code requirements. Throughout the code, the use of a sprinkler system "in accordance with Section 903.3.1.1" is referenced. In addition, where an automatic sprinkler system is required by the code with no direct reference to Section 903.3.1.1, the use of an NFPA 13 system is required.

903.3.1.1.1
Exempt locations. It is the intent of sprinkler protection that sprinklers be installed throughout the structure, including basements, attics, and all other locations specified in the appropriate standard. It is also the intent of the IBC that when an automatic sprinkler system is required throughout, the same meaning is implied. One of the reasons for requiring protection throughout is the possibility of a fire in an unprotected area gaining such a foothold that the automatic sprinkler system would be overpowered. However, over the years, certain areas, locations, or conditions have shown that they require special consideration, and the omission of sprinklers is permitted. In this section, the code itself provides the rationale for the omission of sprinklers.

903.3.1.2
NFPA 13R sprinkler systems. Although residential sprinkler systems installed in accordance with NFPA 13R may be used to satisfy the requirements of specific institutional and residential occupancies, they are not always recognized as full sprinkler protection for the purposes of exceptions or reductions permitted by other code requirements. However, where specifically mentioned through a reference to this section, such systems may be considered acceptable. Where the code indicates that a benefit can be derived from a sprinkler system installed "in accordance with Section 903.3.1.2," it intends that an NFPA 13R system can be used for the benefit. An important point is that an NFPA 13R sprinkler system is only permitted in residential-type buildings up to four stories in height.

903.3.1.2.1
Balconies and decks. Experience has shown that numerous fires in apartment buildings have started from grilling or similar activities on the balconies and patios. Because the NFPA 13R sprinkler standard does not mandate sprinklers in such locations, the code requires such sprinkler protection. The provision is applicable only to dwelling units in buildings of Type VA or VB construction. The automatic sprinkler protection is only required where there is a roof, deck, or balcony directly above a balcony, deck, or patio below. These areas will typically be protected by sidewall-orientation automatic sprinklers. If there is no horizontal element located directly above an exterior balcony, deck, or ground-floor patio, the additional sprinkler protection is not required. See Figure 903-13.

9 Fire Protection Systems

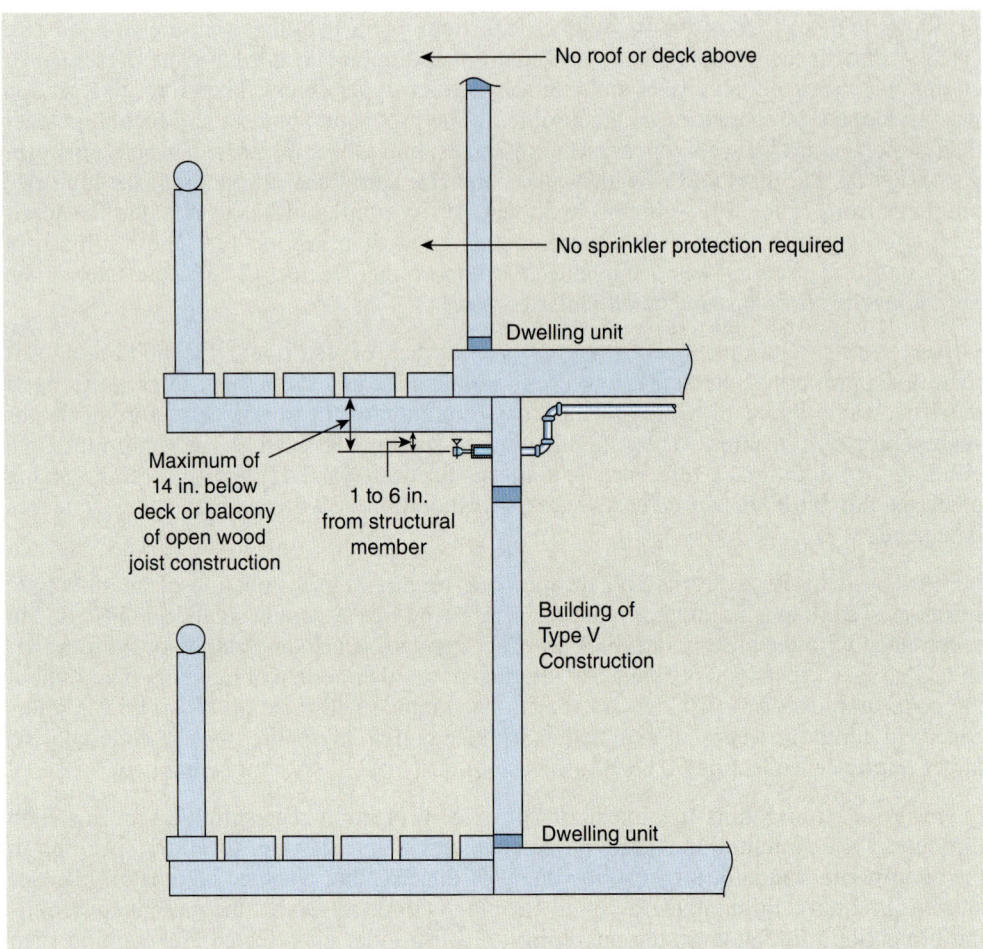

Figure 903-13 Protection of residential deck.

903.3.2 Quick-response and residential sprinklers. Based on the timely performance of quick-response and residential automatic sprinklers, the code requires that they be installed in those occupancies where response or evacuation may not be immediate because of the condition of the occupants. Therefore, all spaces within a Group I-2 smoke compartment containing care recipient sleeping units and all dwelling units and sleeping units in Group R and I-1 occupancies are to be provided with these types of sprinklers. Similarly, ambulatory care facilities must be provided with quick-response or residential sprinklers in all smoke compartments that contain one or more treatment rooms. Such sprinklers are also required in light-hazard occupancies, where the quantity or combustibility of contents is low. Light-hazard occupancies included places of worship, education facilities, office buildings, museums, and seating areas of restaurants and theaters.

Section 904 *Alternative Automatic Fire-Extinguishing Systems*

The code permits the use of automatic fire-extinguishing systems other than automatic sprinkler systems for those circumstances approved by the fire code official. However, the use of an alternative system in lieu of a sprinkler system does not gain the benefit of exceptions or reductions in code requirements. Only those buildings or areas protected by automatic sprinkler systems can take advantage of the allowances provided throughout the code.

The installation of automatic fire-extinguishing systems shall be in compliance with this section, which for the most part refers to the appropriate test standard and listing for each of the various types of systems. Those types addressed include wet-chemical, dry-chemical, foam, carbon dioxide, halon, and clean-agent systems.

The inspection and testing of the system is emphasized because of the importance of a fully operating system. Specific items are identified for inspection, including the location, identification, and testing of the audible and visible alarm devices. More specific criteria are also present for the installation and operation of a fire-extinguishing system for a commercial cooking system.

Section 905 *Standpipe Systems*

A standpipe system is a system of piping, valves, and outlets that is installed exclusively for fire-fighting activities within a building. Standpipes are not considered a viable substitute for an automatic fire-sprinkler system. They are needed in buildings of moderate height and greater, and when used by trained personnel provide an effective means of fighting a fire.

Required installations. This section provides the scoping criteria for when a standpipe system must be provided. Figure 905-1 provides the basic requirements for when a standpipe system is required in a building. Building height is the primary consideration for the installation of a standpipe system. The general requirement calls for Class III standpipe systems where the vertical distance between the highest floor level in the building and the lowest level of fire-department vehicle access exceeds 30 feet (9,144 mm). See Figure 905-2. For measurement purposes, it is not necessary to consider any level of fire-department vehicle access that, because of topographic features, makes access to the building from that point impractical or impossible. Such a condition typically occurs on steeply sloping sites and is illustrated in Figure 905-3. Several exceptions allow the use of Class I rather than Class III standpipes. Additional standpipe requirements may apply to assembly occupancies, covered and open mall buildings, and stages.

905.3

	Class I	Class II	Class III
Buildings where highest floor level is located more than 30 feet above lowest point of fire-department vehicle access			x[1]
Buildings where lowest floor level is located more than 30 feet below highest point of fire-department vehicle access			x[1]
Nonsprinklered Group A buildings with an occupant load >1,000	x		
Covered and open mall buildings	x		
Stages greater than 1,000 square feet			x[2]
Underground buildings	x		
Marinas and boatyards	See IFC Chapter 36		

For SI: 1 foot = 304.8 mm, 1 square foot = 0.093 m^2.
[1]Exceptions allow for Class I in sprinklered buildings, open parking garages, and sprinklered basements.
[2]Only a 1$^1/_2$-in. hose connection is required where buildings or area sprinklered and connection installed per NFPA 13 or NFPA 14 for Class II or III standpipes.

Figure 905-1 Standpipes required.

9 Fire Protection Systems

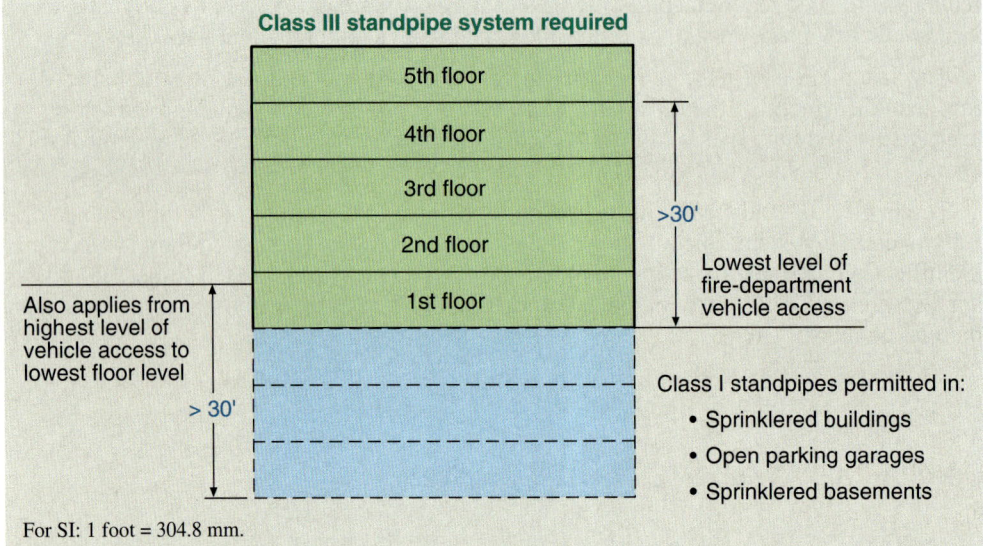

Figure 905-2 Class III standpipe systems.

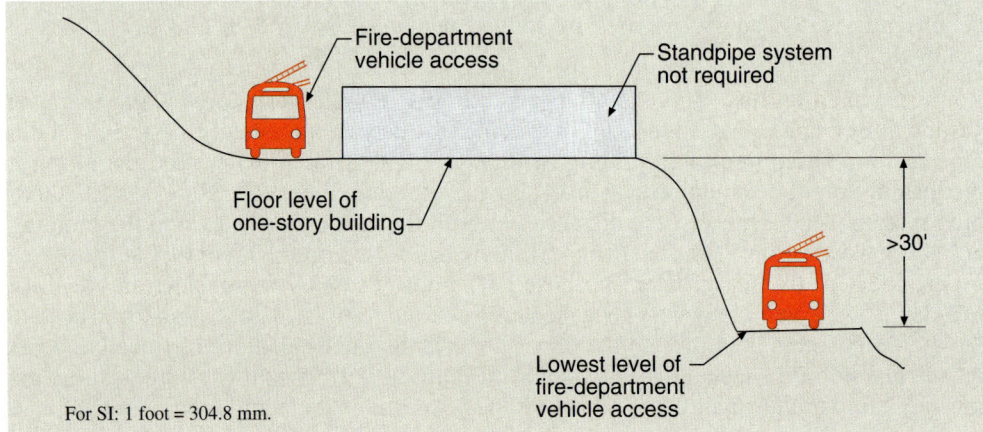

Figure 905-3 Fire-department vehicle access.

905.4 Location of Class I standpipe hose connections. The code intends that Class I standpipes are for the use of the fire department to fight fires within a building. Thus, the code requires that standpipe outlets be at every required stairway. The connections are to be located at every intermediate floor level landing of those stairways required by the code. As an alternative, the hose connections are permitted to be located at the floor level landing, but only where specifically allowed by the fire code official. However, the installation of hose connections at intermediate landings is typically preferred in order to avoid congestion at the stairway door. With these locations for standpipe outlets, the fire-department personnel can bring a hose into the stair enclosure and make a hookup to outlets in a relatively protected area. For this reason, standpipe connections are typically not required for stairways permitted to be unenclosed by the exceptions to Section 1009.3, as unenclosed stairways provide no protection for fire-department personnel. However, it is very seldom that an unenclosed stairway is allowed to be used as a required means of egress in a building that requires a standpipe system.

Because a horizontal exit provides a barrier having a minimum fire-resistance rating of 2 hours, it is a logical location for Class I standpipe outlets. Such outlets are to be provided on both sides of the horizontal exit wall adjacent to the egress doorways through the horizontal exit wall, regardless of whether or not egress is provided from both directions. An exception permits the omission of the standpipe hose connection at the horizontal exit opening where there is a limited distance between the opening and the stairway hose connection. The elimination of the hose connection is permitted on one side of the horizontal exit, as depicted in Figure 905-4, or on both sides, provided

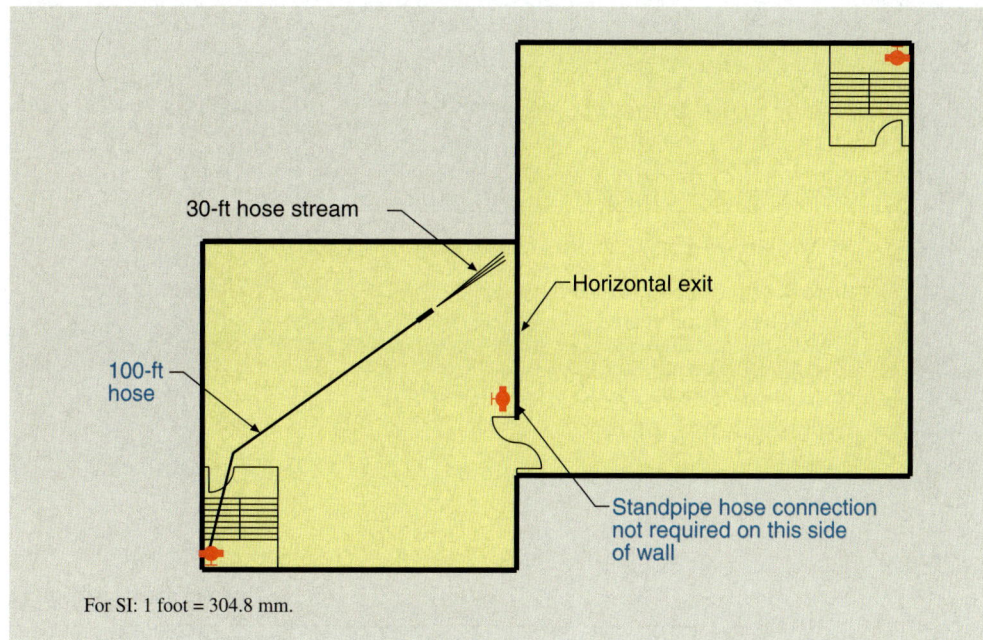

Figure 905-4
Connections at a horizontal exit.

the "100-foot plus 30-foot" distance is not exceeded. The application of the exception is most common where the horizontal exit is provided to allow for the termination of a fire-resistance-rated corridor at an intervening room, or where necessary to address an inadequate number of exits or insufficient egress width. In those cases where the horizontal exit is provided because of a problem with travel distance, the omission of the standpipe connections will seldom be permitted. It should be noted that a standpipe system is not required simply because a horizontal exit occurs within the building. This provision addressing required hose connection locations is only applicable where a horizontal exit is provided in a building described in Section 905.3 as requiring a Class I or Class III standpipe system.

Under unique circumstances, standpipe outlets located in interior exit stairways or at horizontal exit doorways may be a great distance from some portions of the building. Under these circumstances, additional standpipe outlets must be provided in approved locations when required by the fire code official. Figure 905-5 depicts the required locations of Class I standpipe connections. Standpipe connections required for stages exceeding 1,000 square feet (93 m^2) are to be located on each side of the stage, per Figure 905-6.

The enclosure for an interior exit stairway also provides protection for the standpipe and piping system. In those cases where the risers and laterals are not within interior exit stairways, the code requires that they be protected by equivalent fire-resistant construction. The exception to this requirement assumes that the automatic fire-sprinkler system will keep the risers and laterals cool enough so that they will not be damaged by fire.

9 Fire Protection Systems

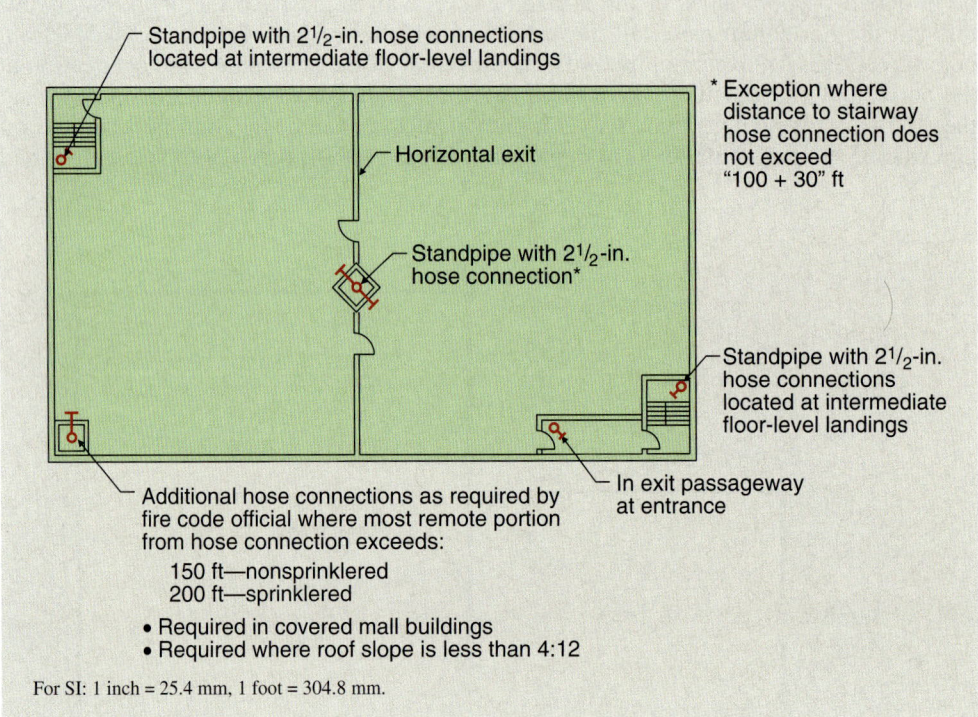

Figure 905-5 Standpipe connection locations.

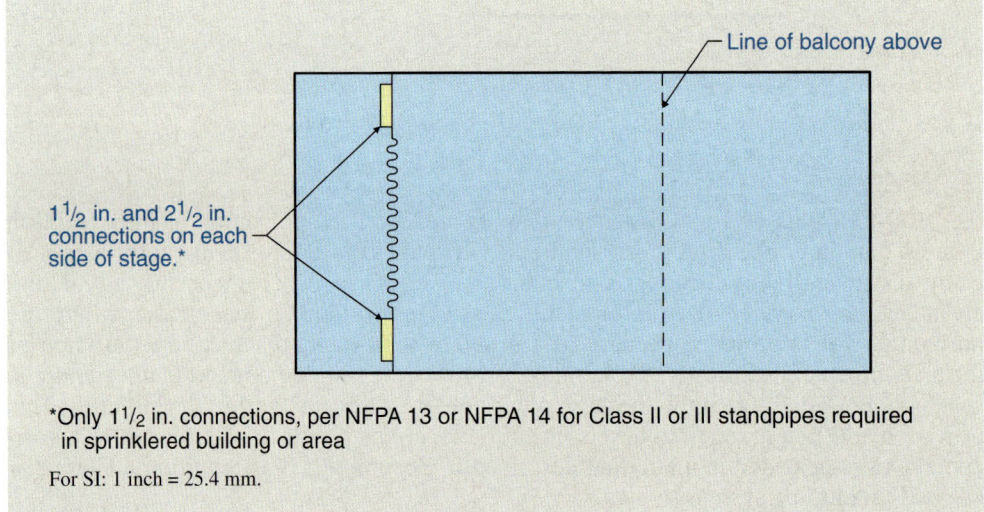

Figure 905-6 Class II standpipe locations at stages > 100 sq ft.

On those roofs that have a flat enough slope for fire fighters to move about, the code requires at least one roof outlet so that exposure fires can be fought from the roof.

The interconnection of the standpipe risers at the bottom for multiple standpipe systems is intended to increase the reliability.

905.5 Location of Class II standpipe hose connections. It is the intent of the code to require the location of hose cabinets for Class II standpipes at intervals ensuring that all portions of a building will be within 30 feet (9,144 mm) of a nozzle attached to 100 feet (30,480 mm) of hose. In plan review, this would necessitate allowing for pulling the hose down corridors and through rooms such that several right-angle turns may be necessary before the hose stream can be placed on the fire. Therefore, judgment is necessary in the determination of standpipe locations. One method to account for this type of partitioning in a building where the future location of partitions is unknown is to subtract 30 feet (9,144 mm) from the straight-line distance between the hose cabinet and the remote location and then multiply the remainder by 1.4. If the result is more than 100 feet (30,480 mm), an additional standpipe connection will be required. Figure 905-7 illustrates the location of a Class II standpipe in a building where an office floor has a central corridor with offices on each side. In this particular arrangement, it is obvious from the layout that the one standpipe will suffice.

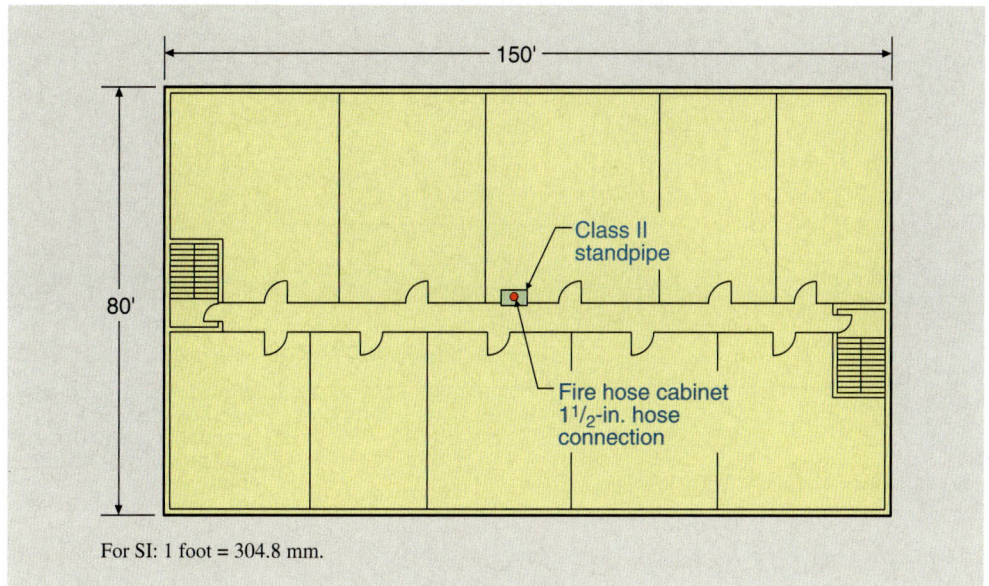

Figure 905-7
Class II connections.

As there are no scoping provisions in Section 905.3 for a Class II or III standpipe system in a sprinklered Group A occupancy, other than the stage requirements of Section 905.3.4, the provisions of Section 905.5.1 have no application. Therefore, the required locations for Class II standpipe hose connections are limited to those buildings or areas required to have Class III standpipe systems.

Because Class II standpipe systems are charged with water, the code does not require fire-resistive protection. The water within the system is considered adequate to keep the pipe cool enough to prevent damage.

Location of Class III standpipe hose connections. Because Class III standpipes are a combination of the benefits of Class I and II systems, containing both 1$\frac{1}{2}$-inch (38-mm) outlets for occupant use and 2$\frac{1}{2}$-inch (64-mm) outlets for fire department use, it is only logical that they be located so as to serve the building as required for both Class I and II standpipes. Figure 905-8 shows the typical arrangement for a Class III standpipe in a building. Usually, the hose rack for 1$\frac{1}{2}$-inch (38-mm) outlets and 2$\frac{1}{2}$-inch (64-mm) hose outlets are both located within the stair enclosure. Where the coverage requirements for Class II standpipes are such that interior exit stairway locations will not cover the entire building, laterals are usually run to other locations from hose cabinets in order to provide for the required coverage. In this case, the laterals are not required to be protected, as they are charged with water.

905.6

Class III standpipe systems and their risers and laterals in sprinklered buildings are not required to have fire-resistive protection for the same reason as discussed for Class I standpipes.

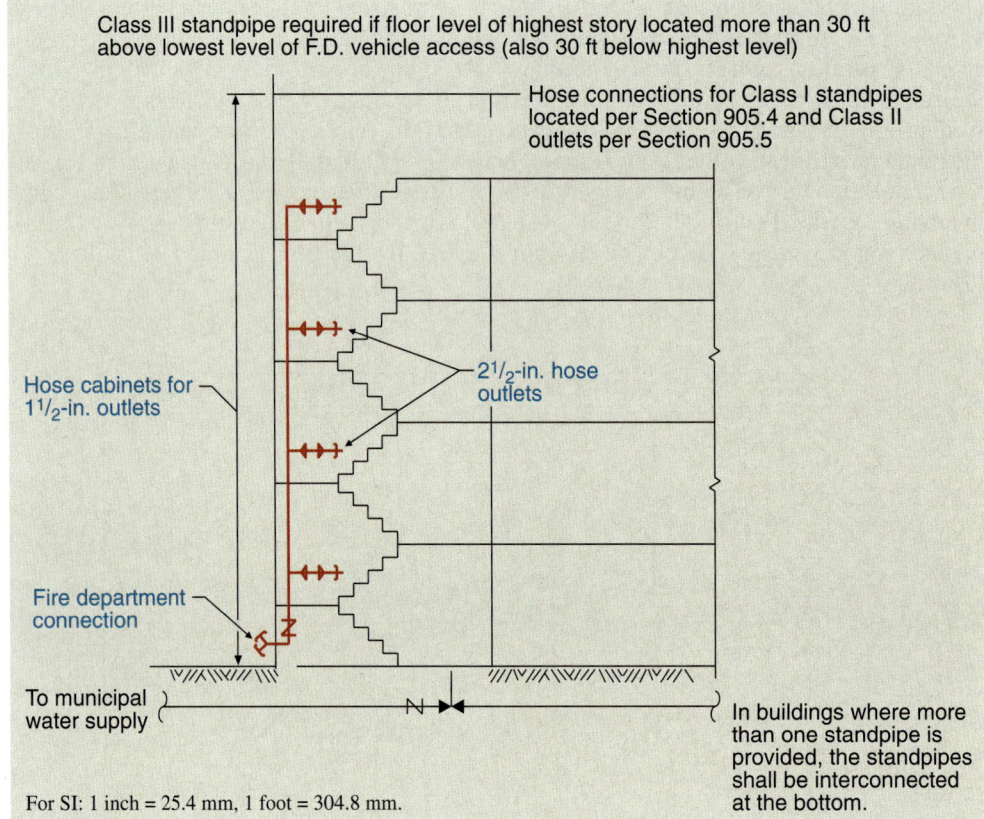

Figure 905-8 Class III standpipe connections.

Section 907 Fire Alarm and Detection Systems

One of the most effective means of occupant protection in case of a fire incident is the availability of a fire alarm system. An alarm system provides early notification to occupants of the building in the event of a fire, thereby providing a greater opportunity for everyone in the building to evacuate or relocate to a safe area. This section covers all aspects of fire alarm systems and their components.

Unlike most of the provisions of Section 903.2 addressing the required installation of an automatic sprinkler system, those requirements mandating a fire alarm system are typically not applied based on the fire area concept. Manual fire alarm systems are most often required based on the occupant load of the occupancy group under consideration, including any occupants of the same occupancy classification that may be identified in another fire area. Fire areas are not to be used in the application of this section regarding fire alarm systems except where specifically addressed in Sections 907.2.1 (multiple Group A occupancies) and 907.2.2.1 (ambulatory care facilities).

907.2 Where required—new buildings and structures. Approved fire alarm systems, either manual, automatic, or both manual and automatic, are mandated in those occupancies and areas identified by this section. Where automatic fire detectors are mandated, smoke detectors are to be provided unless normal operations would cause an inaccurate activation of the detector. All automatic fire-detection systems are to be installed in accordance with NFPA 72.

Fire Alarm and Detection Systems

Group A. Where an assembly occupancy has a sizable occupant load, the safe egress of the occupants becomes an even more important consideration. When the occupant load reaches 300, the code mandates a manual alarm system to provide early notification to the occupants. The IBC also requires portions of a Group E educational occupancy that are occupied for accessory assembly purposes, such as a lunchroom or library, to have alarms as required for the Group E use, rather than based on the less restrictive requirements mandated for Group A occupancies. In addition, the Group E fire alarm provisions are applicable to those areas within a Group E school building that may be classified as Group A occupancies, such as a gymnasium and/or auditorium. As is the case in many other occupancy classifications, manual alarm boxes are not required in those Group A occupancies where an automatic sprinkler system is installed that will immediately activate the occupant notification appliances (horn/strobes) upon water flow.

907.2.1

Where there are two or more Group A occupancies within the same building, the occupant loads of all such occupancies shall be used in evaluating whether or not a manual fire alarm system is required unless multiple fire areas are created in accordance with Section 707.3.10. Where the fire area concept is used, the alarm requirement is based on the occupant load within each individual fire area. Only those fire areas having a Group A occupant load of 300 or more are required to be provided with a manual fire alarm system.

In assembly occupancies containing much larger occupant loads (1,000 or more people), activation of the required fire alarm system must initiate a prerecorded announcement. The emergency voice/alarm communications system, upon approval of the building official, may also be used for live voice emergency announcements originating from a constantly attended location.

Group B. Larger business occupancies classified as Group B require the installation of a fire alarm system. Where the total occupant load exceeds 499 persons, or where more than 100 persons occupy Group B spaces above and/or below the lowest level of exit discharge, a manual fire alarm system shall be installed. See Figure 907-1. Similar to exceptions for other occupancies, the manual fire alarm boxes are not required in a sprinklered building where sprinkler water flow activates the notification devices.

907.2.2

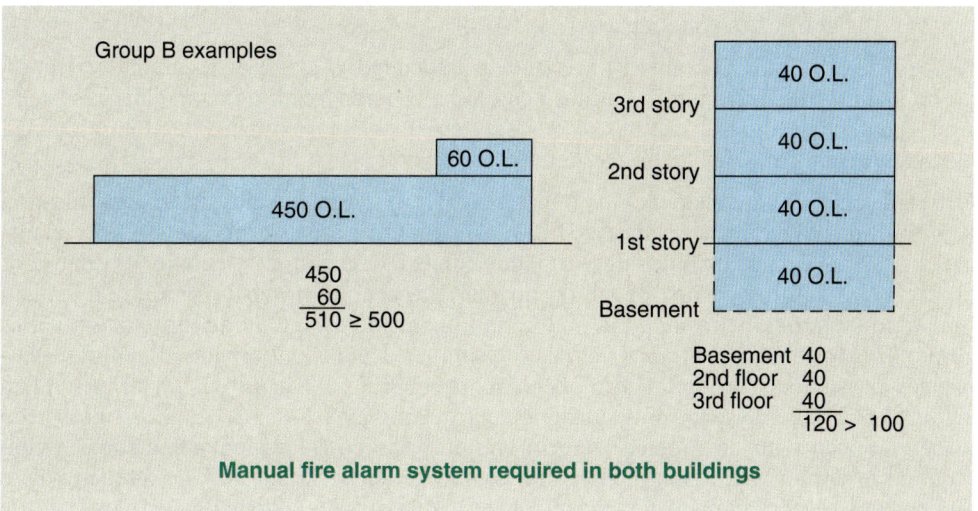

Figure 907-1
Required fire alarm system.

Ambulatory care facilities must be provided with a manual fire alarm system as well as an electronically supervised automatic smoke-detection system. The manual fire alarm system must be provided within the fire area in which the ambulatory care facility is located, while the smoke-detection system must serve the care facility as well as the adjoining public areas.

In both cases, the presence of an automatic sprinkler system modifies the requirements where occupant notification appliances will activate upon water flow.

907.2.3 Group E. The IBC follows the philosophy of society in general that our children require special protection when they are not under parental control. Therefore, in addition to the other life-safety requirements for educational occupancies, manual fire alarm systems are required whenever the occupant load of any Group E occupancy is more than 30. In addition, when the building is provided with a smoke-detection or sprinkler system, such systems shall be connected to the building's fire alarm system. However, the more probable reason for such a low occupant load threshold being established is the exceptional value of such a system in an educational use. Students tend to react quickly and efficiently at the first notification of the alarm system, making safe egress possible. In addition, periodic fire drills reinforce the appropriate egress activity. Where the three conditions are met as listed in Exception 2, manual alarm boxes are not required. Exception 3 also exempts manual fire alarm boxes where, in a sprinklered building, the sprinkler water flow activates the notification appliances and manual activation is possible from a normally occupied location.

The fire alarm system for a Group E occupancy must use an emergency voice/alarm communication system. This type of system will not only provide the necessary notification of occupants under possible fire conditions, but it is also valuable in ensuring the necessary level of life safety inside of the building during a lockdown situation. Because of concerns of school campus safety serving kindergarten through 12th-grade students, it is beneficial to provide an effective means of communication between the established central location and each secured area.

907.2.4 Group F. In multistory manufacturing occupancies, a fire alarm system is mandated where an aggregate occupant load of 500 or more is housed above and/or below the level of exit discharge. Similar to several other occupancies, the alarm system is necessary where there are very large occupant loads and where those occupants must travel vertically to exit. Application of the provision where occupants are located both above and below the lowest level of exit discharge is similar to that for Group B and M occupancies. In Group F occupancies, a manual fire alarm system is mandated where the aggregate occupant load of those levels, other than the exit discharge level, exceeds 499. Where the building is sprinklered, manual alarm boxes are not required if the water flow of the sprinkler system activates the notification appliances.

907.2.5 Group H. Only those Group H occupancies associated with semiconductor fabrication or the manufacture of organic coatings need to be provided with a manual alarm system. Areas containing highly toxic gases, organic peroxides, and oxidizers shall be protected by an automatic smoke-detection system.

907.2.6 Group I. As patients, residents, or inmates of Group I occupancies are asleep during a large portion of the day, Section 907.2.6 provides for early warning of the occupants and staff, thus enhancing life safety. In addition, early response is beneficial because of the lack of mobility of many of the occupants. The general provisions require a manual fire alarm system to be installed in all Group I occupancies, including those classified as Group I-4. However, an automatic smoke-detection system is only required in those occupancies classified as Group I-1, I-2, or I-3. Where a Group I-1 facility is provided with an automatic sprinkler system, only corridors and waiting areas open to such corridors need to be equipped with the smoke-detection system. The corridor smoke-detection system required in Group I-2 nursing homes and detoxification facilities may be omitted where the patient sleeping rooms have smoke detectors that provide a visual display on the corridor side of each patient room, as well as a visual and audible alarm at the appropriate nursing station. Another exception exempts the requirement for corridor smoke detection where sleeping-room doors are equipped with automatic door closers having integral smoke detectors on the room side that performs the required alerting functions.

Because of their special nature, in Group I-3 occupancies the provisions for fire alarm systems are greatly expanded. The manual and automatic fire alarm system is to be

Fire Alarm and Detection Systems

designed to alert the facility staff. Actuation of an automatic fire-extinguishing system, an automatic sprinkler system, a manual fire alarm box, or a fire detector must initiate an automatic fire alarm signal, which automatically notifies staff. For obvious reasons, manual fire alarm boxes need only be placed at staff-attended locations having direct supervision over the areas where boxes have been omitted. Under certain conditions, the installation of an approved smoke-detection system in resident housing areas is required.

Group M. The threshold at which a manual fire alarm system is required for a Group M occupancy is the same as that for a Group B occupancy. During that portion of time when the building is occupied, the signal from the fire alarm box or water-flow switch may be designed to only activate a signal at a constantly attended location, rather than provide the customary visual and audible notification. At this location, the use of an emergency voice/alarm communications system can be used to notify the customers of the emergency conditions. This provision is helpful in eliminating those nuisance alarms that may occur because of the presence of fire alarm boxes. — **907.2.7**

Group R-1. When asleep, the occupants of residential buildings will usually be unaware of a fire, and it will have an opportunity to spread before being detected. As a result, a majority of fire deaths in residential buildings have occurred because of this delay in detection. It is for this reason that the IBC requires fire alarm systems in addition to smoke detectors in certain residential structures. — **907.2.8**

In hotels and other buildings designated as Group R-1, the general provisions mandate that both a manual fire alarm system and an automatic fire-detection system be installed. There is an exception that eliminates the requirement for a manual alarm system for such occupancies less than three stories in height where all guestrooms are completely separated by minimum 1-hour fire partitions and each unit has an exit directly to a yard, egress court, or public way. This exception is based on the compartmentation provided by the separations between units and by the relatively rapid means of exiting available to the occupants. Where guestrooms are limited to egress directly to the exterior, early notification, although important, is not critical. A second exception requires the alarm system, but does not mandate the installation of fire alarm boxes throughout buildings that are protected throughout by an approved supervised fire-sprinkler system. There is, however, a need for at least one manual fire alarm box installed in a location approved by the building official. In addition, sprinkler flow must activate the notification appliances. The automatic smoke-detection system required by this section need only be provided within all corridors that serve guestrooms. An exception eliminates the requirement for the automatic fire-detection system in buildings where egress does not occur through interior corridors or other interior spaces.

Group R-2. Group R-2 buildings such as apartment houses are to be provided with a fire alarm system based on the number of dwelling units and sleeping units, as well as the location of any such units in relationship to the level of exit discharge. Where more than 16 dwelling units or sleeping units are located in a single structure, or where such units are placed at a significant distance vertically from the egress point at ground level, it is beneficial that a detection and notification system be provided. If any one of the three listed conditions exists, the alarm system is required unless exempted or modified by one of the three exceptions. Exceptions similar to those permitted for Group R-1 occupancies apply to Group R-2 buildings as well. — **907.2.9**

Provisions addressing the installation of an automatic smoke-detection system in Group R-2 college and university buildings are essentially the same as those required for Group R-1 occupancies. The single difference is that the detection system must also be provided in laundry rooms, mechanical equipment rooms, and storage rooms, as well as all common areas located outside of the individual sleeping units or dwelling units. It is intended that these provisions be applicable only to those buildings that are owned by a university or college, not privately owned facilities.

Single- and multiple-station smoke alarms. As indicated in the introduction to the residential fire alarm provisions, residential fire deaths far exceed those of any other building classification. Furthermore, more than one-half of the fire deaths in residential buildings — **907.2.11**

occurred because of a delay in detection that is due to the occupants being asleep at the time of the fire. Thus, the IBC requires smoke alarms in all residential buildings and in certain institutional occupancies. In Group R-1 occupancies, single- or multiple-station smoke alarms are to be installed in all sleeping areas, in any room along the path between the sleeping area and the egress door from the sleeping unit, and on each story within the sleeping unit. In all other residential occupancies, the code requires that smoke detectors be located in the sleeping rooms and on the ceiling or wall of the corridor or area giving access to the sleeping rooms. In addition, at least one smoke alarm shall be installed on each story of a dwelling unit, including basements. Where split levels occur in guestrooms or dwelling units, a smoke alarm need only be installed on the upper level, provided there is no intervening door between the adjacent levels. See Figure 907-2 for illustrations of these provisions.

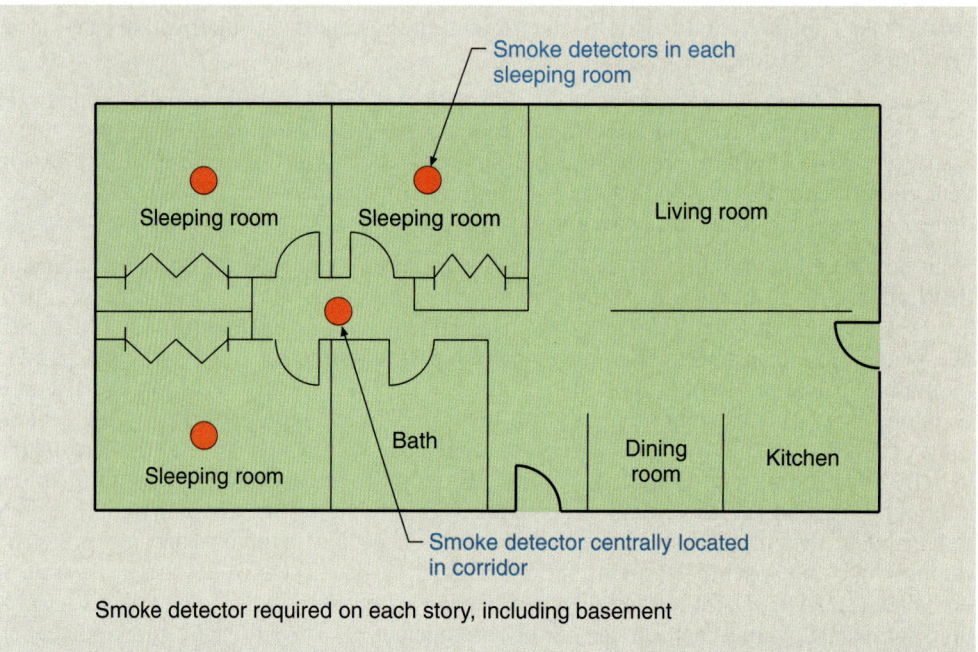

**Figure 907-2
Location of smoke detectors.**

Unless an automatic fire-detection system is provided in compliance with Section 907.2.6.1, single- or multiple-station smoke alarms are to be installed in those sleeping areas of Group I-1 occupancies. It is not the intent of the exception to Section 907.2.11.2 to allow the omission of smoke alarms by referencing the provisions of Section 907.2.6, but rather to allow smoke detectors as a part of the fire alarm system to be installed in the sleeping rooms in lieu of single- or multiple-station smoke alarms. The hazards addressed by the residential smoke alarms also exist to some degree in institutional occupancies. In any new construction, the code requires that smoke alarms receive their power from the building wiring with a battery backup. See Figure 907-3.

In order to notify occupants throughout the dwelling unit or sleeping unit of a potential problem, multiple smoke alarms need to be interconnected. Therefore, when activation of one of the alarm devices takes place, activation of all the alarms must occur. This requirement is also applicable where listed wireless alarms are installed in lieu of physically interconnected smoke alarms. The intent of the code is that the alarms be audible throughout the dwelling, particularly in all sleeping rooms.

907.4.2 **Manual fire alarm boxes.** The IBC often requires a manual fire alarm system because of the special occupants or hazards that exist within the building. Manual fire alarm boxes, defined as manually operated devices used to initiate an alarm signal and often referred to as pull stations, are used in many situations as a means for occupants to notify others of a potential fire emergency. This section identifies the proper locations for the installation of these alarm boxes.

Fire Alarm and Detection Systems

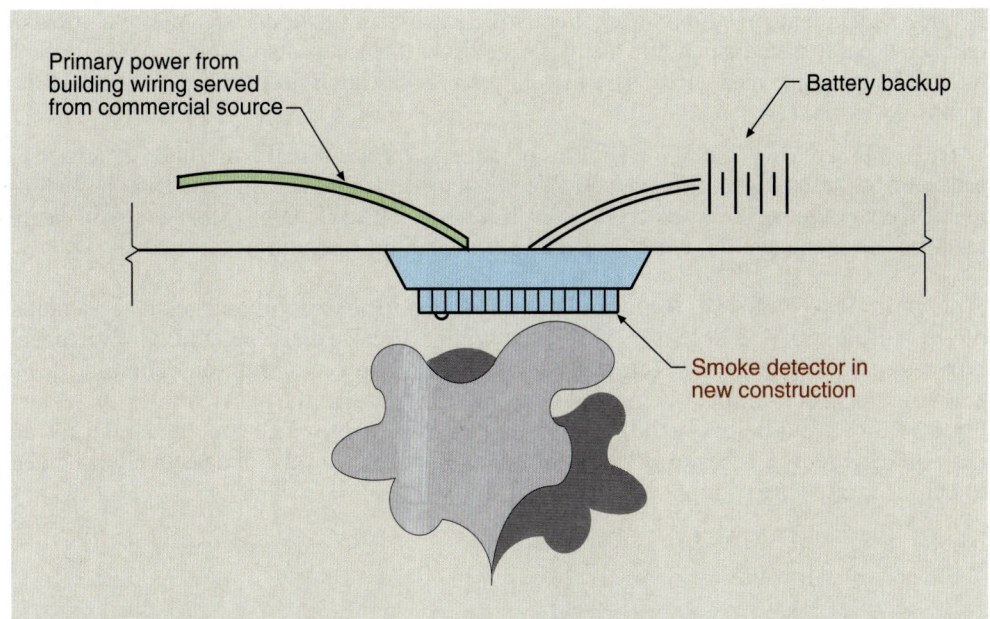

Figure 907-3
Smoke detectors.

In order that manual fire alarm boxes are readily available and accessible to all occupants of the building, they are to be located in close proximity to the point of entry to each exit. This would include placement within 5 feet (1,524 mm) of exterior exit doors, as well as doors entering interior exit stairways and exit passageways, as well as those doors accessing exterior exit stairways and horizontal exits. By placing the boxes adjacent to the exit doors, they will be available to occupants using any of the available exit paths. Additional boxes may be required in extremely large structures, as the maximum travel distance to the nearest alarm box cannot exceed 200 feet (60,960 mm). A manual fire alarm box, required to be red, must be located in a position so that it can be easily identified and accessed. The maximum height of 48 inches (1,372 mm), measured from the floor to the activating lever or handle, is based on the high-end-reach range limited by the accessibility provisions of ICC A117.1. The minimum height of 42 inches (1,067 mm) keeps the activating mechanism in a position readily viewed and providing ease of manipulation. See Figure 907-4.

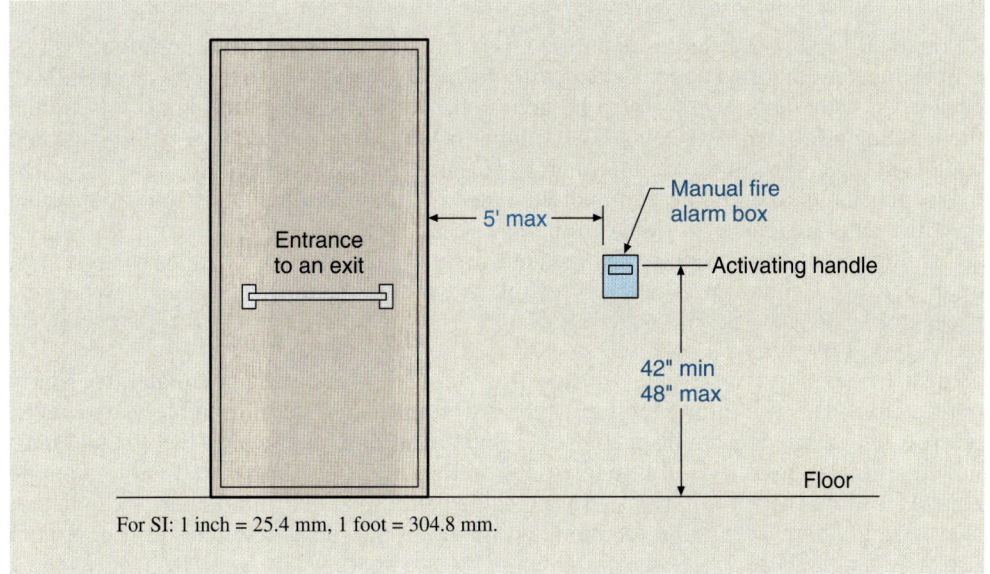

Figure 907-4
Manual fire alarm boxes.

Unless the fire alarm system is monitored by a supervising station, a sign must be installed on or adjacent to each manual fire alarm box advising the occupants to notify the fire department. Where a supervised alarm system is in place, notification becomes automatic and the sign is not necessary.

Often it is necessary to develop a means of reducing the accidental or intentional damage or activation of the alarm-initiating device. Therefore, the code gives the fire code official authority to accept protective covers placed over the listed boxes. The alarm box should remain easily identifiable when covered, with adequate instructions for operation.

907.5.2.2 Emergency voice/alarm communication systems. This section requires that operation of any automatic fire detector, sprinkler water-flow device, or manual fire alarm box result in an alert tone followed by appropriate voice instruction. The alert tone and voice instruction must be provided on the floor of the alarm, as well as the floors directly above and below the alarming floor as applicable. The tone and instruction may be directed on a general basis to the required floor levels, or selectively provided to only the following paging zones:

1. Elevator groups
2. Exit stairways
3. Each floor
4. Areas of refuge

907.5.2.3 Visible alarms. Visible alarms are intended to alert hearing-impaired individuals to a fire emergency. Of those locations where audible alarm systems are required, the IBC identifies the specific conditions under which visible alarm-notification appliances are also mandated. In those portions of the building deemed to be public areas or common areas, visual alarms shall be provided in addition to audible alarms. For example, in an office building, the lobby, public corridors, and public restrooms would be considered public areas, whereas the corridors, toilet rooms, break areas, and conference rooms inside an office suite would be considered common use. On the other hand, the individual offices of each employee would be considered private use. Although such offices would not require the installation of visible alarm notification appliances, wiring must be in place for future installation of the alarms as necessary. The potential of additional visible alarm notification appliances is taken into account by requiring at least 20 percent spare capacity for the appliance circuits. Keep in mind that visual alarms are only required in those occupancies where an alarm system is first required by Section 907.2.

Table 907.5.2.3.3 is used to determine the number of sleeping units in Group R-1 or I-1 occupancies that must be provided with visible alarm notification appliances. In both occupancies, the appliances are to be activated by both the in-room smoke alarm and the building's fire alarm systems. The number of units required to have both visual and audible alarms is based on the total number of units in the building. Not strictly limited to sleeping units, the visible and audible alarm requirements are also applicable where dwelling units are located within a Group R-1 occupancy. In Group R-2 apartment houses and similar occupancies required by the code to have a fire alarm system, provisions must be made for the future installation of visible alarm notification devices as they become necessary.

907.6.3 Zones. In order to quickly identify the fire vicinity, zones must be created for alarm notification. The minimum requirement is to zone each floor separately. However, where a floor exceeds 22,500 square feet (1,860 m^2) in floor area or 300 feet (91,440 mm) in length in any direction, additional zones on the floor must be created with no single zone exceeding these limitations. An indicator panel installed in an approved location shall visually signal the zone location and maintain the identification until the system is reset.

Section 908 *Emergency Alarm Systems*

In addition to those alarms designed to provide early notification of a fire condition, emergency alarm systems are necessary in specific high-hazard-type occupancies to address other identifiable concerns. A reference is made to IBC Section 414.7, which requires a manual emergency alarm system in specific Group H areas used for the storage of hazardous materials. The alarm system is intended to notify the other occupants of the building that there is an emergency situation involving hazardous materials. For the same reason, an alarm or communication method must also be provided where corridors, exit passageways, or interior exit stairways are used for the transportation of highly hazardous materials.

Other conditions call for the installation of a gas-detection system designed to detect the presence of unacceptable levels of hazardous gas and perform additional safety functions. Areas identified in this section as requiring gas-detection systems include HPM facilities, facilities involving the storage and use of toxic and highly toxic gases, ozone gas-generator rooms, and repair garages for nonodorized-gas-fueled vehicles. The activation of a local alarm, audibly distinct from other alarms, is used to provide a warning both inside and outside the area where the gas is detected. In addition, the detection system must initiate the automatic shutdown of the gas supply to the area where the leak is detected.

Section 909 *Smoke-Control Systems*

The provisions of this section are applicable to the design, construction, testing, and operation of mechanical or passive smoke-control systems only when they are required by other provisions of the IBC. Section 909 specifically exempts smoke- and heat-venting requirements that appear in Section 910 and are discussed in the next section of this handbook. Also, this section states that mechanical smoke-control systems are not required to comply with Chapter 5 of the *International Mechanical Code*® (IMC®) for exhaust systems unless their normal use would otherwise require compliance.

The provisions of this section establish minimum requirements for the design, installation, and acceptance testing of smoke-control systems, but nothing within the section itself is intended to imply that a smoke-control system is to be installed. Some sections that specifically reference Section 909 are the requirements for atriums (Section 404.5), underground buildings (Section 405.5), and windowless buildings housing Group I-3 occupancies (Section 408.9). Smoke-control systems are intended to provide a protective environment in areas outside that of fire origination to allow for the evacuation or relocation of occupants in a safe manner. The provisions are not designed to protect the contents from damage or assist in fire-fighting activities.

Much of this section is based on the American Society of Heating, Refrigerating, and Air-Conditioning Engineers publication, *Design of Smoke Control for Buildings*; NFPA publication 92-A, *Recommended Practice for Smoke Control Systems Utilizing Barriers and Pressure Differences*; and the companion publication 92-B, *Technical Guide for Smoke Control Systems in Malls, Atriums, and Large Areas*.

Although this section covers both passive and active smoke-control systems, the majority of the material presented addresses the three mechanical methods—pressurization (Section 909.6), airflow (Section 909.7), and exhaust (Section 909.8)—with other sections addressing related subjects such as the design fire; equipment, including fans, ducts, and dampers; power supply; detection and control systems; and the fire fighter's smoke-control panel.

An important segment of Section 909 addresses acceptance testing of the smoke-control system. Smoke-control system installation requires special inspection per Section 909.18.8 to be performed during erection of ductwork and prior to concealment. These inspections are intended for the purpose of testing for leaks, as well as for recording the specific device locations. The latter creates, in effect, as-built drawings for the system. Additional testing and verification prior to occupancy is also mandated. Section 909.18.8.3 requires the work of the special inspector to be documented in a final report. The report shall be reviewed by the responsible registered design professional who is required to certify the work. The final, designer-approved report, together with other information addressed in Section 909.18.9, shall be provided to the fire code official, and a copy shall be maintained on file at the building.

909.20 Smokeproof enclosures. The provisions of this section identify the methods for complying with Section 1022.10 for the construction of smokeproof enclosures. Smokeproof enclosures are required by Sections 403 and 405 for high-rise buildings and underground buildings, respectively. There are two methods for construction of a ventilated smokeproof enclosure, both of which use an enclosed interior exit stairway. Either an exterior balcony or a ventilated vestibule can be used as the buffer between the floor of the building and the exit stairs. In addition, pressurization of the stair shaft is a permitted alternative.

Unless the pressurization provisions of Section 909.20.5 are used where a smokeproof enclosure is required, the exit path to the stair shall include a vestibule or an open exterior balcony. The minimum size of the vestibule is illustrated in Figure 909-1. A minimum 2-hour fire-resistance-rated fire barrier separates the smokeproof enclosure from the remainder of the building and also separates the stairway from the vestibule. The only openings permitted into the enclosure are the required means of egress doors. Construction of an open exterior balcony is based on the required fire-resistance rating for the building's floor construction, which would typically be 2 hours.

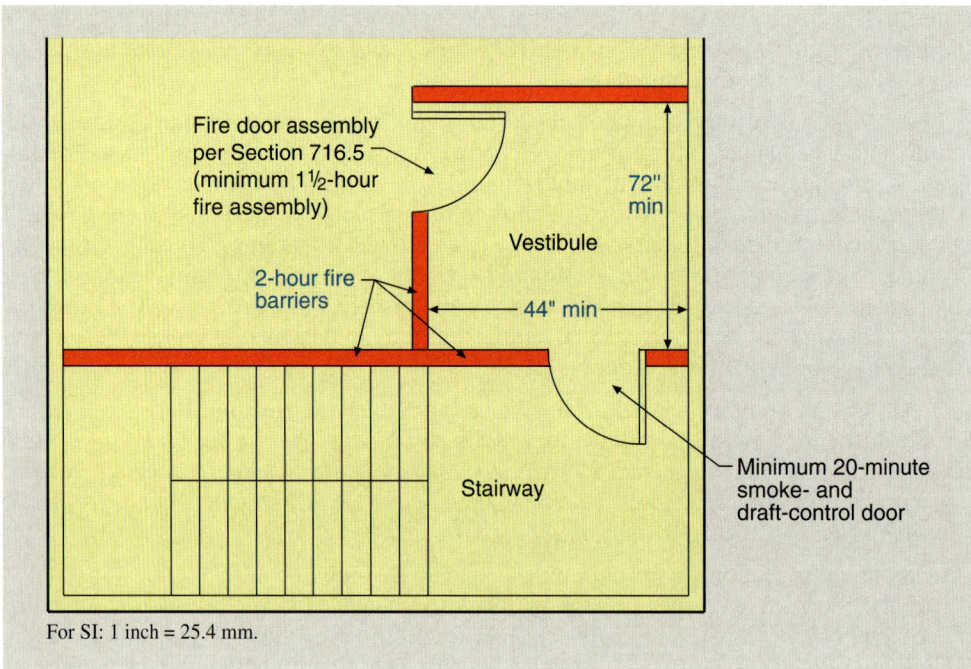

Figure 909-1 Ventilated vestibule.

Smoke-Control Systems

Natural ventilation alternative. In this section, the code provides the details of construction where natural ventilation is used to comply with the concept of a smokeproof enclosure. Where an open exterior balcony is provided, fire doors into the stairway shall comply with Section 716.5. In a vestibule scenario, a similar complying fire door is required between the floor and the vestibule. Between the vestibule and the stairway, the door assembly need only have a 20-minute fire-protection rating. The necessary vestibule ventilation is to be provided by an opening in the exterior wall at each vestibule. Facing an outer court, yard, or public way at least 20 feet (6,096 mm) in width, the exterior wall opening must provide at least 16 square feet (1.5 m²) of net open area. See Figure 909-2.

909.20.3

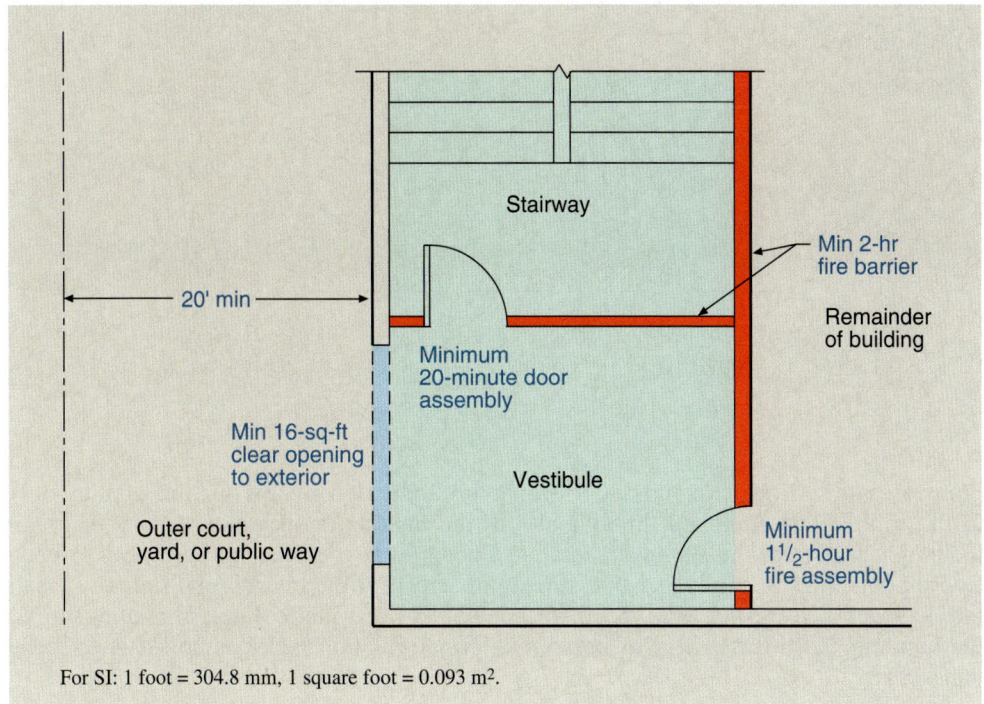

For SI: 1 foot = 304.8 mm, 1 square foot = 0.093 m².

Figure 909-2
Natural ventilation.

Mechanical ventilation alternative. Smokeproof enclosures may also be ventilated by mechanical means. As for naturally ventilated vestibules, a minimum 1^1/$_2$-hour fire door as mandated by Table 716.5 is required between the building and the vestibule, whereas the door between the vestibule and the stairway may have a 20-minute fire-protection rating. The minimum 1^1/$_2$-hour fire door assembly must also meet the criteria of Section 716.5.3 in order to minimize air leakage between the building and the vestibule.

909.20.4

Individual tightly constructed ducts are used to supply and exhaust air from the vestibule. Air is supplied near the floor level of the vestibule and exhausted near the top. The locations of the supply and exhaust registers are illustrated in Figure 909-3, as is the location for the smoke trap. It is important that doors in the open position do not obstruct the duct openings. The code also allows the use of a performance-based engineered vestibule ventilation system per Section 909.20.4.2.1. In addition to ventilation of the vestibule, air shall be provided and relieved from the stair shaft as well. By supplying an adequate amount of air while providing a dampered relief opening, a minimum positive pressure of 0.10 inch (29 Pa) of water column shall be maintained in the shaft relative to the vestibule with all doors closed.

2012 International Building Code Handbook

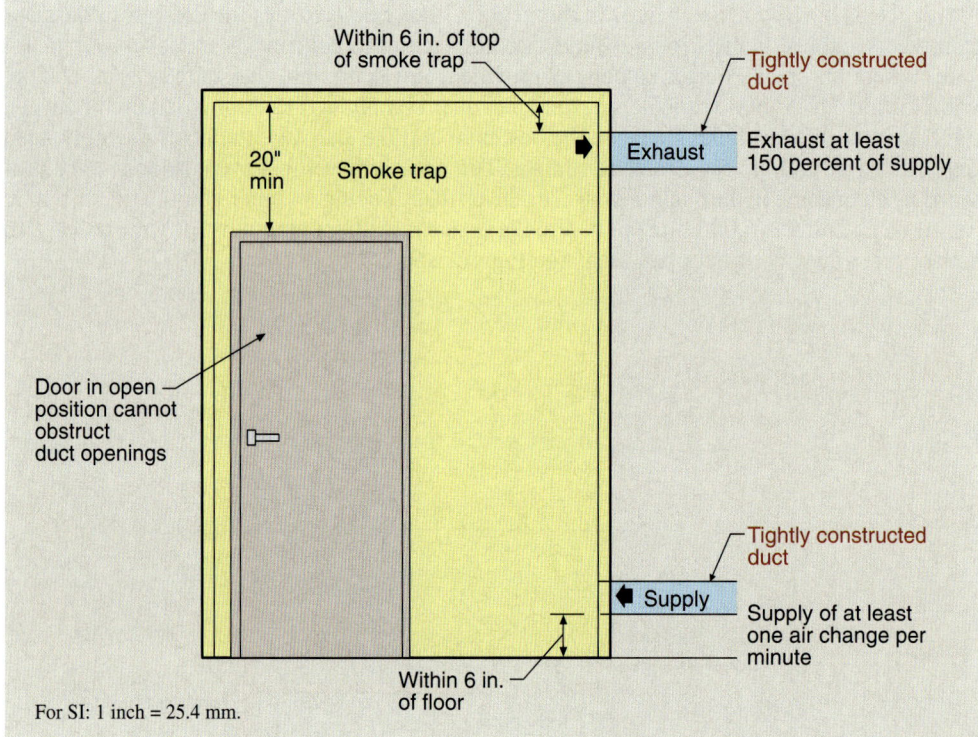

Figure 909-3
Vestibule ventilation.

909.20.5 Stair pressurization alternative. The method addressed in this section applies only to those buildings equipped throughout with an automatic sprinkler system. Through the pressurization of the stair shaft to a prescribed level, the need for vestibules or open exterior balconies is eliminated. Pressurization levels for the interior exit stairways shall fall between 0.10 inch (25 Pa) and 0.35 inch (87 Pa) of water column in relationship to the building. With this extent of pressurization, ventilation methods are deemed to be unnecessary.

Section 910 Smoke and Heat Vents

Smoke and hot gases created by a fire rise to the underside of the roof structure above and then build up so as to cause reduced visibility to the point where fire fighting is relatively ineffective. Also, as the hot gases accumulate near the roof structure, the unburned products of combustion become superheated, and if a supply of air is introduced, these hot, unburned products of combustion will ignite violently. Thus, it has been found that it is imperative that industrial and warehouse-type occupancies be provided with smoke and heat vents in the roofs. Although this section is typically applied to one-story buildings, as well as one-story portions of multistory buildings, the use of smoke and heat vents must also be considered for the top story of multistory structures. In addition, there may be conditions where the space requiring smoke and heat vents is not located at the upper story of the building. In such cases, a mechanical smoke exhaust system can be provided in lieu of the required smoke and heat vent approach.

Smoke and heat vents must typically be installed in conjunction with draft curtains, which are sometimes referred to as curtain boards. Draft curtains installed at code-specified intervals confine the smoke and hot gases so they are not diluted and, thus, increase the effectiveness of automatic vents. Without the confinement of the smoke and hot gases by draft curtains, the vents would be relatively ineffective because of delay in operating, or even because of nonoperation.

In the case of buildings with automatic fire-sprinkler systems, the draft curtains confine the smoke and hot gases so that when sprinklers are actuated they are in a relatively confined area. Without draft curtains, the hot gases would spread laterally and possibly activate so many sprinklers that the sprinkler system may be overcome on account of excessive water-flow demand. An exception to this general requirement permits the elimination of draft curtains within the area where early suppression, fast response (ESFR) sprinklers are installed. However, the application of this provision is not necessary as the installation of ESFR sprinklers eliminates any initial requirement for smoke and heat vents.

Where required. The IBC requires smoke and heat venting in specified industrial buildings and warehouses and in any occupancy, as required by the IFC, where high-piled combustible stock or rack storage is provided. The intent is that those occupancies that have a potential to include large areas of combustible materials be provided with the means to rid the building of hot gases and smoke during a fire. 　　910.2

For those buildings containing high-piled combustible storage, the requirements for smoke and heat venting are addressed in IBC Section 413 (Combustible Storage) and in the IFC.

It is not uncommon for such large areas to be protected by ESFR sprinklers. Such sprinklers are designed to extinguish a fire, rather than control a fire, through the quick application of large amounts of water. Where ESFR sprinklers are installed, smoke and heat vents are not required.

Vent operation. The code requires that smoke and heat vents be approved, labeled, and capable of manual and automatic operation. Required to be located in the roof, smoke and heat vents shall operate automatically in order to release the smoke and hot gases. Where the building is provided with an automatic sprinkler system, it is important to overall fire operations that the smoke and heat vents not operate automatically prior to activation of the sprinkler. 　　910.3.2

Vent dimensions. Aerodynamic studies of flow through rectangular openings have shown that the minimum cross-sectional area for smoke and heat vents should be 16 square feet (1.5 m^2), with a minimum dimension of 4 feet (1,219 mm). Thus, the code has this same requirement. The code also permits projections such as ribs and rain gutters to project into the required 4 feet (1,219 mm), as long as the total width for all projections does not exceed 6 inches (152 mm). 　　910.3.3

Vent locations. As smoke and heat vents are intended to release smoke and hot gases from a fire within the building, the code requires that they be placed a minimum of 20 feet (6,096 mm) from fire walls and from any lot line in order to reduce exposure to adjacent property. Vents shall also be located at least 10 feet (3,048 mm) from fire barriers. Such conditions as roof pitch, curtain location, sprinkler head location, and structural members shall be considered in the location of vents. 　　910.3.4

Draft curtains. In order to be effective, draft curtains are required to be of the types of construction that prevent the passage of smoke through the draft curtain. Therefore, the code requires that they be constructed of sheet metal, lath and plaster, gypsum wallboard, or of another approved material that prevents the passage of smoke. Also, the joints and connections are required to be smoketight. Figure 910-1 illustrates the method for determining draft curtain depth and spacing using the requirements of Table 910.3. In those areas where ESFR sprinklers are provided, draft curtains need only be installed at the point separating the ESFR sprinklers and the conventional sprinklers. 　　910.3.5

9 Fire Protection Systems

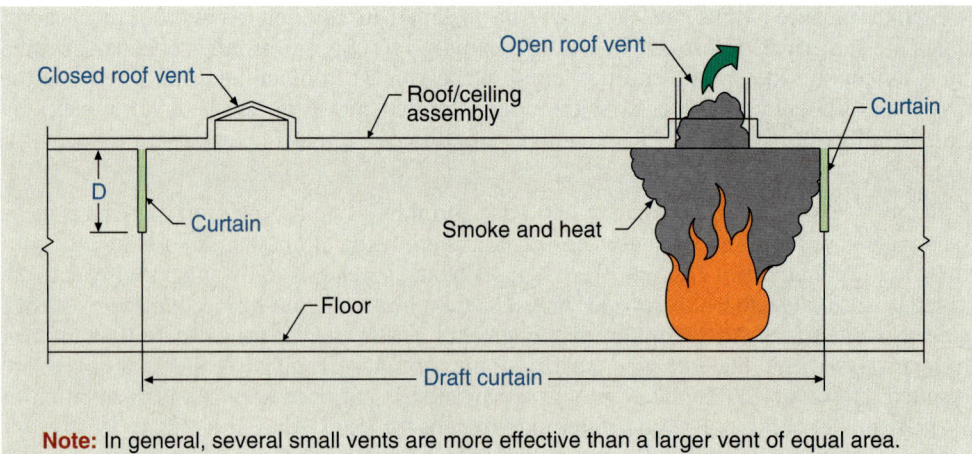

Figure 910-1
Roof vents and curtain boards.

910.4 Mechanical smoke exhaust. An engineered mechanical smoke-exhaust system may be accepted by the fire code official in lieu of smoke and heat vents. Fans located within each draft curtain area can be used for exhaust purposes. Each fan shall be manually controlled, as well as automatically activated, by either an automatic sprinkler system or heat detectors. Supply air for the exhaust system shall be provided uniformly around the perimeter of the area served. It is the intent of the code that the engineered system provide results equivalent to the smoke- and heat-vent method.

Section 911 *Fire Command Center*

The fire command center is the heart of the fire- and life-safety systems in a complex building. The IBC mandates a fire command center be provided in only very special structures such as high-rise buildings regulated by Section 403. The purpose of the command center is to provide a central location where fire personnel can operate during a fire incident or other emergency. Located as determined by the fire department, the fire command center shall be isolated from the remainder of the building by a minimum 1-hour fire-resistance-rated fire barrier. The code lists those system units, controls, display panels, indicators, devices, furnishings, and plans that are to be contained in the command center.

The provisions of Section 911 are only applicable where some other provision of the code specifically mandates that a fire command center be provided. As an example, Section 909.16 mandates that the fire fighter's smoke-control panel required for buildings provided with a smoke-control system must be located in a fire command center if the smoke-control system is used to address smoke-protected assembly seating conditions. For other smoke-control applications, a fire command center is not required to be provided.

Section 914 *Emergency Responder Safety Features*

Section 914 is intended to provide correlation to the current requirements in the IFC for the identification of shaftway hazards and the location of fire protection systems. These requirements are located in Sections 316.2 and 509.1 of the IFC. Section 101.3 of the IBC states that the safety of emergency responders is part of its scope and intent. This new section reinforces the intent by specifying that interior and exterior shaftway hazards be identified as well as the location of fire protection systems, such as fire alarm control units or automatic sprinkler risers.

KEY POINTS

- Automatic sprinkler systems are typically installed because they are mandated by the code, or because they are to be used as equivalent protection to other code requirements.
- Because of the potentially high occupant load and density anticipated in Group A occupancies, coupled with the occupants' probable lack of familiarity with the means of egress system, large assembly uses must be protected by an automatic sprinkler system.
- Most school buildings must be sprinklered throughout unless complying compartmention is provided.
- Large manufacturing buildings and warehouses, when containing combustible goods or materials, must be sprinklered to limit the size of a fire.
- The IBC requires sprinkler protection for all Group H occupancies owing to local hazards within the building and the potential for presenting a high level of hazard to the surrounding properties.
- Because the mobility of the occupants of Group I occupancies is greatly diminished, the code requires automatic fire suppression.
- On account of their fire record, hotels, apartment buildings, assisted-living facilities, and all other residential occupancies must always be sprinklered.
- Adequate openings must be provided in exterior walls for fire department access, or a sprinkler system must be installed.
- Most buildings exceeding 55 feet (16,764 mm) in height are required to be equipped with an automatic sprinkler system throughout.
- Certain occupancies and uses are required to have standpipe protection.
- The locations of Class I standpipe connections are specifically identified in the code.
- The locations of some Class II standpipe hose cabinets are based on the distances that the fire hose can reach throughout the building.
- One of the most effective means of occupant protection in case of a fire incident is the availability of a fire alarm system.
- Pressurization, airflow, and exhaust are the three methods of mechanical smoke control.
- A ventilated smokeproof enclosure uses either an exterior balcony or a ventilated vestibule, whereas pressurization of the stair shaft is a permitted alternative.
- Smoke and heat venting is required in large, open areas of manufacturing, warehouse, and hazardous occupancies, as well as retail sales with high-piled stock.
- Curtain boards are used to divide the area below the roof into zones for smoke and heat venting.
- Special provisions intended to increase the efficiency and safety of fire-department personnel during emergency operations are addressed, including emergency responder safety features, the fire department command center, fire department connections, and fire-pump rooms.

CHAPTER 10

MEANS OF EGRESS

Section 1001 Administration
Section 1002 Definitions
Section 1003 General Means of Egress
Section 1004 Occupant Load
Section 1005 Means of Egress Sizing
Section 1006 Means of Egress Illumination
Section 1007 Accessible Means of Egress
Section 1008 Doors, Gates, and Turnstiles
Section 1009 Stairways
Section 1010 Ramps
Section 1011 Exit Signs
Section 1012 Handrails
Section 1013 Guards
Section 1014 Exit Access
Section 1015 Exit and Exit Access Doorways
Section 1016 Exit Access Travel Distance
Section 1017 Aisles
Section 1018 Corridors
Section 1019 Egress Balconies
Section 1020 Exits
Section 1021 Number of Exits and Exit Configuration
Section 1022 Interior Exit Stairways and Ramps
Section 1023 Exit Passageways
Section 1024 Luminous Egress Path Markings
Section 1025 Horizontal Exits
Section 1026 Exterior Exit Stairways and Ramps
Section 1027 Exit Discharge
Section 1028 Assembly
Section 1029 Emergency Escape and Rescue
Key Points

10 Means of Egress

This chapter establishes the basic approach to determining a safe exiting system for all occupancies. It addresses all portions of the egress system and includes design requirements as well as provisions regulating individual components, which may be used within the egress system. The chapter specifies the methods of calculating the occupant load that are used as the basis of designing the system and, thereafter, discusses the appropriate criteria for the number of exits, location of exits, width or capacity of the egress system, and the arrangement of the system. This arrangement is treated in terms of remoteness and accessibility of the egress system. The accessibility is handled both in terms of the system's usability by building occupants and in terms of it being available within a certain maximum distance of travel. After having dealt with general issues that affect the overall system or multiple zones of the system defined as the exit access, exit, and exit discharge, the chapter then establishes the design requirements and components that may be used to meet those requirements for each of the three separate zones.

In interpreting and applying the various provisions of this chapter, it would help to understand the four fundamental concepts on which safe exiting from buildings is based:

1. A safe egress system for all building occupants must be provided.

2. Throughout the system, every component and element that building occupants will encounter in seeking egress from the building must be under the control of the person wishing to exit.

3. Once a building occupant reaches a certain degree or level of safety, as that occupant proceeds through the exiting system, that level of safety is not thereafter reduced until the occupant has arrived at the exit discharge, public way, or eventual safe place.

4. Once the exit system is subject to a certain maximum demand in terms of the number of persons, that system must thereafter (throughout the remainder of the system) be capable of accommodating that maximum number of persons.

Egress for individuals with physical disabilities is to be provided under the provisions of this chapter, primarily through the design of an accessible means of egress system. Because many of the elements composing the egress system (doors, landings, ramps, etc.) may also form part of the accessible routes as required by Chapter 11, such requirements must be referenced where applicable.

> This chapter includes the three-part approach to the "means of egress." The three-part system, or zonal approach as it is now used, was introduced by the National Fire Protection Association (NFPA) in 1956 and was incorporated over the years into all of the legacy model codes. This approach has established terms that are used throughout the design and enforcement communities to deal with the means of egress system. The three parts of the means of egress system are the exit access, the exit, and the exit discharge. For conceptual ease, the exit access is generally considered any location within the building from where you would start your egress travel, and continues until you reach the door of an exit. The exit access would include all the rooms or spaces that you would pass through on your way to the exit. This may be the room you are in; an intervening room; a corridor; an exterior egress balcony; and any doors, ramps, unenclosed stairs, or aisles that you use along that path. An exit is the point where the code considers that you have obtained an adequate level of safety so that travel distance measurements are no longer a concern. Exits will generally consist of fire-resistance-rated construction and opening protection that will separate the occupants from any problem within the building. Elements that are considered exits include exterior exit doors at ground level, interior exit stairways, interior exit ramps, exit passageways, exterior exit stairways, exterior exit ramps, and horizontal exits. Exterior exit doors, exterior exit stairways, and exterior exit ramps will not provide the fire-protection levels that the

> other elements provide, but insofar as the occupant will be outside the building, they will provide a level of safety by removing the occupants from the problem area. The last of the three parts is the exit discharge. The *International Building Code*® (IBC®) will generally view exterior areas at ground level as the exit discharge portion of the exit system. Therefore, the exit access will be the area within the building that gets the occupants to an exit, whereas the exit discharge will be the exterior areas at grade where the occupants go upon leaving the building in order to reach the public way.

Section 1001 *Administration*

This section requires that every building or portion thereof comply with provisions of Chapter 10. In dealing with portions of buildings, it is important to understand that the code intends this chapter to apply to all portions that are occupiable by people at any time. Therefore, areas such as storage rooms and equipment rooms, although often unoccupied, will still be regulated under the provisions of the chapter.

In order to provide an approved means of egress at all times, it is critical that the exiting system be maintained appropriately. Section 1030 of the *International Fire Code*® (IFC®) regulates maintenance of the means of egress for the life span of the building. Should there be alterations or modifications to any portion of the building, Section 1001.2 mandates that the number of existing exits not be reduced, nor the capacity of the means of egress be decreased, below that level required by the IBC. Section 1104 of the IFC also provides a limited number of specific provisions addressing the means of egress in existing buildings.

A reference to the IFC provisions addressing fire safety and evacuation plans recognizes the need to provide consistent and effective fire- and life-safety operations during emergency conditions. Section 404 of the IFC requires the fire-safety and evacuation plans in certain A, B, E, F, H, I, R, and M occupancies and in high-rise or underground buildings as well as in specific covered mall buildings and buildings with an atrium. These plans are required to include or address a number of different types of issues that may affect the egress of occupants from the building. Along with other items, these include the identification of potential hazards, exits, primary and secondary egress routes, and occupant assembly points, as well as establishing procedures for assisted rescue for people who are unable to use the general means of egress unassisted.

Section 1002 *Definitions*

Section 1002 lists a number of defined terms that are especially important to this chapter. As with most other terms defined in the code, the terms identified in Section 1002.1 are specific to the IBC and typically differ from their ordinarily accepted meanings. Further information on many of these means of egress terms is found in Chapter 2 as part of the discussion of those specific terms defined in the IBC.

Section 1003 *General Means of Egress*

The requirements and topics addressed in this section are used as basic provisions and are to be applied throughout the entire egress path as applicable. Examples of the types of general issues that are found here include ceiling height, protruding objects, floor surface, elevation change, and egress continuity. For consistency purposes, the provisions for ceiling height and protruding objects are identical to the accessibility criteria of ICC A117.1.

1003.2 Ceiling height. In order to provide an exit path that maintains a reasonable amount of headroom clearance for the occupants, this section requires the means of egress to have a minimum ceiling height of 7 feet 6 inches (2,286 mm). The intent of the provision is to address all potential paths of exit travel that can be created based on multiple directions of egress and the layout of the room or space insofar as furniture, equipment, and fixtures are concerned. Any portion of the floor area of the building that can reasonably be considered a possible exit path should be provided with a minimum 7-foot 6-inch (2,286-mm) clear height, unless reduced by exceptions permitted for sloped ceilings, dwelling and sleeping units in residential occupancies, stairway and ramp headroom, door height, and protruding objects. Additional exceptions reduce the minimum required clear height to 7 feet 0 inches (2,134 mm) in parking garage vehicular and pedestrian traffic areas as well as above and below floors considered as mezzanines.

1003.3 Protruding objects. Limitations are placed on the permitted projection of protruding objects for two purposes. First, to maintain an egress path that is essentially free of obstacles. Second, to provide a circulation path that is usable by all occupants, including those individuals with sight-related disabilities. For this reason, provisions regulate the accessibility concerns regarding protruding objects as well as the egress concerns. Note that projections into the required egress width and the minimum clear width of accessible routes are also limited by other provisions of the code.

1003.3.1 Headroom. Consistent with the allowance for stair headroom and doorway height to be reduced below the required egress height of 7 feet 6 inches (2,286 mm), other portions of the egress system may likewise be reduced to a minimum height of 80 inches (2,032 mm). The reduction for signage, sprinklers, decorative features, structural members, and other protruding objects is limited to 50 percent of the ceiling area of the egress path. See Figure 1003-1. Though projections at an 80-inch (2,032-mm) height are not unusual to building occupants, it is necessary to maintain a majority of the egress system at 7 feet 6 inches (2,286 mm) or higher. Passage through a doorway may be further reduced in height to 78 inches (1,981 mm) at the door closer or stop. This reduction at doors is also permitted for accessibility purposes by Section 307.4 of ICC A117.1. Where a vertical clearance of 80 inches (2,032 mm) cannot be achieved, the reduced-height portion of such floor area cannot be used as a portion of the

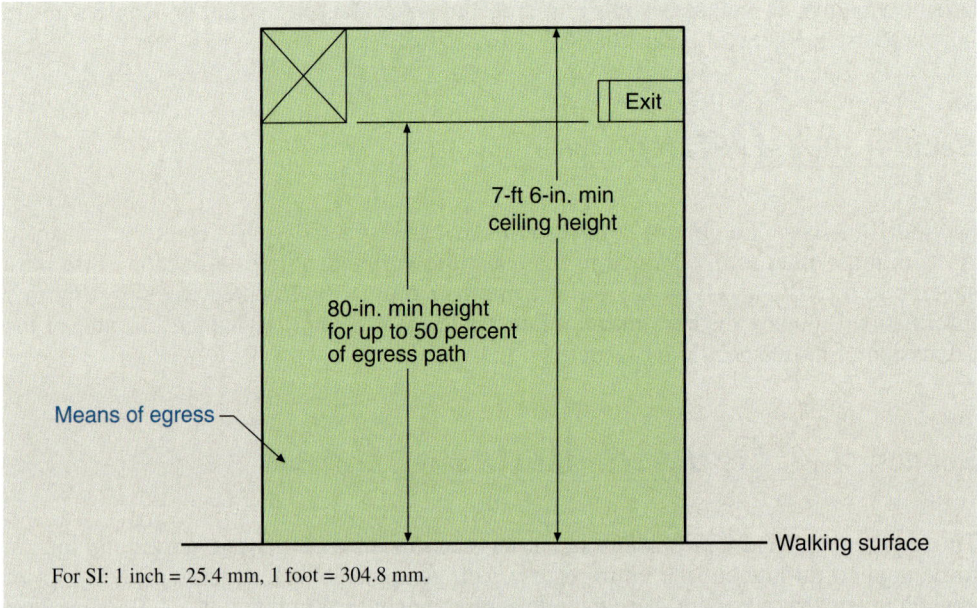

Figure 1003-1 Means of egress headroom.

10 General Means of Egress

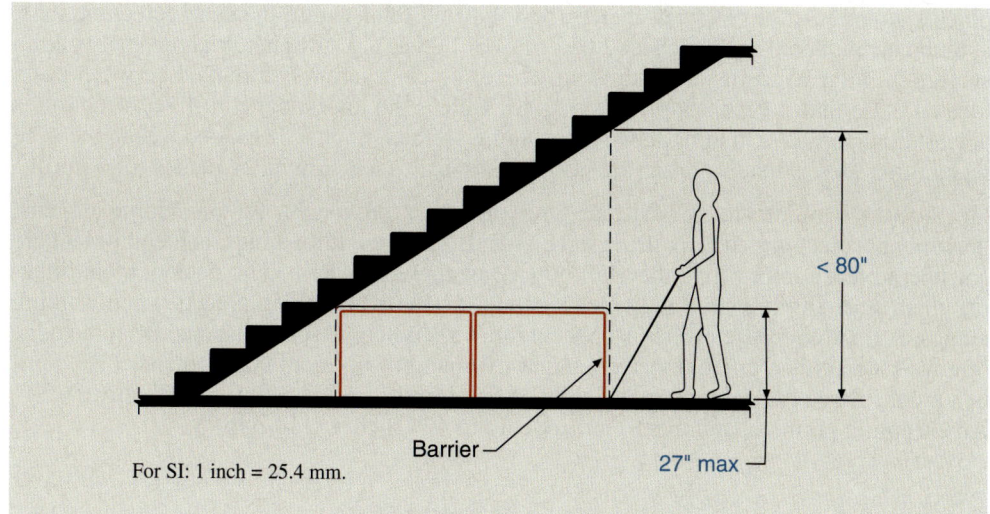

Figure 1003-2
Reduced vertical clearance.

means of egress system. It is also necessary to provide some type of barrier that will prohibit the occupant from approaching the area of reduced height. This is of particular importance where the occupant is sight impaired, with no method other than a barrier to identify the presence of an overhead protruding object. The mandated barrier is to be installed so that the leading edge is no more than 27 inches (686 mm) above the walking surface, as shown in Figure 1003-2. By limiting the height of the barrier edge, it will be located in a manner so that a sight-impaired individual using a long cane will detect the presence of an obstruction and maneuver to avoid the hazard.

Post-mounted objects. Free-standing objects mounted on a post or pylon that are located along or adjacent to the walking surface are potential hazards, particularly to a sight-impaired individual. Objects such as signs, directories, or telephones that are mounted on posts or pylons are, therefore, limited to an overhang of 4 inches (102 mm) maximum if located more than 27 inches (686 mm), but not more than 80 inches (2,032 mm), above the floor level. By limiting the overhang to 4 inches (102 mm), a cane will hit the post or pylon prior to the individual impacting the mounted object. See Figure 1003-3. Free-standing

1003.3.2

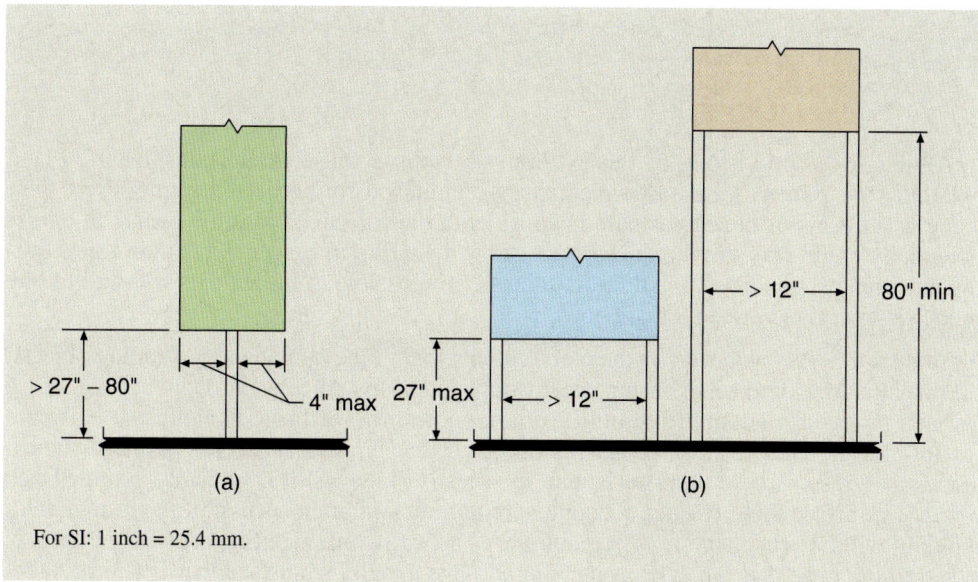

Figure 1003-3
Post-mounted protruding objects.

objects mounted at or below 27 inches (686 mm) will fall within the cane-detection zone, and objects mounted at 80 inches (2,032 mm) or higher are sufficiently above the walking surface. Similar concerns are addressed where the obstruction is mounted between posts located more than 12 inches (305 mm) apart. Unless the lowest edge of the obstruction is at least 80 inches (2,030 mm) above the walking surface, it must be located within the cane recognition area extending from the walking surface to a height of 27 inches (686 mm).

1003.3.3 Horizontal projections. Consistent with the other provisions for protruding objects, horizontal projections such as structural elements, fixtures, furnishings, and equipment are considered hazardous where they fall outside of the area where cane detection can identify them. Visually impaired individuals cannot detect overhanging objects when walking alongside them. Because proper cane techniques keep people some distance from the edge of a walking surface or from walls, a slight overhang of no more than 4 inches (102 mm) is not considered hazardous. An example of this provision is illustrated in Figure 1003-4. An exception permits handrails to protrude up to $4^1/_2$ inches (114 mm).

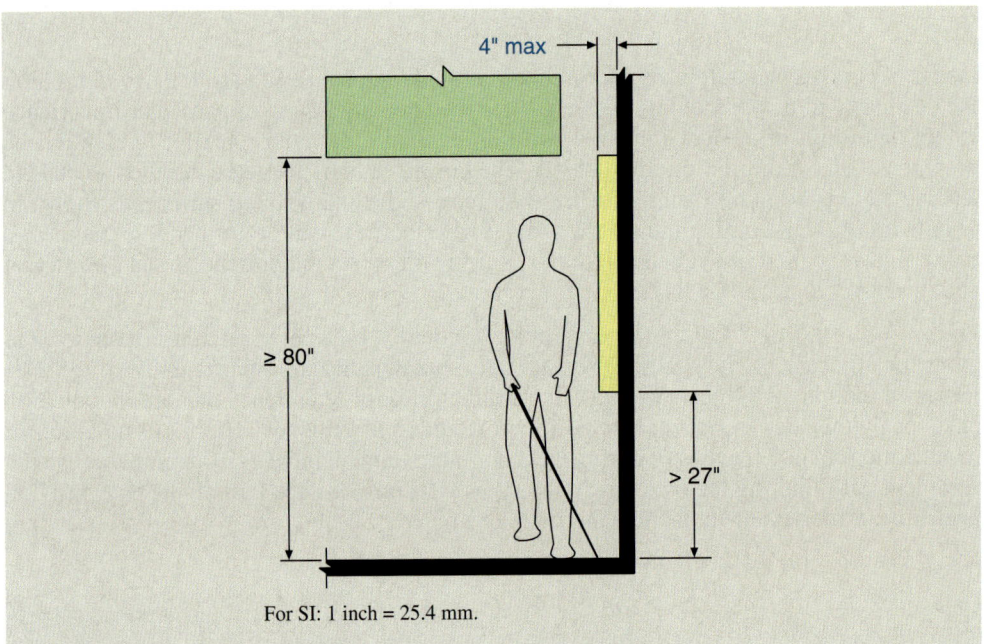

Figure 1003-4
Limits of protruding objects.

Although the provisions of this section, as well as those in Sections 1003.3.1 and 1003.3.2, are primarily based on clearances established for accessibility purposes, their value to all users of the egress path is considerable. Projections into the means of egress potentially could result in a reduced travel flow, resulting in longer evacuation times during emergency conditions. In addition, injuries are possible to individuals who fail to pay proper attention to where they are going.

1003.4 Floor surface. As evidenced by the requirements for ceiling height and protruding objects, the potential for exit travel to be impeded by obstructions is addressed throughout Chapter 10. Various provisions attempt to eliminate the opportunity for hazards along the exit path to slow travel. This section recognizes one area that is often taken for granted when it comes to egress—the walking surface of the means of egress. It is typically assumed that a walking surface that provides adequate circulation, and often accessibility, throughout a building will be acceptable for egress purposes as well. Although this is usually true, it is stated in the code that the egress path should have a walking surface that is slip resistant and securely attached so there is no tripping or slipping hazard that would result in an obstruction of the exiting process. Although the regulation of floor surfacing materials is typically recognized for interior walking surfaces, the provision is also applicable to exterior egress paths. The performance criteria of this code section provide a basis for determining the appropriateness of any questionable exit discharge elements.

Elevation change. The code is concerned that along the means of egress there is no change in elevation along the path of exit travel that is not readily apparent to persons seeking to exit under emergency conditions. Therefore, along the means of egress, any change in elevation of less than 12 inches (305 mm) must be accomplished by means of a ramp or other sloping surface. A single riser or a pair of risers is not permitted. See Figure 1003-5. Steps used to achieve minor differences in elevation frequently go unnoticed and as a consequence can cause missteps or accidents.

1003.5

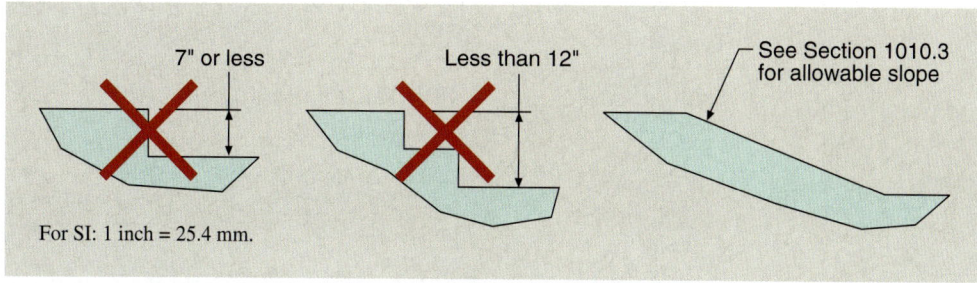

**Figure 1003-5
Longitudinal section through corridor or other exit path.**

This limitation on the method for a change of elevation, however, does not apply in certain locations. Where exterior doors are not required to be accessible by Chapter 11, a single step of 7 inches (179 mm) or less in height is permitted by Exception 1 at such exterior doors in Groups F, H, R-2, R-3, S, and U. See Figure 1003-6. A second exception allows, under specific conditions, a stair with a single riser or with two risers and a tread at those locations not required to be accessible by Chapter 11. In this case, the risers and

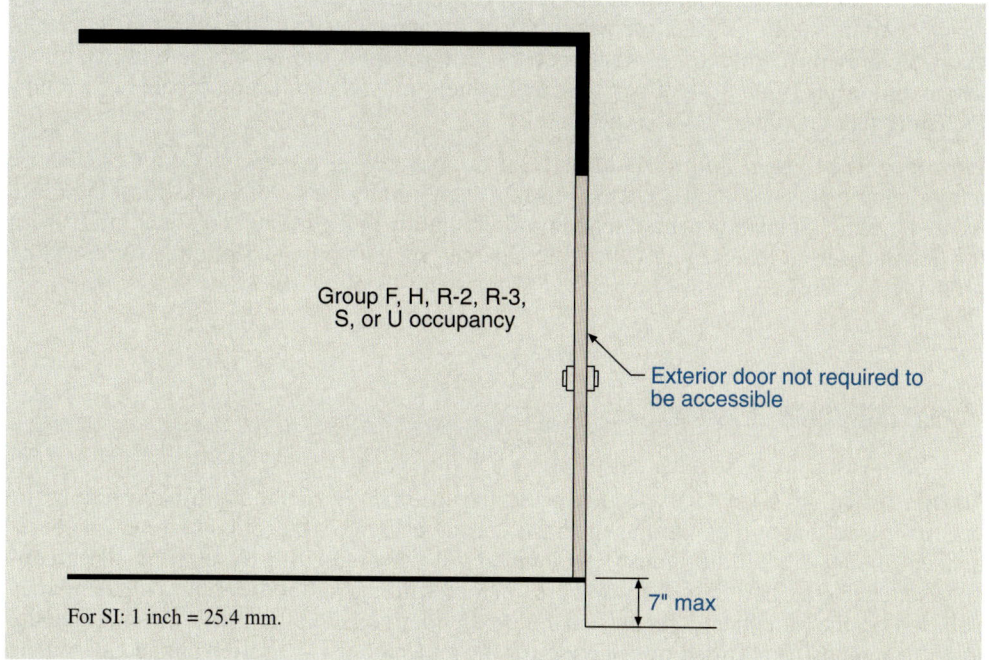

**Figure 1003-6
Single step at exterior door.**

treads must comply with Section 1009.7, the tread depth must be at least 13 inches (330 mm), and a minimum of one complying handrail must be provided within 30 inches (762 mm) of the centerline of the normal path of egress travel on the stair. See Figure 1003-7. A third exception applies to seating areas not required to be accessible. Risers and treads may be used on an aisle serving the seating where a complying handrail is provided.

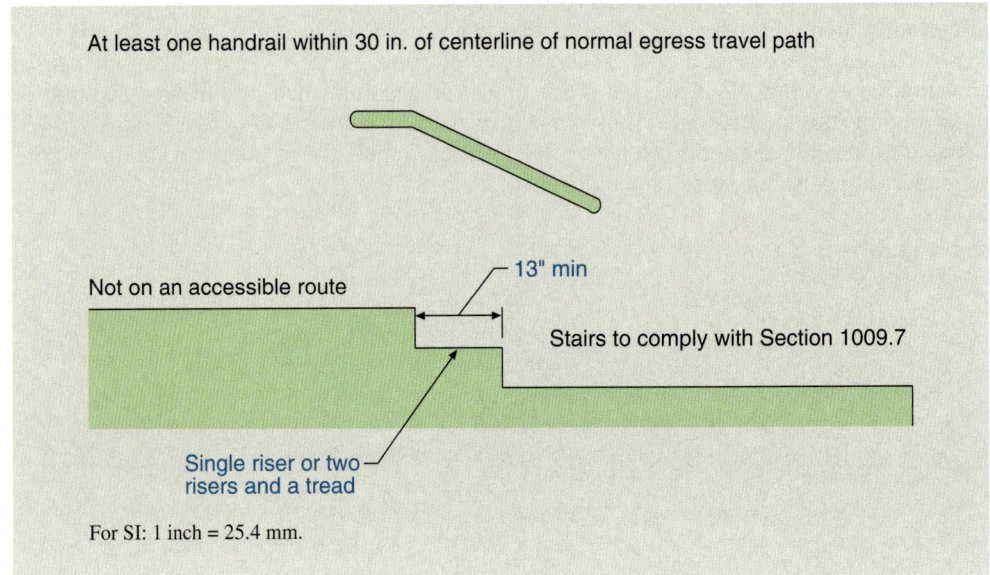

Figure 1003-7
Elevation change.

1003.6 Means of egress continuity. This section emphasizes that wherever the code imposes minimum widths on components in an exiting system, such widths are to be clear, usable, and unobstructed. Nothing may project into these required widths so as to reduce the usability of the full dimension, unless the code specifically and expressly states that a projection is permitted. Two notable examples of permitted projections are doors, either during the course of their swing or in the fully open position, and handrails. The limitations on the amount of such projections are specified within the appropriate sections of the chapter. Additionally, this section places into code language one of the four basic concepts that were previously discussed—that once the exit system is subject to a certain maximum demand in terms of number of persons, that system must thereafter be capable of accommodating that maximum number of persons.

1003.7 Elevators, escalators, and moving walks. For a variety of reasons, elevators, escalators, and moving walks are not to be used to satisfy any of the means of egress for a building. These building components are intended for circulation purposes and do not conform with the detailed egress requirements found in Chapter 10. The only exception is for elevators used as an accessible means of egress as addressed in Section 1007.4.

Section 1004 *Occupant Load*

1004.1 Design occupant load. This section prescribes a series of methods for determining the occupant load that will be used as the basis for the design of the egress system. The basic concept is that the building must be provided with a safe exiting system for all persons anticipated in the building. The process for determining an appropriate occupant load is based on the anticipated density of the area under consideration. Because the density factor is already established by the code for the expected use, variations in occupant load are simply a function of the floor area assigned to that use. It is apparent that in many situations the occupant load as calculated is conservative in nature. This is appropriate because of the extent that the means of egress provides for life-safety concerns. The egress system should be designed to accommodate the worst-case scenario, based on a reasonable assumption of the building's use.

Cumulative occupant loads. This provision mandates that the occupant loads are to be cumulative as the occupants egress through intervening spaces. Under the conditions of Section 1014.2, the path of travel through the intervening space must be discernable to allow for a continuous and obvious egress path. Egress travel is permitted to pass through complying adjoining rooms provided the design occupant load is increased to account for those potential occupants who are assigned to that specific egress path. See Figure 1004-1. Another common application occurs as users of the means of egress merge at aisles, corridors, or stairways, as shown in Figure 1004-2. Where alternative means of egress are provided, only the number

1004.1.1

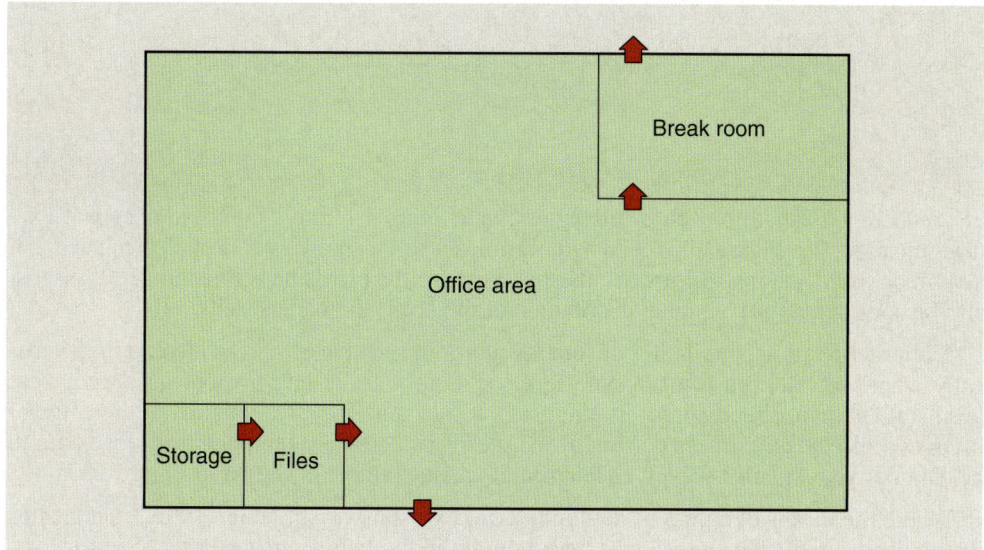

Figure 1004-1
Combination occupant loads.

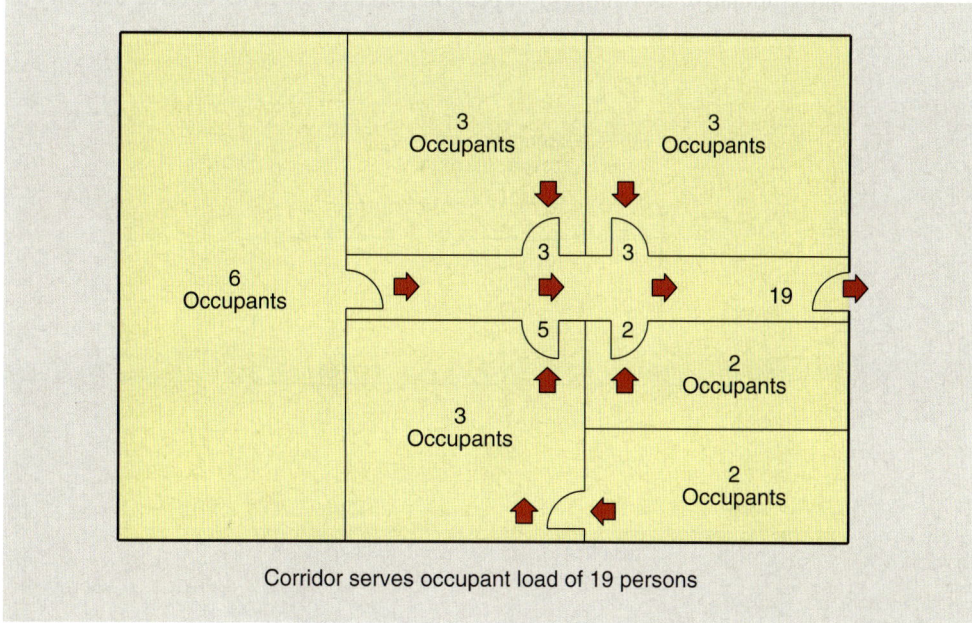

Figure 1004-2
Number of combination.

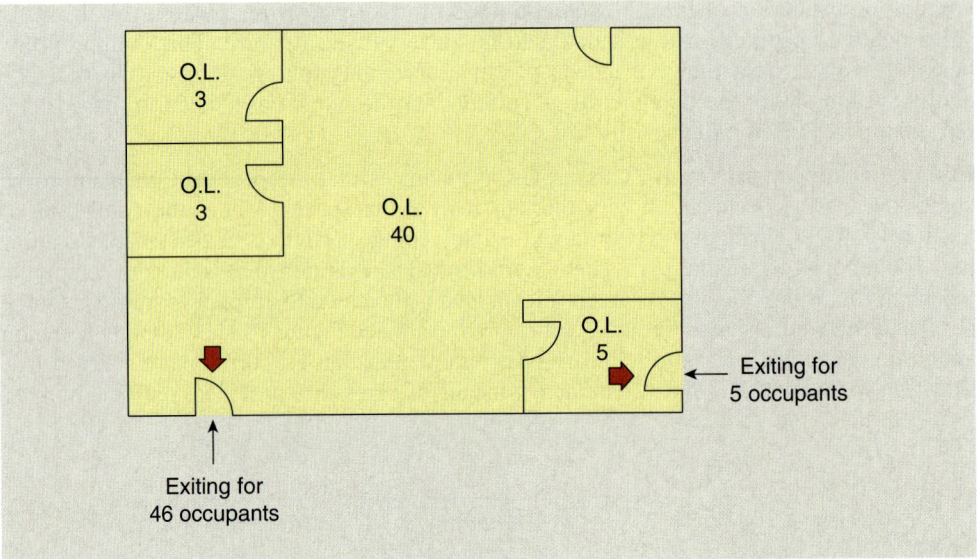

Figure 1004-3
Assignment of occupant load.

of occupants assigned to each of the egress paths is used in the cumulative occupant load determination. See Figure 1004-3. Where significant occupant loads are anticipated such that a room or space requires at least two means of egress, the provisions of Section 1005.5 must also be considered for the proper distribution of the occupant load capacity.

The practice of accumulating occupants along the means of egress also applies vertically where travel from a mezzanine level or a story leads into a room on an adjacent level, rather than directly into an enclosure for an interior exit stairway. Where travel occurs within the exit access portion of the means of egress system, occupant loads are to be cumulative vertically as well as horizontally. An example is shown in Figure 1004-4.

Where the means of egress occurs on interior exit stairways, Section 1005.3.1 indicates that the capacity of the exitways be based on the individual occupant loads of each story. In other words, the number of persons for which the capacity of the stairway is designed is not based on any cumulative total number of persons, but rather on the required capacity of

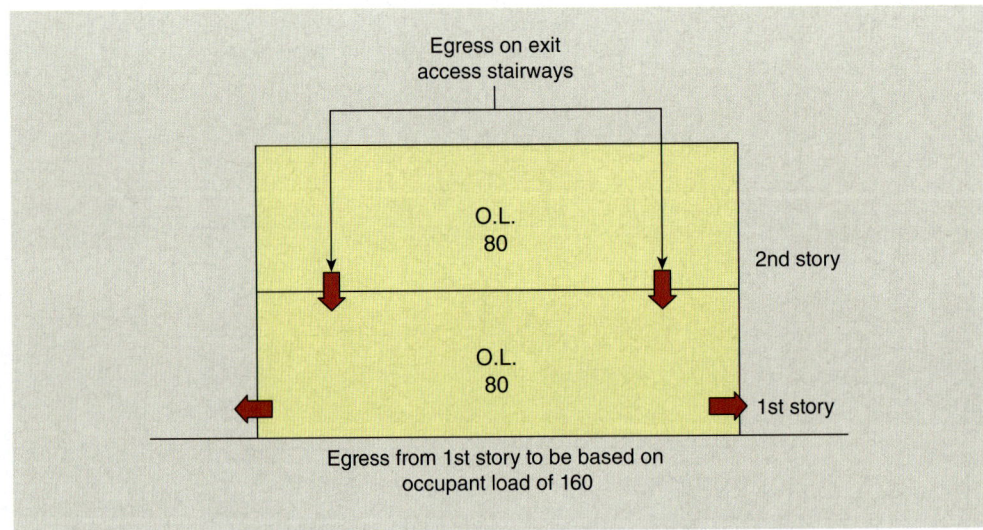

Figure 1004-4
Vertical accumulation of occupant load.

the exits at each particular floor. In no case, however, shall exit capacity decrease along the path of egress travel. A more in-depth discussion of this issue is found in Section 1005.1 and is illustrated in Figure 1005-2. Where exiting occurs from a mezzanine, the provisions of Section 1004.1.1.2 apply rather than those of Section 1005.3.1. Section 1004.1.1.2 is also applicable where exit access stairways serve adjacent stories or floor levels.

Areas without fixed seating. The vast majority of buildings contain uses that do not use fixed seating. Unlike auditoriums, theaters, and similar spaces, in most instances the maximum probable number of occupants may not be known. Therefore, the code provides a formula for determining an occupant load that constitutes the minimum number of persons for which the exiting system must be designed. As a consequence, the code refers to the number obtained by the formula as the design occupant load. Egress systems for all buildings or building spaces must be designed to accommodate at least this minimum number. Basic examples of the use of Table 1004.1.2 are illustrated in Figure 1004-5.

1004.1.2

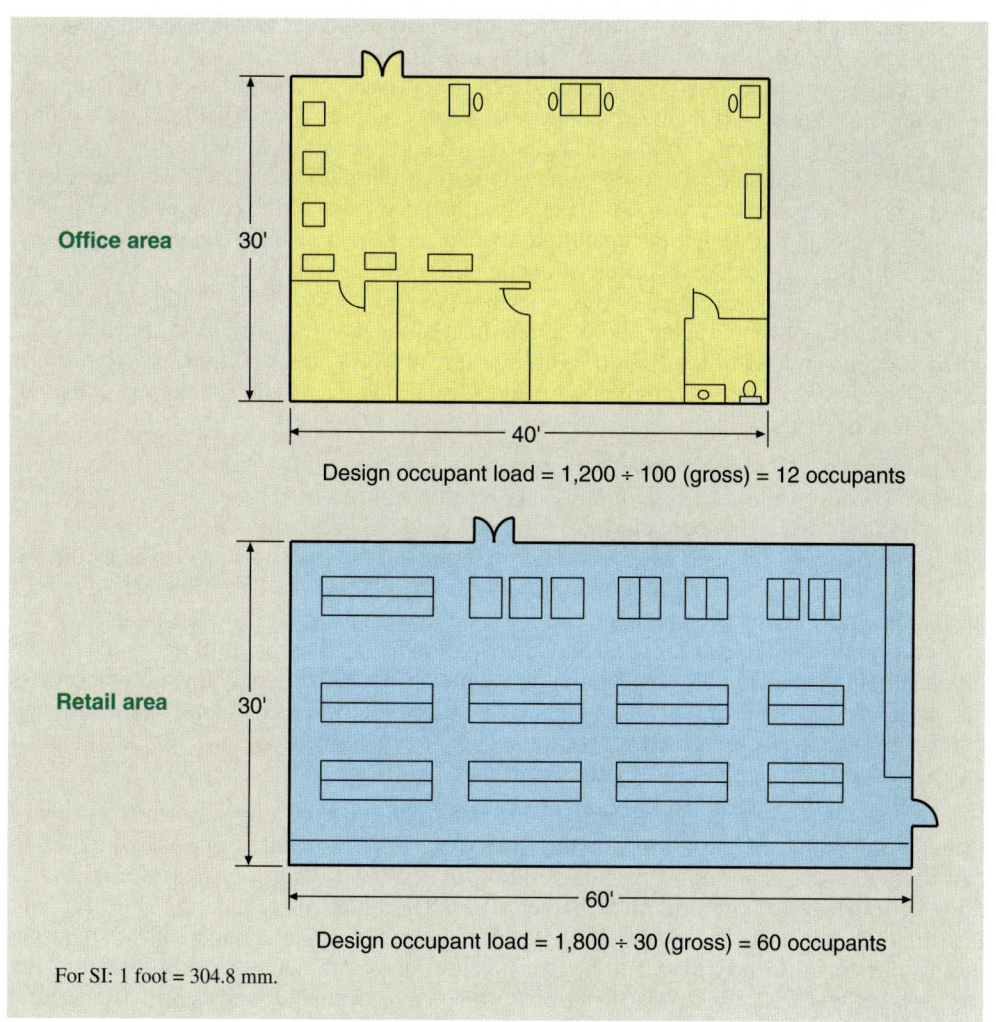

Figure 1004-5
Design occupant load.

As the person responsible for interpreting and enforcing the code, the building official will be called on to make decisions regarding the categories listed in Table 1004.1.2. Although Table 1004.1.2 contains occupant load factors that will serve the code user under most conditions, there will be occasions when either the table does not have an occupant load factor appropriate for the intended use or the occupant load factor contained in the table will not have a realistic application. In such instances, the building official has the authority to establish an appropriate occupant load factor or an appropriate occupant load for those special circumstances and those special buildings.

It may be meaningful to point out that the first column of Table 1004.1.2 is headed "Function of Space." The categories listed in the first column are not the specific groups identified in Chapter 3 for the purpose of assigning an occupancy classification but are the basic generic uses of building spaces. It has been pointed out in the discussion of the various occupancy groups that it is possible to have a classroom classified as a Group E occupancy, a Group B occupancy, or, possibly, a Group A occupancy. In terms of occupant density, however, a classroom is a classroom, and it is reasonable to expect the same density of use in a classroom regardless of the occupancy group in which that classroom might be classified. Therefore, the table specifies that when considering classroom use, one must assume there is at least one person present for each 20 square feet (1.86 m^2) of floor area.

In specifying how the occupant load is to be determined, the code intends that it is to be assumed that all portions of a building are fully occupied at the same time. It may be recognized, however, that in limited instances not all portions of the building are, in fact, fully occupied simultaneously. An example of this approach for support uses might include conference rooms in various occupancies or minor assembly areas such as lunch rooms in office buildings or break rooms in factories. It is important to note that the code does not provide for a method to address such conditions; thus, full occupancy should always be assumed. Only under rare and unusual circumstances should the building official ever consider reducing the design occupant load because of the nonsimultaneous use concept. In such situations, he or she must determine that there are support spaces that ordinarily are used only by persons who at other times occupy the main areas of the building; therefore, it is not necessary to accumulate the occupant load of the separate spaces when calculating the total occupant load of the floor or building. It is always necessary, however, to provide each individual space of the building with egress as if that individual space were fully and completely occupied.

Another type of support area that must be considered in occupant-load calculation includes corridors, closets, toilet rooms, and mechanical rooms. These uses are typical of most buildings and are to be included by definition in the gross floor area of the building. A quick review of Table 1004.1.2 will show that most of the uses listed are to be evaluated based on gross floor area, with no reduction for corridors and the like. However, a few of the listings indicate the use of the net floor area in the calculation of the occupant load. An example would be the determination of an occupant load in a school building. The building official should calculate the occupant load in such buildings using only the administrative, classroom, and assembly areas. It is generally assumed that when corridors, restrooms, and other miscellaneous spaces are occupied, they are occupied by the same people who are at other times occupying the primary use spaces.

The occupant load that can be expected in different buildings depends on two primary factors—the nature of the use of the building space and the amount of space devoted to that particular use. Different types of building uses have a variety of characteristics. Of primary importance is the density characteristic. Therefore, in calculating the occupant load of different uses, by means of the formula, the minimum number of persons that must be assumed to occupy a building or portion thereof is determined by dividing the area devoted to the use by that density characteristic or occupant load factor. The second column of Table 1004.1.2 prescribes the occupant load factor to be used with respective corresponding uses listed in the first column. The occupant load factor does not represent the amount of area that is required to be afforded each occupant. The IBC does not limit, except through the provisions of Section 1004.2, the maximum occupant load on an area basis. Rather, the occupant load factor is that unit of area for which there must be assumed to be at least one person present. For example, when the code prescribes an occupant load factor of 100 gross for business use, it is not saying that each person in an office must be provided with at least 100 square feet (9.29 m²) of working space. Rather, it is saying that, for egress purposes, at least one person must be assumed to be present for each 100 square feet (9.29 m²) of floor area in the business use. It is important to note that the floor area to be used in the application of Table 1004.1.2, both net and gross, is to include counters and showcases in retail stores, furniture in dwellings and offices, equipment in hospitals and factories, and similar furnishings. The floor areas occupied by furniture, equipment, and furnishings are taken into account in the occupant load factors listed in the table.

The numbers contained in the second column of Table 1004.1.2 represent those density factors that approximate the probable densities that can usually be expected in areas devoted to the respective functions listed. For this purpose, the occupant load factors are really a means of estimating the probable maximum density in the varying function areas. They have been developed over a period of years and, for the most part, have been found to consistently represent the occupant/furnishing densities that one might expect in building spaces devoted to the respective uses. See Application Examples 1004-1 and 1004-2 for two methods of occupant load determination.

Application Example 1004-1

GIVEN: A building assembly area and business areas as shown.
DETERMINE: The design occupant load of the building.
SOLUTION: The occupant load is simply 282, the combination of the assembly and business spaces. It is not necessary to consider the corridor, toilet rooms, and other small accessory spaces that serve the entire building. Note that within the office area itself, such circulation and accessory areas would be included in the calculation.

```
        Assembly
        O.L. = 240          Toilet
                            room
    ← ─── Corridor ─── →
        Offices             Mechanical
        O.L. = 42
```

OCCUPANT LOAD DETERMINATION

Application Example 1004-2

GIVEN: A 1,600-square-foot conference room in a hotel.
DETERMINE: The design occupant load of the room.
SOLUTION: Because a variety of assembly activities can occur within the room, the use creating the largest occupant load would be evaluated.
  (1) Conference/seminar use with tables and chairs
      1 person per 15 sq ft = 106.67 = 106 occupants
  (2) Conference/seminar use with chairs only (auditorium-style seating)
      1 person per 7 sq ft for seating = 228.57 = 228
  Therefore, a design occupant load of 228 shall be designated. Note that other potential uses of the room (dining, receptions, dances, etc.) would also utilize these factors.

OCCUPANT LOAD DETERMINATION

The exception to this section allows for a reduction in the calculated design occupant load on a very limited case-by-case basis. The building official is granted authority for the discretionary approval of lesser design occupant loads than those established by calculation. Although the provision allows the building official to be accommodating by recognizing the merits of the specific project, its use should be limited to very unique situations such as extremely large manufacturing or warehousing operations. See Application Example 1004-3. Where the exception is enacted in order to reduce the occupant load, the building official will typically impose specific conditions to help ensure compliance. It is critical that the reasoning for the occupant load reduction be justified and documented.

Application Example 1004-3

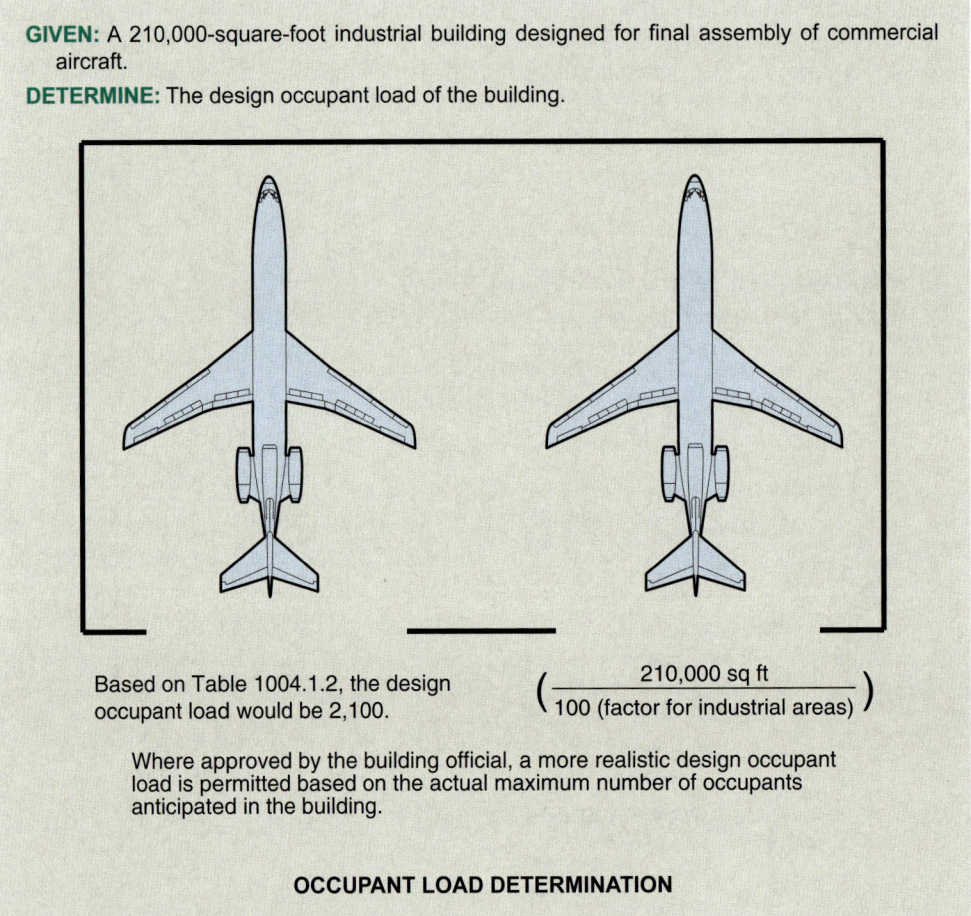

GIVEN: A 210,000-square-foot industrial building designed for final assembly of commercial aircraft.

DETERMINE: The design occupant load of the building.

Based on Table 1004.1.2, the design occupant load would be 2,100. $\left(\dfrac{210{,}000 \text{ sq ft}}{100 \text{ (factor for industrial areas)}}\right)$

Where approved by the building official, a more realistic design occupant load is permitted based on the actual maximum number of occupants anticipated in the building.

OCCUPANT LOAD DETERMINATION

1004.2 Increased occupant load. The provisions of Section 1004.1.2 specify the method to be used in determining the anticipated occupant load for areas without fixed seating. The occupant load determined by this method is the minimum number of persons for which the exiting system must be designed. The provisions do not, as previously pointed out, intend that the maximum permitted occupant load be regulated or controlled on a floor-area basis other than in the manner described by this section.

The provisions of this section specify how the maximum permitted occupant load in a building or portion of the building is to be determined. Here, the approach is taken that the occupant load determined as previously provided may be increased where the entire egress system is adequate, in all of its parts, to accommodate the increased number. In no case, however, shall the occupant load be established using an occupant load factor of less than 7 square feet (0.65 m^2) of floor space per person. See Application Example 1004-4.

Occupant Load 10

> **Application Example 1004-4**
>
> **GIVEN:** A restaurant where the occupant load of the dining area is calculated at 135, based on Table 1004.1.2 (2,025 sq ft/15). The restaurant's owner would like to establish a higher occupant load.
>
> **DETERMINE:** The maximum permitted occupant load of the dining area.
>
> **SOLUTION:** The absolute maximum occupant load per Section 1004.2 appears to be 289 (2,025/7). However, it is obviously impossible for such an occupant load to safely occupy the space, even if adequate exit doors were provided. If tables and chairs were provided to seat 289 customers, there would be inadequate aisle accessways and aisles. If addition, the potential for egress obstruction would be significant. The appropriate maximum occupant load would be approved by the building official on a case-by-case basis, relying on the specific design of the space, the furniture and/or equipment layout, and the egress patterns created.
>
> **MAXIMUM OCCUPANT LOAD**

In order to analyze any increased occupant load, the building official must carefully review all aspects of the arrangement of space as well as the details of the total egress system, not only from the immediate space but continuously through all other building spaces that might intervene. In many cases, a diagram will be required indicating the approved furnishing and equipment layout.

Although it is critical that the building's means of egress system be designed to accommodate the increased occupant load, all other code requirements that are based on the number of occupants must also be reviewed based on the increased number. For example, if it is intended to increase the calculated occupant load of 258 in a Group A-3 conference facility to 340 occupants, all code requirements shall be applied based on the occupant load of 340. This would include the provisions of Section 903.2.1.3 that require an automatic sprinkler system, those of Section 907.2.1 mandating a manual fire alarm system, and the main exit requirements of Section 1028.2. An additional occupant-load-based provision that must be considered is that for plumbing fixtures.

1004.3 **Posting of occupant load.** Where a room or space is to be used as an assembly occupancy, this section requires the posting of a sign indicating the maximum permitted occupant load. This sign serves as a reminder to the occupants of the space, as well as building employees, that any larger occupant load would create an overcrowded condition. In order to be effective, the sign must be conspicuously located near the main exit of the room or space, and must be permanently maintained. An example of an occupant load sign is shown in Figure 1004-6. Where multiple uses causing varying occupant loads are anticipated, it is appropriate to designate the maximum occupant load for each use, as shown in Figure 1004-7.

1004.4 **Fixed seating.** The method of calculating occupant load discussed to this point—that is, the formula that divides an appropriate occupant load factor into the amount of space devoted to a specific function—is used when dealing with building spaces without fixed seating. Where fixed seats are installed, the code specifies that the occupant load be determined simply by counting the number of seats. Although the code does not define the term *fixed seats*, it is intended by this term that the seats provided are, in fact, fastened in position, not easily movable, and maintained in those fixed positions on a more or less permanent basis. A primary example of a fixed-seat facility would be a performance theater. In determining the occupant load for this type of facility, only the number of fixed seats is used because the code also requires that the space occupied by aisles may not be used for any purpose other than aisles and, therefore, may not be used for accommodating additional persons. The aisle system within a fixed-seating facility is, in fact, the exiting system for those fixed seats and, as such, must remain unobstructed. Therefore, the code does not assume any occupancy in the areas that make up the aisles.

Under varying circumstances, fixed-seating assembly spaces may include other assembly areas capable of being occupied. Such areas could include wheelchair spaces, waiting

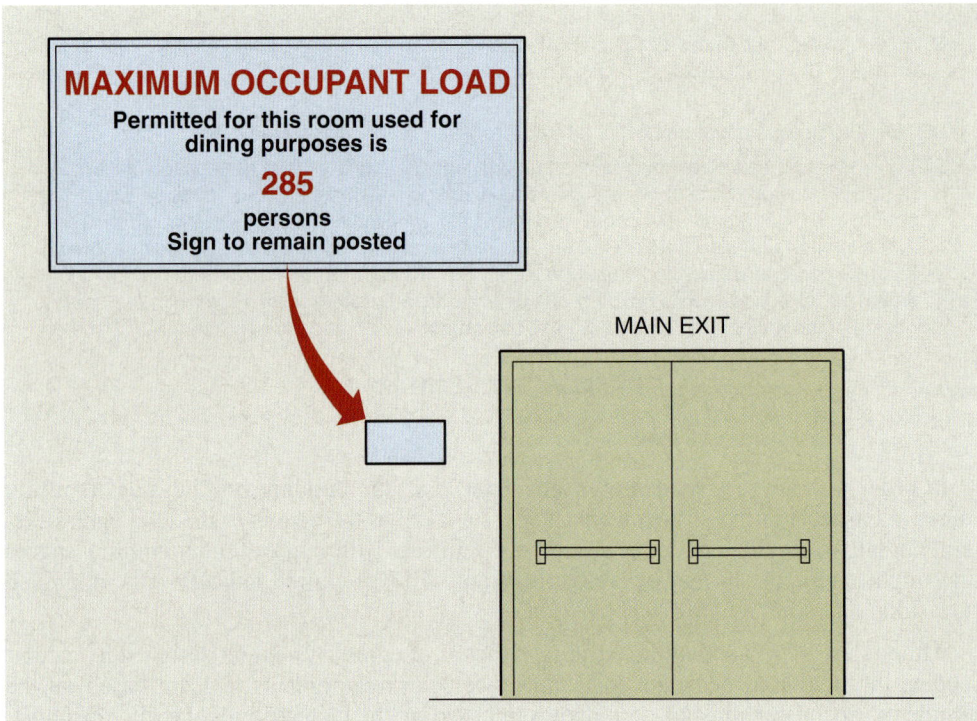

Figure 1004-6
Occupant load sign.

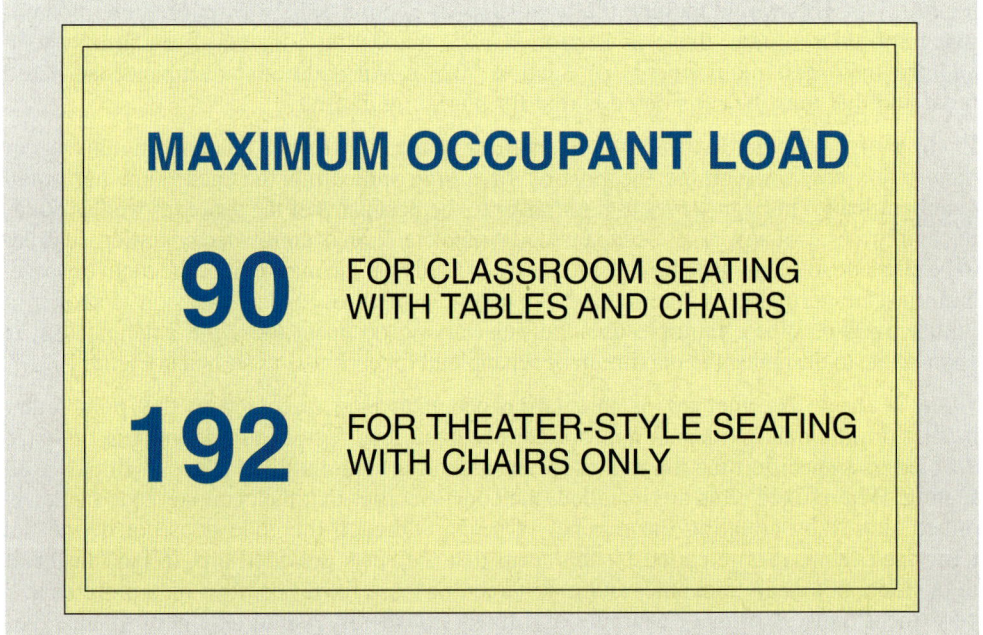

Figure 1004-7
Posting of occupant load.

areas, and/or standing room. Performance areas and similar spaces would also be evaluated and assigned an appropriate occupant load. The occupant load of all such areas must be added to that established for the fixed seating in the calculation of the total occupant load. An example is shown in Figure 1004-8. The inclusion of these additional occupiable areas provides for a more accurate determination of the potential number of persons who could occupy the room or space.

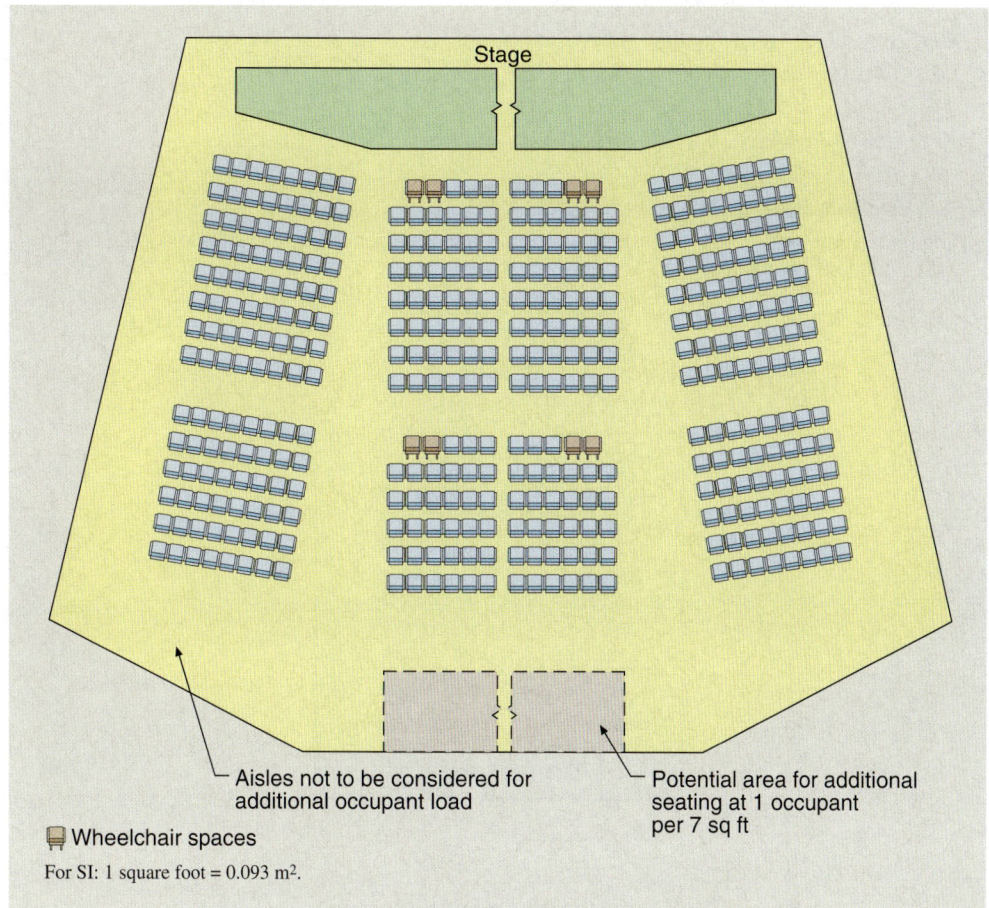

Figure 1004-8
Occupant load determination for fixed seating.

In addition to those fixed-seating arrangements where the seating is provided by a chair-type seat, there will be those that use continuous seating surfaces such as benches and pews. When this type of seating is provided, it is necessary to assume at least one person present for each 18 inches (457 mm) of length of seating surface. Where seating is provided by use of booths, as is frequently done in restaurants, it must be assumed that there is a person present for each 24 inches (610 mm) of booth-seating surface. If the booth seating is curved, the code specifies that the booth length be measured at the backrest of the seating booth. Where seating is provided without dividing arms, such as for benches and booths, it is reasonable to base the occupant load individually to each bench or booth. Similarly, it is appropriate to round the calculated occupant load down to the lower value, as this section only regulates each full 18 inches or 24 inches of width. See Application Example 1004-5.

The method for determining occupant load in a small restaurant is depicted in Application Example 1004-6.

1004.5 Outdoor areas. Occupiable yards, patios, and courts that are used by occupants of the building must be provided with egress in a manner consistent with indoor areas. This provision is applicable to outdoor areas, including building rooftops, that are occupied for a variety of uses, but is primarily applied to outdoor dining at restaurants and cafés. The building official shall assign an occupant load in accordance with the anticipated use of outdoor areas. If an area's occupants need to pass through the building to exit, the cumulative total of the outdoor area and the building shall be used to determine the

Application Example 1004-5

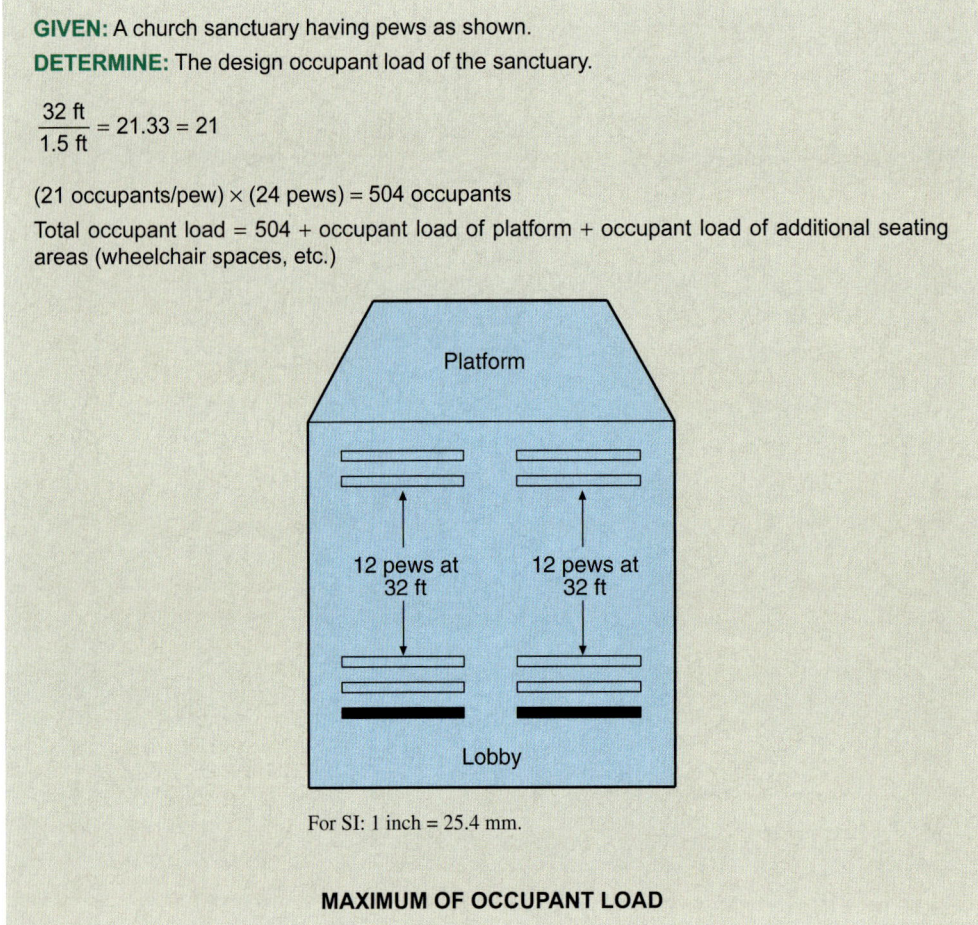

GIVEN: A church sanctuary having pews as shown.
DETERMINE: The design occupant load of the sanctuary.

$$\frac{32 \text{ ft}}{1.5 \text{ ft}} = 21.33 = 21$$

(21 occupants/pew) × (24 pews) = 504 occupants

Total occupant load = 504 + occupant load of platform + occupant load of additional seating areas (wheelchair spaces, etc.)

For SI: 1 inch = 25.4 mm.

MAXIMUM OF OCCUPANT LOAD

exiting requirements. This concept is consistent with the provisions of Section 1004.1.1. See Figure 1004-9.

Another example that is becoming more common is that of secured exterior areas serving nursing homes (Group I-2) or assisted living facilities (Group I-1, R-3, or R-4). Such exterior spaces are often provided to enhance the livability of the facilities by providing outdoor spaces where the patients or residents are free to roam without individual supervision. When evaluating these spaces for egress purposes, there are several issues to consider. If the secured yard is provided with a means of egress independent of the facility, the gates must comply with all of the requirements for egress doors. Special locking arrangements or delayed egress devices installed in accordance with Section 1008.1.9.6 or 1008.1.9.7, as applicable, would be permitted as a means of addressing occupant safety for these areas. Without compliant gates, the means of egress must be designed for travel back through the facility. The facility must also egress independent of the secured yard unless all means of egress from the secured exterior area comply with the code.

The judgment of the building official is very important to the application of these provisions because the building official must determine exactly what occupant load should be considered and to what degree the area is accessible and usable by the building occupants in order to establish the egress requirements. Some cases that will require judgment include large spaces that might have a very limited anticipated occupant load such as areas that are primarily for the service of the building. Where a portion of the required means

Application Example 1004-6

GIVEN: Information as shown in illustration.

DETERMINE: The occupant load for the small restaurant.

SOLUTION: Section 1004.4 states that where booths are used in dining areas, the occupant load shall be based on one person for each 24 inches (610 mm) of booth length. Based on this requirement, each 4-foot 6-inch booth would have an occupant load of four, or a total occupant load for the booth area of 16.

The fixed seats at the counter number eight, which, based on the first paragraph of Section 1004.4, would establish an occupant load of eight.

The open dining area (tables and chairs), having a floor area of 600 square feet (55.7 m^2) and using an occupant load factor of 15 square feet (1.39 m^2) per occupant, as set forth in Table 1004.1.2, would have an occupant load of 40.

The cooking area, having a floor area of 200 square feet (18.6 m^2) and using an occupant load factor of 200 square feet (18.6 m^2) per occupant, as set forth in Table 1004.1.2, would have an occupant load of one.

The total occupant load is as follows:

Booths 16
Counter 8
Dining area (tables and chairs) 40
Cooking area 1

Tolal occupants: 65

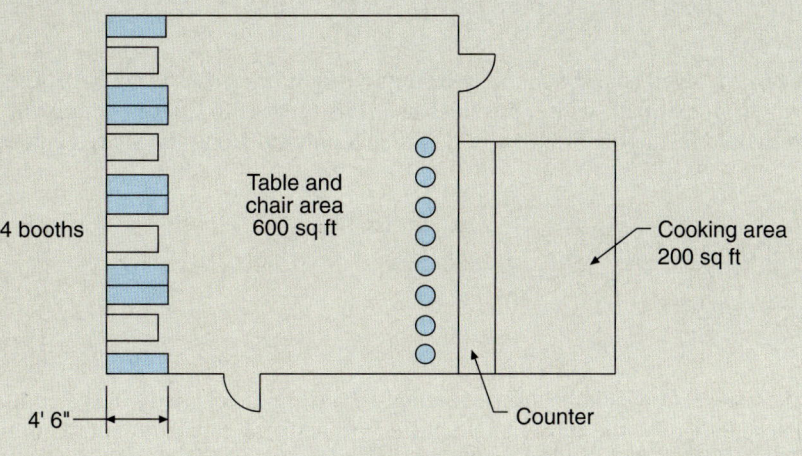

For SI: 1 inch = 25.4 mm, 1 foot = 304.8 mm, 1 square foot = 0.0929 m^2.

DETERMINATION OF OCCUPANT LOAD

of egress for the outdoor area is provided independent of travel back through the building, or where all of such required egress must pass through the building, the applicable provisions would be similar to those for travel through intervening spaces. The distribution of the occupant load from the outdoor area will depend on how many exits are required and how many means of egress paths are available.

Multiple occupancies. In many buildings there are two or more occupancies. Quite often, one or more of the egress paths from an individual occupancy will merge with egress paths from other occupancies. Within each individual occupancy, the means of egress shall be designed for that specific occupancy. However, where portions of the means of egress serve two or more different occupancies, the more restrictive requirements of the

1004.6

10 Means of Egress

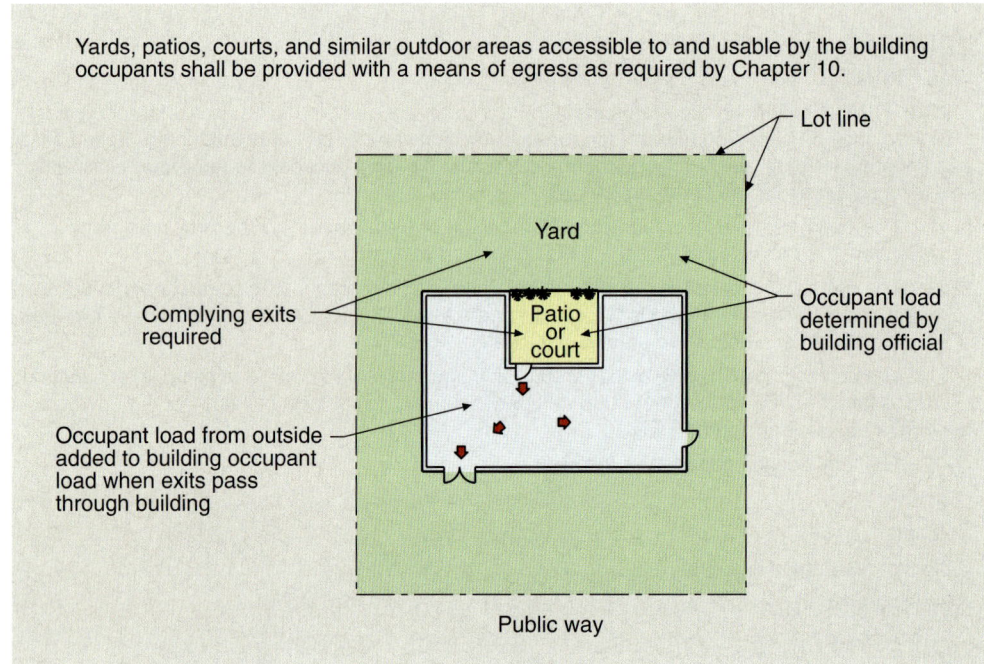

Figure 1004-9 Outdoor areas.

occupancies involved shall be met. An example might be where a sizable assembly occupancy shares an exit path with a business use. The more restrictive requirement for panic hardware would be applicable for any doors encountered along the shared egress route.

Section 1005 *Means of Egress Sizing*

1005.1 General. This section establishes the method for sizing the capacity of the egress system, and more specifically, the minimum required capacity of each individual component in that system. It also establishes the method for distributing egress capacity to various egress paths where multiple means of egress are provided.

There are two methods established by this section for the determination of the minimum required width/capacity of the means of egress all along the various egress paths. These methods are typically referred to as "component" width and "calculated" width. The greater width or capacity required for each egress element based on component width and calculated width is to be applied in the design of the means of egress system. Component width, addressed in Section 1005.2, is specified throughout the code based on the specific means of egress component under review. As an example, the minimum required component width of a corridor is based on the provisions of Section 1018.2. Calculated width, addressed in Section 1005.3, is determined based on the appropriate formula as established in Section 1005.3.1 for stairways and Section 1005.3.2 for egress components other than stairways. It is important to note again that the greater required width as established by component width and calculated width is to be provided. See Application Example 1005-1.

Section 1028 provides special means of egress requirements for those rooms, spaces, and areas used for assembly purposes. Where an assembly use contains seats, tables, displays, equipment, or similar elements, the requirements set forth in Section 1028 are to be applied in addition to the general means of egress requirements of Chapter 10.

Means of Egress Sizing

Application Example 1005-1

GIVEN: Various egress components and the occupant load served by each component.
DETERMINE: The minimum required width for each component in a Group B oocupancy.

Egress Component	Occupant Load Served	Minimum Required Calculated Width	Minimum Required Component Width	Minimum Required Width
Aisle	64	64 (0.2) = 12.8"	36" Sec. 1017.3	36"
Corridor	130	130 (0.2) = 26"	44" Sec. 1018.2	44"
Stairway	200	200 (0.3) = 60"	44" Sec. 1009.4	60"
Door	180	180 (0.2) = 36"	32" Sec. 1008.1.1	36"

For SI: 1 inch = 25.4 mm.

1005.2

Minimum width based on component. Egress components all have a minimum width established by other provisions in the code. It should be noted that where the component width is the appropriate method for determining egress width, in many situations the required component widths may lessen along the egress path. For example, in an office building with an occupant load of 68 persons, a corridor required to be at least 44 inches (1,118 mm) in width by Section 1018.2 may lead to an exit door with a minimum clear width of 32 inches (813 mm) as regulated by Section 1008.1.1. In this example, the minimum component widths for the corridor and the exit door provide for greater widths than required by Section 1005.3.2 addressing calculated width.

1005.3

Required capacity based on occupant load. The formula for means of egress capacity based on occupant load is very succinct. It states that the total required width of the means of egress shall not be less than that obtained by multiplying the total occupant load served by an egress component by the appropriate factor as set forth in Section 1005.3.1 or 1005.3.2, as applicable. It should be noted that the calculation of egress width in most assembly occupancies is not regulated by this section, but rather is governed by Section 1028.6. Where an assembly space contains seats, tables, displays, equipment, or other fixtures or furnishings, it must comply with the exiting provisions of Section 1028.

In designing the means of egress system, it is first necessary to determine the occupant load that must be accommodated through each individual portion of the system. The occupant load anticipated to be served by each individual component is the basis for sizing each component. The design occupant load is to be used when determining both the component width and the calculated width. Continuing to use a corridor as an example, Table 1018.2 requires a minimum component width of 44 inches (1,118 mm) where serving an occupant load of 50 or more, but only 36 inches (914 mm) where serving an occupant load less than 50. For calculated width, multiplying the occupant load by the appropriate factor in Section 1005.3.1 or 1005.3.2 will result in the minimum required width in inches (mm) necessary to accommodate the occupant load. An example is shown in Application Example 1005-2.

Application Example 1005-2

GIVEN: A home-improvement center has an occupant load of 3,180. The building is fully sprinklered and has an emergency voice alarm communication (EVAC) system.
DETERMINE: The total required egress width from the building at the exit doors.
SOLUTION: For a Group M occupancy in a sprinklered building, the exception to Section 1005.3.2 indicates a width factor of 0.15 inches per occupant for egress components other than stairways. 3,180 (0.15) = 477 inches of clear door width to be distributed among available exit doors.

Means of Egress

It cannot be emphasized too strongly that when the code discusses width in terms of an egress system or component, it is referring to the clear, unobstructed, usable width afforded along the exit path by the individual components. Therefore, if it is determined, for example, that a means of egress must have a width of at least 3 feet (914 mm), it shall be arranged so that it is possible to pass a 36-inch-wide (914 mm) object through that egress path and each of its components. Unless the code specifically states that a projection is permitted into the required width by Section 1005.7, nothing may reduce the width of the component required to provide the necessary exit capacity.

Egress width in assembly spaces. As previously mentioned, where the provisions of Section 1028 are applicable for assembly uses, egress width shall be determined based on such provisions. The requirements of Section 1028 regulate all assembly spaces containing seats, tables, displays, equipment, or other material. Thus, it is typical that the egress width requirements of Section 1028.6 are to be followed rather than those of Section 1005.1. Section 1028.6 addresses egress width for assembly uses, based primarily on whether or not smoke-protected assembly seating is provided. A further discussion of this subject is provided in the analysis of Section 1028.

1005.3.1 Stairways. This provision establishes the method for determining the required capacity of an egress stairway, based on the occupant load assigned to the stairway. As a basic requirement, the capacity of a stairway is based on the occupant load assigned to the stairway multiplied by 0.3 inches per occupant. Where the building is provided with both an automatic fire sprinkler system and an emergency voice/alarm communication system, a factor of 0.2 inches per occupant is to be used. The presence of these two fire-protection systems allows for a 33 percent reduction in the minimum required calculated egress width. Again, the minimum required component width must also be addressed. This reduction to 0.2 inches per occupant is not permitted in Group H and I-2 occupancies.

Provided the occupants are traveling in the same direction, there is no need to combine the loads from adjacent stories. The IBC assumes that in exiting multistory buildings there will be occupants feeding into the exit stairs at various stories. This approach recognizes that using the cumulative occupant load for vertical travel often results in stairway widths of significant size. The reduced sizes established by the floor-by-floor method may result in some increase in occupant evacuation time; however, the slower evacuation rate is deemed acceptable since the travel occurs within a fire-resistance-rated enclosure. This is in contrast to the methodology applied to horizontal travel where cumulative occupant loads must be addressed due to the fact that such travel is typically in unprotected portions of the exit access. The capacity factors are substantially higher for stairways than for horizontal egress components, primarily because the speed of exiting on a stair is substantially less than the speed of exiting on level or nearly level surfaces. On stairways there is a forced reduction in normal stride, as the length of stride on a stair must coincide with the stair's run. A study of this difference shows that a stairway requires an increase in width of approximately 50 percent above that for horizontal travel in order to maintain equivalent flow rates.

At one time, building codes addressed a cascading effect when analyzing and determining the required width of stairs, but this concept is no longer applicable. The required width of a stair is calculated on a story-by-story application of the formula.

In the design of multistory buildings, it is quite common that different stories have different occupant loads. Thus, the occupant load calculated for each story must be considered. As a matter of fact, it is not uncommon for buildings to have assembly uses on the top story. As a consequence of that configuration, it is entirely possible that the top story of a building will have an occupant load greater than any other. Under such a condition, it will be necessary to calculate the required stairway width based only on the occupant load of the uppermost story. This required width must be maintained through the successive stories until it serves a greater occupant load from a lower story or until the occupants have reached the public way or ultimate safe place. Figure 1005-1 illustrates this requirement.

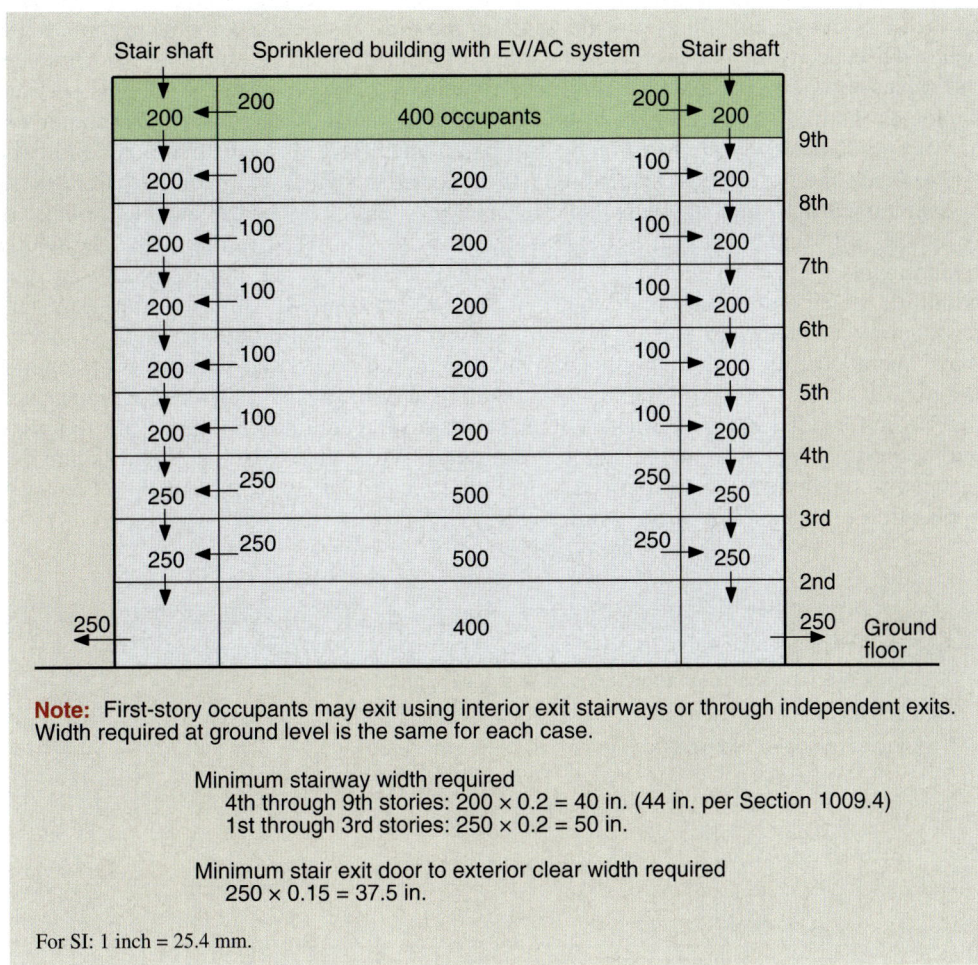

Figure 1005-1 Width of exits—multistory buildings.

It should be noted that the same concept of determining stairway width that is discussed above applies where building occupants exit upward through the stairway. This is the situation in buildings with basements and sub-basements. Occupants of those below-grade floors must exit up the stairway, onto the landing on the ground floor, and then out through the exterior exit door from that landing. The largest capacity calculated from any floor level will be the controlling factor of the exit at ground level, as well as the exit doorway from the stairway. Of course, it is possible that occupants on the ground floor will exit through an enclosure of an interior exit stairway. If so, the required width determined based on that condition may govern. However, the ground floor usually has adequate width of exits independent of any paths through the stairway enclosures. Thus, the occupant load of the ground floor is usually not an issue in the determination of the exterior exit doorway width from the stairway enclosure. In any case, the occupant load of the ground floor would not be added to any occupant load of floors above for determining egress width from the building. Where upper and lower floors converge at an intermediate level, see the discussion of Section 1005.6 on egress convergence.

1005.3.2 Other egress components. A means of egress capacity factor of 0.2 inches per occupant applies to those egress components other than stairways, such as ramps, aisles, and corridors. The factor may be reduced in all occupancies other than Groups H and I-2, provided the building is fully sprinklered and provided with an emergency voice/alarm communication system. Where both of these fire-protection features are provided, the capacity factor may be reduced to 0.15 inches per occupant.

1005.4 Continuity. As stated earlier, it is the width of the most restrictive component that establishes the capacity of the overall exit system. To ensure that a design does not reduce the capacity at some point throughout the remainder of the egress system, this section stipulates that the capacity may not be reduced and that the design must accommodate any accumulation of occupants along the path. Therefore, once the required width is determined for any story (in fact, from any room or other space), that required width must be maintained until the occupants have reached the public way or ultimate safe place. It is important to remember that because different factors are used to determine the width requirements for stairways than for all other components, it is really the capacity, not the width, of the egress system that is not permitted to be reduced. An aisle, door, corridor, passageway, or other horizontal egress component located at the bottom of any stairway may generally be reduced in width from what is required for the stairway. See Application Example 1005-3. This is simply due to the width factor of the stairway being greater than the factor for other egress elements. It must again be mentioned that only the required width (capacity) needs to be maintained. The actual width of the means of egress may be reduced throughout the travel path as long as the required width is provided. A common application of this concept is shown in Figure 1005-2.

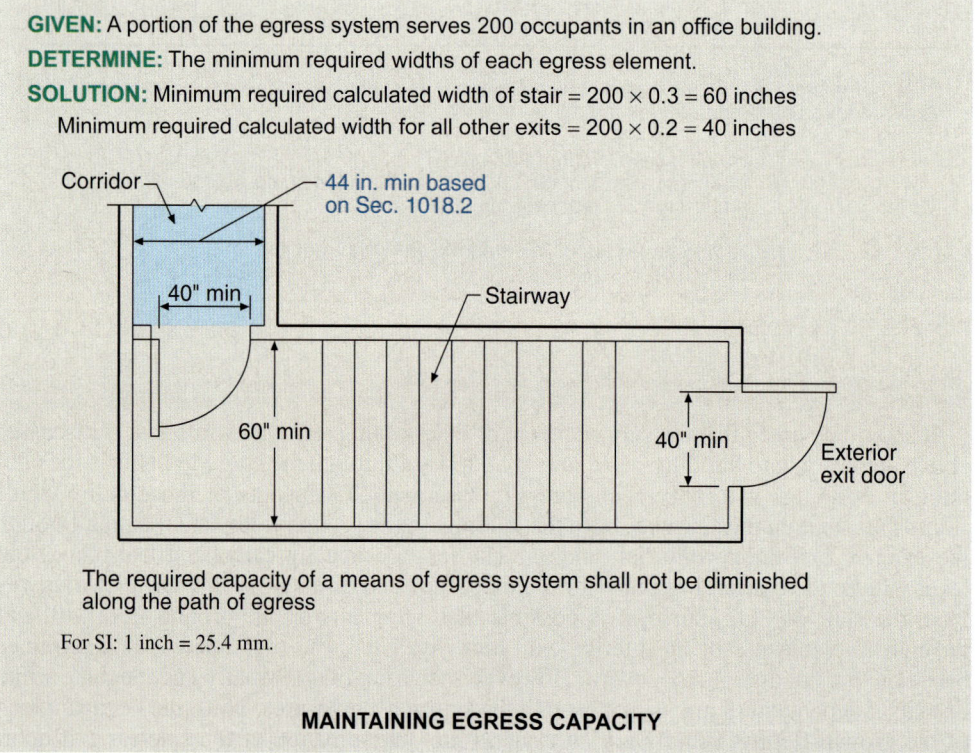

Application Example 1005-3

GIVEN: A portion of the egress system serves 200 occupants in an office building.
DETERMINE: The minimum required widths of each egress element.
SOLUTION: Minimum required calculated width of stair = 200 × 0.3 = 60 inches
Minimum required calculated width for all other exits = 200 × 0.2 = 40 inches

The required capacity of a means of egress system shall not be diminished along the path of egress

For SI: 1 inch = 25.4 mm.

MAINTAINING EGRESS CAPACITY

1005.5 Distribution of egress capacity. When the required egress capacity of the system has been determined by the use of one of the formulas, and the required number of exit access doorways or exits has been determined in accordance with the provisions of Sections 1015.1 and 1021.2, the required egress capacity can be divided among the number of required means of egress. In fact, where additional complying means of egress are provided above the number required by the code, they too can be used for distribution purposes. The manner of distribution shall be such that the loss of any one means of egress will not reduce the available capacity to less than 50 percent of that required. Thus, after the loss of one means

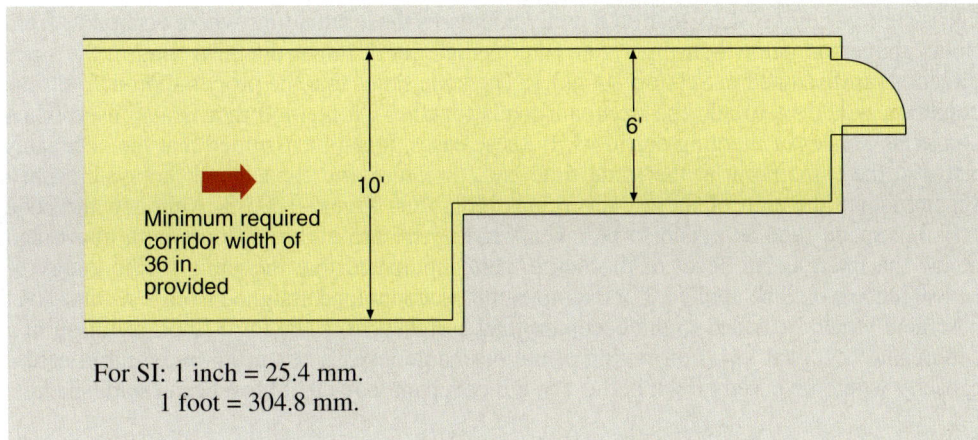

For SI: 1 inch = 25.4 mm.
1 foot = 304.8 mm.

Figure 1005-2
Reduction in actual width.

of egress, at least one-half of the required capacity must be available. See Application Example 1005-4. It is the intent of this section that there be reasonable distribution of the egress capacity necessary to serve a given occupant load.

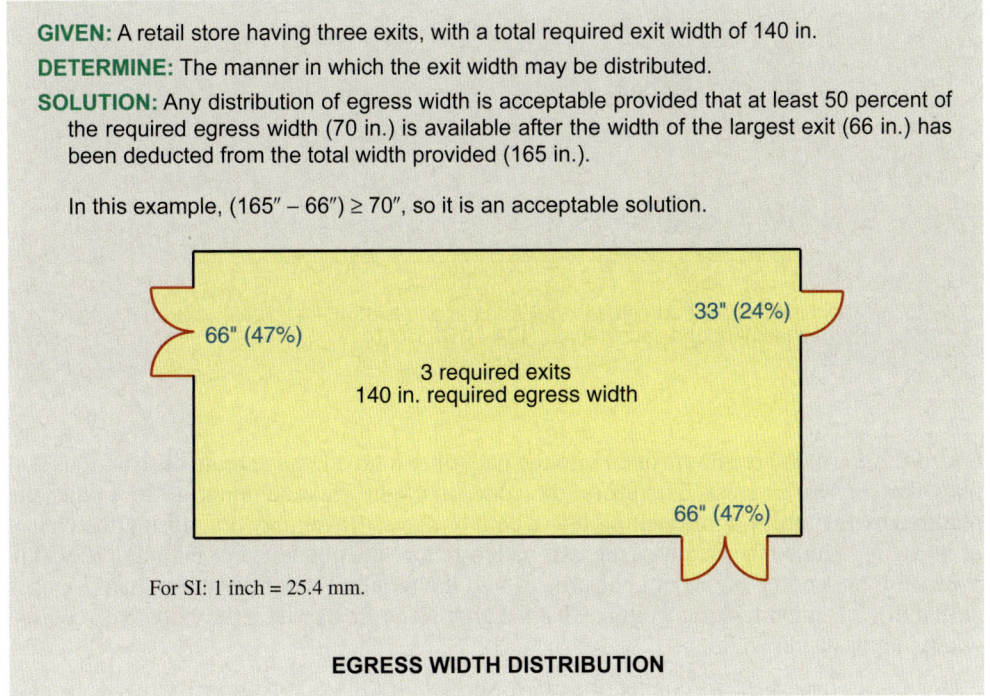

GIVEN: A retail store having three exits, with a total required exit width of 140 in.
DETERMINE: The manner in which the exit width may be distributed.
SOLUTION: Any distribution of egress width is acceptable provided that at least 50 percent of the required egress width (70 in.) is available after the width of the largest exit (66 in.) has been deducted from the total width provided (165 in.).

In this example, (165″ − 66″) ≥ 70″, so it is an acceptable solution.

For SI: 1 inch = 25.4 mm.

EGRESS WIDTH DISTRIBUTION

Application Example 1005-4

The primary reason for requiring multiple egress paths is the fact that in a fire or other emergency, it could be possible that at least one of the routes will be unavailable or blocked by fire. If the egress capacity was concentrated at one exit point, it could very easily be that path that affords the greatest portion of the egress capacity that would be lost. The resulting limitation on the occupants' ability to exit a building or portion thereof is simply unacceptable. In addition, the presence of two or more means of egress allows for a distribution of occupants, which should provide for more efficient and orderly egress under emergency conditions. A third benefit of egress distribution is the potential reduction in the distance occupants must travel to reach an exit or exit access doorway.

1005.6 Egress convergence. This section directly addresses those situations where occupants from floors above and below converge at an intermediate level, rather than traveling in the same direction, as discussed in Section 1005.3.1. The code states that the proper approach for this condition would be to add the occupant loads together—a method that is also used when converging aisles or merging corridors. In these cases, it can be assumed that the occupants arrive at the same point at the same time and, therefore, the capacity of the system must accommodate the sum of these converging floors. See Figure 1005-3. Although the code does not specify the approach to be taken where there are multiple floors both above and below the intermediate level of discharge, it is anticipated that the same methodology of convergence would be applied. For example, the occupant load assigned from a second-level basement would be added to the occupant load assigned from the third floor, resulting in a congregate occupant load converging at the discharge level. The minimum required egress capacity would be based on the highest of the occupant loads that have been established.

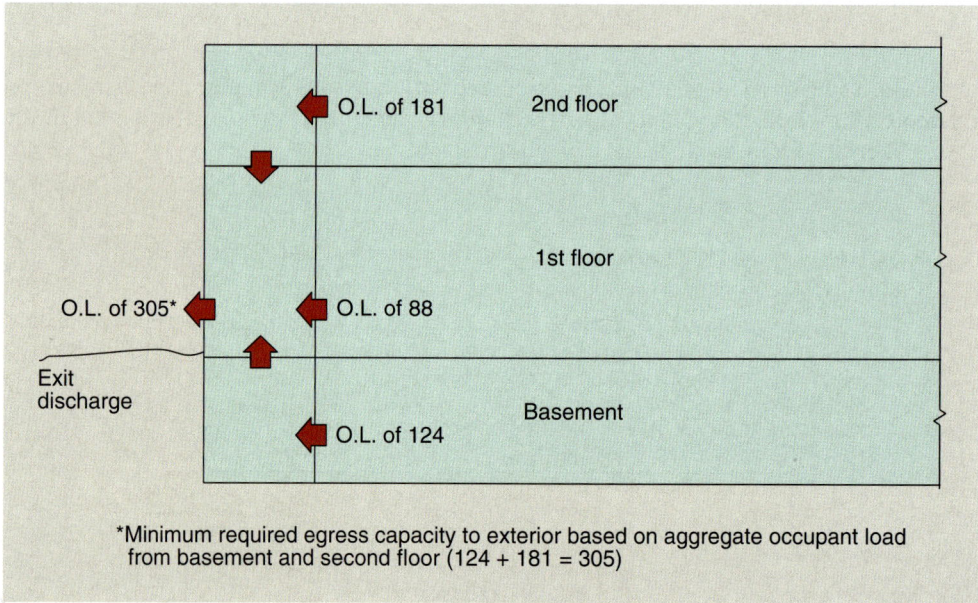

Figure 1005-3 Egress convergence.

*Minimum required egress capacity to exterior based on aggregate occupant load from basement and second floor (124 + 181 = 305)

1005.7 Encroachment. Where doors open into the path of exit travel, they create obstructions that may slow or block egress. Therefore, the code limits the encroachment of doors into the required exit width. A door opening into a path of egress travel may not, during the course of its swing, reduce the width of the exit path by more than one-half of its required width. When fully open, the door may not project into the required width by more than 7 inches (178 mm). It is important to recognize that the provisions are based on the exitway's required width, not its actual width.

Again, as discussed in connection with Section 1008.1.6 for doors swinging over a stairway landing, it might be better to think of the permitted obstruction of a door during the course of its swing from a positive viewpoint. So stated, each door, when swinging into an egress path such as an aisle or a corridor, must leave unobstructed at least one-half of the required width of the path of travel during the entire course of its swing. At least one-half of the required width must always be available for use by the building occupants. When the door is in its fully open position, the required egress width, minus 7 inches (178 mm), must be available. See Figure 1005-4.

In applying the requirements for projections, the code imposes these limitations on a door-by-door basis. It is desirable that doors be arranged so as not to have two doors directly opposing each other on opposite sides of the exit path. Better design would avoid

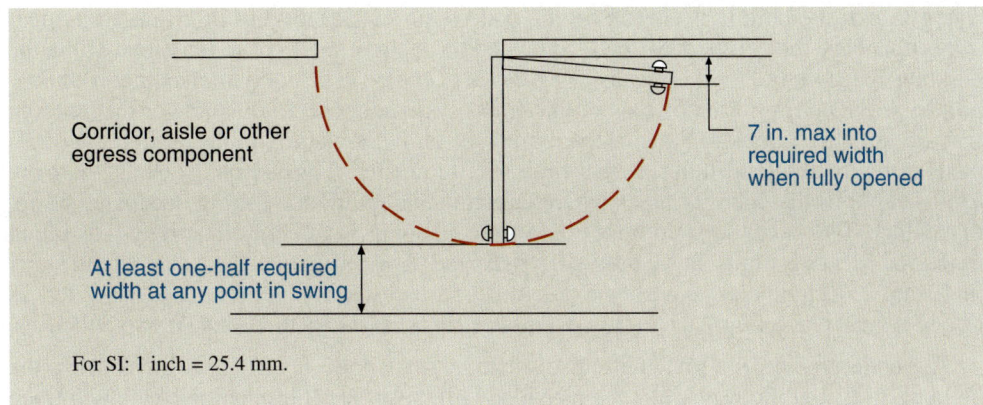

Figure 1005-4
Egress obstruction due to door swing.

this arrangement. The intent of the code is that at least one-half of the required width of the exitway be available for use by the building occupant as illustrated in Figure 1005-5. The restrictions on door swing do not apply to doors within dwelling units and sleeping units of Groups R-2 and R-3.

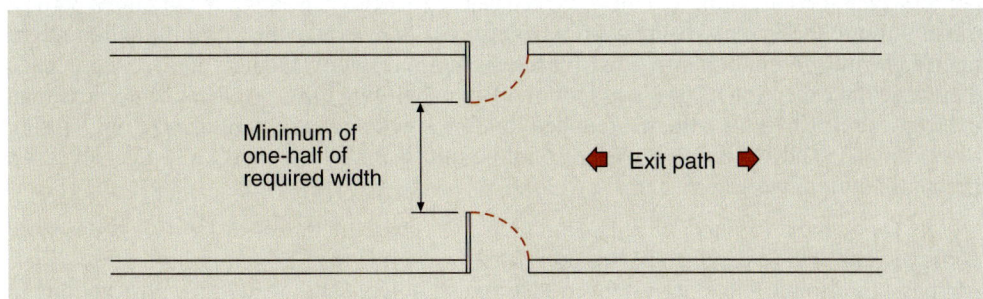

Figure 1005-5
Doors swinging into egress path.

Where nonstructural projections other than doors, such as trim and similar decorative features, extend into the required width of a means of egress component, the limit on their projection into the required width of egress components is $1^1/_2$ inches (38 mm) unless such projection is specifically prohibited by the code. In reviewing the provisions for aisles in Section 1017.1, corridors in Section 1018.3, and exit passageways in Section 1023.2, the encroachments established in this section are specifically permitted. Handrail projections are also permitted provided such projections do not extend beyond the limitations established in Section 1012.8.

Section 1006 *Means of Egress Illumination*

Illumination required. In order for the exit system to afford a safe path of travel and for the building occupant to be able to negotiate the system, it is necessary that the entire egress system be provided with a certain minimum amount of illumination. Without such lighting, it would be impossible for building occupants to identify and follow the appropriate path of travel. The lack of adequate illumination would also be the cause of various other concerns, such as an increase in evacuation time, a greater potential for injuries during the egress process, and most probably an increased level of panic to those individuals

1006.1

trying to exit the building. Therefore, the code requires that, except in a limited number of occupancies, the egress paths be illuminated throughout their entire length any time the building space served by the means of egress is occupied. The code intends that illumination be provided for those portions of the egress system that serve the parts of the building that are, in fact, occupied. Parts of the exiting system that would not be serving the occupants of the building need not, at that time, be illuminated. For obvious reasons, there are four exceptions that identify areas where continuous illumination during occupancy is not mandatory. Two exceptions address uses where sleeping is a common activity—dwelling units and sleeping units in Group R-1, R-2, and R-3 occupancies, and sleeping units in Group I occupancies. Another exception addresses utility structures designated as Group U, whereas a fourth exception exempts aisle accessways in Group A assembly uses.

The code emphasizes that the exit discharge—that portion of egress travel from the building to the public way—also be provided with adequate illumination. Although there are often numerous light sources at a building's exterior, such as lighting for landscaping, parking lots, city streets, and adjacent buildings, it is important that the illumination be effective and reliable for use under this provision. It should also be noted that the requirements of this section are simply for general illumination of the entire egress system, and are not the higher-level conditions for emergency lighting as mandated in Section 1006.3.

1006.2 Illumination level. Such illumination must be capable of producing a light intensity of not less than 1 foot-candle (11 lux) at the walking surface throughout the entire path of travel through the system. An exception recognizes that such levels of illumination might interfere with presentations in such places as motion picture theaters and concert halls; therefore, the exception allows a reduction in such building uses to a level of not less than 0.2 foot-candle (2.15 lux). Such a reduced lighting level, however, is permitted only during a performance and would be brought up to the minimum 1 foot-candle (11 lux) level if a fire alarm system were activated.

One foot-candle (11 lux) of light on a surface is not a great deal of light. It is probably not sufficient light to enable a person to read. However, it is sufficient light to allow a person passing through the exit system to distinguish objects and to identify obstructions in the actual path of travel. The light cast by a full moon on a clear night might approximate the 1 foot-candle (11 lux) light level. When the amount of light intensity is in doubt, it may be necessary to measure it with a light meter.

1006.3 Emergency power for illumination. Normally, the power for illumination of the egress path is provided by the premises' wiring system. However, where the potential life-safety hazard is sufficiently great, it is considered inadequate to solely provide the illumination of the exit system by such a system. In these cases, it is necessary that emergency power—a completely separate source of power—automatically provide illumination of the exitways. In fundamental terms, separate sources of power are required in all occupancies in which two or more means of egress are required. Therefore, any space, area, room, corridor, exterior egress balcony, or other portion of the egress system requiring access to at least two exits or exit access doors is to be provided with emergency lighting. An example of this application is depicted in Figure 1006-1. Also included are exit stairways, both interior and exterior, and exterior landings at exit doors in buildings requiring a minimum of two means of egress.

Where emergency power systems are required, they are to be supplied by storage batteries, unit equipment, or an on-site generator. It is the intent that this power source be automatically available even in the event of the total failure of the public utility system. Therefore, a separate, independent source is generally required. Installation of the emergency power system is regulated by referring to the requirements of NFPA 70.

1006.3.1 Illumination level under emergency power. The initial illumination provided by emergency power along the path of egress at floor level shall average at least 1 foot-candle (11 lux), with a minimum level of illumination required to be 0.1 foot-candle (1 lux). Illumination levels are permitted to decline over the required 90-minute duration of the

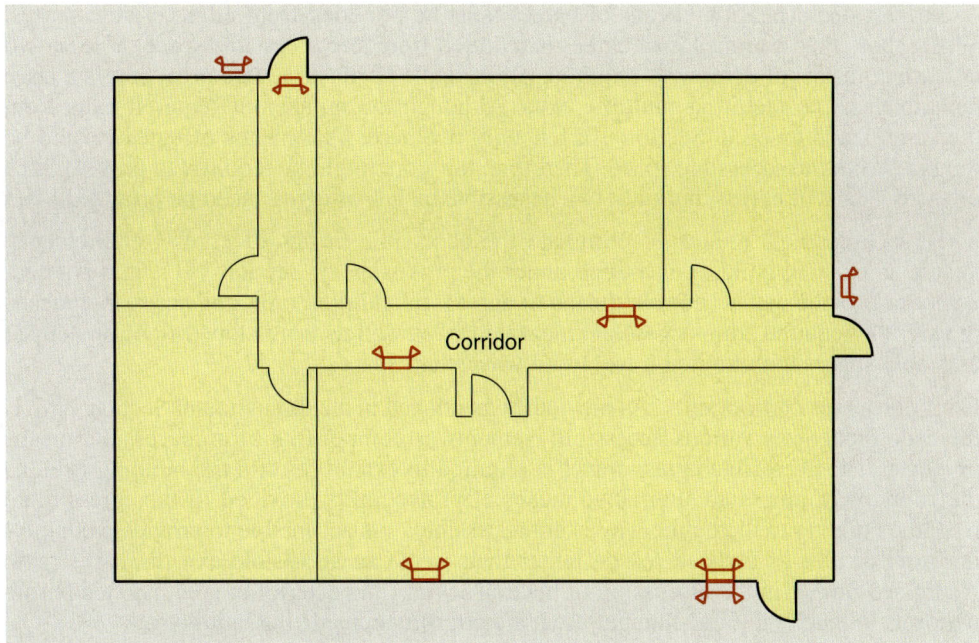

Figure 1006-1
Emergency power for egress illustration.

emergency power source to an average of 0.6 foot-candle (6 lux) with a minimum at any point of 0.06 foot-candle (0.6 lux). Recognizing the variation in light levels throughout the exit path, only the average illumination level needs to be determined; however, an absolute minimum level of illumination must be attained. In no case shall the illumination uniformity ratio between the maximum light level and the minimum light level exceed 40 to 1.

Section 1007 *Accessible Means of Egress*

Accessible means of egress required. In addition to the access to buildings required by the provisions of Chapter 11, it is important that safe egress for physically disabled individuals is provided. Therefore, the code requires that accessible spaces be provided with accessible means of egress consisting of one or more of the following components as set forth in Section 1007.2: **1007.1**

1. Accessible routes complying with Section 1104.
2. Interior exit stairways complying with Sections 1007.3 and 1022.
3. Exit access stairways complying with Sections 1007.3 and 1009.3 (unless connecting levels in the same story).
4. Exterior exit stairways complying with Sections 1007.3 and 1026 (where serving floor levels other than the level of exit discharge).
5. Elevators complying with Section 1007.4.
6. Platform lifts complying with Section 1007.5.
7. Horizontal exits complying with Section 1025.
8. Ramps complying with Section 1010.
9. Areas of refuge complying with Section 1007.6.
10. Exterior areas for assisted rescue complying with Section 1007.7.

At least one accessible means of egress must be provided from all accessible spaces. Where more than one means of egress is required from any accessible space, at least two accessible means of egress are required. An example to illustrate this provision is a large department store requiring multiple exits. Although the number of required exits from the store is addressed in Section 1021.1, only two accessible means of egress would be required from the accessible space. Therefore, the store might be required to provide three or more means of egress, but only two accessible means of egress need be provided.

Three exceptions reduce or eliminate the accessible means of egress requirements. Where an existing building is altered under the provisions of Section 3411, it is not necessary to provide any accessible means of egress. In addition, only one accessible means of egress is required from accessible mezzanines, as well as from sloped-floor or stepped assembly spaces with limited travel to all wheelchair spaces.

1007.2 Continuity and components. As previously mentioned in the discussion of Section 1007.1, the code recognizes various accessible elements as components of an accessible means of egress. The accessible egress travel is required to extend beyond the building itself to the public way, unless an alternative means of protection is provided. If the egress route from the building to the public way is not accessible, it is acceptable to provide a complying exterior area of assisted rescue rather than create an accessible exit discharge path. Addressed further in the discussion of Section 1007.7, the exterior area of assisted rescue performs in much the same manner as an area of refuge inside the building.

1007.2.1 Elevators required. Unlike the general provision found in Section 1003.7 that specifically prohibits considering an elevator as an approved means of egress, this section requires an elevator for rescue purposes under certain conditions. In buildings where a required accessible floor is four or more stories above or below a level of exit discharge, ramps and stairs cannot adequately serve as egress for individuals with a mobility impairment. Therefore, at least one elevator must be provided as an accessible means of egress. The elevator is not required to conform with Section 1007.4 in sprinklered buildings on those floors provided with a conforming ramp or horizontal exit.

In the application of this provision, the second story of a typical building is considered the first story above the level of exit discharge. Accordingly, the building's fifth story is typically viewed as four stories above the level of exit discharge. See the discussion of Section 202 for *level of exit discharge* and Figure 1007-1. Under such conditions, a minimum of one accessible means of egress must be a complying elevator.

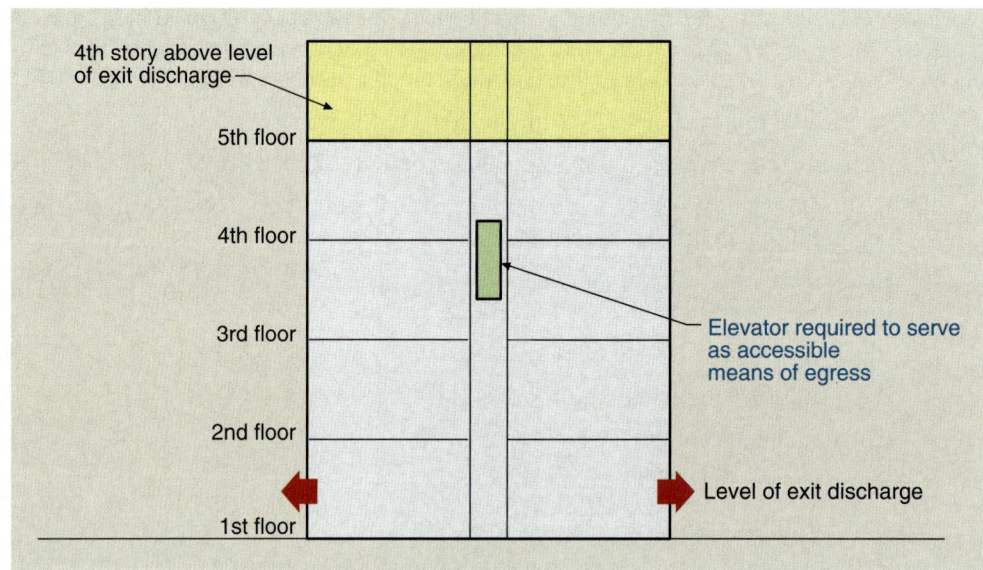

Figure 1007-1
Elevator as accessible means of egress.

1007.3 Stairways. Exit stairways are typically used as one or more of the required accessible means of egress in a multistory building. Increasing the width of stairs to 48 inches (1,219 mm) between the handrails allows for the minimum amount of space needed to assist persons with disabilities in the event of a building evacuation. The provisions for an area of refuge or horizontal exit address the increased time needed for egress. The last sentence of the section is intended to address exit access steps in assembly seating or split-level buildings where an individual awaiting assistance or rescue may be required to wait in the middle of a building and not be able to travel toward an exit. There is no intent to prohibit the use of exit access stairways as part of an accessible means of egress from mezzanines and balconies.

A number of exceptions eliminate the general requirement for an area of refuge. Areas of refuge are not mandated at exit stairways serving open parking garages. In addition, areas of refuge are not mandated in Group R-2 occupancies and in smoke-protected assembly seating areas as regulated by Section 1028.6.2. Another two exceptions remove the requirement for at least 48 inches (1,219 mm) of clear width between handrails. The 48-inch (1,219-mm) width is not mandated in sprinklered buildings, nor is it required for stairways accessed from a horizontal exit.

The most commonly applied exception permits the omission of areas of refuge in buildings protected throughout by an automatic sprinkler system. The purpose of an area of refuge is to provide an area "where persons unable to use stairways can remain temporarily to await instructions or assistance during emergency evacuation." Much of the reasoning in exempting fully sprinklered buildings from the requirement for areas of refuge comes from NISTIP 4770, a report issued by the National Institute of Standards and Technology (NIST) in 1992 titled *Staging Areas for Persons with Mobility Impairments*. The primary conclusion of the report was that the operation of a properly designed sprinkler system eliminates the life threat to all occupants regardless of their individual abilities and can provide superior protection for persons with disabilities as compared to staging areas. The ability of a properly designed and operational automatic sprinkler system to control a fire at its point of origin and to limit production of toxic products to a level that is not life threatening to all occupants of the building, including persons with disabilities, eliminates the need for areas of refuge.

1007.4 Elevators. Although an elevator may be used as an accessible means of egress component in all multilevel facilities, it is only required as such in buildings regulated by Section 1007.2.1. Elevators used as accessible means of egress must comply with the operation and notification criteria of ASME A17.1, Section 2.27. In addition, standby power is required in order to maintain service during emergencies. The general requirement is that any elevator used as an accessible means of egress be accessed from an area of refuge or horizontal exit. However, the area of refuge and horizontal exit are not required in fully sprinklered buildings, open parking garages, smoke-protected assembly seating areas, and where elevators are not required to be protected by shaft enclosures. Additional information is provided in the discussion of Section 1007.3. In such cases, the elevator is still considered an accessible means of egress when in compliance with the other criteria of this section.

1007.5 Platform lifts. Except in limited applications, a platform lift is specifically excluded as an acceptable element of a means of egress. The maintenance of the lift as well as the complexity and delay in using a platform lift are considered substantial obstacles in providing acceptable means of egress for persons in wheelchairs. Section 1109.7 specifically sets forth the few instances where platform lifts are permitted for access and, with the exception of Item 10, egress purposes. Where a complying platform lift is used as an accessible means of egress component, it must be provided with standby power, much in the same manner as an elevator used for the same purpose.

1007.6 Areas of refuge. By definition, an area of refuge is an area "where persons unable to use stairways can remain temporarily to await instructions or assistance during emergency evacuation." Unfortunately, the term *temporary* is not defined, so a number of provisions are applied to an area of refuge to increase the level of protection for anyone using it. These provisions include a size large enough to accommodate wheelchairs without reducing exit width, smoke barriers designed to minimize the intrusion of fire and smoke, two-way communications systems, and instructions on the use of the area under emergency conditions. The two-way communications system is intended to allow a user of the area of refuge to identify his or her location and needs to a central control point. Obviously, it is important that someone be available to answer the call for help when a two-way communications system is provided. The system shall have a timed automatic telephone dial-out capability to a monitoring location or 911 that can be used to notify the emergency services when the central control point is not constantly attended. Each area of refuge shall be identified by a sign with the international symbol of accessibility stating that it is an area of refuge.

The area of refuge must be located along the path of an accessible means of egress from any accessible spaces it may serve. The length of the travel path is limited to the maximum travel distance permitted for the occupancy in accordance with Section 1016.1. An area of refuge may be incorporated into an enlarged floor-level landing of an enclosed exit access stairway or an interior exit stairway. Access is also acceptable from an area of refuge directly, either to a stairway complying with the provisions of Sections 1007.3 or to an elevator complying with Section 1007.4.

1007.6.1 Size. Each required area of refuge shall be sized to accommodate at least one wheelchair space not less than 30 inches by 48 inches (762 mm by 1,219 mm). Where the occupant load of the refuge area and the areas served by the refuge area exceeds 200, additional wheelchair spaces must be provided. Because wheelchair spaces are not permitted to reduce the required exit width and should be located so as to not interfere with access to and use of the fire department hose connections and valves, the designer needs to consider access to fire protection equipment and exit width when placing wheelchair spaces in the area of refuge.

1007.6.2 Separation. The primary concern for individuals awaiting assistance in an area of refuge is the intrusion of smoke and toxic gases into the refuge area. Therefore, the code requires a physical separation between an area of refuge and the remainder of the building. The separation is to be a smoke barrier complying with the provisions of Section 709. The smoke barrier is not required where the area of refuge is located within an enclosure for an exit access stairway or an interior exit stairway due to the inherent protection provided by such an egress enclosure. It is also permissible to create an area of refuge through the use of a horizontal exit.

1007.6.3 Two-way communication. Individuals awaiting assistance in an area of refuge must be provided with a communication means in order to contact a central control point. Where the central control point is not constantly attended, the area of refuge must be provided with a complying telephone with dial-out capability. Both audible and visible signals shall be provided. The requirements for such communication systems are established in Sections 1007.8.1 and 1007.8.2.

1007.7 Exterior area for assisted rescue. Item 10 of Section 1007.2 identifies an exterior area for assisted rescue as an acceptable portion of an accessible means of egress provided it complies with this section. The primary use of an exterior area for assisted rescue is to provide an exterior refuge area for those occupants unable to complete their egress travel due to the lack of an exterior accessible route from the building to the public way. Where exterior steps or other inaccessible site elements are present between the building's discharge level and the public way, it is permissible to use an exterior area for assisted rescue as an alternative to a fully accessible exterior path. Sized in accordance with Section 1007.6.1, the exterior area for assisted rescue must be separated from the interior of the building in a manner similar to that addressed for egress courts and exterior exit stairways. Where the exterior area for assisted rescue is located within 10 feet (3,048 mm) horizontally of the building's interior, a minimum 1-hour fire-resistance-rated wall with any openings protected for at least $^3/_4$ hour shall be provided from the ground level to a point at least 10 feet (3,048 mm) above the floor level of the exterior area for assisted rescue. Such protection need only extend to the roof line if it is less than 10 feet (3,048 mm) vertically above the floor level of the exterior area for assisted rescue. See Figure 1007-2. The required extent of an exterior area for assisted rescue is not described in the code, requiring an individual

Accessible Means of Egress **10**

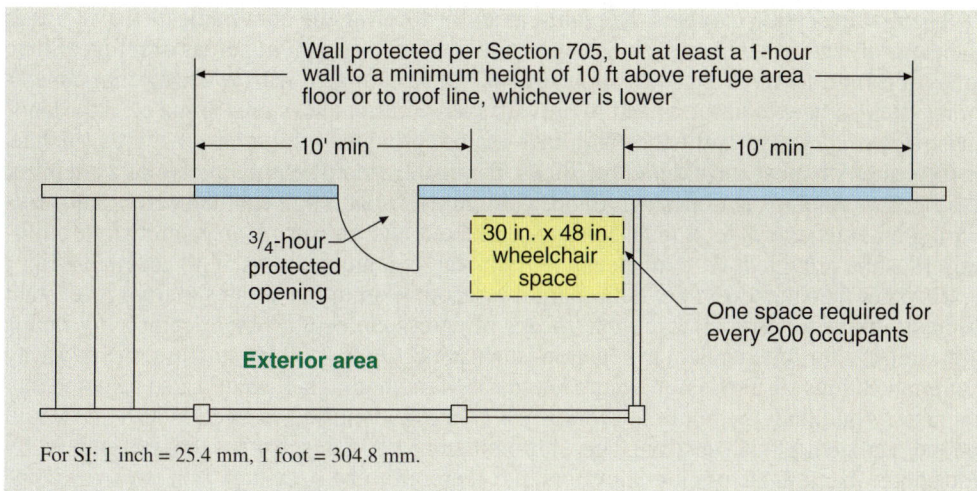

Figure 1007-2
Exterior area for assisted rescue.

evaluation where the size of the exterior area for assisted rescue becomes quite large. For an outdoor space to be considered an exterior refuge area, it must be at least 50-percent open so that toxic gases and smoke do not accumulate. Where an exterior stairway serves the exterior area for assisted rescue, an adequate distance between the handrails is mandated. In addition, complying signage is required per Section 1007.9, Item 2, to identify the exterior area as an appropriate refuge location.

Two-way communications. Unless provided in areas of refuge in multistory buildings, two-way communications systems must be located at the elevator landing of each accessible floor level other than the level of exit discharge. The system is intended to offer a means of communication to disabled individuals who need assistance during an emergency situation. Such a system can be useful not only in the event of a fire but also in the case of a natural or technological disaster by providing emergency responders with the location of individuals who will require assistance in being safely evacuated from floor levels above or below the discharge level. See Figure 1007-3.

1007.8

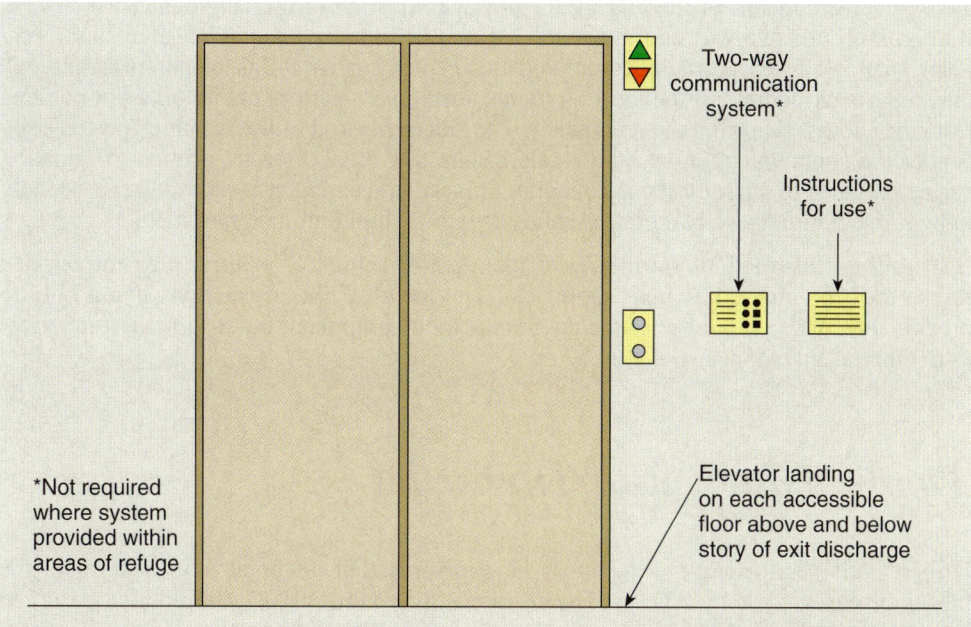

Figure 1007-3
Two-way communication system at elevator landing.

2012 International Building Code Handbook **357**

The first exception exempts the requirement for locating the communication systems at the elevator landings where the building is provided with complying areas of refuge. Since areas of refuge are required by Section 1007.6.3 to be equipped with two-way communication systems, there is limited need to provide such additional systems at the elevator landings. However, where multistory buildings are not provided with areas of refuge, such as is the case with most sprinklered buildings, the installation of communications systems at the elevator landings is important to those individuals unable to negotiate egress stairways during an emergency. As a result, most sprinklered and nonsprinklered multistory buildings must be provided with the means for two-way communications at all accessible floor levels other than the level of exit discharge. A second exemption applies to floor levels that use exit ramps as vertical accessible means of egress elements. Where complying ramps are available for independent evacuation, such as occurs in a sports stadium, the two-way communications system is not required at the elevator landings. It should also be noted that multistory buildings without elevators, such as those identified in Section 1104.4, would not be regulated by this section. Thus, all multistory buildings, except those exempted by Exception 2 and those without elevators, are required to be provided with two-way communication systems.

The arrangement and design of the two-way communication system are specified in Section 1007.8.1. In addition to the required locations specified in Section 1007.6.3 for areas of refuge or Section 1007.8 for elevator landings, a communication device is also required to be located in a high-rise building's fire command center or at a central control point whose location is approved by the fire department. The term "central control point" is not a defined term. However, given the intent and function of the two-way communication system, a central control point is a location where an individual answers the call for assistance and either provides aid or requests aid for an impaired person. A central control point could be the lobby of a building constantly staffed by a security officer, a public safety answering point such as a 9-1-1 center, a central supervising station, or possibly a nurses' station in a Group I occupancy. The key functions at the central control point are that an individual is always available to answer the call for assistance and can either provide assistance or is capable of requesting assistance. In addition, the communication system provides visual signals for the hearing impaired and audible signals to assist the vision impaired.

Guidance to the users of the two-way communication system is also specified. Operating instructions for the two-way communication system must be posted and the instructions are to include a means of identifying the physical location of the communication device. If a signal from a two-way communication system terminates to a public safety answering point, such as a fire department communication center, current 9-1-1 telephony technology only reports the address of the location of the emergency—it does not report a floor or area from the address reporting the emergency. The "identification of the location" posted adjacent to the communication system should ensure that most discrete location information can be provided to the central control point. This will aid emergency responders, especially in high-rise buildings or corporate campuses with multiple multistory structures.

1007.11 Instructions. Instructions on the use of the area of refuge or exterior area for assisted rescue must be provided where applicable. The intent of the instructions is not only to provide directions on the use of the communication equipment, but also to alert the users as to other available means of egress.

Section 1008 *Doors, Gates, and Turnstiles*

1008.1 Doors. This section applies to doors or doorways that occur at any location in the means of egress system. The provision found in Section 1020.2 should also be noted insofar as it will require that at least one exterior door that meets the size requirements of Section 1008.1.1 be provided from every building used for human occupancy. As doors pose a potential obstruction to free and clear egress, they are highly regulated.

Additional doors. The IBC establishes criteria for all egress doors, including those that are not required by Chapter 10. Such additional egress doors must comply with all the provisions of Section 1008.1 for exit doors. Where the doors are installed for egress purposes, whether or not required by the IBC, the building occupant would probably assume that they are a part of the means of egress system. Because the building occupant would then expect the door to provide a safe path from the space, it is imperative that such doors and doorways conform to all applicable code requirements of Section 1008. Two examples are shown in Figures 1008-1 and 1008-2.

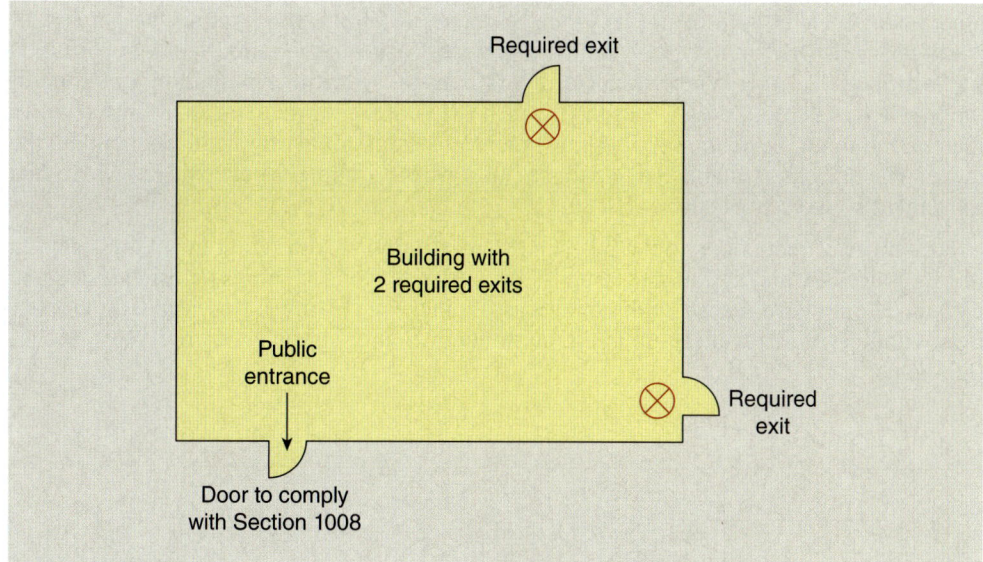

Figure 1008-1
Additional door.

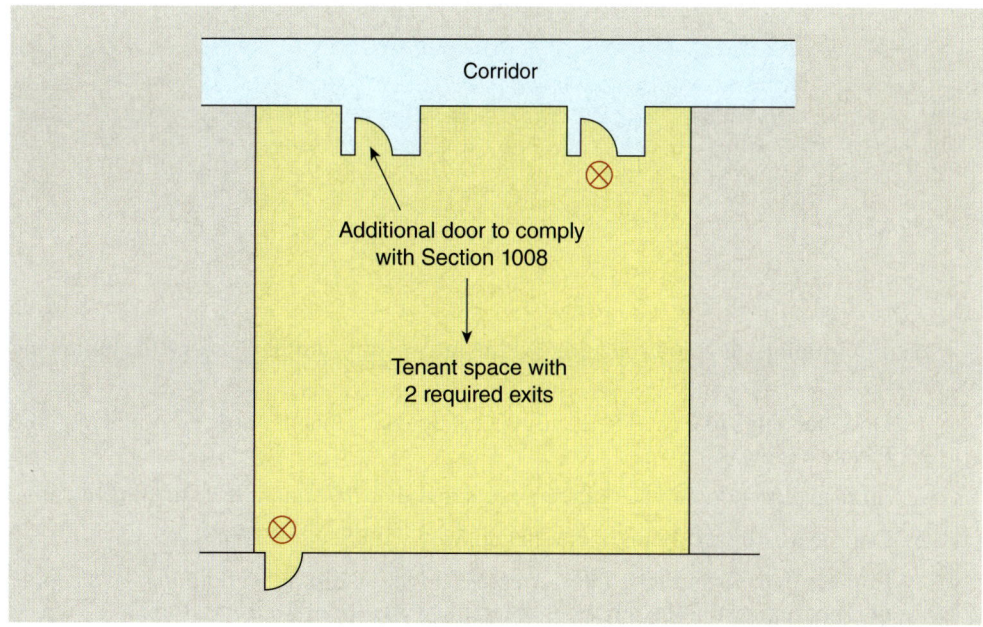

Figure 1008-2
Additional door.

Door identification. The primary gist of the provisions on door identification is that egress doors should be installed so that they are readily recognized as egress doors and are not confused with the surrounding construction or finish materials. It is important that they be easily discernible as doors provided for egress purposes. The corollary of this requirement is that exit doors should not be concealed. In other words, they should not be covered with drapes or decorations, nor should they be provided with mirrors or any other material or be arranged in a way that could confuse the building occupants seeking an exit.

1008.1.1 Size of doors. Every door used for egress purposes must comply with the width and height provisions of this section. It specifies that every required means of egress door opening be of such a size as to provide a clear width of at least 32 inches (813 mm), as illustrated in Figure 1008-3, with a minimum door height of 80 inches (2,032 mm). Again, the code requires that the net dimension of clear width be provided by the exit component. Thus, when a swinging door is opened to an angle of 90 degrees (1.57 rad), it must provide a net unobstructed width of not less than 32 inches (813 mm) and permit the passage of a 32-inch-wide (813-mm) object, unless a projection into the required width is permitted by Section 1008.1.1.1. Where a pair of doors is installed without a mullion, only one of the two leaves is required to meet the 32-inch (813-mm) requirement. As a final requirement, a minimum 41^1/$_2$-inch (1,054-mm) means of egress doorway width to facilitate the movement of beds is mandated for those portions of Group I-2 occupancies where bed movement is likely to occur.

A number of reductions to the 32-inch (813-mm) door-width requirement are found in the exceptions to this section. In Group I-3 occupancies, door openings to resident sleeping units need only have a clear width of at least 28 inches (711 mm). Accessible door openings within Type B dwelling units are permitted a minimum clear width of 31^3/$_4$ inches (806 mm).

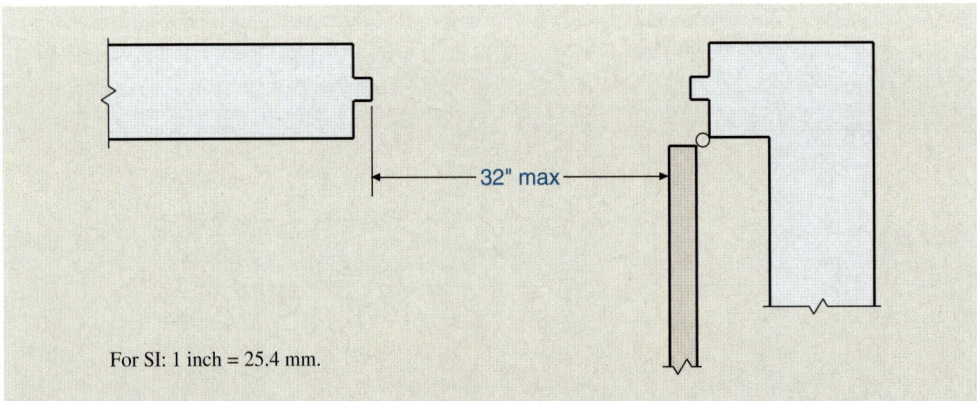

Figure 1008-3
Minimum clear width of egress door.

In addition, minimum door-opening widths are totally unregulated in the following locations:

1. Door openings in Group R-2 and R-3 occupancies that are not part of the required means of egress.

2. Storage closet doors where the closet is less than 10 square feet (0.93 m^2) in area.

3. Door leaves in revolving doors that comply with Section 1008.1.4.1.

4. In other than Group R-1, interior egress doors within a dwelling unit or sleeping unit not required to be an Accessible unit, Type A unit or Type B unit.

Throughout the rest of the code, the intent in specifying means of egress dimensions is to provide only minimum width. This particular section is at variance with that general approach, insofar as it limits the maximum width of any single swinging door leaf in a required egress doorway. As shown in Figure 1008-4, no such leaf may exceed 48 inches (1,219 mm) in width. The reason for this is that doors often do not receive the maintenance necessary to ensure their continued proper operation. The issue being addressed is that door leaves should be reasonably limited in width because wide doors require substantially greater maintenance to ensure reasonable opening effort, and this maintenance is not often provided. The limitation on maximum door width does not apply to complying revolving doors, or to doors in Group R-2 or R-3 occupancies that are not a portion of the required means of egress. It is also not applicable to nonswinging doors, such as overhead doors.

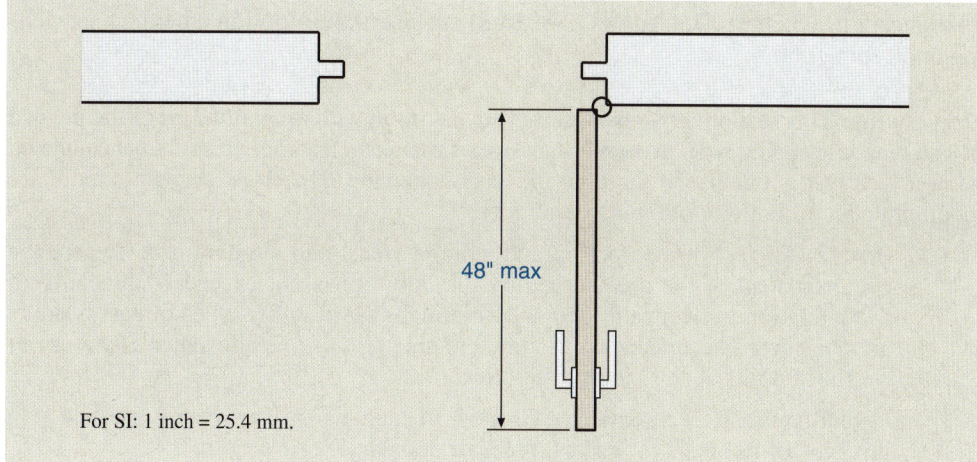

**Figure 1008-4
Maximum door leaf size.**

The required clear width of door openings shall be maintained up to a height of at least 34 inches (864 mm) above the floor or ground. Projections may then encroach up to 4 inches (102 mm) for a height between 34 inches (864 mm) and 80 inches (2,032 mm). See Figure 1008-5. The maximum 4-inch (102-mm) limitation is based partially on those

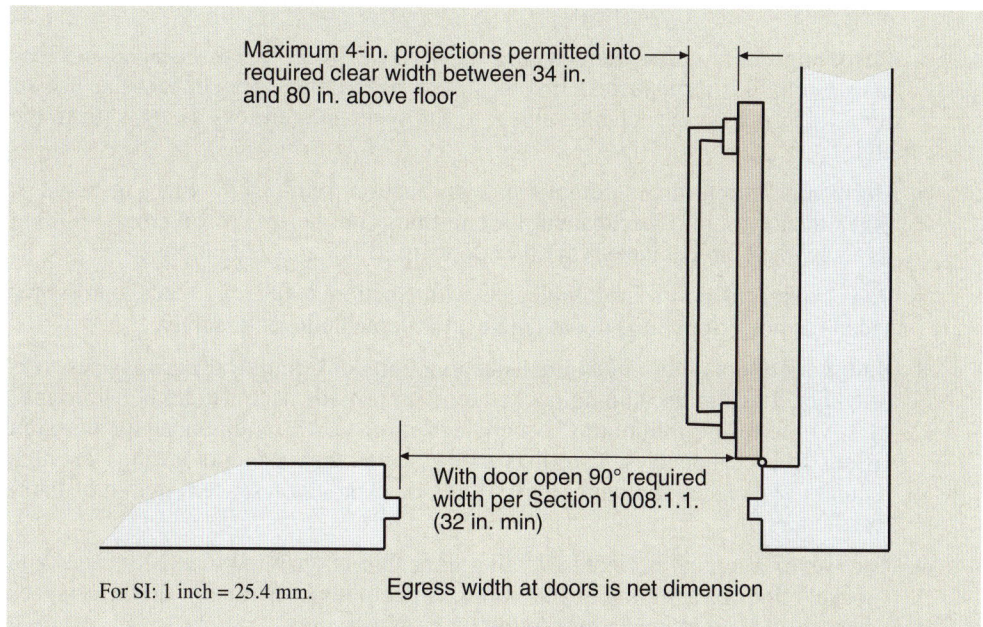

**Figure 1008-5
Egress door width.**

Means of Egress

accessibility provisions regarding protruding objects. Its application allows for the intrusion of panic hardware, or similar door-opening devices, into the required clear width. At a height of 80 inches (2,032 mm) or more above the walking surface, the projection is not regulated.

Although the general requirement for door height is a minimum of 80 inches (2,032 mm), the exceptions to Sections 1003.3.1 and 1008.1.1.1 permit door closers and stops to encroach into this clear height, provided a headroom clearance of at least 78 inches (1,981 mm) is maintained. Door openings at least 78 inches (1,981 mm) in height must be provided within a dwelling unit or sleeping unit. A minimum height of 76 inches (1,930 mm) is required for all exterior door openings in dwelling units other than the required exit door. Only Exception 5 is applicable to the height reduction at a required exit door; therefore, required means of egress door openings must have a minimum height of 80 inches (2,032 mm), 78 inches (1,980 mm) at closers and stops, in other than a dwelling or sleeping unit.

1008.1.2 Door swing. This section requires that every egress door, with exceptions, be of the pivoted or side-hinged swinging type. In most instances, it is necessary that the egress door encountered be of a type that is familiar to the user and easily operated. Therefore, swinging doors are required under all but the following conditions:

1. Private garages, office, factory and storage areas, and similar spaces where the occupant load of the area served by the doors does not exceed 10. Because of the limitation in occupant load and potential hazard, other types of egress doors are considered acceptable. A common application of this allowance is the use of overhead doors at self-storage facilities.

2. Detention facilities classified as Group I-3 occupancies. The security necessary in this type of use calls for special types of doors.

3. Critical care or intensive care patient room within suites of health-care facilities. In these areas, it is preferable to use sliding glass doors to allow for visual observation and the efficient movement of equipment.

4. Within or serving an individual dwelling unit in Group R-2 and R-3 occupancies. Because of the limited occupant loads involved, and the familiarity of the occupants with the doors encountered, door types other than swinging doors are permitted.

5. Revolving doors conforming with Section 1008.1.4.1, where installed in other than Group H occupancies. In other than hazardous occupancies, the use of revolving doors is acceptable subject to the special conditions as set forth in the code.

6. Horizontal sliding doors complying with Section 1008.1.4.3, where installed in other than Group H occupancies. Conditions for the use of horizontal sliding doors make them equivalent to other doors used in egress situations.

7. Power-operated doors in compliance with Section 1008.1.4.2. Safeguards provided for power-operated doors create an acceptable level of safety.

8. Bathroom doors within individual sleeping units of Group R-1 occupancies. It is often beneficial to use sliding pocket doors to provide access to hotel bathrooms, mostly due to the minimum 32-inch (813-mm) width requirement for doors in Group R-1 occupancies. Conflicts often occur between door swings and the required clearances for plumbing fixtures or clear floor space required at bathroom doors.

9. The use of a typical horizontal sliding door that is operated manually, such as a "pocket" door or a sliding "patio" door, is deemed acceptable in those instances where the occupant load served by the door is very low.

In addition, any exit door serving an area or room with an occupant load of 50 or more, or those serving any Group H occupancy, shall swing with the flow of egress travel. In 1942, 492 people died in the Cocoanut Grove fire in Boston. One of the significant contributing factors to that loss of life was the fact that the exterior exit doors swung inward. As a consequence, it was not possible to open the doors because of the press of the crowd attempting to exit the building. This incident was identified as the primary reason for changing building codes to require that, under certain circumstances, exit doors must swing in the direction of exit travel.

1008.1.3 Door opening force. Interior side-swinging doors other than fire doors must have a maximum opening force of 5 pounds (22 N). For doors that are sliding or folding, the door latch shall release when subjected to a 15-pound (66-N) force. This limitation to a 15-pound (66-N) force level also applies to all exterior swinging doors and interior swinging fire doors. In order to set the door in motion, a maximum force of 30 pounds (132 N) is mandated. The door shall swing to a fully open position when subjected to a force not greater than 15 pounds (66 N). These forces are applied to the latch side of the door. Most doors are openable with forces less than these maximum limits. However, when in doubt, the actual force required can be easily measured by use of a spring scale.

1008.1.4 Special doors. Based on the provisions in Section 1008.1.2, the code generally requires that doors in exiting systems be of the pivoted or side-hinged swinging type. In this section, four different types of doors are identified that may be used under very specific conditions.

1008.1.4.1 Revolving doors. Revolving doors continue to be used at building entrances. Where once used primarily in cold climates, they are now being installed in all regions, primarily as an energy-conservation measure. The use of revolving doors is specifically permitted as an alternative to enclosed vestibules where such vestibules are required by International Energy Conservation Code (IECC) at the entrances to commercial buildings. Exception 5 of Section 1008.1.2 permits the installation and use of revolving doors in all occupancies other than Group H when complying with this section. However, it is not permissible to use revolving doors to supply more than 50 percent of the required egress capacity, nor be assigned a capacity greater than 50 persons. Where used, the door must be an approved revolving door and comply with the specific requirements listed.

When revolving doors are installed, they must be of a type where the door leaves will collapse under opposing pressures to a book-fold position with the resulting parallel exit paths providing at least 36 inches (914 mm) of aggregate width. Location of the door in relationship to the foot or top of stairs or escalators is regulated, as is the maximum number of revolutions per minute. At least one conforming exit door shall be located in close proximity to the revolving door. In such an arrangement, the adjacent swinging door can be used to satisfy exit capacity requirements. The maximum force levels required to collapse a revolving door vary based on whether or not the door is to be used as an egress component.

1008.1.4.2 Power-operated doors. Power-operated sliding or swinging doors are often used at the main entry of a building, particularly in mercantile and business occupancies. The same doors are also typically an important aspect in the overall exiting system for the building. There are a number of different types of doors that use a power source to open a door or assist in the manual operation of the door. This may include doors with a photoelectric-actuated mechanism to open the door upon the approach of a person, or doors with power-assisted manual operation. Where such doors are used as a portion of the means of egress, they must be installed in accordance with this section. The main criterion concerns the capability of the door being opened manually in the event of a power failure. Essentially, doors shall have the capability of swinging, and they must be designed and installed to break away from any position in the opening and swing to the fully open position when an opening force not exceeding 50 pounds (222 N) is applied at the normal push-plate location. Doors that are fully power-operated must comply with BHMA A156.10, whereas BHMA A156.19 applies to power-assisted and low-energy doors.

1008.1.4.3 Horizontal sliding doors. Used as smoke and/or fire separation elements, these doors are normally in a fully open position and hidden from view. Closing only under specific conditions, they typically are part of an elevator lobby or similar protected area. Eight provisions are identified in the IBC that regulate the use of horizontal sliding doors as a component of means of egress. Fundamental to the use of horizontal sliding doors is that manual operation of the normally power-operated doors must be possible in the event of a power failure, and no special or complex effort or knowledge should be necessary to open the doors from either side.

1008.1.4.4 Security grilles. Because of the concern of exit doors being obstructed or even completely unusable, the use of security grilles is strictly regulated. By their nature, security grilles are difficult to operate under emergency conditions. Used frequently at the main entrances to retail sales tenants in a covered mall building, such grilles are also permitted in other Group M occupancies as well as Groups B, F, and S. Security grilles, either horizontal sliding or vertical, are only permitted at the main entrance/exit. During periods of time when the space is occupied, including those times where occupied by employees only, the grilles must be openable from the inside without the use of a key or special knowledge. They must be secured in the fully open position during those times where the space is occupied by the general public. Where two or more means of egress are required from the space, a maximum of 50 percent of the exits or exit access doorways are to be equipped with security grilles.

1008.1.5 Floor elevation. The purpose of this section is to avoid any surprises to the person passing through a door opening, such as a change in floor level. Therefore, it is necessary that a floor or landing be provided on each side of a doorway. It is further intended that such a floor or landing should be at the same elevation on both sides of the door. A variation up to $1/2$ inch (12.7 mm) is permitted because of differences in finish materials. See Figure 1008-6. Landings are required to be level, except exterior landings may have a slope of not more than $1/4$ inch per foot (6.4 mm per m) for drainage purposes.

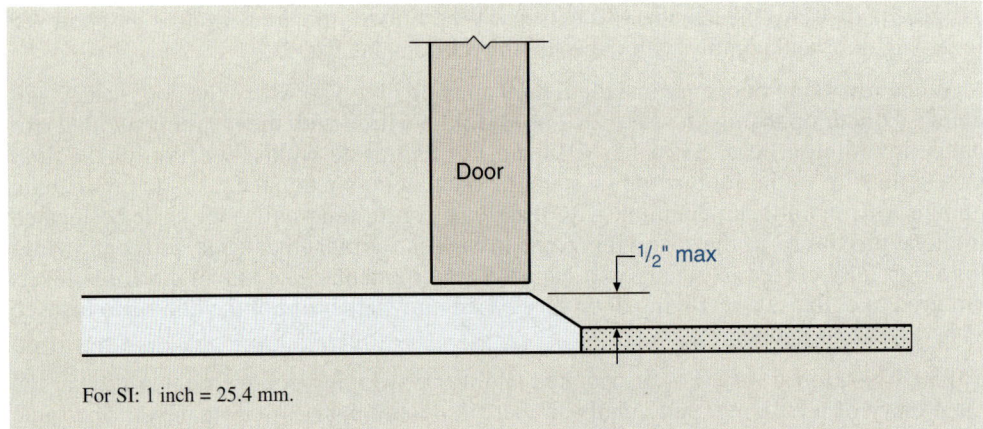

Figure 1008-6 Floor elevation.

Exceptions for individual dwelling units of Group R-2 and R-3 occupancies. An allowance is provided for individual dwelling units where it is permissible to open a door at the top step of an interior flight of stairs, provided the door does not swing out over the top step. The reason for permitting this type of arrangement in dwelling units is that as a building occupant approaches such a door from the nonstairway side, he or she must back away from the door in order to open it. This creates the need for a minimum landing to be traversed before the occupant can proceed to step down onto the stairs. In this situation, with minimal occupant load and familiarity with the unusual condition, the opening may occur at the top of the stairs, but the door must swing toward the person descending the stairs. In an ascending situation, the stair user should have little difficulty in opening

Doors, Gates, and Turnstiles 10

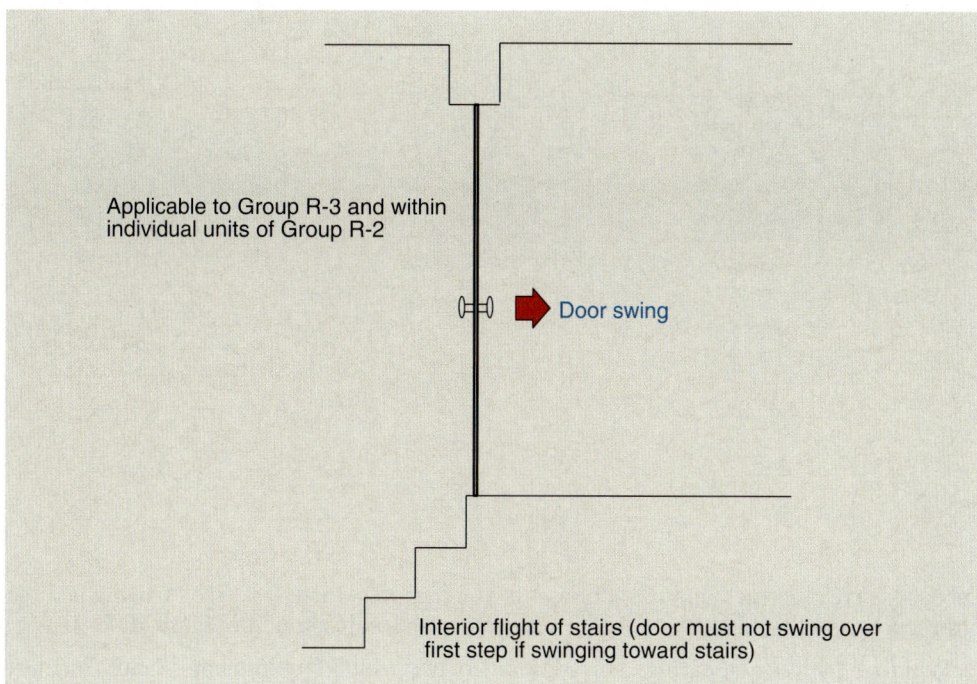

Figure 1008-7
Floor level at doors.

the door while standing on the stair treads, insofar as the door swings in the direction of travel. See Figure 1008-7. Also, in such occupancies it is permissible when screen doors or storm doors are installed, especially on the same jamb as the egress door, to swing them over stairs or landings.

In a Type B dwelling unit as provided in Chapter 11, a maximum drop-off of 4 inches (102 mm) is permitted between the floor level of the interior of the unit down to an exterior deck, patio, or balcony. This limited elevation change is consistent with the level of accessibility provided for travel throughout a Type B dwelling unit.

Exception for exterior doors. Reference is made to the first exception of Section 1003.5 regarding exterior doors in Group F, H, R-2, R-3, S, and U occupancies. Where such exterior doors are not required to be accessible, a single step having a maximum riser height of 7 inches (178 mm) is permitted. The reference to Section 1020.2 is extraneous information and it is not applicable in regard to the exception.

Landings at doors. This section contains the dimensional criteria for landings. It deals only with those landings where there is a door installed in conjunction with the landing. Landings at stairways and ramps are regulated by Sections 1009.8 and 1010.7, respectively.

1008.1.6

Required width of landings. The minimum required width of a landing is determined by the width of the stairway or the width of the doorway it serves. Figure 1008-8 depicts these relationships. The requirement is that the minimum width of the landing be at least equal to the width of the stair or the width of the door, whichever is greater. The code is concerned that doors opening onto landings should not obstruct the path of travel on the landing. In this regard, the code establishes two limitations. The first states that when doors open onto landings, they shall not project into the required dimension of the landing by more than 7 inches (178 mm) when the door is in the fully open position. Second, whenever the landing serves an occupant load of 50 or more, doors may not reduce the dimension of the landing to less than one-half its required width during the course of their swing. Stated from the positive direction, it requires that doors swinging over landings must leave at least one-half of the required width of the landing unobstructed. Although the obstruction of one-half of the required width of the landing might seem excessive, it must be remembered that when

10 Means of Egress

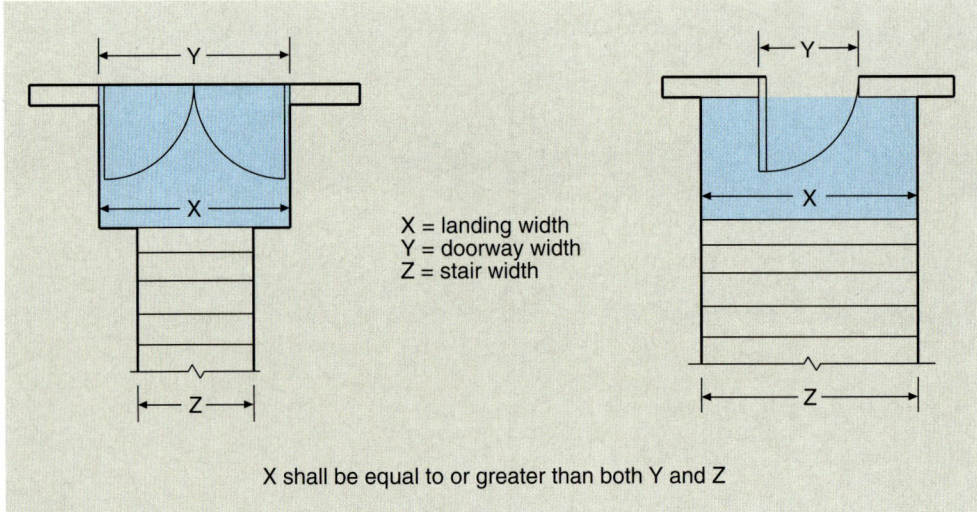

Figure 1008-8
Width of landing at doors.

the door is creating such an obstruction, it is in a position where it is free to swing and the obstruction is not fixed in place. These requirements are illustrated in Figure 1008-9.

Required length of landings. In addition to the width requirements, landings must generally have a length of at least 44 inches (1,118 mm) measured in the direction of travel. Where the landing serves Group R-3 and U occupancies, as well as landings within individual units of Group R-2, the length need only be 36 inches (914 mm). These code requirements are illustrated in Figure 1008-10.

It should be noted that these minimum dimensions for landings in both width and length will be modified by the provisions in Chapter 11 where the door or doorway is a portion of the accessible route of travel.

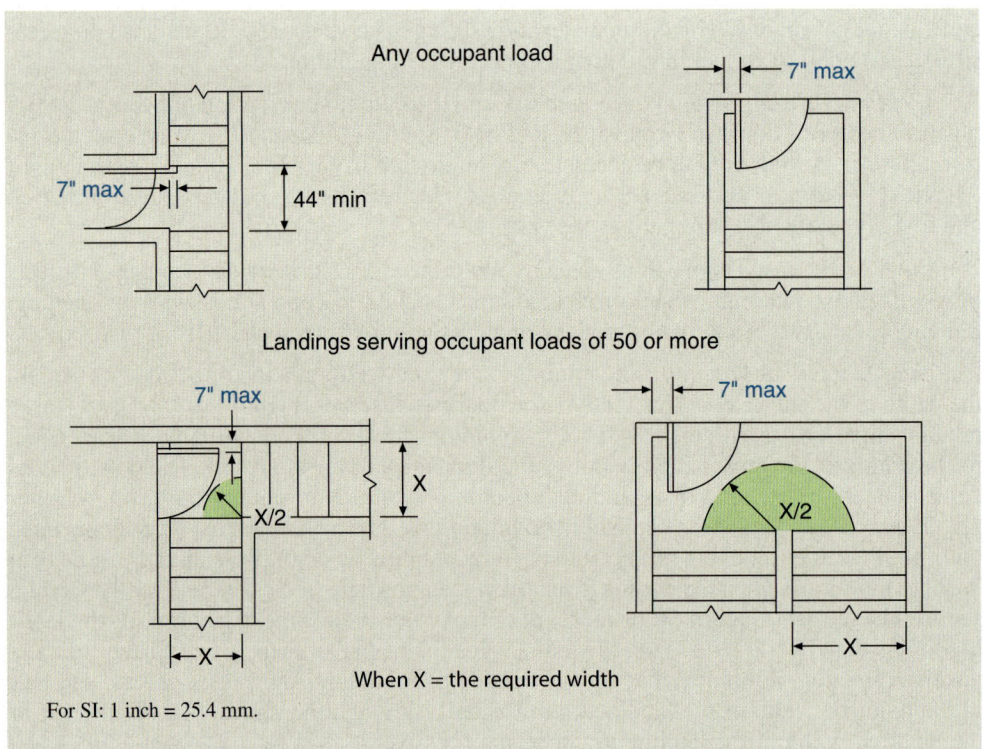

Figure 1008-9
Doors at landings.

366 2012 International Building Code Handbook

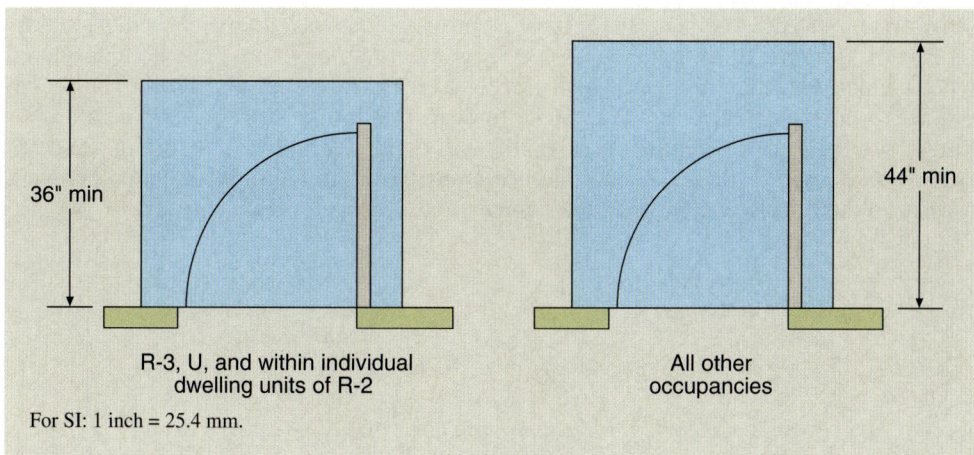

Figure 1008-10
Length of landings at doors.

Thresholds. Raised thresholds make using doors more difficult for people with disabilities. In addition, thresholds with abrupt level changes present a tripping hazard. As a general rule, raised thresholds should be eliminated wherever possible. Where thresholds are provided at doorways, it is necessary to limit their height to provide easy access through the doorway. Changes in floor level and raised thresholds are limited to $1/2$ inch (12.7 mm) in height above the finished floor or landing. Where raised thresholds or changes in floor level exceed $1/4$ inch (6.4 mm), the transition shall be achieved with a beveled slope of 1 unit vertical to 2 units horizontal (1:2) or flatter. See Figures 1008-11 and 1008-12. For a sliding door serving a dwelling unit, a maximum $3/4$-inch (19.1 mm) threshold is permitted. The threshold height at exterior doors may be increased to a maximum height of $7^3/_4$ inches (197 mm) in Group R-2 and R-3 occupancies, but only where such doors are not a required means of egress door, are not on an accessible route, and are not part of an Accessible unit, Type A unit or Type B unit.

1008.1.7

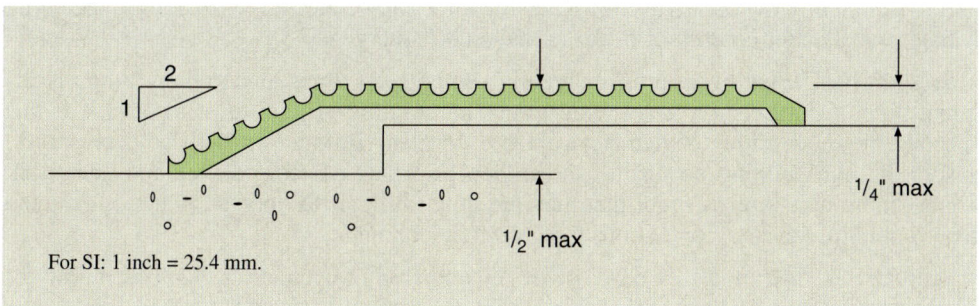

Figure 1008-11
Threshold height.

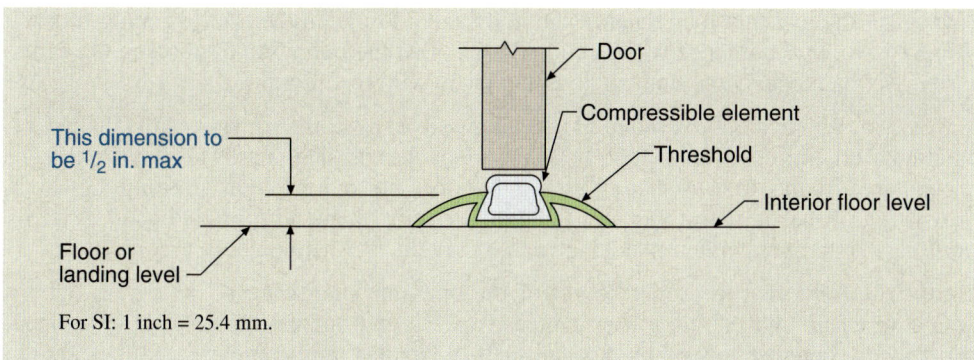

Figure 1008-12
Threshold height.

1008.1.8 Door arrangement. Adequate space must be provided between doors in a series to allow for ease of movement through the doorways. In other than dwelling units not considered Type A units, a minimum clear floor space of at least 48 inches (1,219 mm) in length is sized for a wheelchair user to negotiate through the door arrangement. Where a door swings into the floor space, the clear length shall be increased by the width of the door. As shown in Figure 1008-13, doors in a series must swing in the same direction or swing away from the floor space between the doors.

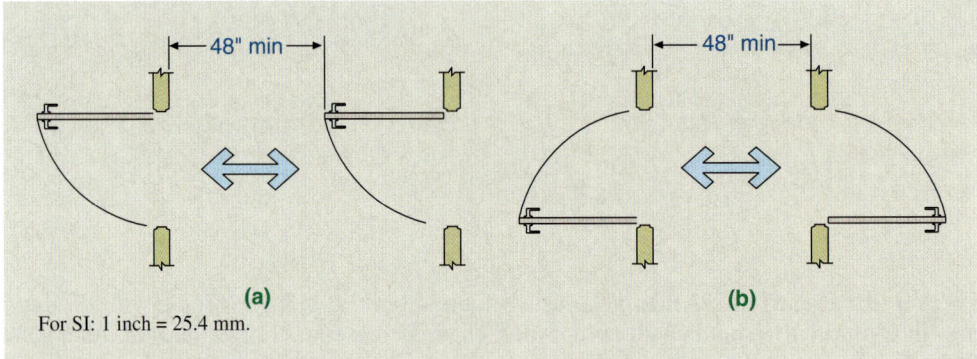

Figure 1008-13
Two doors in series.

1008.1.9 Door operations. This section, along with Section 1008.1.2, is particularly focused on the concept of ensuring that everything in the path of travel through the exit system, particularly doors, shall be under the control of and operable by the person seeking egress. Therefore, as a general statement, this section states that all doors in the egress system are required to be operable from the side from which egress is sought, without the need of a key or any special knowledge or special effort. If a key or special knowledge or effort is required, in all probability the door could not be readily openable by many building occupants. Such devices as combination locks are also prohibited on doors in exiting systems. Essentially, the code intends that the hardware installed be of a type familiar to most users—something that is readily recognizable under any condition of visibility, including darkness, and under conditions of fire or any other emergency.

In addition, the hardware must be readily operable. At times, one will encounter a different type of device such as a thumb turn. The building official must determine if this special type of operating device is acceptable. In many instances, it will be necessary to ensure that the building occupant can, in fact, grip the operating device and operate it. Some thumb turns are so small that they are quite difficult to operate, while others may require multiple twisting operations to achieve unlocking.

Another consideration that needs to be remembered in evaluating the acceptability of operating hardware is the fact that this hardware is going to be in place and in use over a substantial period of time. Unfortunately, doors and their operating hardware do not always get the constant maintenance that they should to keep them in operating order. It is imperative, in accordance with Section 3401.2, that the operation of doors in the egress system be maintained continuously in compliance with this section.

1008.1.9.1 Hardware. Where a door is required to be accessible under the provisions of Chapter 11, additional criteria come into play. It is important that all door hardware intended to be encountered by the door user be of a type that does not require tight grasping, pinching, or twisting of the wrist for operational purposes. Individuals with limited hand dexterity must be able to operate the unlatching or unlocking device without any special effort.

1008.1.9.2 Hardware height. The proper height of the operating hardware on an egress door is critical to ensure that the door user can easily reach and operate the unlatching or opening device. Therefore, operating devices such as handles, pulls, latches, and locks are to

Doors, Gates, and Turnstiles | 10

be installed at no less than 34 inches (864 mm) and no more than 48 inches (1,219 mm) above the finished floor. These limitations are not to be imposed on locks installed strictly for security reasons. An example might be a locking device installed on the bottom rail of an entrance/egress door of a convenience store or similar retail tenant.

Locks and latches. In regard to locking and latching door hardware, the IBC allows five significant exceptions to the general provision. In allowing these exceptions, it permits certain locking conditions that would appear to conflict with the basic requirement. However, in allowing these exceptions, the code often imposes certain compensating safeguards when the exit doors are to be locked or are provided with noncomplying hardware. It is the intent of the code that if the conditions are satisfied, the arrangement then essentially affords an equivalent level of safety as would be provided if the door were, in fact, readily openable at all times without the use of a key or any special knowledge or special effort. **1008.1.9.3**

1. This exception for locking hardware primarily concerns Group I-3 occupancies such as prisons, jails, correctional facilities, reformatories, and similar uses, where individuals are restrained or secured. By their nature, it is necessary that the occupants in these facilities be limited in their movement. Therefore, the IBC allows alternative locks or safety devices when it is necessary to forcibly restrain the personal liberties of the inmates or patients.

   Where a portion of a building is used for the restraint or security of five or fewer individuals, it is not considered a Group I-3 occupancy. Rather, it is anticipated that the secured area would simply be classified the same as the occupancy to which it is accessory. For example, up to five individuals could be restricted in an area such as a merchandise viewing room for customers in a jewelry store. In such situations, the allowance provided by Item 1 would be applicable for the egress door from the viewing room. A lock or latch that would prevent the expected unlocking or unlatching operation of the egress door would be acceptable. Thus, it is not necessary that the room or space be classified as a Group I-3 occupancy in order to apply this provision.

2. The provisions of this exception apply to the main exterior door or doors in Group A (having an occupant load of 300 or less), B, F, M, and S occupancies, and in all places of religious worship. It permits the main exterior entrance/exit doors to be equipped with a key-operated locking device if several conditions are satisfied. In the occupancies listed, it is reasonable to assume that if the building is occupied, the main entrance/exit will, in all probability, be unlocked. The first condition states that the locking device be readily distinguishable as locked. The use of an indicator integral to the locking device may assist in determining when an unsafe condition is present. A second condition requires that there be a sign that is readily visible, permanently maintained, and located on or adjacent to the door. This sign is required to read THIS DOOR TO REMAIN UNLOCKED WHEN BUILDING IS OCCUPIED. The letters must be at least 1 inch (25 mm) in height and placed on a contrasting background. Both of these requirements are for legibility. Although the language of the sign appears to apply without exception, there will be obvious situations where it should not be taken literally. For example, where an employee is working after hours and may be the lone occupant in the building, it is not the intent that the main doors remain in an unlocked position.

   Obviously, the sign or the presence of the sign is not going to ensure that the door is unlocked. However, it does advise the occupant that whenever the space is occupied, the law does require, in the interest of reasonable fire safety, that the door be unlocked. In the event the door is not unlocked, the occupant is advised that his or her life may be at risk. The occupant should seek to alleviate that situation. The limitation imposed by this provision is typically applied only when the public is involved. For example, it is not reasonable to assume that the subject door be unlocked during a time period the occupancy is limited to a janitor or other personnel. Note that the use of this exception may be revoked by the building official for due cause.

3. Where egress doors are used in pairs, it is anticipated that each leaf in the pair of doors should be provided with its own operating hardware. This exception permits a special arrangement, however, when the pair of doors is equipped with automatic flush bolts that are designed so that the act of releasing one of the leaves of the pair releases both leaves. It is critical that the door leaf be provided with the automatic flush bolts and have no door handle or other surface-mounted hardware. To ensure the immediate and reliable operation of the pair of doors, the unlatching of either leaf in the pair must be accomplished by not more than one operation.

4. The fourth exception refers to exit doors from individual dwelling or sleeping units of Group R occupancies. The general requirement of Section 1008.1.9.5 essentially prohibits the use of dead bolts or other security devices that would be installed in addition to the complying door hardware, as such an arrangement would require multiple operations to unlatch the door. This exception does permit, however, the use of a dead bolt, a security chain, or a night latch when the occupant load is 10 or less, on the condition that the device be openable from the inside without the use of a key or any special tool. It follows from the basic requirement of this section, however, that the device must not require any undue effort in order to unlatch the door and gain egress.

5. The listed test procedures for a fire door include the disabling of the door operation mechanism. This exception clarifies that once the minimum elevated temperature has disabled the door's unlatching mechanism, the resulting prevention of the door's operation is acceptable.

1008.1.9.4 Bolt locks. This section specifically prohibits the use of manually operated flush bolts or surface bolts insofar as these clearly do not conform with the intent of Section 1008.1.9. The use of such latching and/or locking hardware on means of egress doors is typically prohibited due to the inability of users to quickly identify and operate such devices under emergency conditions. Exception 1 permits the use of these types of locking devices on doors in individual dwelling units and sleeping units, provided the doors are not required for egress purposes. The second exception recognizes that in certain instances, doorway widths are dictated by the need to pass equipment through the openings. As a consequence, doorways, such as those to a storage room or equipment room, are frequently larger than would be required for exiting purposes alone. Therefore, where that is the case for a normally unoccupied space, manually operated bolts may be used on the inactive leaf. As the space is not normally occupied, this exception presents no significant hazard to life safety. The other side of this coin, however, is that any door leaf that is part of the required egress width must comply with all the requirements that apply to exit doors.

Pairs of doors are often desired in commercial occupancies to allow for the movement of furnishings, equipment, and machinery. Automatic flush bolts and removable center posts can be easily damaged and difficult to maintain in areas of frequent door usage. Exceptions 3 and 4, applicable to Group B, F, and S occupancies, address building functionality while maintaining a high degree of occupant safety. In these moderate-hazard occupancies, the occupants are typically very familiar with the building and the means of egress system. It is expected that they are aware of the operational limits of the inactive door leaf and efficiently use the active leaf. In both exceptions, it is mandated that the inactive leaf not be provided with any hardware, such as levers or panic devices that might cause the user to assume the door is an active egress door. The presence of door hardware on the active leaf will provide the necessary expectation to the building occupants, as occupants will naturally approach the active leaf having the appropriate hardware. If the building is sprinklered throughout with an NFPA 13 system, there is no limit on occupant load assigned to the pair of egress doors other than that based on the clear width of the active leaf. The inactive leaf cannot be assumed to provide for any required egress width. This ensures that the occupants have a fully complying door available for means of egress purposes.

The mandate that the building be fully sprinklered further enhances occupant safety and provides recognition from a general perspective of the value of a fire suppression system. An allowance is also provided for pairs of doors that serve relatively small numbers of people in the occupancies identified. The limit of 49 occupants is consistent with various other means of egress requirements that allow for a reduced level of protection where the occupant load does not exceed 50. Under this exception, the building is not required to be sprinklered. An overview of the provisions is shown in Figure 1008-14.

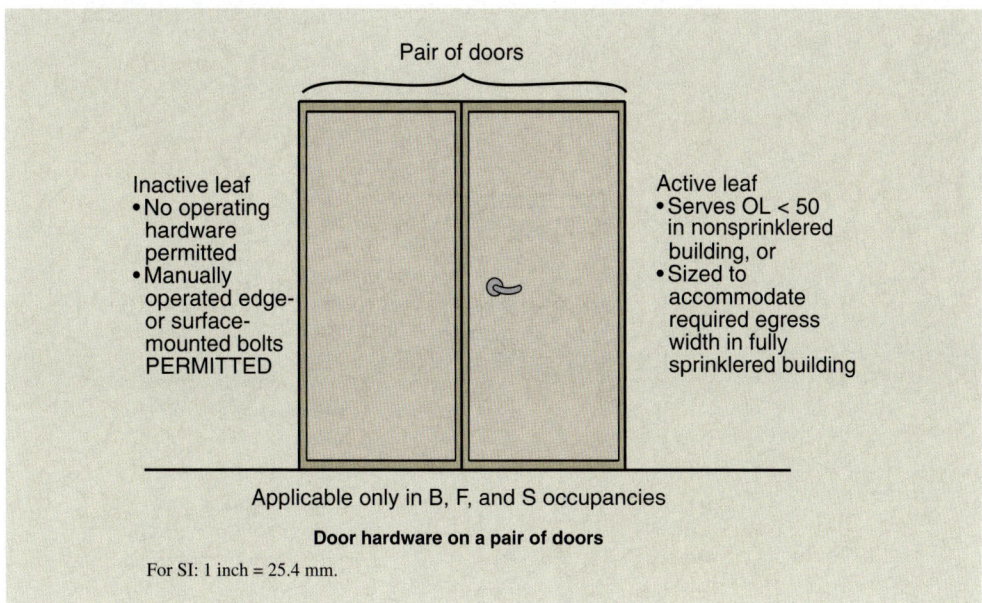

Figure 1008-14
Door hardware on a pair of doors.

A fifth exception addresses those patient room doors in Group I-2 occupancies where additional clear width is needed to allow for the efficient movement of patients and equipment. Where a pair of doors is used to provide the increased opening size, self-latching edge- or surface-mounted bolts may be installed on the inactive leaf. In order to distinguish that the inactive leaf is not an egress door, no lever device or other type of operating hardware is permitted. In addition, the inactive leaf cannot account for any of the minimum required egress width of $41^{1}/_{2}$ inches (1,054 mm).

Unlatching. The installation of multiple devices, or hardware requiring multiple operations, is inappropriate as well. Special effort and special knowledge is often necessary to open a door where more than one operation is required to unlock or unlatch the door. As a result, the multiple operations will typically result in an unacceptable delay in the egress efforts. Four exceptions set forth applications where multiple unlatching or unlocking operations are acceptable.

1008.1.9.5

Special locking arrangements in Group I-2. Many Group I-2 facilities house dementia and Alzheimer's patients. In order to balance the needs of the facility with the life safety of the occupants, the limitations on locking devices in these types of uses must allow for a safe and secure environment for these patients within the means of egress concepts of the code. Locks are permitted on means of egress doors that serve Group I-2 patients whose movement is restrained provided a number of conditions are met. Many of the conditions are similar to those set forth in Section 1008.1.9.7 for delayed egress locks. The building must be fully sprinklered or provided with an approved smoke or heat detection system. The doors must unlock upon actuation of the sprinkler or fire-detection system, the loss of power to the lock or lock mechanism, or by a signal from the nursing station or other approved location. The staff must also have the means to unlock the doors when necessary.

1008.1.9.6

As a further condition, it is mandated that occupants need only pass through one door equipped with this special egress device prior to entering an exit element.

Where patients with mental disabilities are housed, it is often necessary that they be restrained or contained for their own safety. In such cases, the level of restraint must be maintained even if the fire protection systems are activated or the power to the lock fails. However, it is still important that the emergency preparedness plan be developed and the clinical staff has the ability to monitor and enable the evacuation. See Figure 1008-15.

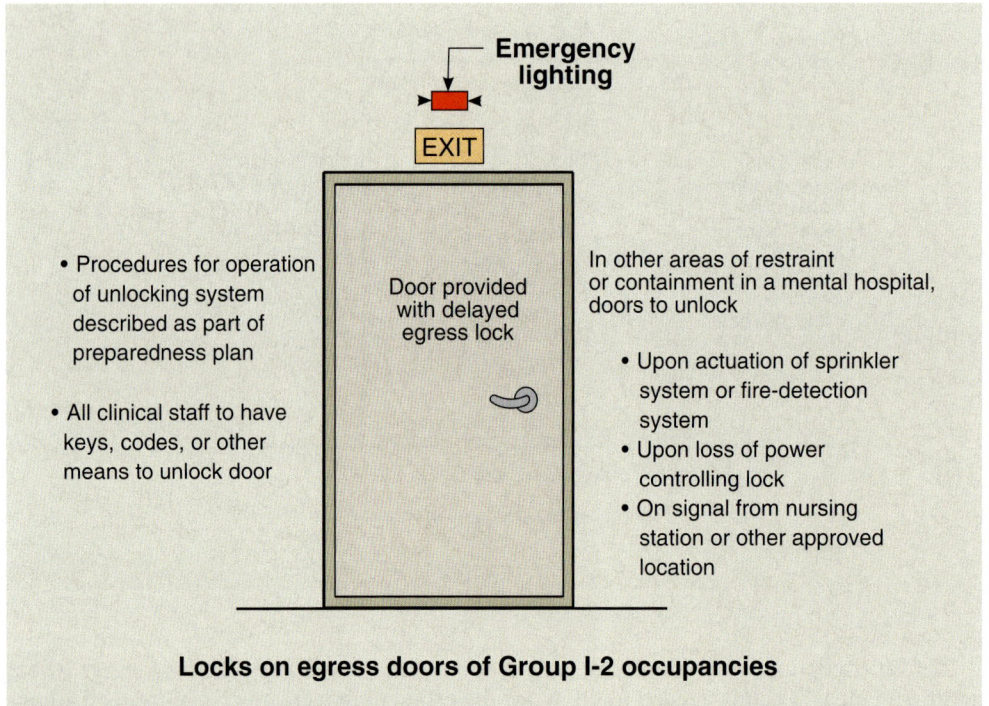

Figure 1008-15
Locks on egress doors of Group I-2 occupancies.

1008.1.9.7 Delayed egress locks. The building code provides for a degree of security to egress doors serving all occupancies other than Groups A, E, and H. This section allows the use of a door that has an approved, listed egress-control device with a built-in time delay under specific conditions. These devices were introduced in the code to resolve the problem of an exit door being illegally blocked by building operators desperate to stop the theft of merchandise through unsupervised, secondary exits. Institutional and residential occupancies are included because it is perceived that they also have security problems that need to be addressed. The devices are sometimes needed in nursing homes or group-care where facility operators must restrict patient egress while still maintaining viable exit systems.

It must be emphasized that under the conditions imposed by this section, and within the reliability of the automatic systems required, there will be no delay whatsoever at the exit in an actual fire emergency—the door will be immediately openable.

Delayed egress locking conditions. Several conditions should be emphasized at the outset:

1. Such devices may be used only in connection with the specifically listed occupancies.

2. The entire building in which the delayed egress locking system is installed must be completely protected throughout by either an approved automatic sprinkler system or an approved automatic smoke- or heat-detection system.

3. A building occupant shall not be required to pass through more than one door equipped with a delayed egress lock before entering an exit.

4. The door shall unlock in compliance with the following criteria: The device must immediately and automatically deactivate on activation of the sprinkler system or fire-detection system, and on the loss of electrical power to the egress lock. There must also be a way of manually deactivating the device by the operation of a signal from the fire command center where provided. Where the operating device is activated, it must initiate an irreversible process that will cause the delayed egress lock to deactivate whenever a manual force of not more than 15 pounds (66 N) is applied for a minimum period of 1 second to the operating hardware. The irreversible process must achieve the deactivation of the device within a time period of not more than 15 seconds from the time the operating hardware is originally activated. Where approved by the building official, a delay of not more than 30 seconds is permitted. Upon activation of the operating hardware, an audible signal shall be initiated at the door so that the person attempting to exit the building will be aware that the irreversible process has been started.

A sign must be installed on the door above and within 12 inches (305 mm) of the operating hardware so that the person seeking egress can be informed as to the type and nature of the egress lock. The sign must read PUSH UNTIL ALARM SOUNDS. DOOR CAN BE OPENED IN 15 (30) SECONDS. An additional requirement requires emergency lighting to be provided at the door where the delayed egress lock is used.

Delayed egress lock reactivation. The code emphasizes that, regardless of the means of deactivation, relocking of the device shall only be by manual means at the door. This requirement ensures that to relock the delayed egress lock, someone must go to the door itself, verify that the emergency no longer exists, and only then relock the door by manual means.

Access-controlled egress doors. Security concerns have prompted the need to provide controlled access at entrance doors to buildings or tenant spaces of certain occupancies. Therefore, this section permits the use of an approved entrance and egress access-control system at the main entrance of a Group A, B, E, I-2, M, R-1, or R-2 occupancy. Access-controlled egress doors will typically be locked from the exterior at all times. Locking from the egress side is also permitted under certain conditions. In order to ensure that the door is fully operable during a fire incident or other emergency, a number of criteria have been developed. For the most part, when a problem situation is identified, the door operates in a manner like any other egress door. Activation of the building fire alarm, automatic sprinkler system, or fire detection system will automatically unlock the door. Unlocking must also be possible from a location adjacent to the door, or occur when there is loss of power to the access-control system. Where assembly, business, educational, or mercantile buildings use this type of door at the main entrance, it must remain unlocked from the inside at any time the building is open to the general public. Very limited in application, this provision is potentially useful where there is a desire to have the entrance doors locked from the inside of the building during those periods of time where the public is not present. It is important to note that the security device must be listed in accordance with UL 294, *Access Control System Units*. **1008.1.9.8**

Electromagnetically locked egress doors. As a general rule, means of egress door hardware must be operable by manual operation to provide for occupant control of the egress system. Locking devices are typically prohibited, as they can interfere or prevent efficient egress through the door during an emergency situation. However, owner concerns that must be considered sometimes require a greater degree of security. In specific occupancies, doors in the means of egress are permitted to be electromagnetically locked if equipped with listed hardware that incorporates a built-in switch that interrupts the power supply to the electromagnetic lock and unlocks the door. The use of this type of locking system provides for a greater degree of security than that offered by other methods addressed in the code, including delayed egress locking systems and access control egress systems. **1008.1.9.9**

The allowance for electronically locked egress doors is limited to low- and moderate-hazard occupancies where security can be a major concern. The listed hardware that incorporates a built-in switch has been tested by UL under Special Locking Arrangements FWAX.SA6635. When the occupant prepares to use the door hardware, the method of operating the hardware must be obvious, even under poor lighting conditions. The operation shall be accomplished through the use of a single hand. This is consistent with the general requirement that the door be readily openable without the use of special knowledge or effort. The unlocking of the door must occur immediately on the operation of the hardware by interrupting the power supply to the electromagnetic lock. As an additional safeguard, the loss of power to the hardware shall automatically unlock the door.

This special type of locking device is permitted to be used in conjunction with panic hardware and fire exit hardware, but only where operation of such hardware also releases the electromagnetic lock.

1008.1.9.11 Stairway doors. The general requirement for interior stairway doors is that they be openable from both sides without the use of a key or special knowledge or effort. Such conditions allow for immediate access from the stairway enclosure to the adjacent floor area for emergency responders. In addition, in the unlikely event that the stairwell becomes untenable during evacuation procedures, occupants may reenter a floor level as an alternative means of egress. However, five exceptions are provided to modify this requirement.

1. Those doors that provide egress from the stair enclosure, discharging directly to the exterior or to an egress component leading to the exterior, are permitted to be locked only on the side opposite the direction of egress travel.

2. In high-rise buildings, stairway doors, other than exit discharge doors, may be locked from the stairway side. Under these conditions, such doors must be capable of being unlocked simultaneously without unlatching upon a signal from the fire command center. Although this exception is not limited to use in high-rise buildings, it is most commonly applied in such situations. When used in buildings that are not considered high-rise, the criteria of Section 403.5.3 may also be applied.

3. Doors are permitted to be locked from the side opposite the egress side for stairways serving four or fewer stories, where emergency personnel have the ability to simultaneously unlock the door. This action must be accomplished upon a signal from a single interior location at the building's main entrance, with the specific location for the actuating device likely approved by the fire code official. Where the building is provided with a fire command center, the signal must be actuated from within the center. It is important that the unlocking signal not deactivate the latching devices of the stairway doors. The doors must remain latched in order to maintain their integrity as fire door assemblies.

4. Applicable only to two-story buildings housing a Group B, F, M, or S occupancy, this exception permits the locking of the stairway door from the stairway side provided the only interior access to the tenant space is a single exit stair. Under specified conditions in Section 1021.2, a single exit is permitted from the basement or second story of the moderate-hazard occupancies scoped in this exception. Where a single exit stair is acceptable due to a limited occupant load and travel distance, it is permissible to lock the door to the tenant space from the stairway side.

5. This exception is similar in application to Exception 4.

1008.1.10 Panic and fire exit hardware. Basically, panic hardware is an unlatching device that will operate even during panic situations, so that the weight of the crowd against the door will cause the device to unlatch. This provision of the code, in harmony with the need to swing the door in the direction of egress travel, is intended to prevent the type of disaster experienced in the Cocoanut Grove fire in Boston in 1942. When a panic hardware device is installed on a door leaf, the press of the crowd, which prevented the opening of the door in the Cocoanut Grove fire, will ensure the automatic opening of the door.

Doors, Gates, and Turnstiles

Where panic hardware is provided, it is necessary that the activating member of the device extend for at least one-half of the width of the door leaf. This minimum length ensures that the unlatching operation will take place when one or more individuals impact the door. In addition, the device must be arranged so that a horizontal force not exceeding 15 pounds (66 N), when applied in the direction of exit travel, will unlatch the door.

Where required. Because of the large concentration of people in an assembly occupancy that may reach an exit door at about the same time during an emergency, the Group A occupancy is one of those occupancy groups that requires the installation of panic hardware listed in accordance with UL 305. Such hardware is also required in Group E occupancies for essentially the same reason. Therefore, in these two occupancies, any egress door provided with a latching or locking device that serves a room or area having an occupant load of 50 or more must be provided with panic hardware. In addition, in all Group H occupancies, panic hardware is required on every egress door regardless of the occupant load served. The potential life-safety hazard in these Group H occupancies is such that when it is necessary to evacuate this type of use, exiting from such spaces must be almost immediate. To facilitate the rapid escape from these occupancies, egress doors shall not be provided with a latch or lock unless it is panic hardware. Again, it should be noted that if a door is used in a situation where panic hardware might otherwise be required, it is not necessary to install panic hardware if the door has no means for locking or latching. If the door is free to swing at all times, there is no need to install panic hardware to overcome a lock or latch. Panic hardware is also not mandated on those doors considered as the main exit in a Group A occupancy having an occupant load of 300 or less where the provisions of Section 1008.1.9.3, Item 2, are applied, permitting the use of key-operated locking devices.

Panic hardware on balanced doors. Special care must be taken when installing panic hardware on balanced doors. In this instance, push-pad-type panic hardware is to be used, and it must be installed at one-half of the door width nearest the latch side to avoid locating the pad too close to the pivot point.

Section 1008.1.2 requires that exit doors should be side-hinged or pivoted swinging doors. The typical pivoted door has its top and bottom pivot points located near the edge of the door frame opposite the latch side. Balanced doors are nothing more than a specialized type of pivoted door in which the pivot point is located some distance inboard from the door edge, creating a counter-balancing effect. The length of the panic bar is limited for balanced doors because the door cannot be opened if the opening force is applied too close to the pivot point or beyond. Limiting the panic bar length to one-half the door width ensures that those who use the door will apply opening pressure at a distance sufficiently removed from the pivot point to allow the door to open. See Figure 1008-16.

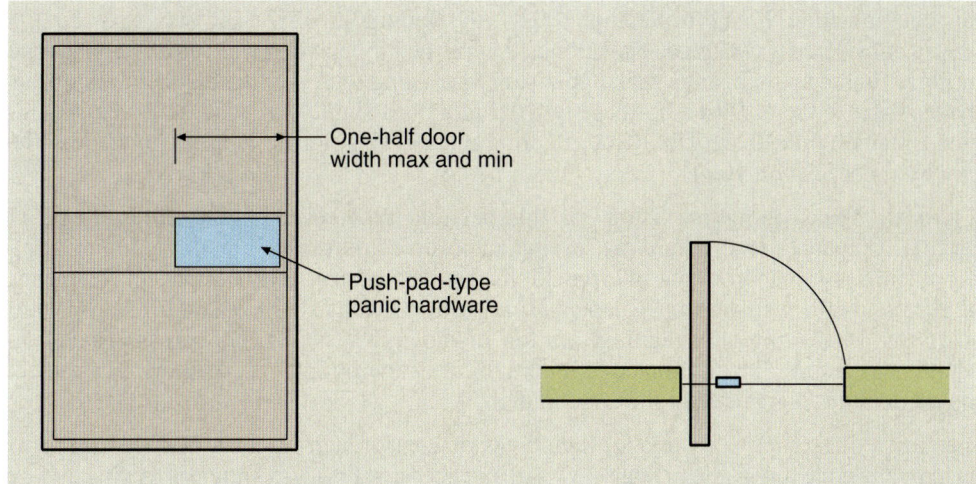

Figure 1008-16
Panic hardware on balanced doors.

10 Means of Egress

1008.2 Gates. This section serves as a reminder that gates within the means of egress system must comply with all of the requirements for doors. The single exception, applicable only to fences and walls surrounding stadiums, overrides the general door requirement that the doors must swing, and that the width of any leaf cannot exceed 4 feet (1,219 mm).

1008.2.1 Stadiums. Facilities such as stadiums may be enclosed by fencing or similar enclosures. The requirement for panic hardware does not apply to gates through such enclosures, provided that the gates are under constant and immediate supervision while the stadium is occupied. However, there must be a safe dispersal area of a size sufficient to accommodate the occupant load of the stadium based on 3 square feet (0.28 m^2) per person, and located between the stadium and the fence or other enclosure. Such a dispersal area must not be less than 50 feet (15,240 mm) from the stadium it serves.

1008.3 Turnstiles. In order to address safety concerns that are due to the use of turnstiles or similar devices that may be placed along the path of exit travel, this section regulates their use when located in a manner to restrict travel to a single direction. The general rule requires that each turnstile be credited with a maximum capacity of 50 occupants, and then only when specific conditions are met. When primary power is lost, the device shall turn freely in the direction of egress travel. Release shall also occur upon manual operation by an employee in the area. When determining the overall egress capacity, turnstiles may only be considered for up to 50 percent of the required capacity. Each device is limited to 39 inches (991 mm) in height and must have a minimum clear width of 16$^1/_2$ inches (419 mm) at and below the height of 39 inches (991 mm). Above the 39-inch (991-mm) height, a clear minimum width of 22 inches (559 mm) is necessary. Obviously, variations in these requirements are necessary where the turnstile is located along an accessible route of travel. Where turnstiles exceed 39 inches (991 mm) in height, they are regulated in a manner consistent with revolving doors.

To address the concern for use of such devices in large occupancies, Section 1008.3.2 requires a side-hinged swinging door for devices other than portable turnstiles. Required at a point where the occupant load served exceeds 300, a swinging door must be located within 50 feet (15,240 mm) of each turnstile. Portable turnstiles are designed to be moved out of the way for large occupancies such as sporting events.

Section 1009 *Stairways*

In order to apply the requirements of this section on stairways in an appropriate manner, the scope of the provisions must be determined. The definition of a stairway is critical for this determination. Found in Section 1002.1, the definition consists of two parts. First, a stair is considered a change of elevation accomplished by one or more risers. Second, one or more flights of such stairs make up a stairway, along with any landings and platforms that connect to them. Based on these two definitions, a single step would also be considered a stairway under the IBC. Both interior and exterior stairways are regulated by the provisions of Section 1009.

1009.1 General. The scoping provision for this section indicates that the requirements of Section 1009 apply to any stairway serving an occupied portion of a building, eliminating the potential for inappropriate interpretations that view stairways not required as a means of egress as not regulated by Chapter 10 nor the provisions of Section 1009. Whether stairways are serving as a required portion of the egress system or simply installed in additional numbers beyond the code minimum, it is appropriate for all stairways to meet the minimum safeguards that the code intends.

1009.2 Interior exit stairways. There are two types of interior stairways established and regulated by the code: interior exit stairways and exit access stairways. An interior exit stairway is defined in Section 202 as "an exit component that serves to meet one or more means of egress design requirements, such as required number of exits or exit access travel distance, and provides for a protected path of egress travel to the exit discharge or public way." Interior exit stairways are to be used in all cases where vertical travel must be within an "exit" component. The specific provisions addressing interior exit stairways are found in Section 1022.

Because travel within the exit access portion of the means of egress system is limited, it is necessary that occupants reach an exit within a prescribed distance. In multistory buildings, this would typically result in travel to an interior exit stairway. In addition, Section 1021.1 requires that each story above the second story of a building be provided with at least one interior exit stairway (or exterior exit stairway). The use of one or more interior exit stairways provides protected egress travel for occupants as they proceed to the public way.

Exit access stairways. By definition, an exit access stairway is "an interior stairway that is not a required interior exit stairway." Any stairway that is not protected by a fire-resistive enclosure as regulated by Section 1022 and used as an "exit" component in the means of egress is to be regulated by this section. Located within the exit access portion of the means of egress, an exit access stairway has no inherent fire protection. Travel on an exit access stairway is considered to be unprotected travel. The base mandate that floor openings created by exit access stairways be enclosed is based on the concept of protection of vertical openings as established in Section 712. The required enclosure is intended to restrict the spread of fire, smoke, and gases vertically through a multistory building and is not intended for the protection of occupants during the exiting process. Multiple exceptions are provided, which either eliminate the requirement for any type of enclosure or provide some other means of addressing openings between stories.

1009.3

Exception 1 emphasizes a fundamental concept of the IBC that is applicable under most conditions—two stories may be open to each other provided such stories do not atmospherically communicate with any other stories. The intent of Exception 1 is that when any two stories are open to one another, neither may be open to yet another story. This is to prevent the formation of an unprotected vertical shaft through more than two stories. This does not mean that the stories under consideration cannot have access to other floors. They can, but complying enclosures must be provided in order to do so. However, this provision does not allow for the enclosure of an interior exit stairway to be interrupted by unenclosed flights. This exception is consistent with the allowance established in Section 712.1.8 addressing vertical opening protection in that it is not applicable in Group I-2 and I-3 occupancies.

Exceptions 3 and 4, applicable only in fully sprinklered buildings, address limited-size exit access stairway openings that are protected by sprinklers and draft curtains rather than fire-resistance-rated enclosures. By limiting the size of the openings, along with the installation of draft curtains and closely spaced sprinklers, the code assumes that the vertical openings in sprinklered buildings do not present an untenable condition. The purpose of the required draft curtain is to trap heat so that the sprinklers will operate and cool the gases that are rising. The curtain is not a fabric; it is constructed of materials consistent with the type of construction of the building. In other than mercantile and business occupancies, the use of this method is limited to openings connecting four or fewer stories. Although required exit access travel is permitted on stairways meeting Exception 3 or 4, such travel is limited to only one adjacent story as set forth in Section 1021.3.1.

Construction. The construction of enclosures for exit access stairways, where enclosure is required, is to be consistent with that required for shaft enclosures as established in Section 713. Fire barriers and/or horizontal assemblies are to be used, and openings, penetrations, joints, and air openings are to be protected, provided such elements are permitted in the stairway enclosure.

1009.3.1

Means of Egress

1009.4 Stairway width. The provisions concerning width of stairways are analogous to the provisions relating to corridors discussed in Section 1018.2. If the stair is subject to use by a sufficiently large occupant load, the minimum required width of the stair is determined by using the formula stated in Section 1005.1. Otherwise, the minimum stairway width cannot be less than the width established by this section. In general terms, the minimum required width of any stair must be at least 44 inches (1,118 mm). In the event the stairway serves an occupant load of 49 or less, the required minimum width of the stairway is only 36 inches (914 mm). The entire occupant load of the floor served by the stairway is considered, rather than divided by all available stairways. See Application Example 1009-1. Other modifications to the width requirements apply to spiral stairways as addressed in Section 1009.12, aisle stairs regulated under the provisions of Section 1028, and stairways that are provided with an incline platform lift or stairway chairlift. Generally, when the code specifies a required width of a component in the egress system, it intends that width to be the clear, net, usable unobstructed width. However, handrails and other projections are permitted to encroach into the required width of a stairway as established by the provisions of Section 1012.8.

Application Example 1009-1

GIVEN: A two-story building with two stairways serving the second floor. The occupant load of the second floor is 68.

DETERMINE: The minimum required width of the stairway.

SOLUTION: Because both of the stairways are required to serve the floor, the minimum required width of each stairway is 44 inches. The occupied load is not divided (as in the case of distributing calculated width) between the two stairways.

For SI: 1 inch = 25.4 mm.

1009.5 Headroom. A minimum headroom clearance of 6 feet 8 inches (2,032 mm) is required in connection with every stairway. Such required clearances shall be measured vertically from the leading edge of the treads to the lowest projection of any construction, piping, fixture, or other object above the stairs, and shall be maintained for the full width of the stairway and landing. See Figure 1009-1. This specific height requirement overrides the general means of egress ceiling height requirement found in Section 1003.2 and is modified for spiral stairways by Section 1009.12.

1009.7.2 Riser height and tread depth. This section provides for a maximum riser height of 7 inches (178 mm), a minimum riser of 4 inches (102 mm), and a minimum tread run of 11 inches (279 mm) for each step on any stairway. These limiting dimensions are identified in Figure 1009-2. Variations in the requirements for treads and risers apply to alternating tread devices, spiral stairways, and stairs serving as aisles in assembly seating areas. These variations are discussed elsewhere in this chapter. Another exception allows $7^{3}/_{4}$ inches (197 mm) maximum and 10 inches (254 mm) minimum for rise and run, respectively, for stairways in Group R-3 occupancies, within dwelling units in Group R-2 occupancies, and in Group U occupancies accessory to a Group R-2 or R-3 dwelling unit.

The 7-inch (178-mm) rise and 11-inch (279-mm) run figures for the steps are based primarily on safety in descending the stairs and are the result of much research. Probably at no prior time in the history of codes has the proportionate of stairs enjoyed a better foundation in research.

As one descends a stairway, balance is essential for safety. Therefore, the tread run must be of such a dimension as to permit the user to balance comfortably on the ball of the foot. The appropriate combination of riser height and tread run provides the proper geometry to enable the user to accomplish the necessary balance to descend the stairway with reasonable safety. Consistent with the importance of the tread dimension, the method of measurement of the tread is expressly stated. Specifically, tread depth (or run) is that

Stairways 10

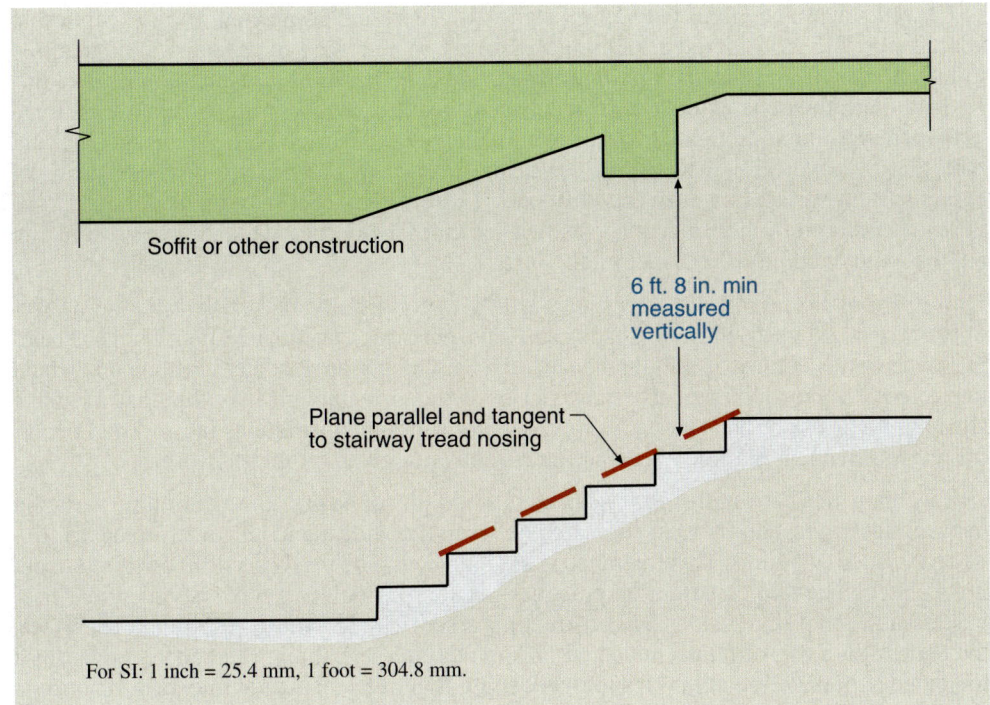

Figure 1009-1
Stairway headroom clearance.

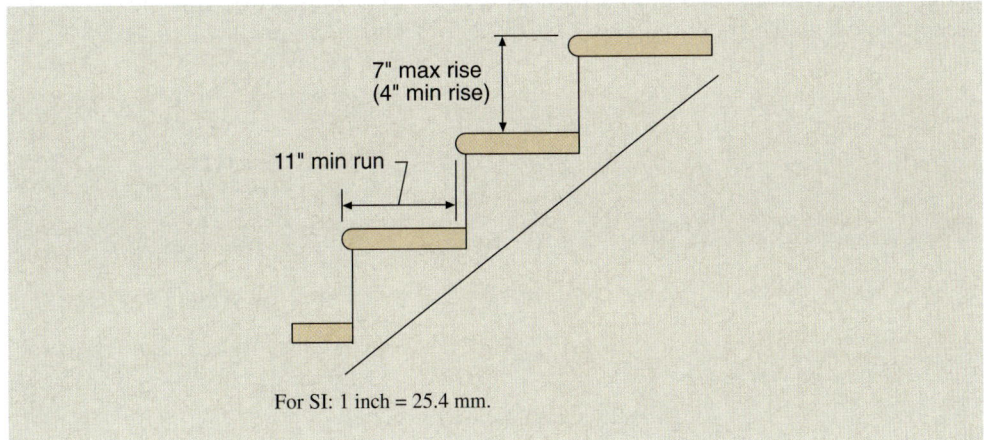

Figure 1009-2
Rise and run.

distance measured horizontally between vertical planes passing through the foremost projections of adjacent treads. As such, the tread dimension is the net gain in the run of the stair. Tread dimension is measured in this manner because any tread surface underneath the overhang of a sloping riser or nosing on the tread above is not available to the person descending the stair. Because descending is the more critical direction, proper dimension of the tread is of paramount importance.

Studies of people traveling on stairways have shown that probably the greatest hazard on a stair is the user. Inattention has been identified as the single factor producing the greatest number of missteps, accidents, and injuries. Inattention frequently results from the user being overly familiar with the stair and its surroundings. It often results from a variety of distractions. It is critical to stair safety that the stair user be attentive to the stair, although attentiveness cannot be codified or dictated. However, stair design and geometry, which usually trigger human error, can be controlled.

Curved stairways, along with spiral stairways, represent somewhat of an exception to what is normally considered a traditional stairway. Where the typical stairway is required to have treads of a consistent and uniform size and shape, these two stairs may have different dimensional characteristics from adjacent treads and vary from one end of the tread to the other. Alternating tread devices are another type of stair whose design is inconsistent with that of a typical stairway. The use of these stairways as a portion of the means of egress system varies; however, the only one of these stairs that may be used as a part of the means of egress in all occupancies and locations is the curved stair. These stairs are addressed in Sections 1009.11 through 1009.13.

1009.7.4 Dimensional uniformity. A significant safety factor relative to stairways is the uniformity of risers and treads in any flight of stairs. The section of a stairway leading from one landing to the next is defined as a flight of stairs. It is very important that any variation that would interfere with the rhythm of the stair user be avoided. Although it is true that adequate attention to the use of the stair can compensate for substantial variations in risers and treads, it is all too frequent that the necessary attention is not given by the stair user.

To obtain the best uniformity possible in a flight of stairs, the maximum variation between the highest and lowest risers and between the widest and narrowest treads is limited to $^3/_8$ inch (9.5 mm). This tolerance is not intended to be used as a design variation, but it does recognize that construction practices make it difficult to get exactly identical riser heights and tread dimensions in constructing a stairway facility in the field. Therefore, the code allows the variation indicated in Figure 1009-3. Although the code allows for a tolerance in both the tread depth and riser height, this tolerance is not intended to permit a reduction in the minimum tread depth or increase in the maximum riser height established by the IBC. For example, a tread depth of $10^5/_8$ inches (270 mm) is not permitted if a minimum of 11 inches (279 mm) is required, nor is a riser height of $7^3/_8$ inches (188 mm) acceptable where the code limits riser height to 7 inches (178 mm).

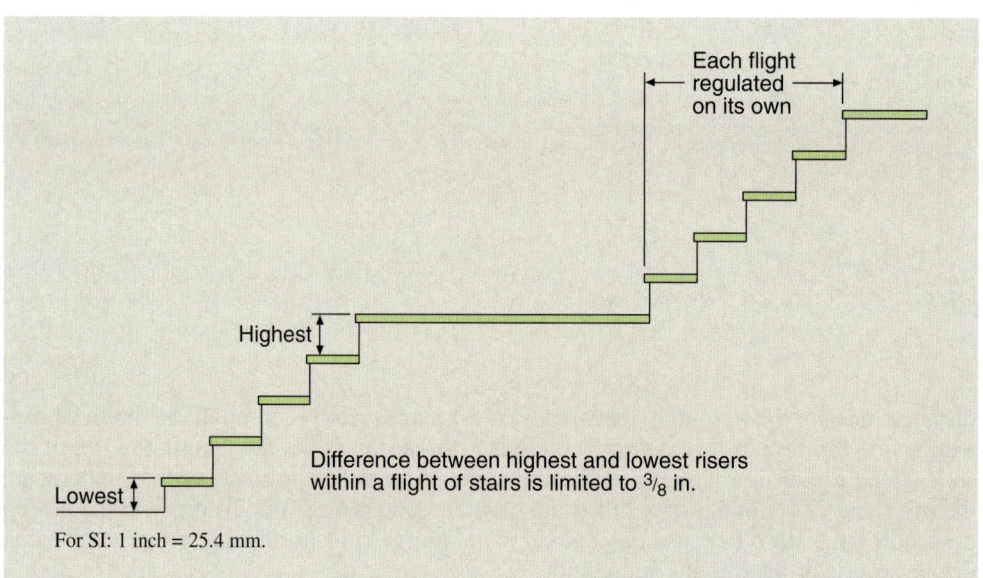

Figure 1009-3
Stair tolerance.

Under the provisions of Section 1028.11.2, riser height nonuniformity is permitted for aisle stairs where changes in the gradient of the adjoining assembly seating area are necessitated in order to maintain adequate lines of sight. Another exception permits the transition between a typical straight run stairway and consistently shaped winders under the conditions of Section 1009.7.

With respect to variation, it is recognized that stairs occasionally descend or rise to areas where the ground or the finished surface is sloping. Where this occurs on private property, the code anticipates that the landing of the stairs be level so that there will not be any variation in the riser height across the width of the stair at that point. However, from time to time, stairs will land on spaces that are not under the control of the property owner, such as a public sidewalk. Therefore, a certain degree of slope across the width of the stair is permitted, resulting in a variation of the height of the riser from one side of the stair to the other. Where this occurs, the height of such a riser may be reduced along the slope to less than 4 inches (102 mm), and the maximum permitted slope shall not exceed 1 unit vertical in 12 units horizontal (8.3-percent slope). Figure 1009-4 shows this condition. It should be clarified that the sloping surface is intended to be an established grade, such as a walkway, public way, or driveway.

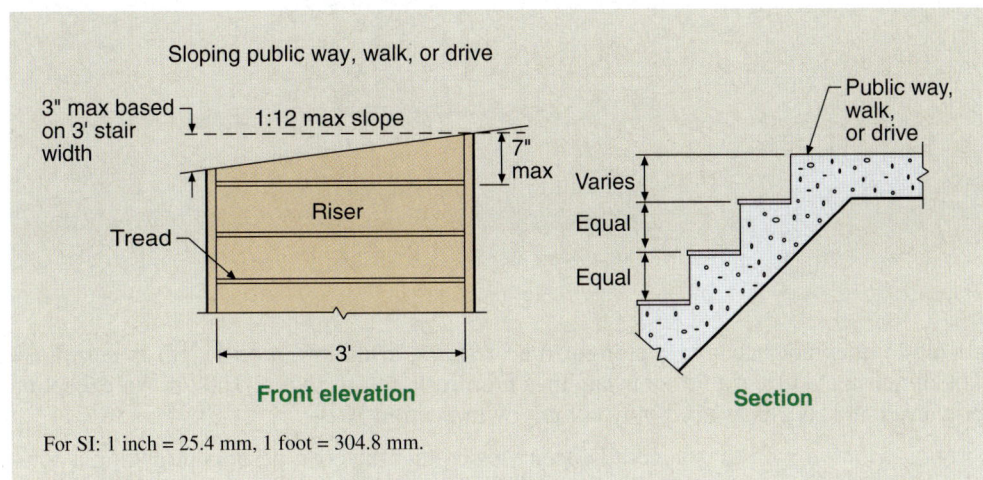

Figure 1009-4
Sloping landing.

1009.7.5 Nosing and riser profile. Tread nosings are limited to a maximum radius or beveling of $^9/_{16}$ inch (14.3 mm) and shall not extend more than $1^1/_4$ inches (32 mm) beyond the tread below. Nosing projections are to be consistent throughout the stair flight, including where the nosing occurs at the floor at the top of a flight. Risers are to be solid and, if sloped, slope no more than 30 degrees (0.52 rad) from the vertical. See Figure 1009-5 for an overview of these provisions.

There are four exceptions that exempt the requirement for solid risers. First, the risers need not be solid where the stairway does not need to comply with the provisions of Section 1007.3 for a stairway used as an accessible means of egress. However, some method of construction must be used to limit the openings between treads to less than 4 inches (102 mm). Second, solid risers are not required in nonpublic areas of Group F, H, I-3, and S occupancies. The third and fourth exceptions indicate that solid risers are not required for complying spiral stairways and alternating tread devices.

1009.8 Stairway landings. Landings are discussed to some extent under Section 1008.1.6, which covers landings that are used with adjoining doors. This section covers landings associated with stairs in creating a stairway. The basis for determining the required dimensions of landings is simple; every landing must be at least as wide as the stair it serves. It must also have a dimension measured in the direction of travel not less than the width of the stairway. However, in those instances where the stair has a straight run, the landing length need not be more than 48 inches (1,219 mm) measured in the direction of travel. As a stairway changes direction at a landing, it is important that the actual width of the stairway be maintained throughout the travel, even if the actual width is greater than the required width. Where the stairway reaches capacity across its width during egress, a landing of reduced size will create an obstruction to the flow pattern that has been established. Because this

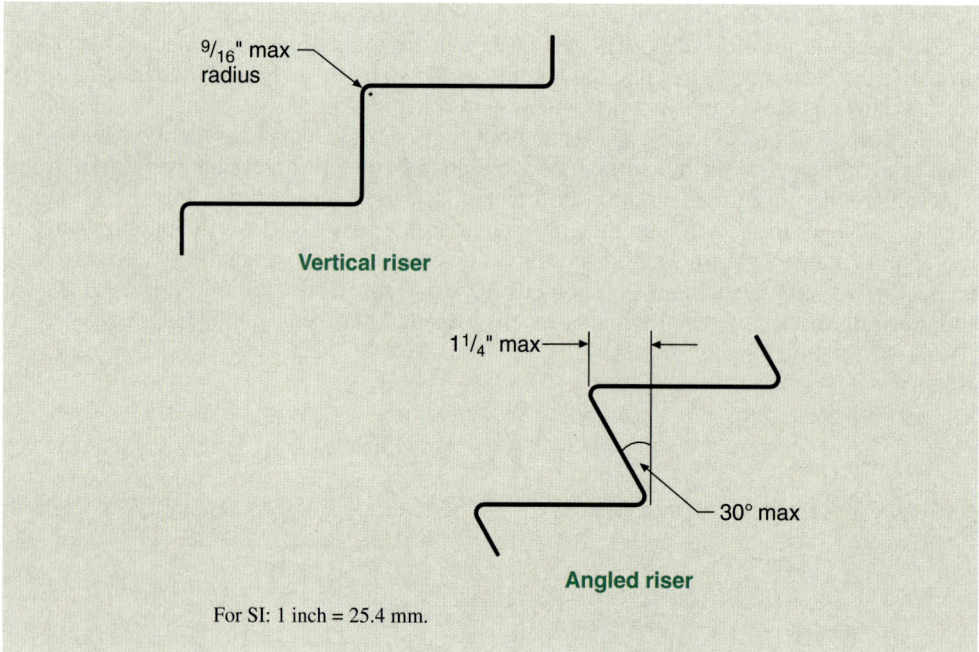

Figure 1009-5 Stair nosing.

condition does not occur in a straight run of stairs, a limitation on length is permitted. The dimensional criteria for stair landings are illustrated in Figure 1009-6. An exception provides that aisle stairs need only comply with Section 1028.

Shape of landing. It has generally been viewed that the code permits providing a complete curve with the radius equal to the width of the stairway, as shown in Figure 1009-7. This viewpoint recognizes that egress travel through the landing would not generally extend into the corners. Application of this approach would provide for the presence of

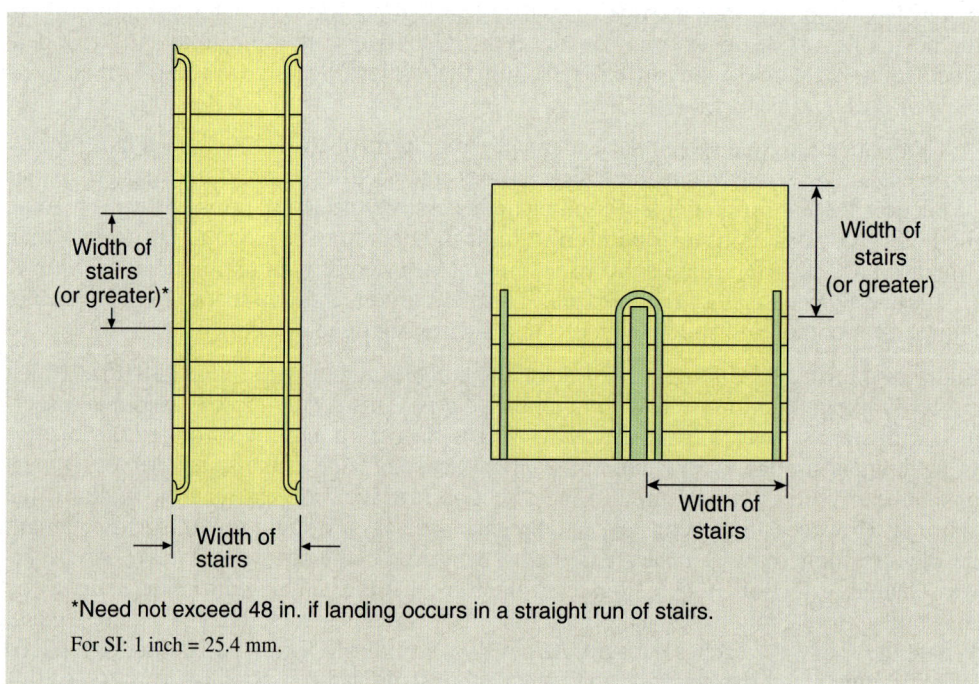

Figure 1009-6 Landing dimensions.

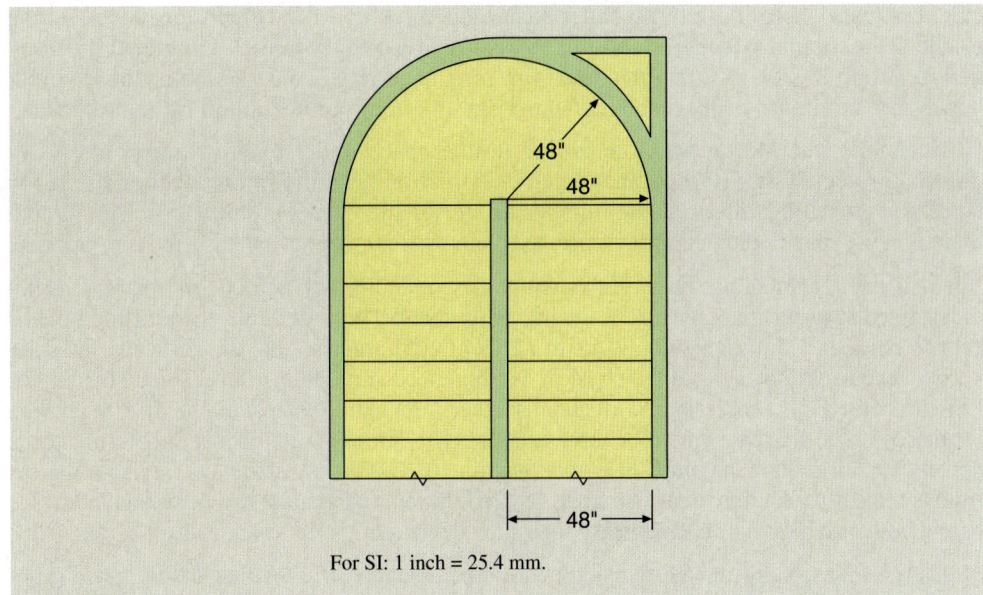

Figure 1009-7
Alternative shape of landing.

structural columns, standpipes, and similar building components within the nonusable portions of the landing.

Stairway construction. Materials used in the construction of stairways are to be consistent with those types of materials permitted based on the building's type of construction. In other words, stairways in Type I and II buildings are to be constructed of noncombustible materials. In buildings classified as Type III, IV, or V construction, combustible or noncombustible materials may be used for stairway construction. These requirements are applicable to the structural elements of the stairway, including stringers and treads. In all types of construction, wood stairway handrails are permitted. Finish materials on stairways are regulated under the provisions of Section 804.

1009.9

It is important that both ascending and descending portions of stairway flights end at relatively level landings, having maximum slopes of 1:48. In addition, the treads themselves are limited to a slope no steeper than 1:48 in any direction. For increased safety and ease of use, the treads and landings are to have a substantially solid surface. However, the use of gratings or similar open-type walking surfaces is permitted for use in treads and landing platforms in manufacturing buildings, warehouses, and hazardous occupancies. Stairways in such uses are not typically accessible to the public. The extent of the openings in the walking surfaces of such stairs and landings is limited to where a sphere with a diameter of $1^1/_8$ inches (29 mm) cannot pass through. In all occupancies, small openings are also permitted provided they do not permit the passage of a $^1/_2$-inch (12.7-mm) sphere. Where the opening is elongated, the long dimension must be placed perpendicular to the direction of stairway travel.

Because of the hazards associated with accumulated water on exterior stairways, outdoor stairs and outdoor approaches to such stairs are to be designed in a manner so that water will not stand on the walking surfaces.

To protect the integrity of this very important element of the exit system, any enclosed usable space under either an unenclosed or enclosed stairway shall be protected by fire-resistance-rated construction. For this section to be applicable, both conditions must exist: the space under the stairway must be enclosed, and it also must be accessible, such as by a door or access panel. The level of protection of the walls and soffits within the enclosed space shall be equivalent to the fire-resistance rating of the stairway enclosure, but in no case less than 1 hour. In order to better ensure the integrity of the enclosure, access to the usable space cannot occur from within the stair enclosure. Within an individual dwelling unit in Group R-2 or R-3, the mandated fire-resistance-rated construction is not required; however, a minimum of $^1/_2$-inch (12.7-mm) gypsum board protection must be provided.

Much like interior stairways, enclosed usable space under exterior egress stairways must be protected by fire-resistance-rated construction. The minimum level of fire resistance is 1 hour, regardless of the number of stories the exterior stairway serves. Where there is open space below exterior stairways, such space is not to be used for any purpose.

1009.10 Vertical rise. Negotiating stairs can sometimes become difficult, particularly for persons not accustomed to using stairs, or for the elderly or disabled. The code limits the maximum vertical rise between landings serving stairs to 12 feet (3,658 mm) so that this difficulty does not become excessive. Aside from the physical exertion necessary, stairs of exceptional height can be intimidating. It is necessary at vertical intervals not exceeding 12 feet (3,658 mm) to provide for places in the stairway where the user can rest. The limitation of 12 feet (3,658 mm) does not apply to aisle stairs in compliance with Section 1028 and spiral stairways serving as egress from technical production areas. Up to 20 feet (6,096 mm) of vertical rise between landings is permitted for alternating tread devices when such devices are used as a means of egress.

The 12-foot (3,658-mm) dimension is not unreasonable. In most instances, it will more than accommodate a single-story height so that in most buildings a single flight of stairs could be used, if desired, to negotiate travel from one floor to the next adjacent floor.

Even though a single flight of stairs is permitted by the code, stairs having an intermediate landing between floors are used in the majority of buildings.

1009.11 Curved stairways. Curved stairs are one of the special types of stairways that the code allows as an alternative to the typical straight stair. Figure 1009-8 depicts this type of

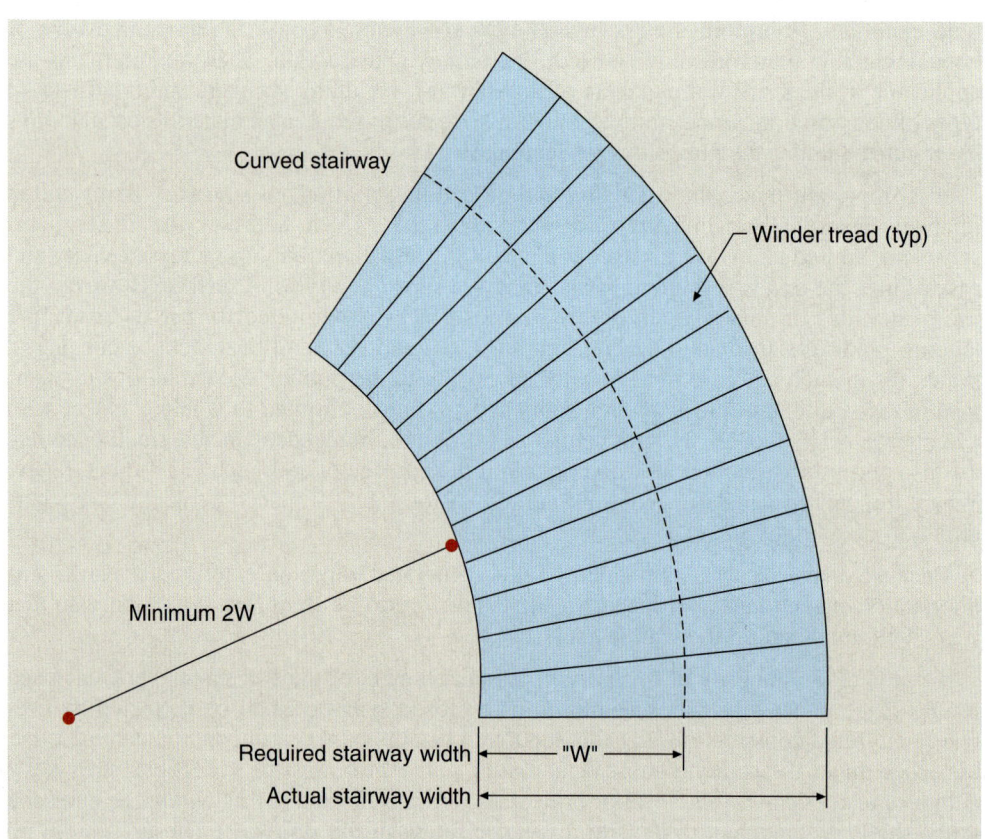

Figure 1009-8
Curved stairways.

alternative stair. It is essentially circular in configuration. The basic requirement is that the inside, or least, radius should be at least twice the required width of the stair. This rule ensures a certain limited degree of curvature deemed acceptable in circular stairs. The only other criterion specified for circular stairs is that treads comply with the winder tread provisions of Section 1009.7. In designing the curved stair and relating the inside radius to the width of the stair, it is important that only the required width of the stair need be used. Because of the stair's geometry, this is a fairly comfortable stairway to use. The geometry of the curved stairway for Group R-3 occupancies and within dwelling units of Group R-2 would only be regulated by the winder tread provisions of Section 1009.7.

Spiral stairways. Spiral stairways are a special type of stair that the code permits in limited applications. As a means of egress component, spiral stairways may be used within dwelling units, or from spaces having an occupant load of five or less and not exceeding 250 square feet (23 m^2) in area. Spiral stairways are also permitted as egress elements from galleries, catwalks, gridirons, and other technical production areas. **1009.12**

A spiral stairway is one where the treads radiate from a central pole. Such a stair must provide a clear width of at least 26 inches (660 mm) at and below the handrail. Each tread must have a minimum dimension of 7$^1/_2$ inches (191 mm) at a point 12 inches (305 mm) from its narrow end. The stair must have at least 6 feet 6 inches (1,981 mm) of headroom measured vertically from the leading edge of the tread. The rise between treads can be as much as, but not more than, 9$^1/_2$ inches (241 mm). The required dimensions of a spiral stairway are depicted in Figure 1009-9.

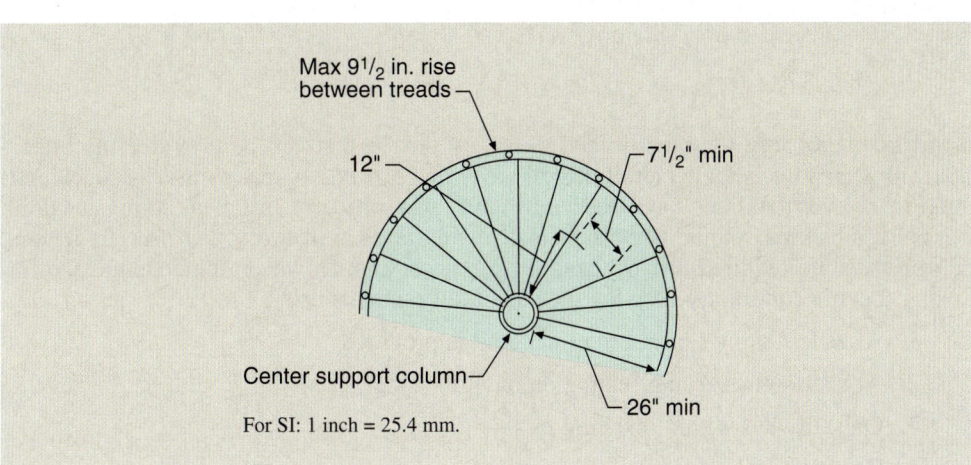

Figure 1009-9
Spiral stairways.

Alternating tread devices. An alternating tread device is a unique type of stairway that also has some characteristics of a ladder. Because it is considered difficult to use for egress purposes, an alternating tread device may only be used as a means of egress in a limited number of occupancies. In factories, warehouses, and high-hazard occupancies, this device can only be used for egress from a small mezzanine serving a limited number of occupants. In Group I-3 occupancies, an alternating tread device may be the egress path from a small guard tower or observation area. An alternation tread device can also be used for access and egress from an unoccupied roof. See Figure 1009-10. **1009.13**

Alternating tread devices must have complying handrails on both sides. In addition, treads are regulated based on whether or not the alternating tread device is used as a means of egress component. Where it is not used for egress purposes, but rather is for convenience only, the minimum tread depth may be reduced and the maximum riser height may be increased.

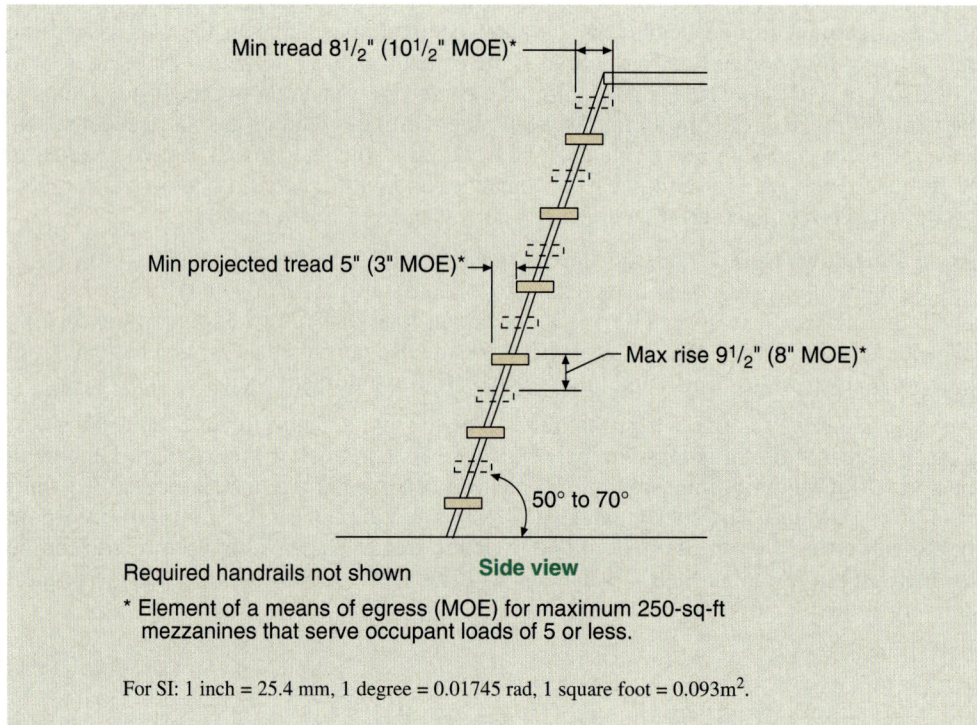

Figure 1009-10 Alternating tread device.

1009.15 Handrails. Probably the most important safety device that can be provided in connection with stairways is the handrail. It will never be known how many missteps, accidents, injuries, or even fatalities have been prevented by a properly installed, sturdy handrail. Basically, a handrail should be within relatively easy reach of every stair user. In general, all stairways are required to have handrails on each side. However, the code has the following specific conditions allowing the use of only one handrail:

1. At aisle stairs where a center handrail is provided
2. At aisle stairs serving seating on only one side
3. On stairways within dwelling units
4. On spiral stairways

Handrails may be omitted under four conditions. Where a single change in elevation occurs at a deck, patio, or walkway, handrails are not required, provided a complying landing area is present. In Group R-3 occupancies, handrails are not required at a single riser serving an entrance door or egress door. In individual dwelling units and sleeping units of Group R-2 and R-3 occupancies, it is also permissible to provide a change in room elevation of three or fewer risers and not install a handrail. Handrails may also be omitted in assembly seating areas as established in Section 1028.13.

1009.16 Stairway to roof. To provide for easy access to roof surfaces and to facilitate fire fighting in buildings four or more stories in height, at least one stairway is required to extend to the roof. This stairway to the roof must comply with all code requirements for stairways, unless the roof is considered unoccupied. Access to an unoccupied roof may be accomplished by an alternating tread device as described in Section 1009.13. However, the stairway or alternating tread device to the roof is not a requirement on steeper roofs where the slope exceeds 4 units vertical in 12 units horizontal (33-percent slope).

Roof access. In buildings having an occupied roof, a penthouse complying with the provisions of Section 1509.2 must be provided for stairway access to the roof. Where the roof is considered unoccupied, access need only be provided through a roof hatch of the minimum size prescribed.

1009.16.1

Protection at roof hatch openings. Roof hatches are permitted as a means of access to unoccupied roofs where such access is required in buildings four or more stories in height. In addition, roof-hatch openings are often provided in low-rise buildings for varied purposes, including access to rooftop equipment. This provision addresses the hazard created where the roof hatch is located very close to the roof edge. See Figure 1009-11. It is necessary that persons accessing the roof by a roof hatch be protected from falling off the roof as a result of a trip or misstep. Occasionally, these roof accesses are used during inclement weather, emergency situations, or times of darkness. It is during these conditions when the hazard level is even higher.

1009.16.2

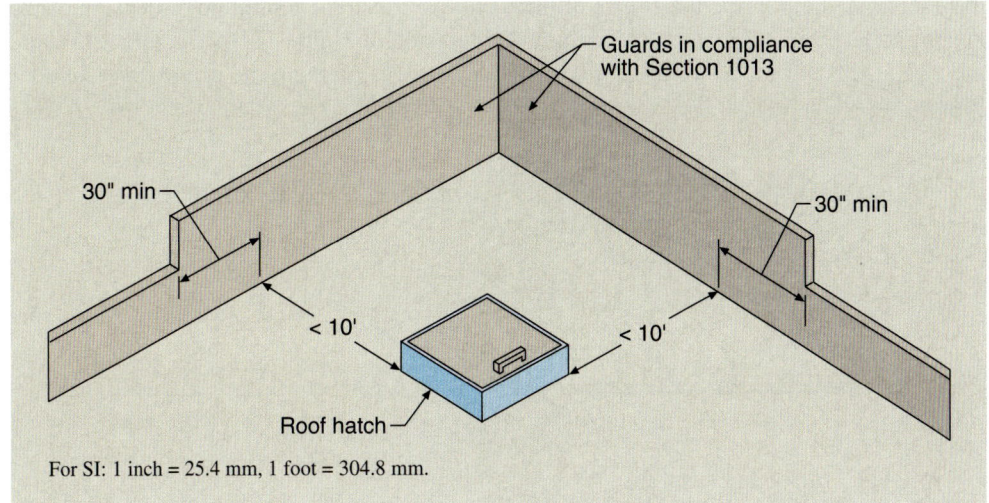

Figure 1009-11
Protection at roof-hatch openings.

Stairway to elevator equipment. Where elevator equipment is located on the roof or in a rooftop penthouse and such equipment must be accessed for maintenance purposes, a stairway must be provided to the roof or penthouse. Consistent with the provisions of ASME A17.1, *Safety Code for Elevators and Escalators*, the requirement mandates a code-complying stairway. As a result, the use of alternating tread devices or ladders as the only means to access the elevator equipment is prohibited.

1009.17

Section 1010 *Ramps*

Scope. With the exception of some ramped aisles in assembly rooms, curb ramps, and vehicle ramps in parking garages used for pedestrian exit access travel, whenever a ramp is used as a component anywhere in a means of egress it is necessary that the ramp comply with the provisions of this section. Note that where ramps are used as part of the egress system, every ramp must meet these standards, regardless of the occupant load served. Ramped aisles in assembly occupancies, other than those to and from accessible wheelchair spaces, need only conform to the provisions of Section 1028.11.

1010.1

In addition to those egress provisions, all ramps located within an accessible route of travel shall be made to conform to the requirements found in ICC A117.1.

1010.3 Slope. Whenever any ramp is used in an exit system, the slope of such a ramp shall not be steeper than 1 unit vertical in 12 units horizontal (8.3-percent slope). Those pedestrian ramps not considered a portion of the means of egress may have a greater slope, but may not be steeper than a slope of 1 unit vertical in 8 units horizontal (12.5-percent slope). For areas of assembly seating, Section 1028.11 allows a maximum 1:8 slope for means of egress ramps not on an accessible route.

1010.6 Minimum dimensions. Because requirements for the widths of ramps are identical to those requirements for widths of corridors, see discussion of Section 1018.2. The net clear width between the handrails is to be no less than 36 inches (914 mm), differing from the stairway provisions that allow for handrail projections into the required width. In fact, all projections into the minimum width requirements of the ramp and its landings are prohibited. Where the minimum required width of an egress ramp is established, it shall not be diminished at any point in the direction of travel. Where doors enter onto or swing over ramp landings, the doors may not, during the course of their swing, reduce the minimum dimension of the landing to less than 42 inches (1,067 mm). As depicted in Figure 1010-1, in all cases there must be at least 42 inches (1,067 mm) of width available on the ramp even while the door is swinging.

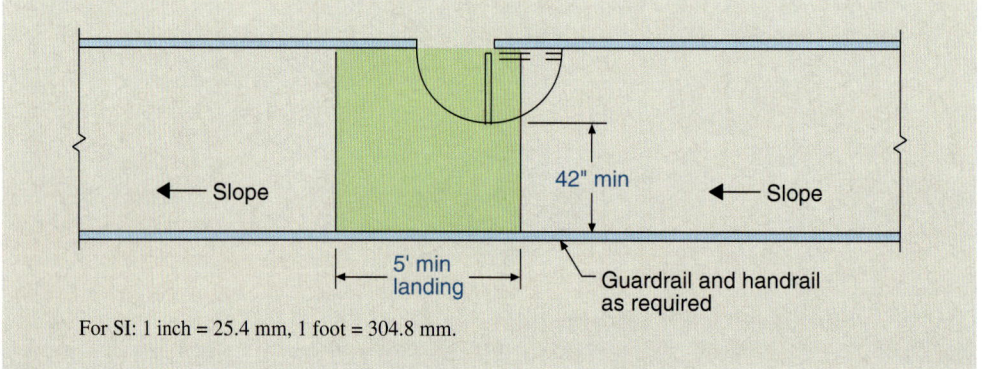

Figure 1010-1 Intermediate ramp landings.

1010.7 Landings. Any ramp, defined as having a slope steeper than 1 unit vertical in 20 units horizontal (5-percent slope), must have landings at the top and bottom, at any changes of direction, at points of entrance or exiting from the ramp, and at doors. All landings are required to have a dimension measured in the direction of the ramp run not less than 5 feet (1,524 mm). These larger-than-normal dimensions are required to reasonably ensure that disabled persons in wheelchairs will have sufficient space to maneuver on any intermediate landings as well as in the area at the top and bottom adjacent to the ramp. See Figure 1010-2. Exceptions permit

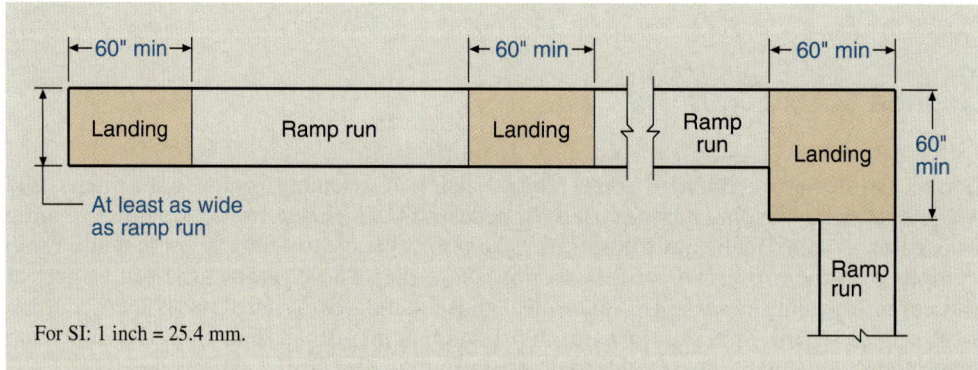

Figure 1010-2 Intermediate ramp landings.

a reduction in landing lengths to 36 inches (914 mm) in nonaccessible Group R-2 individual dwelling units and nonaccessible R-3 occupancies, whereas ramps not on an accessible route in other occupancies must only be 48 inches (1,220 mm) in length.

Ramp construction. Consistent with the provisions for stairway construction found in Section 1009.9, ramps shall be built of materials consistent with the building's type of construction. In addition, exterior ramps and their approaches should be designed in a manner to avoid water accumulation. **1010.8**

Because a ramp has a sloping surface, the potential for accidents resulting from slips and falls is greatly increased. It is critically important that the surface of the ramp be made slip-resistant to help prevent such accidents. It is imperative that very careful attention be given to the selection of materials for ramp surfaces and the methods of finishing such surfaces. Certainly, slick-finished materials should be avoided on ramps, unless it can be ensured that they have been made slip-resistant by some process.

There are a number of products on the market available for treating ramp surfaces that are not sufficiently slip-resistant. In many instances, the use of these materials has not proven to be completely satisfactory. One method is to install carborundum strips across the width of the ramp at appropriate intervals. Other available products can be installed on the ramp surface with an adhesive. It is essential that slip-resistant treatments and materials be of reasonably permanent nature and securely attached so they can perform the function for which they were installed, while at the same time not becoming a potential tripping hazard. Therefore, the slip resistance should be proven by tests and be an integral part of the ramp surface.

Handrails. Whenever any ramp in the means of egress has a rise greater than 6 inches (152 mm), the ramp shall be provided with handrails on both sides. The detailed requirements for the handrails are found in Section 1012. There are no general provisions that allow a handrail on only one side of a ramp, nor is there any requirement for intermediate handrails. It should be remembered that handrails for ramped aisles in assembly occupancies are regulated by Section 1028.13. **1010.9**

Edge protection. In order to further protect the user of a ramp, safeguards are required along ramp edges as well as along each side of ramp landings. Two different methods, as shown in Figure 1010-3, are available for providing compliant edge protection at ramp runs and ramp landings. Either method may be used to provide the necessary level of protection. It is acceptable to use a curb with a minimum height of 4 inches (102 mm) to **1010.10**

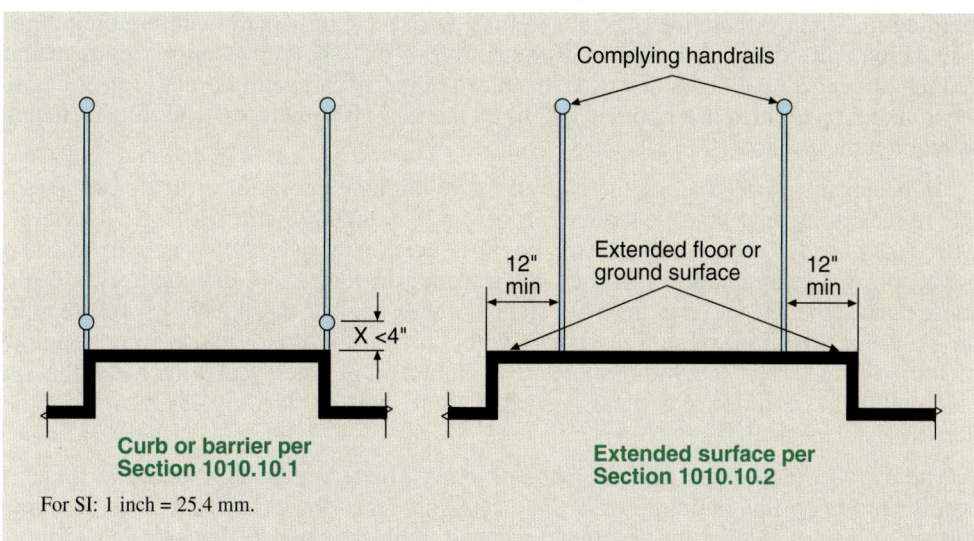

Figure 1010-3
Ramp edge protection.

satisfy the first method. A rail located just above the ramp or landing surface will also suffice, as will a wall or similar barrier. As an alternative methodology, the surface of the ramp or landing may extend a minimum distance of 12 inches (305 mm) beyond a vertical plane established from the inside face of a complying handrail. Both methods are deemed adequate for preventing a ramp user from an unacceptable drop-off at the landing or ramp edge.

Section 1011 *Exit Signs*

1011.1 Where required. To properly identify the egress path through a building, it is necessary to provide exit signs. Although somewhat vague, these provisions are probably more performance oriented than most requirements. The basic provision is that signs are required so that the exiting path is clearly indicated. By combining this requirement with Exception 1, it will mean that, in general, when two or more means of egress are required from any portion of a building, such as from a room, area, floor, or other space, exit signs must be installed at all required exit and exit access doorways, and at any other location throughout the exiting system where deemed necessary by the building official to clearly identify the path of travel to the exit.

Care must be taken in the placement of the exit signs to ensure they are properly located and properly oriented to the direction of travel in the egress system. They need to be easily read by persons seeking the exit. Particular care should be taken to see that nothing occurs in the means of egress that might tend to obscure or screen the exit sign, or that might cause confusion in identifying the exit sign. It is not uncommon in many building plans to find that exit signs are shown at what appear to be the appropriate locations, only to later find they are actually installed behind ducts, equipment, or other building elements in such a manner that they are not really visible to building occupants, or near or over doors that are not the exit. Additionally, the presence of banners, signs, and other movable elements may obstruct the view of required exit signs. In the placement of exit signs, the use of the space and its impact on clear sight lines to exit signs must be considered.

In certain instances, the exit signs may be present, but they are not oriented to the path of travel and to the approaching building occupant. Therefore, they are not visible and are certainly not legible to the persons seeking the exit. It is strongly advised that, as one of the last points of inspection, and before approving the occupancy of any building, the building inspector or building official carefully walk the egress path to ensure the proper installation and effective location and orientation of exit signs. It is also important that ceiling-suspended exit signs or exit signage extending from a wall do not project below the minimum permitted headroom height of 80 inches (2,032 mm) for projecting elements as required by Section 1003.3.1.

As occupants proceed along the path of travel through a corridor or exit passageway, they typically assume that their direction of travel is taking them to an exit. However, where the corridor or exit passageway length is quite extensive, the users may, at some point, make a determination that they are traveling in the wrong direction. This could cause the occupants to reverse their direction and seek another travel path. For this reason, the provisions require exit sign placement within a corridor or exit passageway at points no more than 200 feet (60,960 mm) apart. This method of placement will ensure that no point within the corridor or exit passageway is more than 100 feet (30,480 mm) from the nearest visible sign.

The use of exit signs to identify paths of egress travel are usually limited to the exit access portions of the building. Once the occupants reach the exits, such signs are typically unnecessary as the paths are often direct and single directional. However, in buildings with more complicated means of egress systems, it is possible that egress travel within the exits may not be immediately apparent to the occupants. For this reason, the mandate for exit signs is extended to those portions within exits, such as exit passageways, where such signs are necessary to provide clear egress direction for the occupants. Evacuees may be hesitant or even confused when traveling within an exit that involves transition from a vertical to a horizontal direction and horizontal extension that includes turns and intervening doors within the path of egress. Where travel direction is not clear within an exit, it creates uncertainty and causes delays in evacuations under threatening conditions. Therefore, the direction of egress travel should be identified by exit signs where such direction may not be clearly understood, and all intervening means of egress doors within the exits must also be provided with complying exit signs. An example of the appropriate use of exit signs within exits is shown in Figure 1011-1.

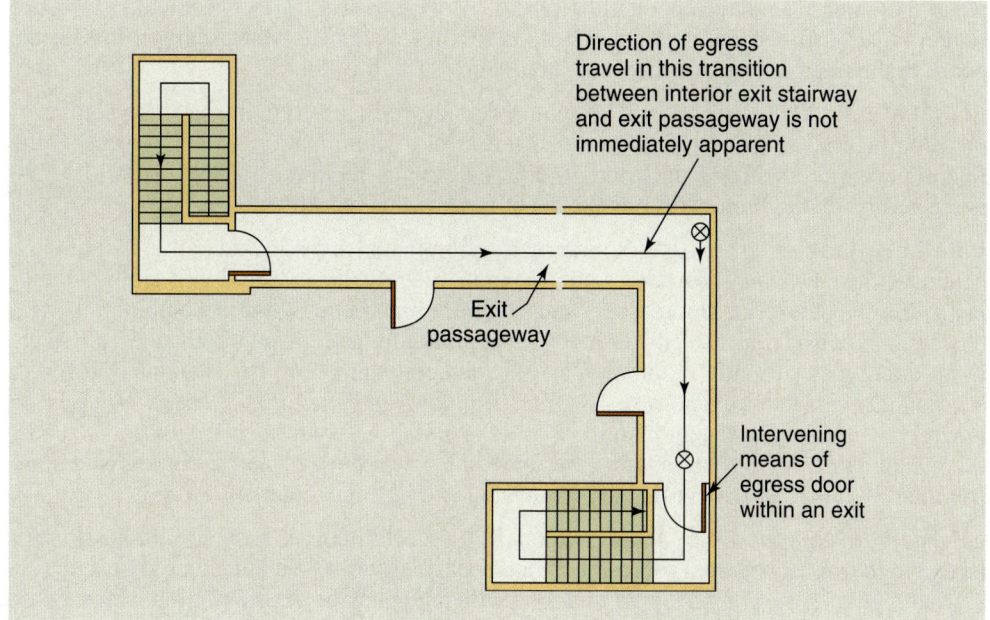

Figure 1011-1
Exit sign locations.

Omission of exit signs. Although the installation of exit signs is of paramount importance in most instances, there will be cases where the means of egress is abundantly obvious to any building occupant. Where the exit point is clearly identifiable, such as at main exterior entrance/exit doors, the exit is essentially its own identifying sign. Where approved by the building official, it may not be necessary to install an exit sign at such an obvious exit. In addition, exit signs may be omitted in Group U occupancies, and within individual sleeping and dwelling units in Group R-1, R-2, and R-3 occupancies. Because of the residential nature of such uses, it is felt that exit signs are unnecessary. Exit signs may also be omitted in day rooms, sleeping rooms, and dormitories of a Group I-3 occupancy, as requiring exit signs for such an occupancy is considered unnecessary. The first exception, which eliminates the need for exit signs in areas that need access to only one exit or exit access element, has application to a great number of areas within most buildings. It is assumed that the exit path is obvious because of its use as the typical entrance to the area or room. It should be noted that the exception applies even where multiple means of egress are provided, provided only one of them is required. A final exception applies to assembly seating arrangements that occur in Groups A-4 and A-5. Exit signs may be omitted in the seating area, provided that the direction to the concourse area is easily identified and adequately illuminated. Once reaching the concourse, exit signs identifying the exit path and egress doors must be provided.

1011.2 Floor-level exit signs in Group R-1. To help guide transient occupants of Group R-1 guest rooms to the exits during emergency conditions, additional exit signs are required within the egress system serving the guest rooms of a Group R-1 occupancy. Limiting the application to the egress system serving the guest rooms of hotels and other Group R-1 occupancies recognizes that the occupants are transient and not familiar with their surroundings. If a corridor or other egress component serving guest rooms were to fill with smoke, the general exit signs that are located higher in the space could quickly be obscured by the rising smoke. As the space fills with smoke, the evacuees are forced to crawl on the floor to reach the nearest exit. They will be confronted with many doors, all looking the same, and will not know which door is the exit. The installation of these low-level exit signs assists these persons in safely exiting the building when signs at the higher levels are obscured as the smoke layer develops at the ceiling.

Low-level exit signs also serve to increase fire-fighter safety while on the fire scene. In their efforts to evacuate the building occupants, the fire fighters will be in the building when the smoke has developed. Although they rely on several other techniques, the fire fighters may also become dependent on this low-level signage while trying to locate the doors to the interior exit stairway and safely egress the fire floor.

1011.3 Illumination. Where exit signs are required, they must be illuminated. The source of the illumination is not material, provided the level of illumination at the sign meets the minimum requirements of the code. Tactile exit signs, required by Section 1011.4, are exempted from the illumination requirement for obvious reasons.

1011.5 Internally illuminated exit signs. Internally illuminated signs include all exit signs that generate their own luminosity in some manner. Electrically powered exit signs, including LED, incandescent, fluorescent, and electroluminescent signs, in combination with those signs considered as self-luminous and photoluminescent, represent the full range of product types currently in the market. All such exit signs must be listed and labeled in accordance with the provisions of UL 924, *Standard for Safety Emergency Lighting and Power Equipment.* Consistent with most other installation requirements found in the code, the manufacturer's instructions must be followed. The value of visible exit signs is demonstrated by the requirement that exit signs must be illuminated at all times.

1011.6 Externally illuminated exit signs. Although it is uncommon to provide illumination for an exit sign from an external source, there are occasions where such methods are used. This section regulates the graphics, illumination, and emergency power supply for such signs.

Although no particular color is specified for exit signs, it is required that the color and design of the signs, the lettering, the arrows, and other symbols on the sign, provide good contrast so as to increase legibility. The letters of the word *exit* are required to be at least 6 inches (152 mm) in height and have a width of stroke of not less than $3/4$ inch (19.1 mm). By specifying the letter spacing, the code ensures that the letters are not placed too close together and that the signs will be legible and more effective as a result. Additionally, when a larger sign is provided, it is important the lettering be increased proportionally.

When the illumination of the sign is from an external source, that source must be capable of producing a light intensity of at least 5 foot-candles (54 lux) at all points over the face of the sign. Measurement of the 5 foot-candles (54 lux) from an external source can be made by a light meter. The measurement of the equivalent luminance, however, from an internal source is more difficult. That is why the code relies on a testing laboratory that has examined, certified, and labeled the exit sign. In labeling such a sign, the testing agency is certifying, among other things, that the light level is at least that required by the code.

Externally illuminated exit signs are required to be illuminated at all times and must provide continued operation for a minimum of one- and one-half hours after the loss of the normal power supply. This backup power supply shall be from a complying storage battery system, unit equipment, or other approved on-site, independent source. The exception addresses illumination that is provided independent of external power supplies and, therefore, does not require compliance with the emergency power provisions.

Section 1012 *Handrails*

A handrail is defined in Section 202 as "a horizontal or sloping rail intended for grasping by the hand for guidance or support." The IBC mandates, with limited exceptions, that handrails be provided to assist the users of stairways and ramps during normal travel conditions. In addition, a handrail must be available for support in case of a misstep or other occurrence that might cause the user to stumble and fall. Numerous criteria for the design and installation of effective handrails are set forth in this section.

Height. In past building codes, the height of handrails was traditionally established within a range of at least 30 inches (762 mm) and not more than 34 inches (864 mm) vertically above the leading edge of treads of the stairway that the handrails serve. However, research has shown that handrails better serve stair users if located in a range higher than this traditional location. Higher handrails can be more readily reached by adult stair users and, interestingly enough, handrails at the higher elevation are also more usable by very small persons, including toddlers. Where handrails are required, they must be located at least 34 inches (864 mm) and not more than 38 inches (965 mm) vertically, measured from the nosing of the stair treads to the top of the rail. See Figure 1012-1 for an illustration of IBC requirements for stairs, including handrail height. A reduction in the handrail height range is provided for alternating tread devices and ship ladders due to the steepness of travel.

1012.2

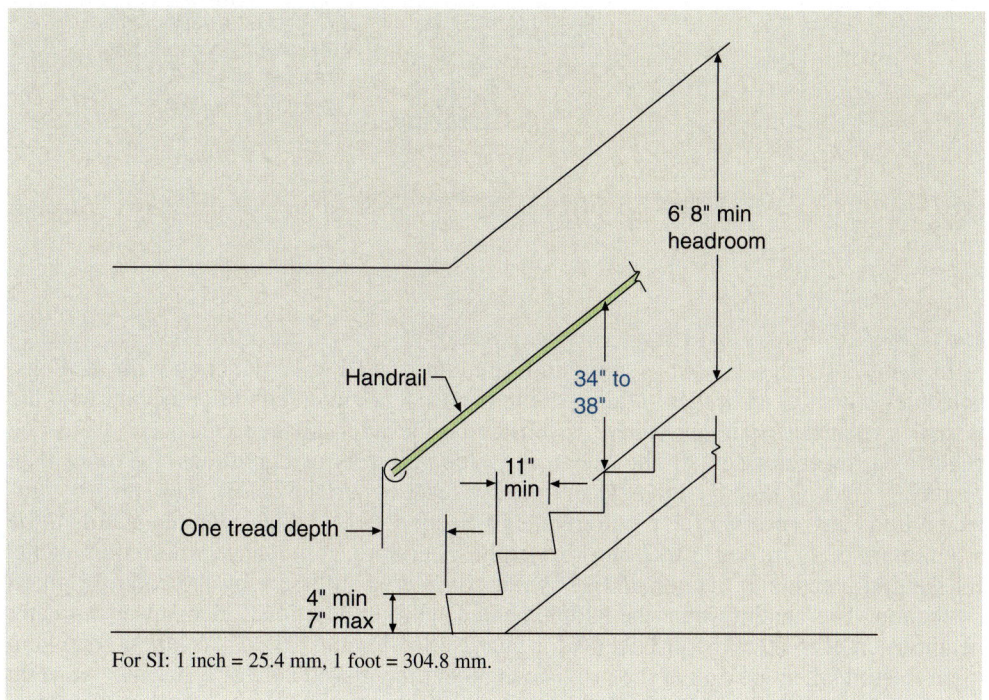

**Figure 1012-1
Handrail and stair detail.**

Where a handrail transitions from one flight of stairs to another at a dog-leg or switchback stair landing, the base code requirement for handrail extensions or for handrail continuity often creates the need for some type of transition. Exception 1 allows for a more gradual variation in the height even though it will allow for portions of the handrail to exceed the normal 38-inch maximum—the belief being that a "continuous" handrail is probably more important than staying within the height limitation. Exception 2 is more comprehensive in that it also addresses the transition from a handrail to a guard; however, its application is limited to residential occupancies. See Figure 1012-2.

10 Means of Egress

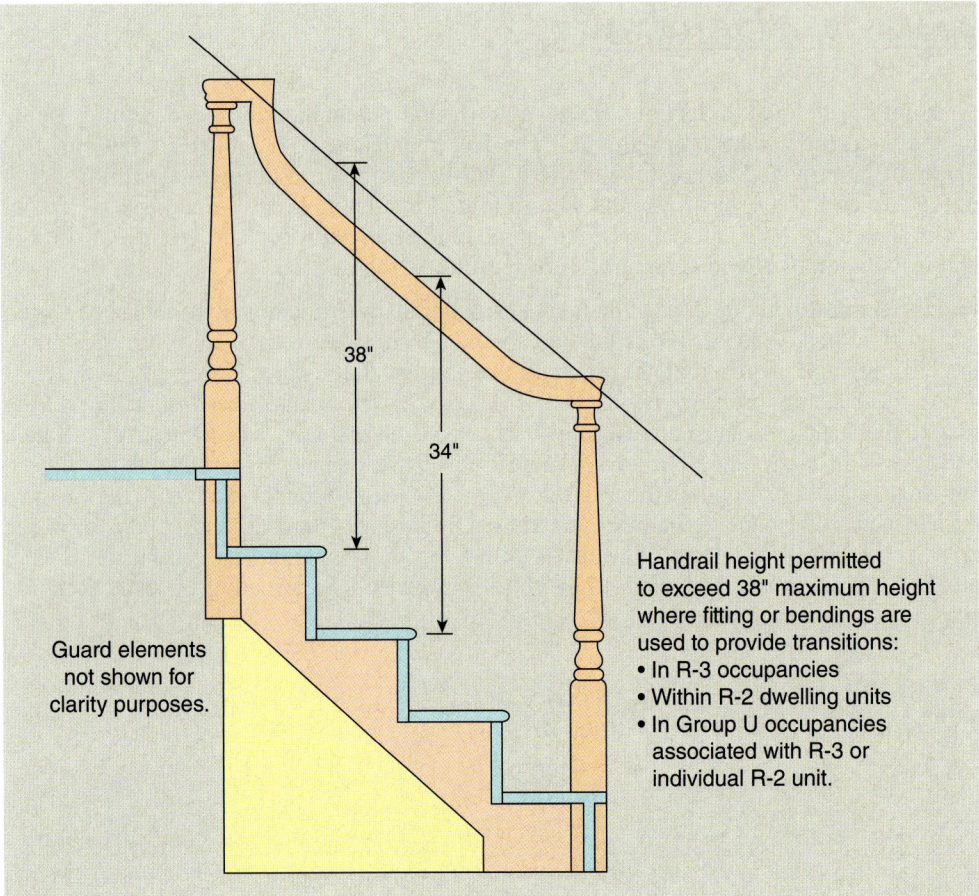

Figure 1012-2
Transition pieces on stair handrail.

In many instances, stairways and ramps are constructed in conjunction with landings, balconies, porches, and other building components. A common condition occurs with the typical switchback stairway, where a substantial elevation change can occur from one flight to a lower adjacent flight. When those components or conditions are more than 30 inches (762 mm) above the adjacent ground level or surface below, they must be protected by guards meeting the requirements of Section 1013, which requires guards to be a minimum of 42 inches (1,067 mm) in height. Therefore, compliance with requirements for handrail heights in Section 1012.2 is not considered sufficient by the code for guard protection. The handrail must be supplemented with an additional element at a greater height in order to meet the provisions for guards. See Figure 1012-3. As a note, one item often missed when looking at the provisions is that the glazed side of a stairway must be protected by a guard unless the glazing meets the strength and attachment requirements of Section 1607.8.

1012.3 Handrail graspability. To be truly effective, handrails must be graspable. The code establishes specific criteria for handrail shape intending to define "graspability" in prescriptive terms. A performance alternative is also provided indicating that the configuration of the handrail can be such that it provides an equivalent, graspable shape. In prescriptive terms, the handgrip portion of the handrail must not be less than $1\frac{1}{4}$ inches (32 mm) or more than 2 inches (51 mm) in a circular cross-sectional dimension, as shown in Figure 1012-4, or where the cross section of the handrail is not circular, an alternative shape is described by the code and illustrated in Figure 1012-5. Such handrails are considered as Type I

Handrails 10

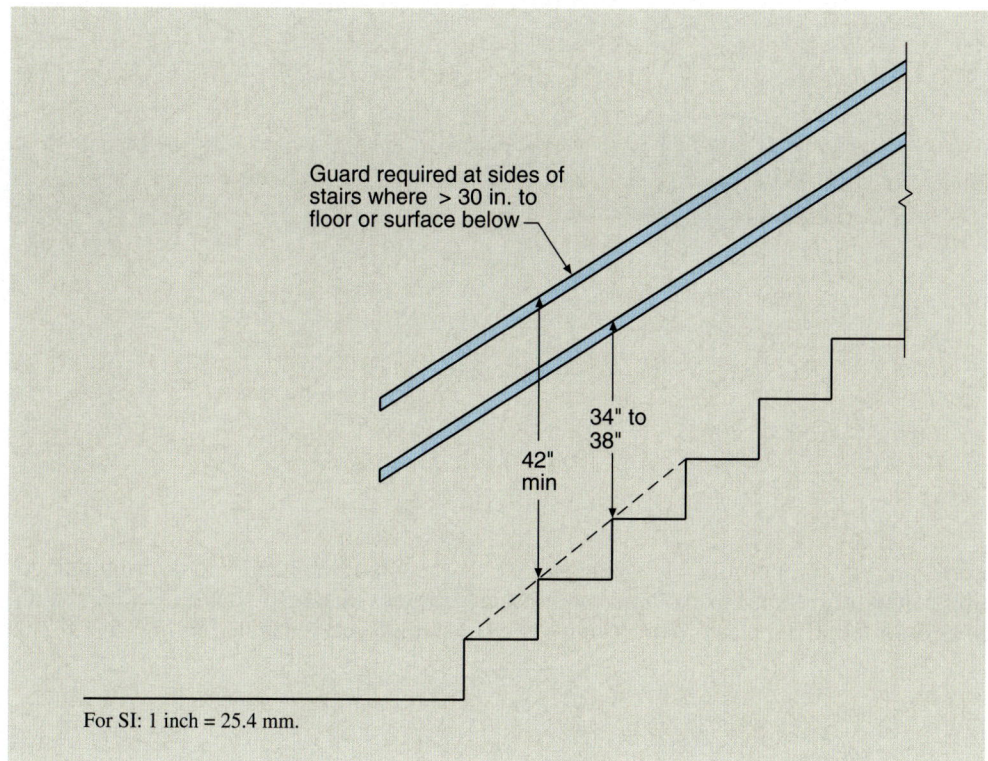

Figure 1012-3
Guard and handrail.

handrails and are permitted for use in all occupancies. Handrail profiles having a perimeter dimension greater than 6 1/4 inches (160 mm), identified as Type II handrails, are also acceptable where installed in specified residential applications.

Research has shown that Type II handrails have graspability that is essentially equal to or greater than the graspability of handrails meeting the long-accepted and codified shape and size defined as Type I. The key features of the graspability of Type II handrails are graspable finger recesses on both sides of the handrail. These recesses allow users to firmly

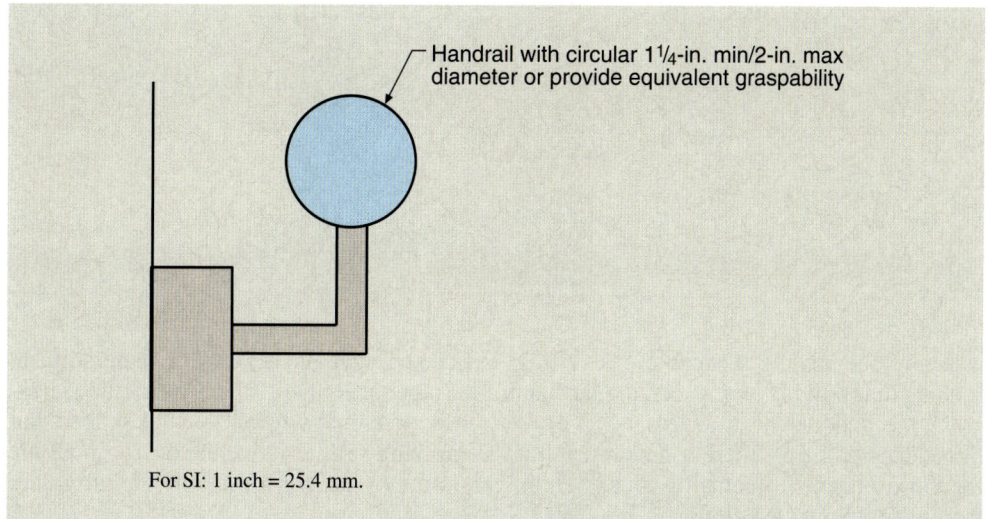

Figure 1012-4
Circular handrail.

2012 International Building Code Handbook 395

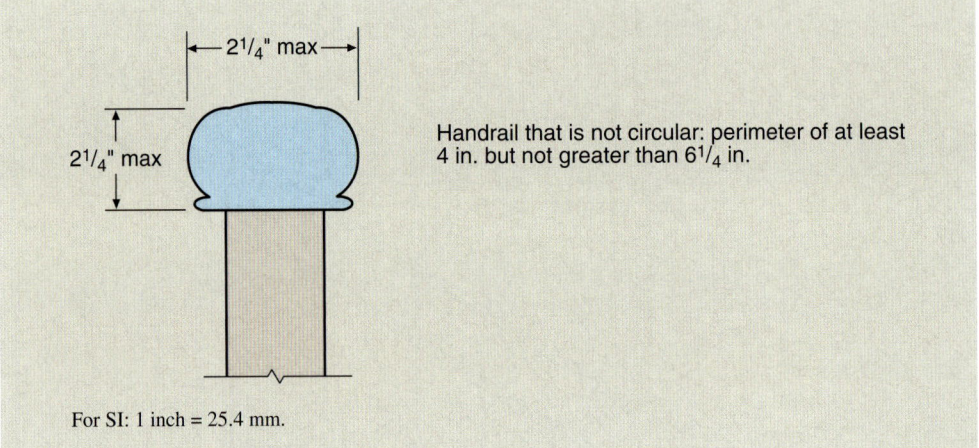

Figure 1012-5
Noncircular handrail.

grip a properly proportioned grasping surface on the top of the handrail, ensuring that the user can tightly retain a grip on the handrail for all forces that are associated with attempts to arrest a fall. Examples of complying Type II handrails are illustrated in Figure 1012-6.

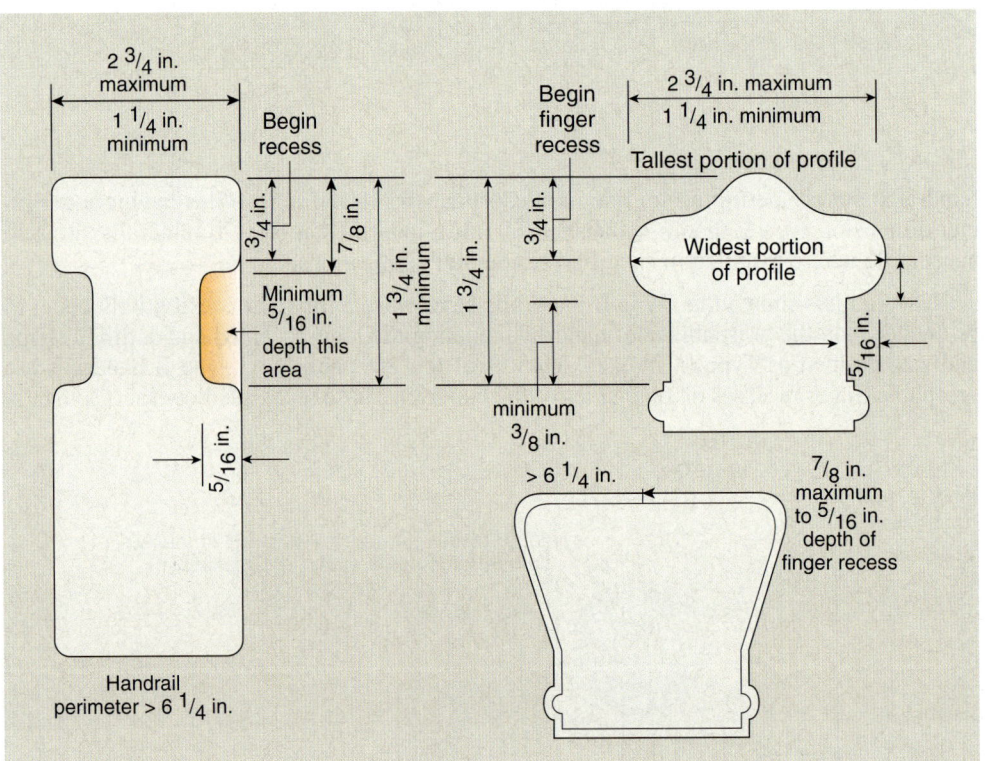

Figure 1012-6
Dimensional properties of Type II handrails.

Many persons are incapable of exerting sufficient finger pressure on a plane surface, such as that provided by a rectangular handrail. To get adequate support or to adequately grasp the handrail, it is necessary to provide those persons with one of the shapes that the code specifies. Where a handrail with a complying profile is installed, it is possible for those persons to actually wrap their fingers around that portion of the handrail and, thereby, obtain better support.

Handrails | 10

Continuity. The handrails must be continuous for the full length of the ramp or the flight of stairs, thereby affording the user support throughout the entire flight. Within dwelling units, handrails are permitted to terminate at a starting newel or ramp or stair volute that is located on the first tread. In addition, a newel post may interrupt the handrail continuity at a stair landing. These types of terminations have been found in residences for years without a record of accidents or lawsuits for an unsafe condition. | **1012.4**

Handrail brackets or balusters are not considered obstructions, provided they are attached to the bottom surface of the handrail and do not project horizontally beyond the sides of the handrail within $1^1/_2$ inches (38 mm) of the bottom of the rail, as shown in Figure 1012-7. This provision is used to regulate the method of support so that the handrail is graspable at any point along its length. A lesser vertical clearance at the handrail's bottom surface is permitted where the perimeter of the handrail exceeds 4 inches (102 mm). As the perimeter of the handrail increases, the vertical clearance may be reduced.

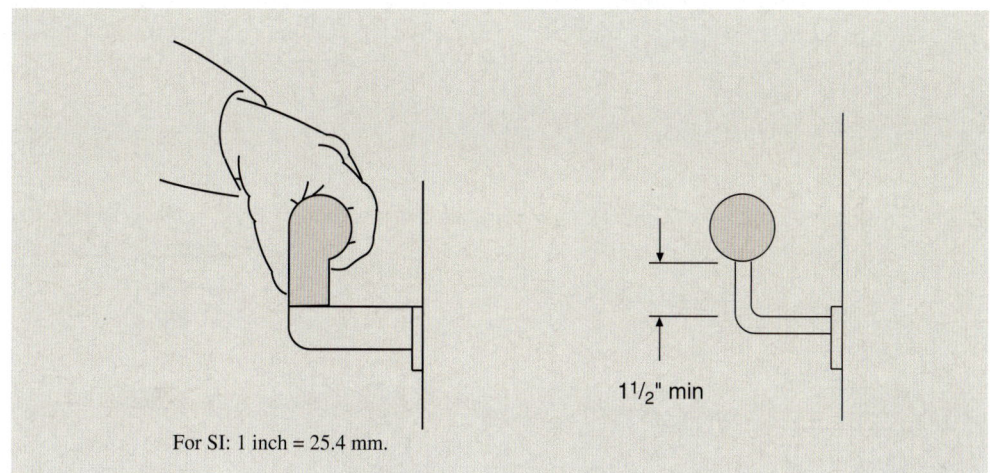

Figure 1012-7
Handrail continuity.

Handrail extensions. Other than locations where handrails are continuous from one flight of stairs to the next, the handrails must extend horizontally not less than 12 inches (305 mm) beyond the top riser. In addition, under such conditions they must also continue to slope beyond the bottom riser for a distance equal to the depth of one tread. See Figure 1012-8. The purpose of the extension of the handrail at the top and bottom of the stairs is to provide minimal additional facility in assisting the stair user. As such, the handrail must extend in the same direction as the travel along the stairway or ramp. This extension is not required for aisle handrails in rooms or spaces used for assembly purposes, for handrails within a dwelling unit that is not required to be accessible, as well as handrails for alternating tread devices and ship ladders. Somewhat similar conditions are applicable where ramp handrails do not continue from one ramp run to the next. For such noncontinuous handrails, a minimum extension of 12 inches (305 mm) is required, as shown in Figure 1012-9. Handrails for both stairs and ramps must return to a wall, guard, or the walking surface unless they continue and connect to a handrail of an adjacent stair flight or ramp run. The reason for requiring that the ends of handrails be returned to the wall, floor, or landing, or end at a guard or similar safety terminal, is to avoid the possibility of loose clothing or other articles being caught on the projection of the handrail. | **1012.6**

Clearance. In view of the desire to make the handrail graspable, and considering the requirement that the handrail be continuous, it is necessary to provide a clear space of at least $1^1/_2$ inches (38 mm) between the handrail and any abutting construction to avoid injury to fingers. See Figure 1012-10. | **1012.7**

10 Means of Egress

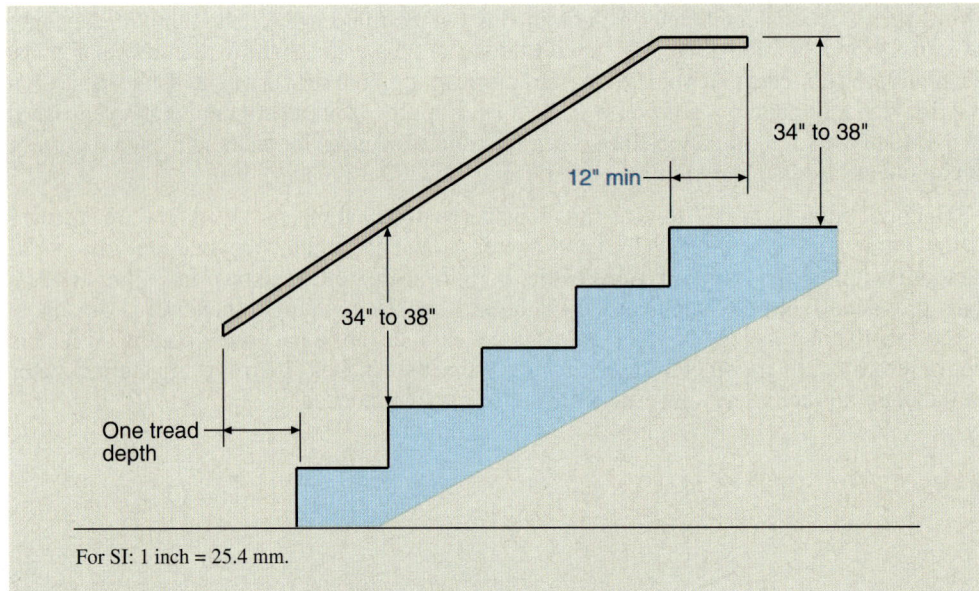

**Figure 1012-8
Stairway handrail extensions.**

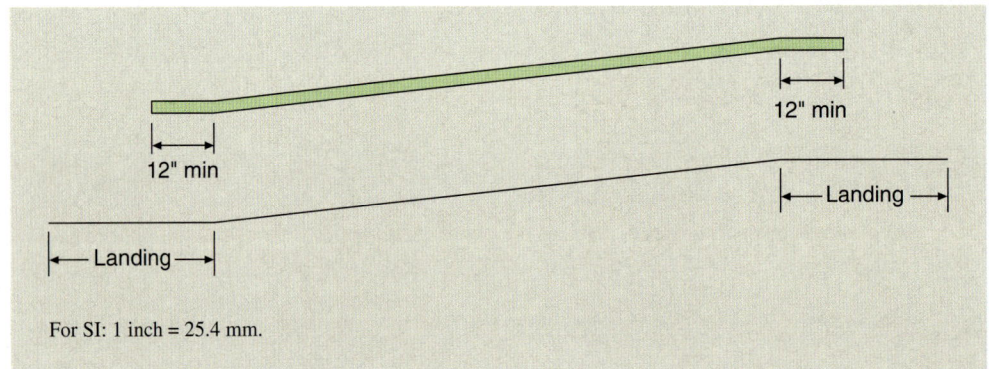

**Figure 1012-9
Ramp handrail extensions.**

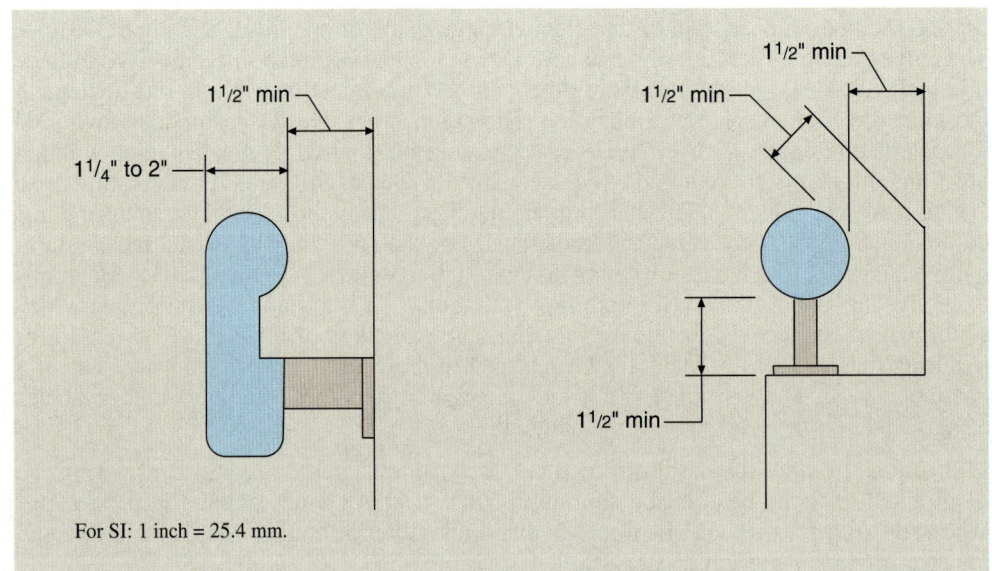

**Figure 1012-10
Handrail clearance.**

Projections. As stated earlier, when the code specifies a required width of a component in the egress system, it generally intends that width to be clear and unobstructed. One permitted projection is for handrails, which may project a maximum distance of $4^1/_2$ inches (114 mm) from each side of the ramp or stair, as shown in Figure 1012-11. In addition, stringers, trim, and other features are permitted to project into the required width up to $4^1/_2$ inches (114 mm) at or below the handrail height. Again, unless a projection into the minimum required width of the stairway is expressly allowed, none is permitted.

1012.8

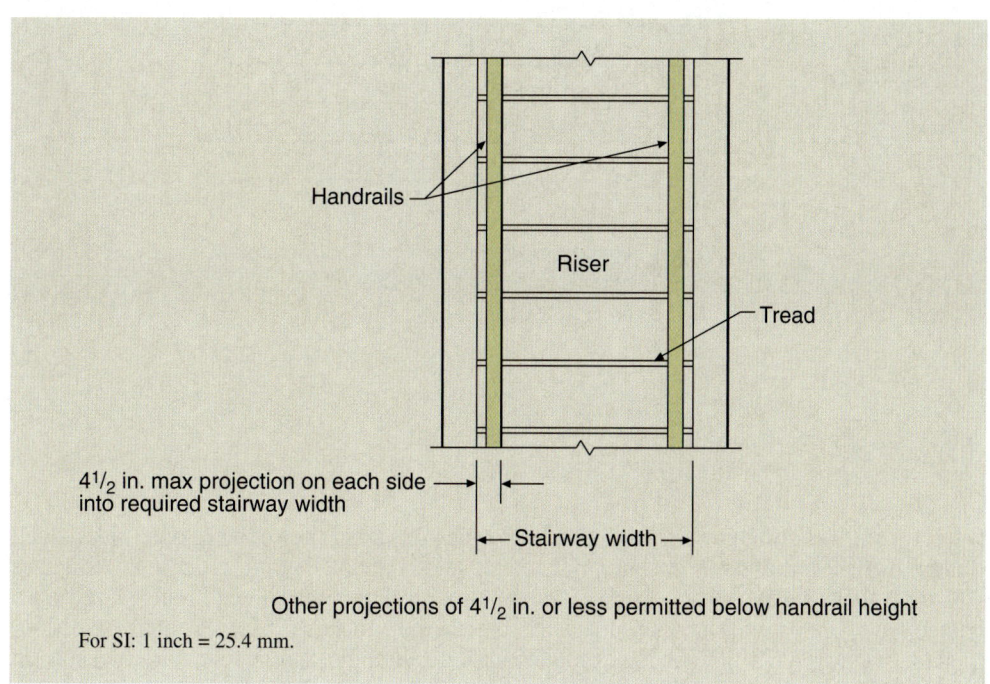

Figure 1012-11 Projections into stairway width.

Unlike the permissible projections of handrails into the required stairway width, such projections into the required ramp width are limited. In no case may the distance between ramp handrails be less than 36 inches (914 mm).

Intermediate handrails. Where the occupant load served by a stairway becomes significant, additional handrails may be necessary to assist stair users. The requirement is based on the required width of the stair established by Section 1005.3.1, not the actual width, and mandates that at no point shall the required stairway width be more than 30 inches (762 mm) from a handrail. See Figure 1012-12. It is difficult to determine the exact point at which intermediate handrails are required, as the handrail projection into the required width can vary from one design to the next. It should be noted that the measurement is to be taken in regard to the handrail location, which is permitted to extend a maximum of $4^1/_2$ inches (114 mm) into the required width. Where the maximum encroachment occurs on each side of the stairway, an intermediate handrail must be provided where the required width exceeds 69 inches (1,752 mm). A lesser required width would apply where the handrails do not extend the full $4^1/_2$ inches (114 mm) into the minimum required stairway width. As an additional safeguard for wide monumental stairs, the handrails must be located along the anticipated travel path of the stair users.

1012.9

10 Means of Egress

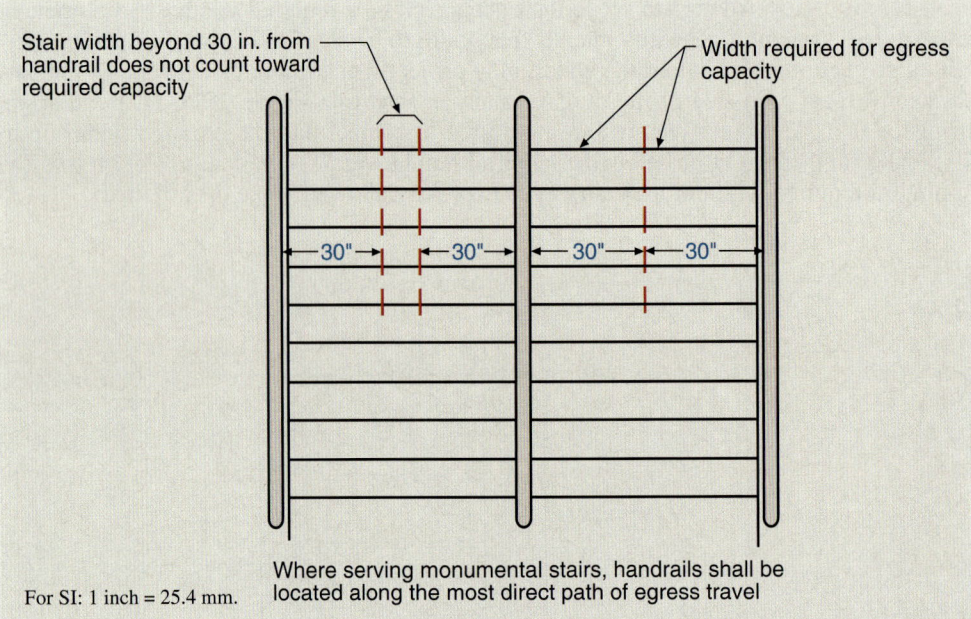

Figure 1012-12
Intermediate handrails.

Section 1013 Guards

1013.2 Where required. In this section, the code provides for guard protection at open sides along walking surfaces, mezzanines, stairways, ramps, and landings that are more than 30 inches (762 mm) above grade or a floor surface below. Also, protection is required for the glazed sides of stairways, ramps, and landings located more than 30 inches (762 mm) above the floor or grade below, unless the glazing complies with the strength and attachment provisions in Section 1607.8 for live loads. The need for guards in these circumstances is evident, although the arbitrary limit of 30 inches (762 mm) is subject to conjecture. Nevertheless, in the case of the IBC, it is assumed that the maximum height differential of 30 inches (762 mm) without guard protection does not create a significant safety hazard.

In the determination of the difference in elevation, an objective method is established for measuring the height of the walking surface above the grade below. Rather than taking this measurement to the ground level or floor directly below the edge of the walking surface, the code requires the height of the walking surface above grade to be based on the lowest point within 36 inches (914 mm) horizontally from the edge of the deck, porch, or other elevated element. This approach recognizes that a sloped site sometimes occurs adjacent to a deck, porch, or similar walking surface, increasing the level of hazard. This method of measurement is illustrated in Figure 1013-1.

There are seven exceptions that identify locations or situations where guards complying with Section 1013 are not required. These exceptions fall into essentially three categories where, for obvious reasons, guards would be inappropriate:

1. Commercial and industrial applications. Guards are not required on the loading side of loading docks or piers, nor along vehicle-service pits inaccessible to the public. Such guards would severely restrict the work that takes place in these areas.

2. Stage and platform areas. Because of the nature of activities involving performance stages or platforms, it is impractical to provide guards in various locations.

Guards 10

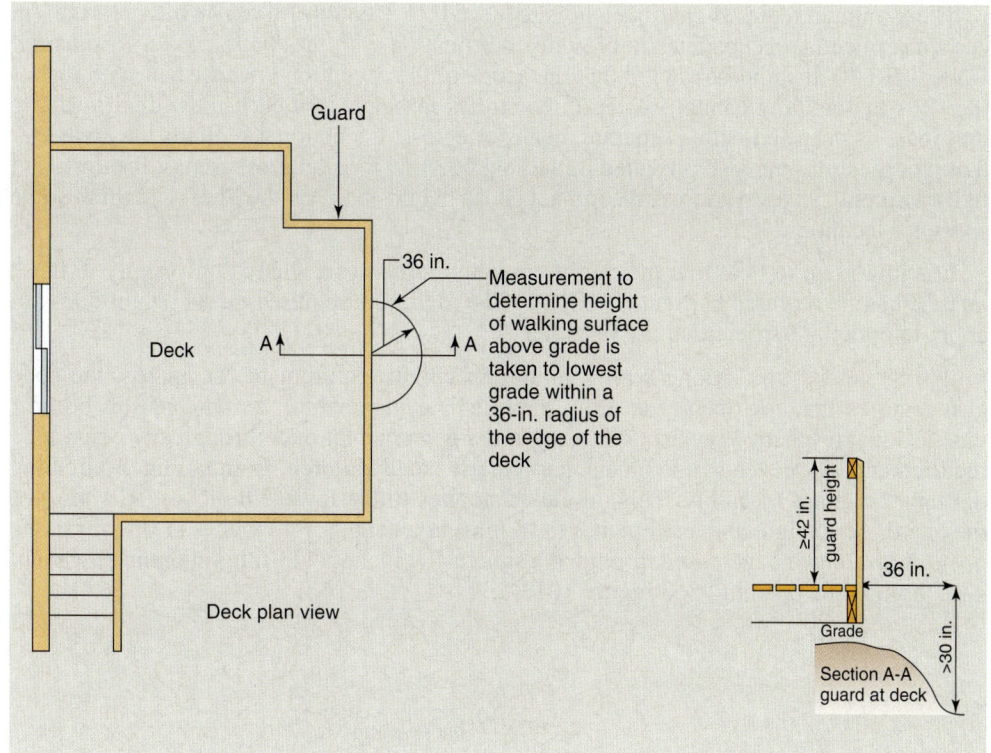

Figure 1013-1
Determination of guard applicability.

Guards may be omitted on the audience side of stages and raised platforms, at runways or side stages used for presentations, at any vertical openings in the performance area, and at elevated walking surfaces used to access or use lighting or equipment. Guards also may be omitted along any steps that may lead from the auditorium area to a stage or platform; however, one or more handrails must still be provided under the provisions of Section 1012.

3. Assembly seating areas. In order to achieve adequate lines of sight for the assembly audience seated in a tiered configuration, it is often necessary to provide for guards of a height lower than required by the provisions of Section 1013. Therefore, reduced guard heights are permitted under the provisions of Section 1028.14.

Height. Where guard protection is required, the guard must be of adequate height to prevent someone from falling over the edge of the protected area. Therefore, the code establishes 42 inches (1,067 mm) as a minimum height that is acceptable for guard protection. In the case of a guard adjacent to a stairway, the minimum height of 42 inches (1,067 mm) is measured vertically above the leading edge of the tread. A number of exceptions to the minimum required guard height address two different conditions: guards adjacent to stairs and guards adjacent to other walking surfaces. **1013.3**

Exception 3 allows the top rail of a stairway handrail in Group R-3 occupancies and within individual dwelling units in Group R-2 occupancies to be considered an adequate guard height, provided the handrail is located between 34 inches (864 mm) and 38 inches (965 mm) measured vertically from the leading edge of the stair tread nosing. A separate guard element with a minimum height of 34 inches (864 mm) is also permitted under similar conditions as indicated in Exception 2. Exception 5 allows alternating tread devices and ship ladders to have guards with top rails serving as handrails located at a reduced height above the tread nosings.

10 Means of Egress

The minimum required guard height is reduced by Exception 1 to 36 inches (914 mm) in certain residential occupancies to provide coordination with the *International Residential Code®* (*IRC®*). It should be noted that the scope of the exception is consistent with that of the IRC, the building cannot exceed three stories in height and each individual dwelling unit must be provided with a separate means of egress. Exception 4 reminds the code user that modifications are also provided in Section 1028.14 for stadiums, arenas, theaters, and other assembly areas to address the impact guard heights have on the lines of sight in some spectator locations.

It is important to note that the minimum height criteria for guards only apply to those guards that are required by Section 1013.1. The minimum required guard height does not apply to optional barriers that are installed.

1013.4 Opening limitations. Along with a minimum height requirement for guards, the code also requires that for open-type rails, intermediate members or ornamentation be provided so that a sphere 4 inches (102 mm) in diameter cannot pass through any opening: a requirement that prevents individuals, particularly small children, from falling or climbing through the guard assembly. This limitation applies to the lower 36 inches (914 mm) of the guard. At a height above 36 inches (914 mm) to a height of 42 inches (1,067 mm), the opening size may be increased, provided a sphere $4^3/_8$ inches (111 mm) in diameter cannot pass through the opening. See Figure 1013-2.

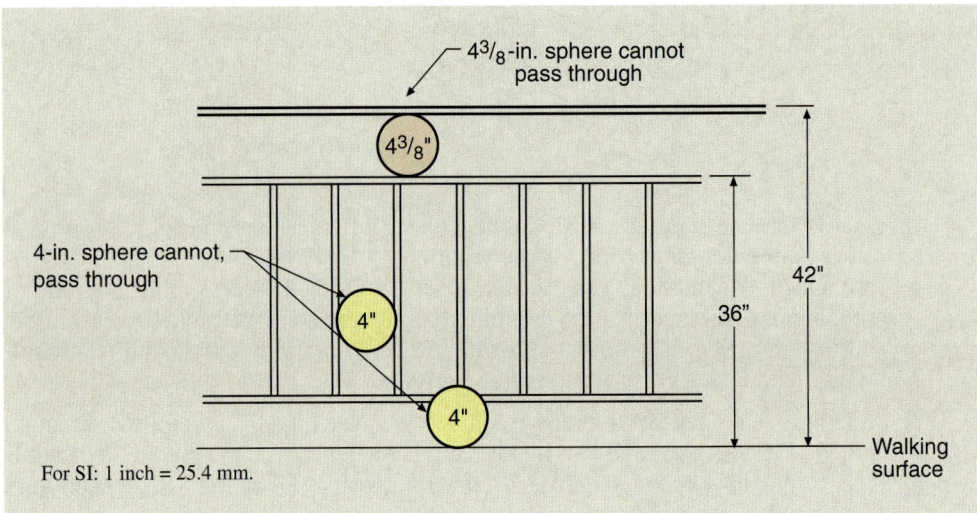

Figure 1013-2 Guard opening limitations.

Several exceptions increase the maximum opening size permitted in an open guard. The triangular area formed by the tread, riser, and bottom rail of the guard is limited in size to where a sphere of 6 inches (152 mm) in diameter cannot pass through the opening. Because of the unusual configuration of a stairway and its required guard, as well as the location of the triangular opening, an increased size is deemed reasonable. This configuration is shown in Figure 1013-3.

In commercial, industrial, and security uses where the public is not invited (therefore, the guard is typically not subject to children crawling or falling through), open guards may have intermediate members or ornamentation spaced so that a 21-inch (533-mm) diameter sphere cannot pass through. This exception applies to those elevated walking surfaces used for access to areas containing electrical, mechanical, or plumbing systems or equipment, as well as platforms provided for the use of such systems or equipment. It also applies to elevated areas in occupancies of Group I-3, F, H, or S where access to the public is not available.

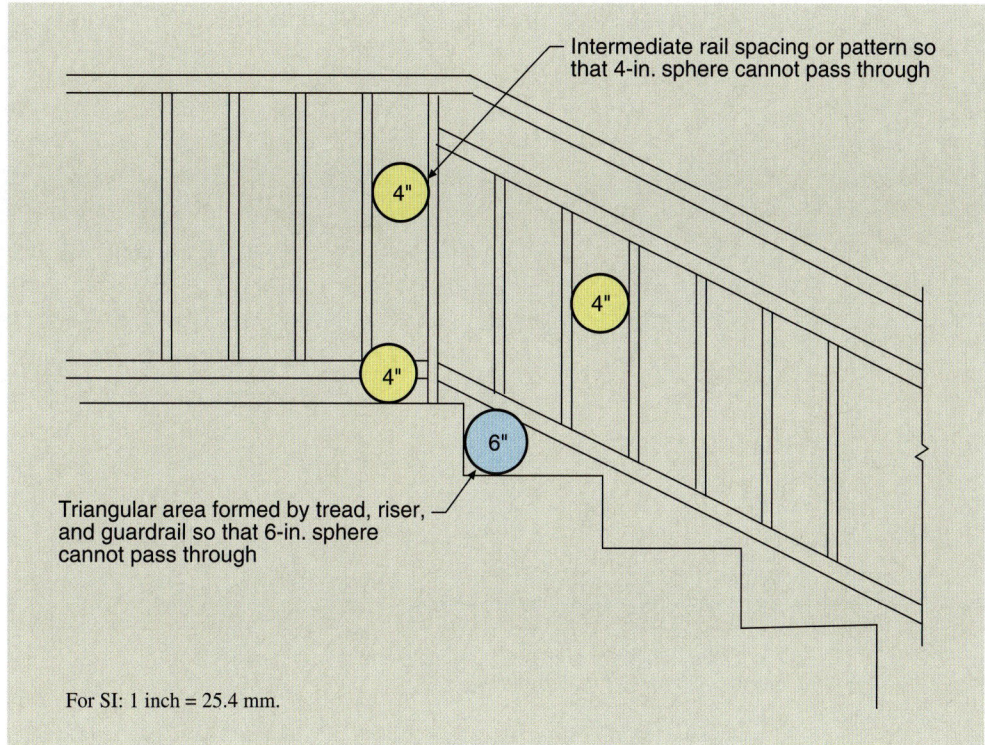

Figure 1013-3
Guardrail openings at stairs.

In order to significantly improve the lines of sight for rows located immediately behind the guard in assembly areas, an exception reduces the amount of infill provided in the guard at the end of the aisles. From a height of 26 inches (660 mm) to the top of the rail, the opening need only be small enough that a sphere 8 inches (203 mm) in diameter cannot pass through. See Figure 1013-4.

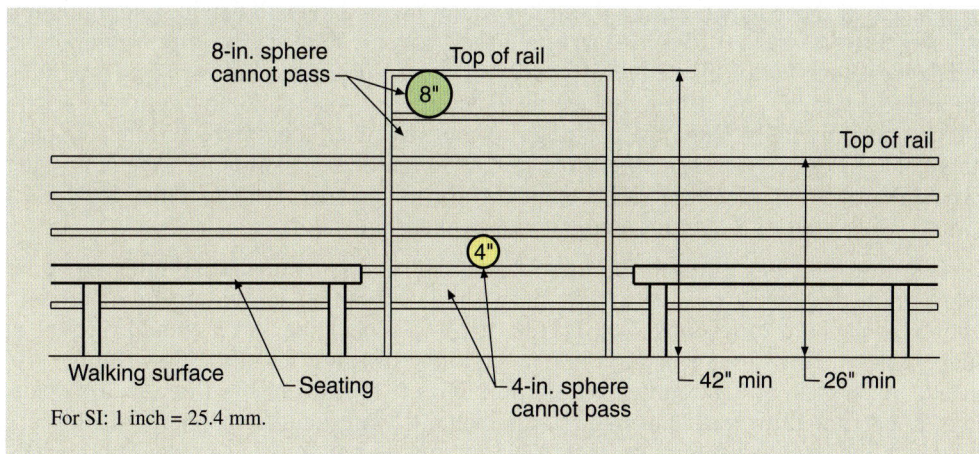

Figure 1013-4
Assembly seating guards.

Applicable only within dwelling units and sleeping units in Groups R-2 and R-3, a slightly larger opening is permitted in guards at the open sides of stairs. See Figure 1013-5. The maximum opening size of $4^3/_8$ inches (111 mm) will permit the installation of two balusters, rather than three, on 10-inch (254-mm) treads without compromising significant safety in regard to infants crawling through the openings. The greater hazard that cannot be addressed in the code is that of infants falling down the stairs.

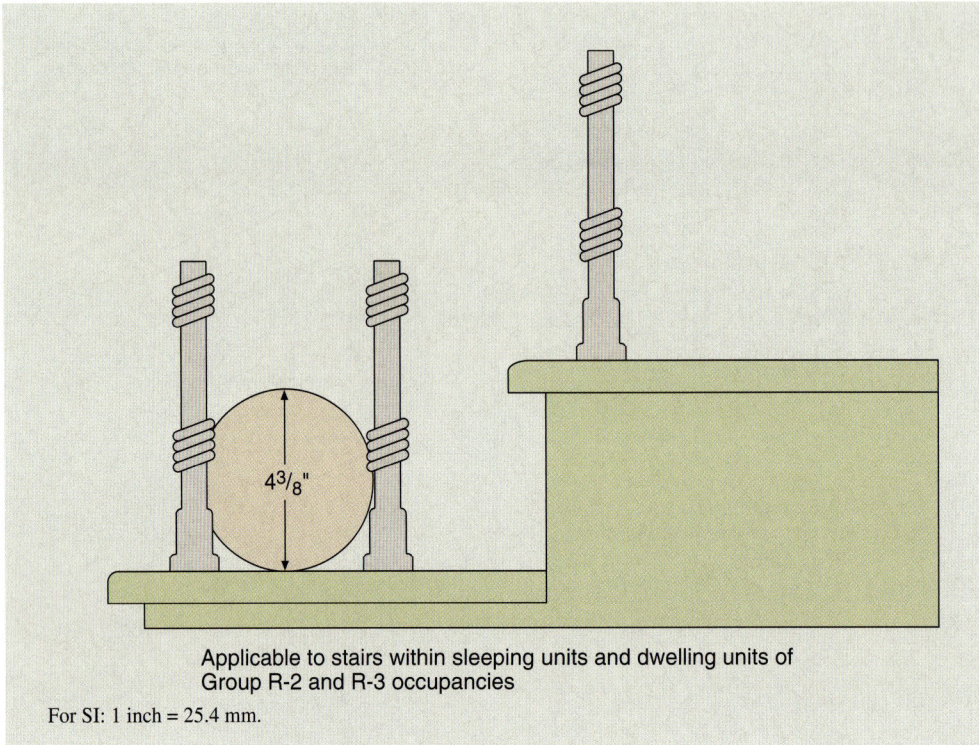

Figure 1013-5 Guard opening limitations for Group R-2 and R-3 occupancies.

1013.8 Window sills. Historical data have shown that each year a considerable number of children fall from windows in residential buildings. It has been estimated that a sizable percentage of those falls occurred through windows with a low sill height. The minimum sill height of 36 inches (915 mm) was established not only to place the lowest point of the window opening above the center of gravity of most children to address accidental falls through the window, but also to reduce the potential for small children to climb onto the sill and possibly fall through the opening. It should be noted that the requirement is only applicable to those buildings where the hazard to children is common—dwelling units classified as Group R-2 or R-3 occupancies. Sleeping units, such as those located in Group R-2 dormitories, are not regulated under these provisions.

The measurements on both the interior and exterior sides of the building are to be taken from the lowest portion of the clear window opening, providing for consistent application of the provisions. See Figure 1013-6. Where the lower window panel is inoperable, the measurements are to be taken to the lowest point of the lowest operable panel. Fixed glazing is permitted within 36 inches (915 mm) of the floor, as are openings that do not allow for the passage of a 4-inch (102-mm) diameter sphere. In addition, openings that are sufficiently protected by complying window fall prevention devices and windows provided with windows opening control devices are not regulated for the minimum sill height.

The requirement for window sill height does not affect the application of the IBC provision addressing emergency escape and rescue windows. Where a window regulated by this section occurs in a sleeping room and is designated as the required emergency escape and rescue opening, it will have both a minimum and a maximum required sill height.

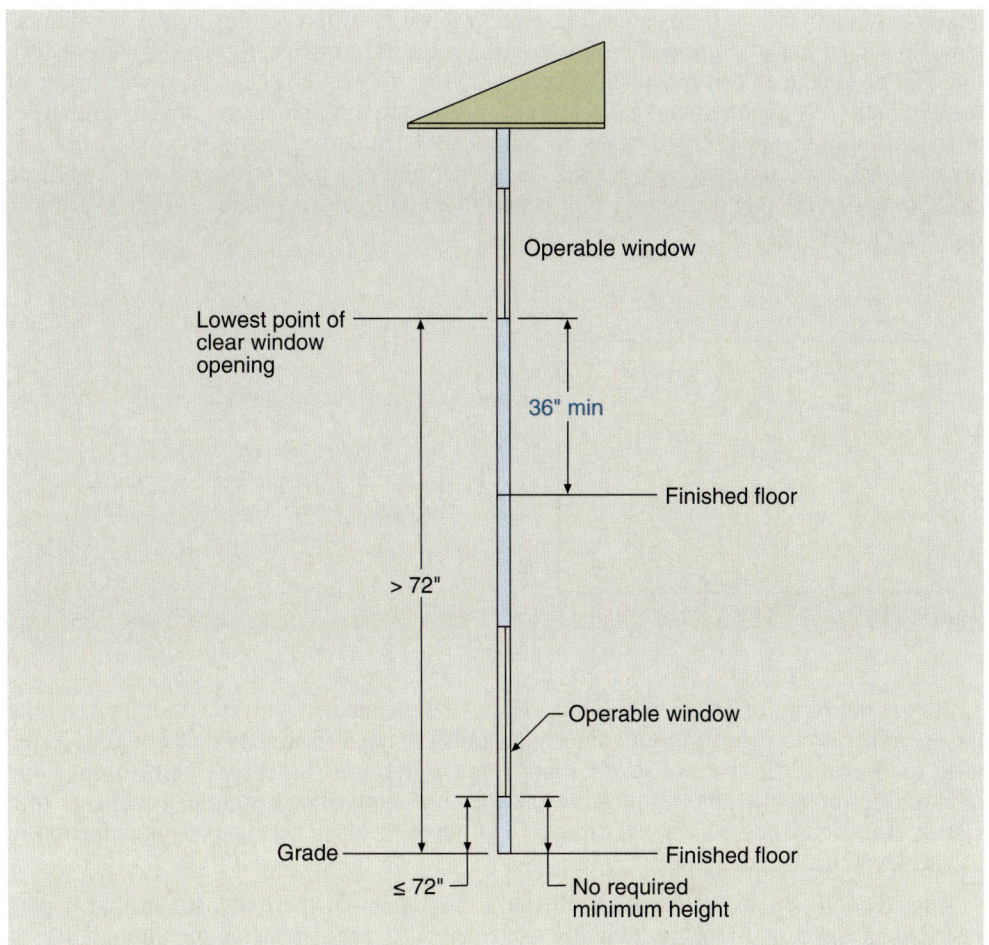

Figure 1013-6
Minimum window sill height.

Section 1014 *Exit Access*

The exit access is identified as the initial component of the means of egress system—that portion between any occupied point in a building and an exit. Leading to an exterior exit door at the level of exit discharge, an exterior egress stairway or ramp, or the door of an exit passageway, interior exit stairway, interior exit ramp, or horizontal exit, the exit access makes up the vast majority of any building's floor area.

General. The key design provisions related to means of egress are typically found in this section. Issues relating to the number and arrangement of exit paths are addressed, as is travel through intervening spaces and travel distance. The concept of a common path of egress travel is also presented. **1014.1**

Egress through intervening spaces. Basically, the code intends that access to exits should be direct from the room or area under consideration. This section, however, makes some modifications where, under certain circumstances, exit paths may be arranged through adjoining rooms or spaces rather than directly into corridors or exit elements, such as interior exit stairways or exit passageways, or through exterior doors. **1014.2**

It is permissible to provide egress through an adjoining room or space, provided the adjacent rooms or spaces are accessory to each other. In this context, the term *accessory* describes an interrelationship between the adjoining spaces based on their use, and not necessarily their size. Where egress must occur through an intervening space, it is important that such a space be under the same control as the initial space. Egress through such an intervening area to an exit must be direct and obvious so that the occupant is well aware of the exit path. It is assumed that a discernible egress path through an area that essentially is an extension of the area served poses no significant hazard to exiting, provided travel does not enter a hazardous area. Egress travel is not permitted to pass through a room or space classified as a Group H occupancy unless permitted under the conditions of the exception. See Figure 1014-1.

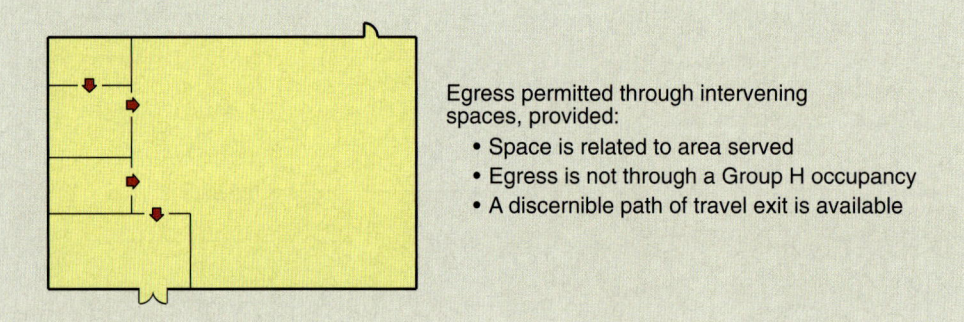

Figure 1014-1
Egress through intervening.

Where the room of origin is a Group H, S, or F occupancy, an exception permits the means of egress to pass through adjoining rooms or spaces designated as equal or lesser hazards. Because the hazard level is not increased along the egress path through the intervening room, the general prohibition does not apply. For example, occupants of a Group H-2 occupancy may travel through a Group H-3 space because of the reduction in hazard level that is anticipated.

This section is also concerned with the arrangement of the exit path in that it puts restrictions on certain spaces that are considered to present an undue probability of obstruction to free egress travel. Therefore, the code prohibits the exit path from passing through kitchens, store rooms, closets, and spaces used for similar purposes where the probability of things obstructing the path of travel is substantially greater. An exception permits travel through a stockroom provided four conditions are met, as illustrated in Figure 1014-2.

As is so frequently the case throughout the code, exceptions are made for rooms within dwelling and sleeping units. Although egress from dwelling units or sleeping units shall not lead through other sleeping areas or through toilet rooms or bathrooms, it is permitted through a kitchen area within the same unit.

An even greater concern is the potential for the egress system to be completely unusable. Therefore, the code specifies that exit access cannot pass through any room or area that can be locked to prevent egress.

1014.2.1 Multiple tenants. Where any floor of a building is occupied by more than one tenant, it is critical that access to all of the required exits be accomplished without passing through any adjacent tenant spaces. This condition also applies to dwelling units and sleeping units. Because it is almost always impossible to have control over what occurs in the tenant spaces of others, it is important that the exit system not be reliant on an egress path through any neighboring tenant spaces. The exception allows relatively small tenant spaces to egress through larger tenant spaces under specified conditions. Independent egress from the smaller tenant spaces is not required where such spaces are limited in size, provided they are classified to an occupancy group similar to that of the main occupancy and their means of egress cannot be locked. An example might be a branch bank or fast-food restaurant tenant that is located within a large retail store.

Exit Access 10

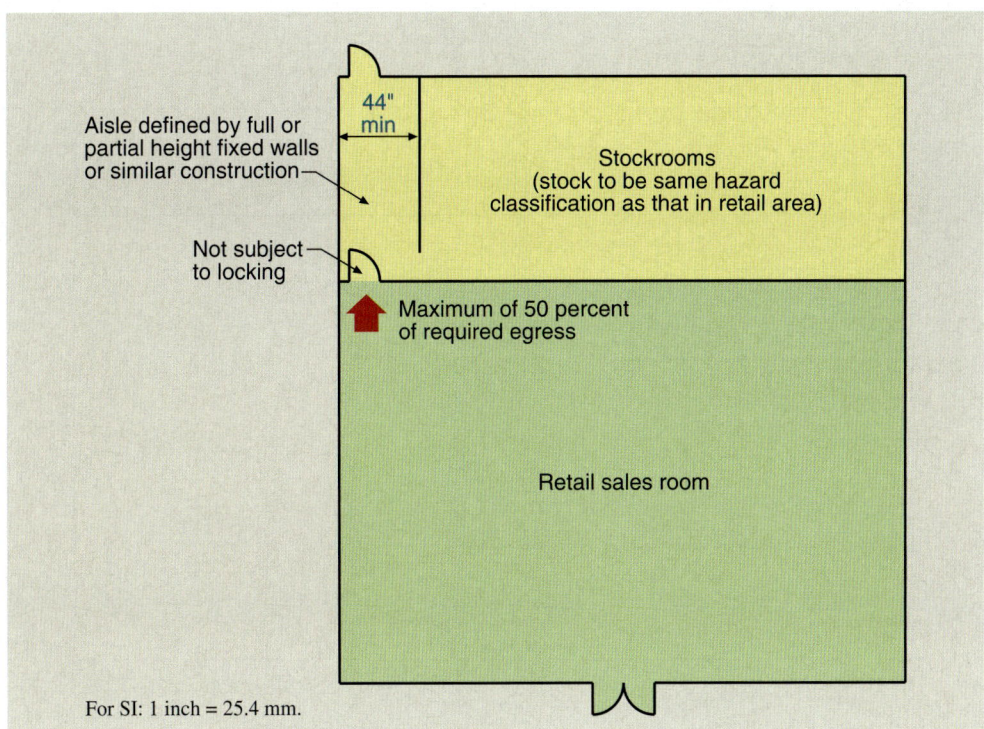

Figure 1014-2
Egress through stockrooms.

Common path of egress travel. The definition of a common path of egress travel is found in Section 202. Described as "that portion of exit access which the occupants are required to traverse before two separate and distinct paths of egress travel to two exits are available," a common path of travel is that portion of the exit system where no egress options are available to the occupant. The length of the common path is measured from the most remote point of a room or area to the nearest location where multiple exit paths to separate exits are available. See Figure 1014-3.

1014.3

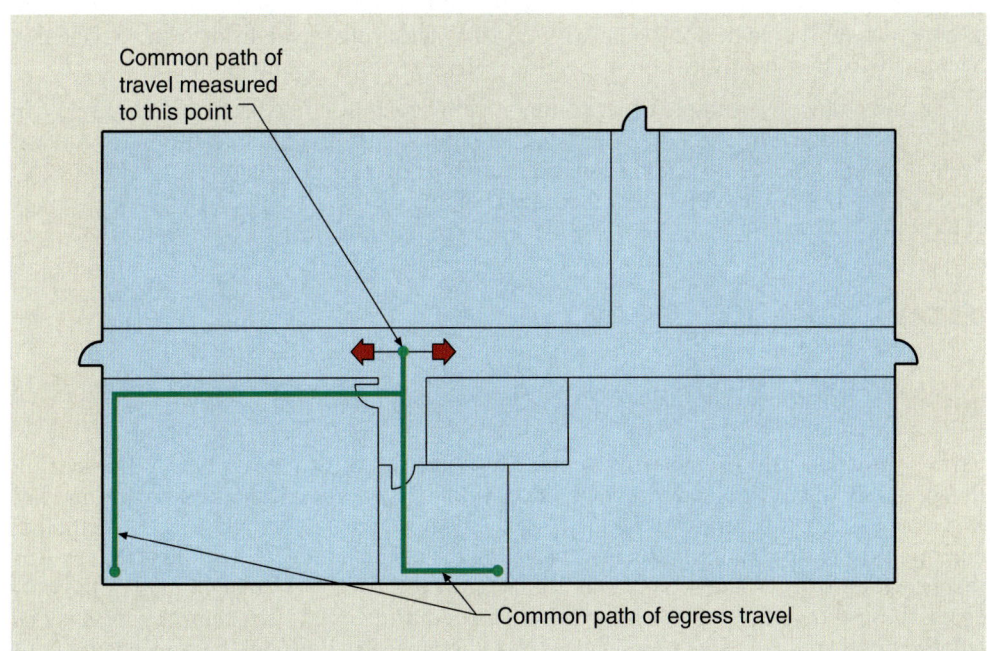

Figure 1014-3
Common path of egress travel.

2012 International Building Code Handbook

10 Means of Egress

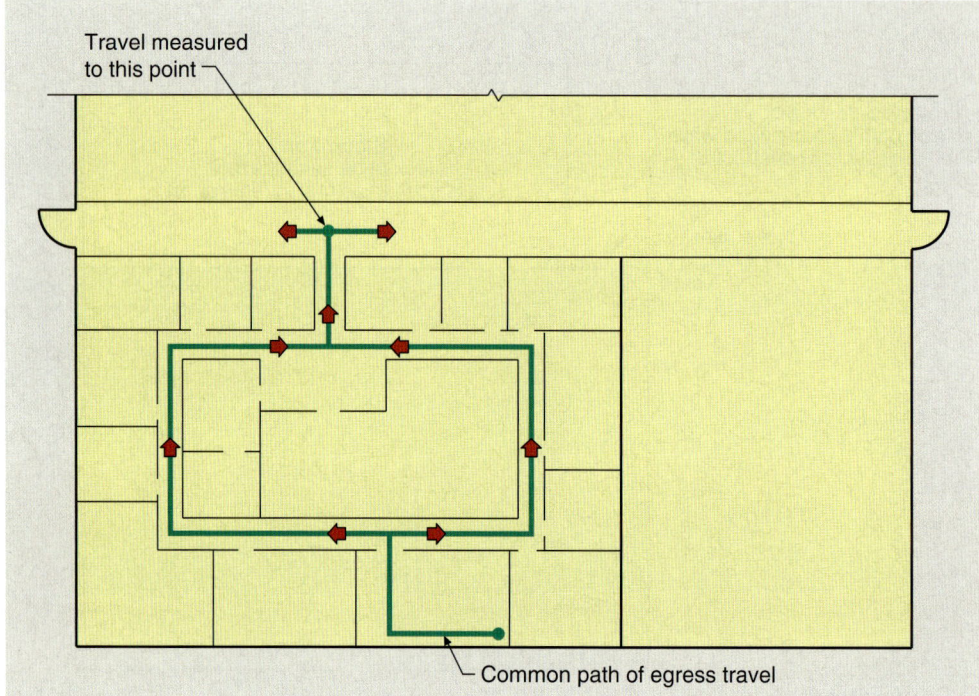

**Figure 1014-4
Merging paths of travel.**

The definition of a common path of egress travel also indicates that exit paths that merge are also considered common paths of egress travel. An example of such a situation is shown in Figure 1014-4.

As a general rule, the maximum length of a common path of egress travel is 75 feet (22,860 mm), as shown in Table 1014.3. Because of unique risks potentially encountered in a high-hazard occupancy, the common path of travel is limited to 25 feet (7,620 mm) in Groups H-1, H-2, and H-3. In certain occupancies and under specific conditions set forth in the table, an increase in the common path of travel to a distance of 100 feet (30,480 mm) or 125 feet (38,100 mm) is permitted. The method of measurement of the common path of egress travel should be consistent with that for measuring travel distance as regulated in Section 1016. See the discussion of Section 1016.3 for further guidance.

The most obvious example of a "common path" condition is a room with a single exit or exit access doorway. Although there are numerous paths that may lead to the doorway, eventually they all end up at the same point. Where two or more complying exits or exit access doorways are provided, common path conditions do not exist.

Section 1015 *Exit and Exit Access Doorways*

1015.1 Exit or exit access doorways from spaces. It would seem obvious that every occupied portion of a building must be provided with access to at least one exit or exit access doorway. It is assumed that if buildings are occupied, then the occupants obviously have a method of entering the various building spaces. Therefore, that same entrance is available to serve as the means of egress. Under many conditions, however, the use of the entrance as the only egress point is insufficient. The basic reason for requiring multiple means of egress is that in a fire or other emergency, it is very possible that the entry door will be obstructed by the fire and, therefore, not be usable for egress purposes. A second exit or exit access doorway can provide an alternative route of travel for occupants of the room or area. However, it is often unreasonable to require multiple egress paths from small spaces or areas with limited occupant loads. It is also seldom beneficial because of the relatively close proximity in which such exits or exit access doorways must be located. Therefore, the code does not require a secondary egress location from all rooms, areas, or spaces.

In this section, the code deals with the number of exits or exit access doorways that are going to be required to accommodate the occupant load to be served. In addition, where the distance of travel to a single exit or exit access doorway is excessive, an alternative egress route is also required. The IBC establishes two basic criteria for providing adequate egress for occupants from any space within the building. First, at least two exits or exit access doorways must be provided when the occupant load of the space exceeds the values set forth in Table 1015.1. Second, two or more egress doorways are required when the common path of egress travel exceeds the limitations found in Section 1014.3. See Application Examples 1015-1 and 1015-2. Third, the provisions of Sections 1015.3 through 1015.6 may mandate a minimum of two egress doorways for specific mechanical equipment areas.

Application Example 1015-1

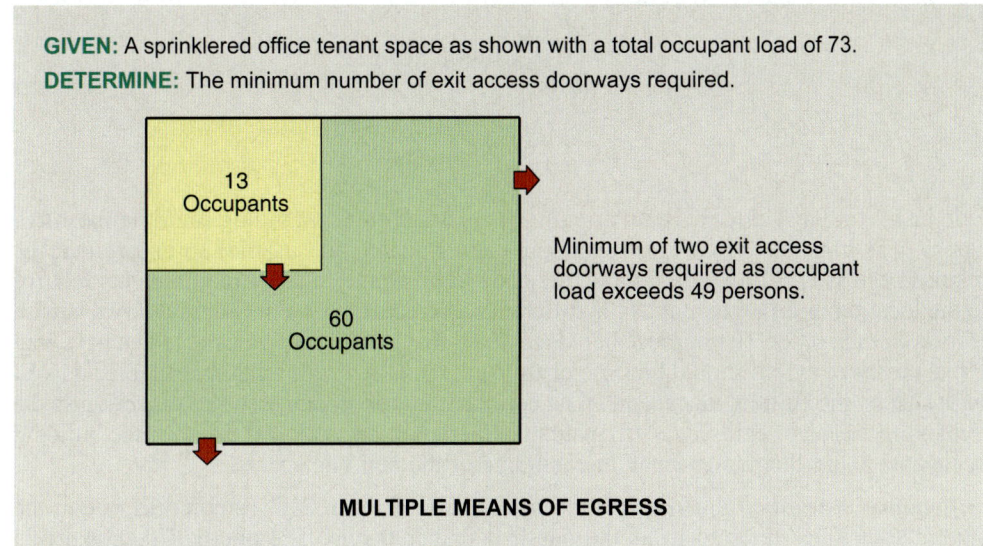

As seen in Table 1015.1, the threshold for requiring a second exit or exit access doorway from a space varies based on the occupancy designation of the space. The variations are caused by conditions associated with the specific uses that occur within the room or area. Factors that contribute to the differences in occupant load include the concentration of occupants, occupant mobility, and the presence of hazardous materials. There is a specific exception that allows a single means of egress from individual dwelling units within and from Group R-2 and R-3 occupancies in fully sprinklered buildings for occupant loads of 20 or less. An additional exception references the provisions of Section 407.4.3 for Group I-2 care suites.

1015.1.1 Three or more exits or exit access doorways. Where any room, space, or other floor area has an occupant load exceeding 500, at least three exits or exit access doorways are mandated. A minimum of four exits or exit access doorways are required where the occupant load exceeds 1,000. These thresholds are consistent with those established in Section 1021.2.4 addressing the minimum number of exits, or exit access stairways, required from an individual story.

Application Example 1015-2

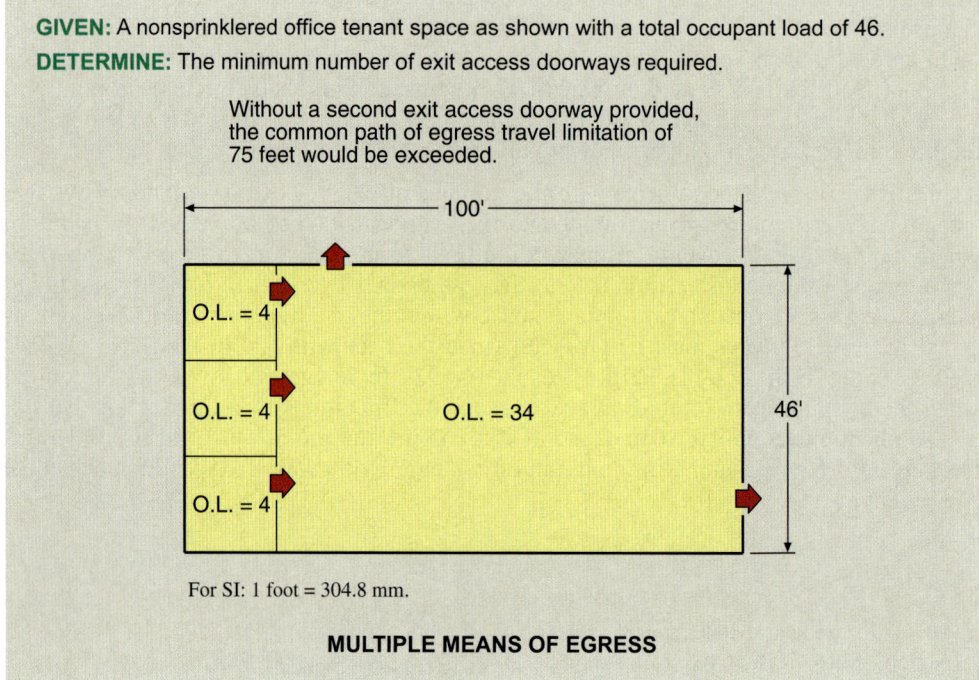

GIVEN: A nonsprinklered office tenant space as shown with a total occupant load of 46.
DETERMINE: The minimum number of exit access doorways required.

Without a second exit access doorway provided, the common path of egress travel limitation of 75 feet would be exceeded.

For SI: 1 foot = 304.8 mm.

MULTIPLE MEANS OF EGRESS

1015.2 Exit or exit access doorway arrangement. In addition to providing multiple means of egress, it is imperative that egress paths remain available and usable. To ensure that the required egress is sufficiently remote, the code imposes rather strict requirements relative to the location or arrangement of the different required exits or exit access doorways with respect to each other. The purpose here is to do all that is reasonably possible to ensure that if one means of egress should become obstructed, the others will remain available and will be usable by the building occupants. As a corollary, this approach assumes that because the remaining means of egress are still available, there will be sufficient time for the building occupants to use them to evacuate the building or the building space.

Required separation of two exits or exit access doorways. This remoteness rule in the IBC is sometimes referred to as the one-half diagonal rule. The one-half diagonal rule states that if two exits or exit access doorways are required, they shall be arranged and placed a distance apart equal to not less than one-half of the maximum overall diagonal of the space, room, story, or building served. Such a minimum distance between the two means of egress, measured in a straight line, shall not be less than one-half of that maximum overall diagonal dimension. See Figure 1015-1 for examples of the application of this rule.

The code does not specifically state the manner in which the straight-line measurement should be taken. In practice, different building officials determine the length of the straight-line measurement for egress separation in different ways. In some instances, building officials measure that distance from the near edge of one egress door to the near edge of the other. Other building officials apply the rule as measuring the distance between the center lines or far edges of the two required egress doors. Based on the lack of specifics in the code, it would seem logical to consider the distance to be measured as between the center lines of exits or exit access doorways. It should be noted that, by definition, the term *exit access doorway* includes any point of egress where the occupant has a single point that must be reached prior to continued travel to the egress door. See Figure 1015-2.

Exit and Exit Access Doorways

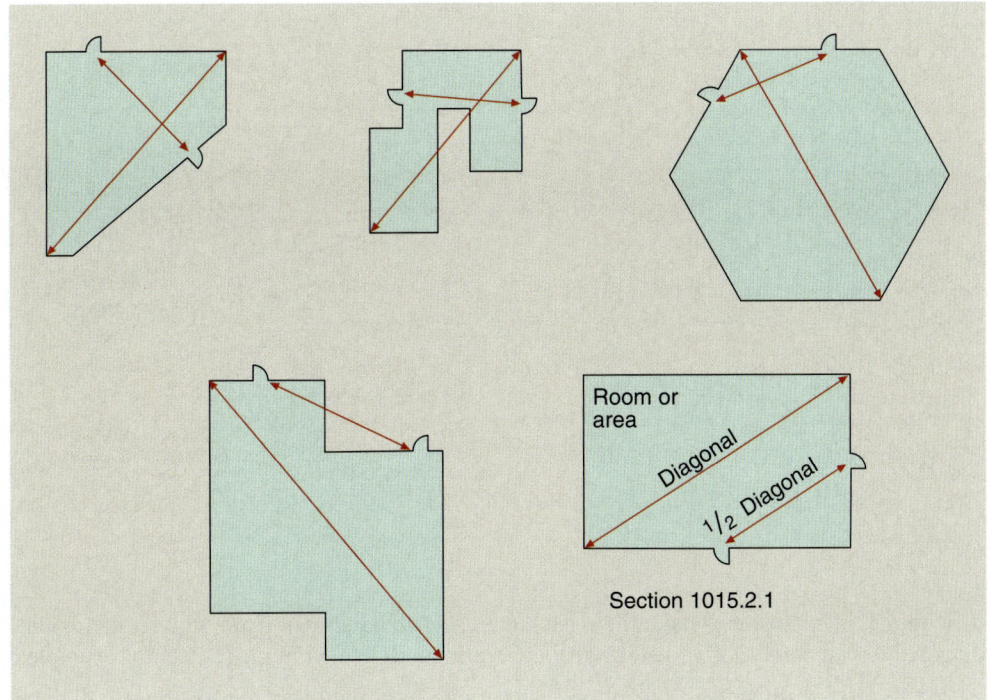

Figure 1015-1
Separation of exits or exit-access doorways.

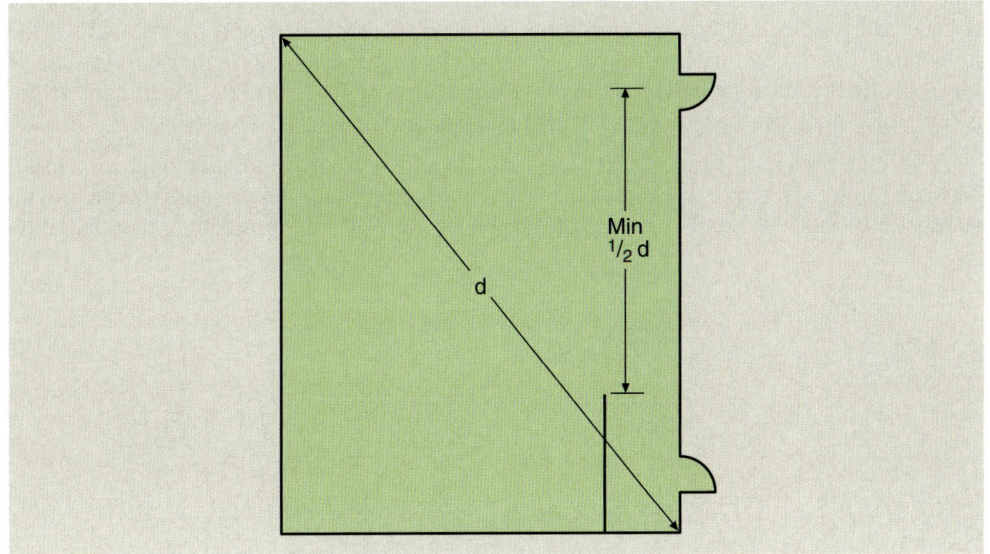

Figure 1015-2
Egress separation.

The use of the one-half diagonal rule has been beneficial to code users for many years. It quantifies the code's intent when the code requires that separate means of egress be remote. It does not leave the building official with a vague performance-type statement that can, in many instances, result in a situation where egress separation would be dictated more by the design or desired layout of the building rather than by a consideration for adequate and safe separation of the means of egress.

In applying the one-half diagonal rule to a building constructed around a central court with an egress system consisting of an open balcony that extends around the perimeter of

10 Means of Egress

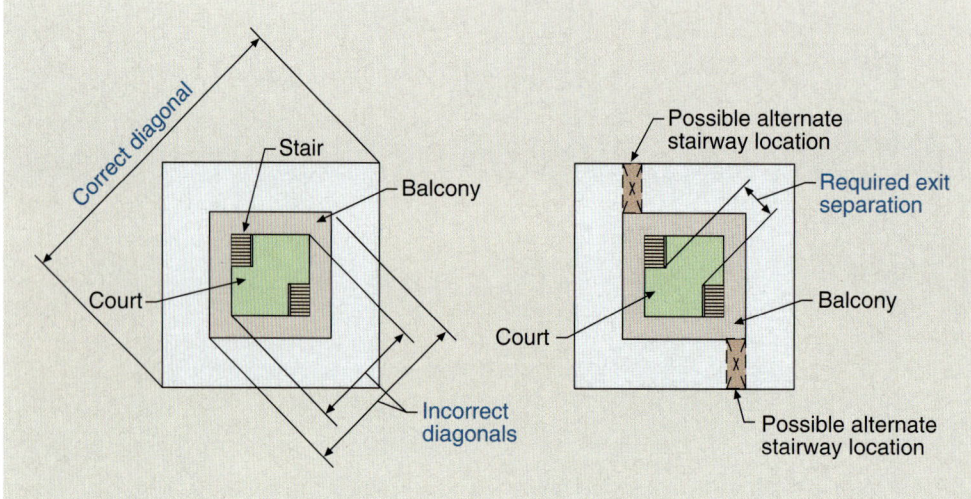

Figure 1015-3
Required egress separation.

the court, it is important to take the measurement of the diagonal from which the one-half diagonal dimension is derived at the proper locations. Refer to Figure 1015-3 for examples.

Figure 1015-4 illustrates Exception 1 to the one-half diagonal rule for those buildings, such as core buildings, where the means of egress are sometimes arranged in rather close proximity. The code recognizes the benefits of such a floor arrangement and makes a specific exception in the event there is such a design. If the exits or exit access doorways are connected by a fire-resistance-rated corridor, the distance determined by one-half of the maximum overall diagonal of the space served may be measured along the shortest path of travel inside the corridor between the two exits. It is specific that the connecting corridor is to be of 1-hour fire-resistant construction.

A second exception reduces the minimum length of the overall diagonal dimension between remote exits or exit access doorways in those buildings equipped throughout with automatic sprinkler systems. Because of the presence of sprinkler protection, the separation

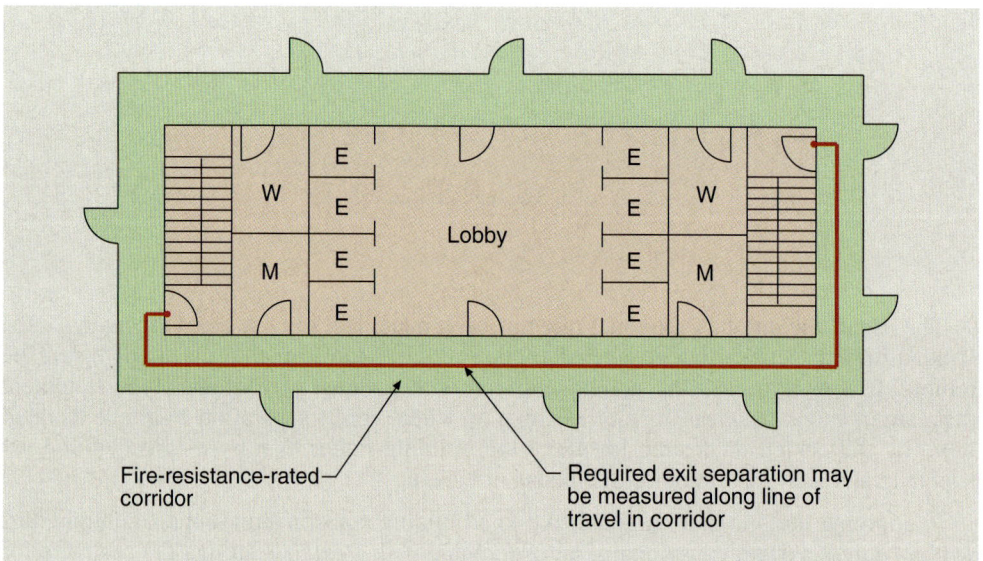

Figure 1015-4
Core arrangement of interior exit stairways.

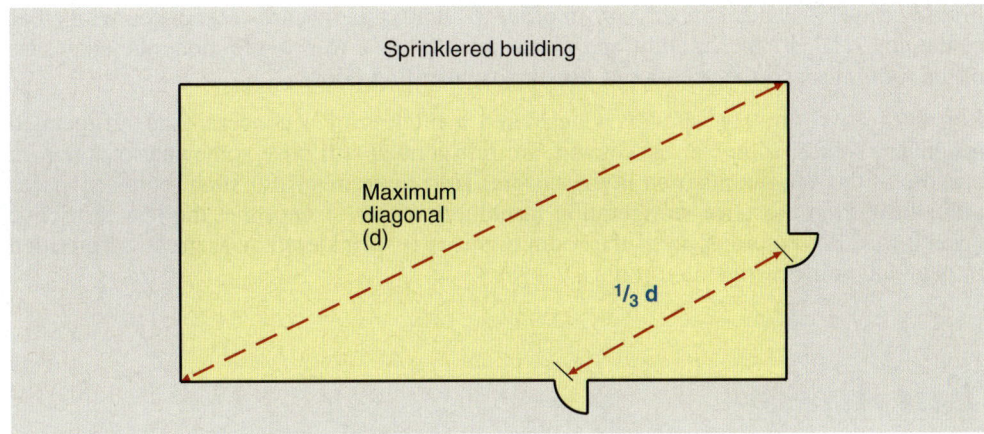

Figure 1015-5
Exit separation—sprinklered building.

distance need only be one-third of the length of the overall diagonal dimension. The use of this exception results in a reduction of the required distance between exits or exit access doorways by $33^1/_3$ percent. See Figure 1015-5.

Three or more exits or exit access doorways. When more than two means of egress are required, the remoteness rule takes on more of a performance character. In such an instance, at least two of the required exits or exit access doorways shall be arranged to comply with the one-half diagonal rule (one-third in fully sprinklered buildings). Although not specifically stated, the other means of egress should be arranged at a reasonable distance from the other egress points so that if any one of the required exits or exit access doorways becomes blocked by a fire or any other emergency, the others will be available. Obviously, this decision will require some very careful evaluation and judgment on the part of the building official. There may be a sufficient basis for applying that same rule when considering each possible pair of egress components in a multi-exit situation. The code is silent in this particular aspect, and proper code administration does require substantial, careful evaluation and judgment on the part of the building official in ensuring that the number of means of egress required is sufficiently remote so that it is not likely that the use of more than one access to exit will be lost in any fire incident.

1015.2.2

Boiler, incinerator, and furnace rooms. Depending on the capacity of the fuel-fired equipment located in a large boiler room or similar area, it may be necessary to provide a secondary egress doorway because of the potential hazards. At least two exit access doorways are required where the room exceeds 500 square feet (46 m^2) and any single piece of fuel-fired equipment such as a furnace or boiler exceeds 400,000 Btu (422,000 KJ) input capacity. The requirement is based on the size of a single piece of equipment, not the aggregate total of all fuel-fired equipment in the room. Access to one of the two exit access doorways may be accomplished through the use of a ladder or an alternating stair device, as described in Section 1009.13. As with other multiple-exit situations, the two doorways must be adequately separated, in this case by a horizontal distance no less than one-half the maximum overall diagonal.

1015.3

Refrigeration machinery rooms. The *International Mechanical Code*® (IMC®) mandates when refrigeration systems must be contained in a refrigeration machinery room. It can be based on the type of refrigerant, the amount of refrigerant, the type of equipment, or other factors. Once a refrigeration machinery room is required by the IMC, the IBC provides the exiting criteria. Where larger than 1,000 square feet (92.9 m^2), the room must have at least two exit access doorways, accessed and separated in the same manner as described for boiler and furnace rooms. Egress doors from the room must be tight-fitting, self-closing, and swing in the direction of travel. Travel distance is also restricted to 150 feet (45,720 mm). It is evident that the presence of multiple exitways and a more

1015.4

limiting travel distance is necessary in order to address the hazards associated with areas containing refrigerants. In addition, provisions are made to provide moderate enclosure of the room through the use of self-closing, tight-fitting doors.

1015.5 Refrigerated rooms or spaces. Considered a bit less of a concern than refrigerated machinery rooms, rooms or spaces that are refrigerated still pose somewhat of a hazard because of the refrigerants used in the system. The requirements for such rooms or spaces differ little from those for refrigeration machinery rooms, except that the less restrictive general travel distances apply if the room or space is sprinklered and egress is permitted through adjoining refrigerated rooms.

Section 1016 *Exit Access Travel Distance*

1016.1 General. In this section, the code is concerned that the means of egress be accessible in terms of their arrangement so that the distance of travel from any occupied point in the building to an exit is not excessive. The IBC, therefore, establishes maximum distances to exits from any occupiable point of the building. This distance is referred to as the travel distance. Travel distance is the distance that a building occupant must travel from the most remote, occupiable portion of the building to either the door of an interior exit stairway or ramp, an exit passageway, or a horizontal exit; to an exterior egress stairway or exterior egress ramp; or to an exterior exit door located at the level of exit discharge. A travel limit is only imposed to the nearest exit component, not to all required exits from the room, floor, or building.

Each of these exit elements in a means of egress system is considered to represent a sufficient level of safety such that the code is no longer concerned about the distance that the building occupant must travel to reach the eventual safe place. On the top floor of the world's tallest buildings, the distance to exits, referred to as the travel distance, is measured on that floor from the most remote occupiable point to the point where the building occupant enters the interior exit stairway. The fact that the building occupant then has dozens of floors of stairway to traverse before exiting the building is not a consideration in dealing with distance to exits. Indirectly, in establishing a maximum distance of travel to a point of reasonable safety, the code is imposing a time factor on the ability of the building occupant to travel from the point of occupancy to a relatively safe place either outside or within the building.

1016.2 Limitations. Basically, the code states that travel distance to an exit may not exceed the distances found in Table 1016.2. For most occupancies, the travel limitation is 200 feet (60,960 mm) in nonsprinklered buildings and 250 feet (76,200 mm) in buildings provided throughout with an automatic sprinkler system. However, there are numerous modifications to the general requirements. Fully sprinklered Group B occupancies are permitted travel distances of 300 feet (91,440 mm). In low-hazard factories and warehouses, as well as utility buildings, the maximum travel distance is also 300 feet (91,440 mm) and 400 feet (121,920 mm) where the building is fully sprinklered. Travel distance varies for high-hazard occupancies from a low of 75 feet (22,860 mm) in a Group H-1 to a high of 200 feet (60,960 mm) in a Group H-5. Those institutional occupancies classified as Groups I-2, I-3, and I-4 are permitted a maximum travel distance of 200 feet (60,960 mm) in a sprinklered building.

To alert code users to the fact that there are also other code sections that affect travel distance requirements, the notes to Table 1016.1 identify several locations where special requirements may be found, including:

1. Section 402.8. Travel distance within mall tenant spaces or within the mall itself. The section limits travel distance within the mall to 200 feet (60,960 mm). The travel distance within individual tenant spaces is also limited to 200 feet (60,960 mm).

2. Section 404.9. Travel distance within the atrium is limited to 200 feet (60,960 mm) on balconies or other egress paths not located on the lowest level of an atrium.

3. Section 1028.7. Travel distance within an assembly building having smoke-protected assembly seating or open-air seating. Where smoke-protective assembly seating is provided, the travel distance from every seat to the nearest concourse entrance is limited to 200 feet (60,960 mm). Travel from the concourse entrance to a stair ramp or walk on the exterior of the building is also limited to 200 feet (60,960 mm). Where the seating is open to the exterior, the maximum travel distance from each seat to the building's exterior shall not exceed 400 feet (121,920 mm). In such buildings of Type I or II construction, travel distance is unlimited.

4. Section 1021.2. Travel distance in buildings requiring only a single exit. Where buildings are permitted to have only one exit, the maximum travel distance is regulated by Tables 1021.2(1) and 1021.2(2).

5. Section 3103.4. Travel distance in temporary structures. The maximum exit access travel distance permitted in a temporary structure is 100 feet (30,480 mm).

1016.2.1 Exterior egress balcony increase. As depicted in Figure 1016-1, the travel limitations specified in Section 1016.2 may be increased by an additional 100 feet (30,480 mm) if the increased travel distance is the last portion of the travel distance on an exterior egress balcony. As an example, for a Group B occupancy, either the 200 feet (60,960 mm) in a nonsprinklered building or the 300 feet (91,440 mm) in a sprinklered building may be increased to 300 feet and 400 feet (91,440 mm and 121,920 mm), respectively, if in each instance the last 100 feet (30,480 mm) of travel distance is on an exterior egress balcony. Simply stated, all travel, up to a maximum of 100 feet (30,480 mm), that occurs beyond that permitted by Section 1016.1 must occur on an exterior egress balcony.

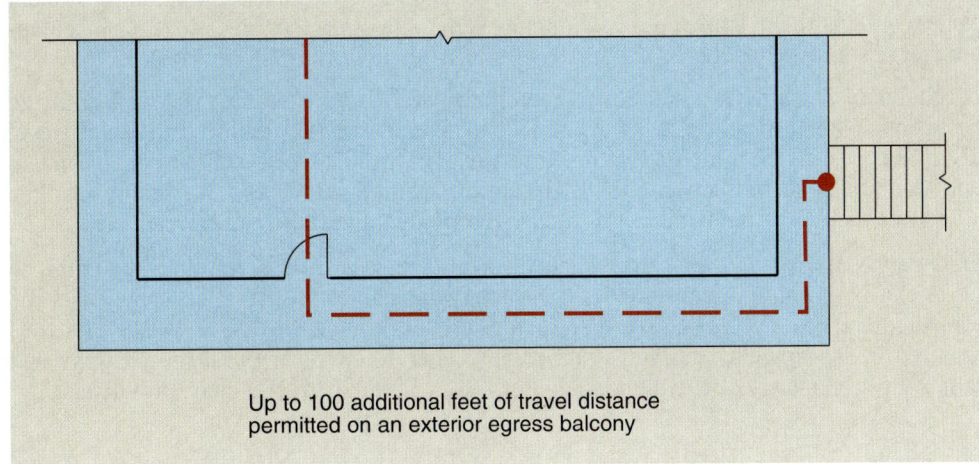

Up to 100 additional feet of travel distance permitted on an exterior egress balcony

Figure 1016-1 Egress balcony travel distance increases.

1016.3 Measurement. Travel distance is one of the most difficult features of the egress system to determine in either the design or the plan review stage. Travel distance is intended to be measured along the natural, unobstructed path available to the building occupant. See Figure 1016-2. That path is often determined by the location of partitions, doors, furniture, equipment, and similar objects. Many of these objects are reasonably portable and, as a consequence, the actual path available is frequently and easily altered. Although it is obvious that travel is to be measured around permanent construction and building elements, how to measure travel in areas with tables, chairs, furnishings, cabinets, and similar temporary or movable fixtures is debatable.

10 Means of Egress

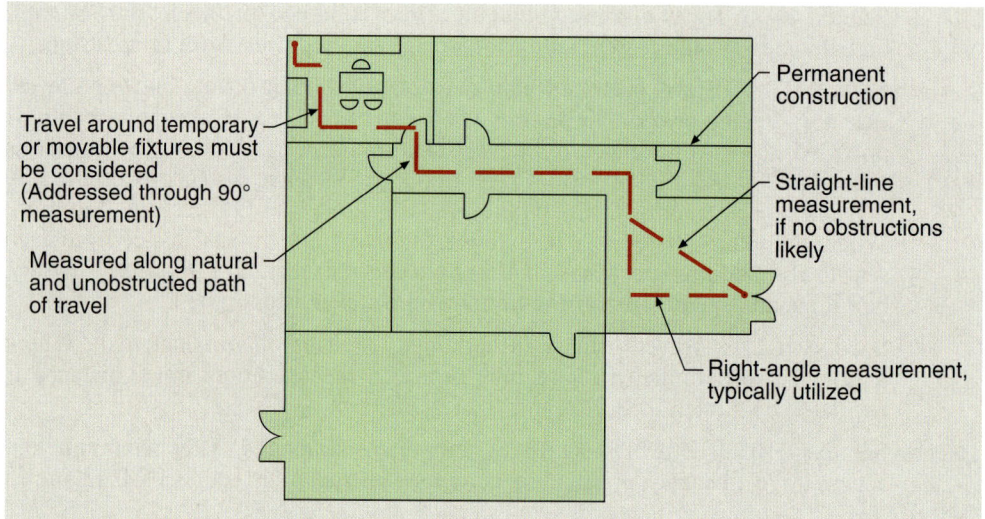

Figure 1016-2
Measurement of travel distance.

The preferred approach, conservative in nature, would dictate using the right-angle method for measuring travel distance. This method recognizes that obstacles such as desks, shelving, modular furnishings, and so on would cause travel to negotiate around such objects. The increased distance determined by right-angle measurement would account for such travel. On the other hand, the straight-line method would assume there are no obstructions along the measured travel path. Although such conditions at times exist, it is seldom that a straight-line path of travel will always be available. Although within private offices and similar small areas the straight-line approach would seem to cause no great concern, the method of measurement could be critical in larger spaces such as home improvement centers or department stores. Care should be taken to measure the travel distance in a manner that best represents the actual means of egress travel through the space.

Although the general requirement requires travel distance to be measured until the occupant's path of travel reaches the entrance to an exit, in open parking garages the travel distance need only be measured to the closest riser of an exit access stairway. Coordinated with the provisions of Section 1021.1, Exception 1, and Section 1009.3, Exception 6, this allowance recognizes that stairways in open parking garages need not be enclosed.

1016.3.1 Exit access stairways and ramps. Since travel distance is measured until the entrance of an exit is reached, travel on exit access stairways and ramps must be included in the measurement. Exit access stairways and ramps do not provide the degree of fire resistance required of interior exit stairways and ramps. Therefore, travel on such stairways and ramps must be included in evaluating compliance with the travel distance provisions.

Section 1017 *Aisles*

As a portion of the exit access, aisles are primarily regulated for minimum width purposes. Aisles typically serve occupant travel from adjoining aisle accessways and often merge into a main aisle. Aisles are commonly found in those buildings that contain furniture, fixtures, or equipment.

1017.1 General. This section requires that aisles and aisle accessways be provided in all occupied portions of the exit access that contain seats, tables, furnishings, displays, and similar fixtures or equipment. Primarily, this would have application to occupied-use areas or rooms where it is necessary to provide a circulation system so that building occupants will have reasonable means for moving around in the occupied spaces, as well as have access to corridors and other components of the egress system. It is customary to think of aisles in such facilities as theaters where an aisle system is installed to serve the fixed-seating areas. However, this section does not apply to such uses. Aisles in assembly areas, including those with seating at tables, such as in restaurants, are regulated solely under the provisions of Section 1028. The provisions of this section apply to circulation systems through open office areas, retail sales rooms, manufacturing areas, warehouse facilities, and other spaces with similar features. As mentioned, aisles serving assembly areas, including grandstands and bleachers, are to comply with Section 1028. Also, all aisles located within an accessible route of travel must also comply with Chapter 11.

The minimum required width of aisles may vary according to the occupancy in which the aisles are located; the nature of the use area that the aisles serve; the occupant load served by the aisles; and even, in some instances, the type of occupant served. In public areas of Group B and M occupancies, such as open offices, retail sales areas, and similar spaces, the minimum required clear width of any aisle is 36 inches (914 mm). In nonpublic areas, where the number of employees or other people served by the aisle is less than 50 and the aisle is not required to be accessible, a minimum required width of only 28 inches (711 mm) is mandated. Except for aisles serving assembly uses, a minimum of 36 inches (914 mm) in width is also required for all other occupancies. All aisles are to be unobstructed, except for those permitted projections such as nonstructural trim and doors. In all cases, the minimum width of an aisle cannot be less than the width needed to satisfy the calculated width requirements of Section 1005.3.

It is important to realize that the first element of any means of egress system is typically an aisle accessway, which then leads to an aisle. The code regulates aisle accessways to some degree in Group M occupancies, as well as in assembly spaces per Section 1028, but it is silent for all other occupancy classifications. The building official must determine at what point an aisle accessway becomes an aisle and must be regulated for minimum width in accordance with this section.

Aisle accessways in Group M. Aisle accessways are regulated within merchandise pads of Group M mercantile occupancies. A merchandise pad is defined as the merchandise display area that contains multiple counters, shelves, racks, and other movable fixtures. Every element within a merchandise pad must adjoin a minimum 30-inch-wide (762-mm) aisle accessway on each side. Travel within a merchandise pad is also limited, with a maximum common path of travel of 30 feet (9,144 mm). This limitation is extended to 75 feet (22,880 mm) in those areas serving a maximum occupant load of 50.   **1017.4**

Section 1018 *Corridors*

The IBC contains a definition of the term *corridor*. Defined as "an enclosed exit access component that defines and provides a path of egress travel," the determination as to when a corridor exists is essentially left to the building official.

For the purpose of the code, a corridor is typically a space where the building occupant has very limited choices as to paths or directions of travel. The available path is restricted and is usually bordered by other occupied-use spaces. As a consequence, it is potentially exposed to fires that might occur in those enclosed spaces unknown to anyone in the corridor. Generally speaking, in a building space of this type, the occupant has only two choices as far as direction of travel through the exiting system is concerned. For that reason, it is sometimes necessary for the building official to evaluate the planned layout of an area and determine whether the space presents a potential fire-hazard exposure to building occupants, as any regular, well-defined corridor might. If the determination is that the fire-exposure potential is the same, the building space should be made to comply with requirements for corridors.

To provide for a greater degree of consistency in the identification or determination of corridors, some jurisdictions have established a set of guidelines that expand on the definition in the IBC. An excellent example addresses four common characteristics of corridors as regulated in the code.

1. It is a space formed by enclosing walls or construction over 6 feet (1,828 mm) in height,
2. It has a length-to-width ratio greater than 3 to 1,
3. Its primary function is for the movement of occupants in the means of egress system, and
4. It has a length greater than that permitted for a dead-end condition.

All four conditions must be present for the element to be considered a corridor. There are many rooms that meet the first, second, and fourth conditions, but their primary use is for something other than the movement of occupants in the egress system. An open office system may have spaces that meet Conditions 2, 3, and 4; however, the walls are not over 6 feet (1,828 mm) in height. In such a space, the egress paths would be regulated as aisles or aisle accessways. It also makes little sense to designate a space as a corridor where it conforms to Conditions 1, 2, and 3 but is very limited in length, as the travel time through the space will be quite short. This approach is just one method for providing uniformity in defining a corridor; there are undoubtedly others that also help in the application of the code's intent. Of course, the definition of a corridor is not as critical where it is not required to be fire-resistance rated.

1018.1 Construction. The thresholds found in Table 1018.1 indicate at what point a fire-resistance-rated corridor must be provided. Where required by the table, walls of corridors shall be considered fire partitions, and as such shall be regulated by the provisions of Section 708. The provisions in Section 708 also address the lid of the corridor in relationship to the continuity of the fire partitions. Examples of appropriate corridor fire-resistance-rated construction are illustrated in the discussion of Section 708.

In a fully sprinklered building, a corridor must be fire-resistance rated only in Groups H, R, I-1, and I-3. In Groups H-1, H-2, H-3, I-1, and I-3, the protection is required regardless of the occupant load served by the corridor. Where a corridor serves an occupant load of more than 10 in a sprinklered Group R occupancy, it must have a minimum $^1/_2$-hour rating. Where a corridor serves an occupant load of more than 30 in a Group H-4 or H-5 occupancy, the walls of the corridors are required to be of not less than 1-hour fire-resistance-rated construction.

In buildings not equipped with an automatic sprinkler system throughout, corridors are required to be fire-resistance rated in all occupancies, based on the occupant load served. When a corridor serves an occupant load greater than 30 in a Group A, B, E, F, M, S, or U occupancy, 1-hour fire-resistance-rated corridor walls are required. Where the occupant load reaches the levels specified, it is appropriate to afford those persons in the means of egress corridor some additional protection from potential fire occurring in the enclosed spaces bordering the corridor. Therefore, a minimum separation for the corridor of 1-hour fire resistance is deemed necessary.

Fire-resistance-rated corridor exceptions. A series of exceptions to the 1-hour corridor requirement addresses a number of special circumstances where it is felt that the fire-resistance-rated separation is not necessary. The first applies to corridors in Group E occupancies. Such corridors need not comply with the 1-hour fire-resistance requirement when:

1. Every classroom that the corridor serves has at least one egress door leading directly to the outside at ground level, and

2. Rooms that are served by the corridor and are used for assembly purposes have at least one-half of their required egress doors leading directly to the outside at ground level.

Exception 2 exempts corridors within dwelling units or sleeping units of Group R occupancies from the fire-resistance rating. Exceptions 3 and 4 indicate that fire-resistance-rated corridors are not required in open parking garages, nor are they required in spaces of Group B occupancies requiring only a single means of egress. Exception 5 is an extension of the exception to Section 708.5 recognizing that corridor walls that are also exterior walls need to comply only with the wall and opening requirements of Section 705.

Corridor width. The minimum required width of a corridor is regulated like any other component of the means of egress. The calculated width, determined by Section 1005.3.2, must be compared with the minimum component width (in this case, established in Table 1018.2). The greater of the two required widths must be available for egress purposes. **1018.2**

For component width purposes, a corridor is required to be at least 44 inches (1,118 mm) in width. A number of exceptions reduce or increase this minimum width to the following dimensions:

1. Twenty-four inches (610 mm) is permitted in those nonpublic areas where access is necessary to service or use electrical, mechanical, or plumbing systems or equipment. These areas are seldom occupied, and then typically only by a single occupant.

2. Thirty-six inches (914 mm) is adequate where the corridor serves as a means of egress for 49 occupants or fewer, or where the corridor is located in a dwelling unit. Small occupant loads can easily egress through corridors of this minimum width. It also allows adequate width for circulation purposes and the movement of furnishings and equipment.

3. Seventy-two inches (1,829 mm) is the minimum width required for school corridors serving at least 100 students and for corridors in surgical health-care areas where the movement of gurneys is anticipated. The increased width for Group E occupancies provides for better circulation during peak usage, whereas the increase in surgical areas allows the limited movement of nonambulatory patients and equipment.

4. Ninety-six inches (2,438 mm) is required in Group I-2 occupancies for the movement of patients in standard hospital beds. This width is necessary to safely accommodate this kind of use.

As stated, this section provides for the minimum required widths of corridors. The code is not totally clear on how the required corridor width is determined when the occupant load is large enough to require corridor widths in excess of these minimums. A common approach to determining the required minimum widths of corridors is that the width should be related to the required width of the exit to which the corridor leads. The most logical procedure for determining corridor widths is to determine the required width of the exits at the end of the corridor and to size the corridor to provide that same minimum width. See Figure 1018-1. The controlling concept is to mutually equate the required width of the corridors and the required width of the exits that the corridor serves. However, the minimum required width of 44 inches (1,118 mm) must be maintained for the corridor unless an exception is applicable.

Dead ends. This section establishes the limitations on dead-end corridors. Basically, wherever more than one exit or exit access doorway is required, the exit access shall be arranged so that occupants can travel in either direction from any point to a separate exit. However, dead ends in corridors are permitted up to a maximum length of 20 feet (6,096 mm), as illustrated in Figure 1018-2. In the event that the conditions of building **1018.4**

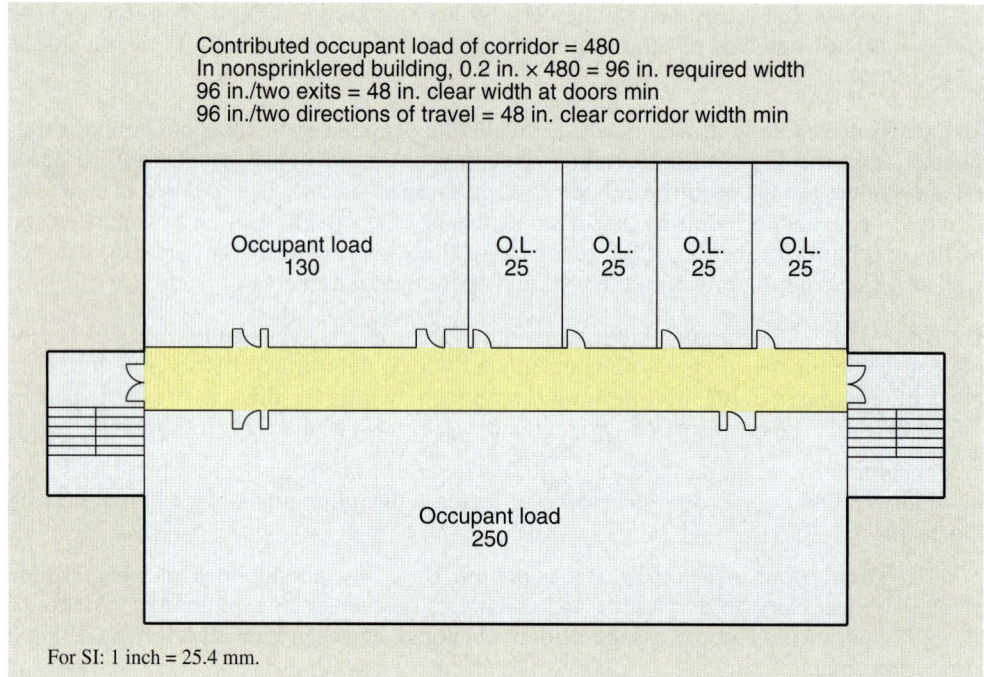

Figure 1018-1 Corridor width.

occupancy permit access to only a single exit, the dead-end limit does not apply. When the basic requirement is for travel in only a single direction to one exit, travel in the opposite direction is not considered to be in violation of the dead-end requirements. However, the maximum distance of travel will be regulated by the common path provisions of Section 1014.3.

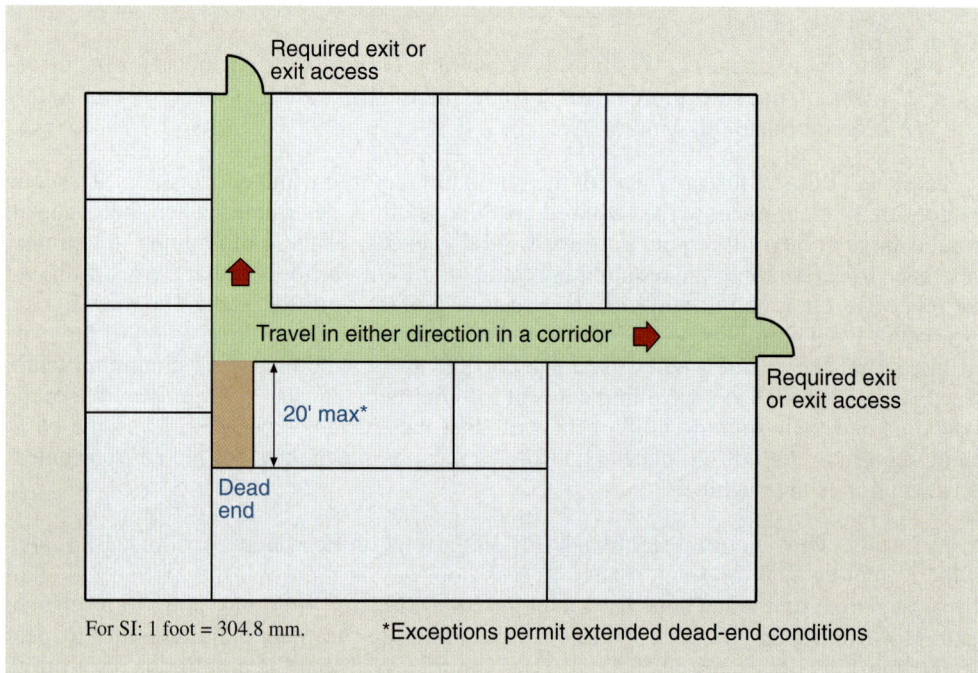

Figure 1018-2 Dead-end corridors.

Two exceptions permit dead ends of a length up to 50 feet (15,240 mm). In Group I-3 occupancies where free movement is allowed to some degree (Conditions 2, 3, or 4), and in low-hazard and moderate-hazard occupancies housed in fully sprinklered buildings, the increased length of the dead-end condition has been found to be acceptable. An additional exception permits a dead end of unlimited length, provided that such length of the dead-end corridor does not exceed two- and one-half times the least width of the dead-end portion of the corridor. See Figure 1018-3.

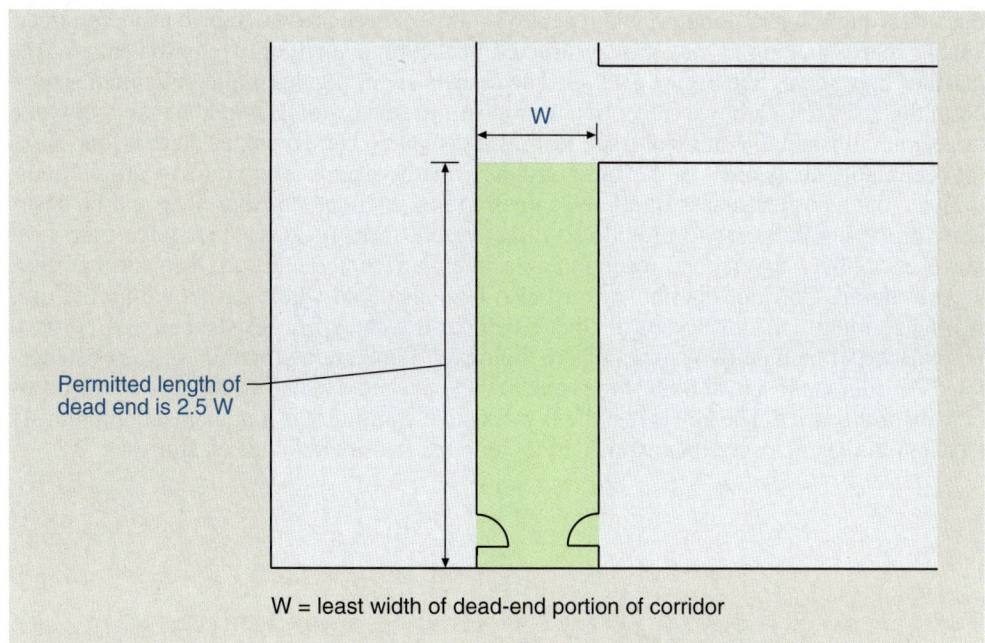

Figure 1018-3
Dead-end limits.

The limitation on dead ends is directed toward avoiding portions of the corridor system that could result in entrapment of the building occupant by creating a situation where the building occupant, following a proper path of travel through the means of egress system, might, under fire conditions, take a wrong turn into a portion of the system from which there is no outlet. In such a situation, it is possible that the building occupant will have to proceed all the way to the end of the dead end before learning that there is no way out and, thereafter, will have to retrace steps to arrive at an exit.

Air movement in corridors. Corridors are not to be used as a portion of the heating, ventilating, and air-conditioning system by serving as plenums or ducts. They also may not be used for relief or exhaust venting purposes. As corridors are relied on to be an important component of the egress system, it is not appropriate to provide potential avenues for the spread of smoke and gases. Exceptions to the general restriction are provided where the corridor is a source of make-up air for exhaust systems in adjacent spaces, a corridor in a dwelling unit, and a corridor within a tenant space not exceeding 1,000 square feet (93 m^2) in area. It should be noted that this provision is applicable to all corridors, not just those that are required to be fire-resistance rated. However, there appears to be inconsistency in the code insofar as adjoining spaces are permitted to be open to a nonrated corridor. Where a lack of physical separation occurs between the corridor and surrounding spaces, how can air be prevented from moving between them? In application, the provision merely is intended to require that those spaces open to the corridor be mechanically designed independent of the corridor (supply, return, exhaust, and make-up air) and not rely on the corridor to move air to or from the space.

1018.5

1018.6 Corridor continuity. It is required that fire-resistance-rated corridors should not be interrupted by intervening rooms, and that they be continuous to the exits. This provision carries out the basic concept, which states that once a building occupant progresses to a certain level of safety—in this case, the safety afforded by a fire-resistance-rated corridor—that level of safety is not thereafter reduced as the building occupant proceeds through the remainder of the means of egress. Therefore, the building occupant, having once reached such a corridor, is not thereafter brought out of the corridor and introduced into another occupied use space of the building; the corridor must be continuous. However, the code emphasizes that it is possible to permit fire-resistance-rated corridors to be conducted through foyers, lobbies, and reception rooms. These are not to be considered intervening spaces as long as they are constructed in accordance with the requirements for the corridor they serve. See Figure 1018-4. The code is silent regarding the maximum size or permitted uses for a lobby or reception room. An extreme example would be the lobby of a large hotel. It is not uncommon to see seating areas, piano bars, breakfast buffets, and similar occupiable areas open to the fire-resistance-rated corridor. The building official must evaluate the appropriateness of all areas open to the extended corridor. It would be desirable for exiting purposes to provide all direct egress from fire-resistance-rated corridors; however, lobbies, foyers, and reception areas are an accepted entrance element that must be considered. Corridor continuity must also be considered where egress within the fire-resistance-rated corridor system includes vertical travel. A fire-resistance-rated corridor may extend to include two stories where the travel is maintained within a fire-resistance-rated corridor enclosure at both stories and at the exit access stairway connecting the corridors on both stories. The key is that the fire-resistive continuity is maintained continuously throughout the enclosure that extends to include the two interconnected stories.

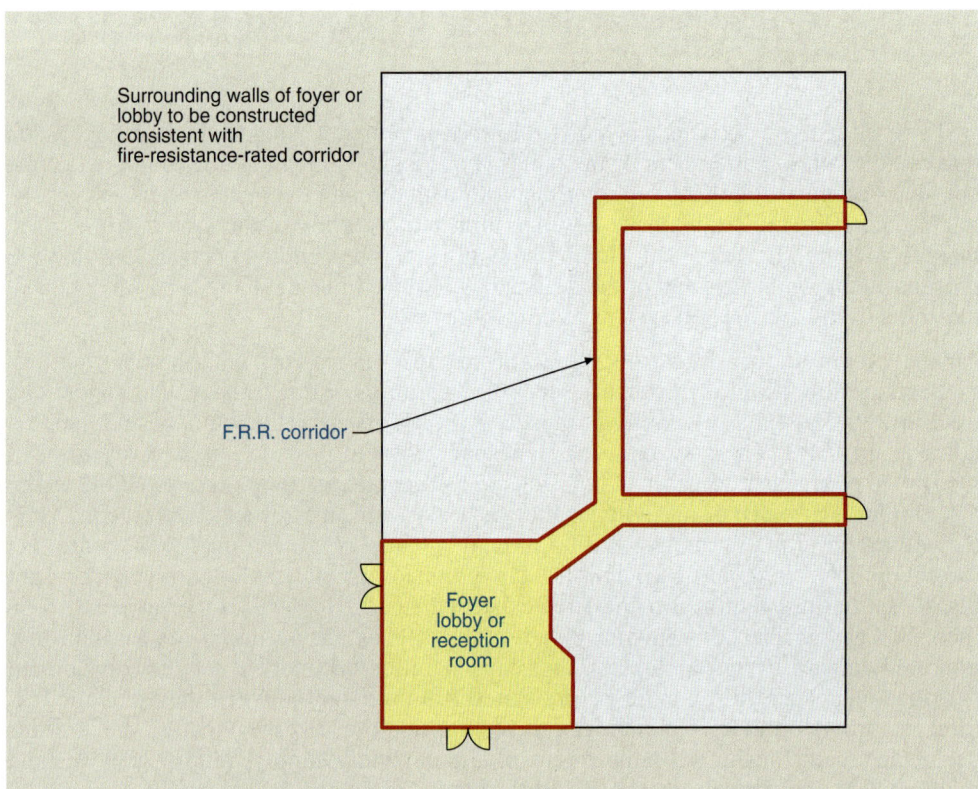

Figure 1018-4
Corridor continuity.

Section 1019 *Egress Balconies*

Balconies used for egress purposes are similar in many ways to corridors and, as such, are required to meet the corridor provisions addressing width, headroom, dead ends, and projections.

Wall separation. Because there is a potential exposure to a fire condition for occupants using an exterior egress balcony during evacuation of a building, a minimum level of fire protection is mandated. This section requires that the balcony be separated from the interior of the building in a manner consistent with the separation requirements for corridors. However, where two stairs serve the egress balcony, and travel from a dead-end condition to a stair does not occur past an unprotected opening, the separation is not required. **1019.2**

Openness. One of the features of any egress balcony is its openness to the atmosphere, limiting the amount of smoke and toxic-gas accumulation. To qualify as part of the exit system, the balcony must be at least 50 percent open on the long side. It is also necessary that the open areas above the guardrail be distributed to allow for adequate natural ventilation. **1019.3**

Location. Because of exposure potential from adjacent property, exterior balconies are prohibited within 10 feet (3,048 mm) of an adjacent lot line. Such components must also maintain a minimum 10-foot (3,048-mm) separation from other buildings on the same lot unless the exterior walls and openings of the adjacent building are protected in accordance with Section 705 based on fire separation distance. See Figure 1019-1. **1019.4**

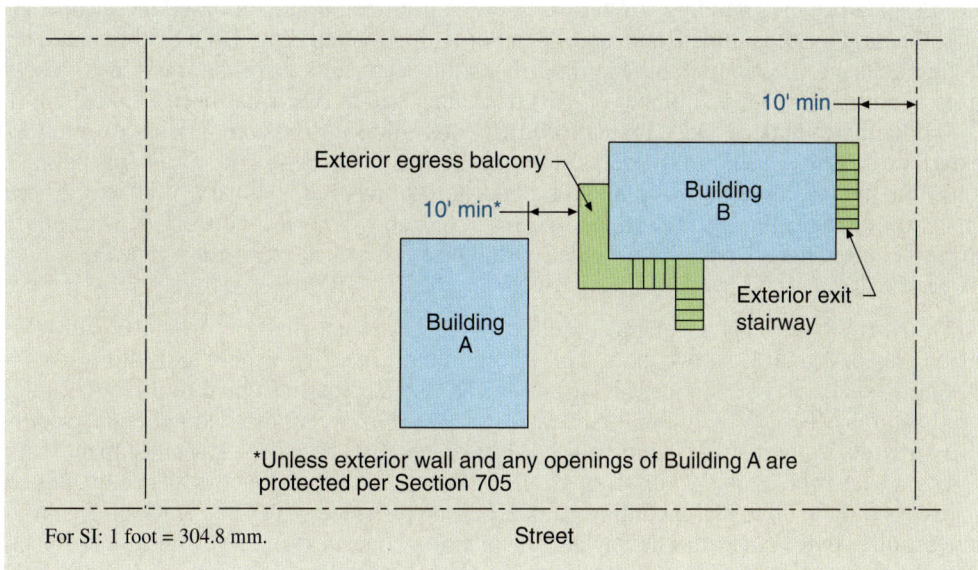

Figure 1019-1 Egress balconies and stairways.

Section 1020 *Exits*

Exits constitute those portions of the means of egress where the occupant first achieves a significant level of fire protection. An exit is expected to provide a mandated level of protection, and that level cannot be reduced until arrival at the exit discharge. Travel within exits is not limited and is typically single directional. Therefore, care is taken to ensure that the exit component is available for use by the building occupant during a fire or other emergency. This section makes it clear that the primary function of an exit component is to provide egress, and any other use cannot interfere with that function. Many of the design requirements for exits are addressed under the general provisions of Sections 1003 through 1013; however, provisions specific to the exit portion of the means of egress are found in Sections 1020 through 1026.

1020.2 Exterior exit doors. Once the provisions of Section 1008 for doors have been reviewed, there are relatively few exterior exit door requirements remaining to address. It should be noted that at least one exterior door that meets the size requirements of Section 1008.1.1 must be provided for every building used for human occupancy.

Section 1021 *Number of Exits and Exit Configuration*

This section is a continuation of the provisions established in Section 1015.1 relating to the minimum number of exits required throughout the means of egress system. Section 1015.1 is to be applied within rooms or spaces within a story and identifies those conditions under which one, two, three, and four means of egress are required from such rooms or spaces. Section 1021 regulates the minimum number of exits required from each story of a building. The allowances for a single exit, as well as the requirements for multiple exits, are established for each individual story, as well as for any occupied roof.

1021.1 General. The section establishes the minimum number of exits, or access to exits, required from each story of a building. As few as one exit may be permitted, based on occupancy classification, occupant load, and story location within the building. On the other hand, up to four exits may be required based primarily on the occupant load of the story. In all cases, once a minimum required number of exits or exit access stairways has been established by the code, that number cannot be reduced. It is possible to use a combination of interior/exterior exit stairways and exit access stairways for means of egress purposes. It is permissible that the means of egress from the second story of a building consist of all exit access stairways as regulated by Section 1009.3. However, above the second story, at least some of the required egress must be provided through the use of interior exit stairways and/or exterior exit stairways, as described in Sections 1022 and 1026, respectively.

1021.2 Exits from stories. As a general requirement, every story shall be provided with two exits, or access to at least two exits. This requirement is also applicable to occupied roofs. A single means of egress is permitted only where the story or occupied roof is within the limits of Table 1021.2(1) or Table 1021.2(2), or where an exception to this section specifically permits a single exit. Section 1021.2.4 indicates that, based solely on occupant load, three, or even four, exits may be required. Any story of a building that has an occupant load in excess of 500, up to and including 1,000, shall be provided with three exits, or access to at least three exits. Any story having an occupant load in excess of 1,000 must be provided with four exits, or access to not less than four exits. Under no circumstances does the IBC require more than four exits for any building or portion thereof based on the number of persons present. It must be noted, however, that additional exits will sometimes be required to satisfy the other egress requirements of Chapter 10. An additional stairway may also be required in high-rise buildings over 420 feet (128 m) in height per Section 403.5.2.

As addressed under the discussion of Section 1015.1, it is desirable that a minimum of two means of egress be provided to building occupants in order to provide for a more reliable and efficient evacuation process. The code, however, recognizes that there are instances where the life-safety risk is so minimal that it is reasonable to permit a single means of egress. Under limited conditions, this allowance extends from a single-level basement to a three-story-above-grade-plane condition. Tables 1021.2(1) and 1021.2(2) identify those stories where a single exit, or access to a single exit, is permitted. The tables

Number of Exits and Exit Configuration

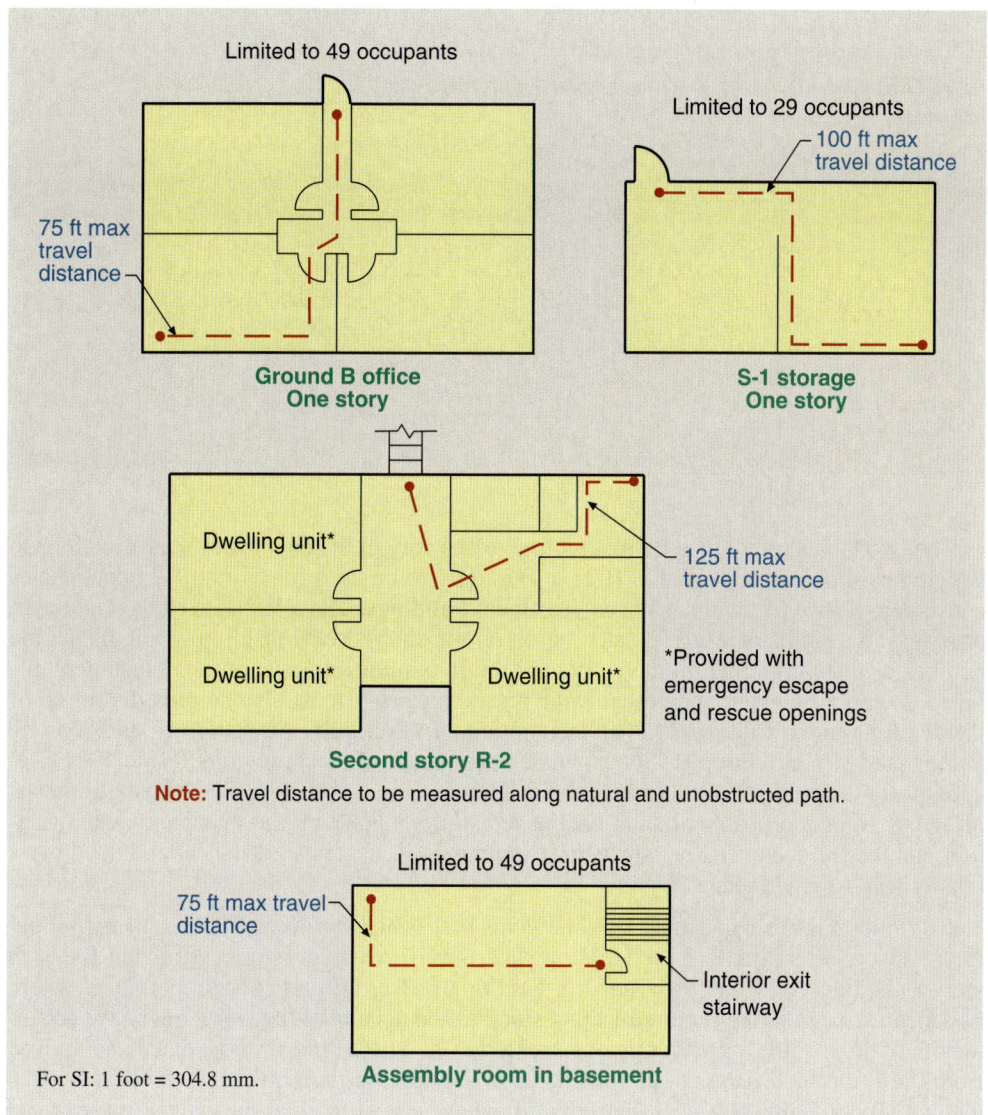

Figure 1021-1
Stories with one exit.

are based on varying criteria, including occupancy classification, number of stories above the grade plane, occupant load, number of dwelling units, and travel distance. Examples illustrating the use of these tables are shown in Figure 1021-1. It should be noted that special consideration is given in the footnotes to certain Group B, F, and S occupancies. All Group R-3 buildings may have a single exit, as well as single-level buildings complying with Section 1015.1 as a space with one means of egress.

Only one exit is required from a story where permitted by Table 1021.2(1) or Table 1021.2(2), regardless of the number of exits required from other stories in the building. For example, a Group B occupancy on the second floor of a multistory building is only required to have one exit from the story provided its occupant load does not exceed 29 and the maximum travel distance to an exit does not exceed 75 feet. The number of occupants and travel distances on the other stories do not affect the determination of the second story as a single-exit story. See Application Example 1021-1. Where applicable, other stories are also regulated independently as to the number of exits.

2012 International Building Code Handbook   425

Application Example 1021-1

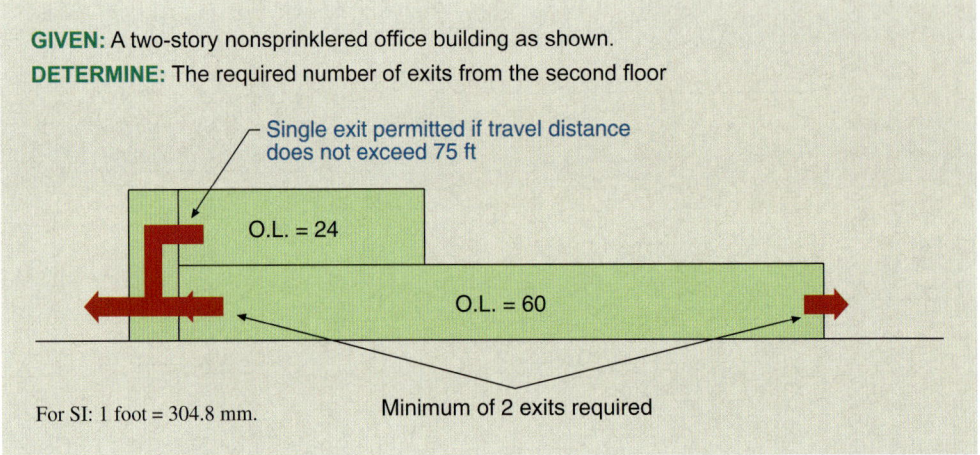

GIVEN: A two-story nonsprinklered office building as shown.
DETERMINE: The required number of exits from the second floor

For SI: 1 foot = 304.8 mm.

The two tables referenced by this section differ only in the occupancy classifications to which they are applicable. Table 1021.2(1) is only to be used for Group R-2 occupancies consisting of dwelling units, such as apartment buildings. The allowance for a single exit, or access to a single exit, is regulated based on the story's relationship to grade plane, the number of dwelling units on the story, and the maximum exit access travel distance provided. Complying emergency escape and rescue openings must also be provided in all of the dwelling units. Table 1021.2(1) does not apply to dormitories, fraternity, and sorority houses, and similar Group R-2 occupancies comprised of sleeping units. Table 1021.2(2) applies to all occupancies other than those addressed by Table 1021.2(1). A single means of egress from each story of such occupancies is regulated by the occupancy classification, number of occupants per story, maximum exit access travel distance, and the story's relationship to grade plane.

It is important to again note that the code first looks at exiting from each individual room, space, or area within the building. The means of egress is then regulated for each story as evidenced by the provisions of Section 1021.2. Of particular importance, where two means of egress are required from a story, all occupants of the story must have access to both of the required egress components. However, where three or more exits are required from the story, such access to all three exits is not necessarily required. Exception 7 states that not all occupants of a story are required to have access to all of the exits provided from that story. This exception recognizes that in some building arrangements, an exit may serve a specific area, such as being within an individual tenant space, and may not be usable by other occupants on that story. Having these isolated exits is not a problem, provided the three requirements are met. These requirements ensure that (1) the entire floor is provided with the proper number of exits, (2) each space on the floor has access to the required number of exits that space needs, and (3) all spaces on the floor have access to the minimum number of required exits, but not less than two. An example is illustrated in Figure 1021-2.

1021.2.1 Mixed occupancies. Where multiple tenants or occupancies are located on a specific story, they are to be regulated independently for single-exit determination. The provisions can be applied to specific portions of the story, rather than the story as a whole. As an example, the second story of a building houses two office tenants, each with its own independent means of egress. Each tenant would be permitted a single, but separate, means of egress provided each had an occupant load of less than 30 and a travel distance not exceeding 75 feet. This portion-by-portion philosophy also applies to a mixed-occupancy condition provided each of the individual occupancies does not exceed the limitations of Table 1021.2. See Figure 1021-3.

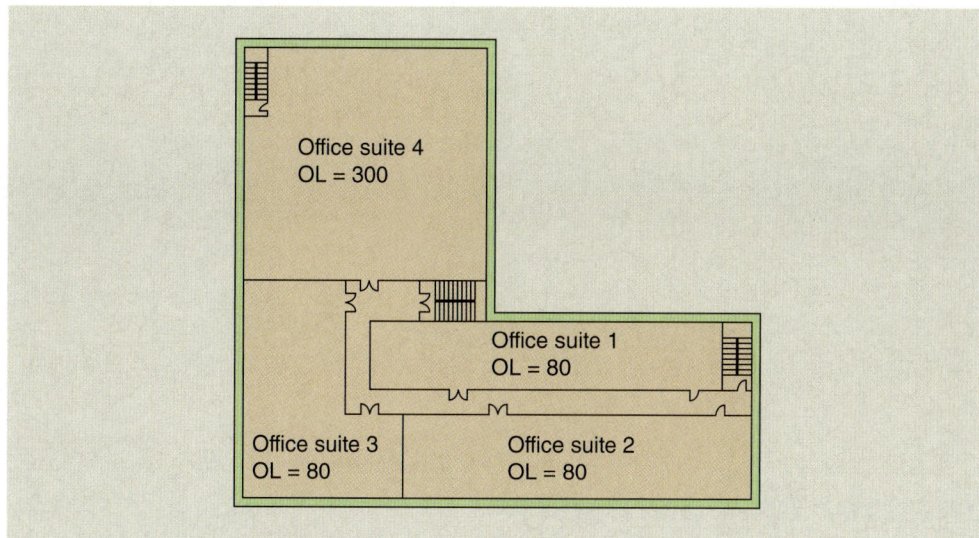

**Figure 1021-2
Exits serving specific spaces or areas.**

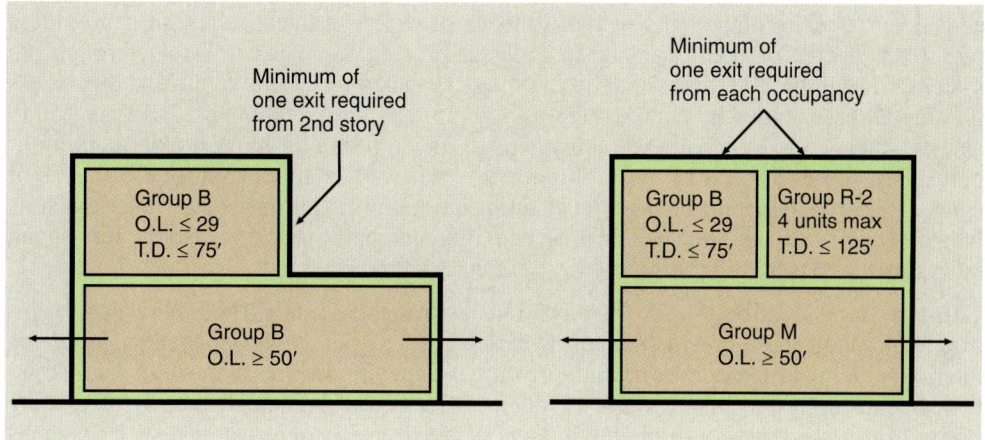

**Figure 1021-3
Stories with one exit.**

Basement. Also illustrated in Figure 1021-1 is a permitted condition for a basement where a single exit is permitted. Conditions of this section limit the basement to a single level below the first story. Travel distance and occupant load are also both limited by Table 1021.2(1) or Table 1021.2(2) as applicable.

1021.2.2

Access to exits at adjacent levels. Where two means of egress are required from a third story above grade plane or higher, Section 1021.1 only requires one interior exit stairway or ramp, or exterior stairway or ramp. The other required means of egress is permitted to be an exit access stairway complying with Section 1009.3. Vertical travel on an exit access stairway must be considered in the evaluation of travel distance and such travel is only available to a single adjacent story. The use of an exit access stairway as a required means of egress component is limited to one story of travel, at which point the occupants must use an interior exit stairway or other exit element. See the example in Figure 1021-4.

1021.3.1

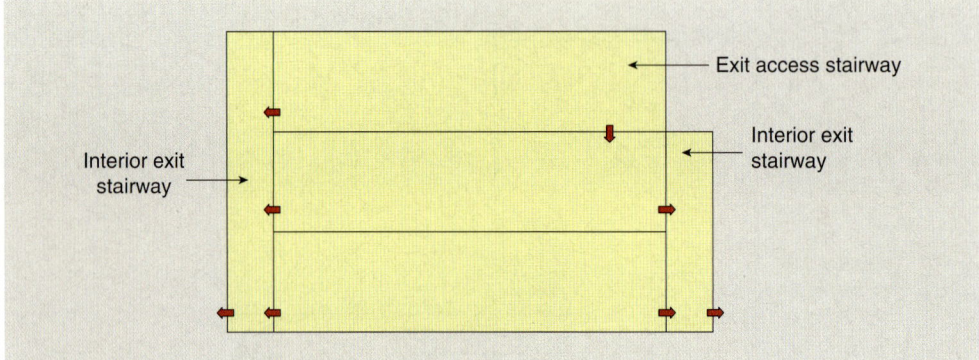

Figure 1021-4
Access to exits at adjacent levels.

Section 1022 Interior Exit Stairways and Ramps

1022.1 General. The interior exit stairways and interior exit ramps addressed in this section are considered as exit components within a means of egress. As such, they must provide a prescribed degree of fire resistance in order to protect occupants traveling through the system. In addition, interior exit stairways must provide direct travel to the exterior unless interrupted or extended by exit passageways complying with Section 1023. Travel within an exit passageway is considered equivalent to travel within an interior exit stairway or ramp. Due to the importance of an interior exit stairway as an element of the means of egress, it is critical that the stairway enclosure be used for no purpose that would lessen its integrity as an exit component. The presence of equipment, furnishings, stored items, and other obstructions within the stairway system is strictly prohibited.

1022.2 Construction. Consistent with the provisions addressing the enclosure of vertical openings with shaft enclosures as set forth in Section 713, all vertical openings for every interior stairway or ramp must be similarly enclosed within fire-resistance-rated construction. Because vertical openings provide the most readily available paths for fire spreading upward from floor to floor through buildings, it is extremely important that such openings be adequately enclosed. This enclosure is also required in order to protect and separate the vertical exitway from potential fire and products of combustion in other spaces of the building to allow for a usable egress path. Other than where the Group I-3 special provisions of Section 408.3.8 are applied, there are no exceptions to the fire-resistance-rated enclosure provisions of this section. Unenclosed stairways and ramps are permitted, but only under the limitations established in Section 1009.3 for exit access stairways and ramps.

The degree of fire resistance required for enclosures for interior exit stairways and ramps is dependent on the number of stories connected by the stairways. Where four or more stories are connected, the enclosing construction must be at least 2-hour fire-resistance-rated construction. In all other instances, required enclosures of interior exit stairways and ramps must be a minimum of 1-hour fire-resistance-rated construction. For the determination of the required fire resistance of the enclosure, the number of stories also includes any basements that may exist within the building, but not mezzanines.

The fire-resistance rating of the enclosure for an interior exit stairway or ramp is also regulated in a manner consistent with shaft enclosures in regard to the rating of the floor(s) being penetrated by the enclosure. Where the floor construction penetrated by the enclosure has a fire-resistance rating, the enclosure must have the same minimum rating.

For example, an interior exit stairway enclosure that penetrates a 2-hour floor assembly must have a minimum fire-resistance rating of 2 hours, regardless of the number of stories the enclosure connects. The fire-resistance rating of an interior exit stairway or ramp enclosure need never exceed 2 hours. If the floor assembly penetrated requires a minimum 3-hour fire-resistance rating, the enclosure rating is only required to be 2-hour fire-resistance rated.

Openings. Because enclosures for interior exit stairways and ramps are so fundamental to the safety of building occupants and their ability to safely exit a multistory building during a fire emergency, the code is careful to protect the integrity of these vertical enclosures in every way possible. In addition to the fire-resistance-rated construction required of the enclosure, the openings that are permitted to penetrate the fire-resistant enclosure are narrowly limited. This section very clearly establishes that only those openings necessary to provide exit facilities for occupants of the building spaces and allowed openings in the exterior walls are permitted. **1022.4**

Because the exterior walls of an interior exit stairway or ramp enclosure are not protecting the exit from other building spaces, openings through the exterior wall into the atmosphere are permitted. In fact, in buildings that are located on the lot so that there would be no requirement for the fire-resistance-rated construction of the exterior wall or for the protection of the openings in such walls, the exterior wall of an enclosure could be eliminated entirely. However, such openings must comply with the proper requirements of the code relating to the location of those openings with respect to lot lines or to other potential exterior fire exposures.

This provision also makes it clear that it is the intent of the code to prohibit openings from typically unoccupied spaces directly into the interior exit stairway or ramp enclosure. Therefore, it is not permitted to provide openings from such spaces as store rooms, toilet rooms, equipment rooms, machinery rooms, electrical rooms, and similar rooms directly into the enclosure. In addition, elevators are specifically prohibited from opening into an enclosure for interior exit stairways and ramps.

Where openings for exit doorways are provided, it is necessary that they be protected with a fire-rated assembly. Fire-door assemblies and other opening protectives permitted under this section are addressed in Section 716. Where an exit passageway is used as an extension to the exterior of the interior exit stairway or ramp enclosure, Section 1022.3.1 indicates that a fire door is required to separate the vertical enclosure from the exit passageway. See Figure 1022-1. Such a fire door shall have a fire-protective rating in accordance with Section 716.5.

Penetrations. Penetrations into an interior exit stairway or ramp are prohibited unless necessary to service or protect the exit component. Acceptable penetrations include sprinkler piping, standpipes, and electrical conduits serving the stairway or ramp enclosure. All such penetrations shall be made in a manner that will maintain the structural and fire-resistance integrity of the enclosure. Under no circumstances shall there be communicating openings or penetrations between adjacent interior exit stairways. The exception permits complying membrane penetrations, but only on the outside of the enclosure. There is no limit on the type or purpose of the penetration other than limiting the location to the exterior membrane and requiring the proper protection. **1022.5**

Ventilation. Equipment and ductwork necessary for independent pressurization of an enclosure for an interior exit stairway or ramp is permitted under specific conditions. Where ventilation of the enclosure is desired, the ventilation systems shall be independent and isolated from the other building ventilation systems. There are three methods set forth to regulate the installation of the ventilation equipment and ductwork. The equipment and ductwork may be located at the exterior of the building, within the enclosure, or within the building. In all cases, the provisions of Section 713 for shaft enclosures will regulate the separation of the equipment and ductwork from the remainder of the building. **1022.6**

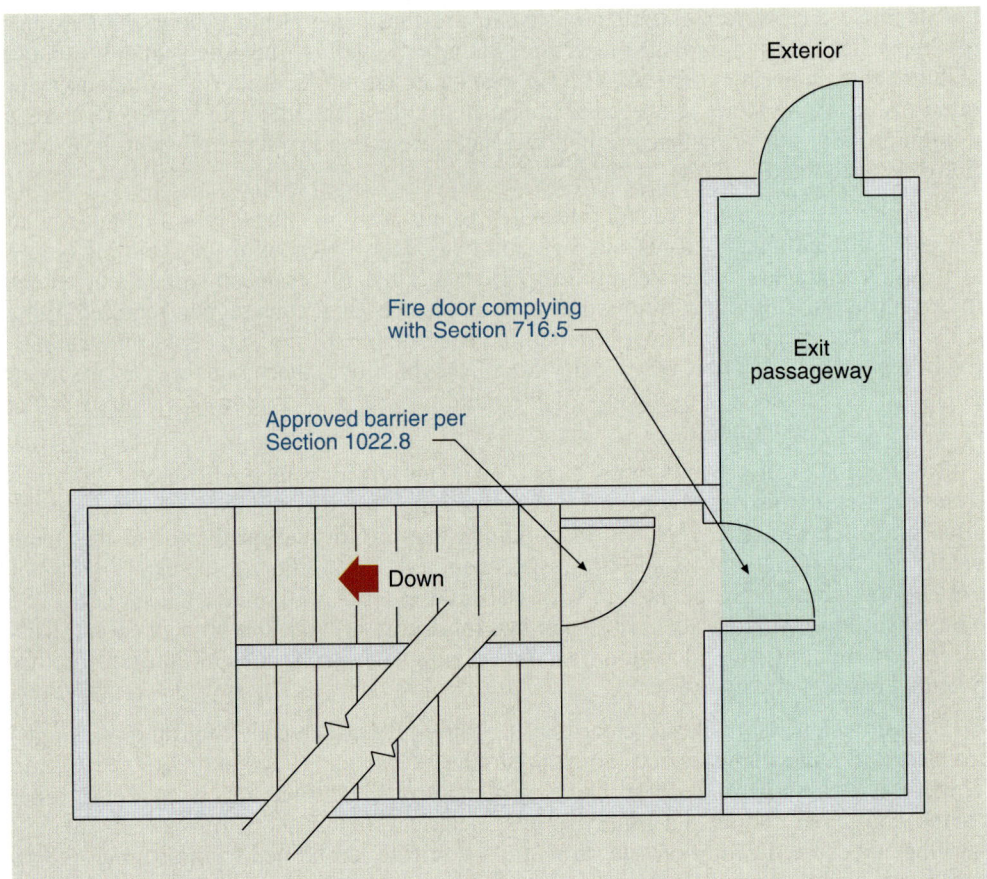

Figure 1022-1 Extent of enclosure.

1022.7 Interior exit stairway and ramp exterior walls. Whenever a stairway or ramp is installed as a component in an exiting system, it is important to protect the stair or ramp user from potential exposure to any fire that might occur in the building. Therefore, where exterior walls of interior exit stairway and ramp enclosures are nonrated and openings in such walls are unprotected, the location and protection of adjacent portions of the building must be considered. Only those walls and openings within 10 feet (3,048 mm) horizontally and located at an angle less than 180 degrees (3.14 rad) from the enclosure walls need to be protected. Such walls shall have a minimum fire-resistance rating of 1 hour and any openings shall be protected at least $^3/_4$ hour. The extent of the protected construction shall be from the ground to a point at least 10 feet (3,048 mm) above the topmost landing of the stairway or ramp, or to the roof line if it is lower than 10 feet (3,048 mm). An example of this provision is shown in Figure 1022-2. In all cases, the exterior walls of the enclosure must comply with the provisions of Section 705 based on fire separation distance.

This provision is based on the enclosure exterior walls being of nonrated construction with any openings left unprotected. Because the main issue is to protect the egress path, any adjacent building construction that would present an exposure hazard to occupants using the stair or ramp enclosure must be fire-resistance-rated construction. On the other hand, the enclosure walls and openings could be fully protected on all sides, including those on the exterior of the building, under the provisions of Section 1022.2 and, therefore, the requirements of this section would not be applicable. Whether the protection is at the enclosure or at the location of the hazard is ultimately immaterial.

Interior Exit Stairways and Ramps 10

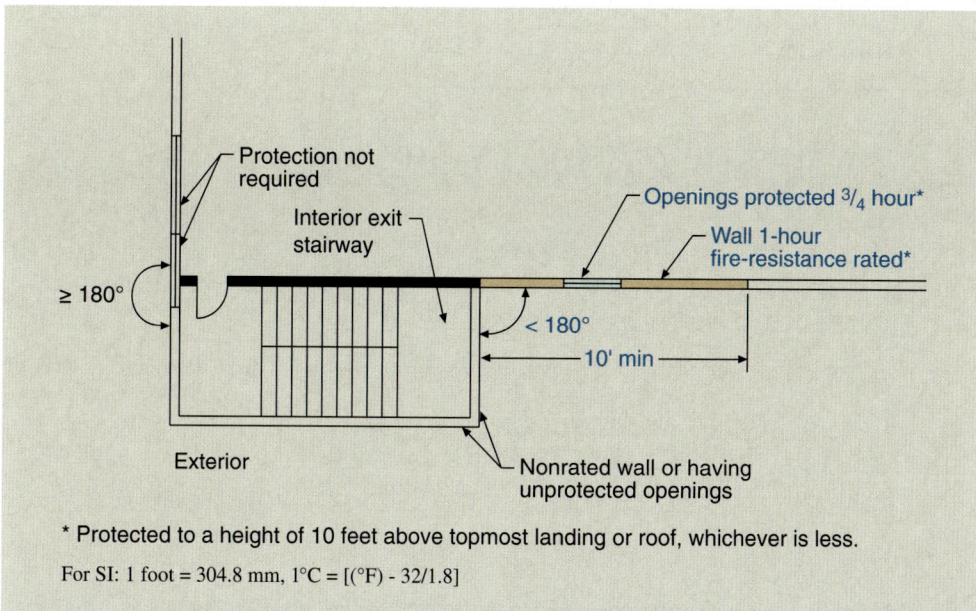

Figure 1022-2
Interior exit stairway exterior walls.

Discharge identification. The barrier required by this section is intended to prevent persons from accidentally continuing into the basement. The design and location details of the barrier must be approved by the building official. Directional exit signs as specified in Section 1011 are also required, in addition to the physical barrier. **1022.8**

Stairway identification signs. This section specifies a system whereby any persons, particularly fire fighters, inside an interior exit stairway or ramp enclosure in a building will be provided with information telling them where they are in the building and where the stairway or ramp leads to both above and below that point. This required sign can be critically important for fire-fighting purposes and is frequently useful to other building occupants. Enclosures connecting three stories or fewer are exempt from the identification requirements. **1022.9**

As set forth in this provision, this sign is to be positioned 5 feet (1,524 mm) above the floor landing in such a manner that it is readily visible whether the door is open or shut. Information to be provided on the sign includes:

1. The floor level.
2. The top terminus of the enclosure.
3. The bottom terminus of the enclosure.
4. The identification or location of the stairway or ramp.
5. The story of exit discharge.
6. The direction of exit discharge.
7. The availability of roof access from the stairway.

A sample sign complying with these requirements is shown in Figure 1022-3. In addition, tactile signage in compliance with ICC A117.1 must be provided to identify the floor level.

Smokeproof enclosures and pressurized stairways and ramps. Under certain conditions of building occupancy, protection of the vertical exitways over and above that required in the more usual situations is warranted. In such situations, it is necessary that the vertical exitway be constructed not only with a vertical enclosure, but also with either **1022.10**

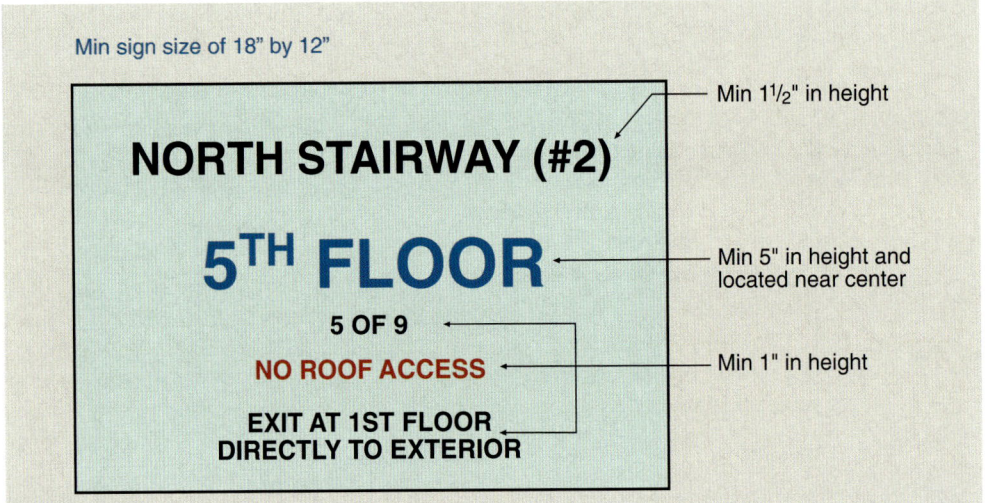

Figure 1022-3
Typical stairway floor number sign.

an outside balcony, by a ventilated vestibule, or by pressurization meeting the requirements of Section 909.20. A smokeproof enclosure is a special arrangement of the vertical exitway to minimize, if not prevent, the infiltration of smoke and other products of combustion into the actual stairway. Therefore, a smokeproof enclosure is one where the building occupant does not enter the stair enclosure directly from the occupied space of the building, unless the stair pressurization alternative addressed in Section 909.20.5 is used.

Only those buildings regulated by Section 403 (high-rise buildings) or Section 405 (underground buildings) are required to use smokeproof enclosures in their means of egress. It is required that all exits in such buildings be smokeproof enclosures or pressurized stairways for each of the exits that serve stories where the floor surface is located more than 75 feet (22,860 mm) above the level of fire-department vehicle access or more than 30 feet (9,144 mm) below the level of exit discharge serving such floor levels. Details of the various features of smokeproof enclosures are found in the discussion of Section 909.20. Similar requirements apply to airport traffic control towers, as established in Section 412.1.

Where a smokeproof enclosure or pressurized stairway reaches the level of exit discharge, it is to exit directly to a public way or to a yard or open space providing access to a public way. As an option, egress is permitted to travel through areas on the level of exit discharge if the conditions of Exception 1 or 2 of Section 1027.1 are met. Where the enclosure is not located at the exterior wall, an exit passageway may be used to maintain the protection of the exitway. The exit passageway must be of 2-hour fire-resistance-rated construction and have no openings that could adversely affect the integrity of the enclosure. An exception permits openings where they are adequately fire protected and the exit passageway is pressurized in the same manner as the smokeproof enclosure. Pressurized stairways are offered the same exception. As previously mentioned, the access to a stairway within a smokeproof enclosure must be made through a vestibule or an open exterior balcony. The one exception is the use of a pressurized system complying with the provisions of Section 909.20.

Section 1023 *Exit Passageways*

In many ways, the role of an exit passageway is identical to that of an interior exit stairway (or ramp). Therefore, the provisions regulating exit passageways are very similar to those governing interior exit stairways. The code specifies that both components are not to be used for any purpose other than as a means of egress. Once a building occupant is inside an exit passageway or an interior exit stairway, there is no subsequent limitation on travel distance. In most instances, the exit component must be continuous to the exit discharge or a public way.

The width of an exit passageway is regulated in the same manner as for a corridor. Those passageways serving an occupant load of less than 50 are permitted to have a minimum width of 36 inches (914 mm). Where the occupant load is 50 or more, such a width must be at least 44 inches (1,118 mm). In no case, however, shall the width be less than that determined by calculation in Section 1005.3.2. Other than the permitted projections for doors, trim, and similar decorative features, the required width of exit passageways is to be clear and unobstructed.

A minimum 1-hour fire-resistance rating is required for the walls, floors, and ceilings enclosing exit passageways. Where the exit passageway is provided as an extension of an interior exit stairway, the required fire-resistance rating of the exit passageway shall not be less than that required for connecting the interior exit stairway. For example, an exit passageway serving as a horizontal extension of stairway travel in a 2-hour fire-resistance-rated interior exit stairway must also be 2-hour enclosed.

Openings and penetrations that occur in the enclosure elements of an exit passageway are strictly controlled. In fact, the limitations are consistent with provisions regulating openings into and penetrations through interior exit stairways. See the discussion of Sections 1022.4 and 1022.5 for an analysis of the provisions that are applicable to both exit passageways and interior exit stairways.

Uses of exit passageways. Exit passageways are commonly used in several different exiting situations. In many cases, it is required that interior stairways be enclosed and that the enclosure extend completely to the exterior of the building, including, if necessary, an exit passageway on the floor of the level of exit discharge. When used in this configuration, the exit passageway assumes the same fire-protection requirements as for the stairway enclosure it serves. An exit passageway is also required to maintain the protective continuity of an interior exit stairway at horizontal travel above the discharge level. Where travel on a story is required from the termination point of one interior exit stairway to the entrance to another interior exit stairway in order to provide a continuous exit path, the connecting element must be an exit passageway.

An ongoing use of exit passageways is in connection with covered mall buildings. An example of the use of exit passageways in covered mall buildings results from the fact that travel distance in the mall is limited to a maximum of 200 feet (60,960 mm). It will occasionally be necessary between major exits from the mall to introduce an exit passageway or provide an additional exit to satisfy the requirements limiting travel distance. By the use of such exit passageways, it is possible to locate the main entrance/exit points to the mall building at substantially greater intervals. In addition, the use of an exit passageway can potentially eliminate dead-end conditions that occur where back-of-tenant egress is provided along the same path as mall egress.

A historic use of exit passageways is in buildings that have very large floor areas. In such buildings, it is sometimes not possible to get the building occupants to the exits at the building's perimeter within the limitations of the permitted travel distance. Therefore, an exit passageway is used to literally bring the exit to the interior of the spaces and to the building occupant so that it is possible for any building occupant to reach and enter the exit passageway within the permitted travel distance. This type of exit passageway is frequently accomplished by constructing, in effect, a special type of fire-resistive corridor. It can also be accomplished by constructing either an overhead, fire-resistant, enclosed passageway, or a tunnel. By these latter means, it is possible to avoid manufacturing processes and other functions at the floor level within the building.

Another use of exit passageways occurs when a separation of multiple exit doors is insufficient. By extending the points of egress through the use of one or more exit passageways it is possible to relocate the exit doors to the point where they comply with Section 1015.2. The exit separation distance would then be measured in a straight line between the exit doors, which could each enter into an exit passageway.

Again, it should be noted that once in an exit passageway, the building occupant is considered to be in a relatively safe location and travel distances within the exit passageway are not limited, just as travel distances within interior exit stairways are not limited.

In most cases, the primary difference between an exit passageway and an exit corridor lie in their respective requirements for opening protection and the fact that the passageway requires a complete enclosure, including ceiling and floor, of at least 1-hour fire-resistance-rated construction. Permitted openings and penetrations in exit passageways are also much more limited than what is allowed into a corridor.

Section 1024 *Luminous Egress Path Markings*

Improving the visibility of stair treads and handrails under normal and emergency conditions is a significant factor in raising the level of occupant safety for individuals negotiating stairs during egress of a high-rise building. A second source of emergency power for exit illumination, exit signs, and stair shaft pressurization systems in smokeproof enclosures is mandated for high-rise buildings. In the event of an emergency that disconnects utility power, the emergency power source should engage, causing the stair shaft to be illuminated and maintained smoke-free by the pressurization system. Unfortunately, such systems can fail under demand conditions. The mandate for luminous egress path markings adds an additional level of safety to the egress activity. The installation of photoluminescent or self-illuminating marking systems that do not require electrical power and its associated wiring and circuits provide an additional means for ensuring that occupants can safely egress a building via exit stairs, even if the emergency power supply and system fails to operate.

The use of photoluminescent or self-illuminating materials to delineate the exit path is required in Group A, B, E, I, M, and R-1 occupancies having occupied floors more than 75 feet (22,860 mm) above the lowest level of fire-department vehicle access. In such high-rise buildings, the required use of these markings is limited to interior exit stairways and ramps, as well as exit passageways. The selected materials must meet the requirements of UL 1994, *Luminous Egress Path-Marking Systems*, or ASTM E 2072, *Standard Specification for Photoluminescent (Phosphorescent) Safety Markings*.

All markings are required to be solid and continuous stripes. A key requirement for marking systems is that their design must be uniform. The placement and dimensions of markings shall be consistent throughout the same stairway or ramp enclosure. By specifying standard marking dimensions, the requirements ensure that the marking is visible during dark conditions and provides consistent and standard application in the design and enforcement of exit path markings. Markings installed on stair steps, perimeter demarcation lines, and handrails must have a minimum width of 1 inch (25 mm). For stair steps and perimeter demarcation lines, their maximum width cannot exceed 2 inches (51 mm). The provisions for stair steps, perimeter demarcation lines, and handrails allow the width of the marking to be reduced to less than 1 inch (25 mm) when marking stripes are listed in accordance with UL 1994.

Markings are required along the entire length of the leading edge of each stair step and along the leading edge of stair landings. Markings are also required along the perimeter of stair landings and other floor areas within the enclosure. These demarcation lines serve to identify the transition from the stair steps to the landing, which is important to minimize the risk of a fall inside of a stairway or ramp enclosure that is not illuminated. In order to discern the transition from the stair to the floor, the demarcation line is located either across the bottom of the door or on the floor in front of the door.

Selected materials used in the construction of the luminous egress path markings must comply with either UL 1994 or ASTM E 2072. ASTM E 2072 allows the use of paints and coatings, which can be useful because it avoids a potential tripping hazard, especially in locations where the surface substrate may not be even. The luminescence of the selected marking system must provide an illumination of at least 1 foot-candle (11 lux), which is consistent with the requirement in IBC Section 1006.2 for the general illumination of walking surfaces. This degree of illumination must be provided for at least 60 minutes.

Analogous to rechargeable batteries, many photoluminescent and self-illuminating egress path markings require exposure to light to perform properly. Thus, luminous egress path markings must be exposed to a minimum 1-foot candle (11 lux) of light energy at the walking surface for at least 60 minutes prior to the building being occupied. The charging rate for luminous egress path markings is based on the wattage of lamps used to provide egress path illumination. Therefore, it is important to verify that the specified lamps have sufficient wattage to meet the specified time period.

Section 1025 *Horizontal Exits*

The horizontal exit may well be the least understood and most under-used component in the means of egress. It can be a very effective method for providing adequate required exiting capacity while at the same time realizing some very substantial construction cost and space savings.

A horizontal exit consists essentially of separating a story into parts by dividing it with construction having a fire-resistance rating. The construction of one or more horizontal exit walls divides the floor into fire compartments. A horizontal exit may also be located between two buildings where travel occurs from one building to an area in another building at approximately the same level. This would include travel through a fire wall. The concept of the horizontal exit is to permit each of these fire compartments to serve as an area of refuge for occupants in one or more of the fire compartments in the event of a fire emergency. Building occupants in the compartment of fire origin may then pass through the fire-resistance-rated horizontal exit into the compartment of refuge, thereby gaining sufficient protection and sufficient time for either the extinguishment of the fire and elimination of the fire threat, or the orderly use of the remaining exits from the compartment serving as the area of refuge.

The horizontal exit is used effectively in hospitals, as well as in detention and correctional facilities, where total evacuation from the building may present numerous physical and other problems. If in a health-care facility it is necessary to move patients from their rooms in a fire emergency, it is desirable to avoid the need for moving them vertically by stairs. Therefore, if an arrangement can be provided whereby patients would only be subject to horizontal movement, the safety of the building occupant can be far more easily achieved. Although horizontal exits are most frequently used only where the more traditional forms of egress design are not satisfactory, they can often be used quite effectively in many situations. In fact, in those instances where fire walls having a fire-resistance rating of not less than 2 hours are provided, the resulting arrangement is often tailor-made for use as a horizontal exit.

Horizontal exits. When properly constructed, a horizontal exit may serve as a required exit. It may, in fact, be substituted on a one-for-one basis for other types of exits. However, in only very limited situations may horizontal exits be used as the only exit from a portion **1025.1**

10 Means of Egress

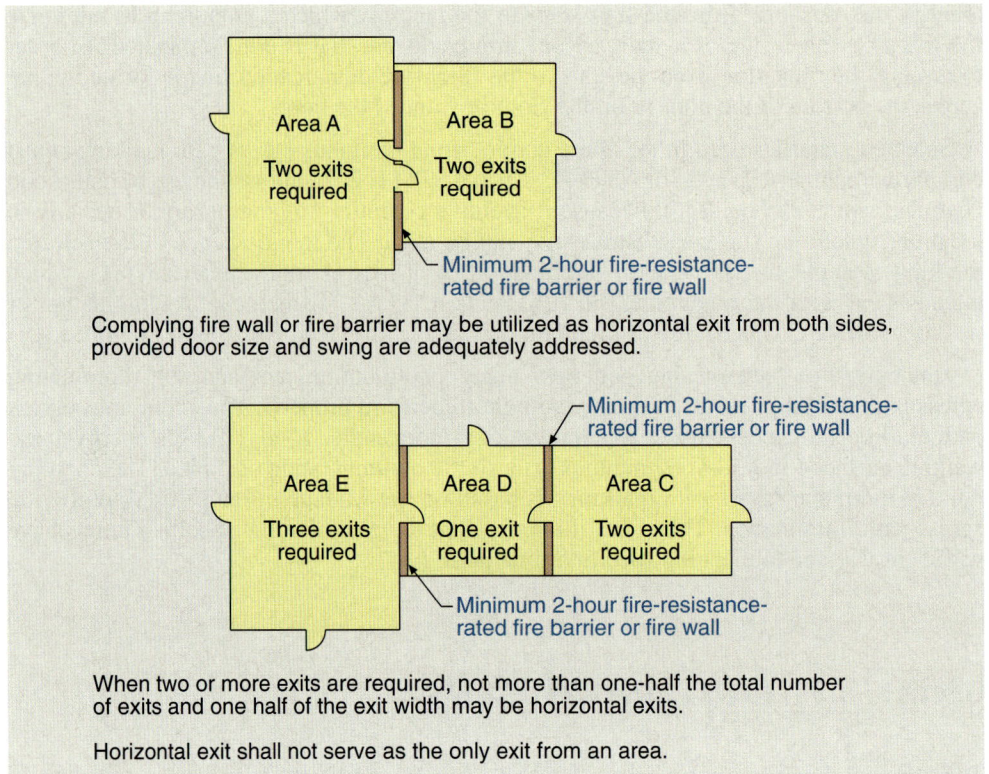

Figure 1025-1
Horizontal exits.

of a building, or to provide more than one-half the total required number of exits from any building space. See Figure 1025-1 for the general rule. In a Group I-2 occupancy, horizontal exits may comprise up to two-thirds of the required exits from any building or floor area, and Group I-3 occupancies may use horizontal exits for 100 percent of the required egress.

In the design of a horizontal exit system, it must be emphasized that it is not necessary to accumulate the total occupant load of the two compartments and then provide exit capacity for that total occupant load from each of the compartments. Were that the case, the horizontal exit would often provide little or no benefit. In designing a horizontal exit arrangement, one simply treats the separate compartments as if they were, in fact, separate buildings. Each compartment is provided with an exit system that is in compliance with all of the various criteria for a total exit system from a building. The main difference is that in this configuration, one of the exits from the separate compartments is a horizontal exit into the compartment of refuge. Figure 1025-2 depicts possible arrangements for horizontal exits.

1025.2 Separation. As previously mentioned, a fire barrier and fire wall are the only two construction methods acceptable for the separation provided by a horizontal exit. In both cases, a minimum fire-resistance rating of 2 hours is required. A fire wall provides a natural horizontal exit. For a fire barrier, acting as a horizontal exit, to completely divide a floor of a building into two or more separate refuge areas, the fire barrier walls must be continuous from exterior wall to exterior wall. In addition, a fire barrier used as the horizontal exit must extend vertically through all levels of the building, unless 2-hour fire-resistance-rated floor assemblies are provided with no unprotected openings. This method of isolation affords safety in the refuge area from fire and smoke in the area of incident origin.

Horizontal Exits 10

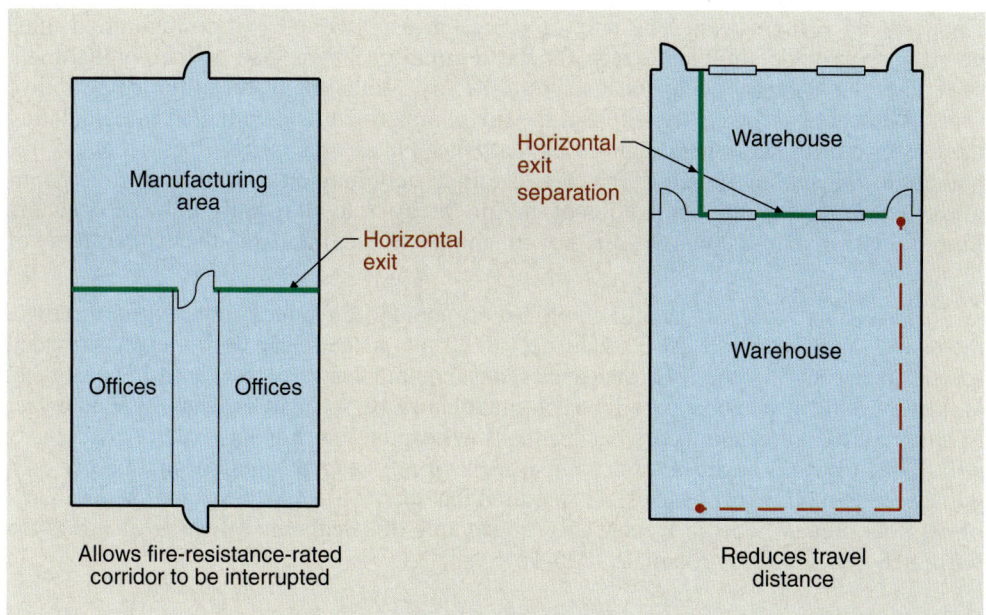

Figure 1025-2
Horizontal exit uses.

Opening protectives. Openings through a horizontal exit are required to be protected, and thus they must have a fire-protection rating, consistent with the fire-resistance rating of the wall, as required by Section 716. Where installed through a horizontal exit wall, door openings must be self-closing, or smoke-detector-actuated automatic-closing assemblies must be installed in accordance with Section 716.5.9.3. In fact, when a horizontal exit is installed across a corridor, or when a corridor terminates at a horizontal exit, smoke-detector-actuated automatic-closing assemblies must be used. As is the case with any exit door, doors in horizontal exits must swing in the direction of exit travel when serving an occupant load of 50 or more, as provided in Section 1008.1.2. When the horizontal exit is to be used for exiting in both directions, it will often be necessary to provide separate exit doors for the separate directions to satisfy this requirement. Most likely, this will require the use of one or more pairs of opposite-swinging doors. See Figure 1025-3.

1025.3

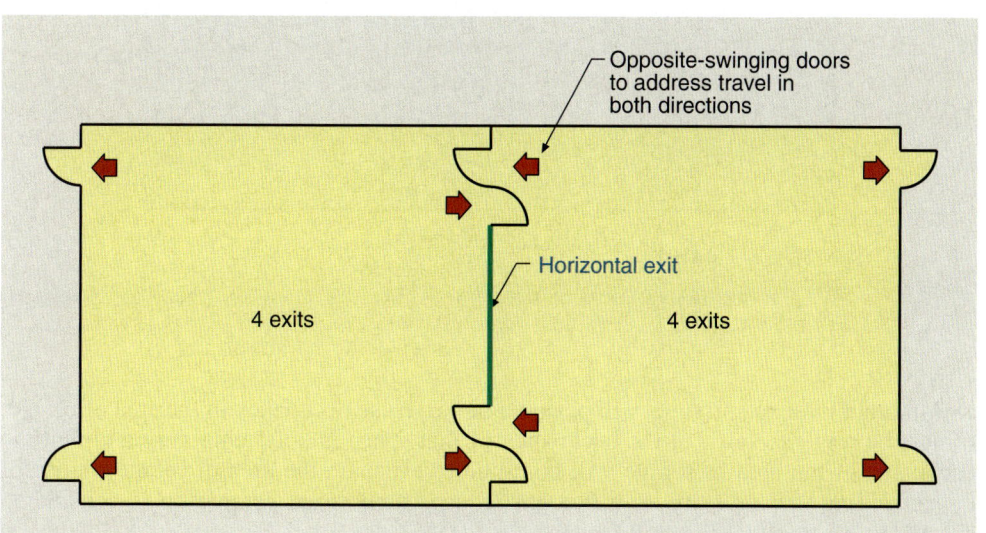

Figure 1025-3
Opposite-swinging doors.

2012 International Building Code Handbook

10 Means of Egress

1025.4 Capacity of refuge area. The area of refuge in a horizontal exit configuration must be sized to provide sufficient space for the original occupant load of the compartment of refuge plus a partial occupant load from the fire compartment for a limited period of time. The occupant load assigned from the fire compartment is determined by calculating the capacity of the horizontal exit doors that enter the area of refuge. It is necessary for building occupants to remain in the area of refuge only long enough to permit the extinguishment of the fire and elimination of the fire threat, or to allow the combined occupant load of the two compartments to use the remaining exit facilities from the compartment of refuge.

To reasonably accommodate the combined occupant loads in the compartment of refuge, the code requires that at least 3 square feet (0.28 m^2) of net clear floor area be provided for each occupant. In Group I-2 occupancies, it is required that there be at least 15 square feet (1.4 m^2) provided per occupant for each ambulatory person and at least 30 square feet (2.8 m^2) for each nonambulatory occupant. As in hospitals, such nonambulatory occupants will frequently be brought into the compartment of refuge either in a hospital bed or on a gurney. In Group I-3 occupancies, it is required that there be at least 6 square feet (0.56 m^2) of net floor area provided per occupant. An example of calculating refuge area capacity is illustrated in Application Example 1025-1.

Application Example 1025-1

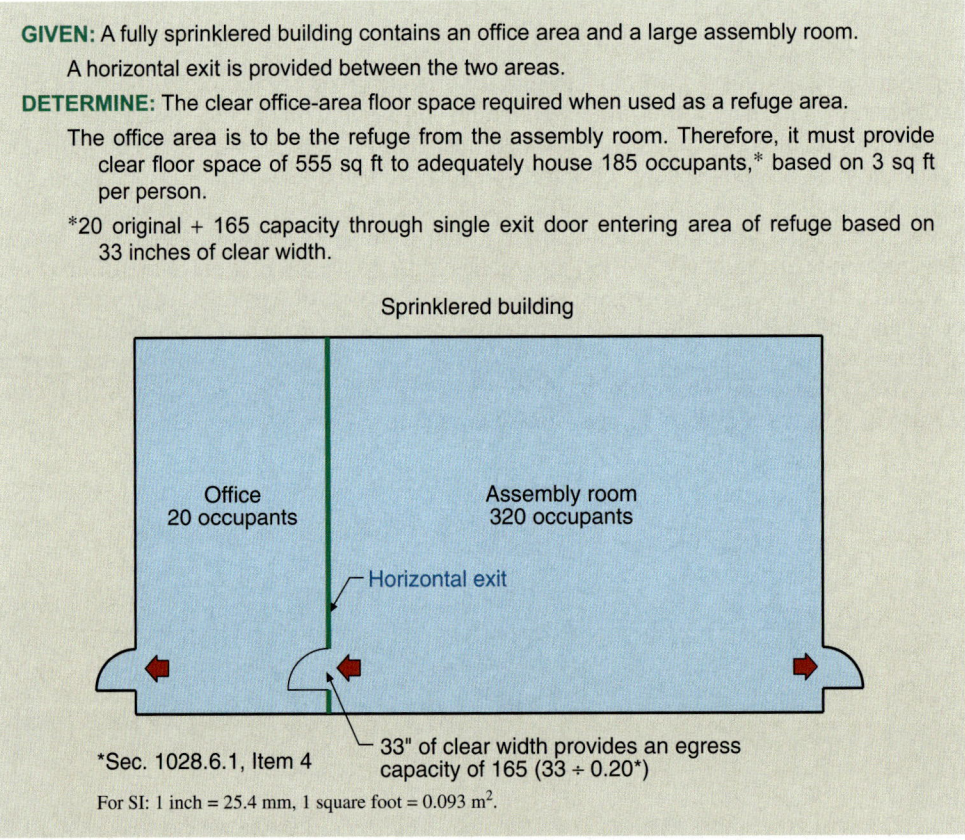

GIVEN: A fully sprinklered building contains an office area and a large assembly room. A horizontal exit is provided between the two areas.

DETERMINE: The clear office-area floor space required when used as a refuge area.

The office area is to be the refuge from the assembly room. Therefore, it must provide clear floor space of 555 sq ft to adequately house 185 occupants,* based on 3 sq ft per person.

*20 original + 165 capacity through single exit door entering area of refuge based on 33 inches of clear width.

Sprinklered building

Office 20 occupants

Assembly room 320 occupants

Horizontal exit

*Sec. 1028.6.1, Item 4

33" of clear width provides an egress capacity of 165 (33 ÷ 0.20*)

For SI: 1 inch = 25.4 mm, 1 square foot = 0.093 m^2.

Although these area figures will permit a rather dense occupancy in the area of refuge, it must be remembered that the occupancy of that space is only temporary and that the occupants of the area of refuge will continue to evacuate the area of refuge by use of the remaining exit or exits. In a fire emergency, such space per person is considered adequate for a short period of time.

It is also important to provide such a required refuge area for occupants in spaces that will, in fact, be available to the occupants of the building as they enter the compartment of refuge. Such spaces can be provided in corridors, lobbies, and other public areas, as long as they are sufficient to accommodate the total occupant load at the appropriate rate of area per person. Spaces in the refuge area occupied by the same tenant as in the area of fire origin are also permitted.

Section 1026 *Exterior Exit Stairways and Ramps*

Use in a means of egress. An exterior exit stairway or ramp may serve as an exit component in the means of egress system in all occupancies other than Group I-2. The use of an exterior exit stairway or ramp as a required means of egress element is limited to non-high-rise buildings with a maximum of six stories above grade plane. **1026.2**

Open side. To be classified as an exterior stair, it must be open on at least one side. The open side must then adjoin open areas such as yards, courts, or public ways. In order to qualify as an open side, there must be at least 35 square feet (3.3 m²) of aggregate open area adjacent to each floor level and at the level of each intermediate landing. In addition, the required open area must be at least 42 inches (1,067 mm) above the adjacent floor or landing level. See Figure 1026-1. By limiting the amount of enclosure by the exterior walls of the building, an exterior stair will be sufficiently open to the exterior to prevent accumulation of smoke and toxic gases. Any stairway that does not comply with these criteria is considered an interior stairway. **1026.3**

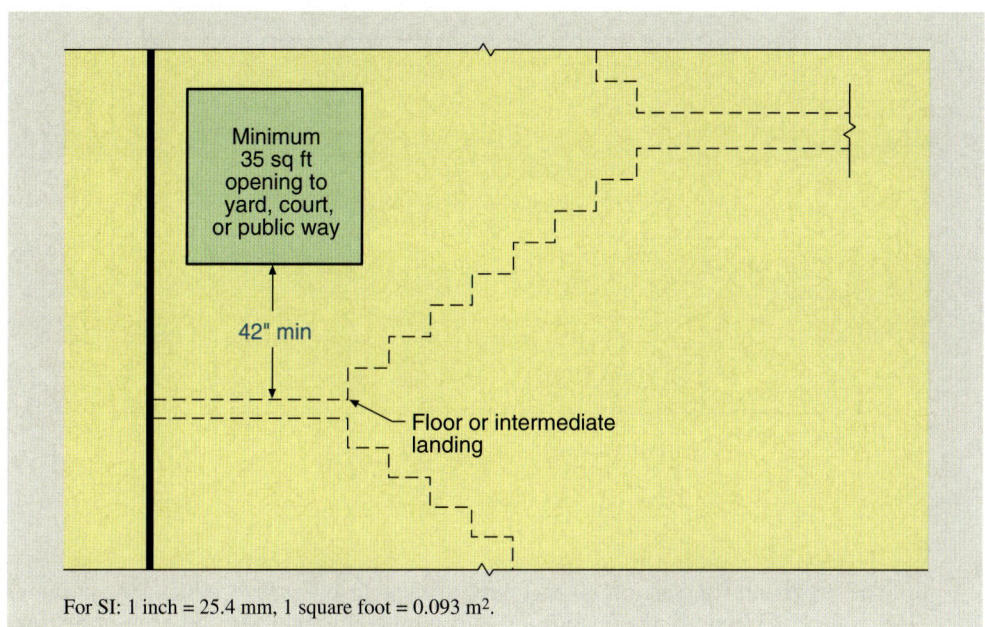

For SI: 1 inch = 25.4 mm, 1 square foot = 0.093 m².

Figure 1026-1 Exterior exit stairways.

Location. Because of exposure potential from adjacent property, exterior stairways and ramps are prohibited within 10 feet (3,048 mm) of an adjacent lot line. Such components must also maintain a minimum 10-foot (3,048-mm) separation from other buildings on the same lot unless the exterior walls and openings of the adjacent building are protected in accordance with Section 705 based on fire separation distance. See Figure 1019-1. **1026.5**

1026.6 Exterior ramps and stairway protection. In order to adequately protect the building occupants as they travel an exterior stairway or ramp during egress, the exterior exit path must be adequately separated from the interior of the building. With exceptions, this section requires that an exterior stairway or ramp be provided with protection in the same manner and to the same degree as is provided by an interior exit stairway. See the discussion of Section 1022. Consistent with these provisions, openings are not permitted unless necessary for egress from normally occupied spaces to the exterior stair. There are four situations where the separation between the exterior exit ramp or stairs and the building's interior is not required:

1. In buildings no more than two stories above grade plane housing other than Group R-1 and R-2 occupancies, separation is not required where the level of exit discharge is at the first story. In such a scenario, only one story of vertical exit travel would be required. This would limit the exposure to a degree that protection is deemed unnecessary.

2. If an open exterior balcony provides access to at least two remote exterior stairways or other exits, the separation is not required. See Figure 1026-2. To be considered open, the balcony must be open to the exterior for at least 50 percent of its perimeter. This length must then be open vertically at least 50 percent of the wall height, with openings extending at least 7 feet (2,134 mm) above the balcony. Where the building occupant has an alternative choice of exit travel, the occupant is not forced to use the stairway or ramp adjacent to the hazard. Therefore, wall and opening protection is not required.

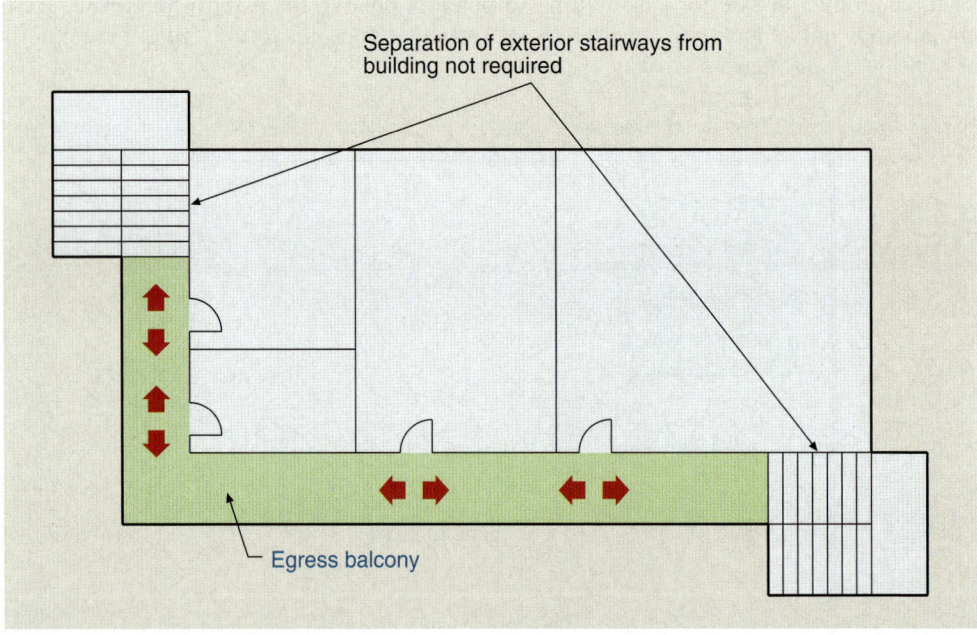

Figure 1026-2
Exterior stairway protection.

3. The provisions of Section 1009.3 permit unenclosed exit access stairways under certain conditions. If the conditions warrant the elimination of any fire protection for a stairway inside the building, there should be no reason to require fire protection for a similar exterior situation.

4. An open-ended corridor is simply a corridor that is open to the outside at the exterior of the building, leading directly to an exterior stairway or ramp at each end with no intervening doors or enclosures at the exterior wall. Where open-ended corridors are used, such as in a breezeway design, the code identifies

under what conditions the fire separation is not necessary between the building interior and the exterior stairway or ramp. First, the building, including the corridor and stairs, must be sprinklered throughout. Second, the corridor shall meet all of the corridor provisions of Section 1018. Third, the exterior stairways at the ends of the open-ended corridor are to comply with the general exterior exit stairway and ramp provisions of Section 1026. Fourth, exterior walls and permitted openings adjacent to the exterior exit stairway or ramp must be protected when required by Section 1022.7. Fifth, where a change in direction of more than 45 degrees (0.79 rad) occurs in the corridor, an exterior stairway, exterior ramp, or openings to the exterior shall be provided. The openings shall be such that the accumulation of smoke and toxic gases will be minimized, but in no case less than 35 square feet (3.25 m²) in area. See Figure 1026-3.

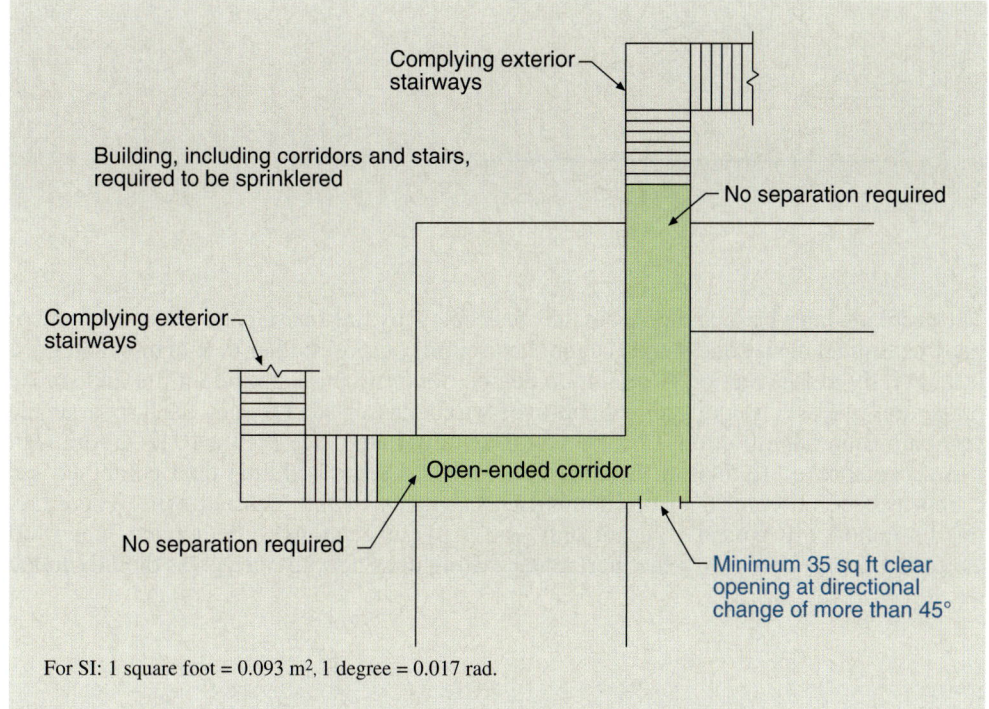

Figure 1026-3
Open-ended corridor.

Section 1027 *Exit Discharge*

General. Exits are intended to discharge directly to the exterior of the building. Three exceptions permit the exit path to include a portion of the building beyond the exit component. An exception to the requirements for the continuity of interior exit stairways (and ramps) is permitted where a maximum of 50 percent of the exits pass through areas on the level of exit discharge. The path of travel to the exterior must be unobstructed and easily recognized. Sprinkler protection is required for the egress path between the termination of the interior exit stairway to the building's exterior, as is fire-resistance-rated construction isolating any areas below the discharge level. See Figure 1027-1. Effectively, all portions of the discharge level that provide access to the egress path must be sprinklered as well. A second exception permits egress from an interior exit stairway to enter a vestibule where limited to 50 percent of the number and capacity of the total stairways provided.

1027.1

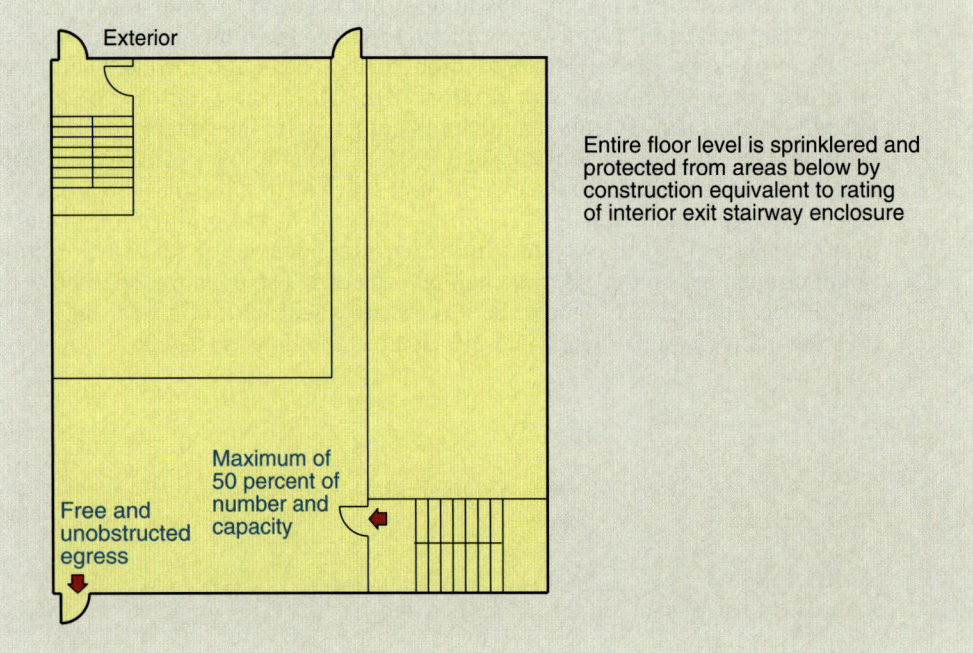

Figure 1027-1
Exit discharge through building.

The vestibule must be separated from the areas below by fire-resistance-rated construction, must be limited in size and shape, cannot be used for purposes other than egress, and must lead directly to the exterior. A minimum degree of construction providing fire and smoke protection, at least the equivalent of approved wired glass in steel frames, shall separate the vestibule from other portions of the level of exit discharge. See Figure 1027-2. Although it is acceptable to use both of these exceptions in the same building, their combined use cannot exceed 50 percent of the number and capacity of the required exits. Therefore, this limitation will typically permit only one of the exceptions to be applied. The third exception simply recognizes that horizontal exiting does not provide egress directly to the outside of a building.

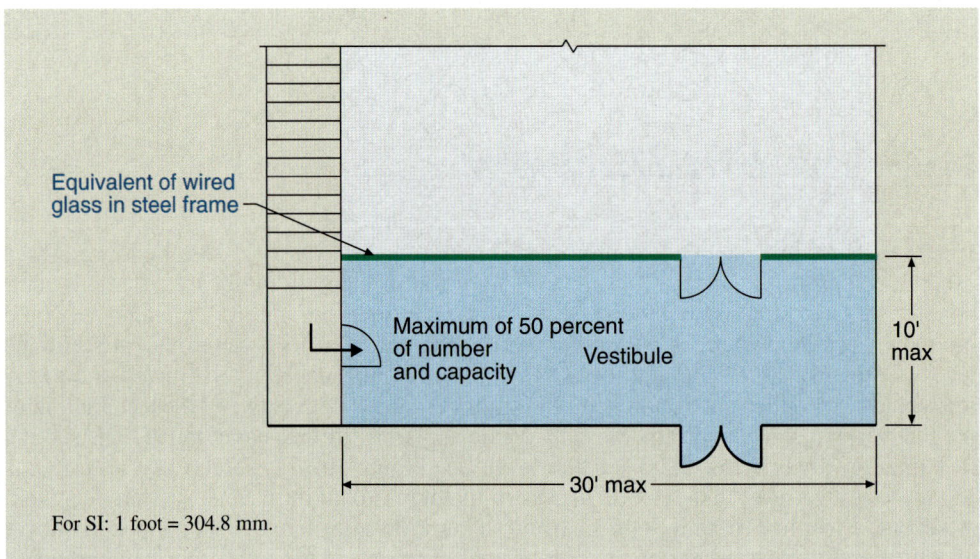

Figure 1027-2
Exit discharge through vestibule.

Exit Discharge

Exit discharge is prohibited from reentering a building to ensure that the user does not go back into any portion of the building once he or she has reached the exit discharge. This prohibition can be viewed as excluding the reentry into interior exit stairways and ramps, exit passageways, or any portion that would be considered a component of the exit.

Exit discharge components. The general concept of the exit discharge portion of the means of egress is that the components be sufficiently open to the exterior to prevent the accumulation of smoke and toxic gases. As occupants reach the exterior of the building at grade level, they expect to have arrived at a point of relative safety. Where adequate natural cross ventilation is available to disperse any smoke or gases that may be present, one of the major fire- and life-safety concerns is assumed to have been addressed. It is common that roofed areas be provided at the exterior side of exterior exit doors in order to provide weather protection and/or decoration. Such areas are typically considered as a portion of the exit discharge where they are adequately open on the sides to provide for adequate natural ventilation.

1027.3

Egress courts. An egress court is defined as any court or yard that provides access to a public way for one or more required exits. As such, the code requires that every egress court discharge into a public way. See Figure 1027-3.

1027.4

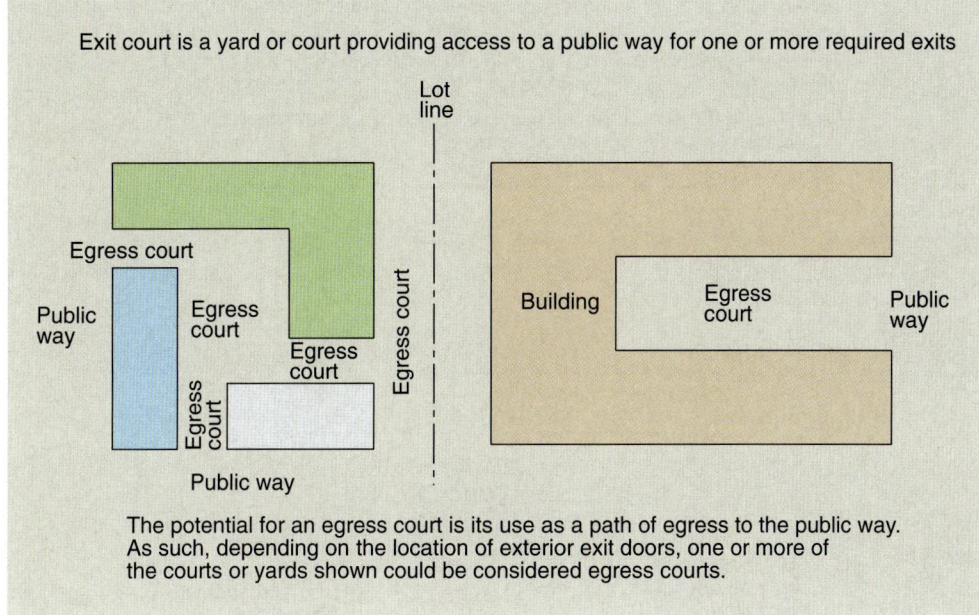

Figure 1027-3
Egress courts.

Width. The minimum required width of an egress court is determined in a similar manner to that of a corridor. An egress court must provide at least 44 inches (1,118 mm) of clear width in all occupancies except Groups R-3 and U, where the width may be reduced to 36 inches (914 mm). When egress courts are subject to use by a sufficiently large occupant load, the required width may be wider than 44 inches (1,118 mm). Such a greater width would be determined in accordance with the applicable provisions of Section 1005.1 based on the occupant load served. As is the case for most other egress components, limited encroachments into the required width are permitted for doors and handrails. Whatever the required width of the egress court, there must be at least 7 feet (2,134 mm) of unobstructed headroom.

1027.4.1

In no event may the minimum required width be less than that required by the above paragraph. However, should the actual width of an exit court be greater than the minimum width required, and it becomes necessary to reduce the width of the court as the building occupants proceed toward the public way, such a reduction cannot be an abrupt change. The changes must be made by making an angle of not more than 30 degrees (0.52 rad) with the axis of the egress court so that the reduction in width is a gradual one.

1027.4.2 Construction and openings. Because an egress court is a component in a means of egress system, building occupants using that component must be afforded sufficient protection to reasonably ensure that they will reach the safety of the public way. Therefore, in other than a Group R-3 occupancy, any time an egress court serves an occupant load of 10 or more and is less than 10 feet (3,048 mm) in width, the walls of the exit court must be of 1-hour fire-resistance-rated construction for a minimum height of 10 feet (3,048 mm) above the floor of the court. By this means, a fire-resistant separation is maintained between the persons in the court and the occupied-use spaces of the building. Should any openings occur in the portion of the egress court wall required to be fire-resistance-rated construction, those openings must be protected by assemblies having a fire-protection rating of not less than $^3/_4$ hour. See Figure 1027-4.

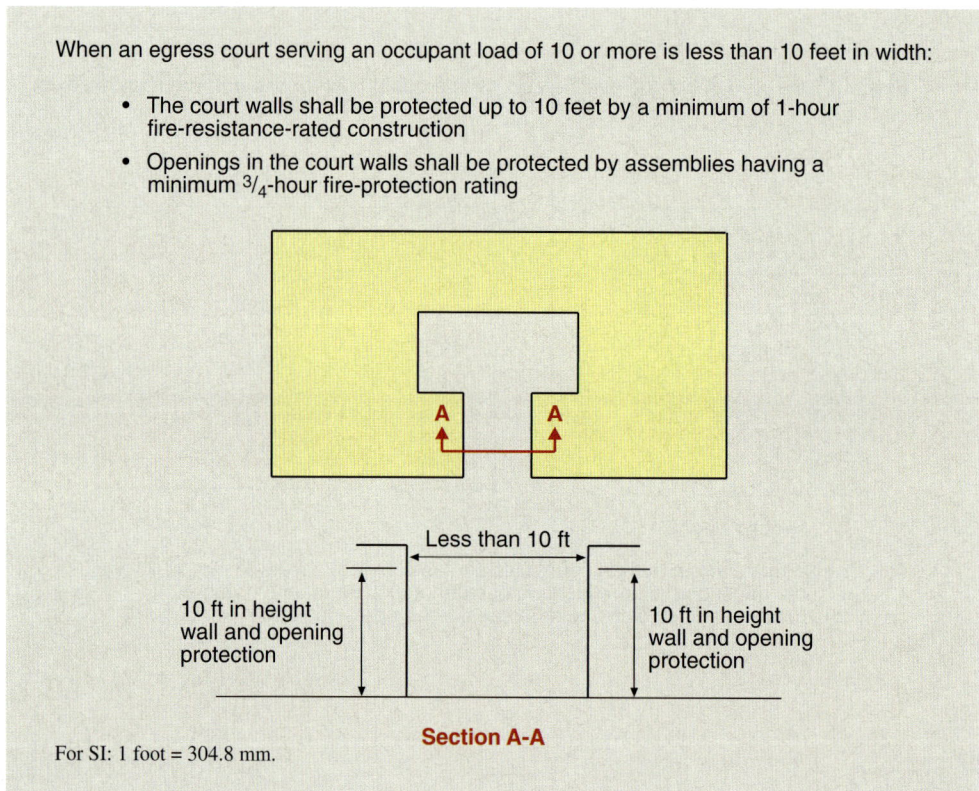

Figure 1027-4 Egress court construction.

Section 1028 *Assembly*

Because of the potential for large occupant loads in concentrated areas, assembly uses are regulated for egress a bit differently than other occupancies. This section addresses a variety of issues that are specific to buildings and portions of buildings that are used for assembly purposes, including exit capacity, aisle widths, and smoke-protected assembly seating.

Although the provisions of this section do not apply to all assembly-type uses, they are applicable to most. The means of egress for any assembly room or space containing seats, tables, displays, shelving, equipment, fixtures, or similar elements must comply with this section.

Section 1028 is applicable to assembly uses regardless of their occupancy classification. Although the provisions are most often applied to Group A occupancies, they would also address other assembly uses, such as those assembly rooms accessory to Group E occupancies. Although the classification of such areas may be based on the educational function of the building, it is important to recognize that from a means-of-egress perspective the large assembly uses still function as assembly spaces. All of the hazards involved with assembly-type functions are present in school assembly rooms, regardless of whether the room is classified as Group A or E. Therefore, the specific means of egress provisions of Section 1028 are to be applied to all assembly rooms or spaces, not just those classified as Group A.

Bleachers, grandstands, folding seating, and telescoping seating are not regulated by Section 1028, but rather by ICC 300, *Bleachers, Folding and Telescoping Seating, and Grandstands*. This standard, developed by the ICC Consensus Committee on Bleacher Safety, addresses the means of egress for such special types of seating arrangements. Simply an extension of the IBC, ICC 300 also addresses structural design and construction features.

Spaces under grandstands and bleachers. In order to protect the assembly seating in bleachers and grandstands from an exposure to hazards, a minimum separation of 1-hour fire-resistance-rated construction is required to protect the seating area from the spaces below. This separation requirement applies where the space beneath the seating is used for any purpose other than restrooms of any size and limited-size ticket booths, even if the spaces are protected by an automatic sprinkler system in accordance with Section 903.2.1.5. Conceptually, this requirement for protection on the underside of the seating is similar to the required protection of enclosed usable space beneath a stairway (see Section 1009.9.3). It is intended that the provision apply to storage rooms, concession stands, and similar areas where there is a concern due to the amount of combustible loading directly beneath an assembly seating area. It should be assumed that the requirement for a protective covering need not be applied where the means of egress passes below the seating area due to the absence of combustible materials. **1028.1.1.1**

Assembly main exit. This section provides special exiting requirements for assembly buildings, rooms, and spaces. The first specification is that every assembly use having an occupant load greater than 300 must be provided with a main exit. The minimum required width of this main exit is determined by two criteria—calculated width and component width. The rule that would produce the larger required width for the main exit is the one that would govern the required minimum size. **1028.2**

First, in any assembly use having such a sizable occupant load, the designated main exit must have sufficient width to accommodate at least one-half of the total occupant load. Second, its width shall not be less than the total required width of all means of egress such as aisles, corridors, exit passageways, and stairways that lead to the main exit. The main exit must always be connected by appropriate exit components to ensure that the occupants have continuous and unobstructed access to a public way.

Basically, the requirement that the main exit be adequate to accommodate 50 percent of the total occupant load takes into account a characteristic of human nature. The majority of the occupants are, in all probability, not completely familiar with the facility and its exit system. As a consequence, in the event of an emergency it is typical that people will attempt to reach the exit through which they entered. This natural tendency could put an unduly large load on the main exit if it were not sized according to this requirement.

An exception permits distribution of exits where there is no well-defined main exit or where multiple main exits are provided. In no case, however, may the total width of egress be less than 100 percent of the required width. Should there be multiple points of entry into the building, it would be unreasonable to size each point as if it were a main exit. However, where a main exit is specifically identified, the 50 percent rule is applicable.

1028.3 Assembly other exits. In addition to requiring that the main exit accommodate 50 percent of the total occupant load, it is also necessary that every assembly use with an occupant load of more than 300 be provided with additional means of egress. They are required to be of sufficient capacity to accommodate at least 50 percent of the total occupant load served by that level. As expected, these additional means of egress must also comply with Section 1015.2 addressing doorway arrangement.

1028.4 Foyers and lobbies. In Group A-1 occupancies, typically theaters and similar uses, persons often occupy a lobby to await a presentation in the major-use area. It is important that while the lobby is occupied, the required clear egress width from the major-use area is maintained through the lobby to the exit from the building.

1028.5 Interior balcony and gallery means of egress. Consistent with the basic requirement for the number of exits in most assembly uses as reflected in Table 1015.1, every balcony, gallery, and press box that has an occupant load of 50 or more must be provided with a minimum of two remote, separate means of egress. In some assembly uses, more than one such feature is provided. These requirements for exits apply to any and all balconies, galleries, and press boxes in the building. The code indicates that at least one of the required means of egress must lead directly to an exit. However, it is intended that the means of egress may include unenclosed exit access stairways and ramps between the balcony, gallery, or press box and main assembly floor in uses such as theaters, churches, and auditoriums.

1028.6 Width of means of egress for assembly. The method for calculating egress widths has two distinct categories. These are buildings with and buildings without smoke-protected assembly seating. Thus, the first thing that must be established is the category in which the egress must be analyzed. Section 202 defines smoke-protected assembly seating as "seating served by means of egress that is not subject to smoke accumulation within or under a structure." Section 1028.6.2 regulates smoke-protected assembly seating in regard to smoke control, roof height, and automatic sprinklers.

If the reduced width requirements of Table 1028.6.2 are applied in a building that has smoke-protected assembly seating, a Life Safety Evaluation complying with NFPA 101 must be conducted. An acceptable evaluation would include such criteria as pedestrian flow rates, movement characteristics of persons using exit systems, the nature of the means of egress system beyond the aisles, location of potential hazards, staff-response capabilities, facility preplanning and training, and other items relating to occupants' safety. Should it be determined that a building is not smoke protected, or that no life-safety evaluation has been performed, the aisle width will need to be calculated using the more stringent non-smoke-protected conditions.

1028.6.1 Without smoke protection. For egress on stairs, two modifications to the basic width factor of 0.3 inch per occupant (7.6 mm) are to be considered. The general requirement of 0.3 inches (7.6 mm) is based on stairs having riser heights of no more than 7 inches (178 mm). Where the riser height exceeds 7 inches (178 mm), at least 0.005 inch (0.127 mm) of additional stairway width for each occupant shall be provided for each 0.10 inch (2.5 mm) of riser height above 7 inches (178 mm). The following formula represents this method of width increase:

$$W = 0.3 + 10(R - 7.0)(0.005)$$

where:

W = Required width in inches per occupant
R = Riser height in inches

Formula 10-1

The second potential modification applies only where egress is accomplished on the stair in a descending fashion. Under such a condition, an increase of at least 0.075 inch (1.9 mm) of additional width per occupant must be provided where no handrail is provided within a horizontal distance of 30 inches (762 mm). This egress-width increase is illustrated by the following formula:

$$W = 0.3 + (0.75) = 0.375$$

where:

W = Required width in inches per occupant

Formula 10-2

It is also very possible that both increases will be applied to the same aisle stairs. Where a stair has a riser height exceeding 7 inches (178 mm), and a portion of the required stair width is more than 30 inches (762 mm) from a handrail, both modifications to the basic requirement of 0.3 inch (7.6 mm) per occupant are applicable. The increase in required egress width can be shown as:

$$W = 0.3 + 10(R - 7.0)(0.005) + (0.075)$$

where:

W = Required width in inches per occupant
R = Riser height in inches

Formula 10-3

Where a ramp is used as a portion of the means of egress, the difference in calculated width is based on the slope of the ramp. Ramps having a slope steeper than 1 in 12 (1:12) shall have a minimum clear width based on 0.22 inch (5.6 mm) per occupant served. Note that this condition would only apply to ramps not regulated by Chapter 11 for accessibility. The minimum clear width of level or ramped egress paths having a slope of 1 in 12 (1:12) or less is based on 0.20 inch (5.1 mm) for each occupant.

The required increases in egress width are all based on one of the primary IBC concepts regarding means of egress: where it is anticipated that users of the egress system may be slowed in their egress travel, the path is required to be widened to compensate for the reduced travel speed. By increasing the width of the exit path, the occupants are expected to continue to travel at a relatively consistent rate through the exit system. For stair travel, a riser height over the customary maximum of 7 inches (178 mm) creates an uncomfortable condition for most stair users. In such cases, it is natural for persons to slow their travel in order to more safely use the stairway. Occupants also tend to travel more slowly when descending a stairway where a handrail is not within easy reach. Without the ability to easily grasp a handrail in case of a misstep, the stair user tends to be a bit more cautious. On the other hand, the presence of an adjacent handrail provides a degree of security that encourages faster stair travel. In ramp travel, a slope steeper than that normally encountered also creates a small degree of hesitancy, requiring a greater width to compensate for the reduction in travel speed.

The increases in required egress width prescribed by this section are only intended to apply where the minimum width is calculated based on the number of occupants served by the egress path. The width increases are not applicable to the component widths established in Section 1028.9.1. For example, a minimum width of 36 inches (914 mm) is mandated for aisle stairs with seating on only one side. This condition would typically allow for a single handrail, which would be located more than 30 inches (762 mm) horizontally from a portion of the required aisle width. It is not appropriate to increase the minimum 36-inch (914-mm) required width based on the factors established in Item 3 of this section. Additionally, if the aisle stair included risers that exceeded a height of 7 inches (178 mm), the adjustment found in Item 2 would also not be applied to the minimum component width as established in Section 1028.9.1.

1028.6.2 Smoke-protected seating. Calculation of egress width in an assembly space located in a smoke-protected environment is based on Table 1028.6.2. As the total number of seats in the smoke-protected assembly occupancy increases, egress-width requirements continue to decrease until reaching an end point of 25,000 or more seats. As addressed earlier, a Life Safety Evaluation must be done for a facility using the reduced width requirements of Table 1028.6.2. Application Example 1028-1 shows how the calculated width of exit

Application Example 1028-1

GIVEN: A 10,000-seat arena with seating sections as shown.
  Case I—Smoke-protected assembly seating for which a life-safety evaluation has been provided.
  Case II—Nonsmoke-protected assembly seating.
DETERMINE: Required aisle width for both cases.
SOLUTION:
  Case I—Use Table 1028.6.2. For a 10,000-seat arena, the width equation for stairs is 0.130 inch per seat served, and the required width of aisle becomes:

$$400 \times 0.130 = 52 \text{ Inches or } 4.33 \text{ feet } (1321 \text{ mm})$$

Since 52 inches (1321 mm) exceeds the minimum required width of 48 inches (1219 mm), it becomes the governing width.

Case II—For nonsmoke-protected assembly seating, the width is based on 0.3 inch per seat served. However, since a center handrail will exceed the 30-inch rule, an increase of 0.075 inch is required. Thus, the factor is 0.375 inch per seat and the required width of aisle becomes:

$$400 \times 0.375 = 150 \text{ inches}$$

Thus, the aisle required for nonsmoke-protected assembly seating is almost three times as wide as the aisle required for smoke-protected assembly seating.

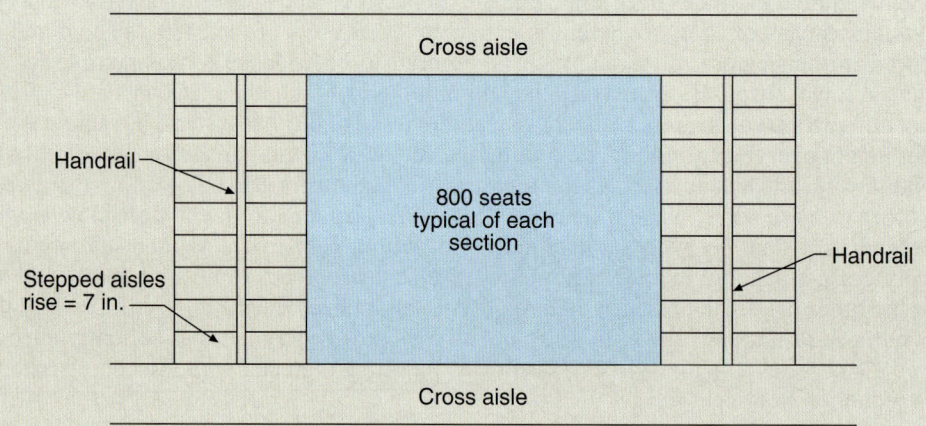

For SI: 1 inch = 25.4 mm, 1 foot = 304.8 mm.

provisions would be applied for smoke-protected seating and seating without smoke protection. The permitted reduction in egress width applies to all elements in the means of egress system (vomitories, concourses, stairways, etc.), but only to the extent that they too are smoke protected. Any egress elements that are not provided with complying smoke protection are subject to the greater widths established for areas that are not smoke protected.

1028.6.2.1 **Smoke control.** To maintain an essentially smoke-free means of egress system, a smoke-control system complying with Section 909 must be provided. As stated in Section 909.1, a smoke-control system should be designed to provide a tenable environment for the evacuation or relocation of occupants. When it can be satisfactorily demonstrated to the building official, a design incorporating a natural venting system is permitted. The natural ventilation must be designed to maintain the smoke level at a point 6 feet (1,829 mm) or more above the floor level of any portion of the means of egress within the smoke-protected assembly seating area.

1028.6.2.2 **Roof height.** Whenever smoke-protected assembly seating is covered by a roof, a minimum clearance of 15 feet (4,572 mm) is required between the highest aisle or aisle accessway to the lowest portion of the roof. In an outdoor stadium, the roof canopy need only be 80 inches (2,032 mm) or more above the highest aisle or aisle accessway, provided there are no projections or obstructions below the 80-inch (2,032-mm) level. By providing an adequate roof height above the occupiable portion of the building or structure, a smoke-containment area is created. Smoke control or removal would then limit smoke migration into the egress environment.

1028.6.2.3 **Automatic sprinklers.** Another condition for the use of the liberal egress width provisions for smoke-protected assembly seating areas is the installation of an approved automatic sprinkler system. In general, the sprinkler system is required in all areas enclosed by walls and ceilings in buildings or structures containing smoke-protected assembly seating. However, three exceptions identify locations where sprinklers may be omitted. Where the area on the assembly room floor is used for low fire-hazard uses such as performances, contests, or entertainment, sprinklers may be omitted, provided the roof construction is more than 50 feet (15,240 mm) above the floor level. Small storage facilities and press boxes under 1,000 square feet (92.9 m^2) in area are also exempt. A third exception clarifies that sprinkler protection is not required in the seating area of outdoor facilities where the means of egress for the seating area is essentially open to the outside.

1028.6.3 **Width of means of egress for outdoor smoke-protected assembly.** Where the facilities are outdoors, such as in a stadium, and the egress system for the assembly seating is considered smoke protected owing to the natural ventilation available, the clear width is based on one of two factors. Where the egress is by aisles and stairs, the width factor is 0.08 inch per occupant, whereas it is 0.06 inch for ramps, corridors, tunnels, or vomitories. An example of this calculation is shown in Application Example 1028-2. If, however, the width calculated through the use of Table 1028.6.2 for all types of smoke-protected seating is determined to be a lesser width, such a lesser width is acceptable. As indicated in the exception to Section 1028.6.2, the point where Table 1028.6.2 applies for outdoor seating is 18,000 occupants.

Application Example 1028-2

GIVEN: An outdoor smoke-protected assembly seating area having stairs serving an occupant load of 200. The total number of seats in the facility is 2,600.

DETERMINE: The required aisle width.

CASE 1: Using Table 1028.6.2, the width per person is 0.280* inch. 200 persons (0.28 inch) = 56 inches.

CASE 2: Using Section 1028.6.3, the width per person is 0.08 inch. 200 persons (0.08 inch) = 16 inches < 48 inches minimum required for seating on both sides.

*Interpolated between 0.3 inch and 0.2 inch.

∴ Minimum required width is 48 inches.

1028.7 Travel distance. Travel distance in assembly occupancies is regulated under one of three conditions: seating without smoke protection, smoke-protected seating, and open-air seating. The measurement of this distance shall be along the line of travel, including along the aisles and aisle accessways, from each seat to the nearest exit door. It is improper to measure this travel distance over or on the seats. Inside a building without the benefit of smoke protection, travel distance is limited to 200 feet (60,960 mm) in nonsprinklered buildings and 250 feet (72,200 mm) in sprinklered buildings. In smoke-protected assembly seating, a maximum travel distance of 200 feet (60,960 mm) is permitted from each seat to the nearest entrance to an egress concourse. Up to another 200 feet (60,960 mm) of travel is permitted from the egress concourse entrance to an egress stair, ramp, or walk at the building exterior. Where the assembly seating is located in an outdoor facility and all portions of the means of egress are open to the outside, the maximum travel distance is 400 feet (121,920 mm). This distance is measured from each seat to the exterior of the building. When the outdoor seating facilities are of Type I or II noncombustible construction, the travel distance may be unlimited.

1028.8 Common path of egress travel. By providing persons in an assembly occupancy a choice of travel paths, egress opportunities will be enhanced. The distance an occupant must travel prior to reaching a point where two paths are available is more limited than permitted by the general provisions of Section 1014.3. In most assembly seating areas, the common path of travel is limited to 30 feet (9,144 mm). In smoke-protected areas, up to 50 feet (15,240 mm) of common path travel is permitted. Where the seating area has a limited occupant load, additional travel is also permitted.

1028.8.1 Path through adjacent row. Because the common path of travel is most often limited to 30 feet (9,144 mm), single-access seating areas are often required to egress through rows of the adjoining seating area to reach another aisle in order to comply. Such seating areas are served by an aisle on only one side, and the aisle is single directional (top loading or bottom loading only). Where this condition occurs, the code (1) limits the maximum number of seats in the adjoining row to 24, and (2) increases the minimum required width of the aisle accessway serving the row. A similar concept is applied in Section 1028.9.5 for increasing the maximum permitted length of dead-end aisles.

1028.9 Assembly aisles are required. This section requires that aisles be provided in all occupied portions of any assembly occupancy that contains seats, tables, displays, and similar fixtures or equipment. Although the intent of the aisle provisions is to provide safe access to components of the egress system such as exits or exit access doorways, the provisions would also have application to occupied-use areas where it is necessary to provide a circulation system so that building occupants will have reasonable means for moving around in the occupied spaces. Egress travel within restaurants, classrooms, and similar uses with seating at tables is regulated by Section 1028.10.1.

1028.9.1 Minimum aisle width. Where seating is arranged in rows, the clear width shall not be less than the following, while still conforming to the calculated aisle-width provisions:

1. Forty-eight inches (1,219 mm) for aisle stairs where seating is provided on both sides of an aisle. A 36-inch (914-mm) aisle is permitted where the aisle serves less than 50 seats.

2. Thirty-six inches (914 mm) for aisle stairs where seating is provided only on one side of an aisle. Only 23 inches (584 mm) of aisle width is required between an aisle stair handrail and the nearest seat where an aisle serves no more than five rows on only one side.

3. Twenty-three inches (584 mm) between an aisle stair handrail and the nearest seat where the rail is within the aisle.

4. Forty-two inches (1,067 mm) for level or ramped aisles where seating is provided on both sides of an aisle. A 36-inch (914-mm) aisle is permitted where the aisle serves less than 50 seats, and a minimum of 30 inches (762 mm) is allowed where serving 14 seats or less.

5. Thirty-six inches (914 mm) for level or ramped aisles where seating is provided on only one side of an aisle. Only 30 inches (762 mm) of aisle width is required where the aisle serves no more than 14 seats.

This section must be used in concert with the other provisions of Section 1028 to completely define and arrange an assembly seating area.

Aisle width. The occupant load served by an aisle is the determining factor in establishing its minimum required width. In this determination, it is assumed that the egress travel is distributed evenly among the adjacent travel paths. The tributary occupant load would be assigned to each aisle proportionally, based on the arrangement of the means of egress. — **1028.9.2**

Converging aisles. Where aisles converge to form a single aisle, the capacity of that single aisle shall not be less than the combined required capacity of the converging aisles. There is no penalty for providing aisles that are wider than the minimum code requirement. — **1028.9.3**

Uniform width. Where egress is possible in two directions, the shape of an aisle cannot be of an hourglass configuration. The clear width shall be uniform throughout. A tapered aisle is allowed only for dead-end aisle conditions. — **1028.9.4**

Assembly aisle termination. In the arrangement of a seating area, all aisles serving the seating area must end in a cross aisle, foyer, doorway, vomitory, or concourse. In large facilities, it is not uncommon to find a number of aisles leading to a number of cross aisles. The required egress capacity of a cross aisle is the same as that for converging aisles, the combined required capacities of the aisles leading to the cross aisle. Although dead-end aisles are allowed, their length is limited to 20 feet (6,096 mm) except where the seats served by the dead-end aisle are not more than 24 seats from another aisle measured along a row having a minimum clear width of 12 inches plus 0.6 inch (305 mm plus 15 mm) for each additional seat over seven in a row. For smoke-protected assembly seating, up to 21 rows are permitted in a dead-end vertical aisle except where the seats served by the dead-end aisle are not more than 40 seats from another aisle measured along a row having a minimum clear width of 12 inches plus 0.3 inch (305 mm plus 7.6 mm) for each additional seat over seven in a row. — **1028.9.5**

Aisle accessways. The provisions of this section are specific to aisle accessways serving two different conditions: seating at tables and seating in rows. — **1028.10**

Seating at tables. Because it is often difficult to determine the exact dimensions of chairs and other seating that may encroach into an aisle or aisle accessway, a measurement of 19 inches (483 mm) is used to account for any potential obstruction caused by the seating. The 19-inch (483-mm) distance, applicable for movable seating, is measured perpendicular to the edge of the table or counter where the seating is provided. Where fixed seating is provided, the width is to be measured to the back of the seat. Examples of the use of this provision are illustrated in Figure 1028-1. Where other side boundaries are present, the clear width, other than that permitted for handrail projections, is to be measured to walls, tread edges, seating edges, or similar elements. — **1028.10.1**

Aisle accessway width for seating at tables. The clear width of aisle accessways serving various arrangements of seating at tables or counters is based on 12 inches (305 mm) plus 0.5 inch (12.7 mm) of additional width for each 1 foot (305 mm), or fraction thereof, beyond 12 feet (3,658 mm) of aisle accessway length. This length is measured from the center line of the seat farthest from the aisle. — **1028.10.1.1**

Minimum clear width = 12 inches + 0.5 inch (x − 12 feet),

> **WHERE:**
>
> x = the distance in feet between the aisle and the center line of the most remote seat.

10 Means of Egress

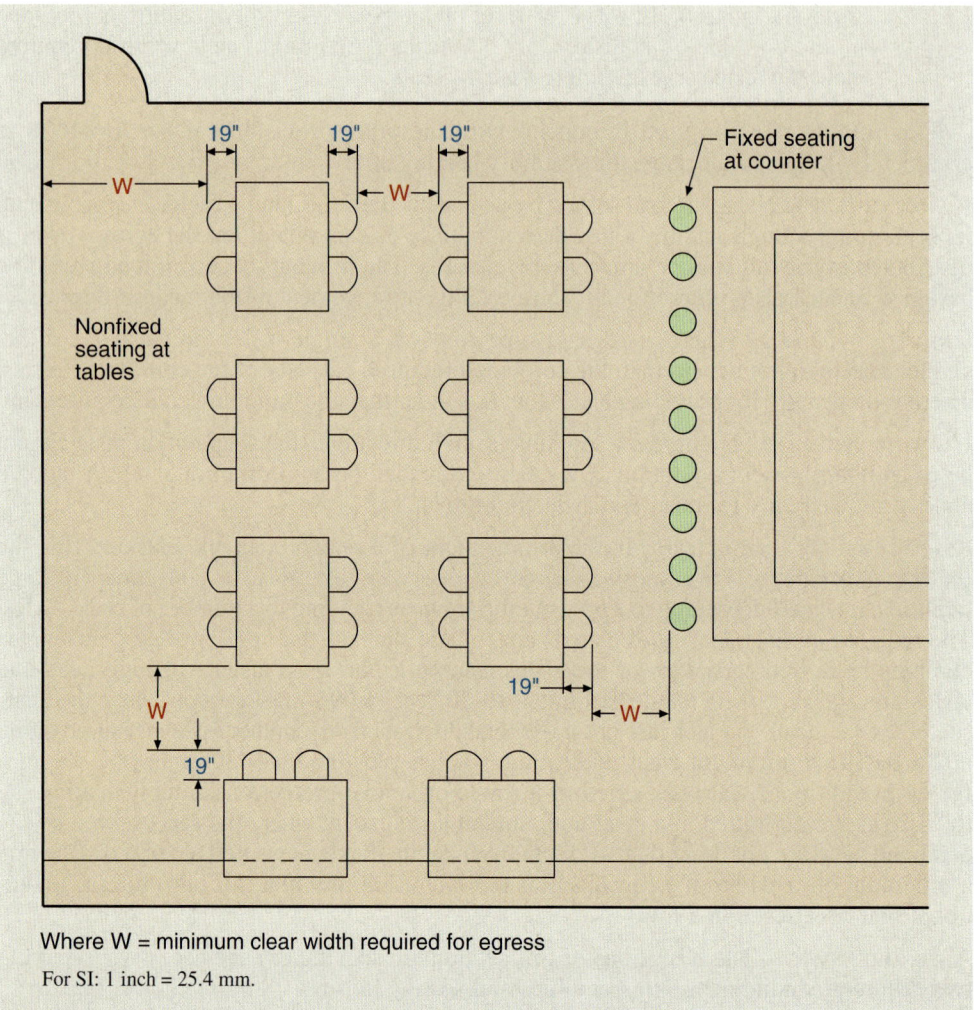

Figure 1028-1
Seating at tables and counters.

It should be remembered that in no case should the width be less than that based on the capacity requirements of Section 1005.1. On the other hand, there is no minimum width required for those portions of aisle accessways serving four or fewer persons and 6 feet (1,829 mm) or less in length.

1028.10.1.2 Seating at table aisle accessway length. Consistent with other provisions for means of egress, single-directional travel along aisle accessways is also limited in length. The distance from any seat to the point where two or more paths of egress travel are available to separate exits is limited to 30 feet (9,144 mm). In application, the 30-foot (9,144-mm) limitation would be the maximum length of a dead-end aisle accessway. Within that distance, the aisle accessway must reach an aisle or other means of egress element where travel would be limited based on the common path provisions of Section 1014.3. Various examples of the width provisions for aisle accessways serving seating and tables are shown in Figure 1028-2.

1028.10.2 Clear width of aisle accessways serving seating. The minimum clear width between rows of seats is 12 inches (305 mm), measured between the rearmost projection of the seat in the forward row and the foremost projection of any portion of the seat in the row behind, where the rows have 14 or fewer seats. If automatic or self-rising seats are used (such as in movie theaters), the minimum clear width shall be measured with the seat up. If any chair in the row

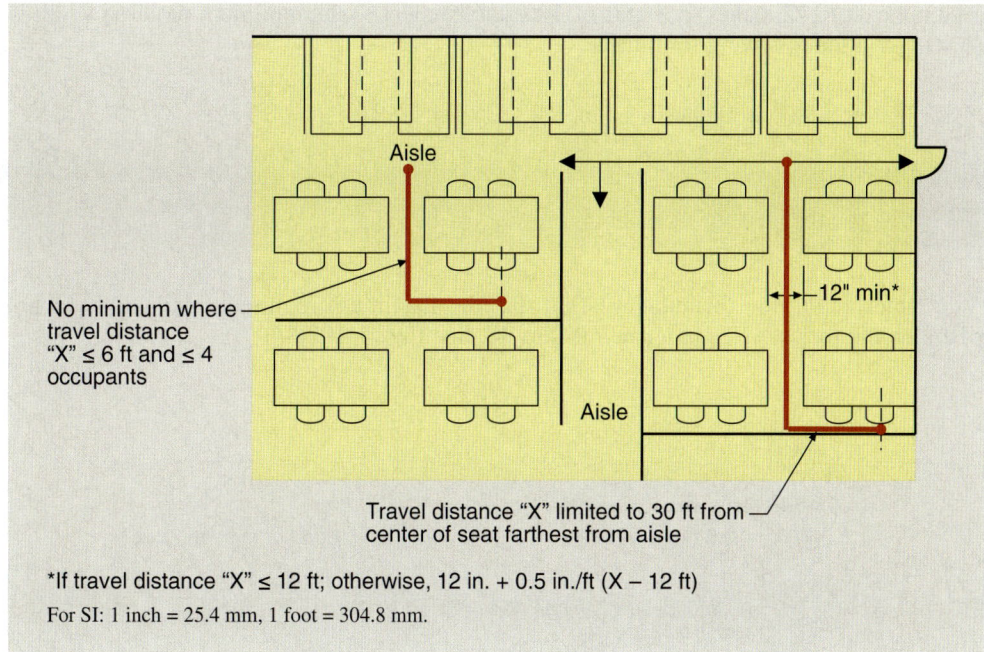

Figure 1028-2
Egress width at tables.

does not have an automatic- or self-rising seat, the measurement must be made with the chair in the down position. See Figure 1028-3. Seats with folding tablet arms are regulated in a special manner. Unless the tablet arm is of a type that automatically returns to the stored position by gravity when raised manually to the vertical position, the row spacing measurement must be taken with the arm in the position in which it is used.

Such clear width must be increased whenever the number of seats in a row exceeds 14, but in no case shall the number of seats in any row exceed 100. The increased width when seating rows are served by aisles or doorways at both ends of a row is the minimum 12 inches plus 0.3 inch (305 mm plus 7.6 mm) for each additional seat over 14; however, the width

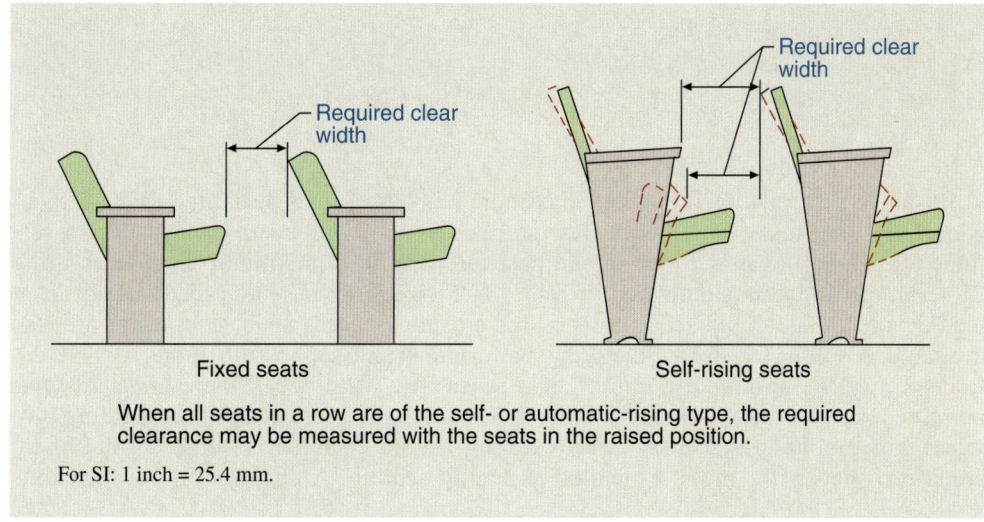

Figure 1028-3
Aisle accessway width.

need not exceed 22 inches (559 mm). The establishment of the minimum required width is illustrated in the following formula:

> Clear width, in inches = 12 inches (305 mm) + 0.3 inch (7.6 mm) (x − 14 seats)
>
> where:
>
> x = the number of seats in a row.

By interpreting the formula, we find that the maximum required clear width occurs when the number of seats in a row reaches 48. See Figure 1028-4.

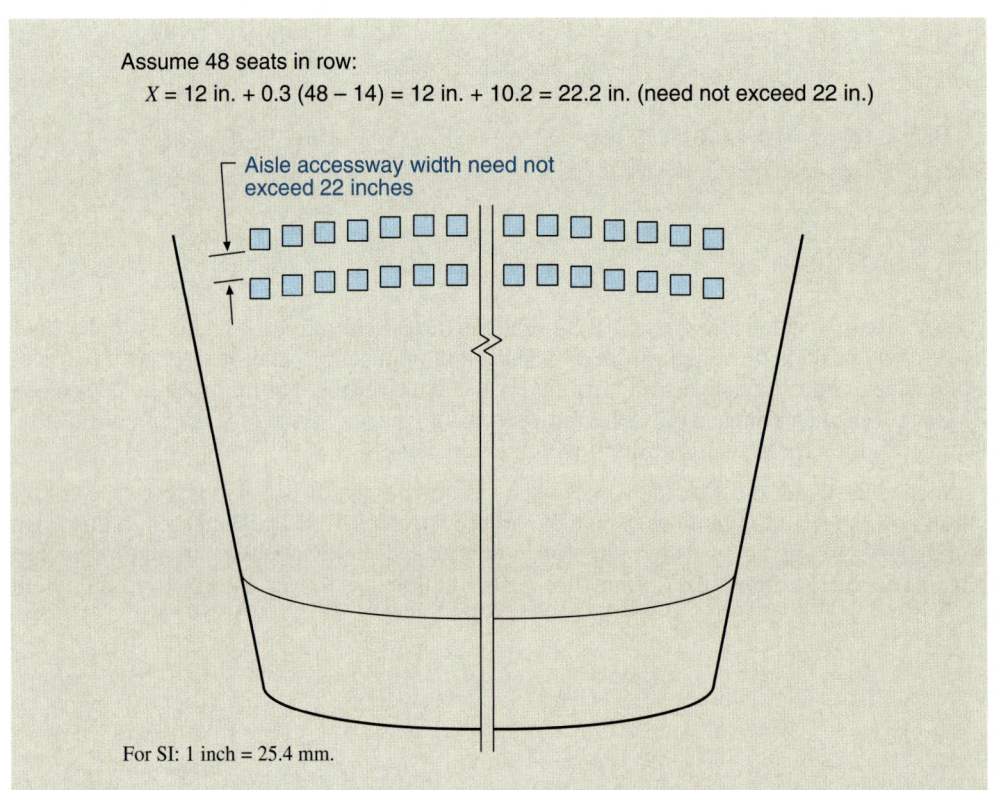

Figure 1028-4
Maximum required aisle accessway width.

In smoke-protected facilities, the maximum number of seats permitted for a 12-inch-wide (305-mm) aisle accessway with dual access is permitted to exceed the 14-seat limit for seating areas without smoke protection. With a maximum that varies up to 21 seats, the limitations indicated in Table 1028.10.2.1 are based on the total occupant load devoted to assembly seating. Increases on single-access aisle accessways are also available. A similar version of the previous formula can be used with Table 1028.10.2.1 to determine the minimum required aisle accessway width when the number of seats permitted for a 12-inch-wide (305-mm) aisle accessway is exceeded in a building provided with smoke-protected assembly seating. As an example, assume a 16,000-seat arena provided with smoke-protected assembly seating. A dual-access row contains

34 seats. The minimum required aisle accessway (row) width is determined by the following formula:

> 12 inches + [(0.3 inch)(x − y)]
>
> where:
>
> x = the number of seats in the row (34 in this example), and
> y = the maximum number of seats permitted with a 12-inch aisle accessway (19 in this example)
>
> 12 + [(0.3)(34 − 19)] = 12 + 4.5 + 16.5 inches minimum width

For rows of seating served by an aisle or doorway at only one end of a row, the formula for the clear width is the minimum 12 inches plus 0.6 inch (305 mm plus 15 mm) for each seat over seven, but the clear width need not exceed 22 inches (559 mm). This is similar to the previous provision with one major difference: a maximum 30-foot (9,144-mm) path of travel is permitted from the occupant's seat to a point where there is a choice of two directions of travel to an exit. This can be to the adjacent two-way aisle or along a dead-end aisle to a cross-aisle or doorway. See Figure 1028-5. Where one of the two paths of travel is across the aisle through a row of seats to another aisle, the maximum of 24 seats rule described previously is in effect. For smoke-protected assembly seating, single direction travel distance can be increased per Table 1028.10.2.1.

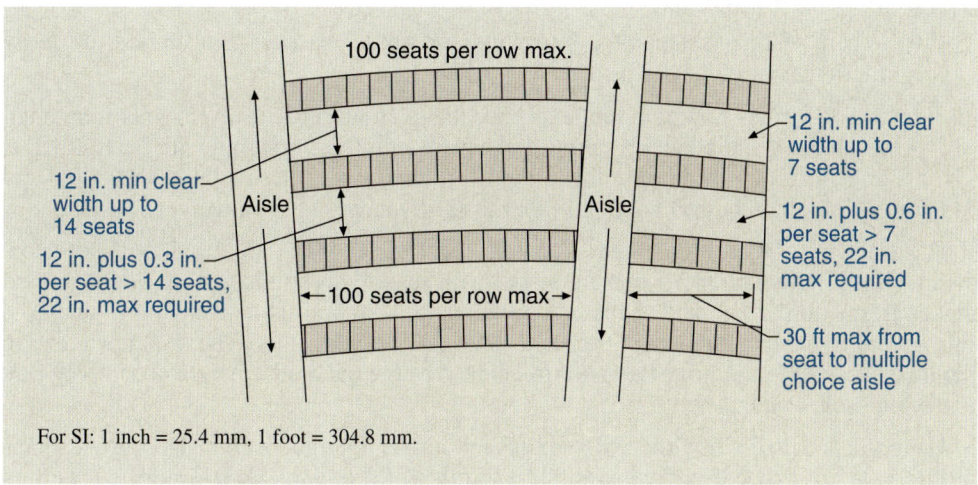

For SI: 1 inch = 25.4 mm, 1 foot = 304.8 mm.

Figure 1028-5
Row spacing.

Assembly aisle walking surfaces. Where aisles have a slope of 1 unit vertical in 8 units horizontal (12.5-percent slope) or less, steps in the aisles are prohibited because occupants in low-slope aisles have a tendency to not notice steps as readily as they would in the steeper aisles. Continuous surfaces are safer surfaces. **1028.11**

Where an aisle has a slope steeper than 1 unit in 8 units horizontal (12.5-percent slope), a series of risers and treads must be used. These risers and treads shall extend the entire width of the aisle, shall have a rise of no more than 8 inches (203 mm) nor less than 4 inches (102 mm), and shall be uniform for the entire flight. The tread shall not be less than 11 inches (279 mm) and shall be uniform throughout the flight. Variations in run or height between adjacent treads or risers shall not exceed $3/16$ inch (4.8 mm).

One provision that helps the user of an aisle notice a step is the provision requiring a contrasting strip or other approved marking on the leading edge of each tread. Designed to identify the edge of each tread when viewed in descent, this marking strip may be omitted where it can be shown that the location of each tread is readily apparent.

An exception permits variations in rise or height to exceed $^3/_{16}$ inch (4.8 mm) between risers, provided the exact location of such a variation is clearly identified with a marking strip at the nosing or leading edge adjacent to the nonuniform risers. This edge marking strip, having a width between 1 inch and 2 inches (25 mm and 51 mm), shall be distinctively different from the contrasting marking strip required on each tread. In another exception to the riser height provision, riser heights may be increased to 9 inches (229 mm) where it can be demonstrated that lines of sight would otherwise be impaired.

1028.12 Seat stability. Because of the potential obstructions to the paths of egress travel caused by loose seating, this section requires seats in assembly uses to be securely fastened to the floor. The following six exceptions identify conditions under which the securing of seating is either impractical or unnecessary:

1. Where 200 or fewer seats are provided on a flat floor surface
2. Where seating is at tables and the floor surface is flat
3. Where more than 200 seats are fastened together in groups of three or more on a flat floor surface
4. Where seating flexibility is critical to the function of the space, and 200 or fewer seats are provided on tiered levels
5. Where level seating is separated by railings or similar barriers into groupings of 14 seats or fewer
6. Where seating is separated by railings or similar barriers and limited to use by musicians or other performers

1028.13 Handrails. All aisles having a slope steeper than 1 unit vertical in 15 units horizontal (6.7-percent slope), and all aisles stairs shall have handrails complying with Section 1012. The handrails can be placed on either side of or down the center of the aisle served and can project into the required width no more than $4^1/_2$ inches (114 mm).

Handrails may be omitted where the slope of the aisles is not greater than 1 unit vertical in 8 units horizontal (12.5-percent slope) with seating on both sides, or where a guard that conforms to the size and shape requirements of a handrail is located at one side. The first exception intends the seating to be a substitute for handrails. The second exception permits a graspable top rail of a guard to be used as the handrail where a drop-off occurs on the one side of the aisle.

Handrails located within the aisle width shall not be continuous, but shall provide gaps at intervals not exceeding five rows. The width of these gaps should not be less than 22 inches (559 mm), nor more than 36 inches (914 mm). This is to provide access to seating on either side of the rail and to facilitate the flow of users on the aisle. An intermediate handrail located 12 inches (305 mm) below the main handrail is required to prevent users from ducking under the handrail and hindering flow. Also, it provides a handrail for toddlers who may be using the aisle.

1028.14 Assembly guards. The code requires minimum 26-inch-high (660-mm) guards between aisles parallel to seats (cross aisles) and the adjacent floor or grade below where an elevation change of 30 inches (762 mm) or less occurs. An exception exists where the backs of seats on the front of the cross aisle project 24 inches (610 mm) or more above the adjacent floor of the aisle. See Figure 1028-6. Where the elevation change adjacent to a cross aisle exceeds 30 inches (762 mm), the general guard height requirements of Section 1013.2 shall apply.

The intent of this provision is to provide a certain degree of protection from falls that may occur while occupants are using a cross aisle adjacent to a drop-off. Even if the drop

Emergency Escape and Rescue

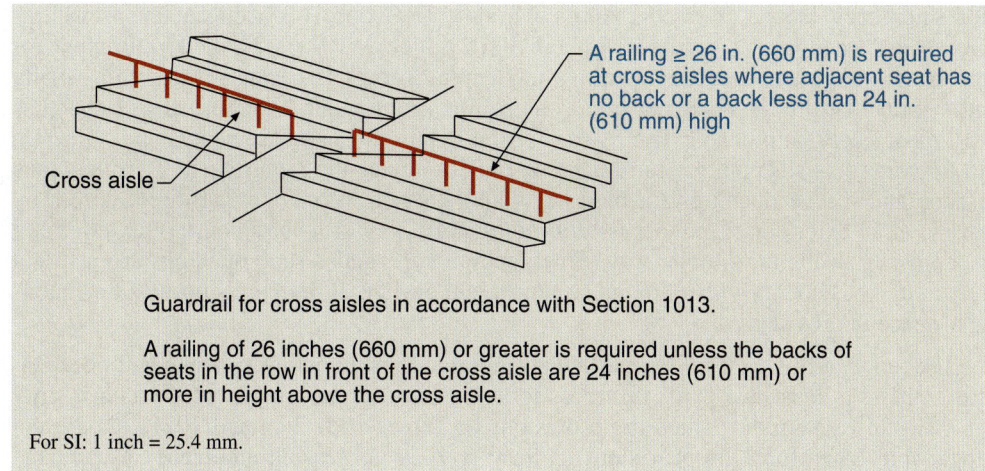

Figure 1028-6
Guards at cross aisles.

is minimal, the conditions of egress from an assembly use, particularly in low light, dictate the need for an increased level of safety. In addition, where the top of the seat backs are less than 24 inches (610 mm) above the aisle floor, an unintentional impact of the seat back could cause a fall over the seats.

In order to provide for proper viewing in auditoriums, theaters, and similar assembly uses where the floor or footboard elevation is more than 30 inches (762 mm) above the floor or grade below, a guard in front of the first row of fixed seats, and that is not at the end of an aisle, may be 26 inches (660 mm) in height. Under such conditions, a guard height of at least 36 inches (914 mm) high shall extend the full width of the aisle at the foot of the aisle. In addition, the top of the guard shall be located at least 42 inches (1,067 mm), measured diagonally, from the nosing of the nearest tread, as depicted in Figure 1028-7.

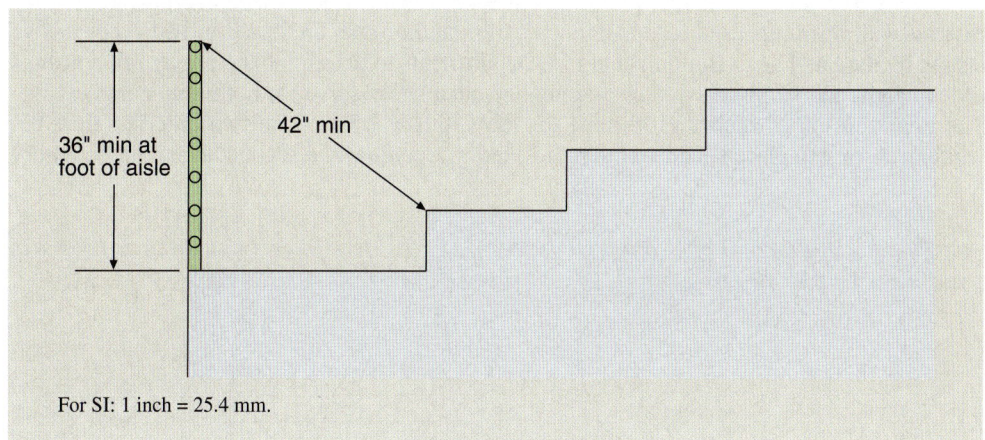

Figure 1028-7
Guard heights.

Section 1029 *Emergency Escape and Rescue*

General. Because so many fire deaths occur as the result of occupants of residential buildings being asleep at the time of a fire, the IBC selectively requires that basements and all sleeping rooms below the fourth story have windows or doors that may be used

1029.1

for emergency escape or rescue. Applicable only to Group R-3 occupancies, as well as Group R-2 occupancies regulated by Table 1021.2(1) or 1021.2(2), the requirement for emergency escape and egress openings in sleeping rooms is because a fire will usually have spread before the occupants are aware of the problem, and the normal exit channels will most likely be blocked. The reason for the requirement in basements is that they are so often used as sleeping rooms. An exception eliminates the requirement for emergency escape and rescue openings for basements and sleeping rooms having direct access by means of an exit door or exit access door to a public way or a yard, court, or exterior exit balcony that leads to a public way. Emergency escape and rescue openings are also not required in basements with a limited ceiling height or a small floor area, provided no habitable space is provided.

The scope of this section is of particular importance as it applies to Group R-2 occupancies. Where at least two exits, or access to at least two exits, are provided on each story of a Group R-2 building, then the provisions of Tables 1021.2(1) and 1021.2(2) are not applicable. Therefore, the provisions of Section 1029 addressing emergency escape and rescue openings also do not apply. However, where the allowances of Table 1021.2(1) or 1021.2(2) permitting a single means of egress are used, then the Group R-2 dwelling units must be provided with complying emergency escape and rescue openings. In those situations where, in multistory buildings, one or more stories may have access to two or more means of egress and there are other stories with access to only one exit, the requirements of this section would only be applied to those stories with access to just one exit.

The code intends that the openings required for emergency escape or rescue be located on the exterior of the building so that rescue can be affected from the exterior or, alternatively, so that the occupants may escape from that opening to the exterior of the building without having to travel through the building itself. Therefore, where openings are required, they shall open directly onto a public street, public alley, yard, or court. This provision ensures that continued egress can be accomplished after passing through the emergency escape and rescue opening.

1029.2 Minimum size. The dimensions prescribed in the code, and as illustrated in Figure 1029-1 for exterior wall openings used for emergency egress and rescue, are based in part on extensive testing by the San Diego Building and Fire Departments to determine the proper relationships of the height and width of window openings to adequately serve for both rescue and escape. The minimum of 20 inches (508 mm) for the width was based on two criteria—first, the width necessary to place a ladder within the window opening, and second, the width necessary

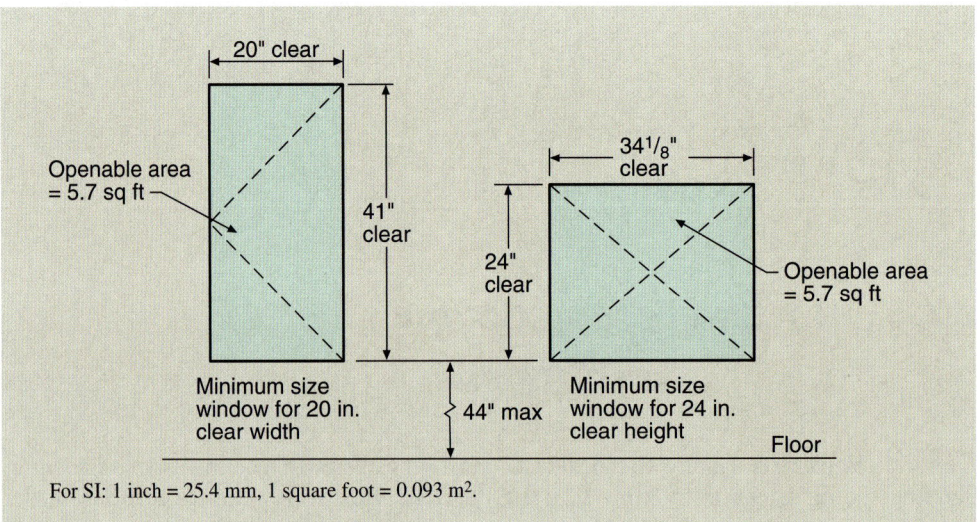

Figure 1029-1
Emergency escape and rescue window.

to admit a fire fighter with full rescue equipment. The minimum 24-inch (610-mm) height dimension was based on the minimum necessary to admit a fire fighter with full rescue equipment. By requiring a minimum net clear opening size of at least 5.7 square feet (0.53 m^2), the code ensures that an opening of adequate dimensions is provided. Where the opening occurs at grade level, the opening need only be 5 square feet (0.46 m^2) because of the increased ease of access from the exterior.

Maximum height from floor. In order to be relatively accessible from the interior of the sleeping room or basement, the emergency escape and rescue opening cannot be located more than 44 inches (1,118 mm) above the floor. The measurement is to be taken from the floor to the bottom of the clear opening. **1029.3**

Operational constraints. As stated in the code, these openings used for emergency escape or rescue must be operational from the inside of the room. Where windows are used, the intent is that they be of the usual double-hung, horizontal sliding, or casement windows operated by the turn of a crank. Although there is no specified limit on the type of window that may be used, Section 1029.2.1 requires that the net clear opening dimensions must be provided through normal operation of the window. The building official should evaluate special types of windows other than those just described based on the difficulty of operating or removing the windows. If no more effort is required than that required for the three types of windows just enumerated, they could be approved as meeting the intent of the code as long as no keys or tools are required. **1029.4**

The ever-increasing concern for security, particularly in residential buildings, has created a fairly large demand for security devices such as grilles, bars, and steel shutters. Unless properly designed and constructed, the security devices over bedroom windows can completely defeat the purpose of the emergency escape and rescue opening. Therefore, the IBC makes provisions for security devices, provided the release mechanism has been approved and is operable from the inside without the use of a key, tool, or force greater than that which is required for normal operation of the escape and rescue opening. Furthermore, in this case, the code requires that the building be equipped with smoke detectors in accordance with Section 907.2.11. Fire deaths have been attributed to the inability of the individual to escape from the building because the security bars prevented emergency escape.

The very essence of the requirement for emergency escape openings is that a person must be able to effect escape or be rescued in a short period of time because the fire will have spread to the point where all other exit routes are blocked. Thus, time cannot be wasted in figuring out means of opening rescue windows or obtaining egress through them. Therefore, any impediment to escape or rescue caused by security devices, inadequate window size, difficult operating mechanisms, and so on, is not permitted by the code.

Window wells. Window wells in front of emergency escape and rescue openings also have minimum size requirements. These provisions address those emergency escape windows that occur below grade. Obviously, just providing the standard emergency escape window criteria to these windows will get occupants through the window, but the window well may actually trap them against the building without providing for their escape from the window well or providing for fire-fighter ingress. **1029.5**

The minimum size requirements in cross section are similar in intent to the emergency escape and opening criteria; that is, to provide a nominal size to allow for the escape of occupants or ingress of fire fighters. See Figure 1029-2. The ladder or steps requirement is the main difference.

Emergency escape openings below the fourth story are not required to have an escape route down to grade; however, those openings below adjacent grade are so required. When the depth of a window well exceeds 44 inches (1,118 mm), a ladder or steps from the window well are required. The details for construction of steps are not identified in the provisions; however, the design of the ladder is specifically addressed. Rungs are to have a minimum interior width of 12 inches (305 mm), shall project at least 3 inches (76 mm)

10 Means of Egress

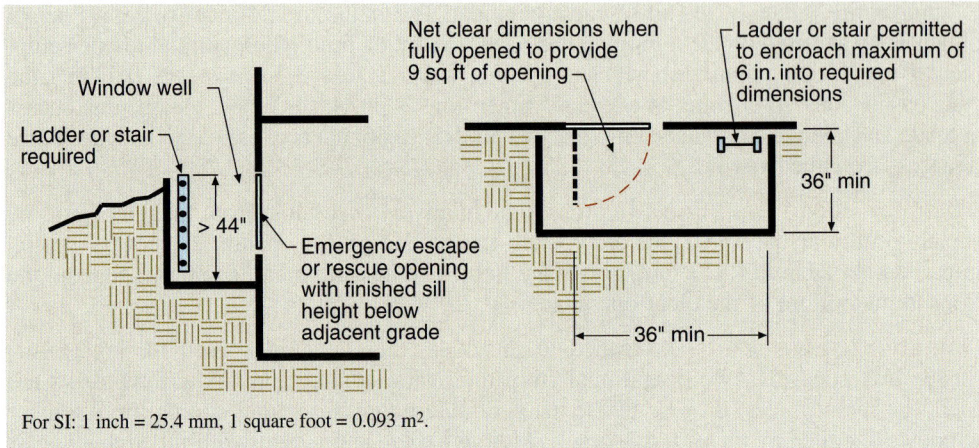

Figure 1029-2 Window wells.

from the wall, and shall be spaced no more than 18 inches (457 mm) on center vertically for the full height of the window well. Because ladders and steps in window wells are provided for emergency use only, they are not required to comply with the provisions for stairways found in Section 1009.

KEY POINTS

- The means of egress is an exiting system that begins at any occupied point in a building and continues until the safety of the public way is reached.
- Three distinct elements compose the means of egress—the exit access, the exit, and the exit discharge.
- Occupant load, the driving force behind the design of an exiting system, must be determined for the expected use or uses of a building.
- Components along the path of egress travel must be sized to accommodate the expected occupant load served by the components.
- Specific minimum component widths, such as those provided for doors, aisles, corridors, and stairways, often dictate the capacity of the means of egress.
- Where multiple complying exitways are provided, the calculated width may be dispersed among the various exits or exit-access doorways.
- With limited exceptions, the means of egress must have a minimum clear height of 7 feet 6 inches (2,286 mm) throughout the travel path.
- The code regulates the means of egress so that there is no change in elevation along the path of exit travel that is not readily apparent to persons seeking to exit under emergency conditions.
- As a general rule, exit signs are required from rooms or areas requiring access to two or more paths of exit travel.
- Requiring continuous illumination, exit signs must be provided with a secondary source of power.
- In those rooms or areas requiring access to at least two exitways, a second source of power is required for maintaining illumination to the exit path.
- Guards must be designed to reduce the probability of falls from one level to a lower level that exceeds 30 inches (762 mm) in elevation difference.

Key Points

- Guards are typically required to be at least 42 inches (1,067 mm) in height.
- In all public areas, guards must have limited openings to prevent individuals from falling or climbing through the guard assembly.
- In addition to providing proper access to and through a building, an accessible means of egress must be provided.
- Areas of refuge are mandated for certain buildings where stairs and/or elevators occur along the accessible means of egress.
- Doors are highly regulated in the IBC because of their potential for obstructing the means of egress.
- The use of revolving, overhead, and sliding doors for egress purposes is strictly limited.
- Doors swinging toward the direction of egress travel are mandated for all hazardous uses, as well as areas in other uses having an occupant load of 50 or more.
- Criteria for an acceptable latching or locking device on an egress door are very basic in that no key, special effort, or special knowledge is necessary to open the door.
- Where security issues are as important as those addressing fire and life safety, the IBC permits the installation of delayed egress locks.
- Panic hardware is mandated in Group A and E occupancies having an occupant load of 50 or more, as well as in all Group H occupancies.
- Gates located in the means of egress are regulated in a manner similar to doors.
- A stair is considered a change of elevation accomplished by one or more risers, whereas one or more flights of such stairs make up a stairway, along with any landings that connect to them.
- There are two different stairways addressed in the IBC—exit access stairways and interior exit stairways.
- Exit access stairways may be unenclosed where connecting no more than two stories.
- Treads and risers must be appropriately sized and uniform throughout the stair flight.
- Spiral stairways, curved stairways, and alternating tread devices are limited in their use because of the uniqueness of their configurations.
- Handrail design is regulated for height, size, shape, and continuity.
- Ramps must be designed for egress purposes as well as for accessibility.
- Exit access describes the vast majority of a building's floor area that provides the access necessary to reach a protected area (an exit).
- Access to at least two exits is typically required from floor levels above the first story, and rooms or areas having sizable occupant loads or excessive travel distance.
- Multiple exit paths must be arranged in order to minimize the risk of a single fire blocking all of the exit ways.
- Egress from a room through a nonhazardous accessory area is permitted, provided there is a discernible egress path that is direct and obvious.
- Travel distance is limited within the exit access portion of the means of egress; however, once an exit is reached, travel distance is no longer regulated.

10 Means of Egress

- In rooms where seating is at tables, additional limitations are placed upon the aisles and aisle accessways used for egress purposes.
- Corridors are intended to be used for circulation and egress purposes, and at times must be constructed as a protected element for use as a path for egress travel.
- The exit is the portion of the means of egress that provides a degree of occupant protection from fire, smoke, and gases.
- Horizontal exits, exit passageways, interior exit stairways and ramps, exterior exit doors at the level of exit discharge, exterior exit ramps, and exterior exit stairways are the exit components addressed in the IBC.
- Interior exit stairways are to be constructed of either 1-hour or 2-hour fire-resistance-rated construction with protected openings.
- Openings and penetrations into an interior exit stairway or ramp are strictly limited because of the hazards involved with vertical egress.
- An exit passageway is similar to a corridor but is built to a higher level and limited in much the same manner as a vertical exit enclosure.
- The concept of a horizontal exit is the creation of a refuge area to be used by occupants fleeing the area of fire origin.
- The use of an exterior exit stairway as a required means of egress element is limited to buildings not exceeding six stories or 75 feet (22,860 mm).
- In high-rise buildings, luminous egress path markings are required in exit enclosures and exit passageways of Group A, B, E, I, M, and R-1 occupancies.
- Egress travel outside of the building at grade level is considered exit discharge, continuing until the public way is reached.
- An egress court, open so that smoke and toxic gases will not accumulate, is an exit discharge component.
- Egress courts of limited width must be provided with a minimum level of fire protection in order to protect occupants as they pass through the egress court.
- Larger auditoriums, theaters, and similar assembly spaces are uniquely regulated as to the design of the egress system.
- In assembly spaces, the method for calculating aisle widths is modified where smoke-protected assembly seating is provided.
- Grandstands and bleachers, although similar in many aspects to typical assembly seating, are regulated by unique provisions found in ICC 300.
- Emergency escape and rescue openings are only required in Group R-3 occupancies and certain Group R-2 occupancies.

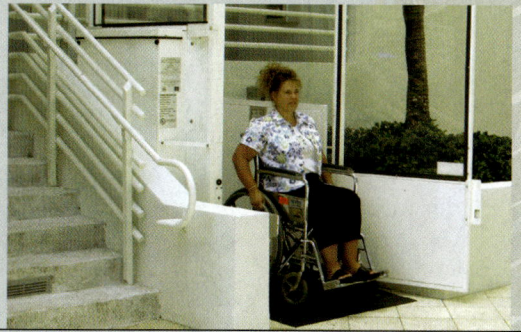

CHAPTER 11

ACCESSIBILITY

Section 1101 General
Section 1102 Definitions
Section 1103 Scoping Requirements
Section 1104 Accessible Route
Section 1105 Accessible Entrances
Section 1106 Parking and Passenger
            Loading Facilities
Section 1107 Dwelling Units and Sleeping Units
Section 1108 Special Occupancies
Section 1109 Other Features and Facilities
Section 1110 Signage
Key Points

11 Accessibility

This chapter addresses accessibility and usability of buildings and their elements for persons with physical disabilities. Where a facility is designed and constructed in accordance with this chapter and other related provisions throughout the *International Building Code®* (IBC®), it is considered accessible.

A historical perspective. In 1961, the American National Standards Institute (ANSI) published ANSI Standard A117.1. The President's Committee on Employment of the Handicapped and the National Easter Seal Society were designated as the secretariat for the standard. Since that time, a number of historic events have occurred that have brought accessibility and usability issues to the forefront of not only building code enforcement, but of society as a whole.

In the early 1970s, all three United States legacy model code groups approved code changes to make buildings more accessible and usable for people with disabilities. These independent developments resulted in confusion in the regulatory design and construction community. As a result, the Council of American Building Officials (CABO) requested that the Board for the Coordination of Model Codes (BCMC) review the regulations and suggest provisions to all of the model codes that would result in uniformity. In addition, ANSI requested that CABO become the secretariat for ANSI A117.1.

In October 1987, BCMC began its assignment to provide regulations that set forth when, where, and to what degree access must be provided (commonly referred to as scoping) for persons with disabilities. The ANSI A117.1-1986 standard contained design specifications intended to provide buildings and facilities accessible to and usable by people with disabilities but did not specify scoping provisions. Authorities who chose to employ ANSI A117.1 found it necessary to adopt amendments to establish when, where, and to what degree its provisions applied. During the BCMC work on accessibility, it became apparent that safe egress for people with disabilities was essential if access to buildings was to be increased. Therefore, the final BCMC report addressed both access and egress for people with disabilities. While BCMC was working on scoping provisions for ANSI A117.1-1986, the ANSI A117.1 committee continued to study revisions to their standard for public review.

In 1988, the United States Congress passed the Fair Housing Amendments Act to cover multifamily housing of four units or more on a site. On July 26, 1990, President George Bush signed the Americans with Disabilities Act (ADA), which set forth comprehensive civil rights protection to individuals with disabilities in the areas of employment, public accommodations, state and local government services, and telecommunications. One of the reasons legislators supported the ADA was the recognized inadequacy, limited application, and nonuniformity of existing protection for individuals with disabilities. One year later, on July 26, 1991, the United States Department of Justice (DOJ) issued its final rules, the Americans with Disabilities Act Accessibility Guidelines (ADAAG), that provided for access and usability for disabled persons in public accommodations and commercial facilities. Both acts were born from the Civil Rights Act of 1964. The ADA set forth statutory deadlines for when certain requirements became effective. One of these requirements was that new facilities designed and constructed for first occupancy after January 26, 1993, must be accessible.

The public review draft of revisions to the ANSI A117.1 standard dated January 24, 1992, was submitted to the DOJ with a request for technical assistance. A staff comparison of ANSI A117.1 and ADAAG yielded only a few areas in which ADAAG was deemed to provide greater accessibility. Generally, the differences found between ADAAG and ANSI A117.1 indicated that the ANSI standard provided for greater overall accessibility. At the BCMC meetings in May and June of 1992, the committee reviewed suggestions that would incorporate ADA guidelines into the ANSI A117.1 standard and other regulations. At the BCMC meeting on June 8, 1992, the committee finalized its report. From June 9 through 11, 1992, the ANSI A117.1 committee finalized its standard.

The final BCMC report of June 8, 1992, and the final draft of the ANSI A117.1 standard were soon adopted by all of the model code groups as their accessibility requirements. The final BCMC report of June 8, 1992, and the CABO/ANSI A117.1-1992 standard were submitted to the DOJ for a technical review. The resulting letter received in November 1995 described nine general problems, many of which were differences in philosophy. Noncode items such as laboratory equipment, automated teller machines, and telephones were also a concern of the DOJ. During the DOJ review, a joint task force was established to make recommendations for changes to both the CABO/ANSI A117.1, *Accessible and Usable Buildings and Facilities*, and ADAAG. This harmonization effort over many months resulted in suggested revisions to both documents. In an effort to reduce conflict, confusion, and frustration among all users, the results of the harmonization report were accepted in July 1996 by the ADAAG Review Advisory Board. That board presented its final report to the Architectural and Transportation Barriers Compliance Board, which in turn resulted in additional rulemaking by the Access Board.

The CABO/ANSI revisions were finalized and subsequently approved by ANSI's Board of Standards Review on February 13, 1998. Because CABO was incorporated into the International Code Council (ICC) in November 1997, the resulting standard was retitled ICC A117.1-1998. The 2009 edition of the standard is now referenced in the 2012 IBC, Chapter 11, for the design and construction of accessible buildings and facilities. The 2009 edition of ICC A117.1 was finalized in late 2009, and its technical provisions and format closely parallel those technical requirements in what at that time was called the proposed "new" ADA/ABA AG (Americans with Disabilities Act and Architectural Barriers Act Accessibility Guidelines). These new federal regulations were ultimately approved by DOJ and are now known as the *2010 ADA Standards for Accessible Design*.

These updated federal requirements were the first major rewrite of ADAAG since 1990 and were published in the Federal Register on July 23, 2004. The updates resulted in the reconciliation of most differences that occurred between the A117.1 standard and the original ADAAG. Although the coordination effort is not fully completed, it is expected that the efforts of all parties will result in a system that greatly enhances compliance with the recently released *2010 ADA Standards for Accessible Design*. These new federal standards became an allowable alternative to the original ADAAG in March 2011 and ultimately became mandatory on March 15, 2012, when they replaced the original ADAAG.

On another front, efforts were initiated addressing the differences between those provisions of the IBC, ICC A117.1, and the federal Fair Housing Accessibility Guidelines (FHAG). In 2000, the Department of Housing and Urban Development (HUD) reviewed the IBC and ICC A117.1-1998 for compliance with FHAG. Based on HUD's report, a series of modifications were proposed as part of the 2000 code change cycle. The proposed modifications were accepted by the voting members and were incorporated into the 2001 Supplement to the IBC. As a result of these changes, HUD issued a press release that stated that the 2000 IBC with the 2001 Supplement and ICC A117.1-1998 could be considered "safe harbor" for anyone wanting to comply with the FHAG requirements. The 2003 and 2006 editions of the IBC incorporated those provisions, along with other appropriate modifications, and they have both been recognized as a safe harbor in compliance with the Fair Housing requirements. At the time of this publication, HUD is reviewing the 2009 and 2012 IBCs to determine if they will also be granted "safe harbor" status. Given the previous approvals and the way HUD and ICC have worked together to coordinate the IBC, it is anticipated that the 2009 and 2012 editions of the IBC will also be determined to be equivalent when addressing the housing accessibility requirements.

11 Accessibility

Section 1101 *General*

1101.2 Design. This section adopts ICC A117.1, more specifically the 2009 edition, as the adopted design standard to be used to ensure that buildings and facilities are accessible to and usable by persons with disabilities. With this section providing the accessible design and construction standards for buildings, the remaining sections of the chapter provide the scoping provisions that set forth when, where, and to what degree access must be provided.

The importance of this section and the requirements it imposes should not be overlooked. As previously stated, buildings must be designed and constructed to the minimum provisions of Chapter 11, along with the other applicable provisions of the IBC and ICC A117.1, to be considered accessible. Therefore, prior to applying the code provisions for accessibility, it is important for the code user to review the technical requirements found in ICC A117.1. Although these items are not completely addressed in the IBC, they are an important part of making a building accessible. Such elements include, but are not limited to, space allowance and reach ranges, accessible route, protruding objects, and ramps.

Space requirements can vary greatly depending on the nature of the disability, the physical functions of the individual, and the skill or ability of the individual in using an assistive device. However, it is generally accepted that spaces designed to accommodate persons using wheelchairs will be functional for most people.

It is important to note that not all portions of ICC A117.1 are referenced by the IBC. For example, the criteria set forth in Section 504 for stairways as well as the stair handrail provisions of Section 505 have no application insofar as accessible stairways are not specifically scoped by IBC Chapter 11. As a result, the provisions of IBC Sections 1009 and 1012 solely regulate all stairways and associated handrails for accessibility purposes. As another example, accessible telephones and automatic teller machines are addressed in ICC A117.1, Sections 704 and 707, respectively. However, these elements are only regulated in the IBC by Appendix E, which must be specifically adopted to be in effect.

Section 1102 *Definitions*

It is important to note that the defined terms listed in this section are specifically for this code and may have different meanings from definitions in other accessibility provisions or regulations, such as ADAAG. Although the accessibility terms defined specifically in the IBC are shown in Section 1102, their definitions are all located in Section 202. Additional definitions that apply to these provisions can be found in Section 106 of ICC A117.1.

The term *accessible* takes on a very broad meaning by requiring compliance with Chapter 11. Space requirements must be addressed for all portions of the building to be considered accessible or to provide an accessible route within a building. Space requirements apply to adequate maneuvering space, clearance width for doors and corridors, and height clearances.

Location of controls, switches, and other forms of hardware becomes a function of forward and parallel reach ranges if they are to be considered accessible. To assist persons with limited dexterity, controls and other forms of hardware need to be operable without tight gripping, grasping, or twisting of the wrist.

People with visual impairments are provided with accessible routes, by the inclusion of provisions for clear and unobstructed routes that are free of protrusions created by benches, overhanging stairways, poles, posts, and low-hanging signs. Many hazards can be eliminated by using different textural surfaces in the accessible route to alert sight-impaired persons.

Visually impaired or partially sighted persons can be assisted with proper signage of the correct size, surface, and contrast. Signage is also provided for the hearing impaired. Directional signage should always be clear, concise, and appropriately placed.

An accessible route is defined in both the IBC and ICC A117.1, and is described in Section 402 of the latter document. Accessible routes have a number of components, such as walking surfaces with a slope not steeper than 1:20, marked crossings at vehicular ways, clear floor spaces at accessible elements, access aisles, ramps, curb ramps, and elevators. Each component must comply with specific applicable standards and requirements.

Several key elements to review when addressing accessible routes are the requirements related to protruding objects and ramps. An accessible route may contain a number of protruding objects that can affect its use. These objects include ordinary building elements such as telephones, water fountains, signs, directories, and automatic teller machines.

Protruding objects and other such obstructions are regulated by Section 307 of ICC A117.1. Similar provisions governing the means of egress system are located in IBC Section 1003.3. Ramps with proper slopes may serve as acceptable means of egress, as well as part of an accessible route. A sloped surface that is steeper than 1:20 is considered to be a ramp and is required to comply with both the requirements of Section 1010 for egress paths and Section 405 in ICC A117.1. To be considered acceptable, ramps shall have a slope not steeper than 1:12, with a maximum rise for any ramp not to exceed 30 inches (762 mm) and a minimum width of at least 36 inches (914 mm).

Three types of dwelling units and sleeping units are defined: Accessible, Type A, and Type B. The Accessible units are deemed to be fully accessible and must be in compliance with the IBC and Section 1002 of ICC A117.1. Type A units are considered dwelling units and sleeping units that are designed and constructed for accessibility in accordance with Section 1003 of ICC A117.1. Designed and constructed in accordance with Section 1004 of ICC A117.1, a Type B dwelling unit or sleeping unit is intended to be consistent with the technical requirements of HUD's Fair Housing Guidelines. A Type A unit is considered to provide a significant degree of accessibility, whereas a Type B unit provides for only a minimum level.

Several other definitions in Section 202 assist in clarifying the intent of the provisions. An employee work area is limited to those spaces used directly for work activities and does not include those common use areas such as toilet rooms and break rooms. Public use areas are identified as those spaces used by the general public and may be interior or exterior.

Section 1103 *Scoping Requirements*

In general, access to persons with physical disabilities is required for all buildings and structures, whether temporary or permanent. There are, however, certain conditions under which sites, buildings, facilities, and elements are exempt from the provisions where specifically addressed. Various modifications to the general requirements for accessibility are also found throughout other areas of Chapter 11. For example, an exception to Section 1104.4 eliminates the requirement for an accessible route of travel to levels having relatively small floor areas, applicable in all but a few types of uses. There are also a number of exceptions that apply only within dwelling units.

Where the building under consideration is existing, only the requirements found in Section 3411 apply. Intended to include historic buildings, the provisions apply to the maintenance, alteration, or change in use of an existing structure. An exception relating to

11 Accessibility

Type B dwelling units and sleeping units states that they need not be provided in existing buildings unless substantial alterations or a change of occupancy takes place.

In commercial applications, individual employee workstations are not required to be fully accessible. They must, however, comply with the appropriate provisions for visible alarms, accessible means of egress, and common use circulation paths. In addition, such work areas shall be located on an accessible route in order to provide access to, into, and out of the work area. Modifications to each individual workstation can then be made in order to address the specific needs of the employee. None of these limited provisions are applicable for small, elevated work areas where the area must be elevated because of the nature of the work performed. A number of other employee-use areas are also considered spaces that need not be provided with accessibility. Raised areas used for security or safety purposes, limited access spaces, equipment spaces, and single-occupant structures have been identified as those types of areas where it seems unreasonable, if not impossible, to provide full access.

Observation galleries, prison-guard towers, fire towers, lifeguard stands, and similar elevated observation areas need not be accessible or served by an accessible route. Ladders, catwalks, crawl spaces, freight elevators, very narrow passageways or tunnels, and any other space deemed to be nonoccupiable require no access. Spaces such as elevator pits and penthouses; mechanical, electrical, or communications equipment rooms; equipment catwalks; water or sewer treatment pump rooms and stations; electric substations in transformer vaults; and any other areas accessed solely by personnel for maintenance, repair, or monitoring of equipment are not required to be accessible. In addition, single-occupant structures where access occurs only by passageways elevated above grade or buried below grade are not required to be accessible. An example would be a toll booth accessed only by an underground tunnel, or a bank teller booth reached from an overhead enclosure. Accessibility is also not required to walk-in coolers and freezers, provided they are strictly employee access.

Section 1103.2.6 recognizes that some activities directly associated with construction projects will not be safe for persons with certain physical disabilities. Therefore, structures, sites, and equipment used or associated with construction are not required to be accessible. The limited scope of this provision is important insofar as the accessibility provisions generally do apply to the temporary buildings. Although accessibility would not be required to a construction trailer, it must be provided to sales trailers that are common in new subdivisions or multifamily projects. It is also necessary to know that where pedestrian protection is required by Chapter 33 and Table 3306.1, such walkways are required to be accessible.

For the most part, occupancies designated as Group U are exempt from the provisions of Chapter 11. One exception requires agricultural buildings that are associated with the general public to be provided with access to those public areas. An example might be a produce stand set up just inside the entry of the agricultural greenhouse. In addition, all paved work areas for agricultural buildings must be provided with access to such areas. A second exception mandates that where accessible parking is provided in private garages or carports, such parking structures shall be accessible.

A live/work unit, primarily regulated by Section 419, is considered to be a dwelling unit or sleeping unit in which a significant portion of the space includes a nonresidential use operated by the tenant. Although the entire unit is classified as a Group R-2 occupancy in the IBC, for accessibility purposes it is viewed more as a mixed-use condition. The residential portion of the unit is regulated differently for accessibility purposes than the nonresidential portion. The floor area of the dwelling unit or sleeping unit that is intended for residential use is regulated under the provisions of Section 1107.6.2 for Group R-2 occupancies. The requirements for an Accessible Type A or B unit would be applied based on the specific residential use of the unit and the number of units in the structure. The exceptions for Type A and B units set forth in Section 1107.7 would also exempt such units where applicable. In the nonresidential portion of the unit, full accessibility would be required based on the intended use. For example, if the nonresidential area of the unit is used for hair care services, all elements related to the service activity must be accessible. This would include site parking where provided, site- and building-accessible routes, the public entrance, and applicable service facilities. In essence, this portion of the live/work unit would be regulated in the same manner as a stand-alone commercial occupancy.

Section 1104 *Accessible Route*

An accessible route is defined as a continuous unobstructed path that complies with the provisions in Chapter 11. This route connects all accessible elements or spaces of the building or facility, including corridors, aisles, doorways, ramps, elevators, lifts, and clear floor space at fixtures. Code users should review the accessible route provisions in Chapter 4 of ICC A117.1. In addition, exterior portions of accessible routes must be evaluated and may include parking access aisles, curb ramps, crosswalks at vehicular ways, walks, ramps, and lifts.

Within a site. The clear intent of these provisions is to provide, on sites with single or multiple buildings or facilities, access to each accessible element from all parking areas, as well as from one accessible element to another on the same site. Where the only means of access between accessible facilities on a site is a vehicular way, an accessible route is not required between such facilities. Should a sidewalk, walking path, or similar circulation route be provided connecting the site elements, the route is to be designed and constructed as an accessible route. **1104.2**

Connected spaces. Those portions of a building that are required to be accessible must be connected by at least one accessible route of travel. Such route shall connect to all accessible entrances and lead to accessible walkways connecting other accessible site elements and potentially the public way. One exception clarifies that in assembly areas with fixed seating an accessible route must only be provided to the accessible seating areas. The other exception modifies the maneuvering clearance requirements of ICC A117.1 for doors to Group I-2 sleeping units. **1104.3**

Employee work areas. Although employee work areas are typically exempt from the accessibility provisions of the code, the circulation paths within such areas that are used by multiple employees are regulated. The paths must be designed and constructed as a complying accessible route unless exempted by one of the three exceptions. In those cases where an exception is applicable, it is still necessary to connect the work area to an accessible route such that physically disabled persons can approach, enter, and exit the area. For the purpose of applying any of the three exceptions, it is important to note that *common use* is defined as nonpublic areas shared by two or more individuals. An example is shown in Figure 1104-1. **1104.3.1**

Multilevel buildings and facilities. In addition to providing an accessible route to each portion of a building, each accessible level in a multistory facility must be connected via at least one accessible route of travel. The first exception waives the requirement for an accessible route to floors above and below any accessible level, provided such inaccessible levels have an aggregate floor area of not more than 3,000 square feet (278.7 m²). This allowance is not permitted, however, where the vertically inaccessible level contains offices of health-care providers, passenger transportation facilities, or multitenant sales facilities. **1104.4**

This exception eliminates the requirement for an elevator or other means of vertical access; however, it does not reduce or eliminate the obligation of compliance with other provisions for accessibility. As an example, a toilet room on an inaccessible level permitted

11 Accessibility

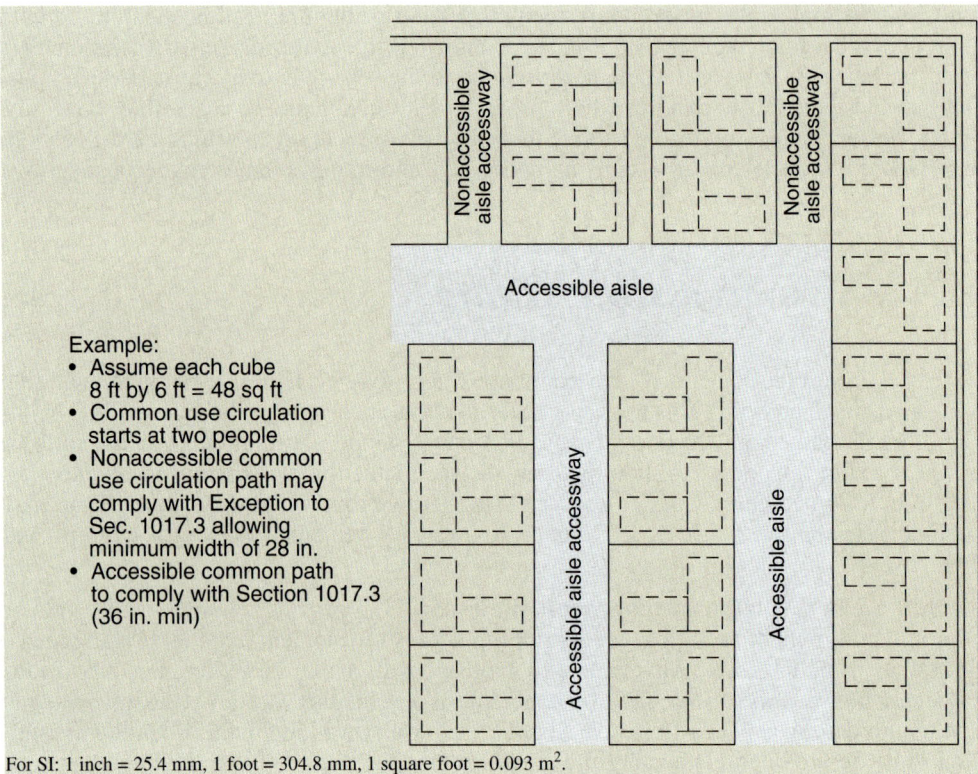

Figure 1104-1 Employee work areas.

by the exception would still be required to comply with all of the provisions for accessible toilet rooms. In addition, whereas the floor either above or below the accessible grade-level floor would not be required to meet the applicable accessible route provisions, any facilities located on these floors would also be required to be provided on the accessible floor. For example, if toilet facilities are located on the floor either above or below the accessible floor, then the same facilities are required on the accessible floor and shall be constructed as accessible facilities.

1104.5 Location. Where an interior route of travel is provided between floor levels within a building, any required accessible route provided between such levels shall also be interior. The intent of the provisions is to provide equal means of access. For example, an interior stairway and exterior ramp fails the equality test. The first exception applies solely to parking garages within and serving Type B dwelling units.

Section 1105 *Accessible Entrances*

In general, at least one public entrance, but not less than 60 percent of all such entrances to a building or individual tenant space, must be accessible.

In Figure 1105-1, two entrances (Doors A and B) are considered to be public entrances to the entire building. As such, both entrances must be accessible. If Doors C and D are public tenant entrances, they must be accessible in addition to the accessible entrances provided into each tenant space from the common lobby. One easy requirement to remember is that if there are only one or two entrances into a building or tenant, those entrances must be accessible. In addition to the general provisions for public entrances, the code mandates

Accessible Entrances 11

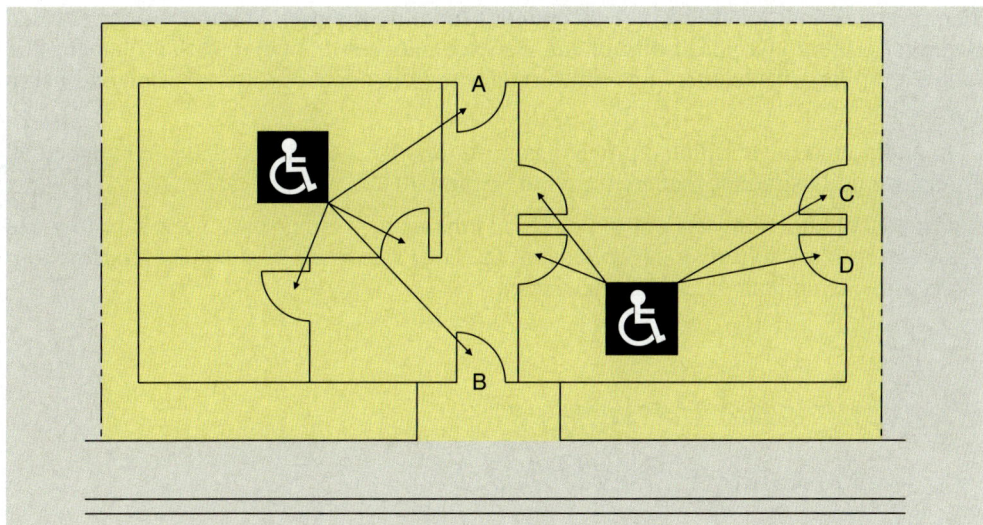

**Figure 1105-1
Accessible entrances.**

accessible entrances under other conditions, such as between a parking garage and the building served by the garage. Even though the minimum number of accessible entrances has been provided at other locations, the direct access between the parking garage and the building must be accessible. Exceptions are provided for entrances used exclusively for loading and service, as well as entrances to spaces not required to be accessible.

An important aspect of accessibility and accessible entrances is obviously the door assembly and its related components, including the threshold, hardware, closers, and opening force. Important requirements that affect doors and their accessibility are in IBC Section 1008 and in Section 404 of ICC A117.1.

Landings on both sides of doorways are also important accessibility features. When access for persons with disabilities is required, a floor or landing shall not be more than $1/2$ inch (12.7 mm) [$3/4$ inch (19.1 mm) for sliding doors] lower than the threshold of the doorway according to Section 1008.1.7 of the IBC. Where the level change or threshold exceeds $1/4$ inch (6.4 mm), it is required to be beveled at a slope of one unit vertical in two units horizontal (1:2) or less. Similar provisions are found in Sections 303 and 404.2.4 of the A117.1 standard.

Door hardware, including handles, pulls, latches, locks, or any other operating device on accessible doors, is required to be of a shape that is easy to grasp with one hand and does not require a tight pinching, tight grasping, or twisting of the wrist to operate. Many individuals have great difficulty operating door hardware that does not include push-type, U-shaped handles, or lever-operated mechanisms.

Door closers with delayed action capability are also important and allow a person more time to maneuver through a door. These closers are required to be adjusted so that from an open position of 90 degrees (1.57 rad), the time required to move the door to an open position of 12 degrees (0.21 rad) will be a minimum of 5 seconds. Door closers are required to have minimum closing forces in order to close and latch the door; however, for other than fire doors and exterior doors, maximum force levels are set to limit the force levels for pushing or pulling open doors. Opening forces and the methods used to measure them are specified in Section 404.2.8 of ICC A117.1.

Minimum maneuvering clearances at doors, other than those that are for automatic or power-assisted doors, are based on a combination of forward- and side-reach limitations, the direction of approach, and minimum clear width required for wheelchairs. They also permit enough space that a slight angle of approach can be gained, which then provides additional leverage or opening force by the user. Without these required clearances, there is a possibility of interference between the edge of the door and the footrest on the wheelchair. This could render the door inaccessible to someone using a wheelchair.

In addition to the traditional provisions for doors, ICC A117.1 contains many detailed provisions in illustrations that are found in Section 404.

Care should be exercised when considering the maneuvering space necessary to make doors accessible. An inadvertent reversal of the latch side to hinge side may render a door inaccessible to an individual in a wheelchair.

Section 1106 *Parking and Passenger Loading Facilities*

The number of accessible parking spaces required on a site varies by the total number of spaces provided. For other than specific residential and medical uses, Table 1106.1 is used as the basis for calculating the required number of spaces. Rehabilitation facilities, as well as those facilities providing outpatient physical therapy, require a larger percentage of accessible spaces than addressed in the table. Obviously, this is due to the much higher probability of individuals with a mobility impairment visiting the facility. On the other hand, the required number of accessible parking spaces for Groups R-2 and R-3 is typically reduced from Table 1106.1.

For every six accessible parking spaces, at least one of the spaces must be an accessible van space. As an example, consider a parking lot with 23 total parking spaces. According to Table 1106.1, at least one accessible parking space is to be provided, and it must be designed and constructed to be van accessible. As another example, a parking lot with 202 parking spaces must be provided with a minimum of seven accessible spaces. At least two of those seven spaces shall be van accessible.

An indicator of the critical nature of site development is the requirement for the shortest accessible route of travel. On a site with multiple buildings and a number of requirements for each, accessibility may pose complex site-design problems related to direct routes from parking areas. Where the parking facilities do not necessarily serve a particular building, the provisions require the accessible parking spaces to be located as closely as possible to the accessible pedestrian entrance to the parking facility. Early involvement or early attention to the location of multiple accessible elements during site development may tend to eliminate most, if not all, related problems.

Where multiple distinct parking facilities are provided on a site, such as a parking garage and a surface parking lot, the provisions of Section 1106.1 require that the total minimum required number of accessible parking spaces be determined individually. See Application Example 1106-1. However, Section 1106.6 allows accessible spaces to be relocated from remote lots to locations near accessible building entrances. This addresses a concern that in a large facility, the dispersion of accessible parking spaces into remote lots may result in decreased access for persons with disabilities.

If a passenger loading zone is provided, it shall have an adjacent access aisle that is part of the accessible route to the building. In accordance with ICC A117.1, the space shall have a vertical clearance of at least 98 inches (2,490 mm) at the zone and along the vehicle access route on the site. There are only two conditions under which the code mandates the installation of a passenger loading zone, where valet parking services are provided, and at accessible entrances of specified medical facilities.

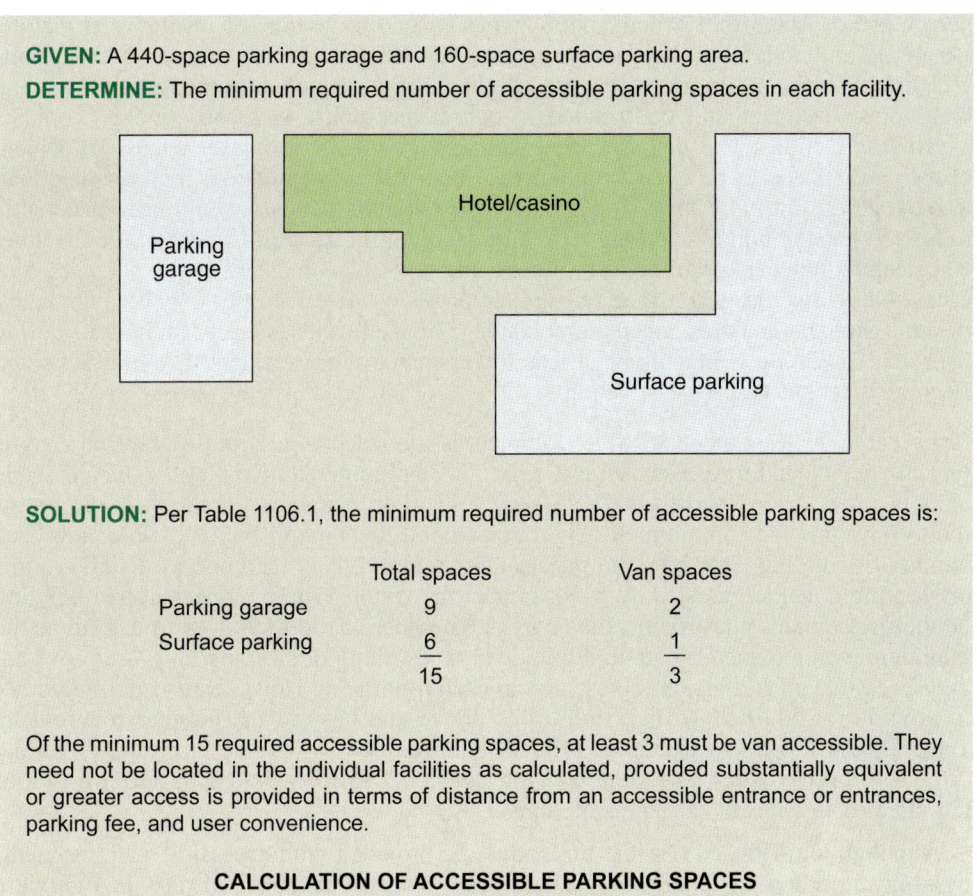

Application Example 1106-1

GIVEN: A 440-space parking garage and 160-space surface parking area.
DETERMINE: The minimum required number of accessible parking spaces in each facility.

SOLUTION: Per Table 1106.1, the minimum required number of accessible parking spaces is:

	Total spaces	Van spaces
Parking garage	9	2
Surface parking	6	1
	15	3

Of the minimum 15 required accessible parking spaces, at least 3 must be van accessible. They need not be located in the individual facilities as calculated, provided substantially equivalent or greater access is provided in terms of distance from an accessible entrance or entrances, parking fee, and user convenience.

CALCULATION OF ACCESSIBLE PARKING SPACES

Section 1107 *Dwelling Units and Sleeping Units*

This section is limited to the accessibility provisions related to dwelling units and sleeping units as defined in Chapter 2. It provides guidance as to the conditions under which Accessible units, Type A units, and Type B units are mandated. These scoping provisions generally indicate where some degree of accessibility is required, and to what extent.

Design. As addressed under the discussion of Section 1102, there are three types of dwelling and sleeping units that provide varying degrees of accessibility. Accessible units are provided with the most comprehensive accessibility requirement and are required to comply with those applicable provisions of Section 1002 of ICC A117.1. Type A and B units, regulated by ICC A117.1 Sections 1003 and 1004, respectively, not only provide a reasonable degree of accessibility, but also allow for the use of adaptive features. This section reflects how it is always acceptable to design and construct to a higher degree of accessibility than that required by the code.

1107.2

Group I. Dwelling units and sleeping units in a variety of Group I occupancies, including nursing homes, hospitals, nurseries, assisted-living facilities, group homes, and care facilities, are regulated for accessibility. In Group I-1 occupancies, at least 4 percent of the dwelling and sleeping units shall be Accessible units. In the Group I-2 category, however, the

1107.5

percentage of Accessible patient rooms varies based on the type of institutional facility. Hospitals and rehabilitation facilities that specialize in the treatment of conditions affecting mobility are required to have all patient rooms, including the toilet rooms and bathrooms, designed and constructed as Accessible units. General-purpose hospitals, psychiatric facilities, and detoxification facilities are required to have at least 10 percent of their patient rooms be Accessible units for their patient population. It is assumed that, in most cases, not more than 10 percent of the facility's patients would need accessible rooms. In nursing homes and long-term care facilities, at least 50 percent of the dwelling and sleeping units are required to be Accessible units. This increase in the percentage of accessible rooms recognizes that the care recipients of these facilities may be ambulatory on admission, but may become nonambulatory or have further mobility limitations during their stay. In Group I-3 facilities, at least two percent of the resident dwelling units and sleeping units must be Accessible units.

1107.6.1 Group R-1. Unless intended to be occupied as a residence, a Group R-1 occupancy is typically regulated for Accessible units only. The minimum required number of Accessible units is easily determined from Table 1107.6.1.1. For example, a hotel with 185 guestrooms must provide a minimum of eight guestrooms that comply as Accessible units. The number of buildings in which the guestrooms, referred to in the code as dwelling units or sleeping units, are located does not impact the result. Where more than one building on the site contains guestrooms, the aggregate number of rooms is used to determine the minimum requirements. Where multiple types of dwelling or sleeping units are provided, the Accessible units must be represented in each room type. However, it is not necessary to provide for additional Accessible units above and beyond the number required by Table 1107.6.1.1. As an example, a motel requiring two Accessible units and providing three room types (such as a double, king, and king suite) would only require two of the three room types to be Accessible units.

Although all of the Accessible units must be provided with accessible bathing facilities, only a portion are required to have roll-in showers. The table indicates the minimum number of Accessible units that must contain roll-in showers as described in Section 608 of ICC A117.1. The code mandates that such shower facilities be provided with a permanently mounted folding shower seat to address various disabilities. The minimum required number of units provided with roll-in showers is complemented by requiring a minimum required number of Accessible units without roll-in showers. The intent is to provide persons with physical disabilities a range of options equivalent to those available to other persons served by the facility. If the standard rooms have bathtubs, then some of the Accessible units should also be provided with bathtubs, and a small percentage of rooms should incorporate roll-in showers. Likewise, if all of the standard rooms have shower compartments, then most of the Accessible units should have transfer showers, with again a small percentage being provided with roll-in showers. It has been shown that accessible bathtubs are preferred by many people with mobility impairments for both security when sitting and the therapeutic relief from a warm bath.

1107.6.2 Group R-2. The provisions for Group R-2 occupancies are divided into two general categories: those applicable to typical apartment buildings and those applicable to dormitories, fraternity houses, and sorority houses. Apartment houses, along with monasteries and convents, are regulated for Type A and B units. On the other hand, dormitories and similar congregate living facilities do not need to contain Type A units, but do require one or more Accessible units.

The mandate for Type A units in apartment buildings applies where 21 or more dwelling units or sleeping units are contained within the building. If two or more buildings are located on the same site, the aggregate number of units is used to determine the minimum required number of Type A units. Assuming a site contains four apartment buildings, each containing 30 units, the provisions are based on the sum total of 120 units. Using the

Dwelling Units and Sleeping Units 11

2-percent rule, at least three of the dwelling units are required to be designed and constructed as Type A units. Assuming that throughout the four buildings three types of units (studio, one bedroom, and two bedroom) are represented, a minimum of one unit of each type shall be a Type A unit. All three Type A units are permitted in the same building, and all may be located on the same floor level, which would typically be at grade. Those units that are not required to be Type A must be designed and constructed as Type B units unless exempted by the provisions of Section 1107.7.

In dormitories, sorority houses, fraternity houses, and boarding houses, Accessible units are required in the same manner as Group R-1 occupancies. At least one of the dwelling or sleeping units within the building must be an Accessible unit, with additional Accessible units as required by Table 1107.6.1.1. As with other Group R-2 occupancies, the remaining units must be designed and constructed as Type B units unless allowed to be reduced by Section 1107.7.

General exceptions. The required number of Type A and B dwelling units and sleeping units required in Section 1107 may be reduced under the provisions of this section. However, there is no provision allowing the reduction or elimination of Accessible units mandated by Section 1107.6.1.1, 1107.6.2.2.1, or 1107.6.4.1. Five exceptions to these requirements are included in this section. The first four exceptions relate to buildings with limited or no elevator service and are relatively self-explanatory as worded in the code. The fifth exception relates to buildings that are required to have raised floors at the primary entrances as measured from the elevations of the vehicular and pedestrian arrival points in order to accommodate the base-flood elevation. Where no such arrival points are within 50 feet (15,240 mm) of the primary entrance, the closest arrival points shall be used.

1107.7

As noted, most of the exceptions are only applicable to buildings with no elevator service. It is expected that once an elevator is installed to access floors above the grade level, the additional measures to make the units more accessible are appropriate. However, the provisions do not mandate the installation of an elevator simply to make the upper floors accessible. Several of the applications of this section are shown in Figures 1107-1 through 1107-3.

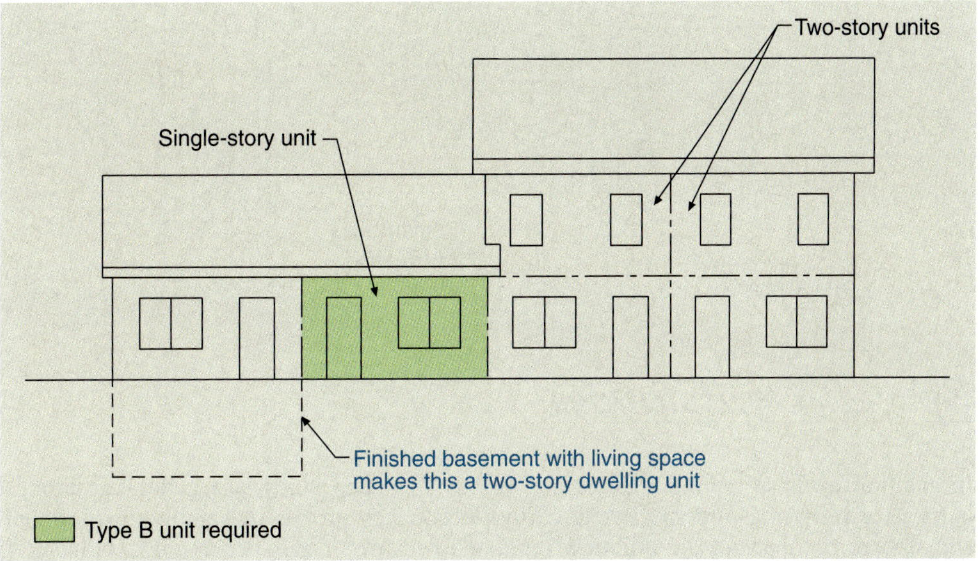

Figure 1107-1
One-story and multistory units.

11 Accessibility

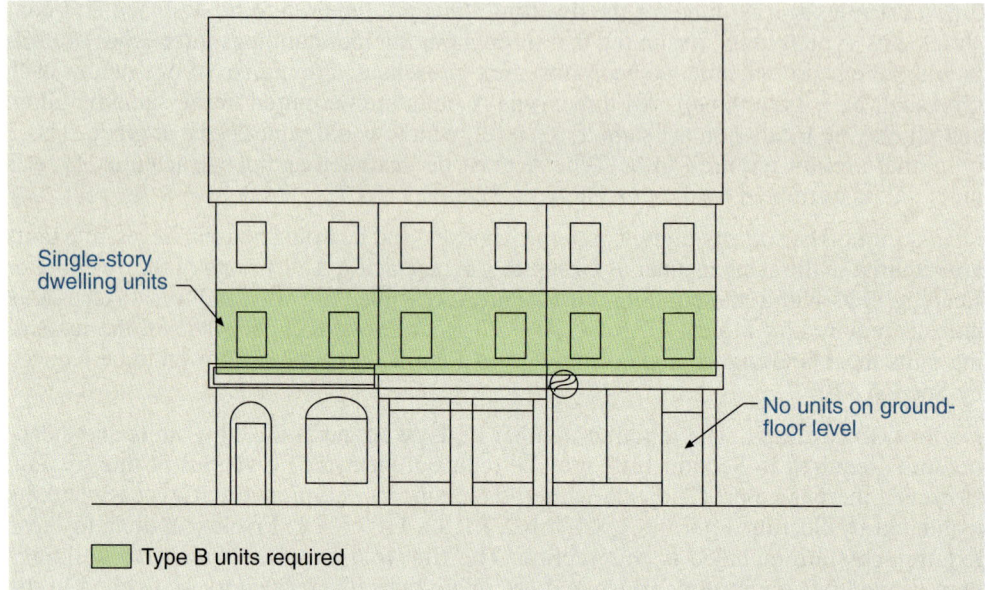

Figure 1107-2
One story of Type B units required.

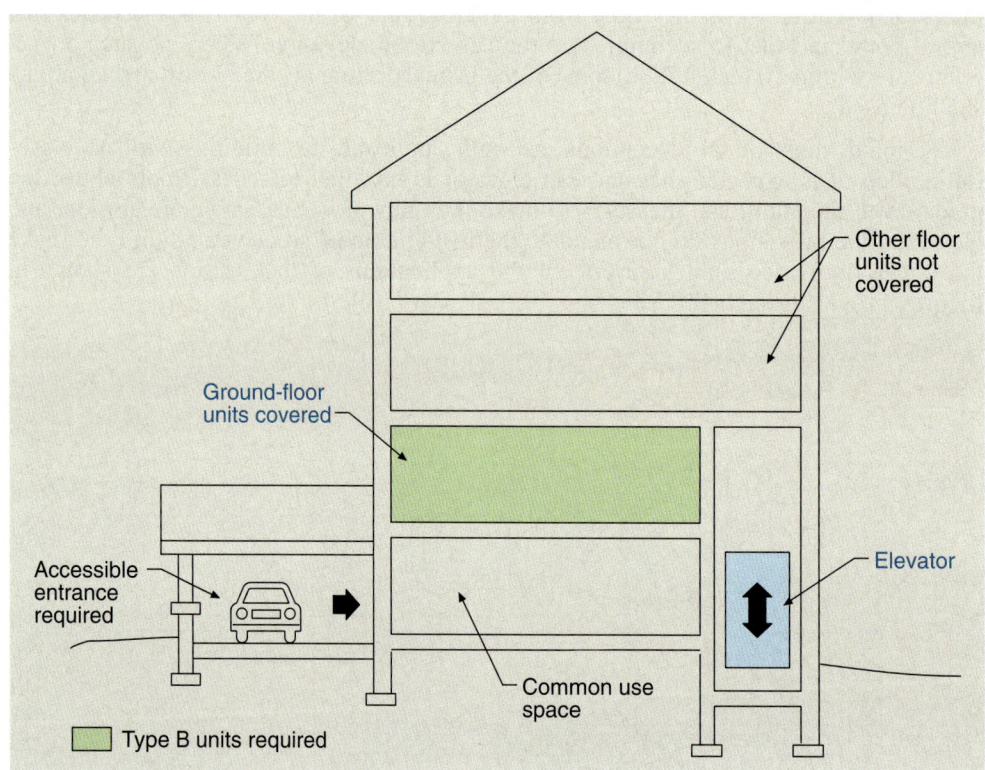

Figure 1107-3
Elevator service to lowest story with units.

Section 1108 *Special Occupancies*

In addition to those requirements that apply to all occupancies in a general fashion, a number of provisions are specific to certain occupancy groups. Assembly, institutional, and storage occupancies are uniquely regulated because of the special characteristics of their use.

Special Occupancies

1108.2 Assembly area seating. In stadiums, theaters, auditoriums, and similar assembly seating areas with fixed seats, the number of accessible wheelchair spaces to be provided is addressed in Table 1108.2.2.1. Unlike parking requirements where each lot is to be considered separately in the calculation of the minimum required number of accessible parking spaces, the minimum required number of wheelchair spaces for grandstands and bleachers serving a single function should be based on the aggregate number of seats. For example, the total number of seats provided in a high school football facility would be the basis for determining the minimum required number of wheelchair spaces even though the bleachers may be located on opposite sides of the playing field. However, the accessible spaces must be dispersed in an appropriate manner to accommodate individuals on both sides of the field.

The requirements for wheelchair locations can be found in Section 802 of ICC A117.1, including a provision mandating that at least one seat be provided beside each wheelchair space to allow for companion seating. The required dispersion of wheelchair spaces is comprehensively addressed in the 2009 edition of ICC A117.1. Wheelchair spaces should be an integral feature of any seating plan to provide individuals with physical disabilities with a choice of admission prices and a line of sight comparable to that provided to the general public. There are two exceptions to the general rule for the location of wheelchair spaces in multilevel facilities that permit all wheelchair spaces to be located on the main level where the second-floor or mezzanine level is limited in capacity.

1108.2.5 Designated aisle seats. The intent of providing designated aisle seats is to permit individuals who might find it difficult to negotiate an aisle accessway the opportunity to use a seat directly adjacent to the aisle. The requirements of ICC A117.1 call for signs to identify the designated aisle seats. In addition, where armrests are installed, they must be retractable or folding on the aisle side of the seat.

1108.2.7 Assistive listening systems. Assistive listening systems are required to be installed where audible communications are a necessary part of the assembly room's use. If the type of assembly use does not necessitate the installation of an audio-amplification system, an assistive listening system is not required. The number of assistive listening receivers is based on the total seating capacity of the assembly area and determined by Table 1108.2.7.1. As an example, an 800-seat theater would require at least 30 receivers, of which at least eight must be hearing-aid compatible, as shown in the following calculation:

20 receivers + 1 receiver [(800 seats − 500 seats)/33 seats per receiver] = 29.1 = 30 receivers.

25% of 30 receivers = 7.5 = 8 hearing-aid compatible receivers.

These systems are intended to augment a standard public address or other audio system by providing signals that are free of background noise to individuals who use special receivers or their own hearing aids. Further provisions found in Section 706 of ICC A117.1 require that individual fixed seating served by an assistive listening system be located to provide a complete view of the stage, playing area, or cinema screen. AM and FM radio-frequency systems, infrared systems, induction loops, hard-wired earphones, and other equivalent devices are all permitted.

1108.2.9 Dining and drinking areas. The general rule for accessibility in dining rooms is that the total floor area be accessible. This requirement includes both interior dining/drinking spaces as well as those that are outdoors. The first exception to this section eliminates the requirement for an elevator or ramp system to a mezzanine seating area in a dining room under strict conditions. Caution should be exercised when determining the same services mentioned in this exception. It is generally accepted that the same services not only address actual food and drink service but also include decor, views and ambiance, and so on, and it is equally important that a specific accessible space or area not be set aside and restricted for use only by people with disabilities.

For spaces at accessible fixed tables or counters, at least one table in the facility shall be provided for wheelchairs. When more than one is required by this section, then they shall be equally distributed around the facility so as not to isolate an accessible area from the rest of the establishment.

1108.3 Self-service storage facilities. The number of individual storage spaces that must be accessible at self-service storage facilities is determined from Table 1108.3. Assuming a total of 280 spaces in the facility, at least 12 accessible storage spaces must be provided as shown:

10 spaces + [2% (280 spaces −200 spaces)] = 11.6 = 12 storage spaces.

Where different sizes or classes of storage space are available, the individual accessible spaces shall be appropriately dispersed among the sizes or classes available, but only up to the total number of accessible spaces that are required. All accessible storage spaces complying with this section are permitted to be located in a single building.

1108.4 Judicial facilities. Courtrooms, holding cells, and visitation areas of judicial facilities are required to be accessible to the degree mandated by this section. Courtrooms typically have elements unique to their use, such as witness stands, jury boxes, judge's benches, and similar areas. Provisions are in place to provide for a limited degree of accessibility to such areas. The spectator area of a courtroom is regulated under the provisions for assembly seating. Workstations are regulated differently based on their use. Employee workstations, such as the judge's bench, bailiff's station, and court reporter's station, need to be on an accessible route, but the portion of the route leading to such elevated work areas is not required at the time of construction. It is only necessary to provide adequate space and support such that a complying ramp, platform lift, or elevator can be installed in the future without extensive reconstruction. Workstations for other than employees, such as the litigant's station, counsel station, and lectern, must provide full accessibility.

Section 1109 Other Features and Facilities

This section provides scoping provisions to ensure that certain elements or areas within buildings, where specific services are provided or activities are performed, are made accessible. Certain components in ICC A117.1 or other accessibility regulations have not been included in this section. Items such as telephones, automatic-teller machines, and fare machines are referenced in Appendix E, insofar as these are not typically considered items regulated by the building code.

1109.2 Toilet and bathing facilities. As a general rule, all toilet rooms and bathing facilities shall be accessible. The term *accessible* applies to the doors, fixtures, clear floor space, operable parts, towel dispensers, and mirrors, among other elements. Typically, the main components of toilet facilities are the water closet, the toilet stall, and the lavatory; and the main accessibility issues are the clear space, door swing, transfer capability, height of fixtures, grab bars, and controls. In bathing facilities, a number of items related to the shower or bathtub and its location, controls, and grab bars need to be considered. The ICC A117.1 standard contains details and many illustrations that depict these requirements.

A number of exceptions revise, reduce, or eliminate the requirements for toilet rooms and bathing facilities under specific conditions. Within the facility, at least one of each type of fixture, element, control, or dispenser must be accessible. There is an allowance for a nonaccessible urinal where only a single urinal is located within a toilet room.

Other Features and Facilities 11

Exception 6 permits toilet facilities to be designed using the children's size provisions of the A117.1 standard and to still be considered as being accessible. In facilities or portions of buildings that are primarily designed for children's use, the general accessibility requirements may result in the elements not being usable by the major portion of the occupants. This exception provides the "scoping" requirement that will allow designers to use the children's height requirements of the A117.1 standard when determining the accessibility requirements for areas that are primarily for children's use. The A117.1 standard defines *children's use* as "spaces and elements specifically designed for use primarily by people 12 years old and younger." Examples of where the adults are in the minority and the space is "primarily for children's use" include most areas of an elementary school, preschool, or kindergarten; a children's library; or a children's museum. Adult-dimensioned fixtures should be provided in other areas or spaces where there is a mix of all ages or where the space serves staff, parents, older students, or the general public. Therefore, if a restroom is provided in the staff area of an elementary school, that toilet room should meet the adult requirements and not attempt to use the reduced-size children's provisions in that location. So even though the code permits children's-sized accessible elements to substitute for accessible adult-sized provisions, the intent is to clearly limit the application to the areas that the children occupy and the fixtures that they would be using.

Family or assisted-use toilet and bathing rooms. Specific provisions are provided in this section on accessibility requirements for family or assisted-use bathing and toilet rooms. Such bathing and toilet rooms are required to comply with this section and ICC A117.1. The primary issue relative to family or assisted-use toilet/bathing facilities is that some people with disabilities may require the assistance of persons of the opposite sex and, therefore, require a toilet or bathing facility that accommodates both persons. **1109.2.1**

Bathing facility requirements for recreational facilities require that an accessible family or assisted-use bathing room be provided where separate-sex bathing facilities are provided. If each separate-sex bathing facility has only one shower fixture, family or assisted-use bathing facilities will not be required. Accessible family or assisted-use toilet room requirements for Group A and M occupancies are mandated by this section where an aggregate of six or more male and female water closets are required. These occupancies generally have high occupant loads with a minimum stay of approximately 1 hour for the occupants; thus, a high probability exists that there will be occupants who will need the use of such facilities. In determining the total number of fixtures required in a building as mandated by Chapter 29, those fixtures located in family or assisted-use facilities may be included in the total fixture count.

Family or assisted-use bathing rooms are to be provided with a water closet and lavatory in addition to the single shower or bathtub fixture. Family or assisted-use toilet rooms are limited to a single water closet, lavatory, and optional urinal. A complying family or assisted-use bathing room may be considered the family or assisted-use toilet room. In order to provide the appropriate privacy for a family or assisted-use facility, doors to a family or assisted-use toilet room or bathing room must be capable of being secured from the inside.

Family or assisted-use toilet and bathing rooms shall be located on an accessible route. Family or assisted-use toilet rooms are to be located not more than one story above or below separate-sex toilet facilities. The accessible route from any separate-sex toilet room to a family or assisted-use toilet room must not exceed 500 feet (152,400 mm). Additionally, in passenger transportation facilities and airports, the accessible route from separate-sex toilet facilities to a family or assisted-use toilet room shall not pass through security checkpoints. The restriction regarding crossing through security checkpoints at airports and similar facilities is intended to eliminate any potential delays that may cause missed flights and/or connections.

Water closet compartment. At least one wheelchair-accessible compartment shall be provided in all toilet rooms or bathing facilities where compartments are installed. Also, an ambulatory-accessible water closet compartment must be provided in addition to the **1109.2.2**

wheelchair-accessible compartment in those toilet rooms having an aggregate total of six or more water closet compartments and urinals. The ambulatory-accessible compartment benefits those individuals who, although not wheelchair users, have physical limitations or impairments that make it difficult to use other types of toilet compartments. Both wheelchair-accessible and ambulatory-accessible compartments are to be in compliance with ICC A117.1. It should be noted that the threshold for this provision is based on a room-by-room evaluation, not the aggregate number of fixtures throughout the facility.

1109.2.3 Lavatories. The provisions specific to lavatories are a subsection to the other provisions for toilet and bathing facilities. The requirement for a minimum of 5 percent of the lavatories to be accessible results in a single mandated accessible lavatory in almost every toilet room and bathing room. Only where more than 20 lavatories are provided must additional accessible lavatories be provided.

Additional accessibility features are required where the total number of lavatories in a toilet room or bathing room exceeds five. Where six or more lavatories are provided, a minimum of one lavatory must be provided with enhanced reach ranges. It is permissible for the lavatory with the enhanced reach range to serve as the required accessible lavatory. The technical requirements for such lavatories with enhanced reach ranges are found in ICC A117.1 Section 606.5. These types of lavatories are usable by individuals who have a limited obstructed reach depth. Such individuals can often only reach faucets and soap dispensers up to a reach depth of 11 inches (279 mm) in lavatories with a height of 34 inches (864 mm). The maximum 11-inch (279-mm) depth is possible by locating the faucet controls to the side of the bowl while leaving the spout toward the back, mounting the faucet on a sidewall, installing the faucet on the side of the bowl, or other potential locations that will provide the necessary access.

1109.5 Drinking fountains. This section ensures that all of the drinking fountains provided are made accessible. This is another of the many provisions where it is important to focus on the word *provided* and note that the provision does not require drinking fountains to be installed.

Drinking fountains must be accessible based on use from a wheelchair while still providing access to people with a limited ability to bend or stoop. The technical provisions set forth in Section 602 of ICC A117.1 address drinking fountains designed for wheelchair access, as well as those intended for use by standing persons. Other than the requirement for clear floor space, the provisions of Section 602 are applicable to both heights of fountains. Where a single water fountain (combination *hi-lo* unit) can comply with the requirement for both user groups, it is permitted to be used in lieu of two individual fountains.

The minimum required number of drinking fountains is established by the *International Plumbing Code*® (*IPC*®). At times, drinking fountains are installed even though they are not required by the code. It is also not uncommon for the actual number of fountains provided to exceed the minimum required. Under such conditions, one-half of the number provided must comply with the provisions of ICC A117.1 for wheelchair users, and the remainder shall accommodate standing persons. If an odd number of fountains are installed, an additional drinking fountain is not required in order to meet the 50-percent criteria. The remaining water fountain after the 50/50 split is permitted to be of either type. For example, if three drinking fountains are provided on the first floor of a building, at least one, but not more than two, is to be wheelchair accessible.

Exception 2 to Sections 1109.5.1 and 1109.5.2 addresses drinking fountains used primarily by children. The current height of 38 to 43 inches (965 mm to 1,090 mm) specified in the A117.1 for standing-height drinking fountains is too high for small children. The exceptions override the standard and will allow the height to be reduced to 30 inches (762 mm).

1109.8 Lifts. The provisions of this section address the limited acceptance of platform (wheelchair) lifts as a portion of an accessible route. To use platform lifts in lieu of an elevator or ramp, they must comply with Section 408 of ICC A117.1 and be installed in one of the 10 listed locations. A companion provision, Section 1007.5, allows platform lifts to be used as an accessible means of egress for the first nine locations established in this section. However, the platform lift cannot be used as a portion of an accessible means of egress where the lift is used along the accessible route because of existing exterior site constraints as described in Item 10. Because of the slow operation of a platform lift, it is considered inappropriate for egress purposes where there is the potential for a large number of lift users.

1109.10 Detectable warnings. Detectable warnings must be provided on edges of passenger transit platforms where a drop-off occurs, unless platform screens or guards are present. Not applicable to bus stops, detectable warnings should be standardized to assist in the universal recognition and reaction of persons using these platforms. Detectable and tactile warning devices also serve guide dogs and should contrast visually with the adjoining surfaces in a light-on-dark or dark-on-light application.

1109.12 Service facilities. This section provides nominal accessibility provisions that are intended to allow persons with disabilities to use these facilities without much assistance. Other features commonly found in these uses would have to comply with the other provisions in this chapter.

1109.13 Controls, operating mechanisms, and hardware. To be able to operate controls, operating mechanisms, and hardware, a clear floor space and reach ranges for either forward or parallel approaches must be provided. This section also requires that, in other than kitchens or bathrooms, an accessible operable window be provided in those rooms in Accessible units and Type A units where operable windows are located.

1109.15 Recreational and sports facilities. Where recreational facilities are provided serving Type A or B units in Group R-2 or R-3 occupancies, at least 25 percent but not less than one of each type of such facilities shall be accessible. All recreational facilities of each type on a site shall be considered to determine the total number of each type that is required to be accessible. This requirement recognizes that not all recreational facilities need to be accessible, nor would such a requirement be feasible. However, at least one of every four recreational facilities shall be available to someone with a disability. Where multiple residential buildings are on a site, each type of recreational facility, such as a basketball court, a handball court, a weight room, an exercise area, a game room, or a television room serving each building, must be fully accessible. Each type of recreational facility on a site, such as a racquetball court in an apartment complex, is considered to determine the total number required to be accessible. Thus, if there were 12 racquetball courts on the site, at least three of the courts would need to be accessible. If they were evenly distributed in two separate buildings, one building would need to contain at least one accessible court, and another would need at least two accessible courts to meet the overall 25-percent rule. If the courts were located in four separate buildings, each building would require at least one accessible facility, even if the total exceeded the overall 25-percent requirement.

Section 1110 *Signage*

This section lists those required accessible elements and locations that must be identified with appropriate signage. Elements such as accessible parking spaces and loading zones, accessible areas of refuge, accessible dressing rooms, and family or assisted-use toilet rooms shall be provided with the International Symbol of Accessibility, as shown in Figure 1110-1. Directional signage, including the International Symbol of Accessibility, is to be used to indicate the nearest route to a like accessible element.

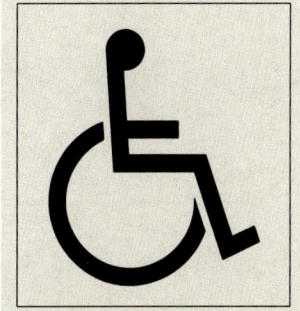

Figure 1110-1 **International symbol of accessibility.**

11 Accessibility

Directional signage must be provided at inaccessible building entrances, inaccessible public toilets and bathing facilities, elevators not serving an accessible route, separate-sex toilet and bathing facilities (to indicate the location of the nearest family or assisted-use facility), and exits and exit stairways serving a space required to be accessible but not providing an accessible means of egress. Additional signage will be needed where assistive-listening systems are available for assembly occupancies: at doors to egress stairways, exit passageways, and exit discharge; at areas of refuge; at areas of rescue assistance; and at two-way communication systems.

> **KEY POINTS**
> - Chapter 11 of the IBC provides the scoping provisions for accessible spaces and elements within buildings.
> - ICC A117.1-2009 is the referenced standard for regulating the facilities and elements of buildings that are required to be accessible by IBC Chapter 11.
> - Virtually all buildings, both temporary and permanent, must be designed and constructed for accessibility and usability.
> - Accessible routes must be provided to connect all accessible elements within a building, as well as all accessible elements on the site.
> - At least 60 percent of all public building entrances must be accessible.
> - The number of accessible parking spaces required on a site varies by the total number of spaces provided.
> - All occupancies require some degree of accessibility, with special provisions for assembly, institutional, residential, and utility uses.
> - Assembly areas are further regulated for wheelchair spaces and assistive-listening systems.
> - Both Type A and B dwelling units may be required in a large apartment building.
> - A variety of building elements are regulated as facilities required to be accessible, including toilet rooms, bathing facilities, drinking fountains, elevators, stairs, platform lifts, fixed or built-in seating, storage, customer-service facilities, controls, operating mechanisms, and alarms.
> - Under specific conditions, accessible family or assisted-use bathing rooms and toilet rooms must be provided in addition to the accessible separate-sex facilities.
> - Appropriate signage is mandated in order to properly identify the various accessible elements.

CHAPTER 12

INTERIOR ENVIRONMENT

Section 1203 Ventilation
Section 1204 Temperature Control
Section 1205 Lighting
Section 1206 Yards or Courts
Section 1207 Sound Transmission
Section 1208 Interior Space Dimensions
Section 1209 Access to Unoccupied Spaces
Section 1210 Toilet and Bathroom Requirements
Key Points

12 Interior Environment

> This chapter is designed to address those issues related to the interior environment aspects of a building's use, such as ventilation, lighting, temperature control, yards and courts, sound transmission, and room dimensions.

Section 1203 *Ventilation*

Ventilation in buildings is regulated based on the ventilating method used. This section addresses the use of natural ventilation, whereas the use of mechanical ventilation is regulated by the *International Mechanical Code®* (IMC®).

1203.2 Attic spaces. During cold weather, condensation is deposited on cold surfaces when, for example, warm, moist air rising from the interior of the building and through the attic comes in contact with the roof deck. This alternative wetting and drying that is due to condensation creates dry rot in the wood, and preventive measures are required. In attic areas of noncombustible construction, it is also important to ventilate the area, particularly in light-gauge steel construction. Therefore, enclosed attics and enclosed rafter spaces formed where ceilings are applied directly to the underside of roof-framing members, such as in cathedral ceiling applications, are to have cross ventilation for each separate space. Ventilation of the attic prevents moisture condensation on the cold surfaces and, therefore, will prevent dry rot on the bottom surfaces of shingles or wood roof decks. Figure 1203-1 provides three examples of attic ventilation. In the areas where the moisture condensation is a particular problem or where the normal requirement for attic ventilation cannot be provided, ventilation of the attic by mechanical exhaust fans may be required. Exhaust fans are particularly beneficial in all cases because of the extra movement of the air provided.

The method and arrangement of providing ventilated openings is an important aspect in the proper ventilation of attic spaces. It is critical that any such openings be protected against the entrance of rain and snow. In addition, blocking and bridging that is installed must be located so as to not interfere with the movement of air. At least 1 inch (25 mm) of air space must be provided between the insulation and the roof sheathing, as shown in Figure 1203-2. The net free ventilating area must be at least 1/150 of the area of the space ventilated. Two exceptions permit a reduction in the amount of ventilating area to 1/300 of the attic area being ventilated, provided specific criteria are met. Exception 1 is illustrated in Figure 1203-3. A third exception allows the building official to determine that attic ventilation is not required due to atmospheric or climatic conditions.

Something often overlooked when sizing attic vents is that the code requires that the area provided be the net free area. The net free area can be as much as 50 percent less than the gross area. For example, one manufacturer's 24-inch (610-mm) square gable vent [gross area equals 576 square inches (0.37 m^2)] is listed in their catalog as having a net free area of 308 square inches (0.20 m^2), which is about 53 percent of the gross area. The manufacturer's literature for the specific vents being used needs to be consulted in order to obtain accurate free area information.

1203.2.1 Openings into attic. Exterior ventilation openings are required by the code to be screened in order to prevent entry of birds, squirrels, rodents, and other similar creatures. A mesh size between 1/16 inch (1.6 mm) and 1/4 inch (6.4 mm) is required to address the problems of both smaller openings being blocked by debris and spider webs, and larger openings permitting access to small rodents. In addition to the use of corrosion-resistant-wire cloth screening, it is also permissible to use hardware cloth, perforated vinyl, or any other similar material that will prevent unwanted entry. A cross-reference is also provided to IMC Chapter 7 to remind users that there are special requirements where combustion air is obtained from the attic area.

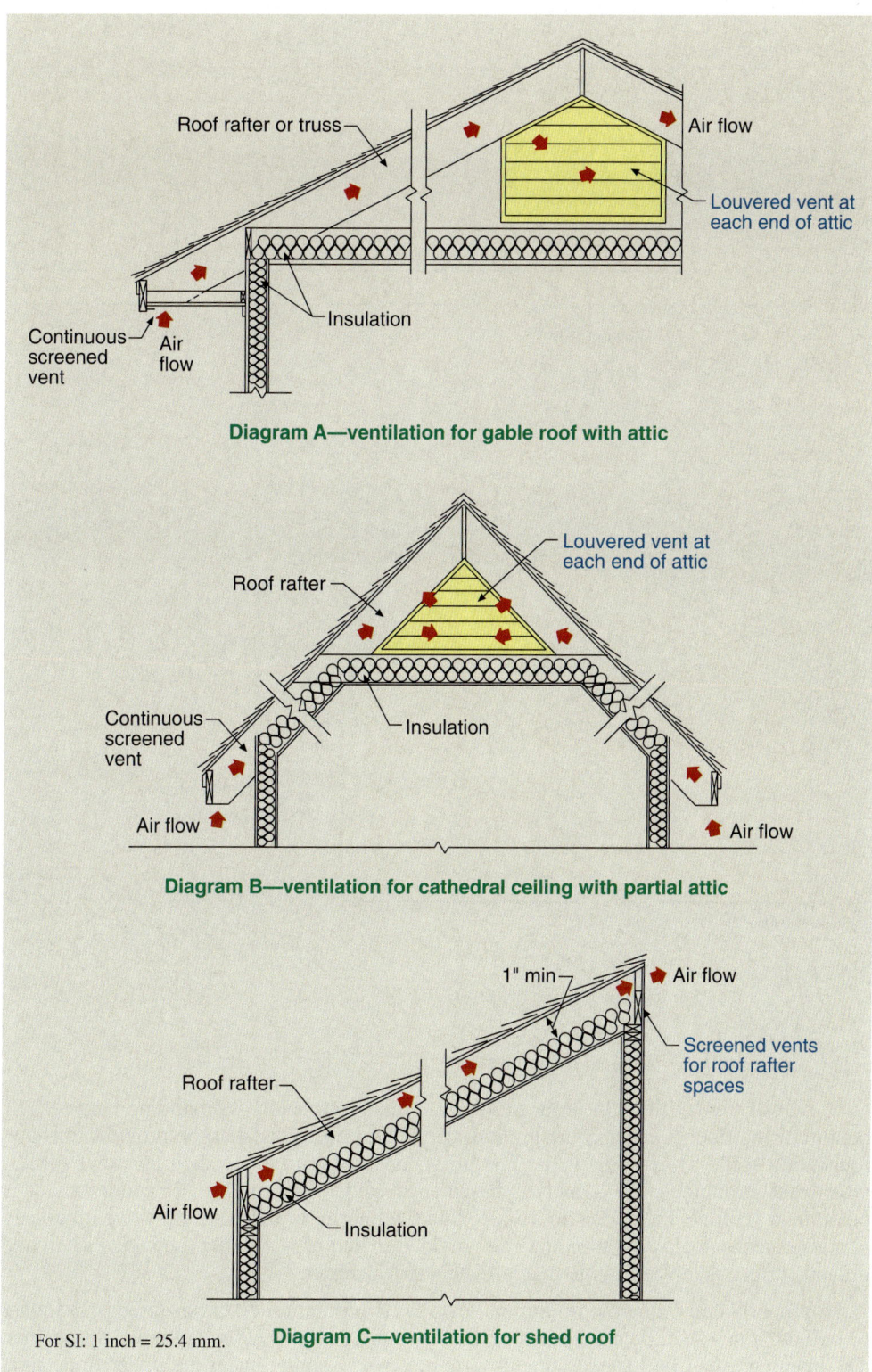

Figure 1203-1
Attic ventilation.

12 Interior Environment

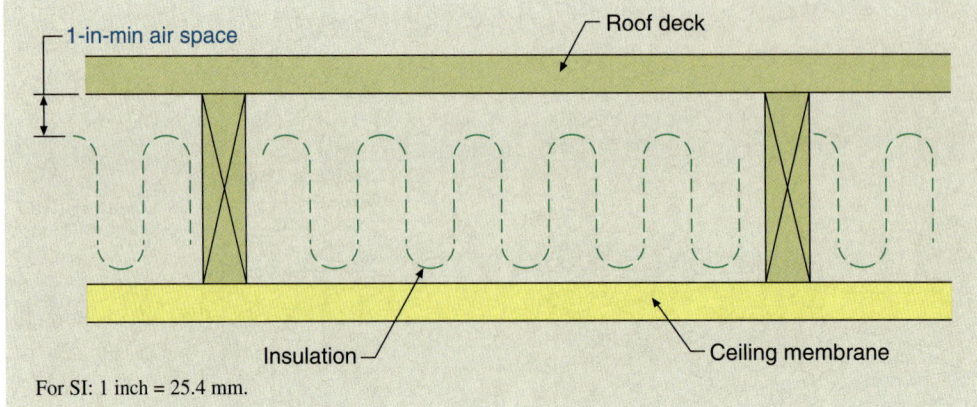

Figure 1203-2
Attic ventilation—air space.

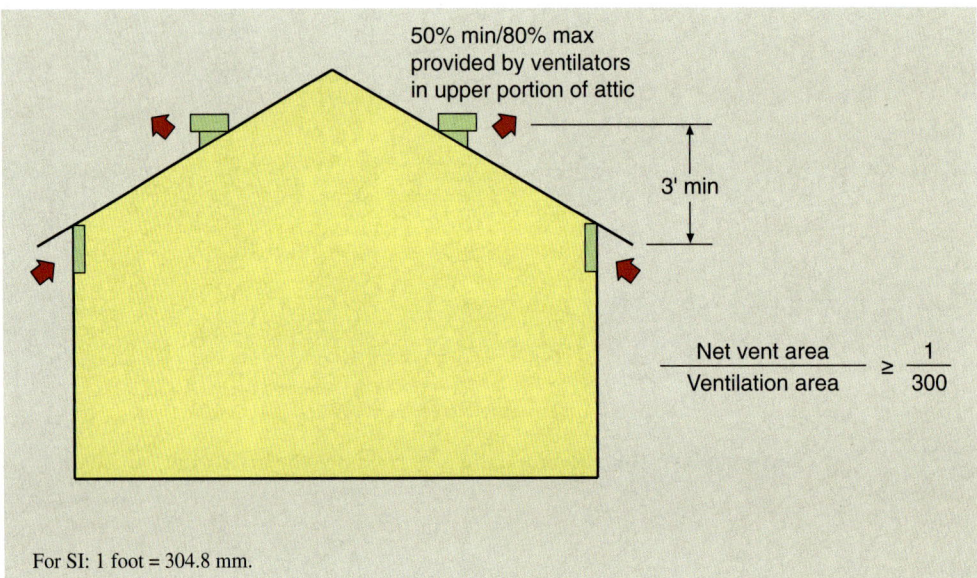

Figure 1203-3
Attic ventilation—calculations.

1203.3 Under-floor ventilation. In order to ventilate the space below the building between the bottom of the floor joists and the ground, ventilation openings shall be provided through foundation walls or exterior walls. The provisions apply to areas such as crawl spaces, rather than occupiable areas such as basements. Under certain climatic conditions, it is possible to ventilate the under-floor space into the interior of the building, or continuously operated mechanical ventilation may be provided in lieu of ventilation openings where the ground surface is covered with an approved vapor retarder.

To properly determine the minimum net area of ventilation openings, at least 1 square foot (0.0929 m^2) shall be provided for each 150 square feet (13.9 m^2) of crawl space area. The openings shall be located so as to provide cross ventilation in the under-floor area. See Figure 1203-4. An exception permits a dramatic reduction in the amount of ventilation opening area, provided the ground surface is treated with Class I vapor-barrier material. In this case, the total area of ventilation openings need not exceed 1/1,500 of the under-floor area, provided such openings are located to provide for adequate cross ventilation of the under-floor space. In this case, the vents may have operable louvers.

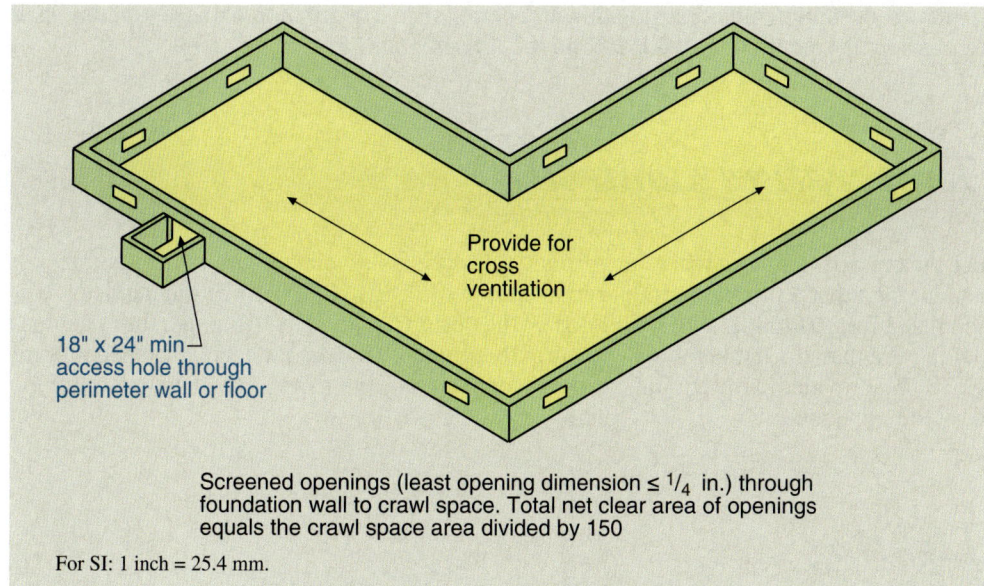

Figure 1203-4
Under-floor ventilation.

It is critical that ventilation openings be completely covered with a substantial material to prevent the entrance of insects and animals. Corrosion-resistant wire mesh, with the least dimension not exceeding 1/8 inch (3.2 mm), is one of six materials identified by the code to address this concern.

Where the under-floor space is conditioned, it is unnecessary to provide ventilation openings. It has been shown that by insulating the perimeter walls, covering the ground surface with a Class I vapor barrier, and conditioning the space in accordance with the *International Energy Conservation Code*® (IECC®), unvented crawl spaces outperform vented under-floor areas.

1203.4 Natural ventilation. Where buildings are not provided with adequate mechanical ventilation as specified in the IMC, natural ventilation through openings directly to the exterior must be provided. In order to determine the amount of ventilation air required, the minimum openable area to the outdoors is based on 4 percent of the floor area being ventilated.

In those cases where rooms or spaces do not have direct openings to the exterior, it is still necessary to ventilate the interior space, often through an adjoining room. In this case, the opening to the adjoining room should be unobstructed and have an area not less than 8 percent of the floor area of the interior room or space. In no case should the opening between the rooms be less than 25 square feet (2.3 m^2). Where the intervening room is a thermally isolated sunroom or patio cover, the minimum openable area of 8 percent is still applicable; however, the opening need only be 20 square feet (1.86 m^2).

As previously mentioned, in calculating the total openable area to the outdoors, such opening area shall not be less than 4 percent of the total floor area being ventilated. In those conditions where openings that provide the natural ventilation are located below grade, the outside horizontal clear space measured perpendicular to the opening is required to be at least one- and one-half times the depth of the opening.

Where rooms contain bathtubs, showers, spas, and similar bathing fixtures, natural ventilation is not an acceptable method. Because of the common reluctance to open exterior windows, particularly in cold-weather conditions, a mechanical system provides for more consistent ventilation. Therefore, bathrooms and similar spaces must be mechanically ventilated in accordance with the IMC. Where flammable and combustible hazards or other

12 Interior Environment

contaminant sources are present within an interior space, ventilation-exhaust systems shall be provided as required by the IMC and the *International Fire Code®* (IFC®).

Section 1204 *Temperature Control*

For those interior spaces where the primary purpose is associated with human comfort, it is important that a minimum indoor temperature of 68°F (20°C) can be maintained, measured at 3 feet (914 mm) above the floor on the design heating day. Although the code does not require that this temperature be constantly provided, it does mandate that such interior spaces be provided with equipment or systems having the capability of maintaining the desired temperature.

Section 1205 *Lighting*

Almost every occupancy requires some level of lighting that is due to its use. Means of egress illumination is also required by Section 1006. In spite of the obvious need for interior light as a necessary part of a building's function, the code mandates that some degree of lighting, whether artificial or natural, be provided to every occupiable space.

1205.2 Natural light. Where glazing to the exterior is used as the method for providing natural light, the exterior openings shall open directly onto a public way, yard, or court in compliance with Section 1205. Exterior wall openings used to provide natural light must have an area that is computed based on the net glazed area for windows and doors, and not the nominal size of the opening. Where a room is not located on an exterior wall, the provisions of this section permit the borrowing of light from an adjoining room. Figure 1205-1 illustrates the requirements for this condition.

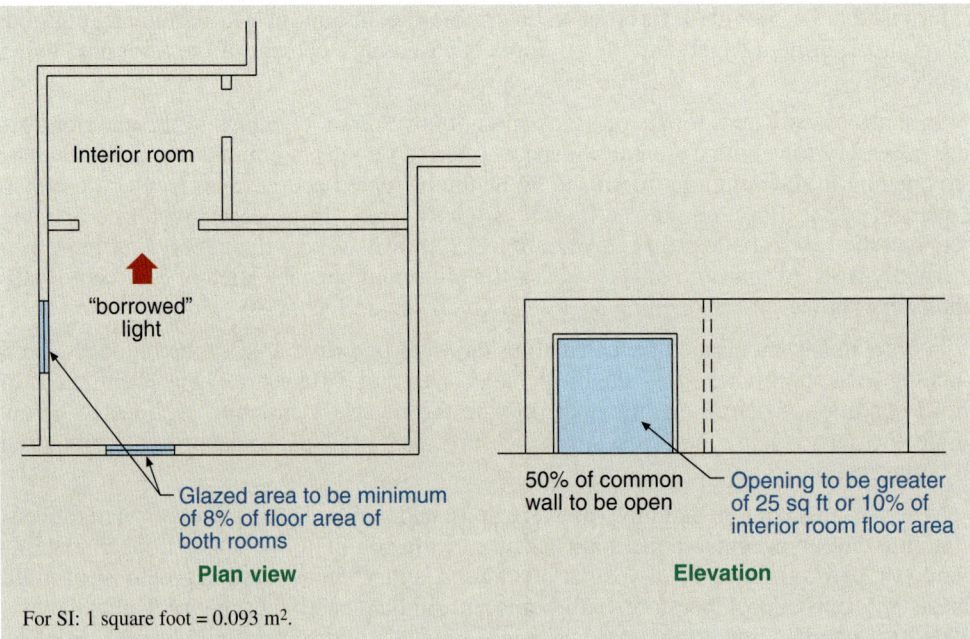

Figure 1205-1 Borrowing natural light.

Section 1206 *Yards or Courts*

The *International Building Code®* (IBC®) contains provisions for yards and courts where they are used to provide the required light and ventilation to exterior openings in the building. Most modern-day zoning ordinances also have requirements for yards and courts, and quite often these are more than adequate to gain the lighting and ventilation required by the code. In addition, where the alternatives of artificial light and mechanical ventilation as provided by Sections 1205.3 and 1203.1, respectively, are used in lieu of natural light and ventilation, the provisions of Section 1206 are not applicable.

To be considered providing adequate natural light and ventilation, each yard or court must have a minimum width of 3 feet (914 mm). In addition, these yards and courts must be increased in width, depending on the height of the building. See Figure 1206-1. The intent for the tall building is to have an increased court width so that light coming into the court will be able to reach the lower stories of the building, and this is only possible where the width of the court is in proper relationship to the height of the court. The requirements in the code are an obvious compromise between optimum light at the bottom of a court and the need to build as much building area on the lot as possible for economic purposes.

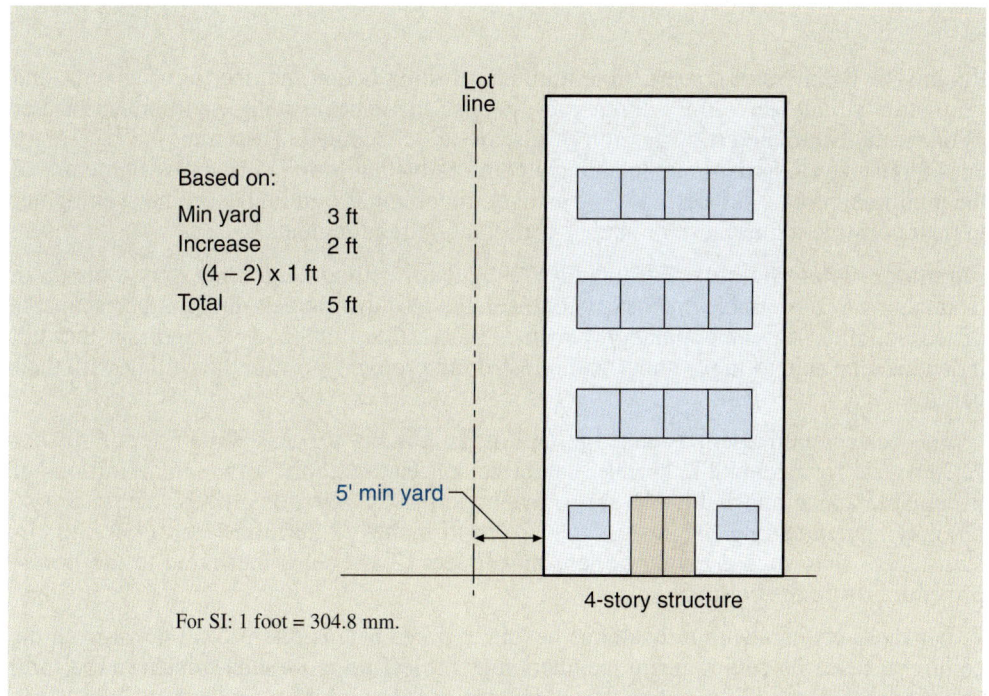

Figure 1206-1
Yards and courts.

Inner courts that are enclosed by the walls of the buildings, sometimes referred to as light wells, obviously need some means to remove accumulated trash at the bottom, provide for adequate drainage, and provide for circulation of air for ventilation purposes. In keeping with this intent, the code requires that an air intake be provided at the bottom of courts for buildings more than two stories in height, that grading and drainage be addressed, and that all courts be provided with access for cleaning.

12 Interior Environment

Section 1207 Sound Transmission

Applicable only to buildings containing dwelling units, this section is intended to provide regulations covering sound transmission control in residential buildings. It pertains to wall and floor/ceiling assemblies separating dwelling units from each other and from public space, such as interior corridors, stairs, or service areas. These must be provided with airborne-sound insulation for the walls and both airborne- and impact-sound insulation for the floor/ceiling assemblies.

For airborne-sound insulation, the separating walls and floor/ceiling assemblies must be provided with insulation equal to that required for a Sound Transmission Class (STC) of 50 (45 when field tested) as defined by ASTM E 90. As an alternative for concrete masonry and clay masonry assemblies, the sound transmission class may be calculated per TMS 0302. Penetrations or openings through the assemblies must be sealed, lined, insulated, or otherwise treated to maintain the required ratings. Dwelling unit entrance doors from corridors only need to be tight fitting to the frame and sill. Floor/ceiling assemblies between separate dwelling units must also provide impact-sound insulation equal to that required to meet an impact insulation class of 50 (45 when field tested), as defined in ASTM E 492.

Section 1208 Interior Space Dimensions

Room size, tightness of construction, minimum ceiling height, number of occupants, and ventilation all interact with each other to establish the interior living environment insofar as odors, moisture, and transmission of disease are concerned. Therefore, the IBC regulates room sizes to assist in maintaining a comfortable and safe interior environment, and the minimum room sizes become increasingly important as buildings become even tighter in their construction because of energy-conservation requirements.

1208.2 Minimum ceiling heights. Section 1208.2 regulates ceiling height, not only to assist in maintaining a comfortable indoor environment, but also to provide safety for the occupants of the building. As our population becomes increasingly taller, it is important that tall individuals be able to move about without striking projections from the ceiling with their heads.

The basic requirement is that the ceiling height be not less than 7 feet 6 inches (2,286 mm) for occupiable spaces, habitable spaces, and corridors (see definitions of occupiable space and habitable space in Section 202). Kitchens, halls, baths, and so on, may have a ceiling height less than 7 feet 6 inches (2,286 mm), but under no circumstances may such a height be less than 7 feet (2,134 mm) measured to the lowest projection from the ceiling.

For those ceilings within dwellings having exposed beams that project down from the ceiling surface, the ceiling beam members may project no more than 6 inches (152 mm) below the required ceiling height, provided the beams or girders are spaced at not less than 4 feet (1,219 mm) on center.

For rooms with sloped ceilings, the code requires only that the prescribed ceiling height be maintained in one-half the area of the room. However, no portion of the room that has a ceiling height of less than 5 feet (1,524 mm) shall be used in the computations for floor area. In the case of a room with a furred ceiling, the code requires the prescribed ceiling height in two-thirds of the area and, as in all cases for projections below the ceiling, the furred area may not be less than 7 feet (2,134 mm) above the floor.

Efficiency dwelling units. This section of the code provides for a specific type of dwelling unit—a dwelling unit consisting of only one habitable room. Many of the requirements in this section are redundant, as this chapter already requires many of these provisions. However, there are some requirements that are unique to the efficiency dwelling unit:

1208.4

1. The living room (which also serves as a bedroom and kitchen) is to have not less than 220 square feet (20.4 m^2) of floor area. It is the intent of the code that this floor area be the total gross floor area, less the area occupied by built-in cabinets and other built-in appliances that are not readily removed and that preclude any other use of the floor space occupied by the built-in cabinets and fixtures.
2. The minimum room size shall be increased by 100 square feet (9.29 m^2) of floor area for each intended occupant in the unit in excess of two.
3. A closet is required.
4. A kitchen sink, cooking appliance, and refrigeration facilities are required, each providing a clear working space of not less than 30 inches (762 mm) in front.
5. A separate bathroom containing a water closet, lavatory, and bathtub or shower is required.

Section 1209 *Access to Unoccupied Spaces*

Access to crawl spaces and attic spaces is regulated by this section. Though typically unoccupied, it is sometimes necessary that these normally concealed areas be accessed for various reasons.

Crawl spaces. This section of the code mandates that under-floor areas be accessible by a minimum 18-inch by 24-inch (457-mm by 610-mm) access opening. Where the access opening opens to the exterior of the building, the code intends it to be screened or covered to prevent the entrance of insects and animals. Also, it is the intent of the code that all portions of the under-floor area be accessible and access be provided beneath or around obstructions created by pipes, ducts, and so on.

1209.1

Attic spaces. Because enclosed attics provide an avenue for the undetected spread of fire in a concealed space, the code requires that access openings be provided into the attic so that fire-fighting forces may gain entry to fight the fire. To be of any value, the access openings must be of sufficient size to admit a fire fighter with fire-suppression gear and must also have enough headroom so that entry into the attic may be secured. Although not specified, the access should be located in a readily accessible location. A public hallway is the best location for attic-access openings. Fire department personnel will not then have to open private offices, apartments, or hotel rooms in order to enter the attic. Attic access may be provided through a wall as well as through a ceiling. In split-level buildings with multiple attics, an attic-access opening must be provided to each attic space.

1209.2

Attic access is only required by the IBC for those attic areas having a clear height greater than 30 inches (762 mm). Where such conditions occur, an attic-access opening of not less than 20 inches by 30 inches (559 mm by 762 mm) shall be provided. However, if the attic contains mechanical equipment, the opening may need to be enlarged to gain compliance with the IMC.

12 Interior Environment

Section 1210 Toilet and Bathroom Requirements

The primary thrust of this section is to provide easily cleanable, sanitary, and water-resistant surfaces in toilet rooms and shower areas.

1210.2.1 Floors and wall bases. Except for dwelling units, the code requires toilet room and bathing room floors to have a smooth, hard, nonabsorbent surface. Finishes such as concrete and ceramic tile are certainly acceptable, as are other approved materials that may also be used. It is the intent of the code that the building official determine the suitability of the proposed floor surface insofar as cleanability and water resistance are concerned.

Although the materials used for the floor covering and the wall base are to all have a smooth, hard, nonabsorbent surface, they can be of different materials. For example, a top set rubber base extending at least 4 inches (102 mm) up the wall is generally viewed as acceptable in conjunction with a floor covering of vinyl composition tiles.

Toilet and bathing room floor-finish requirements apply to all uses and occupancies except for dwelling units. Because motel and hotel rooms are typically not dwelling units, toilet room flooring in these uses must often comply with this section.

1210.2.2 Walls and partitions. Walls and partitions within 2 feet (610 mm) of the front and sides of urinals, water closets, and service sinks are required to have a smooth, hard, nonabsorbent finish. Required finishes shall extend to a height of at least 4 feet (1,219 mm) above the floor and shall be of a type not adversely affected by moisture. Water-resistant gypsum backing board may be used as a base for tile or wall panels, provided it satisfies the limitations set forth in Section 2509. Although the code does not specifically state that concrete walls must be sealed, concrete is not, strictly speaking, a nonabsorbent material and needs some type of surface treatment. Sealing is particularly important for concrete block walls.

Special wall finishes at water closets and urinals are not required for dwelling units, sleeping units, and toilet rooms not accessible to the public that contain only one water closet. Note that the exceptions for floor finishes and wall finishes are not the same. A private toilet room containing one water closet would be required to have flooring that complies with Section 1210.2.1, but there would be no special requirements for wall finishes.

In all occupancies, including dwelling units, penetrations of the water-resistant surfacing of the walls for the installation of accessories such as grab bars, towel bars, and so on, are required to be sealed to protect the structural elements from moisture. The intent of the code is that because the structural elements are not required to be moisture resistant, the penetrations should be sealed to protect the structural elements. Although sealing is required for all walls in a toilet room, sealing is obviously most critical in areas of water splash such as in showers or behind or adjacent to lavatories.

1210.2.3 Showers. All showers must have floor and wall finishes that are smooth, nonabsorbent, and not affected by moisture. Wall finishes must extend not less than 70 inches (1,778 mm) above the drain inlet.

KEY POINTS

- The method and arrangement of providing ventilated openings is an important aspect in the proper ventilation of attic spaces.
- Under-floor ventilation is to be provided through foundation walls or exterior walls in order to adequately ventilate the space below the building.
- Ventilation of interior spaces may be accommodated through exterior openings or by a mechanical system.
- The IBC mandates that some degree of lighting, whether artificial or natural, is to be provided to every occupiable space.
- Wall and floor/ceiling assemblies separating dwelling units in residential buildings must be provided with sound insulation.
- In specific areas of a building, floors and walls are required to have a smooth, hard, nonabsorbent finish.

ENERGY EFFICIENCY

CHAPTER 13

13 Energy Efficiency

> This chapter of the *International Building Code®* (IBC®) references the *International Energy Conservation Code®* (IECC®) for provisions regulating the design and construction of buildings for energy efficiency. The IECC sets forth minimum requirements for new and existing buildings by regulating their exterior envelopes. In addition, it addresses the selection of heating, ventilating, and air-conditioning; service water-heating; electrical distribution; and illumination systems and equipment in order to provide for efficient and effective energy usage.
>
> The provisions of this chapter apply to both residential and commercial buildings; however, specific energy requirements are provided in the *International Residential Code®* (IRC®) to address structures regulated by the IRC.

CHAPTER 14

EXTERIOR WALLS

Section 1402 Definitions
Section 1403 Performance Requirements
Section 1404 Materials
Section 1405 Installation of Wall Coverings
Section 1406 Combustible Materials on the Exterior Side of Exterior Walls
Section 1407 Metal Composite Materials
Section 1408 Exterior Insulation and Finish Systems (EIFS)
Key Points

14 Exterior Walls

> This chapter establishes the basic requirement for exterior wall coverings, namely that they shall provide weather protection for the building at its exterior. The other important item necessary for complete weather protection, the roof system, is addressed in the next chapter. Chapter 14 presents general weather-protection criteria and special requirements for veneer. In addition, exterior wall coverings, balconies, and similar architectural appendages constructed of combustible materials are regulated in Section 1406.

Section 1402 *Definitions*

The code intends by the definition of veneer that it be a nonstructural facing of masonry, concrete, metal, plastic, or similar approved material. It is intended to provide ornamentation, protection, or insulation. On this basis, face bricks that are laid with common bricks so as to provide a composite structural assembly are not considered veneer, as they act structurally along with the common bricks. To be considered veneer, a material must not act structurally with the backing insofar as the consideration of structural strength of the assembly is concerned. However, in many cases the veneer does act structurally with the backing, and problems can result if not constructed properly.

Veneer can be either adhered or anchored and either exterior or interior, depending on its method of attachment to the backing and whether or not it is applied to a weather-exposed surface. The definitions for all of the specific terms listed in this section are located in Chapter 2.

Section 1403 *Performance Requirements*

This section provides specifications for the protection of the interior of the building from weather; therefore, there are requirements for:

1. Protection of the interior wall covering from moisture that penetrates the exterior wall covering.

2. Flashing of openings in, or projections through, exterior walls.

3. Vapor retarders used to resist the transmission of water vapor through the exterior envelope.

It is also necessary to provide for the removal of any water that may accumulate behind the exterior veneer. The code requires a means for draining water to the exterior that enters the assembly. The intent of the provision is to merely utilize the felt building paper and flashing to drain the water to the outside, rather than provide an air space between the siding and the water-resistant barrier. Thus, the practice of placing siding directly over the building paper is acceptable, provided flashing is correctly installed to direct water within the wall assembly to the exterior.

The code exempts the need for a weather-resistant wall envelope over concrete or masonry walls designed in accordance with Chapters 19 and 21, respectively. The *International Building Code®* (IBC®) requirements for weather coverings, flashings, and drainage are also not applicable where the exterior wall envelope has been tested in accordance with ASTM E 331 and shown to resist wind-driven rain. In addition, where an exterior insulation and finish system (EIFS) is installed in accordance with Section 1408.4.1, the requirements of Section 1403.2 need not be followed.

The provisions of Section 1405.3 are referenced in regard to protection against condensation in exterior walls. Where framed walls, floors, and ceilings are not ventilated in a manner that will allow moisture to escape, it is mandated that a complying vapor barrier be installed on the warm-in-winter side of the insulation. A vapor barrier is not required for masonry or concrete exterior walls insofar as the provision is only applicable to framed walls. In addition, vapor barriers are not required in warmer climatic areas, specifically Climate Zones 1, 2, 3, and 4 (other than Marine 4) and where the value of such barriers is negligible because of the lack of damage any moisture will create. Vapor barriers may also be omitted in basement walls and any other portions of walls located below grade.

Testing has shown that the addition of a combustible water-resistive barrier can cause an exterior wall system to fail the NFPA 285 test, even if the wall had successfully passed the test prior to the addition of the barrier. Small-scale testing has shown that these types of materials can provide significant amounts of combustible fuel loading to a wall assembly. Therefore, exterior walls of those buildings required to have noncombustible exterior walls (Types I, II, III, and IV) must be tested in accordance with NFPA 285 where they contain a combustible water-resistive barrier and are over 40 feet (12,192 mm) in height. The use of this testing method is necessary to address the potential vertical and lateral flame spread that can occur either on or within exterior wall systems that contain combustible barriers.

Section 1404 *Materials*

As most exterior wall coverings are permeable by moisture, the code requires that there be a water-resistive barrier placed under the exterior wall covering and over the sheathing. The purpose of the barrier is to prevent moisture infiltration to the sheathing and subsequently the interior wall surfaces. The required water-resistive barrier is identified as at least one layer of No. 15 asphalt felt complying with ASTM D 226 for Type I felt. All necessary flashing must be installed as described in Section 1405.4 in order to complete the water-resistant envelope.

Section 1405 *Installation of Wall Coverings*

Exterior wall coverings are to be installed in accordance with the provisions of this section. Of primary concern are the types and thicknesses of the weather protection provided by the wall coverings. In addition, it is important that flashing be installed in those areas prone to leakage. Various types of veneer are addressed, including wood, masonry, stone, metal, glass, vinyl, fiber, and cement. The specifics for each veneer type are described in detail.

Weather protection. A broad range of exterior wall-covering materials is listed in Table 1405.2 as being satisfactory when utilized as protection against the elements of nature. Covering types addressed include wood and particleboard siding, wood shingles, stucco, anchored and adhered masonry veneer, aluminum and vinyl siding, stone, and structural glass. For most of the 36 types listed, the minimum required thickness for weather protection is mandated. In the case of particleboard and plywood without sheathing, reference is made to Chapter 23.

1405.2

1405.3 Vapor retarders. In order to protect against condensation in framed exterior wall assemblies of buildings constructed in the specified Climate Zones, complying vapor retarders are mandated. Table C301.1 in the *International Energy Conservation Code®* (IECC®) establishes the appropriate Climate Zones in the United States based on counties. Those counties designated as Zone 5, 6, 7, and 8, as well as those classified as Zone 4 Marine must be provided, with exceptions, with either a Class I or II vapor retarder on the interior side of any exterior framed walls. The use of Class III vapor retarders is also acceptable provided such retarders are set forth in Table 1405.3.1 for the appropriate Climate Zones. In addition to the use of assemblies tested and identified as Class I, II, or III vapor retarders, Section 1405.3.2 also provides a prescriptive listing of materials considered as compliant.

1405.4 Flashing. The code requires that all points subject to the entry of moisture be appropriately flashed. Roof and wall intersections, as well as parapets, are especially troublesome, as are exterior wall openings exposed to the weather, and particularly, those exposed to wind-driven rain. Even though the code may not cover every potential situation that might occur, it intends that the exterior envelope of the building be weatherproofed so as to protect the interior from the weather. Furthermore, for buildings of human occupancy, the interior must be sanitary and livable. Therefore, whether prescribed in the code or not, any place on the envelope of the building that provides a route for admission of water or moisture into the building is required to be properly protected.

1405.5 Wood veneers. The use of wood veneer in buildings is limited when installed on buildings of other than Type V construction. The height of the veneer cannot exceed 40 feet (12,190 mm), or 60 feet (18,290 mm) where fire-retardant-treated wood is used. The backing must be noncombustible, with any open or spaced veneer projecting no more than 24 inches (610 mm) from the building wall.

1405.6 Anchored masonry veneer. In addition to complying with the applicable provisions of Chapter 14, anchored masonry veneer must also comply with Sections 6.1 and 6.2 of TMS 402/ACI 530/ASCE 5. Anchored masonry veneer must typically be supported by noncombustible corrosion-resistant structural framing, with the exception that this section does permit the use of wood construction supports when in conformance with four criteria. In addition to the structural considerations, the deflection limitation and joint requirements are intended by the code to limit the differential interaction of the veneer with its backing so as to prevent cracking, spalling, and buckling of the veneer surfaces. Figure 1405-1 shows examples of anchored masonry veneer.

1405.7 Stone veneer. Two methods are detailed for the application of stone veneer having a maximum thickness of 10 inches (254 mm). One method is applicable where the stone is anchored directly to masonry or concrete construction, whereas the second is to be used for stud construction. Figure 1405-2 details several of the requirements for both methods of attachment.

1405.8 Slab-type veneer. When limited in thickness to 2 inches (51 mm), slab-type veneer units are to be anchored directly to concrete, masonry, or stud construction. The maximum size of each unit is limited to 20 square feet (1.9 m^2). Attachment shall be made with dowels and ties that are corrosion resistant in a manner prescribed by this section. An illustration of the application of slab-type veneer units is shown in Figure 1405-3.

1405.9 Terra cotta. Where anchored terra cotta or ceramic units are used as exterior veneer, they must be at least $1^5/_8$ inches (41 mm) in thickness. They shall be anchored directly to masonry, concrete, or stud construction in a manner consistent with the requirements of this section, as shown in Figure 1405-4.

1405.10 Adhered masonry veneer. Adhered masonry veneer must comply with the requirements of TMS 402/ACI 530/ASCE 5 as well as the provisions of Section 1405.10. Figure 1405-5 provides an example of the application of adhered veneer.

1405.11 Metal veneers. To help ensure the integrity and durability of the exterior wall covering material, metal veneer must be manufactured of corrosion-resistant materials. As an alternative, the

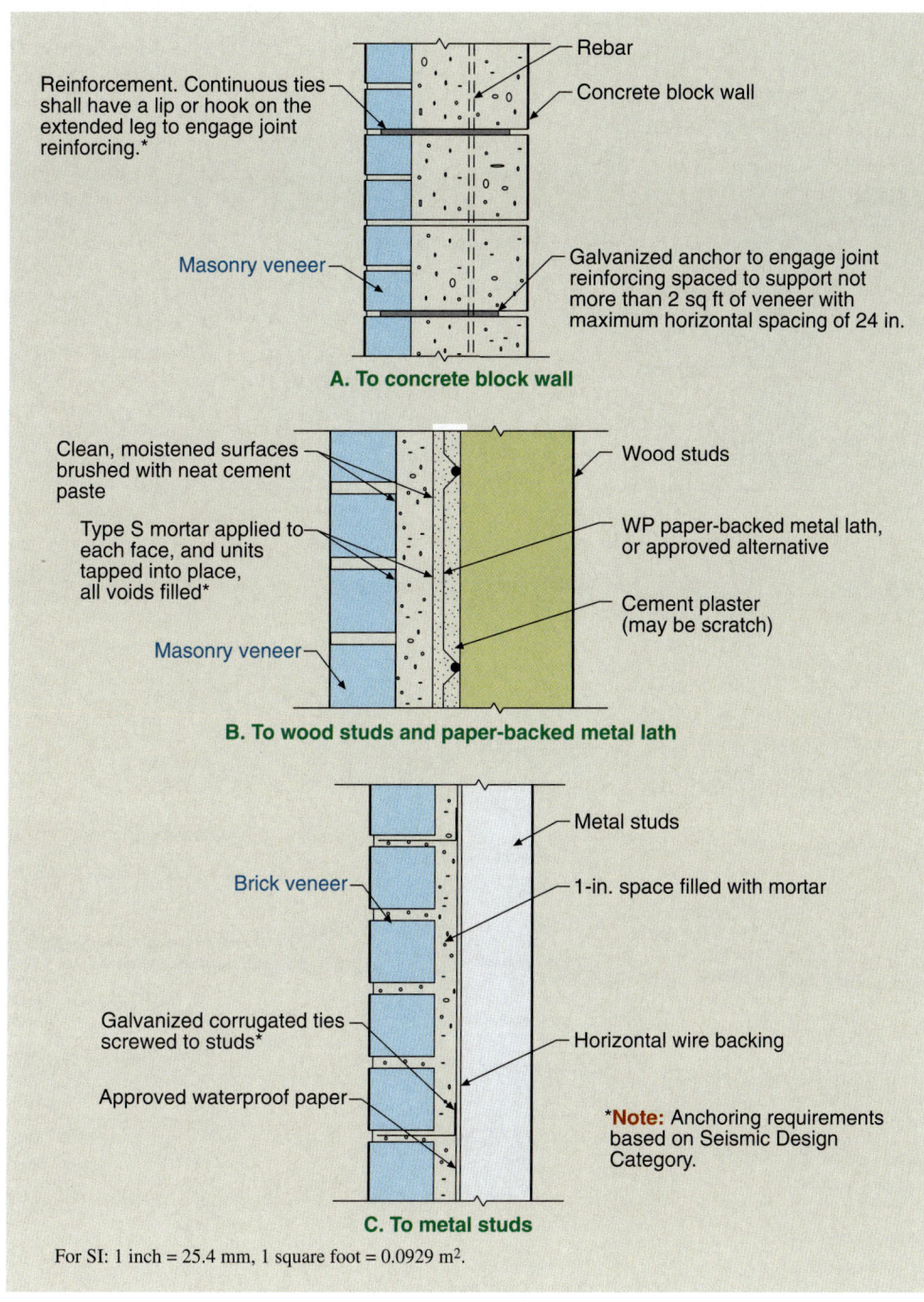

Figure 1405-1 Generic application of anchored masonry veneer (5-inch maximum thickness).

metal panels may be completely encased in porcelain enamel or treated in some other manner so that the metal is resistant to corrosion. The nominal thickness of sheet steel-metal veneer is to be at least 0.0149 inch (0.378 mm), with the veneer to be mounted on furring strips or approved sheathing.

As would be expected, the fasteners, metal ties, or other attachment devices must also be corrosion resistant. The fasteners are to be spaced at no more than 24 inches (610 mm) vertically and horizontally, with at least four attachments for veneer panels exceeding 4 square feet (0.4 m^2) in area. It is also important that some means of weather protection be provided for the exterior metal veneer, including the use of pressure-treated wood.

14 Exterior Walls

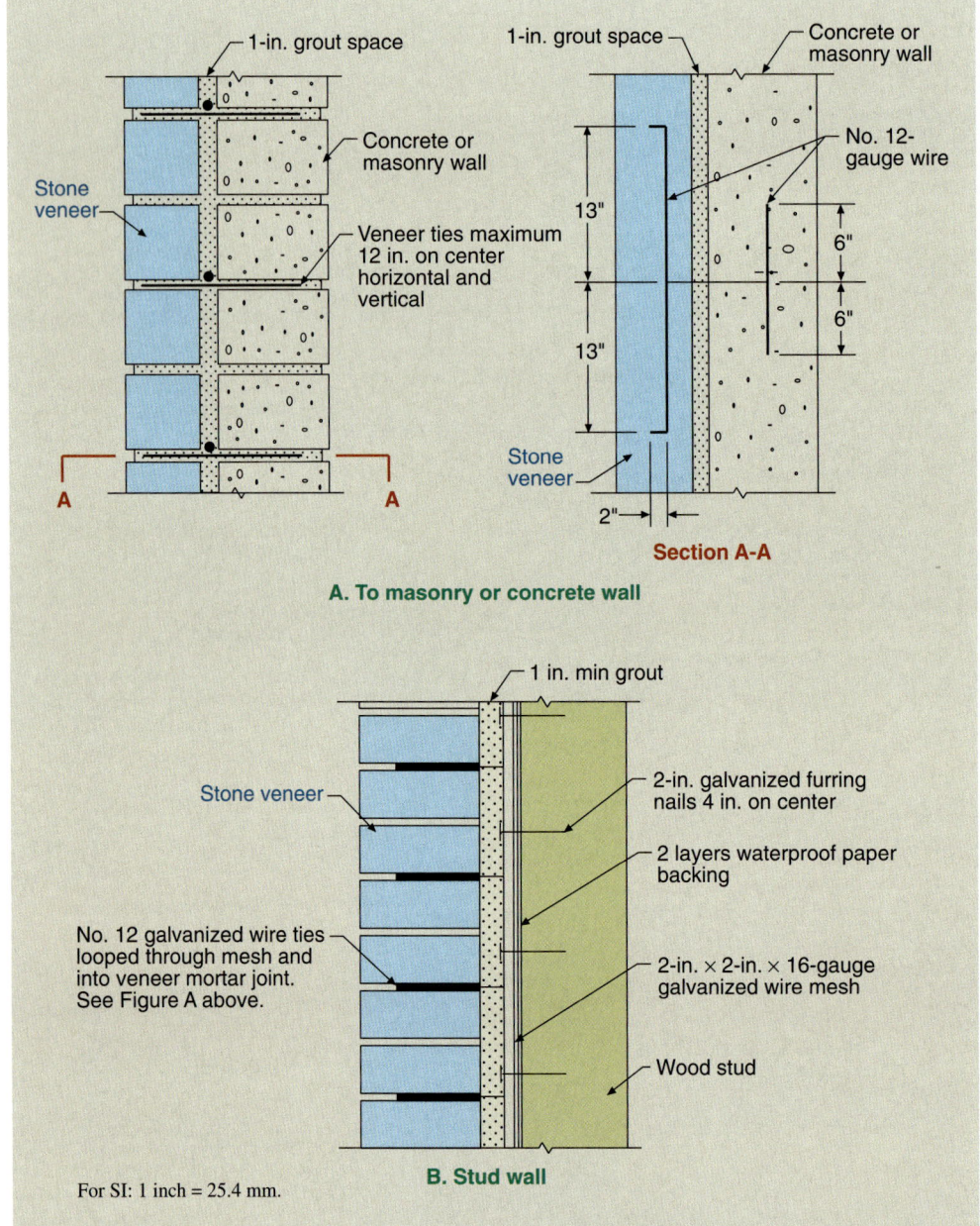

Figure 1405-2
Application of anchored stone veneer units up to 10-inch maximum thickness.

1405.12 Glass veneer. For safety reasons, the use of glass veneer as an exterior wall covering is highly regulated. Any single piece of veneer is limited to 6 square feet (0.56 m^2) in area, unless located within 15 feet (4,572 mm) of the sidewalk level or grade directly below, in which case it may be up to 10 square feet (0.93 m^2) in area. The maximum height or width of any section is limited to 48 inches (1,219 mm), with the thickness to be no less than 0.344 inch (8.7 mm).

The application method requires at least one-half of the area of each veneer panel to be directly bonded to the backing with an approved mastic cement. Metal moldings shall be used for any veneer that extends to the sidewalk surface, with shelf angles required for additional support under certain conditions. At a height over 12 feet (3,658 mm) or over show windows, additional support shall be provided by fasteners at either the four corners or the vertical or horizontal edges of the glass veneer. Design of the fasteners must allow

Installation of Wall Coverings

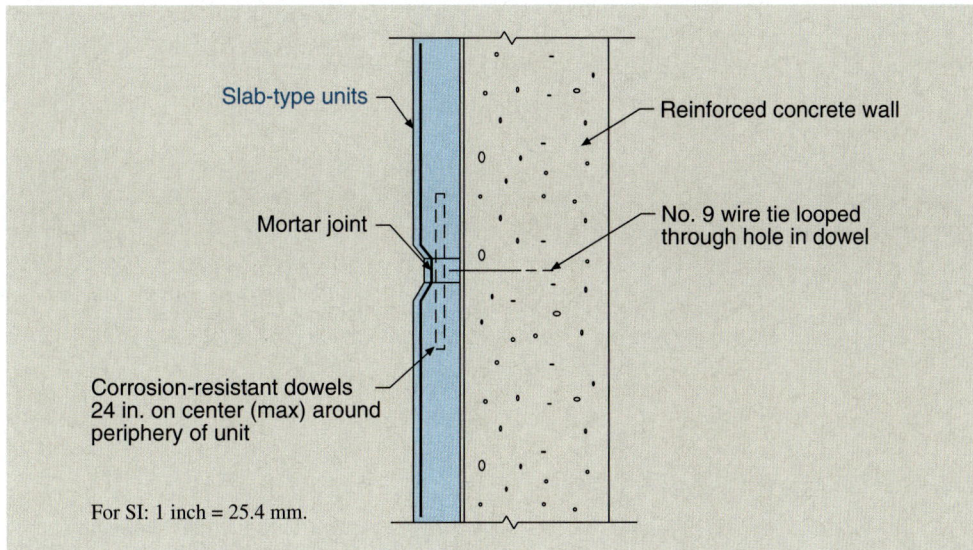

Figure 1405-3
Application of slab-type units (2-inch maximum thickness).

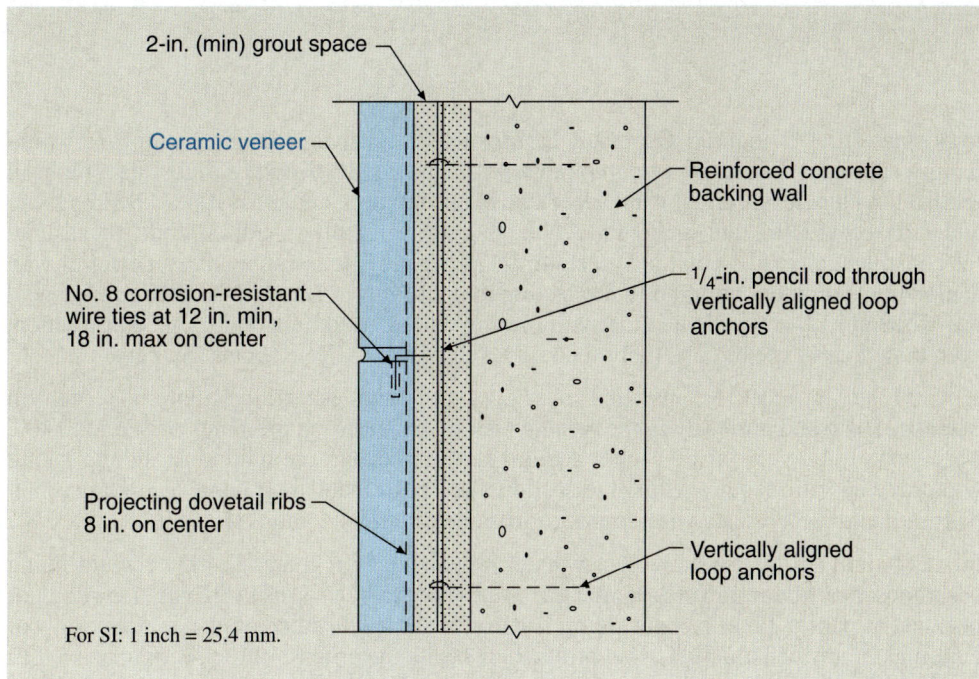

Figure 1405-4
Application of terra cotta or ceramic units.

for the independent support of the veneer. Exposed edges must also be adequately flashed to prevent the entrance of moisture between the backing and the glass veneer.

Vinyl siding. ASTM D 3679 establishes requirements and testing methods for extruded single-wall siding manufactured from rigid polyvinyl chloride compound. It is important that the building official determine if a particular siding material is code complying by referring to the packaging for information as to compliance with the material standard.

Siding regulated by this section is typically limited to installation on the exterior walls of buildings of Type V construction located in areas where the wind speed, according to Chapter 16, does not exceed 100 miles per hour (45 m/s) and the building height is limited

1405.14

14 Exterior Walls

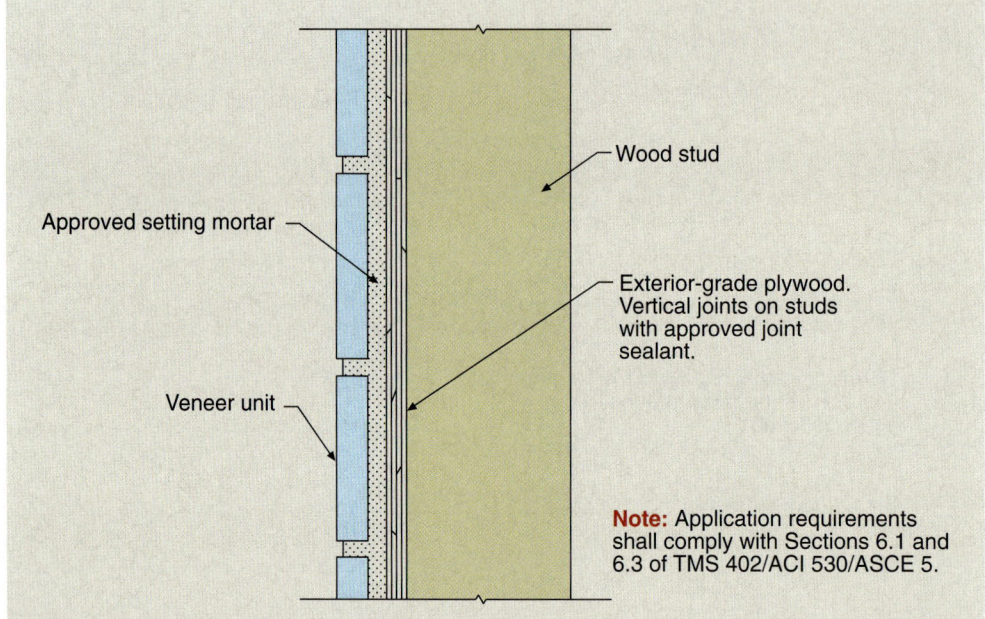

**Figure 1405-5
Application of adhered masonry veneer.**

to 40 feet (12,192 mm) in Exposure C. More severe wind or height conditions would require submittal data showing compliance with Chapter 16. However, most geographical areas fall within the specified wind speed and exposure limit, and most Type V buildings are under 40 feet (12,192 mm) in height. Although not specifically stated, the code intends that the construction type be nonrated unless test data are provided to show that an exterior wall using vinyl siding has undergone fire testing to qualify as at least a 1-hour fire-resistance-rated assembly. It is anticipated that vinyl siding will be used on wood-frame construction, even though any type of construction could be classified as Type V construction.

Vinyl siding must be installed over a wood-based sheathing material listed in Section 2304.6 and must satisfy the weather-resistive barrier requirements of Section 1403. This section also sets forth specific requirements regarding the nailing of siding. Along with these specific requirements, the section also states that "siding and accessories shall be installed in accordance with approved manufacturer's instructions."

1405.16 **Fiber cement siding.** A relatively new exterior wall covering material, fiber cement siding is composed of fiber-reinforced cement. It is designed to be used as horizontal lap siding or installed as panels. Fiber cement siding is permitted to be installed on the exterior walls of buildings of any construction type, provided it meets the criteria of the section, as well as the manufacturer's installation requirements.

1405.18 **Polypropylene siding.** Siding made of polypropylene can spread a fire by several means. Polypropylene that has not been appropriately fire retarded will release a greater amount of heat (about four times greater) than non-fire-retarded polypropylene or other combustible sidings that are permitted by the code such as wood siding or vinyl (PVC) siding. This greater heat release and resultant radiant heat can help spread the fire to other adjacent structures. In addition, the flaming droplets from the plastic can also contribute to the fire spreading by igniting grass, mulch, and debris found near the building. Therefore, the use of this material is strictly regulated under the provisions of this section and those of Section 1404.12. Except for the fire testing, the provisions are similar to those dealing with vinyl siding and will provide a way for the polypropylene to be used safely. A major limitation is that polypropylene siding is limited to use on buildings of Type V construction where combustible exterior wall construction is permitted.

Section 1406 Combustible Materials on the Exterior Side of Exterior Walls

Exterior walls of buildings of Types I, II, III, and IV are typically required to be of noncombustible construction. However, it is common for some limited combustible elements to be installed on the exterior side of such exterior walls. Type V buildings are permitted to use combustible materials for the entire wall construction. Where the exterior walls of a building include combustible materials such as wall coverings, trim, balconies, and similar appendages, the materials must be in compliance with this section. The only exception is for plastics, which are regulated by Chapter 26. The installation of such materials in compliance with the code does not negatively impact the desired level of fire safety.

The general requirements address the ignition resistance of the combustible wall coverings, based on the distance between the exterior wall and the lot line. Where the fire separation distance is 5 feet (1,524 mm) or less, the combustible wall covering must be subjected to the test method of NFPA 268 and exhibit no sustained flaming. As the exterior wall is moved farther and farther from the lot line, the tolerance to radiant heat energy need not be as great, allowing the use of alternative materials. Table 1406.2.1.1.2 identifies the tolerable level of incident radiant heat energy based on fire separation distance.

Section 1406.2.1 provides for the use of combustible wall coverings on otherwise noncombustible exterior walls, provided the building is limited to 40 feet (12,192 mm) in height above grade plane. The amount of such materials is also limited where the fire separation distance is 5 feet (1,524 mm) or less. The reductions from the general requirements are granted because of the restricted height and, in some cases, the limited amount of combustible materials. At a height exceeding 40 feet (12,192 mm), the use of combustible materials or supports is prohibited, except where the use of wood veneer and fire-retardant-treated wood is specifically allowed.

Balconies and similar projecting elements in buildings of Type I and II construction are to be constructed of noncombustible materials, unless the building is no more than three stories in height above grade plane. Under this condition, fire-retardant-treated wood may be used where the balcony or similar element is not used as a required egress path. Unless constructed of complying heavy-timber members, a combustible balcony or similar combustible projection must have a minimum fire-resistance rating equivalent to the required floor construction.

In Type III, IV, and V buildings, balcony construction may be of any material permitted by the code, combustible or noncombustible. Where a fire-resistance rating is mandated by the code, it must be maintained at the projecting element unless sprinkler protection is provided or it is of Type IV construction.

In addition to these types of construction limitations, the aggregate length of all projections cannot exceed 50 percent of the building perimeter at each floor unless sprinkler protection is extended to the balcony areas. In all cases, the use of untreated wood is permitted for pickets, rails, and similar guard elements when limited to a height of 42 inches (1,067 mm).

Section 1407 Metal Composite Materials

Metal composite materials (MCM) consist of a thin, extruded plastic core encapsulated within metal facings. As the use of MCM continues to increase, it is important to provide

14 Exterior Walls

detailed requirements addressing this unique building element. Many of the provisions are based on requirements from elsewhere in this chapter and Chapter 26. Although having some of the same characteristics as foam-plastic insulation, light-transmitting plastic, plastic veneer, and combustible construction, MCM used as exterior wall coverings are specifically regulated by this section.

Where installed on exterior walls of buildings, MCM systems are regulated for surface-burning characteristics. In buildings of other than Type V construction, the flame-spread index cannot exceed 75 and the smoke-developed index is limited to 450. In such buildings required to have noncombustible exterior walls, the use of a thermal barrier is necessary to separate the MCM from the interior of the building unless specifically approved by appropriate testing. Alternatively, the installation of MCM to a limited height in compliance with Section 1407.11 modifies the general requirements.

Section 1408 *Exterior Insulation and Finish Systems (EIFS)*

Exterior Insulation and Finish Systems (EIFS) are non-load-bearing exterior wall coverings that are used extensively throughout North America, Europe, and the Pacific Rim. The provisions of Section 1408 are primarily intended to reference the applicable ASTM standards that are specific to EIFS. Reference is made to E 2273 *Standard Test Method for Determining the Drainage Efficiency of Exterior Insulation and Finish Systems (EIFS) Clad Wall Assemblies,* E 2568 *Standard Specification for PB Exterior Insulation and Finish Systems (EIFS),* and E 2570 *Standard Test Method for Evaluating Water-Resistive Barrier (WRB) Coatings Used Under Exterior Insulation and Finish Systems (EIFS) for EIFS with Drainage.* In addition to the several ASTM standards previously identified, reference is also made to various provisions found elsewhere in the IBC that are applicable to EIFS. Current ICC ES Acceptance Criteria further establish requirements for EIFS and related components, and numerous EIFS manufacturers hold evaluation reports to demonstrate code compliance.

KEY POINTS

- The interior of a building must be protected with a weather-resistant envelope, including wall coverings, flashing, and drainage methods.
- The IBC regulates numerous veneer materials for exterior applications.
- A limited amount of combustible materials such as wall coverings, trim, balconies, and similar appendages are permitted on exterior walls of Type I, II, III, and IV buildings.
- The applicable referenced standards are identified for the installation of exterior insulation and finish systems (EIFS).

CHAPTER 15

ROOF ASSEMBLIES AND ROOFTOP STRUCTURES

Section 1502 Definitions
Section 1503 Weather Protection
Section 1504 Performance Requirements
Section 1505 Fire Classification
Section 1506 Materials
Section 1507 Requirements for Roof Coverings
Section 1508 Roof Insulation
Section 1509 Rooftop Structures
Section 1510 Reroofing
Key Points

15 Roof Assemblies and Rooftop Structures

> In addition to the requirements for roof assemblies and roof coverings, this chapter regulates roof insulation and rooftop structures. Rooftop structures include such elements as penthouses, tanks, cooling towers, spires, towers, domes, and cupolas. Roofing materials and components are regulated for quality as well as installation.
>
> The provisions in Chapter 15 for roof construction and roof covering are intended to provide a weather-protective barrier at the roof and, in most circumstances, to provide a fire-retardant barrier to prevent flaming combustible materials such as flying brands from nearby fires from penetrating the roof construction. The chapter is essentially prescriptive in nature and is based on decades of experience with the various traditional roof-covering materials. These prescriptive rules are very important to ensuring the satisfactory performance of the roof covering, even though the reason for a particular requirement may be lost. The provisions are based on an attempt to prevent observed past unsatisfactory performances of the various roofing materials and components.
>
> Those measures that have been shown by experience to prevent past unsatisfactory performance generally are included in the manufacturer's instructions for application of the various roofing materials. In many cases, the manufacturer's instructions are incorporated in this code by reference. The code intends, then, that they be followed as if they were part of the code.
>
> The overriding safety need of roofs is resistance to external fire factors. In this regard, the enforcement of this chapter is driven by Table 1505.1, as well as the appropriate standards for fire-retardant roof assemblies and roof coverings, including ASTM E 108, UL 790, and ASTM D 2898. Typically, a roof covering by itself cannot be a listed fire-retardant roof. Therefore, the regulations clearly separate assemblies and coverings to enforce construction of listed roof assemblies to the level at which listed wall and floor/ceiling assemblies are regulated.

Section 1502 *Definitions*

As with other industries supplying specialty building products, the roof-covering industry has a language of its own. In order to properly understand and apply the provisions of the *International Building Code®* (IBC®), the unique terms employed in the IBC must be understood. The roofing industry publishes several publications containing excellent glossaries of the terms of their industry. This section identifies those terms specifically defined in Chapter 2 that are related to roofs, roof assemblies, roof coverings, and roof structures.

Section 1503 *Weather Protection*

In all cases, a roof must be designed to provide protection from the elements. To ensure that the roof will adequately perform this function, it must be designed in accordance with this chapter. In addition, and just as important, the approved manufacturer's installation instructions must be adhered to.

This section requires flashing where the roof intersects vertical elements such as walls, chimneys, dormers, plumbing stacks, plumbing vents, and other penetrations of the weather-protective barrier. Coping of parapet walls shall be accomplished with noncombustible, weather-proofed materials. Figures 1503-1 through 1503-4 depict some examples of roof flashing details at vertical surfaces.

Performance Requirements 15

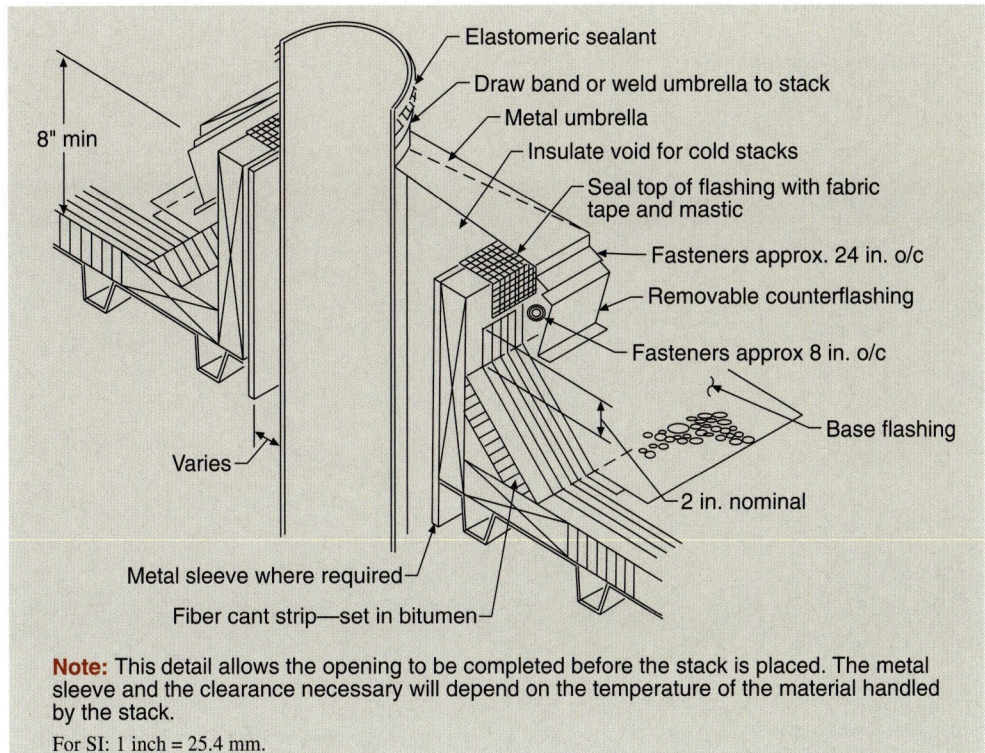

**Figure 1503-1
Stack flashing.**

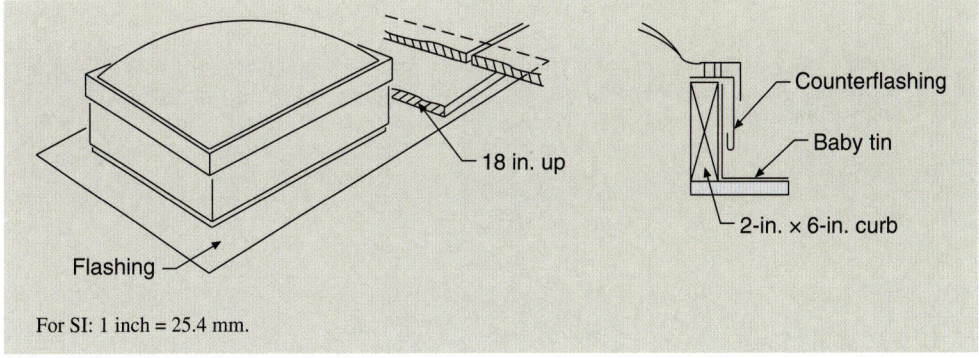

**Figure 1503-2
Flashing at skylight (wood roof).**

Section 1504 *Performance Requirements*

Roof decks and roof coverings must be able to withstand the effects of nature in a satisfactory manner. This section of the code regulates the performance of a roof against three concerns: wind, weathering, and impact. For wind resistance, roofs must comply with this section and Chapter 16. Low-slope roofs must demonstrate that they are resistant to both weathering and impact damage by complying with the appropriate standards.

The use of aggregate as a roof-covering material and aggregate, gravel, or stone as ballast is prohibited in specified locations in an effort to reduce property loss that is due to high winds. Field assessments of damage to buildings caused by high-wind events

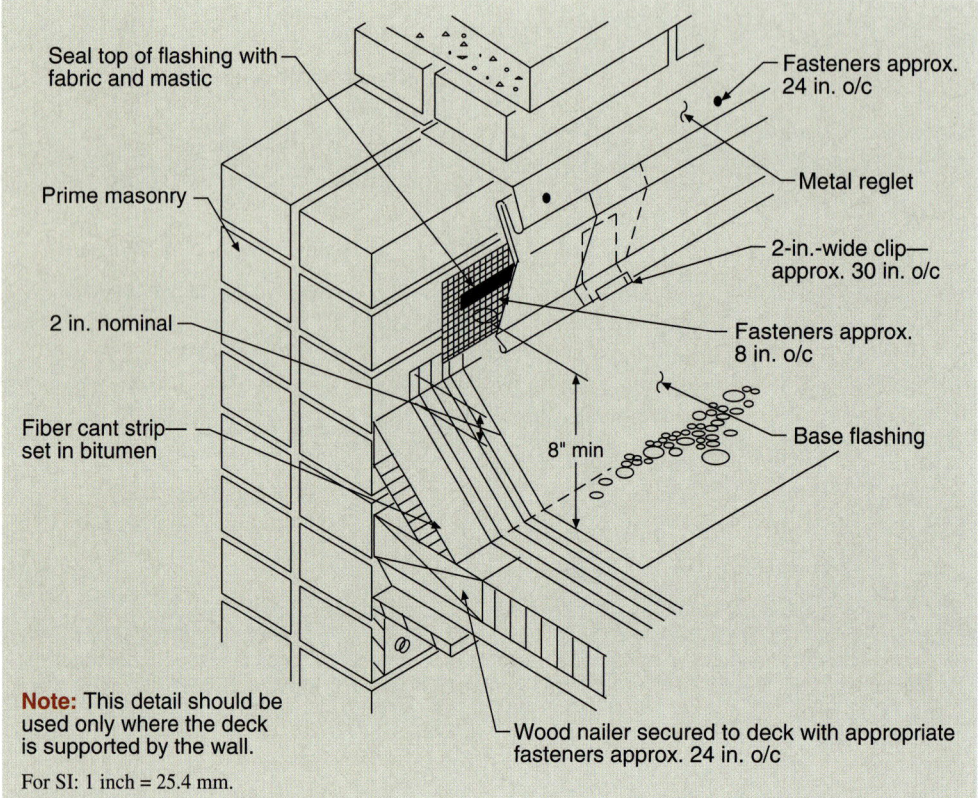

Figure 1503-3
Base flashing at bearing wall.

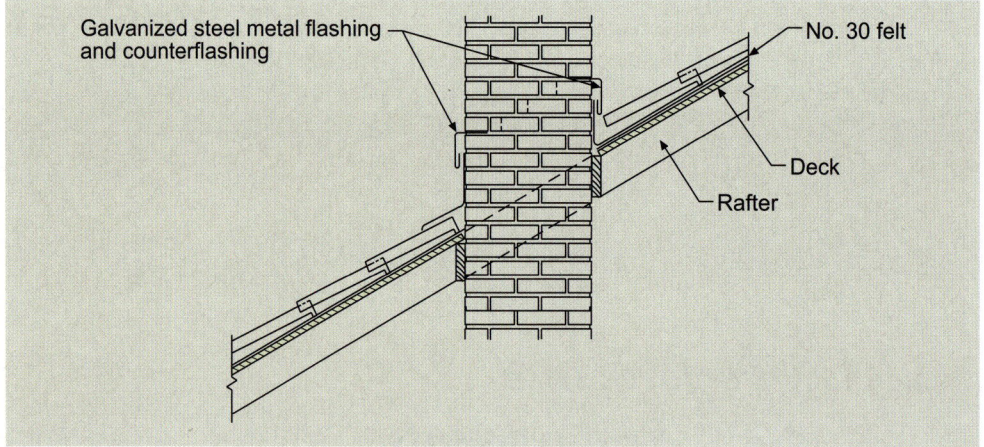

Figure 1503-4
Chimney flashing detail.

have shown that gravel or stone blown from the roofs of buildings has exacerbated damage to other buildings because of breakage of glass. The code prohibits the use of aggregate, gravel, or stone on roofs of buildings in hurricane-prone regions. These regions are defined as areas along the United States Atlantic Ocean and Gulf of Mexico coasts where the basic wind speed exceeds 90 miles per hour (40 m/s), and also include the islands of Hawaii, Puerto Rico, Guam, Virgin Islands, and American Samoa. Aggregate, gravel, and stone roof-covering materials are also prohibited on those buildings where the mean height exceeds that allowed by Table 1504.8 on the basis of the basic wind speed and exposure category. Under these conditions, there is a great enough potential for gravel stone, debris, or other unsecured objects to become airborne and possibly break glass in buildings downwind.

Section 1505 *Fire Classification*

As a minimum, the IBC generally requires Class B or C roof coverings for most buildings. These are roof coverings that provide protection of the roof against moderate and light fire exposures, respectively. The various sizes of brands used for testing are shown in Figure 1505-1. These exposures are external and are generally created by fires in adjoining structures, wild fires (brush fires and forest fires, for example), and fire from the subject building that extends up the exterior and onto the top surface of the roof. Wild fires and some structural fires create flying and flaming brands that can ignite nonclassified roof coverings. With regard to clay tile roofing, which is defined in Section 1505.2 as a Class A roof assembly, it is of interest to note that the Spanish missionaries shipped clay roofing tile to North America to protect their mission buildings from fire caused by flaming arrows shot onto the roofs.

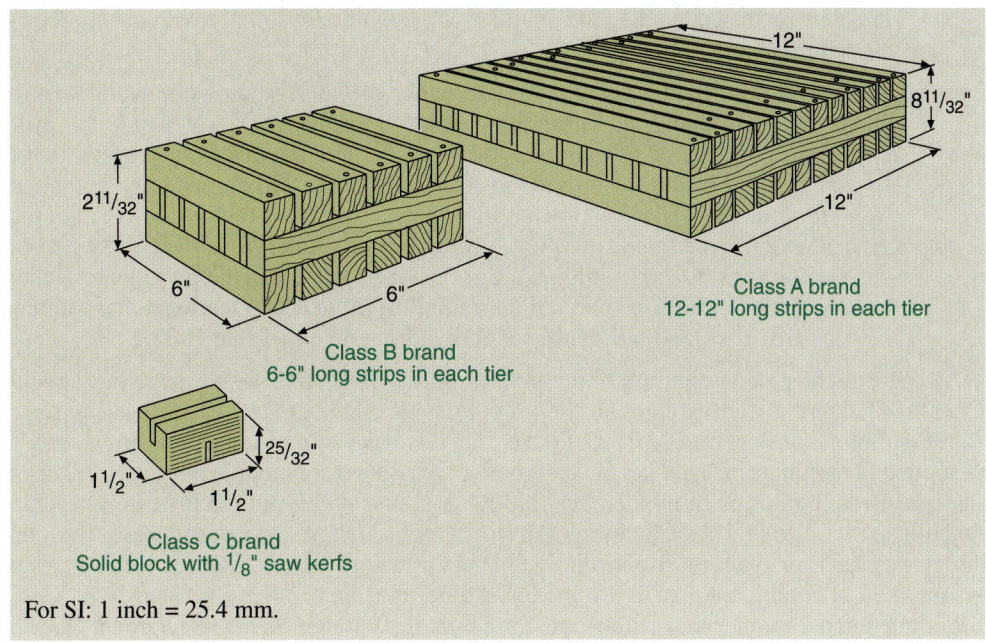

Figure 1505-1 Brands for Class A, B, and C tests.

The roof assembly classifications required by the code, which are related primarily to type of construction, are delineated in Table 1505.1.

Section 1505 defines the following roof assemblies and roof coverings:

1. Class A roof assemblies. Roof assemblies recognized as Class A are effective against severe fire exposures. Roof coverings of brick, masonry, slate, clay or concrete roof tile, exposed concrete roof deck, and metal, ferrous or copper shingles or sheets are all considered Class A roof assemblies, as well as any roof assembly or roof covering tested and listed as Class A. Class A roof assemblies are permitted for use on all buildings, regardless of the building's construction type.

2. Class B roof assemblies. Class B roof assemblies are effective against moderate fire exposures and are considered appropriate for all types of construction. There are no prescriptive roof assemblies or roof coverings considered as Class B, as they must be listed as such. Consistent with the universal acceptance of Class A assemblies, Class B roof assemblies are also permitted for use on all buildings.

3. Class C roof assemblies. Buildings not required to be of fire-resistant construction—Types IIB, IIIB, and VB—are permitted to use Class C roof assemblies. Such assemblies are effective against light fire exposures.

4. Fire-retardant-treated wood shingles and shakes. Where fire-retardant-treated wood shakes and shingles are components of Class A, B, or C roof assemblies, they must comply with the criteria of Section 1505.6. AWPA C1, *All Timber Products-Preservative Treatment by Pressure Processes*, is referenced as the standard for regulating the pressure treatment for fire-retardant purposes. In addition to the required markings that identify the shakes or shingles and their manufacturer, a label is mandated to identify the appropriate Class A, B, or C classification.

5. Nonclassified roofing. Roof coverings that are considered nonclassified roof coverings are approved for use by the IBC on Group R-3 and U buildings where the roof is located at least 6 feet (1,829 mm) from all lot lines. These roof coverings have been shown by experience to provide the necessary resistance to weather as intended by the code when the qualities of the materials comply with the appropriate requirements.

6. Special-purpose roofs. These roofs are either of wood shingles or wood shakes and are applied with a minimum 5/8 inch (15.9 mm) Type X water-resistant gypsum backing board or gypsum sheathing panel. The intent of the provisions for special-purpose roofs is to provide a roof covering that, although it may be ignited by flying brands, will not burn through to the interior of the building. Also, the special underlayment tends to prevent fires from the interior of the building from burning through to and igniting the roof covering, which helps prevent flying brands. Special-purpose roofs are permitted in limited applications on buildings of Type IIB, IIIB, and VB construction by Footnote c to Table 1505.1.

Because of the inconsistent use of the terms *roof assembly* and *roof covering* throughout Chapter 15, there is confusion as to the proper use of combustible materials at the roof. Based on Item 4 of Section 603.1, roof coverings that have an A, B, or C classification are permitted in buildings of Type I or II construction. This does not include the structural deck materials, which must be of noncombustible construction or fire-retardant-treated wood in compliance with Item 1.3 of Section 603.1. It would, however, permit the use of wood structural panels or foam plastic insulation boards as a part of the classified roof covering where used in combination with a noncombustible roof deck. Where a Class A, B, or C roofing assembly includes a combustible structural deck, other than fire-retardant-treated wood where permitted, it is limited to use on a building of Type III, IV, or V construction.

Section 1506 *Materials*

Certainly, roofing materials must comply with quality standards embodied in the IBC for Chapter 15. Furthermore, identification of the roofing materials is mandatory in order to verify that they comply with quality standards. In addition to bearing the manufacturer's label or identifying mark on the materials, roof-covering materials are required by the code to carry a label of an approved agency having a service for inspection of materials and finished products during manufacture.

Section 1507 Requirements for Roof Coverings

The following sections in the IBC contain the basic requirements for the installation of the more common roof-covering materials:

Section 1507.2 Asphalt shingles

Section 1507.3 Clay and concrete tile

Section 1507.4 Metal roof panels

Section 1507.8 Wood shingles

Section 1507.9 Wood shakes

Section 1507.10 Built-up roofing materials

Asphalt shingles. Figure 1507-1 shows a typical installation of asphalt shingles on roofs with a slope of 4:12 as required by the IBC. However, the IBC also permits application on a roof that has a slope as shallow as 2:12 when installed in accordance with Section 1507.2.8.

Where the roof slope is less than 4:12, water drainage from the roof is slowed down and has a tendency to back up under the roofing and cause leaks. Also, the effect of ice dams at eaves is more pronounced on low-slope roofs and, as a result, special precautions are necessary to ensure satisfactory performance of the roofing materials. Thus, the code requires that the underlayment be laid with two layers of shingled felt, which provides two thicknesses of the underlayment at any point. Figure 1507-2 shows the method of underlayment application, and Figure 1507-3 shows the method for shingle application for low-slope roofs. The code requires that underlayment of one layer of approved felt be provided under asphalt-shingle roof coverings with a slope of 4:12 or greater.

Clay and concrete tile. Tile is among the oldest of roofing materials, and clay tile was used by the builders of ancient Greece and Egypt. Clay and concrete tile come in two

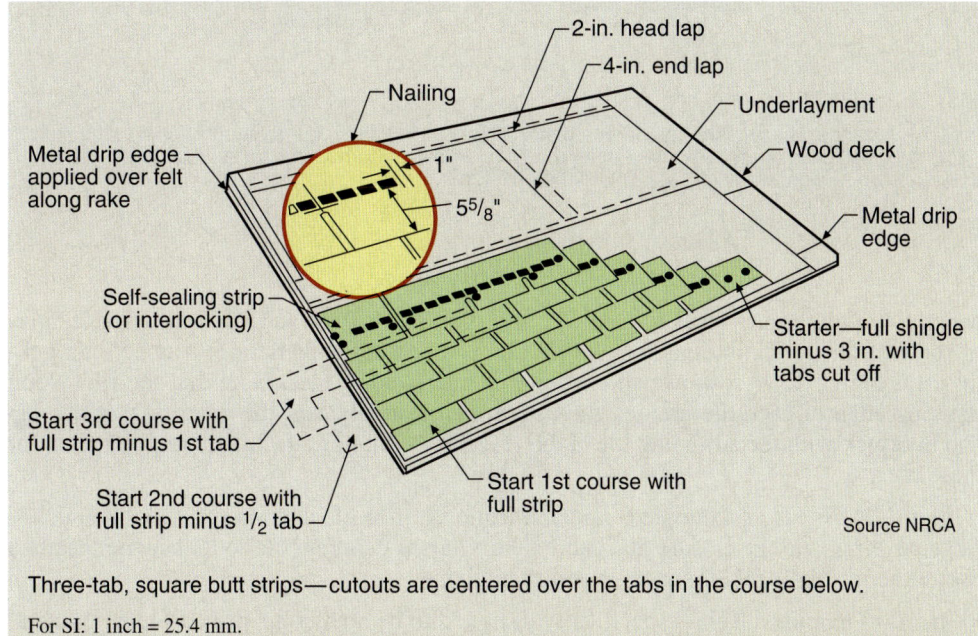

Figure 1507-1 Asphalt roofing shingle application on high slope (4:12 minimum).

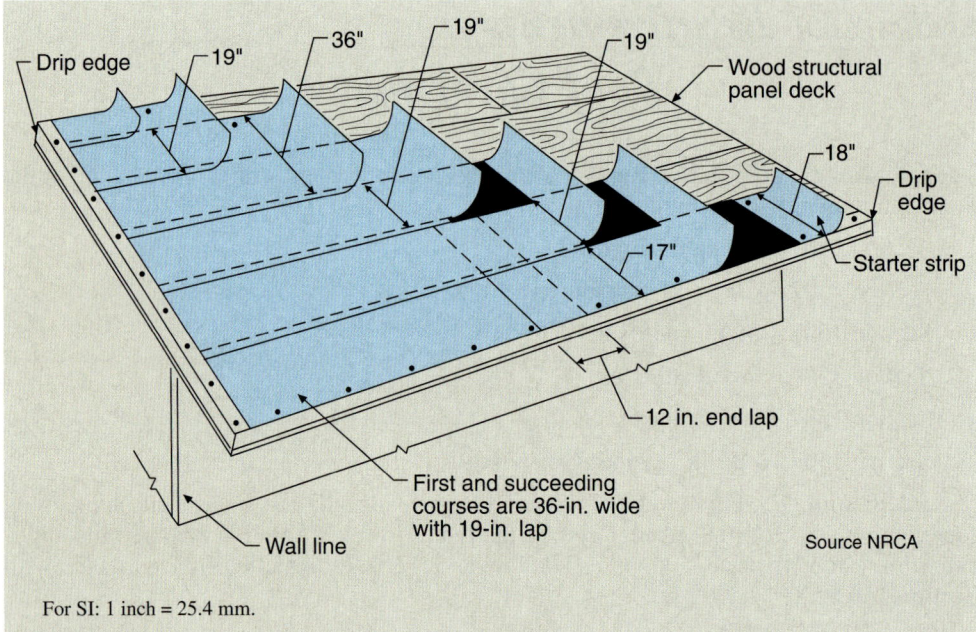

Figure 1507-2
Low slope (less than 4:12) underlayment application.

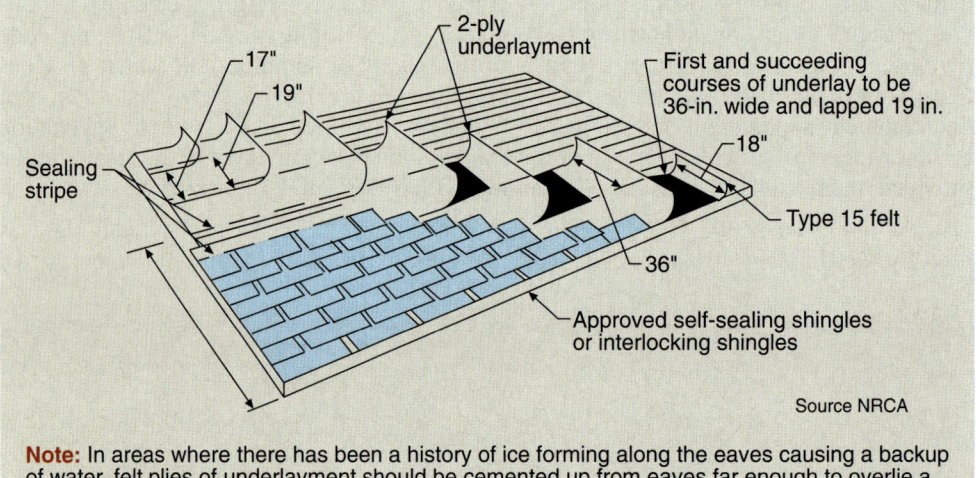

Figure 1507-3
Application of asphalt shingle on slopes between 2:12 and 4:12.

Note: In areas where there has been a history of ice forming along the eaves causing a backup of water, felt plies of underlayment should be cemented up from eaves far enough to overlie a point 24 in. inside the inside wall line of the building.

For SI: 1 inch = 25.4 mm.

generic configurations—roll tile and flat tile. Figures 1507-4 and 1507-5 give examples of some of the configurations for roofing tile. Either roll tile or flat tile may be interlocking, and Figure 1507-5 shows an example of interlocking tiles. Note that the ribs along the long edge of each tile are designed such that each adjacent tile has ribs that overlap and interlock with the adjoining tile. Table 1507.3.7 contains attachment requirements for these tiles.

Figure 1507-6 is an example of the application of roll tile and also provides details for ridge covering and the closure at a gable rake. Clay or concrete tile roofs may be installed over either a solid deck or spaced wood sheathing.

Metal roof panels. Unless specifically designed to be applied to spaced supports, metal roof panel roof coverings must be applied to a solid or closely fitted deck. Depending on

Requirements for Roof Coverings 15

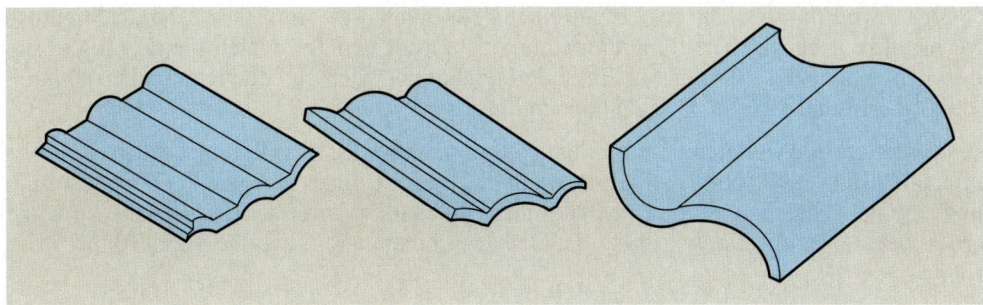

Figure 1507-4
Clay roll tile.

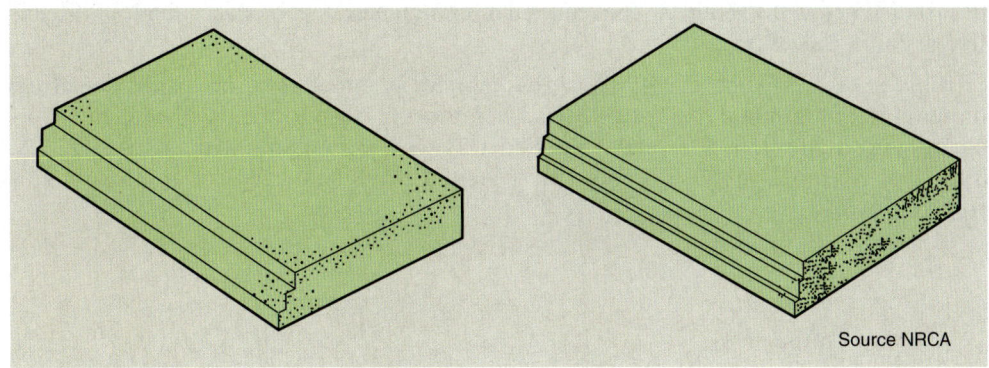

Source NRCA

Figure 1507-5
Concrete flat tile.

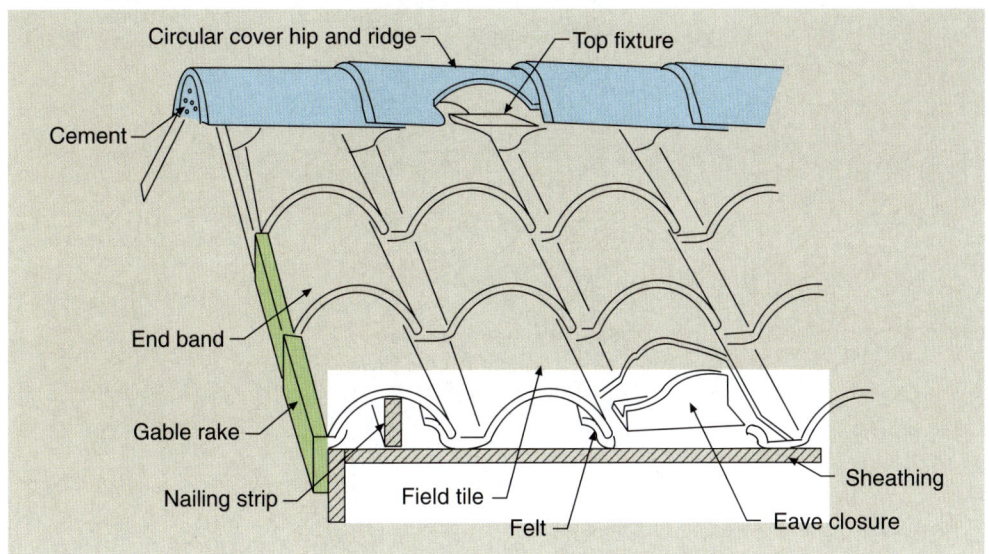

Figure 1507-6
Application of roll tile.

the type of metal-roof system, the minimum slope of the roof deck varies between $1/4$:12 and 3:12. The attachment of metal roofing should be based on the recommendations of the manufacturer. Where such recommendations are not provided, stainless-steel fasteners shall be used. Galvanized fasteners are permitted for galvanized roofs, and hard copper or copper-alloy fasteners shall be used for copper roofs. Table 1507.4.3(1) identifies the application rate or thickness for metal-sheet roof coverings installed over structural decking.

2012 International Building Code Handbook 515

15 Roof Assemblies and Rooftop Structures

Metal roof shingles. The deck requirements for metal roof shingles are identical to those for metal roof panels. The minimum slope of the roof deck on which metal shingles are installed is to be 3:12. Provisions for underlayment and flashing are very similar in nature to those for other types of shingle applications.

Mineral-surfaced roll roofing. Limited to application on solidly sheathed roofs only, mineral-surfaced roll roofing shall not be applied on roof slopes less than 1:12. Conforming underlayment shall be provided, and additional precautions must be taken in severe climate areas. Several publications provide detailed recommendations for the application of roll roofing.

Slate shingles. The application of slate shingles is also limited to solidly sheathed roofs. Because of the nature of these roofing materials, a deck slope of at least 4:12 is required, and underlayment complying with ASTM D 226, Type I or ASTM D 4869 must be installed below the shingles. The required minimum head lap for slate shingles is identified in Table 1507.7.6.

Wood shingles. Wood shingles may be applied on either solid or spaced sheathing on roofs with a minimum slope of 3:12 when the exposure to the weather is in accordance with Table 1507.8.6, and with underlayment complying with ASTM D 226, Type I or ASTM D 4869. Figures 1507-7 and 1507-8 show typical applications of wood shingles.

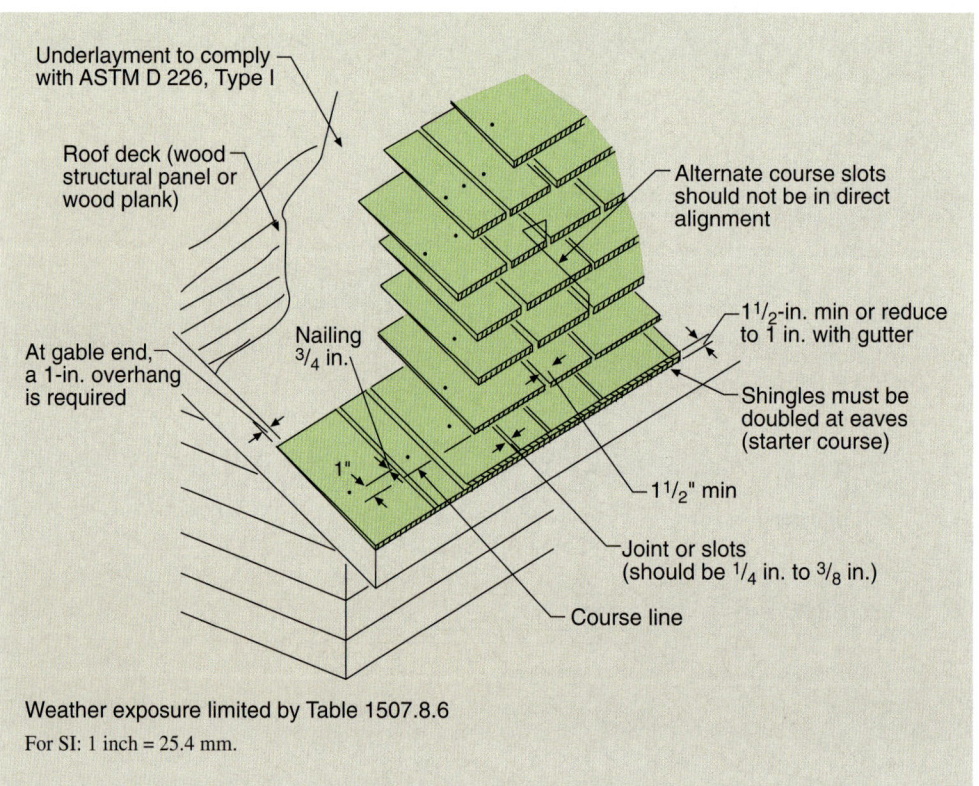

Figure 1507-7
Wood shingle application.

Wood shakes. Wood shakes are to be applied on roofs with a minimum slope of 4:12 on either spaced or solid sheathing. In addition to the required underlayment, felt interlayment is required to be shingled between the courses of wood shakes. As with other special requirements for shingle and tile roofs laid in areas of severe cold weather, extra precautions are required to prevent the backup of water on the roof and under the shingles

Requirements for Roof Coverings

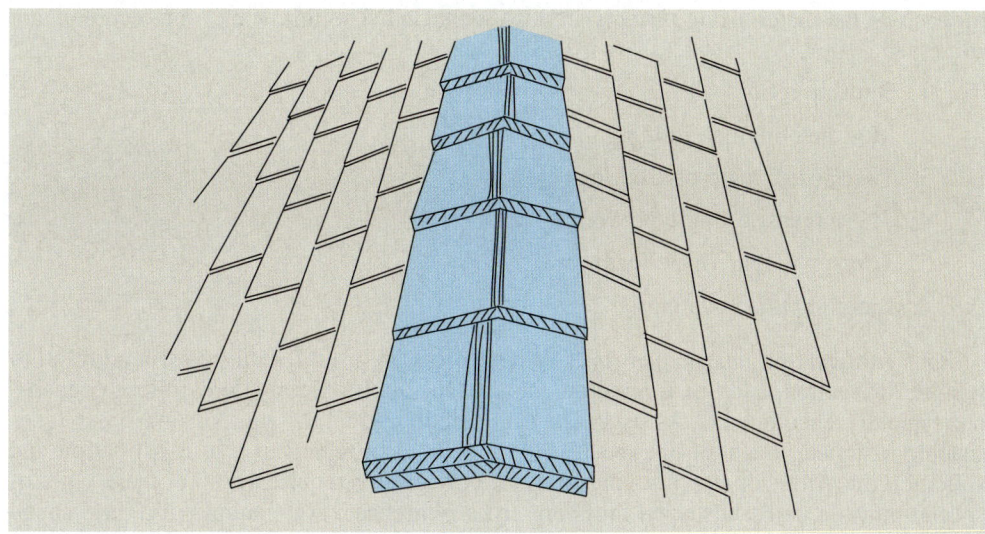

Figure 1507-8
Application of wood shingles at ridges.

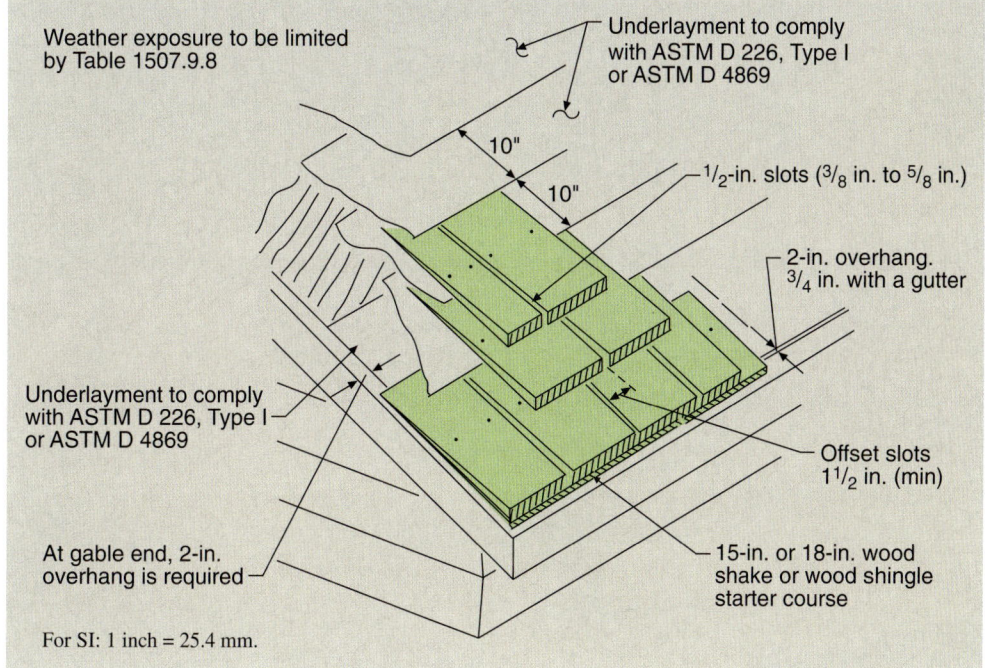

Figure 1507-9
Wood shake application.

with resulting leaks into the building. Figure 1507-9 illustrates an example of wood-shake application.

Both wood shingles and wood shakes are required by the code to have certain maximum exposures to the weather, and in the case of shingles, this exposure is dependent on the roof slope. These exposures for roofs with slopes of 4:12 or greater provide a minimum of three thicknesses of shingle with an overlap of $1^1/_2$ inches (38 mm) and have been shown through experience to be the maximum exposure that will provide a leak-free roof.

Other roof coverings. Sections 1507.10 through 1507.15 address those roof covering methods typically applicable to flat roofs. For most installations, a minimum roof slope of one vertical in 48 units horizontal ($^1/_4$:12) is specified. The appropriate material standards for each of the materials or systems are also identified. The following roof coverings are addressed:

1. Built-up roofs.
2. Modified bitumen roofing.
3. Thermoset single-ply roofing.
4. Thermoplastic single-ply roofing.
5. Sprayed polyurethane foam roofing.
6. Liquid-applied coatings.

Roof gardens and landscaped roofs. Where roofs are used for purposes in addition to weather protection, additional requirements may be applicable to address those concerns not typically encountered. As established in Section 1507.16, gardens and other landscaping installed on a roof are specifically addressed in regard to roof construction and structural integrity. For structural purposes, the Chapter 16 requirements for these types of special-purpose and landscaped roofs are cross-referenced. An example of a landscaped roof is shown in Figure 1507-10.

Figure 1507-10
Roof garden.

The addition of rooftop vegetation or landscaping can provide a number of benefits in building construction, such as reducing the heat-gain and cooling demands, helping to control storm water runoff, and simply providing pleasant areas or sites within a community. However, these landscaped roofs also have a potential to place combustible vegetation in an area where it is less accessible to the fire department in the event of an emergency or to increase the potential fire exposure of either the building itself or buildings nearby. Section 317 of the IFC, as referenced in this section, has requirements that limit the size of a single landscaped roof area, require a minimum separation and protection between adjacent areas, and address maintenance issues and separation from other combustible rooftop elements such as penthouses or mechanical equipment.

Section 1508 Roof Insulation

Because of an increased level of energy consciousness, the use of roof insulation has become more and more prevalent, as it has distinct benefits not only in energy conservation but also in building occupant comfort. Insulation also provides a smooth uniform substrate for application of the roofing materials. The code requires that above-deck thermal insulation be covered by an approved covering and be in compliance with FM 4450 or UL 1256 when tested as an assembly. For foam plastic insulation used under roof coverings, see the commentary under Section 2603.

Section 1509 Rooftop Structures

Penthouses and other roof structures are regulated by the IBC as if they were appurtenances to the building rather than occupiable portions.

In fact, if a penthouse is used for any purpose other than shelter of mechanical equipment or shelter of vertical shaft openings, the code requires that it be considered an additional story of the building.

As intended by the IBC, roof structures are equipment shelters, equipment screens, platforms that support mechanical equipment, water-tank enclosures, and other similar structures generally used to screen, support, or shelter equipment on the roof of the building. This section also regulates towers and spires, which are addressed separately in Section 1509.5.

The IBC regulates penthouses, roof structures, tanks, towers, and spires to prevent hazardous conditions that are due to internal and external fire concerns or structural inadequacy, and to ensure their proper use as equipment shelters.

1509.2 **Penthouses.** The code does not regulate the height of penthouses and roof structures on Type I buildings. However, for buildings of other construction types, the code limits the height of penthouses and roof structures to 18 feet (5,486 mm) above the height of the roof, except where the penthouse is used to enclose a tank or elevator. In such cases, a maximum height of 28 feet (8,534 mm) is permitted. As this section also limits the aggregate area of all penthouses and roof structures to one-third the area of the roof, the additional height permitted for penthouses and roof structures does not pose any significant fire- and life-safety hazard that is due to other restrictions that this section places on construction. See Figure 1509-1.

Penthouses and other rooftop structures are also regulated by the provisions of Section 504.3, where the requirements are based on the effect of a rooftop structure on the allowable height permitted for the entire building. Where applicable, the provisions from both Chapters 5 and 15 are in effect, and where there is a conflict, the most restrictive condition will apply.

As the code has reduced requirements for construction of penthouses and roof structures, it is logical that their use should be limited as specified in this section. Thus, if other uses are made of penthouses or other roof structures, it also seems appropriate that they should be constructed as would be required for an additional story of the building. It is

15 Roof Assemblies and Rooftop Structures

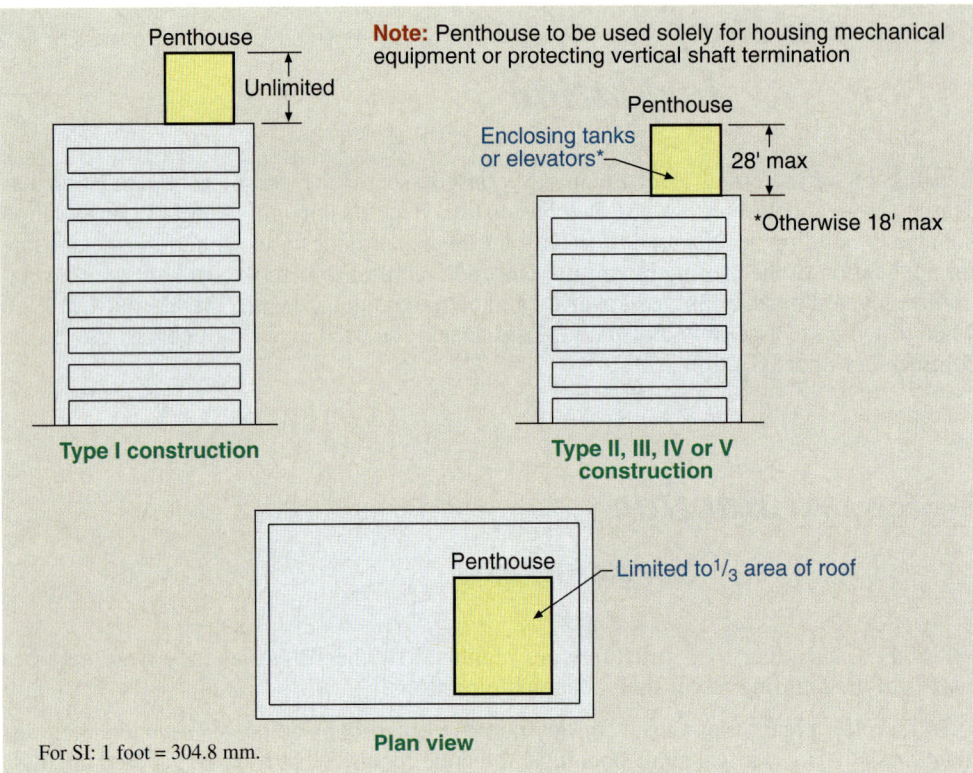

Figure 1509-1 Penthouse limitations.

the intent of the code that a penthouse or roof structure complying with this section not be considered to create an additional story above that permitted by Section 503.

If a rooftop structure qualifies as a penthouse, it is intended that the floor of the penthouse is only required to meet the roof provisions of Table 601. As an extension of this recognition, any fire-resistance-rated shaft that extends to or through the penthouse floor does not need to be protected at the floor line. Under both conditions, the floor of the penthouse is solely regulated as the building's roof. Regarding any required means of egress from the penthouse, the provisions for an occupied floor are not applicable, nor are the requirements of Section 1021.1 for an occupied roof. It would seem appropriate that the access provisions of the *International Mechanical Code*® (IMC®) for rooftop equipment also provide for adequate egress.

1509.2.5 Type of construction. The intent of the code is that penthouses be constructed with the same materials and the same fire resistance as required for the main portion of the building. However, because of the nature of their use, the code does permit exceptions for exterior walls, roofs, and interior walls. Where the exterior walls of penthouses are at least 5 feet (1,524 mm) [or in some cases 20 feet (6,096 mm)] from the lot line, reductions in any required fire resistance are typically permitted. See Figure 1509-2.

1509.5 Towers, spires, domes, and cupolas. The IBC intends that towers, spires, domes, and cupolas be considered separately from penthouses and other roof structures. The towers contemplated in this section are towers such as radio and television antenna towers, church spires, and other roof elements of similar nature that do not support or enclose any mechanical equipment and that are not occupied. As with penthouses and other roof structures, the code intends to obtain construction and fire resistance consistent with that of the building to which they are attached. Under a variety of conditions, however, towers and similar elements are required to be constructed of noncombustible materials, regardless of the building construction.

Reroofing

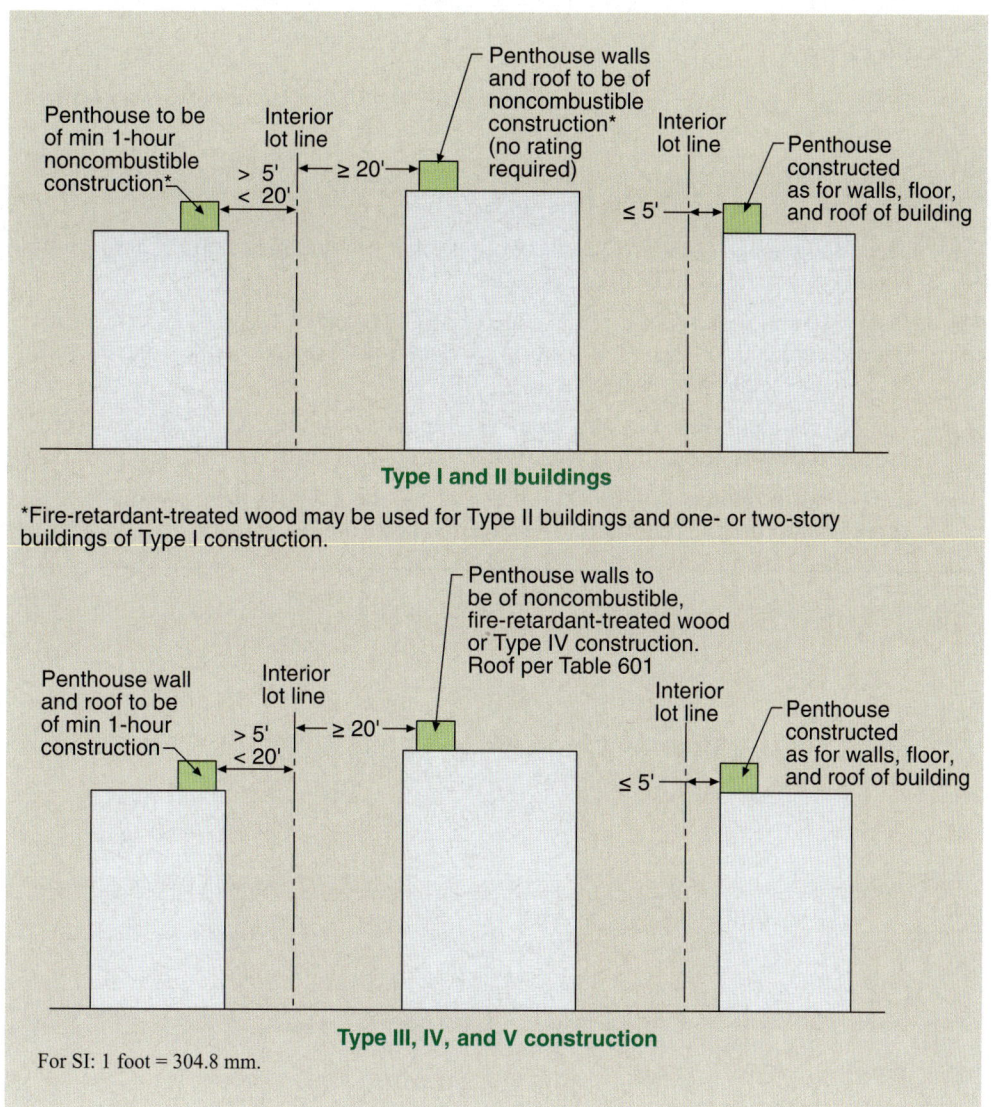

Figure 1509-2
Penthouse construction.

Section 1510 *Reroofing*

This section addresses the problems associated with unregulated reroofing operations. The intent of these provisions is to ensure that when an existing building is reroofed, the existing roof is structurally sound and in a proper condition to receive the new roofing. With exceptions, the replacement of existing roof coverings is required where any one of the following conditions occurs:

1. The existing roof or roof covering is water soaked or has deteriorated such that it is an inadequate base for additional roofing.

2. The existing roof covering consists of wood, slate, clay, cement, or asbestos-cement tile.

3. There are currently two or more applications of roof covering on the existing roof.

15 Roof Assemblies and Rooftop Structures

> **KEY POINTS**
>
> - Roofs and roof coverings are addressed for both weather protection and fire retardancy.
> - The performance of a roof for weather protection is regulated against three concerns—wind, weathering, and impact.
> - The IBC generally requires roof coverings that provide protection for the roof against moderate or light fire exposures.
> - Roof assemblies and roof coverings are classified as either Class A, B, or C roofing assemblies; nonclassified roofing; fire-retardant-treated wood shingles and shakes; or special-purpose roofs.
> - Penthouses and other roof structures are regulated by the code as if they were appurtenances to the building rather than occupiable portions.
> - Reroofing is limited to those buildings where the existing roof is structurally sound and in proper condition to receive the new roofing.

INTRODUCTION TO THE STRUCTURAL PROVISIONS

Introduction to the Structural Provisions

This introduction provides a brief background on the structural provisions in building codes in the United States, including the development of the *International Building Code®* (IBC®). For a more detailed discussion of the history of structural loads and seismic design provisions in the IBC, refer to the *2009 IBC Handbook—Structural Provisions*[1] published by ICC.

There was a difference in the way standards were adopted into the UBC and the other two model codes. The BOCA/NBC[2] and SBC[3] adopted standards by reference, reproducing only such portions as were likely to be frequently used by building officials, making few, if any, modifications in the process. The UBC, on the other hand, generally adopted standards by transcribing provisions from the standards into the body of the code itself. Frequently used standards such as material design standards were generally reproduced in the code itself, while the other standards were transcribed as UBC Standards. Modifications to the structural provisions, sometimes significant ones, were often made during the code development process.

The topic of adoption of standards by reference rather than transcription was the subject of lengthy debates within the IBC Structural Subcommittee. Primary arguments against transcribing standards have been that the practice increases the length of the code and that, more importantly, provisions of the referenced standards become subject to change through the code change process. Legal ramifications of transcribing only portions of a standard into the code have also been pointed out: code users may rely on the code alone for what they need when they should be relying on the standard. In the end, the ICC chose to emulate the BOCA and SBCCI practice of adopting standards by reference. Thus, the IBC adopts many standards by reference. From the 2000 IBC to the current 2012 IBC, many structural provisions were deleted from the code in favor of referencing standards. An example is Section 2305, which has been significantly reduced in favor of a reference to the AF&PA SDPWS standard.

The current 2012 edition of the IBC references several hundred national standards, which are listed alphabetically in Chapter 35. Note that the year or edition of the standard is only shown in Chapter 35, not in the body of the code. For example, Section 1613 references ASCE 7 for determination of seismic load effects, but does not indicate what edition. Chapter 35 shows the edition (year) of the standard being referenced. The main structural standards referenced in the 2012 IBC for loads and materials are shown in Table I-1.

Table I-1. **Key Referenced Structural Standards in the 2012 IBC**

Subject	IBC Chapter	Referenced Standard in 2012 IBC
Structural loads	16	ASCE/SEI 7-10
Concrete	19	ACI 318-11
Aluminum	20	ADM1-2010
Masonry	21	TMS402-11/ACI 530-11/ASCE 5-11 TMS602-11/ACI 530.1-11/ASCE 6-11
Structural steel	22	AISC 360-10 AISC 341-10
Cold-formed steel	22	AISI S100-07/S1-10[1]
Wood	23	AWC NDS-2012 AF&PA SDPWS-08

[1]See Chapter 35 for complete list of AISI standards referenced in the IBC.

Structural Design (Chapter 16): Nonseismic

Chapter 16 of all three legacy model codes as well as the IBC address general design requirements and prescribe various types of structural loads and, also very importantly, design load combinations.

The 2000 IBC used the 1998 edition of ASCE 7[4] as the basis of all provisions related to nonseismic loads, making only relatively rare exceptions. The 2003 IBC updated to ASCE 7-02 and the 2006 IBC updated to ASCE 7-05. In the 2006 IBC, much of the text was dropped from various sections of Chapter 16 in favor of provisions covered by ASCE 7-05. Simplified wind design provisions, for instance, were to be found only in Chapter 6 of ASCE 7-05. The 2009 IBC also referenced ASCE 7-05 because the 2010 edition of ASCE 7 was still under development when the 2009 IBC was published. The 2012 edition of the IBC references ASCE 7-10 for structural loads. Many of the code changes to structural loads in the 2012 IBC were done in order to coordinate the load provisions in the IBC with the corresponding provisions in ASCE 7-10.

Structural Design (Chapter 16): Seismic

Until the very beginning of the 1990s, seismic design provisions in U.S. building codes followed a certain pattern of development. Earthquake design provisions were first proposed by the Structural Engineers Association of California (SEAOC) in its *Recommended Lateral Force Requirements*[5] (commonly referred to as the "SEAOC Blue Book"), then incorporated by ICBO into its *Uniform Building Code* (UBC). The UBC provisions were then adopted (often with modifications) by the American National Standards Institute (ANSI) Standard A58.1 (later to become ASCE 7), which, in turn, was adopted by the other two model codes: the BOCA/NBC and the SBC. Thus, the seismic design provisions of all three model codes were based on the SEAOC *Blue Book* provisions.

A departure from the above pattern was initiated in 1972 when the National Science Foundation and the National Bureau of Standards (now the National Institute of Standards and Technology) jointly sponsored a Cooperative Program in Building Practices for Disaster Mitigation. Under that program, the Applied Technology Council (ATC) developed a document entitled *Tentative Provisions for the Development of Seismic Regulations for Buildings.*[6] This document, published in 1978 and commonly referred to as ATC 3-06, underwent a thorough review by the building community in ensuing years. Trial designs were conducted to establish the technical validity of the new provisions and to assess their impact. A new entity, the Building Seismic Safety Council (BSSC), was created under the auspices of the National Institute of Building Sciences (NIBS) to administer and oversee the trial design effort. The trial designs indicated the need for certain modifications to the original ATC 3-06 document, and modifications were made. The resulting document was the first edition, dated 1985, of the National Earthquake Hazards Reduction Program (NEHRP) *Recommended Provisions for the Development of Seismic Regulations for New Buildings*[7] (*and Other Structures* was added to the title starting with the 1997 edition). Under continued federal funding from the Federal Emergency Management Agency (FEMA), this document has been updated every three years; the 1988, 1991, 1994, 1997, 2000, and 2003 editions of the NEHRP Provisions have been issued by the Building Seismic Safety Council.

The seismic design provisions of the IBC were treated separately from the rest of the structural provisions during the code development process. In 1996, the IBC Structural Code Committee agreed in concept for the IBC to be based on the 1997 edition of the NEHRP *Provisions*, which was being developed at the time the last edition of the UBC (1997) was published. A Code Resource Development Committee (CRDC), funded by FEMA, was formed under the direction of the BSSC to generate seismic code provisions based on the 1997 edition of the NEHRP *Provisions*, for incorporation into the 2000 IBC. This effort was successful and the CRDC submittal was accepted by the IBC Structural Code Committee for inclusion in the IBC. The seismic design provisions of the IBC Working Draft were thus based on the 1997 NEHRP *Provisions*, but also included a number of features from the 1997 UBC that were not included in the 1997 NEHRP *Provisions*. Many changes were made to the provisions of the IBC Working Draft through the two sets of code development hearings. The BSSC's Code Resource Support Committee (CRSC), a successor group to the CRDC, played an active role in this development by sponsoring changes of their own and by taking positions on other changes submitted at the code development hearings. Thus, the seismic design provisions of the first (2000) edition of the IBC were based on the 1997 NEHRP *Provisions*, with some of the features of the 1997 UBC also included.

The 2003 IBC references the seismic design provisions of ASCE 7-02, which were based on the 2000 NEHRP *Provisions*. The 2003 IBC gave the designer two specific options: seismic design entirely by ASCE 7-02, or seismic design by ASCE 7-02, as modified by the 2003 IBC. This approach proved to be cumbersome and confusing.

The 2006 IBC references the seismic design provisions of ASCE 7-05, which are based on the 2003 NEHRP *Provisions*. Unlike previous editions of the standard where the seismic provisions were entirely contained in Chapter 9, the seismic design provisions in ASCE 7-05 were reorganized and broken into various chapters. The provisions in ASCE 7-05 and ASCE 7-10 are now found in Chapters 11 through 23, and Appendices 11A and 11B. In addition, the seismic design provisions of the 2006, 2009, and 2012 IBC are contained in just one section, 1613. That section adopts, by reference, Chapters 11, 12, 13, and 15 through 23 of ASCE 7 as well as Appendix 11B (Existing Building Provisions). Chapter 14, Material Specific Requirements, is not adopted because the design of concrete, aluminum, masonry, wood, and steel structures must be done in accordance with the material Chapters 19 through 23 of the IBC, respectively, and not by Chapter 14 of ASCE 7. Also not adopted is Appendix 11A, which covers inspection and testing, because inspection and testing of a structure designed under the IBC must be done in accordance with Chapter 17 of the code and not by Appendix 11A of ASCE 7. The 2012 IBC references ASCE 7-10, which has essentially retained the same organization as ASCE 7-05 in regard to seismic design provisions.

The IBC uses seismic design categories (SDC) to determine permissible structural systems, limitations on height and irregularity, the type of lateral force analysis method that must be performed, the level of detailing for structural members and joints that are part of the seismic-force-resisting system, and for the components that are not. The SDC is a function of the nature of the occupancy (previously called Seismic Use Group in the 2000 and 2003 IBC and the 1997, 2000, and 2003 NEHRP *Provisions*) and of soil-modified seismic risk at the site of the structure.

In 1978, ATC 3-06, the predecessor document to the NEHRP *Provisions*, made the level of detailing (and other restrictions concerning permissible structural system, height, irregularity, and analysis procedure) a function of occupancy, which was continued in all the NEHRP *Provisions* through the 1994 edition. Now, in the IBC and the 1997 and subsequent NEHRP *Provisions*, the level of detailing and the other restrictions have been made a function of the soil characteristics at the site of the structure by virtue of the site class that affects the seismic design category of the structure.

FEMA found that the NEHRP *Provisions*' updates became preoccupied with maintaining congruence with the provisions in ASCE 7, which led to a major philosophical change in the 2009 edition of the NEHRP *Provisions* (FEMA P-750). By adopting the latest (2005) edition of ASCE 7 as the reference standard to be updated in the 2009 *Provisions*, instead of revising the previous (2003) edition of the *Provisions*, the developers of FEMA P-750 intend the *Provisions* to resume its role as the resource for introducing new knowledge, innovative concepts, and design methods with the goal of improving national seismic standards and codes. Unlike previous editions, the 2009 NEHRP *Provisions* is a single volume organized into three parts. Part 1 proposes specific modifications to ASCE 7-05 and provides a new set of national seismic design maps based on the U.S. National Seismic Hazard Maps released by USGS in 2008. Part 2 focuses on how to use ASCE 7-05 to design seismic-resistant structures. Part 3 comprises 13 resource papers on special topics in seismic design. Thus, the 2009 edition of the NEHRP *Provisions* not only represents a change in the three-year cycle on which this document has been published in the past, but also represents a change in the role it plays in the seismic codes and standards arena.

Structural Design (Chapter 16): Load Combinations

The design load combinations in all three of the legacy model codes progressively became inordinately complex, primarily because of an emphasis on modernizing the seismic requirements in each of those codes over their last decade of existence. Advancements in the understanding of seismic structural response brought about changes in design philosophy, which in turn resulted in changes in load combinations that are used when earthquake load effects are considered. The design load combinations of ASCE 7, if and when adopted, were subject to so many modifications that confusion was commonplace. A reader seeking clarity may benefit from consulting References 9–11 and Appendix 1 of the *2009 IBC Handbook—Structural Provisions* published by ICC.

The complexity involving load combinations continued into the 2000 IBC but since has dissipated with subsequent editions of the IBC and ASCE 7. The current state of design load combinations in the IBC and ASCE 7 is discussed in some detail under Section 1605.

REFERENCES

1. International Code Council, *2009 IBC Handbook, Structural Provisions*, Country Club Hills, IL, 2009.

2. Building Officials & Code Administrators International, *The BOCA National Building Code*, Country Club Hills, IL, 1993, 1996, 1999.

3. Southern Building Code Congress International, *Standard Building Code*, Birmingham, AL, 1994, 1997, 1999.

4. American Society of Civil Engineers, *ASCE Standard Minimum Design Loads for Buildings and Other Structures*, ASCE 7, Reston, VA, 2000, 2002, 2005, 2010.

5. Seismology Committee, Structural Engineers Association of California, *Recommended Lateral Force Requirements and Commentary*, San Francisco (later Sacramento), CA, 1974, 1988, 1996, 1999.

6. Applied Technology Council, *Tentative Provisions for the Development of Seismic Regulations for Buildings*, ATC Publication ATC 3-06, NBS Special Publication 510, NSF Publication 78-8, U.S. Government Printing Office, Washington, DC, 1978.

7. Building Seismic Safety Council, *NEHRP (National Earthquake Hazards Reduction Program) Recommended Provisions for the Development of Seismic Regulations for New Buildings (and Other Structures)*, Washington, DC, 1994 (1997, 2000, 2003, 2009).

CHAPTER 16

STRUCTURAL DESIGN

Introduction
Section 1601 General
Section 1602 Definitions and Notations
Section 1603 Construction Documents
Section 1604 General Design Requirements
Section 1605 Load Combinations
Section 1606 Dead Loads
Section 1607 Live Loads
Section 1608 Snow Loads
Section 1609 Wind Loads
Section 1610 Soil Lateral Loads
Section 1611 Rain Loads
Section 1612 Flood Loads
Section 1613 Earthquake Loads
Section 1614 Atmospheric Ice Loads
Section 1615 Structural Integrity
Key Points
References
Example 16-1 Design Axial Force, Shear Force, and Bending Moment for Shear Wall Due to Lateral and Gravity Loads (Strength Design)
Example 16-2 Design Axial Force, Shear Force, and Bending Moment for Shear Wall Due to Lateral and Gravity Loads (Allowable Stress Design Using Basic Load Combinations)
Example 16-3 Design Axial Force, Shear Force, and Bending Moment for Shear Wall Due to Lateral and Gravity Loads (Allowable Stress Design Using Alternate Basic Load Combinations)
Example 16-4 Calculations of Live Load Reduction

16 Structural Design

Introduction

This chapter explains and provides background on the development of the structural design requirements of Chapter 16 of the 2012 *International Building Code*® (IBC®). Significant portions of Chapter 16 provisions related to the determination of snow, wind, and seismic loads are included only by reference to ASCE 7 Standard *Minimum Design Loads for Buildings and Other Structures*.[1] Portions of these provisions that remain in the IBC relate to the local geologic, terrain, or other environmental conditions that many building officials will wish to specify when adopting the model code by local ordinance. In addition, the IBC includes a simplified alternative wind design provision that is not found in ASCE 7. The seismic provisions of the IBC are derived from two primary sources: ASCE 7 and the NEHRP *Provisions*.[2] Fortunately, both of these documents come with detailed commentaries, portions of which have at times been paraphrased in an attempt to make the discussion in this handbook reasonably self-contained.

Numerical examples have been included where they serve to illustrate the design requirements.

Section 1601 General

Chapter 16, Structural Design, governs the structural design of IBC-regulated buildings, nonbuilding structures, and portions thereof. A building is defined in Section 202 as any structure used or intended for supporting or sheltering any use or occupancy.

Chapter 16 provides requirements for minimum structural loads as well as criteria or methods of load application to be used in the design of buildings and other structures. The various types of structural loads specified by Chapter 16 are either gravity loads or lateral loads. Gravity loads specifically addressed are dead loads, live loads, snow loads, and rain loads. Lateral loads specifically dealt with are those due to wind, earthquakes, soil pressure, or flood. Loading conditions, such as uniformly distributed and concentrated live loads, impact loads, and most important, the design load combinations, are also regulated by the provisions of Chapter 16. Section 1601 presents the scope of the chapter. Section 1602 lists defined terms that are used in the structural design requirements. The actual definitions of the terms are provided in Section 202. Section 1603 specifies the minimum information that must be provided on the construction documents. Section 1604 gives general design requirements and specifically addresses strength criteria and serviceability criteria, including deflection limitations, structural analysis, risk categories, and load tests. Section 1605 addresses the vital topic of how to apply design load combinations. Strength design load combinations as well as allowable stress design load combinations are given. Section 1606 specifies design dead loads. Section 1607 specifies the minimum uniformly distributed live loads and minimum concentrated live loads for various types of uses and occupancies. This section also permits a reduction of design live loads under certain conditions. Section 1608 specifies the design snow loads, largely by reference to Chapter 7 of ASCE 7. Section 1609 contains structural design requirements for wind loads, largely by reference to Chapters 26 through 30 of ASCE 7. Section 1609.6 contains an alternative wind design method that may be used where applicable. Sections 1610, 1611, and 1612 treat soil lateral loads, rain loads, and flood loads, respectively. The rain load provisions of Section 1611 are by reference to Chapter 8 of ASCE 7. Section 1613 is devoted to the seismic design of buildings and other structures and references

Chapters 11 through 23 (excluding Chapter 14) of ASCE 7. Section 1614 references Chapter 10 of ASCE 7 for atmospheric ice loads, and Section 1615 includes provisions to provide a minimum level of structural integrity in certain types of high-rise buildings.

Section 1602 *Definitions and Notations*

Section 1602 lists structural terms that are used in Chapter 16.

Although some notations are given in Section 1602, symbols and notations are typically defined throughout the IBC following the equation(s) in which they are used. The ICC Structural Subcommittee thought this to be more user friendly than requiring the reader to refer back to the beginning of a chapter every time an equation appears.

Section 1603 *Construction Documents*

This section details the items to be shown on construction documents.

Construction documents are a part of the submittal documents required by Section 107, and the term is defined in Section 202. Note that the loads are not required to be on the construction drawings, but must be included within the construction documents in such a way that the design loads are clear for all parts of the structure. Of course, the indicated loads are required to be equal to or greater than the minimum loads required by the code. The information required to be included in the construction documents is useful to the building official in performing plan review and field inspection. It is also an extremely useful piece of information when additions or alterations are made to a structure at a later date. Each of the items indicated in Section 1603.1.4 is an important parameter in the determination of the wind loads that are required in the design of the structural system of the building.

The exception to Section 1603.1 simplifies the structural design information required for wood buildings constructed to the conventional light-frame construction provisions of Section 2308. A registered design professional is not required for the design of such buildings. However, many of the requirements of Sections 1603.1.1 through 1603.1.9 clearly would require the services of such a professional in order to provide the specified design data. The requirements in the exception are intended to provide adequate information for the building official to verify the structural design basis of wood-frame buildings built to these conventional construction provisions.

Section 1603.1.7 provides a pointer to Section 1612.5, which is entitled "Flood Hazard Documentation." Section 1612.5 requires statements to be included on the construction documents if certain flood-related situations exist. By including this pointer in Sections 1603.1.7 to 1612.5, the likelihood that these statements will be included on the plans is enhanced.

Section 1603.1.9 goes beyond items to be shown on construction documents. To ensure that earthquake-resistant systems and components are properly constructed and/or installed, special inspection for seismic resistance is required in Section 1705.11. To do this inspection, approved construction documents for these systems and components are required.

16 Structural Design

Section 1604 *General Design Requirements*

1604.1 General. This section requires buildings and other structures and all portions thereof to be designed and constructed in accordance with the general strength and serviceability requirements contained in Section 1604.

1604.2 Strength. The basic strength requirement is that buildings and structural systems be capable of supporting the factored loads without exceeding the applicable material strength limit. For structural elements designed using nominal rather than factored loads, the actual design stresses are not to exceed the applicable allowable stress levels.

1604.3 Serviceability. The requirements for serviceability mean that structural systems and members must have adequate stiffness to limit deflection and lateral drift to an appropriate degree based on the intended use. Specific requirements are given in Sections 1604.3.1 through 1604.3.6. Table 1604.3 contains deflection limits of structural members as a function of span and load type.

The general statement is adapted from Section 1.3.2 of ASCE 7, which reads as follows: "Structural systems, and members thereof, shall be designed to have adequate stiffness to limit deflections, lateral drift, vibration, or any other deformations that adversely affect the intended use and performance of buildings and other structures." Note that the IBC excludes any reference to vibration or any other deformations that have an adverse impact on intended use and performance of the structure because:

1. The code has no objectively defined standard for structural vibration. Acceptable vibration limits are frequently subjective and highly dependent on the specific requirements of occupants of a building. This information is not necessarily available to the building official.

2. It is impossible for the building official to anticipate everything that can "adversely affect the intended use and performance" of a building. Sections 1604.3.1 through 1604.3.6 provide objectively defined deflection limits that are deemed to suffice for a wide range of structures. Limits more restrictive than these should be a matter of the design professional understanding the client's needs and goals, but they are not typically part of a minimum life-safety building code. For example, there are situations in which sensitive computer, optical, or mechanical equipment requires extraordinary measures to limit their movement or vibration. These measures are often very complex and well beyond the life-safety requirements of most structures.

1604.4 Analysis. The first two paragraphs are reproduced with minor modifications from Section 1.3.4 of ASCE 7. The third paragraph can be traced back to 1997 *Uniform Building Code* (UBC)[3] Section 1605.2, Rationality. Similarly, the fourth paragraph is mostly from 1997 UBC Section 1605.2.1, Distribution of horizontal shear, and the last paragraph is from UBC Section 1605.2.2, Stability against overturning.

Structural analysis is to be based on fundamental principles of structural mechanics. These principles are equilibrium, stability, geometric compatibility, and material properties. Although the code in general does not intend to specify the design method used by the engineer, it does intend that the design method be rational and in accordance with well-established principles of mechanics. Departures from this latter requirement can still be made based on the provisions of Section 104.11 when approved by the building official. For example, the structural adequacy of a building may not admit to a rational analysis; a program of full-scale testing may be the only reasonable way to determine its structural behavior. If the testing program shows that a certain building can safely resist the loads required by the code, the building official may approve the construction of the building based on test results.

The requirement for stability against overturning makes reference to the following specific sections related to lateral loads: 1609 for wind, 1610 for lateral soil loads, and 1613 for earthquakes.

It is important to note that the requirement that provisions be made for the increased forces induced in resisting elements of the structural system, resulting from torsion that is due to eccentricity between the center of application of the lateral forces and the center of rigidity of the lateral-force-resisting system, does not apply "where diaphragms are flexible, or are permitted to be analyzed as flexible," because flexible diaphragms cannot transmit torsion.

Risk category. This section and Table 1604.5 make a combined presentation of the risk categories of buildings and other structures. The risk category reflects the relative anticipated seriousness of consequence of failure from lowest hazard to human life (Risk Category I) to highest (Risk Category IV), and is used to relate the criteria for maximum environmental loads and distortions specified in the code to the consequence of the loads being exceeded for the structures and their occupants.

1604.5

The term "occupancy category" was changed to "risk category" in the 2012 IBC to better reflect the meaning and to coordinate the terminology used in ASCE 7-10. The term "occupancy category" was felt to be misleading because it implies something about the nature of the building occupants, and the word "occupancy" relates primarily to the non-structural fire- and life-safety provisions, not the risks associated with structural failure. In fact, some of the structures regulated by the IBC and IEBC are not even occupied, but have an occupancy category assigned because their failure could pose a substantial risk to the public. Although the terminology changed, the classifications continue to reflect the progression of the consequences of failure from the lowest risk category to the highest risk category. A detailed discussion of the risk categories is contained in Section C1.5 of the ASCE 7-10 commentary.

Risk Category I contains buildings and other structures that represent a low hazard to human life in the event of failure, either because they have a small number of occupants or because they have a limited period of exposure to extreme environmental loading.

Risk Category II contains all occupancies other than those in Risk Categories I, III, or IV, and are sometimes referred to as *ordinary* for the purpose of risk exposure.

Risk Category III contains those buildings and other structures that have large numbers of occupants, that are designed for public assembly, or in which physical restraint or other incapacity of occupants hinders their movement or evacuation. Therefore, these structures represent a substantial hazard to human life in the event of a failure. Risk Category III also includes important infrastructure structures such as power-generating stations, water treatment facilities, and so on, where a failure may not create an unusual life-safety risk, but can cause large-scale economic impact and/or mass disruption of day-to-day civilian life.

For further clarity, Table 1604.5, in some cases, defines the risk categories based on the occupancy classifications defined in Chapter 3 of the code. For instance, hospitals and health-care facilities are referred to as Group I-2 occupancies to specify the requirements of Risk Categories III and IV in order to make it clear that the care facilities in Group I-1 classification are not intended to be assigned to Risk Category III or IV.

Note that the footnote to Table 1604.5 makes a distinction between the calculation of occupant load for the purpose of determining whether a building qualifies for Risk Category III because the occupant load exceeds 5,000. The calculation of the occupant load for the purpose of Table 1604.5 can be based on the net floor area even when Table 1004.1.2 requires gross floor area to be used. The net floor area does not include corridors, stairways, elevators, closets, accessory areas, structural walls and columns, and so on. This is because of the basic difference in the purposes the two tables serve. Table 1004.1.2 provides minimum occupant loads for the purpose of egress design, and egress design is determined on a floor-by-floor basis, assuming that the maximum occupant load can occur on a certain floor at a given time. However, for the purpose of Table 1604.5, occupant loads are calculated for the whole building, and it is very unlikely that all floors of a building would have the maximum occupant load at the same time. This is similar to the reasoning that is behind permitting live load reductions in the code.

Risk Category IV contains buildings and other structures that are designated as essential facilities, and are intended to remain operational in the event of extreme loading such as hospitals, fire stations, and so on. Also included are the structures that are supplementary to Risk Category IV structures, which are required for the operation of Risk Category IV facilities during an emergency, for example, facilities to maintain water pressure for fire suppression. Furthermore, structures holding extremely hazardous materials are also included in Risk Category IV because of the potentially adverse effect of a release of those materials in the environment.

1604.5.1 **Multiple occupancies.** In the case where there are multiple occupancies in a structure, the highest (or most restrictive) risk category is to be assigned to the structure unless the portions are structurally separated. In other words, when a lower risk category impacts a higher risk category, the higher risk category must either be structurally independent of the other, or the two must be in one structure designed to the requirements of the higher risk category. In cases where the two uses are structurally independent but are functionally dependent, both portions are required to be assigned to the higher risk category. To be a structural separation, there must be a physical separation (not just an expansion joint) that conforms to the requirements of ASCE 7 Section 12.12.3. If portions do not meet the minimum separation distance, then they are not structurally separated and must be designed and interconnected to act as one unit.

1604.6 **In situ load tests.** Whenever there is reasonable doubt as to the stability or load-bearing capacity of a completed building, structure, or portion thereof for the expected loads, an engineering assessment may be required by the building official. The engineering analysis may involve either a structural analysis or an in situ load test, or both. See IBC Section 1709 for more details.

1604.7 **Preconstruction load tests.** In evaluating the physical properties of materials and methods of construction that are not capable of being designed by approved engineering analysis, or that do not comply with applicable material design standards listed in Chapter 35, the structural adequacy must be predetermined based on the load test criteria given in Section 1710.

1604.8 **Anchorage.** This section addresses the anchorage of the various components of a building to resist the uplift and sliding forces that result from the application of the code-prescribed lateral forces. It intends that all members be tied together or anchored to resist the uplift and sliding forces. The section differentiates between the uplift and sliding forces to be resisted in general (Section 1604.8.1) and the lateral support required for structural walls (Section 1604.8.2). Anchorage requirement of decks attached to an exterior wall is provided in Section 1604.8.3.

Many observed failures of concrete or masonry walls in the 1971 San Fernando and the 1994 Northridge earthquakes were attributable to inadequate anchorage between the walls and the roof system. Although code requirements for anchorage to prevent the separation of heavy masonry or concrete walls from floors or roofs have been common in areas of high seismicity, they have been minimal or nonexistent in most other parts of the country. While a minimum wall anchorage requirement alone may not provide complete earthquake-resistant design, observations of earthquake damage indicate that it can greatly increase the earthquake resistance of buildings and reduce hazards in those localities where earthquakes may occur but are rarely damaging.[2] Note that IBC Section 1604.8.2 refers to walls of all materials, not just concrete and masonry walls, as was the case in earlier editions of the IBC.

Structural walls. These requirements pertain to "structural walls," which are essentially defined within this section to include load-bearing walls and shear walls. This section establishes the minimum wall anchorage capacity, based on the structure's seismic design category (SDC). In accordance with the provisions of ASCE 7, Section 1.4.5 is referenced for the determination of the minimum anchorage force in structures that are classified as SDC A. The referenced ASCE 7 provision provides minimum connection forces based on the consideration of a minimum level of structural integrity, and in the context of Section 1604.8.2, it means that wall anchorage needs to be adequate to carry a horizontal force equal to 0.2 times the wall weight tributary to the anchor, but not less than 5 psf (see Figure 1604-1). For all other SDCs, the more stringent requirements in Section 12.11 of ASCE 7 are referenced.

1604.8.2

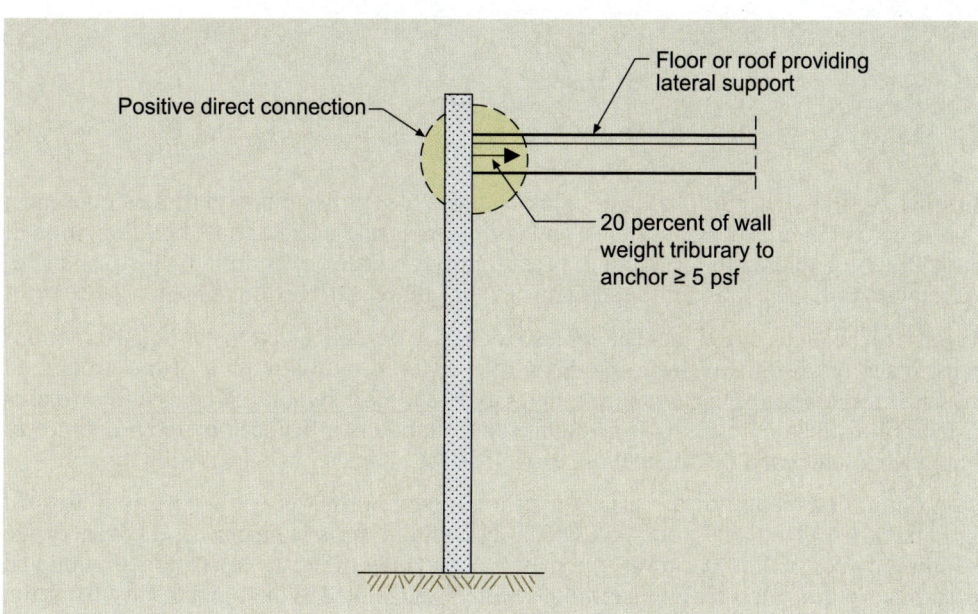

Figure 1604-1
Minimum anchorage of structural walls in SDC A.

If the wall being laterally supported is of hollow-unit masonry or a cavity wall, the required anchors must be embedded in a reinforced grouted structural element of the wall.

Decks. When a deck is attached to an exterior wall, such attachment needs to be capable of resisting the effects of dead load as well as live load (as specified in IBC Table 1607.1) or snow load. Such connections need to be verified during inspection. Otherwise, the deck needs to be capable of supporting itself without any attachment to a wall.

1604.8.3

When a deck is supported by an exterior wall at one end as well as by an intermediate support from a framing member (Figure 1604-2) such that a portion of the deck cantilevers beyond the intermediate support, two load cases need to be considered for the purpose of designing the attachment to the wall. In the first case, in addition to the dead load, the full span of the deck is subjected to the live load or the snow load, whichever is higher.

This load case provides the maximum possible downward reaction in the wall attachment. In the second load case, only the cantilevered portion of the deck is subjected to the live load or the snow load in order to obtain the maximum uplift force in the wall attachment. In both cases, loads need to be combined in accordance with Section 1605 of the IBC.

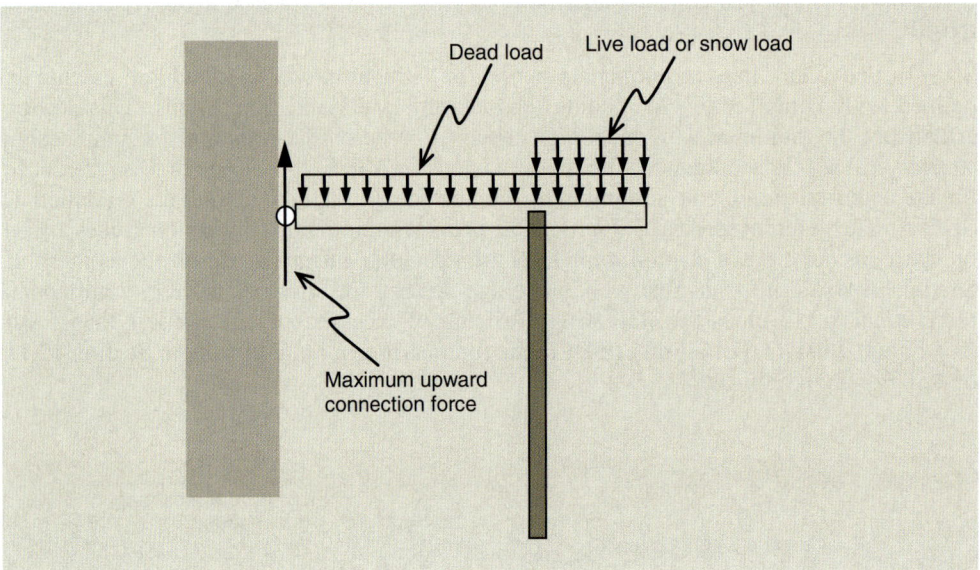

Figure 1604-2
Application of live/snow load on a cantileverd deck.

1604.9 Counteracting structural actions. This contains a very basic and general requirement that structural members and systems, and components and cladding in a building or other structure, be anchored to resist wind- or earthquake-induced overturning, uplift, and sliding, and to provide continuous load paths for those forces to the foundation.

1604.10 Wind and seismic detailing. The forces that a structure subjected to earthquake motions must resist result directly from the distortions induced by the motion of the ground on which it rests. The response (i.e., the magnitude and distribution of forces and displacements) of a structure resulting from such a base motion is influenced by the properties of both the structure and the foundation, as well as the character of the exciting motion.

A simplified picture of the behavior of a building during an earthquake is illustrated in Figure 1604-3. As the ground on which the building rests is displaced, the base of the building moves with it. However, the inertia of the building mass resists this motion and causes the building to suffer a distortion (greatly exaggerated in the figure). This distortion wave travels along the height of the structure in much the same manner as a stress wave in a bar with a free end. The continued shaking of the base causes the building to undergo a complex series of oscillations.

It is important to draw a distinction between forces that are due to wind and those produced by earthquakes. Occasionally, even engineers tend to think of these forces as belonging to the same category just because codes specify design wind as well as earthquake forces in terms of equivalent static forces. Although both wind and earthquake forces are dynamic in character, a basic difference exists in the manner in which they are induced in a structure. Whereas wind loads are external loads applied and, therefore, proportional to the exposed surface of a structure, earthquake forces are essentially inertia forces. The latter results from the distortion produced by both the earthquake motion and inertial resistance of the structure. Their magnitude is a function of the mass of the structure rather than its exposed surface. Also, in contrast to the structural response to essentially static gravity loading or even to wind loads, which can often be validly treated as static loads, the dynamic character of the response to earthquake excitation can seldom be ignored. Thus, although

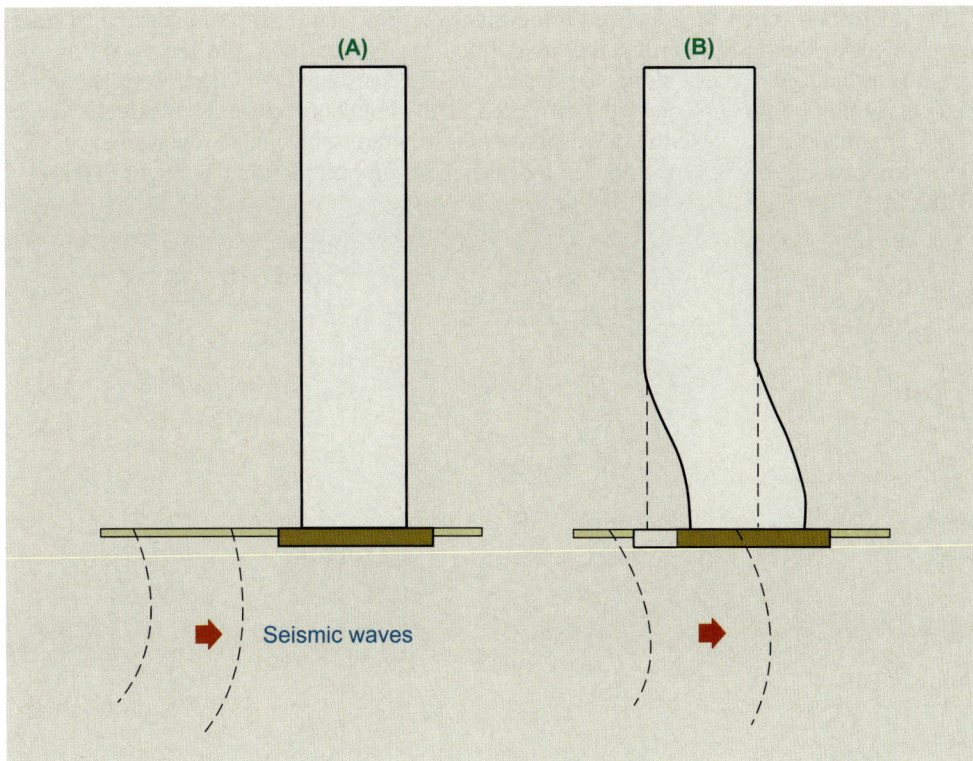

Figure 1604-3
Behavior of building during an earthquake.

in designing for static loads one would feel greater assurance about the safety of a structure made up of members of heavy section, in the case of earthquake loading, the stiffer and heavier structure does not necessarily represent the safer design.

When a structure responds elastically to ground motions during a severe earthquake, the maximum response accelerations may be several times the maximum ground acceleration and may depend on the mass and stiffness of the structure and the magnitude of the damping. It is generally uneconomical and also unnecessary to design a structure to respond in the elastic range to the maximum earthquake-induced inertia forces. Thus, the design seismic horizontal force recommended by codes is generally less than the elastic response inertial forces induced by a major earthquake (the design earthquake ground motion is defined in Section 1613).

Experience has shown that structures designed to the level of seismic horizontal forces required by current codes can survive major earthquake shaking. This is because of the ability of well-designed structures to dissipate seismic energy by inelastic deformations in certain localized regions of certain members. Decrease in structural stiffness caused by accumulating damage and soil-structure interaction also helps at times. It should be evident that the use of the level of seismic design forces recommended in our codes implies that the critical regions of inelastically deforming members should have sufficient inelastic deformability to enable the structure to survive without collapse when subjected to several cycles of loading well into the inelastic range. This means avoiding all forms of brittle failure and achieving adequate inelastic deformability by the yielding of certain localized regions of certain members (or of connections between members) in flexure, shear, or axial action. This is precisely why the materials chapters (Chapters 19, 21, 22, and 23) provide detailing requirements and other limitations that go hand in hand with the code-prescribed seismic forces. The design earthquake forces (Chapter 16) and the detailing requirements and other restrictions of the materials chapters form an integral package for seismic design.

Figure 1604-4 shows the idealized force-displacement relationship of a particular structure subject to the design earthquake, as defined in Section 1613. On the y-axis are the earthquake-induced forces; along the x-axis are the earthquake-induced displacements. The curve may be thought of as the envelope or the backbone curve of hysteretic force-displacement loops that describe the response of a structure subjected to reversed cyclic displacement histories of the type imposed by earthquake ground motion (see Figure 1604-5).[4]

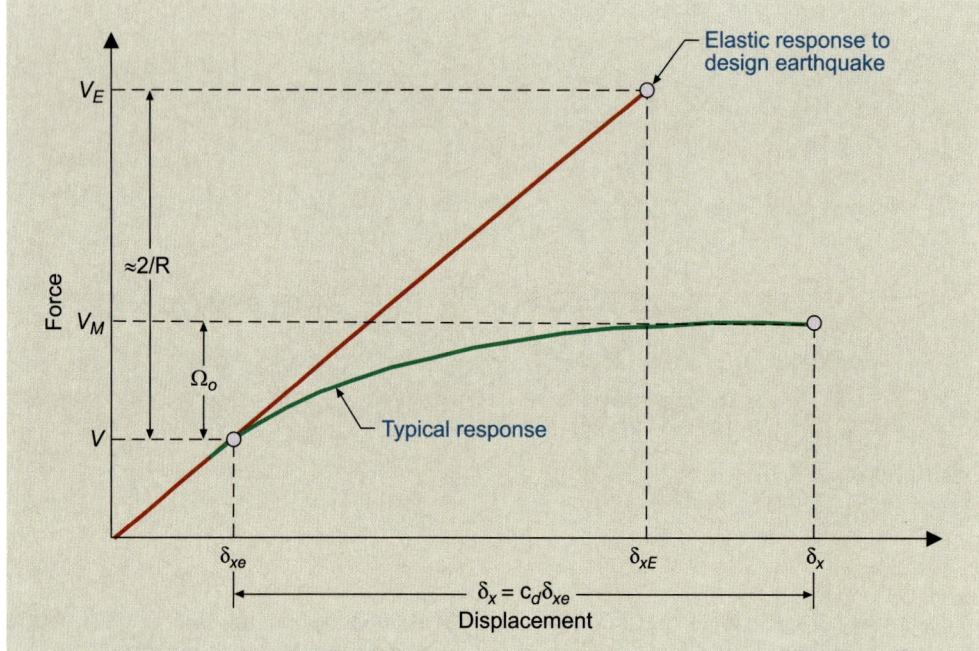

Figure 1604-4
Idealized force-displacement relationship of a building subjected to the IBC design earthquake.

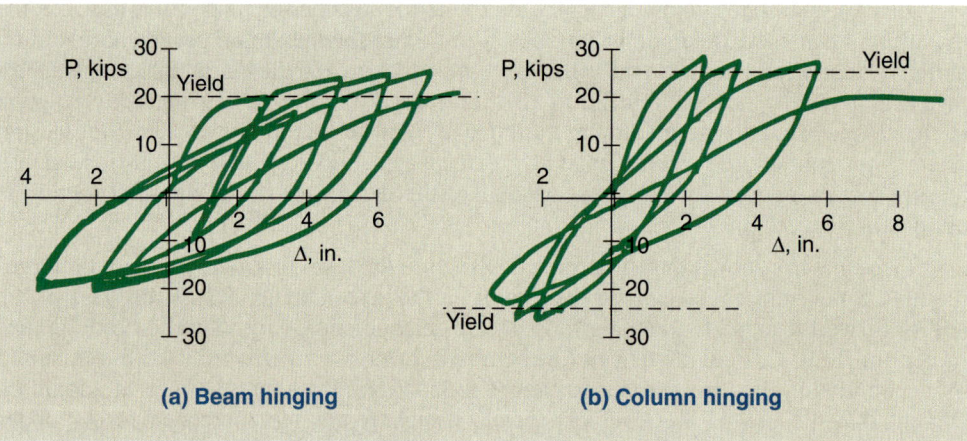

Figure 1604-5
Load-deflection curves of structural subassemblies subjected to reverse cyclic displacements.

It should be obvious from the above that as long as seismic design is done using code-prescribed forces, which are reduced below the level that would have produced elastic structural response to the design earthquake, the detailing requirements of the materials chapters must also be met, irrespective of how high the wind effects might be in comparison with the earthquake effects. Wind and earthquake effects are not considered to occur simultaneously on the structure in U.S. design practice. Section 1605 contains gravity load combinations, combinations of gravity and wind loads, as well as combinations of gravity loads and earthquake forces. Design of every critical section of every structural member must be done considering all of these load combinations. If the gravity and wind load combinations produce demands that are closer to the design strength of a section than do the combinations of gravity loads and earthquake effects, then wind rather than earthquakes may be thought of as governing the design of that section. If the same happens for every critical section of a structure, then wind may be thought of as governing the design of that entire structure. However, this fact has no bearing on the necessity to comply with the detailing requirements of the materials chapters. Theoretically, it could be argued that if wind effects were larger than unreduced earthquake effects (earthquake effects corresponding to the elastic response of the structure to the design earthquake), then the detailing requirements of the code could be dispensed with. However, even that would not be allowed by the IBC. Totally irrespective of the severity of wind effects, the SDCs defined in Section 1613.3.5 would determine the applicability of the detailing requirements. The SDCs are used in the code, irrespective of the severity of wind effects, to determine permissible structural systems, limitations on height and irregularity, those components of the structure that must be designed for seismic resistance, and the type of lateral force analysis that must be performed.

Section 1605 *Load Combinations*

General. This section requires that buildings and other structures and portions thereof be designed to resist combined load effects as given by the strength design or LRFD load combinations of Section 1605.2 or the ASD load combinations of Section 1605.3, and the load combinations specified in Chapters 18 through 23. In addition, this section refers to ASCE 7 Section 12.4.3.2 for load combinations with overstrength factors (previously referred to as "special seismic load combinations" in earlier editions of the IBC) that are required to be used by Sections 12.2.5.2 (cantilever column systems), 12.3.3.3 (elements supporting discontinuous walls or frames), and 12.10.2.1 (collector elements requiring load combinations with overstrength factor for SDCs C through F) of ASCE 7. Reference is made to ASCE 7 Section 12.14.3.2 for load combinations with overstrength factors required in the simplified design of ASCE 7 Section 12.14. **1605.1**

Collectors and elements supporting discontinuous shear walls are designed for magnified forces (the estimated maximum axial forces that can realistically develop in these elements in an earthquake situation) so that they will not fail before the vertical resisting elements (Figure 1605-1). This enables these components to deliver earthquake forces to or support the vertical resisting elements so that the vertical elements are able to dissipate energy through inelastic deformation. An analogy is provided by an electrical circuit in which the wire (collector) is sized to safely carry more current than the capacity of the fuse (shear wall) to ensure that the fuse blows before the wire melts. The forces in the affected elements are magnified by multiplying by system overstrength factor Ω_0 given in ASCE 7 Table 12.2-1, which mostly varies between 2 and 3.

ASCE 7 Sections 12.4.3.2 and 12.14.3.3 provide load combinations with overstrength effects for use in strength design or allowable stress design, respectively. These provisions are actually clarifications of how the seismic load effect with overstrength is to be used in the ASCE 7 load combinations (or the corresponding IBC load combinations).

Load combinations using strength design or load and resistance factor design. The basic strength design load combinations of the IBC are adapted from the strength design load combinations of ASCE 7.[1] **1605.2**

Note that differences exist between the load combination equations in ASCE 7 and those in the IBC. Factors f_1 and f_2 are used with live load effect (L) and snow load effect (S) in IBC Equations 16-3, 16-4, and 16-5. Although ASCE 7 accomplishes the same effect as factor f_1 through an exception to the load combinations, a substantial

16 Structural Design

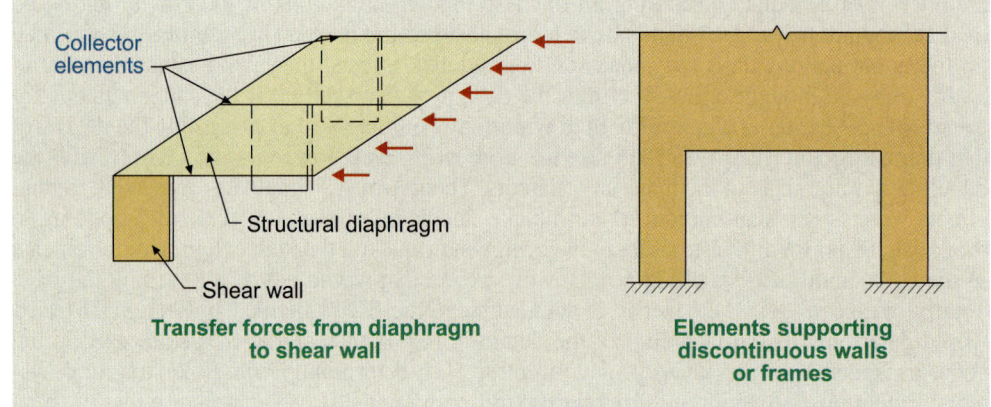

Figure 1605-1 Collector elements for diaphragms and discontinuous shear walls.

difference exists in the case of snow load. Instead of applying a fixed factor of 0.2, the IBC snow load factor, f_2, equals 0.7 for roof configurations (such as saw tooth) that do not shed snow off the structure and 0.2 for other roof configurations.

Where F_a (flood load) is to be considered in design, the load combinations of Section 2.3.3 of ASCE 7 are to be used.

1605.3 Load combinations using allowable stress design. The IBC contains two alternative sets of ASD load combinations. The basic ASD load combinations are the same as those in ASCE 7, whereas the alternative basic ASD load combinations are adapted from the legacy model codes.

The basic ASD load combination equations include F (load due to fluids) and H (load due to lateral pressure of soil and water in soil). Where self-straining force, T (arising from contraction or expansion resulting from temperature changes, shrinkage, moisture change, creep in component materials, movement due to differential settlement, or combinations thereof) or atmospheric icing is considered, the corresponding requirements in ASCE 7 are referenced. Also, where F_a (flood load) is to be considered in design, the load combinations of Section 2.4.2 of ASCE 7 are to be used.

In the alternative basic ASD load combinations, load effects F, H, and T are dealt with under Section 1605.3.2.1, whereby 1.0 times each applicable load (F, H, or T) must be added to the combinations specified in Section 1605.3.2.

The two sets of ASD load combinations of the IBC are based on different philosophies and are not intended to be equivalent to each other. The basic set of ASD load combinations, adapted from ASCE 7, is based on the premise that the design strength resulting from the allowable stress method should, in general, not be less than that resulting from the strength design method. The alternate basic set of ASD load combinations is based on the premise that designs comparable to those permitted under the legacy model codes should be permitted under the IBC.

The IBC specifically states that increases in allowable stresses "specified in the appropriate materials section of this code or referenced standard" shall not be used with the basic ASD load combinations, except that increases shall be permitted in accordance with Chapter 23. Chapter 23 of the IBC simply adopts AF&PA NDS by reference. On the other hand, when using the alternate basic load combinations that include wind or seismic loads, allowable stresses are permitted to be increased or load combinations reduced, "where permitted by the material section of this code or referenced standard."

In the alternative ASD load combinations that include the counteracting effects of dead loads and wind loads, only two-thirds of the minimum dead load that is likely to be in place during a design wind event shall be used.

1. For evaluating sliding, overturning, and soil bearing at the soil–structure interface, using the alternative ASD load combinations, the reduction of foundation overturning from ASCE 7 Section 12.13.4 is not permitted.

2. For the purpose of proportioning foundations for seismic loadings, using the alternative ASD load combinations, the vertical seismic load effects, E_v, are permitted to be taken as zero.

The ASD load combinations are subject to an important exception. Flat roof snow loads of 30 psf or less as well as roof live loads of 30 psf or less need not be combined with seismic loads. Where flat roof snow loads exceed 30 psf, only 20 percent of the flat roof snow load is required to be combined with seismic loads. In the case of roof promenades, roof gardens, and similar uses where the roof live load exceeds 30 psf, it needs to be combined with seismic loads. The elimination or reduction of snow loads is consistent with the definition of W, the effective seismic weight of the structure, in Section 12.7.2 of ASCE 7, Effective Seismic Weight. Also, roof live loads up to 30 psf are primarily representative of maintenance work, and the probability of maintenance work on a roof occurring at the same time as the design earthquake is considered to be very low for a given structure.

Section 1606 Dead Loads

According to the definition, dead loads consist of the weight of materials of construction incorporated into the building, including but not limited to walls, floors, roofs, ceilings, stairways, built-in partitions, finishes, cladding and other similarly incorporated architectural and structural items, and fixed service equipment, including the weight of cranes.

To establish uniform practice among designers, the ASCE 7 Commentary provides an extended list of weights for commonly used building materials that can be useful to the designer and building official alike. While special cases will inevitably arise, authority is granted to the building official to deal with such cases.

Engineers, architects, and building owners are also cautioned in the ASCE 7 Commentary to consider factors that result in differences between actual and calculated loads. Conditions have been encountered in the past that, if not considered in design, may reduce the future utility of a building or reduce its margin of safety. The ASCE 7 Commentary points out two such conditions:

1. There have been numerous instances in which the actual weights of members and construction materials have exceeded the values used in design. Care is advised in the use of tabular values. Also, allowances should be made for such factors as the influence of formwork tolerances as well as support deflections on the actual thickness of a concrete slab of prescribed nominal thickness.

2. Allowance should be made for the weight of future wearing or protective surfaces where there is a good possibility that such may be applied. Special consideration should be given to the likely types and position of partitions, as insufficient provision for partitioning may reduce the future utility of a building.

Section 1607 Live Loads

1607.1 General. Live loads are defined as loads produced by the use and occupancy of the building or other structure, and do not include construction or environmental loads such as wind load, snow load, rain load, earthquake load, flood load, or dead load.

1607.2 Loads not specified. For occupancies and uses not specifically included in Table 1607.1, the method of determination of the design live load is subject to the approval of the building official.

Extremely valuable information is provided in the commentary to ASCE 7, Tables C4-1 and C4-2, concerning the determination of design live loads for occupancies not listed in Table 1607.1.

1607.3 Uniform live loads. This section charges the designer to use the unit live loads set forth in Table 1607.1 and specifies that these loads must be considered minimum live loads. In other words, floors must be designed for the maximum live loads to which they are likely to be subjected during the life of the building based on its intended use, but in no case should the design loads be less than those given in Table 1607.1. The commentary to ASCE 7 advises that in selecting the occupancy and use for the design of a building, the owner should consider the possibility of later changes of occupancy involving loads heavier than originally contemplated. The lighter loading appropriate to the first occupancy should not necessarily be selected, when an owner has reason to anticipate different uses for the building in the future. The owner's planning should also consider the possibility of temporary changes in the use of a building as in the case of clearing a portion of dormitory for a dance or other recreational activity.

Under one- and two-family dwellings (see Item 25 in Table 1607.1), Footnotes i, j, and k provide the criteria that determine whether an attic should be designed for storage, or when it must be considered a habitable space. The storage condition is based on the clearance available within the attic. Figure 1607-1 illustrates the clearance requirements that are stated in Footnotes i and j for truss and joist construction.

The commentary to ASCE 7 provides background to the development of the tabulated live loads. Many surveys of live loads in buildings, particularly office buildings, have been conducted over the years.[5–8] Buildings must be designed to resist the maximum live loads they are likely to be subjected to during some reference period, frequently taken as 50 years. Table C4-2 of the ASCE 7 Commentary briefly summarizes how load survey data are combined with a theoretical analysis of the load process for some common occupancy types, and illustrates how a design load might be selected for an occupancy that is not included in the live load table. To help prevent floors from being overloaded, Section 106 of the code provides requirements for the posting of live loads. In addition, Section 106.3 makes it unlawful to place a load greater than the code permits on any floor or roof of a building.

1607.4 Concentrated live loads. Many uses are susceptible to the movement of equipment, files, machinery, and so on. Therefore, the code requires that floors for these uses, which are listed in Table 1607.1, be designed for the indicated concentrated load placed on a space $2^1/_2$ feet square whenever this load, on an otherwise unloaded floor, produces stresses greater than those caused by the uniform loads required by the code. As this concentrated load can take many forms in the real world, and as the design structural engineer usually does not know in advance what form the load will take and how it will be applied, the best compromise to cover most situations is to consider the concentrated load to be applied through a rigid base $2^1/_2$ feet square.[9]

Footnote a to Table 1607.1 requires garages and other areas where motor vehicles are stored to be designed for concentrated loads or, alternatively, the uniform loads as specified in Table 1607.1. In the case of garages for storage of private pleasure-type vehicles, the note prescribes a single 3,000-pound load over a 4.5-inch by 4.5-inch area. For mechanical parking structures without slab or deck for the storage of passenger cars only, a load of 2,250 pounds per wheel is specified.

For all the concentrated loads specified in this section, the intent of the code is that each concentrated load shall be placed on the floor in such a position as to create maximum load effects in the structural members. Likewise, the loading condition, either uniform or

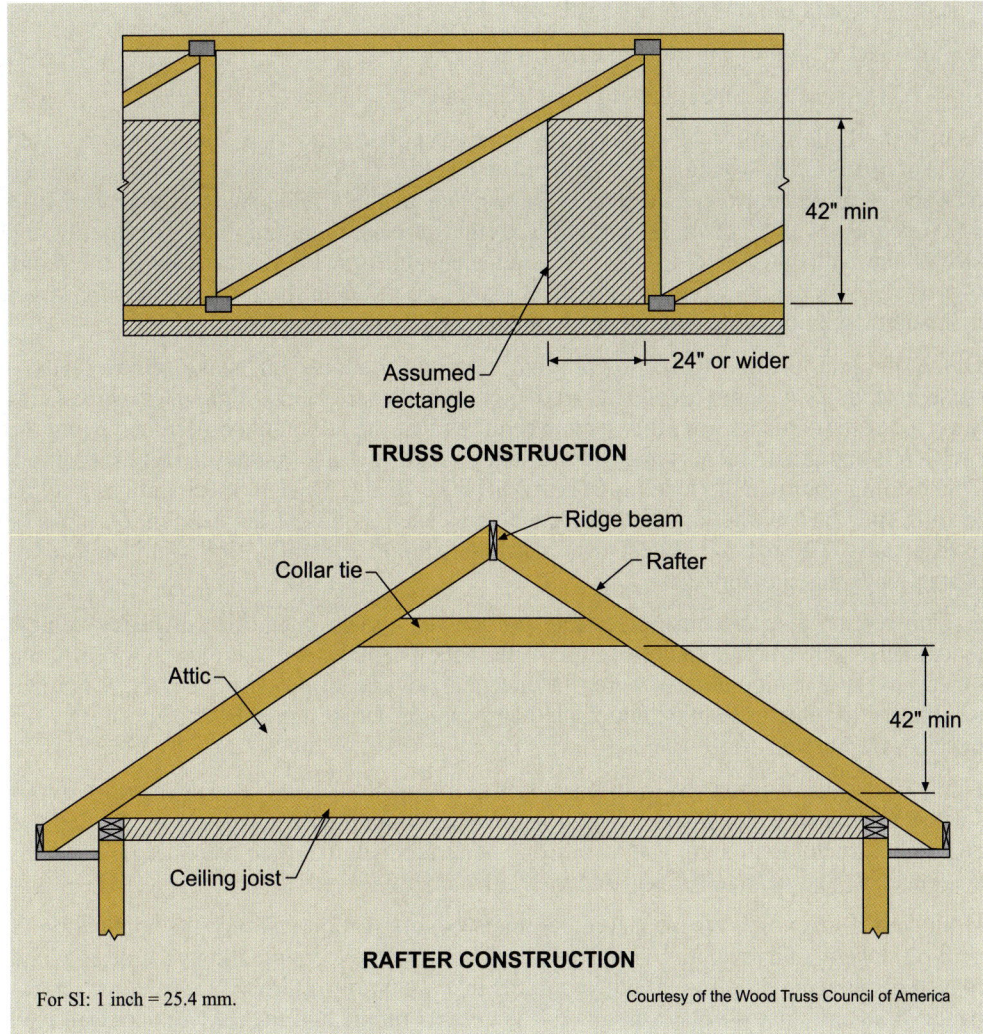

Figure 1607-1
Uninhabitable attics with storage.

concentrated, that produces the greatest load effects in the structural members would be used for the design of those members. Note that the term "load effects" refers not only to member forces and the resulting stresses, but also to member deformations (see deflection criteria in Section 1604.3).

Partition loads. In those uses where partition arrangements are periodically reconfigured such as office buildings and flexible-plan school buildings, the code requires that the floor system be designed to support a partition live load of 15 psf. This requirement is irrespective of whether partitions are shown on the construction documents. The 15 psf value was arrived at by assuming 10-foot-high partition walls of wood or steel stud wall construction with $1/2$-inch gypsum board on each side, and arranged in a square grid of 10-foot sides. This assumption was thought to provide a fairly conservative estimate of partition loading. However, the ASCE 7 Commentary also advises designers to consider a larger partition load if a higher density of partitions is anticipated. It should be noted that the uniformly distributed load of 15 psf is considered by the code to be a live load. Thus, it should be included with other live loads in the load combinations.

1607.5

Helipads. In this section, the IBC establishes minimum design loading criteria for helicopter landing surfaces, which include:

1607.6

1. A minimum uniform live load.
2. The weight of the helicopter.
3. The landing impact effect of the helicopter.

In Item 1, the uniform live load of 40 psf is applied when the landing area is used by helicopters with a take-off weight not exceeding 3,000 pounds. A study indicated that about 56 percent of all registered helicopters in the United States weigh less than 3,000 pounds, which is comparable to the weight of a small automobile. Considering the size of the helicopter landing area, the equivalent uniform load is actually in the range of only 2.1–4.3 psf, meaning that the live load of 40 psf is fairly conservative. For larger helicopters, the 60-psf design live load would apply.

1607.7 Heavy vehicle loads. This section provides criteria for addressing heavy vehicle loads—those with gross weights exceeding 10,000 pounds. Where heavy highway-type vehicles have access onto a structure, this section requires that the structure be designed using the same code and requirements that are applicable to roadways and bridges in that jurisdiction. This loading could be the loading from AASHTO's Bridge Design Specification,[10] or the loading specified by the jurisdiction for elements such as lids of large detention tanks or utility vaults. The registered design professional should consult with the jurisdiction for design loads that are applicable.

This section also establishes loading criteria for certain categories of heavy vehicle loads such as fire trucks and other similar emergency vehicles—in addition to forklifts and other moveable equipment. As a precaution against overloading a structure, the maximum weight of the vehicles that are anticipated and used in the design should be posted by the owner—see Section 106.1.

1607.8 Loads on handrails, guards, grab bars, seats, and vehicle barriers. The majority of this section is adapted from ASCE 7. These requirements are intended to provide an adequate degree of structural strength and stability to handrails, guards, grab bars, accessible seats and benches, and vehicle barriers and their attachments.

1607.8.1 Handrails and guards. The basic requirement of this section calls for the application of a 50-plf design load to handrails and guards. The second exception allows the design load to be reduced from 50 plf to 20 plf for guards that are within a Group I-3, F, H, or S occupancy in an area that is not accessible to the general public and that has an occupant load of less than 50. These are the same occupancies listed in Exception 4 to Section 1013.4, which are permitted to have larger openings in the guard.

Also, in the first exception, the IBC clearly states that while one- or two-family dwellings are exempted from the requirement for a minimum distributed load of 50 plf, the requirement for a single concentrated load of 200 pounds specified in Section 1607.8.1.1 still applies.

The commentary of Section C4.5.1 of ASCE 7 points out that loads that can be expected to occur on handrail and guardrail systems are highly dependent on the use and occupancy of the protected area. It further points out that for cases in which extreme loads can be anticipated, such as long, straight runs of guardrail systems against which crowds can surge, appropriate increases in loading need to be considered.

1607.8.2 Grab bars, shower seats, and dressing room bench seats. The components listed in this section are those that are typically required in providing accessibility in accordance with Chapter 11. Where accessibility is required, conformance with ICC A117 *Accessible and Usable Buildings and Facilities*[11] is mandated in Section 1101.2. ICC A117 determines where elements such as grab bars or shower seats must be provided. The design loading on these elements that is specified herein is also consistent with *ADA Accessibility Guidelines for Building and Facilities* (ADAAG).[12]

1607.8.3 Vehicle barriers. This section requires the application of a 6,000-pound design load that accounts for impact. Because of differing vehicle bumper heights, barrier configurations, and anchorage methods, the specified load must be applied over a range of heights so that the maximum load effects are determined for design of the barrier.

Impact loads. These provisions are adapted from ASCE 7. Where unusual vibration or impact forces are likely to occur, their effect may be to produce additional stresses and deflections in the structural system. This section requires that the structural design takes these effects into account. Typically, the dynamic effects are approximated through the application of a static load equal in effect to the dynamic loads. In most cases, this is sufficient. However, in certain situations, a dynamic analysis may be necessary to properly consider the natural frequencies of vibration of a structure.

1607.9

Reduction in uniform live loads. The live load reduction provisions of Section 1607.101 are based on ASCE 7, while the alternative floor live load reduction provisions are based on legacy model codes, such as the 1997 *Uniform Building Code* and the 1997 *Standard Building Code*.[13] Small portions of a floor are more likely to be subjected to the full uniform load than larger floor areas. Unloaded or lightly loaded areas tend to reduce the total load on the structural members supporting those floors. In recognition that the larger the tributary area of a structural member, the lower the likelihood that the full live load will be realized, the specified uniformly distributed live loads from Table 1607.1 are permitted to be reduced.

1607.10

The alternative floor live load reduction provisions, based on tributary floor area (Figure 1607-2), represent the *original* live load reduction provisions that used to be in older editions of the ANSI A58.1 standard (predecessor to ASCE 7) and in all three legacy model codes. The alternative floor live load reduction equation is used for reducing floor live loads only, whereas in prior model codes, it was used to reduce roof live loads as well. Roof live load reductions are covered in Section 1607.12.2.

The concept of, and method for, determining member live load reductions as a function of a loaded member's influence area, *AI*, was first introduced into ANSI A58.1 in 1982 and was the first such change since the concept of live load reduction was introduced over 40 years ago. The revised method was the result of more extensive survey data and theoretical analyses. Figure 1607-3, reproduced from the commentary to ASCE 7-95, illustrates the influence area concept. The influence area is considered to be the floor area over which the influence surface for structural effects is significantly different from zero.

In the basic uniform live load approach of Section 1607.10.1, the permitted reduction is a function of the tributary area, *AT*, multiplied by the element factor, K_{LL}. This factor is the ratio of the influence area (*AI*) of a member to its tributary area (*AT*), i.e., $K_{LL} = AI/AT$.

Table 1607.10.1 has established K_{LL} values (derived from calculated K_{LL} values) to be used in Equation 16-23 for a variety of structural members and configurations. K_{LL} values vary for column and beam members having adjacent cantilever construction, and the Table 1607.10.1 values have been set for these cases to result in live load reductions that are slightly conservative (Figure 1607-4). For unusual shapes, the concept of significant influence for structural effect needs to be applied.

See Example 16-4 of this chapter.

Basic uniform live load reduction. Reductions in the minimum uniformly distributed live load are permitted, based on an influence area, $K_{LL} A_T$, of 400 square feet or more. Essentially, the influence area of a structural element is the total floor area surrounding the element from which it derives any of its loads. The basis for the permitted reduction is that in the design of structural elements with large influence areas, it is highly unlikely that the floors will be fully loaded over their entire area. Note that Footnote m in Table 1607.1 prohibits the reduction of certain live loads unless a specific exception applies—see Sections 1607.10.1.2 and 1607.10.1.3.

1607.10.1

Live load reduction in excess of 50 percent is not permitted for columns or other structural elements (such as bearing walls) that support the loads of a single floor. In essence,

16 Structural Design

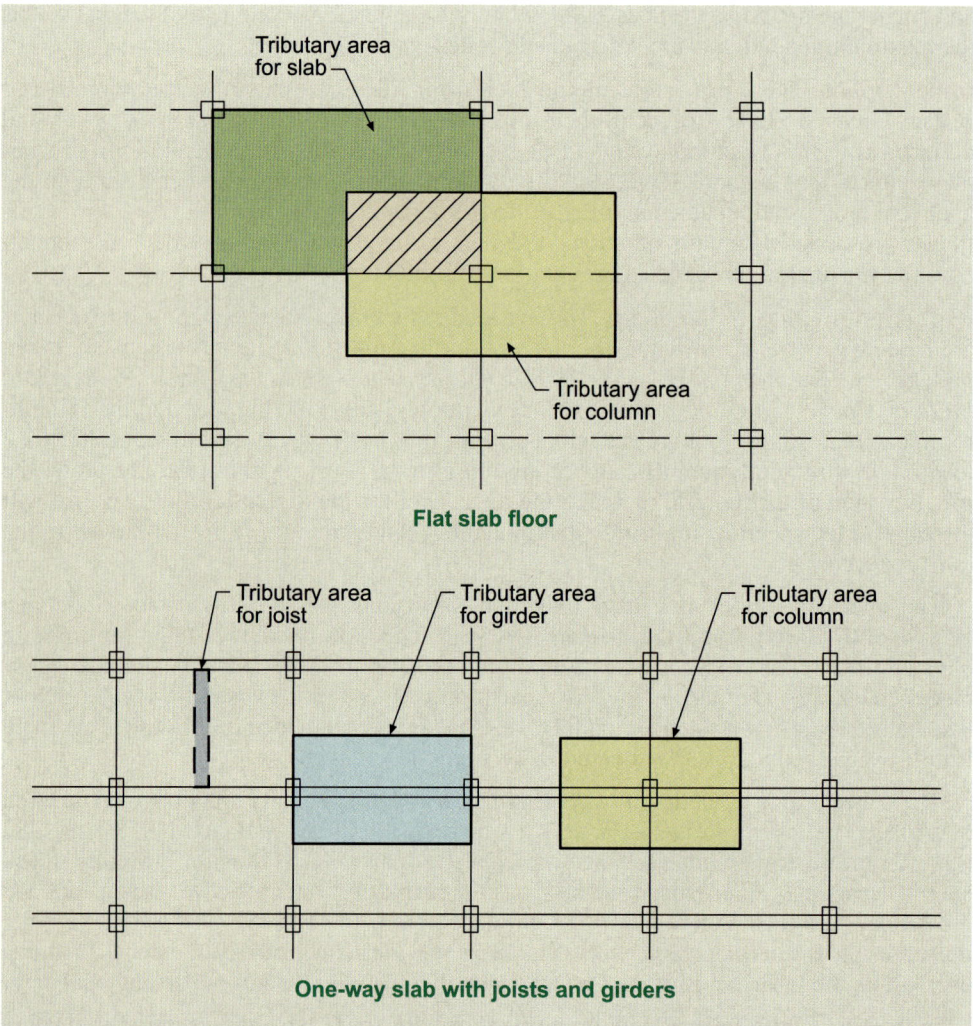

Figure 1607-2
Tributary areas.

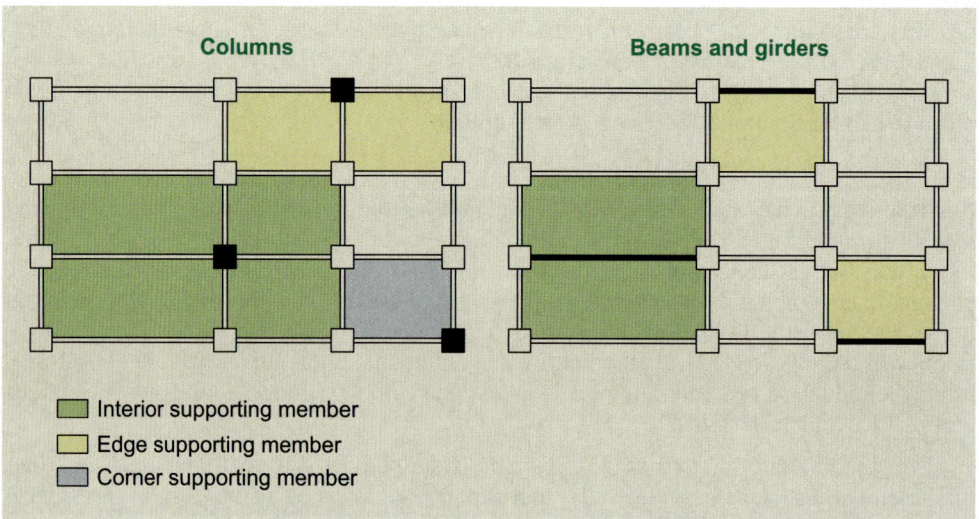

Figure 1607-3
Influence areas.

Live Loads 16

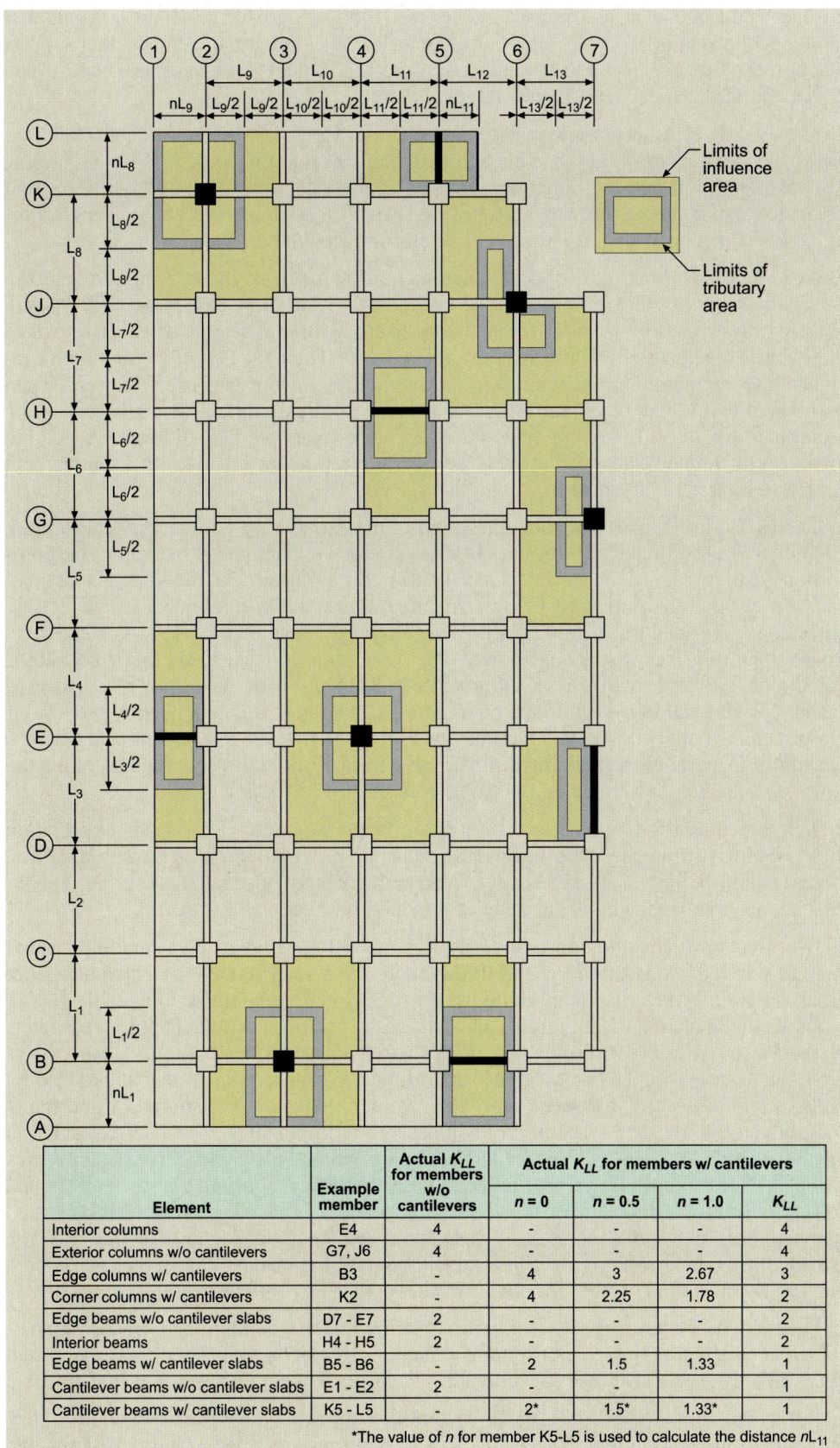

Element	Example member	Actual K_{LL} for members w/o cantilevers	Actual K_{LL} for members w/ cantilevers			K_{LL}
			$n = 0$	$n = 0.5$	$n = 1.0$	
Interior columns	E4	4	-	-	-	4
Exterior columns w/o cantilevers	G7, J6	4	-	-	-	4
Edge columns w/ cantilevers	B3	-	4	3	2.67	3
Corner columns w/ cantilevers	K2	-	4	2.25	1.78	2
Edge beams w/o cantilever slabs	D7 - E7	2	-	-	-	2
Interior beams	H4 - H5	2	-	-	-	2
Edge beams w/ cantilever slabs	B5 - B6	-	2	1.5	1.33	1
Cantilever beams w/o cantilever slabs	E1 - E2	2	-	-	-	1
Cantilever beams w/ cantilever slabs	K5 - L5	-	2*	1.5*	1.33*	1

*The value of n for member K5-L5 is used to calculate the distance nL_{11}

Figure 1607-4
Typical tributary and influence areas.

this means that the influence area may not exceed 3,600 square feet (334.4 m²) in calculating the reduced unit floor live load. For columns or other structural elements that support two or more floors, the sum of the reduced live loads from all floors must not be less than 40 percent of the sum of the unreduced live loads.

1607.10.1.1 One-way slabs. This section allows live load reduction in one-way slabs—see Table 1607.10.1 for K_{LL} value of a one-way slab. Although both one-way and two-way slabs have the same K_{LL} value of 1.0, this section imposes an upper limit on the slab width that can be used to calculate AT of a one-way slab as 1.5 times the span of the slab. This restriction is indicative of the lower redundancy in one-way slabs when compared to two-way slabs.

1607.10.1.2 Heavy live loads. In the case of occupancies requiring relatively heavy minimum uniform live loads, such as storage buildings, several adjacent floor panels can be fully loaded. Field surveys indicate that rarely is any story loaded with an average actual live load of more than 80 percent of the average design live load.[1] Thus, the ASCE 7 committee concluded that the minimum uniform live load should not be reduced for the floor and beam design, but that it may be reduced a flat 20 percent for the design of members supporting more than one floor. In Exception 1, the IBC further qualifies this allowance to require that the reduction be calculated in accordance with Section 1607.10.1, with the maximum reduction limited to 20 percent.

The IBC also includes a second exception to the prohibition on reduction, permitting additional live load reduction for uses other than storage when the registered design professional can provide an acceptable substantiation for doing so. The rationale is that there can be uses other than storage where the maximum design live loads may exceed 100 psf, but the average load on members with large tributary areas may be less. For example, floors supporting heavy machinery may have very high uniform loads that are concentrated mostly over a small part of the floor area. This provision will allow the registered design professional to present to the building official a rational load reduction proposal if those scenarios apply. Since there are no specific criteria stated and the reduction is subject to the approval of the building official, this exception is very much like reiterating the concept of an alternative method of design as described in Section 104.11.

1607.10.1.3 Passenger vehicle garages. There are no significant variations in the loads imposed on these facilities, which are often fully loaded, thus the prohibition on live load reductions. An exception permits a 20-percent live load reduction for members supporting two or more floors. The reasoning is the same as provided for Section 1607.10.1.2.

1607.10.2 Alternative uniform live load reduction. This section establishes a minimum tributary area of 150 square feet as the threshold for live load reductions computed from Equation 16-24, which is plotted in Figure 1607-5. Note that Footnote m in Table 1607.1 prohibits the reduction of certain live loads unless a specific exception applies—see items one and two. Live loads in excess of 100 psf may not be reduced with two exceptions. First, the design live loads on columns supporting two or more floors may be reduced by 20 percent, and second, for usage other than storage, reduction is permitted when found acceptable by a registered design professional through rational analyses, as explained in Section 1607.10.1.2 above. Also, reduction is not permitted for passenger-vehicle garages except for a maximum 20-percent reduction for columns supporting two or more floors. An upper limit is also specified for the tributary width of a one-way slab for reduction calculations as 0.5 times the span of the slab, as explained in Section 1607.10.1.1. The maximum live load reduction permitted is 40 percent for members receiving loads from one level only and 60 percent for other members (such as columns or transfer girders).

Equation 16-24 was derived so that if a structural member supporting a tributary area of sufficient size to qualify for the maximum reduction allowed by the equation were subjected to the full design live load over the entire area, the overstress would not exceed 30 percent.[14]

It may be noted from Equation 16-25 that the maximum live load reduction is proportional to the ratio of dead load to live load. Therefore, for heavy framing systems, the

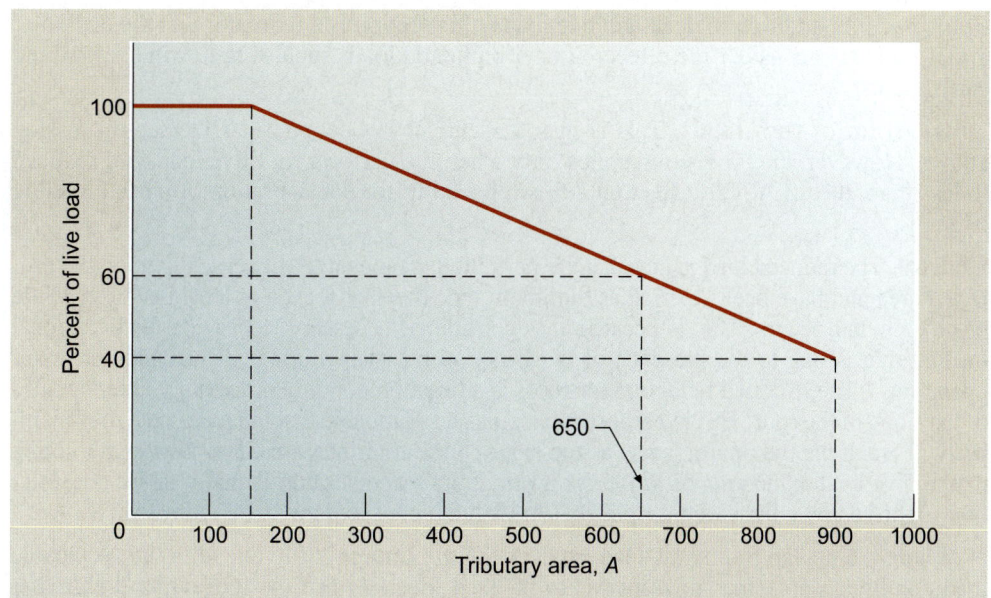

**Figure 1607-5
Live load reduction Equation 16-24.**

reduction is permitted to be greater than it would be for lighter framing systems. This reflects the thinking that for a given magnitude of overload on a structural system, the system with the heavier dead load is overstressed proportionately less than one with a lighter dead load. For example, if a floor system weighing 30 pounds per square foot and designed for a live load of 40 pounds per square foot were subjected to a 20-pounds-per-square-foot overload, the amount by which the structural system would be overloaded is about 30 percent, assuming the system was designed to support just the minimum design live load of 40 pounds per square foot. If this floor had a dead load of 60 pounds per square foot, the overload would be only 20 percent, again assuming that the system was designed to support just the 40-pounds-per-square-foot live load.

Distribution of floor loads. Where loads are uniformly distributed on continuous structural members, they shall be arranged so as to create maximum bending moment in any given critical section. This may require a design to consider so-called skip loading or alternate span loading, as shown in Figure 1607-6.

1607.11

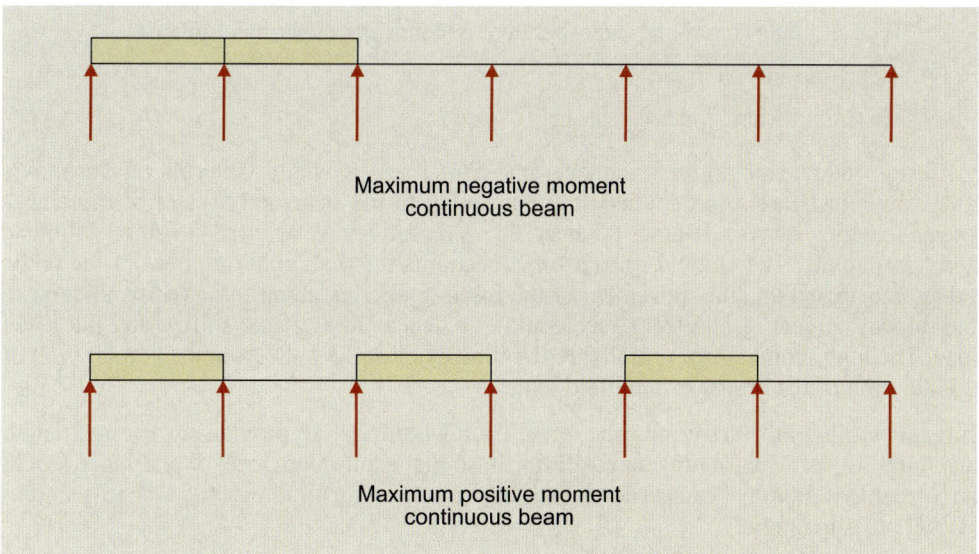

**Figure 1607-6
Alternate span loading for continuous beams.**

1607.12 Roof loads. In addition to dead and live loads, this section is a reminder that the design of a roof must also consider the effects of environmental loads such as rain, wind, snow, and earthquakes.

1607.12.1 Distribution of roof loads. This is nearly identical to Section 1607.11 (see discussion above). However, the provision applies only when the uniform roof live loads are reduced to less than 20 psf. ASCE 7 Section 7.5 is referred to for consideration of partial loading of snow.

1607.12.2 General. The reduced load values that are permitted are meant to act vertically upon the projected area and have been selected as minimum roof live loads, even in localities where little or no snowfall occurs. This is because it is considered necessary to provide for occasional loading that is due to the presence of workers and materials during maintenance or repair operations.[1] The live load reduction for roofs is a function not only of tributary area, but also of the slope of the roof. This is because it becomes less probable that the loads on a roof member will reach the maximum levels as the slope of the roof increases. It is also worth noting that no live load reduction is allowed for awnings and canopies consisting of fabric construction supported by a lightweight rigid skeleton structure—see Item 26 of Table 1607.1.

Where the design roof snow load exceeds the minimum roof live load in the applicable load combination(s), the snow load is to be used for design of the roof. Only the greater of the roof load value established by the minimum roof live load or the design snow load determined as indicated above is required to be applied to the roof. It should be noted that for the roofs covered in this section the absolute minimum roof live load is 12 psf, as indicated by the limits on Equation 16-26.

1607.12.3 Occupiable roofs. Since roofs that are occupiable are designed for live loads that are commensurate with the intended use—see Item 26 of Table 1607.1—these roof live loads may be reduced in accordance with the live load reductions allowed by Section 1607.10.[8]

1607.12.3.1 Landscaped roofs. Designers need to consider any additional dead loads that may be imposed by saturated landscaping materials.

1607.12.4 Awnings and canopies. For the design of awnings and canopies, snow loads and wind loads, as specified in Sections 1608 and 1609, need to be considered in addition to the live loads specified in Item 26 of Table 1607.1.

1607.13 Crane loads. All craneways and supporting construction must be designed and constructed in compliance with this section, which parallels the crane load criteria of ASCE 7. The crane live loads depend on:

- Type of crane (monorail, cab-operated, pendant-operated, hand-geared)
- Rated capacity of crane
- Maximum wheel loads

Design lateral, longitudinal, and vertical forces are provided in terms of the above. These live loads are to be applied simultaneously to the structural system of the craneway, including runway beams, connections, support brackets, cross-bracing, columns, and foundations. The vertical impact force accounts for the vibration effect of the crane bridge movement and the movement of the lifted load. The lateral force (perpendicular to the runway girder) results from acceleration or deceleration of the trolley and the lifted load. The longitudinal force (parallel to the runway girder) results from the acceleration or deceleration of the bridge or the lifted load.

1607.14 Interior walls and partitions. The intent of this section is to provide sufficient strength and durability of wall framing and wall finish, so that a minimum level of resistance would be available to nominal impact loads that commonly occur in the use of a facility and to HVAC pressurization.

Fabric partitions are defined in Section 202 as follows:

FABRIC PARTITIONS. A partition consisting of a finished surface made of fabric, without a continuous rigid backing, that is directly attached to a framing system in which the vertical framing members are spaced greater than 4 feet on center.

The definition clearly differentiates them from other more traditional-type partitions, which include partial-height office partitions that contain rigid panels finished with fabric and attached to a rigid frame. In the case of a fabric partition, there is no rigid panel to which the fabric is attached. The fabric simply spans the open space between the rigid frame over which it is stretched and attached. In the definition, it states that the vertical framing members are spaced greater than 4 feet on center, again to differentiate this type of partition from a more traditional partition where the vertical framing members are spaced at 4 feet on center or less. Typically, these partitions are not intended to be full (ceiling) height, so they would normally not be attached directly to the ceiling. They are usually supported by the floor, except under conditions where, because of the layout and the height of the ceiling, it may be more appropriate to hang the partition from the ceiling grid or use a combination of floor supports and ties to the ceiling grid or a special structural ceiling grid designed to support such partitions. Figure 1607-7 shows some examples of fabric partitions.

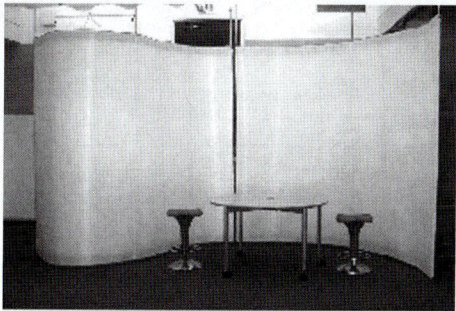

**Figure 1607-7
Applications of fabric partitions.**

The intent behind the concentrated load of 40 pounds was to ensure that the partition does not tip over if someone were to inadvertently lean up against the frame or the fabric, and that a person inadvertently leaning against the fabric would not cause the fabric to tear and the person to fall abruptly. In the case of the 5 psf horizontal load, it was decided that the 5 psf is to be distributed by calculating the total load based on the area of the fabric and then having that load distributed proportionally over the horizontal and vertical structural framing members of the partition. Thus, the framing system, in effect, will be resisting the total horizontal distributed load of 5 psf even though the 5 psf is not applied over the field of the fabric between the supports.

16 Structural Design

Section 1608 Snow Loads

The snow load provisions reference Chapter 7 of ASCE 7. Only the provisions regarding determination of ground snow loads in the contiguous United States, Alaska, and Hawaii are contained in the code, which many building officials will wish to specify when adopting the model code by local ordinance. Based on the provisions in the IBC and ASCE 7, Figure 1608-1 has been developed to show the organization of the design snow load determination. The variables that must be determined for the calculation of the design snow loads include ground snow load, exposure factor, thermal factor, and importance factor. Other considerations include a rain-on-snow surcharge, partial loading, and ponding instability from melting snow or rain on snow. The discussions about these provisions are presented in the same order as shown in Figure 1608-1.

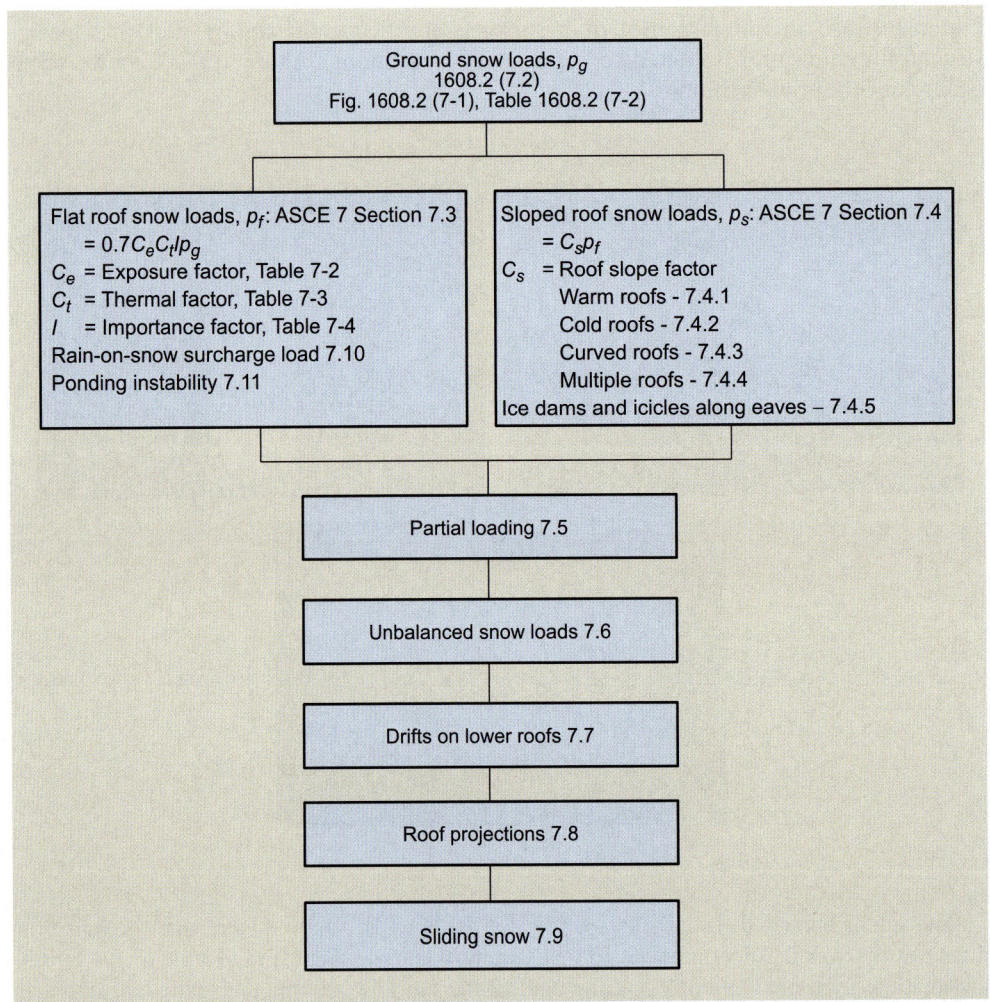

Figure 1608-1 Snow load provisions.

1608.1 General. The IBC stipulates in this section that design snow loads be determined in accordance with Chapter 7 of ASCE 7. Requiring that the design roof load shall not be less than the roof live loads that are determined in Section 1607 merely states what is certainly required by the load combinations of Section 1605.

1608.2

Ground snow loads. Figure 1608.2 and Table 1608.2 are reproductions of Figure 7-1 and Table 7-1 of ASCE 7, respectively. Figure 1608.2 for the contiguous United States and Table 1608.2 for Alaska give the ground snow loads to be used in determining the design snow loads for roofs. The IBC states: "Snow loads are zero in Hawaii, except in mountainous regions as approved by the building official." The ground snow loads on the map and the table are generally based on snow depths recorded over a period approaching 50 years. The snow loads on the maps have a 2-percent probability of being exceeded (a 50-year mean occurrence interval). The mapped values indicate the ground snow loads in pounds per square foot (psf). In mountainous areas, the map indicates the highest elevation up to which it is appropriate to use a given snow load. For elevations higher than indicated on the map, a site-specific case study is necessary to determine the appropriate snow load. Areas identified as CS on the map require site-specific case studies irrespective of elevation. Assistance in the determination of an appropriate ground snow load for these areas may be obtained from the U.S. Department of Army Cold Regions Research and Engineering Laboratory in Hanover, New Hampshire.

ASCE 7 Section 7.3 Flat Roof Snow Loads. This section converts the ground snow load, p_g, to flat roof snow load, p_f. This calculation utilizes an exposure factor, a thermal factor, as well as the importance factor. In Section 7.4, the flat roof snow load is used to determine the sloped roof, or balanced, snow load. It does so by applying the roof slope factor C_s. When the value of C_s is 1.0, then the snow load that applies is essentially the flat roof snow load.

ASCE 7 Section 7.3.1 Exposure Factor, C_e. The roof exposure factor depends on the wind exposure category (or terrain category) as well as the exposure of the roof, as described in the footnotes to Table 7-2. The roof snow load is higher for a sheltered site such as a wooded area than it is for an exposed site that is flat and open. Attention should be paid to the footnotes to Table 7-2, which require that the terrain category and roof exposure condition chosen shall be representative of the anticipated conditions during the life of the structure. An exposure is required to be determined for each roof of a structure.

ASCE 7 Section 7.3.2 Thermal Factor, C_t. The thermal factor accounts for the heat that is transmitted from the interior of the structure, which tends to reduce the snow depth on the roof. Thermal conditions considered in the determination of the thermal factor, C_t, from Table 7-3 are required to be representative of the anticipated conditions during winters for the life of the structure.

ASCE 7 Section 7.3.3 Importance Factor, I_s. The snow load importance factor attempts to address the need to relate design loads to the consequences of failure. Roofs of most structures having normal occupancies and functions are designed with an importance factor of 1.0, which corresponds to unmodified use of the statistically determined ground snow load for a 2-percent annual probability of being exceeded (50-year mean recurrence interval). Lower- and higher-risk situations are established using the importance factors for snow loads, which range from 0.8 to 1.2. The factor of 0.8 corresponds to an annual probability of being exceeded of about 4 percent (about a 25-year mean recurrence interval). The factor of 1.2 is nearly that for a 1-percent annual probability of being exceeded (about a 100-year mean recurrence interval).[1] This factor has been explained further under the ASCE 7 Commentary Section C7.3.3.

ASCE 7 Section 7.3.4 Minimum Snow Load for Low-Slope Roofs, p_m. This section accounts for a number of situations where the ground-to-roof conversion factor of 0.7 as well as the factors C_e and C_t may underestimate the snow load, such as the load resulting from a single storm in an area where the ground snow load p_g is less than 20 psf. This section also includes the clarification that it is not necessary to combine the minimum snow load with drift load and sliding snow load, as well as unbalanced or partial snow loads. When this minimum snow load is applicable, it is considered a separate snow load case.

ASCE 7 Section 7.4 Sloped Roof Snow Loads, p_s. The design snow load for a sloped roof is equal to the flat roof snow load multiplied by a roof slope factor that is given by Section 7.4.1 for warm roofs, Section 7.4.2 for cold roofs, Section 7.4.3 for curved roofs, and Section 7.4.4 for multiple folded plate, sawtooth, and barrel vault roofs. The standard notes, parenthetically, that this sloped roof snow load is the "balanced" snow load. Section 7.6 further clarifies that the balanced and unbalanced snow loads are analyzed separately.

ASCE 7 Section 7.5 Partial Loading. In many situations, a reduction in snow load on a portion of a roof by wind scour, melting, or snow removal operations may simply reduce the stresses in the supporting members. However, in other cases, removal of snow from an area may induce heavier stresses in the roof structure than can occur when the entire roof is loaded. Cantilevered roof joists are a good example; removing half the snow load from the cantilevered portion increases the bending stress and deflection of the adjacent continuous span. In some situations, adverse stress reversals may result.[1] This section requires consideration to be given in design to those adverse situations.

ASCE 7 Section 7.6 Unbalanced Roof Snow Loads. Snow on the roof of a building rarely accumulates evenly. The code intends that the designer investigate conditions of imbalance by requiring that roof designs consider unbalanced snow loading. One case would be the loading of one slope of a gable roof with snow while the other slope is unloaded (Figure 1608-2). If this creates a less favorable condition in the design of the roof structural members, then it is the loading to be considered. This type of loading covers the case where the snow may have been removed from one side of the roof because of sliding or melting.

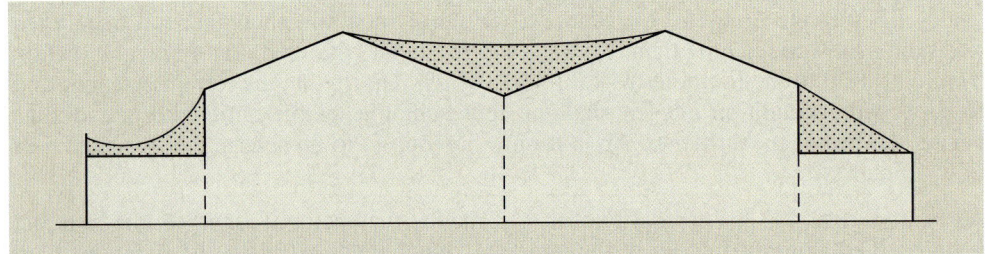

Figure 1608-2 Potential snow accumulation, gable roofs with sidesheds.

ASCE 7 Section 7.7 Drifts on Lower Roofs (Aerodynamic Shade). It is extremely important to consider localized drift loads in designing roofs, because drifts onto lower roofs are a common cause of roof failures after a heavy snow (Figure 1608-2). This section separately addresses windward drift that forms on the windward side of a high-bay wall area, and leeward drift that forms on the leeward side of a high-bay wall area. It provides the formulations for determining the intensity of the drift load and prescribes which portion of the lower roof is subject to the drift loading.

ASCE 7 Section 7.8 Roof Projections and Parapets. Solar panels, mechanical equipment, parapet walls, and penthouses are examples of roof projections that may cause windward drifts on the roof around them. The drift-load provisions of this section cover most of these situations adequately.

ASCE 7 Section 7.9 Sliding Snow. The snow that slides from a higher sloped roof imposes loads that are in addition to the snow load already on the lower roof. This section requires consideration of such additional loads in roof design.

ASCE 7 Section 7.10 Rain-on-Snow Surcharge Load. This load accounts for the increased weight of snow after it rains on the snow. It has been shown by O'Rourke and Downey[15] that the major portion of the rain-on-snow surcharge load comes from the rainwater flowing along the slope of the roof by percolating through the snow layer immediately above the roof. Thus, the surcharge load increases with larger ridge-to-eave span W and smaller roof slope.

ASCE 7 Section 7.11 Ponding Instability. Because of roof deflection caused by the weight of the snow, positive slope to the roof drains may be lost, thereby causing ponding. This section requires the roof framing to be stiff enough to maintain positive slope to the roof drains, so as to prevent potential instability from progressive deflection caused by ponding.

Section 1609 *Wind Loads*

Applications. The following basic requirement is specified: Wind loads must not be decreased considering the effect of shielding by other structures. As the ASCE 7 Commentary points out, the one possible exception to this rule would be the use of wind tunnel tests in accordance with Chapter 31 of ASCE 7. **1609.1**

Determination of wind loads. In general, wind pressures must be assumed to come from any horizontal direction and to act normal to the surfaces considered. In addition to referring to Chapters 26 through 30 of ASCE 7 for the determination of wind forces, the IBC also includes the Alternate All-Heights Method in Section 1609.6 applicable to rigid simple diaphragm buildings (refer to discussion of Section 1609.6). Exception 6 permits the use of wind tunnel tests in accordance with Chapter 31 of ASCE 7. There are three exceptions for building structures: **1609.1.1**

1. Subject to the limitations spelled out in Section 1609.1.1.1, the ICC 600 *Standard for Residential Construction in High Wind Regions*[16] is permitted to be used for applicable Group R2 and R3 buildings. ICC 600 provides a set of prescriptive requirements that is consistent with the wind design requirements in the IBC and ASCE 7.

2. Subject to the limitations of Section 1609.1.1.1, the AF&PA *Wood Frame Construction Manual for One- and Two-Family Dwellings*[17] is permitted to be used. This manual gives engineering and prescriptive requirements for the design and construction of one- and two-family dwellings of wood-frame construction.

3. Subject to the limitations of Section 1609.1.1.1, residential structures are permitted to use the provisions of AISI S230, *Standard for Cold-Formed Steel Framing—Prescriptive Method for One- and Two-Family Dwellings*,[18] as an acceptable alternative for the calculation of wind loads. This edition of the standard is based on the ASCE 7 wind design provisions. Because the standard addresses wind speeds up to 150 mph, it is also included by reference in the new ICC 600, *Standard for Residential Construction in High Wind Regions*.

Note that Section 1609.1.1.1 imposes topographic restrictions on the use of ICC 600, AF&PA WFCM, and AISI 230.

The ASCE 7 Commentary on the wind design provisions is fairly extensive. The basic steps of the simplified design procedure are discussed below to serve as a guide through the wind load provisions.

The wind load provisions of ASCE 7-10 have been reorganized and broken into six new chapters. Chapter 26 covers general wind load design requirements, Chapters 27 and 28 cover the directional and envelope procedures for design of main wind-force-resisting systems (MWFRSs); Chapter 30 covers design of components and cladding; and Chapter 31 covers requirements for the wind tunnel procedure. Chapter 27 (directional procedure) applies to enclosed, partially enclosed, and open buildings of all heights. Chapter 28 (envelope procedure) applies to low-rise buildings. Part 2 of Chapter 28 is a projected area method that applies to enclosed simple diaphragm buildings.

Simplified provisions for low-rise buildings. The simplified provisions in ASCE 7 Chapter 28, Part 2 and Chapter 30, Part 2 are applicable to the design of MWFRSs and components and claddings of buildings that satisfy all the following conditions:

1. Enclosed building (see Section 26.10 of ASCE 7).
2. Flat roof, gabled roof, or hipped roof with $\theta \leq 45°$.
3. Mean roof height is less than or equal to 60 feet in accordance with definition of "low-rise building" in Section 26.2 of ASCE 7.
4. The building does not have response characteristics making it subject to across wind loading, vortex shedding, and instability that is due to galloping or flutter. Also, the building is not to be located on a site for which channeling effects or buffeting in the wake of upwind obstructions warrant special consideration.
5. The building is regular shaped (see definition in Section 26.2 of ASCE 7).

In addition, the building must meet further requirements for its MWFRS to qualify for the simplified design procedure, including simple diaphragm buildings with no expansion joint or structural separation, exempted from torsional load cases, and not classified as flexible building. Note that in case only the components and cladding of a building qualify to be designed by the simplified procedure, then the MWFRSs need to be designed by the analytical procedure or wind-tunnel procedure.

Some examples of simple diaphragm buildings are houses with plywood shear walls, typical CMU (concrete masonry unit) wall buildings, concrete-frame buildings, and steel-frame buildings with vertically spanning walls and floor and roof diaphragms. Metal buildings with horizontal girts that span between frames are not simple diaphragm buildings.

Much of the simplicity of the procedure derives from the fact that internal pressures are not involved in the design pressures for the MWFRS. Because the wind forces are delivered to the MWFRS via floor and roof diaphragms, and the building, by definition, is enclosed, the internal pressures simply do not come into play (or cancel out).

Step 1: Mapped wind speed. The ultimate design wind speed is to be determined from IBC Figures 1609A, 1609B, and 1609C, which are reproductions of ASCE 7 Figures 26.5-1A, 26.5-1B, and 26.5-1C. The wind speed maps provide ultimate wind speeds for the contiguous United States, Alaska, and other selected locations. The wind speeds correspond to 3-second gust speeds at 33 feet (10 m) above ground level for exposure category C.

Although the mapped wind speeds are valid for most regions of the country, there are special regions in which wind-speed anomalies are known to exist. Some of these "special wind regions" are noted on the maps. Winds blowing over mountain ranges or through gorges or river valleys in these special regions can develop speeds that are substantially higher than the values indicated on the map. When determining wind speeds in these special regions, the IBC requires conformance with ASCE 7 Section 26.5.3. ASCE 7 advises the use of regional climatic data and consultation with a wind engineer.

When referenced documents, such as those named in the exceptions in Section 1609.1.1, are based on service level design wind speeds, the mapped ultimate wind speeds, V_{ult}, must be converted to a nominal wind speed, V_{asd}, using Table 1609.3.1 or Equation 16-33. This conversion is also necessary for various code provisions that use wind speed as a trigger or threshold (e.g., see Sections 2308.2 and 2308.2.1).

Step 2: Exposure category. IBC Section 1609.4 requires that an exposure category that adequately reflects the characteristics of ground surface irregularities be determined at the site for each wind direction considered.

The IBC exposure category definitions are structured similarly to the ASCE 7 provisions. A Surface Roughness Category is first determined for the building site, and the Exposure Category of the building is then determined based on the Surface Roughness Category prevailing over specified distances in the upwind direction. The definitions for the three types of surface roughness categories (B, C, and D) and the three types of exposure categories (B, C, and D) are identical to those given in ASCE 7.

Once a surface roughness category has been established for a given wind direction, an exposure category is determined from the provisions of Section 1609.4.3. Exposure B applies where Surface Roughness B prevails in the upwind direction for a distance of at least 1,500 feet for buildings with mean roof heights less than or equal to 30 feet. When the mean roof height is greater than 30 feet, the upwind distance is 2,600 feet or 20 times the height of the building, whichever is greater. Exposure D is applicable when Surface Roughness D prevails in the upwind direction for a distance of at least 5,000 feet or 20 times the building height, whichever is greater. This exposure is to extend into downwind areas of Surface Roughness B or C for a distance of 600 feet or 20 times the building height, whichever is greater. Exposure C applies where Exposures B and D do not apply. Commentary Section C26.7 of ASCE 7 contains a method to determine surface roughness categories and exposure categories in cases where a more detailed assessment is required or desired.

In IBC Section 1609.4.1, clear guidelines are provided for determining the governing exposure category whereby two upwind sectors, each enclosing an angle of 45° from the selected wind direction (Figure 1609-1), are to be considered separately, and the exposure resulting in the highest wind load governs the design for the selected wind direction. This is particularly helpful in establishing a more objective determination of exposure category for buildings located in a transition zone of two types of exposures.

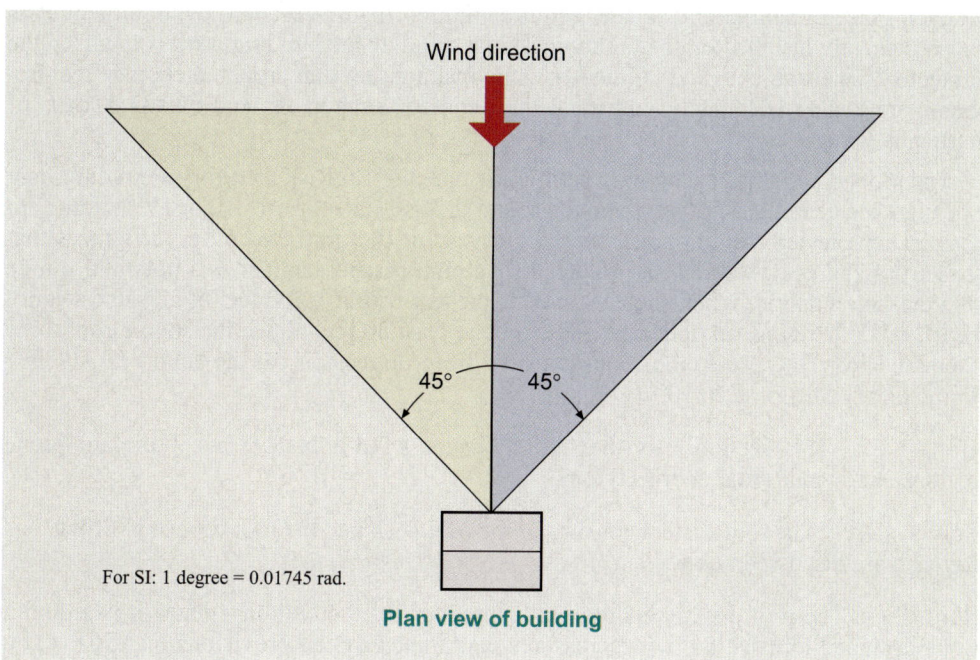

Figure 1609-1 Upwind sectors for determination of governing exposure category.

Step 3: Topographic factor. Topographic factor, K_{zt}, accounts for the increase in wind speeds due to an abrupt change in the upwind terrain such as the presence of an isolated hill, ridge, or escarpment.

However, the isolated feature needs to make a significant difference in the general terrain features to have an effect on the design wind speed. ASCE 7 Section 26.8.1 provides the detailed requirements for determining the topographic factor. The value of K_{zt} can be calculated by using ASCE 7 Figure 26.8-2 and Equation 26.8-1. More discussion can be found in ASCE 7 Commentary Section C26.8.

Step 4: Design wind pressure. Simplified design wind pressure is calculated by using ASCE 7 Equation 28.6-1 for MWFRSs. Figure 28.6-1 provides the simplified design wind pressures on different pressure zones, designated as A through H, for Exposure Category B and mean roof height of 30 feet. An adjustment factor, λ, is also provided in the same tables to adjust the tabulated wind pressure values to those for the actual exposure category and mean roof height of the structure being considered.

Minimum design wind loads are specified in Section 28.6.4. A minimum pressure of 16 psf is to be applied over Zones A and C along with a pressure of 8 psf applied at Zones B and D, while having zero pressure on Zones E through H.

1609.1.2 Protection of openings. This requires that in wind-borne debris regions, glazing in buildings be made impact resistant or be protected with an impact-resistant covering in compliance with the following: (1) glazed openings located within 30 feet of grade meet requirements of the Large Missile Test of ASTM E 1996,[16] and (2) glazed openings located more than 30 feet above grade meet the provisions of the Small Missile Test of ASTM E 1996. The mandatory protection requirements are meant to maintain the integrity of the building envelope and thereby prevent damage to the building due to wind and water and the consequent financial as well as functional losses.

An exception to the above requirement provides the builder and/or owner of one- or two-story buildings that are classified as Group R-3 or R-4 occupancy with a low-cost alternative to the installation to permanent shutters or laminated glass meeting the requirements of ASTM E 1996. This exception requires the builder to provide panels precut to fit each glazed opening and further requires that corrosion-resistant panel attachments be permanently installed on the building. This is to allow the building owners to attach the protective panels quickly and efficiently while making sure that proper anchorage requirements are met to adequately withstand the required components and cladding loads of hurricane winds.

The second and third exceptions permit any openings in Risk Category I buildings and openings located at least 60 feet above ground in Risk Category II, III, or IV buildings to remain unprotected. Additionally, the second part of the third exception recognizes that loose roof aggregate that is not protected by a high parapet can act as a potential source of wind-borne debris. When such a source is present within 1,500 feet of a Risk Category II, III, or IV building, an opening needs to be at least 30 feet above the source roof to be exempted from the protection requirements. More discussion can be found in ASCE 7 Commentary Section C26.10.

1609.2 Definitions. This section lists the definitions of wind-related terms, hurricane-prone regions, and wind-borne debris regions.

1609.5 Roof systems. The roof system consists of the roof deck and the roof covering. Provisions for each are discussed below.

1609.5.1 Roof deck. The roof deck is a structural component of the building. Thus, it is required to be designed to resist the wind pressures determined by the provisions of ASCE 7, by ICC 600,[17] by the AF&PA *Wood Frame Construction Manual for One- and Two-Family Dwellings*,[18] or by AISI S230.[19] The last three procedures have limits on their applicability.

1609.5.2 Roof coverings. If the roof deck is relatively impermeable, the wind pressures will act through the roof deck to the building framing system. If the roof covering is also relatively impermeable and fastened to the roof deck, the two components will be subject to

Wind Loads 16

the same wind pressures. If the roof covering is air permeable, wind pressures are able to develop on both the top of the roof covering and underneath the roof covering. This venting action of the roof negates some of the wind pressures on the roof covering.

Rigid tile. As explained above, in certain types of installations, the roof covering is not subject to the same wind pressures as the roof deck. Such installations include concrete and clay tile roof coverings. The gaps occurring at the joints allow some equalization of pressure between the inner and outer faces of the tiles, leading to reduced pressures. The equation given in this section has been developed by research for the determination of wind loads (uplift moments) on loose-laid and mechanically fastened roof tiles when laid over sheathing with an underlayment. The procedure is based on practical measurements on real tiles to determine the effect of air being able to penetrate the roof covering.

1609.5.3

Alternate all-heights method. In a major addition to the wind provisions, a simplified wind design procedure is included in the IBC. This approach has broader applicability than the simplified method for enclosed simple diaphragm low-rise buildings found in Chapter 28, Part 2 of ASCE 7. This method is an alternative to the Directional Procedure of ASCE 7 and it is somewhat similar to the wind design procedure that was in the 1997 UBC.

1609.6

This method is based on the Directional Procedure of ASCE 7, but its application is restricted to rigid buildings to allow simplifying of assumptions in addition to simplifying the computation of design parameters. These simplifications typically lead to a more conservative design load compared to that of the complete Directional Procedure. The restrictions and the basis of the development of the alternative are described below.

In addition to the limitations already required for the application of the Directional Procedure of ASCE 7, three additional restrictions are imposed on the building for the application of the Alternate All-Heights Method:

1. The building or other structure is less than or equal to 75 feet (22,860 mm) in height, with a height-to-least-width ratio of 4 or less, or the building or other structure has a fundamental frequency greater than or equal to 1 hertz.

2. The building shall meet the requirements of a simple diaphragm building as defined in ASCE 7 Section 26.2, where wind loads are only transmitted to the MWFRS at the diaphragms.

3. For open buildings, multispan gable roofs, stepped roofs, sawtooth roofs, domed roofs, roofs with slopes greater than 45°, solid free-standing walls and solid signs, and rooftop equipment, apply ASCE 7.

The restriction on the building dimensions (or the fundamental frequency) effectively requires that the building must be a rigid building in order to take advantage of this alternative. This allows the use of the Gust Effect Factor, G, to be simply taken as 0.85, as is permitted for rigid structures (see Section 26.9.1 of ASCE 7). This eliminates the need to go through long and complicated calculations necessary to determine the Gust Effect Factor of a flexible building. When the fundamental frequency of the building is greater than or equal to 1 hertz, it is a rigid building by definition (ASCE 7 Section 26.2). It has also been shown that when a building is up to 75 feet high with a height-to-least-width ratio 4 or less, the fundamental frequency is generally greater than 1 hertz. The procedure is derived from the provisions included in ASCE 7 Chapter 27, and the special types of structural configurations mentioned in the limitations are not included.

The alternative procedure is derived from the provisions of ASCE 7 Chapter 27 as follows:

In the ASCE 7 Directional Procedure, the wind pressure for the MWFRS as well as for Components and Cladding of a building is given by

$$p = qGC_p - q_i(GC_{pi}) \text{ lb/ft}^2 \qquad (1)$$

where q is equal to q_z, evaluated at height z above the ground, for the windward wall and q_h, evaluated at the mean roof height h, for the leeward wall, sidewalls, and the roof. The internal velocity pressure, q_i, is taken simply equal to q_h, evaluated at the mean roof height.

In the first step of simplification, the new procedure assumes q_i to be equal to q as defined above. Because, for the windward wall, q is equal to q_z, which is always less or equal to q_h, this simplification results in smaller absolute values of the internal pressure below the roof. Although this does not have any effect on the final wind forces on the MWFRS because the internal pressures on the opposing walls cancel each other, this does tend to push the design wind pressures on the components and cladding to the unconservative side, when compared to those obtained by Chapter 30 of ASCE 7. However, this effect is relatively small and is only apparent in the lower stories of the building.

Thus, Equation 1 above can be rewritten as

$$p = q(GC_p - GC_{pi}) \text{ lb/ft}^2 \quad (2)$$

Substituting for q using Equation 27.3-1 of ASCE 7, the equation above can be further rewritten as

$$p = 0.00256 V^2 K_z K_{zt} K_d (GC_p - GC_{pi}) \text{ lb/ft}^2 \quad (3)$$

The quantity $K_d(GC_p - GC_{pi})$ is called the Net Pressure Coefficient, C_{net}, and is provided in IBC Table 1609.6.2. Thus, the wind pressure in the IBC Alternate All-Heights Method is given by Equation 16-35:

$$p = 0.00256 V^2 K_z C_{net} K_{zt} \text{ lb/ft}^2 \quad (4)$$

For the calculation of C_{net} values, the Directionality Factor K_d is assumed to be a constant 0.85 (ASCE 7 Table 26.6-1), while the Gust Effect Factor G is also taken as constant 0.85 as the method is applicable to rigid buildings (ASCE 7 Section 26.9.1). Thus, the Net Pressure Coefficient is calculated as

$$C_{net} = 0.85 \times (0.85 \times C_p - GC_{pi}) \text{ lb/ft}^2 \quad (5)$$

The values of C_p and GC_{pi} are taken directly from the pertinent sections of ASCE 7 with occasional simplifications. A few examples of the derivations of C_{net}, as presented in IBC Table 1609.6.2, are given below.

MWFRS: Windward Wall ($G = 0.85$, $K_d = 0.85$)

From ASCE 7 Figure 27.4-1, $C_p = 0.8$ and $GC_p = (0.85)(0.8) = 0.68$

From ASCE 7 Table 26.11-1, for enclosed buildings, $GC_{pi} = \pm 0.18$

Thus, $C_{net} = 0.85 \times (0.68 - 0.18) = \mathbf{0.43}$ for positive internal pressure

and $C_{net} = 0.85 \times [0.68 - (-0.18)] = \mathbf{0.73}$ for negative internal pressure

MWFRS: Leeward Wall ($G = 0.85$, $K_d = 0.85$)

In ASCE 7 Figure 27.4-1, three values of C_p are provided for different ratios of length-to-width dimensions (L/B) of a building. Only the numerical maximum of these values is conservatively considered for the alternative procedure for all L/B ratios. Thus,

$$C_p = -0.5 \text{ and } GC_p = (0.85)(-0.5) = -0.43$$

From ASCE 7 Table 26.11-1, for enclosed buildings, $GC_{pi} = \pm 0.18$

Thus, $C_{net} = 0.85 \times (-0.43 - 0.18) = \mathbf{-0.51}$ for positive internal pressure, and $C_{net} = 0.85 \times [-0.43 - (-0.18)] = \mathbf{-0.21}$ for negative internal pressure.

Note that Footnote b to Table 1609.6.2 allows the calculation of more precise (and less conservative) C_{net} values when the coefficients are grouped together for the sake of simplicity, as is the case here. Thus, one is permitted to calculate lower C_{net} values based on the L/B ratio of the building being considered.

MWFRS: Parapets ($G = 0.85$, $K_d = 0.85$)

ASCE 7 Section 27.4.5 provides the combined net pressure coefficients for parapets that include pressures from the front and the back parapet surfaces.

For windward parapet, $GC_{pn} = 1.5 => C_{net} = K_d GC_{pn} = 0.85 \times 1.5 = \mathbf{1.28}$

For leeward parapet, $GC_{pn} = -1.0 => C_{net} = K_d GC_{pn} = 0.85 \times (-1.0) = \mathbf{-0.85}$

MWFRS: Roof with $\theta < 10°$ ($G = 0.85$, $K_d = 0.85$)

In ASCE 7 Figure 27.4-1, pairs of C_p values are provided for different height-to-length ratios (h/L) of a building as well as for different horizontal distances from the windward edge. For the alternative procedure, only the numerical maximum pair of these values is conservatively considered for all cases. Thus,

$C_{p,max}$(Condition 1) $= -1.3$ and $GC_p = (0.85)(-1.3) = -1.105$

From ASCE 7 Table 26.11-1, for enclosed buildings, $GC_{pi} = \pm 0.18$

Thus, $C_{net} = 0.85 \times (-1.105 - 0.18) = \mathbf{-1.09}$ for positive internal pressure, and $C_{net} = 0.85 \times [-1.105 - (-0.18)] = \mathbf{-0.79}$ for negative internal pressure.

As was the case with the leeward wall above, Footnote b to Table 1609.6.2 allows the calculation of more precise C_{net} values when the coefficients are conservatively grouped together for the sake of simplicity. Thus, one is permitted to calculate lower C_{net} values based on the h/L ratio of the building as well as the horizontal distances from the windward edge.

Note also that ASCE 7 Figure 27.4-1 allows a reduction of $C_p = 1.3$ based on the roof area over which it is applicable. Even though the tabulated C_{net} value is based on this C_p, the reduction is not included under the alternative method.

Component and Claddings: Walls in Zone 4 with $h > 60$ ft ($G = 0.85$, $K_d = 0.85$)

From ASCE 7 Figure 30.6-1, for positive pressure in Zone 4 on an effective wind area less or equal to 20 square feet, $GC_p = 0.9$

From ASCE 7 Table 26.11-1, for enclosed buildings, $GC_{pi} = \pm 0.18$

For maximum effect, only the *negative* internal pressure is considered.

Thus, $C_{net} = 0.85 \times [0.9 - (-0.18)] = \mathbf{0.92}$

Similarly, for *positive* pressure in Zone 4 on an effective wind area greater or equal to 500 square feet, $GC_p = 0.6$

Thus, $C_{net} = 0.85 \times [0.6 - (-0.18)] = \mathbf{0.66}$

Note that Footnote a to Table 1609.6.2 permits linear interpolation between these two C_{net} values for intermediate values of the effective wind area. This is different from how these values are graphed in ASCE Figure 30.6-1, where the x-axis shows the log of effective wind area. For instance, with an effective wind area of 100 square feet, the graphed GC_p value is 0.75, which would result in a C_{net} value of $0.85 \times [0.75 - (-0.18)] = 0.79$. However, from the direct linear interpolation between the values in IBC Table 1609.6.2 as calculated above, C_{net} for effective wind area of 100 square feet would be

$$C_{net} = 0.92 + \frac{(0.66 - 0.92)}{(500 - 20)}(100 - 20) = 0.88$$

It is clear that a direct linear interpolation, as permitted by Footnote a to Table 1609.6.2, would produce a more conservative result when compared to that obtained from ASCE 7 Figure 30.6-1.

16 Structural Design

Section 1610 *Soil Lateral Loads*

1610.1 General. Table 1610.1, Lateral Soil Load, is a modified version of Table 3-1 of ASCE 7. For example, Footnotes c and d to Table 3-1 are incorporated into the last column in Table 1610.1 entitled "At-rest pressure."

Table 1610-1, in addition to showing soil lateral loads from the IBC and the ASCE 7 Standard, also shows soil lateral load values using the calculation procedure that formed the basis of the values included in *The BOCA National Building Code*.[20] The calculated soil lateral loads are presented for both moist and saturated conditions. Table 1610-1 shows that for gravels and sands, IBC, ASCE, and the calculated soil lateral loads for moist conditions closely agree. For silts and silt-clay mixtures, the IBC values tend to agree more closely with the calculated soil lateral loads for moist conditions, whereas ASCE 7 values are closer to the calculated soil lateral loads for saturated conditions. Footnote a to IBC Table 1610.1 as well as ASCE 7 Table 3-1 states that the design lateral soil loads are given for moist soil conditions, which appears to be the case more so for the IBC soil loads.

Table 1610-1. **Soil Lateral Loads—Calculated Versus Code and Standard Values**

Soil Description	Unified Soil Classification	Design Lateral Soil Load (pounds per square foot per foot of depth)		Calculated Lateral Soil Load — Active Lateral Pressure (pounds per square foot per foot of depth)	
		IBC	ASCE 7	Moist	Saturated
Well-graded clean gravels, gravel-sand mixes	GW	30	35	34	81
Poorly graded clean gravel, gravel-sand mixes	GP	30	35	34	80
Silty gravels, poorly graded gravel-sand mixes	GM	40	35	40	84
Clayey gravels, poorly graded gravel-sand-clay mixes	GC	45	45	44	86
Well-graded clean sand, gravelly sand mixes	SW	30	35	32	80
Poorly graded clean sands, sand-gravel mixes	SP	30	35	32	79
Silty sands, poorly graded sand-silt mixes	SM	45	45	38	82
Sand-silt clay mix with plastic fines	SM-SC	45	85	40	84
Clayey sands, poorly graded sand-clay mixes	SC	60	85	42	85
Inorganic silts and clayey silts	ML	45	85	39	82
Mixture of inorganic silts and clay	ML-CL	60	85	40	83
Inorganic silts and silt-clay, medium plasticity	CL	60	100	46	86
Organic silts and silt-clays, low plasticity	OL	Unsuitable	Unsuitable	—	—
Inorganic clayey silts, elastic silts	MH	Unsuitable	Unsuitable	—	—
Inorganic clays or high plasticity	CH	Unsuitable	Unsuitable	—	—
Organic clays and silty clays	OH	Unsuitable	Unsuitable	—	—

According to the exception to Section 1610.1, foundation walls (basement walls) extending not more than 8 feet below grade and laterally supported at the top by flexible diaphragms are permitted to be designed for active pressure. Note that ASCE 7 uses the term *light floor system* rather than *flexible diaphragm*. Examples of light floor systems supported on shallow basement walls, given in the ASCE 7 Commentary, are floor systems with wood joists and flooring, and cold-formed steel joists without cast-in-place concrete floors attached.

Expansive soils are found in many regions of the United States. Without special design considerations, expansive soil can cause serious damage to basement walls.[1] Footnote b of Table 1610.1 prohibits the use of expansive soils as backfill because of the potential for very high lateral pressures acting against walls. In this case, it is preferable to excavate expansive soils and backfill with suitable nonexpansive material such as sands and gravels.

Section 1611 *Rain Loads*

This section is similar to Chapter 8 of ASCE 7. The IBC includes Figure 1611.1, which provides the 100-year hourly rainfall rates, and Section 1611.1 specifically refers to these maps (or other approved local weather data) for the purpose of determining the design rain loads. These maps are taken from the maps in IPC Figure 1106.1. Section 1611.2, Ponding Instability, simply references the provisions in ASCE 7 Section 8.4. Note that Figure 1611.1(2) in the 2012 IBC Commentary[21] and Figure 1101.7(1) of the 2012 IPC Commentary[22] provide flow rates in gallons per minute and the corresponding hydraulic heads for various types of drainage systems. This is necessary information when determining the water depth, d_h, due to the hydraulic head. For example, a 6-inch wide by 4-inch high scupper with 3 inches of hydraulic head will discharge 90 gallons per minute.

Design rain loads. Each portion of a roof is required to be designed to sustain the load of rainwater that will accumulate on it if the primary drainage systems for that portion are blocked, plus the uniform load caused by water that rises above the inlet of the secondary drainage system at its design flow. It should be obvious from the design load combinations of Sections 1605.2 and 1605.3 that the design rain loads would affect design only if they are larger than the design snow loads and the roof live loads. **1611.1**

Section 1612 *Flood Loads*

General. Section 1612.1 simply states that buildings and structures in flood hazard areas as established in Section 1612.3 must be designed and constructed to resist the effects of flood hazards and flood loads, as prescribed in this section. It also requires that a building that is located in more than one flood hazard area be designed in accordance with the provisions of the most restrictive flood hazard area. This requirement addresses the situation where a building site is not subjected to high velocity wave action, but is affected by an area designated as a floodway. **1612.1**

Much of the impetus for flood-resistant design has come from initiatives of flood insurance and flood-damage mitigation sponsored by the federal government. The National Flood Insurance Program (NFIP) is based on an agreement between the federal government and participating communities that have been identified as being flood prone. The Federal Emergency Management Agency (FEMA), through the Federal Insurance Administration, makes flood insurance available to the residents of communities, provided that the community adopts and enforces adequate flood-plain management regulations that meet the minimum requirements. Included in the NFIP requirements, found under Title 44 of the U.S. Code of Federal Regulations, are minimum design and construction standards for buildings located in special flood hazard areas.

Special flood hazard areas are those identified by FEMA's Mitigation Directorate as being subject to inundation during the 100-year flood (flood having a 1-percent chance of being equaled or exceeded in any given year), special flood hazard areas are shown on flood insurance rate maps that are produced for flood-prone communities. Special flood hazard areas are identified on such maps as A Zones (A, AE, A1-30, A99, AR, AO, or AK) or V Zones (V, VE, VO, or V1-30). The special flood hazard areas are those in which communities must enforce NFIP-compliant, flood-damage-resistant design and construction practices.

Ensuring that the provisions of the IBC and the manner in which they are administered are consistent with those of the NFIP is accomplished through adoption by reference of the ASCE 24 *Flood Resistant Design and Construction Standard*.[23] The provisions of ASCE 24 meet or exceed the building science provisions of the NFIP and present the consensus state-of-the-art approach to flood-resistant construction.

ASCE 24 requires that the design of structures within flood hazard areas be governed by the loading provisions of ASCE 7. That standard requires that the structural systems of buildings be designed, constructed, connected, and anchored to resist flotation, collapse, and permanent lateral movement due to the action of wind loads and loads from flooding associated with the design flood. Wind loads and flood loads may act simultaneously at coastlines, particularly during hurricanes and coastal storms. This may also be true in some other situations cited in the ASCE 7 Commentary Section C5.3.1. Flood loads are the loads or pressures on the surfaces of buildings and structures caused by the presence of floodwaters. These loads are of two basic types—hydrostatic and hydrodynamic. Impact loads result from objects transported by floodwaters striking against structures or parts of structures. Wave loads are considered a special type of hydrodynamic load. Hydrostatic loads, hydrodynamic loads, wave loads, and impact loads are specified in ASCE 7. Wave loads may be determined by using an analytical procedure outlined by more advanced numerical modeling procedures, or by laboratory test procedures or physical modeling.

ASCE 24 requires that the design and construction of structures located in flood-hazard areas consider all flood-related hazards, including hydrostatic loads, hydrodynamic loads and wave action, debris impacts, alluvial fan flooding, flood-related erosion, ice flows or ice jams, or mudslides in accordance with the requirements of that standard if specified or, if not specified in that standard, then in accordance with the requirements approved by the authority having jurisdiction. Design documents must identify, and take into account, flood-related and other concurrent loads that will act on the structure. Design documents must include, but not be limited to, the applicable conditions listed below:

1. Wave action
2. High-velocity floodwaters
3. Impacts due to debris in the floodwaters
4. Rapid inundation by floodwaters
5. Rapid drawdown of floodwaters
6. Prolonged inundation by floodwaters
7. Wave- and flood-induced erosion and scour
8. Deposition and sedimentation by floodwaters

1612.2 Definitions. The particular definitions to be noted are Design Flood, Flood-Hazard Area, Flood-Hazard Area Subject to High-Velocity Wave Action, and Special Flood Hazard Area.

Establishment of flood-hazard areas. Flood-hazard areas are established by the local jurisdiction through adoption of a flood-hazard map and supporting data. Minimum requirements imposed on the jurisdiction are spelled out. Two key pieces of information that are not part of the minimum requirement are the design flood elevations and designation of floodways (floodways are areas along riverine bodies that convey the bulk of floodwaters). A large percentage of areas that are mapped as special flood hazard areas by the National Flood Insurance Program do not have flood elevations and do not have floodway designations. In case of such an absence, the building official can require that the design flood elevations be either obtained from any pertinent data available from the federal, state, or other sources, or determined by a registered design professional in accordance with accepted hydrologic and hydraulic engineering practices.

1612.3

Development in riverine floodplains can increase flood levels and loads on other properties, especially if it occurs in areas known as floodways that must be reserved to convey flood flows. The floodway, by definition, is the area along riverine waterways that "must be reserved in order to discharge the base flood without cumulatively increasing the water surface elevation more than a designated height." For the purpose of this section, the designated height is 1 foot.

Design and construction. This section requires the design and construction of buildings and structures located in flood-hazard areas, including flood-hazard areas subject to high-velocity wave action, to be in accordance with the requirements of Chapter 5 of ASCE 7 and ASCE 24.

1612.4

Flood-hazard documentation. This lists a number of documents that must be furnished to the building official. The requirements are different for construction in flood-hazard areas not subject to high-velocity wave action and for construction in flood-hazard areas subject to high-velocity wave action.

1612.5

Section 1613 *Earthquake Loads*

Overview. In the IBC, the seismic load provisions are largely adopted through reference to ASCE 7. The earthquake provisions of ASCE 7 are mainly derived from the NEHRP *Provisions*. For information regarding the application of the IBC seismic provisions, one should refer to the NEHRP *Provisions* Part 2—Commentary. In addition, the ASCE 7 Commentary provides some background on the seismic requirements. One other valuable reference is the *2009 IBC SEAOC Structural Seismic Design Manual, Volume 1, Code Application Examples*,[24] which is a compilation of sample calculations developed by the Structural Engineers Association of California.

Using these provisions, restrictions on building height and structural irregularity, choice of analysis procedures that form the basis of seismic design, as well as the level of detailing required for a particular structure are all governed by a structure's SDC, a classification that considers the nature of the occupancy along with the soil-modified ground motion at the site of the structure. In other words, when designing for earthquakes under the IBC, the required detailing as well as other seismic system limitations are affected by the soil characteristics at the site of a structure. This is a major departure from prior model codes.

Also see Section 1604.10 discussion on building behavior under earthquake ground motions.

Scope. Section 1613 incorporates the seismic load provisions of ASCE 7. Chapter 14 of ASCE 7 is excluded because the IBC materials Chapters 19 through 23 must be used. Appendix 11A of ASCE 7 is also excluded because the inspection and testing requirements in Chapter 17 of the IBC must be used. Thus, the reference to ASCE 7 earthquake requirements has the net effect of adopting Chapters 11, 12, 13, and 15 through 23 of ASCE 7.

1613.1

After having stated that every structure and portions thereof shall, at a minimum, be designed and constructed to resist the effects of earthquake motions and assigned an SDC in accordance with Section 1613 or ASCE 7, the IBC permits a number of important exceptions:

1. Detached one- and two-family dwellings in SDC A, B, or C or located where S_S is less than $0.4g$ are exempt from seismic design requirements. The last part of this exception removes the need to determine the SDC, which in turn requires the determination of site class, in low-to-moderate hazard areas. Figure 1613-1 shows areas of the contiguous United States that have S_S less than $0.4g$.

2. The seismic-force-resisting systems of wood-frame buildings that conform to the provisions of Section 2308 (Conventional Light-Frame Construction) need not be analyzed as required by Section 1613.1. Based on the limitations in IBC Sections 2308.2 and 2308.11.1, this exception, in effect, increases the allowable height of a wood-frame building from two to three stories when the building is assigned to SDC A or B.

3. Agricultural storage structures intended only for incidental human occupancy are exempt from all seismic design requirements. This is because of the very low risk to life that is involved.

4. Certain special structures such as "vehicular bridges, electrical transmission towers, hydraulic structures, buried utility lines and their appurtenances, and nuclear reactors" are placed outside the scope of Section 1613. These structures require special consideration of their response characteristics and environment, and need to be designed in accordance with other regulations.

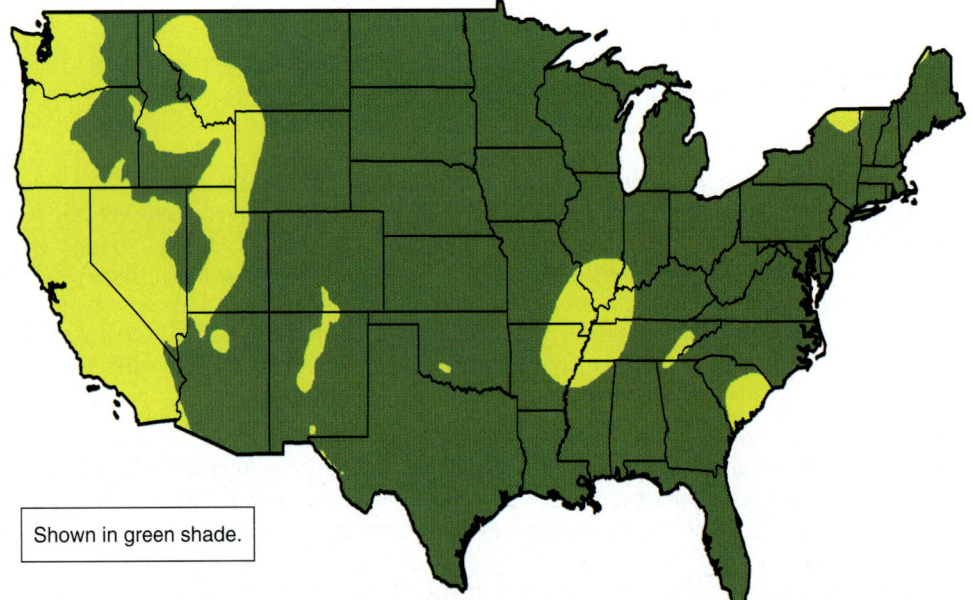

Figure 1613-1 Areas of the contiguous United States where $S_S < 0.4g$. (Map provided courtesy of S. K. Ghosh Associates.)

Shown in green shade.

1613.3 Seismic ground motion values. The procedure for determining the maximum considered earthquake design spectral response accelerations (SD_S and SD_1), site class, and SDC are given in this section.

1613.3.1 Mapped acceleration parameters. The mapped maximum considered earthquake spectral response accelerations at short periods (S_S) and at 1-second period (S_1) for a particular site are to be determined from Figures 1613.3.1(1) through 1613.3.1(6). Where a site is between contours, straight-line interpolation or the value of the higher contour may be used.

The IBC maps for S_S and S_1 are based on USGS maps that are available on the USGS website at http://earthquake.usgs.gov/research/hazmaps. The USGS has prepared an Internet calculation tool for obtaining seismic design parameters using the same data that were used to prepare the ground motion maps published in the IBC and ASCE 7. Along with the mapped values, the user can also view a design spectrum for a site specified by latitude-longitude or ZIP code. Site coefficients other than Site Class B may be included in the calculations.

The mapped MCE spectral response accelerations S_S and S_1 of the 2009 IBC are mapped on Site Class B, soft rock of the Western United States. This happens to be equivalent to Soil Profile Type S_B of the 1997 UBC.

IBC Section 1613.3.1 assigns SDC A to any structure regardless of its risk category when mapped short-period and 1-second-period spectral response accelerations, S_S and S_1, at its site are less than or equal to 0.15g and 0.04g, respectively. See Figure 1613-2.

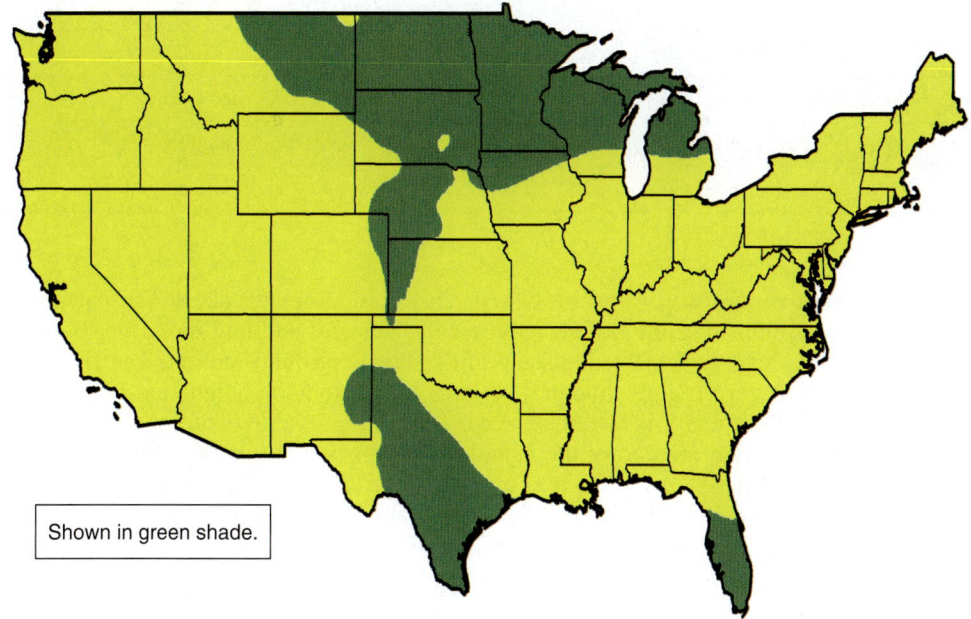

Figure 1613-2
Areas of the contiguous United States where $S_S \leq$ 0.15g and $S_1 \leq$ 0.04g. (Map provided courtesy of S. K. Ghosh Associates Inc.)

Shown in green shade.

1613.3.2 Site class definitions. The site class definitions of the code date back to the 1994 NEHRP *Provisions*,[2] in which extensive modifications were made to the consideration of side effects.

1613.3.3 Site coefficients and adjusted maximum considered earthquake spectral response acceleration parameters. S_{MS} is obtained by multiplying the mapped MCE spectral response acceleration S_S (at short periods) by F_a, the acceleration-related site coefficient [see Table 1613.3.3(1)]. S_{M1} is similarly obtained by multiplying the mapped MCE spectral response acceleration S_1 (at 1-second period) by F_v, the velocity-related site coefficient [see Table 1613.3.3(2)].

Site coefficients are determined by the closest distance from a site to known seismic sources that are capable of generating large-magnitude earthquakes and the types of those seismic sources. The short- and long-period amplification factors implied by the Loma Prieta strong-motion data and related calculations for the same earthquake by Joyner et al.,[25] as well as modeling results at the 0.1g ground acceleration level, provided the basis for these site coefficients F_a and F_v, which first appeared in the 1994 NEHRP *Provisions*.

Commentary guidance is provided in the 2000 NEHRP *Provisions* on how to determine the response of Type F soils for which values of site factors F_a and F_v were not tabulated in the IBC and for which the note to the tables requires that "site-specific geotechnical investigation and dynamic site response analyses shall be performed."

1613.3.4 Design spectral response acceleration parameters. Five-percent damped design spectral response accelerations at short periods, S_{DS}, and at 1-second period, S_{D1}, are equal to two-thirds S_{MS} ($= F_a S_s$) and two-thirds S_{M1} ($= F_v S_1$), respectively. In other words, the design ground motion is 1/1.5 or two-thirds times the soil-modified maximum considered earthquake ground motion. This is in recognition of the inherent margin contained in the NEHRP *Provisions* that would make collapse unlikely under one and one-half times the design level ground motion. Table 1613-1 summarizes the derivation of the design quantities S_{DS} and S_{D1}.

Table 1613-1. **Design Ground Motion of the IBC**

S_S = MCE spectral acceleration in the short-period range for Site Class B.
S_1 = MCE spectral acceleration at 1-second period for Site Class B.
$S_{MS} = F_a S_S$, MCE spectral acceleration in the short-period range adjusted for site class effects.
$S_{M1} = F_v S1$, MCE spectral acceleration at 1-second period adjusted for site class effects.
$S_{DS} = 2/3\ S_{MS}$, spectral acceleration in the short-period range for the design ground motion.
$S_{D1} = 2/3\ S_{M1}$, spectral acceleration at 1-second period for the design ground motion.

1613.3.5 Determination of seismic design category. This section creates six design categories that are key to establishing the design requirements for any building based on its occupancy and on the level of expected soil-modified seismic ground motion. The IBC uses SDCs to determine permissible structural systems, limitations on height and irregularity, the type of lateral force analysis that must be performed, the level of detailing for structural members, and joints that are part of the lateral-force-resisting system and for the components that are not. The SDC is a function of occupancy and of soil-modified seismic risk at the site of the structure in the form of the design spectral response acceleration at short periods, S_{DS}, and the design spectral response acceleration at 1-second period, S_{D1}.

A structure located where $S_1 \geq 0.6g$ is assigned to SDC E if it is in Risk Category I, II, or III and to SDC F if it is in Risk Category IV. For structures not assigned to SDC E or F, the seismic design category needs to be determined twice—first as a function of S_{DS} by Table 1613.3.5(1) and a second time as a function of S_{D1} by Table 1613.3.5(2); the more severe category governs.

SDC A applies to structures, irrespective of their occupancy, in regions where anticipated ground motions are minor, even for very long return periods.

SDC B includes Risk Category I, II, or III structures in regions where moderately destructive ground shaking is anticipated.

SDC C includes Risk Category IV structures where moderately destructive ground shaking may occur as well as Risk Category I through III structures in regions with somewhat more severe ground-shaking potential.

SDC D includes structures of Risk Category IV structures located in regions of severe seismicity and above, and Risk Category I through III structures located in regions expected to experience destructive ground shaking.

However, as noted above, structures located close to major active faults where the mapped value of S_1 is greater than or equal to 0.75 are directly assigned to SDC E or F. SDC E includes Risk Category I, II, or III structures in regions located close to major active faults, and SDC F includes Risk Category IV structures in those locations. See Figure 1613-3.

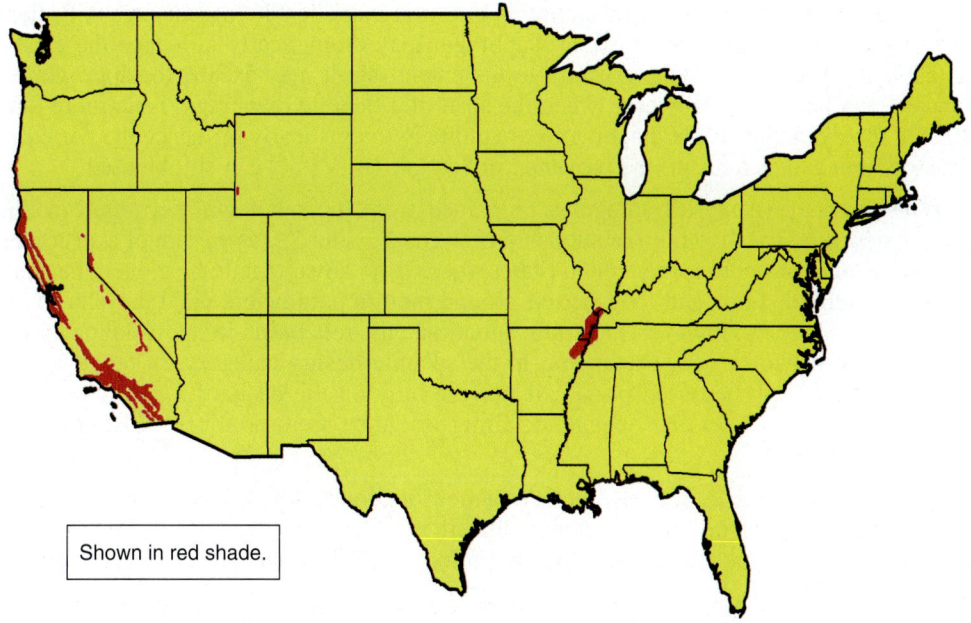

Figure 1613-3
Areas of the conterminous United States where $S_1 \geq 0.75g$. (Map provided courtesy of S. K. Ghosh Associates Inc.)

Shown in red shade.

1613.3.5.1

Alternative seismic design category determination. In view of the fact that it is unnecessary and wasteful to require that the seismic design category of a short-period structure be based on long-period ground motion, the IBC permits the SDC determination to be based on S_{DS} alone [Table 1613.3.5(1)], provided:

1. 1-second mapped spectral response acceleration, S_1, is less than 0.75g;
2. $T_a < 0.8T_{S5}$ where T_a is the approximate fundamental period of the structure and $T_S = S_{D1}/SD_S$;
3. Fundamental period of the structure used to calculate story drift is $< T_S$;
4. Upper-bound design base shear is used in design [e.g., $V = SD_SW/(R/I)$]; and
5. Diaphragm is not flexible, or when the distance between vertical elements of the seismic-force-resisting system does not exceed 40 feet.

The first criterion means that SDC determination cannot be based on S_{DS} alone for structures in the vicinity of known faults that can generate large earthquakes (so-called near-fault structures).

The period $T_S = S_{D1}/SD_S$ is the period at which the short-period or constant-acceleration part of the design spectrum transitions into the long-period or velocity-governed part of the spectrum. It is the dividing line between short-period and long-period response. By requiring in Item 2 above that the approximate fundamental period, T_a, be less than $0.8T_S$ rather than T_S itself, the intent of the code is to minimize the possibility that the actual fundamental period, T, might equal or exceed T_S, even though T_a is less than T_S. However, this requirement appears to become superfluous in the presence of the requirement in Item 3. ASCE 7 does not impose an upper limit on the period used to calculate story drift, and as a result, it probably is a more realistic representation of the actual period of the structure than T_a.

Item 4 above requires that the upper-bound design base shear, as given by the constant-acceleration or *flat-top* part of the design spectrum, be used in the design of a structure utilizing the above exception. This requirement is intended to impose a design force penalty on a structure for which T may equal or exceed T_S, while T_a is less than $0.8T_S$.

The last criterion makes the relaxation in question inapplicable to structures with flexible diaphragms, because the flexible diaphragm may dramatically influence the elastic fundamental period of such a structure to the extent that it may exceed the approximate fundamental period, T_a. However, when the span of a flexible diaphragm is no more than 40 feet, the dynamics of the diaphragms are unlikely to drastically influence the period of the structure, and as a result, the provision in Section 1613.3.5.1 can still be used.

Whether there is any advantage to be gained from the relaxation permitted in this section depends strictly on the relationship between S_{DS} and S_{D1} at the site of a structure. For many locations across the United States, there is no advantage to be gained, because the soil-modified short- and long-period ground motion parameters yield the same seismic design category. However, in certain situations, the relaxation in question may yield a one- or even two-category reduction in the seismic design category. In other words, while the SDC may be based on S_{D1}, it may be only on the basis of S_{DS}. In that case, only intermediate, rather than special, detailing would be required for the seismic-force-resisting system.

1613.4 Alternatives to ASCE 7. The IBC includes a modification to ASCE 7 in Section 1613.4.1. This modification is optional, meaning that it may be employed by the engineer if desired; otherwise, the unmodified corresponding ASCE 7 provisions are to be used.

1613.4.1 Additional seismic-force-resisting systems for seismically isolated structures. An exception is included in ASCE 7 Section 17.5.4.2, whereby restrictions placed on the use of Ordinary Steel Concentrically Braced Frames (Steel OCBFs) and Ordinary Steel Moment Frames (OMFs of steel) in a seismically isolated SDC D, E, or F structure are relaxed significantly. Due to their poor performance during a large earthquake, ASCE 7 Table 12.2-1 appropriately limits the use of these systems in SDC D, E, or F structures. While these limitations are valid for buildings with fixed bases where large ductility demands are expected, seismically isolated structures warrant a different approach, as the ductility demand on them is much smaller, owing to the much smaller value of the response modification factor R. Thus, a significant savings is possible by not requiring the use of special detailing associated with special moment frames and braced frames. This section in the IBC permits OCBFs and OMFs of steel in a building classified as SDC D, E, or F up to a height of 160 feet, provided the value of RI is taken as 1 and the structural systems are designed in accordance with AISC 341, *Seismic Provisions for Structural Steel Buildings*.[26]

Section 1614 *Atmospheric Ice Loads*

This section provides charging text in the IBC, which references the technical provisions of ASCE 7 for atmospheric ice loads. This section relies on the determination of which structures are ice sensitive in order to determine the need to comply with the applicable provisions of ASCE 7. An "ice-sensitive structure" is defined in Section 202 and provides the technical basis for determining which structures are ice sensitive.

Section 1615 *Structural Integrity*

This section aims to provide a minimum level of structural integrity by establishing minimum requirements for tying together the primary structural elements. The provision has been developed by a broad coalition of industry members involved in steel, concrete, masonry, and wood structures, and applies to applicable structures of all materials.

Although the structural integrity requirements already included in ACI 318, coupled with the common design and construction practices adopted in the United States, are generally adequate to provide a satisfactory level of structural integrity to most concrete structures, this section expands on those provisions in order to include other building materials as well.

The provisions apply to high-rise buildings that are classified as Risk Category III and IV. The code defines a high-rise building as one that has an occupied floor located more than 75 feet above the lowest level of fire department vehicle access. Detailing requirements are provided for concrete- and steel-frame structures, whereas minimum requirements for vertical, longitudinal, transverse, and perimeter ties are provided for bearing-wall structures.

KEY POINTS

- Chapter 16 prescribes minimum structural loading requirements for the design of buildings and other structures regulated by the IBC.

- In addition to the general requirements for construction documents given in Section 107, Section 1603 contains specific requirements for construction documents pertaining to structural loads.

- The American Society of Civil Engineer's *Minimum Design Loads for Buildings and Other Structures* (ASCE/SEI 7-10) is the referenced standard for structural loads in the 2012 IBC.

- General construction requirements are provided in Section 1604 for strength, serviceability, and risk category, as well as permissible design methods such as allowable stress design, load and resistance factor design, empirical design, and prescriptive conventional construction.

- How various structural loads are required to be combined is prescribed in Section 1605.

- Buildings and other structures must resist all prescribed loads and be stable against overturning and sliding.

- In addition to dead and live loads, buildings and other structures must resist the effects of environmental loads such as snow, wind, earthquake, soil lateral, rain, flood, and ice loads as required by Chapter 16 and ASCE 7.

- High-rise buildings assigned to Risk Category III or IV must also meet minimum structural integrity requirements of Section 1615 based on whether they are frame structures (gravity loads supported by columns) or bearing-wall structures (gravity loads supported by walls) as defined in Section 202.

REFERENCES

1. American Society of Civil Engineers, *ASCE Standard Minimum Design Loads for Buildings and Other Structures*, ASCE 7-93, ASCE 7-95, ASCE 7-98, New York, NY, 1993, 1995, 1998, and 2000, respectively, ASCE 7-02, ASCE 7-05, ASCE 7-10, Reston, VA, 2002, 2005, and 2010, respectively.

2. Federal Emergency Management Agency, *NEHRP (National Earthquake Hazards Reduction Program) Recommended Provisions for Seismic Regulations for New Buildings and Other Structures*, Washington, DC, 1985, 1988, 1991, 1994, 1997, 2000, 2003.

3. International Conference of Building Officials, *Uniform Building Code*, Whittier, CA, 1997.

4. Jirsa, J.O., "Behavior of Elements and Subassemblages—R.C. Frames," *Proceedings of a Workshop on Earthquake-Resistant Reinforced Concrete Building Construction*, Berkeley, July 1977, Vol. III, pp. 1196–1214.

5. Peir, J.C., and Cornell, C.A., "Spatial and Temporal Variability of Live Loads," *Journal of the Structural Division*, ASCE, Proceedings Vol. 99, No. ST5, May 1973, pp. 903–922.

6. McGuire, R.K., and Cornell, C.A., "Live Load Effects in Office Buildings," *Journal of the Structural Division*, ASCE, Proceedings Vol. 100, No. ST7, July 1974, pp. 1351–1366.

7. Culver, C.G., *Survey Results of Fire Loads and Live Loads in Office Buildings*, NBS Building Science Series 85, National Bureau of Standards, Washington, DC, May 1976.

8. Ellingwood, B.R., and Culver, C.G., "Analyses of Live Loads in Office Buildings," *Journal of the Structural Division*, ASCE, Proceedings Vol. 103, No. ST8, pp. 1551–1560, August 1977.

9. International Conference of Building Officials, *Handbook to the Uniform Building Code: An Illustrative Commentary*, Whittier, CA, 1998.

10. American Association of State Highway and Transportation Officials, *LRFD Bridge Design Specifications*, Fourth edition, 2007.

11. International Code Council, *Accessible and Usable Buildings and Facilities*, A117.1, Washington, DC, 2009.

12. ADA, *Accessibility Guidelines for Building and Facilities (ADAAG)*. 28 CFR, Part 36 revised, July 1, 1994.

13. Southern Building Code Congress International, *Standard Building Code*, Birmingham, AL, 1997, 2000.

14. The National Bureau of Standards, *Live Loads on Floors and Buildings, Building Materials and Structures Publication No. 133*, Washington, DC, 1952.

15. O'Rourke, M., and Downey, C., "Rain-on-Snow Surcharge for Roof Design," *Journal of Structural Engineering*, ASCE, Vol. 127, No. 1, 2001, pp. 74–79.

16. American Society for Testing and Materials, *Specification for Performance of Exterior Windows, Glazed Curtain Walls, Doors and Impact Protective Systems Impacted by Windborne Debris in Hurricanes*, ASTM E 1996-06, West Conshohocken, PA, 2006.

17. International Code Council, *Standard for Residential Construction in High Wind Regions*, ICC 600-08, Washington, DC, 2008.

18. American Forest and Paper Association, *Wood Frame Construction Manual for One- and Two-Family Dwellings*, ANSI/AF&PA WFCM-2001, 2001.

19. American Iron and Steel Institute, *Standard for Cold-Formed Steel Framing—Prescriptive Method for One- and Two-Family Dwellings*, with Supplement 2, AISI 230-07, Washington, DC, 2007.

20. Building Officials and Code Administrators International, *The BOCA National Building Code*, Country Club Hills, IL, 1993, 1996, 1999.

21. International Code Council, *2012 International Building Code Commentary*. Washington, DC: International Code Council, 2012.

22. International Code Council, *2012 International Plumbing Code Commentary*. Washington, DC: International Code Council, 2012.

23. American Society of Civil Engineers, *Flood Resistant Design and Construction Standard*, Reston, VA, 1998, 2005.

24. Structural Engineers Association of California, *2009 IBC SEAOC Structural Seismic Design Manual*, Volume 1, Code Application Examples, Sacramento, CA, 2012.

25. Joyner, W.B., Fumal, T.E., and Glassmoyer, G., "Empirical Spectral Response Ratios for Strong Motion Data from the 1989 Loma Prieta, California Earthquake," *Proceedings of the NCEER/SEAOC/BSSC Workshop on Site Response during Earthquakes and Seismic Code Provisions*, November 18–20, 1992, University of Southern California, Los Angeles, Martin, G.M., Ed., 1994.

26. American Institute of Steel Construction, *Seismic Provisions for Structural Steel Buildings*, AISC 341-05, Chicago, IL, 2010.

Example 16-1 Design Axial Force, Shear Force, and Bending Moment for Shear Wall Due to Lateral and Gravity Loads (Strength Design)

Load Effect	Symbol	Axial Force (kips)	Shear Force (kips)	Bending Moment (ft-kips)
Dead load effect	D	5,381	0	0
Live load effect	L	837	0	0
Wind load effect	W	0	462	29,938
Effect of horizontal design earthquake forces	Q_E	0	1,571	118,596

Code Formula	Combination	Axial Force (kips)	Shear Force (kips)	Bending Moment (ft-kips)
(16-1)	$1.4D$[1]	7,533	0	0
(16-2)	$1.2D + 1.6L$[2]	7,796	0	0
(16-4)	$1.2D + 1.0W + 0.5L$[3]	6,876	462	29,938
(16-5)	$1.2D + (pQ_E + 0.2S_{DS}D) + 0.5L$[4]	7,952	1,571	118,596
(16-6)	$0.9D + 1.0W$[5]	4,843	4,62	29,938
(16-7)	$0.9D + (pQ_E - 0.2S_{DS}D)$[6]	3,767	1,571	118,596

Notes:
(1) The effect of fluid load (F) not considered here.
(2) The effect of fluid load (F), lateral earth pressure (H), roof live load (L_r), snow load (S), or rain load (R) not considered here.
(3) The live load was assumed to be less than 100 psf so that $f_1 = 0.5$; the effect of fluid load (F), lateral earth pressure (H), roof live load (L_r), snow load (S), or rain load (R) not considered here.
(4) The live load was assumed to be less than 100 psf so that $f_1 = 0.5$; $\rho = 1.0$ and $S_{DS} = 1.0$; the effect of fluid load (F), lateral earth pressure (H), or snow load (S) not considered here.
(5) The effect of lateral earth pressure (H) not considered here.
(6) The effect of fluid load (F) or lateral earth pressure (H) not considered here; $\rho = 1.0$ and $S_{DS} = 1.0$.

16 Structural Design

Example 16-2 *Design Axial Force, Shear Force, and Bending Moment for Shear Wall Due to Lateral and Gravity Loads (Allowable Stress Design Using Basic Load Combinations)*

Load Effect	Symbol	Axial Force (kips)	Shear Force (kips)	Bending Moment (ft-kips)
Dead load effect	D	5,381	0	0
Live load effect	L	837	0	0
Wind load effect	W	0	462	29,938
Effect of horizontal design earthquake forces	Q_E	0	1,571	118,596

Code Formula	Combination	Axial Force (kips)	Shear Force (kips)	Bending Moment (ft-kips)
(16-8)	D[1]	5,381	0	0
(16-9)	$D + L$[2]	6,218	0	0
(16-13)	$D + 0.75(0.6)W + 0.75L$[3]	6,009	208	13,472
(16-14)	$D + 0.75(0.7)(\rho Q_E + 0.2 S_{DS} D) + 0.75L$[4]	6,574	825	62,263
(16-15)	$0.6D + 0.6W$[5]	3,229	277	17,963
(16-16)	$0.6D + 0.7(\rho Q_E - 0.2 S_{DS} D)$[6]	2,475	1,100	83,017

Notes:
(1) The effect of fluid load (F) not considered here.
(2) The effect of fluid load (F) and lateral earth pressure (H) not considered here.
(3) The effect of fluid load (F), lateral earth pressure (H), roof live load (L_r), snow load (S), or rain load (R) not considered here.
(4) The effect of fluid load (F), lateral earth pressure (H), or snow load (S) not considered here; $\rho = 1.0$ and $S_{DS} = 1.0$.
(5) The effect of lateral earth pressure (H) not considered here.
(6) The effect of fluid load (F) and lateral earth pressure (H) not considered here.

Example 16-3 Design Axial Force, Shear Force, and Bending Moment for Shear Wall Due to Lateral and Gravity Loads (Allowable Stress Design Using Alternate Basic Load Combinations)

Load Effect	Symbol	Axial Force (kips)	Shear Force (kips)	Bending Moment (ft-kips)
Dead load effect	D	5,381	0	0
Live load effect	L	837	0	0
Wind load effect	W	0	462	29,938
Effect of horizontal design earthquake forces	Q_E	0	1,571	118,596

Code Formula	Combination	Axial Force (kips)	Shear Force (kips)	Bending Moment (ft-kips)
(16-17)	$D + L$ [1]	6,218	0	0
(16-18)	$D + L + \omega 0.6W$ [2]	6,218	360	23,352
(16-18) [3]	$0.67D + L + \omega 0.6W$	4,424	289,360	23,352
(16-20)	$D + L + \omega 0.6W/2$ [4]	6,218	180	11,676
(16-21)	$D + L + (\rho Q_E + 0.2 S_{DS} D)/1.4$ [5]	6,987	1,122	84,711
(16-22)	$0.9D + (\rho Q_E - 0.2 S_{DS} D)/1.4$ [5][6]	4,073	1,122	84,711

Notes:

(1) The effect of roof live load, snow load, or rain load not considered here.

(2) $\omega = 1.3$ as wind forces are calculated per ASCE 7.

(3) In IBC Section 1605.3.2, there is a statement that reads, "For load combinations that include the counteracting effects of dead and wind loads, only two-thirds of the minimum dead load likely to be in place during a design wind event shall be used." This is obviously the same as multiplying the dead load by 0.67, which is not very different from 0.6 (IBC Equation 16-15, see Example 2). The live load effect will help in counteracting the wind load effect when using the "alternative basic" ASD load combinations, and as a result, they are somewhat less conservative than the "basic" load combinations.

(4) The effect of snow load not considered here.

(5) The effect of snow load not considered here; $\rho = 1.0$ and $S_{DS} = 1.0$.

(6) In seismic design, the alternative basic load combinations are likely to result in a more economical structure, because 90 percent, rather than 60 percent, of the design dead load effects can be counted upon to counteract service-level earthquake effects.

16 Structural Design

Example 16-4 Calculations of Live Load Reduction

The computation of live load reduction for roof and floor per Sections 1607.10 and 1607.12 is illustrated in the following example.

Given: Plan Dimensions (as shown)
Story Height = 12 ft
Live load: Floors = 50 psf (for a typical office building, see Table 1607.1)
Roof = 20 psf (assumed)
Dead load: Floors = 120 psf (assumed)
Roof = 90 psf (assumed)

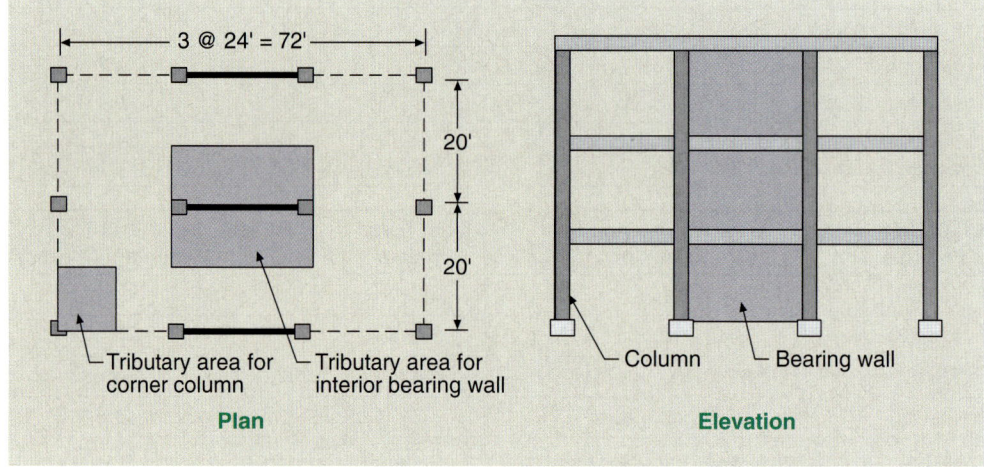

Figure 16-E4-1 Tributary roof area.

Roof Live Load, L_r, Reduction						
L_r for Corner Column						
A_T (ft²)	L_o (psf)	R_1	R_2	L_r (psf)	$R = 1 - L/L_o$ Reduction (%)	L_r (kips)
120	20	1	1	20	0	2.4
L_r for Interior Bearing Wall						
A_T (ft²)	L_o (psf)	R_1	R_2	L_r (psf)	$R = 1 - L/L_o$ Reduction (%)	L_r (kips)
480	20	0.72	1	14.4	28	6.9

576 2012 International Building Code Handbook

Example 16-4

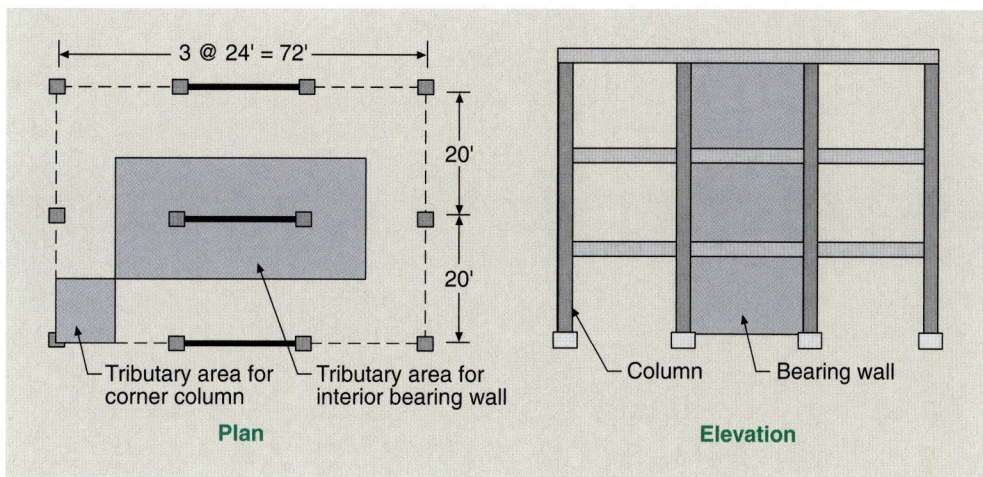

Figure 16-E4-2
Tributary floor area.

Floor Live Load, *L*, Reduction										
L for Corner Column										
Floor Level	A_T (ft²)	D (psf)	D (kips)	L_o (psf)	K_{LL}	$K_{LL} \times A_T$	R (%)	R (max) (%)	Reduction (%)	L (psf)	L (kips)
3rd	120	120	14.4	50	4	480	6.5	50	6.5	46.8	5.6
2nd	240	120	14.4	50	4	960	27	60	27	36.5	8.8

L for Corner Column (Alternate Method)										
Floor Level	A_T (ft²)	D (psf)	D (kips)	L_o (psf)	R	R (max)*	R (max)**	Reduction (%)	L (psf)	L (kips)
3rd	120	120	14.4	50	0	60	78.5	0	50	6.0
2nd	240	120	14.4	50	7.2	60	78.5	7.2	46.4	11.1

*Equation 16-24.
**Equation 16-25.

L for Interior Bearing Wall										
Floor Level	A_T (ft²)	D (psf)	D (kips)	L_o (psf)	K_{LL}	$K_{LL} \times A_T$	R (%)	R (max) (%)	Reduction (%)	L (psf)	L (kips)
3rd	960	120	115.2	50	2	1920	41	50	41	29.5	28.3
2nd	1920	120	115.2	50	2	3840	51	60	51	24.6	47.2

L for Interior Bearing Wall (Alternate Method)										
Floor Level	A_T (ft²)	D (psf)	D (kips)	L_o (psf)	R	R (max)*	R (max)**	Reduction (%)	L (psf)	L (kips)
3rd	960	120	115.2	50	64.8	60	78.5	60	20	19.2
2nd	1920	120	115.2	50	141.6	60	78.5	60	20	38.4

*Equation 16-24.
**Equation 16-25.

CHAPTER 17

STRUCTURAL TESTS AND SPECIAL INSPECTIONS

Introduction
Section 1701 General
Section 1702 Definitions
Section 1703 Approvals
Section 1704 Special Inspections, Contractor Responsibility, and Structural Observations
Section 1705 Required Verification and Inspection
Section 1706 Design Strengths of Materials
Section 1707 Alternate Test Procedures
Section 1708 Test Safe Load
Section 1709 In Situ Load Tests
Section 1710 Preconstruction Load Tests
Section 1711 Material and Test Standards
Overview of Chapter 17
Key Points
References

17 Structural Tests and Special Inspections

Introduction

The primary goal of Chapter 17 of the *International Building Code®* (IBC®) is to improve the quality and workmanship of certain structural systems by requiring structural testing, special inspection, and structural observation. Many of the requirements are specifically intended to improve the quality of the lateral-force-resisting system when buildings are subjected to the design wind or seismic event. To accomplish these goals, the chapter sets forth provisions for quality of materials, workmanship, testing, and labeling of materials incorporated into the construction of buildings or structures regulated by the code. Generally, all components and materials used in new buildings must conform to the requirements in the code, or the applicable standards referenced by the code.* Specific tests and standards are referenced in other parts of the code. Chapter 35 contains a complete alphabetical list of all the testing and material standards referenced in the IBC.

This chapter provides the requirements for special inspection and verification of the construction at various stages, special inspection for wind and seismic resistance, special testing and qualification for seismic resistance, structural observation by the registered design professional, and alternative methods to establish test procedures for products that do not have applicable standards.

From a historical perspective, several successful code changes to Chapter 17 of the 2009 IBC clarified the provisions to make them more useable. Further reorganization and consolidation was done in the 2012 IBC, which will be discussed in some detail in the appropriate section. The most significant of these changes are as follows, where the section numbers refer to the 2009 IBC:

- Section 1704.1—The previous exemption from special inspection for R-3 occupancies was eliminated. Modern structural systems in single-family residential buildings can be as complex and challenging as commercial structures, especially in large custom homes. Engineered seismic-force-resisting systems are common in Seismic Design Categories D, E, and F where there are often a variety of structural systems and components that require special inspection. A companion code change clarified and expanded the requirements pertaining to special inspector qualifications. Another code change clarified that the registered design professional acting as an approved agency can provide special inspection for work designed by them, provided they demonstrate to the building official that they are qualified to perform the special inspections involved.

- Section 1704.4, Table 1704.4—Continuous special inspection is required for cast-in-place bolts installed in concrete where the strength design procedure is used. This was subsequently changed to periodic special inspection in Table 1705.3 of the 2012 IBC because cast-in-place bolts are similar to placement of reinforcement, which does not require the inspector to provide continuous inspection. Periodic special inspection is also required for post-installed anchors in hardened concrete. It should be noted that the 2015 IBC will likely require continuous special inspection for post-installed adhesive anchors in hardened concrete where the anchors are designed to resist sustained tension loads.

- Sections 1704.3.4 and 1704.6.2—Two new sections were added to require periodic special inspection of temporary and permanent bracing for cold-formed steel and metal-plate-connected wood trusses spanning 60 feet or greater.

- Section 1706—A new section was added requiring special inspection for build-

---

*Items not specifically addressed by the code or a referenced standard can be approved under Section 104.11, Alternative materials, design and methods of construction. ICC Evaluation Services Reports (ICC ESRs) are often used as the basis for approving items that are not specifically addressed by the code or referenced standards.

ings sited in high-wind areas based on exposure category and wind speed. The requirements apply to wood-frame structures and cold-formed steel structures, and include roof and wall cladding.

- Sections 1704.7 through 1704.10—Because Chapter 18 was revised entirely in the 2009 IBC, editorial changes were made to Section 1704.7 Soils, Section 1704.8 Driven deep foundations, Section 1704.9 Cast-in-place deep foundations, Section 1704.10 Helical pile foundations, and Section 1704.11 Vertical masonry foundation elements. Continuous special inspection is required during installation of helical pile foundations, which were added to Chapter 18.

The most significant change to Chapter 17 in the 2012 IBC was the reorganization of provisions pertaining to special inspection, the statement of special inspections, contractor responsibility, structural observations, and required verifications. All of the administrative requirements related to special inspection, the statement of special inspections, contractor responsibility, and structural observations, are now contained in Section 1704. All of the verifications required for various systems, elements, and materials, including special inspection requirements for wind and seismic, are now contained in Section 1705. In the 2009 IBC, Section 1704 covered what specific items required special inspection, and Section 1705 covered what is required to be included in the statement of special inspections. In the 2009 IBC, there are a variety of conflicts and inconsistencies between the two sections. The charging sentence of Section 1705 stated that where special inspection or testing was required by Section 1704, 1707, or 1708, the registered design professional in responsible charge must prepare a statement of special inspections in accordance with Section 1705. For example, suspended ceilings in Seismic Design Category D are required to be included in the statement of special inspections, yet Sections 1704, 1707, and 1708 did not specifically require special inspection for suspended ceilings. Additionally, items that required special inspection or testing by Section 1704, 1707, or 1708 were not all covered in the requirements for the statement of special inspections in Section 1705. In other words, not all items that require special inspection under Section 1704, 1707, or 1708 were listed in Section 1705, and not all items required to be in the statement of special inspection in Section 1705 require special inspection in Section 1704, 1707, or 1708. To resolve these issues, the special inspection and statement of special inspections provisions were reorganized in an effort to clarify the intent and improve proper application and enforcement.

Section 1701 *General*

This section sets forth the scope for Chapter 17 and the general requirements for both new and used construction materials. Because Chapter 17 deals with construction documents and submittals, some discussion of the subject is warranted. The term *construction documents* is defined in Section 202 as "written, graphic and pictorial documents prepared or assembled for describing the design, location and physical characteristics of the elements of a project necessary for obtaining a building permit." Construction documents are a part of the submittal documents required by Section 107. It should be noted that there are also specific requirements for construction documents in Section 1603 for structural loads, as well as in the structural material chapters, such as Section 1901.3 for concrete structures and Section 2101.3 for masonry structures. There are also requirements for specific types of submittals, which are considered part of the construction documents. For example, Section 1803.6 for geotechnical reports, Section 2207.4 for steel joist drawings, Section 2211.3.1 for cold-formed steel truss drawings, and Section 2303.4.1.1 for wood truss drawings.

17 Structural Tests and Special Inspections

1701.2 New materials. Testing is required for all materials that are not specifically provided for in the code or referenced standard. For example, a composite wood material that is not listed in Chapter 23 would be required to follow the procedures set forth in this chapter. A similar provision for acceptance of alternative materials, systems, or methods for which the standards are not adopted in the code is set forth in Section 104.11. This section restates that alternative or new materials and methods may be used if it can be established by tests or other means that the performance of the new material or method will equal that required by the code for the replaced product. As noted, ICC Evaluation Service Reports are often used by building officials as the basis for approving items that are not specifically addressed by the code or referenced standards.

1701.3 Used materials. Materials may be reused, provided they meet *all* the code requirements for new materials. Note, however, that Section 104.9.1 specifically restricts the use of used equipment and devices unless approved by the building official. One should always exercise caution in approving reuse of materials. The applicable material or design standards should be consulted to determine if reuse of materials is allowed or prohibited. For example, reuse of high-strength A490 structural bolts is prohibited by the AISC RCSC[1] specifications. Even a piece of used structural steel should be carefully checked for conformance to the design specifications, applicable standards, and code requirements.

Section 1702 *Definitions*

1702.1 General. Definitions of various terms help in the understanding and application of code requirements. A list of the terms used in Chapter 17 that are defined in Chapter 2 is provided in this section so that the reader is aware of specific terminology used in Chapter 17. The actual definitions are provided in Section 202.

APPROVED AGENCY. The definition of this term is needed in order to effect the requirements of Section 1703.1. The word *approved* means "acceptable to the building official or authority having jurisdiction" (see the definition of *approved* in IBC Section 202). The basis for approval of an agency for a particular activity by the building official includes the competence or technical capability to perform the work in accordance with Section 1703.

APPROVED FABRICATOR. An approved fabricator is a qualified person, firm, or corporation that is approved by the building official to perform specified work without special inspection because they have approved quality-control procedures and are subject to periodic auditing of fabrication practices by an approved special inspection agency. Approved fabricators issue certificates of compliance for their work product. See discussion of Section 1704.2.5.

CERTIFICATE OF COMPLIANCE. An *approved* fabricator is required to submit a Certificate of Compliance for work performed without special inspection. See Section 1704.2.5 for more detailed discussion of fabricator approval. Figure 1703-3 shows an example of an AITC certificate of compliance for a glue-laminated timber.

DESIGNATED SEISMIC SYSTEM. The designated seismic system consists of those architectural, electrical, and mechanical systems and their components that require design in accordance with Chapter 13 of ASCE 7 for which the component importance factor, I_p, is greater than 1.0, as prescribed in Section 13.1.3 of ASCE 7. Section 13.1.3 of ASCE 7 lists four components that have an importance factor of 1.5, which include components required to function after an earthquake including fire sprinkler systems and egress stairways that are not an integral part of the building structure; components containing hazardous, toxic, or explosive materials; and components in Risk Category IV structures that are necessary for continued operation of the facility. All other components are assigned a component importance factor of 1.0. Risk Category IV structures are described in detail in IBC Table 1604.5. Note that Section 13.1.4 of ASCE 7 lists those nonstructural components that are exempt from the seismic design requirements of Chapter 13 and are therefore not considered to be part of the designated seismic system.

FABRICATED ITEM. Fabricated items are materials assembled prior to installation in a building and are referred to in Section 1704.2.5. The definition is provided to clarify the intent of the code, as the term *fabricated items* could easily be interpreted to mean items for which special inspection is not intended by the code. An example of a fabricated item for which special inspection is required is roof trusses not manufactured in accordance with the in-plant quality control requirements of the TPI 1 standard. The section also describes elements that are not considered fabricated items, such as rolled structural steel, reinforcing bars, masonry units, and wood structural panels that are fabricated under specific quality control standards. Because special inspections are analogous to the quality control (QC) programs required by some standards referenced in the code, a change in the 2009 IBC was made that items produced in accordance with standards listed in Chapter 35 that have quality control by a third-party quality-control agency are not considered "fabricated items." This change makes it clear that the code does not intend to impose a redundant special inspection requirement for such items.

INSPECTION CERTIFICATE. An inspection certificate is an identification applied to a product indicating that the individual product has been inspected by an approved agency. The inspection certificate is obligatory for components subject to special inspection requirements for seismic resistance (see Section 1705.11). Note that the requirements for an inspection certificate differ from the requirements for a label. An inspection certificate is issued for the specific piece of a product when it is inspected and is an ongoing process, whereas a label requires only that a representative sample of a product be periodically tested.

MAIN WIND-FORCE-RESISTING SYSTEM. The main wind-force-resisting system (MWFRS) is one of the structural systems that must be designed to resist wind loads. The other system that is designed to resist wind loads is components and cladding. The MWFRS comprises those structural elements that provide lateral support and stability for the overall structure. In general, these are out-of-plane walls, diaphragms, chords, collectors, vertical lateral-force-resisting elements such as shear walls, braced frames, and so on, and foundations. In general, the MWFRS receives wind loading from more than one surface. In contrast, components and cladding are generally on the exterior envelope of the building and receive loading directly from the wind, such as roof and wall covering.

SPECIAL INSPECTION. That category of field inspection for which special knowledge, special attention, or both are required. For example, inspection of complete penetration welds requires both special knowledge and special attention to ensure that the requirements of the codes and standards are met. Note that the special inspector does not have the same authority as that of the jurisdiction inspector. The role of the special inspector is to report discrepancies between the construction in the field and the approved construction documents to the contractor. If uncorrected, discrepancies should be reported to the design professional in responsible charge and to the building official.

SPECIAL INSPECTION, CONTINUOUS. Continuous full-time inspection is required where compliance of the work or product cannot be determined after incorporation into the building or structure. For example, one cannot determine whether a multipass fillet weld is in compliance with the code requirements unless each pass of the weld is inspected during the welding process. Whether a particular special inspection is continuous or periodic is specified in the various inspection and verification tables in Section 1705.

SPECIAL INSPECTION, PERIODIC. Intermittent or part-time inspection, which may be allowed when the compliance of the work or product can be determined after being incorporated into the structure. For example, compliance with the design nailing requirements of a

17 Structural Tests and Special Inspections

wood shear wall can be determined after construction of the wall (but before closure); hence, verification by periodic special inspection is adequate in this case. Another good example is placement of reinforcing bars. The special inspector need not be continuously present during placement. Whether a particular special inspection is periodic or continuous is specified in the various inspection and verification tables in Section 1705.

STRUCTURAL OBSERVATION. Structural observation is intended to ensure general conformance with the design intent, and not necessarily specific conformance with the code. For example, in a building under construction in a high seismic area, the registered design professional acting as the structural observer may focus on crucial elements of the seismic-force-resisting system. Like the special inspector, the structural observer does not have the same authority as that of the jurisdiction inspector. The role of the structural observer is to report discrepancies between the construction in the field and the approved design documents to the contractor. If unresolved, discrepancies should be reported to the building official. Note that structural observation by the registered design professional does not replace or substitute for any of the requisite special inspections required by Sections 1705 or the jurisdiction inspections required by Section 110.

Several definitions that were previously located in Section 1702, such as "label," "manufacturers designation," and "mark," are no longer listed in Chapter 17. They were deleted in the 2009 IBC and relocated to Section 202 because these terms are more general in nature. As noted, all definitions in the 2012 IBC are now found in Section 202 only.

Section 1703 *Approvals*

1703.1 Approved agency. The word *approved* means "acceptable to the building official or the authority having jurisdiction" (see the definition of "Approved" in IBC Section 202.1). The basis for approval of an agency for a particular activity by the code or building official may include the capacity or technical capability and expertise necessary to perform the work in accordance with Section 1703. Special inspection agencies, testing laboratories, and the inspection agencies that provide quality assurance of concrete and steel fabricators should be accredited by a recognized Accreditation Body. For example, testing laboratories should be accredited to ISO/IEC 17025 *General requirements for the competence of testing and calibration laboratories.*

Accreditations through the International Accreditation Service (IAS), a subsidiary of the International Code Council (ICC), is the most straightforward method for the building official to assure that an agency provides quality services and meets all of the requirements of the building code. Quality standards for agency approvals through IAS are defined by the following standards:

- Special inspection agencies should be accredited to IAS AC291.
- Fabricators of reinforced and precast/prestressed concrete should be inspected by an approved special inspection agency, or by an inspection agency, which is accredited to IAS AC157.
- Fabricators of structural steel should be inspected by an approved special inspection agency, or by an inspection agency, which is accredited to IAS AC172.
- Fabricators of metal building systems should be inspected by an agency, which is accredited to IAS AC472.
- Fabricators of cold-formed steel structural and nonstructural components not requiring welding should be inspected by an agency, which is accredited to IAS AC473.

Other accreditation programs are also available that may be approved by the building official, such as the AISC Certification Program for Structural Steel Fabricators, the Plant Certification Program for steel joists by the Steel Joist Institute, or the Plant Certification Program for precast concrete products by the Precast/Prestressed Concrete Institute.

Accreditation helps to ensure that the agency has the necessary equipment and employs qualified persons. For example, if an agency wishes to be approved for the special inspection of structural welds, the agency should submit evidence that its welding inspector is certified in accordance with applicable International Code Council (ICC), American Welding Society (AWS), or American Society of Nondestructive Testing (ASNT) requirements.

Independence. The agency should have objectivity as well as competence. Objectivity can be measured by the agency's financial and fiduciary independence. The agency should be independent from the contractor responsible for the work and have no financial ties to the organization it inspects. For example, a testing laboratory checking concrete strength should have no financial ties to the contractor, its subcontractors, or the concrete supplier. **1703.1.1**

Equipment. If any agency is not accredited as described above, the building official should evaluate the agency to confirm compliance with the requirements of applicable standards and that it has the appropriate equipment to perform required tests or inspections. **1703.1.2**

Personnel. If the agency is not accredited, the building official should evaluate both the experience and qualifications of personnel. An agency may have personnel with the appropriate certifications but not the necessary experience. Supervisory and inspection personnel should have the appropriate certifications as well as the requisite experience and/or education. **1703.1.3**

If the services being provided by an inspection or testing agency come under the purview of the professional registration laws of the state or jurisdiction, the building official should request evidence that personnel are qualified to perform the work in accordance with the requisite professional registration.

Written approval. A written approval by the building official is required for all material, appliances, equipment, or systems incorporated into the work in order to have a documented record of approval and the basis for approval. **1703.2**

Approved record. Records must be kept for all approvals, including conditions and limitations of approval. The approvals must be kept on file and available for public inspection. The records must demonstrate compliance with the applicable code requirements of any material, appliance, equipment, or system incorporated into the structure. **1703.3**

Performance. When conformance with the code is predicated on the performance and quality of materials or products, the building official must require the submission of test reports from an approved agency establishing this conformance. In the absence of such reports, the building official should require specific data that show compliance with the intent of the applicable code requirements in accordance with Chapter 16 (see IBC Sections 1604.6 and 1604.7). For example, core tests of in situ concrete could be used to determine compliance with design strength if the sample cylinders required by ACI 318[2] were destroyed. **1703.4**

Materials and products must be subjected to various levels of quality control and identification in order to determine that the material complies with the requirements of the code. The degree of quality control and identification to which a material must be subjected is based on its relative importance to the structure's performance and function. As noted above, the terms "label," "manufacturers designation," and "mark" were deleted in the 2009 IBC and relocated to Section 202 because these terms are more general in nature. See Section 202 for definitions of the terms *mark*, *manufacturer's designation*, *label*, and *inspection certificate*. By use of these terms, the code establishes a hierarchy of quality control and identification as follows:

Level 1: Manufacturer identifies the material or product with the name of the material or product, the manufacturer's name, and the intended usage (see *Mark*).

Level 2: Manufacturer identifies the material or product as in Level 1 and also certifies compliance with a given standard or set of rules (see *Manufacturer's designation*).

Level 3: Manufacturer identifies the material or product as in Level 1. An approved quality control agency performs periodic audits of the manufacturer's facilities and QA/QC procedures.

Level 4: Each batch of material or individual product is inspected by an approved quality control agency (see *Inspection Certificate*).

The term *mark* is an identification applied on a product by the manufacturer indicating the name of the manufacturer and the function of a product or material. An example of the APA trademark for rated wall sheathing is shown in Figure 1703-1.

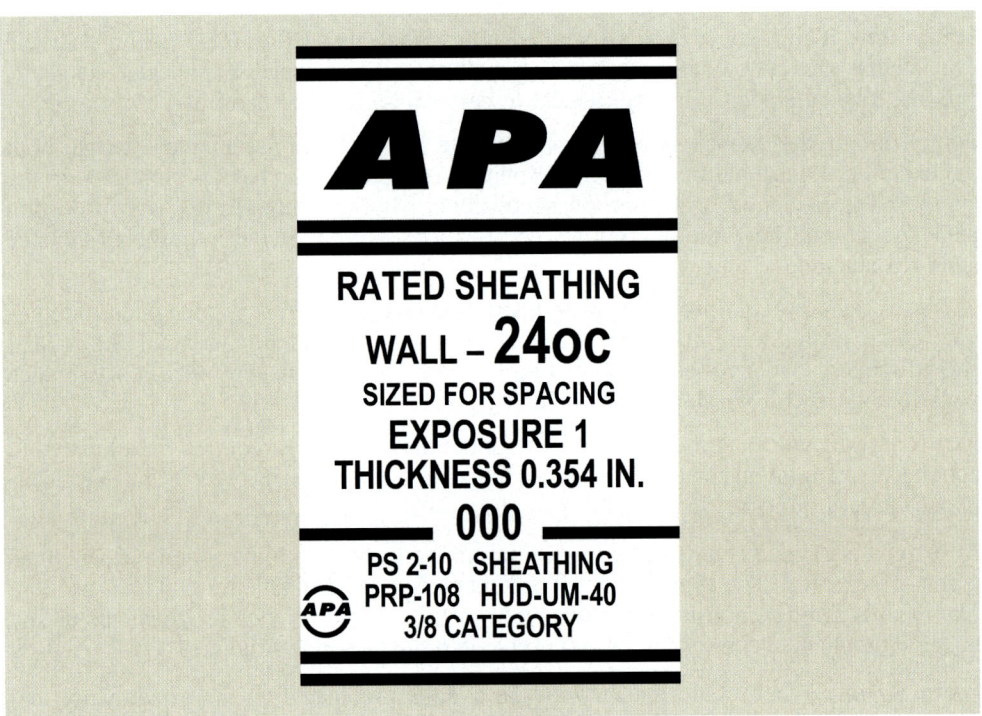

Figure 1703-1
APA-rated sheathing.

The term *manufacturer's designation* refers to an identification applied on a product by the manufacturer indicating that a product or material complies with a specified standard or set of rules. An example of a manufacturer's designation would be the designation on ASTM A706 reinforcing bars where the "W" indicates the bars conform to the ASTM A706 standard for weldable rebar. See Figure 1703-2.

The term *inspection certificate* refers to an identification applied to a product by an approved agency containing the name of the manufacturer, the function, and the performance characteristics, and indicates that the product or material has been inspected and evaluated by an approved agency. An example of an inspection certificate would be a Certificate of Conformance for a glue-laminated beam that meets the ANSI/AITC A190.1 specification. The specification requires that each glue-laminated timber be marked with the AITC Quality Mark (see "Mark") or be accompanied by an AITC Certificate of Conformance. Where appearance is important, such as exposed glue-laminated beams, a certificate of conformance is used. See Figure 1703-3.

Approvals 17

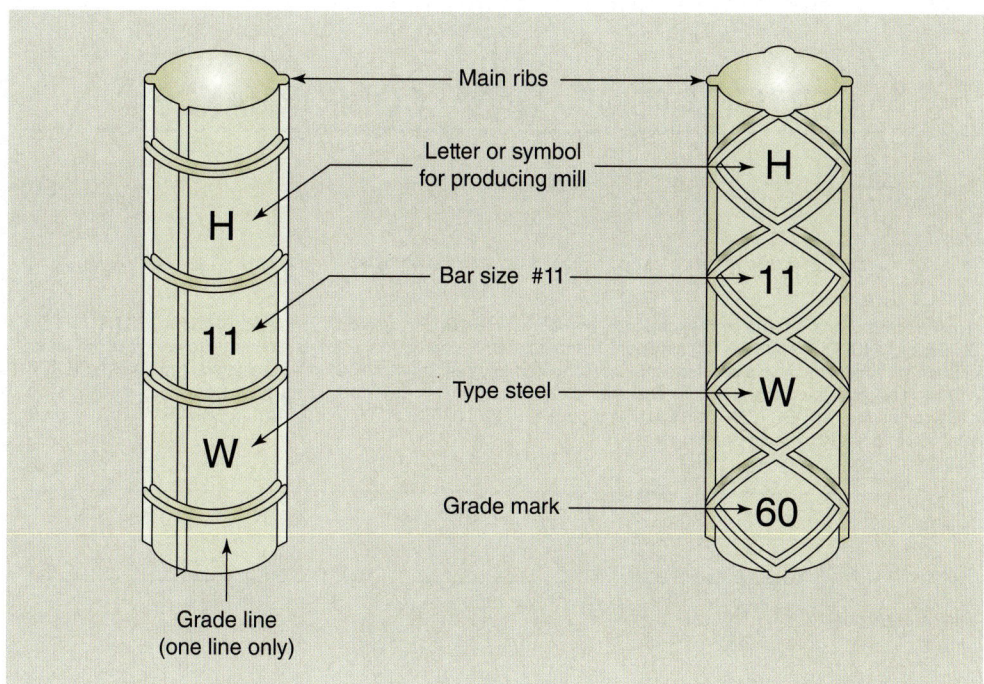

**Figure 1703-2
Required marking for ASTM A706 reinforcing bars.**

1703.4.1 **Research and investigation.** This section is intended to implement the requirements of Section 104.11 for use of innovative or alternative materials, design, and methods of construction. For example, an innovative prestressed concrete system that does not emulate the performance of cast-in-place concrete could be evaluated in accordance with this section. Note that the section requires that the costs of collecting data and preparing reports are to be borne by the applicant.

1703.4.2 **Research reports.** This section is identical to Section 104.11.1 regarding the use of innovative or alternative materials, design, and methods of construction. Evaluation reports prepared by approved agencies, such as those published by the organizations affiliated with the model code groups, may be accepted as part of the data needed by the building official to form the basis of approval of a material or product not specifically covered by the code. ICC Evaluation Service Reports are considered by many building officials to be the most straightforward method for the building official to review products and systems that are not specifically covered by the code. Such reports supplement the resources of the building official and eliminate the need for the official to conduct detailed analysis on every new product that is not covered by the code or a referenced standard. Because evaluation reports issued by the model code-affiliated organizations are advisory in nature, the building official is not mandated to accept or approve them. Technically, such products and reports are approved under the alternative materials and methods of construction provisions of Section 104.11.

1703.5 **Labeling.** When materials or assemblies are required to be labeled by the code, such as wood structural panels, rated sheathing, lumber, fire doors, and so on, the labeling must be in accordance with the procedures outlined in this section and its subsections. Labeling of materials or assemblies is an indication that the materials or assemblies have been subjected to testing, inspection, and/or other operations by the labeling agency. The presence of a label does not necessarily indicate compliance with code requirements. For example, use of plywood siding, although labeled, would not comply with the code if Structural I sheathing was specified to resist the design lateral forces. The installation of labeled products must comply with the specific requirements and limitations of the label. For example, the actual span in the field for wood structural panel roof sheathing must meet the panel

CERTIFICATE OF  CONFORMANCE

THE UNDERSIGNED MANUFACTURER HEREBY CERTIFIES that the products identified below and on attached sheets are marked with the Collective Mark of the AMERICAN INSTITUTE OF TIMBER CONSTRUCTION (AITC) and were manufactured in conformance with applicable provisions of the latest revision of American National Standard for wood products – Structural Glued Laminated Timber, ANSI/AITC A190.1, and that such manufacture occurred at our plant in _____, which plant has a quality control system approved by the INSPECTION BUREAU OF THE AMERICAN INSTITUTE OF TIMBER CONSTRUCTION, and is audited periodically by such Bureau.

Job Name: _____

Job Location: _____

Customer's Order No. _____ Date: _____ Mfgr's Order No. _____

Order Description:

Signature: _____ Company: _____

Title: _____ Address: _____ Date: _____

AITC HEREBY CERTIFIES that the said company at its said plant is licensed by the AMERICAN INSTITUTE OF TIMBER CONSTRUCTION to use the AITC Collective Mark in respect of products which comply with applicable provisions of said Standard, that the adequacy of the quality control system in effect at said plant is periodically audited and verified by the AITC INSPECTION BUREAU, and that in the judgment of such Bureau, said company is capable of complying with applicable manufacturing and testing provisions of said Standard in respect of products manufactured in said plant. Conformance with the Standard in respect of any specific or particular product is the sole responsibility of the manufacturer; AITC's guarantee hereunder being only that the said company is qualified to produce a product meeting the said Standard and that its plant is periodically audited and verified by the AITC INSPECTION BUREAU.

AITC Certificate No. _____

Void without Label

AMERICAN INSTITUTE OF TIMBER CONSTRUCTION
© 2010 AMERICAN INSTITUTE OF TIMBER CONSTRUCTION

Figure 1703-3 Certificate of Conformance for a glue-laminated beam.

span rating shown on the label (stamp). See Figure 1703-4 for examples of typical lumber grade labels. Another example is a fire-rated door that is labeled with the hardware requirements specified on the label. The building inspector must ensure that the hardware used in the installation of the door meets the labeling requirements to ensure that the door complies with the code.

See Section 202 for the definition of the term *label*.

Approvals 17

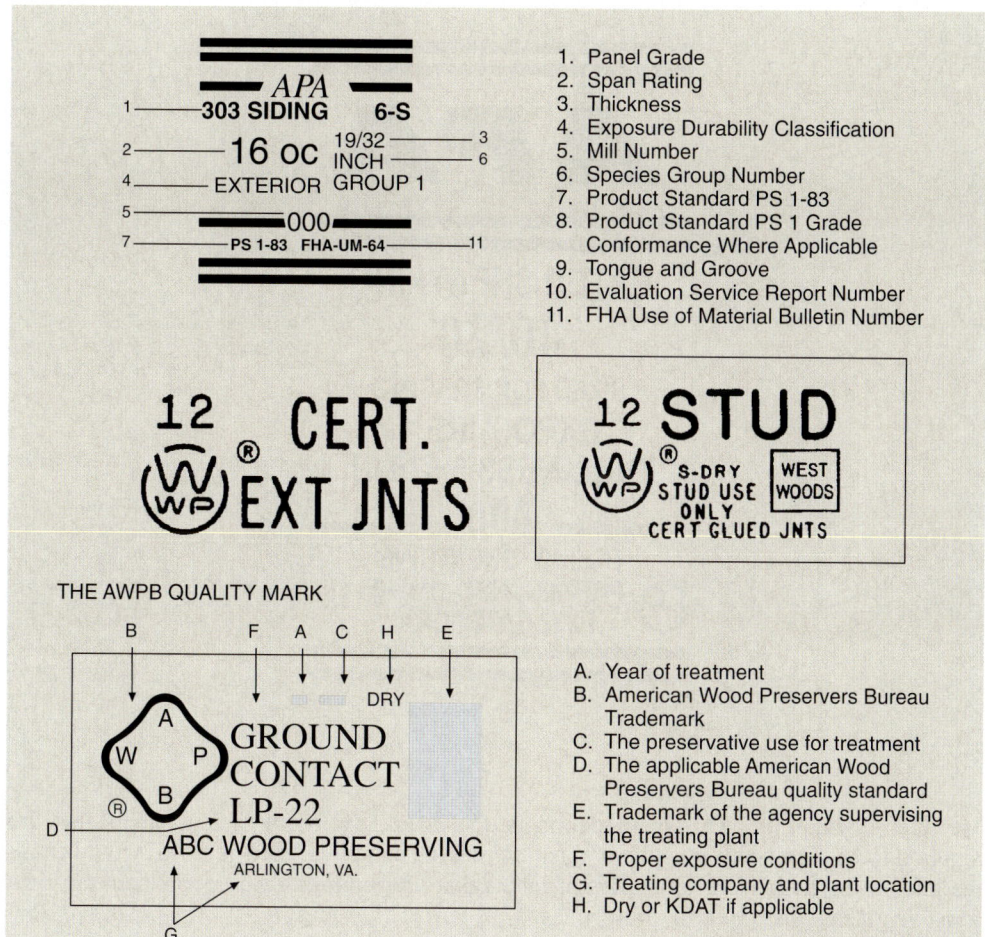

Figure 1703-4
Examples of lumber grade labels.

Testing. For a material or product to be labeled, the labeling agency is required to perform **specific** testing on representative samples of the material or product in accordance with the applicable standards referenced by the code. For example, quality-control testing for strength and stiffness of machine stress-rated lumber to ensure that the products meet structural requirements in accordance with the American Lumber Standard system. Another example is factory-built fireplace assemblies that must be tested in accordance with the requirements of UL 127 (see Chapter 35). **1703.5.1**

Inspection and identification. The approved agency whose label is applied to a material or product must perform periodic inspections of the manufacture of the material or product to determine that the manufacturer is indeed producing the same material or product as tested and labeled. For example, if the labeling agency had tested $1/_2$-inch C-C plywood sheathing made with five plies but the manufacturer was now making the plywood with only three plies, the agency would need to withdraw the use of its label and listing. **1703.5.2**

Label information. This section specifies the minimum information necessary on a label for the building inspector to determine that the installed material conforms to the approved plans. See Figure 1703-4 as an example of typical lumber grade labels. Figure 1703-5 shows an actual label for APA Performance Rated Panel. The manufacturer (Georgia Pacific), the performance characteristics (Span Rating 40/20, Exposure 1), and approved agency (APA—The Engineered Wood Association) are shown. **1703.5.3**

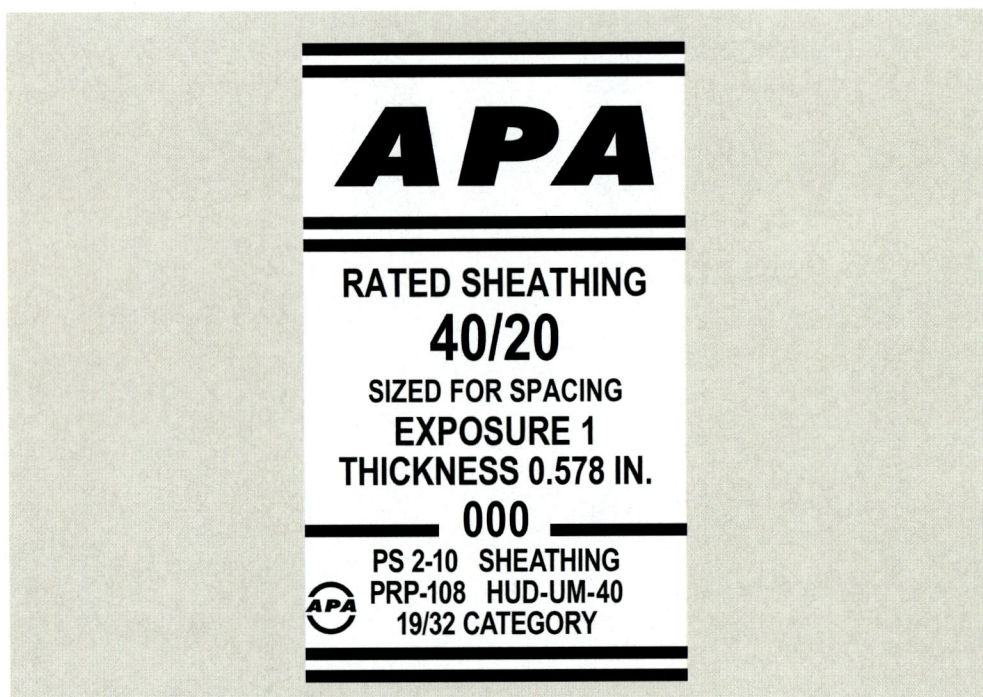

Figure 1703-5
APA Performance Rated Panel label.

1703.5.4 Method of labeling. This new section in the 2012 IBC requires that labels that are required to be permanent must be of a nature that once applied, their removal would cause the label to be destroyed. Examples of labels that are required to be permanent are fire-resistive glazing and opening protectives.

1703.6 Evaluation and follow-up inspections. This provision applies where a structural component cannot be inspected after completion of a prefabricated assembly. An example might be a prefabricated shear wall consisting of wood structural panel sheathing over a welded steel frame; the welding must be inspected prior to the installation of the sheathing and the entire assembly must be inspected after it is fabricated. The testing and inspection should follow the provisions described in Section 1703.

Previous editions of the IBC had a section on *Approval of heretofore approved materials* that was deleted in the 2009 IBC because the use of existing material is covered in Section 3401.4.

Section 1704 *Special Inspections, Contractor Responsibility, and Structural Observations*

This section provides administrative requirements for special inspections (1704.2), preparation and submittal of the statement of special inspections (1704.3), the contractor's statement of responsibility (1705.4), and structural observation by the registered design professional (1704.5). These requirements, along with the provisions for required verification and testing, were reorganized in the 2012 IBC so that requirements for special inspections, the statement of special inspections, contractor's statement of responsibility, and structural observation are now entirely contained in Section 1704, and all the required inspection and verification requirements are contained in Section 1705. The reorganization of Sections 1704 and 1705 in the 2012 IBC removed several ambiguities and is a significant improvement.

One of the oldest mechanisms for providing quality assurance in construction is the process known as *special inspection*. The purpose of special inspection is to ensure proper fabrication, installation, and placement of components or materials that require special knowledge or expertise, such as welding of structural steel or placement of grouted masonry. The knowledge and duties of a special inspector differ from that of the jurisdiction building inspector in that the special inspector's expertise is narrower in scope, such as that of a structural steel welding or prestressed concrete special inspector.

The concept of special inspection dates back to the 1927 edition of the *Uniform Building Code* (UBC), where it existed under the designation *Special Engineering Supervision*. The first special inspection provisions similar to those presently used appeared in the 1943 UBC under the designation *Registered Inspectors*. The requirements in the 1943 UBC contained the following essential elements that are also in the current special inspection provisions:

1. The particular types of work requiring special inspection were specified.
2. The special inspector had to be qualified and demonstrated his or her qualifications to the building official.
3. The requisite special inspections were in addition to those performed by the jurisdiction building inspector.
4. The special inspector was to be employed by the owner or design professional, not the contractor, in order to avoid any conflict of interest.

The special inspection provisions in the 1997 UBC continued with essentially the same elements. Although similar, the special inspection provisions in the IBC were expanded to be much more extensive than those in the UBC.

The special inspection requirements in the IBC address three areas:

1. Adequacy or quality of materials, such as concrete strength
2. Adequacy of fabrication, such as verification of manufacturer's certified test reports
3. Adequacy of construction techniques, such as proper placement of reinforcing

Special inspection is that category of inspection requiring special knowledge, special attention, or both. The knowledge required is generally more specialized than that required by a general building inspector. An individual with a high degree of specialized knowledge is generally required; hence, the term *special inspection*.

Most building departments do not have the staff necessary to do detailed inspections on large or complex structures, nor do the permit fees allow the level of inspection necessary for the types of construction where extra care in quality control must be exercised to ensure compliance with the code. In some cases, the special inspection must be continuous during a particular operation, such as pretensioning high-strength bolts or making complete penetration groove welds. Hence, there is a need for a special inspector with specialized expertise who can be there continuously during a particular operation.

ICC offers a publication titled *Model Program for Special Inspection*[3] that provides the building official with guidance on the administration and implementation of the special inspection requirements of the IBC. The guidance is based on recommended practices and the consensus of building officials and design professionals, as well as inspection and testing agencies. The duties and responsibilities of the building official, special inspector, project owner, engineer or architect of record, contractor, and building official are covered in the guide. Suggested forms are also included that can be easily adapted to the specific needs of the jurisdiction.

A good example of a checklist document for special inspection is included in the *Model Program for Special Inspection* (ICC). Copies of this document are also available from any of the member organizations of the ICC.

Another good source of information for special inspection and testing for seismic resistance is the *Structural Construction and Special Inspection Manual: A Companion to the 2006 IBC Structural/Seismic Design Manuals* developed by the Structural Engineers Association of California (SEAOC). The examples in this manual are based on design examples in the 2006 *IBC Structural/Seismic Design Manuals* developed by SEAOC.

Note that in addition to the general special inspection requirements in Section 1705, there are additional special inspection requirements in Section 1705.10 that specifically apply to wind resistance, and in Section 1705.11 that specifically apply to seismic resistance.

1704.1 General. This section outlines the general requirements for special inspections, the statement of special inspections, contractor's statement of responsibility, and structural observations. The actual inspections and verifications required are contained in Section 1705.

1704.2 Special inspections. The owner is responsible for the employment of special inspectors meeting the approval of the building official and all costs associated with the employment of special inspectors. Note that the special inspectors must be employed by the owner, or the responsible registered design professional acting as the owner's agent, not by the contractor or builder. This ensures independence of the special inspector and avoids any potential conflict of interest that could occur if the special inspector were employed by the contractor or builder. Note that the special inspections required by Chapter 17 are in addition to, not in lieu of, the jurisdiction inspections required by Section 110.

There are exceptions to the requirement for special inspections for minor work or work not required to be designed or sealed by a registered design professional, Group U occupancies accessory to a residence, and prescriptive cold-formed steel or wood light frame construction. The exemption from special inspection R-3 occupancies was deleted in the 2009 IBC because the structural systems in modern single-family residential buildings can be as complex and challenging as commercial structures, especially large custom homes. Engineered seismic-force-resisting systems are very common in residential structures in Seismic Design Categories D, E, and F. Group R-3 occupancies often have components that require special inspection such as high-load diaphragms, high-strength concrete, structural steel frames, high-strength bolting, complete penetration groove welds, engineered masonry, and deep (pile) foundation systems. The exemption for Group U occupancies accessory to a residential occupancy is for those structures that are typically not required to be designed by a registered design professional or those designed and constructed in accordance with the *International Residential Code*® (IRC)®. Exception 1 does not necessarily mean that inspections are not required, only that they are not required to be made by a special inspector. However, the above comments regarding structural systems used in the R-3 occupancies could also apply to U occupancies. It is not inconceivable that a large private garage could have structural components that would require special inspection. Exception 1 refers to "conditions in the jurisdiction" as a possible exception. The primary conditions envisioned by the code in this case refer either to the jurisdiction having the resources and skill level necessary to perform the requisite special inspection tasks, thus obviating the need for a special inspector hired by the owner, or the work is of a minor nature in the opinion of the building official. Note that this exception for special inspection cannot be invoked by the owner. One purpose of the exception is to allow jurisdictions to perform special inspections if the jurisdiction so desires. Exception 3 waives special inspection for prescriptive light-frame construction of cold-formed steel or conventional wood structures. Section 2211.7 applies to prescriptively framed detached one- and two-family dwellings and townhouses less than or equal to three stories constructed in accordance with AISI S230.

Special Inspections, Contractor Responsibility, and Structural Observations

1704.2.1 Special inspector qualifications. Special inspectors are required to be qualified and demonstrate competence to the satisfaction of the building official. Code changes to the 2009 IBC clarified and expanded the requirements pertaining to special inspector qualifications and also clarified that the registered design professional acting as an approved agency may provide special inspection for work designed by them, provided they demonstrate to the building official that they are qualified to perform the special inspections involved.

1704.2.2 Access for special inspection. Prior to the 2012 IBC, there were no specific requirements for providing access to construction for special inspectors other than the general requirement in Section 110.1. This new section requiring access to the construction specifically applies to special inspection.

1704.2.3 Statement of special inspections. The permit applicant must submit a detailed statement outlining the required special inspections and designate those who will perform the special inspections. In general, the responsible registered design professional is required to prepare the statement of special inspections, because the special inspections relate directly to the design and construction documents required by Sections 107.1 and 1603. The statement of special inspections must conform to the detailed requirements described in Section 1704.3. The exception in Section 1704.2.3 allows the statement of special inspections to be waived for prescriptive light-frame construction of cold-formed steel or conventional wood structures. The exception in Section 1704.3 allows the statement of special inspections to be prepared by a qualified person instead of a registered design professional where approved by the building official for construction that is not designed by or required to be designed by a registered design professional.

Although a statement of special inspections is not required for cold-formed steel structures constructed in accordance with the provisions of Section 2211.7 or conventional wood frame structures constructed in accordance with the prescriptive provisions of Section 2308, it should be noted that Sections 2308.1.1 and 2308.4 permit portions and elements of an otherwise conventional building to be designed in accordance with the engineering provisions of the code; therefore, these engineered elements and portions could require special inspection. For example, an otherwise conventional wood frame residence that uses a steel moment frame to resist lateral forces at the clear story entrance foyer. Such an element is outside of the conventional construction provisions, requires engineering under Section 2308.4, and could require special inspection under Section 1705.2.

1704.2.4 Report requirement. Records of each inspection must be kept and submitted to the building official so as to document compliance with the code. The records must include all inspections made and compliance with the code requirements, as well as all discrepancies or violations. A final report must show that all required special inspections have been made and that discrepancies have been resolved before a certificate of occupancy can be issued by the building official. It is the responsibility of the special inspector to document and submit inspection records to the building official and to the registered design professional in responsible charge. The final special inspection report documenting resolution of discrepancies must be submitted at a time agreed upon by the permit applicant and the building official prior to the commencement of work.

1704.2.5 Inspection of fabricators. Unless structural items are fabricated in an approved fabricator's shop, special inspection is required during the fabrication process. This section should be used in conjunction with Section 1703.6 relating to evaluation and follow-up reports. The use of a special inspector does not relieve the fabricator from his or her own required quality-control procedures and personnel. See Section 1704.2.5.2 regarding fabricator approval.

1704.2.5.1 Fabrication and implementation procedures. The special inspector is required to verify that the fabricator maintains quality control and review the fabricator's procedures to achieve code compliance for the scope of work. The special inspector acts in the quality assurance role in this instance by verifying that the quality control procedures that the fabricator has in place will ensure compliance and that the fabricator follows the requisite quality control procedures. For example, welding of structural steel is to be done in accordance with the requirements of AWS D1.1.[4] AWS D1.1 requires that the fabricator have its own quality control procedures and personnel. The fabricator's QC personnel are responsible for determining conformance of the welds with the requirements of AWS D1.1 by applicable visual and nondestructive testing (NDT) methods. The special inspector acts as the owner's agent for verification by auditing the fabricator's QC program. The fabricator's welding inspector is responsible for making sure that all welders are qualified in accordance with D1.1, and that proper weld procedure specifications (WPS) are used for all welds and for keeping documentation for each weld made in the fabrication shop. The special inspector, on the other hand, is responsible for reviewing the qualification records of the welders, for determining that the WPS were suitable for the specified weld and were properly qualified, for reviewing NDT procedures and records, and for observing a representative number of welds to ensure that the fabricator's QC program is adequate and being followed.

The exception applies to the approved fabricators. Unless fabrication is done in an approved fabricator's shop in accordance with Section 1704.2.5.2, special inspection is required during the fabrication process. In other words, special inspection is not required where fabrication is done on the premises of an approved fabricator's shop. What constitutes an approved fabricator is described in detail in Section 1704.2.5.2.

1704.2.5.2 **Fabricator approval.** Special inspection at a fabrication plant is not required where a fabricator has been specifically approved by the building official. This approval may be based on available third-party accreditations, previously listed ("approved agency"), or by the building official's own review of the fabricator's written procedures and quality assurance/quality control program. The fabricator should be periodically audited by an independent, approved special inspection agency. This section is intended to apply to fabricators accredited by organizations such as International Accreditation Service (IAS), the AISC Certification Program for Structural Steel Fabricators, the Plant Certification Program for steel joists by the Steel Joist Institute, or the Plant Certification Program for precast concrete products by the Precast/Prestressed Concrete Institute. Some low-rise metal building manufacturers meet both the AISC Certification Program for Structural Steel Fabricators and the IAS Fabricator Inspection Program and are recognized by most building officials as having the personnel, organization, experience, knowledge, and equipment to produce the quality required for structural steel buildings.

A minor change to the language of this section was made in the 2009 IBC to avoid a possible misinterpretation. The 2009 IBC states, "Special inspections required by Section 1704 are not required where ..." In previous editions of the IBC, the section stated, "Special inspections required by this code are not required where ..." The previous language could be interpreted to mean that all special inspections, including that required for seismic and wind resistance, could be waived when work is done by an approved fabricator. Note that special inspections for wind and seismic resistance are not included in Section 1704, but are in their own specific Sections 1706 and 1707. However, with the reorganization of Chapter 17 in the 2012 IBC, the special inspection requirements for wind and seismic resistance are now contained in Section 1705. The implication of this is that all special inspections, including those for wind and seismic resistance, are now waived where fabrication is done in an approved fabricator's shop.

1704.3 **Statement of special inspections.** There were several code changes to the 2006 IBC that reorganized and clarified the special inspection provisions of Chapter 17 to more clearly convey the intent. Beginning with the 2006 IBC, the term *statement of special inspections* is used instead of the previous terms, *special inspection program* and *quality assurance plan*. Provisions that were formerly covered in the quality assurance plan requirements in the 2003 IBC are now included as part of the statement of special inspections when required by the seismic or wind criteria. In addition, the statement of contractor responsibility requirements were consolidated in the 2006 IBC and put in a separate section, eliminating the redundant language of the 2000 and 2003 IBC, where they appeared in both the

Special Inspections, Contractor Responsibility, and Structural Observations

wind and seismic quality assurance plans. These code changes, in addition to the changes to the 2009 IBC and the further reorganization in the 2012 IBC, make the special inspection provisions in the current edition of the IBC much easier to understand and enforce.

When special inspection, special inspection for wind or seismic resistance, or structural testing and qualification for seismic resistance is required by Section 1705, the registered design professional is required to prepare a detailed statement of special inspections. The statement of special inspections must be submitted by the permit applicant and is a condition of permit issuance as prescribed in Section 1704.2.3. The statement of special inspection must identify ordinary special inspections required by the various subsections in Section 1705, special inspection requirements for wind resistance prescribed in Section 1705.10, special inspection requirements for seismic resistance prescribed in Section 1705.11, and the seismic testing and certification requirements covered by Section 1712.

In general, the responsible registered design professional is required to prepare the statement of special inspections because the special inspections relate directly to the design and construction documents required by Sections 107.1 and 1603. The exception allows the statement of special inspections to not be prepared by the registered design professional if it is prepared by a qualified person approved by the building official for construction that is not designed by or required to be designed by a registered design professional.

Content of statement of special inspections. The statement of special inspections must identify the work requiring special inspection or testing. It should include the type and extent of each special inspection or test and indicate whether the inspections are to be continuous or periodic. Where required, the statement also includes the additional special inspection or testing requirements for wind or seismic resistance as prescribed by Sections 1705.10, 1705.11, and 1705.12. — **1704.3.1**

Seismic requirements in the statement of special inspections. The statement of special inspections must include elements of the seismic-force-resisting system, the designated seismic system and additional systems, components listed in Section 1705.11, and the seismic testing and qualifications required by Section 1705.12. The terms "seismic-force-resisting system" and "designated seismic system" are defined in Section 202. — **1704.3.2**

Wind requirements in the statement of special inspections. The statement of special inspections must identify the elements of the main wind-force-resisting system and components and cladding (that are subject to special inspection) where the nominal design wind speed, v_{asd}, is greater than or equal to 120 mph in wind Exposure Category B, or where the nominal design wind speed, v_{asd}, is greater than or equal to 110 mph in wind Exposure Category C or D. The nominal design wind speed, v_{asd}, is a new term introduced into the 2012 IBC as a result of the new ultimate wind speed maps. See discussion under Section 1609 for descriptions of the nominal design wind speed and Exposure Categories. — **1704.3.3**

Contractor responsibility. In the 2006 IBC, the statement of contractor responsibility requirements were consolidated and relocated in a separate section, eliminating the redundant language of the 2000 and 2003 IBC, where it appeared in both the wind and seismic quality assurance plans. The intent of the section is to require the contractor responsible for construction of the main wind-force-resisting system, the seismic-force-resisting system, the designated seismic system, and those wind-resisting components listed in the statement of special inspections to submit a statement of responsibility to the owner and building official. The statement of responsibility is to be submitted prior to commencement of work on a particular structural system or component, and must acknowledge awareness of special requirements contained in the statement of special inspections. The provisions in the 2006 IBC listed four items that essentially required the contractor to acknowledge that control would be exercised to achieve conformance with construction documents, outline the procedures for exercising quality control, specify the methods and frequency of reporting, and identify and provide qualifications of key personnel responsible for implementing the requirements contained in the statement of special inspections. A code change to the 2009 IBC proposed to delete the statement of contractor responsibility — **1704.4**

17 Structural Tests and Special Inspections

requirement entirely. However, this statement of responsibility is considered a central part of a long-standing national effort to improve construction quality for seismic and wind resistance by stressing the contractor's role in providing adequate quality control during the construction process. The final language was consolidated in the 2009 IBC and simplified by eliminating Items 2, 3, and 4 that appeared in previous editions of the IBC. In the 2012 IBC, the provision was relocated under Section 1704 along with structural observation. The intent of the statement of responsibility is to ensure that the contractor acknowledges awareness of special requirements contained in the statement of special inspections related to the lateral-force-resisting systems for wind and seismic loads.

1704.5 Structural observations. Structural observation requirements first appeared in the 1988 UBC and were applicable to buildings in areas of high seismic risk. The purpose of structural observation is to ensure that critical elements of the lateral-force-resisting system are constructed in general conformance with the design as shown in the approved structural drawings and specifications. Because the registered design professional is most familiar with the design and the details of the lateral-force-resisting system, he or she is the most appropriate person to execute the requisite observation. Note that structural observation is in addition to, not in place of, special inspection, and it does not replace or waive any of the inspections by the jurisdiction inspector required by Section 110.

Structural observation consists of visual observation of the structural systems by the registered design professional for general conformance with the approved construction documents at significant construction stages and at the completion of the structural system. As noted above, structural observation does not include or waive the inspections performed by the jurisdiction as required by Section 110 or the special inspections required by Section 1705. See the definition of "Structural observation" in Section 202.

Structural observation is triggered by the seismic risk category of the building and the nominal design wind speed at the site. Section 1704.5.1 requires structural observation for buildings assigned to Seismic Design Category D, E, or F when any of the five conditions listed in Section 1704.5.1 exists. Structural observation is also required for buildings where the nominal design wind speed exceeds 110 mph when any of the four conditions listed in Section 1704.5.2 exists.

Prior to performing structural observations, the observer is required to notify the building official in writing of the frequency and extent of structural observations. At the conclusion of work, the structural observer is to submit a written statement that the site visits have been made and identify any unresolved deficiencies.

It is important to note that the last item in both the code sections (for seismic and wind) allows either the registered design professional or the building official to require structural observation at their discretion.

Section 1705 *Required Verification and Inspection*

In the 2012 IBC, all items requiring special inspection and verification have been consolidated in Section 1705. All general administrative requirements related to special inspections, the statement of special inspections, contractor responsibility, and structural observations are contained in Section 1704. Other than the reorganization of Chapter 17 in the 2012 IBC, the most significant change is the deletion of the specific requirements for special inspection of structural steel and masonry structures. In both cases, the code now refers to the respective referenced standard for quality assurance requirements related to structural steel and masonry construction.

1705.1.1 Special cases. This section pertaining to special inspection in special cases was relocated from the end to the beginning of the section in the 2012 IBC. Special inspection is required for proposed work that is unique or unusual in nature and products or systems that are not specifically addressed in the code or in standards referenced by the code. This section is used to apply special inspection requirements to items that are not specifically covered in the code but are approved under the alternate design and methods of construction provisions in Section 104.11. Many ICC Evaluation Service Reports for structural products require special inspection in accordance with this section. For example, ESR 2302 for the Hilti Kwik Bolt 3 requires continuous special inspection during installation.

Steel construction. This section sets forth the special inspection requirements for the fabrication and erection of steel structures. Prior editions of the IBC contained detailed requirements for special inspection of structural steel in Table 1704.3. The table contained verification and inspection requirements, whether the frequency of inspection is required to be continuous or periodic, the appropriate referenced standard, and the applicable IBC code section. A code change to the 2012 IBC by AISC deleted the special inspection requirements for structural steel and replaced them with a reference to the quality assurance and inspection requirements in AISC 360-10.[5] The special inspection requirements that remain in the code pertain to steel structures other than structural steel, such as cold-formed steel, sheet steel, and reinforcing.

1705.2

The exception eliminates the need for special inspection of steel structures in certain cases. Special inspection is not required if the fabricator does not alter the properties of the parent material by welding, thermal cutting, or heating operations. For example, if the members being fabricated were cut by mechanical means, such as a band saw, and punched or drilled for bolt holes, with no application of heat, special inspection would not be required. But if the same members were cut with an oxy-acetylene torch, special inspection *would* be required. Even if special inspection is not required, the fabricator must provide evidence that his tracking procedures are adequate to verify that the material used to fabricate any member meets the required specification, is of the proper grade, and has an associated mill test report.

Structural steel. As noted above, the section references the quality assurance provisions of AISC 360-10 for special inspection of structural steel. Substantial portions of the special inspection requirements for structural steel were deleted from the 2012 IBC because the 2010 edition of ANSI/AISC 360, *Specification for Structural Steel Buildings*, incorporates a new Chapter N, which includes comprehensive quality control and quality assurance requirements for structural steel construction. AISC 360, Chapter N, covers quality control requirements pertaining to the structural steel fabricator and erector, as well as quality assurance requirements pertaining to the owner's inspecting and/or testing agencies. The requirements in ANSI/AISC 360-10 are similar to those that were incorporated into AISC 341-05,[6] Appendix Q. AISC 360-10, Chapter N, provides the foundation for the quality control and quality assurance requirements for general structural steel construction, along with AISC 341-10, Chapter I, thereby extending specific requirements to high-seismic applications. The inspection requirements in AISC 360-10 of the Quality Assurance Inspector are purported to be equivalent to those specified for the special inspector in IBC Chapter 17.

1705.2.1

Section 1704.3 of the 2009 IBC addressed all forms of steel construction, but the majority of the requirements in the section and Table 1704.3 pertained to structural steel construction and therefore have been deleted. However, some items apply to cold-formed steel construction and rebar welding, which are not covered by AISC 360. Requirements for special inspection of other forms of steel construction are in a separate section and in a reduced table titled Steel Construction Other Than Structural Steel. The exception in Section 1705.2 has been retained but modified to clarify the requirement. In practice, the "representative mill test reports" are supplied as described in the AISC Code of Standard Practice, so the added sentence in the exception on mill test reports allows traceability when required by the construction documents and defers to AISC 360 in other cases. For a correlation between the provisions that were deleted from 2009 IBC that are covered in AISC 360-10, Chapter N, refer to code change S121-09/10 in the Code Changes Resource Collection: 2012 Edition, available from ICC.

1705.2.2 Steel construction other than structural steel. Special inspection requirements for steel structures other than structural steel are provided in Table 1705.2.2. The table covers three items: material verification of cold-formed steel decks, welding of cold-formed steel decks, and welding of reinforcing. It should be noted that Table 1705.3, which contains special inspection requirements for concrete construction, references back to Table 1705.2.2 for welding of reinforcing steel.

1705.2.2.1 Welding. The two subsections related to welding reference standards for requirements related to welding inspection and welder inspector qualifications for cold-formed steel roof and floor decks and reinforcing steel. Section 1705.2.2.1.1 references AWS D1.3 for cold-formed steel roof and floor decks, and Section 1705.2.2.1.2 references D1.4 and ACI 318 for reinforcing steel. Note that AISC 360-10 references AWS D1.3-2008 and ACI 318-11 references AWS D1.4-2011. The 2012 IBC references AWS D1.3-98 and AWS D1.4-98.

1705.2.2.2 Cold-formed steel trusses spanning 60 feet or greater. This is a new section that was added to the 2009 IBC requiring special inspection for cold-formed steel trusses spanning more than 60 feet. Because long-span trusses have significant loads and support reactions, proper installation and bracing is of critical importance. The special inspector must verify that the temporary and permanent truss bracings are installed in accordance with the approved truss engineering and submittal package. A similar requirement was added for wood trusses in Section 1705.5.2.

Table 1705.2.2 Required Verification and Inspection of Steel Construction Other Than Structural Steel. This table presents the requirements for special inspection of certain steel elements other than structural steel. As noted above, a code change to the 2012 IBC deleted the specific requirements for special inspection of structural steel and replaced them with a reference to the quality assurance and inspection requirements contained in AISC 360-10. The special inspection requirements that remain in Table 1705.2.2 are for material verification of cold-formed steel decks, welding of cold-formed steel decks, and welding of reinforcing. The table describes the specific item requiring verification, the inspection frequency required (whether continuous or periodic), and the applicable referenced standard. For example, Item 1a references ASTM standards for materials and Item 2 references AWS D1.3 for welding floor and roof decks.

1705.3 Concrete construction. This section provides the special inspection and verification requirements for concrete construction such as buildings, foundations, and other elements. Detailed requirements are covered in Table 1705.3, which sets forth the specific verification and inspection requirements, whether the frequency of inspection is to be continuous or periodic, the referenced standard, and/or the applicable IBC code section. Exceptions for special inspection are isolated spread footings for buildings three stories or less in height, lightly loaded elements such as slabs on grade and concrete foundations such as conventional foundations for light-frame construction and nonstructural concrete flatwork such as sidewalks, patios, and driveways. Where certain criteria are met, Exceptions 1 and 2 exempt conventional foundations, either spread (pad) footings or continuous footings, for buildings three stores or less in height. Under Exception 2, Item 2.3 exempts continuous concrete footings from special inspection where the design is based on a specified concrete strength of 2,500 psi, even if a higher specified compressive strength is provided for other reasons such as durability. This allows engineers to specify a higher concrete strength as a matter of preference where a 2,500-psi design strength is otherwise acceptable. Note that Section 1808.8.1 requires a specified concrete strength of 3,000 psi for foundations in Seismic Design Category D, E, or F. This means that foundation concrete in Seismic Design Category D or higher must be designed for a specified concrete strength of 3,000 psi concrete, and therefore special inspection is required. However, Table 1808.8.1 allows a specified concrete strength of 2,500 psi for Group R or U of light-frame construction two stories or less in height for foundations in Seismic Design Category D, E, or F. Therefore, 2,500 psi concrete is permitted for these structures, and special inspection would not be required.

Concrete basement or foundation walls constructed in accordance with the prescriptive provisions in Section 1807.1.6.2 are not required to be engineered and are exempt from special inspection.

Materials. Constituent materials for concrete such as aggregate, cement, admixtures, and water must conform to the requirements set forth in Section 1903 and the standards of Chapter 3 of ACI 318. When sufficient documentation is not available to verify that constituent materials conform to these requirements, the building official should require testing of the materials. Where welded reinforcing does not conform to ASTM A706, the reinforcing must comply with Section 3.5.2 of ACI 318, which requires the material properties to conform to the requirements of AWS D1.4.

1705.3.1

Table 1705.3 Required Verification and Inspection of Concrete Construction. Table 1705.3 presents the requirements for special inspection of concrete structures in a concise format along with the inspection frequency, referenced standard, and/or applicable code section. The table summarizes the required inspections and test samples necessary to verify that the in-place concrete meets code requirements. Refer to ACI 318[2] and the applicable ASTM standards for more details on constituent material tests, sampling of fresh concrete, testing of slump or air content, casting of test specimens, and other test requirements.

Item 1. Inspections of reinforcing should verify that the reinforcement is of the correct size and grade, as required by the approved drawings and specifications, and is properly placed prior to placement of concrete. Proper placement of reinforcement has a significant impact on the integrity and strength of reinforced concrete. The reinforcement should be placed within the tolerances set forth in ACI 318 Section 7.5 and ACI 117, *Standard Tolerances for Concrete Construction and Materials*. Additional requirements of ACI 318, such as surface conditions of reinforcement (Section 7.4), spacing limitations (Section 7.6), and concrete protection for reinforcement or cover (Section 7.7), must also be checked.

Item 2. Note that Item 2, inspection of reinforcing steel welding, refers to Table 1705.2.2, Item 2b, for welding reinforcing steel.

Item 3. Item 3 comes into effect when cast-in-place anchor bolts are designed for the higher load allowed by special inspection (see Section 1908.5) or where the strength design procedure is used to design the anchorage. Most jobs that require special inspection for concrete will be designed to take advantage of the higher allowable bolt loads with special inspection. Hence, if special inspection is required for the concrete, most likely it will also be required for the anchor bolts. Proper placement and embedment of anchor bolts is of extreme importance. If an anchor bolt is set too low for proper thread engagement, there are few satisfactory methods to remedy the situation. The common practice of placing a puddle weld in the nut is of questionable value, as neither the bolt nor the nut may actually be weldable material. The chemistry of nuts is not controlled by the ASTM requirements, and the chemistry allowed for an ordinary A307 bolt is such that it may or may not be weldable. In either case, the weld is not a prequalified weld in accordance with AWS D1.1, and a Procedure Qualification Record should be developed to qualify the welding procedure. Note also that if the anchor bolt is a high-strength bolt such as an ASTM A325 or ASTM A449, the bolt is quenched and tempered, and application of heat by welding may destroy the strength of the bolt and make the bolt brittle.

Prior to the 2009 IBC, this table required continuous special inspection of bolts installed in concrete prior to placement of concrete where allowable loads have been increased but did not require special inspection for bolts designed in accordance with the strength design procedure of ACI 318 Appendix D, anchors installed in hardened concrete (e.g., expansion and undercut anchors), and anchors designed to resist seismic loads. Two new requirements were added to the 2009 IBC. Item 3 requires continuous special inspection for bolts cast in concrete where the strength design procedure is used, and Item 4 requires periodic special inspection for anchors installed in hardened concrete. For post-installed anchors, the imposition of continuous special inspection was felt to be a hardship for projects involving the installation of large numbers of anchors where simultaneous inspection of all installations is impractical. In the 2012 IBC, Item 3 was changed to require periodic inspection because inspecting cast-in-place anchors is really no different than inspecting placement of reinforcement. Thus, the anchors in both Items 3 and 4 only require periodic special inspection. The lack of special inspection for bolts designed in accordance with ACI 318 Appendix D (which includes all bolts and anchors designed to resist seismic loads) was considered an oversight. Table 1705.3 now requires periodic special inspection for bolts cast in concrete where strength design is used, which also includes anchors designed to resist seismic loads by virtue of the restrictions imposed by Sections 1908 and 1909.

Item 4. As noted above, Item 4 requires periodic special inspection for anchors post-installed in hardened concrete. It should be noted that the 2015 IBC will likely require continuous special inspection for post-installed adhesive anchors in hardened concrete where the anchors are designed to resist sustained tension loads. The new Footnote b has been added in the 2012 IBC to account for post-installed anchors approved through the alternate methods of construction provisions of Section 104.11, such as anchors installed in accordance with ICC Evaluation Service Reports. It is also intended to distinguish between the requirements for special inspection of anchors designed to comply with the IBC alone versus those qualified by approved research reports in accordance with ACI 355.2, Qualification of Post-Installed Mechanical Anchors in Concrete. Typically, items requiring special inspection that are approved under Section 104.11 are covered by Section 1705.1.1, Special Cases. Where special inspection requirements are not provided in a research report, the special inspection requirements must be specified by the registered design professional, who would indicate whether inspections are continuous or periodic, and be approved by the building official prior to commencement of the work.

Item 5. Item 5 is a particularly important verification on larger jobs that may have many required mix designs with differing strength requirements and aggregate sizes.

Item 6. Sampling of fresh concrete for making specimens for strength tests is extremely important for proper quality control. Properly sampled and prepared specimens are necessary to determine that the concrete will meet or exceed the design strength. The frequency of sampling should be in accordance with ACI 318 Section 5.6.2—one set of specimens for each class of concrete not less than once per day, once per 150 cubic yards, or once per 5,000 square feet of slab or wall. Sampling should be done in accordance with ASTM C 172 to ensure representative samples for determining compressive strength. The tests specified in Item 6 may be supplemented by other tests such as unit weight or air content.

Item 7. Observation of the actual placement is important to determine that the fresh concrete is properly handled so that it does not segregate during placement and that the concrete is properly consolidated by vibration. The mixing requirements (see ACI 318 Section 5.8), conveying requirements (see ACI 318 Section 5.9), and depositing requirements (see ACI 318 Section 5.10) should be strictly enforced to ensure proper placement with adequate consolidation and without segregation. Concrete voids or *rock pockets* can adversely affect design strength and are unattractive.

Item 8. Maintenance of proper cure is essential to obtaining quality concrete that will reach the design strength. Concrete that is not properly cured often will be below design strength and may suffer degradation at the surface from use or from environmental effects much earlier than properly cured concrete.

Item 9. The inspections required by Item 9 are of extreme importance as the strength of a prestressed member is highly dependent on proper prestressing. When checking the application of prestressing force, both the force applied to the tendon and the tendon elongation should be checked simultaneously to ensure that the tendon has not been hung up in the tendon sheath.

Item 10. Criteria for the erection procedures of precast concrete must be provided on the design drawing by the design engineer. The drawings should identify each panel to be cast and should specify: dimensions and thickness of panels, reinforcement grade, size and location, location of inserts, and minimum concrete strength at lifting and in-service. The special inspector should ensure that the erection process complies with the approved procedures.

Item 11. Concrete strength must be verified prior to stressing tendons used in post-tensioned concrete. Form supports for prestressed concrete should not be removed until sufficient prestressing has been applied. Forms and shoring for conventionally reinforced concrete members such as beams and structural slabs should not be removed until adequate concrete strength is achieved.

Item 12. Inspection of concrete forms for proper shape, dimensions, and location is essential for adequate performance on concrete members such as beams, columns, walls, and structural slabs.

Masonry construction. The masonry standard is developed by a joint committee of TMS, ACI, and ASCE, called the Masonry Standards Joint Committee (MSJC), and consists of *Building Code Requirements for Masonry Structures*, TMS 402/ACI 530/ASCE 5,[7] *Specifications for Masonry Structures*, TMS 602/ACI 530.1/ASCE 6,[8] and commentaries to the code and specifications. These documents are often referred to as the MSJC Code and MSJC Specification. Prior editions of the IBC had two tables for verification and inspection of masonry structures. Level 1 inspection applies to engineered masonry structures in Occupancy Category I, II, and III and empirically designed masonry structures in Occupancy Category IV. Level 2 inspection applies to engineered masonry structures in Occupancy Category IV. A code change to the 2012 IBC by TMS deleted these special inspection requirements and tables and replaced them with a reference to the quality assurance and inspection requirements in the 2011 edition of the MSJC Code and Specification. The level of quality assurance required by Section 1.19 of the 2011 MSJC Code is driven by the type of masonry used and the risk category of the structure. As previously noted, the term "Occupancy Category" was changed in the 2012 IBC to "Risk Category" to better reflect the meaning. Table 1705-1 reflects the level of quality assurance required based on the type of masonry and risk category of the structure. **1705.4**

There are three exemptions from special inspection for masonry structures. Empirically designed masonry, glass unit masonry, or masonry veneer in Risk Category I, II, or III structures are exempt from special inspection when constructed in accordance with Sections 2109, 2110, or Chapter 14. Note that the MSJC Code requires Level A quality assurance for these structures, which requires no specific testing and only consists of verification of approved submittals for materials in accordance with MSJC Specification Article 1.5. The other two exceptions are for masonry foundation walls constructed in accordance with the prescriptive tables of Section 1807.1.6, which do not require engineering, and masonry fireplaces, chimneys, and heaters constructed in accordance with Sections 2111, 2112, or 2113.

Empirically designed masonry, glass unit masonry, and masonry veneer in Risk Category IV. Special inspection for empirically designed masonry, glass unit masonry, and masonry veneer in Risk Category IV structures must comply with TMS 402/ACI 530/ASCE 5 **1705.4.1**

Table 1705-1. **Level of Quality Assurance Required by MSJC Code Section 1.19**

Type of Masonry Structure	Risk Category	Level of Quality Assurance Required
Empirical[a] Veneer Glass Unit	I, II, III	Level A
Veneer Glass Unit	IV	Level B
Masonry *other than*: Empirical Veneer	I, II, III	Level B
Masonry *other than*: Empirical Veneer Glass Unit	IV	Level C

[a]Empirically designed masonry is not permitted in Risk Category IV.

Level B Quality Assurance requirements. See Table 1.19.2 of the 2011 MSJC Code for specific details.

1705.4.2 Vertical masonry foundation elements. The intent of this section is that the special inspection requirements for vertical masonry foundation elements are as noted in Section 1705.4 and its subsections. In other words, vertical masonry foundation elements may be required to comply with quality assurance and inspection provisions of the MSJC Code and Specification, or they may be required to only comply with the Level B Quality Assurance requirements prescribed in Section 1705.4.1, or they may be exempt from special inspection in accordance with one of the exceptions. See Section 1808.9 for requirements for vertical masonry foundation elements.

1705.5 Wood construction. This section requires special inspection for certain types of wood construction. The requirement is for inspection of prefabricated elements such as wood trusses and refers to Section 1704.2.5 regarding special inspection of fabricators. For portions of wood structures designated as the seismic-force-resisting system, see also the special inspection requirements for wind and seismic resistance discussed in the analysis of Sections 1705.10 and 1705.11. Special inspection is required for site-built assemblies such as high-load diaphragms and wood trusses spanning 60 feet or more.

1705.5.1 High-load diaphragms. When lateral loads become significant, designers often use high-load diaphragms, which have multiple rows of fasteners. Proper construction of high-load diaphragms is critical so this section requires periodic special inspection. Design values for high-load diaphragms fastened with nails are in Table 4.2B of the SDPWS-08, and high-load diaphragms fastened with staples are in Table 2306.2(2). The special inspector is required to inspect the diaphragm for proper sheathing grade and thickness; proper size, species, and grade of framing members; and proper fastener type, size, spacing, and edge distance from sheathing and framing members.

1705.5.2 Metal-plate-connected wood trusses spanning 60 feet or greater. This is a new section that was added to the 2009 IBC requiring special inspection for wood trusses spanning over 60 feet. Because long-span trusses have significant loads and support reactions, proper installation and bracing is of critical importance. The special inspector must verify that the temporary and permanent truss bracings are installed in accordance with the approved truss engineering and submittal package. A similar requirement was added for cold-formed steel trusses spanning over 60 feet in Section 1705.2.2.2.

Required Verification and Inspection — 17

Soils. The table for verification and inspection of soils was added to the 2006 IBC to clarify the specific inspection tasks that are required for soils. Prior to the 2006 IBC, the inspections were presumed to be whatever was required by the geotechnical engineer. This section covers special inspection requirement soils such as site preparation, engineered fills, and materials supporting load-bearing foundations. The load-bearing capacity of the site soil and any fill has a significant impact on the structural integrity of a building supported by the fill. For example, settlements in an improperly compacted fill can cause significant structural distress. Differential settlements of $1/4$ inch in a 20-foot grade beam can induce stresses that exceed the yield limits. Hence, fills should be engineered and compaction carefully controlled. The special inspection tasks outlined in Table 1705.6 must be performed to verify compliance with the approved construction documents and the geotechnical report required by Section 1803.2. What is required to be included in geotechnical reports is covered in Section 1803.6. The special inspector is required to (1) verify that materials below footings are adequate to achieve the design bearing capacity; (2) verify that excavations are extended to the proper depth and have reached proper bearing material; (3) perform classification and testing of controlled fill materials; (4) verify use of proper materials, densities, and lift thicknesses during placement and compaction of controlled fill material; and (5) verify that the site and subgrade have been properly prepared prior to placement of controlled fill.

1705.6

The exception in previous editions of the code that exempted placement of controlled fills 12 inches or less in depth from having to comply with the special inspection requirements was changed in the 2009 IBC to require that the special inspector verify a minimum of 90-percent compaction.

There were editorial changes to this section in the 2009 IBC to update the terminology to current practice. The term "geotechnical report" is used to be consistent with changes to Chapter 18. The term "shallow foundation" is used instead of "footing" to be consistent with the new terminology used in Chapter 18 regarding types of foundations.

Driven deep foundations. Special inspection is required for installation and testing of driven deep (pile) foundations. The special inspection tasks outlined in Table 1705.7 are required to verify compliance with the approved construction documents and the geotechnical report required by Section 1803.2. What is required to be included in geotechnical reports is covered in Section 1803.6. The term *pile* was changed in the 2009 IBC to "driven deep foundation element" to be consistent with changes to Chapter 18. Table 1705.7 requires continuous special inspection to (1) verify that pile materials, sizes, and lengths comply with the requirements; (2) determine capacities of test piles and conduct additional load tests; (3) observe driving operations and maintain complete and accurate records for each pile; and (4) verify placement locations and plumbness, confirm type and size of hammer, record the number of blows per foot of penetration, determine required penetrations to achieve design capacity, record tip and butt elevations, and document any pile damage.

1705.7

For driven deep foundations, records should, at a minimum, contain for each pile a driving log showing the number of hammer blows for each foot (and the energy) and the resistance at final penetration in blows per inch or blows per foot as applicable. The inspector should verify that the hammer throttle is set correctly to give the desired energy. The driving log should also give information on the duration and cause of any delays, the depth of any preexcavation, and the elevation of the tip and the cutoff. Any known or suspected pile damage, as well as any observation of pile drift or heave, should be noted.

For specialty piles, the special inspector must perform additional inspections in accordance with the registered design professional's recommendations.

The table now has seven items. Item 8 was deleted from the table in the 2009 IBC because augered piles and caissons are cast in place. Special inspection requirements for these piles are covered in Section 1705.8.

1705.8 Cast-in-place deep foundations. The terminology related to foundations in Chapter 18 was changed in the 2009 IBC. Chapter 18 now has two general types of foundations—shallow foundations and deep foundations. What were previously referred to pier foundations are now covered under the more general category of cast-in-place deep foundations. Continuous special inspection is required for installation and testing of cast-in-place deep foundations. The special inspection tasks outlined in Table 1705.8 are required to verify compliance with the approved foundation and geotechnical report required by Section 1803.2. The table requires the special inspector to observe drilling operations and maintain complete and accurate records for each pile, and verify placement locations and plumbness, and confirm pile diameters, bell diameters (if applicable), lengths, embedment into bedrock (if applicable), and adequate end-bearing strata capacity. The volume of concrete or grout placed is often the first indicator of potentially significant problems during construction. Item 2 requires the special inspector to record the volumes of concrete or grout placed as a critical diagnostic tool. For concrete cast-in-place deep foundations, Item 3 of the table refers to Table 1705.3 for special inspection tasks related to concrete construction.

Augered piles are drilled and then grouted in place. For each drilled pile, data should include a drilling log showing the types of soils encountered for each foot and the material stratum at the required tip elevation. The inspector should verify that the soil at the required tip elevation is the correct soil. The drilling log should also give information on the duration and cause of any delays, data on the rebar cage and concreting procedures, casing or other procedures necessary to prevent intrusion of ground water, and results of any concrete strength tests.

1705.9 Helical pile foundations. Helical piles were added to Chapter 18 of the 2009 IBC. Helical piles are deep foundation systems used for new construction, for foundation repair to mitigate structure settlement, or for tiebacks in shoring applications. Helical piles can be installed with a column of high-strength grout that encases the helical shaft to provide additional compressive and lateral load capacity. The section requires continuous special inspection during installation of helical pile systems. The installation procedure must be in accordance with the recommendations of the registered design professional and the approved geotechnical report.

1705.10 Special inspection for wind resistance. This new section about requiring special inspection for wind resistance was added to the 2009 IBC. In areas of high seismic risk (i.e., Seismic Design Categories C, D, E, and F), the IBC requires special inspection of seismic-force-resisting systems in buildings of light-frame construction (wood framing and cold-formed steel framing). Similar risks exist in high-wind areas and special inspection of main wind-force-resisting systems and components, and cladding in buildings of light-frame construction was deemed to be warranted. Damage to buildings due to high-wind forces often begins with failure of the cladding system, which often exposes the main wind-force-resisting system to damage from wind-driven rain and other forces that the wind-force-resisting system is typically not designed to resist. Unless specifically exempted in the general requirements in Section 1704.2, special inspection for wind resistance is required for buildings sited in Exposure B where the nominal design wind speed is 120 mph or more, and Exposure C or D where the nominal design wind speed is 110 mph or more. The specific requirements apply to structural wood, cold-formed light-frame steel structures, and exterior cladding. The inspections are continuous or periodic depending on the type of structural elements involved. Table 1705-2 summarizes the requirements in Sections 1705.10.1 (wood), 1705.10.2 (cold-formed steel), and 1705.10.3 (cladding).

1705.10.1 Structural wood. Continuous special inspection is required for any field gluing of the main wind-force-resisting system (see Section 202 for definition of main wind-force-resisting system). Periodic special inspection is required for fastening (nailing, bolting, anchoring) of elements of the main wind-force-resisting system such as shear walls, diaphragms, chords, collectors (drag struts), braces, and hold-downs. The exception applies where the fastener spacing of the sheathing is more than 4 inches on center.

Cold-formed steel light-frame construction. Continuous special inspection is required for any welding of the main wind-force-resisting system. Periodic special inspection is required for fastening (screw attachment, bolting, anchoring) of elements of the main wind-force-resisting system such as shear walls, braces, diaphragms, chords, collectors (drag struts), and hold-downs. The first exception applies when the sheathing is gypsum board or fiberboard because of their relatively low load capacity. Similar to wood framing, the second exception applies where the fastener spacing of the sheathing is more than 4 inches on center. **1705.10.2**

Wind-resisting components. Damage to buildings in high-wind events often begins with failure of the cladding system. Thus, periodic special inspection is required for roof and wall cladding. **1705.10.3**

Special inspection for seismic resistance. Unless specifically exempted in the general requirements in Section 1704.2, special inspection for seismic resistance is required in addition to the general special inspection requirements covered in Section 1704. The requirements are triggered by the seismic design category of the building. Special inspection is required for the seismic-force-resisting system, the designated seismic system, and architectural, mechanical, and electrical components depending on the seismic design category. Other items such as storage racks and base isolation systems require periodic special inspection. **1705.11**

The seismic-force-resisting system consists of those structural elements and systems that provide lateral stability of the structure and are specifically designed to provide resistance to the anticipated seismic forces. The designated seismic system consists of those architectural, electrical, and mechanical systems and components that require design in accordance with Chapter 13 of ASCE 7 for which the component importance factor, I_p, is greater than 1.0, as prescribed in Section 13.1.3 of ASCE 7 (see Section 202). Section 13.1.3 of ASCE 7 lists four classes of components that have a component importance factor of 1.5. These are components required to function after an earthquake including fire-sprinkler systems and egress stairways that are not an integral part of the building structure, components containing or conveying toxic materials, components in Risk Category IV structures that are deemed to be necessary for continued operation of the facility, and components containing or conveying hazardous materials. Refer to Section 13.1.3 of ASCE 7 for complete descriptions of these systems. Egress stairways was added to Item 1 of ACE 7-10 Section 13.1.3, and Table 13.5-1 gives amplification/response factors for egress stairways that are not a part of the building structure.

In addition to the general exceptions in Section 1704.2, there are three specific exceptions where special inspection for seismic resistance is not required: Light-frame structures not over 35 feet in height and in which S_{DS} does not exceed 0.5; buildings with reinforced

Table 1705-2. **Special Inspection for Wind Resistance**

	Type of Special Inspection	
Type of Structural Element	Continuous	Periodic
Wood–Field gluing of MWFRS elements	X	
Wood frame–Nailing, bolting, anchoring, fastening components of MWFRS such as shear walls, chords, collectors, hold-downs[a]		X
Cold-formed steel light frame–Welding of MWFRS elements		X
Cold-formed steel light frame–Screws, bolting, anchoring, fastening components of MWFRS such as shear walls, chords, collectors, hold-downs[a]		X
Roof cladding		X
Wall cladding		X

[a]Note that exceptions for both wood frame cold-formed steel do not require special inspection where the fastener spacing of the sheathing is more than 4 inches on center.

masonry or reinforced concrete seismic-force-resisting systems not over 25 feet and in which S_{DS} does not exceed 0.5; and detached one- or two-family dwellings not exceeding two stories that do not have the horizontal or vertical irregularities listed.

1705.11.1 Structural steel. For special inspection requirements for structural steel, the IBC references the quality assurance requirements of AISC 341. A new exception added to the 2009 IBC permits steel structures not detailed for seismic resistance that use an *R* factor of 3 or less in Seismic Design Category C to be exempt from special inspection for seismic resistance. The exception does not apply to cantilever column systems due to their limited ductility and redundancy. Table 12.2-1 of ASCE 7 assigns "steel systems not specifically detailed for seismic resistance, excluding cantilever column systems," a response modification coefficient, *R*, equal to 3 and limits their use to Seismic Design Category A, B, or C. For these building systems, the seismic response coefficient of 3 reflects their inherent lack of ductility. Thus, these structures are permitted to be designed using only AISC 360 and are not required to be detailed in accordance with the additional provisions of AISC 341. The details and connections are the same as typical steel buildings following AISC 360, and no additional special inspection or testing is required beyond that ordinarily applied to typical steel buildings.

1705.11.2 Structural wood. The seismic special inspections required for wood structures are primarily to ensure continuity of load path within the seismic lateral-force-resisting system. The walls must transfer their inertial loads to the diaphragms, which in turn transmit the inertial loads through the lateral-force-resisting system to the foundation. Periodic inspection is allowed except in the case of field gluing operations, which are to be continuous. The section requires periodic special inspection for fastening such as nailing, bolting, and anchoring components of the seismic-force-resisting system, which includes chords, collectors (struts), braces, and hold-downs. Particular care should be given to the nailing of diaphragms and shear walls. Common nails are often specified in the design, but smaller-diameter sinkers or power-driven (gun) nails are often substituted in the field because the smaller-diameter nails have a lower lateral resistance. For example, the lateral resistance of a 0.131-inch-diameter power-driven nail used as a replacement for a 10d common nail in Douglas Fir-Larch is only 76 pounds, whereas the value for the 10d common nail is 90 pounds. Of additional importance is the connection of collectors to shear walls and the proper installation and tightening of hold-down bolts in shear walls. The exemption from special inspection applies where diaphragm and shear panel construction has a fastener spacing of more than 4 inches on center. Where the fastener spacing is greater than 4 inches, there is lower demand and less potential for splitting, and special inspection is not required.

1705.11.3 Cold-formed steel light-frame construction. Similar to wood framing, the seismic special inspections for cold-formed steel structures are to ensure continuity of load path within the seismic lateral-force-resisting system; that is, the walls must transfer their inertial loads to the diaphragms, which in turn transmit the inertial loads through the lateral-force-resisting system to the foundation. The section requires periodic special inspection of welding and fastening such as screw attachment, bolting, and anchoring components of the seismic-force-resisting system including chords, collectors (struts), braces, and hold-downs. Particular care should be given to connections of braces and hold-downs. Wood and cold-formed steel light-frame constructions have similar requirements for their lateral-force-resisting systems. Therefore, the exception for structural wood also applies, with the appropriate adaptation, to cold-formed steel light-frame construction. The 4-inch spacing for wood construction roughly translates to a minimum capacity of 380 plf. Thus, where fastener spacing for wood structural panel sheathing or steel sheets is greater than 4 inches, there is lower demand and special inspection is not required. The exception for gypsum board and fiberboard shear walls is based on the fact that the capacity of these materials is below the demand threshold.

1705.11.4 Designated seismic systems. As noted above, the designated seismic system consists of those architectural, electrical, and mechanical systems and components that require design in accordance with Chapter 13 of ASCE 7 for which the component importance factor, I_p, is greater than 1.0, as prescribed in Section 13.1.3 of ASCE 7 (see Section 202). Section 13.1.3

of ASCE 7 lists four classes of components that have a component importance factor of 1.5. Refer to Section 13.1.3 of ASCE 7 for complete descriptions of these systems. The special inspector must verify that elements of the designated seismic system that require certification according to Section 1705.12.3 are properly labeled and anchored or mounted in accordance with the certificate of compliance. See discussion of Section 1705.12.3.

1705.11.5 **Architectural components.** Exterior cladding and veneers can be a serious safety hazard if they become detached from the structure during seismic ground motion as well as potentially blocking required exit paths. The code requires periodic special inspection for exterior cladding, nonbearing walls and partitions, and veneer in Seismic Design Category D, E, or F. The exemption from special inspection applies to these elements when they are relatively low or lightweight. Thus, special inspection is not required when these elements are 30 feet or less above grade or the walking surface, are light-weight cladding and veneer weighing 5 psf or less, or light-weight partitions weighing 15 psf or less.

1705.11.5.1 **Access floors.** Access floors consist of a system of panels and supports that create a raised floor above the actual structural floor system. The space between the raised floor and the structural floor contains components like wiring for power, voice, and data. The space may also be used for HVAC distribution either as a plenum or with ductwork. Because failure of the floor system can pose a threat to the occupants in high seismic areas, periodic special inspection is required for anchorages of access floors in structures in Seismic Design Category D, E, or F.

1705.11.6 **Mechanical and electrical components.** Inspection is necessary for components that must function in post-earthquake conditions, such as emergency electrical systems, or for anchorage of mechanical equipment, piping, and ducting using or carrying flammable or hazardous materials. Periodic special inspection is required for the following items in structures in Seismic Design Category C, D, E, or F: (1) during the anchorage of electrical equipment for emergency or standby power systems; (2) during installation of piping systems intended to carry hazardous contents and their associated mechanical units; (3) during the installation of HVAC ductwork that contains hazardous materials; and (4) during installation of vibration isolation systems in structures where the design requires ¼ inch or less between the equipment support frame and restraint. For structures in Seismic Design Category E or F, periodic special inspection is required for the installation of anchorage systems of all other electrical equipment.

1705.11.5.7 **Storage racks.** Tall storage racks such as those typically found in large "big box" building supply stores can pose a threat to the public in high seismic areas. Because proper anchorage is critical to keep tall storage racks from overturning during a seismic event, periodic special inspection is required for storage racks 8 feet or greater in height in Seismic Design Category D, E, or F.

1705.11.8 **Seismic isolation systems.** The performance of seismic isolators is critical to the performance of isolation systems and energy dissipation devices. Periodic special inspection is required during fabrication and installation of these devices. See Section 1705.12.4, which refers to Section 17.8 of ASCE 7 for testing of seismic isolation systems.

1705.12 **Testing and qualification for seismic resistance.** Sections 1705.12.1 through 1705.12.4 contain special testing and qualification requirements for the seismic-force-resisting system, elements of the designated seismic system, architectural, mechanical, and electrical components, as well as seismic isolation systems. The requirements are driven by the seismic design category of the structure and, with the exception of seismic isolation systems, apply to structures in Seismic Design Categories C through F. The following is a summary of the requirements:

- Concrete reinforcement and structural steel in the seismic-force-resisting systems must meet the requirements of Sections 1705.12.1 and 1705.12.2.
- Elements of the designated seismic system that are subject to the certification requirements of ASCE 7 Section 13.2.2 must meet the certification requirements in Section 1705.12.3.

- Architectural, mechanical, and electrical components where the requirements of ASCE 7 Section 13.2.1 are met by submittal of manufacturer's certification in accordance with Item 2 therein must meet the certification requirements in Section 1705.12.3.

- The isolation system in seismically isolated structures must be tested in accordance with Section 1705.12.4, which references ASCE 7 Section 17.8.

1705.12.1 Concrete reinforcement. When reinforcing steel conforming to ASTM A615 is used to resist seismic forces in special moment-resisting frames, in the boundary elements of shear walls (special structural walls) or coupling beams in Seismic Design Categories B through F, the steel yield and ultimate strengths must be tested and in accordance with Chapter 21 of ACI 318.

Additionally, if A615 reinforcing steel is to be welded, the chemical composition must be determined. The weldability of the steel is determined by its chemical content. AWS D1.4, *Structural Welding Code—Reinforcing Steel*, bases the welding requirements, including preheat and interpass temperature, on the carbon equivalent, which is based on the steel chemistry and the bar size. Testing in accordance with Chapter 3 of ACI 318 is required. Certified mill test reports are required for each shipment of reinforcing.

Changes were made to this section in the 2009 IBC to coordinate the requirements with ASCE 7 and ACI 318. The title of the subsection was changed from "Reinforcing and prestressing steel" to "Concrete reinforcement" because it applies to the reinforcement requirements in ACI 318. The requirements apply to ASTM A615 reinforcement used to resist earthquake forces in special moment frames, special structural walls, and coupling beams in structures assigned to Seismic Design Category B and above.

1705.12.2 Structural steel. Prior to the 2009 IBC, the section required testing as set forth in AISC 341 supplemented by the following requirements: (1) All complete joint penetration (CJP) and partial joint penetration groove welds subject to net tension forces must be tested by nondestructive methods; (2) the acceptance criteria for nondestructive testing are either those for static loading or dynamic loading, as designated by the registered design professional; (3) base metal thicker than $1^1/_2$ inches subject to the through-thickness weld shrinkage strains. For example, a T-joint or the column flange in a typical beam-column connection must be tested for discontinuities such as lamellar tearing after testing. See Figure 1705-1 for an example of a weld susceptible to weld shrinkage strains. Between the 2003 and 2006 editions of the IBC, the terminology was changed from "quality assurance plan" to "statement of special inspection." The change in terminology was not in the 2005 edition of AISC 341. For simplification and clarity, the section in the 2012 IBC provides a direct reference to the quality assurance plan requirements in AISC 341, which provide the user with the minimum acceptable requirements for the construction of welded joints, bolted joints, and other details in the seismic-force-resisting system. Where appropriate, AISC 341 references AWS D1.1 for specific acceptance criteria. The paragraph on ultrasonic testing of base metal that may be subject to lamellar tearing or have laminations present was deleted because AISC 341 addresses this by specifying when nondestructive testing (NDT) is needed. After joint completion, base metal thicker than $1^1/_2$ inches loaded in tension in the through-thickness direction in tee and corner joints, where the connected material is greater than $3/_4$ inches and contains CJP groove welds, AISC 341 requires ultrasonic testing for discontinuities behind and adjacent to the fusion line of such welds. Any base metal discontinuities found within $t/4$ of the steel surface shall be accepted or rejected on the basis of criteria of AWS D1.1 Table 6.2, where t is the thickness of the part subjected to the through-thickness strain.

Referenced in AISC 341, AWS D1.1 Table 6.2 provides the acceptance criteria for ultrasonically tested joints when statically loaded. The criteria are similar to that used prior to adoption of the language in the 2000 IBC, which used the term "larger reflector criteria" from the UBC. The "larger reflector criteria" is a term used in the 1970s that is now identified as a "Class A" discontinuity in Table 6.2. By referencing only Table 6.2, and not referencing

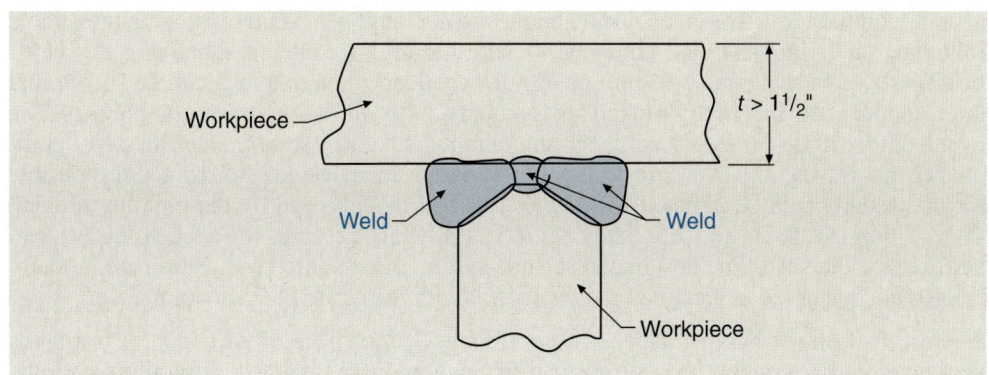

**Figure 1705-1
Weld susceptible to shrinkage stresses.**

Class A, the additional considerations of flaw length and reflector height are made. Finally, the direct references to ASTM A435 and ASTM A898 in previous editions of the IBC are no longer needed because the AISC 341 criteria have been made more restrictive regarding permitted flaws, and more properly reflects the angle-beam ultrasonic methodology used for post-welding examinations. The prior reference to ASTM A435 and A898 were straight-beam ultrasonic tests to detect laminations in base metal prior to welding, and have been deemed inadequate for post-welding lamellar tearing checks.

Table 12.2-1 of ASCE 7 assigns "steel systems not specifically detailed for seismic resistance, excluding cantilever column systems" a response modification coefficient, R, equal to 3 and limits their use to Seismic Design Category A, B, or C. For these building systems, the seismic response coefficient of 3 reflects their inherent lack of ductility. Thus, these structures are permitted to be designed using only AISC 360 and are not required to be detailed in accordance with the additional provisions of AISC 341. The details and connections are the same as typical steel buildings following AISC 360, and no additional special inspection or testing is required beyond that ordinarily applied to typical structural steel buildings. A new exception in the 2009 IBC permits steel structures not detailed for seismic resistance that use an R factor of 3 or less in Seismic Design Category C to be exempt from special inspection for seismic resistance. The exception does not apply to cantilever column systems due to their limited ductility and redundancy.

Note that for ordinary moment frames, only demand-critical complete joint penetration (CJP) groove welds are required to have UT and mag particle testing. This is consistent with the recognition that ordinary moment frames have minimal inelastic straining. Therefore, only demand-critical welds are of concern with regard to potential flaws in CJP groove welds.

Seismic certification of nonstructural components. The seismic certification for nonstructural components and designated seismic systems in structures assigned to *Seismic Design Category* C, D, E, or F applies to those listed in Items 2 and 3 of Section 1705.12. These are elements of the designated seismic system that are subject to the certification requirements of ASCE 7 Section 13.2.2 and architectural, mechanical, and electrical components where ASCE 7 Section 13.2.1 is met by submittal of manufacturer's certification in accordance with Item 2 of Section 13.2.1. Section 13.2.1 of ASCE 7 applies to architectural, mechanical, and electrical components, and Section 13.2.2 applies to special certification required for the designated seismic systems (see Section 202 for the definition of components that are considered to be designated seismic systems). The registered design professional is to provide the applicable seismic qualification requirements on the construction documents. Section 13.2 of ASCE 7 permits justification of components by project-specific design or certification by the manufacturer through analysis, testing, or experience data using historical data that demonstrate acceptable seismic performance. The special certification required by Section 13.2.2 of ASCE 7 for design seismic systems

1705.12.3

is only required for active mechanical and electrical components that must remain operable following an earthquake, and components with hazardous contents. Obtaining this certification requires shake-table testing or use of experience data unless it can be shown that the component is "inherently rugged" by comparison to similar qualified components. The manufacturer of designated seismic system components must submit a certificate of compliance that is to be reviewed and accepted by the registered design professional responsible for the design of the designated seismic system and be approved by the building official. This section is referenced from Section 1705.11.4, which requires the special inspector to verify that elements of the designated seismic system that require certification are properly labeled and anchored or mounted in accordance with the certificate of compliance.

1705.12.4 **Seismic isolation systems.** This section refers to Section 17.8 of ASCE 7 for testing of seismic isolation systems. See discussion of Section 1705.11.8, which requires periodic special inspection during fabrication and installation of these devices.

1705.13 **Sprayed fire-resistant materials.** This section provides requirements for special inspection of spray applied fire-resistant materials (SFRM). For an SFRM to perform as intended, its application must be within the proper range for certain parameters as determined by the system manufacturer. These include temperature at time of application, substrate conditions, thickness, bond strength, and density. These parameters must be checked prior to and during installation of the SFRM.

1705.13.1 **Physical and visual tests.** The special inspector must verify the five conditions listed. The substrate condition, thickness, density, bond, and finished application must conform to the applicable listing for the fire rating required.

1705.13.2 **Structural member surface conditions.** The integrity of an SFRM system primarily depends on the condition of the surface of the steel member to which it is applied. For proper performance, the system must be fully adhered to the surface. The prepared surface must be inspected prior to application of SFRM.

1705.13.3 **Application.** During the application and curing of SFRMs, several parameters must be controlled, including ambient temperature, temperature of the substrate, and the temperature of the applied SFRM. Control of these temperatures is necessary so that the chemical reactions needed to make the SFRM bond to the steel substrate properly occur. The minimum or maximum temperatures needed for proper bond and cure depend on the particular system.

1705.13.4 **Thickness.** The SFRM must be applied at the correct thickness for the system to provide the required fire resistance. The sampling and thickness determinations are based on ASTM E 605. This standard also provides testing methods commonly used by industry. See Reference 10 for additional information on inspection of SFRM. Section 1705.13.4.1 specifies minimum thickness as a function of the design thickness, and Sections 1705.13.4.2 through 1705.13.4.9 cover sampling requirements for various structural members.

1705.13.5 **Density.** The SFRM must be applied at the correct density in order for the system to provide the required design fire resistance. The sampling and density determinations are based on ASTM E 605. A sampling rate for density testing is one sample for each 2,500 square feet at each story. For floor, roof, and wall assemblies, the sampling area is at least one sample for every 2,500 square feet of the sprayed area at each story and for structural members such as beams, girders, trusses, and columns, the sampling area is at least one sample for each type of structural member for each 2,500 square feet of floor area at each story.

1705.13.6 **Bond strength.** The bond of the SFRM to the steel substrate is essential for the SFRM to provide the required fire resistance. The sampling and bond strength determinations are based on ASTM E 736. The minimum cohesive/adhesive bond strength of 150 pounds per square foot is based on the American Institute of Architects (AIA) Master Specifications. The sampling rates for bond strength, as set forth in IBC Sections 1704.11.5.1 and 1704.11.5.2, are lower than those for thickness and density, as the determinations for thickness and density give an indirect indication of bond. See Section 1705.14 for discussion of mastic and intumescent fire-resistant coatings.

Mastic and intumescent fire-resistant coatings. Where fire-resistant coatings are used, the special inspector must verify that they are applied to structural elements and decks in accordance with the Association of the Wall and Ceiling Industry Technical Manual 12-B[9] (AWCI 12-B), *Standard Practice for the Testing and Inspection of Field Applied Thin Film Intumescent Fire-Resistive Materials*, based on the required fire resistance as shown in the approved construction documents.

1705.14

Exterior insulation and finish systems (EIFS). Special inspection is required for exterior insulation and finish systems (EIFS) based on the approved research report and manufacturer's installation instructions. Critical areas necessary for adequate EIFS performance are proper installation of the waterproofing membrane and installation of flashings at windows, doors, joints, eaves, corners, and penetrations. The Association of the Wall and Ceiling Industry (AWCI) offers training and certification programs for performing proper special inspection of EIFS systems. The exceptions from requiring special inspection are for EIFS installed over a water-resistive barrier, or concrete or masonry walls.

1705.15

Water-resistant barrier coating. Where a water-resistive barrier coating is applied between the EIFS and the sheathing, special inspection is required for the barrier coating. The coating must be properly applied to provide additional protection to the building from incidental moisture intrusion that may occur through the building envelope.

1705.15.1

Fire-resistant penetrations and joints. This is a new section in the 2012 IBC that requires special inspection for fire-resistant penetrations and joints in high-rise buildings or buildings in Risk Category III or IV. Through-penetration and membrane-penetration firestop systems, as well as fire-resistant joint systems and perimeter fire barrier systems, are critical to maintaining the fire-resistive integrity of fire-resistance-rated construction elements, including fire walls, fire barriers, fire partitions, smoke barriers, and horizontal assemblies. The proper selection and installation of such systems must be in compliance with the code and/or appropriate listing. Where these systems are used in two types of buildings considered as "high risk," they must be included in the special inspection program.

1705.16

Although the proper application of firestop and joint system requirements is very important in all types and sizes of buildings, the requirement for special inspection is limited to specific building types that represent a substantial hazard to human life in the event of a system failure or that are considered to be essential facilities. Inspection conforming to ASTM E 2174 for penetration firestop systems and ASTM E 2393 for fire-resistant joint systems increases the level of quality assurance.

Penetration firestops. A primary method of addressing a penetration of a fire-resistance-rated wall assembly is through the use of an approved firestop system installed in accordance with ASTM E 814 or UL 1479. The system must have an F rating that is not less than the fire-resistance rating of the wall being penetrated. It is critical that the firestop system be appropriate for the penetration being protected. The choice of firestop systems varies based on the size and material of the penetrating item, as well as the construction materials and fire-resistance rating of the wall being penetrated. Special inspection of the firestop system is intended to verify that the appropriate system has been specified and the installation is in conformance with its listing.

1705.16.1

Fire-resistant joint systems. A joint is defined as a "linear opening in or between adjacent fire-resistance-rated assemblies that is designed to allow independent movement of the building in any plane caused by thermal, seismic, wind or any other loading." The joint creates an interruption of the fire-resistant integrity of the wall or floor system, requiring the use of an appropriate fire-resistant joint system. The code mandates general installation criteria for such systems and requires them to be tested in accordance with ASTM E 1966 or UL 2079. Much like the inspection of penetration firestop systems, the proper choice and installation of fire-resistant joint systems can be verified through a comprehensive special inspection process. Although regulated under the provisions for fire-resistant joint systems, a second type of system is technically not a joint, but rather an extension of protection afforded by a horizontal assembly. The void created at the intersection of an

1705.16.2

exterior curtain wall assembly and a fire-resistance-rated floor or floor-ceiling assembly must be filled in a manner that maintains the integrity of the horizontal assembly. The system utilized to fill the void must be in compliance with ASTM E 2307 and able to resist the passage of flame for a time period equal to that of the floor assembly. Special inspection is necessary to ensure that the appropriate joint system is chosen and properly installed.

1705.17 Special inspection for smoke control. This section, although related to mechanical systems rather than structural or architectural systems, is required by the code because the mechanical ductwork and signaling devices are likely to be concealed during the building construction, and the ductwork needs to be leakage tested prior to concealment.

1705.17.1 Testing scope. The special inspection applies to leakage testing of the ductwork prior to concealment and overall system performance of the completed system prior to occupancy.

1705.17.2 Qualifications. The special inspection agencies that perform the testing are required to have specific expertise in fire protection or mechanical engineering and air balancing. The Associated Air Balance Council (AABC) is the certifying association for air balance testing. AABC has certifications for technicians and equipment commissioning.

Section 1706 Design Strengths of Materials

This section requires the strength of structural materials to conform to the applicable design standards, or in the absence of such standards, must conform to accepted engineering practice.

1706.2 New materials. Materials not explicitly covered by the code are allowed, subject to testing that demonstrates adequate performance. Section 1707 provides alternative test procedures, and Section 104.11 provides administrative means of accepting alternative methods of design and construction.

Section 1707 Alternate Test Procedures

Test reports from approved agencies may be used as the basis for approval of materials not specifically covered by the code or approved standards. New materials not covered in the code are also mentioned in Section 1701.2. This section references Section 104.11 in regard to the use of new, innovative, or alternative materials. The building official has the authority to accept such materials based on reports from an approved agency that is independent from the material supplier. ICC Evaluation Service Reports are the most straightforward method used by building officials to review products and systems that are not specifically covered by the code. The cost associated with performing testing and issuing research reports is to be borne by the applicant.

Section 1708 Test Safe Load

Testing to determine a safe load is required when a structure or component cannot be designed or analyzed in accordance with accepted engineering principles and practices, or where the construction design method does not fully comply with the respective material design standards referenced by the code. In this case, the code refers to Section 1710 for preconstruction load test requirements and procedures.

Section 1709 *In Situ Load Tests*

This section covers structural analysis or load testing for existing structures or portions thereof where the load-bearing capacity or stability is in doubt.

General. When there is doubt as to the structural integrity, load-carrying capacity, or stability of an existing structure or portion thereof, the building official has the option of requiring a structural analysis or a load test, or both. If an engineering analysis is required, it should be based on the as-built conditions and actual material properties. If the structural analysis shows that the structure is not capable of safely carrying the code-design loads, the structure must be load tested. If the structure is found to be inadequate or unstable, the structural system must be modified to provide adequate structural integrity. — **1709.1**

Test standards. When load-test procedures are given in a referenced standard, those procedures should be followed. An example would be the test procedures in the SJI standard specification.[11] When the referenced standard lacks a load test procedure or there is no referenced standard for the structure or portions thereof, a registered design professional should develop a test procedure and test protocol that simulate the actual loading conditions and displacements that the structure is expected to sustain. — **1709.2**

In situ load tests. In situ tests fall into two categories: procedures specified and procedures not specified. Load testing must simulate the loading required by Chapter 16 and be performed under the supervision of the registered design professional. — **1709.3**

Load test procedure specified. When the applicable standard has a test procedure and acceptance criteria specified, the standard should be applied. — **1709.3.1**

Load test procedures not specified. When there is no applicable load test procedure in a referenced standard, the structure or portion thereof should be tested with a loading protocol that simulates the actual loads and deformations, both lateral and vertical, that the structure is expected to receive. For the gravity system, the vertical loads should be twice the unfactored design loads. Dead loads should include expected partition loads. — **1709.3.2**

Section 1710 *Preconstruction Load Tests*

This section applies to materials or methods of construction that are not capable of being designed by conventional engineering analysis or do not comply with referenced standards.

General. When the load-carrying capacity and the physical properties of materials or methods of construction are not amenable to analysis by accepted engineering methods, or the material or method does not comply with applicable standards, the structural load capacity and physical properties must be determined by tests. This section is applicable to components, assemblies, and elements of structures, for example, windows. This section is not applicable to load testing of existing structures (see discussion of Section 1714). — **1710.1**

Load test procedures specified. When an applicable standard has a test procedure and acceptance criteria, the standard should be applied. — **1710.2**

Load test procedures not specified. When there is no applicable load test procedure in a referenced standard, the structure or portion thereof should be tested with a loading protocol that simulates the applicable loading conditions and deformations, both lateral and vertical, that the structure is expected to sustain. For components, assemblies, and elements that are not part of the seismic-force-resisting system, the loading and acceptance criteria are set forth in Section 1710.3.1. — **1710.3**

1710.3.1 Test procedure. The loading procedure and acceptance criteria use commonly accepted engineering practices to test the adequacy of the component or assembly to resist structural failure at the design loads. The test load is 2 times the design load for 24 hours, and must recover at least 75 percent of the maximum deflection within 24 hours after the test. The test assembly is then reloaded until failure occurs or 2½ times the load corresponding to the maximum allowable deflection is reached (see Section 1604.3) or 2½ times the design load is reached. This procedure should be used only for loads for which the upper bounds can be established with a reasonable degree of accuracy such as dead and live loads. This procedure should not be used for earthquake loads and should be used with care for wind loads.

1710.3.2 Deflection. Deflection under design load is limited to the allowable limits set forth in Section 1604.3.

1710.4 Wall and partition assemblies. Walls and partitions must be tested for simultaneous vertical and lateral loads, both with and without door and window framing.

1710.5 Exterior window and door assemblies. Door and window assemblies are generally qualified by the tests specified in Section 1710.5.1, which requires testing and labeling in accordance with the AAMA/WDMA/CSA standard, or tested in accordance with Section 1715.5.2 and the referenced ASTM and DSMA standards. The exception permits allowable wind pressures for smaller units made of identical components, including glass thickness, to be higher as determined by engineering analysis, provided that an additional test is done on the assembly with the highest pressure to validate the analysis.

1710.5.1 Exterior windows and doors. Exterior windows and sliding doors must be tested and labeled as conforming to the American Architectural Manufacturers Association/Window and Door Manufacturers Association standard, AAMA/WDMA/CSA101/I.S.2/A440. Side-hinged exterior doors can comply with the standard or meet the requirements of Section 1710.5.2, as discussed below. The products tested and labeled in accordance with the AAMA/WDMA/CSA standard need not comply with the analysis and testing requirements of Section 2403.2 and the deflection limitations specified in Section 2403.3.

1710.5.2 Exterior windows and door assemblies not provided for in Section 1715.5.1. Exterior window and door assemblies not covered by Section 1715.5.1 must be tested in accordance with ASTM E 330. The test load is equal to 1.5 times the design pressure, as determined per Chapter 16, applied for 10 seconds. Exterior window and door assemblies covered by this section and containing glass are required to conform to the requirements of Section 2403. Specific requirements for structural performance of garage doors were added to the 2009 IBC by reference to ASTM E 330 and ANSI/DASMA 108 for sectional garage doors and rolling doors.

1710.6 Skylights and sloped glass. A tubular daylighting device (TDD) is typically field-assembled from a manufactured kit, unlike a unit skylight, which is typically shipped as a factory-assembled unit. The dome of a TDD is not necessarily constructed out of a single panel of glazing material. Thus, separate definition of a TDD was added to Chapter 2 based on the definition in AAMA/WDMA A440. The section refers to Section 2405 for sloped unit skylight and TDD requirements.

1710.7 Test specimens. Test specimens should be representative of what is actually used in practice. Tests must be conducted or witnessed by an independent approved agency.

Section 1711 *Material and Test Standards*

This section sets forth the test standards and acceptance criteria for joist hangers and concrete and clay roof tiles.

Material and Test Standards

1711.1 Joist hangers. Testing requirements for joist hangers are given in this section based on ASTM D1761 and the AWC NDS.

1711.1.1 General. Because the criteria limit the specific gravity of the species of wood between 0.49 and 0.55, testing is typically performed using specimens of Douglas Fir-Larch (specific gravity 0.50) or Southern Pine (specific gravity 0.55) in accordance with ASTM D 1761. The torsional moment capacity test uses a 24-inch long joist. The vertical load capacity test has a minimum joist length but no upper limit on the length of the joist, so the exception provides that the joist length need not exceed 24 inches.

1711.1.2 Vertical load capacity for joist hangers. This section sets forth the test and acceptance criteria for testing joist hangers for vertical loads. The allowable vertical load is the lowest value obtained from the five items listed. Note that Items 1 and 2 are the strength limit states of the hanger, Item 3 is a deflection limit state, and Items 4 and 5 are design load limit states for the fasteners and wood, respectively.

1711.1.2.1 Design value modifications for joist hangers. Only the design values based on the allowable design load for the fasteners or wood member (Item 4 or 5 in Section 1711.1.2) are allowed to be increased for the duration of load factors specified in the NDS, but may not exceed the capacity determined by Item 1, 2, or 3. No duration of load increase is allowed for capacities determined by ultimate load tests or limited by deflection.

1711.1.3 Torsional moment capacity for joist hangers. The allowable torsional moment capacity is determined by testing an assembly of three hangers in accordance with ASTM D 1761 based on a rotation corresponding to $1/8$-inch lateral movement of the top or bottom of the joist.

1711.2 Concrete and clay roof tiles. Concrete and clay roof tiles are required to be designed to resist overturning from component and cladding wind pressures.

1711.2.1 Overturning resistance. This section references ICC's SBCCI SSTD 11 *Test Standard for Determining Wind Resistance of Concrete or Clay Roof Tiles*, and Chapter 15 (see Section 1504.1), which references Section 1609.5. Section 1609.5 requires the roof deck to be capable of resisting component and cladding pressures determined by ASCE 7. Note that SBCCI SSTD 11 is an ICC standard as shown in Chapter 35.

1711.2.2 Wind tunnel testing. Roof tiles that do not meet the requirements in Section 1609.5.2 for rigid tiles require wind tunnel tests in accordance with the SBCCI SSTD 11 standard to determine performance and response characteristics.

Overview of Chapter 17

Chapter 17 is designed to improve the construction quality of structural systems through the following requirements:

- Special inspection is required as prescribed in Section 1704. The various types of structural materials, elements, and systems that require verification are listed in Section 1705.
- The registered design professional must provide a detailed statement of special inspections in accordance with Section 1704.3.
- Additional special inspections for wind resistance may be required in accordance with Section 1705.10.
- Additional special inspections for seismic resistance may be required in accordance with Section 1705.11.
- Specific structural testing and qualifications for seismic resistance may be required in accordance with Section 1705.12.

17 Structural Tests and Special Inspections

- The contractor responsible for constructing the wind or seismic-force-resisting system or component listed in the statement of special inspection must provide a statement of responsibility in accordance with Section 1704.4.

- Structural observation by a registered design professional may be required in accordance with Section 1704.5.

- Special inspection and verification for various types of structural elements and systems prescribed in Section 1705 include steel (Section 1705.2), concrete (Section 1705.3), masonry (Section 1705.4), wood construction (Section 1705.5), soils (Section 1705.6), driven deep foundations (Section 1705.7), cast-in-place deep foundations (Section 1705.8), and helical pile foundations (Section 1705.9). In addition, special inspection is required for special cases such as alternate materials and unusual designs, as prescribed in Section 1705.1.1.

- The requirements for special inspection for wind resistance are triggered by the nominal design wind speed and exposure category of the building. The requirements for special seismic inspection and seismic testing are triggered by the seismic design category of the building.

- The requirements for structural observation are triggered by the seismic design category of the building and the assigned nominal design wind speed.

Example Problem

A new fire station building is proposed in Grass Valley, California. The building will be constructed of fully grouted reinforced concrete masonry unit (CMU) walls with a steel truss roof system. What are the specific quality assurance requirements for the masonry construction required by Chapter 17 of the 2012 IBC?

Solution: Because many of the special inspection, structural observation, and structural testing requirements of the IBC are based on the nominal design wind and speed seismic design category, the first step is to determine these parameters for the proposed building. In order to determine the proper wind speed map to use, we first have to determine the risk category of the building.

- **Risk category.** Table 1604.5 indicates that a fire station is in Risk Category IV.

- **Nominal design wind speed.** Figure 1609B for Risk Category IV indicates that the ultimate 3-second-gust wind speed for Grass Valley, California, is 115 mph. Using Table 1609.3.1 and interpolating, the nominal design wind speed for the site is 89 mph. Therefore, there are no special requirements in Chapter 17 based on wind speed.

- **Seismic Design Category.** The latitude and longitude of the building site are determined from the Grass Valley USGS quadrangle map as follows:

    Latitude = 39.219 degrees, Longitude = −121.060 degrees.

- From the USGS website program for the 2012 IBC/ASCE 7-10 at http://earthquake.usgs.gov/hazards/designmaps/, the mapped short- and long-period mapped spectral accelerations are:

    The mapped spectral acceleration for short periods, $S_S = 0.588$

    The mapped spectral acceleration for 1-second period, $S_1 = 0.243$

- These mapped spectral acceleration values are for Site Class B soils. Since S_1 is less than $0.75g$, the building is not in Seismic Design Category F (see Section 1613.3.5). Assuming Site Class D without a geotechnical report that establishes the site class is permitted under Section 1613.3.2. Because the existing soil at the site is firm, rocky soil, the seismic design category will be determined assuming Site Class D as the default soil profile. From Tables 1613.3.3(1) and 1613.3.3(2), the soil factors F_a and F_v for Site Class D and S_S and S_D are found:

  $F_a = 1.33$

  $F_v = 1.914$

- The design spectral accelerations S_{DS} and S_{D1} are $2/3$ of the soil modified spectral acceleration values as follows:

  $S_{DS} = 2/3\ (F_a\ S_S) = 2/3 \times 1.33 \times 0.588 = 0.521$

  $S_{D1} = 2/3\ (F_v\ S_1) = 2/3 \times 1.914 \times 0.243 = 0.310$

Refer to the 2003 NEHRP Commentary[12] (FEMA 450) for a discussion of the $2/3$ factor.

Entering Table 1613.3.5(1) with $S_{DS} = 0.521$ and Risk Category IV, we find the building is assigned to Seismic Design Category D. Entering Table 1613.3.5(2) with $S_{D1} = 0.310$ and Risk Category IV, we find the building is assigned to Seismic Design Category D. The building is therefore assigned to Seismic Design Category D.

Permissible design methods. Section 2109 references Section 5.1.2 of the MSJC Code, which specifically prohibits the use of the empirical design method for buildings in Seismic Design Category D, E, or F. Therefore, the proposed fire station building must be designed by the engineering provisions prescribed by either the allowable stress design procedure of Section 2107 or the strength design procedure of Section 2108. The seismic design of the building must conform to Section 2106, which references Section 1.18 of the MSJC Code.

Special inspection. Special inspection is required for all engineered masonry structures. The masonry must be inspected and verified in accordance with the Level C quality assurance requirements of Section 1.19 of TMS 402/ACI 530/ASCE 5 and TMS 602/ACI 530.1/ASCE 6.

Special inspection for seismic resistance. Although the height of the structure does not exceed 25 feet, S_{DS} exceeds 0.5 seconds so the building does not meet exception 2 of Section 1705.11, and special inspection for seismic resistance is required.

Testing and qualification for seismic resistance. Because the building is in Seismic Design Category D, elements of the seismic-force-resisting system, designated seismic system, and architectural, mechanical, and electrical components must comply with the applicable requirements of Section 1705.12. The designated seismic system consists of those nonstructural components that require design in accordance with Chapter 13 of ASCE 7 with a component importance factor, I_p, greater than 1 in accordance with Section 13.1.3 of ASCE 7. Reinforcement and structural steel must conform to Sections 1705.12.1 and 1705.12.2, respectively.

Contractor responsibility. The contractor responsible for construction of the seismic-force-resisting system and components listed in the statement of special inspections must submit a statement of responsibility to the owner and building official in accordance with Section 1704.4.

Structural observation. Because the building is classified as Risk Category IV and assigned to Seismic Design Category D, Section 1704.5.1 requires structural observation as defined in Section 202.

17 Structural Tests and Special Inspections

Summary of Requirements

- The proposed fire station is in Risk Category IV in accordance with Table 1604.5.
- The seismic design category of the building was determined to be Seismic Design Category D based on assumed Site Class D soil conditions.
- The concrete masonry structural system must be designed in accordance with the engineering provisions prescribed by the allowable stress design procedure of Section 2107 or the strength design procedure of Section 2108.
- As a condition of permit issuance, the registered design professional must provide a statement of special inspections to be submitted by the permit applicant in accordance with Section 1704.3.
- Special inspection must be provided for the masonry construction in accordance with Level C quality assurance requirements of Section 1.19 of the MSJC Code.
- Special inspection for seismic resistance is required for the seismic-force-resisting system, the designated seismic system, and architectural, mechanical, and electrical components in accordance with Sections 1705.11
- The seismic-force-resisting system, the designated seismic system, and architectural, mechanical, and electrical components listed in the statement of special inspections must conform to the testing and qualification for seismic resistance requirements in accordance with Section 1705.12.
- Structural observation for seismic resistance in accordance with Section 1704.5 is required.
- The contractor responsible for construction of the main wind and seismic-force-resisting systems, the designated seismic system, and components listed in the statement of special inspections must submit a statement of responsibility to the owner and building official prior to commencement of work on a particular structural system or component in accordance with Section 1704.4.

KEY POINTS

- Chapter 17 provides requirements for quality assurance for construction of buildings and other structures regulated by the IBC through special inspection and verification, structural testing, and structural observation.
- Specific provisions for approvals, approved agencies, records, labeling, and testing are provided in Section 1703.
- General requirements for special inspections, contractor responsibility, structural observation, fabricator approval, and the statement of special inspections are provided in Section 1704.
- Specific items that require verification by the special inspector are outlined in Section 1705 including special cases, steel, concrete, masonry, wood, soils, and deep foundations.
- Specific verifications are required for wind resistance based on the nominal design wind speed at the site and for seismic resistance based on the seismic design category of the building.
- Special testing and qualification for seismic resistance is required for certain structural elements and architectural, mechanical, and electrical components based on the seismic design category of the building.
- Requirements for alternative test procedures, safe load testing, in situ load testing, preconstruction load testing, and testing requirements for joist hangers are provided.

REFERENCES

1. RCSC-04, *Specification for Structural Joints Using A325 or A490 Bolts (June 30, 2004)*, Research Council on Structural Connections, Chicago, IL, 2004.

2. ACI 318-11, *Building Code Requirements for Structural Concrete*, American Concrete Institute, Farmington Hills, MI, 2011.

3. *Model Program for Special Inspection*, International Code Council, Washington, DC.

4. AWS D1.1-04, *Structural Welding Code—Steel*, American Welding Society, Miami, FL, 2004.

5. AISC 360-10, *Specification for Structural Steel Buildings*, American Institute for Steel Construction, Inc., Chicago, IL, 2010.

6. AISC 341, *Seismic Provisions for Structural Steel Buildings*, American Institute of Steel Construction, Chicago, IL, 2010.

7. TMS 402/ACI 530/ASCE 5, *Building Code Requirements for Masonry Structures*, American Concrete Institute, Farmington Hills, MI, 2011.

8. TMS 602/ACI 530.1/ASCE 6, *Specification for Masonry Structures*, American Concrete Institute, Farmington Hills, MI, 2011.

9. Technical Manual 12-B, 2nd Edition; *Standard Practice for the Testing and Inspection of Field Applied Thin Film Intumescent Fire-Resistive Materials*; an Annotated Guide, The Association of the Wall and Ceiling Industries International, Falls Church, VA, 1998.

10. Technical Manual 12-A, 3rd Edition; *Standard Practice for the Testing and Inspection of Field Applied Sprayed Fire-Resistive Materials*; an Annotated Guide, The Association of Wall and Ceiling Industires International, Falls Church, VA, 2012.

11. SJI, *Standard Specification for Joist Girders, Open Web Steel Joists (K-Series), Longspan Steel Joists (LH Series) and Deep Longspan Steel Joists (DLH Series)*, Steel Joist Institute, Myrtle Beach, SC, 2005.

12. NEHRP, *NEHRP (National Earthquake Hazard Reduction Program) Recommended Provisions for New Buildings and Other Structures (FEMA 450)*, Building Seismic Safety Council, Washington, DC, 2003.

CHAPTER 18

SOILS AND FOUNDATIONS

Introduction
Section 1801 General
Section 1802 Definitions
Section 1803 Geotechnical Investigations
Section 1804 Excavation, Grading, and Fill
Section 1805 Dampproofing and Waterproofing
Section 1806 Presumptive Load-Bearing
 Values of Soils
Section 1807 Foundation Walls, Retaining Walls,
 and Embedded Posts and Poles
Section 1808 Foundations
Section 1809 Shallow Foundations
Section 1810 Deep Foundations
Key Points
References
Bibliography

18 Soils and Foundations

Introduction

Chapter 18 was completely reorganized in the 2009 *International Building Code*® (IBC®) to facilitate proper application of the provisions. Some notable elements of the restructured Chapter 18 format include categorization of foundations into two main types of shallow and deep foundations, the collection of the common provisions for piles and piers, such as installation, under specific sections for deep foundations and neatly organizing other previously scattered requirements, such as material strength requirements for concrete and grout and stresses, in two new tables rather than being scattered in various sections of the code. Overall the reorganization of Chapter 18 is a big improvement over previous editions of the IBC and should result in better understanding of the intent of the code. Relatively few significant changes were made to the 2012 IBC. Changes were made to Sections 1803.5.11 and 1803.5.12 regarding seismic requirements for geotechnical reports in areas of high seismic risk and to Section 1810.3.11.2 related to seismic design of pile caps.

Satisfactory performance of the building structural system is critical in the overall performance of any building during its life cycle. The 2012 IBC structural provisions in Chapters 16, 17, 18, 19, 20, 21, 22, and 23 work together to provide for the overall performance and safety of the structural system. Section 1604.4, Analysis, requires that all structural systems have a continuous load path from the point of origin to the resisting element. The resisting element for buildings and other structures is the foundation that ultimately transfers the loads to the supporting soil. Therefore, the satisfactory performance of the foundation system is critical to the satisfactory performance of the overall structure. Most building structures are designed assuming a fixed unyielding base that is not subjected to large total or differential settlements or displacements. Shallow foundations on firm soils will generally perform satisfactorily if the requirements of these provisions are followed.

Foundation design, however, becomes a significant factor for large structures, embedded structures such as a tall building constructed over a multilevel basement garage, structures on soft soils, structures supporting rotating or reciprocating equipment, and structures sensitive to differential displacements. It is important to have a good knowledge of the behavior of the various foundation types, including their limitations. In addition, for structures subject to high-wind forces or seismic ground motion, special consideration must be given to the lateral load path, and, in the case of deep foundation supported structures, the ability of the deep foundation to survive the displacements and curvatures imposed on the pile by seismic ground motion.

Sufficient understanding of the behavior and limitations of the various deep and shallow foundation systems is necessary to determine that the foundation and the supported structure will provide the intended serviceability. It is important to determine whether the estimated total and differential settlements of the foundation are compatible with the selected structure type. For example, a stiff bearing-wall structure with openings may be more sensitive to differential settlements than a more flexible light-frame structure. When considering seismic ground motion, sufficient knowledge of ground-shaking effects on the foundation is important, particularly in soft soils.

Section 1801 General

1801.1 Scope. The requirements of this chapter apply to all building and foundation systems of any type and any location. For example, buildings located in beach-front properties that would be subject to wave run up and inundation during hurricanes can still be designed using applicable loads from ASCE 7 (Minimum Design Loads for Buildings and Other Structures) and ASCE 24 (Flood Resistant Design and Construction) with proper geotechnical investigation and engineering analysis. ASCE 7 Chapter 5 and ASCE 24 are referenced in Section 1612.4 for the design and construction of buildings and structures in flood hazard areas, including flood hazard areas subject to high velocity wave action.

Design basis. Bearing pressures, stresses, and lateral pressures used in this chapter are allowable pressures or stresses, not strength level values. These allowable foundation pressures are to be used with the allowable stress design load combinations set forth in Section 1605.3 unless noted otherwise. **1801.2**

Section 1802 *Definitions*

Definitions. Specific definitions applicable to Chapter 18 are referenced in this section and include the following terms: Deep Foundation, Drilled Shaft, Socketed Drilled Shaft, Helical Pile, Micropile, and Shallow Foundation. Definitions of terms commonly used in foundation design can be found in various foundation engineering references.[2–6] All chapters of the IBC will now show, in italics, any term that is defined if the definition is applicable to the topic being presented. All definitions are located in Chapter 2 in the 2012 IBC. **1802.1**

For example, terms such as Approved, Building Official, and Allowable Stress Design will be shown in italics.

Section 1803 *Geotechnical Investigations*

General. A geotechnical investigation must be conducted when required by the building official. In general, the investigation should be required unless the foundation is designed and constructed in accordance with the presumptive allowable foundation pressures and lateral-bearing pressures set forth in Section 1806. Some minimal knowledge of soil classification at the bearing elevation of the foundation is required by Section 1806. A registered design professional should be used as required by the professional practice laws of the state in which the jurisdiction is located. The practice of geotechnical or soils engineering is a branch of civil engineering and is generally regulated by the various states. **1803.1**

Investigation required. A foundation and soils investigation is required for any of the adverse subsurface conditions listed in Sections 1803.5.2 through 1803.5.12. **1803.2**

In certain cases, an exception allows the building official some flexibility in requiring a foundation and soils investigation when the soil conditions of the site are already known from other soils reports. For example, if the site is located in an area where there are reasonably uniform and horizontal soil strata and a soils report is available for the adjacent parcels, then a new soils report should not be necessary. This exception does not apply to some very specific conditions such as analysis of lateral pressure, liquefaction potential, and ground stabilization techniques in Seismic Design Categories (SDCs) D, E, and F. Other conditions where the exception does not apply and a geotechnical investigation is mandated are excavation near foundations, compacted fill materials, and controlled low-strength materials (CLSMs).

Soils and Foundations

1803.3 Basis of investigation. The investigation cannot be solely based on theoretical analysis and research; rather, analysis and observation through tests of materials by borings, test pits, or other subsurface explorations in appropriate locations is needed. The number and types of tests, equipment used, type of site inspections, and other issues relevant to the scope of the investigation are the responsibility of and must be under the supervision of a registered design professional experienced in soils exploration. Additional studies shall also be made as necessary for conditions such as slope stability, soil strength, position and adequacy of load-bearing soils, the effect of moisture variation on soil-bearing capacity, compressibility, liquefaction, and expansiveness.

1803.4 Qualified representative. Whenever the allowable bearing capacity is in doubt or a geotechnical investigation is necessary, exploratory borings are necessary to determine the soil characteristics and the load-bearing capacity. The investigation procedures must be outlined and identified by the registered design professional in accordance with the accepted engineering practice and acceptable standards. The apparatus and equipment used in the investigation must be calibrated and perform as expected for accurate results. Special inspection requirements for soil-related activities are provided in Chapter 17 of the IBC. Qualified representatives must be present on site during the boring or sampling operations. Qualifications of investigative or inspection agencies and individuals are typically established by their experience, certification, and accreditation through an approved accreditation body such as the International Accreditation Service (IAS).

1803.5 Investigated conditions. Those cases where geotechnical investigations are required have been provided in Sections 1803.5.2 through 1803.5.12. It should be emphasized that the exception discussed in Section 1803.2 that allows the building official to waive the investigation can be used in most of these conditions except those that are critical or are site specific and cannot be determined based on existing geotechnical investigations of surrounding areas (lateral pressure, liquefaction potential, and ground stabilization techniques in SDCs D, E, and F, excavation near foundations, compacted fill materials, and CLSMs). Soil materials are required to be classified in accordance with the Unified Soil Classification System found in ASTM D 2487 (Practice for Classification of Soils for Engineering Purposes) (Unified Soil Classification System). IBC Table 1610.1 provides the Unified Soil Classification for some backfill materials. This table establishes the minimum soil lateral loads for the design of foundation walls and retaining walls.

1803.5.2 Questionable soil. Where the classification, strength, or compressibility is uncertain, or where bearing capacity of soil in excess of the presumptive value is claimed, the code allows the *building official* to obtain a geotechnical report. In this section, two cases trigger a foundation soils investigation and report:

1. Where the design load-bearing value is greater than the presumptive allowable foundation pressures and lateral-bearing pressures set forth in Section 1806.
2. Where the type of soil, the bearing capacity, or the stiffness of the soil is questionable, such as in areas subject to liquefaction from strong ground shaking, in areas containing soft or sensitive clays such as bay muds, or in areas with unconsolidated or improperly consolidated fills.

1803.5.3 Expansive soil. Expansive soils are those that shrink and swell appreciably because of changes in soil moisture content. Reference should be made to Section 1808.6 for mitigation methods and design for expansive soils. Frost heave is not considered in this section. Expansive soils are present or prevalent in all 50 states of the United States and in many other countries around the globe. Soils must meet all four of the criteria to be classified as expansive, not just a high Plasticity or Expansion Index.

The section allows two different ways to identify expansive soils. The first option involves meeting four criteria: (1) Plasticity Index (PI) of 15 or greater determined by ASTM D 4318; (2) more than 10 percent of the soil particles pass a No. 200 sieve determined by ASTM D 422; (3) more than 10 percent of the soil particles are less than 5 micrometers in size as determined by ASTM D 422; and (4) an Expansion Index greater than 20 according to ASTM D 4892. The second option is to determine if the Expansion Index (according to ASTM D 4892) is greater than 20. If the Expansion Index is determined, the first three tests need not be conducted. In other words, Expansion Index > 20 is both a necessary and sufficient condition by itself.

1803.5.4 **Ground-water table.** This section may cause significant changes to the conventional approach to foundation design and construction for light-frame buildings with subsurface floors, either a basement or a hillside building with a floor cut into the hillside. A foundation and soils investigation is required to show that the ground-water table is at least 5 feet below the elevation of the lowest floor level unless waterproofing is provided in accordance with Section 1805.

1803.5.5 **Deep foundations.** The purpose of a geotechnical investigation is to define the general subsurface stratifications of soil and rock materials, determine the soil and rock profiles, and locate the ground-water table. Such information will help in selecting the type of deep foundations and in estimating deep element lengths. Furthermore, a geotechnical investigation is often required to render data on specific soil properties, such as shear strength, relative density, compressibility of the soil, and other such findings that will help in analyzing subsurface conditions for determining design loads, type of deep foundations, driving criteria, suitable bearing strata, and probable durability of foundation materials relative to the particular soil conditions found at the construction site. It is generally not economical or feasible to use deep foundations without a geotechnical investigation and report.

Subsurface information is normally obtained by means of test borings that yield suitable samples of soils and rock and give the depths from which they are obtained. Sometimes, certain in situ tests are also conducted, but these should not be made without first making test borings.

This section outlines the kinds of information to be derived from a geotechnical investigation and report. In addition to the items listed in the code provision, the investigation could include other valuable and applicable data that would help in the evaluation, such as:

- Information on existing construction at the site or at adjacent sites, including the type and condition of these structures, age, types of foundations used, performance data, and so on.

- Information on the existence of deleterious substances in the soils or other conditions that could seriously affect the durability and structural performance of the piles.

- Information on the geologic conditions at the site, which could include such items as the existence of mines, earth cavities, underground streams or other adverse water conditions, history of seismic activity, and so on.

1803.5.6 **Rock strata.** If a rock stratum is being used for bearing, the characteristics of the layer must be known in sufficient detail to classify the rock per Section 1803.5.1.

1803.5.7 **Excavation near foundations.** The intent of this section is clear in that lateral and subjacent support of any foundation must not be removed unless an investigation is conducted to identify the potential problems that might be created and how to provide the needed lateral support. For example, the area shown in Figure 1803-1 should not be excavated without conducting an investigation to address support for the foundation.

1803.5.8 **Compacted fill material.** When compacted fill of more than 12 inches (305 mm) in depth is used for shallow foundation support, the geotechnical investigation required in Section 1803 must also contain the seven items listed in this section.

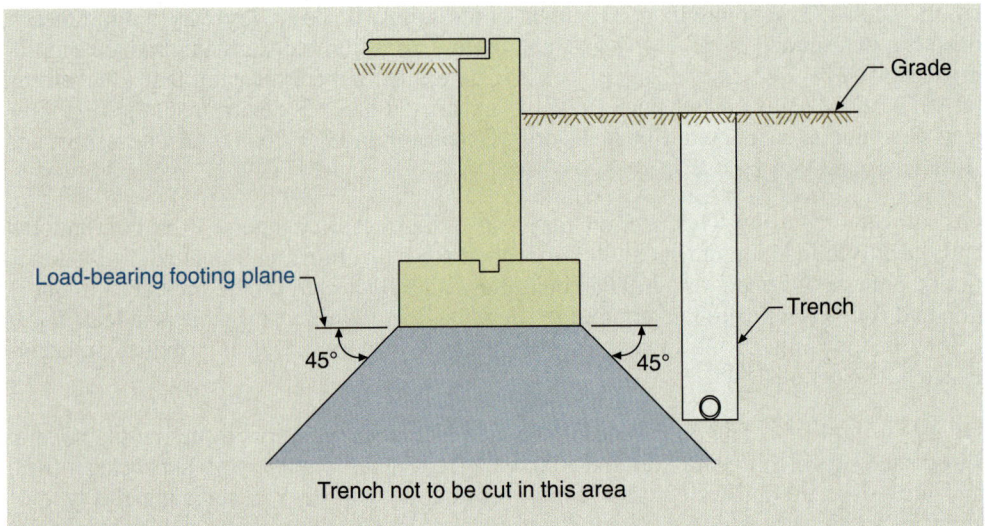

Figure 1803-1 Lateral support.

1803.5.9 Controlled low-strength material (CLSM). CLSM was introduced into the 2003 IBC as an acceptable backfill material that need not be compacted. Prior to the 2003 IBC, CLSM would need to be approved under the alternative materials, design, and methods of construction provisions in Section 104.11. It is a self-compacted, cementitious material used as backfill instead of compacted backfill. It has also been referred to as "flowable fill" and "lean mix backfill." CLSM has compressive strength of 1200 psi or less. Most CLSM applications require unconfined compressive strengths of 200 psi or less to allow for future excavation of CLSM. CLSM is composed of water, portland cement, aggregate, and fly ash. It is a fluid material with typical slumps of 10 inches or more and has a consistency similar to a milk shake. Although there is no referenced standard for CLSM in the IBC, there is a national report promulgated by the ACI Committee 229, entitled "ACI 229R-99 Controlled Low-Strength Materials." The ACI standard was reapproved in 2005 and a new edition is expected to be published in 2012.

1803.5.10 Alternate setback and clearance. This alternate procedure allows the building official to approve alternate setbacks and clearances from slopes, provided that the intent of Section 1808.7 is met. This section gives the building official the authority to require a geotechnical investigation by a qualified geotechnical engineer to establish that the intent of Section 1808.7 is met and specifies the minimum parameters to be investigated.

1803.5.11 Seismic Design Categories C through F. For all structures in SDCs C through F, a geotechnical investigation is required to evaluate liquefaction, slope stability, total and differential settlement, and surface displacement caused by faulting or lateral spreading. Significant ground motion can occur even in areas with moderate seismic risk. Liquefaction typically occurs at sites with loose sands and high water tables. Surface rupture generally occurs with large-magnitude earthquake events. Lateral spreading generally occurs adjacent to waterways where a saturated soil has a free edge.

1803.5.12 Seismic Design Categories D through F. In addition to the investigation required for SDCs C through F by Section 1803.5.11, the investigation must evaluate the additional lateral pressures on basement or retaining walls from ground shaking, the detrimental effects and consequences of liquefaction or soil strength loss, and mitigation measures. The potential for liquefaction and soil strength loss must be evaluated using the peak acceleration based on a site-specific study and including the effects of soil amplification.

Liquefaction causes loss of bearing capacity with resulting large differential or total settlements. Structures with high height-to-width ratios that have liquefaction occur under a portion of the structure are subject to overturning. Many of the structures in various Japanese earthquakes were damaged by liquefaction.

Mitigation measures for liquefaction include:

1. In situ densification of the loose sands subject to liquefaction.

2. Use of pile foundations penetrating through the liquefying layers. The pile capacity must be developed through bearing or skin friction in the soils below the liquefying layers.

3. Use of rigid raft foundations that can minimize the effects of settlement caused by loss of bearing capacity.

Soil strength loss is associated with "sensitive" or "quick" clays that are sensitive to remolding effects. These soft clays typically occur in marine (or former marine) environments and are often called "*bay muds*." The clays lose significant strength when remolded. Remolding and subsequent strength loss occurs when a pile foundation through these clays is subjected to strong ground shaking. As a consequence, lateral support for the pile is lost.

Mitigation measures for soils susceptible to strength loss such as "sensitive" or "quick" clays include:

1. Replacement of the soft clay.

2. In situ consolidation of the soft clays by preloading or water removal.

3. Use of pile foundations penetrating through the soft clay layers. The pile capacity must be developed through bearing or skin friction in the soils below the soft clay layers.

When using pile foundations, the effects of the layers of liquefaction or soils susceptible to strength loss on the curvature (and hence, moment) demands on the pile must be investigated. Often these soil-induced curvatures place a much higher moment demand on the piles than would be determined from conventional lateral-force P-y analyses. The curvatures are significantly increased at the interface between the soft soils and stiffer soils. The curvature demands increase as the ratio of stiffness of the stiff layer to the soft layer increases.

The requirement that geotechnical reports address earthquake loads on foundation walls and retaining walls in SDCs D, E, and F (see Item 1) has been modified in the 2012 IBC so that it only applies to those walls supporting more than 6 feet of backfill. In the previous editions of the IBC, there was no exemption based on the height of the wall or the amount of soil supported by the wall. This was deemed to be overly restrictive for light-frame foundation walls, small retaining walls, and swimming pools. Evidence from recent earthquakes and recent experimental research results, including work recently completed at the University of California–Berkeley, has demonstrated that retaining wall structures must move in order to develop the failure wedge postulated in the so-called Mononobe and Okabe method. However, the postulated condition can only occur when the wall has already failed due to other causes. The current body of field evidence does not provide any evidence for the existence of this mechanism of failure. It was determined that the requirement in the 2009 IBC and ASCE 7-05 imposed an unjustifiable burden on the permit applicant to investigate a site for small retaining structures such as foundation walls, retaining walls, and swimming pools that support no more than 6 feet of backfill.

Reporting. The objective of a geotechnical investigation is to produce all the necessary information for design and construction of a structure's foundation. To ensure that the report of the foundation and soils investigation meets this objective and provides enough data to ensure compliance with the code and a safe foundation, the report must contain at a minimum the items enumerated in this section. See also report requirements for compacted

1803.6

18 Soils and Foundations

fill in Section 1803.5.8 and for CLSM in Section 1803.5.9. Other data such as results of consolidation tests, compaction curves, sieve analyses, Atterberg Limits tests, Plasticity and Expansion Index tests, and so on should be included in the report where available.

Section 1804 Excavation, Grading, and Fill

1804.1 **Excavations near foundations.** See discussion under Section 1803.5.7.

1804.2 **Placement of backfill.** Backfill should be performed in accordance with an approved soils report. If no soils report is needed, the backfill should be placed in maximum 6-inch layers free from any rocks or cobbles larger than 4 inches and compacted to a minimum 90-percent relative density (Modified Proctor per ASTM D 1557) using the appropriate compaction equipment. The 2003 IBC introduced the use of CLSM as an acceptable backfill material that need not be compacted. See further discussion under Section 1803.5.9.

1804.3 **Site grading.** The general requirement is that surface water must drain away from foundations. Minimum slope is 5 percent for a distance of 10 feet. An alternate method is permitted if physical obstructions or lot lines prohibit the 5-percent slope for a minimum of 10 feet horizontally. In this case, swales or impervious surfaces must have a minimum 2-percent slope where located within 10 feet of the building foundation.

1804.4 **Grading and fill in flood hazard areas.** In general, grading is not permitted in designated flood hazard areas. Changes in the configuration or shape of floodways by grading or fill can divert erosive flows and increase wave energies that could increase forces and adversely affect adjacent buildings and structures. The restrictions in the code are consistent with provisions related to fill in ASCE 24, *Flood Resistant Design and Construction.* The exceptions permit grading in flood hazard areas, provided an engineering analysis demonstrates that the proposed grading will not increase flood levels or otherwise adversely affect the design flow. The last exception allows grading in flood hazard areas that are not designated floodways, provided the overall effect of the encroachment does not increase the design flood elevation by more than 1 foot at any point.

1804.5 **Compacted fill material.** See discussion under Section 1803.5.8.

1804.6 **Controlled low-strength material (CLSM).** See discussion under Section 1803.5.9.

Section 1805 Dampproofing and Waterproofing

This section covers the requirements for waterproofing and dampproofing those parts of substructure construction that need to be provided with moisture protection. Sections 1805.1 through 1805.3 identify the locations where moisture barriers are required and specify the materials to be used and the methods of application. The provisions also deal with subsurface water conditions, drainage systems, and other protection requirements.

Dampproofing requirements are outlined in Section 1805.2, and waterproofing requirements are covered in Section 1805.3. Although both terms are intended to apply to the installation and the use of moisture barriers, dampproofing does not furnish the same degree of moisture protection as does waterproofing.

Dampproofing generally refers to the application of one or more coatings of a compound or other materials that are impervious to water, which are used to prevent the passage of water vapor through walls or other building components, and which restrict the flow of water under slight hydrostatic pressure. Waterproofing, on the other hand, refers to the application of coatings and sealing materials to walls or other building components to prevent moisture from penetrating in either a vapor or liquid form, even under conditions of significant hydrostatic pressure. Hydrostatic pressure is created by the presence of water under pressure. This pressure can occur when the ground-water table rises above the bottom of the foundation wall, or the soil next to the foundation wall becomes saturated with water caused by uncontrolled storm water runoff.

1805.1 General. This section is an overall requirement specifying that waterproofing and dampproofing applications are to be made to horizontal and vertical surfaces of those below-ground spaces where the occupancy would normally be adversely affected by the intrusion of water or moisture. Moisture or water in a floor below grade can cause damage to structural members such as columns, posts, or load-bearing walls, as well as pose a health hazard by promoting growth of bacteria or fungi. Moisture can adversely affect any mechanical and electrical appliances that may be located at that level. It can also cause a great deal of damage to goods that may be located or stored in that lower level. These vertical and horizontal surfaces include foundation walls, retaining walls, underfloor spaces, and floor slabs. Waterproofing and dampproofing are not required in locations other than residential and institutional occupancies where the omission of moisture barriers would not adversely affect the use of the spaces. An example of a location where waterproofing or dampproofing would not be required is in an open parking structure, provided the structural components are individually protected against the effects of water. Waterproofing and dampproofing are not permitted to be omitted from residential and institutional occupancies where people may be sleeping or services are provided on the floor below grade. A person waking in a flooded basement may find themselves in a very hazardous situation particularly if the possibility exists of an electrical charge in the water caused by electrical service at that level.

Section 1805.1.1 addresses the type of problem faced when a portion of a story is above grade, whereas Section 1805.1.2 limits any infiltration of water into crawl spaces so as to protect this area from potential water damage and prevent ponding of water in the crawl space. These sections reference other applicable sections of Chapter 18.

1805.1.1 Story above grade plane. The provisions of this section require that where a basement is deemed to be a story above grade plane, the section of the basement floor that occurs below the exterior ground level and the walls that bound that part of the floor are to be dampproofed in accordance with the requirements of Section 1805.2.

The use of dampproofing, rather than waterproofing, is permitted here because high hydrostatic pressure will not tend to develop against the walls if the basement is a story above grade plane and the ground level adjacent to the basement wall is below the basement floor elevation for not less than 25 percent of the basement perimeter.

Any water pressure that may occur against the walls below ground or under the basement floor would be relieved by the water drainage system required in this section. The drainage system would be installed at the base of the wall construction in accordance with Section 1805.4.2 for a minimum distance along those portions of the wall perimeter where the basement floor is below ground level.

Because of the relationship of grade to the basement floor and the inclusion of foundation drains, the potential for hydrostatic pressure buildup is not significant. Therefore, a ground-water table investigation, waterproofing, and a basement floor gravel base course are not required.

18 Soils and Foundations

The objective of this section is to prevent moisture migration in basement spaces. In story-above-grade plane construction that meets the requirements of this section, the basement floor would be only partly below ground level (sometimes a small part) and the need for section-required moisture protection is unnecessary. Dampproofing of the floor slab would be required, however, in accordance with Section 1805.2.1.

1805.1.2 **Under-floor space.** Essentially, the requirements of this section are designed to prevent any ponding of water in under-floor areas such as crawl spaces. Crawl spaces are particularly susceptible to ponding of water as they are usually uninhabitable spaces that are infrequently observed. Water can build up in these spaces and remain for an extended period of time without being noticed by the building occupants. This is also to prevent water from ponding under the structure if it is flooded. Under-floor spaces of Group R-3 buildings located in flood hazard areas that meet the requirements of FEMA/FIA-TB-11 need not comply with the requirements in this section.

Stagnant water collected under a building can result in a serious health concern. Water buildup in a crawl space can also damage the structural integrity of the building. Wood exposed to water can deteriorate and rot, and concrete and masonry exposed to water can deteriorate with a loss of strength.

Steel exposed to water or high humidity can eventually rust to the extent that effective structural capability is jeopardized. Water buildup in a crawl space can also damage mechanical or electrical appliances located in the space by causing corrosion of electrical parts or metal skins and deterioration of insulation used to protect heating elements.

Where it is known that the water table can rise to within 6 inches of the outside ground level, or where there is evidence that surface water cannot readily drain from the site, then the finished ground surface in under-floor spaces is to be set at an elevation equal to the outside ground level around the perimeter of the building unless an approved drainage system is provided. For the drainage system to be approved, it must be demonstrated that the system will be adequate to prevent the infiltration of water into the under-floor space. This is done by determining the maximum possible flow of water near the foundation wall and footing, and designing the drainage system to remove that flow of water as it occurs, thereby preventing the buildup of water at the foundation wall.

To prevent the ponding of water in the under-floor space from a rise in the ground-water table, or from storm water runoff, the finished ground level of an under-floor space is not to be located below the bottom of the foundation footings.

Dampproofing, waterproofing, or providing subsoil drainage is not necessary if the ground level of the under-floor space is as high as the ground level at the outside of the building perimeter, as the foundation walls do not enclose an interior space below grade.

Compliance with Sections 1805.2, 1805.3, and 1805.4 would be required where the finished ground surface of the under-floor space is below the outside ground level.

1805.1.3 **Ground-water control.** After completion of building construction, it is necessary to maintain the water table at a level that is at least 6 inches below the bottom of the lower floor to prevent the flow or seepage of water into the basement. Where the site consists of well-draining soil and the highest point of the water table occurs naturally at or lower than the required level, there is no need to provide any kind of a site drainage system specifically designated to control the ground-water level. Where the soil characteristics and the site topography are such that the water table can rise to a level that will produce a hydrostatic pressure against the basement structure, a site drainage system may be installed to reduce the water level. When ground-water control in accordance with this section is provided, waterproofing in accordance with Section 1805.3 is not required.

There are many types of site drainage systems that can be employed to control ground-water levels. The most commonly used systems may involve the installation of drainage ditches or trenches filled with pervious materials, sump pits and discharge pumps, well point systems, drainage wells with deep-well pumps, sand-drain installations, and so on. This section requires that all such systems be designed and constructed using accepted engineering principles and practices based on considerations that include the permeability of the soil, amount and rate at which water enters the system, pump capacity, capacity of the disposal area, and other such factors that are necessary for the complete design of an effective drainage system.

Dampproofing. Where a ground-water table investigation has established that the high water table will occur at such a level that the building substructure will not be subjected to significant hydrostatic pressure, then dampproofing in accordance with this section and a subsoil drain in accordance with Section 1805.4 are sufficient to control moisture in the floor below grade.

1805.2

Wood foundation systems specified in Section 1807.1.4 are to be dampproofed as required by the American Forest and Paper Association Permanent Wood Foundation (AF&PA PWF) standard. AF&PA PWF-2007 replaces the previous AF&PA Technical Report No. 7. The new PWF Design Specification was written as part of an effort to update design recommendations and procedures in the wood industry's design aides, such as Technical Report 7: The Permanent Wood Foundation System (1987) and the *Permanent Wood Foundation System: Design, Fabrication and Installation Manual* (1987).

Floors. Floors requiring dampproofing in accordance with Section 1805.2 are to employ materials specified in Section 1805.2.1. The dampproofing materials must be placed between the floor construction and the supporting gravel or stone base, as shown in Figure 1805-1. Even if a floor base is not required, dampproofing should be placed under the slab.

1805.2.1

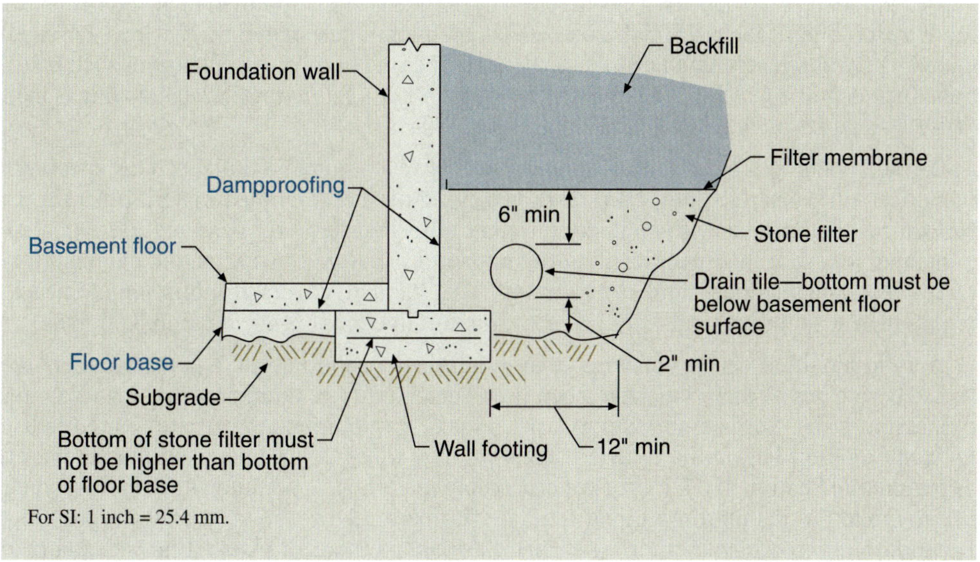

Figure 1805-1
A foundation drainage system.

The installation is intended to provide a moisture barrier against the passage of water vapor or seepage into below-ground spaces.

The dampproofing material most commonly used for underslab installations consists of a polyethylene film not less than 6 mil in thickness, which is applied over the gravel or stone base required in Section 1805.4.1. Care must be used in the installation of the material over the rough surface of the base and during the concreting operation so as not to puncture the polyethylene. Joints must be lapped at least 6 inches.

Dampproofing materials can also be applied on top of the base concrete slab if a separate floor is provided above the base slab, because the dampproofing is provided to prevent moisture infiltration of the interior space, and not the concrete slab.

Materials commonly used for dampproofing floors are listed in Table 1805-1.

Table 1805-1. **Dampproofing Materials**

Material	Specification
Asphalt	ASTM D 449
Asphalt primer	ASTM D 41
Coal-tar	ASTM D 450
Concrete and masonry oil primer (for coal-tar applications only)	ASTM D 43
Treated glass fabric	ASTM D 1668

1805.2.2 **Walls.** Walls requiring dampproofing in accordance with Section 1805.2 are first to be prepared as required in Section 1805.2.2.1, then coated with any of the bituminous materials listed in Table 1805-1 or by other approved materials and methods of application. Approved materials are those that will prevent moisture from penetrating the foundation wall.

Coatings are applied to cover prepared exterior wall surfaces extending from the top of the wall footings to slightly above ground level so that the entire wall that contacts the ground is protected. Surfaces are usually primed to provide a bond coat and then dampproofed with a protection coat of asphalt or tar pitch.

Dampproofing materials for walls may be any of the materials specified in Section 1805.3.2 for waterproofing. Table 1805-1 gives a list of bituminous materials that can be used, including the applicable standards that may be used as the basis of acceptance of such materials. Included in Table 1805-1 is ASTM D 1668 for glass fabric that is treated with asphalt (Type I), coal-tar pitch (Type II), or organic resin (Type III).

Surface-bonding mortar complying with ASTM C 887 may be utilized. This specification covers the materials, properties and packaging of dry, combined materials for use as surface-bonding mortar with concrete masonry units that have not been prefaced, coated, or painted. Because this specification does not address design or application, the manufacturer's recommendations should be followed. This standard covers proportioning, physical requirements, sampling, and testing. The minimum thickness of the coating is $1/8$ inch.

Acrylic-modified cement coatings may be utilized at the rate of 3 pounds per square foot. These types of materials have been used successfully as dampproofing materials for foundation walls. Surface-bonding mortar and acrylic-modified cement are limited in use to dampproofing. The ability of these two types of products to bridge nonstructural cracks, as required in Section 1805.3.2 for waterproofing materials, is not known. Therefore, their use is limited to dampproofing and they are not permitted to be used as waterproofing. Dampproofing may also include other materials and methods of installation acceptable to the building official.

1805.2.2.1 **Surface preparation of walls.** Before applying dampproofing materials, the concrete must be free of any holes or recesses that could affect the proper sealing of the wall surfaces. Air trapped beneath the dampproofing coating or membranes can cause blistering. Rocks and other sharp objects can puncture membranes. Irregular surfaces can also create uneven layering of coatings, which can result in vulnerable areas of dampproofing. Surface irregularities commonly associated with concrete wall construction can be sealed with bituminous materials or filled with portland cement grout or other approved materials.

Dampproofing and Waterproofing

Unit masonry walls are usually parged (plastered) with a $1/2$-inch-thick layer of portland cement and sand mix (1:2$1/2$ by volume) or with a Type M mortar proportioned in accordance with the requirements of ASTM C 270 and applied in two $1/4$-inch-thick layers. In no case is parging to result in a final thickness of less than $3/8$ inch. The parging is to be coved at the joint formed by the base of the wall and the top of the wall footing to prevent the accumulation of water at that location.

The moisture protection of unit masonry walls provided by the parging method may not be required where approved dampproofing materials such as grout coatings, cement-based paints, or bituminous coatings can be applied directly to the masonry surfaces.

Waterproofing. Waterproofing installations are intended to provide moisture barriers against water seepage that may be forced into below-ground spaces by hydrostatic pressure. **1805.3**

Where a ground-water table investigation has established that the high water table will occur at such a level that the building substructure will be subjected to hydrostatic pressure, and where the water table is not lowered by a water control system, as described in the discussion to Section 1805.4.2, all floors and walls below ground level are to be waterproofed in accordance with Sections 1805.3.1 and 1805.3.2.

Floors. Because floors required to be waterproofed are subjected to hydrostatic uplift pressures, such floors must, for all practical purposes, be made of concrete and designed and constructed to resist the maximum hydrostatic pressures possible. It is particularly important that the floor slab be properly designed, as severe cracking or movement of the concrete would allow water seepage into below-ground spaces. The ability of the waterproofing materials to bridge cracks is limited. Concrete floor construction is to comply with the applicable provisions of Chapter 19. **1805.3.1**

Materials used for waterproofing below-ground floors are to conform to the requirements of Section 1805.3.1.

Below-ground floors subjected to hydrostatic uplift pressures are to be waterproofed with membrane materials placed as underslab or split-slab installations, including such materials as rubberized asphalt, butyl rubber, and neoprene, or with polyvinyl chloride or polybutylene films not less than 6 mil in thickness, lapped at least 6 inches. All membrane joints are to be lapped and sealed in accordance with the manufacturer's instructions to form a continuous, impermeable waterproof barrier. There are many proprietary membrane products available that are specifically made for waterproofing floors and walls (i.e., polyethylene sheets sandwiched between layers of asphalt), which may be used when approved by the building official. Products that have an ICC-ES evaluation report are acceptable in most jurisdictions if the requirements of the evaluation report are followed. ICC-ES reports are intended to address the technical aspects and requirements of new and innovative products that are approved by the building official under the alternative materials and methods of construction provisions of Section 104.11. All ICC-ES evaluation reports are posted online at www.icc-es.org. A sample ICC-ES report is shown in Figure 1805-2.

Walls. Walls that are required to be waterproofed in accordance with Section 1805.3 must first be prepared as required in this section and then waterproofed with the required membrane-type installations. **1805.3.2**

The walls must be designed to resist the hydrostatic pressure anticipated at the site, as well as any other lateral loads to which the wall will be subjected, such as soil pressures or seismic loads. As with the floors required to be waterproofed, it is particularly important that the walls required to be waterproofed be properly designed to resist all anticipated loads, as cracking and other damages would allow water seepage into below-ground spaces. Concrete or masonry construction must comply with the applicable provisions of Chapters 19 and 21, respectively.

ICC-ES Evaluation Report

ESR-4802

Issued March 1, 2011
This report is subject to re-examination in one year.

www.icc-es.org | (800) 423-6587 | (562) 699-0543   A Subsidiary of the International Code Council®

DIVISION: 07-THERMAL AND MOISTURE PROTECTION
Section: 07410-Metal Roof and Wall Panels

REPORT HOLDER:

ACME CUSTOM-BILT PANELS
52380 FLOWER STREET
CHICO, MONTANA 43820
(808) 664-1512
www.customblltpanels.com

EVALUATION SUBJECT:

CUSTOM-BILT STANDING SEAM METAL ROOF PANELS: CB-150

1.0 EVALUATION SCOPE

Compliance with the following codes:

- 2006 *International Building Code®* (IBC)
- 2006 *International Residential Code®* (IRC)

Properties evaluated:

- Weather resistance
- Fire classification
- Wind uplift resistance

2.0 USES

Custom-Bilt Standing Seam Metal Roof Panels are steel panels complying with IBC Section 1507.4 and IRC Section R905.10. The panels are recognized for use as Class A roof coverings when installed in accordance with this report.

3.0 DESCRIPTION

3.1 Roofing Panels:

Custom-Bilt standing seam roof panels are fabricated in steel and are available in the CB-150 and SL-1750 profiles. The panels are roll-formed at the jobsite to provide the standing seams between panels. See Figures 1 and 3 for panel profiles.

The standing seam roof panels are roll-formed from minimum No. 24 gage [0.024 inch thick (0.61 mm»)] cold-formed sheet steel. The steel conforms to ASTM A 792, with an aluminum-zinc alloy coating designation of AZ50.

3.2 Decking:

Solid or closely fitted decking must be minimum $^{15}/_{32}$-inch-thick (11.9 mm) wood structural panel or lumber sheathing, complying with IBC Section 2304.7.2 or IRC Section R803, as applicable.

4.0 INSTALLATION

4.1 General:

Installation of the Custom-Blt Standing Seam Roof Panels must be in accordance with this report, Section 1507.4 of the IBC or Section R905.10 of the IRC, and the manufacturer's published Installation Instructions. The manufacturer's installation instructions must be available at the job site at all times during installation.

The roof panels must be installed on solid or closely fitted decking, as specified in Section 3.2. Accessories such as gutters, drip angles, fascias, ridge caps, window or gable trim, valley and hip flashings, etc., are fabricated to suit each job condition. Details must be submitted to the code official for each installation.

4.2 Roof Panel Installation:

CB-150: The CB-150 roof panels are installed on roofs having a minimum slope of 2:12 (17 percent). The roof panels are installed over the optional underlayment and secured to the sheathing with the panel clip. The clips are located at each panel rib side lap spaced 6 inches (152 mm) from all ends and at a maximum of 4 feet (1.22 mm) on center along the length of the rib, and fastened with a minimum of two No. 10 by 1-inch pan head corrosion-resistant screws. The panel ribs are mechanically seamed twice, each pass at 90 degrees, resulting in a double locking fold.

4.3 Fire Classification:

The steel panels are considered Class A roof coverings in accordance with the exception to IBC Section 1505.2 and IRC Section R902.1.

4.4 Wind Uplift Resistance:

The systems described in Section 3.0 and installed in accordance with Sections 4.1 and 4.2 have an allowable wind uplift resistance of 45 pounds per square foot (2.15 kPa).

5.0 CONDITIONS OF USE

The standing seam metal roof panels described in this report comply with, or are suitable alternatives to what is specified in, those codes listed in Section 1.0 of this report, subject to the following conditions:

5.1 Installation must comply with this report, the applicable code, and the manufacturer's published installation instructions. If there is a conflict between this report and the manufacturer's published Installation Instructions, this report governs.

5.2 The required design wind loads must be determined for each project. Wind uplift pressure on any roof area must not exceed 45 pounds per square foot (2.15 kPa).

6.0 EVIDENCE SUBMITTED:

Data in accordance with the ICC-ES Acceptance Criteria for Metal Roof Coverings (AC166), dated October 2007.

7.0 IDENTIFICATION

Each standing seam metal roof panel is identified with a label with a label bearing the product name, the material type and gage, the Acme Custom-Bilt Panels name and address, and the evaluation report number (ESR-4802).

ICC-ES Evaluation Reports are not to be construed as representing aesthetics or any other attributes not specifically addressed, nor are they to be construed as an endorsement of the subject of the report or a recommendation for its use. There is no warranty by ICC Evaluation Service, LLC, express or implied, as to any finding or other matter in this report, or as to any product covered by the report.

Copyright © 2010

Figure 1805-2 Sample ICC-ES evaluation report.

Table 1805-2. **Waterproofing Materials**

Material	Specification
Asphalt-saturated asbestos felt	ASTM D 250
Asphalt-saturated burlap fabric	ASTM D 1327
Asphalt-saturated cotton fabric	ASTM D 173
Asphalt-saturated organic felt	ASTM D 226
Coal-tar-saturated burlap fabric	ASTM D 1327
Coal-tar-saturated cotton fabric	ASTM D 173
Coal-tar-saturated organic felt	ASTM D 227

Table 1805-2 lists materials commonly used for the installation of moisture barriers in wall construction and the related standards that may be used as a basis for acceptance of such materials.

Asphalt and coal-tar products are not compatible and should not be used together.

Waterproofing installations are to extend from the bottom of the wall to a height not less than 12 inches above the maximum elevation of the ground-water table. The remainder of the wall below ground level (if the height is small) may be either waterproofed as a continuation of the installation or must be dampproofed in accordance with the requirements of Section 1805.2.

This section requires that waterproofing must consist of two-ply hot-mopped felts. The practice of the waterproofing industry is to select the number of plies of membrane material based on the hydrostatic head (height of water pressure against the wall). As a general practice, if the head of water is between 1 and 3 feet, two plies of felt or fabric membrane are used; between 4 and 10 feet, three-ply construction is needed; and between 11 and 25 feet, four-ply construction is necessary.

Waterproofing installations may also use polyvinyl chloride materials of not less than 6-mil thick, 40-mil polymer-modified asphalt, or 6-mil polyethylene. These materials have been widely recognized for their effectiveness in bridging nonstructural cracks. Other approved materials and methods may be used, provided that the same performance standards are met. All membrane joints must be lapped and sealed in accordance with the manufacturer's instructions.

Surface preparation of walls. Before applying waterproofing materials to concrete or masonry walls, the wall surfaces must be prepared in accordance with the requirements of Section 1805.2.2.1, which requires the sealing of all holes and recesses. Surfaces to be waterproofed must also be free of any projections that might puncture or tear membrane materials that are applied over the surfaces.

1805.3.2.1

Joints and penetrations. This section requires that all joints occurring in floors and walls and at locations where floors and walls meet, as well as all penetrations in floors and walls, be made watertight by approved methods. Sealing the joints and penetrations in the waterproofing is of primary importance to ensure the effectiveness of the waterproofing. If the joints or penetrations are not sealed properly, they can develop leaks, which become a passageway for water to enter the building. Because the remainder of the foundation is wrapped in waterproofing, moisture can actually become trapped in the foundation walls or floor slab, and serious damage to these structural components can occur.

1805.3.3

Methods may involve the use of construction keys between the base of the wall and the top of the footing, or, if there is a hydrostatic pressure, floor and wall joints may require the use of manufactured waterstops made of metal, rubber, plastic, or mastic materials.

Floor edges along the walls and floor expansion joints may employ any of a number of preformed expansion joint materials, such as asphalt, polyurethane, sponge rubber, self-expanding cork, cellular fibers bonded with bituminous materials, and so on, which all comply with applicable ASTM standards or other approved specifications. A variety of sealants may be used together with the preformed joint materials. Gaskets made of neoprene and other materials are also available for use in concrete and masonry joints. The National Roofing Contractor's (NRCA) Roofing and Waterproofing Manuals provide details for the reinforcement of membrane terminations, corners, intersections of slabs and walls, through-wall and slab penetrations, and other locations.

Penetrations in walls and floors may be made watertight with grout or manufactured fill materials and sealants made for the purpose.

1805.4 **Subsoil drainage system.** Subsoil drainage systems are required to drain the area adjacent to basement walls to eliminate hydrostatic loads.

This section covers subsoil drainage systems in conjunction with dampproofing (see Section 1805.2) to protect below-ground spaces from water seepage. Such systems are not used where basements or other below-ground spaces are subject to hydrostatic pressure, because they would not be effective in disposing of the amount of water anticipated if hydrostatic pressure conditions exist. Ground-water tables may be reduced to acceptable levels by methods described in the discussion of Section 1805.1.3.

The details of subsoil drainage systems are covered in the requirements of Sections 1805.4.1 through 1805.4.3.

1805.4.1 **Floor base course.** This section requires that floors of basements, except for story-above-grade-plane construction, must be placed on a gravel or stone base not less than 4 inches thick. Not more than 10 percent of the material is to pass a No. 4 sieve to provide a porous installation and provide a capillary break. Material that passes a No. 4 sieve would be silt or clay that does not permit the free movement of water through the floor base, but allows for upward migration by capillary action.

This requirement serves three purposes. The first is to provide an adjustment to the irregularities of a compacted subgrade so as to produce a level surface upon which to cast a concrete slab. The second purpose is to provide a capillary break so that moisture from the soil below will not rise to the underside of the floor. Finally, where required, the porous base can act as a drainage system to expel underslab water by means of gravity, or the use of a sump pump or other approved method.

The exception allows for the omission of the floor base when the natural soils beneath the floor slab consist of well-draining granular materials such as sand, stone, or mixtures of these materials. Some caution, however, is justified in the use of this exception. If the granular soils contain an excessive percentage of fine materials, the porosity and the ability of the soil to provide a capillary break may be considerably diminished. The exception should be applied only if the natural base is equivalent to the floor base otherwise required by this section.

1805.4.2 **Foundation drain.** This section describes in considerable detail the materials and features of construction required for the installation of foundation drain systems.

This type of drain system is suitable where the water table occurs at such elevation that there is minimal hydrostatic pressure exerted against the basement floor and walls and where the amount of seepage from the surrounding soil is so small that the water can be readily discharged by gravity or by mechanical means into sewers or ditches. The objective is to combine the protection afforded by the dampproofing of walls and floors (see Section 1805.2) and that given by the perimeter drains to maintain below-ground spaces in a dry condition.

A foundation drain system usually consists of the installation of drain tiles made of clay or concrete or of drain pipes of corrugated metal or nonmetallic pipes surrounded by crushed stone or gravel and a filter membrane material (filter fabric). The foundation drain is set adjacent to the wall footing and extends around the perimeter of the building. Drain tiles are placed end to end with open joints to permit water to enter the system. Metallic and nonmetallic drains are made with perforations at the invert (bottom) section of the pipe and are installed with connected ends. Where drain tile or perforated drain pipe is used, the invert must not be set higher than the basement floor line so that water conveyed by the drain does not seep into the filter material and then create a hydrostatic pressure condition against the foundation wall and footing. The inverts should not be placed below the bottom of the adjacent wall footings to avoid carrying away fine soil particles whose loss, in time, could possibly undermine the footing and cause settlement of the foundation walls.

Tile joints or pipe perforations should be covered with an approved filter membrane material to prevent them from becoming clogged and to prevent fine particles that may be contained in the surrounding soil from entering the system and being carried away by water.

The filter material around the drain tiles or pipes (not to be confused with filter membrane material) should consist of selected gravel and crushed stone containing not more than 10 percent of material that passes a No. 4 sieve. The filter materials should be selected to prevent the movement of particles from the protected soil surrounding the drain installation into the drain. Filter material is to be placed in the excavation so that it will extend out from the edge of the wall footing a distance of at least 12 inches, with the bottom of the fill being no higher than the bottom of the base under the floor (see Section 1805.4.1) and the top of the footing.

Requiring the bottom of the foundation drain to be no higher than the bottom of the floor base is necessary so that if the water table rises into the floor base, it will also be able to rise unobstructed into the foundation drain. The foundation drain will then drain the water away from the building, as required by Section 1805.4.3. The top of the filter fill material must be covered with an approved filter membrane to allow water to pass through to the perimeter drain tile or pipes without allowing fine soil materials to enter the drainage system.

Drain tiles or pipes are to be installed in the filter bed and should be seated on at least 2 inches of filter material and covered with at least 6 inches of filter material to maintain good water flow into the drain tile or pipe.

1805.4.3 **Drainage discharge.** This section references the *International Plumbing Code*® (IPC®) for requirements for installing piping systems for the disposal of water from the floor base and the foundation drains. Chapter 11 of the IPC considers the piping materials, applicable standards, and methods of installation of subsurface storm drains to facilitate water discharge either by gravity or by mechanical means.

Where the soil at the site consists of well-drained granular materials such as gravel or sand-gravel mixtures to prevent the occurrence of hydrostatic pressure against the foundation walls and under the floor slab, the use of a dedicated drainage system as prescribed in the IPC is not required, because the site soils would permit natural drainage.

Section 1806 *Presumptive Load-Bearing Values of Soils*

1806.1 **Load combinations.** The presumptive bearing values in the code are allowable stress values, not strength level values. The format of IBC Table 1806.2 shown in Table 1806-1 comes from the legacy 1997 UBC. However, some of the footnotes from the UBC version

Soils and Foundations

have been moved into the text of the code. Each of the legacy model codes used a different approach to generate their allowable values. Hence, there was a disparity between the values shown in the model codes. The tabular values for allowable foundation pressure cannot be increased for width or depth as was allowed by the UBC. Note, however, that the allowable foundation pressures are permitted to be increased by one-third with the alternative ASD load combinations of Section 1605.3.2 for combinations including wind or earthquake so as to be consistent with previous editions of the legacy model codes.

Table 1806-1. **Presumptive Load-Bearing Values**

Class of Materials	Vertical Foundation Pressure (psf)	Lateral Bearing Pressure (psf/ft below natural grade)	Lateral Sliding Resistance	
			Coefficient of friction[a]	Cohesion (psf)[b]
1. Crystalline bedrock	12,000	1,200	0.70	—
2. Sedimentary and foliated rock	4,000	400	0.35	—
3. Sandy gravel and/or gravel (GW and GP)	3,000	200	0.35	—
4. Sand, silty sand, clayey sand, silty gravel, and clayey gravel (SW, SP, SM, SC, GM, and GC)	2,000	150	0.25	—
5. Clay, sandy clay, silty clay, clayey silt, silt, and sandy silt (CL, ML, MH, and CH)	1,500	100	—	130

For SI: 1 pound per square foot = 0.0479 kPa, 1 pound per square foot per foot = 0.157 kPa/m.
a. Coefficient to be multiplied by the dead load.
b. Cohesion value to be multiplied by the contact area, as limited by Section 1806.3.2.

1806.2 **Presumptive load-bearing values.** The presumptive allowable bearing and lateral pressures must be used unless a geotechnical investigation substantiates higher values. The term "unprepared fill" refers to fill that was not placed and compacted in accordance with an approved soils report.

The classifications in Table 1806.2 are from the Unified Soil Classification System. Most foundation-bearing strata can be classified into one of the classifications in the table. The allowable bearing pressures and lateral-bearing values are based on long experience with the behavior of these materials. However, the presumptive value for CL and CH may not be conservative for "soft" clays, depending on the degree of consolidation of the clay. The selection of an allowable bearing pressure should take into account the strength of weaker underlying soil strata so that the pressure in any weaker stratum does not exceed the allowable pressure, particularly in clay soils. Because of this, it is important to know the soil profile and classifications of the different strata.

1806.3 **Lateral load resistance.** Sections 1806.3.1 through 1806.3.4 deal with lateral load calculations when using the presumptive load-bearing values of Table 1806.2.

The classifications for lateral bearing in Table 1806.2 are from the Unified Soil Classification System. The lateral-bearing values are based on long experience with the behavior of these materials. The limitation on frictional resistance for silts and clays is intended to provide structural stability and improve serviceability.

The formulae for lateral bearings employing posts and poles (found in IBC Section 1807.3, embedded posts and poles) were originally for outdoor advertising structures. For these structures, deflections of $1/2$ inch at the surface do not affect serviceability. Thus, the allowance of two times the tabular value for these structures is permitted.

Foundation Walls, Retaining Walls, and Embedded Posts and Poles

Section 1807 *Foundation Walls, Retaining Walls, and Embedded Posts and Poles*

1807.1 **Foundation walls.** Foundation walls of materials such as concrete and masonry, rubble stone, or wood are addressed in Section 1807. Lateral soil loads that must be considered in foundation wall design are covered in Section 1610. If a drainage system is not placed behind the wall to drain ground water away from the wall, hydrostatic pressures, which can easily exceed the lateral pressures from the retained soil, will occur. For example, the active lateral pressure from a well-graded granular soil may be in the range of 30 to 35 pounds per square foot, whereas the hydrostatic pressure is 62.4 pounds per cubic foot. Hence, the equivalent fluid pressures on a wall that was designed as drained, but constructed without an effective drainage system, could be subjected to pressures approximately three times the design pressure.

Unbalanced backfill height and its method of measurement are presented in Section 1807.1.2 and rubble stone foundation walls in Section 1807.1.3. Because rubble masonry uses rough stones of irregular shape and size, a larger thickness, as compared to hollow unit masonry or concrete, is required for adequate bonding of the stone and mortar. Rubble stone foundation walls are not permitted in SDC C, D, E, or F.

1807.1.4 **Permanent wood foundation systems.** The requirements set forth in AF&PA Technical Report PWF must be rigidly followed. The wood foundation system is an assembly similar to a fire assembly—no substitution of materials or methods is allowed. All lumber and plywood must be treated in accordance with AWPA U1 (Commodity Specification A, Use Category 4B, Section 5.2) and must be identified and labeled in accordance with Section 2303.1.8.1. ICC in partnership with AWPA publishes all 24 AWPA standards that are referenced in the IBC.

All hardware and fasteners must be corrosion resistant. Metals in contact with the preservative salts will corrode at a much faster rate than normal because of the influence of the salts. Hence, only corrosion-resistant fasteners made of silicon bronze, copper, or Type 304 or 316 stainless steel may be used, except that hot-dipped galvanized nails may be used when installed under the specific conditions set forth in the technical report for surface treatment of the nails and moisture protection of the foundation.

1807.1.5 **Concrete and masonry foundation walls.** Foundation walls must be designed within the applicable provisions of IBC Chapters 19 and 21. However, if the foundation wall is laterally supported at the top and bottom, such as a basement wall laterally supported by a floor diaphragm at the top and a basement floor slab at the base, the wall may be constructed in accordance with the prescriptive provisions of Section 1807.1.6 and its associated tables. These tables allow the use of unreinforced and plain (lightly reinforced) concrete or masonry walls that have been used in low or very low seismic risk areas.

Walls in moderate-to-high seismic risk areas will be subjected to ground shaking and ground displacements of unknown magnitude, and the walls will have an additional lateral load caused by the seismic ground motion. In the 2000 IBC, there were no specific requirements based on regional seismic considerations. Code changes were made to the 2003 IBC that were designed to address these concerns by the addition of a new section that has seismic requirements based on SDC. Specific seismic requirements for concrete and masonry foundation walls are now covered in Sections 1807.1.6.2.1 (concrete) and 1807.1.6.3.2 (masonry), based on the SDC of the building.

Additionally, if the prescriptive provisions are used, sufficient soil investigation should be done to properly classify the retained soils as indicated in the tables in accordance with the Unified Soil Classification Method (see IBC Section 1803.5.1).

1807.1.6 Prescriptive design of concrete and masonry foundation walls. The requirements and provisions of Sections 1807.1.6.1 through 1807.1.6.3.2 are applicable to the prescriptive design of concrete and masonry foundation walls that are laterally supported at the top and bottom. The minimum wall thicknesses are specified in the appropriate sections, based on the thickness of the supported wall, soil loads, unbalanced backfill height, and overall height of the wall. Rubblestone walls cannot be less than 16-inches thick where permitted (Section 1807.1.3). These minimum thickness provisions are to facilitate support of the wall above. These thickness provisions are empirical and have been used successfully in low or very low seismic risk areas. Additional seismic requirements for concrete and masonry foundation walls are covered in Sections 1807.1.6.2.1 (concrete) and 1807.1.6.3.2 (masonry), based on the SDC of the building.

1807.1.6.2 Concrete foundation walls. This section specifies the material requirements for walls constructed in accordance with the prescriptive tables. Note the effective depth, "d," in Section 1807.1.6.2(3). Placement of reinforcing at the prescribed "d" is critical to develop adequate flexural strength necessary to resist the combined vertical and lateral soil loads.

The concrete section contains seven specific requirements to prescriptively select a foundation wall.

Concern has been expressed that the prescriptive foundation wall provisions do not impose a limitation on the maximum axial loads that the walls should support. To resolve this concern, a conservative maximum unfactored axial load of $1.2tf'_c$ for concrete and $1.2tf'_m$ for masonry are included in the requirements. The maximum unfactored axial load is based on a compressive stress on the outside face of the wall that is due to the axial load and bending moment induced by the backfill that is well below that permitted by ACI 318 or TMS 402/ACI 530/ASCE 5. Although this axial load limitation has merit, it requires a calculation to determine actual maximum axial load acting on a given wall. Table 1807-1 shows the maximum unfactored allowable axial load for the typical concrete foundation wall.

Table 1807-1. Maximum Permissible Axial Load for Concrete Walls Based on $1.2tf'_c$ in Pounds per Foot of Wall

Wall Thickness (inches)	f'_c = 2000 psi	f'_c = 3000 psi
7.5	22,500	27,000
9.5	28,000	34,200
11.5	34,500	41,400

1807.1.6.2.1 Seismic requirements. The 2000 IBC had no specific seismic requirements for the prescriptive foundation wall provisions. This was of particular concern in the western states where earthquakes are relatively frequent and destructive. In the 2003 IBC, these concerns were addressed by adding specific seismic-related requirements in a new section, which covered seismic requirements based on SDC. These same specific seismic requirements for concrete and masonry foundation walls are now covered in Sections 1807.1.6.2.1 and 1807.1.6.3.2, based on the SDC of the building. The requirements are summarized below for concrete foundation walls.

Seismic requirements for concrete foundation walls constructed in accordance with Table 1807.1.6.2 are as follows:

1. SDCs A and B—One #5 bar is required at a minimum around window and door openings, which must extend beyond the corners of the openings or be anchored so as to develop f_y in tension at the corner of openings (reference Section 1909.6.3).

2. SDCs C, D, E, and F—The prescriptive tables are not allowed to be used except as permitted for plain concrete members in accordance with Section 1908.1.8, which modifies ACI 318, Section 22.10. The modification states that structural plain concrete members are not permitted in SDC C, D, E, or F except for structural plain concrete basement, foundation, or other walls below the base in detached one- and two-family dwellings three stories or less in height constructed with stud-bearing walls. Additional restrictions apply to dwellings in SDC D or E, where the walls cannot exceed 8 feet in height, cannot be less than 7.5-inches thick, and can retain no more than 4 feet of unbalanced fill. The last requirement states that the walls must be reinforced in accordance with ACI Section 22.6.6.5.

Masonry foundation walls. This section contains requirements for masonry foundation walls, both plain and with reinforcement. Masonry is required to be solid in order to distribute the concentrated force, or hollow units must be solidly grouted as noted in Footnote c of the plain masonry foundation wall Table 1807.1.6.3(1). The masonry section contains 10 specific requirements for prescriptive selection of a masonry foundation wall. See Section 2104.2 if corbelling is necessary or desired to match the width of a masonry cavity wall above the foundation wall. **1807.1.6.3**

Table 1807-2 shows the maximum unfactored allowable axial load for the typical masonry foundation wall. Also see the discussion under Section 1807.1.6.2, Concrete foundation walls.

Table 1807-2. Maximum Permissible Axial Load for Masonry Walls Based on $1.2tf'_m$ in Pounds per Foot of Wall

Wall Thickness (inches)	$f'_m = 1500$ psi	$f'_m = 2000$ psi
7.625	13,725	18,300
9.625	17,325	23,100
10.625	19,125	25,500
11.625	20,925	27,900

Alternative foundation wall reinforcement. The code permits equivalent cross section of reinforcing, provided the spacing does not exceed 72 inches and the bar size does not exceed #11. If alternative reinforcement is used, it is preferable to reduce bar size and spacing rather than increase bar size and spacing. In development of the reinforcing, the bar size must be small enough that the reinforcing, and any splices, can be adequately developed. Development refers to the embedment of the reinforcing to adequately develop the bond between the reinforcing and the grout. A good rule of thumb to prevent splitting of concrete masonry is that the bar size number should not exceed $t - 1$, where t is the nominal thickness of the wall in inches. **1807.1.6.3.1**

Seismic requirements (masonry). See discussion under Section 1807.1.6.2.1, Seismic requirements. Table 1807-3 summarizes requirements for masonry foundation walls based on seismic design category and gives applicable sections of the MSJC Code. **1807.1.6.3.2**

Table 1807-3. Seismic Requirements for Masonry Foundation Walls

Seismic Design Category	TMS 402-11/ACI 530-11/ASCE 5-11 Section
C	1.18.4.3
D	1.18.4.4
E, F	1.18.4.5

Seismic requirements for masonry foundation walls constructed in accordance with Tables 1807.1.6.3(1) through 1807.1.6.3(4) are as follows:

1. SDCs A and B—No additional requirements apply.

2. SDC C—Additional requirements cover discontinuous members that are part of the lateral-force-resisting system, such as columns, pilasters, and beams that support reactions from walls or frames, but no specific requirements for foundation walls. Refer to Section 1.18.4.3 of TMS 402/ACI 530/ASCE 5 for other requirements.

3. SDC D—Must conform to the requirements of SDC C, as well as Section 1.18.4.4 of TMS 402/ACI 530/ASCE 5.

4. SDCs E and F—Must conform to the requirements of SDCs C and D, as well as Section 1.18.4.5 of TMS 402/ACI 530/ASCE 5.

1807.2 **Retaining walls.** Although the legacy codes 1999 BOCA/NBC and 1999 SBC, and ASCE 7-98 contained some requirements for retaining walls, they were very limited in scope. The legacy 1997 UBC had more detailed requirements for retaining walls, which are essentially the same as the provisions in the IBC. The IBC requires retaining walls to be designed to resist overturning, sliding, and excessive foundation-bearing pressure with a safety factor of at least 1.5 against lateral sliding and overturning using allowable stress design loads. (See discussion under Section 1610 and Table 1610.1 for soil lateral loads.) A keyway incorporated into the retaining wall base extending into the soil is considered to enhance the ability of the retaining wall against sliding.

There has been considerable debate in the structural engineering community whether both the passive pressure resisting the slide and the active pressure acting on the driving side of the key should be considered in the analysis. While some believe considering soil lateral pressure on both sides of the keyway is too conservative, there are others who believe not considering the active pressure will be too unconservative. Hence, the code now requires the consideration of all lateral pressures as is required to do in a free-body diagram. See Figure 1807-1.

1807.3 **Embedded posts and poles.** The design criteria for the use of poles or posts embedded in the ground, or in concrete footings in the ground and unconstrained at the ground surface, were developed for the Outdoor Advertising Association of America, Inc. (OAAA). The research was conducted at Purdue University from 1938 to 1940, and continued in 1947 at the University of Notre Dame. The results of this research were used by OAAA for the design of outdoor advertising structures, which had previously used trussed A-frame supporting systems. Charts and a monograph were developed, which the association used for the design of poles as cantilever uprights for support of its outdoor advertising structures. These data were subsequently submitted through one of the ICC legacy organizations, International Conference of Building Officials' (ICBO), code change process and were incorporated into the 1964 edition of the *Uniform Building Code* (UBC).

The criteria relate to lateral bearing and apply to a vertical pole considered a column embedded in either earth or in a concrete footing in the earth and used to resist lateral loads. In order for the pole to meet the conditions of research that resulted in the code formula, the code requires that the backfill in the annular space around a column that is not embedded in a concrete footing be either of 2,000 psi concrete or of clean sand thoroughly compacted by tamping in layers not more than 8 inches in depth.

The original design criteria established for the Outdoor Advertising Association of America, Inc., resulted in a $1/2$-inch lateral pole deformation at the surface of the ground. These criteria were also based on field tests conducted in a range of sandy and gravelly soils and silts and clays.

Foundation Walls, Retaining Walls, and Embedded Posts and Poles 18

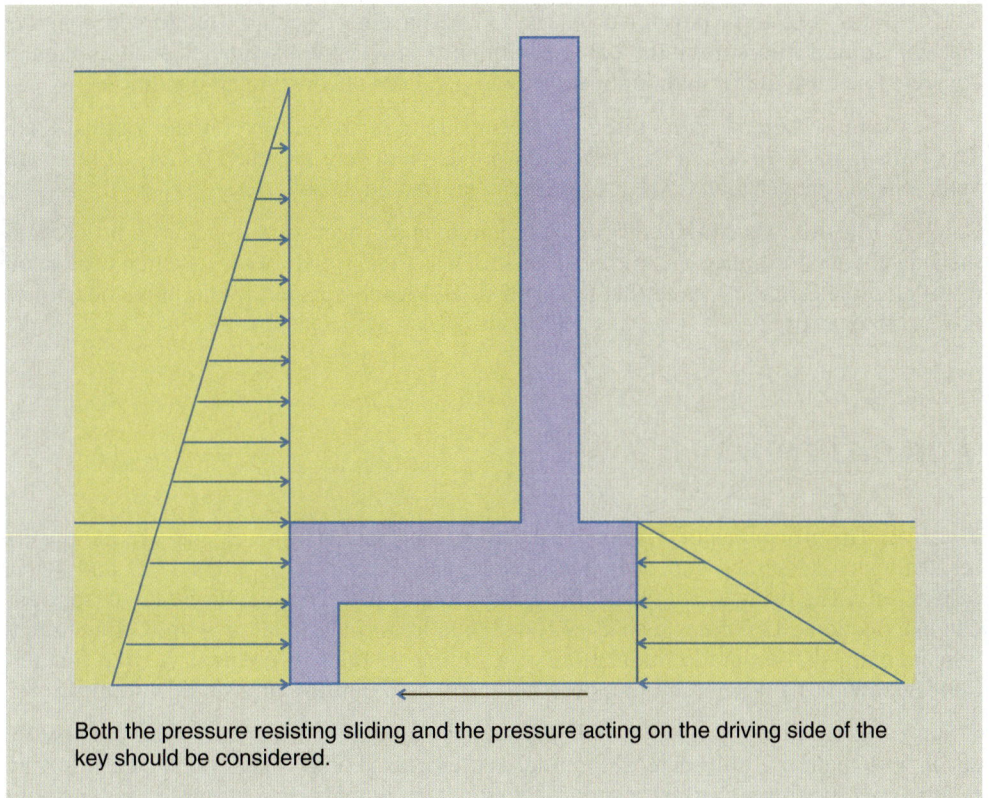

Both the pressure resisting sliding and the pressure acting on the driving side of the key should be considered.

Figure 1807-1 Retaining wall keyway in soil.

The IBC employs allowable lateral-bearing stresses in IBC Table 1806.2, which are considerably lower than those developed for the Outdoor Advertising Association of America. Consequently, Section 1806.3.4 permits a doubling of the lateral-bearing values for isolated poles and poles supporting structures that can safely tolerate the $1/2$-inch movement at the ground surface.

Limitations. The limitations imposed by this section are intended for both structural stability and serviceability. The limitation of the frictional resistance for silts and clays is consistent with the UBC, which also limited the sliding resistance to one-half the dead load. **1807.3.1**

The limitations on the types of construction that use the lateral support of poles are based on the brittle nature of the materials. To prevent excessive distortions that would cause the cracking of these brittle materials, the code limits the use of the poles unless some type of rigid cross-bracing is provided to limit the deflections to those that can be tolerated by the materials.

Wood poles must be treated in accordance with AWPA U1. Sawn timber posts are Commodity Specification A, Use Category 4B, and round timber posts are Commodity Specification B, Use Category 4B.

Design criteria. See IBC Section 1806.3 for allowable values of lateral bearing. Note that the two-time increase allowed per Section 1806.3.4 may only be used for structures where deflection of $1/2$ inch at the surface is tolerable, for example, signs, flagpoles, and light poles. **1807.3.2**

Nonconstrained. See Section 1807.3 discussion for the empirical basis of the formula. This formula should be used only for minor foundations of moderate size, which will fit within the constraints of the data from which the formula was developed. For large-sized piers, that is, more than 2 feet in diameter, a more appropriate method should be used. See Winterkorn et al.[2] **1807.3.2.1**

1807.3.2.2 Constrained. The term *pavement* means a rigid pavement such as reinforced concrete that will form a fulcrum for the column. Columns in flexible pavements such as asphalt concrete must use the formula in Section 1807.3.2.1 for unconstrained conditions.

1807.3.2.3 Vertical load. There is no requirement to consider a combined lateral and vertical load. The vertical loads for which the formulae were derived were less than $0.1FA_g$. If there are vertical loads greater than $0.1FA_g$, these formulae should not be used.

1807.3.3 Backfill. Backfill in accordance with the requirements is necessary to achieve the strength predicted by the formulae. The required backfill was used as part of the research conducted to develop the formulae. Note that the sand should be compacted to a relative density of at least 85 percent.

Section 1808 *Foundations*

1808.1 General. The provisions of Section 1808 apply to all foundations. Specific requirements for shallow foundations and deep foundations are located in Sections 1809 and 1810, respectively. The two general types of foundations are shallow foundations and deep foundations. Section 202 defines a shallow foundation as an individual or strip footing, a mat foundation, a slab-on-grade foundation, or a similar foundation element. A deep foundation is defined as a foundation that does not meet the definition of a shallow foundation.

1808.2 Design for capacity and settlement. Footings should be designed for approximately equal settlements to minimize differential settlements. For footings on sands, this may require unequal footing pressures to affect equal settlements. For example, see Terzaghi et al.[1] Expansive soils are addressed in Section 1808.6.

1808.3 Design loads. Footings are to be designed using full dead load (including overlying fill materials), floor and roof live loads, snow load, wind or seismic forces, and any other loads required by Section 1605 that will produce the most severe loading. Live loads acting at the foundation may be reduced based on the reduced probability of simultaneous occurrence of maximum live loads. This section specifically permits live load reduction as specified in Sections 1607.10 and 1607.12 for the foundation design.

1808.3.1 Seismic overturning. When strength design loads are used to proportion the foundations, the seismic overturning effects are permitted to be reduced in accordance with ASCE 7 Section 12.13.4. This maximum recognizes that the seismic forces determined in accordance with the ASCE 7 standard are based on strength design, not allowable stress design (ASD). Foundations proportioned in accordance with ASD procedures have historically performed satisfactorily. Because of expected deviation from the results from the equivalent lateral force method, which assumes a fixed base of the building, overturning effects at the foundation are permitted to be reduced 25 percent for structures other than inverted pendulum or cantilever column systems when designed by the equivalent lateral force procedure. Overturning effects at the foundation are permitted to be reduced 10 percent for structures designed by the modal analysis method because of the higher degree of accuracy of the procedure. Note that these reductions cannot be used with the alternative basic ASD load combinations of Section 1605.3.2.

1808.4 Vibratory loads. Footings supporting equipment should be designed to minimize the transmission of vibratory loads to the soils. The dynamic interaction of the footing, equipment, and soil mass should be analyzed, and the footing "tuned" to minimize the transmission. As a rough rule of thumb, footings for rotating or reciprocating equipment should have a mass that is at least four times the mass of the equipment.

Vibratory loads from equipment foundations that are transmitted to the soil can cause significant and damaging settlements. The transmitted vibration will cause densification of granular materials, particularly loose or medium dense sands. The reduction in volume can cause large settlements depending on the initial density of the sands. In saturated granular materials, such as loose or medium dense sands with a high water table, the transmitted vibrations can cause a buildup of pore pressure and liquefaction of the sands, with resulting loss of bearing capacity and settlements. In saturated clays, the vibrations can enhance the drainage of water from the pores and increase long-term settlements.

Shifting or moving soils. For example, loose sands.   **1808.5**

Design for expansive soils. The requirements to mitigate the effects of expansive soils are set forth in this section. In addition to mitigation by foundation design, the effects of expansive soils may also be mitigated by removal of the expansive soils or stabilization by chemical means, pre-saturation, or dewatering. Expansive soils are cohesive soils, typically high plasticity clays, with a high Plasticity Index and a high Swell Index.   **1808.6**

Foundations. The large volume changes in expansive soils caused by changes in the soils' water content can cause significant differential deflections in a building if not uniform. In a typical building on expansive soils, the soils at the perimeter of the building will have seasonal moisture changes, whereas the soils at the interior of the building will remain at a fairly constant moisture content. The perimeter foundations will rise and fall with the seasonal volume changes in moisture content, whereas the soil at the interior footings or slab will not have any volume changes, because of constant moisture content. The resulting differential displacements between the interior and exterior footings can cause significant structural distress. Hence, the requirements that the foundation be designed to resist the differential volume changes and to minimize racking or differential displacements in the structure.   **1808.6.1**

Slab-on-ground foundations. The slab-on-ground or raft foundation design methods cited in this section result in a raft that has sufficient stiffness to bridge differential displacements caused by the volume changes in the supporting soil.   **1808.6.2**

Design moments, shears, and deflections are to be determined in accordance with WRI/CRSI *Design of Slab-on-Ground Foundations* or PTI *Standard Requirements for Analysis of Shallow Concrete Foundations on Expansive Soils*. Once the design moments, shears, and deflections are determined from the applicable standard, then conventionally reinforced (nonprestressed) foundations on expansive soils must be designed in accordance with WRI/CRSI *Design of Slab-on-Ground Foundations*, and post-tensioned foundations on expansive soils must be designed in accordance with PTI *Standard Requirements for Design of Shallow Post-Tensioned Concrete Foundations on Expansive Soils*.

The code also permits alternative methods of analysis, provided the methodology is rational and the basis for the analysis and design parameters are available for peer review.

Removal of expansive soil. Removal of the expansive soil is an acceptable mitigation method and is the preferred method if the stratum of expansive soil is near the surface and reasonably thin. This method may also be the least expensive method if the expansive soil is at the surface.   **1808.6.3**

Stabilization. Expansive soils may be stabilized so that the moisture content does not change; hence, there will be no volume changes to cause differential displacements. Stabilization can be by chemical methods, by pre-saturating the soils to a maximum swell and capping the expansive layer to keep the moisture content constant, or by dewatering to a minimum shrinkage and providing drainage to keep the moisture content constant.   **1808.6.4**

Foundations on or adjacent to slopes. The provisions of this section apply only to buildings placed on or adjacent to slopes steeper than 1 vertical to 3 horizontal.   **1808.7**

Building clearance from ascending slopes. This setback requirement is intended to provide protection to the structure from shallow slope failure (sloughing) and protection for erosion and slope drainage. The setback space also provides access around the structure and helps to create a light and open environment. IBC Figure 1808.7.1 depicts the criterion   **1808.7.1**

for the setback or clearance. Figure 1808-1 also depicts the criteria set forth in this item for determination of the toe of the slope when the slope exceeds 1:1.

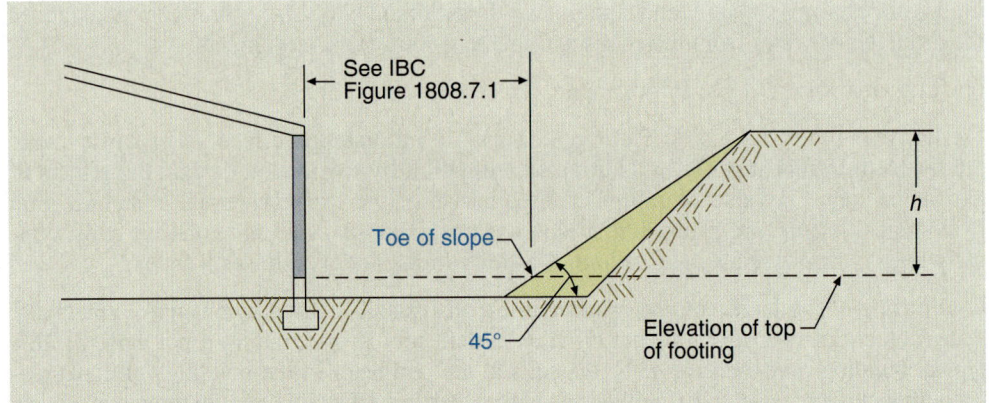

Figure 1808-1
Buildings adjacent to ascending slope exceeding 1:1.

1808.7.2 **Foundation setback from descending slope surface.** The setback requirement at the top of slopes is intended to provide vertical and lateral support for the foundations and minimize the possibility of shallow bearing failure of the foundation because of lack of lateral support. The setback also provides space for drainage away from the slope without creating too steep a drainage profile, which could cause erosion problems. The setback space also provides access around the structure and helps to create a light and open environment. IBC Figure 1808.7.1 depicts the criterion for the setback or clearance. Figure 1808-2 herein depicts the criteria set forth in this item for determination of the toe of the slope when the slope exceeds 1:1.

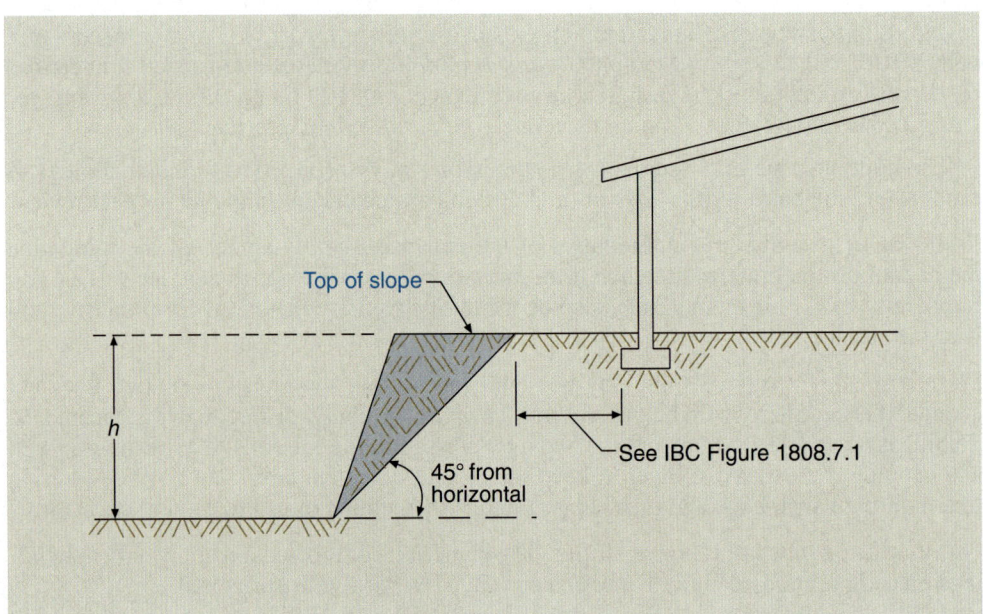

Figure 1808-2
Buildings adjacent to descending slope exceeding 1:1.

It is possible to locate a structure closer to the slope than indicated in IBC Figure 1808.7.1. The footing of the structure may be located on the slope itself, provided that the depth of embedment of the footing is such that the face of the footing at the bearing plane is set back from the edge of the slope at least H/3.

Pools. Figure 1808-3 depicts the criteria for the design of swimming pool walls near the top of a descending slope. The wall must be sufficient to resist the hydrostatic water pressure without support from the soil to protect against failure of the pool wall should localized minor slope movement or sloughing occur. The pool setback should be established as one-half of the setback required by IBC Figure 1808.7.1.

1808.7.3

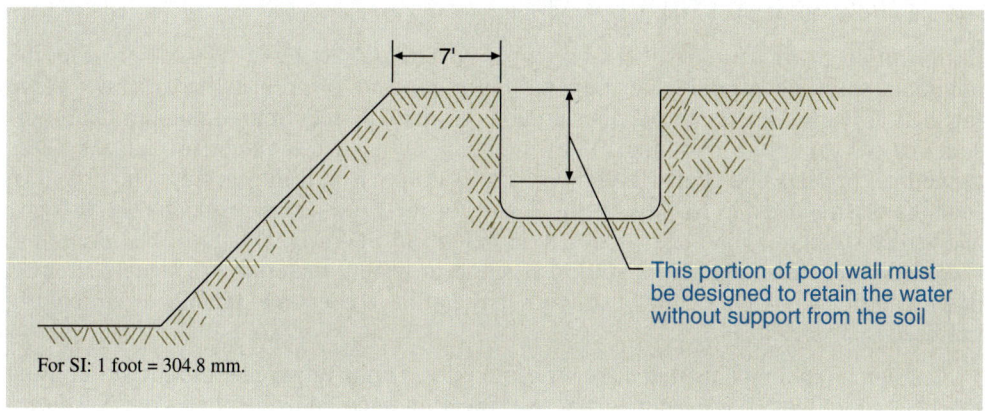

Figure 1808-3
Swimming pool adjacent to descending slope.

Foundation elevation. Figure 1808-4 depicts the criteria from this section for the elevation of the exterior foundations relative to the street, gutter, or point of inlet of a drainage device.

1808.7.4

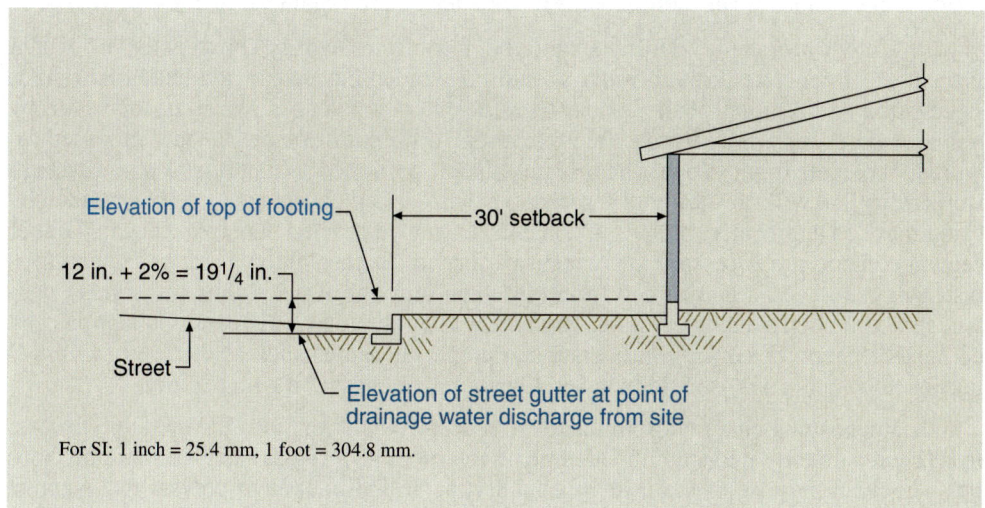

Figure 1808-4
Footing elevation on graded sites.

The elevation of the street or gutter shown is that point at which drainage from the site reaches the street or gutter. This requirement is intended to protect the structure from water encroachment in the case of heavy or unprecedented rains. This requirement may be modified if the building official finds that positive drainage slopes are provided to drain water away from the building and that the drainage pattern is not subject to temporary flooding from clogged drains, landscaping, or other impediments.

Alternate setback and clearance. This alternative procedure allows the building official to approve alternative setbacks and clearances from slopes, provided that the intent of Section 1808.7 is met. This section gives the building official the authority to require a

1808.7.5

geotechnical investigation by a qualified geotechnical engineer to establish that the intent of Section 1808.7 is met. The parameters for such geotechnical investigation are established in Section 1803.5.10 and include consideration of material, height of slope, slope gradient, load intensity, and erosion characteristics of the slope material.

1808.8 **Concrete foundations.** Footings may be designed, or the requirements of Table 1809.7 may be used, for structures with light-framed walls of conventional construction, where frost heave or expansive soils are not a problem. Table 1809.7, which originated with the UBC, is based on anticipated dead and live loads from the floors and roof and an assumed soil classification of ML, MH, CL, or CH.

1808.8.1 **Concrete or grout strength and mix proportioning.** A new Table 1808.8.1 provides the minimum specified compressive strength, f'_c for concrete or grout to be used in specific foundation types. Where the previous editions of the IBC required a minimum 2,500 psi compressive strength for footings, the new table now provides values for various SDCs as well as for piles and shafts. The minimum 2,500 psi is still the correct value for most footings of structures in SDC A, B, or C and for the footings of residential light-frame and utility structures, one or two stories in height, in SDCs D, E, and F. The minimum specified concrete strength of 2,500 psi is set to provide a material of adequate strength and durability. Concrete of lower strength may not have adequate durability, particularly in freeze-thaw areas.

The slump requirements stated are for cased piles. Slump requirements must be adjusted for other conditions. For example, concrete placed in uncased drilled holes needs to be in the 6- to 8-inch range so that the concrete flows readily into the irregularities in the drilled hole. Use of superplasticizers will provide the desired slump while keeping the water to cementitious material ratio low.

1808.8.2 **Concrete cover.** Cover requirements are set to provide protection and minimize steel corrosion. All concrete cover requirements of Chapter 18 are organized in Table 1808.8.2.

1808.8.3 **Placement of concrete.** Holes should be free from debris, loose soils, or water. Placement of concrete through water should be avoided because of the increased risk of segregation and dilution of the concrete paste. When concrete is placed under water, by tremie or other approved method, the mix must be different from the standard mix used for ordinary concrete foundations. The mix must be proportioned so that it is plastic with high workability and will flow without segregation. The desired consistency can be obtained by using rounded aggregates, high sand contents, entrained air, and superplasticizers. Higher cement contents are necessary to compensate for the increase in the water to cementitious materials ratio caused by dilution from placement through water. Minimum cement content should be 600 pounds per yard. Placement from the top of the deep foundations can cause segregation of concrete mix also, and proper measures such as the use of a funnel hopper (elephant trunk) should be taken to avoid the potential for segregation.

When depositing concrete from the top of a deep foundation, the IBC requires concrete to be chuted directly into smooth-sided pipes and tubes or through a centering funnel hopper. The main purpose of the centering funnel for drilled piles is to prevent the concrete from encountering the soil at the perimeter of the hole, which is generally not a problem for pipes and tubes. The term smooth sided is included in the code to prevent possible segregation from the ridges or corrugations if nonsmooth pipes or tubes are used.

1808.8.4 **Protection of concrete.** Concrete footings should not be placed during rain, sleet, snow, or freezing weather without protection against either freezing or increase in water content at the surface from rain while plastic. If concrete placement is undertaken under such conditions without adequate protection, numerous complications can be expected. There have been many cases where a project was forced to come to a halt while concrete core samples, testing, analysis, and investigation had to be performed on the hardened concrete to determine suitability of the deposited concrete and of the structural elements. See ACI 306R and ACI 306.1 for cold-weather concrete operations.

Forming of concrete. The soil should have sufficient strength and cohesion that the shape, dimensions, and vertical sides of the excavation can be maintained without sloughing prior to and during the concreting operations. Excavations in loose granular materials must be formed. — **1808.8.5**

Seismic requirements. Specific requirements for foundations in SDC C, D, E, or F are contained in Section 1905, which contains the modifications to ACI 318. In SDC D, E, or F, concrete must have a specified compressive strength of not less than 3,000 psi, except that 2,500 psi concrete strength is permitted in Group R or U occupancies of light-frame construction two stories or less in height. — **1808.8.6**

Buildings in SDCs D, E, and F are required to comply with ACI 318, Sections 21.12.1 through 21.12.4, except for detached one- and two-family dwellings of light-frame construction two stories or less in height. ACI 318 Section 21.12.1 covers foundation requirements in general, and Sections 21.12.2 through 21.12.4 cover requirements for footings, mat foundations, pile caps, grade beams, and slabs on grade.

Note that plain concrete is either unreinforced or lightly reinforced concrete that contains less reinforcing than required to meet the minimum reinforcement requirements set forth in ACI 318, Section 10.5.

Section 1809 Shallow Foundations

General. Foundations are divided into two major categories of shallow foundations and deep foundations. Shallow foundations are the individual or strip footings, the bottom of which is typically close to the surface such as mat foundations, slab on grade, or similar foundation types. Shallow foundations are regulated in Sections 1890.2 through 1809.13. Deep foundations are those that are not classified as shallow foundations. — **1809.1**

Supporting soils. Because the supporting soil for shallow foundations is close to the surface, in order to minimize differential settlement, shallow foundations must be constructed on undisturbed native soil, compacted fill material, or CLSM. Where constructed on fill, the material must be properly placed and compacted to achieve adequate density in accordance with Section 1804.5. CLSM must be placed and tested in accordance with Section 1804.6. — **1809.2**

Stepped footings. Footings are required to be stepped when the slope of the bearing surface exceeds 1 in 10. No recommendations or restrictions are provided. Figure 1809-1 schematically represents a satisfactory stepped foundation. The figure shows a recommended — **1809.3**

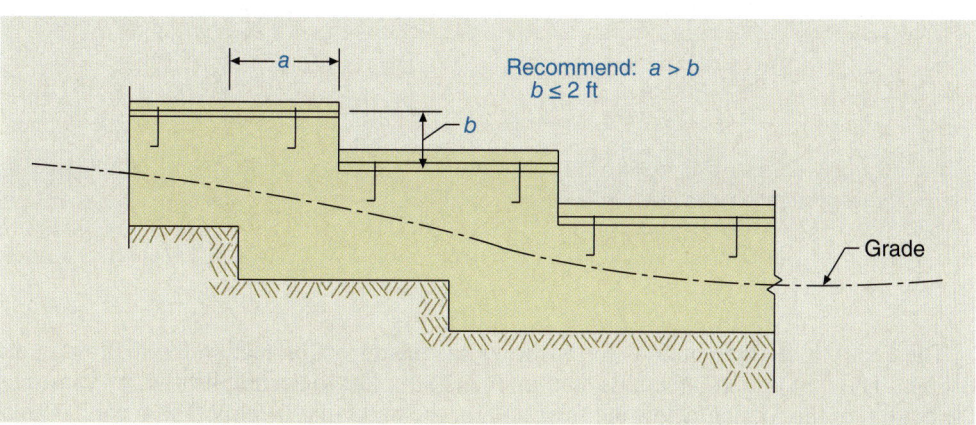

Figure 1809-1 Stepped foundations.

horizontal overlap of the top of the foundation wall beyond the step in the foundation to be larger than the vertical step in the foundation wall at that point. This is recommended to keep any crack propagation approximately at a 45-degree angle. To keep this cracking to a minimum, it is also recommended that the height of each step not exceed 1 to 2 feet. Other measures to protect against cracking, such as special reinforcing details, may be needed.

1809.4 Depth and width of footings. Footings should always be placed a minimum of 12-inches deep on either firm, undisturbed earth or properly compacted fill, and be at least 12 inches in width.

1809.5 Frost protection. To prevent frost heave during winter and subsequent settlement upon thawing, foundations and building supports should be placed on a stratum with adequate load-bearing resistance that is below the frost line. Frost heave occurs because of the increased soil volume from the freezing of pore water in the soil. Clay soils, particularly saturated clays, are most susceptible to frost heave. Well-drained sands and gravels will not be susceptible to significant movement. If the foundations are built on soils that can freeze, the resulting frost heave, which is rarely uniform, can cause serious damage from differential settlements.

The frost line is defined as the lowest depth below the ground surface to which a temperature of 32°F extends. The factors governing the depth of the frost line are air temperature, the length of time the air temperature is below freezing (32°F), and the soil's thermal conductivity. Frost lines vary significantly throughout the country from no penetration in southern Florida to 100 inches in the northern regions of Michigan and Maine. Data on frost penetration are available from the U.S. Department of Commerce Weather Bureau. See Figure 1809-2.

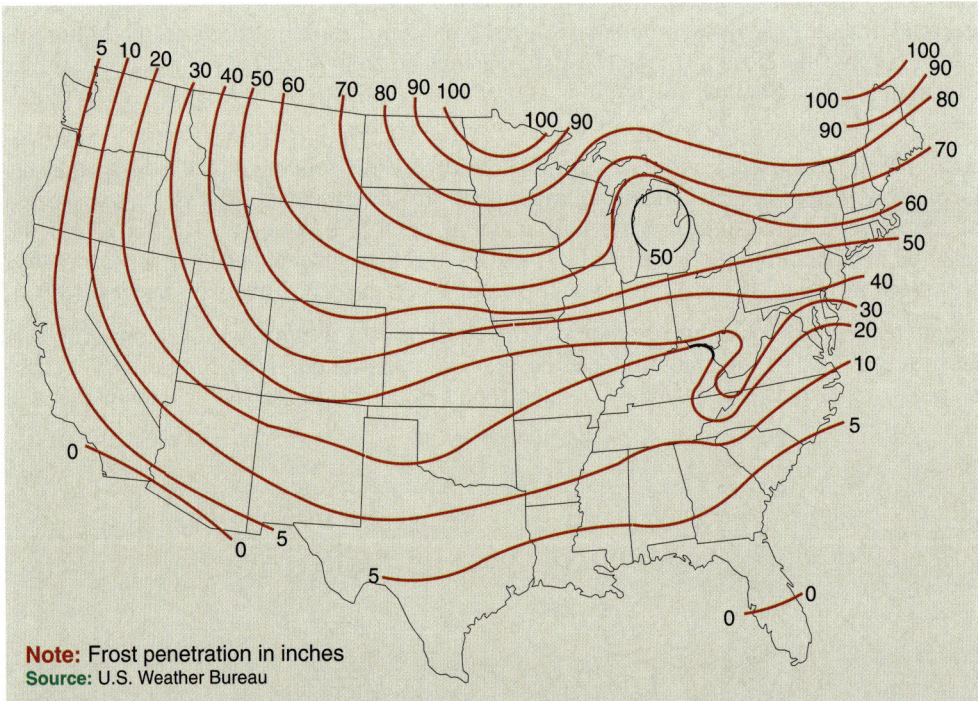

Figure 1809-2
Frost penetration depths.

Note: Frost penetration in inches
Source: U.S. Weather Bureau

The code offers three options for ensuring adequate frost protection for shallow foundations. The first option—the most common and simplest to accomplish—is to construct the bottom of the footing below the frost line for the particular locality. The second option is to construct the footing in accordance with the referenced standard, ASCE 32, *Design and Construction of Frost-Protected Shallow Foundation*. The third option, which is often encountered in areas where bedrock is prevalent, is to construct the footing on solid rock.

Note the exception where frost-protected foundation is not required: free-standing buildings classified in Risk Category I (see Section 1604.5), floor area of 600 square feet or less for light-frame construction or 400 square feet or less for other than light-frame construction, and eave height of 10 feet or less. Note that the term *light-frame construction* is defined in Section 201 as a system that uses repetitive wood or light-gauge steel-framing members.

The code prohibits footings from bearing directly on frozen soil unless the soil is permanently frozen. Permafrost may not meet this condition as permafrost is considered soil that remains in a frozen state for more than two years in a row.

Location of footings. This restriction is intended to minimize the influence of vertical and lateral loads from footings at a higher elevation on footings at a lower elevation. See Figure 1809-3. **1809.6**

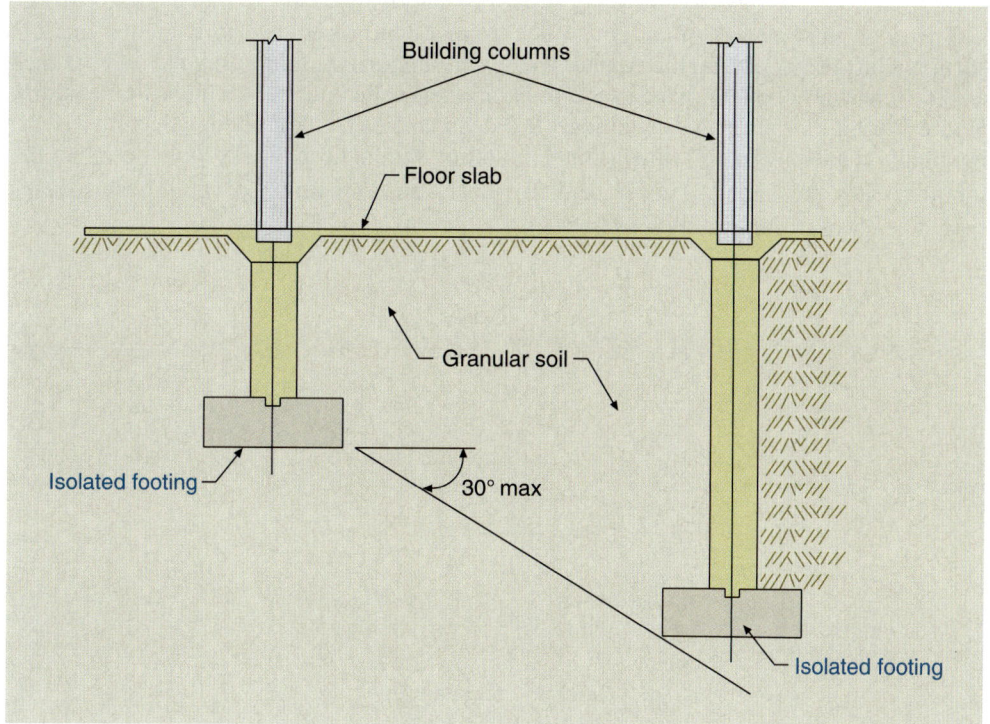

Figure 1809-3 Isolated foundation.

Prescriptive footings for light-frame construction. Footings of concrete or masonry unit may be designed, or the requirements of Table 1809.7 may be used, for structures with light-frame walls of conventional construction, where frost heave or expansive soils are not a problem. IBC Table 1809.7 shown herein as Table 1809-1, which originated with the UBC, is based on anticipated dead and live loads from the floors and roof and an assumed soil classification of ML, MH, CL, or CH. **1809.7**

Plain concrete footings. In compliance with ACI 318, Section 22.7.4, the edge thickness of plain concrete footings in other than light-frame construction must not be less than 8 inches. In accordance with ACI 318, Section 22.4.7, the thickness or depth used to compute footing stresses (flexure, combined axial load and flexure, or shear) should be 2 inches less than the actual thickness of the footing for footings cast against soil. This is done to allow for unevenness of excavation and contamination of the concrete adjacent to the soil. **1809.8**

18 Soils and Foundations

Table 1809-1. **Prescriptive Footings Supporting Walls of Light-Frame Construction**[a,b,c,d,e] (Formerly Table 1809.7)

Number of Floors Supported by the Footing[f]	Width of Footing (inches)	Thickness of Footing (inches)
1	12	6
2	15	6
3	18	8[g]

For SI: 1 inch = 25.4 mm, 1 foot = 304.8 mm.
a. Depth of footings shall be in accordance with Section 1809.4.
b. The ground under the floor shall be permitted to be excavated to the elevation of the top of the footing.
c. Interior stud-bearing walls shall be permitted to be supported by isolated footings. The footing width and length shall be twice the width shown in this table, and footings shall be spaced not more than 6 feet on center.
d. See Section 1908 for additional requirements for concrete footings of structures assigned to Seismic Design Category C, D, E, or F.
e. For thickness of foundation walls, see Section 1807.1.6.
f. Footings shall be permitted to support a roof in addition to the stipulated number of floors. Footings supporting a roof only shall be as required for supporting one floor.
g. Plain concrete footings for Group R-3 occupancies shall be permitted to be 6-inches thick.

The edge thickness of plain concrete footings can be reduced to 6 inches for R-3 occupancies, provided that the edge distance (projection) of the footing beyond the face of the stem wall does not exceed the thickness (6 inches depth = 6 inches extension). Figure 1809-4 illustrates this condition. (R-3 occupancies are described in Section 310.1.) For lightly loaded walls, this dimensional limitation should keep the flexural stresses in the footing below the limit of $5\phi\sqrt{f'_c}$ and the shear stresses below $2\phi\sqrt{f'_c}$. These stresses should be checked for heavily loaded walls.

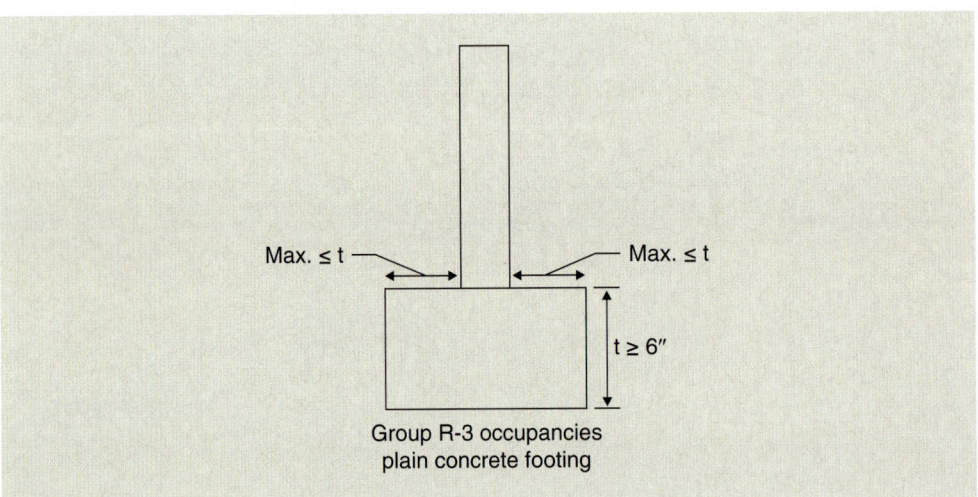

Figure 1809-4
Plain concrete footing.

1809.9 **Masonry-unit footings.** Masonry footings were widely used until the middle of the last century when they were replaced by steel or wood grillage footings, which in turn were replaced by more economical plain or reinforced concrete footings. Masonry footings were often built of stone cut to a specific size, or rubble masonry of random-sized stones bonded with mortar. Although seldom used, masonry footings may be constructed of hard-burned brick set in cement mortar to support lightweight buildings.

1809.9.1 **Dimensions.** Type M mortar is suitable for unreinforced masonry below grade or with earth contact. Type S should be used for reinforced footings. Projections of the footing beyond a wall or pier should not exceed one-half of the footing depth to keep the shear and flexural stresses in the footing within safe limits. For example, a footing with a 12-inch depth should project no more than 6 inches beyond the face of the wall.

Offsets. The stepping back, or racking, of successive courses of the foundation wall supported by a masonry footing must not exceed $1^1/_2$ inches for a single course or 3 inches for a double course. Where wide footings are necessary for bearing, the wall must be stepped back to keep the footing projection within the limits of Section 1809.9.1. See Figure 1809-5.

1809.9.2

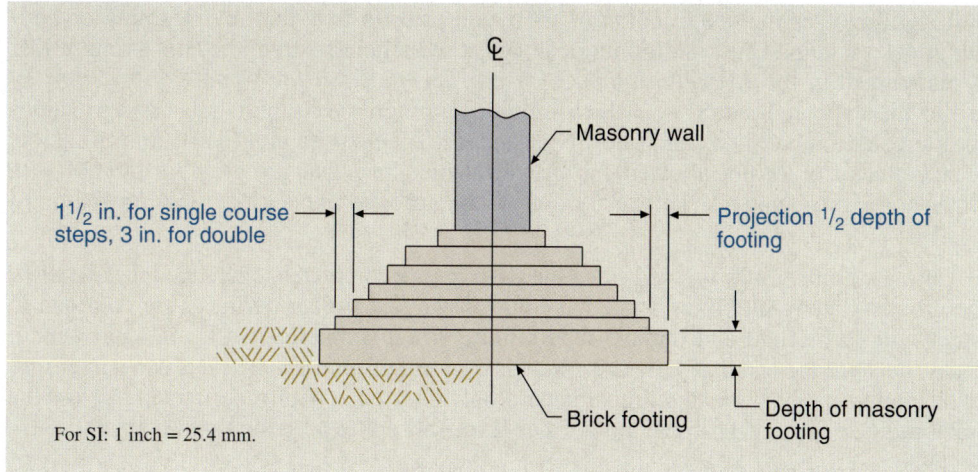

**Figure 1809-5
Brick footing wall offsets.**

Pier and curtain wall foundations. Pier foundations may not be used to support structures assigned to SDCs D, E, and F because seismic detailing requirements for higher SDCs have not been developed. These foundation elements have been popular particularly in the Eastern United States, and under this section are allowed only for the support of light-frame construction of not more than two stories above grade plane. "Story above grade plane" is defined in Section 202 as any story having its finished floor surface entirely above grade plane, or in which the finished surface of the floor next above is:

1809.10

1. More than 6 feet above grade plane; or
2. More than 12 feet above the finished ground level at any point.

Steel grillage footings. Steel grillage footings were used extensively during the latter half of the last century, but the development of reinforced concrete made grillage footings obsolete, except for underpinning work.

1809.11

A typical grillage consists of two or more tiers of steel beams, with each tier placed at right angles to the tier below. The beams in each tier are usually held together by a system of bolts and pipe spacers. For construction of new grillage footings, the beams should be clean and unpainted and the entire grillage system filled with and encased in concrete with at least 4 inches of cover. The grillage should be placed on a concrete pad at least 6-inches thick to distribute the load evenly to the soil.

Timber footings. Use of timber footings is allowed only for Type V structures. (Type V structures are described in Section 602.5 as that type of construction in which the structural elements, exterior walls, and interior walls are of any material permitted by this code. Buildings of Type V construction are limited in area and height in accordance with Chapter 5 of the code.) Timber must be pressure treated to American Wood Protection Association (AWPA) U1 (Commodity Specification A, Use Category 4B) standards, except for foundations permanently below the ground-water table. The pressure preservative treatment protects the timber from decay, fungi, and harmful insects. The AWPA Use Category System is based on the end use hazard, similar to other international standards for wood treatment. The Use Category System (UCS) is used to specify the wood treatment based on the desired wood species and the environment of the intended end use. There are six use categories, which describe the exposure conditions that wood may be subject to in service.

1809.12

18 Soils and Foundations

ICC in partnership with AWPA publishes all 24 AWPA standards that are referenced in the IBC. Stronger preservatives are necessary to prevent marine borers when timber foundations are used in coastal brackish or marine environments.

Preservative treatment by the pressure process within the limitations of the AWPA standards should not significantly affect the strength of the wood. Part of the process, however, involves the conditioning of the wood prior to treatment by steaming or boiling under vacuum. This conditioning can cause reduction in strength. This strength loss is recognized in the AF&PA *National Design Specification for Wood Construction* by use of the untreated factor, C_u, which provides an increase in the tabular design values for untreated timber poles and piles. Note that the NDS states that load duration factors greater than 1.6 are not allowed for structural members that are pressure treated with water-borne preservatives. This restriction would apply to impact loads that have a duration factor of 2.0.

Untreated timber may be used when the footings are completely embedded in soil below the ground-water table. Experience has long shown that timber permanently confined in water will stay sound and durable indefinitely. Wood submerged in fresh water cannot decay, because the necessary air is excluded. Because ground-water levels can sometimes change appreciably, untreated timber should only be used at depths sufficiently below the water table so that small drops in the water level will not expose the timbers to air.

1809.13 **Footing seismic ties.** Interconnection of individual spread footings is required for structures in SDCs D, E, and F sited on soils in Site Class E or F. The footings must be interconnected with ties capable of transmitting a force equal to the larger footing load multiplied by the short period response acceleration, S_{DS}, divided by 10 or 25 percent of the smaller footing design gravity load, whichever is smaller. The intent of this requirement is to minimize differential movement or spreading between the footings during ground shaking, and have the individual footings act as a unit. If slabs on grade, or beams within slabs on grade, are used to meet the tie requirement, the load path from footing to slab or beam/slab and across joints in the slab or beam/slab should be checked for continuity. The slab or beam must be reinforced for the design tension load. In addition, the slab should be checked for buckling under the required compression load using an assumed slab width of no more than six times the slab thickness.

Section 1810 Deep Foundations

1810.1 **General.** Foundations are divided into two major categories of shallow foundations and deep foundations. Shallow foundations are the individual or strip footings, the bottom of which is typically close to the surface such as mat foundations, slab on grade, or similar foundation types. Deep foundations are those that are not classified as shallow foundations. Deep foundations are regulated in Sections 1810.1 through 1810.4 and are those typically referred to as piles, piers, or caissons that transfer load from a superstructure such as a building to the underlying soils or rock that provides support for the superstructure. Deep foundations are generally used when the loads are too high to be supported by shallow footings or mats. Load resistance is provided by skin friction between soil and the sides of the pile and by end bearing at the tip. Settlement has to be limited within acceptable levels and may control the pile capacity for design purposes.

Deep foundations can be subjected to compression, tension, and lateral loads. The loads can be static and dynamic forces resulting from soil pressure, wind pressure acting on the building, or earthquake load effects from ground motion. In many cases, there are groups of piles, and group interaction must be considered. Closely spaced deep foundation elements have lower capacities.

Deep Foundations 18

Deep foundations can be constructed of cased or uncased concrete, precast-prestressed concrete, timber, steel pipe or H sections, or other special types such as micropiles and helical piles. Piles can be either driven in place or drilled in place. The installation method for a particular project depends on a variety of factors such as the geotechnical engineer's experience with local conditions, settlement limitations, acceptable noise levels, and so on.

The term *caisson* refers to deep foundations that are required when very heavy loads are supported such as high-rise building towers. Caissons are large-diameter concrete elements installed with a casing that can be left in place. The caisson bottom can be expanded, called a bell bottom, to provide more bearing area or it may include a rock socket. Trump Tower in Chicago is a 92-story building supported on 230 caissons bearing on rock and hardpan. The core caissons have a record-breaking 500-ksf bearing capacity with rock sockets up to 10 feet in diameter.[1]

Within Section 1810 provisions are found for seismic design and detailing for various types of deep foundations in SDCs C through F. These mostly reflect the concern of the NEHRP Code Resource Development Committee that significant ground motions can occur in SDC C. Additional requirements are imposed on deep foundation elements in SDCs D through F, including a requirement that the upper portion of piles be detailed as special moment-resisting-frame columns, to prevent failure of the piles under severe ground motions. These provisions intend to include the deep foundation element bending and curvatures resulting from horizontal ground movement during an earthquake in the structural design. The reinforcement in the deep element, required to resist the curvature effect, increases ductility of the foundation such that bending or shear failure is precluded.

Geotechnical investigation. See Section 1803.5.5 for discussion on geotechnical investigations. — **1810.1.1**

Use of existing deep foundation elements. This section allows the reuse of existing deep foundations where sufficient information is submitted to the building official to demonstrate they are adequate. This introduces flexibility for both the building designer and the building owner to make use of existing materials where it makes sense to do so. — **1810.1.2**

Deep foundations remaining after structures that are demolished should not be used for the support of new loads, unless evidence shows them to be adequate. This is because of the lack of soil data or detailed information on the piling material, and because of the unavailability of the pile-driving records or pier construction records made during the construction of these older buildings or structures. As such, the true condition of the deep foundation elements is unknown and, over time, they may have deteriorated, or their load capacity may have been reduced. Such deep foundation elements may be used, however, if they are load tested or in the case of piles, if they are retracted and redriven to verify their capacities. Only the lowest allowable load capacity as determined by test data or redriving information should be used in the design.

Deep foundation elements classified as columns. (Column Action.) The code requires that deep foundation elements standing unbraced in air, water, or material not capable of providing adequate lateral support shall be designed in accordance with the column formulas of the code. Obviously, water and air do not provide lateral support. On the other hand, most soils do provide lateral support, although exceptionally loose and unconsolidated fills, liquified sands, and remolded clays are inadequate to provide lateral support. Cast-in-place elements can be designed as a "pedestal" under conditions where unsupported height to least horizontal dimension is three or less. — **1810.1.3**

Special types of deep foundations. Deep foundations such as pile or pier types are basically classified in accordance with the structural material used, such as concrete, steel, or wood. They can also be categorized in accordance with the method of construction or installation. There are many variations of foundation types used in the construction of deep foundations, including some special or proprietary types beyond — **1810.1.4**

the scope of the code. Section 1810 includes only those basic pile types commonly used in the construction practices of today.

Special deep foundation types that are not specifically included in the provisions of the code are not precluded from use, provided that adequate information covering test data, calculations, structural properties, load capacity, and installation procedures is submitted and accepted by the building official.

1810.2 **Analysis.** Sections 1810.2.1 through 1810.2.5 are related to the analysis methodology for deep foundations and apply to all deep foundations in all locations except where a specific analysis is called for in higher seismic active areas. The discussion of lateral support, stability, settlement, lateral loads, SDCs D through F, and group effects discussed below is related to analysis procedures and requirements.

1810.2.1 **Lateral support.** This section provides needed guidance to the designer and building official on what constitutes adequate lateral support for deep foundations. It specifies that any soil other than fluid soil is allowed to be considered to provide lateral support to prevent buckling of deep foundation elements. Liquefaction causes loss of lateral-bearing capacity with resulting loss of support for deep foundation elements. Loss of lateral support can also occur from the soil strength loss associated with sensitive or quick clays that are sensitive to remolding effects. These clays lose significant strength when remolded, as might occur when a pile foundation is moved through muds by seismic-induced displacements.

Portions of deep foundation elements standing unbraced in air, water, or material not capable of providing adequate lateral support are permitted to be considered laterally supported at a depth of 5 feet in stiff soils and at a depth of 10 feet in soft soils, as determined by the geotechnical investigation.

1810.2.2 **Stability.** A group of deep foundation elements such as piles designed to support a common load or to resist horizontal forces must be braced or rigidly tied together to act as a single structural unit that will provide lateral stability in all directions. Deep foundation elements connected by a rigid, reinforced concrete pile cap are deemed to be sufficiently braced to meet the intent of this provision.

This section clarifies that for pile or pier groups to be considered to provide lateral stability, they must meet the radial spacing requirements defined herein. Three or more deep foundation elements are generally used to support a building column load or other isolated, concentrated load. In a three-element group, lateral stability is assured by requiring that the elements are located such that they will not be less than 120 degrees apart as measured from the centroid of the group in a radial direction.

For stability of deep foundation elements in a group supporting a wall structure, the elements are braced by a continuous, rigid footing and are alternately staggered in two lines at least 1 foot apart and symmetrically located on each side of the center of gravity of the wall. Other approved deep foundation element arrangements may be used to support walls, provided the elements are adequately braced and lateral stability of the foundation construction is assured.

Exception 1 allows isolated cast-in-place deep foundation elements without lateral bracing where the minimum dimension is 2 feet, and the length must be less than or equal to 12 times the least dimension of the pier.

Exception 2 allows one- and two-family dwellings of lightweight construction, such as R-3 buildings, not exceeding two stories above grade plane or 35 feet in height, a single row of piles located within the width of the wall.

1810.2.3 **Settlement.** The purpose of a settlement analysis is to provide the data needed to design a deep foundation system that will maintain the stability and structural integrity of the supported building or structure. The load-bearing stratum of every soil must support the loads transferred through the deep foundation system, as well as the weight of all soil above. The capability of the strata underlying the deep foundation element to support additional loads without detrimental settlement can often be determined by analytical procedures. For example, serious settlements in a pile foundation system, particularly differential settlements, can cause great structural damage to the supported structure and the foundation itself.

Although the settlement analysis of an individual deep foundation element is complex, the analysis of a group of elements is significantly more complicated because of the overlapping soil stresses caused by closely spaced elements. Analytical procedures vary with the type of deep foundation element and especially with the soil conditions. Settlement analysis would generally include cases involving point-bearing piles on rock, and in granular soils and hard clay. It would also involve friction piles in sand and gravel soils, and in clay materials.

Load tests are often used to aid in the analysis. In the case of pile foundations in clay soils, however, there are no practical ways to determine long-term settlement from load tests, and therefore only approximations of settlement may be derived from laboratory tests.

Lateral loads. Deep foundation moments, shears, and deflections must be based on nonlinear soil-deep foundation element interaction. If using deep foundations in soils with lenses of soft clays, lenses subject to liquefaction, or soils susceptible to strength loss from remolding, the effects of these layers on the curvature and, hence, moment demands on the deep foundations should be investigated. Often these soil-induced curvatures place a much higher moment demand on the deep foundations than would be determined from conventional lateral force P-y analysis. If the ratio of the depth of embedment of the deep foundation element to the element diameter or width is less than or equal to six, the deep foundation element may be assumed to be rigid. See Winterkorn et al.[2] for methods of analysis of rigid piers.

1810.2.4

Seismic design categories D through F. In addition to the requirements for SDC C, moments, shears, and deflections must be based on nonlinear soil-pile interaction. If using pile foundations in soils with layers of soft to medium stiff clays, layers subject to liquefaction, or soils susceptible to strength loss from remolding (see Site Class E and F), the effects of these layers on the curvature and, hence, moment demands on the pile must be investigated. Often these soil-induced curvatures place a much higher moment demand on the piles than would be determined from conventional lateral force P-y analysis. In addition, at the interfaces of the layers described above and stiffer layers, plastic hinging may occur. Hence, confinement reinforcement per the concrete special-moment frame provisions must be provided at these interfaces and at the pile-to-cap connection for concrete piers and piles. See ASCE 7 Chapter 12. The curvature capacity requirements are considered to have been met without the analysis outlined in this section under two conditions. One condition is for precast-prestressed concrete piles where the transverse reinforcement detailing complies with Section 1810.3.8.3.3. These provisions were developed specifically for precast-prestressed piles to meet the curvature requirements. The other condition is for cast-in-place concrete deep foundation elements that meet a minimum longitudinal reinforcement ratio of 0.005 the full length of the element and detailed in accordance with ACI 318, Sections 21.6.4.2 through 21.6.4.4.

1810.2.4.1

Group effect. Where deep foundation elements are spaced far apart, they are considered as individual elements and analyzed and designed accordingly based on various requirements of the code. As spacing between the elements is reduced, the loads and stresses from one element may affect surrounding deep foundation elements. This is the spacing at which group action must also be considered. The spacing for group action consideration in analysis of lateral loads is a center-to-center spacing of less than eight times the least horizontal dimension of an element, and for axial loads is a center-to-center spacing of less than three times the least horizontal dimension of a deep element. See Figure 1810-1.

1810.2.5

18 Soils and Foundations

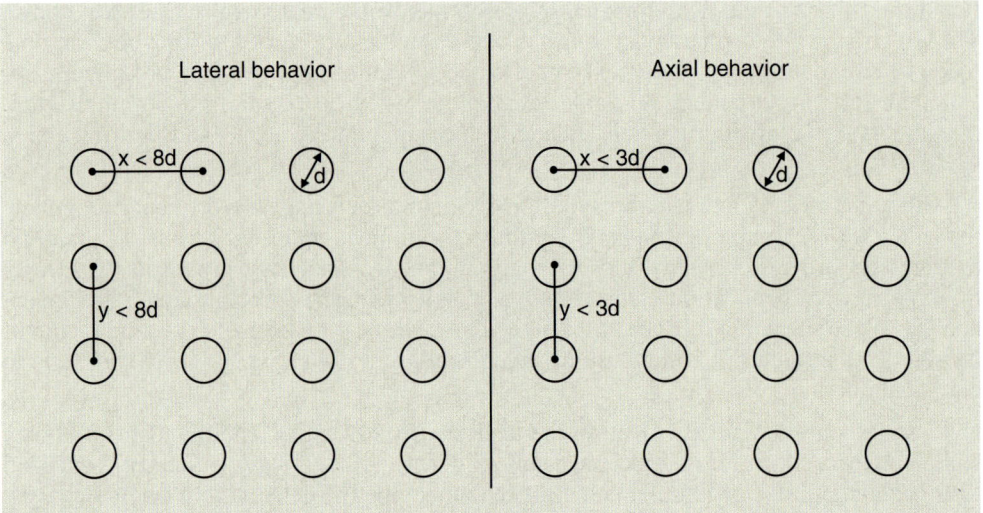

Figure 1810-1
Group effect.

Determination of the proper spacing of a pile group in relation to the type of pile foundation employed and the soil conditions encountered is a matter of design. The spacing of piles must be such that the loads transferred to the load-bearing strata do not exceed the safe load-bearing values of the supporting strata as determined by test borings, field load tests, or other approved methods.

1810.3 **Design and detailing.** Sections 1810.3.1 through 1810.3.12 are related to the design and detailing of deep foundations and address general requirements as well as special seismic requirements where applicable. The subsections of Section 1810.3 are listed in Table 1810-1.

Table 1810-1. **Deep Foundation Seismic Detailing Requirements**

1810.3.1	Design conditions	1810.3.8	Precast concrete piles
1810.3.2	Materials	1810.3.9	Cast-in-place deep foundations
1810.3.3	Determination of allowable loads	1810.3.10	Micropiles
1810.3.4	Subsiding soils	1810.3.11	Pile caps
1810.3.5	Dimensions of deep foundation elements	1810.3.12	Grade beams
1810.3.6	Splices	1810.3.13	Seismic ties
1810.3.7	Top of element detailing at cutoffs		

Driven piles and helical piles are required to be designed and manufactured in accordance with accepted engineering practice to consider handling, driving, and service conditions and are addressed briefly in Sections 1810.3.1.5 and 1810.3.5.3.3.

1810.3.1.3 **Mislocation.** Because of subsurface obstructions or other reasons, it is sometimes necessary to offset deep foundation elements a small distance from their intended locations so that they are not driven out of position. In such cases, the load distribution in a group of deep elements may be changed from the design requirements and cause some of them to be overloaded. To prevent major problems from displacements, the foundation system and the superstructure are required to be designed to resist the effects of deep foundation mislocation of at least 3 inches. This section requires that the maximum compressive load on any pile caused by mislocation not exceed 110 percent of the allowable design load. Deep elements such as piles exceeding this limitation must be extracted and redriven in the proper location or other approved remedies applied, such as installing additional piles to balance the group.

1810.3.1.4 Driven piles. Except for steel H-piles, driven piles covered in Chapter 18 are the displacement type. That is, as the pile is driven, a volume of soil is displaced by the pile volume, resulting in compaction of the surrounding soils. Piles must be designed to resist driving and handling stresses in addition to anticipated service loads. In long piles, tensile stresses resulting from driving may govern the design. In shorter piles, the handling loads may dominate. Driven uncased piles are displacement piles and are constructed by driving a temporary casing, removing the soils from the casing, and placing concrete in the hole as the casing is removed. The casing is driven with a closed end, thereby displacing and compacting adjacent soil during driving. The casing is kept closed either by a detachable tip, which is left in place when the casing is withdrawn, or by a mandrel that closes off the casing tip during driving.

1810.3.1.5 Helical piles. "Helical pile" is a manufactured steel deep foundation element consisting of a central shaft and one or more helical bearing plates. These are piles that are installed by rotating into the ground very much like screwing in a rotating plate. Provisions for helical piles are found in IBC Sections 202.1 (Helical Pile definition), 1810.3.1.5 (Helical Pile-design condition), 1810.3.3.1.9 (Helical Pile-allowable axial load), 1810.3.5.3.3 (Helical Pile-dimensions), 1810.4.11 (Helical Pile-installation), and 1810.4.12 (Helical Pile-special inspection).

1810.3.1.6 Casings. Casings are used in many deep foundation systems. The steel-cased pile is the most widely used type of cast-in-place concrete pile. This pile type is characterized by a thin steel shell and is a displacement pile. This pile type consists of a closed-end light-gauge steel shell or a thin-walled pipe driven into the soil and left permanently in place, reinforced when required for uplift, lateral bending, or seismic-induced curvatures, and filled with concrete. The shell or pipe is usually driven with a removable mandrel. The shell is either a constant section or a tapered shape. Steel-encased piles are generally friction piles. The steel shell must have sufficient strength to remain watertight and not collapse from ground pressure when the mandrel is removed.

1810.3.2 Materials. Deep foundations of various materials such as concrete, steel, and timber are covered in this section including seismic hook requirements and other relevant issues such as protection of materials and allowable stresses.

1810.3.2.1 Concrete. Concrete used for the bulb type of pile must have a zero slump to be stiff enough to be compacted by the drop weight. The maximum sized aggregate allowed is $3/4$ inch to allow proper compaction and prevent segregation. To prevent the pull-out of the hoops, spirals, and ties in higher seismic areas of SDCs C through F, seismic hoops as defined in ACI 318 should be used at the end of such hoops, spirals, and ties.

1810.3.2.2 Prestressing steel. Prestressing steel used in deep foundations must comply with ASTM A 416 *Specification for Steel Strand, Uncoated Seven-Wire for Prestressed Concrete.* Table 1808.8.1 lists when higher-strength concrete is used in prestressed piles to reduce volume changes, which reduce prestress losses, and to provide a more dense concrete to reduce cover requirements.

1810.3.2.3 Structural steel. Structural steel piles, pipes, and fully welded steel piles shall conform to the appropriate and applicable ASTM standard referenced in this section of the code. H-piles are typically available in ASTM A 36 and A 572 steel. Pipe piles are fabricated from ASTM A 252 or A 283 plate. ASTM A 252 is a specification specifically for welded pipe piles. Piles using ASTM A 690 and A 992 are used in common practice. The Pile Driving Contractors Association (PDCA) *Installation Specification for Driven Piles* (PDCA 102-07) contains both these material specifications for steel piles.

1810.3.2.4 Timber. Timber piles, although not having the high load capacities of steel or concrete piles, are the most commonly used type of pile, mainly because of their availability and ease of handling. Timber piles are shaped from tree trunks and are tapered because of the natural taper of the trunk. Round timber piles are generally made from Southern Pine in lengths up to 80 feet or Pacific Coast Douglas Fir in lengths up to 120 feet. Other species that are used are red oak and red pine in lengths up to 60 feet.

Untreated timber piles that are embedded permanently below the ground-water level (fresh water only, not brackish or marine conditions) may last indefinitely. If embedded above the water table, the piles are subject to decay, and if above ground are also subject to insect attack. Hence, piles should be preservative treated.

Timber piles may be either end-bearing or friction piles. ASTM D 25 sets forth the minimum circumference at the butt and the tip based on pile taper, as well as quality of the wood and tolerances on straightness, knots, twist of grain, and other requirements.

Piles must be pressure treated to prevent decay and insect attack. The IBC references AWPA U1 (Commodity Specification E, Use Category 4C) for round timber piles and AWPA U1 (Commodity Specification A, Use Category 4B) for sawn timber piles. See also discussion in Section 1809.12 for timber footings.

1810.3.2.5 **Protection of materials.** Unless properly protected, deep foundation elements may deteriorate because of biological, chemical, or physical actions caused by particular conditions that exist or that may later develop at the site. The durability of deep foundation elements will be long lasting if care is taken in the selection and protection of element materials.

Some of the problems associated with deep foundation element durability are:

- Untreated timber piles may be successfully used if they are entirely embedded in earth and their butts (cutoffs) are below the lowest ground-water level or are submerged in fresh water. The risk, however, is in situations where unexpected lowering of the water table occurs and exposes the upper parts of piles to decay and insect attack. Such conditions may occur where the water table is significantly lowered by pumping or deep drainage. There is also the remote possibility that wood piles will be damaged by the percolation of ground water heavily charged with alkali or acids.

- Wood piles extending above the water table or exposed to air or saltwater are subject to decay and attack by insects and marine borers. The piles need to be pressure treated with preservatives conforming to the requirements of AWPA U1.

- Steel piles that are driven and embedded entirely in undisturbed soil are generally not significantly affected by corrosion caused by oxidation, regardless of soil types or soil properties. The reason for this is that undisturbed soil is so deficient in oxygen at levels only a few feet below the ground line or below the water table that progressive corrosion is inhibited. However, where upper portions of steel piles protrude above ground into the air, where piles are placed in corrosive soils, or where ground water contains deleterious substances from sources such as coal piles, alkali soils, active cinder fills, chemical wastes from manufacturing operations, or other sources of pollutants, the steel may be subject to corrosive action. Under such conditions, steel piles may be protected by a concrete encasement or a suitable coating, extending from a level slightly above ground to a depth below the layer of disturbed earth. For piles above ground level that are exposed to air and subject to rusting, the steel should be protected by being painted, as any other type of structural steel construction would be protected.

- Steel piles installed in saltwater or exposed to a saltwater environment are subject to corrosion, and therefore should be protected with approved coatings or encased in concrete.

- Concrete deep elements, plain or reinforced, that are entirely embedded in undisturbed earth are generally considered permanent installations. The level of the water table does not normally affect the durability of concrete deep elements. Ground water that readily flows through either granular materials or disturbed soil and contains deleterious substances can have deteriorating effects. Concrete piles embedded in impervious clay materials will not generally suffer from ground water containing harmful substances. The primary deleterious substances that attack concrete are acids and sulfates. In the case of acids, it is best to use an alternative pile material if the acid attack is potentially destructive as coating piles may be ineffective because of soil abrasion during driving. Concrete can be attacked, however, by exposure to soils with high sulfate content. In the case of high alkaline soils with sulfate salts, Type V portland cement may be used. Where the exposures are only moderate, a Type II portland cement will usually be adequate. If the piles are in a marine environment, Type II or V portland cement is also indicated to provide the necessary sulfate resistance.

The conditions of the underground environment should be ascertained so as to protect deep foundation elements against possible corrosion of either the concrete or exposed load-bearing steel. Corrosion by oxidation is generally very minor and often disregarded. Corrosion caused by electrolytic action or by destructive chemicals on load-bearing steel can be protected by suitable coating, concrete encasement, and cathodic protection. This also applies to bare steel piles previously discussed. Concrete can be protected from chemical attack by the use of special cements, dense concrete mixtures, and special coatings.

Deep foundation elements installed in saltwater, such as for buildings or other structures in waterfront construction, are subject to chemical action on concrete materials coming from polluted waters, frost action on porous concrete, spalling of concrete, and rusting of steel reinforcement. Spalling may become particularly serious under tidal conditions where alternate wetting and drying occurs coupled with cycles of freezing and thawing. Spalling can be minimized or prevented by providing additional cover over the reinforcement; by the use of rich, dense concrete; by air entrainment and suitable concrete admixtures; and, in the case of precast piles, by careful handling to minimize stresses and avoid cracking during placement. See discussion of Chapter 19 and ACI 318 for more detailed information on concrete quality and concrete materials.

Allowable stresses. Allowable stresses for various types of deep foundations have been summarized in Table 1810.3.2.6. The allowable stress limits are set to provide an adequate margin of safety. **1810.3.2.6**

For precast-prestressed concrete, the term $0.33f'_c$ is the same as conventionally reinforced piles. Because prestressing places additional compressive stresses on the pile, this stress must be subtracted from the allowable compressive stress; hence, the subtractive term $-0.27f_{pc}$. The term f_{pc} is the effective prestress on the gross area, which is the prestressing force remaining after losses have occurred.

For H-piles, the stresses allowed consider stability. Tests have shown that for H-piles driven to refusal in rock through soils that provide full lateral support, the stresses at failure can approach the yield stress of the material. Hence, higher stresses are allowed if a soils investigation and load tests are performed.

For concrete cast in place without a permanent steel casing, pile capacity must be based on concrete strength alone without consideration of soil capacity, and the area of the internal cross section of the pile, that is, casing inside diameter. The allowable stresses for drilled uncased piles are the same as cast-in-place piles for which holes are formed by machine drilling with auger or bucket-type drills, with or without temporary casing. Concrete is placed by conventional methods including tremies or funnel hoppers. In augered piles, concrete is injected through the hollow stem auger as the auger is withdrawn. Reinforcement, if placed without lateral ties, is also placed through ducts in the hollow stem auger. Concrete drilled or augered cast-in-place uncased piles have an allowable stress limit of 33 percent of the 28-day specified compressive strength (f'_c), and an allowable compressive stress in the reinforcement of 40 percent of the yield strength of the steel (or 30,000 psi). These limits are consistent with ASCE *Standard Guidelines for the Design and Installation of Pile Foundations* (ASCE 20-96). For steel-cased deep concrete foundation elements, the allowable stress is $0.33 f'_c$ as for other concrete piles because the steel shell is too thin to act as a composite pile but does act as confinement reinforcement. The allowable stress is $0.40 f'_c$ if the shell meets the confinement conditions in Section 1810.3.2.7.

1810.3.2.7 **Increased allowable compressive stress for cased cast-in-place elements.** The conditions discussed in Section 1810.3.2.6 and Table 1810.3.2.6 where the allowable concrete compressive stress is allowed to be increased consist of six conditions. Note that the shell thickness is not considered load carrying, the shell must be seamless or spirally welded, the yield strength of steel normally used for the casings is 30 ksi, and the maximum diameter is set so that the volumetric ratio of shell to concrete is sufficient to provide confinement for the concrete.

1810.3.2.8 **Justification of higher allowable stresses.** In those sections of Chapter 18 that specifically deal with the types of elements most commonly used in the construction of deep foundations, there are limitations placed on the stresses that can be used in the deep foundation element design. In most cases, the allowable stresses are stated as a percentage of some limiting strength property of the deep foundations material. For example, in the case of piles made of steel materials, the allowable stresses are prescribed as a percentage of the yield strength of the several grades of steel that can be used for pile construction. For concrete, the allowable stress is stated as a percentage of compressive strength of the material. The allowable stresses for timber piles are based on tabulations of already reduced stresses. The reduced stresses are based on the strength values of different species of wood and reductions in strength caused by preservative treatment.

The allowable design stresses designated in Chapter 18 for each of the different types of deep foundations are intended to provide a factor of safety against the dynamic forces of deep foundation element driving that may cause damage to the element, and to avoid overstresses in the element under the design loads and other loads that may be induced by subsoil conditions.

This section allows the use of higher allowable stresses when evidence supporting the values is submitted and approved by the building official. The data submitted to the building official should include analytical evaluations and findings from a foundation investigation as specified in Section 1803, and the results of load tests performed in accordance with the requirements of Section 1810.3.3.1.2. The technical data and the recommendation for the use of higher stress values must come from a registered design professional that is knowledgeable in soil mechanics and experienced in the design of deep foundations. This registered design professional must supervise the deep foundation design work and witness the installation of the deep foundation so as to certify to the building official that the construction satisfies the design criteria. In any case, the use of greater design stresses should not result in design loads that are larger than one-half of the ultimate axial load capacity (see Section 1810.3.3.1.2).

1810.3.3 **Determination of allowable loads.** The IBC specifies that the determination of allowable loads shall be based on one of three methods:

1. An approved driving formula
2. Load tests
3. Foundation analysis

In most cases, the allowable loads will be determined by a combination of Items 2 and 3. However, there may be circumstances where the soil conditions, such as granular soils, and the type of deep foundation selected are such that the use of an approved dynamic deep foundation driving formula can be an aid to a qualified practitioner in establishing reasonable but safe allowable loads for the foundation system. Nevertheless, some literature indicates that "the use of a complicated formula is not recommended since such formulas have no greater claim to accuracy than the more simple ones."[7]

The dynamic pile-driving formula included in the 1970 and earlier editions of the UBC was dropped from the code because of its unreliability for cohesive soils. It is interesting to note that the earlier editions of the UBC utilized the so-called Engineering News formula, $R = 12WH/S+c$, which is the most simple of the dynamic pile-driving formulas. In 1937, the Pacific Coast formula was adopted into the UBC until its deletion prior to the 1973 edition of the code. This was one of the more complex dynamic pile-driving formulas and was based on a dynamic pile-driving formula developed by Terzaghi. However, as stated previously, in the hand of a qualified practitioner, a dynamic pile-driving formula does have some utility even though the IBC no longer provides such a formula.

There are two general considerations for determining capacity as required for the design and installation of deep foundations. The first consideration involves the determination of the underlying soil or rock characteristics. The second is the application of approved driving formulas, load tests, or accepted methods of analysis to determine the pile capacities required to resist the applied axial and lateral loads, as well as to provide the basis for the proper selection of pile-driving equipment.

Allowable axial load determination is further addressed in Sections 1810.3.3.1.1 through 1810.3.3.1.9.

Driving criteria. Deep foundation elements must be of a size, strength, and stiffness capable of resisting without damage: **1810.3.3.1.1**

- Crushing caused by impact forces during driving.
- Bending stresses during handling.
- Tension from uplift forces or from rebound during driving.
- Bending stresses caused by horizontal forces during driving.
- Bending stresses caused by deep element curvatures.

Additionally, the deep element must be capable of transmitting dynamic driving forces to mobilize the required ultimate deep element capacity within the soil without severe elastic energy losses. Driveability depends on the deep element stiffness, which is a function of deep element length, cross-sectional area, and modulus of elasticity. Yield strength does not affect stiffness. Thus, caution should be observed in the use of high-yield-strength steels for high loads on smaller cross sections requiring high dynamic driving energy. For allowable loads greater than 40 tons, a wave equation method of analysis reflects deep element stiffness or driveability. The selection of deep foundation types and dimensional requirements for driveability is a function of soil characteristics.

For many decades, it has been the practice to try to predict the capacity of a deep foundation element from its resistance to driving. The usual procedure has been to make such determinations by the application of pile-driving formulas, none of which have been completely dependable. The singular premise used in the development of these formulas is simple and is best expressed by R.B. Peck as follows: The greater the resistance of a pile to driving, the greater the pile's capacity to support load. With complex engineering problems, however, occasionally there are special circumstances under which there will be exceptions to general statements of this kind.

There are many pile-driving formulas. The simplest and most widely used formula in the United States is the Engineering News formula. This particular expression and other formulas in common use today have all generally shown poor correlations with load test results. Such comparisons are considerably better, however, when they are applied to the determination of deep foundation element capacity in soils consisting of free-draining, coarse-grained materials such as sand and gravel. In soils such as silt, clay, and fine sand, the water cannot escape fast enough during driving operations to not have an adverse influence on the frictional resistance of the piles. As a consequence, information may be unreliable.

This section limits the allowable compression load on a deep foundation element as established by an approved driving formula to a maximum of 40 tons. Generally, the use of pile-driving formulas to determine pile capacity should be avoided except, perhaps, in cases involving small jobs where the piles are to be driven in well-drained granular soils and load testing cannot be economically justified.

1810.3.3.1.2 Load tests. This section specifies the standards to be used to load test deep foundation elements, where higher compressive loads than allowed in other sections of the code are exceeded or where cast-in-place deep foundation elements have an enlarged base formed either by compacting concrete or by driving a precast base. See the discussion under Section 1810.3.3.1.3 for test evaluation methods.

Questions were raised at the public hearings as to whether or not ASTM D 4945, which is a dynamic test, is sufficient by itself to verify the pile capacity. Many standards, including ASTM D 4945, indicate that a dynamic test may not be sufficient without a static test (ASTM D 1143) to calibrate the results, but leave it up to the registered design professional to decide if the dynamic test is sufficient. Other standards require a static load test to calibrate the dynamic test.

The safest method for determining deep foundation element capacity is by load test. The load-bearing capacity of enlarged base piles is specifically required to be determined from load tests. A load test should be conducted wherever feasible and used where the deep foundation element capacity is intended to exceed 40 tons per element (see Section 1810.3.3.1.2). Test deep elements are to be of the same type and size as intended for use in the permanent foundation and installed with the same equipment, by the same procedure, and in the same soils intended or specified for the work.

Load tests are to be conducted in accordance with the requirements of ASTM D 1143, which covers procedures for testing vertical or batter foundation piles, individually or in groups, to determine the ultimate pile load (pile capacity) and whether the pile or pile group is capable of supporting the loads without excessive or continuous settlement. Recognition, however, must be given to the fact that load-settlement characteristics and pile capacity determinations are based on data derived at the time and under the conditions of the test. The long-term performance of a pile or group of piles supporting actual loads may produce behaviors that are different than those indicated by load test results. Judgment based on experience must be used to predict pile capacity and expected behavior.

The load-bearing capacity of all deep foundation elements, except those seated on rock, does not reach the ultimate load until after a period of rest. The results of load tests cannot be deemed accurate or reliable unless there is an allowance for a period of adjustment. For piles driven in permeable soils such as coarse-grained sand and gravel, the waiting period may be as little as two or three days. For test piles driven in silt, clay, or fine sand, the waiting period may be 30 days or longer. The waiting period may be determined by testing (i.e., by redriving piles) or from previous experience.

This section also requires that at least one deep foundation element be tested in each area of uniform subsoil conditions. The statement should not be misconstrued to mean that the area of test is to have only one uniform stratum of subsurface material, but rather that the soil profile, which may consist of several layers (strata) of different materials, must represent a substantially unchanging cross section in each area to be tested.

The allowable deep foundation element load to be used for design purposes is not to be more than one-half of the ultimate deep element capacity, as determined by the load test in which the net settlement of the test element is not to exceed 0.01 inch per ton or more than a total of $^3/_4$ inch. The rate of penetration of permanent deep foundation elements must be equal to or less than that of the test element(s).

All production deep foundation elements should be of the same type, size, and approximate length as the prototype test elements, as well as installed with the same or comparable equipment and methods. They should also be installed in soils similar to those of the test element.

Load test evaluation methods. Three specific methods are given that are acceptable for performing deep foundation load tests. Other methods are permitted at the discretion of the building official. **1810.3.3.1.3**

Allowable frictional resistance. Resistance that is due to skin friction is limited to a maximum of 500 psf unless a greater value is permitted by the building official based on recommendations of an approved geotechnical investigation or a greater value is substantiated by load test methods described in Section 1810.3.3.1.2. **1810.3.3.1.4**

Uplift capacity of a single deep foundation element. This section gives both the designer and building official needed guidance on criteria to use for design of a single deep foundation element for uplift. The IBC requires an approved method of analysis with a safety factor of 3 or a test in accordance with ASTM D 3689. The maximum allowable uplift load cannot exceed the ultimate load capacity determined by the methods described in Section 1810.3.3.1.2 divided by a factor of safety of two. **1810.3.3.1.5**

When deep foundation elements are designed to withstand uplift forces, they act in tension and are actually friction elements. The amount of tension that can be developed not only depends on the strength properties of the element, but also on the frictional or cohesive properties of the soil. The uplift or tensile resistance of a deep element is not necessarily a function of its load-bearing capacity under compressive load. For example, the tensile resistance of a friction pile in clay will usually be about the same value as its load-bearing capacity, as the skin friction developed in such soils is very large. In contrast, a friction pile in sand or in other granular materials will develop a tensile resistance considerably less than its load-bearing capacity.

Where the properties of the soil are known, the ultimate uplift resistance value of a pile can be determined by approved analytical methods. This section requires that where the ultimate tensile value is determined by analysis, a safety factor of 3 must be applied to establish the allowable uplift load of the deep foundation element.

The best way to determine the response of a vertical or batter pile to a static tensile load (uplift force) applied axially to the pile is by applying an extraction test in accordance with the requirements of ASTM D 3689. The maximum allowable uplift load is not to be more than one-half of the total test load. This section of the code gives a limitation on the upward movement of the pile in compliance with the provisions of the ASTM D 3689 test method. The measurements of pile movement in the standard test procedure, however, are time-dependent incremental measurements and should be adhered to in determining allowable pile load.

To be effective in resisting uplift forces as tension members of a foundation system, deep foundation elements must be well anchored into the cap by adequate connection devices. In turn, the cap must be designed for the uplift stresses. Deep foundation elements must also be designed to take the tensile stresses imposed by the uplift forces. For example, concrete piles must be reinforced with longitudinal steel to take the full net uplift. Special consideration needs to be given in the design of pile splices that are intended to act in tension. When design uplift is due to wind or seismic loading, the factor of safety for the analytical method requires a factor of safety of 2 while the load test method requires a factor of safety of 1.5.

1810.3.3.1.6 Uplift capacity of grouped deep foundation elements. The allowable uplift load on a group of deep foundation elements is to be reduced from the value obtained on a single element as described in Section 1810.3.3.1.5 of the code and in compliance with comprehensive analytical methods. In the 2012 IBC, the capacity of deep foundation groups is also limited to two-thirds of the weight of the group and the soil contained in the group plus two-thirds of the ultimate shear resistance along the soil block. This is consistent with requirements in other sections in the code on uplift and overturning, where the dead load resistance is limited to two-thirds of the weight. Previous editions of the IBC have allowed two-thirds of the effective weight of a pile group and the weight of the soil contained within the block defined by the perimeter of the group, but did not include an allowance for the shear resistance of the soil block. This was unreasonably conservative because not only the weight of the soil within the pile group resists uplift, but also the shear resistance developed contributes to the resistance to uplift of the pile group. The 2012 IBC now allows the use of two-thirds of the effective weight of the pile group, two-thirds of the weight of the soil contained within a block defined by the perimeter of the group and the length of the piles, plus two-thirds of the ultimate shear resistance along the soil block. Where the center-to-center spacing of deep foundation elements is at least 2.5 times the least horizontal dimension of the largest single element, the allowable working uplift load for the group must be calculated by an approved method of analysis.

1810.3.3.1.7 Load-bearing capacity. The load-bearing capacity of a deep foundation element is determined as a deep element–soil system. For example, the load-bearing capacity of a single pile is the function of either the structural strength of the pile or the supporting strength of the soil. The load-bearing capacity of the deep foundation element is controlled by the smaller value obtained in the two considerations. The load-bearing capacity of a deep element group may be greater than, equal to, or less than the capacity of a single element multiplied by the number of elements in the group, depending on deep element spacing and soil conditions.

Because the supporting strength of the soil generally controls the load-bearing capacity of a deep foundation element, this section requires that the ultimate load-bearing capacity of an individual element or a group of elements be at least twice the design load capacity of the supporting load-bearing strata.

Sometimes, weaker layers of soil underlie the soil load-bearing strata supporting a pile foundation and may cause damaging settlements. Under such subsurface conditions, it must be determined by an approved method of analysis that the safety factor has not been reduced to a figure less than 2. Otherwise, the piles are to be driven to deeper load-bearing soils to obtain adequate and safe support, or the design capacity is to be reduced and the number of piles increased.

1810.3.3.1.8 Bent deep foundation elements. Deep foundation elements that are discovered to have sharp or sweeping bends because of obstructions encountered during the driving operations or for any other cause are to be analyzed by an approved method, or a representative deep element is to be load tested to determine its load-carrying capacity. Otherwise, the deep foundation elements could be used at some reduced capacity as determined by test or analysis; or, if necessary, they can be abandoned and replaced.

1810.3.3.2 Allowable lateral load. Because of wind loads, unbalanced building loads, earth pressures, and seismic loads, it is inevitable that individual deep foundation elements or groups of vertical elements supporting buildings or other structures will be subjected to lateral forces. The distribution of these lateral forces to the deep elements largely depends on how the loads are carried down through the structural framing system and transferred through the supporting foundation to the deep foundation elements. The amount of lateral load that can be taken by the deep foundation element is a function of (1) the type of deep foundation element used; (2) the soil characteristics, particularly in the upper 10 to 30 feet of the deep foundation element; (3) the embedment of the deep foundation element head (fixity); (4) the magnitude of the axial compressive load on the deep foundation element; (5) the nature of the lateral forces; and (6) the amount of horizontal deep foundation element movement deemed acceptable.

The degree of fixity of the deep foundation element head is an important design consideration under very high lateral loading unless some other method, such as the use of batter piles, is employed to resist lateral loads. The fixing of the deep foundation element head against rotation reduces the lateral deflection. In general, pile butts are embedded 3 to 4 inches into the pile cap (see Section 1810.1.4) with no ties to the cap. These pile heads are neither fixed nor free, but somewhere in between. Such construction is satisfactory for many loading conditions, but not for high seismic loads.

The magnitude of friction developed between the surfaces of two structural elements in contact with each other is a function of the weight or load applied. The larger the weight, the greater the frictional resistance developed. In the design of deep foundation elements, frictional resistance between the soil and the bottom of the deep element caps (footings) should not be relied on to provide lateral restraint, because the vertical loads are transmitted through the deep foundation elements to the supporting soil below and to the ground immediately under the deep element caps. Only the weights of the caps can supply some frictional resistance insofar as such footings are constructed by placing fresh concrete on the soil, thus providing a positive contact. The weight of the caps in comparison to the magnitude of loads and lateral forces transmitted to the deep foundation elements is nominal and not significant from a structural design standpoint. Also, in rare occurrences, soils have been known to settle under caps, leaving open spaces and thus eliminating the development of any frictional restraint.

Where vertical deep elements are subjected to lateral forces exceeding acceptable limitations, the use of batter piles may be required. Lateral forces on many structures are also resisted by the embedded foundation walls and the sides of the deep foundation element caps.

The allowable lateral-load capacity of a single deep foundation element or group of such elements is to be determined either by approved analytical methods or by load tests. Load tests are to be conducted to produce lateral forces that are twice the proposed design load; however, in no case is the allowable deep foundation element load to exceed one-half of the test load, which produces a gross lateral element movement of 1 inch as measured at the ground surface or the top of foundation element, whichever is lower. This criterion can be exceeded if it can be shown that the predicted lateral movement will not cause any harmful distortion of or instability in the structure and that no element will be loaded beyond its capacity.

Subsiding soils. Where deep foundation elements are driven through subsiding soils and derive their support from underlying firmer materials, the subsiding soils cause an additional load to the deep foundation elements through so-called negative friction. This negative friction is actually a downward friction force on the deep foundation elements, which increases the axial load on such elements. The code permits an increase in the allowable stress on the deep foundation elements if an analysis of the geotechnical investigation indicates that the increase is justified. **1810.3.4**

Dimensions of deep foundation elements. Deep foundation elements must have minimum dimensions as described in Sections 1810.3.5.1 through 1810.3.5.3 for precast, cast-in-place cased, and cast-in-place uncased deep foundations. **1810.3.5**

Precast. Eight inches is the minimum practical dimension to accommodate reinforcement. **1810.3.5.1**

1810.3.5.2 Cast-in-place or grouted-in-place. Eight inches is the minimum practical dimension to accommodate reinforcement for cased cast-in-place deep foundation elements. For uncased cast-in-place deep foundation elements, the minimum 12-inch diameter is for inspection purposes. The length-to-diameter ratio is based on construction and stability considerations.

1810.3.5.2.3 Micropiles. A micropile is defined as a bored, grouted-in-place deep foundation element that develops its load-carrying capacity by means of a bond zone in soil, bedrock, or a combination of soil and bedrock. The maximum outside diameter of a micropile is 12 inches. This dimension was originally part of the definition of a micropile when first introduced in the 2006 IBC. The 12-inch dimension is no longer in the definition but is now used in the section as the technical criterion by which a micropile is identified.

1810.3.5.3.1 H-piles. Structural steel piles are characterized by high axial load capacity. Piles may be H-piles or steel pipe piles.

H-piles are usually used as deep end-bearing piles because they are essentially nondisplacement-type piles that can readily penetrate solid strata to reach rock or other suitable hard-bearing strata such as dense gravels. Ideally, steel H-piles are driven to hard or medium hard rock.

H-piles are proportioned to withstand the impact stresses from hard driving. The flange and web thicknesses are usually equal. The flange widths are proportioned such that the section modulus, S_y, in the weak axis is approximately one-third of S_x.

1810.3.5.3.2 Steel pipes and tubes. Driven pipe piles are displacement-type piles if driven closed, and are nondisplacement-type piles if driven open. Pipe piles are made of seamless or welded pipes and are frequently filled with concrete after driving. Pipe piles conforming to ASTM A 252 are used in both friction and end-bearing applications. Pipe piles may be driven open ended or closed ended. Open-ended pipe piles are generally used when the geotechnical investigation shows rock or a suitable end-bearing stratum close to the ground surface, especially if the loads to be supported are large. The pipe is driven to bearing, the soils forced into the pipe during driving are cleaned out, and the pipe is filled with concrete. Closed-end piles are generally used as friction piles when a suitable bearing stratum is not available at suitable depths. There are several proprietary closed-end pipe piles available.

When steel pipes are driven open ended, minimum thickness to diameter is related to hammer energy by requiring a minimum area per kip-foot of energy. Open-ended pipes require a minimum area of 0.34 square inches to resist each 1,000 ft-lbs. This requirement equates to a wall thickness of 0.27 inch for a 10-inch pipe driven with a hammer energy of 25 kip-feet. Note that if the wall thickness is less than 0.179 inch, a driving shoe is required to prevent local buckling at the tip from hard driving, regardless of diameter or hammer energy. The 0.179-inch thickness originally entered in the 2003 IBC to be consistent with the most common minimum thickness for closed-end pipe piles.

Concrete-filled steel pipes or tubes in structures assigned to SDC C, D, E, or F shall have a wall thickness of not less than $^3/_{16}$ inch. The pipe or tube casing for socketed drilled shafts shall have a nominal outside diameter of not less than 18 inches and a wall thickness of not less than $^3/_8$ inch. The pipe is a welded or seamless pipe conforming to ASTM A 252. ASCE 20-96 *Standard Guidelines for the Design and Installation of Pile Foundations* as well as the recommendations of the Driven Pile Committee of the Deep Foundations Institute list 0.179 inches as the minimum wall thickness.

Concrete-filled steel piles are either seamless or welded pipe, or closed-end tubular piles with either straight or tapered sections that are driven into the soil. The piles may be installed as either friction or end-bearing piles. This pile type is characterized by a steel shell that is thicker than the thin shell used in some steel-cased piles, hence both the concrete and steel shell are assumed to carry load compositely. If driven open ended, the earth core is removed from the shell prior to concreting. The shell may be driven with an internal mandrel.

Splices. This section specifies the requirements for splicing of deep foundation elements. The 50-percent requirement provides more strength where the bending moments are low, insofar as it is based on the capacity of the deep foundation element, not the design loads. The 2009 IBC addressed splices of the same type of deep foundation elements and splices of deep foundation elements of different materials or different types. Splices of deep foundation elements of different materials or types are required to develop the full compressive strength and not less than 50 percent of the tension and bending strength of the weaker section. Although it is physically and economically better to drive piles in one piece, site conditions sometimes necessitate that piles be driven in spliced sections. For example, when the soil or rock-bearing stratum is located so deep below the ground that the leads on the driving equipment will not receive full-length piles, it becomes necessary to install the piles in sections or, where possible, to take up the extra length by setting the tip in a preexcavated hole (see discussion, Section 1810.4.4). When piles are installed in areas such as existing buildings with restricted headroom, they are also required to be placed in spliced sections. There are a number of other reasons for field-splicing piles, such as restrictions on shipping lengths or the use of composite piles. **1810.3.6**

This provision requires that splices be constructed to provide and maintain true alignment and position of the deep foundation element sections during installation. Splices must be of sufficient strength to transmit the vertical and lateral loads on the deep foundation elements, as well as to resist the bending stresses that may occur at splice locations during the driving operations and under long-term service loads. Splices are to develop at least 50 percent of the value of the deep foundation element in bending. Consideration should be given to the design of splices at locations where the deep foundation elements may be subject to tension. Splices that occur in the upper 10 feet of pile embedment are to be designed to resist the bending moments and shears at the allowable stress levels of the pile material, based on an assumed pile load eccentricity of 3 inches, unless the pile is properly braced. Proper bracing of a spliced pile is deemed to exist if stability of the pile group is furnished in accordance with the provisions of Section 18.10.2.2, provided that other piles in the group do not have splices in the upper 10 feet of their embedded length.

There are different methods employed in splicing deep foundation elements depending on the materials used in the deep element construction. For example, timber piles are spliced by one of two commonly used methods. The first method uses a pipe sleeve with a length of about four to five times the diameter of the pile. The butting ends of the pile are sawn square for full contact of the two pile sections, and the spliced portions of the timber pile are trimmed smoothly around their periphery to fit tightly into the pipe sleeve. The second splicing method involves the use of steel straps and bolts. The butting ends of the pile sections are sawn square for full contact and proper alignment, and the four sides are planed flat to receive the splicing straps. This type of splicing can resist some uplift forces.

Splicing of precast concrete piles usually occurs at the head portions of the piles. After the piles are driven to their required depth, pile heads are cut off or spliced to the desired elevation for proper embedment in the concrete pile caps. Any portion of the pile that is cracked or shattered by the driving operations or cutting off of pile heads should be removed and spliced with fresh concrete. To cut off a precast concrete pile section, a deep groove is chiseled around the pile exposing the reinforcing bars, which are then cut off (by torch) to desired heights or extensions. The pile section above the groove is snapped off (by crane) and a new pile section is freshly cast to tie in with the precast pile.

Steel H-piles are spliced in the same manner as steel columns, usually by welding the sections together. Welded splices may be welded-plate or bar splices, butt-welded splices, special welded splice fittings, or a combination of these. Spliced materials should be kept on the inner faces of the H-pile sections to avoid forcing a hole in the ground larger than the pile, causing at least a temporary loss in frictional value and lateral support that might result in excessive bending stresses.

Steel pipe piles may be spliced by butt welding, sometimes using straps to guide the sections and to provide more strength to the welded joint. Another method is to use inside sleeves having a driving fit, with a flange extending between the pipe sections. By applying bituminous cement or compound on the outside of the ring before driving, a water-tight joint is obtained.

1810.3.6.1 Seismic Design Categories C through F. Splices of deep foundation elements in SDCs C through F are addressed in this section and require that the splices develop the lesser of two elements: (1) the nominal strength of the deep foundation element and (2) the axial and shear forces and moments from the seismic load combinations including overstrength factor of ASCE 7, Section 12.4.3 or 12.14.3.2. For a discussion of overstrength factor, see Section 1810.3.11.2.

1810.3.7 Top of element detailing at cutoffs. The requirements of this section are intended to account for conditions where a deep foundation element encounters refusal at a shallower depth than anticipated and a portion of the deep element is cut off. It is imperative that the required reinforcement be provided at the top of the deep foundation element when the excess deep element length is cut off.

1810.3.8 Precast concrete piles. Precast concrete piles are manufactured as conventionally reinforced concrete or as prestressed concrete. Both types can be formed by bed casting, centrifugal casting, slipforming, or extrusion methods. Piles are usually square, octagonal, or round, and either solid or hollow. Precast piles, which may be either friction or end-bearing piles, are of the displacement type and are driven into place.

1810.3.8.1 Reinforcement. The closely spaced spirals or ties at the ends are to accommodate radial tensile principal stresses from driving.

1810.3.8.2.1 Minimum reinforcement. (Of precast-nonprestressed piles). Four bars are the practical minimum for placement.

1810.3.8.2.2 Seismic reinforcement in Seismic Design Categories C through F. Minimum longitudinal reinforcing steel and transverse tie or spiral confinement reinforcing is required to provide some ductility. This is required for precast-nonprestressed piles in all SDCs except SDCs A and B.

1810.3.8.2.3 Additional seismic reinforcement in Seismic Design Categories D through F. This section contains additional transverse reinforcement requirements for buildings in SDCs D, E, and F. These are in addition to the longitudinal and transverse reinforcement requirements provided in Section 1810.3.8.2.2. Spirals or ties are spaced closer to provide for the higher ductility requirements in SDC D and higher. The details of this additional transverse requirement are found in Section 1810.3.9.4.2.

1810.3.8.3 Precast-prestressed piles. Minimum prestress is set to minimize cracking from handling and driving stresses. The purpose of prestressing piles is to place the concrete under a compressive stress so that hairline cracks caused by any subsequent tensile stress that may occur from handling, driving, superimposed loads, or seismic imposed curvatures, and which are larger than the prestressed compression stress, will close when the tensile stresses are removed. This is easily achievable for handling and driving loads, but may not be feasible for seismic-imposed curvatures.

Prestressed piles can be either pretensioned or post-tensioned. Pretensioned piles are generally cast full length in a casting bed at a manufacturing plant and often contain only prestressing steel reinforcement. Post-tensioned piles may be plant cast or site cast, and generally contain mild steel reinforcing to resist handling stresses.

Seismic reinforcement in Seismic Design Category C (precast-prestressed piles). **1810.3.8.3.2**
Spiral transverse (confining) reinforcement is required to mitigate the effects of soil-induced curvatures from seismic ground displacements. The volumetric requirement is the same as that for columns in ductile frames. (See ACI 318, Section 21.6.2.)

Seismic reinforcement in Seismic Design Categories D through F. These transverse **1810.3.8.3.3** reinforcing requirements are based on testing of prestressed piles in New Zealand and the subsequent recommendations by the Prestressed Concrete Institute (PCI). The requirements result in prestressed piles with good ductility without creating construction problems from reinforcement congestion. In Item 2, the "distance from the underside of the pile cap to the point of zero curvature" is determined as in Section 1810.3.9.4.1 below. See Sheppard[8] and Joen et al.[9]

Cast-in-place deep foundations. Cast-in-place deep foundations are covered in **1810.3.9** Sections 1810.3.9.1 through 1810.3.9.4.2.2. Cast-in-place (CIP) concrete piles are installed by placing concrete into holes preformed by drilling or by driving a temporary or permanent casing to the required bearing depth. Drilled or augered piles are also known as cast-in-drill-hole or CIDH piles. Drilled or augured uncased piles are nondisplacement piles and are installed by drilling or augering a hole and filling the uncased hole with concrete, either during or after withdrawing the auger. CIP piles may be either cased or uncased. Uncased piles are difficult to construct when below the ground-water table. Except for enlarged base piles, the concrete in CIP piles is not subjected to driving forces, only the forces imposed by the service loads and downdrag from settlement. One advantage of drilled CIP piles is that the tip elevation can easily be adjusted to have the tip on the correct bearing stratum. Reinforcement is installed during the concreting operation. CIP-drilled piles are of the nondisplacement type, that is, the soil is not displaced or compacted by the drilling operation. CIP piles constructed by first driving a closed-end shell are displacement piles, where the soil surrounding the shell is displaced and compacted during the driving operation. CIP piles constructed by driving an open-ended casing without a mandrel or temporary tip closure are nominally a nondisplacement pile, although some compaction around the shell may occur in cohesive soils, and densification may occur from driving in granular soils.

The steel-cased pile is the most widely used type of cast-in-place concrete pile. This pile type is characterized by a thin steel shell and is a displacement pile. This pile type consists of a closed-end light-gauge steel shell or a thin-walled pipe driven into the soil and left permanently in place, reinforced when required for uplift, lateral bending, or seismic-induced curvatures, and filled with concrete. The shell or pipe is usually driven with a removable mandrel. The shell is either a constant section or a tapered shape. Steel-encased piles are generally friction piles.

Design cracking moment. The design cracking moment that is established as nominal **1810.3.9.1** moment capacity multiplied by the factor is determined from the equation

$$\varphi M_n = 3\sqrt{f'_c S_m}$$

The design cracking moment in SI is: $\varphi M_n = 0.25\sqrt{f'_c S_m}$.

The equation and the reinforcement requirements of Section 1810.3.9.2 clarify the present requirements, making the IBC consistent with the requirements of ACI 318, and allow elimination of the definition for flexural length. For both uncased and cased cast-in-place deep foundation elements (but not concrete-filled pipes and tubes), reinforcement must be provided where moments exceed a reasonable lower bound for the capacity of the plain concrete section, known as the cracking moment.

1810.3.9.2 Required reinforcement. See Section 1810.3.9.1.

1810.3.9.3 Placement of reinforcement. With three exceptions, reinforcing steel must be assembled into a cage and placed in the hole or casing prior to concreting, not stabbed after concreting. One exception is for dowels less than 5 feet in length, and the other exception is for auger-injected piles, which are placed by injecting the concrete through a hollow stem auger. The third exception is for smaller residential and utility buildings (Group R-3 and U occupancies) where the method of reinforcement placement after concrete placement must be approved by the building official.

1810.3.9.4 Seismic reinforcement. There are four specific cases where prescriptive provisions are provided for special cases to comply with the seismic reinforcement requirements in SDCs C through F. Other than these four exceptions, cast-in-place deep foundations must comply with Section 1810.3.9.4.1 for seismic reinforcement in SDC C and Section 1810.3.9.4.2 for seismic reinforcement in SDCs D, E, and F.

1810.3.9.4.1 Seismic reinforcement in Seismic Design Category C. Minimum steel requirements are established to provide some ductility. The minimum reinforcement must be continued throughout the flexural length of the pile. The term flexural length was added to the 2003 IBC as the "length of the pile from the first point of zero lateral deflection to the underside of the pile cap or grade beam." The point of zero lateral deflection can be determined from the P-y analysis. In the 2009 IBC, the term "flexural length" and its definition were removed and replaced with "minimum reinforced length." Minimum reinforced length is not a defined term in the 2012 IBC; rather, in this section there are four criteria to determine the length.

1810.3.9.4.2 Seismic reinforcement in Seismic Design Categories D through F. The requirements in SDC D, E, or F are similar to those in SDC C, except that the minimum reinforcement ratio is higher and extends over a longer length to improve ductility. In addition, closed ties or spirals are required to provide confinement in regions of plastic hinging, that is, at the pile-cap interface, at the interface of soft to stiff layers, and in liquefaction zones. Confinement reinforcing should also be used for bay muds and sensitive clays. See Sections 1803.5.11 and 1803.5.12. The term "flexural length" has been replaced with "minimum reinforced length," which is determined in this section by four criteria.

1810.3.9.4.2.1 Site Classes A through D. Transverse confinement reinforcement requirements for stiffer soil and rock sites are in this section, which references applicable provisions in ACI 318.

1810.3.9.4.2.2 Site Classes E and F. Transverse confinement reinforcement requirements for soft or sensitive soil sites are in this section, which references applicable provisions in ACI 318.

1810.3.9.5 Belled drilled shafts. Bells are designed to increase the bearing surface on which the loads will be transferred. Bells are typically designed in more cohesive soils so that there will not be any collapsing of the bell walls or roof.

1810.3.9.6 Socketed drilled shafts. Socketed drilled shaft deep foundations are what were previously called the caisson pile or more commonly known as a drilled-in caisson. They are installed as a special type of high-load-capacity pile and are characterized by a structural steel core, an upper-cased section extending to bedrock, and a lower uncased tip that is socketed into rock.

The socketed drilled shaft deep foundation element is a cased cast-in-place concrete pile that is formed by (1) driving a heavy-wall open-ended pipe down to bedrock, (2) cleaning out the soil materials within the pipe, (3) drilling an uncased socket into the bedrock, (4) inserting a structural steel core into the pipe, and (5) filling the entire pipe and drilled socket with concrete.

The core material is usually made of hot-rolled structural steel wide-flange or I-beam sections, or steel rails. This section specifies that the steel core is to extend full length from the base of the drilled socket to the top of the steel pipe or, as an alternative and depending on design requirements, the steel core may extend halfway up the pipe or as a stub core to a distance in the pipe at least equal to the depth of the socket. The strength of the deep foundation element is developed in combined friction and end-bearing of the rock socket.

Micropiles. Micropiles are bored, grouted-in-place deep foundation elements that develop their load-carrying capacity by means of a bond zone in soil, bedrock, or a combination of soil and bedrock. Prior to their inclusion in the code, the use of micropiles had to be approved under the alternative materials, design, and methods of construction provisions in Section 104.11. The provisions are based on the recommendations of the ADSC/DFI (International Association of Foundation Drilling/Deep Foundations Institute) Committee on Micropiles, and are intended to provide a uniform standard for micropiles, and eliminate inconsistencies in their design and installation. The IBC provisions are based primarily on the *Massachusetts Building Code* (MBC) with additional changes and modifications. **1810.3.10**

Pile caps. Pile caps are to be of reinforced concrete and designed in accordance with the requirements of ACI 318. For footings (pile caps) on piles, computations for moments and shears may be based on the assumption that the load reaction from any pile is concentrated at the pile center. See ACI 318 for loads and reactions of footings on piles. **1810.3.11**

The soil immediately under the pile cap should not be considered to provide any support for vertical loads. For a more detailed explanation of this requirement, see the discussion on Section 1810.3.3.2, allowable lateral load.

The heads of all piles are to be embedded not less than 3 inches into pile caps, and the edges of the pile caps are to extend at least 4 inches beyond the closest sides of all piles. The degree of fixity between a pile head and the concrete cap depends on the method of connection required to satisfy design considerations.

Seismic Design Categories C through F. Deep foundation elements in SDCs C through F must have a positive connection to the pile cap for sliding or uplift purposes. This is achieved by connecting the deep foundation element to the pile cap by either embedding the deep element reinforcement in the pile cap or by field-placed dowels anchored into the element and extended into the pile cap for a distance equal to the dowel development length in accordance with ACI 318. **1810.3.11.1**

Piles in structures subject to seismic ground shaking are likely to be subjected to uplift (tension) forces, either by design or because of insufficient resistance to overturning forces by gravity loads. Hence, concrete piles and concrete-filled steel pipe piles must be able to develop the strength of the pile (in tension) in the connection to the cap. This is accomplished by the requirement that the pile be embedded in the cap by a distance equal to the development length. The development length may not be reduced by the ratio $A_{required}/A_{supplied}$. Alternative means, such as increasing concrete confinement, may be used to reduce the development length.

Similarly, the various types of steel piles are required to develop the strength in tension and to transmit this strength to the cap by positive means other than bond to the bare steel; for example, welded studs or welded reinforcement must be used.

Splices must develop the full strength of the pile, both tension and compression, for all pile types.

Seismic Design Categories D through F. Anchorage of piles or piers into pile caps must consider the combined effects of uplift and pile fixity. The anchorage must develop at least 25 percent of the strength of the pile in tension. For piles subject to uplift or required to provide rotational restraint, the anchorage must develop the lesser of the nominal tensile strength **1810.3.11.2**

of the longitudinal reinforcement in a concrete element, the nominal tensile strength of a steel element, and 1.3 times the frictional force developed between the element and the soil. Because of the large variability in soils, it would be prudent to design these piles for the full tensile capacity rather than 1.3 times the uplift capacity (frictional force). Exceptions allow the use of ASCE 7 Section 12.4.3 or 12.14.3.2 for design of the anchorage to resist axial tension forces or, in the case of rotational restraint, the design of the anchorage to resist axial and shear forces, and moments.

Batter piles have performed poorly in past earthquakes. This is because the batter piles are laterally stiff relative to vertical piles and resist most of the seismic-induced inertial forces. The piles are not usually designed to resist the actual forces, but are designed to resist an inertial force reduced by an assumed ductility. However, the batter piles are axially stiff and generally not detailed for ductility; hence the failures. To preclude this type of failure in batter piles, the piles and their connections must be designed to resist the anticipated maximum earthquake forces from the load combinations with overstrength factor in Section 12.4.3 or 12.14.3.2 of ASCE 7.

The load combinations with overstrength only apply where specifically required by the seismic provisions. They constitute an additional requirement that must be considered in the design of specific structural elements to account for the maximum earthquake load effect, E_m, which considers "system overstrength." This system characteristic is accounted for by multiplying the effects of the lateral earthquake load by the overstrength factor, Ω_0, for the seismic-force-resisting system involved. It represents the upper bound system strength for purposes of designing nonyielding elements for the maximum expected load. Under the design earthquake ground motions, the forces generated in the seismic-force-resisting system can be much greater than the prescribed seismic design forces. If not accounted for, the system overstrength effect can cause failures of structural elements that are subjected to these forces. Because system overstrength is unavoidable, design for the maximum earthquake force that can be developed is warranted for certain elements. The intent is to provide key elements with sufficient overstrength so that inelastic (ductile) response/behavior appropriately occurs within the vertical resisting elements. It should be noted that these load combinations are only to be applied where specified in the earthquake load provisions or in other structural chapters of the code. The requirement for using load combinations with overstrength in Section 1810.3.11.2 is such a case.

1810.3.12 Grade beams. In SDC D, E, or F, grade beams must be designed as ductile in accordance with the provisions of Section 21.12.3 of ACI 318 unless the beam is strong enough to resist the anticipated maximum earthquake force as set forth in the load combination with overstrength factor of Section 12.4.3 or 12.14.3.2 of ASCE 7. That is, grade beams must be either strong or ductile. For a detailed discussion of overstrength factor, see Section 1810.3.11.2.

1810.3.13 Seismic ties. Interconnection of piles and caissons (2009 IBC replaced the term caissons with socketed drilled shaft deep foundations, which are covered in Section 1810.3.9.6) is necessary to prevent differential movement of the components of the foundation during an earthquake. It is well known that a building must be thoroughly tied together if it is to successfully resist earthquake ground motion. These provisions apply to SDCs C through F.

Individual piles, piers, or pile caps required for structures in SDCs C through F must be interconnected with ties capable of transmitting the lesser of a force equal to the larger pile cap or column load times the short-period response acceleration, S_{DS}, divided by 10 and 25 percent of the smaller pile cap or column load. The intent of this requirement is to minimize differential movement or spreading between the footings during ground shaking. If slabs on grade, or beams within slabs on grade, are used to meet the tie requirement, the load path from footing to slab or beam/slab and across joints in the slab or beam/slab should be checked for continuity. The slab or beam must be reinforced for the design tension load. In addition, the slab should be checked for buckling under the required compression load using an assumed slab of no more than six times the slab thickness.

1810.4 **Installation.** Provisions of the code dealing with the details of installations for various deep foundation types were collected and relocated to Section 1810.4 of the 2009 IBC, which is the last section of Chapter 18.

Care must be used during installation to prevent damage during handling and driving. The proper cushion must be used at the driving end. Precast concrete pile recommendations for design, manufacture, and installation are given in ACI 543R. Damage to piles can be classified into four types:

1. Spalling at the butt or head (driving end) caused by high or irregular compressive stress concentrations. The spalling may be caused by insufficient cushioning, pile butt not square with the pile longitudinal axis, hammer and pile not aligned, reinforcing steel not flush or below the top of the pile allowing the hammer force to be transmitted through the steel, or insufficient transverse reinforcement.

2. Spalling at the tip, which is usually caused by an extremely high driving resistance such as when the tip is bearing on a rock.

3. Breaking or transverse cracking. This is caused by the rarefaction wave reflected from the tip. When the hammer strikes the cushion or head, a compression wave is produced that travels down the pile. The wave can be reflected from the tip as a rarefaction (tension) wave or a compression wave depending on the soil stiffness. Rarefaction waves usually occur when the soil at the tip is soft with very little resistance to penetration, causing tension waves that can cause significant tensile damage. This phenomenon usually occurs only in long piles exceeding 50 feet. Prestressed piles have more resistance to rarefaction damage than do conventionally reinforced piles. The hammer energy should be reduced when driving long piles through soft soils.

4. Spiral or transverse cracking may be caused by a combination of torsional stress and rarefaction stress. Torsion is usually caused by excessive restraint in the leads.

In the case of precast-prestressed piles, because of the precompression, less care is needed in the handling and driving than for conventionally reinforced piles, and prestressed piles are, in general, more durable than conventionally reinforced precast concrete piles.

If a deep foundation element consists of two or more sections of various materials or different types of deep foundation elements spliced together, each section is required to satisfy the applicable installation requirements of this section. Generally referred to as "composite piles," these refer to deep foundation elements placed in series, such as a cast-in-place concrete deep foundation element placed over a submerged wood pile.

1810.4.1 **Structural integrity.** Deep foundation elements can be exposed to damage or could potentially cause damage to surrounding areas, especially piles that are generally installed by either driving, vibration, jacking, jetting, direct weight, or a combination of such methods. Most types of piles are exposed to some degree of damage during placement. However, with knowledge of soil conditions and the proper selection of equipment, installation methods, and techniques, damage may be prevented or minimized.

Due care must be exercised during pile installation to avoid interference with adjacent piles or other structures so as to leave their strength and load capacity unimpaired. If any pile is damaged during installation so as to affect its structural integrity, the damage must be satisfactorily repaired or the pile rejected.

Displacement piles have their own special issues. As displacement piles are driven within a group, progressive compaction of the surrounding soil occurs, particularly where it involves closely spaced piles. This can cause piles to be deflected off-line because of the buildup of unequal soil pressures around the piles. Soil compaction during driving operations can cause extreme variations in pile lengths within a group, with some piles failing to reach specified load-bearing material. Ground heave is another effect of soil compaction (see discussion, Section 1810.4.6).

To prevent or significantly reduce the problems associated with soil compaction, the driving sequence of pile installations becomes an important consideration. For example, if the outer piles of a group are driven first so that the inner piles, because of soil compaction, fetch up to specified sets (hammer blows) at much higher elevations than the outer piles, the total load-bearing value of the group will be adversely affected. As another example, starting pile driving at the edge of a group makes the piles progressively more difficult to drive and results in a one-sided bearing group. The general driving practice is to work from the center of a group outward. For large groups consisting of rows of widely spaced piles, driving can be done progressively from one side to the other.

The provisions within Section 1810.4 are to ensure pile structural integrity by adhering to proper installation procedures. The code cannot cover all possibilities, however, thus establishing the need for the general nature of this section.

1810.4.1.1 **Compressive strength of precast concrete.** Handling and driving forces will likely govern pile strength requirements. Through the use of steam cure and Type III cements, 75 percent of specified strength can be achieved relatively quickly.

1810.4.1.2 **Casing.** These requirements are intended to result in a satisfactory pile. The construction of drilled piles is fraught with problems: caving, ground water, and other issues.[10] Most of these problems generally relate to soil conditions, including soil or rock debris accumulating at the base of the pile or occurring in the pile shaft, reductions in the shaft cross section caused by the necking of soil walls because of soft materials or earth pressures, discontinuities in the deep foundation shaft, hollows on the surface of the shaft, and other problems related to the drilling operations.

1810.4.1.3 **Driving near uncased concrete.** These requirements are intended to prevent damage to uncured concrete in adjacent deep foundation elements from the soil displacements caused by driving adjacent elements. The spacing requirements should not be construed to mean that the center-to-center spacing must not be closer than six average diameters of a cased element in granular soils nor within one-half the pile depth in cohesive soils; just that deep elements cannot be driven at that spacing within 48 hours of placement of concrete.

1810.4.1.4 **Driving near cased concrete.** The restrictions on driving within four and one-half average diameters of a cased element filled with concrete less than 24 hours old are primarily to avoid damage to uncured concrete in adjacent deep foundation elements.

1810.4.1.5 **Defective timber piles.** Damage to the pile, including breakage, should be suspected when there is a sudden drop in penetration resistance while driving that cannot be explained by the soil profile. The pile should be withdrawn for examination. If penetration resistance should suddenly increase, driving should be stopped to avoid possible damage. A significant problem encountered during installation of timber piles is damage from overdriving. Overdriving can cause failure by bending, brooming of the tip, crushing, brooming at the butt end, or splitting or breaking along the pile section. See Figure 1810-2.

1810.4.2 **Identification.** All deep foundation materials must be identified for conformity to the code requirements and construction specifications. Information such as strength (species and grade for timber piles), dimensions, and other pertinent information is required. Such identifications must be provided for all deep foundation elements, whether they are taken from manufacturers' stock or made for a particular project. Identifications are to be maintained from the point of manufacture through the shipment, on-site handling, storage, and installation of the piles. Manufacturers, upon request, usually furnish certificates of compliance with construction specifications. In the absence of adequate data, piles must be tested to demonstrate conformity to the specified grade.

Deep Foundations 18

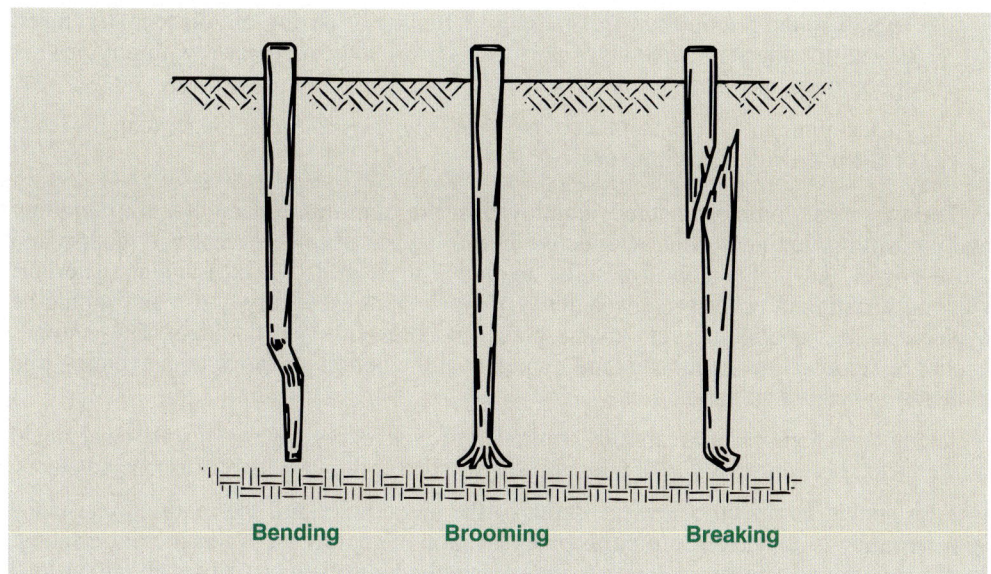

Figure 1810-2
Effect of overdriving timber piles.

In addition to mill certificates (steel piles), identification is made through plant manufacturing or inspection reports (precast concrete and timber piles) and delivery tickets (concrete). Timber piles are stamped (labeled) with information such as producer, species, treatment, and length.

Identification is essential when high-yield-strength steel is specified. Frequently, pile cutoff lengths are reused and pile material may come from a jobber, a contractor's yard, or a material supplier. In such cases, mill certificates are not available and the steel should be tested to see if it complies with the code requirements and the project specifications.

Location plan. A plan clearly showing the designation of all deep foundation elements on a project by an identification system is to be filed with the building official before the installation is started. The inspector (see discussion in Chapter 17 on special inspection) must keep piling logs and other records and submit written reports based on this identification system. The use of such a system becomes particularly important at sites where the variations in soil profiles are so extensive that it becomes necessary to manufacture piles of different lengths to satisfy bearing conditions. **1810.4.3**

The building official should also be furnished copies of all modifications to the original deep foundation location plan that may be necessary as the work proceeds (as-built drawings). This would show elements added, eliminated, or relocated. Such records would facilitate the use of existing deep foundations in the future (see Section 1810.1.2) if the structure is altered or another structure is built on the site.

Preexcavation. There are several important reasons for the use of preexcavation to facilitate the installation of foundation piles. Some of these purposes are: **1810.4.4**

- To install piles through upper strata of hard soil.
- To penetrate through subsurface obstructions, such as timbers, boulders, rip rap, thin stone strata, and the like.
- To reduce or eliminate the possibility of ground heave that could lift adjacent structures or piles already driven.
- To reduce ground pressures resulting from soil displacement during driving and to prevent the lateral movement of adjacent piles or structures.
- To reduce the amount of driving required to seat the piles in their proper load-bearing strata.

- To reduce the possibility of damaging vibrations or jarring of adjacent structures, as well as reduce the amount of noise, all of which are associated with pile-driving operations.
- To accommodate the placement of piles that may be somewhat longer than the leads of the pile-driving equipment.

The two most common methods employed in preexcavation operations are prejetting and predrilling. Jetting is usually effective in most types of soils, except very coarse and loose gravel and highly cohesive soils. Jetting is most effective in granular materials. Generally, jetting in cohesive soils is not very practical or especially useful and should be avoided in soils containing very coarse gravel, cobbles, or small boulders. These stones cannot be removed by the jet and tend to collect at the bottom of the hole, preventing pile penetration below that depth.

Jetting operations must be carefully controlled to avoid excessive loss of soil, which could affect the load-bearing capacity of piles already installed or the stability of adjacent structures.

Piles should be driven below the depth of the jetted hole until the required resistance or penetration is obtained. Before this preexcavation method is used, consideration should be given to the possibility that jetting, unless strictly controlled, can adversely affect load transfer, particularly as it involves the placement of nontapered piles.

Predrilling or coring before driving is effective in most types of soils and is a more controllable method of preexcavation than jetting. The risk of adversely affecting the structural integrity of adjacent piles or structures or the frictional capacity of piles is considerably less than jetting.

Predrilling can be performed as a dry operation or as a wet rotary process. Dry drilling can be done by using a continuous-flight auger or a short-flight auger attached to the end of a drill stem or kelly bar. Wet drilling requires a hollow-stem continuous-flight auger or a hollow drill stem employing the use of spade bits. When the wet rotary process of predrilling is used, bentonite slurry or plain water is circulated to keep the hole open. As in the case of jetting, piles should be driven with tips below the predrilled hole. This is necessary to prevent any voids or very loose or soft soils from occurring below the pile tip.

There are other methods used for preexcavation purposes, such as the dry tube method and spudding, but such procedures are seldom used. In any case, the methods to be employed for preexcavation are subject to the approval of the building official.

1810.4.5 Vibratory driving. The use of vibratory drivers for the installation of piles is not applicable to all types of soil conditions. They are effective in granular soils with the use of nondisplacement piles, such as steel H-piles and pipe piles driven open ended. Vibratory drivers are also used for extracting piles or temporary casings employed in the construction of cast-in-place concrete deep foundation elements.

Vibratory drivers, either low or high frequency, cause the pile to penetrate the soil by longitudinal vibrations. Although this type of pile driver can produce good results in the installation of nondisplacement piles under favorable soil conditions, the greatest difficulty is the lack of a reliable method of estimating the load-bearing capacity. After the pile has been installed with a vibratory driver, pile capacity can be determined by using an impact-type hammer to set the pile in its final position.

One method to determine pile capacity is to calibrate the power consumption in relation to the rate of penetration. Nonetheless, the use of a vibratory driver is only permitted where the pile load capacity is established by load tests in accordance with the requirements of Section 1810.3.3.1.2.

1810.4.6 Heaved elements. Piles that are driven into saturated plastic clay materials can often displace a volume of soil equal to that of the piles themselves. When this happens, the soil displacement sometimes occurs as ground heave and may lift adjacent deep foundation elements already driven. Under such conditions, heaved deep foundation elements may no longer be properly seated and a loss of pile capacity occurs. Heaved piles must be redriven to firm bearing to again develop the required capacity and penetration. If heaved piles are not redriven, their capacity must be verified by load tests made in accordance with the requirements of Section 1810.3.3.1.2.

This section applies only to piles that can be safely redriven after installation. Heaved uncased cast-in-place concrete deep foundation elements or sectional piles with joints that cannot take tension should be abandoned and replaced. When redriving heaved piles, a comparable driving system or the same as that of the initial driving should be employed. It should be noted that in redriving concrete-filled pipe piles, the driving characteristics of the pile have been altered and the pile is substantially stiffer than when the empty pipe was initially driven. In such cases, the required driving resistance would be less than originally required.

One method used to prevent or reduce objectionable soil displacement is to remove some of the soil in the spaces to be occupied by the piles. This is done by predrilling the pile holes (see discussion, Section 1810.4.4).

Enlarged base cast-in-place elements. Enlarged base elements are intended to be end-bearing-type deep foundation elements that spread the bearing load over a larger area than a prismatic element, thereby increasing capacity. Enlarged base deep elements may be either cased or uncased. Enlarged base elements are used only in granular soils, which, because of the voids between soil particles, allow densification of the soils around the deep element tip without creating excessive pressures. One type is the compacted base type, which consists of a bulb-shaped footing formed after driving the shaft casing to its final depth. Another type is the concrete-pedestal type, in which a truncated cone- or pyramid-shaped precast concrete tip larger than the steel casing diameter is driven into the soil with the casing. See Figure 1810-3.

1810.4.7

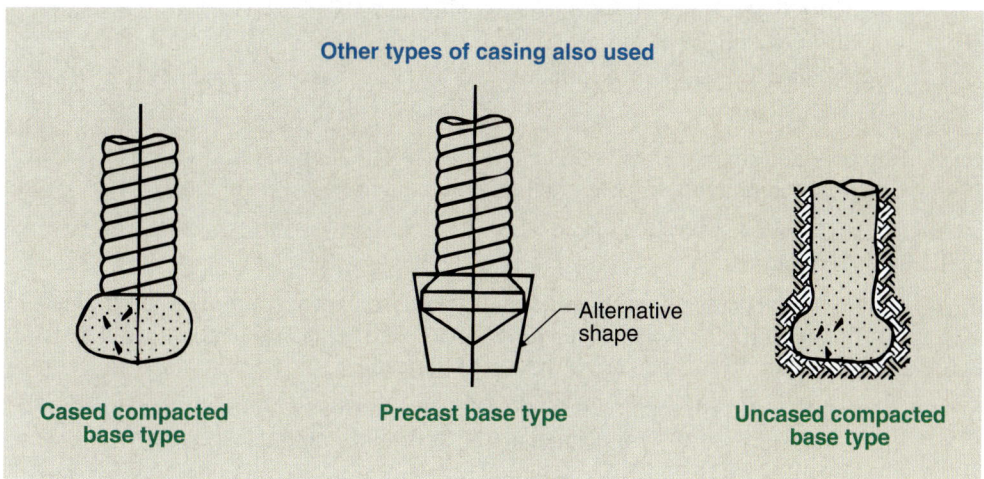

Figure 1810-3
Enlarged base pile cased or uncased shafts.

Installation must employ the same methods used to install the load test piles. The compacted base pile is usually installed by driving a steel casing. A zero slump concrete plug is placed at the tip of the casing and impacted with a heavy drop weight, thereby driving the casing and plug. Sometimes a gravel plug is used rather than zero slump concrete. When the casing and plug have been driven to the required depth, the plug is driven out and the bulb is formed by progressively compacting additional layers of zero slump concrete. A welded reinforcing steel cage is added where required. If the deep foundation element is to be uncased, the shaft is formed by compacting zero slump concrete in small lifts as the casing is withdrawn. If the deep foundation element is to be cased, as would be required for piles through peats or other organic soils, the shaft is formed by inserting a steel shell inside the drive casing after forming the bulb, withdrawing the drive casing, and filling the shell with conventional concrete. The precast base-type pile is installed with the precast base placed at the tip of a mandrel-driven steel shell.

A problem that occurs with the precast base type and the cased bulb-type piles is that an annular space between the casing and the soil remains. Either the pile must be designed as a slender reinforced concrete column governed by buckling, or the annular space must be filled to provide the requisite lateral support. The usual practice is to fill the annular space by pumping grout.

1810.4.10 **Micropiles.** Micropile boreholes are typically advanced by either rotary drilling or rotary percussive drilling. Installation requirements differ based on whether a steel casing is permanent or temporary or not provided (IBC Section 1810.4.10 Items 6, 1, and 2, respectively).

1810.4.11 **Helical piles.** See Section 1810.3.1.5.

1810.4.12 **Special inspection.** See analysis and discussions in Chapter 17 regarding special inspection of deep foundations.

> **KEY POINTS**
> - Chapter 18 provides requirements for design and construction of foundation systems for buildings and other structures regulated by the IBC.
> - Geotechnical investigations are required based on the SDC of the building or where certain site conditions exist.
> - The requirement for a geotechnical investigation may be waived by the building official in some cases for buildings in SDCs A and B.
> - Geotechnical investigations are required for buildings in SDCs C through F and where other conditions exist such as compacted fill more than 12 inches deep, foundations bearing on CLSM, and excavations that may compromise lateral support of existing foundations.
> - Section 1803.6 specifies the minimum content required to be included in geotechnical reports.
> - Requirements for site grading related to foundations are provided in Section 1804. General requirements for site grading are contained in Appendix J.
> - Foundations must be dampproof and waterproof where required in accordance with Section 1805.
> - Where a geotechnical investigation and report does not specify load-bearing values for soils, presumptive values for vertical bearing, lateral bearing, and lateral sliding friction resistance are provided in the code.
> - Prescriptive requirements for concrete and masonry foundation walls are provided based on the properties of the wall and unbalanced backfill.
> - Requirements for the design of retaining walls and embedded poles are provided.
> - Foundations are divided into two distinct groups: shallow foundations (footings) and deep foundations (piles), with general requirements for both and specific requirements for each.
> - Section 1808 contains general requirements for all foundations, Section 1809 covers shallow foundations, and Section 1810 covers deep foundation requirements.
> - Where a specific design is not provided, prescriptive requirements for constructing footings supporting light-frame construction are given in Table 1809.7.
> - The IBC now contains three tables that provide minimum compressive strength for concrete and grout, minimum concrete cover for reinforcement used in foundations, and allowable stresses for materials used in deep foundation elements.

REFERENCES

1. Terzaghi, Karl, and Peck, Ralph B., *Soil Mechanics in Engineering Practice*, John Wiley and Sons, 1967.
2. Winterkorn, Hans F., and Fang, Hsai-Yang, *Foundation Engineering Handbook*, Van Nostrand Reinhold Company, Inc., New York, NY, 1975.
3. Bowles, J. E., *Foundation Analysis and Design*, 5th Edition, McGraw Hill, Inc., 1996.
4. Johnson, S. M., and Kavanagh, T. C., *The Design of Foundations for Buildings*, McGraw Hill, Inc., 1968.
5. Liu, C., and Evett, J. B., *Soils and Foundations*, 2nd Edition, Prentice-Hall, Inc., 1980.
6. Sowers, G. B., and Sowers, G. F., *Introductory Soil Mechanics in Engineering Practice*, Macmillan Publishing Co., 1970.
7. Committee on the Bearing Value of Pile Driving Foundations, *Pile-Driving Formulas—Progress Report*, Proceedings, May, American Society of Civil Engineers, New York, NY, 1941.
8. Sheppard, David A., "Seismic Design of Prestressed Concrete Piling," *PCI Journal*, March-April, 1993.
9. Joen, Pam Hoat, and Park, Robert, "Simulated Seismic Load Tests on Prestressed Concrete Piles and Pile-Cap Connections," *PCI Journal*, November-December, 1990.
10. ICC, *The BOCA National Building Code—Commentary*, Volume 2, International Code Council, Inc., Washington, DC, 2006.

BIBLIOGRAPHY

ACI 229R, *Controlled Low-Strength Materials*, American Concrete Institute, Farmington Hills, MI, 1999.

ACI 318-11, *Building Code Requirements for Structural Concrete*, American Concrete Institute, Farmington Hills, MI, 2011.

Building Code Requirements and Specifications for Masonry Structures (TMS 402/ACI 530/ ASCE 5) and (TMS 602/ACI 530.1/ASCE 6), American Concrete Institute, Farmington Hills, MI, 2011.

Federal Highway Administration, *Micropile Design and Construction Guidelines, Implementation Manual*, FHWA Publication No. FHWA-SA-97-070, McLean, Virginia, June 2000.

ICC, *Handbook to the Uniform Building Code: An Illustrative Commentary*, International Code Council, Washington, DC, 1998.

Joint Micropile Committee of The Deep Foundations Institute (DFI) and The International Association of Foundation Drilling (ADSC), *Guide to Drafting a Specification for High Capacity Drilled and Grouted Micropiles for Structural Support*, Dallas, Texas, 2004.

NEHRP, *NEHRP (National Earthquake Hazard Reduction Program) Recommended Provisions for the Development of Seismic Regulations for New Buildings (and Other Structures)*, Building Seismic Safety Council, Washington, DC, 1994 (1997).

SEAOC, *Recommended Lateral Force Requirements and Commentary*, Structural Engineers Association of California, Sacramento, CA, 1996.

UBC-IBC Structural Comparison & Cross Reference, International Code Council, Washington, DC, 2000.

Zeevaert, L., *Foundation Engineering for Difficult Soil Conditions*, 2nd Edition, Van Nostrand Reinhold Company, Inc., New York, NY, 1982.

CHAPTER 19

CONCRETE

Introduction
Section 1901 General
Section 1902 Definitions
Section 1903 Specifications for Tests and Materials
Section 1904 Durability Requirements
Section 1905 Modifications to ACI 318
Section 1906 Structural Plain Concrete
Section 1907 Minimum Slab Provisions
Section 1908 Anchorage to Concrete—Allowable Stress Design
Section 1909 Anchorage to Concrete—Strength Design
Section 1910 Shotcrete
Section 1911 Reinforced Gypsum Concrete
Section 1912 Concrete-Filled Pipe Columns
Key Points
References

19 Concrete

Introduction

Chapter 19 of the 2012 IBC references the 2011 edition of ACI 318, Building Code Requirements for Structural Concrete. Chapter 19 of the 2000 *International Building Code®* (IBC®) was based on the 1999 edition of the ACI 318 Standard (ACI 318-99). Portions of Chapters 2 through 7 of ACI 318-99 were reproduced in the 2000 IBC, with a small number of amendments printed in italics. The remainder of ACI 318-99 was adopted by reference, subject to eleven amendments that were listed in Section 1908. This was in keeping with the Building Officials and Code Administrators International (BOCA) and Southern Building Code Congress International (SBCCI) practice of adopting standards by reference. The reproduction of much of Chapters 2 through 7 was for the convenience of the building inspector who, particularly in smaller jurisdictions, may not have ready access to the ACI 318 Standard. The engineering provisions of ACI 318 (Chapter 8 and subsequent chapters) were not reproduced because it was assumed that access to the ACI 318 Standard would be readily available to design professionals and structural plan reviewers.

Chapter 19 of the 2000 IBC included substantive provisions in addition to those of the ACI 318 Standard. This material was contained in Section 1911: Minimum Slab Provisions; 1912: Anchorage to Concrete—Allowable Stress Design; 1913: Anchorage to Concrete—Strength Design; 1914: Shotcrete; 1915: Reinforced Gypsum Concrete; and 1916: Concrete-Filled Pipe Columns. All these sections (except 1913) were from one or more of the legacy model codes. Therefore, Section 1913 of the 2000 IBC requires some explanation.

It had been anticipated that ACI 318-99 would include a new Appendix D, Fastening to Concrete, which would have been applicable to the strength design of anchors cast in concrete as well as to post-installed anchors (i.e., anchors installed in hardened concrete). The provisions for post-installed anchors were dependent on the completion of a standard test method (ACI 355.1) for evaluating the strengths of such anchors; however, the development of that standard ran into difficulties and it was not completed in time to be referenced in ACI 318-99. Therefore, ACI 318-99 was issued without any provisions for fastening to concrete.

The concrete industry then asked ACI for permission to submit the provisions of the proposed Appendix D on cast-in-place anchors to the IBC as a replacement for the strength design provisions of IBC Section 1913. ACI agreed to permit the submission of this copyright-protected material for inclusion in the IBC, provided it was understood that ACI retained exclusive copyright to the material.

The provisions of Appendix D appeared for the first time in ACI 318-02, which was referenced in the 2003 IBC. Thus, Section 1913 of the 2003 IBC referenced Appendix D, Anchoring to Concrete, and made one significant amendment to its provisions. With this standard referenced in the code, the anchorage provisions appearing in Section 1913 of the 2000 IBC were then deleted. A brief background on the development of Appendix D is given in the discussion of Section 1909 below.

Section 1901 *General*

1901.1 Scope. Chapter 19 provides minimum requirements governing the materials, quality control, design, and construction of structural concrete elements of any structure erected under the requirements of the 2012 IBC.

1901.2 **Plain and reinforced concrete.** This section requires that structural concrete be designed and constructed in accordance with the requirements of Chapter 19 and ACI 318 as amended in Section 1905. Chapter 35 specifies that the 2011 edition of ACI 318 in particular must be used. This section states that Chapter 19 does not govern the design and construction of soil-supported slabs (i.e., slabs on grade), unless the slab transmits vertical loads or lateral forces from other portions of the structure to the soil, although Sections 1904, Durability Requirements, and 1907, Minimum Slab Provisions, do apply. A similar statement is to be found in Section 1.1.7 of ACI 318, although that statement does not make ACI 318 Chapter 4, Durability Requirements, applicable to slabs on grade. As mentioned, the equivalent of Section 1907 does not exist in ACI 318.

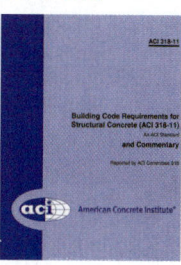

1901.3 **Construction documents.** This section reproduces from ACI 318, Section 1.2.1, the list of items to be shown on design drawings and specifications. Two items on the ACI list are omitted from the IBC. These are (1) the name and date of issue of the code and supplement to which the design conforms, and (2) live load and other loads used in design. See Section 1603 for additional structural-related requirements pertaining to construction documents.

1901.4 **Special inspection.** This section provides a cross-reference to the special inspection requirements of IBC Chapter 17. Section 1705 includes special inspection requirements for structural concrete. ACI 318, in general, does not contain provisions for special inspection of structural concrete, except that Section 1.3.5 requires continuous inspection of the placement of the reinforcement and concrete in special moment frames designed in accordance with ACI 318 Chapter 21.

Section 1902 *Definitions*

This section simply references Chapter 2 of ACI 318 for the definitions of terms related to concrete construction used in Chapter 19. No definition is actually contained in this section. Amendments are made to a number of ACI 318 definitions in Section 1905.1.1. Therefore, Section 1902 refers the user to ACI 318 as modified by Section 1905.1.1.

Section 1903 *Specifications for Tests and Materials*

1903.1 **General.** This section specifies that materials required to produce concrete and the testing of such materials must be in compliance with the provisions of Chapter 3 of ACI 318. The referenced provisions include those regarding cement, aggregates, water, steel reinforcement, admixtures, and storage of materials. The corresponding ACI 318 provisions are discussed below under ACI 318 Sections 3.2 through 3.8.

Chapter 17 of the IBC contains special inspection and testing requirements that are not included in ACI 318, except that Section 1.3.5 does require continuous inspection of reinforcement and concrete placement in special moment frames designed in accordance with ACI 318 Chapter 21. Chapter 17 also contains test requirements that are not included in ACI 318. Section 1903.1 makes these IBC Chapter 17 requirements applicable whenever required by the provisions of that chapter.

ACI 318 Section 3.2 Cementitious materials. ACI 318 requires cements to conform to ASTM C150, Specification for Portland Cement[6]; ASTM C 595, *Specification for*

Blended Hydraulic Cements[7] [excluding Type IS (≥ 70), which are not intended as principal cementing constituents of structural concrete]; ASTM C 845, *Specification for Expansive Hydraulic Cement*[8]; and ASTM C 1157, *Performance Specification for Hydraulic Cement*.[9] ACI 318 also requires other cementitious materials to conform to ASTM C 618, *Standard Specification for Coal Fly Ash and Raw or Calcined Natural Pozzolan for Use in Concrete*[10]; ASTM C 989, *Standard Specification for Ground Granulated Blast Furnace Slag for Use in Concrete and Mortars*[11]; or ASTM C 1240, *Standard Specification for Silica Fume Used in Cementitious Mixtures*.[12] ACI 318 Section 3.2.2 requires that "cement used in the Work shall correspond to that on which selection of concrete proportions was based." Refer to ACI 318 Commentary Section R3.2.2 for explanation of this requirement.

ACI 318 Section 3.3 Aggregates. ACI 318 requires concrete aggregates to conform to either ASTM C 33, *Specification for Concrete Aggregates*,[13] or ASTM C 330, *Specification for Lightweight Aggregates for Structural Concrete*.[14] However, aggregates conforming to these ASTM specifications are not always economically available and, in many instances, noncomplying materials have a long history of satisfactory performance. Such nonconforming materials are permitted with the approval of the building official when acceptable evidence of satisfactory performance is provided.

ACI 318 Section 3.3.2 specifies size limitations on aggregates to ensure the proper encasement of reinforcement and to minimize honeycombing. These requirements are waived if, in the judgment of the engineer, workability and methods of construction are such that concrete can be placed without honeycombs or voids.

ACI 318 Section 3.4 Water. Water used in mixing concrete is required to conform to the relatively recent ASTM Specification C 1602.[15] Almost any water that is drinkable and has no pronounced taste or odor can be used as mixing water for making concrete. However, some waters that are not fit for drinking may still be suitable for mixing concrete.

Excessive impurities in mixing water may not only affect setting time and concrete strength, but also cause efflorescence, staining, corrosion of reinforcement, volume instability, and reduced durability. ASTM C 1602[15] as well as Reference 16 advise that certain optional limits may be set on chlorides, sulfates, alkalis, and solids in the mixing water, or that appropriate tests may be performed to determine the effect that the impurities have on various properties. Some impurities may have little effect on strength and setting time, yet they may adversely affect durability and other properties.

According to Reference 16, water containing no more than 2,000 parts per million (ppm) of total dissolved solids can generally be used satisfactorily for making concrete. Water containing more than 2,000 ppm of dissolved solids should be tested for its effects on strength and time of set.

ASTM C 1602 allows the use of drinkable water as mixing water for concrete without testing. It also includes methods for qualifying nondrinkable sources of water with consideration of effects on setting time and strength. Testing frequency is established to ensure that water quality is monitored on a continuing basis.

Water of questionable suitability can be used for making concrete if mortar cubes (ASTM C 109[17]) made with it have 7-day and 28-day strengths equal to at least 90 percent of the corresponding strengths of companion specimens made with drinkable or distilled water. Reference 16 advises that ASTM C 191[18] tests should additionally be made to ensure that impurities in the mixing water do not adversely shorten or extend the setting time of the cement.

Concern over high chloride content in mixing water is chiefly due to the possible adverse effect of chloride ions on the corrosion of reinforcing steel or prestressing strand. In fact, prestressing strand and aluminum embedments are singled out as areas of concern in ACI 318, Section 3.4.2. Chloride ions attack the protective oxide film formed on the steel or aluminum by the highly alkaline (pH > 12.5) chemical environment present in concrete. Chlorides can be introduced into concrete with the separate mix ingredients—admixtures, aggregates, cement, and mixing water—or through exposure to deicing salts, seawater, or salt-laden air in coastal environments. Placing an acceptable limit on chloride content for any one ingredient, such as mixing water, is difficult, considering the several possible sources of chloride ions in concrete. An acceptable limit depends primarily on the type of structure and the environment to which it is exposed during its service life.[16] ACI 318 Chapter 4 devotes particular attention to this topic.

ACI 318 Section 3.5 Steel reinforcement. This section regulates reinforcement as well as welding of reinforcement to be placed in concrete.

Reinforcement is required to be deformed reinforcement, except that plain reinforcement is allowed for spirals or prestressing steel. Reinforcement consisting of structural steel, steel pipe, or steel tubing is permitted as specified in ACI 318. Tables 1903-1, 1903-2, and 1903-3 show the different reinforcement types recognized by ACI 318. Other metal elements such as inserts, anchor bolts, or plain bars for dowels at isolation or contraction joints are not normally considered to be reinforcement under the provisions of ACI 318.

Welding of reinforcing bars is required to conform to "Structural Welding Code—Reinforcing Steel," AWS D1.4 of the American Welding Society.[33] This document covers aspects of welding reinforcing bars, including criteria to qualify welding procedures.

The type and location of welded splices and other required welding of reinforcing bars must be indicated on the design drawings or in the project specifications. Much of this is also required under Section 1.2.1 of ACI 318.

Weldability of steel is based on its chemical composition or carbon equivalent. The AWS D1.4 Welding Code establishes preheat and interpass temperatures for a range of carbon equivalents and reinforcing bar sizes. Carbon equivalent is calculated from the chemical composition of the reinforcing bars. For bars other than ASTM A 706, *Specification for Low-Alloy Steel Deformed and Plain Bars for Concrete Reinforcement*,[20] the producer of the reinforcing bars does not routinely provide the chemical analysis required to calculate the carbon equivalent. For welding reinforcing bars other than ASTM A 706 bars, the design drawings or project specifications need to specifically require that results of the chemical analysis be furnished.

The ASTM A 706 specification covers low-alloy steel reinforcing bars intended for applications requiring controlled tensile properties or welding. Weldability is accomplished in the ASTM A 706 specification by limits or controls on chemical composition or carbon equivalent. The producer is required by the ASTM A 706 specification to report the chemical composition and carbon equivalent.

The AWS D1.4 Welding Code requires the contractor to prepare written welding procedure specifications conforming to the requirements of the Welding Code. Appendix A of the Welding Code contains a suggested form that shows the information required for such a specification.

It is often necessary to weld to existing reinforcing bars in a structure when no mill test report of the existing reinforcement is available. This condition is particularly common in alterations or additions to existing buildings. The AWS D1.4 Welding Code provides guidance concerning this situation.

AWS D1.4 does not cover welding of wire to wire, and of wire or welded wire reinforcement to reinforcing bars or structural steel elements. Some guidance on this subject is available in ACI 318 Commentary Section R3.5.2.

Note that ASTM A 616, *Specification for Rail-Steel Deformed and Plain Bars for Reinforcement*, and ASTM A 617, *Specification for Axle-Steel Deformed and Plain*

Table 1903-1. **Deformed Reinforcement Recognized in ACI 318**

Product	ASTM Specification	Grade or Type		Minimum Yield Strength ksi	Minimum Tensile Strength ksi
Reinforcing bars	A 615[19]	40[f]		40	60
		60		60	90
		75[f]		75	100
	A 706[20]	60		60 (78 max)	80
	A 955[21]	40[f]		40	70
		60		60	90
		75[f]		75	100
	A 996[e, 22]	Type R	50	50	80
			60	60	90
		Type A	40	40	70
			60	60	90
	A 1035[23]	100		100[g]	150
		120		120[h]	150
Bar mats[a]	A 184[24]	—		—	—
Wire, deformed	A 496[25]	—		75	85
Welded wire reinforcement, plan (W 1.2 and larger)	A 185[26]	—		65	75
Welded wire reinforcement, deformed	A 497[27]	—		70	80
Galvanized reinforcing bars[b]	A 767[28]	—		—	—
Epoxy-coated reinforcing bars[b]	A 775[29] A 934[30]	—		—	—
Epoxy-coated wires[c] and welded wire reinforcement[d]	A 884[31]	—		—	—
Stainless steel wire and welded wire reinforcement, deformed	A 1022[32]	—		—	—

[a]Same as reinforcing bars, except only available with Grade 40 bars (ASTM A 615) and Grade 60 bars (ASTM A 615 or ASTM A 706).
[b]Same as reinforcing bars.
[c]Same as wires.
[d]Same as welded wire reinforcement.
[e]Bars are furnished only in sizes 3 through 8.
[f]Grade 40 bars are furnished only in sizes 3 through 6. Grade 75 bars are furnished only in sizes 6 through 18.
[g]0.2% offset stress. Stress corresponding to extension under load of 0.0035 in./in. is 80 ksi.
[h]0.2% offset stress. Stress corresponding to extension under load of 0.0035 in./in. is 90 ksi.

Bars for Concrete Reinforcement, have been replaced by ASTM A 996, *Specification for Rail-Steel and Axle-Steel Deformed Bars for Concrete Reinforcement.*[22] Reinforcing bars conforming to ASTM A 996 are generally unavailable in all but a few areas of the United States ASTM A 996 bars can be Type Rail Symbol (with a rail symbol stamped on the bars), Type R (with the letter R stamped on the bars), or Type A (with the letter A stamped on the bars). The first two types are rail steel, and the third type is axle steel. ACI 318 requires rail-steel reinforcing bars to conform to the provisions for Type R bars, which are required to meet more restrictive provisions for bend tests.

Specifications for Tests and Materials

Table 1903-2. **Plain Reinforcement and Prestressing Tendons Recognized in ACI 318**

Product	ASTM Specification	Grade or Type	Minimum Yield Strength ksi	Minimum Tensile Strength ksi
Reinforcing bars	A 615[19]	40	40	60
		60	60	90
		75	75	100
	A 706[20]	60	60	80
Wire, plain	A 82[34]	—	70	80
Prestressing wire	A 421[35]	—	199.75-212.5	235-250
Prestressing wire, low-relaxation	A 421[35] + supplement	—	199.75-212.5	235-250
Prestressing strand, stress-relieved	A 416[36]	250	212.5	250
		270	229.5	270
Prestressing strand, low-relaxation	A 416[36]	250	225	250
		270	243	270
Prestressing bar	A 722[37]	Type I	127.5	150
		Type II	120	150

Table 1903-3. **Structural Steel, Steel Pipe, or Tubing Recognized in ACI 318 for Use in Composite Compression Member**

Product	ASTM Specification
Structural Steel Used with Reinforcing Bars in Compression Members Meeting Requirements of Section 10.16.7 or 10.16.8	
Carbon structural steel	A 36[38]
High-strength low-alloy structural steel	A 242[39]
High-strength low-alloy columbium-vanadium structural steel	A 572[40]
High-strength low-alloy structural steel with 50 ksi minimum yield strength up to 4-in. thick	A 588[41]
Structural shapes	A 992[42]
Steel Pipe or Tubing for Composite Compression Members Composed of Steel Encased Concrete Core Meeting Requirements of Section 10.16.6	
Seamless and welded black and hot-dipped galvanized steel pipe	A 53 Grade B[43]
Cold-formed welded and seamless carbon steel structural tubing in rounds and shapes	A 500[44]
Hot-formed welded and seamless carbon steel structural tubing	A 501[45]

ACI 318 Section 3.6 Admixtures. ACI 318 Section 3.6.1 requires admixtures used in concrete to be subject to prior approval by the registered design professional.

The admixtures recognized for use in concrete in Section 3.6 of ACI 318 are listed in Table 1903-4. The so-called mineral admixtures—Fly Ash or Other Pozzolans (ASTM C 618), Ground Granulated Blast Furnace Slag (ASTM C 989), and Silica Fume

Table 1903-4. Concrete Admixtures Recognized by ACI 318

Product	ASTM Specification	Desired Effect
Accelerators	C 494, Type C	Accelerate setting and develop early strength
Air-entraining admixtures	C 260	Improve durability in environments of freeze-thaw, deicers, sulfate and alkali reactivity, improved workability
Retarders	C 494, Type B	Retard setting time
Superplasticizers	C 1017, Type 1	Reduce water-cementitious materials ratio Produce flowing concrete
Superplasticizer and retarder	C 1017, Type 2	Produce flowing concrete with retarded set Reduce water
Water reducer	C 494, Type A	Reduce water demand by at least 5%
Water reducer and accelerator	C 494, Type E	Reduce water (minimum 5%) and accelerate set
Water reducer and retarder	C 494, Type D	Reduce water (minimum 5%) and retard set
Water reducer—high range	C 494, Type F	Reduce water demand (minimum 12%)
Water reducer—high range and retarder	C 494, Type G	Reduce water demand (minimum 12%) and retard set

(ASTM C 1240) now appear in Section 3.2, Cementitious materials. They are no longer included under Section 3.6 and are therefore not included in Table 1903-4.

Wire, strands, and bars not specifically listed in ASTM A 421,[35] A 416,[36] or A 722[37] are allowed by ACI 318 Section 3.5.6.2, provided they conform to minimum requirements of these specifications and do not have properties that make them less satisfactory than those listed in ASTM A 421, A 416, or A 722.

ACI 318 Section 3.7 Storage of materials. This section requires proper storage of cementitious materials and aggregates to prevent contamination or deterioration, and prohibits the use of any material so affected in concrete.

ACI 318 Section 3.8 Referenced standards. All the standards referenced in ACI 318 (most of them ASTM standards) are listed in this section.

1903.2 **Glass fiber reinforced concrete.** This section in the IBC, which is not to be found in 1903.2 ACI 318, requires that glass fiber reinforced concrete (GFRC) and the materials used in such concrete conform to PCI MNL 128, *Recommended Practice for Glass Fiber Reinforced Concrete Panels.*[5] This publication contains the latest information on the planning, design, manufacture, and installation of GFRC panels.

1903.3 **Flat wall insulating concrete form (ICF) systems.** This section references ASTM E 2634 for the material used to make forms for insulating concrete form (ICF) systems using molded expanded polystyrene (EPS) insulation panels. ICFs are rigid plastic foam forms used in the construction of cast-in-place concrete structural members. The ICFs remain in place as permanent building insulation for energy-efficient, cast-in-place, reinforced concrete walls, floors, roofs, beams, and columns. The forms are interlocking modular units that are dry-stacked (without mortar) and filled with concrete. ICFs are generally manufactured from polystyrene or polyurethane. Reinforcing steel is placed in the forms and then concrete is pumped into the cavity to form the structural portion of the walls. The forms provide thermal and acoustic insulation, space to run electrical conduit and plumbing, and backing for interior and exterior finishes. ASTM E 2634 applies to ICF systems that consist of molded EPS insulation panels that are connected by cross ties and act as permanent formwork for cast-in-place reinforced concrete structural members such as reinforced concrete beams, lintels, exterior and interior bearing and nonbearing walls, foundations, and retaining walls. ASTM E 2634 lists test methods appropriate for establishing ICF system performance as a permanent concrete forming system.

Section 1904 *Durability Requirements*

Durability requirements are adopted through reference to Chapter 4 of ACI 318, with certain exceptions. The provisions of that chapter emphasize the importance of considering durability requirements before the designer selects f'_c and cover over reinforcing steel. They include exposure categories and classes with applicable durability requirements for concrete in a unified format.

As pointed out in the commentary to Chapter 4 of ACI 318, maximum water-cementitious materials ratios of 0.40 to 0.50 that may be required for concrete exposed to freezing and thawing, sulfate in soils or waters, or for preventing corrosion of reinforcement will typically be equivalent to requiring an f'_c of 5,000 to 4,000 pounds per square inch (psi), respectively. Generally, the required average concrete strengths, f'_{cr}, will be 500 to 700 psi higher than the specified compressive strength, f'_c. Since it is difficult to accurately determine the water-cementitious materials ratio of concrete during production, the f'_c specified should be reasonably consistent with the water-cementitious materials ratio required for durability. This will help ensure that the required water-cementitious materials ratio is actually obtained in the field. Because the usual emphasis in inspection is on strength, test results substantially higher than the specified strength may lead to a lack of concern for quality and production of concrete that exceeds the maximum water-cementitious materials ratio. Thus, an f'_c of 3,000 psi and a maximum water-cementitious materials ratio of 0.45 should not be specified for a parking structure, if the structure will be exposed to deicing salts.

The commentary in Chapter 4 of ACI 318 also points out that the chapter does not include provisions for especially severe exposures, such as acids or high temperatures, and is not concerned with aesthetic considerations such as surface finishes. These items should be covered in project specifications.

Exposure categories and classes. Concrete is required to be assigned to exposure classes in accordance with ACI 318 Section 4.2, based on: **1904.1**

1. Exposure to freezing or thawing in a moist condition or deicer chemicals;
2. Exposure to sulfates in water or soil;
3. Exposure to water where the concrete is intended to have low permeability; and
4. Exposure to chlorides from chemicals, salt, saltwater, brackish water, seawater, or spray from these sources, where the concrete has steel reinforcement.

Concrete properties. This section requires concrete mixtures to conform to the most restrictive maximum water-cementitious materials ratios and maximum specified compressive strength requirements of ACI 318 Section 4.3, based on exposure classes assigned in Section 1904.1. **1904.2**

This section includes an exception to the ACI 318 provisions requiring that in buildings less than four stories in height that house Group R (residential) occupancies, normal-weight concrete subject to weathering (freezing and thawing), as determined from Figure 1904.2, or deicer chemicals, must comply with the requirements of Table 1904.2. Neither Figure 1904.2 nor Table 1904.2 is part of ACI 318. The exception, the table, and the figure are adopted into the IBC from the *International Residential Code®* (IRC®).[46]

19 Concrete

Table 1904.2 mandates a minimum specified compressive strength as a function of the concrete element and exposure, but no maximum water-cementitious materials ratio. Figure 1904.2 shows the geographic regions within the U.S. mainland that are classified as having negligible, moderate, and severe weather exposures for the purposes of Table 1904.2. It is clearly indicated in a note to the figure that the boundary lines defining the weather regions are only approximate. Building officials in communities near a boundary line must establish applicable exposure classifications.

Section 1905 *Modifications to ACI 318*

1905.1 General. A summary of the modifications to the provisions of ACI 318 can be found in Table 1905-5. A more detailed discussion of each modification is provided below.

1905.1.1 ACI 318 Section 2.2. Modifications are made to existing definitions and new definitions are added to those in ACI 318 Section 2.2. The definition of Design Displacement is modified to reference Section 12.8.6 of ASCE 7.[47] Definitions of *Detailed Plain Concrete Structural Wall, Ordinary Precast Structural Wall, Ordinary Reinforced Concrete Structural Wall, Ordinary Structural Plain Concrete Wall, Special Structural Wall*, and *Wall Pier* are added to Section 2.2.

The definition of *Special Structural Wall* from ACI 318 is modified in order to make it compatible with ASCE 7. In ACI 318, special cast-in-place walls and special precast walls are defined under a common name of *Special Structural Wall*. Because Table 12.2-1 of ASCE 7-05, listing the values of seismic design parameters and height limitations for different structural systems, has only one entry of *Special Reinforced Concrete Shear Wall*, it needs to be clarified that the entry corresponds to *Special Structural Wall* in ACI 318, and as a result, the corresponding values of seismic design parameters and height limitations apply to both special cast-in-place and special precast structural walls.

Note that *Shear Walls* are referred to as *Structural Walls* to make the building code consistent with ACI 318, although the term *Shear Wall* continues to be used in Table 12.2-1 of ASCE 7. These terms should be treated as interchangeable as far as concrete structural systems are concerned.

Table 1905-5. **2012 IBC Modifications to ACI 318**

Item	IBC Section	Subject	ACI 318 Section Modified
1	1905.1.1	Additional definition of terms	Section 2.2
2	1905.1.2	Clarification of applicability of the provisions of ACI 318 Chapter 21	Section 21.1.1
3	1905.1.3	Requirements for intermediate precast structural walls	Section 21.4
4	1905.1.4	Wall piers and wall segments	Section 21.9
5	1905.1.5	Special precast structural walls	Section 21.10
6	1905.1.6	Seismic design of concrete foundations	Section 21.12.1.1
7	1905.1.7	Requirements for detailed plain concrete structural walls	Section 22.6
8	1905.1.8	Application of plain concrete in SDC C, D, E, and F structures	Section 22.10
9	1905.1.9	Anchors in SDC C, D, E, and F structures	Section D.3.3
10	1905.1.10	Anchors larger than 2 inches in diameter and/or with embedment depths exceeding 25 inches	Section D.4.2.2

ACI 318 Section 21.1.1. The modifications to Section 21.1.1 provide clarity of these provisions. The modifications unambiguously state exactly which ACI 318 requirements are to be complied with for structures in different seismic design categories, and make it clear that the *ordinary structural walls* mentioned in Item (b) of Section 21.1.1.7 include both cast-in-place and precast walls. The use of plain concrete is also specifically forbidden in structures assigned to SDC C or higher, other than those permitted by IBC Section 1905.1.8.

1905.1.2

ACI 318 Section 21.4. This section modifies ACI 318 Section 21.4, Intermediate precast structural walls, by first renumbering Section 21.4.3 to Section 21.4.4 and then adding four new sections (21.4.3, 21.4.5, 21.4.6, and 21.4.7).

1905.1.3

- Section 21.4.3 specifies a minimum ductility requirement for a steel element used in a connection between wall panels or between wall panels and the foundation, so that yielding can be allowed in that element. This section is adopted from the 2003 NEHRP *Provisions*,[48] and requires that a yielding steel connection element retain at least 80 percent of its design strength at the deformation level corresponding to the design displacement of the structure. Thus, a clear distinction is made between a ductile steel element and a nonductile one. ACI 318 Section 21.4.2 allows yielding in steel reinforcement as well as in steel elements present in the connection, while the 2003 IBC modified this provision to restrict yielding in the reinforcement only. This was done because no formal distinction between ductile and nonductile steel elements was available at the time, and it was felt that allowing yielding in a nonductile element was not desirable. However, with a proper definition of ductile elements in place, the 2003 IBC modification to ACI 318 Section 21.4.2 was omitted in the 2006 IBC.

- Section 21.4.5 clarifies that wall piers in buildings that are assigned to Seismic Design Category D, E, or F must comply with Section 1905.1.4.

- Section 21.4.6 specifies a minimum transverse reinforcement requirement for intermediate precast wall piers when not designed as part of a moment frame, to prevent a premature shear failure. Similar requirements already existed for wall piers meant to be used in SDC D, E, or F structures (IBC Section 1905.1.4). This provision, adopted from the 2003 NEHRP *Provisions*,[48] expands the scope to SDC C structures as well. In a second addition, Section 21.4.7 stipulates that wall piers with a horizontal length-to-thickness ratio less than 2.5 shall be designed as columns.

ACI 318 Section 21.9. This modification specifies transverse reinforcement requirements for wall piers, defined in Section 1905.1.1 (Figure 1905-1), when they are not designed as part of a special moment frame. The requirements were first added to the 1991 edition of the UBC out of concern that thin column-like elements between openings in shear walls were being designed without proper transverse reinforcement. There are two important exemptions to the requirements of Section 1905.1.4. A wall pier satisfying the requirement of deformation compatibility with the lateral-force-resisting system need not be detailed by Section 21.9.8.2. Wall piers laterally supported by much stiffer shear walls along the same line within a story also need not be detailed by Section 21.9.8.2.

1905.1.4

ACI 318 Section 21.10. ACI 318 Section 21.10.2 requires that precast special structural walls satisfy all the requirements for cast-in-place special structural walls as specified in Section 21.9, as well as the requirements for precast intermediate structural walls specified in Sections 21.4.2 and 21.4.3. However, supplemental requirements are added for precast intermediate walls by 2009 IBC Section 1905.1.3 (discussed earlier). As a result of that modification, this section adds the supplemental ACI 318 Section 21.4.4 to the requirements for special precast structural walls.

1905.1.5

19 Concrete

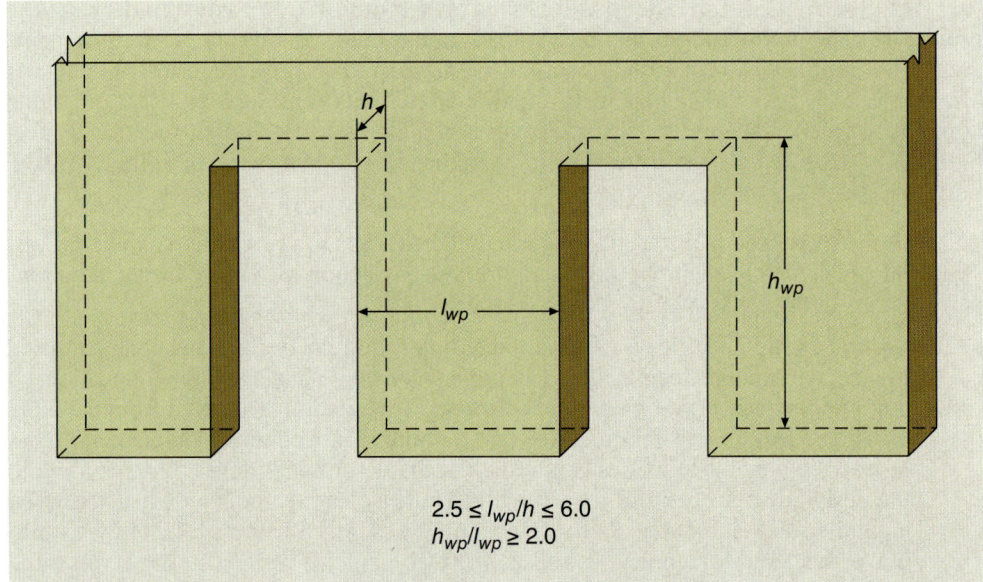

**Figure 1905-1
Wall piers and wall segments.**

$2.5 \leq l_{wp}/h \leq 6.0$
$h_{wp}/l_{wp} \geq 2.0$

1905.1.6 ACI 318 Section 21.12.1.1. IBC requires that the seismic design of foundations comply with relevant provisions in ACI 318, unless they are modified by Chapter 18, Soils and Foundations, of the 2009 IBC. In general, the provisions of IBC Chapter 18 are more stringent than those found in ACI 318 Section 21.12.

1905.1.7 ACI 318 Section 22.6. This modification adds new Sections 22.6.7 through 22.6.7.2 to specify the reinforcement requirement that detailed plain concrete structural walls need to conform to in addition to that required for ordinary plain concrete structural walls. Note that detailed plain concrete structural walls are not recognized in ACI 318.

1905.1.8 ACI 318 Section 22.10. This modification replaces ACI 318 Section 22.10 in its entirety, while serving essentially the same purpose, that is, to prohibit the use of plain concrete elements in structures assigned to Seismic Design Category C, D, E, or F, with several specific exceptions that permit the use of plain concrete in foundations. This modification describes these exceptions in more detail than ACI 318 Section 22.10.

1905.1.9 ACI 318 Section D.3.3. This modification to Appendix D of ACI 318 is based on the 2008 edition of the standard. It provides flexibility in designing anchors for earthquake forces by providing exceptions to the ACI 318 provision. The basic premise of D.3.3 is that anchorage design is controlled by the strength of a "ductile steel element," which is defined in the ACI 318 appendix.

Section D.3.3.5 includes four exceptions. Exception 1 applies to concrete wall anchorage that is designed for maximum expected seismic forces. Under ASCE 7, higher wall anchorage force levels are required to protect against brittle failure. This exception clarifies that these special wall anchorage design forces need not be compounded with the ACI 318 Appendix D anchorage ductility requirement.

Exception 2 applies to the anchorage of wood sill plates to concrete foundations in light-frame construction. Based on light-frame shear wall testing, the wood sill plate controls the ductile behavior of the anchorage assembly. If the anchor meets the requirements of the exception, then the anchor need not meet the requirements of Section D.3.3.5. Allowable in-plane shear capacity of the anchor bolt is determined in accordance with NDS Table 11E rather than ACI 318 Appendix D.

Exception 3 applies to the anchorage of cold-formed steel track to concrete foundations in light-frame construction. Based on light-frame shear wall testing, the cold-formed steel track controls the ductile behavior of the anchorage assembly. If the anchor meets the requirements of the exception, then the anchor need not meet the requirements of Section D.3.3.5. Allowable in-plane shear capacity of the anchor bolt is determined from AISI S100 Section E3.3.1 rather than ACI 318 Appendix D.

There are five criteria to meet in order to use Exception 2 or 3 for the design of anchor bolts connecting wood sill plates of shear walls of light-frame structures. The maximum nominal diameter of the anchor is $5/_8$ inches (15.9 mm); anchors must be embedded a minimum of 7 inches into the footing and located a minimum of $1^3/_4$ inches (44 mm) from the edge of the concrete parallel to the wood sill plate or steel track, and a minimum of 15 diameters from the edge of the concrete perpendicular to the wood sill plate or steel track; the wood sill plate must be a 2X or 3X, and the track must be within a thickness range of 33 mil to 68 mil.

For light-frame construction, Exception 4 allows use of provision D.3.3.8 rather than D.3.3.5. Section 202 defines light-frame construction as a system where vertical and horizontal structural elements are primarily formed by repetitive wood or cold-formed steel framing members.

Section D.3.3.6 allows design of the anchor attachment to control the anchorage assembly design. Yielding of the attachment must begin before the anchor begins to yield. ACI 318 Appendix D defines "attachment" as the structural assembly external to the surface of the concrete that transmits loads to or from the anchor. As steel today has a higher expected yield strength than its specified yield strength, the attachment must have an expected yield strength no larger than the design strength of the anchor.

There are two exceptions to this provision. Exception 1 deals with nonstructural components designed for earthquake loading in accordance with ASCE 7 Section 13.4.2. That provision imposes additional nonductile anchor force increases on anchors in structures assigned to Seismic Design Category C and higher. The exception clarifies that it is not intended for nonductile anchor force increase to be compounded with this ACI 318 anchorage requirement. Exception 2 applies to concrete wall anchorage that is designed for maximum expected seismic forces. Under ASCE 7, higher wall anchorage force levels are required to protect against brittle failure. This exception clarifies that these special wall anchorage design forces need not be compounded with the ACI 318 Appendix D anchorage ductility requirement.

Section D.3.3.7 allows the design strength of an anchor to be set as 0.4 multiplied by the steel strength of the element as determined in accordance with ACI 318 Appendix D.5.1 for steel tensile strength or D.6.1 for steel shear strength. This provision was initially added to ACI 318-08 to provide a capacity for nonductile steel anchors. In ACI 318-11, anchors are assumed to be ductile steel elements.

Section D.3.3.8 refers to ACI 318 D6.2.1(c) where shear loads are examined parallel to the length of the sill plate or track. This provision allows the use of ductile or nonductile anchors in wood sill plates and tracks without the decrease in design strength required in Section D.3.3.7. This practice is considered acceptable since wood plate or steel track failure occurs before the anchor reaches its design strength.

ACI 318 Section D.4.2.2. This modification revises ACI 318 Section D.4.4.2 to make the provision more specific regarding the applicability of ACI 318 Sections D.5.2 and D.6.2 for computing concrete breakout strengths under tension and shear, respectively. The way Section D.4.2.2 is phrased leads one to believe that the limitations on both the maximum anchor diameter (2 inches) and the maximum embedment depth (25 inches) need to be satisfied in order to be able to compute the concrete breakout strengths in accordance with Sections D.5.2 and D.6.2. This is not correct. The diameter limitation pertains only to concrete breakout in shear (Section D.6.2), whereas the limitation on the embedment depth pertains only to concrete breakout in tension (Section D.5.2). As a result, not meeting any one limitation should not disqualify a designer from using the provisions related to the other limitation. In addition, it has also been clarified that even when the anchor embedment depth

1905.1.10

is more than 25 inches, the concrete breakout strength under tension can still be computed in accordance with ACI 318 Section D.6.2 as long as Equation D-8 is not used.

Inconsistencies with ACI 318-11. The 2012 IBC was published prior to the publication of ACI 318-11. As noted above, the modifications to ACI 318 in the 2012 IBC are based on ACI 318-08, which is the referenced standard for concrete in the 2009 IBC. Chapter 19 of the 2012 IBC references ACI 318-11. Because of this, some of the modifications in Section 1905 are inconsistent with the 2011 edition of the ACI 318 Standard. To remedy this, two code changes (S340-12 and S215-12) were proposed and approved for the 2015 IBC. These two code changes were brought to the attention of designers and adopting agencies to address technical and inspection requirements for anchors in concrete and properly correlate the provisions of the 2012 IBC with those of ACI 318-11. Perhaps most critical is the requirement for continuous special inspection for post-installed adhesive anchors in hardened concrete where the anchors are designed to resist sustained tension loads. For a complete discussion of this matter, refer to the article by S.K. Ghosh, "The Concrete Provisions of the 2012 IBC and ACI 318-11," published by ICC in the February 2013 issue of *Building Safety Journal Online.* It has been suggested by Ghosh that jurisdictions that have not adopted the 2012 IBC should make amendments in their adopting ordinance, and those that have previously adopted the 2012 IBC could approve the amendments under the alternative materials, design, and methods of construction provisions in Section 104.11.

Section 1906 *Structural Plain Concrete*

This section refers to Chapter 22 of ACI 318 as modified in Sections 1905.1.7 and 1905.1.8.

An exception is included for Group R-3 occupancies as other occupancies that are less than two stories in height and of light-frame construction. The required edge thickness of 8 inches in ACI 318 is permitted to be reduced to 6 inches, provided the footing does not extend more than 4 inches on either side of the supported wall.

Section 1907 *Minimum Slab Provisions*

Section 1905.0 of the 1999 BOCA/NBC, Section 1900.4.4 of the 1997 UBC, and Section 1909.1 of the 1999 SBC contained identical minimum thickness requirements for floor slabs supported directly on the ground. These are included in Section 1907 of the IBC. In addition, provisions aimed at retarding vapor transmission through the floor slab are included. These provisions are essentially the same as those found in the 1999 edition of the BOCA/NBC (Section 1905.1) and the 1999 edition of the SBC (Section 1909.2). BOCA and SBCCI published explanations of their provisions,[49,50] which are used as source material for the following discussion.

It is specifically stated in ACI 318 Section 1.1.6 that it does not govern the design and construction of soil-supported slabs, unless the slab transmits vertical loads or lateral forces from other portions of the structure to the soil. This does not preclude the application of any ACI 318 provision that would ensure the proper strength, durability, and abrasion resistance of concrete slabs on ground. Slabs on ground may be plain concrete without any reinforcement or may contain shrinkage and temperature reinforcement in the form of reinforcing bars or wire mesh.

1907.1 General. The minimum thickness of $3^1/_2$ inches is based on the fact that the largest aggregate size typically used in the construction of slabs on grade is 1 inch. ASTM C 33[13] allows 5 percent of 1-inch graded material to be larger than 1 inch, but smaller than $1^1/_2$ inches in size. Therefore, in accordance with the requirement of ACI 318 Section 3.3.2 that the nominal size of coarse aggregate must be no larger than one-third the depth of a slab, the minimum depth becomes $3^1/_2$ inches (3 times 1 inch plus a $^1/_2$ inch allowance for larger stones).

Although good-quality concrete is practically impermeable to the passage of water (that is not under significant pressure), concrete is not impervious to the passage of water vapors. If the surface of the slab is not sealed, water vapor will pass through the slab. If a floor finish such as linoleum, vinyl tile, wood flooring or any type of covering is placed on top of the slab, the moisture is trapped in the slab. Any floor finish adhering to the concrete may eventually loosen or buckle or blister.

Many of the moisture problems associated with enclosed slabs on ground can be prevented or minimized by installing vapor retarders, such as polyethylene sheeting or other approved materials, between the slab and the ground. Such retarders are needed under slabs in habitable spaces. Where garages, utility areas, and similar spaces are not to be heated or occupied, the use of vapor barriers is generally not necessary. If moisture migration is not expected to be a problem based on the occupancy of a building, Exception 3 permits the vapor retarder to be omitted. The vapor retarder provisions of Section 1907 did not change practice in jurisdictions that had previously adopted the UBC since the soil investigation report typically required such vapor retarders.

Section 1908 *Anchorage to Concrete—Allowable Stress Design*

The IBC provides two distinct methods for designing anchorage to concrete. Section 1908 presents an allowable stress design (ASD) method of limited applicability. Section 1909 presents a strength design method of broader applicability.

1908.1 Scope. The ASD provisions of this section are adapted from the 1997 edition of the *Uniform Building Code*. They are based on limited test data on headed bolts cast in normal-weight concrete subjected to static loading. It is, therefore, not applicable to hooked (J or L) bolts, lightweight aggregate concrete, post-installed anchors (meaning anchors installed in hardened concrete), or when load combinations include earthquake effects.

1908.2 Allowable service load. Table 1908.2 provides allowable service loads for headed bolts or stud anchors cast in normal-weight concrete, which are subject to tension only or shear only. The table values are for a particular range of bolt sizes ($^1/_4$ to $1^1/_4$ inches) and concrete strengths (2,500, 3,000, and 4,000 psi). When using the tabulated values, minimum edge distances and minimum spacing between bolts, specified in the table, must be carefully maintained. Edge distances are measured from the face of the bolt to the sides and ends of concrete members. Section 1908.3 provides for decreases of up to 50 percent in edge distances and/or spacing, with corresponding decreases in allowable service loads.

For a headed bolt or stud anchor subjected to combined tension and shear loading, Section 1908.2 requires that the applied service-level load in tension and the applied service-level load in shear satisfy the interaction formula given in that section, which involves the allowable values given by Table 1908.2.

1908.3 Required edge distance and spacing. The purpose of this section is to expand the applicability of Table 1908.2 by providing for situations where bolt locations cannot meet the required minimum edge distances and/or the spacing requirements of the table.

Both the minimum edge distance and the minimum spacing requirements of the table may be reduced by up to 50 percent for a corresponding 50-percent reduction in the allowable service loads. Where reduction in edge distance and spacing is less than 50 percent, the allowable service loads may be determined by linear interpolation.

1908.4 **Increase in allowable load.** This section allows a one-third increase in the tabulated values of allowable service loads of Table 1908.2 when service loads include those that are due to wind, in addition to dead loads, live loads, and other loads such as snow. Wind loads are transitory in nature. It is considered highly unlikely that two or more variable loads such as wind and live loads will attain their design values simultaneously. This is the rationale behind the one-third increase in allowable stresses permitted when the alternative basic load combinations of Section 1605.3.2 are used in design, and load combinations include the effects of wind or seismic loads.

1908.5 **Increase for special inspection.** The provisions of this section have been taken from Footnotes 5 and 6 to Table 19-D (Allowable Service Load on Embedded Bolts) of the 1997 UBC, which is essentially identical to Table 1908.2. These footnotes first appeared in the 1976 edition of the UBC. Footnote 5 applies to allowable service loads in tension and states that the values shown are for work without special inspection; where special inspection is provided, values may be increased 100 percent. Footnote 6 applies to allowable service loads in shear and states that the values shown are for work with or without special inspection.

Section 1909 *Anchorage to Concrete—Strength Design*

This section references the provisions of Appendix D, Anchoring of Concrete of ACI 318 by reference. A brief discussion on the development of Appendix D, first introduced in the 2002 edition of ACI 318, follows.

As of the late 1990s, the primary sources of design information for connections to concrete using cast-in-place anchors were Appendix B of ACI 349[51] and the PCI Design Handbook.[52] The design of connections using post-installed anchors typically was performed using information from the anchor manufacturers.

ACI Committee 318, with the support of ACI Committee 355 (Anchorage to Concrete) and ACI Committee 349 (Concrete Nuclear Structures), took the lead in developing provisions for both cast-in-place and post-installed mechanical anchors. During the code cycle leading to ACI 318-99, ACI Committee 318 approved a proposed Appendix D to ACI 318 dealing with the design of anchorages to concrete. At the same time, ACI Committee 355 was in the process of developing a test method for evaluating the performance of post-installed anchors in concrete. Final adoption of Appendix D in ACI 318-99 depended on the approval of this test method.

The test method for evaluating the performance of post-installed anchors was not completed in time to meet the publication deadline of ACI 318-99. A subsequent attempt was made to process Appendix D with provisions for only cast-in-place anchors, but this move failed to garner support within ACI. Thus, ACI 318-99 was issued without any provisions for fastening to concrete.

The concrete industry then asked ACI for permission to submit the provisions of the proposed Appendix D on cast-in-place anchors to the IBC as a replacement for the strength design provisions of what was then IBC Section 1913. ACI agreed to permit the submission of this copyright-protected material for inclusion in the IBC, provided it was understood that ACI retained exclusive copyright to the material.

During the code cycle leading to the 2003 IBC, a code change was submitted and subsequently approved to remove the IBC anchorage provisions and to reference the new provisions in Appendix D of ACI 318-02.

Scope. The provisions of this section govern the strength design of anchors installed in concrete that transmit structural loads between elements. Appendix D of ACI 318 now contains provisions for both cast-in-place and post-installed mechanical anchors. Two modifications are made to these provisions in Sections 1905.1.9 and 1905.1.10, as described previously. The modification in Section 1905.1.10 pertains to the concrete breakout strength requirements of Section D.4.2.2 for single anchors exceeding 2 inches in diameter and/or 25 inches of embedment depth. Anchors that are not within the scope of Appendix D are to be designed by an approved procedure. As previously noted, the modifications in Section 1905 were based on ACI 318-08. See the discussion of Section 1905 for more detail.   **1909.1**

Section 1910 *Shotcrete*

This section contains provisions very similar to those in Section 1911 of the BOCA/NBC, Section 1924 of the UBC, and Section 1915 of the SBC (latest editions). BOCA and SBCCI published commentaries on their shotcrete provisions,[49,50] from which the material here is largely drawn.

General. Shotcrete is pneumatically projected concrete or mortar. Other terms such as spraycrete, sprayed concrete, and gunite are also associated with shotcrete construction. Shotcrete needs to conform to Chapter 19 requirements for plain or reinforced concrete, unless specifically exempted by Section 1910.   **1910.1**

Proportions and materials. Proportions of shotcrete mixtures should be determined prior to the beginning of construction by trial applications on test specimens. The test specimens should be representative of the in-place application (flat, vertical, overhead), and the shotcrete should be applied using the same materials and equipment that will be used for construction.   **1910.2**

Aggregate. For construction applications in which the shotcrete will be several inches thick, coarse aggregate may be used in the mixture. In those cases, the aggregate size is limited to $^3/_4$ inch, to minimize the effects of rebounding during placement and the creation of voids in the shotcrete. Rebound refers to shotcrete that ricochets off the receiving surface. See Section 1910.6.   **1910.3**

Reinforcement. The size and the spacing of the reinforcement are required to be such as to minimize interference with the high-velocity placement of the shotcrete and to ensure that the reinforcement is completely covered. The clearance between the form and the reinforcement may vary depending on whether concrete or mortar is used for the shotcrete. The use of reinforcing bars or welded wire reinforcement can also affect the minimum clearance from the form. The exception allowing reduction of required clearances, subject to approval of the building official and preconstruction testing, should be noted.   **1910.4**

Noncontact lap splices are preferred, to minimize the creation of weak sections in the shotcrete. Where possible, at least 2 inches should separate lapped bars. Welded wire reinforcement should be lapped by one square in all directions.[49] When adequate encasement can be shown, contact lap splices are permitted with the approval of the building official.

Preconstruction tests. The preconstruction tests provided for in this section are at the discretion of the building official. They were part of the UBC provisions, but not part of the BOCA/NBC or the SBC provisions. The requirements are quite explicit and self-explanatory.   **1910.5**

1910.6 Rebound. As mentioned earlier, rebound is shotcrete that ricochets off the receiving surface. The position of the work (flat, vertical, or overhead), layer thickness, discharge pressure, cement content, water content, size and gradation of aggregate, and type and amount of reinforcement can affect the amount of rebounding that occurs.[49] Rebounded material may not be reused or worked back into the construction and must be removed from the surface prior to placement of additional layers of shotcrete.

1910.7 Joints. Construction joints are generally tapered to a thin edge over a width of approximately 12 inches. Square construction joints should be avoided, except as specifically permitted by this section.

1910.8 Damage. After placement, any shotcrete that lacks uniformity or that exhibits segregation, honeycombing, or delamination, or which contains dry patches, slugs, voids, or sand pockets (porous areas low in cement content), or that sags or sloughs must be removed and replaced.

1910.9 Curing. As in most construction involving cementitious materials, proper curing practices need to be followed from the time of completion of the shotcrete application, as outlined in this section.

1910.10 Strength tests. The BOCA/NBC and the SBC required strength tests of shotcrete to be made in accordance with the quality assurance provisions of ACI 506.2.[53] Test specimens were required to be obtained from the in-place shotcrete or from a test panel that was representative of the work, and tested in accordance with ASTM C 42, *Test Method for Obtaining and Testing Drilled Cores and Sawed Beams of Concrete*.[54] The IBC, like the UBC, spells out strength test provisions, including sampling requirements and acceptance criteria.

Section 1911 Reinforced Gypsum Concrete

Provisions on reinforced gypsum concrete were included in Section 1925 of the 1997 UBC and Section 1914 of the 1999 SBC.

The specifications of Section 1911 cover poured-in-place reinforced gypsum concrete over permanent formboard. Gypsum concrete is used as a structural material in the construction of roof decks or slabs and as a nonstructural material in floor topping.

Gypsum concrete is required to conform to the specifications of ASTM C 317, *Specification for Gypsum Concrete*.[55] The design and application of reinforced gypsum concrete must be in accordance with the requirements of ASTM C 956, *Specification for Installation of Cast-in-Place Reinforced Gypsum Concrete*.[56]

1911.2 Minimum thickness. The minimum thickness of reinforced gypsum concrete is required to be 2 inches, except that this may be reduced to no less than $1^1/_2$ inches if certain conditions given in Section 1911.2 are met.

Section 1912 Concrete-Filled Pipe Columns

These provisions are largely the same as those of Section 1912 of the 1999 edition of the BOCA/NBC. The BOCA Commentary[49] is used as a source for much of this discussion.

1912.1 General. According to Reference 57, steel pipe should be manufactured to the requirements of ASTM A 501[45] for hot-formed welded and seamless carbon steel of round, square, or rectangular shape for general structural purposes. The steel must have a minimum tensile strength of 58,000 psi and minimum yield strength of 36,000 psi.

Cold-formed welded and seamless carbon steel of square or rectangular shape conforming to the Grade B requirements of ASTM A 500,[44] having a minimum tensile strength of 58,000 psi and a minimum yield strength of 36,000 psi, may also be used for concrete-filled pipe columns.

Design. The load-carrying capacity of concrete-filled pipe columns may be computed in accordance with approved rules such as those in Section 10.13 of ACI 318, or may be determined by load tests. **1912.2**

Connections. Conditions for making structural connections to concrete-filled pipe columns are prescribed. If connections require welding to the steel shell, it must be done before the core is filled with concrete, unless it is possible to demonstrate that the concrete will not be damaged from the heat of welding. **1912.3**

Reinforcement. Reinforcement must comply with the requirements of this section as well as with the applicable provisions of ACI 318. A minimum clearance of 1 inch between any such reinforcement and the outer shell must be provided. This is to permit the concrete to flow around the reinforcement and bond to it, which is necessary for composite action. **1912.4**

Fire-resistance-rating protection. Concrete-filled pipe columns are required to be fire-resistance rated in accordance with Table 601. Irrespective of whether the protective cover is of concrete, concrete with a metal encasement, or a metal cover that encases other fire-insulating materials, it must not be considered to contribute to the load-carrying capacity of the column. **1912.5**

This section is revised to clarify that concrete-filled pipe columns in a Type V construction can be 3 inches in diameter provided the building does not exceed three stories above grade plane. Further, the height limitation of 40 feet is clarified to be the "building height," which is defined as "the vertical distance from grade plane to the average height of the highest roof surface."

Approvals. Concrete-filled pipe columns, including their connection details and splices, which are shop-fabricated as pre-engineered items, are subject to certain inspection and approval requirements spelled out in this section. **1912.6**

KEY POINTS

- Chapter 19 contains requirements for the design and construction of concrete buildings and other concrete structures regulated by the IBC.
- The 2012 IBC references ACI 318-11, *Building Code Requirements for Structural Concrete*, and in most cases, refers to specific sections in the ACI Standard for detailed design and construction requirements.
- In addition to the general requirements for construction documents given in Section 107, Section 1901.3 contains specific requirements for construction documents pertaining to concrete structures.
- Durability requirements for concrete structures are covered in Section 1904 and ACI 318 Chapter 4 based on exposure categories F (freezing and thawing), S (sulfate), P (requiring low permeability), C (corrosion protection for reinforcement), and various classes within each category.
- An exception for Group R occupancies appurtenances thereto less than four stories above grade plane permits normal-weight aggregate concrete to comply with the requirements of Table 1904.2 and Figure 1904.2 in lieu of the durability requirements of ACI 318.

- Modifications (amendments) to the provisions in ACI 318-08 are found in Section 1905.
  - Section 1905.1.8 contains important provisions for structural plain concrete foundation walls, isolated footings, and footings supporting light-frame walls.
  - Section 1905.1.9 contains exceptions from requirements of Appendix D for (1) anchors designed to resist wall out-of-plane forces with design strengths equal to or greater than the force determined in accordance with ASCE 7 Equation 12.11-1 or 12.14-10, and (2) anchor bolts attaching wood sill plates or cold-formed steel tracks of bearing or nonbearing walls of light-frame wood structures to foundations or foundation stem walls, provided several conditions are met.
- Provisions for anchorage to concrete by allowable stress design and strength design are contained in Sections 1908 and 1909.
  - Anchorage to concrete by the allowable stress design procedure is not permitted for anchors installed in hardened concrete (post-installed anchors) and anchors resisting earthquake load effects.
  - Anchors installed in hardened concrete (post-installed anchors) and anchors resisting earthquake load effects must be designed by the strength design method and applicable provisions of Appendix D.
- Requirements for concrete elements not specifically covered by ACI 318, such as slabs supported directly on grade, shotcrete, reinforced gypsum concrete, and concrete-filled pipe columns, are provided in Sections 1907, 1910, 1911, and 1912, respectively.

REFERENCES

1. *Uniform Building Code*, International Conference of Building Officials, Whittier, CA, 1997, copyright held by International Code Council.
2. *The BOCA National Building Code*, Building Officials and Code Administrators International, Country Club Hills, IL, 1993, 1996, 1999, copyright held by International Code Council.
3. *Standard Building Code*, Southern Building Code Congress International, Birmingham, AL, 1994, 1997, 1999, copyright held by International Code Council.
4. ACI Committee 318, *Building Code Requirements for Structural Concrete (ACI318–95) and Commentary (ACI 318R–95)*, American Concrete Institute, Farmington Hills, MI, 1995, 1999, 2002, 2005, 2008, 2011.
5. *Recommended Practice for Glass Fiber Reinforced Concrete Panels*, PCI MNL 128, Precast and Prestressed Concrete Institute, Chicago, IL, 2001.
6. *Specification for Portland Cement*, ASTM C 150, ASTM International, West Conshohocken, PA, 2007.
7. *Specification for Blended Hydraulic Cements*, ASTM C 595, ASTM International, West Conshohocken, PA, 2007.
8. *Specification for Expansive Hydraulic Cement*, ASTM C 845, ASTM International, West Conshohocken, PA, 2004.
9. *Standard Performance Specification for Hydraulic Cement*, ASTM C 1157, ASTM International, West Conshohocken, PA, 2008.
10. *Standard Specification for Coal Fly Ash and Raw or Calcined Natural Pozzolan for Use in Concrete*, ASTM C 618, ASTM International, West Conshohocken, PA, 2008.

11. *Standard Specification for Slag Cement for Use in Concrete and Mortars*, ASTM C 989, ASTM International, West Conshohocken, PA, 2009.

12. *Standard Specification for Silica Fume Used in Cementitious Mixtures*, ASTM C 1240, ASTM International, West Conshohocken, PA, 2005.

13. *Specification for Concrete Aggregates*, ASTM C 33, ASTM International, West Conshohocken, PA, 2003.

14. *Specification for Lightweight Aggregates for Structural Concrete*, ASTM C 330, ASTM International, West Conshohocken, PA, 2005.

15. *Standard Specification for Mixing Water Used in the Production of Hydraulic Cement Concrete*, ASTM C 1602, ASTM International, West Conshohocken, PA, 2006.

16. *Design and Control of Concrete Mixtures*, 14th Edition, Portland Cement Association, Skokie, IL, 2002 (revision 2008).

17. *Standard Test Method for Compressive Strength of Hydraulic Cement Mortars (Using 2-in. [or 50-mm] Cube Specimens)*, ASTM C 109, ASTM International, West Conshohocken, PA, 2008.

18. *Standard Test Method for Time of Setting of Hydraulic Cement by Vicat Needle*, ASTM C 191, ASTM International, West Conshohocken, PA, 2008.

19. *Specification for Deformed and Plain Billet-Steel Bars for Concrete Reinforcement*, ASTM A 615, ASTM International, West Conshohocken, PA, 2004.

20. *Specification for Low-Alloy Steel Deformed and Plain Bar for Concrete Reinforcement*, ASTM A 706, ASTM International, West Conshohocken, PA, 2005.

21. *Standard Specification for Deformed and Plain Stainless-Steel Bars for Concrete Reinforcement*, ASTM A 955, ASTM International, West Conshohocken, PA, 2007.

22. *Standard Specification for Rail-Steel and Axle-Steel Deformed Bars for Concrete Reinforcement*, ASTM A 996, ASTM International, West Conshohocken, PA, 2006.

23. *Standard Specification for Deformed and Plain, Low-Carbon, Chromium, Steel Bars for Concrete Reinforcement*, ASTM A 1035, ASTM International, West Conshohocken, PA, 2007.

24. *Standard Specification for Fabricated Deformed Steel Bar Mats for Concrete Reinforcement*, ASTM A 184, ASTM International, West Conshohocken, PA, 2006.

25. *Standard Specification for Steel Wire, Deformed, for Concrete Reinforcement*, ASTM A 496, ASTM International, West Conshohocken, PA, 2007.

26. *Standard Specification for Steel Welded Wire Reinforcement, Plain, for Concrete*, ASTM A 185, ASTM International, West Conshohocken, PA, 2007.

27. *Standard Specification for Steel Welded Wire Reinforcement, Deformed, for Concrete*, ASTM A 497, ASTM International, West Conshohocken, PA, 2007.

28. *Standard Specification for Zinc-Coated (Galvanized) Steel Bars for Concrete Reinforcement*, ASTM A 767, ASTM International, West Conshohocken, PA, 2005.

29. *Standard Specification for Epoxy-Coated Steel Reinforcing Bars*, ASTM A 775, ASTM International, West Conshohocken, PA, 2007.

30. *Standard Specification for Epoxy-Coated Prefabricated Steel Reinforcing Bars*, ASTM A 934, ASTM International, West Conshohocken, PA, 2007.

31. *Standard Specification for Epoxy-Coated Steel Wire and Welded Wire Reinforcement*, ASTM A 884, ASTM International, West Conshohocken, PA, 2006.

32. *Standard Specification for Deformed and Plain Stainless Steel Wire and Welded Wire for Concrete Reinforcement*, ASTM A 1022, ASTM International, West Conshohocken, PA, 2007.

33. *Structural Welding Code—Reinforcing Steel*, ANSI/AWS D1.4–05, American Welding Society, Miami, FL, 2005.

34. *Standard Specification for Steel Wire, Plain, for Concrete Reinforcement*, ASTM A 82, ASTM International, West Conshohocken, PA, 2007.

35. *Standard Specification for Uncoated Stress-Relieved Steel Wire for Prestressed Concrete*, ASTM A 421, ASTM International, West Conshohocken, PA, 2005.
36. *Specification for Steel Strand, Uncoated Seven-Wire for Prestressed Concrete*, ASTM A 416, ASTM International, West Conshohocken, PA, 2006.
37. *Specification for Uncoated High-Strength Steel Bar for Prestressing Concrete*, ASTM A 722, ASTM International, West Conshohocken, PA, 2007.
38. *Specification for Carbon Structural Steel*, ASTM A 36, ASTM International, West Conshohocken, PA, 2008.
39. *Standard Specification for High-Strength Low-Alloy Structural Steel*, ASTM A 242, ASTM International, West Conshohocken, PA, 2004.
40. *Specification for High-Strength Low-Alloy Columbium-Vanadium Structural Steel*, ASTM A 572, ASTM International, West Conshohocken, PA, 2007.
41. *Specification for High-Strength Low-Alloy Structural Steel with 50 ksi (345 MPa) Minimum Yield Point to 4 Inches (100 mm) Thick*, ASTM A 588, American Society for Testing and Materials, West Conshohocken, PA, 2005.
42. *Standard Specification for Structural Shapes*, ASTM A 992, ASTM International, West Conshohocken, PA, 2006.
43. *Standard Specification for Pipe, Steel, Black and Hot-Dipped, Zinc-Coated, Welded and Seamless*, ASTM A 53, Grade B, ASTM International, West Conshohocken, PA, 2006.
44. *Standard Specification for Cold-Formed Welded and Seamless Carbon Steel Structural Tubing in Rounds and Shapes*, ASTM A 500, ASTM International, West Conshohocken, PA, 2007.
45. *Standard Specification for Hot-Formed Welded and Seamless Carbon Steel Structural Tubing*, ASTM A 501, ASTM International, West Conshohocken, PA, 2007.
46. *International Residential Code*, International Code Council, Washington, DC, 2011.
47. *Minimum Design Loads for Buildings and Other Structures*, ASCE 7, American Society of Civil Engineers, New York, 2005, 2010.
48. *NEHRP (National Earthquake Hazards Reduction Program) Recommended Provisions for the Development of Seismic Regulations for New Buildings*, Building Seismic Safety Council, Washington, DC, 2003, 2009.
49. Building Officials and Code Administrators International, *The BOCA National Building Code/1999—Commentary*. Country Club Hills, IL, 1999.
50. Southern Building Code Congress International, *An Illustrated Commentary to the 1999 Edition of the Standard Building Code*, Birmingham, AL, 1999.
51. ACI Committee 349, "Code Requirements for Nuclear Safety Related Concrete Structures," ACI 349–97, *ACI Manual of Concrete Practice*, 1999 (Appendix B—Steel Embedments, pp. 349-76–349-82).
52. *PCI Design Handbook*, 5th Edition, Precast/Prestressed Concrete Institute, Chicago, IL, 1999.
53. ACI Committee 506, *Specifications for Shotcrete*, ACI 506.2-95, American Concrete Institute, Detroit, MI, 1995.
54. *Test Method for Obtaining and Testing Drilled Cores and Sawed Beams of Concrete*, ASTM C 42, ASTM International, West Conshohocken, PA, 2004.
55. *Specification for Gypsum Concrete*, ASTM C 317, ASTM International, West Conshohocken, PA, 2005.
56. *Specification for Installation of Cast-in-Place Reinforced Gypsum Concrete*, ASTM C 956, ASTM International, West Conshohocken, PA, 2004.
57. Muguruma, H., and Watanabe, F. (1990), "Ductility Improvement of High-Strength Concrete Columns with Lateral Confinement," *Proceedings, Second International Symposium on High-Strength Concrete*, SP-121, American Concrete Institute, Detroit, MI, 47-60.

CHAPTER 20

ALUMINUM

Introduction
Section 2002 Materials
Key Points

20 Aluminum

Introduction

2001.1 **Scope.** This chapter covers the requirements for the quality, design, fabrication, and erection of aluminum structures.

Section 2002 Materials

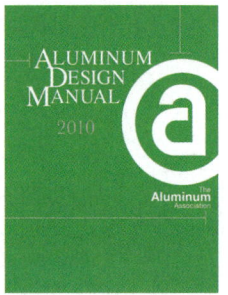

The IBC references two aluminum industry design standards: *Aluminum Design Manual: Part 1—A Specification for Aluminum Structures (ADM1—2010)* and *Aluminum Sheet Metal Work in Building Construction, Fourth Edition (ASM 35—00)*. As with other structural materials (concrete, masonry, steel, and wood), the *International Building Code®* (IBC®) references industry standards rather than transcribing the provisions directly into the code. The advantage to this approach is that the referenced standards are readily updated by revising the year in Chapter 35, Referenced Standards, thereby keeping the code current with the latest available industry standards.

2002.1 **General.** The 2012 IBC references the 2010 edition of the Aluminum Design Manual, Part 1 of which is the Specification for Aluminum Structures, the first unified allowable strength design, and load and resistance factor design aluminum standard. It has design provisions for aluminum structural members and connections and provides minimum strengths for wrought, cast, and welded aluminum products.

The manual includes a commentary that discusses the provisions in the Specification along with useful references. The included design guide addresses structural design issues not included in the Specification, including diaphragms, adhesive bonded joints, aluminum composite material, extrusion design, corrosion prevention, fire protection, sustainability, and design references for aluminum structural components. Material properties include alloy and temper designation systems for wrought and cast aluminum alloys; comparative characteristics of wrought alloys; foreign alloy designations correlated with U.S. alloy designations; and typical mechanical and physical properties, including thermal expansion, electrical conductivity, and density (all in U.S. and SI units). Dimensions and section properties for aluminum channels, I-beams, angles, tees, zees, square and rectangular tube, round tube, pipe, and roofing and siding, as well as sheet metal and wire gauges are provided. Design aids provide buckling constants, allowable stress tables for various alloys, allowable load tables for channels and I-beams, in bending, tread plate, roofing and siding; fastener strengths, minimum bend radii for aluminum sheet and plate, wire, and rod, design stresses for groove and fillet welds, and beam formulas. Illustrative design examples are provided that include structural design calculation examples based on the ADM1 Specification.

The 2010 Specification includes new or revised provisions addressing safety and resistance factors, design for stability and combined stresses, adding 6005A-T61 and 6082-T6, a glossary, shear yield strengths, shear strength of tubes, screw pull-over, screw slot pull-out strength, serviceability, evaluating existing structures, axial compressive strength of complex cross sections, fatigue strength of light pole bases, members subject to torsion, local buckling strength of welded elements, design for fire conditions, and design of braces.

The Aluminum Design Manual is available in printed book form, as a CD-ROM, or as a downloadable PDF file from the Aluminum Association at www.aluminum.org/.

The nominal loads used to design aluminum structures are as required by Chapter 16, which references ASCE 7.

KEY POINTS

- Chapter 20 contains requirements for the design and construction of aluminum structures regulated by the IBC.
- The 2012 IBC references ADM1—2010 *Aluminum Design Manual: Part 1—A Specification for Aluminum Structures* and ASM 35—00 *Aluminum Sheet Metal Work in Building Construction (Fourth Edition)*.
- Design loads and load combinations are to be determined in accordance with Chapter 16 or ASCE 7 as required.

CHAPTER 21

MASONRY

Introduction
Section 2101 General
Section 2102 Definitions and Notations
Section 2103 Masonry Construction Materials
Section 2104 Construction
Section 2105 Quality Assurance
Section 2106 Seismic Design
Section 2107 Allowable Stress Design
Section 2108 Strength Design of Masonry
Section 2109 Empirical Design of Masonry
Section 2110 Glass Unit Masonry
Section 2111 Masonry Fireplaces
Section 2112 Masonry Heaters
Section 2113 Masonry Chimneys
Key Points
References
Bibliography

21 Masonry

Introduction

The consensus standards referenced in the *International Building Code®* (IBC®) for masonry design and construction are produced by the Masonry Joint Standards Committee (MSJC), which is a joint committee of The Masonry Society (TMS), American Concrete Institute (ACI), and the American Society of Civil Engineers (ASCE). These jointly published documents, formally known as TMS 402/ACI 530/ASCE 5, *Building Code Requirements for Masonry Structures*,[1] and TMS 602/ACI 530.1/ASCE 6, *Specification for Masonry Structures*[2] are often referred to as the MSJC Code and the MSJC Specification, respectively. The terms "MSJC Code" and "MSJC Specification" are used in the discussion that follows.

The fact that the IBC now adopts standards by reference means that many of the design provisions, test procedures, and material standards for masonry design and construction are not included in the body of the code but are incorporated by reference. The national consensus standard for masonry referenced in the 2012 IBC is the 2011 edition of the MSJC Code and Specification, TMS 402/ACI 530/ASCE 5, *Building Code Requirements for Masonry Structures*, and TMS 602/ACI 530.1/ASCE 6, *Specification for Masonry Structures*. The seismic provisions in Section 1.8 of the MSJC Code also reference standards, most notably the 2010 edition of ASCE 7. The remaining referenced standards are ASTM standards for the various masonry materials and test methods. See Section 1.4 for standards referenced in the 2011 MSJC Code and Specification.

The discussion in this chapter primarily focuses on the masonry provisions within the IBC itself because the MSJC Code and Specification includes an extensive commentary. If the reader is interested in a historical perspective, extensive information on the origin of the masonry provisions in the 2000 IBC can be found in Table 2101-1 of the *2000 IBC Handbook—Structural Provisions*,[3] which shows the source of the various masonry provisions in the 2000 IBC. For more extensive background on specific provisions, the reader is referred to the MSJC commentary.

During the code development process that led to the 2012 IBC, the trend of adopting standards by reference rather than transcribe standards into the code continued. Accordingly, Chapter 21 decreased from 43 pages in the 2000 IBC to 16 pages in the 2012 IBC. This reduction in the number of pages is due primarily to the increasing reliance on referencing the provisions in the MSJC Code, rather than maintaining the provisions within the code itself.

With the publication of the 2008 edition of the MSJC Code and Specification, TMS became the lead sponsoring organization of the standard, which is charged with reviewing and maintaining the provisions. As such, the title was changed in 2008 from ACI 530/ASCE 5/TMS 402 to TMS 402/ACI 530/ASCE 5, and from ACI 530.1/ASCE 6/TMS 602 to TMS 602/ACI 530.1/ASCE 6.

The 2011 edition of the standard, which is referenced in the 2012 IBC, is divided into four parts:

- Building Code Requirements for Masonry Structures (TMS 402-11/ACI 530-11/ASCE 5-11)
- Specification for Masonry Structures (TMS 602-11/ACI 530.1-11/ASCE 6-11)
- Companion Commentaries to both Code and Specification

As previously noted, the standard contains a complete commentary on the provisions within the standard and specification; thus, the reader is referred to that commentary for a more detailed discussion of the specific provisions contained therein.

In general, the majority of the code changes to Chapter 21 of the IBC were primarily done to make the masonry provisions in the code consistent with the standard. Where duplication existed, provisions in the code were deleted. In some cases, Chapter 21 contains some modifications that amend the provisions in the referenced standard. These modifications are found in the following sections:

Section 2107 Allowable stress design

- Section 2107.2.1 Lap splices
- Section 2107.3 Splices of reinforcement
- Section 2107.4 Maximum bar size

Section 2108 Strength design of masonry

- Section 2108.2 Development length
- Section 2108.3 Splices

Section 2101 *General*

The masonry provisions govern the materials, design, construction, and quality for masonry designed in accordance with allowable stress design, strength design, or empirical design, and for glass masonry and masonry fireplaces and chimneys. Masonry veneer is covered in Chapter 14 of the IBC and Chapter 6 of the MSJC Code. This section contains the basic road map for the user. Section 2101.2 directs the user to specific sections depending on the specific design method or material being used. The requirements for construction documents in Section 2101.3 provide the minimum masonry-related information to be included in the design drawings and specifications. Note that IBC Sections 107 and 1603 also contain specific requirements for construction documents. Table 2101-1 shows the basic layout of Chapter 21 and how the sections relate to the MSJC Code.

Table 2101-1. IBC Chapter 21–MSJC Cross-Reference

Subject	IBC Section(s)	TMS 402/ACI 530/ASCE 5 Section(s)
General design requirements	2101	Chapter 1
Construction documents	2101.3	Section 1.2
Definitions and notation	2102	Sections 1.5, 1.6
Masonry construction materials	2103	Chapter 1
Construction requirements	2104	Chapter 1, Section 1.20
Quality assurance	2105	Section 1.19
Seismic design	2106	Section 1.18
Allowable stress design	2101.2.1, 2107	Chapter 2
Strength design	2101.2.2, 2108	Chapter 3
AAC masonry design	2101.2.2	Chapter 8
Prestressed masonry	2101.2.3	Chapter 4
Empirical design	2102.4, 2109	Chapter 5
Masonry veneer	2101.2.6, 1405	Chapter 6
Glass unit masonry	2101.2.5, 2110	Chapter 7
Masonry fireplaces	2111	—
Masonry heaters	2112	—
Masonry chimneys	2113	—

21 Masonry

2101.2 Design methods. This section basically references the various sections within the code and in the standard depending on the specific design method and material used. Regardless of the design method used, all masonry construction must comply with the general requirements of Sections 2101 through 2104, as well as the applicable seismic design requirements in Section 2106 without amendment. Section 2106 merely references Section 1.18 of the MSJC Code for seismic requirements. Section 2101.2 is broken down into seven categories: allowable stress design, strength design, prestressed masonry, empirical design, glass unit masonry, masonry veneer, and the direct design method. The reference to the new standard, *Direct Design Handbook for Masonry Structures* (TMS 403),[4] was added to the 2012 IBC. In most cases, the code references as well as modifies the provisions within the standard.

2101.2.1 Allowable stress design. Allowable stress design must comply with the seismic requirements in Section 2106 and the general design requirements in Section 2107. The term *working stress design* was changed to *allowable stress design* in the 2006 IBC to be consistent with the new terminology used in the referenced standard and the load combinations in Section 1605.3. Section 2107 essentially references Chapters 1 and 2 of TMS 402/ACI 530/ASCE 5, which covers the general design requirements for masonry, and also modifies portions of Chapter 2 of the standard.

2101.2.2 Strength design. Strength design of masonry must comply with the seismic requirements in Section 2106 and the general strength design requirements in Section 2108. Section 2108 references Chapters 1 and 3 of TMS 402/ACI 530/ASCE 5, which covers the general design requirements for masonry and modifies specific portions of Chapter 3 of the standard.

Autoclaved aerated concrete (AAC) masonry is a relatively new building material in the United States, although its use is continually increasing. AAC masonry is designed by the strength design procedure and must comply with seismic requirements in Section 2106 and the general design requirements in Chapters 1 and 8 of TMS 402/ACI 530/ASCE 5. Prior to its inclusion in the 2006 IBC, AAC masonry was approved under the alternative materials, design, and methods of construction provisions in Section 104.11. Since its inclusion in the building code and the MSJC Code, it has become a more popular masonry construction method. Having been used in Europe for many years, it is considered an effective and efficient building material. In approving the use of AAC in the 2006 IBC, a modification by the committee prohibited the use of AAC in the seismic-force-resisting system of structures classified as Seismic Design Category B, C, D, E, or F. At that time, the restriction on the use of AAC masonry in seismic-force-resisting systems to Seismic Design Category A structures was considered prudent until it could be cyclically tested and its seismic response characteristics evaluated by the Building Seismic Safety Council (BSSC). A code change to the 2009 IBC relaxed the restriction to Seismic Design Category A and added shear wall design coefficients for the seismic-force-resisting system in AAC masonry structures assigned to Seismic Design Categories B and C. For a summary of the seismic design coefficients and limitations found in Section 1613.6.4 of the 2009 IBC, refer to the *2009 IBC Handbook—Structural Provisions*.[5] The 2010 edition of ASCE 7, which is referenced in the 2012 IBC, now includes two types of AAC masonry shear walls. See Table 12.2.1 of ASCE 7.

2101.2.3 Prestressed masonry. Prestressed masonry must be designed in accordance with Chapters 1 and 4 of TMS 402/ACI 530/ASCE 5 and the seismic requirements in Section 2106. Special inspection is required during construction of prestressed masonry as described in Section 1704.5. As noted in the discussion of Chapter 17, the special inspection requirements were deleted in the 2012 IBC and replaced with a reference to the quality-assurance provisions in Section 1.19 of the MSJC Code.

2101.2.4 Empirical design. Empirically designed masonry must comply with the seismic requirements in Sections 2106 and the general design requirements in Section 2109 and Chapter 5 of TMS 402/ACI 530/ASCE 5. Section 2109 was deleted entirely in the 2009 IBC and replaced with provisions that are not covered in the MSJC code. Most notably is

Section 2109.3, which contains extensive provisions for adobe construction. Adobe construction is defined in Section 202 and designed and constructed according to the empirical design method in Chapter 5 of the MSJC Code and the specific requirements in Section 2109.3.

2101.2.5 Glass unit masonry. Glass unit masonry must comply with the provisions of Section 2110 or Chapter 7 of TMS 402/ACI 530/ASCE 5. Section 2110 was deleted entirely in the 2009 IBC and replaced with provisions that are not covered in the MSJC code. These are the fire-resistive and structural limitations in Section 2110.1.1. Glass unit masonry is not permitted to be used in fire-resistive walls or load-bearing walls. Glass unit masonry must be designed and constructed according to Chapter 7 of the standard and the specific requirements in Section 2110.

2101.2.6 Masonry veneer. Masonry veneer must comply with the provisions of IBC Chapter 14 or Chapter 6 of TMS 402/ACI 530/ASCE 5. Because veneer is a wall-covering material, the IBC also contains specific requirements for veneer in Chapter 14. In some cases, the masonry veneer provisions in Chapter 14 reference requirements in the MSJC code.

2101.2.7 Direct design. The 2012 IBC now includes a simplified design method for single-story, concrete masonry buildings based on the new referenced standard TMS 403, *Direct Design Handbook for Masonry Structures*. The methodology is based on the strength design provisions and the factored load combinations for dead, roof live, wind, seismic, snow, and rain loads in accordance with ASCE 7. The new design standard was developed by the masonry industry in response to concerns from the design community that structural loads and design requirements have become too complicated, particularly for relatively small, simple structures. The direct design procedure is a table-based structural design method that permits the user to follow a series of steps to design and specify relatively simple, single-story, concrete masonry bearing-wall structures. The method is simple to implement compared to conventional design approaches, but it limits the design to only those configurations addressed by the standard. It introduces slightly more conservatism compared to conventional design procedures as a result of the conditions and assumptions inherent to the design method. There are limitations based on ground snow load, wind speed, mapped seismic spectral acceleration, the type of masonry walls and roof systems used, and minimum reinforcing requirements. The *Direct Design Handbook for Masonry Structures* is intended to capture many of the simple load-bearing masonry structures commonly designed today.

It should be emphasized that irrespective of the design method used, whether allowable stress design, strength design, empirical design, or prestressed masonry design, masonry structures must also comply with the applicable seismic design requirements prescribed in Section 2106 based on the seismic design category of the building or structure. The seismic design category of a structure is determined in accordance with Section 1613.3.5 based on the potential seismic ground motion at the site (S_{DS} and S_{D1}), soil type (Site Class), and use of the structure. Section 2106 references Section 1.18 of the MSJC code without amendments for seismic design of masonry structures.

2101.3 Construction documents. Construction documents are a part of the submittal documents required by Section 107. The general requirements for construction documents are given in Sections 107.2 and 1603. This section contains specific requirements for construction documents related to masonry structures. Items 5 through 9 were added to the 2009 IBC by the Masonry Alliance for Codes and Standards to coordinate the construction document requirements with Chapter 16 and the other structural materials.

2101.3.1 Fireplace drawings. Fireplace drawings are required to be submitted so that compliance with the requirements for masonry fireplaces (Section 2111) and masonry chimneys (Section 2113) can be resolved during the plan review process and prior to construction.

21 Masonry

Section 2102 *Definitions and Notations*

The terms related to masonry construction are listed in Section 2102 with the complete definitions given in Section 202. As the IBC evolved since the 2000 edition, several new definitions were added during the ICC code development process. Most of the new definitions related to new materials and construction methods or were clarifications of masonry terminology. Most notable were the new definitions for prestressed masonry and the associated types of prestressed masonry shear wall definitions that were introduced into the 2003 IBC. The definition of Autoclaved Aerated Concrete (AAC) masonry was introduced into the 2006 IBC as well.

Some additional definitions were added to be more consistent with definitions in the MSJC Code. For example, the terms *foundation pier*, *glass unit masonry*, and *unreinforced (plain) masonry* were added to the code. Some minor clarifications were also made to the definitions. For example, the term *required strength* was relocated under the general term *strength* so that the three strength design terms *design strength*, *nominal strength*, and *required strength* appear in one place.

In the 2009 IBC, several definitions were deleted because they had become obsolete or they were included in the 2008 edition of the TMS 402/ACI 530/ASCE 5 standard. Several new definitions were added to the 2008 MSJC Code for new materials or to coordinate terminology with the IBC. For example, new definitions for *self-consolidating grout*, *slump flow*, and *visual stability index* were added. The term *architect/engineer* was replaced with *licensed design professional*, which is essentially the same as the term *registered design professional* in the IBC. The terms *periodic inspection* and *continuous inspection* were added to be consistent with special inspection terminology used in Chapter 17.

Section 2103 *Masonry Construction Materials*

This section contains minimum requirements for various masonry construction materials. The 2012 IBC references about 240 ASTM Standards and Specifications, many of which apply to masonry construction materials.

2103.1 Concrete masonry units. Table 2103-1 gives the specifications for concrete masonry units.

Table 2103-1. **Concrete Masonry Unit Specifications**

Type of Masonry Unit	ASTM Standard
Concrete brick	ASTM C 55
Calcium silicate face brick	ASTM C 73
Load-bearing concrete masonry units	ASTM C 90
Prefaced concrete and calcium silicate masonry units	ASTM C 744

2103.2 Clay or shale masonry units. Table 2103-2 gives the specifications for clay or shale masonry units.

2103.3 AAC masonry. AAC is autoclaved aerated concrete masonry, a new material that was first introduced into the 2006 IBC by reference to the 2005 edition of the ACI 530/ASCE 5/TMS 402 standard. It is defined in Section 202 as a low-density cementitious product of calcium silicate hydrates whose material specifications are defined in ASTM C 1386. See Sections 2102.1 and 2101.2 for discussion of design requirements for AAC masonry structures.

Table 2103-2. **Clay or Shale Masonry Unit Specifications**

Type of Masonry Unit	ASTM Standard
Structural clay load-bearing wall tile	ASTM C 34
Structural clay non-load-bearing wall tile	ASTM C 56
Building brick (solid masonry units made from clay or shale)	ASTM C 62
Solid units of thin veneer brick	ASTM C 1088
Ceramic-glazed structural clay facing tile, facing brick, and solid masonry units	ASTM C 126
Structural clay facing tile	ASTM C 212
Facing brick (solid masonry units made from clay or shale)	ASTM C 216
Hollow brick (hollow masonry units made from clay or shale)	ASTM C 652
Glazed brick (single-fired solid brick units)	ASTM C 1405

Stone masonry units. Stone masonry units must conform to the various applicable ASTM standards for exterior marble, limestone, granite, sandstone, and slate building stone. Table 2103-3 gives the specifications for stone masonry units.

2103.4

Table 2103-3. **Stone Masonry Unit Specifications**

Type of Stone Masonry Unit	ASTM Standard
Marble building stone (exterior)	ASTM C 503
Limestone building stone	ASTM C 568
Granite building stone	ASTM C 615
Sandstone building stone	ASTM C 616
Slate building stone	ASTM C 629

Architectural cast stone. The 2012 IBC references ASTM C 1364 for physical requirements, sampling, testing, and visual inspection of architectural cast stone.

2103.5

Ceramic tile. References for ceramic tile mortar were added based on requirements in the legacy model codes. During the development of the IBC, a frequent point of discussion concerned architectural materials such as ceramic tile and whether they belonged in the structural chapters. For lack of a better place, the ceramic tile provisions were placed in Chapter 21, as in the legacy model codes. The code references the American National Standard Specifications for Ceramic Tile, ANSI A137.1.

2103.6

Glass unit masonry. The requirement for treating surfaces of glass block in contact with mortar originated with the 1997 UBC. Section 2103.9 and the MSJC Code require glass unit masonry to be laid in Type S or N mortar conforming to ASTM C270. Reclaimed glass block units are not permitted to be reused.

2103.7

Second-hand units. Some masonry materials may be reused, provided that they meet all the code requirements for new materials. Note, however, that Section 104.9.1 restricts the use of used materials unless specifically approved by the building official. One should exercise caution in approving reuse of materials. The applicable material or design standards should be consulted to determine if reuse of used materials should be permitted. For example, glass block units cannot be reused, as indicated in Section 2103.6.

2103.8

This section allows the use of salvaged or *used* brick. Used bricks are often salvaged from the demolition of old URM buildings. Masonry units manufactured in the past may not have the same quality as masonry made to meet current standards. Caution should also be practiced when specifying used brick as a structural material. Even though the brick may appear clean, the pores in the bedding faces may be filled with cement paste, lime, and other deleterious microscopic particles that may reduce the absorption properties of the brick, thereby adversely affecting the bond between the mortar and the masonry and reducing the mortar strength. Testing for the absorption rate of used masonry units in accordance with ASTM C 67 will give an indication of the bonding qualities of the used units with mortar. Used masonry units are best used as veneers, where the units are not relied upon for structural strength.

2103.9 **Mortar.** Mortar for use in masonry construction must conform to ASTM C 270 and the applicable sections of the MSJC Specifications except for surface bonding mortar, mortar for ceramic tile, and dry set Portland cement mortar, which are covered in Sections 2103.10, 2103.11, and 2103.12.

Two tables for proportioning mortar and mortar properties that were in the previous editions of the IBC were deleted in the 2009 IBC because they are in the ASTM C270 standard. Mortars proportioned in accordance with Table 1 of ASTM C270 should have a cube strength in excess of that required by ASTM C 270 for the various mortar types. Type M mortar is high-strength mortar having a minimum average 28-day compressive strength of 2,500 psi. Type M mortar is suitable for general use, and is recommended where maximum compressive strength is required such as in unreinforced masonry below grade. Type S mortar is recommended where a high lateral strength is required and is specifically recommended for reinforced masonry. The minimum average compressive strength for Type S mortar is 1,800 psi. Type N mortar is a medium-strength mortar having a minimum average compressive strength of 750 psi. Type N mortar may be used where high compressive or lateral strength is not required. Type N is generally used in exposed masonry above grade where exposed to weather. Type O mortar is a low-strength mortar having a minimum average compressive strength of 350 psi. Type O mortar should only be used for nonbearing walls not exposed to severe weathering.

Table 2 in ASTM C270 allows for the proportioning of mortar to meet the minimum requirements for 28-day compressive strength, minimum water retention, and maximum air content.

2103.10 **Surface-bonding mortar.** The specifications for premixed mortar, masonry units, and other materials used in constructing dry-stack, surface-bonded masonry, as well as the construction requirements, are referenced in this section.

2103.11 **Mortars for ceramic wall and floor tile.** The code references ANSI A108.1 for Portland cement mortars used to lay ceramic tile. Table 2103.11 gives compositions of hydrated lime, cement, and sand for different applications.

2103.12 **Mortar for AAC masonry.** Mortar requirements for AAC masonry are covered in the MSJC Specification. The requirements for thin-bed mortar and mortar for leveling courses were deleted in the 2009 IBC because these requirements are in Articles 2.1 C.1 and 2.1 C.2 of the MSJC Specification. Thin-bed mortar is used in AAC masonry construction with joints 0.06 inch or less. Mortar used for the leveling courses of AAC masonry must be Type M or S.

2103.13 **Grout.** Grout proportion requirements are given in Article 2.2 of the MSJC Specification, which references ASTM C 476. The grout proportioning by volume in Table 2103.12 was deleted in the 2009 IBC because it is in Table 1 of ASTM C 476. The minimum grout strength requirement of 2,000 psi set forth in the model codes has been removed from the code provisions because ASTM C 476 permits grout to comply with the minimum strength requirement of 2,000 psi or to comply with the proportions given in Table 1. Grout batched in accordance with the proportions in the table will have strength in excess of 2,000 psi.

2103.14 **Metal reinforcement and accessories.** Requirements for metal reinforcement and accessories used in masonry construction reflect updates made to the national consensus standards. Several subsections were deleted in the 2009 IBC because these provisions are included in Article 2.4 of the MSJC Specification, which references various ASTM standards. The various ASTM standards are listed in Table 2103-4. For a complete list of specifications for reinforcement, prestressing tendons, and metal accessories, see Article 2.4 of the MSJC Specification.

Table 2103-4. **ASTM Standards for Metal Reinforcement and Accessories**

Reinforcement Type	ASTM Standard
Deformed reinforcing bars	A 615 (Billet steel) A 706 (Weldable) A 767 (Zinc coated) A 775 (Epoxy coated) A 996 (Rail and axle steel)
Masonry joint reinforcement	A 951
Deformed reinforcing wire	A 496
Wire fabric	A 185 (Plain steel welded wire fabric) A 497 (Welded deformed steel wire fabric)
Anchors, ties, and accessories	A 36 (Plate and bent-bar anchors) A 1008 (Sheet metal anchors and ties) A 185 (Wire mesh tie) A 82 (Plain steel wire ties and anchors) A 307 Grade A (Header anchor bolts)
Prestressing tendons	A 421 (Wire and low-relaxation wire) A 421 (Low-relaxation wire) A 416 (Strand and low-relaxation strand) A 722 (Bar)

Coatings and corrosion protection requirements for mill galvanized, hot-dipped galvanized, and epoxy coating methods are provided in Section 2.4E of the MSJC Specification. Corrosion protection requirements for tendons are provided in Section 2.4G of the MSJC Specification.

Section 2104 *Construction*

The provisions in IBC Section 2104, together with the requirements in the MSJC Specification, provide minimum requirements to ensure the masonry is properly constructed, consistent with the design methods used. More stringent construction requirements and referenced specifications than those contained in the code may be needed to satisfy aesthetic and architectural design criteria. Such requirements should be included in the construction documents.

Many of the construction requirements contained in IBC Section 2104 of previous editions of the IBC were also contained in the MSJC Code. During development of the 2006 IBC, the structural committee decided to not duplicate many of these requirements in the IBC insofar as they are already contained in the standard. In the 2009 IBC, nearly all the construction provisions were deleted and replaced with references to the MSJC Code and Specification so that duplication and conflicts between the provisions in the IBC and the standard are minimized. In some cases, provisions are in the IBC but not in the standard. For example, the requirements in Section 2404.1.4 for chases and recesses and Section 2104.1.6 (which references Section 2304.12) for masonry supported by wood members.

Masonry

2104.1 Masonry construction. This section cites the applicable code sections and references the applicable construction provisions of TMS 602/ACI 530.1/ASCE 6.

Section 2105 *Quality Assurance*

During the development of the IBC, it was noted that masonry damaged in earthquakes and high wind events often showed poor quality and workmanship, which often contributed to the damage. Many of the masonry structures that failed showed evidence of missing reinforcement, missing grout or missing connectors, and various other improper construction techniques. The best way to improve the quality of construction in the field is through a quality-assurance program that ensures proper materials and construction methods are used.

2105.1 General. Section 2105 requires a quality-assurance program to ensure that masonry structures are constructed in accordance with the construction documents. Historically, the quality-assurance program is achieved through the special inspection and testing requirements covered in Chapter 17 of the IBC. However, the specific inspection and verifications for masonry structures were deleted in the 2012 IBC and replaced by a reference to the quality-assurance provisions in the MSJC Code and Specification. Section 1.19 of the MSJC Code has three levels of quality assurance that are based on the risk category (formerly called occupancy category) of the structure. Prior editions of the IBC had two tables for verification and inspection of masonry structures. Level 1 inspection applies to engineered masonry structures in Occupancy Categories I, II, and III and empirically designed masonry structures in Occupancy Category IV. Level 2 inspection applies to engineered masonry structures in Occupancy Category IV. A code change to the 2012 IBC by TMS deleted these special inspection requirements and tables and replaced them with a reference to the quality-assurance and inspection requirements in the 2011 edition of the MSJC Code and MSJC Specification. The level of quality assurance required by Section 1.19 of the 2011 MSJC Code is driven by the type of masonry used and the risk category of the structure. (The term "Occupancy Category" was changed in the 2012 IBC to "Risk Category" to better reflect the meaning.) See discussion of Section 1705.4 in regard to special inspection for masonry structures and Table 1705-1, which reflects the level of quality assurance required based on the type of masonry structure and risk category.

2105.2 Acceptance relative to strength requirements. Compliance with minimum specified compressive strength is determined by the unit strength method covered in Section 2105.2.2.1 or the prism test method covered in Section 2105.2.2.2. Where these methods are not used, masonry prisms saw cut from constructed masonry can be used to establish compliance in accordance with Section 2105.3.

2105.3 Testing prisms from constructed masonry. When approved by the building official, masonry is permitted to be based on tests of prisms saw cut from the masonry construction. A set of three masonry prisms at least 28 days old are saw cut for each 5,000 square feet of wall area and not less than one set of three masonry prisms for a given project. The testing and transportation of the prisms must comply with the requirements of ASTM C 1314.

Section 2106 *Seismic Design*

In early editions of the IBC, Section 2106 contained the minimum design requirements for masonry structures in areas of seismic risk that were based on requirements in the 1997 *Uniform Building Code*, the MSJC Code, and the NEHRP *Recommended Provisions for*

Seismic Regulations for New Buildings and Other Structures.[6] This section consists of only one sentence because the seismic design requirements for masonry structures are no longer in the code itself but are found in Section 1.18 of the MSJC Code.

Seismic design requirements for masonry. The 2012 IBC references Section 1.18 of TMS 402/ACI 530/ASCE 5 for seismic design requirements without any amendments. **2106.1**

As with other structural materials, seismic load effects are determined from the IBC, which in turn references ASCE 7, and seismic resistance is determined from the specific design provisions in the MSJC Code. Seismic design of reinforced masonry structures is based on the inelastic ductility response of the structure, which reduces the seismic design force level. This is reflected by the R-factor for the particular seismic-force-resisting system. Section 1604.10 and ASCE 7 Section 11.1.1 require that structures meet all of the applicable seismic detailing requirements, even though other loads such as wind may be greater than the seismic load effects.

Prior to the 2006 IBC, it was not clear which masonry walls were considered part of the seismic-force-resisting system. The most common understanding included those walls that were specifically designed to resist seismic forces. A change in the IBC clarified which masonry walls are considered part of the seismic-force-resisting system by stating that all masonry walls, unless they are isolated on three edges from the in-plane motion of the basic structural system, must be considered part of the seismic-force-resisting system and designed accordingly. This clarification was intended to address nonload-bearing walls that are connected to the lateral-force-resisting system and as a result could be subject to seismic forces. The new code language in the IBC intended to prevent the incorporation of masonry elements that are incapable of resisting seismic demands that could be imposed on them during an earthquake. Beginning with the 2008 MSJC Code, elements are classified as participating or nonparticipating in the seismic-force-resisting system. Participating elements must be designed and detailed to resist seismic forces such as shear walls, columns, beams, and coupling elements. Nonparticipating elements can be any masonry element but are not required to be designed and detailed to resist seismic forces. Element classifications are described in Section 1.18.3 of the MSJC Code.

Basic seismic-force-resisting systems. The basic seismic-force-resisting systems are based on concepts from the NEHRP *Provisions*. The 2000 IBC listed five distinct types of shear wall systems based on the expected performance of the walls with various construction techniques. The five types of masonry shear wall systems in the 2000 IBC are ordinary plain (unreinforced) masonry, ordinary reinforced masonry, detailed plain (unreinforced) masonry, intermediate reinforced masonry, and special reinforced masonry shear walls. The various shear wall types are assigned different seismic design coefficients found in ASCE 7, such as response modification coefficient R, based on the expected performance and ductility of the particular shear wall system used. Certain shear wall types are required in some seismic regions, and unreinforced shear wall types are not permitted in intermediate and high seismic-risk regions, based on the seismic design category of the building. See Table 12.2-1 of ASCE 7-10 for seismic design coefficients for the various types of masonry shear wall systems.

In the 2003 IBC, the specific requirements for the shear wall types were removed from the code because the MSJC Code included all of the design requirements for these shear wall systems. Subsequently, all of the seismic design requirements were deleted from the 2009 IBC and replaced with a reference to the MSJC Code. In addition to the five types of conventionally reinforced shear wall systems, three new prestressed masonry shear wall systems were added. These are now covered in Section 1.18.3.2 of the MSJC Code: ordinary plain (unreinforced) prestressed masonry (Section 1.18.3.2.10), intermediate reinforced prestressed masonry (Section 1.18.3.2.11), and special reinforced prestressed masonry shear walls (Section 1.18.3.2.12). All of the various types of masonry shear wall systems are defined under the term *shear wall*, in IBC Section 202 and Section 1.6 of the MSJC Code.

In the 2006 IBC, which references Table 12.2-1 of ASCE 7-05, the three distinct types were all combined into one category so that all prestressed masonry shear wall systems have the same seismic design coefficients and height limitations for the same type of building system. ASCE 7-10 contains only one prestressed masonry shear wall system. See item A12 of ASCE 7-10 Table 12.2-1 for bearing-wall buildings with prestressed masonry shear wall systems, and item B21 of ASCE 7-10 Table 12.2-1 for building-frame buildings with prestressed masonry shear wall systems. Note that prestressed masonry shear wall systems are not permitted in Seismic Design Categories C, D, E, and F.

In addition to the eight shear wall systems described above, the MSJC Code has an empirically designed shear wall and three types of AAC masonry shear walls for a total of 12 different types of masonry shear wall systems. The requirements for the different shear walls are given in Section 1.18.3 of the MSJC Code. Table 2106-1 summarizes the requirements for each of the different masonry shear wall systems described above. It should be emphasized, however, that the limitations and restrictions imposed on the different types of masonry shear wall systems are given in Tables 12.2-1 and 12.14-1 of ASCE 7.

Table 2106-1. **Shear Wall Types and Requirements**[a]

Shear Wall System	Design Methods	Reinforcement Requirements	Permitted in Seismic Design Category
Empirical design masonry shear walls	Section 5.3	None	A
Ordinary plain (unreinforced) masonry shear walls	Section 2.2 or 3.2	None	A, B
Detailed plain (unreinforced) masonry shear walls	Section 2.2 or 3.2	Section 1.18.3.2.3.1	A, B
Ordinary reinforced masonry shear walls	Section 2.3 or 3.3	Section 1.18.3.2.3.1	A, B, C
Intermediate reinforced masonry shear walls[b]	Section 2.3 or 3.3	Section 1.18.3.2.5	A, B, C
Special reinforced masonry shear walls	Section 2.3 or 3.3	Section 1.18.3.2.6	A, B, C, D, E, F
Ordinary plain (unreinforced) AAC masonry shear walls	Section 8.2	Section 1.18.3.2.7.1	A, B
Detailed plain (unreinforced) AAC masonry shear walls	Section 8.2	Section 1.18.3.2.8.1	A, B
Ordinary reinforced masonry AAC shear walls	Section 8.3	Section 1.18.3.2.9	A, B, C, D, E, F
Ordinary plain (unreinforced) prestressed masonry shear walls	Chapter 4	None	A, B
Intermediate reinforced prestressed masonry shear walls	Chapter 4	Section 1.18.3.2.11	A, B, C
Special reinforced prestressed masonry shear walls	Chapter 4	Section 1.18.3.2.12	A, B, C, D, E, F

[a]See ASCE 7 Table 12.2-1.
[b]The maximum spacing for vertical reinforcement in an intermediate reinforced masonry shear wall is 4'–0" c.c.

Anchorage of masonry walls. Section 1.18.2.3 of the MSJC Code references the "legally adopted building code" or ASCE 7 for anchorage of masonry walls. Because of the significant out-of-plane seismic forces that develop in concrete and masonry walls during seismic ground motion, the code requires them to be anchored to roofs and floors that provide lateral support for the wall. The anchorage must provide a positive and direct connection capable of resisting the design seismic forces. Although the term *positive and direct* is not explicitly defined in the code, the intent is to transfer the lateral loads as directly as possible and not by indirect circuitous load paths that are less reliable and more prone to failure during an earthquake.

Early editions of the legacy model codes required connections to floors and roofs parallel and perpendicular to wall to transfer design forces but not less than 200 pounds per lineal foot, which was an allowable stress design level force. The strength level force is determined by 1.4 × 200 plf = 280 plf. The anchorage requirement for masonry walls was deleted in the 2009 IBC, and Section 1604.8.2 was modified to apply to anchorage of structural walls in general, not just concrete and masonry walls. Section 1604.8.2 references Section 1.4.5 of ASCE 7 for walls in structures in Seismic Design Category A and Section 12.11 for walls in structures in other seismic design categories. Section 1.4.5 of ASCE 7 requires the anchorage to develop a design strength equal to 0.2 times the weight of the wall tributary to the anchor but not less than 5 psf. Obviously, where the anchors are used in masonry walls constructed of hollow units or cavity wall systems, the code requires them to be embedded in a reinforced grouted structural element within the wall. See the discussion under Section 1604.8.2.

For buildings in Seismic Design Categories B through F, anchorage of concrete and masonry walls must be designed in accordance with Section 12.11 of ASCE 7, or Section 12.14.7.5 for the simplified seismic design procedure. Section 12.11 requires the anchorage to be designed for the greater of several seismic forces, with specified minimum forces. The structural wall anchorage requirements have been revised and updated in ASCE 7-10. For a discussion of these new requirements for anchorage of structural walls, refer to *Structural Loads: 2012 IBC® and ASCE/SEI 7-10* by David Fanella.[7]

Requirements based on Seismic Design Category. The specific requirements based on the seismic design category were deleted in the 2009 IBC and are now found in Section 1.18.4 of the MSJC Code. Seismic design is not optional, but the permissible systems and the level of detailing required depend on the seismic design category of the building. For example, empirical design is not permitted in Seismic Design Category D, E, or F, and empirically designed masonry elements of the seismic-force-resisting system are only permitted in Seismic Design Category A. Note that the seismic design requirements are cumulative, which means the requirements in Seismic Design Category D include the requirements for Seismic Design Category C, and so on.

Seismic Design Category A. The requirements for Seismic Design Category A are given in Section 1.18.4.1. The interaction of structural and nonstructural elements that may affect the seismic response of the structure must be considered. Structural elements must be classified as nonparticipating and participating elements. Nonparticipating elements must be isolated from the seismic-force-resisting system. Participating elements are elements of the seismic-force-resisting system and must be designed and detailed accordingly. Masonry elements in the seismic load path to the foundation must be capable of resisting the design seismic forces. The MSJC Code does not prescribe anchorage and connection forces, because these are design loads. Thus, the MSJC Code references the legally adopted building code or ASCE 7 for determination of anchorage design forces. Although drift limits must be met in Seismic Design Category A, it is permitted to assume that some of the shear wall types are deemed to meet the drift limits required by ASCE 7. See Section 1.18.2.4. Note that all of the seismic load requirements for Seismic Design Category A are entirely self-contained in Section 11.7 of ASCE 7.

Seismic Design Category B. The 2003 IBC imposed some seismic-related requirements on masonry structures beginning with Seismic Design Category B. The code required masonry partition walls, screen walls, and other masonry elements that are not specifically designed to resist external loads other than the loads produced by their own weight to be isolated from the rest of the structure so that the forces from the structure are not imparted to these elements. Any joints or connections between the structure and these isolated elements must be designed to accommodate the design seismic story drift. These provisions are now the nonparticipating element design requirements of MSJC Code Section 1.18.4.1 and apply to buildings in both Seismic Design Categories A and B. All shear wall types are permitted in Seismic Design Category B except empirically designed masonry shear walls.

Seismic Design Category C. Structures in Seismic Design Category C must conform to the requirements for Seismic Design Category B, as well as the additional requirements in Section 1.18.4.3 of the MSJC Code. Horizontal or vertical reinforcing is required for nonparticipating elements described in Section 1.18.4.3.1. The vertical reinforcing requirement for nonparticipating elements is one No. 4 bar spaced not more than 10 feet on center and within 16 inches of the end of masonry walls. Past earthquakes such as the 1971 San Fernando earthquake have demonstrated that connections to columns are particularly vulnerable. Thus, anchor bolts used to connect horizontal members to columns require lateral ties. Anchorages used to transfer seismic forces from roof and floor diaphragms for AAC masonry shear walls must be embedded in grout. Structural clay wall tiles are not permitted in the seismic-force-resisting system. Although masonry buildings may be designed to use shear walls to resist seismic forces, columns may be incorporated to support gravity loads. The MSJC code permits piers and columns to support lateral loads where the R-factor is not greater than 1.5. In this case, the MSJC Code stipulates that the lateral stiffness at each story level and at each line of lateral resistance be not less than 80 percent of the lateral stiffness provided by the shear walls. Columns that resist loads from discontinuous walls, and beams that resist loads from discontinuous walls or frames, require minimum transverse reinforcement with a ratio of 0.0015. These requirements for strength and toughness are intended to prevent local failure or collapse in elements supporting discontinuous portions of the lateral-force-resisting system. Inherent in the code philosophy is the assumption that the inelastic demands on the structure will be reasonably distributed throughout the lateral-force-resisting system. The value of the response modification factor, R, is based on this assumption, as well as the assumption of sufficient ductility and overstrength to meet the maximum anticipated seismic demands. Elements used to redistribute or transfer the effect of overturning forces and shears from stiff discontinuous elements are susceptible to increased localized or concentrated inelastic demands, which violate the above assumption, and may not achieve the required ductility. The requirement for a minimum amount of transverse reinforcement at a maximum spacing increases the maximum usable masonry strain and ductility in the element so that the element will meet the maximum seismic demand and distribute the overturning forces and shears without failure.

Seismic Design Category D. Structures in Seismic Design Category D must conform to the requirements for Seismic Design Categories B and C and the additional requirements of Section 1.18.4.4 of the MSJC Code. The vertical reinforcing requirement for nonparticipating elements is one No. 4 bar spaced not more than 48 inches on center and within 16 inches of the end of masonry walls. Lateral ties embedded in grout are required for masonry columns. Lateral ties must be anchored with standard 135° or 180° hooks. Type N mortar and masonry cement is not permitted to be used in participating elements of the seismic-force-resisting system.

Special reinforced shear walls. Special reinforced masonry walls are permitted in any seismic design category and are the only type of shear wall allowed in Seismic Design Category D and above. The special reinforced shear wall has the most favorable (highest) R-factor of the shear wall types, and as such requires specific reinforcement to improve and ensure ductility.

The special reinforced shear wall reinforcement requirements are given in Section 1.18.3.2.6 of the MSJC Code:

(a) The maximum spacing of vertical reinforcement is the lesser of one-third the shear wall length or one-third the shear wall height, but not more than 48 inches for running bond and 24 inches for stack bond.

(b) The maximum spacing of vertical reinforcement is the lesser of one-third the shear wall length or one-third the shear wall height, but not more than 48 inches for running bond and 24 inches for stack bond. Horizontal reinforcing must be uniformly distributed and embedded in grout.

(c) The minimum cross-sectional area of vertical reinforcement is one-third the area of required shear reinforcement. The sum of the cross-sectional area of horizontal and vertical reinforcement is at least 0.020 times the cross-sectional area of the wall, and the minimum cross-sectional area of reinforcement in each direction is 0.007 times the cross-sectional area of the wall. For stack bond masonry, the minimum cross-sectional area of horizontal reinforcement is 0.015 times the cross-sectional area of the wall.

(d) Shear reinforcement must be anchored vertical reinforcing with standard hooks.

(e) Masonry laid in stack bond must be hollow open-end units grouted solid or two wythes of solid units grouted.

Special reinforced shear walls designed by the allowable stress design method. Special reinforced shear walls designed by the allowable stress design method are to be designed to resist 1.5 times the code-prescribed in-plane shear force determined from ASCE 7. However, the 1.5 multiplier does not apply to the overturning moment or out-of-plane forces. See Section 1.18.3.2.6.1.2. The intent of this requirement is to ensure that flexural failure dominates in order to ensure ductile performance.

Seismic Design Category E or F. Structures in Seismic Design Category E or F must conform to the requirements specified for Seismic Design Categories B, C, and D and the additional requirements in Section 1.18.4.5 of the MSJC Code. The additional minimum reinforcing requirements for stack bond elements that originated with the 2000 IBC are now part of the standard. Stack bond used in nonparticipating elements (elements that are not a part of the lateral-force-resisting system) is required to have a horizontal reinforcing ratio of at least 0.0015 with a maximum spacing of 24 inches on center and must be constructed of hollow open-end units grouted solid or two wythes of solid units fully grouted. Only special reinforced masonry shear walls are allowed in Seismic Design Categories E and F. Refer to Tables 12.2-1 and 12.14-1 of ASCE 7 for the seismic design coefficients, limitations, and restrictions pertaining to special reinforced masonry shear walls.

Section 2107 *Allowable Stress Design*

2107.1 General. Section 2107 references the general design requirements in Chapter 1 and the allowable stress design method in Chapter 2 of the MSJC Code, with additional modifications that were approved through the code development process. The modifications to the standard are contained in Sections 2107.2 through 2107.4.

The allowable stress design provisions in the standard are based on the use of full design stresses only, assuming that all engineered structural masonry will have some level of special inspection. These minimum levels of special inspection have accordingly been incorporated into Section 1704.5, which references the quality-assurance provision of the MSJC Code. The IBC and MSJC Code do not allow the half stress design without special inspection that was permitted in the UBC. It is important to note that the one-third increase in allowable stresses that was permitted for load combinations that include earthquake and wind forces is no longer permitted in the 2011 MSJC Code. From an IBC perspective, masonry structures designed by the allowable stress design provisions would be designed by either set of load combinations in Section 1605.3.

2107.2 TMS 402/ACI 530/ASCE 5 Section 2.1.7.7.1.1, lap splices. This modification allows the reinforcing-bar lap splice lengths to be consistent for all masonry design methods (allowable stress design and strength design). In the 2009 IBC, this amendment was mandatory in that it modified the lap splice length in MSJC Code Section 2.1.9.7.1.1. In the 2012 IBC, it is a permissible alternative to the lap splice length in MSJC Code

Section 2.1.7.7.1.1. The splice lengths in the MSJC Code are based on developing the allowable stress in the bar by means of an assumed simplified bond mechanism, which is not conservative for large bar sizes.

Both allowable stress and strength design methods assume that the reinforcing will have sufficient strength to resist the imposed forces. Splices of reinforcing bars must meet this same strength test; the same holds true for development length. In seismic regions, the force in any particular bar is indeterminate but may be at the yield strength of the bar. Required development and splice lengths are based on developing the yield strength of the reinforcing bar, including any apparent overstrength, without distress in the masonry, as is done in the strength design method.

The splice length or development length required by the equation in the 2003 IBC, including the f factor, will develop 125 percent of the specified yield strength of the reinforcing bar. This allows for the likely overstrength of reinforcing bars and matches the requirements for welded or mechanical splices, which also must develop 125 percent of the specified yield strength of the bar.

During the development of the 2006 IBC, it was pointed out that the lap splice Equation 21-2 in the 2003 IBC produces lap splices that in some cases are unreasonably long and result in conditions that cannot be constructed in the field. During the development of the 2005 edition of the ACI 530/ASCE 5/TMS 402 standard, over half of the public comments were directed at this issue. A code change modified the lap splice provisions to be essentially the same as the development length provision in the 1997 UBC (which originated with the 1985 UBC). The code change to the 2006 IBC also increased lap splices by 50 percent in areas of high tensile demand (greater than 80 percent of the allowable steel tension stress F_s), which originated with the 1997 UBC.

The amendment in Section 2107.2 is only for lap splices. Although the same length is also needed to develop the strength of the reinforcing bar, the requirement for development length in the MSJC Code Section 2.1.7.3 is not amended in the IBC.

The code language related to epoxy-coated bars was added to the 2006 IBC because IBC Section 2107.3 replaces MSJC Section 2.1.9.7.1.1, which in essence deleted the language related to epoxy-coated bars. The same requirement for increasing developing length of epoxy-coated bars by 50 percent is in Section 2.1.7.3 of the MSJC Code.

2107.3 **TMS 402/ACI 530/ASCE 5 Section 2.1.7.7, splices of reinforcement.** The amendment to the standard is the requirement that bars larger than No. 9 be spliced using mechanical connectors. The reason for the amendment is that the allowable stress design method of the standard allows reinforcing bar sizes up to No. 11. As noted in Section 2107.4, these large bar sizes are very difficult to lap splice without splitting the masonry. Using mechanical connectors can mitigate the splitting problem that occurs in ordinary lap splices. Hence, the code requires mechanical connectors for bar sizes larger than No. 9. Section 2.1.7.7.3 requires the mechanical splices to develop 1.25 times the yield strength of the bar. This requirement is the same for strength design in Section 3.3.3.4 of the standard. The requirements that welded splices must be ASTM A706 steel reinforcement was added to the 2009 IBC.

2107.4 **TMS 402/ACI 530/ASCE 5 Section 2.3.7, maximum bar size.** The amendment adds a new section (Section 2.3.7) to the standard that restricts the size of reinforcing bars used in walls. Placing large bars in masonry walls over-reinforces the section so that a brittle failure of the masonry is likely. It is difficult to develop lap splices for large reinforcing bars in thin masonry walls. Research shows that when the reinforcing bar size number is larger than the nominal thickness of the wall, the splice will fail by splitting of the wall and pullout of the bar before the strength of the bar is developed in the splice. The research shows that a rough rule of thumb for maximum bar size is $(n - 1)$, where n is the nominal thickness of the wall. Hence, the limit on bar size to one-eighth the nominal thickness. This requirement is similar to the requirement for strength design, except that Section 3.3.3.1 also imposes a maximum size limit of No. 9 on reinforcing bars.

Lightly loaded masonry columns. In previous editions of the IBC, Section 2107.4 contained specific provisions for lightly loaded masonry columns in light-frame structures having a maximum area of 450 square feet and located in regions of relatively low seismicity to be reinforced with a single vertical reinforcing bar in each cell, as long as the column can safely support the code-required loads and deformations. In the MSJC Code, columns are defined by geometry, not by applied loads. Thus, masonry members that have certain geometry are classified as columns, although they more closely resemble a flexural element with low axial load. Such columns were required by the MSJC Code to be reinforced with a minimum amount of vertical reinforcement and horizontal ties. The IBC exempted lightly loaded columns such as those used to support light-frame carport roofs, porches, sheds, or similar structures that primarily experience axial tension and flexure in high wind events from the prescriptive requirements of the MSJC Code. The requirements were deleted in the 2009 IBC because they were incorporated into Section 1.14.2 of the 2008 MSJC Code. Lightly loaded columns in Seismic Design Category A, B, or C supporting light-frame structures and no more than 2,000 lbs are permitted if they are a minimum 8-inch nominal side dimension, maximum of 12 feet in height, and 0.2 square inches of reinforcing (one No. 4 bar) centered in the column.

Section 2108 *Strength Design of Masonry*

Section 2108 of the 2000 IBC has detailed requirements for the strength design of masonry that were based entirely on the 1997 UBC because the 1999 edition of the ACI 530/ASCE 5/TMS 402 standard did not include a strength design procedure. The UBC strength design provisions for masonry were developed from 1986 through 1994 by the Masonry Joint Ad Hoc Committee, which was a joint committee of the Structural Engineers Association of California (SEAOC) Code and Seismology Committees. Some of the SEAOC Masonry Joint Ad Hoc Committee members were also members of NEHRP Technical Subcommittee 5 (TS-5). At the time the 2000 IBC was being developed, the IBC structural committee encouraged the MSJC to develop and incorporate strength design provisions within the standard. This was achieved in the 2002 edition of the ACI 530/ASCE 5/TMS 402 standard, which contains comprehensive strength design provisions. The strength design provisions in Chapter 3 of the 2002 MSJC Code were under development by masonry design experts for over 10 years. Many of the provisions in the IBC and in other documents such as NEHRP were based on the draft provisions that were being developed by the MSJC. The strength design provisions were removed from the 2003 IBC, and Section 2108 referenced Chapter 3 of the standard. Section 2108 of the 2012 IBC references Chapter 3 of the MSJC Code, but also includes some modifications to those provisions.

Although the 2000 IBC included masonry wall frame provisions, the MSJC decided to leave provisions for wall frames out of Chapter 3 because of considerable differences of opinion on the requirements for masonry wall frames and because very few masonry wall frame buildings have been constructed using either the 2000 IBC or 1997 UBC provisions.

There are no longer any provisions for designing masonry wall frames in the IBC or the MSJC Code. Therefore, masonry wall frame design would probably be done using Section 2108 of the 2000 IBC and could be approved under the alternative materials, design, and methods of construction provisions in Section 104.11.

For a detailed discussion of the strength design provisions in the 2000 IBC, the reader is referred to the *2000 IBC Handbook—Structural Provisions*.

General. The 2009 IBC references general design requirements in Chapter 1 and the strength design procedure in Chapter 3 of the 2011 edition of the TMS 402/ACI 530/ASCE 5 standard with two modifications. In the 2011 edition of the standard, the design **2108.1**

provisions for AAC masonry structures were relocated from Appendix A to the body of the code in Chapter 8. For the design of AAC masonry, which uses a strength design procedure, the code references Chapters 1 and 8 of the MSJC Code.

2108.2 TMS 402/ACI 530/ASCE 5 Section 3.3.3.3, development. The required development length of reinforcement is determined by Equation 3-16 of the standard but cannot be less than 12 inches and need not be greater than 72 d_b. Equation 3-16 produces lap splices that are highly variable and in some cases are unreasonably long, resulting in conditions that can be difficult to construct in the field. During the development of the 2005 edition of the MSJC Code, 148 public comments emphasized this problem. The 1997 UBC had a nearly identical equation, but the maximum lap splice length was capped at 52 bar-diameters. The 1997 UBC lap splices for Grade 60 reinforcement were determined by ASD in areas of high moment to be 1.5 times 48 bar-diameters, or 72 bar-diameters. This modification to the standard essentially increases the 1997 UBC cap from 52 to 72 bar-diameters to coordinate the ASD and strength design requirements. This modification will likely be removed from the IBC after further research and the provisions are incorporated into the MSJC Code.

2108.3 TMS 402/ACI 530/ASCE 5 Section 3.3.3.4, splices. Welded splices are not permitted in plastic hinge zones of intermediate or special reinforced masonry shear walls. Although the section also mentions "special moment frames of masonry," there is no such system in ASCE 7-10. Where used, welded splices must be of ASTM A706 (weldable) steel reinforcement. This modification to the standard is based on a revision to the 2000 NEHRP *Provisions*. To achieve adequate performance, splices in reinforcing used in the seismic-force-resisting system that are subjected to high seismic strains must be capable of developing the full strength of the reinforcing steel. In order to be welded properly, the chemistry of the steel must have a limited carbon content as well as other elements such as sulfur and phosphorus. If the chemistry of the steel is not carefully controlled, the welding procedures in AWS D1.4, which are based on the steel chemistry, must be carefully adhered to in order to produce welds that develop the strength of the steel. If the carbon equivalent or the sulfur or phosphorus content is too high, the steel may not be weldable. Because the chemistry of reinforcing steel conforming to ASTM A615, A616, and A617 is not controlled and is often unknown, a quality weld that can adequately develop the strength of the steel is not always possible. In contrast, ASTM A706 steel has controlled chemistry and is always weldable. Welded splices are required to be able to develop only 125 percent of the specified yield strength of the spliced bars. However, because A615, A616, A617, as well as A706 bars can have actual yield strengths in excess of 125 percent of the specified yield strength, a code-conforming welded splice may fail before the spliced bars can yield. This would compromise the inelastic deformability of a structural member. Therefore, welded splices are prohibited within the potential plastic hinge region of members in structural systems in buildings assigned to the higher seismic design categories.

Type 1 mechanical splices are not permitted to be used within a plastic hinge zone or within a beam-column joint of intermediate or special reinforced masonry shear walls, because Type 1 splices may not be able to resist the stress levels that develop within the yielding region. However, Type 2 mechanical splices are permitted in any location within a member because Type 2 splices are required to develop the specified tensile strength of the spliced bars. ACI recommends that good detailing precludes the use of splices within regions subjected to potential yielding. However, if splices cannot be avoided, the designer should investigate and document the force-deformation characteristics of the spliced bar and the ability of the splice to meet the expected inelastic demand. See Sections 21.1.6.1 and 12.14.3.2 of ACI 318 for discussion of the types of mechanical splices.

This modification to the standard is based on a revision to the NEHRP *Recommended Provisions*. Reinforcing steel is predominantly produced from remelted steel scrap. Because it is difficult to control the strength of the scrap steel, the resulting products tend to have a strength considerably higher than the specified yield strength. This is similar to the situation that occurred in structural steel where the actual yield strength can be much greater than the specified yield strength. Because there is a lower limit but no upper limit on the yield strength other than for ASTM A706 bars, most reinforcing steel has a higher yield point than specified. Testing by the California Department of Transportation (CALTRANS) has shown that the overstrength can be as much as 60 percent over the specified strength. Splices in reinforcing steel used in the seismic-force-resisting system within plastic hinge zones and in beam-column joints are subjected to high seismic strains. In order to achieve adequate performance, these splices must be able to develop the full strength of the reinforcing. Therefore, Type 1 splices are prohibited within a plastic hinge zone or within a beam-column joint, but Type 2 splices are allowed because they are required to develop the specified tensile strength of the bar. Cyclic tests by CALTRANS of current splices meeting only the 125-percent requirement show that, in many cases, although the splices meet the 125-percent criterion, they cannot survive several excursions in the post yield range imposed by cyclic load testing. ACI 318 now uses the terminology *mechanical splices*, which replaced the term *mechanical connections* used in older editions of the standard, to provide consistency between the IBC and ACI 318 with respect to mechanical splices.

Maximum areas of flexural tensile reinforcement. The 2006 IBC required that the strain in prestressing steel be compatible with a strain in the tension reinforcement equal to five times the strain at reinforcement yield stress for special prestressed masonry shear walls. This requirement is not required in the MSJC Code because the "five times" requirement was too restrictive and walls were practically unbuildable. Special prestressed masonry shear walls are covered in Section 1.18.3.2.12, which references Sections 3.3.3.5 for maximum reinforcement ratios and Section 3.3.6.5 for boundary member requirements.

Section 2109 *Empirical Design of Masonry*

The empirical design procedure for masonry is a prescriptive method of sizing and proportioning masonry structures using rules and formulas that were developed over many years. The procedure is based on experience and predates the engineering design methods. The empirical method was developed for use in smaller buildings with more interior walls and stiffer floor systems than are commonly built today. Gravity loads are assumed to be approximately centered on bearing walls and foundation piers, and the effects of reinforcement are neglected.

General. Section 2109 was deleted entirely in the 2009 IBC and replaced with provisions that are not covered in the MSJC Code such as Section 2109.3, which contains extensive provisions for adobe construction. The requirements for adobe masonry construction originated with and are similar to those contained in the *Standard Building Code* and the *Uniform Code for Building Conservation* (UCBC). Adobe construction is defined in Section 202 as designed and constructed according to the empirical method in Chapter 5 of the MSJC Code and the specific requirements given in Section 2109.3. From a fire- and life-safety standpoint, adobe construction must meet the requirements for Type V construction specified in Chapter 6. **2109.1**

Limitations. The restrictions on the use of empirical design were expanded in the 2006 IBC to make the IBC consistent with the 2005 edition of the MSJC Code. The six limitations on empirical design of masonry listed in the IBC are generally based on the level of lateral load risk, and therefore are driven by seismic design category, basic wind speed, and building height. The limitations are now given in Section 5.1.2 of the MSJC Code. Empirical design is not permitted in buildings in Seismic Design Category D, E, or F and cannot be used for the seismic-force-resisting system in Seismic Design Category B or C. Where empirically **2109.1.1**

designed masonry walls are used in the lateral-load-resisting system (wind or seismic), the height of the building is limited to 35 feet. See Table 5.1.1 of the MSJC Code for limitations based on wind speed. A new provision in the 2012 IBC modifies Section 5.1.2.2 in regard to limitations on wind speed. The empirical design method cannot be used to design or construct masonry buildings or parts of masonry buildings located in areas where the nominal wind speed (V_{asd}) determined in accordance with Section 1609.3.1 exceeds 110 mph. If a building structure exceeds the limitations prescribed in this section or the standard, then the building or structure must be designed in accordance with the engineering provisions covered in Section 2107 for allowable stress design or Section 2108 for strength design. Masonry foundation walls may be constructed in accordance with the prescriptive masonry foundation wall provisions covered in Chapter 18, Section 1807.1.6.

Section 5.1.2.6 prohibits the use of empirical design for AAC masonry, which must be designed in accordance with the strength design procedure in Chapter 8 of the MSJC Code.

Section 2110 Glass Unit Masonry

This section covers the empirical requirements for nonload-bearing glass unit masonry elements used in exterior or interior walls.

Glass unit masonry provisions are similar to those contained in the UBC with additional updates and revisions based on the MSJC Code. Section 2110 was deleted entirely in the 2009 IBC and replaced with provisions that are not specifically covered in the 2008 edition of the MSJC Code. These are the fire-resistive and structural limitations in Section 2110.1.1. Glass unit masonry is not permitted to be used in various types of fire-resistive walls, smoke barriers, or load-bearing walls. See further discussion in the *2012 International Building Code Commentary*.[8] Glass unit masonry is designed and constructed according to Chapter 7 of the standard and the specific requirements in Section 2110.

Section 2111 Masonry Fireplaces

This section covers requirements for masonry fireplaces and their foundations constructed of concrete or masonry. Table 2111.1 and Figure 2111.1, which summarize the fireplace and chimney requirements, were deleted from the 2006 IBC because they were out of date and inconsistent with other provisions in the code. Note that Section 2111 covers requirements for masonry fireplaces, not masonry chimneys. Specific requirements for masonry chimneys are covered in Section 2113.

The masonry fireplace and chimney provisions in Sections 2111 and 2113 originated with the three legacy model codes but were updated and revised to be consistent with the corresponding provisions in the IRC. The format of the provisions, however, is dramatically different than the UBC, and numerous changes were made to update the requirements and achieve consistency between the IBC and the IRC.

In previous editions of the IBC, the seismic reinforcing and anchorage requirements only applied to masonry fireplaces in Seismic Design Category D. A code change to the 2009 IBC extended these requirements to also apply to masonry fireplaces in Seismic Design Category C.

Section 2112 *Masonry Heaters*

This section covers requirements for masonry heaters as defined in Section 2112.1. Section 2112 in the 2006 IBC was revised in its entirety to coordinate the masonry heater provisions in the IBC and IRC and to reference applicable ASTM and UL standards. Although a code change to the 2009 IBC extended the seismic reinforcing and anchorage requirements for masonry fireplaces and chimneys to include Seismic Design Category C, there were no similar changes to Section 2112 for masonry heaters. Seismic anchorage and reinforcing is only required in masonry heaters in Seismic Design Category D. The section references Section 2113.3 for anchorage and reinforcing.

Section 2113 *Masonry Chimneys*

This section applies to masonry chimneys constructed of concrete or masonry. The section covers seismic anchorage and reinforcing, footing support, and general construction requirements for masonry chimneys. The provisions are primarily based on the fireplace and chimney requirements of the three legacy model codes but were subsequently updated and revised to be consistent with the corresponding provisions in the IRC. In previous editions of the IBC, the seismic reinforcing and anchorage requirements only applied to masonry chimneys in Seismic Design Category D. A code change to the 2009 IBC extended these requirements to also apply to masonry fireplaces in Seismic Design Category C.

KEY POINTS

- Chapter 21 contains requirements for the design and construction of masonry buildings and other masonry structures regulated by the IBC.
- The 2012 IBC references the 2011 edition of TMS 402/ACI 530/ASCE 5, *Building Code Requirements for Masonry Structures*, and TMS 602/ACI 530.1/ASCE 6, *Specification for Masonry Structures*, and in many cases, the IBC refers to specific sections in the standard for detailed design and construction requirements.
- These TMS/ACI/ASCE standards are also known as the "MSJC Code" and "MSJC Specification" because they are developed by the Masonry Standards Joint Committee (MSJC).
- Masonry structures may be designed by one of the procedures prescribed in Section 2101.2 for allowable stress design, strength design, prestressed masonry design, empirical design, or the direct design procedure in accordance with TMS 403.
- Allowable stress designed, strength designed, AAC, prestressed, and empirically designed masonry must also meet the applicable seismic design requirements of Section 2106, which references Section 1.18 of the MSJC Code. Seismic requirements are based on the seismic design category of the structure.
- Provisions for glass unit masonry may be designed by provisions in Section 2110 of the code or by Chapter 7 of the MSJC Code.
- Provisions for masonry veneer may be designed by provisions in Chapter 14 of the code or by Chapter 6 of the MSJC Code.

- In addition to the general requirements for construction documents given in Sections 107 and 1603, Section 2101.3 contains specific requirements for construction documents pertaining to masonry structures.

- Masonry materials must comply with the requirements and applicable ASTM and other standards referenced in Section 2103.

- Masonry construction methods must comply with the requirements in Section 2104, which references various provisions in the MSJC Specification.

- A quality-assurance program in accordance with Section 2105 is required for masonry structures to ensure that constructed masonry is in conformance with the construction documents.

- Chapter 17 references the quality-assurance provisions in Section 1.19 of the MSJC Code, which are based on the risk category of the structure.

- Requirements for masonry fireplaces and chimneys including specific seismic reinforcing in Seismic Design Categories C and D are provided in Sections 2111 and 2113, respectively.

- Masonry fireplaces and chimneys in structures in Seismic Design Categories E and F must be designed in accordance with the engineering provisions of Chapter 21 and the MSJC Code.

REFERENCES

1. TMS 402/ACI 530/ASCE 5, *Building Code Requirements for Masonry Structures*, The Masonry Society, Boulder, CO, 2008, 2011.

2. TMS 602/ACI 530.1/ASCE 6, *Specification for Masonry Structures*, The Masonry Society, Boulder, CO, 2008, 2011.

3. *2000 IBC Handbook—Structural Provisions*, International Code Council, Washington, DC, 2001.

4. TMS 403, *Direct Design Handbook for Masonry Structures*, The Masonry Society, Boulder, CO, 2010.

5. *2009 IBC Handbook—Structural Provisions*, International Code Council, Washington, DC, 2009.

6. NEHRP, *NEHRP (National Earthquake Hazard Reduction Program) Recommended Provisions for Seismic Regulations for New Buildings and Other Structures*, Building Seismic Safety Council, Washington, DC, 1997, 2000, 2003.

7. Fanella, David, *Structural Loads: 2012 IBC® and ASCE/SEI 7-10*, International Code Council, Washington, DC, 2012.

8. *2012 International Building Code Commentary,* International Code Council, Washington, DC, 2012.

BIBLIOGRAPHY

ICC, *Handbook to the Uniform Building Code: An Illustrative Commentary*, International Code Council, Washington, DC, 1998.

ICC, *2000 IBC Handbook—Structural Provisions*, International Code Council, Washington, DC, 2001.

ICC, *2006 IBC Handbook—Structural Provisions*, International Code Council, Washington, DC, 2008.

ICC, *2009 IBC Handbook—Structural Provisions*, International Code Council, Washington, DC, 2009.

ICC, *2006 IBC Code Changes Resource Collection*, International Code Council, Washington, DC, June, 2006.

ICC, *2009 IBC Code Changes Resource Collection*, International Code Council, Washington, DC, June, 2009.

SEAOC, *Recommended Lateral Force Requirements and Commentary*, Structural Engineers Association of California, Sacramento, CA, 1996, 1999, 2009.

CHAPTER 22

STEEL

Introduction
Section 2201 General
Section 2202 Definitions
Section 2203 Identification and Protection of Steel for Structural Purposes
Section 2204 Connections
Section 2205 Structural Steel
Section 2206 Composite Structural Steel and Concrete Structures
Section 2207 Steel Joists
Section 2208 Steel Cable Structures
Section 2209 Steel Storage Racks
Section 2210 Cold-Formed Steel
Section 2211 Cold-Formed Steel Light-Framed Construction
Key Points
References
Bibliography

22 Steel

Introduction

This chapter essentially gives the user a roadmap to the various design standards that apply to the design and construction of steel structures. During the code development process since the 2000 *International Building Code*® (IBC®), the trend of adopting steel standards by reference continued to the point where Chapter 22 of the 2012 IBC contains only three pages. Sections 2205 through 2211 contain references to standards for structural steel, composite steel and concrete, steel joists, steel cable structures, steel storage racks, and cold-formed steel.

Section 2201 *General*

2201.1 Scope. Chapter 22 covers requirements for the quality, design, fabrication, and construction of steel structures including structural steel, cold-formed steel, steel bar joists, steel cable structures, and steel storage racks.

Section 2202 *Definitions*

2202.1 Definitions. The terms listed in this section are defined in Chapter 2. The 2012 IBC only has three definitions: cold-formed steel construction, steel joist, and structural steel member. All other definitions were deleted in the 2009 IBC because they are covered in the various referenced standards for steel structures.

Cold-formed steel members are cold formed to shape from sheet or strip steel to make roof and floor deck, wall panels, studs, and joists. The term *structural steel member* is defined in Section 202 and includes rolled steel structural shapes other than cold-formed steel or steel joist members.

The definition of light-frame construction in Section 202 pertains to repetitive wood framing and also includes repetitive cold-formed steel framing members. This definition is important because in some cases the code has exceptions that apply to buildings of light-frame construction, which includes cold-formed steel structures. For example, Section 1809.7 permits concrete footings supporting walls of light-frame construction to be designed in accordance with the prescriptive requirements in Table 1809.7, which is the familiar prescriptive foundation table that gives the width and thickness of continuous footings based on the number of floors supported. The table applies to both wood and cold-formed steel light-frame structures.

Earlier editions of the IBC included a section that defined the steel-related engineering nomenclature (notation and symbols) used in the code. The section is no longer in the IBC because the engineering nomenclature and notation for steel structures are defined in the various referenced steel standards.

Section 2203 *Identification and Protection of Steel for Structural Purposes*

2203.1 Identification. All structural steel used for load-carrying purposes must be properly identified in order to determine conformance with the appropriate specification and grade.

Structural steel must be identified in accordance with AISC 360, Specification for Structural Steel Buildings. Chapter N of AISC 360-10 addresses minimum requirements for quality control, quality assurance, and nondestructive testing for structural steel systems and steel elements of composite members for buildings and other structures. Section N2 requires material identification procedures to comply with the requirements of Section 6.1 of the AISC Code of Standard Practice for Structural Steel Buildings and Bridges (AISC 303-10), and be monitored by the fabricator's quality control inspector (QCI). The identification of cold-formed steel members should be in accordance with the requirements of AISI S100. Section A of AISI S100 addresses the identification of cold-formed steel. Cold-formed steel used in light-frame construction should be identified in accordance with the additional requirements of AISI S200, Section A5. Other types of steel must be identified in accordance with the applicable ASTM standard or other standard referenced in the chapter. The code requires unidentified steel to be tested for conformity to the applicable standard.

Protection. Protection of structural steel and cold-formed steel structural members or panels from corrosion is required. Protection may be by painting or galvanizing. Painting of structural steel is required to comply with the requirements of AISC 360. Painting steel joists must comply with the appropriate Steel Joist Institute (SJI) standard for the particular type of joist or girder. Protection of cold-formed steel members should be in accordance with the general requirements of AISI S100, Section A, which addresses the protection of cold-formed steel from corrosion. Cold-formed steel in light-frame construction should be protected in accordance with the additional requirements of Section A4 of AISI S200. **2203.2**

Section 2204 *Connections*

Early editions of the IBC had separate sections covering requirements for bolted and welded connections. In the 2006 IBC, bolting and welding were consolidated into one single section on connections. The section now references the subsections that apply to the various types of steel (Section 2205 for structural steel, Section 2207 for steel joists, Section 2208 for steel cable structures, Section 2210 for cold-formed steel, and Section 2211 for cold-formed steel in light-framed construction) and the applicable referenced standards.

Welding. This section does not directly reference the *AWS Structural Welding Code*. Rather, it requires that welding be performed in accordance with the applicable steel specification used for the design. The applicable standards are referenced in Section 2205 for structural steel, Section 2207 for steel joists, Section 2208 for steel cable structures, Section 2210 for cold-formed steel, and Section 2211 for cold-formed steel in light-framed construction. This is done because the various referenced standards either reference different editions of an AWS D1.1 specification, reference a different AWS specification, or the referenced standard includes its own welding requirements such as in the SJI standards. For example, AISI S100-07/S2-10 and AISI S200-07 reference AWS D1.1-04 and AWS D1.3-98; the SDI standards, ANSI/NC1.0-10 and ANSI/RD1.0-10 reference AWS D1.1-10 and AWS D1.3-08. Although the IBC does not directly reference AWS D1.1, the 2010 edition of AWS D1.1 is referenced in AISC 360-10. **2204.1**

Both the 2006 and 2009 IBC reference AISC 360-05, *Specification for Structural Steel Buildings*, which is a unified standard that replaced the ASD specification (AISC 335s1), the LRFD specification (AISC 350), the specification for steel hollow structural steel sections (AISC 346), and the specifications for single-angle members (AISC 336 and 351) into one document. The 2009 IBC also references AISC 341-05, *Seismic Provisions for Structural Steel Buildings*, including Supplement No. 1 dated 2006. The 2012 IBC references the new 2010 editions of AISC 360 and AISC 341, which both reference the 2010 edition of the AWS D1.1 structural steel welding code.

Special inspection of welding is required by this section as specified in Section 1705.2.2.1 and Table 1705.2.2. As noted in the discussion of Chapter 17, the specific special inspection requirements for structural steel were deleted from the 2012 IBC, and Section 1705.2.1 now references the quality-assurance and inspection provisions in AISC 360 instead. Chapter 17 now only includes the special inspection requirements for steel construction other than structural steel. Section N5.4 and Tables N5.4-1 through N5.4-3 of AISC 360-10 contain quality-assurance and inspection requirements for structural steel welding in structural steel buildings. It should be noted that the terminology used in the AISC quality assurance provisions differ from Chapter 17 of the IBC. Most notably the terms "special inspection," "continuous," and "periodic" are not used. In AISC 360, the term "quality control" refers to tasks performed by the steel fabricator and erector to verify the quality of construction. The term "quality assurance" refers to inspection tasks performed by organizations other than the steel fabricator and erector such as the project owner's representative, and are intended to ensure that the product meets the project requirements. The terms quality assurance inspector (QAI) and quality control inspector (QCI) are defined in the standard. Thus, the QAI refers to the special inspector as defined in the building code. Chapter N of AISC 360 also defines two inspection levels as either "observe" or "perform" in contrast to the building code, which uses the terms "periodic" or "continuous."

2204.2 **Bolting.** The philosophy of this section is similar to that used for welding requirements, as the design standards reference different editions of the bolting standards. Bolting standards are not referenced directly in the code itself, but are referenced in the particular design standard. The applicable standards are referenced in Section 2205 for structural steel, Section 2207 for steel joists, Section 2210 for cold-formed steel, and Section 2211 for cold-formed steel light-framed construction.

As noted, the 2010 IBC references the 2010 editions of AISC 360 (*Specification for Structural Steel Buildings*) and AISC 341 (*Seismic Provisions for Structural Steel Buildings*), both of which reference the 2009 edition of the RSCS *Specification for Structural Joints Using ASTM A325 or A490 Bolts*.

The requirements for special inspection of the installation of high-strength bolts are no longer provided in Chapter 17 of the 2012 IBC. Section N5.6 and Tables N5.6-1 through N5.6-3 of AISC 360-10 contain quality-assurance and inspection requirements for high-strength bolting in structural steel buildings. High-strength bolting must be installed in accordance with the approved construction documents and the RCSC specifications. Special inspection is required for installation and tightening of high-strength bolts regardless of the tightening method, or whether the bolts are slip critical or snug-tight connections. The only case where special inspection may not be required is where the connections are *designed* to use ordinary mild steel (ASTM A307) bolts but the engineer *specifies* high-strength bolts to be installed.

As noted above, the terminology used in the AISC quality assurance provisions differ from Chapter 17 of the IBC. Most notably, the terms "special inspection," "continuous," and "periodic" are not used. The terms quality assurance inspector (QAI) and quality control inspector (QCI) are defined in the standard. "QAI" refers to the special inspector as defined in the building code. Chapter N of AISC 360 also defines two inspection levels as either "observe" or "perform" rather than "periodic" or "continuous." For example, turn-of-nut installation with matchmarking, installation using twist-off bolts, and installation using direct tension indicators provides visual evidence of a completed installation, therefore "periodic" special inspection is permitted for these methods. In this case, Chapter N states, "The QCI and QAI need not be present during the installation of fasteners when these methods are used by the installer." On the other hand, calibrated wrench installation provides no such visual evidence and therefore "continuous" special inspection is required, and the inspector needs to be on site during installation. In this case, Chapter N states, "The QCI and QAI shall be engaged in their assigned inspection duties during installation of fasteners when these methods are used by the installer." Refer to the Commentary to Chapter N of AISC 360 for more detail.

Anchor rods. Anchor rods must be set in accordance with the approved construction documents. The code specifies the bolt protrusion such that all of the threads of the nut are fully engaged but the end of the threaded portion does extend beyond the surface of the connected part so the nut does not run out of threads. The most common specification for anchor rods is ASTM F1554. The standard covers straight and bent, headed and headless anchor bolts (also known as anchor rods) made of carbon, carbon boron, alloy, or high-strength low-alloy steel, and having specified yield strengths of 36, 55, and 105 ksi. The anchor bolts are furnished in two thread classes, and in various sizes, and are intended for anchoring structural supports to concrete foundations. AISC 360 requires threads on anchor rods to conform to ASME B 18.2.6.

2204.2.1

Section 2205 *Structural Steel*

Section 202 defines structural steel members as any steel structural member of a building or structure consisting of a rolled steel structural shape other than cold-formed steel, or steel joist members.

General. Structural steel buildings must be designed, fabricated, and constructed in accordance with the referenced standard, ANSI/AISC 360-10. AISC 360 is an ANSI-approved standard that provides requirements for the design and construction of structural steel buildings and other structures, and incorporates *both* allowable strength design, and load and resistance factor design methods. The dual-units format provides both U.S. customary and SI units. In addition to the new standard, AISC publishes AISC 325-10, the 14th edition of the AISC *Steel Construction Manual*. The manual contains the following specifications, codes, and standards: 2010 AISC 360 *Specification for Structural Steel Buildings*, 2009 RCSC *Specification for Structural Joints Using ASTM A325 or A490 Bolts*, and 2010 AISC *Code of Standard Practice for Steel Buildings and Bridges*.

2205.1

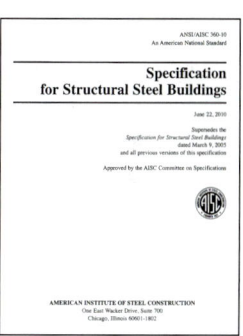

In practice, design engineers use the 14th edition AISC Steel Construction Manual, which is based on and includes ANSI/AISC 360-10. According to AISC, a concerted effort was made in the development of both the standard (specification) and the manual to limit the number of changes. Some of the important changes include:

- All tabular information and discussions have been updated to comply with the 2010 Specification for Structural Steel Buildings, and the standards and other documents referenced therein.

- Updated shape information based on ASTM A6-09.

- Weights of HSS are revised based on industry practice.

- The minimum S_z value for single angles is published based on the heel and toes of the angles.

- Tables for eccentrically loaded single angles are revised to be consistent with the Specification using the unsymmetric shape equations in Specification Section H2.

- Available compressive strength tables for filled HSS now indicate at what length the bare steel member controls the strength.

- The coefficients, C, in the eccentrically loaded weld group tables have been revised for the $a = 0$ case (no eccentricity) to be consistent with Specification Chapter J. Additionally, the tables are supplemented to provide the strength for L-shaped welds loaded from either side.

- The procedure for the design of conventional single-plate shear connections is revised to accommodate the increased bolt shear strengths of the 2010 specification.
- Information is provided to determine if stiffening plates (stabilizers) are required for extended single-plate shear connections.
- The bracket plate design procedure is revised.

AISC 360-10 is available for free downloading on the AISC website: aisc.org/2010spec. Many additional resources are available from aisc.org.

2205.2 Seismic requirements for steel structures. The 2012 IBC references ANSI/AIC 341-10, *Seismic Provisions for Structural Steel Buildings*. The AISC seismic provisions incorporate the latest requirements based on the NEHRP *Provisions*[1] and the significant research results of the SAC Joint Venture investigation,[2,3] which was initiated as a result of the extensive damage to welded steel moment frames in the 1994 Northridge earthquake. AISC 341 contains an extensive commentary and bibliography. The reader is referred to the commentary in AISC 341 for a detailed discussion on the seismic design provisions for steel buildings.

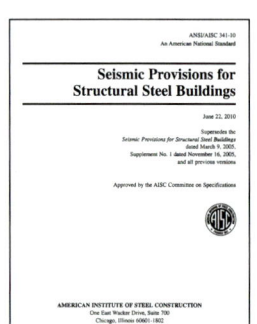

The ANSI-approved specification is a companion to the ANSI/AISC 360-10 that extends coverage to the connection detailing and member design requirements for structural steel and composite systems in high-seismic applications. The 2010 edition of AISC-341 supersedes and is an update of the 2005 edition. A number of major changes were made in this edition, and a cross-reference list following the table of contents provides a reference to the corresponding section of the 2005 standard. Both LRFD and ASD methods of design are incorporated, and dual-units format provides for both U.S. customary and SI units.

In addition to the new seismic design standard, AISC publishes the *Seismic Design Manual*, AISC 327. The AISC *Seismic Design Manual* includes printed versions of AISC 341-10 and AISC 358-10. AISC 358 covers prequalified connections for special and intermediate steel moment frames.

The requirements for design of structural steel structures resisting seismic forces are based on the seismic design category of the building as determined from Section 1613.3.5. Requirements for Seismic Design Category A, B, and C are given in Section 2205.2.1, and requirements for Seismic Design Category D, E, and F are given in Section 2205.2.2. For a detailed discussion on determination of seismic design category, see discussion of Section 1613.

2205.2.1 Seismic design category B or C. All structures in Seismic Design Category A need only comply with the general structural integrity requirements prescribed in Section 1.4 of ACSE 7, which is referenced from Section 11.7. Steel structures assigned to Seismic Design Category B or C may be of any construction permitted in Section 2205. Where a structure is designed using a Response Modification Factor, *R*, from Table 12.2-1 of ASCE 7, the structure must be designed and detailed in accordance with AISC 341. The exception allows structures assigned to Seismic Design Category B or C to be designed as "steel systems not specifically detailed for seismic resistance" using an *R*-factor of 3 using AISC 360 and not AISC 341. See Item H of Table 12.2-1 of ASCE 7. It should be noted that this category does not include cantilever column systems, which are specifically covered under Item G in Table 12.2-1. Although not required, structures in Seismic Design Category B or C may be designed and detailed in accordance with AISC 341 using the appropriate *R*-factor from Table 12.2-1 or Table 12.14-1 for the type of building system involved. In a sense, this is trading ductile detailing for design force level. Using systems with higher *R* values produces lower lateral forces but requires more ductile detailing to ensure ductile response characteristics. In general, the detailing requirements can have a significant impact on the cost to construct the seismic-force-resisting system; therefore, providing the required detailing in order to design for lower lateral forces may not be economical in the lower seismic design categories. Because the 2012 IBC references ASCE 7 for seismic design, the seismic design coefficients (such as the *R*-factor) for the various types of buildings and seismic-force-resisting systems are given in ASCE 7 Table 12.2-1 for the equivalent lateral force procedure or Table 12.14-1 for the simplified seismic design procedure.

Seismic design category D, E, or F. Unlike structural steel systems in Seismic Design Category A, B, or C, steel structures in Seismic Design Category D, E, or F are required to be designed and detailed in accordance with AISC 341 using the appropriate *R*-factor from ASCE 7 Table 12.2-1 or Table 12.14-1 for the building type and seismic-force-resisting system involved. Note that ASCE 7 Table 15.4-1 contains specific detailing reference for nonbuilding structures similar to buildings.

2205.2.2

Section 2206 *Composite Structural Steel and Concrete Structures*

General. The IBC references ANSI/AISC 360 and ACI 318 (excluding Chapter 22, Plain Concrete) for the design of composite structural steel and concrete structures. Seismic design of composite steel and concrete structures is provided in Section 2206.2.

2206.1

Seismic requirements for composite structural steel and concrete construction. The design of composite structural steel and concrete systems resisting seismic forces based on the appropriate *R*-factor from ASCE 7 is required to conform to the requirements of the AISC 341 and ACI 318 standards. The requirement in the 2009 IBC that composite structures in Seismic Design Categories D, E, and F provide substantiating evidence demonstrating that they will perform as intended by Part II of AISC 341 has been deleted because these structures are now addressed in the 2010 edition of AISC 341. AISC 341 contains detailed provisions for testing composite special moment frames, composite partially restrained moment frames, and composite eccentrically braced frames. Unlike structural steel structures regulated by Section 2205, composite structures must be designed and detailed in accordance with AISC 341 regardless of seismic design category. The new section makes no specific reference to Seismic Design Categories D, E, and F; thus, composite systems in buildings in Seismic Design Category B or higher are required to be designed and detailed in accordance with AISC 341. Structures in Seismic Design Category A may be designed in accordance with AISC 360 and need only comply with the general structural integrity requirements prescribed in Section 1.4 of ASCE 7, which is referenced from Section 11.7.

2206.2

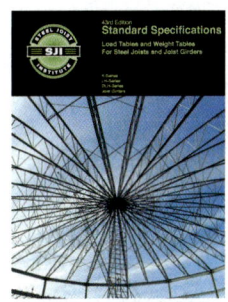

Section 2207 *Steel Joists*

Steel joists are required to be designed, manufactured, and tested in accordance with the various specifications published by the SJI and referenced in Section 2207.1.

Joists are used primarily as gravity-load-carrying members. When joists are incorporated into the lateral-force-resisting system as collectors, chords, and diaphragm ties, care must be taken to ensure that the joists are specifically designed and detailed to properly function as these elements. Normally, joists are designed only for gravity and wind uplift loads. A continuous load path is necessary for chords, collectors, and diaphragm ties, and their connections. The typical seat on a steel joist is eccentric to the chord and is not ordinarily designed to carry the axial force and resulting moment through the seat when functioning as a collector or chord, which are in tension and compression. Generally, special seats must be designed and manufactured, or the load path must be specifically directed through the chords by direct chord-to-chord connection. See Fisher[4] for methods of designing chord-to-chord connections.

A code change during the 2004/2005 code cycle created four new code sections in the 2006 IBC that were intended to specify what is required in the way of submittal documents and clarify the responsibilities of the registered design professional and the joist engineer/specialty structural engineer. The new sections pertain to design, calculations, steel joist construction drawings, and certification. These new provisions require the steel joist industry to meet requirements similar to those for the pre-engineered metal plate-connected wood truss industries; thus, the requirements are similar to the requirements for Section 2303.4 for pre-engineered wood trusses.

The new provisions were intended to clarify the difference between joist placement and layout plans and open web steel joists and joist girder construction drawings. Steel joist placement and layout plans are generally recognized by the industry as not requiring an engineer's signature and seal, whereas steel joist and joist girder construction documents do require an engineer's design, review, and seal. Note that Section 2207.4 specifically states that steel joist placement plans do not require the seal and signature of the joist manufacturer's engineer. The new sections clarify the responsibilities of the registered design professional and the joist engineer/specialty structural engineer based on the American Society of Civil Engineer (ASCE)/Council of American Structural Engineers (CASE) Document 962 D 2003, "National Practice Guidelines for Specialty Structural Engineers." A specialty structural engineer is often retained by the supplier or subcontractor who is responsible for the design, fabrication, and sometimes the installation of engineered structural elements. For example, virtually all engineered wood trusses are designed by specialty engineers who are generally not the registered design professional responsible for the overall building design.

A position paper published in March of 2003 by the SJI regarding signing and sealing of steel joist placement plans pointed out that there is a potential for critical items that can only be designed and detailed by the joist engineer/specialty structural engineer to slip through the cracks because the SJI paper does not require a seal to be placed on the joist placement plans. Critical items that could be overlooked are (1) joist-to-joist or joist-to-girder connections; (2) compression chord bridging design and bridging connection details for cantilevered and uplift conditions; (3) compression chord design and detailing for conditions where the compression chord is not continuously braced; (4) special loading conditions; and (5) special configurations. In approving the code change, the IBC structural committee felt that the code change clarifies steel joist submittal requirements and the roles of the registered design professional and the steel joist manufacturer, and will better serve the engineering community and construction industry. The four sections are summarized as follows:

Section 2207.2 Design—This section describes what specific items must be addressed by the registered design professional and shown on the construction documents. See Section 202 for definitions of the terms "registered design professional" and "construction documents."

Section 2207.3 Calculations—This section requires the steel joist and girder manufacturer to design the system in accordance with the referenced SJI specifications and load tables and any loads specified by the registered design professional described in Section 2207.2.

Section 2207.4 Steel Joist Drawings—Steel placement drawings are used to install the joist system in the field. This section itemizes what is to be included on joist placement drawings. As noted above, the steel joist placement drawings do not require the seal and signature of the joist manufacturer's engineer.

Section 2207.5 Certification—The manufacturer is required to submit a certificate of compliance stating that the joist system conforms to the approved construction documents and the applicable SJI specifications.

2207.1 General. The design, manufacture, and installation of open web steel joists and joist girders are required to conform to one of the four SJI standard specifications listed in this section: Composite Steel Joists - CJ-series (CJ–10), Joist Girders (JG–10), Open Web Steel Joists - K-series (K–10) and Longspan Steel Joists - LH-series, and Deep Longspan Steel Joists - DLH-series (LH/DLH–10). Seismic design must be in accordance with the provisions of Section 2205 or Section 2210.6 for light-framed cold-formed steel construction. A new SJI standard for design of composite steel joists was added to the 2009 IBC, *Standard Specification for Composite Steel Joists, CJ-Series*. Published in 2006 as an ANSI standard, the new SJI specification covers LRFD design, manufacture, and use of simply supported open-web composite steel joists called the CJ-series. The standard includes the requirements deflection limits for composite steel joists. The standard requires the specifying professional to give due consideration to the effects of short- and long-term deflection and vibration in the design and selection of composite joist systems. For steel joists, deflection calculations must account for the inherent flexibility of the open web configuration.

2207.2 Design. This section requires the registered design professional to indicate on the construction documents the steel joist or joist girder designations used as well as layout scheme, end support, anchorages, bridging, bridging termination connections, and bearing connections that resist uplift and lateral loads. The construction documents must also include any special loads or conditions that are listed in the code section. The construction documents are part of the submittal documents required by Section 107. See Section 202 for the definition of construction documents.

2207.3 Calculations. The steel joist/girder manufacturer is required to design the steel joists and/or steel joist girders to support the anticipated loads and load combinations prescribed by Chapter 16 in accordance with the applicable SJI specifications. The registered design professional responsible for the design of the building may require joist/girder calculations to be prepared along with a cover letter bearing the seal and signature of the joist manufacturer's registered design professional. In addition to the standard steel joist or joist girder calculations, other items such as non-SJI standard bridging details and nonstandard connection details, field splices, and joist headers are required to be submitted.

2207.4 Steel joist drawings. Steel joist placement plans that indicate the steel joist products specified on the construction documents are required to be submitted to the building department for review and approval. Joist placement plans are primarily used to install the joist system in the field. This code section is very specific as to what the joist placement plans are required to include. Note that the section specifically states that steel joist placement plans do not require the seal and signature of the joist manufacturer's registered design professional.

2207.5 Certification. The steel joist manufacturer is required to submit a certificate of compliance stating that work was performed in accordance with approved construction documents and the applicable SJI standard specifications. See also Section 1704.2.5 regarding exemption of special inspection where work is done by an approved fabricator. A certificate of compliance from the fabricator is required to be submitted to the building official stating that the work was performed in accordance with approved construction documents and with SJI standard specifications.

Section 2208 *Steel Cable Structures*

2208.1 General. The IBC references ASCE 19, *Structural Applications of Steel Cables for Buildings*, for design and construction of steel cables used in buildings. See the commentary in ASCE 19 for a detailed explanation of the structural requirements for cable structures.

2208.2 Seismic requirements for steel cable. The modifications to the seismic requirements of ASCE 19 that appear in this section are based on recommendations in the NEHRP *Provisions*.

Section 2209 Steel Storage Racks

The 2010 IBC references the 2008 edition of ANSI/MH16.1, the Rack Manufacturers Institute (RMI) *Specification for Design, Testing, and Utilization of Industrial Steel Storage Racks*. Use of the RMI specification is limited to the specific types of storage racks within the scope of the standard that are made of cold-formed or hot-rolled steel structural members, which are the typical low- to moderate-height industrial racks mounted at grade such as those used in warehouses and "big box" retail stores. The standard covers industrial pallet racks, movable shelf racks and stacker racks, push-back racks, pallet-flow racks, case-flow rack pick modules, rack supported platforms, and rack structures that support exterior walls and roofs. It does not cover other types of racks, such as drive-in and drive-through racks, cantilever racks, or portable racks. Where required by ASCE 7, seismic design of steel storage racks must be in accordance with Section 15.5.3 of ASCE 7, which contains modifications to several sections of the RMI standard.

2209.1 Storage racks. During the development of the 2003 NEHRP *Provisions*, the issue of seismic safety of steel storage racks was raised because of the recent proliferation of big box retail stores utilizing tall storage racks. As a result, FEMA funded the report, *Seismic Considerations for Steel Storage Racks Located in Areas Accessible to the Public* (FEMA 460). For a number of years and several code change cycles, RMI worked to harmonize its rack specifications with the IBC and NEHRP *Provisions*. Working with the nonbuilding structures committee of the BSSC and ASCE 7, as well as the FEMA 460 Task Group, the seismic requirements in the current RMI standard have been updated to be consistent with the NEHRP *Provisions*. The code section clarifies the applicability of ASCE 7 Section 15.5.3.

Section 2210 Cold-Formed Steel

2210.1 General. The term "cold-formed steel construction" is defined in Section 202. This section references the 2007 edition of the *North American Specification for the Design of Cold-Formed Steel Structural Members* with the 2010 Supplement (AISI S100/S2-10), published by the American Iron and Steel Institute (AISI). The 1996 edition of the specification combined LRFD and ASD into one specification. The 1999 edition, published as Supplement No. 1 to the 1996 edition, further refined the specification and added some new provisions. Subsequently, a cooperative effort by the AISI Committee on Specifications for the Design of Cold-Formed Steel Structural Members, the Canadian Standards Association (CSA) Technical Committee on Cold-Formed Steel Structural Members (S136), and the Camara Nacional de la Industria del Hierro y del Acero (CANACERO) in Mexico developed the 2001 edition of the *North American Specification for the Design of Cold-Formed Steel Structural Members*. The 2007 edition of the standard with the 2010 Supplement, which has been given the new number designation of AISI S100/S2-10, is the latest edition of the standard and is referenced in the 2012 IBC. Because AISI S100-07 is intended for use in Canada, Mexico, and the United States, it uses a format that allows for requirements particular to each country. This results in a main document—Chapters A through G and

Appendices 1 and 2—that is intended for use in all three countries, and two country-specific appendices (A and B). Appendix A is required for use in both the United States and Mexico, and Appendix B is required for use in Canada. Reflecting its North American scope, AISI S100 provides an integrated treatment of Allowable Strength Design (ASD), Load and Resistance Factor Design (LRFD), and Limit States Design (LSD). This is accomplished by including the appropriate resistance factors (ϕ) for use with LRFD and LSD and the appropriate factors of safety (Ω) for use with ASD. The use of the LSD procedure is limited to Canada, and the use of LRFD and ASD is limited to the United States and Mexico.

AISI S100-07 provides well-defined procedures for the design of load-carrying cold-formed steel members in buildings, as well as other applications, provided that proper allowances are made for dynamic effects. The provisions reflect the results of continuing research to develop new and improved information on the structural behavior of cold-formed steel members. For a detailed discussion of the major changes made in the 2007 edition of AISI S100, refer to the *2009 IBC Handbook—Structural Provisions.*[5] The phrase "Cold-formed steel light-frame construction shall also comply with Section 2211" was added to the 2009 IBC for light-framed construction to clarify that AISI S100 is to be used in conjunction with the AISI standards referenced in Section 2211.

Cold-formed stainless steel structural systems are required to be designed in accordance with ASCE 8, *Standard Specification for the Design of Cold-Formed Stainless Steel Structural Members.*

Where required by ASCE 7, seismic design of cold-formed steel structures must be in accordance with Section 2210.2, which references AISI S100 and ASCE 8. The section references AISI S110 for design of cold-formed steel special-bolted moment frames. See Section 2210 for a detailed discussion of cold-formed steel special-bolted moment frames.

Steel decks. The section covers design and construction of cold-formed steel floor and roof decks. Two new SDI standards on cold-formed steel decks, ANSI/SDI NC1.0 and ANSI/SDI RD1.0, were added to the 2009 IBC. The 2012 IBC references to 2010 editions of these standards. It is intended that designers be permitted to use these documents in lieu of the more formal approach of AISI S100.

2210.1.1

A code change proposal to add the Steel Deck Institute (SDI) standard C1.0-06 for composite steel floor decks to the 2009 IBC as an alternative to ASCE 3 was disapproved because of technical issues regarding the use of fibers to substitute for steel reinforcement required by ACI 318. The section in the 2009 IBC that referenced ASCE 3-91 for the design of composite decks was deleted in the 2012 IBC because the standard is outdated and SDI's standard C1.0-06 was disapproved for the 2012 IBC because of questions raised on its treatment of serviceability and wheel loads as well as the need to clarify the exclusion of fiber reinforcement. As it now stands, there is no specific referenced standard in the IBC for design of composite steel decks. The design approach would be based on the general design requirements in Section 1604.4 and other referenced standards such as AISC 360, AISI S100, and ACI 318, where the steel and concrete capacities are analyzed separately or composite resistance is developed using shear connectors. A more straight forward alternative would be to design and specify a composite deck system from a manufacturer with a current ICC Evaluation Service Report (ESR), which can be accepted by the building official under the alternative design and construction provisions of Section 104.11. The ESRs contain information on composite deck loads and spans, including the need for mid-span shoring where the steel capacity is exceeded during the concrete pours. Typically the deck panel styles described in the reports have embossments that provide shear interlock with the concrete fill. For the purpose of determining the shear bond strength between the concrete and deck panels with embossments, ICC-ES still references the test methods given in the aforementioned ASCE 3. Although not in the IBC, AC43 references SDI's Composite Deck Design Handbook (CDD2) and allows other rational methods of analysis with prior concurrence of the ICC-ES staff. Thus, any ICC-ES report that includes composite deck designs would have used ASCE 3, CDD2, or other approved methods of design.

2210.1.1.1 Noncomposite steel floor decks. The section references ANSI/SDI NC1.0, *Standard for Noncomposite Steel Floor Deck*, which governs the materials, design, and erection of cold-formed noncomposite steel deck used as a form for reinforced concrete slabs. In the 2009 IBC, there was a modification to SDI NC1.0 that was intended to remove any doubt that noncomposite concrete slabs must be designed in accordance with the applicable provisions of ACI 318. By specifically requiring compliance with ACI 318, which does not permit fibers or fibrous admixtures in lieu of steel reinforcement, and removing any mention of fibers or fibrous admixtures in Section 2.4B 1 of SDI NC1.0, the concerns of the IBC Structural Committee were addressed. The 2012 IBC references the 2010 edition of SDI NC1.0 standard, which specifically requires the slab design to comply with ACI 318. Because the amendment is no longer useful, it was deleted from the 2012 IBC.

2210.1.1.2 Steel roof decks. This section references ANSI/SDI RD 1.0, *Standard for Steel Roof Deck*, which governs the materials, design, and erection of cold-formed steel deck used for the support of roofing materials, design live loads, and construction loads. It is intended that designers be permitted to use this standard in lieu of the more formal approach of AISI S100/S2-10.

The SDI standards are available as free downloads from the SDI website at www.sdi.org.

2210.2 Seismic requirements for cold-formed steel structures. This section on seismic design of cold-formed steel structures is new in the 2012 IBC. Where required by ASCE 7, seismic design of cold-formed steel structures is to be in accordance with AISI S100 and ASCE 8. The section references AISI S110 for the design of cold-formed steel special-bolted moment frames. The new AISI S110 standard for seismic design of cold-formed steel special moment frames includes design provisions for a new cold-formed steel seismic-force-resisting system called Cold-Formed Steel–Special-Bolted Moment Frames (CFS-SBMF). The standard was developed by AISI as a result of research conducted at the University of California at San Diego by Professors Chia-Ming Uang and Atsushi Sato. CFS-SBMF systems experience substantial inelastic deformation during a design seismic event, with most of the inelastic deformation occurring at the bolted connections due to slip and bearing. See Figure 2210-1. The CFS-SBMF system was vetted through the Building Seismic Safety Council (BSSC) process for inclusion in the 2009 NEHRP *Provisions* and subsequently incorporated into the ASCE 7-10 standard. Cyclic testing has

Figure 2210-1
Cold-formed steel–special-bolted moment frames (CFS-SBMFs).

shown that the system has large ductility capacity and significant hardening. To develop the expected mechanism, requirements based on capacity design principles are provided in the standard for the design of the beams, columns, and their connections. Table 12.2-1 of ASCE 7-10 includes seismic design parameters for CFS-SBMF system of $R = 3.5$, $\Omega_0 = 3.0$ and $C_d = 3.5$. AISI S110 also includes specific requirements for quality-assurance and quality-control procedures.

Section 2211 *Cold-Formed Steel Light-Framed Construction*

This section requires structural and nonstructural members using cold-formed steel of light-frame construction where the specified minimum base steel thickness is between 0.0179 inches and 0.1180 inches to be designed and installed in accordance with AISI S200 and the applicable standards referenced in Sections 2211.2 through 2211.7. As noted above, cold-formed steel members are distinguished from structural steel members by Section 202, which defines both "cold-formed steel construction" and "structural steel member." Structural steel members are defined as any steel structural member of a building or structure consisting of a rolled steel structural shape other than cold-formed steel, or steel joist members. This section further distinguishes *light-frame* cold-formed steel based on the base metal thickness. As noted, the term "light-frame construction" is also defined in Section 202 because the code has specific requirements that apply to light-frame construction only.

During the code development process from the 2000 edition to the 2006 edition of the IBC, AISI developed several national consensus standards for cold-formed steel used in light-framed construction. The term *light-framed construction* is defined in IBC Section 202 as a type of construction whose vertical and horizontal structural elements are primarily formed by a system of repetitive wood or cold-formed steel framing members. The 2012 IBC references the 2007 editions of the various AISI standards and added one new standard for floor and roof system design. The AISI standards referenced in this section are:

- AISI S200-07/S2-10: *North American Standard for Cold-Formed Steel Framing—General Provisions*
- AISI S210-07: *North American Standard for Cold-Formed Steel Framing—Floor and Roof System Design*
- AISI S211-07: *North American Standard for Cold-Formed Steel Framing—Wall Stud Design*
- AISI S212-07: *North American Standard for Cold-Formed Steel Framing—Header Design*
- AISI S213-07/S1-09: *North American Standard for Cold-Formed Steel Framing—Lateral Design*
- AISI S214-07: *North American Standard for Cold-Formed Steel Framing—Truss Design*
- AISI S230-07: *Standard for Cold-Formed Steel Framing—Prescriptive Method for One- and Two-Family Dwellings*

A new numeric ANSI designation system was initiated with these 2007 edition AISI standards in the hope that this will simplify the referencing of the documents in this growing series of AISI design and installation standards. The words *North American* are now in the titles of many of these standards to emphasize that these documents are intended for adoption and use not just in the United States, but throughout North America. In all such documents, provisions applicable to Canada were added. For a detailed discussion of the major changes to the 2007 AISI standards, refer to the *2009 IBC Handbook—Structural Provisions*.

2211.1 General. This section gives general requirements related to the design, installation, and construction of cold-formed steel structural and nonstructural framing and specifies the range of base steel thicknesses (0.0179 inches to 0.1180 inches) that apply. The design and installation of these members must meet the general requirements of AISI S200 and the applicable requirements in the specific standards referenced in subsequent Sections 2211.2 through 2211.7.

2211.2 Header designs. This section references the *North American Standard for Cold-Formed Steel Framing—Header Design* (AISI S212-07) as an alternative to AISI S100 for the design and installation of cold-formed steel headers. The two-part standard gives design professionals the required tools to design efficient built-up and L-shaped headers.

2211.3 Truss design. This section references the *North American Standard for Cold-Formed Steel Framing—Truss Design* (AISI S214) for the design and installation of cold-formed steel trusses. The standard provides technical information and specifications on cold-formed steel truss construction and applies to the design, quality assurance, installation, and testing of cold-formed steel trusses. During the development of the 2009 IBC, the National Council of Structural Engineers Associations (NCSEA) worked with AISI and the Wood Truss Council of America (WTCA) to coordinate the truss design requirements in the code with AISI S214. AISI agreed to produce a supplement to AISI S214 that addresses both NCSEA's and WTCA's concerns. AISI S214 Supplement 2 to AISI S214-07 was issued on June 9, 2008.

The 2012 IBC references the 2007 edition of S214 with Supplement 2. Several new subsections were added in the 2009 IBC to make the requirements for truss design drawings and submittals similar to the requirements for metal-plate-connected wood trusses.

2211.3.1 Truss design drawings. This section requires truss design drawings to conform to the requirements of Section B2.3 of AISI S214 and to be provided with the shipment of trusses delivered to the job site. Truss design drawings are required to include details of permanent individual truss member restraint/bracing.

2211.3.2 Deferred submittals. This section essentially deletes Section B4.2, because IBC Section 106.3.4.2 covers requirements for deferred submittals. In addition, it is not appropriate for AISI Section B4.2 to list requirements for deferred submittals on other items or materials.

2211.3.3 Trusses spanning 60 feet or greater. AISI S214 covers permanent bracing but not temporary bracing. This section requires the owner to contract with a registered design professional to design the temporary and permanent truss member bracing for trusses with clear spans 60 feet or greater. In addition, for trusses with clear spans 60 feet or greater, special inspection is required. See discussion of Section 1705.2.2.2 for more detail.

2211.3.4 Truss quality assurance. Chapter E of AISI S214 provides requirements for quality criteria for steel trusses that are part of a manufacturing process with quality control. For trusses that are not part of a manufacturing process that includes quality control under the supervision of a third-party quality control agency, the section requires trusses to be manufactured in compliance with Sections 1704.2.5 and 1705.2 as applicable.

2211.4 Wall stud design. This section references the *North American Standard for Cold-Formed Steel Framing—Wall Stud Design* (AISI S211-07) as an alternative to AISI S100 for the design and installation of cold-formed steel wall studs. The standard provides requirements for the design and installation of structural and nonstructural walls in buildings.

2211.5 Floor and roof system design. This section references the *North American Standard for Cold-Formed Steel Framing—Floor and Roof System Design* (AISI S210-07) as an alternative to the more formal approach of AISI S100. Either S210-07 or AISI S100 may be used to design floor and roof systems.

2211.6 Lateral design. This section references the *North American Standard for Cold-Formed Steel Framing—Lateral Design* (AISI S213-07) for the design of cold-formed steel light-framed shear walls and diaphragms used to resist wind and seismic loads. The standard contains design requirements for shear walls, diagonal strap bracing, and diaphragms consisting of light-frame cold-formed steel members.

Prescriptive framing. This section references the *Standard for Cold-Formed Steel Framing—Prescriptive Method for One- and Two-Family Dwellings* (AISI S230-07) for the design and construction of cold-formed steel framing in detached one- and two-family dwellings and townhouses up to three stories in height. This standard is an updated version of previous CABO and IRC prescriptive requirements as well as the previous 2000 edition of the AISI standard. It incorporates latest developments such as the L-header and an efficient design procedure for built-up headers.

2211.7

KEY POINTS

- Chapter 22 contains requirements for the design and construction of structural steel buildings and other steel structures regulated by the IBC.
- The 2012 IBC references the 2010 editions of AISC 360 *Specification for Structural Steel Buildings* and AISC 341 *Seismic Provisions for Structural Steel Buildings*, and various AISI standards for cold-formed steel (CFS) structures.
- In most cases, the IBC refers to specific sections in the referenced standards for detailed design and construction requirements.
- Structural steel members must be properly identified in accordance with AISC 360-10, and CFS structural members must be identified in accordance with AISI S100.
- Steel members used for structural purposes other than structural steel and CFS must be properly identified in accordance with applicable ASTM standards.
- Welded and bolted connections must conform to the applicable referenced specification (standard) that governs their specific application.
- Structural steel structures in Seismic Design Category A need not have a designated seismic-force-resisting system and need only comply with the basic structural integrity requirements of Section 1.4 of ASCE 7-10.
- Structural steel structures in Seismic Design Category B or C may be designed in accordance with AISC 360 using an *R*-factor of 3 and need not conform to the seismic detailing requirements of AISC 341, or may be designed to conform to the detailing requirements of AISC 341 where the applicable *R*-factor from ASCE 7-10 is used.
- Structural steel structures in Seismic Design Category D, E, or F are required to conform to the seismic detailing requirements of AISC 341 using the applicable *R*-factor from ASCE 7-10.
- Steel joist systems must comply with Section 2207, which includes requirements for design, calculations, joist drawings, and certification.
- Steel storage racks must be designed in accordance with the 2008 edition of the RMI standard, ANSI/MH16.1-08 *Specification for Design, Testing and Utilization of Industrial Steel Storage Racks*, and the applicable seismic design requirements of Section 15.5.3 of ASCE 7-10.
- Section 2210 references various AISI standards for the design and construction of CFS structures.
- Section 2211 references various AISI standards for the design and construction of light-frame CFS structures.

REFERENCES

1. NEHRP, *NEHRP (National Earthquake Hazard Reduction Program) Recommended Provisions for Seismic Regulations for New Buildings and Other Structures*, Building Seismic Safety Council, Washington, DC, 1997, 2000, 2003.
2. SAC 95-02, *INTERIM GUIDELINES: Evaluation, Repair, Modification and Design of Steel Moment Frames*, SAC Joint Venture—A Partnership of the Structural Engineers Association of California, Applied Technology Council, and California Universities for Research in Earthquake Engineering, Sacramento, CA, 1995 (also known as FEMA 267).
3. SAC 96-03, *INTERIM GUIDELINES: Advisory No. 1*, SAC Joint Venture—A Partnership of the Structural Engineers Association of California, Applied Technology Council, and California Universities for Research in Earthquake Engineering, Sacramento, CA, 1996 (also known as FEMA 267A).
4. Fisher, James A., West, Michael, and Van de Pas, Julian, *Designing with Steel Joists, Joist Girders, and Steel Deck*, Vulcraft/Nucor, 1991.
5. *2009 IBC Handbook—Structural Provisions*, International Code Council, Country Club Hills, IL, 2009.

BIBLIOGRAPHY

ICC, *2009 IBC Code Changes Resource Collection*, International Code Council, Country Club Hills, IL, 2009.

ICC, *2009 IBC Code and Commentary Volume II*, International Code Council, Country Club Hills, IL, 2009.

ICC, *2012 IBC Code and Commentary Volume II*, International Code Council, Country Club Hills, IL, 2012.

SEAOC, *Recommended Lateral Force Requirements and Commentary*, Structural Engineers Association of California, Sacramento, CA, 1996 (1999).

CHAPTER 23

WOOD

Introduction
Section 2301 General
Section 2302 Definitions
Section 2303 Minimum Standards
 and Quality
Section 2304 General Construction Requirements
Section 2305 General Design Requirements
 for Lateral-Force-Resisting Systems
Section 2306 Allowable Stress Design
Section 2307 Load and Resistance Factor Design
Section 2308 Conventional Light-Frame
 Construction
Key Points
References
Bibliography

23 Wood

Introduction

Chapter 23 of the *International Building Code®* (IBC®) originated as an amalgamation of the wood design provisions in the 1997 *Uniform Building Code* (UBC), the 1996 *BOCA National Building Code* (NBC), and the 1997 *Standard Building Code* (SBC), with the addition of seismic provisions from the 1997 National Earthquake Hazard Reduction Program (NEHRP) *Recommended Provisions for Seismic Regulations for New Buildings and Other Structures.*[1] The provisions in the IBC were selected from these source documents based on their technical merit, and new provisions were added where they did not already exist. The use of seismic design categories instead of seismic zones was the most significant change for code users who were most familiar with the UBC. Seismic design categories are determined based on a combination of use of the structure, as defined by the Risk Category (formerly called Occupancy Category), the soil type, and the potential seismic hazard at the site. See Table 1604.5 for Risk Category descriptions based on the use of the structure. The seismic hazard at the site is a combination of the seismic risk associated with potential ground motion (mapped spectral response accelerations) and the site soil classification (Site Class). See the discussion of Section 1613 for a more detailed analysis of earthquake load effects, site class, and determination of seismic design category.

During the code development process from the 2000 IBC to the current 2012 edition, the most significant change to Chapter 23 consisted of the elimination of substantial code language in Sections 2305 and 2306, in favor of the referenced standards, *National Design Specification (NDS) for Wood Construction*[2] and *Special Design Provisions for Wind and Seismic* (SDPWS).[3] Both the *NDS* and *SDPWS* are dual-format standards that include allowable stress design (ASD) and load and resistance factor design (LRFD) procedures. The NDS and SDPWS are referenced in Sections 2305, 2306, and 2307 without amendments.

In June 2010, the American Wood Council (AWC) was re-chartered, evolving from its predecessor groups. Prior to the founding of the new AWC, the forest products industry was represented by the American Forest & Paper Association (AF&PA), which grew out of the National Forest Products Association (NFPA) and the American Paper Institute (API). The 2012 IBC references the 2012 edition of the *National Design Specification (NDS) for Wood Construction (ANSI/AWC NDS-2012) and Supplement*, which is an AWC standard developed by the AWC's Wood Design Standards Committee.

Section 2301 General

2301.1 Scope. Chapter 23 covers materials, design, construction, and quality of wood buildings and structures. The chapter is formatted into eight major sections that follow a logical format: general requirements (2301), definitions (2302), minimum standards and quality for wood-related construction materials (2303), general construction requirements for wood structures (2304), general design requirements for lateral-force-resisting systems used in wood structures, allowable stress design of wood structures (2305), load and resistance factor design of wood structures (2307), and prescriptive conventional construction requirements for light wood-frame structures.

The 2003 IBC referenced the 1997 *National Design Specification (NDS) for Wood Construction*, published by the American Forest and Paper Association (AF&PA)[2] for allowable stress design (ASD), and ASCE 16-95, *Standard for Load and Resistance Factor Design (LRFD) for Engineered Wood Construction,*[4] for strength design of wood structures. The AF&PA NDS was referenced in Section 2306 (ASD), and ASCE 16-95 was referenced in Section 2307 (LRFD). One significant change to the 2006 IBC is that both Sections 2306 and 2307 reference the 2005 edition of the AF&PA NDS, which is a dual-format standard that includes both ASD and LRFD procedures. The 2009 IBC also referenced the 2005 NDS and Supplement. For lateral design of wood structures, the 2009 IBC references the 2008 edition of the *Special Design Provisions for Wind and Seismic* (SDPWS). The 2012 IBC references the latest 2012 edition of the NDS and 2012 Supplement and the 2008 edition of the SDPWS. The SDPWS will be covered in more detail in the discussion of Section 2305.

Wood products addressed in Chapter 23 include boards, dimension lumber, posts, timbers, glued-laminated members, wood structural panels including plywood, composite panels, oriented strand board, particleboard, fiberboard, hardboard, prefabricated wood I-joists, structural composite lumber, laminated veneer lumber, and parallel strand lumber.

The chapter contains material specifications, quality requirements, and design provisions, as well as empirical and prescriptive provisions for wood-frame construction.

Wood and wood-based products are also regulated in other chapters of the code. For example, Chapter 14 contains provisions for weather coverings for walls, as well as provisions for veneers and exterior trim. Wood and wood-based products used in fire-resistance-rated assemblies must also comply with Chapter 7. Wood and wood-based products used as interior finish on walls, ceilings, and floors must comply with Chapter 8. Wood roof coverings are regulated in Chapter 15.

General design requirements. There are three basic design methods permitted for wood structures—allowable stress design in accordance with Section 2306, load and resistance factor design in accordance with Section 2307, and prescriptive conventional light-frame construction in accordance with Section 2308. Lateral design requirements for wood structures are covered in Section 2305. Regardless of the design method used, the general design and construction requirements in Section 2304 apply as well. As an alternative to the prescriptive conventional construction provisions of Section 2308, an exception allows wood-frame buildings to be designed in accordance with the AF&PA *Wood Frame Construction Manual* (WFCM). The 2012 IBC references the 2012 edition of the WFCM.

2301.2

The first two design methods, ASD under Section 2306 and LRFD under Section 2307, are engineering methods, and the third method under Section 2308 is a prescriptive method. Engineering methods require the structure to be designed and detailed to resist all the applicable loads prescribed in Chapter 16 and ASCE 7. In contrast, the prescriptive provisions are essentially a collection of rules that anyone may follow without calculating any loads. If all of the rules are followed, the resulting structure is deemed to comply with the intent of the code. See Section 2308 for further discussion of the prescriptive conventional wood-frame construction provisions.

The three design methods are discussed in more detail below.

1. **Allowable stress design.** Allowable stress design uses the load combinations of Section 1605.3 for the determination of strength. The designer has two options for ASD load combinations: the basic load combinations in Section 1605.3.1, or the alternative basic load combinations in Section 1605.3.2. It should be noted that all deformations and drifts from seismic load effects are determined using strength level forces in accordance with the requirements of Section 12.12 of ASCE 7. The seismic load effect, E, is not multiplied by 0.7 (or divided by 1.4) for determination of deformation and drift. The so-called "special seismic load combinations" were deleted from the 2009 IBC. They are now called "Seismic Load Effect Including Overstrength Factor" in Section 12.4.3 of ASCE 7. These load combinations are used for collector design for all structures assigned to Seismic Design Categories C through F, except for wood-frame buildings braced entirely by light-frame shear walls. In this case, the collectors must be designed to resist the diaphragm forces in ASCE 7 Section 12.10.1.1. See Exception 2 in Section 12.10.2.1 of ASCE 7.

Refer to the discussion of Section 1605 and ASCE 7 Section 12.4.3 for general application of the seismic load combinations with overstrength factor.

2. **Load and resistance factor design.** The AF&PA/ASCE 16-95 *Standard for Load and Resistance Factor Design for Engineered Wood Construction* was first added to the 1997 UBC as a recognized standard. AF&PA/ASCE 16 was then incorporated into the 2000 IBC as a referenced standard in Section 2307 for LRFD of wood structures. In the 2006 IBC, the LRFD procedure is included as part of the 2005 edition of the NDS, so Section 2307 now references the NDS. The applicable load combinations for LRFD and strength design are in Section 1605.2. Note that wood structures designed using the LRFD procedure are subject to the same general and lateral force design provisions as structures designed using ASD. See the 2012 NDS for requirements, strength properties, and design values related to the LRFD procedure.

3. **Conventional light-frame wood construction.** The prescriptive conventional construction provisions are in Section 2308. Perhaps the most important aspect of the prescriptive conventional construction provisions is the restrictions and limitations in Section 2308.2. A building or any portion of a building that does not conform to the limitations must be designed to resist all applicable loads of Chapter 16 in accordance with one of the engineering methods. See also Sections 2308.1.1 and 2308.4 for design of portions and elements. As with the engineering methods, buildings designed by the prescriptive conventional construction provisions of Section 2308 are also required to comply with the general construction requirements in Section 2304. For clarification, a new definition of conventional light-frame construction was added to the 2009 IBC. The term *light-frame construction* is also defined in Section 202.

Log structures. The IBC references ICC 400 for the design and construction of log structures. Because ICC 400 gives base values and references AF&PA NDS-05 for design, either ASD or LRFD procedures can be used. Section 2303.1.10 references ASTM D 3957 for determining stress grades for structural log members and requires identification by grade stamps or certificates of inspection.

2301.3 Nominal sizes. Nominal lumber sizes, for example, 2 inches by 4 inches, are the sizes usually referred to or specified in this chapter. Actual or net dimensions, which are less than nominal dimensions, are used in structural calculations to determine member section properties, actual stresses, and strength properties. The nominal and actual sizes of dimension lumber are established by Department of Commerce (DOC) Voluntary Product Standard PS 20, *American Softwood Lumber Standard*. The edition referenced in previous editions of the IBC is DOC PS 20-99. The most current edition is DOC PS 20-05, which is referenced in the 2012 IBC. The PS 20 standard is available from the National Institute of Standards and Technology (NIST). The nominal size, dressed size, and section properties for sawn lumber and glued laminated timber are given in Tables 1A, 1B, 1C, and 1D of the NDS Supplement. See Section 202 for the definition of nominal size lumber.

Section 2302 *Definitions*

The specific terms related to wood construction listed in Section 2302 are defined in Section 202 to clarify their meaning. Many of the terms are to clarify terms used to describe elements of the lateral-force-resisting systems.

New definitions in the 2006 IBC for prefabricated wood I-joist, structural composite lumber, laminated veneer lumber, parallel strand lumber, and some modifications to the definition of composite panels, oriented strand board, and plywood were made to coordinate the terms in the IBC with the NDS. Some changes were made to the definition of treated wood in the 2009 IBC. Treated wood is a general term with two specific types defined: *fire-retardant-treated wood* and *preservative-treated wood*. Also, the definition of *termite-resistant wood* was expanded to include Alaska yellow cedar and Western red cedar.

Some specific terms are discussed below.

COLLECTOR. The collector collects shear from the (floor or roof) diaphragm and delivers it to vertical lateral-force-resisting elements such as shear walls. The term *drag strut* is a colloquial expression that means *collector*. Collectors can also be used to transfer forces within a diaphragm. The IBC recognizes both horizontal and sloped (or nearly horizontal) diaphragms and as such the definition of collector applies to both horizontal and sloped diaphragms. The word *horizontal* was used to differentiate diaphragms from vertical lateral-force-resisting elements such as a shear wall. In the legacy codes, the terms *horizontal diaphragm* and *vertical diaphragm* were often used to describe diaphragms and shear walls. The newer terminology in the IBC is to simply use the terms *diaphragm* and *shear wall*.

CONVENTIONAL LIGHT-FRAME CONSTRUCTION. For clarification, a new definition of *conventional light-frame construction* was added to the 2009 IBC. It is a type of construction whose primary structural elements such as walls, floors, and roof are formed by repetitive wood-framing members constructed in accordance with Section 2308. The term *light-frame construction* is also defined in Section 202.

DIAPHRAGM, UNBLOCKED. A diaphragm is a horizontal or nearly horizontal (sloped) structural element that transmits lateral forces to the vertical resisting elements (shear walls or frames) of the lateral-force-resisting system. An unblocked diaphragm has edge nailing at the supported edges only. In an unblocked diaphragm, the continuous panel joint is unblocked. In a blocked diaphragm, all sheathing panel edges are supported by framing members or solid blocking members. Blocked diaphragms have continuity between the sheathing panel edges and therefore have significantly less deflection than unblocked diaphragms because of the stiffness developed by the continuity at the blocked panel edges.

The definition of a diaphragm existed in Chapter 23 of the 2000 IBC but not in Chapter 16. In the 2003 IBC, the definition of a diaphragm was deleted from Chapter 23 and added to Chapter 16 along with the various types and elements of diaphragms. Chapter 23 only has a definition for unblocked diaphragm. See Section 1602 for additional terms pertaining to diaphragms. It is important to note that the definition of diaphragm also includes horizontal bracing systems.

In terms of distribution of seismic or wind forces, the stiffness of the diaphragm relative to the stiffness of the vertical lateral-force-resisting elements (shear walls or frames) is the important parameter by which to classify the diaphragm. Diaphragms can be flexible, rigid, or semi-rigid. Refer to the discussion of Section 12.3 of ASCE 7 for further discussion of diaphragm classifications.

GRADE (LUMBER). References Department of Commerce (DOC) standard PS 20, *American Softwood Lumber Standard*, listed under DOC in Chapter 35. Voluntary Product Standard PS 20 is published by the U.S. Department of Commerce and establishes standard sizes and requirements for development and coordination of the lumber grades of the various species, the assignment of design values when called for, and the preparation of grading rules applicable to each species. The PS 20 standard is available from the National Institute of Standards and Technology (NIST), which administers the Department of Commerce Voluntary Product Standards program or the American Lumber Standards Committee website at www.alsc.org.

NATURALLY DURABLE WOOD. Naturally durable wood is the heartwood of a durable listed species. Naturally durable wood includes decay-resistant woods, which are redwood, cedar, black locust, and black walnut, and termite-resistant woods, which are redwood and

Eastern red cedar. The definition of *termite-resistant wood* was expanded in the 2009 IBC to include Alaska yellow cedar and Western red cedar, which were recently determined to be termite resistant in a study involving the Formosan subterranean termite. Note that only the heartwood of redwood and red cedar are both decay and termite resistant.

PERFORMANCE CATEGORY. A new term "performance category" in the 2012 IBC reflects the latest versions of the DOC PS 1 and PS 2 standards, which use terminologies of bond classification to reference glue type and performance categories to reference the thicknesses tolerance consistent with the nominal panel thicknesses in the IBC. The performance category value is the "nominal panel thickness" or "panel thickness." See Section 2303.1.4.

PREFABRICATED WOOD I-JOIST. This definition was added to the 2006 IBC for structural members manufactured of sawn or structural composite lumber flanges and wood structural panel webs bonded together with exterior exposure adhesives in the form of an "I" cross-sectional shape.

SHEAR WALL. A shear wall is a vertical lateral-force-resisting element that is designed to resist lateral seismic and wind forces parallel to the plane of a wall. The definitions of perforated shear wall and perforated shear wall segment were consolidated under the shear wall definition in the 2006 IBC. A perforated shear wall is a wood structural panel sheathed wall with openings that has not been specifically designed and detailed for force transfer around the openings. Perforated shear wall segments are sections of the shear wall with full-height sheathing that meet the height-to-width ratio limits specified in AF&PA SDPWS Section 4.3.4. Substantial portions of Section 2305 were deleted in the 2009 IBC and the section references the 2008 AF&PA SDPWS, which is also referenced in the 2012 IBC. The detailed requirements for design of shear walls in wood-frame structures are now in the SDPWS standard and not the IBC.

STRUCTURAL COMPOSITE LUMBER. The definition of structural composite lumber (SCL) was added to the 2006 IBC. The two main types are laminated veneer lumber (LVL), which is composed of thin wood veneer sheet elements, and parallel strand lumber (PSL), which is composed of wood strand elements. LVL is the most widely used SCL product. Both LVL and PSL have wood fibers that are primarily oriented along the length of the member. Some common examples of structural composite lumber are laminated veneer lumber (LVL) and parallel strand lumber (PSL) such as Microllam® LVL Beams or Parallam® PSL Beams manufactured by Weyerhaeuser, VERSALAM® LVL manufactured by Boise Cascade, and LP® SolidStart® LVL manufactured by Louisiana Pacific (LP). See also discussion of Section 2303.1.9.

TREATED WOOD. Includes both fire-treated wood and wood treated to resist decay and termites. The definition was reorganized in the 2009 IBC so that treated wood is now a general term with two specific types defined: fire-retardant-treated wood and preservative-treated wood. Additional required information for treated wood was added, the definition of preservative-treated wood was revised, and the definition for fire-retardant-treated wood was added. The ability of the wood to extinguish itself once the source of ignition is consumed or removed is an important element of the material. The definition of preservative-treated wood was not really a definition but a reference to a code section. In addition, preservative-treated wood will not reduce susceptibility to all insects, only those that actually eat the wood. Section 2303.2 requires testing of fire-retardant-treated wood. The test must be continued 20 minutes beyond the 10 minutes required to establish the flame spread without any significant progressive combustion.

WOOD STRUCTURAL PANEL. Wood structural panels are manufactured from veneers, wood strands, or wafers, or a combination thereof, bonded together with waterproof synthetic resins. The terms *composite panels, oriented strand board* (OSB), and *plywood* were added in the 2003 IBC under the definition of *wood structural panel* for clarification. When used for structural purposes such as siding, roof and wall sheathing, subflooring, diaphragms, and built-up members, wood structural panels must conform to the requirements for their type in DOC PS 1 or PS 2. DOC PS 1 covers plywood and DOC PS 2 covers wood-based structural panels. DOC Voluntary Product Standards are developed under procedures published by the Department of Commerce in Title 15 Code of Federal Regulations Part 10. The purpose of these standards is to establish nationally recognized requirements for products and to provide all concerned interests with a basis for common understanding of the characteristics of the products. The National Institute of Standards and Technology (NIST) administers the Voluntary Product Standards program. The DOC PS 1 and PS 2 standards are available from the NIST global standards information program website at www.gsi.nist.gov.

Section 2303 *Minimum Standards and Quality*

General. Structural lumber, wood structural panels, and other wood products are highly variable in strengths and other mechanical properties. The code requires that these materials (defined in the first paragraph of this section) conform to the applicable standards and grading rules specified in the code. Furthermore, the code requires that they be identified by a grade mark or be accompanied by a Certificate of Inspection issued by an approved agency. The grade mark is also required to be placed on the material by an approved agency (see labeling—Chapter 17). The proper use of a wood structural member cannot be determined unless it has been properly identified as to species and grade. Counterfeit grade stamps do occasionally appear on lumber in the field, and it is important that designers and enforcement personnel be familiar with the grade-approved stamps. Examples of grade marks are shown in Figure 2303-1. **2303.1**

Joist hangers are subject to the applicable requirements of this chapter as well as the requirements in Chapters 17 and 22.

Sawn lumber. Lumber references the voluntary standard, *American Softwood Lumber Standard*, PS 20. The 2012 IBC references the 2005 edition of DOC PS 20. The current edition is PS 20-10 and is available from the American Lumber Standard Committee at www.alc.org. The NDS references ASTM D 1990, ASTM D 245, ASTM D 2555, the *Wood Handbook*,[5] and PS20 for the classification, definition, methods of grading, and development of design values for lumber. The NDS also references the various standard grading rule documents such as NLGA, NELMA, NSLB, SPIB, WCLIB, and RIS. **2303.1.1**

In the 1991 NDS, changes in design values for dimension lumber were based on the In-Grade Testing program conducted by the North American forest products industry. The program was carried out over an eight-year period and involved the destructive testing of 70,000 pieces of lumber from 33 different species groups. A new test method standard, ASTM D4761, was also developed for the mechanical test methods used in the program. In addition, the standard practice, ASTM D1990, was developed for procedures used to establish design values for visually graded dimension lumber from test results obtained from in-grade test programs.

Certificate of compliance. Certification is an acceptable alternative to a grade mark from both U.S. and Canadian grading agencies certified by the ALSC. The code allows certain types of structural lumber to have a certificate of inspection instead of a grade mark. A certificate of inspection is acceptable for precut, remanufactured, or rough-sawn lumber and for sizes larger than 3 inches (76 mm) nominal in thickness. It is industry practice to place only one label (grade mark) on a piece of lumber, which may be removed on precut and remanufactured lumber. Each piece of lumber is graded after it has been cut to a standard size. **2303.1.1.1**

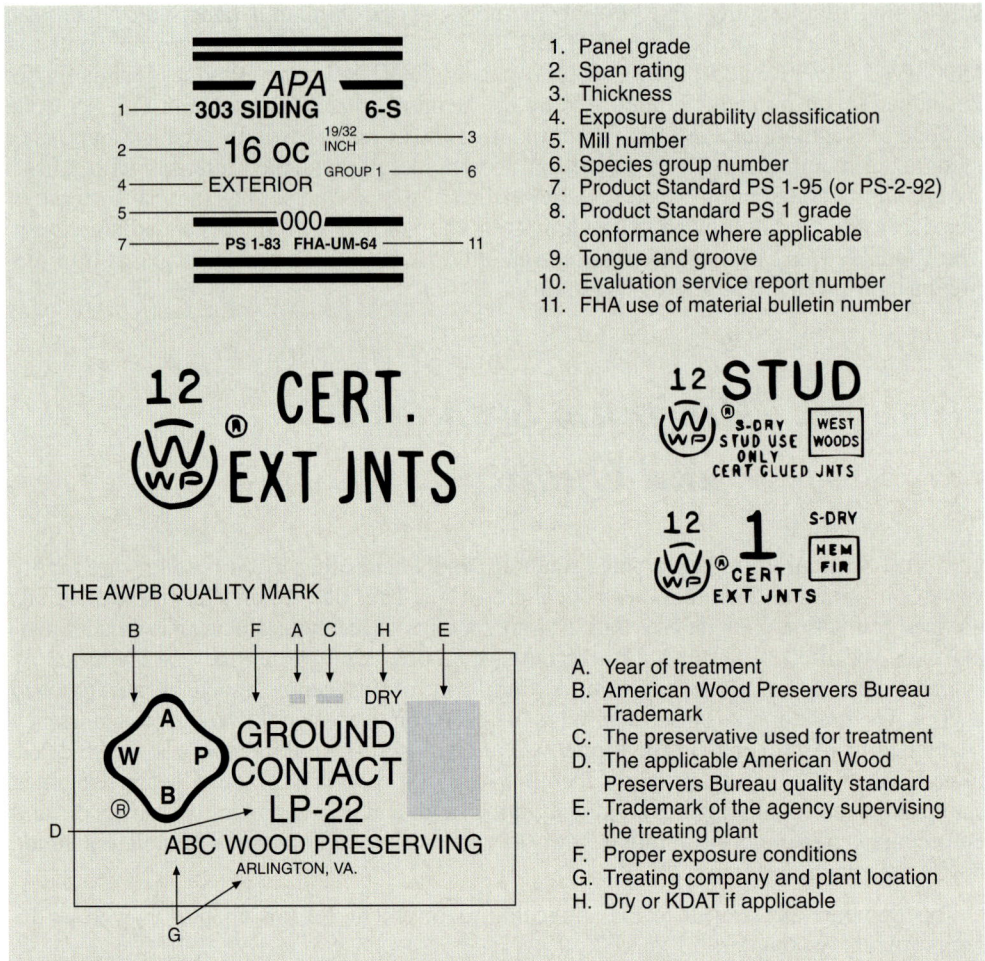

Figure 2303-1
Typical lumber grade stamps.

The grade of the piece is determined based on its size, number, and location of strength-reducing characteristics. Therefore, one log of timber may produce lumber of two or more different grades. It is also industry practice not to label lumber having a nominal thickness larger than 3 inches, or rough-sawn material where the label may be illegible. A certificate of inspection from an approved agency is acceptable instead of the label for these types of lumber. The certificate should be filed with the permanent records of the building or structure. If defects exceeding those permitted for the grade allegedly installed are visible, then a certified grader would be able to determine that the wood is definitely not of a suitable grade. To determine if the wood in question is definitely of a suitable grade, the grader must inspect all four faces of the piece.

2303.1.1.2 End-jointed lumber. Approved end-jointed or edge-glued lumber is presumed to be equivalent to solid-sawn lumber of the same species and grade. Section 4.1.6 of the NDS permits the use of end-jointed lumber of the same species and grade. When finger-jointed lumber is marked "STUD USE ONLY" or "VERTICAL USE ONLY," such lumber is limited to use where bending or tension stresses are of short duration. The use of the term *approved* is intended to convey the need for quality control during the production of these glued products, and also to establish the qualification tests for the type of end joint used. Joints are tested for strength and for durability, and adhesive manufacturers test their products for durability. End-joined lumber can be manufactured in different ways. Finger joints or butt joints are typical methods of joinery. Adhesives used in finger-jointed lumber are of two basic types, depending on whether they are to be used for members with long-duration bending loads like floor joists or short-duration bending and tension loads like wall studs. To add elevated-temperature performance requirements for end-jointed lumber adhesives intended for use in fire-resistance-rated assemblies, end-jointed lumber manufactured with adhesives must meet the new requirements and be designated as "Heat Resistant Adhesive" or "HRA" on the grade stamp.

An example of end-jointed lumber is shown in Figure 2303-2, which illustrates a finger-jointed end joint.

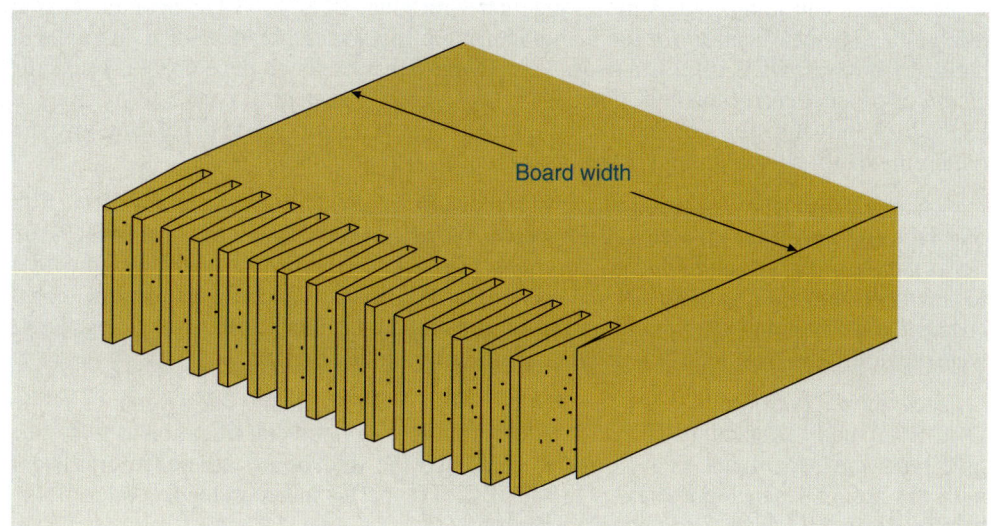

Figure 2303-2
Finger-jointed end joint.

Prefabricated wood I-joists. A new definition for prefabricated wood I-joist was added to the 2006 IBC. Wood "I"-joists are structural members manufactured of sawn or structural composite lumber flanges and wood structural panel webs bonded together with exterior exposure adhesives in the form of an "I" cross-sectional shape. The shear, moment, and stiffness capacities of prefabricated wood I-joists must be established in accordance with ASTM D 5055. This standard also specifies that application details, such as bearing length and web openings, are to be considered in determining the structural capacity. Wood I-joists are manufactured of sawn or structural composite lumber flanges and structural panel webs, and are bonded together with exterior adhesives to form an "I" cross section. Wood I-joists are structural members typically used in floor and roof construction. The standard requires I-joist manufacturers to employ an independent inspection agency to monitor the procedures for quality assurance. The standard specifies that proper installation instructions accompany the product to the job site. The instructions are required to include weather protection, handling requirements and, where required, web reinforcement, connection details, lateral support, bearing details, web hole-cutting limitations, and any special conditions.

2303.1.2

Structural glued laminated timber. Glued laminated timbers are manufactured in accordance with ANSI/AITC A 190.1, which references several other AITC standards. ASTM D 3737 is the standard method for establishing allowable stresses for glued laminated timber. See AITC *Timber Construction Manual*[6] for more information on and construction of glue laminated timber.

2303.1.3

Wood structural panels. Wood structural panels must conform to the Department of Commerce voluntary product standards PS-1 or PS-2. PS-1 is the product standard for Construction and Industrial Plywood, and PS-2 is the Performance Standard for Wood-Based Structural-Use Panels such as oriented strand board (OSB). Plywood, oriented strand board, and composite panels are all considered wood structural panels and are covered under US DOC PS-1 and PS-2. Other than in the definition of wood structural panel in Section 2302, the code does not differentiate between the different types of wood structural

2303.1.4

panels such as composite panels, oriented strand board, or plywood. For example, the shear wall and diaphragm Tables 2306.2(1), 2306.2(2), and 2306.3(1) refer to wood structural panels. Wood structural panels include all-veneer plywood; composite panels consisting of a combination of veneer and wood-based material; and mat-formed panels that contain wood fiber only, such as OSB and waferboard. The primary distinction between OSB and waferboard is that the wood fibers in OSB are generally all oriented in the same direction, whereas the wood fibers in waferboard are oriented randomly in all directions within the plane of the board. Plywood is defined as panels made by cross laminating three or more wood veneers and joining the veneers together with glue. DOC PS-1 has been developed as a guide and specification for the manufacturing of plywood intended for industrial and construction uses. DOC PS-2 is a consensus standard that has been developed as a specification of product performance for various grades of wood structural panels. Provisions in DOC PS-1 and DOC PS-2 define the requirements for structural-grade panels and give the requirements for sheathing and single floor-grade wood structural panels.

A new APA standard for wood siding was added to the 2012 IBC, ANSI/APA PRP 210-8 Standard for Performance-Rated Engineered Wood Siding. Based on APA's PRP-108 *Performance Standards and Policies for Structural-Use Panels*, ANSI/APA PRP-210 provides requirements and test methods for qualification and quality assurance for performance-rated engineered wood siding intended for use in construction applications as exterior siding. There were previously no American National Standards covering these products.

A new term, "performance category," in the 2012 IBC reflects the latest versions of the DOC PS 1 and PS 2 standards, which use terminologies of bond classification to reference glue type and performance categories to reference the thicknesses tolerance consistent with the nominal panel thicknesses in the IBC. The performance category value is the "nominal panel thickness" or "panel thickness."

The user is encouraged to obtain the referenced standards for additional technical information on wood structural panel products. It must be emphasized that the proper fastening of wood structural panels to the supporting structural frame is very important. The nailing schedules contained in the code, referenced standards, and the manufacturer's recommendations must be strictly observed for good performance. The correct nail size and spacing is necessary to achieve the design strength and performance of the wood structural panel system.

Wood structural panels manufactured in accordance with DOC PS 1 and DOC PS 2 are inspected and labeled to certify compliance by an approved agency. Examples are the American Plywood Association and Timber Engineering Company (TECO). The label identifies the grade and span ratings of the product. The inspection agencies maintain a continuous monitoring program designed to produce products that meet or exceed the applicable product standard. A number of tests, including deflection measurements, are required.

Besides the grades cited above, wood structural panels are also classified by exposure type:

1. Exterior—exterior type with a 100-percent waterproof glue line. Only the higher grades of veneers are allowed in exterior grades. Exterior-rated panels are suitable for continuous exposure to weather.

2. Exposure 1—interior type made with waterproof exterior glue. Exposure 1–rated panels are suitable for extended construction exposure. The lower grades of veneers or strands used in the backs and interiors of Exposure 1 panels can affect the glue-line performance and cause delamination/deterioration during continuous exposure to weather.

3. Exposure 2—interior type made with interior glue. Exposure 2–rated panels are not suitable for exposure to weather.

Exposure ratings for APA structural wood panels designated in APA trademarks are based on bond classification. The exposure ratings are Exterior, Exposure 1, Exposure 2, or Interior. Exterior panels have a fully waterproof bond and are designed for applications subject to permanent exposure to the weather or to moisture. Exposure 1 panels have a fully waterproof bond and are designed for applications where long construction delays may be expected prior to providing protection, or where high moisture conditions may be encountered in service. Exposure 1 panels are made with the same exterior adhesives as those used in Exterior panels. However, because other compositional factors may affect bond performance, only Exterior panels should be used for permanent exposure to the weather. Exposure 2 panels (identified as Interior type with intermediate glue under PS 1) are intended for protected construction applications where only moderate delays in providing protection from moisture may be expected. Interior panels or panels that lack further glue-line information in their trademarks are manufactured with interior glue and are intended for interior applications only.

Plywood is manufactured from more than 70 species of wood, which are divided into five groups in accordance with their stiffness and strength characteristics. Construction and industrial-panel grades are generally identified under PS-1 in terms of face veneer grade or by a name indicating an intended use, such as APA-Rated Sturd-I-Floor. The plies may be of any species listed, except for panels designated Structural I and other special-use panels, which use only Group I species. The veneer grade defines the appearance in terms of natural, unrepaired growth characteristics (knots) and the number and size of repairs that may be made during manufacturing. The highest grades are N and A. The lowest grade is D. Grade D veneers may only be used for backs and inner plies of interior-use panels. Panels are also marked as sanded, unsanded, and touchsanded.

OSB is manufactured from several species of wood. PS-2 sets forth performance requirements in terms of strength, stiffness, and durability. OSB is manufactured to meet the strength, stiffness, and durability requirements instead of being manufactured to a prescriptive recipe, as is plywood.

Wood structural sheathing and subflooring panels are classified as:

- Rated Sheathing—Exterior-Rated Sheathing—Exposure 1–Rated Sheathing—Exposure 2
- Structural I-Rated Sheathing—Exterior
- Structural I-Rated Sheathing—Exposure 1

Wood structural panels intended for single-floor construction have limited voids in the inner plies in addition to the solid face veneer to prevent indentation caused by small, concentrated loads.

Single-floor panels intended for use under carpets and resilient flooring are classified as:

- Rated Sturd-I-Floor Exterior
- Rated Sturd-I-Floor Exposure 1
- Rated Sturd-I-Floor Exposure 2

Rated Single Floor Exposure 2 Underlayment panels for use over subflooring are classified as:

- Underlayment Interior
- Underlayment Exposure 1
- C-C Plugged Exterior

Siding is manufactured as panels or as lapped siding and includes:

- 303-OL-MDO Exterior
- 303 Siding Exterior

Note that wood structural panels permanently exposed to weather, such as siding grades, must be exterior type. Panels that are interior type bonded with exterior glue (Exposure 1) are not allowed for siding applications but are allowed on the exposed underside of roof overhangs.

2303.1.5 Fiberboard. Fiberboard is a smooth textured panel made up of natural fibers such as wood or cane. Fiberboard is used primarily as an insulating board and for decorative purposes, but may also be used as wall or roof sheathing under the provisions of this section. Unlike particleboard, the cellulosic components of the fiberboard are broken down to individual fibers and molded to create the bond between the fibers. Other ingredients may be added during processing to provide or improve certain properties such as strength or water resistance, or to achieve specific surface finishes for decorative products. Fiberboard is used in most locations where panels are necessary, including wall sheathing, insulation of walls and roofs, roof decking, doors, and interior finish. Although fiberboard sheathing board may be used for shear walls [see Table 2306.3(2)], fiberboard may not be used for diaphragms. Certification of fiberboard products is performed by an approved agency. The material is generally labeled to indicate an intended use, strength values, and flame resistance where applicable.

Fiberboard may be used as roof or wall insulation, but is not intended for prolonged exposure to sunlight, wind, rain, and snow. Where fiberboard is used as roof insulation, it must be protected with an approved roof covering to prevent water saturation and subsequent delamination, and to avoid decay and destruction of the roofing bond caused by moisture. The section in the IBC pertaining to insulating roof decking was deleted from the 2006 IBC because fiberboard insulating roof decking is no longer manufactured.

Fiberboard is permitted without any fire-resistance treatment in the walls of all types of construction. When used in fire walls and fire separation walls, the fiberboard must be attached directly to a noncombustible base and protected by a tight-fitting, noncombustible veneer that is fastened through the fiberboard to the base. This will prevent the fiberboard from contributing to the spread of fire.

Fiberboard used in building construction must comply with ASTM C 208, *Specification for Cellulosic Fiber Insulating Board*. For several decades, the fiberboard industry supported parallel ASTM and ANSI standards. During the last revision of ASTM C 208, the differences were resolved and the Board of Directors of the American Fiberboard Association voted to discontinue support of the ANSI standard in favor of ASTM C 208. Thus, the previous ANSI/AHA A 194.1 standard was deleted in the 2006 IBC because the standard was withdrawn by ANSI and the fiberboard manufacturers no longer support it.

When used as structural sheathing, fiberboard must be identified by an approved agency. See SDPWS Table 4.3A for nominal unit capacities for wood-based panel shear walls, including structural fiberboard sheathing.

2303.1.6 Hardboard. Hardboard is used as exterior siding and in interior locations for paneling and underlayment. The code references three Composite Panel Association standards for hardboard panels.

2303.1.7 Particleboard. Particleboard is a generic term for construction panels and products manufactured from cellulosic materials, usually wood, in the form of discreet pieces and particles as distinguished from fibers. The particles are combined with synthetic resins and other binders and bonded together under heat and pressure.

Particleboard is used as underlayment, siding, and for shear walls. Particleboard used structurally for siding or shear walls must be stamped (labeled) M-S Exterior or M-2 Exterior. The "M" stands for medium density; the "2" designates the strength grade (grades range from 1 to 3); and "S" designates "special grade." Particleboard designated M-S is medium density and has physical properties between an M-1 and M-2 designation. Both must be made with exterior glue. See SDPWS Table 4.3A for nominal unit capacities for wood-based panel shear walls, including particleboard.

Minimum Standards and Quality

Although similar in characteristics to medium-density Grade 1 particleboard, the particleboard intended for use as floor underlayment is designated "PBU" and has stricter limits on permitted levels of formaldehyde emission than those placed on Grade M particleboard. The particleboard intended for use as floor underlayment is not commonly manufactured with exterior glue, which could emit higher levels of formaldehyde than that permitted for "PSU"-grade floor underlayment by ANSI A 208.1.

Particleboard underlayment is often applied over a structural subfloor to provide a smooth surface for resilient-finish floor coverings or textile floor coverings. The minimum $1/4$-inch thickness is suitable for use over panel-type subflooring. Particleboard underlayment installed over board or deck subflooring that has multiple joints should have a thickness of $3/8$ inch. Joints in the underlayment should not be located over the joints in the subflooring.

All particleboard underlayment with thicknesses of $1/4$ inch through $3/4$ inch should be attached with minimum 6d annular threaded nails spaced 6 inches on center on the edges and 12 inches on center for intermediate supports. See Table 2304.9.1.

Preservative-treated wood. The applicable American Wood Preservers Association (AWPA) standards are cited. Different preservative treatments are used depending on whether the wood is above ground or in contact with the ground. See commentary in Section 2304.11. ICC publishes a book containing all of the AWPA standards referenced in both the 2009 IBC and the IRC. — **2303.1.8**

Identification. All wood required to be preservative treated by Section 2304.11 must be stamped (labeled) with the information listed in the section. There are no exceptions. — **2303.1.8.1**

Moisture content. Preservative treatments used in above-ground locations are waterborne salts. These salts may leach unless the wood is dried below a moisture content of 19 percent (i.e., dry) and covered with a protective material. — **2303.1.8.2**

Structural composite lumber. Structural composite lumber (SCL) is covered in the code because of its increasingly widespread use. Structural properties and strength capacities for SCL are set forth in manufacturers' literature and evaluation reports by ICC. A new definition for structural composite lumber was added to the 2006 IBC for structural members manufactured using wood elements bonded together with exterior adhesives. — **2303.1.9**

LVL is the most widely used SCL product. It is produced by bonding thin wood veneers together. The grain of the veneers is parallel to the long direction of the member. LVL members have enhanced mechanical properties and dimensional stability and offer a broader range in product width, depth, and length than conventional sawn lumber. LVLs are used for a variety of applications such as headers and beams, hip and valley rafters, rim board, scaffold planking, studs, flange material for prefabricated wood I-joists, and truss chords. Some common examples of laminated veneer lumber products are Microllam® LVL Beams manufactured by Weyerhaeuser, VERSALAM® LVL manufactured by Boise Cascade, and LP® SolidStart® LVL manufactured by Louisiana Pacific (LP).

PSL members consist of long veneer strands in parallel formation and bonded together with adhesive to form the finished structural section. PSL members are commonly used for long-span beams, heavily loaded columns, and beam and header applications where high bending strength is needed. A common example of a parallel strand lumber product is Parallam® PSL Beams manufactured by Weyerhaeuser.

Another type of SCL is laminated strand lumber (LSL) and oriented strand lumber (OSL), which are similar to PSL, but are manufactured from flaked wood strands that have a high length-to-thickness ratio. The primary difference between OSL and LSL is the length of strand used to fabricate them. OSL strands are shorter (up to 6 inches) than LSL strands (approximately 12 inches). The strands are combined with adhesive, and are oriented and formed into a large mat or billet and pressed. Although their strength and stiffness properties are somewhat lower than LVL and PSL members, they have good fastener-holding strength and mechanical connector performance. LSL and OSL members are used in a variety of applications, such as beams, headers, studs, rim boards, and millwork. Examples of LSL and OSL products are Ainsworth Engineered Durstrand LSL/OSL and LP® SolidStart® LSL.

Reports prepared by approved agencies or evaluation reports published by the ICC Evaluation Service may be accepted as part of the evidence and data needed by the building official to form the basis of approval of a material or product not specifically covered in the code. Such research reports supplement the resources of the building official and eliminate the need for the official to conduct a detailed analysis on every new product. Note that evaluation reports are approved under the alternative materials, design, and methods of construction provisions of Section 104.11.

2303.1.10 Structural log members. A new Section 2303.1.10 was added to the 2006 IBC to provide structural capacity and grading requirements for logs used as structural members. In the past, the design of log structures could be challenging for both designer and building official because the building code contained no specific provisions that addressed structural capacity and grading requirements for logs used as structural members. All log structures require engineering, and the structural design values were approved under the alternative materials, design, and methods of construction provisions in Section 104.11. This new section provides acceptable methods for establishing structural capacities of logs based on ASTM D3957 *Standard Practices for Establishing Stress Grades for Structural Members Used in Log Buildings* and specifies the requirement for a grading stamp or alternative means of identification. Between publication of the 2006 IBC and the 2009 IBC, the International Code Council (ICC) developed a new standard for log structures known as *Standard on the Design and Construction of Log Structures* (ICC 400-2007). This new standard is now referenced in Section 2301.2 of the 2012 IBC. The goal of the new standard is to provide technical design and performance criteria that will facilitate and promote the design, construction, and installation of safe, reliable structures constructed of log timbers. It is intended to be used by design professionals, manufacturers, constructors, and building and other government officials, and continue as a referenced standard in future building codes. Because ICC 400 gives base values and references AF&PA NDS-05 for design, either ASD or LRFD can be used.

2303.1.11 Round timber poles and piles. This new section referencing the ASTM standards for round timber poles and piles was added to the 2006 IBC to coordinate the requirements with Chapter 6 of the NDS. Chapter 6 of the 2012 NDS has been updated to address changes to the ASTM standards for developing and adjusting round timber pile and round construction pole design values.

2303.2 Fire-retardant-treated wood. Fire-retardant-treated wood (FRTW) is defined as plywood and lumber that has been pressure impregnated with chemicals to improve its flame-spread characteristics beyond that of untreated wood. The principal objective of impregnating wood with fire-retardant chemicals is to produce a chemical reaction at certain temperature ranges. This chemical reaction reduces the release of certain intermediate products that contribute to the flaming of wood, and also results in the increased formation of charcoal and water. Some chemicals are also effective in reducing the oxidation rate for the charcoal residue. Fire-retardant chemicals also reduce the heat release rate of the FRTW when burning over a wide range of temperatures. This section gives provisions for the treatment and use of FRTW. However, the fire-retardant chemicals are generally quite corrosive and corrosion-resistant fasteners may be required with FRTW. See Section 2304.9.5.

The effectiveness of the pressure-impregnated fire-retardant treatment is determined by subjecting the material to tests conducted in accordance with ASTM E 84, with the modification that the test is extended to 20 minutes rather than 15 minutes. Under this procedure, a flame-spread index is established during the standard 10-minute test period. The test is continued for an additional 20 minutes. During this added time period, there must not be any significant flame spread. At no time must the flame spread more than $10\frac{1}{2}$ feet past the centerline of the burners. These criteria have been correlated with large-scale fire tests.

Changes to this section in the 2009 IBC included the addition of three new subsections that clarify the meaning of the phrase *pressure process or other means during manufacture* and provide testing requirements of treatments not impregnated by a pressure process in accordance with Section 2303.2.

Pressure process. Treatment using a pressure process requires a minimum pressure of 50 psi. — **2303.2.1**

Other means during manufacture. This section requires treatment using other means during manufacture to be an integral part of the manufacturing process. — **2303.2.2**

Testing. This section requires added testing of treatments not impregnated by a pressure process. Requiring equivalent performance from all sides of the wood product eliminates any concern over the orientation when it is installed. Only the front and back faces of wood structural panels need to be tested. — **2303.2.3**

Labeling. Each piece of FRTW must be stamped (labeled). The labeling must show the performance of the material, including the 20-minute ASTM E84 test. The labeling must state the strength adjustments, and conformance to the requirements for interior or exterior application. — **2303.2.4**

The FRTW label must be distinct from the grading label to avoid confusion between the two. The grading label gives information about the properties of the wood before it is fire-retardant treated. The FRTW label gives properties of the wood after FRTW treatment. It is imperative that the FRTW label be presented in such a manner that it complements the grading label and does not create confusion over which label takes precedence.

The requirements for labeling fire-retardant-treated lumber and wood structural panels were expanded in the 2003 IBC, giving a list of specific requirements for the labeling.

Strength adjustments. Several factors can significantly affect the physical properties of FRTW. These factors are the pressure treatment and redrying processes used, and the extremes of temperature and humidity that the FRTW will be subjected to once installed. The design values for all FRTW must be adjusted for the effects of the treatment and environmental conditions, such as high temperature in attic installations and humidity. The design adjustment values must be based on an investigation procedure, which includes subjecting the FRTW to similar temperatures and humidities. The procedure must be approved by the building official. The FRTW tested must be identical to that which is produced. The building official reviewing the test procedure should consider the species and grade of the untreated wood and conditioning of wood, such as drying before the fire-retardant-treatment process. A fire-retardant wood treater may choose to have its treatment process evaluated by the ICC Evaluation Services. — **2303.2.5**

The FRTW is required to be labeled with the design adjustment values. These design adjustment values can take the form of factors that are multiplied by the original design values of the untreated wood to determine its allowable stresses, or new allowable stresses that have already been factored down in consideration of the FRTW treatment.

Two subsections were added to the 2003 IBC that prescribe specific strength adjustment requirements for treated wood structural panels and lumber. The effects of treatment and redrying after treatment and exposure to high temperatures and high humidities on the flexural properties of treated plywood and the design properties of treated lumber are required to be determined in accordance with ASTM standards D 5516 and D 5664. The section requires the manufacturer to publish allowable maximum loads and spans for service as floor and roof sheathing and the modification factors for roof framing for its particular treatment process. The section references the ASTM standard D 6841 to be used to evaluate the ASTM D 5664 test data.

Wood structural panels. This section references the test standard developed to evaluate the flexural properties of fire-retardant-treated plywood that is exposed to high temperatures. Note that while the section title refers to wood structural panels, the referenced standard is limited to softwood plywood. Therefore, judgment is required in — **2303.2.5.1**

determining the effects of elevated temperature and humidity on other types of wood structural panels. The product manufacturer is required to publish the allowable maximum loads and spans for floor and roof sheathing for the particular type of treatment.

2303.2.5.2 Lumber. This section references the test standard developed to determine the necessary adjustments to design values for lumber that has been fire-retardant treated including the effects of elevated temperatures. The lumber manufacturer is required to publish the modification factors to design values.

2303.2.6 Exposure to weather, damp, or wet locations. Some fire-retardant treatments are soluble when exposed to the weather or used under high-humidity conditions. When an FRTW product is to be exposed to weather conditions, it must be further tested in accordance with ASTM D 2898. The material is then subjected to the ASTM weathering test and retested after drying. There must not be any significant differences in the performance recorded before and after the weathering test.

2303.2.7 Interior applications. When an FRTW product is intended for use under high-humidity conditions, it must be further tested in accordance with ASTM D 3201. The material must demonstrate that when tested at 92-percent relative humidity, the moisture content of the FRTW does not increase to more than 28 percent. The label must show the test results.

2303.2.8 Moisture content. FRTW contains water-borne salts that are subject to leaching. The FRTW must be dried to the specified moisture contents after treating to minimize leaching. In addition, FRTW chemicals are quite corrosive to metal fasteners. Where the moisture content of the treated wood is too high, the corrosivity of the treated wood is even higher and contributes to greater corrosion of metal fasteners.

For wood that is kiln dried after treatment (KDAT), the kiln temperatures cannot exceed that used to dry the lumber and plywood that was submitted for the tests required by Section 2303.2.2.1 for plywood and Section 2303.2.2.2 for lumber.

2303.2.9 Type I and II construction applications. Use of FRTW in Type I or II construction is limited to nonload-bearing partitions and exterior walls.

2303.3 Hardwood and plywood. Hardwood plywood and decorative plywood are not used for structural purposes. This section references the American National Standard for Hardwood and Decorative Plywood.

2303.4 Trusses. Metal-plate-connected trusses are typically constructed out of nominal dimension lumber with the metal-plate connectors placed on either the narrow or wide dimension of the lumber (4-inch by 2-inch lumber for floor trusses versus 2-inch by 4-inch lumber for roof trusses). The NDS specifies the allowable design stresses for lumber, whereas the Truss Plate Institute (TPI) *National Design Standard for Metal-Plate-Connected Wood Trusses* specifies how the allowable metal-plate design values and maximum stresses in the truss elements are to be determined.

This section was revised in the 2006 IBC to more clearly define the requirements pertaining to metal-plate-connected wood trusses to achieve consistency with current design practice. Additional editorial and organizational changes were made in the 2009 IBC by the Wood Truss Council of America and the Structural Building Components to improve the organization and language of the provisions. The section clarifies the requirements pertaining to metal-plate-connected wood trusses to be consistent with the current industry practice and eliminates confusion regarding truss submittal requirements. The provisions include general design requirements; specific and detailed requirements for truss design drawings; requirements for truss member permanent bracing; a definition of the truss designer and truss design drawing seal and signature requirements; provisions for truss placement diagrams; specific requirements for the requirements for the truss submittal package; and specific truss anchorage requirements. The truss submittal package is part of the construction documents, which are part of the submittal documents required by Section 107.1. The term *construction documents* is defined in Section 202.

Minimum Standards and Quality

Adequate bracing for trusses is critical. Lateral bracing requirements (e.g., brace points, bracing size, or strength and stiffness) should be specified by the truss designer. Methods of permanent bracing are described in Section 2304.1.2. Temporary bracing should be left in place until permanent bracing is installed. All lateral bracing must be installed as assumed in the truss design so that the truss will have the same structural capacity for which it was designed. In any case, the individual truss member continuous lateral bracing locations are to be shown on the truss design drawings. See Section 2303.4.1.1, Item 14.

Permanent bracing must be installed in compliance with the truss industry's permanent bracing standard details that follow sound engineering practice. These details are usually provided by the component manufacturers to the building design professional as the projects are being designed. The Building Component Safety Information (BCSI 1-08) *Guide to Good Practice for the Handling, Installing & Bracing of Metal-Plate-Connected Wood Trusses* is a booklet produced by the Truss Plate Institute (TPI) and the Wood Truss Council of America (WTCA). It is the truss industry's new and improved guide for job-site safety and truss performance and replaces the HIB-91 booklet from TPI. In 2008, the WTCA changed its name to Structural Building Components Association (SBCA). The BCSI 1-08 publication is available from the SBCA website at www.sbcindustry.com.

The truss design drawings and specifications must be prepared by a registered design professional and must be provided to and approved by the building official prior to installation in the structure. In general, the regulations that govern registered design professionals are professional practice acts regulated by state law.

Note that the items listed in the code are the minimum requirements. The intent of adding the terms *truss design drawings* and *truss placement diagram* in the 2006 IBC was to minimize confusion that exists in the construction industry between a variety of terms that may mean the same thing, such as *construction documents*, *shop drawings*, and so on. The term *truss placement diagram* is used by the truss industry and is very specific. These terms are intended to provide better clarity where truss submittals are concerned.

The truss design drawings are required to show permanent bracing, and the truss designer generally is the most knowledgeable concerning the required strength, stiffness, and location of the bracing necessary to prevent buckling of the truss members. Bracing is necessary to resist buckling of the compression webs and chord under maximum gravity loads, as well as the uplift condition of dead load plus wind. Bracing may be necessary on the bottom chord at the first interior panel point to resist the wind uplift. The truss designer should specify either the strength and stiffness of braces, or the member size and grade of the braces (e.g., 2 by 4 DF #2) at the specified locations.

The truss designer should furnish complete calculations substantiating the size of all members and connector plate sizes. The truss calculations should indicate the combined stress index for members subjected to combined stresses from bending and axial compression and tension. The combined stress index should be less than one. The calculations are generally performed with a computer program; therefore, documentation may be required by the building official that substantiates the basis of the computer program used.

The anchorage requirements in Section 2303.4.4 was added in the 2006 IBC to clarify that the transfer of all design loads through the building structure and the connections of the trusses to the supporting structure to resist code-prescribed loads is the responsibility of the registered design professional of the building.

In the 2006 IBC, several sections pertaining to trusses were relocated from Section 2308 to 2303.4 for clarification and consolidation, most notably Section 2303.4.5 pertaining to alterations to trusses. Truss members cannot be altered without concurrence by a registered design professional. In most cases, altering pre-engineered trusses requires additional engineering. Additional loading, such as installation of new mechanical equipment, requires engineering to verify that the trusses can adequately support additional loads.

There were changes to the inspection and quality-assurance requirements for metal-plate-connected wood trusses in the 2009 IBC. Section 2303.4.6 requires metal-plate-connected wood trusses to meet the quality assurance requirements of the 2007 TPI 1 standard and be inspected at the job site under Section 110.4, which gives the building official the authority to accept reports from approved inspection agencies. Trusses not manufactured under TPI 1 or in accordance with a referenced standard that provides quality control under the supervision of a third-party quality control agency must comply with Sections 1704.2.5 and 1705.5, which require an approved fabricator or special inspection during fabrication.

The 2009 and 2012 IBCs reference ANSI/TPI 1–2007, *National Design Standard for Metal-Plate-Connected Wood Trusses*.

2303.4.1 Design. Wood trusses must be designed in accordance with accepted engineering practice, which is generally governed by state laws that regulate professional engineering. Truss members are permitted to be joined by any acceptable method.

2303.4.1.1 Truss design drawings. The section prescribes what information is to be provided on truss drawings. Truss design drawings are to be approved by the building official prior to installation, and the design drawings are to be on the job site. Note that where required by the building official or state law, truss deign drawings do need to be stamped and signed by the truss engineer.

2303.4.1.2 Permanent individual truss member restraint. Where permanent bracing or restraint is required it should be indicated on the truss design drawings and meet the methods prescribed in the section.

2303.4.1.3 Trusses spanning 60 feet or greater. Trusses spanning 60 feet or more require an RDP to design the temporary and permanent bracing. Note that a similar requirement in Section 2211.3.3 applies to cold-formed steel trusses spanning 60 feet or more.

2303.4.1.4 Truss designer. In general, the truss designer is a specialty engineer that works for the truss manufacturer and is not the same as the RDP responsible for the overall building design. Section 2303.4.1.4.1 specifies what documents need to be stamped and signed by the truss engineer.

2303.4.2 Truss placement diagram. The truss placement diagram is used in the field to facilitate proper installation of the trusses. Note that unlike truss design drawings, truss placement diagrams are not required to be stamped and signed by the truss engineer.

2303.4.3 Truss submittal package. The section describes what documents are required to be included in the truss submittal package.

2303.4.4 Anchorage. The requirement in this section is an essential ingredient in providing a complete and continuous load path as required by Section 1604.4. The design of the anchorage required to transfer loads (gravity and lateral) from each truss to the supporting structure is the responsibility of the RDP. This is most critical for lateral loads where forces must be transferred from the roof diaphragm to the collectors and supporting shear walls, but it is also important for gravity loads. To illustrate a simple case, if the trusses are spaced at 24 inches on center and the supporting studs are spaced at 16 inches on center, some trusses will occur between studs. Depending on the roof load, such as where high snow loads occur, the double top plate may not be adequate to span between studs and support the truss reaction. In this case, the simple solution is for the RDP to require a stud under each truss. Another example is multi-ply girder trusses with very high reactions. In this case, the simple solution is for the RDP to require a stud under each ply in the girder truss.

2303.4.5 Alterations to trusses. Obviously truss members should not be cut, notched, drilled, spliced, or altered in any way without approval of the engineer. Any addition of loads from HVAC equipment, piping, additional roofing, and so on should be reviewed and approved by the engineer to ensure that the trusses are capable of supporting additional loading.

TPI 1 specifications. The design, manufacture, and quality assurance of metal-plate-connected wood trusses are to be in accordance with the TPI 1 standard, *National Design Standards for Metal-Plate-Connected Wood Truss Construction*. The 2012 references the 2007 edition of TPI 1, which is available from the Truss Plate Institute at www.tpinst.org.

2303.4.6

Truss quality assurance. The TPI 1 standard includes quality-assurance procedures for truss manufacturers. Trusses not manufactured in accordance with the TPI 1 standard or another approved standard that provides quality control under the supervision of a third-party quality control agency require special inspection unless the manufacturer is an approved fabricator in accordance with Section 1704.2.5.

2303.4.7

Test standard for joist hangers and connectors. Section 1711.1 sets forth the test and acceptance criteria for joist hangers. Note that Section 1711.1.2.1 stipulates that only the design values based on the design limit states for the fasteners or the wood member may be increased for duration of load. If the hanger load is governed by the strength- or deflection-limit states of the steel, the design value may not be increased for duration of load. Evaluation reports and manufacturer's literature must provide the limit state data in order for the structure designer to comply with this requirement.

2303.5

Nails and staples. This section references ASTM F1667 for driven fasteners such as nails, spikes, and staples. Bending yield strength requirements for nails are provided. The bending yield strength requirements are those used in the NDS lateral strength tables, the IBC and IRC fastener schedules, as well as model code evaluation reports. These strengths are standardized within the nail industry for engineered fasteners and are set forth in ASTM F 1667 *Specification for Driven Fasteners: Nails, Spikes and Staples*. A code change to the 2006 IBC added the shank length and diameter in parentheses in the tables for the various types of nails used in wood connections with the intent of emphasizing the nail size, not just the penny weight designation. See a more complete discussion of fasteners for wood construction under Section 2304.9.1.

2303.6

Shrinkage. Because wood shrinkage is a concern, two new sections were added to the 2003 IBC pertaining to shrinkage. Section 2303.7 requires the designer to consider the effects of cross-grain dimensional changes (shrinkage) in the vertical direction that can occur in lumber that was fabricated green. See Sections 2.7 and 2.11 of PS 20 for definitions of dry and green lumber: Dry lumber is lumber of less than nominal 5-inch thickness that has been seasoned or dried to a maximum moisture content of 19 percent. Green lumber is lumber of less than nominal 5-inch thickness that has a moisture content in excess of 19 percent. For lumber of nominal 5-inch or greater thickness (timbers), green is defined in accordance with the provisions of the applicable lumber grading rules. See also Section 2304.3.3 for specific requirements related to shrinkage.

2303.7

Section 2304 *General Construction Requirements*

General. The requirements of Section 2304 are general and apply to all design methods, ASD, LRFD, and the conventional construction provisions.

2304.1

Size of structural members. Net dimensions of structural lumber are set forth in NDS. Refer to Tables 1A, 1B, 1C, and 1D of the NDS Supplement for nominal and dressed lumber sizes and section properties of sawn lumber and glued laminated timber members.

2304.2

Wall framing. Interior and exterior walls must be framed in accordance with conventional construction provisions unless a specific design is provided. Even most engineered

2304.3

wood-frame buildings have portions, such as walls, that are framed in accordance with the prescriptive conventional construction provisions of the code.

2304.3.1 Bottom plates. Sill plates must be at least nominal 2x even if design calculations show that a smaller size is acceptable. This requirement applies both to engineered (designed) structures and those constructed in accordance with the prescriptive conventional construction provisions.

2304.3.2 Framing over openings. Windows, doors, air-conditioning units, and other service equipment require openings to be provided in wood-stud walls and partitions. Loads imposed above these openings must be transferred by a structural element above the opening to supports on both sides of the opening and then to a load-bearing wall or partition. In most wood-frame structures, these structural elements are composed of two pieces of 2-inch dimensional lumber plus a spacer or plate, or solid 4x or 6x members. These elements must be fastened in accordance with Table 2304.9.1. Based on the span and the loading, some openings may require trusses, laminated veneer lumber, glued-laminated members, or even steel beams. Headers may be engineered or selected from Tables 2308.9.5 and 2308.9.6. Although these tables only have multiple 2x members, 4x and 6x headers are more common in the western regions of the United States. Header tables are also published in a number of other technical documents. For example, Western Wood Products Association (WWPA) publishes Tech Note No. 6, Design Load Tables for Solid Sawn Lumber Beams and Headers—Single 4x and 6x. All other headers not designed by span tables must be engineered in accordance with Section 2301.2. In all cases, headers and their supports must be adequate to support the imposed loads (see Figure 2304-1).

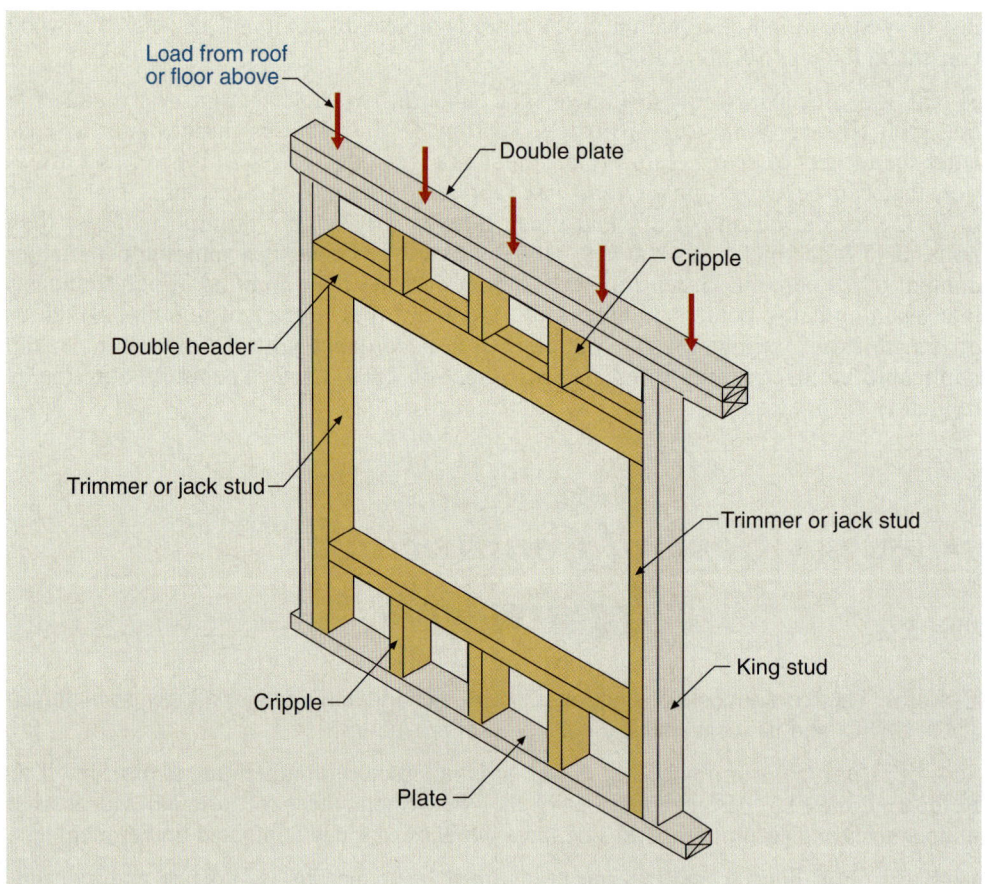

**Figure 2304-1
Headers over wall openings.**

General Construction Requirements

2304.3.3 **Shrinkage.** This new section was added to the 2003 IBC and imposes requirements for considering wood shrinkage. Wood-framed walls and bearing partitions cannot support more than two floors and a roof unless an analysis shows that there will be no adverse effects on the structure that are due to excessive shrinkage or differential movement caused by shrinkage. The provisions in this section were originally in the UBC to take into account the cumulative effects of shrinkage for horizontal wood-frame members (joists and stud wall plates) and the possible adverse effects on plumbing and mechanical systems. There have been cases reported in multistory wood-frame projects where shrinkage was not considered in the building design, and shrinkage of the framing caused plumbing breaks in the stud walls. Although some designers are familiar with the need for shrinkage analysis in multistory wood-frame buildings, many designers may not be aware of the problem. Four- and five-story wood-frame construction is growing in popularity, and in some areas of the country designers may not be aware of problems associated with multistory wood-frame buildings.

In addition, new Section 2303.7 applies to areas where green lumber is used in construction, and in the past has been considered a West Coast issue. The basis of this requirement is that green lumber shrinks more than dry lumber; therefore, shrinkage should be considered in buildings more than two stories in height. See discussion under Section 2303.7 for discussion of dry and green lumber.

2304.4 **Floor and roof framing.** The framing of floors, joists, and roof rafters must be in accordance with the conventional construction provisions or be engineered. Even some engineered wood-frame buildings have portions, such as floor systems, that are framed in accordance with the prescriptive conventional construction provisions of the code.

2304.5 **Framing around flues and chimneys.** Specific wording in the code requires wood framing to be a minimum of 2 inches clear from flues, chimneys, and fireplaces, and 6 inches clear from flue openings. This air space is required because the resistance of wood to ignition is lowered when it is subjected to low levels of heat for extended periods of time. The separations specified in this section and in the *International Mechanical Code* (IMC) are intended to prevent the possibility of the wood being subjected to low heat for extended periods of time. Note that 2 inches of masonry will not provide the same level of thermal resistance as 2 inches of air space. Therefore, adding 2 inches to the required thickness of a fireplace or chimney wall will *not* serve as an acceptable alternative to the 2-inch clear air space requirement. The openings between the wood and the masonry must be adequately fire blocked in accordance with Section 718.

2304.6 **Wall sheathing.** Wall sheathing is intended to:

1. Protect the interior of the building from the weather. As required by Section 1405.2, exterior wall coverings must provide weather protection for the structure. Although not explicitly required by the code, where wood structural panel products are used for exterior siding, such as plywood, particleboard, or hardboard, joints should occur over framing members and be protected with continuous wood battens, tongue-and-groove joints, caulking, flashing, vertical or horizontal shiplaps, or other methods approved by the building official that will make the joints resistant to water penetration. Alternatively, these panels may be applied over lumber and wood structural panel sheathing, in which case the double layers provide for an overlapping of joints and prevent direct penetration of water.

2. Be of sufficient strength to span between the studs or other structural members. The strengths or thicknesses of most of the exterior wall-covering materials listed in Table 2304.6 are based on a stud spacing of 16 inches on center. However, there are a few cases where the thickness is based on a 24-inch spacing.

3. Have sufficient durability to withstand, for the life of the building, the weathering effects of the elements to which they are exposed. Thus, where wood structural panels are used, the code requires that they be rated as exterior type. Other materials must be of a weather-resistant type of material, such as cedar shingles, or the material must be finished with a weather-resistant finish, such as exterior paint.

Provisions for other wall coverings are prescribed in Section 1405, and exterior plaster provisions are set forth in Section 2512.

2304.6.1 Wood structural panel sheathing. Wood structural panel sheathing used as an exterior wall finish must have an Exterior type durability designation. For example, T1-11 siding is an APA 303 siding designed to be used as exterior finish exposed to outdoor conditions and is identified as Exterior. Wood structural panels used under an exterior finish such as roof or wall covering must have an Exterior or Exposure 1 designation. Exposure 1 panels are allowed to be exposed on the underside at eaves. See Section 2303.1.4.

Recent high-wind events including Hurricane Katrina and several thunderstorms associated with tornados have shown that failure of wall sheathing in winds as low as 60 mph has led to significant damage resulting from breaching of the wall envelope. A code change in the 2009 IBC added Table 2304.6.1, giving maximum wind speed for exterior wood structural panel wall sheathing or siding that must resist wind loads perpendicular to the wall. The table gives maximum wind speeds (Section 1609.3) based on the exposure category of the building site (Section 1609.4). For a given wind speed and exposure category, the table gives the minimum nail size, wood structural panel span rating, panel thickness, stud spacing, and nailing schedule. The limitations of the table are: the building must be enclosed (ASCE 7 Section 26.2); the mean roof height must be not greater than 30 feet (ASCE 7 Section 26.2), and topographic factor must be equal to 1.0 (ASCE 7 Section 26.8). Buildings that do not meet these limitations must have the exterior sheathing designed to resist component and cladding wind pressures in accordance with ASCE 7. The table was developed by comparing the component and cladding wind pressures given in ASCE 7-05 Figure 6-3 with the wood structural panel capacities based on U.S. DOC PS 2 standard, engineering calculations, and the APA Panel Design Specification referenced in Section 2306.1. Nail head pull through and withdrawal capacity was also considered in addition to panel stiffness and bending strength. Panel-fastener capacity was based on tributary area to a single critical nail.

2304.6.2 Interior paneling. Interior paneling must be installed in accordance with Table 2304.9.1 and comply with DOC PS-1 or PS-2. Wood structural panels installed in accordance with this section are considered interior wall finishes and cannot be used to resist lateral forces. Prefinished hardboard paneling must conform to the requirements of AHA A135.5, and hardwood plywood must conform to HPVA HP-1.

2304.7 Floor and roof sheathing. Structural floor and roof sheathing meeting the requirements given in the tables in Section 2307 are deemed to comply with the code.

2304.7.1 Structural floor sheathing. Structural floor sheathing serves three purposes:

1. Provides support for and distributes superimposed gravity loads to the floor joists or trusses.
2. Provides lateral support for the top of the floor joists.
3. Serves as the shear-resisting element when the floor acts as a diaphragm to distribute lateral wind or seismic loads to the shear walls or foundation.

The sheathing may either be engineered to meet the loading requirements, or be selected from the tables in the code. The tables are prescriptive and therefore deemed to comply with the minimum code requirements.

The maximum span for wood structural panels is limited by both strength and deflection. Table 2304.7(3) limits deflection to $L/180$ under the tabulated uniform dead and live load, or $L/240$ under the tabulated uniform live load only. Deflections for combination subfloor/underlayment grades in Table 2304.7(4) are based on a deflection of $L/360$ caused by a total uniform load of 100 psf except for $1^1/_8$-inch panels on joists spaced at 48 inches on center, which are rated at a deflection of $L/360$ at a uniform load of 65 psf. When using the tables, *all* the conditions set forth in the table must be met regarding loads, type, grade, species group, plies, blocking, direction of span, and other requirements. Sheathing systems that do not meet *all* the conditions must be engineered.

Wood structural panels in Table 2304.7(3) are designated by the panel span rating (also called the panel identification index), instead of thickness. The rating consists of two numbers such as $^{32}/_{16}$. The first number is the allowable span for use as roof sheathing, and the second is the allowable span for use as floor sheathing. Panels intended only for single floor grades are designated by one number, for example, 16, but the panels may be used for roof sheathing with the spans listed in the table. The single floor grades have higher-grade faces than Structural I panels and may be preferable for aesthetic reasons as roof sheathing where visible from the underside.

Several thicknesses of structural panel are included under a single panel span rating. This is because different grades of interior and face veneers or strands may be used in different combinations. For example, a five-ply plywood panel will have a higher load rating than a four-ply panel with the same thickness. Conversely, the five-ply panel may be thinner than the four-ply panel, yet have the same panel span rating. For example, a $^{15}/_{32}$-inch-thick four-ply panel will support the same total load as a $^{7}/_{16}$-inch-thick five-ply panel.

The $^{5}/_{16}$-inch wood structural panels were deleted from Table 2304.7(3) in the 2009 IBC because $^{5}/_{16}$-inch thickness is a very small fraction of the panels currently produced. Although $^{5}/_{16}$-inch wood structural panel has been the minimum panel thickness specified for many applications over the years, the building industry has shifted away from producing them due to manufacturing efficiencies and marketplace demand. The $^{3}/_{8}$-inch wood structural panel is currently the minimum thickness in the table.

Structural roof sheathing. Roof sheathing must be designed or conform to the provisions of Table 2304.7(1), 2304.7(2), 2304.7(3), or 2304.7(5). Wood structural panel roof sheathing must be bonded by exterior glue. See the discussion under Section 2304.7.1, Structural floor sheathing. **2304.7.2**

Lumber decking. Section 2304.8 was revised to incorporate the lumber deck design provisions of American Institute of Timber Construction (AITC) 112-93 *Standard for Tongue-and-Groove Heavy Timber Decking* directly into the body of the code. The reference to AITC 112-93 in Section 2306.1 was deleted from the 2006 IBC because AITC no longer maintains the standard. The revisions to Section 2304.8 in the 2006 IBC along with the addition of new Section 2306.1.4 incorporated the pertinent provisions of AITC 112-93 directly into the body of the code, eliminating the need to reference AITC 112-93. The title of Section 2304.8 was revised to indicate that the provisions cover all decking, including mechanically laminated and solid sawn decking. The capacity of lumber decking is arranged according to the various layup patterns described in Section 2304.8.2. Section 2306.1.4 gives the design capacity of lumber decking for flexure and deflection according to the formulas given in Table 2306.1.4. In addition to the technical changes that were made to the section in the 2006 IBC, several editorial changes were made to improve the language in the 2009 IBC. **2304.8**

General. The general provisions specify the requirements for square ends, beveled ends, and the orientation of tongue-and-groove decking on roofs. Section 2304.8.2 provides specific requirements for the various layup patterns described in the subsections. Sections 2304.8.3 through 2304.8.5 cover specific requirements for mechanically laminated decking, 2-inch sawn tongue-and-groove decking, and 3- and 4-inch sawn tongue-and-groove decking, respectively. **2304.8.1**

Connections and fasteners. Connectors and fasteners must comply with the applicable provisions of Sections 2304.9.1 through 2304.9.7. The section covers fasteners used to connect wood members, sheathing fasteners, fasteners for joist hangers, and framing hardware, and specific requirements for fasteners used in treated wood. Other fastening methods such as clips, staples, and glue are allowed where approved. **2304.9**

Fasteners requirements. Table 2304.9.1 in the IBC is comparable to the nailing tables in the legacy model codes. Power-driven fasteners, along with the typical sizes used, and **2304.9.1**

staples are included in the table. The size designations in the table are common to all fastener manufacturers. The code intends that Table 2304.9.1 provide minimum requirements for the number and size of fasteners used to connect wood-framing members. This table accommodates the builder of non-designed (conventional) construction, but also provides minimum fastening requirements for designed construction. Details such as end and edge distances and nail penetration are required to be in accordance with the applicable provisions of the NDS. Where required, corrosion-resistant fasteners must be either hot-dipped zinc-coated galvanized steel, stainless steel, silicon bronze, or copper.

A code change to the 2006 IBC added the shank length and diameter in parentheses to the tables for the various types of nails used in wood connections. It has been reported that improper nail sizes have been used in wood-frame building construction because the pennyweight system of specifying nail sizes is not universally understood. Code users sometimes focus on pennyweight (8d - 8 penny, 16d - 16 penny, etc.) and do not pay sufficient attention to the specific type of nail such as common, box, cooler, sinker, finish, and so on. A typical example is substitution of box nails for common nails of the same pennyweight. The specific type of nail is critical because there can be significant differences in strength properties of connections nailed with nails of the same pennyweight but of different nail type. Specifying the nominal dimensions of nails in the fastening tables may help avoid confusion and reduce misapplications in nailed connections. The code change proponent expected some reluctance on the part of some in the building construction community to completely abandon the pennyweight system of designating nail sizes, so the code continues to maintain the pennyweight designations. Because nominal dimensions are not as subject to misinterpretation, the shank length and diameter in parentheses will help prevent confusion and misapplication of the various types of nails used in wood connections.

2304.9.2 Sheathing fasteners. Fasteners should be driven flush with the surface of the sheathing, but not overdriven. Overdriving of fasteners can significantly reduce the shear capacity and ductility of the diaphragm or shear wall. For three-ply material, the strength reduction is significant if the fastener is overdriven through one ply.

If no more than 20 percent of the fasteners around the perimeter of the panel are overdriven by $1/8$ inch or less, then no reduction in shear capacity need be considered. If more than 20 percent of the fasteners around the perimeter are overdriven by any amount, or if any fasteners are overdriven by more than $1/8$ inch, then additional fasteners must be driven to maintain the desired shear capacity, provided that the additional fasteners will not split the substrate. For every two fasteners overdriven, one additional fastener should be driven. Panels with more than 20 percent of the fasteners overdriven greater than $1/8$ inch should be replaced.

Also, if the actual panel thickness is greater than the design panel thickness needed to resist the design shear, the panel shear capacity may be adequate without the driving of additional fasteners. For example, if the design required a $15/32$-inch-thick panel, but a $19/32$-inch-thick panel with all fasteners overdriven by $1/8$ inch is used for the sheathing, the panel is adequate because the net thickness is $15/32$ inch.

2304.9.3 Joist hangers and framing anchors. Most joist hangers and other framing hardware have reports issued by building product evaluation services. Such ICC Evaluation Services and detailed reports are issued for the various products. There are acceptance criteria for the performance of the hardware in terms of the strength limit states of the metal, wood, and fasteners, as well as deflection limit states. For example, ICC AC 13, Acceptance Criteria for Joist Hangers and Similar Devices, is used to evaluate devices used to support or attach wood members, such as joists, rafters, purlins, beams, girders, plates, posts, studs, and headers to wood, metal, concrete, or masonry where the attachment is by means of mechanical fastenings such as nails, spikes, lag screws, wood screws, bolts, and so on. The term "device" refers to one or more pieces or units so arranged as to transfer load vertically or laterally, within safe limits, from the end of a supported member (such as a joist) to a supporting member (such as a header, beam, or girder).

General Construction Requirements

Test requirements for joist hangers are found in Section 1711.1. Only the design values based on the strength limit states for the fasteners or wood member (Item 4 or 5 in Section 1711.1.2) are allowed to be increased for duration of load per the NDS. No increase is allowed for loads limited by steel strength or deflection. ICC Evaluation Service reports and manufacturer's specifications provide the controlling limit strength of the connector in order for the designer to meet this requirement.

Other fasteners. Fasteners not specifically cited in the code may be used but are subject to building official approval. See Section 1703.4 for requirements related to research reports and acceptance of evaluation reports. **2304.9.4**

Fasteners in preservative-treated and fire-retardant-treated wood. The water-borne salts in preservative-treated and fire-retardant-treated wood are corrosive. Fasteners must be corrosion resistant when used with these materials. Corrosion-resistant fasteners are made of type 304 or 316 stainless steel, silicon bronze, copper, or steel that has been hot-dipped or mechanically deposited zinc-coated galvanized with a zinc coating of not less than 1.0 ounce per square foot. **2304.9.5**

The 2006 IBC introduced an alternative method for mechanically galvanizing, which is preferable to hot dipping for some types of fasteners, as indicated in the exception to this section. Class 55 was added to provide an equivalent amount of zinc as would be provided by the hot-dip process in accordance with ASTM A 153. According to the American Galvanizers Association, mechanically plating to a thickness of 55 microns provides an equivalent coating to 1 ounce per square foot of hot-dipped galvanized zinc, which is what is provided for fasteners by ASTM A 153. Class 55 provides 55 microns of thickness. The section references ASTM A 53 or B 695 for coating weight requirements.

Electro-galvanized steel fasteners do not qualify as corrosion resistant; the zinc coating typically is about 0.1 ounce per square foot. Electro-galvanized nails are suitable only for occasionally wet locations such as for nailing composition shingles.

In the 2009 IBC, the section was subdivided into three categories of fasteners in treated wood: fasteners in preservative-treated wood, fasteners in fire-retardant-treated wood in exterior or damp locations, and fasteners in fire-retardant-treated wood in interior locations.

A significant change in the 2009 IBC is the exception for fasteners used in preservative-treated wood in an interior dry environment. There is no documented evidence of any detrimental fastener corrosion when plain steel fasteners are used in SBX/DOT (sodium borate) or zinc borate preservative-treated wood in interior, dry environments. In this case, the exception permits plain carbon steel fasteners to be used.

In the 2012 IBC, the phrase "including nuts and washers" was added to each of the sections to clarify that the corrosion-resistance requirements apply to the fastener as well as nuts and washers.

Load path. The code requires the load path to be continuous from the point of origin to the resisting element, which generally means from the roof to the foundation. A continuous load path for both gravity and lateral loads is necessary for adequate performance of the structure in response to superimposed vertical and lateral loads. This is especially critical in the case of high wind and load effects from earthquake ground motion. For example, visualize what happens to the wind suction load on a low-slope roof. The upward force that is not offset by the dead load of the roof elements must be resisted by dead load elsewhere. Similarly, where the structure is subjected to lateral load effects from earthquake ground motion, the inertial forces from the floor and roof diaphragms must be effectively transferred to the vertical lateral-force-resisting elements, for example, wood-framed shear walls and then to the supporting foundation. A positive, properly detailed continuous load path is essential to ensure the transfer of all gravity and lateral loads from the roof and floors through the structural system down to the foundation and supporting soil. **2304.9.6**

2304.9.7 Framing requirements. Columns must be provided with full end bearing to transfer their loads or connections must be designed to resist the full compressive load. Connections must also be able to resist lateral and net uplift loads.

2304.10 Heavy timber construction. The provisions contain general provisions for column continuity, transfer of loads from beams to columns, other connection criteria, and requirements for structural anchorage and continuity. Minimum element size requirements for beams and columns used in Type IV construction are found in Section 602.4.

2304.10.1 Columns. Columns in heavy timber construction must be continuous throughout all stories by being connected by concrete or metal caps, base plates, timber splice plates, or other approved methods.

2304.10.1.1 Column connections. Girders and beams in heavy timber construction are required to be fitted around columns, and adjoining ends must be adequately tied to each other to transfer horizontal loads across the joints.

2304.10.2 Floor framing. Wall pockets or hangers are required where wood beams, girders, or trusses in heavy timber construction are supported by masonry or concrete walls. Beams supporting floors are required to bear on girders or be supported by ledgers, blocks, or hangers.

2304.10.3 Roof framing. Roof girders and alternative roof beams in heavy timber construction are to be anchored to supporting members with steel or iron bolts designed to resist vertical uplift of the roof.

2304.10.4 Floor decks. Floor decks and floor covering in heavy timber construction cannot extend closer than $1/2$ inch to walls with the space covered by a molding fastened to the wall.

2304.10.5 Roof decks. Where supported by a wall, roof decks in heavy timber construction must be anchored to walls by steel or iron bolts to resist uplift forces. This section in the 2000 IBC requires roof decks supported by a wall to be anchored to walls at intervals not exceeding 20 feet, but did not specify the type or purpose of the anchorage. Code changes to the 2003 IBC added that the anchors must be capable of resisting uplift forces determined in accordance with Chapter 16, and the anchors must be steel or iron bolts of sufficient strength to resist vertical uplift of the roof.

2304.11 Protection against decay and termites. The provisions from the legacy model codes were reorganized in the IBC for clarity. The provisions are grouped into the following areas:

1. Wood used above ground
2. Laminated timbers exposed to weather
3. Wood with ground or fresh water contact
4. Supports for appurtenances
5. Termite protection
6. Retaining walls and cribs

The provisions of this section are intended to protect against decay and termite infestation. The provisions are based on the extensive material on biodeterioration of wood in the *Wood Handbook*.

2304.11.2 Wood used above ground. Wood used above ground, if preservative treated, is usually treated with a water-borne preservative such as ammoniacal copper arsenate (ACA) or chromated copper arsenate (CCA) in accordance with AWPA U1 (Commodity Specifications A or F). The retention rates are lower than those required for ground contact.

2304.11.2.1 Joists, girders, and subfloor. There must be 18 inches clearance to joists and 12 inches clearance to wood girders from exposed ground if they are not of naturally durable or preservative-treated wood. See Figure 2304-2.

General Construction Requirements 23

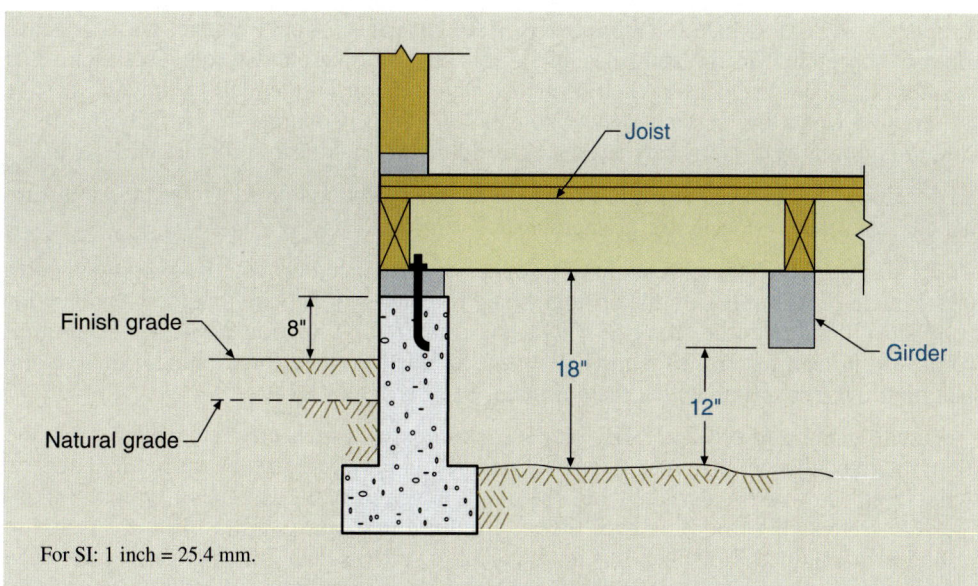

Figure 2304-2
Under-floor clearance.

Wood supported by exterior foundation walls. Framing, including sheathing (not siding), must have 8 inches of clearance from exposed earth if it is not naturally durable or preservative treated. See Figure 2304-3. This section was retitled in the 2006 IBC to clarify that the section applies to wood framing resting on exterior foundation walls. Note that Section 2304.11.2.6 allows a 6-inch clearance for wood siding.

2304.11.2.2

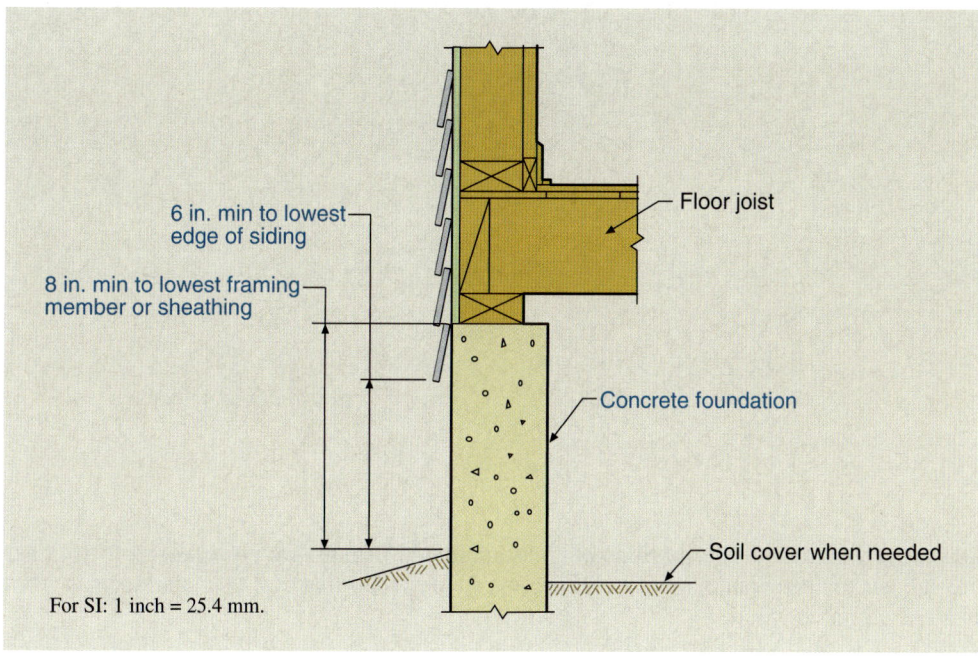

Figure 2304-3
Clearance between wood framing, wood siding, and earth.

Exterior walls below grade. These requirements were put in a separate subsection in the 2006 IBC to clarify that they apply to wood framing attached to the interior side of exterior concrete or masonry foundation walls.

2304.11.2.3

2012 International Building Code Handbook 775

23 Wood

2304.11.2.4 Sleepers and sills. Concrete and masonry slabs that are in direct contact with the earth are very susceptible to moisture because of absorption of ground water. This can occur on interior slabs as well as at the perimeter. This section is intended to prevent the use of untreated wood that may decay under such conditions. Concrete that is fully separated from the ground by a vapor barrier is not considered to be in direct contact with earth.

2304.11.2.5 Girder ends. An airspace is required around the ends of wood girders to reduce moisture that can contribute to decay of the member.

2304.11.2.6 Wood siding. Wood siding must have 6 inches of clearance between the siding and earth unless made of naturally durable or preservative-treated wood. Note that the clearance for wood sheathing under the siding is *8 inches* as required by Section 2304.11.2.2. In other words, siding can extend 2 inches below the foundation plate, or framing, whereas the sheathing must be terminated at the sill plate. See Figure 2304-4.

A code change in the 2009 IBC added a minimum 2-inch clearance between wood siding and an adjacent concrete slab such as a patio or walk. The previous code language should result in wood materials being at least 2 inches from the surface of typical 4-inch-thick concrete walk or porch slab if the required minimum of 6 inches distance from the ground is maintained. However, it is not unusual to see less than 2-inch clearance between wood siding and a concrete slab. Without specifying a minimum clearance, water that collects on the concrete can lead to decay in the wood. The IBC now requires a minimum 2-inch clearance between wood siding and a concrete slab in addition to the 6-inch clearance between the siding and the ground.

2304.11.2.7 Posts or columns. This section is an amalgam of the requirements from the three model codes. Figure 2304-4 illustrates the requirements.

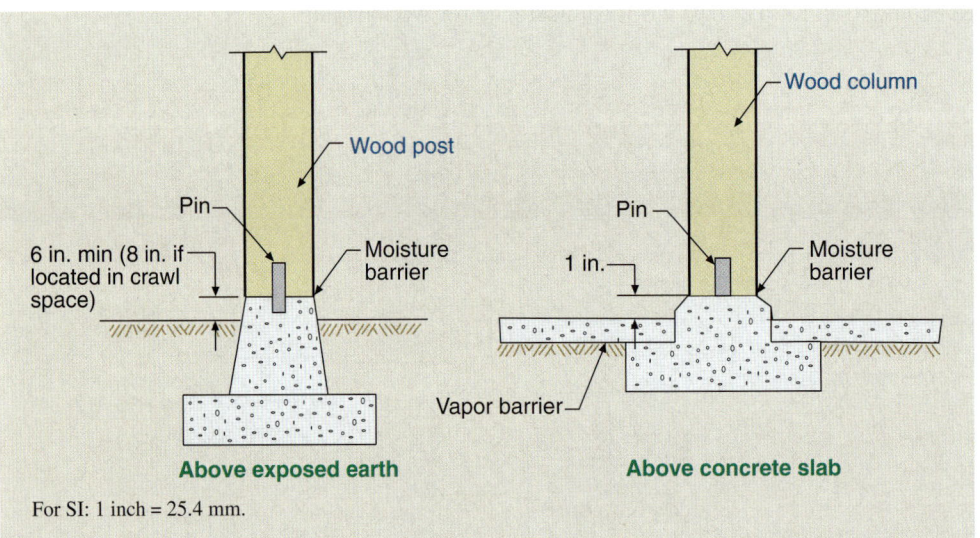

Figure 2304-4 Posts and columns.

2304.11.3 Laminated timbers. The portions of glue-laminated beams directly exposed to weather are subject to decay and should be of preservative-treated material or manufactured from naturally durable wood.

2304.11.4 Wood in contact with the ground or fresh water. Note specifically the limiting adjective *fresh*. This section only applies to wood in contact with the ground or fresh water. The water-borne preservatives used for fresh water are not suitable for brackish or salt water, where attack can also come from marine borers. The preservative retention rates are higher for wood in ground contact than for above-ground uses. The first paragraph of this code section allows wood in direct contact with the earth to be naturally durable. However, this

General Construction Requirements

only applies to wood in contact with the ground, not posts or columns. Posts and columns are required by Section 2304.11.4.1 to be preservative treated.

Posts or columns. Posts or columns embedded in concrete or embedded in earth, such as columns in a pole-supported structure, have no opportunity to dry and are subject to decay. Hence, they must be of preservative-treated wood.

2304.11.4.1

Wood structural members. Where wood framing is used to support floors and roofs that are moisture permeable, such as a concrete slab over a patio or a patio slab over a garage, the framing must be of pressure-treated wood or approved naturally durable species, unless the slab is separated from the wood framing by a waterproof membrane.

2304.11.4.2

Although these wood-framing members are not necessarily in direct contact with the ground, their exposure to moisture is similar to that of wood in direct contact with the ground. Therefore, the wood framing must be of naturally durable wood, or it must be preservative treated in accordance with AWPA C2, C9, or C22.

Supporting member for permanent appurtenances. Balconies, porches, and other appurtenances that are exposed to weather conditions may not have protective overhangs. Water can collect in the joints and on the surfaces, creating alternating cycles of wetting and drying conducive to deterioration and decay. In this case, the wood-framed structural supports must be of naturally durable or preservative-treated wood. The exception can be applied to buildings located in areas where climatic conditions are favorable enough to preclude the use of naturally durable or preservative-treated wood. For example, very dry desert areas such as Death Valley, California, and Yuma Valley, Arizona, have extremely low annual precipitation where the exception could be used with the approval of the building official.

2304.11.5

Termite protection. Where termites are a significant hazard such as in some southern states, floor framing must be preservative treated, naturally durable, or have some other approved method of termite protection. Section 2603.9 has specific restrictions on the use of foam plastics in areas where the probability of termite infestation is very heavy. Section 2603.9 refers to the termite infestation probability map shown in Figure 2603.9. (See Figure 2304-5.) In areas where the probability of termite infestation is very heavy, naturally durable termite-resistant wood or preservative-treated wood must be used.

2304.11.6

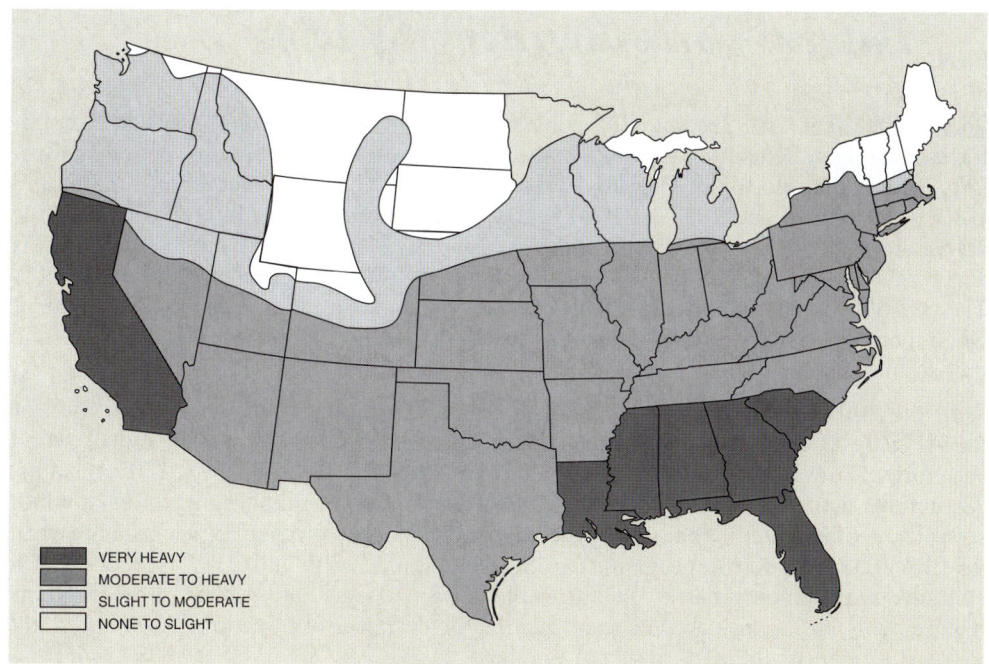

Figure 2304-5 Termite infestation probability map.

2304.11.7 Wood used in retaining walls and cribs. Wood used in retaining walls, crib walls, bulkheads, and other walls that retain or support the earth must be of preservative-treated wood specified as ground-contact-treated wood.

2304.11.8 Attic ventilation. Refer to the discussion under Section 1203.2.

2304.11.9 Under-floor ventilation. Refer to the discussion under Section 1203.3.

2304.12 Long-term loading. Wood structural members are subject to long-term creep, which increases deflection, particularly where a high dead load is present. The 2000 and 2003 IBC do not permit wood members to permanently support the dead load of masonry or concrete (with some exceptions), because additional deflection produced by long-term creep can cause severe cracking in the masonry or concrete. The provision in the IBC was a carryover from the 1997 UBC, which was intended to address concerns with long-term creep in wood members permanently supporting concrete and masonry. The section consisted of a restriction against wood members being used to permanently support dead load of masonry or concrete followed by a list of exceptions to the restriction. The provision was deleted in the 2006 IBC and replaced by a new section that references the design method for limiting long-term deflections in Section 3.5.2 and Appendix F of the NDS. Under sustained loading, wood members exhibit additional time-dependent deformation (creep), which generally develops over long periods of time. The tabulated modulus of elasticity design values, E, in the NDS are intended to be used to estimate the immediate deformation under load. Where dead loads or sustained live loads represent a relatively high percentage of total design load, creep is an appropriate design consideration, which is addressed within the NDS. The total deflection under long-term loading is estimated by increasing the initial deflection associated with the long-term load component by 1.5 for seasoned lumber or 2.0 for unseasoned or wet lumber or glued laminated timber. See Section 3.5.2 and Appendix F of the NDS for design provisions related to time-dependent deformations known as creep.

Section 2305 General Design Requirements for Lateral-Force-Resisting Systems

Prior to the 2009 IBC, Section 2305 contained extensive design requirements for lateral-force-resisting systems used in wood-frame buildings such as diaphragms, chords, collectors, shear walls, and so on. The provisions are primarily based on a combination of 1997 UBC Section 2315 and 1997 NEHRP Sections 12.1 through 12.4. In the 2006 IBC, the 2005 edition of AF&PA *Special Design Provisions for Wind and Seismic* (SDPWS) was added as a permissible alternative to the detailed requirements contained in Section 2305. The SDPWS standard provides complete requirements for design and construction of wood members, fasteners, and assemblies that resist lateral forces from wind and seismic ground motion.

In the 2009 IBC, nearly all of Section 2305 was deleted, and the 2008 edition of the AF&PA SDPWS is a mandatory referenced standard for lateral design of wood structures. The 2012 IBC also references the 2008 SDPWS. Section 2305.1 states, "Structures using wood-framed shear walls or wood-framed diaphragms to resist wind, seismic or other lateral loads shall be designed and constructed in accordance with AF&PA SDPWS and the provisions of Sections 2305, 2306 and 2307." Section 2306 contains requirements based on allowable stress design and Section 2307 contains requirements based on load and resistance factor design. Both Sections 2306 and 2307 reference the 2012 edition of the NDS, which contains both ASD and LRFD design procedures. In the 2012 IBC, Section 2305 has only three subsections: Section 2305.1 is general requirements; Section 2305.1.1 contains a general requirement regarding openings in shear walls that originated with the UBC; Section 2305.2 contains provisions for calculating deflection of stapled diaphragms; and Section 2305.3 contains provisions for calculating deflection of stapled shear walls. All other requirements pertaining to lateral design of wood-framed structures are in the 2008 AF&PA SDPWS. For a complete history of the code change that deleted substantial portions of Section 2305, refer to Code Change S82-06/07 in the ICC publication, 2009 IBC Code Changes Resource Collection. For a detailed table that cross-references provisions in the 2006 IBC and the 2008 SDPWS, refer to the discussion of Section 2305 in the *2009 IBC Handbook—Structural Provisions*. For detailed discussion of the provisions in Section 2305 that were in previous editions of the IBC, refer to the discussion in the *2006 IBC Handbook—Structural Provisions*.

2305.1 General. The section references the 2008 SDPWS for design and construction of wood-frame shear walls or wood-frame diaphragms to resist wind, seismic, or other lateral loads. Provisions in Sections 2305, 2306, and 2307 are also required to the extent that they apply.

2305.1.1 Openings in shear panels. This provision originated with the UBC and predates the perforated shear wall design method. Although the section requires that openings in shear walls that materially affect their strength be detailed and have their edges reinforced, the meaning of the phrase *materially affect their strength* is not defined in the code. It should also be noted that the perforated shear wall design method in SDPWS Section 4.3.5.3 does permit openings in shear walls without force transfer design, provided certain requirements, adjustments, and restrictions are met.

2305.2 Diaphragm deflection. The SDPWS does not address deflection calculations for stapled diaphragms. The code change that deleted most of Section 2305 from the 2009 IBC retained Equation 23-1 and the parameters necessary to calculate deflection of diaphragms fastened with staples. Stiffness properties (apparent shear stiffness, kip/inch) for diaphragms constructed of wood structural panels and lumber are given in SDPWS for purposes of complying with diaphragm classification, drift, and stiffness compatibility requirements specified in ASCE 7. See Section C4.2.2 of the SDPWS Commentary for detailed discussion of calculating diaphragm deflection.

Although there was no rational method to calculate the deflection of unblocked diaphragms, monotonic racking tests indicate that the deflection of an unblocked diaphragm may be on the order of three times the deflection of a blocked diaphragm. The 1999 edition of the SEAOC Blue Book[7] (*Recommended Lateral Force Requirements and Commentary of the SEAOC Seismology Committee*) suggests that the deflection of an unblocked diaphragm at its tabulated allowable shear capacity will be about 2.5 times the calculated deflection of a blocked diaphragm of similar construction and dimensions at the same shear capacity. In the 2008 SDPWS, apparent shear stiffness values are tabulated for each combination of nailing and sheathing thickness for typical blocked and unblocked diaphragms in order to simplify calculation of diaphragm deflection. See Section C4.3.2 of the SDPWS Commentary for detailed discussion of calculating shear wall deflection.

2305.3 Shear wall deflection. As noted above, the SDPWS does not address deflection of stapled shear walls. The code change that deleted most of Section 2305 from the 2009 IBC retained Equation 23-2 and the parameters necessary to calculate deflection of shear walls fastened with staples. Stiffness properties (apparent shear stiffness, kip/inch) for shear walls constructed of wood structural panels and other materials are given in SDPWS for purposes of complying with drift and stiffness compatibility requirements specified in ASCE 7.

23 Wood

Section 2306 *Allowable Stress Design*

In the 2009 IBC, substantial portions of Section 2306 were deleted because many of the provisions are included in the 2008 edition of the AF&PA SDPWS, which is referenced in Section 2306.1. As noted, the 2012 IBC also references the 2008 SDPWS in Sections 2305, 2306, and 2307. For detailed discussion of the provisions in Section 2306 that were in previous editions of the IBC, refer to the *2006 IBC Handbook—Structural Provisions*. For a complete history of the code change that deleted substantial portions of Section 2306, refer to Code Change S83-06/07 in the ICC publication, *2009 IBC Code Changes Resource Collection*.

2306.1 Allowable stress design. The 2012 edition of ANSI/AWC *National Design Specification (NDS) for Wood Construction and Supplement* is adopted by reference without any amendments. As previously noted, the 2012 NDS is now an AWC standard developed by the American Wood Council's (AWC) Wood Design Standards Committee. Many changes were made to the 2012 edition of the NDS that are described in an article entitled *2012 NDS Changes* by John "Buddy" Showalter, et al. The article and other AWC publications and resources related to wood engineering are available at www.awc.org.

AITC 500, *Determination of Design Values for Structural Glued Laminated Timber*, was deleted from the 2006 IBC because methods to determine design values for structural glued laminated timber are included in ASTM D 3737, which is the referenced standard in Section 2303.1.3.

The ANSI standard for Shallow Post Foundation Design, ANSI/ASAE EP486.1, was added to the 2003 IBC. The standard provides a design procedure for shallow post foundations that resist moments and lateral and vertical forces. The standard includes definitions, material requirements, and design equations for designing post foundations. The Standard was developed by the ASAE Post and Pole Foundation Subcommittee, approved by the Structures and Environment Division Standards committee and adopted by ASAE in March 1991. It was revised from 1992 to 1999. The 2012 IBC references the current edition of ASABE EP 486.1-99 (R2005).

2306.1.1 Joists and rafters. Although the section indicates that rafters may be designed using the AF&PA span tables, the intent is that both joists and rafters are permitted to be designed by the *AF&PA Span Tables for Joists and Rafters*. The 2012 IBC references the 2012 edition of the *AF&PA Span Tables for Joists and Rafters* and *Design Values for Joists and Rafters and Span Tables for Joists and Rafters*, which are available online from the American Wood Council at awc.org.

2306.1.2 Plank and beam flooring. Plank and beam flooring may be designed in accordance with the *AF&PA Wood Construction Data No. 4* (WCD No. 4—2003), *Wood Construction Data—Plank and Beam Framing for Residential Buildings*. The WCD4 document is available online from the American Wood Council at awc.org.

2306.1.3 Treated wood stress adjustments. Several factors can significantly affect the physical properties of FRTW. These factors are the pressure treatment and redrying processes used, and the extremes of temperature and humidity that the FRTW will be subjected to once installed. The design values for all FRTW must be adjusted for the effects of the treatment and environmental conditions, such as high temperature and humidity in attic installations. The design adjustment values must be based on an investigation procedure that includes subjecting the FRTW to similar temperatures and humidities, and which has been approved by the building official. The FRTW tested must be identical to that which is produced. The building official reviewing the test procedure must consider the species and grade of the untreated wood, and conditioning of the wood, such as drying before the fire-retardant-treatment process. Fire-retardant-wood treaters may choose to have their treatment process evaluated by ICC Evaluation Service and have an ESR issued that will assist the building official in determining compliance with the code.

The FRTW must be labeled with the design adjustment values. The adjustment factors must be obtained from the company providing the treatment and redrying service. These design adjustment values can take the form of factors that are multiplied by the nominal design values of the untreated wood to determine allowable stresses, or new allowable stresses that have already been factored down in consideration of the FRTW treatment. Other wood design adjustment factors are applicable to FRTW, except that the load duration factor for impact loads does not apply. Section 2.3.4 of the NDS has a similar requirement. Refer to NDS Table 2.3.2 for load duration factors.

Lumber decking. The reference to AITC 112-93, *Standard for Tongue-and-Groove Heavy Timber Decking*, in Section 2306.1 was removed from the 2006 IBC because the American Institute of Timber Construction (AITC) no longer maintains the standard. Revisions to Section 2304.8 along with the addition of new Section 2306.1.4 incorporated the pertinent provisions of AITC 112-93 directly into the body of the code. Section 2304.8 was revised to indicate that the provisions cover all decking including mechanically laminated and solid sawn decking. The capacity of lumber decking is arranged according to the various layup patterns described in Section 2304.8.2. Section 2306.1.4 gives the design capacity of lumber decking for flexure and deflection according to the formulas given in Table 2306.1.4.

2306.1.4

Wind provisions for walls. The provisions for increasing the allowable bending stress for studs subjected to out-of-plane (component and cladding) wind loads were deleted from the 2009 IBC because they are covered in Section 3.1.1 of the SDPWS standard. The allowable bending stress, F_b, in wood studs resisting wind loads may be increased in lieu of using the repetitive member factor (C_r) increase of 1.15. This increase recognizes that the studs and sheathing exhibit some composite action, which results in load sharing when properly designed and the appropriate blocking and fasteners are used. The wind pressures are transferred to the studs based on their relative stiffness to the sheathing. Because the minimum sheathing thickness is constant, the relative stiffness of the sheathing decreases as the stud depth increases; therefore, a higher proportion of load is carried by the stud as the stud depth increases. A minor change in the 2006 IBC clarified that the provision only applies to sawn lumber studs and the sheathing panel joints must occur over studs or blocking. This requirement was incorporated into the provisions in the SDPWS.

Wood-frame diaphragms. The term *diaphragm* refers to horizontal or sloping panel elements that resist and distribute lateral forces. (Vertical panel elements that resist lateral forces are termed *shear walls*.) See Section 202 for definitions of various types of diaphragms and diaphragm components. The title of Section 2306.2 was changed to apply to all wood-framed diaphragms and the section references the SDPWS for their design. The specific sections covering diagonally sheathed lumber diaphragms were deleted in the 2012 IBC because their design provisions are included in the SDPWS.

2306.2

Tables 2306.2(1) and 2306.2(2) contain values for diaphragms fastened with staples because the SDPWS does not address stapled diaphragms. Nails were deleted from the diaphragm tables in the 2012 IBC so the tables only contain design values for staples. The design values given in the tables are based on Douglas Fir-Larch or Southern Pine framing. When the framing lumber is of a species other than Douglas Fir-Larch or Southern Pine, the tabular values must be adjusted to account for the fastener behavior in framing members of other lumber species. The adjustments must be made in accordance with Footnote a of the table.

The high-load diaphragm table was added to the 2003 IBC. High-load diaphragms are required to have special inspection in accordance with Section 1705.5.1, as indicated in Footnote g of the table. The footnote was added to the table as a cross reference to Chapter 17 so that code users are aware of the requirement for special inspection for high-load diaphragms. When the high-load diaphragm table was first added to the IBC, the figures that illustrate the nailing patterns at panel joints were inadvertently left out of the code. These figures have been part of the APA high-load diaphragm table for over two decades. The figures show the various multi-row fastening patterns, required spacing, and staggering. These fastening diagrams are intended to clarify proper application of the table and are essential to develop the diaphragm shear values listed.

The last footnote in the diaphragm and shear wall tables [Table 2306.2(1), Table 2306.2(2), and Table 2306.3(1)] was added in the 2006 IBC to provide the necessary factors to convert table values for shear loads of normal or permanent load duration as defined by the NDS. The factors reflect the fact that the values in the tables have a built-in load duration factor of 1.6. As such, to convert to a normal load duration of 1.00, the tabular value must be multiplied by 0.63 ($1.6 \times 0.63 = 1.00$). To convert to a permanent load duration of 0.90, a factor of 0.56 must be used ($1.6 \times 0.56 = 0.90$).

Shear capacities modifications for wind loads. Section 2306.2 permits allowable shear capacities for wood diaphragms to be increased 40 percent for wind design. The 40-percent increase is permitted in Section 2306.2 for diaphragms and Section 2306.3 for wood-framed shear walls. A similar increase is permitted by the SDPWS for lumber diaphragms and shear walls, as well as particleboard and fiberboard shear walls. The wind loads determined by ASCE 7 have changed over time, and recent research has provided a better understanding of how structures are affected by wind, and code-prescribed wind loads have been refined since the diaphragm and shear wall tables were originally developed. The tables use a 2.8 factor of safety. Because wind loads are essentially monotonic and bounded, a safety factor of 2 is deemed to be sufficient for wind loading. This results in a 40-percent increase in the tabulated values ($2.8 - 2.0/2.0 = 0.4$). In the 2008 SDPWS, the diaphragm shear capacity tables have been separated so that nominal values for wind and seismic loading are given in separate columns. The wind load capacities in the SDPWS tables are 40 percent higher than the seismic load capacities. It should also be noted that the SDPWS has separate tables for blocked and unblocked diaphragms.

Diagonally sheathed lumber diaphragms. The diagonally sheathed lumber diaphragm tables were deleted in the 2009 IBC and replaced with references to the SDPWS. In the 2012 IBC, the sections pertaining to diagonally sheathed lumber diaphragms were deleted entirely. The detailed requirements for diagonally sheathed lumber diaphragms are covered in SDPWS Section 4.2.7. See SDPWS Table 4.2D for nominal unit shear capacities for lumber sheathed diaphragms of single horizontal, single diagonal, and double diagonal sheathing.

2306.2.1 Gypsum board diaphragm ceilings. Section 2508.5 allows the use of gypsum board ceiling diaphragms provided certain requirements and restrictions are met. For example, gypsum board ceiling diaphragms cannot be used to resist lateral forces resulting from concrete or masonry. The maximum diaphragm aspect ratio is $1\frac{1}{2}:1$, and cantilevered diaphragms and diaphragm rotation are not permitted.

2306.3 Wood-frame shear walls. The title of Section 2306.3 was changed to apply to all wood-framed shear wall systems and the section references the SDPWS for their design. The specific sections covering lumber sheathed, particleboard, fiberboard, Portland cement plaster (stucco), and gypsum board shear walls were deleted in the 2012 IBC because design provisions are included in the SDPWS. Where shear wall panels are fastened with staples, the IBC contains tables with allowable shear values for wood structural panels [Table 2306.3(1)], fiberboard [Table 2306.3(2)], and Portland cement plaster and gypsum materials [Table 2306.3(3)]. As noted in Footnote e of Table 2306.3(2), fiberboard shear walls fastened with staples are not permitted to be used in Seismic Design Categories D, E, and F. The SDPWS restricts the use of particleboard and structural fiberboard shear walls in Seismic Design Categories A, B, and C and Portland cement plaster (stucco) and gypsum shear walls in Seismic Design Categories A, B, C, and D. Thus, only wood structural panel shear walls are permitted to be used in Seismic Design Categories E and F.

Allowable Stress Design

See further discussion below under seismic considerations in regard to more specific seismic limitations imposed by ASCE 7.

Prior to the 2009 IBC, the code required all edges of wood structural panel shear walls to be supported by studs or blocking. The 2008 SDPWS has an exception allowing horizontal blocking to be omitted if specific conditions are met and nominal shear capacity is reduced by the unblocked shear wall adjustment factor. See SDPWS Sections 4.3.2 and 4.3.7.

As with diaphragms, the allowable shear capacities for wood structural panel shear walls may be increased 40 percent for wind design only. Because wind loads are essentially monotonic and bounded, a safety factor of 2.0 is deemed sufficient for wind loading. Because code tables use a factor of safety equal to 2.8, this results in a 40 percent increase in the tabulated values for wind load design. It is simply a coincidence that this 1.4 increase factor is numerically the same as the factor used to convert strength level seismic loads to ASD level in the alternative ASD load combinations in Section 1605.3.2. The 1.4 increase factor for wind is to calibrate tabular values to a divisor (safety factor) of 2.0, whereas the 1.4 load factor is used to scale code-prescribed seismic loads from strength level down to service load (ASD) level. In contrast, the load factor for E in the strength design load combinations in Section 1605.2.1 is 1.0 because all code-prescribed seismic load effects are at strength level.

As previously noted, the 2009 and 2012 IBC reference the 2008 SDPWS, which covers design and construction requirements for wood members, fasteners, and assemblies designed to resist wind and seismic forces. Table 4.3A of the SDPWS gives nominal unit shear capacities for wood-frame shear walls of wood-based panels. The nominal unit shear capacities for wind are given in a separate part of the table than the nominal unit shear capacities for seismic loads. The shear capacities in the wind load portion of the table are 1.4 times the values in the seismic load portion of the table because they incorporate the 1.4 increase factor for wind.

Because nails were deleted from the shear wall tables in the 2012 IBC, the tables only contain design values for staples. With the exception of the removal of nails, Table 2306.3(1) is similar to the wood structural panel shear wall table in previous editions of the IBC. The design values given in the table are based on Douglas Fir-Larch or Southern Pine framing. When the framing lumber is of a species other than Douglas Fir-Larch or Southern Pine, the tabular values must be adjusted to account for the fastener behavior in framing members of other lumber species. The adjustments must be made in accordance with Footnote a.

Footnotes d and f require 3-inch nominal or wider framing at adjoining panel edges where fastener spacing at panel edges is 2 inches on center or less, and where panels are applied to both faces, panel edge fastener spacing is less than 6 inches on center and panel joints are not offset from each other.

Footnote g applies to shear walls in Seismic Design Categories D, E, and F. All framing members receiving edge nailing from abutting panels must be 3-inch nominal when the shear design value exceeds 350 pounds per lineal foot (ASD). Note that two 2-inch nominal members fastened together to transfer shear between framing members in accordance with the NDS is an acceptable alternative. Both panel joint and sill plate fasteners must be staggered to prevent splitting. The footnote references SDPWS Sections 4.3.6.1 and 4.3.6.4.3 for sill plate size and anchorage requirements because these requirements were deleted from the 2009 IBC. SDPWS Section 4.3.6.4.3 requires 3-inch square by 0.229-inch thick square plate washers with some exceptions. A standard cut washer can be used where all of the following are met: Anchor bolts are designed to resist shear only; the full-height segment design method is used; hold downs are designed without considering any dead load to resist overturning; the shear wall aspect ratio does not exceed 2:1; and the nominal shear capacity of the shear wall does not exceed 980 plf.

Section 2307 of the 2009 IBC has a requirement similar to Footnote g but applicable to LRFD. The requirement is essentially the same as Footnote g but has a threshold of 490 plf

LRFD, which corresponds to 350 plf ASD. Section 2307.1.1 was deleted from the 2012 IBC. Note that Section 2307 (LRFD) now references the NDS and SDPWS without any modifications to the standards.

Lumber sheathed shear walls. The specific sections covering lumber sheathed shear walls were deleted from the 2012 IBC because various types of lumber shear walls are covered in the SDPWS. See SDPWS Sections 4.3.7.6 and 4.3.7.7 for construction requirements of single and double layer diagonally sheathed shear walls. The SDPWS only permits diagonally sheathed lumber shear walls in Seismic Design Categories A, B, C, and D. As noted, the 40-percent increase in capacity is permitted for lumber sheathed shear walls for wind design.

Particleboard shear walls. The specific section covering particleboard sheathed shear walls was deleted from the 2012 IBC because particleboard shear walls are covered in the SDPWS. The SDPWS only permits particleboard shear walls in buildings in Seismic Design Categories A, B, and C. See SDPWS Section 4.3.7.3 and Table 4.3A. Particleboard used for shear walls must conform to ANSI A208.1. Maximum shear wall aspect ratios for particleboard shear walls are given in SDPWS Table 4.3.4. Note that particleboard shear walls have a maximum aspect ratio of 2:1.

Fiberboard shear walls. The specific section covering structural fiberboard sheathed shear walls was deleted from the 2012 IBC because fiberboard shear walls are covered in the SDPWS. The SDPWS only permits fiberboard shear walls in buildings in Seismic Design Categories A, B, and C. See SDPWS Section 4.3.7.3. Fiberboard used for shear walls must conform to ASTM C 208. SDPWS Table 4.3.4 allows fiberboard shear walls to have an aspect ratio of up to 3.5:1, provided the nominal unit shear capacity is reduced by the Aspect Ratio Factor for wind or seismic. The Aspect Ratio Factor cannot exceed 1.0. The Aspect Ratio Factor for wind and seismic are shown in Table 23-1 for an 8-foot shear wall. See Footnote 3 of SDPWS Table 4.3.4.

Table 23-1. **Fiberboard Shear Wall Aspect Ratio Factors for 8-Feet Shear Wall Height**

Shear Wall Width, b (feet)	Aspect Ratio, h/b	Aspect Ratio Factor (seismic)	Aspect Ratio Factor (wind)
2.28	3.5	0.357	0.775
3.00	2.67	0.438	0.850
4.00	2.00	0.550	0.910
5.00	1.60	0.663	0.946
6.00	1.33	0.775	0.970
7.00	1.14	0.888	0.987
8.00	1.00	1.00	1.00

Fiberboard shear walls fastened with staples are not permitted to resist lateral loads resulting from concrete or masonry walls. See Table 2306.3(2), Footnote a.

Shear walls sheathed with other materials. The specific section covering shear walls composed of cement plaster and gypsum materials was deleted from the 2012 IBC because these shear walls are covered in the SDPWS. The detailed construction requirements for gypsum board, gypsum sheathing, gypsum lath and plaster, and Portland cement plaster shear walls are in SDPWS Section 4.3.7.5 and IBC Chapter 25. The applicable ASTM standards for the various types of materials are given in SDPWS Sections 4.3.7.5.1 through 4.3.7.5.4. The SDPWS permits cement plaster and gypsum shear walls to be used in Seismic Design Categories A, B, C, and D. Shear walls using these materials are considered brittle (nonductile), and as such the strength and stiffness of both plaster and gypsum shear walls decrease significantly under a limited number of cycles when subjected to seismic demands.

Examination of the performance of structures subjected to strong ground shaking from the Northridge earthquake showed that these walls may not exhibit adequate strength or ductility. Therefore, these walls should be used to resist seismic demands only after careful consideration of their performance characteristics.

Seismic considerations. In the 2006 IBC, the technical provisions for seismic design were removed from the code because they are covered in the ASCE 7 standard. The IBC only contains the design criteria for environmental loads such as seismic and wind. Shear walls resisting seismic loads are subject to the limitations in Section 12.2.1 and Table 12.2-1 of ASCE 7. Similar requirements applicable to the simplified seismic provisions are found in Section 12.14 and Table 12.14-1. Tables 12.2-1 and 12.14-1 contain two types of wood-frame shear walls: light-frame (wood) walls sheathed with wood structural panels rated for shear resistance or steel sheets, and light-frame walls with shear panels of all other materials. See Items 15 and 17 of ASCE 7 Table 12.2-1 under bearing wall systems. For light-frame shear walls sheathed with materials other than wood structural panels or steel sheets, Table 12.2-1 assigns a response modification factor of 2 and restricts the height of the building to 35 feet in Seismic Design Category D. Also, these shear walls are not permitted to be used to resist seismic forces in Seismic Design Category E or F. The ASCE 7 and SDPWS standards have somewhat different limitations for shear walls with different materials based on the seismic design category of the building. Table 2306-1 provides a summary comparison of these requirements.

Table 2306-1. **Summary of Different Types of Wood-Frame Shear Wall System Limitations Based on Seismic Design Category**

Shear Wall Type	SDPWS Section	Permissible Seismic Design Categories per SDPWS-08	Permissible Seismic Design Categories per ASCE 7-10 Tables 12.2-1 and 12.14-1
Particleboard	Section 4.3.7.3	A, B, C	A, B, C, D[a]
Structural Fiberboard	Section 4.3.7.4	A, B, C	A, B, C, D[a]
Horizontally Sheathed Single-Layer Lumber	Section 4.3.7.8	A, B, C	A, B, C, D[a]
Vertical Board Siding	Section 4.3.7.9	A, B, C	A, B, C, D[a]
Gypsum Wallboard, Gypsum Base for Veneer Plaster, Water-Resistant Gypsum Backing Board, Gypsum Sheathing Board, Gypsum Lath and Plaster, Portland Cement Plaster	Section 4.3.7.5	A, B, C, D	A, B, C, D[a]
Diagonally Sheathed Lumber - Single Layer Diagonally Sheathed Lumber - Double Layer	Section 4.3.7.6 Section 4.3.7.7	A, B, C, D	A, B, C, D[a]
Wood structural panels over gypsum wallboard or gypsum sheathing	Section 4.3.7.2	A, B, C, D, E, F	A, B, C, D[b], E[b], F[b]
Wood structural panel	Section 4.3.7.1	A, B, C, D, E, F	A, B, C, D[b], E[b], F[b]

[a]Limited to 35 feet building height in Seismic Design Category D.
[b]Limited to 65 feet building height Seismic Design Categories D, E, and F.

23 Wood

Section 2307 *Load and Resistance Factor Design*

In the 2000 and 2003 IBC, this section references the *Load and Resistance Factor (LRFD) Standard for Engineered Wood Construction*, AF&PA/ASCE 16-95, for design of wood structures using the LRFD procedure. The reference to the AF&PA/ASCE 16 LRFD standard was deleted from the 2006 IBC and replaced with a reference to the 2005 edition of the *National Design Specification (NDS) for Wood Construction*, which is a dual format specification that contains up-to-date provisions for both allowable stress design (ASD) and LRFD. The 2009 IBC also references the 2005 edition of the NDS. The 2012 IBC references the current 2012 edition of the NDS and Supplement (*ANSI/AWC NDS-2012*), developed by the AWC Wood Design Standards Committee. Many significant changes for wood member and connection design were introduced during the development of the 1997, 2001, and 2005 editions of the NDS for the ASD procedure, which became part of the LRFD design procedure contained in the 2005 edition of the NDS. Many additional changes were made to the 2012 edition of the NDS, which are described in an article entitled *2012 NDS Changes* by John "Buddy" Showalter, et al. The article and other AWC publications and resources related to wood engineering are available at www.awc.org.

2307.1 General. The section references the 2012 NDS and the 2008 SDPWS for load and resistance factor design of wood structures. Both are dual format standards that permit ASD and LRFD procedures.

Wood structural panel shear walls. Section 2307.1.1 of the 2006 and 2009 IBC contained specific requirements for shear walls in buildings assigned to Seismic Design Category D, E, or F where the LRFD shear exceeds 490 pounds per foot. The requirements are essentially the same as Footnote g of Table 2306.3(1) in regard to framing at abutting panel joints and sill plate anchorage. The section was deleted in the 2012 IBC because the requirements for shear wall framing and anchorage are now in the SDPWS standard. Shear wall framing, sheathing, and anchorage requirements are covered in Section 4.3.6 of the SDPWS. Framing and blocking must be a minimum of 2-inch nominal members, and the shear wall chords must be of adequate size and grade to resist tension and compression forces resulting from overturning. Section 4.3.7.1 and Table 4.3A provide specific requirements for wood structural panel shear walls. A 3-inch nominal framing member and staggered nailing are required at abutting panel joints where (1) nail spacing is 2 inches on center or less, or (2) 10d common nails with spacing of 3 inches on center or less with a penetration of more than $1^1/_2$ inches are used, or (3) where the required nominal shear capacity on either side of the shear wall exceeds 700 plf in Seismic Design Category D, E, or F. SDPWS Section 4.3.6.4.3 requires 3-inch square by 0.229-inch thick square plate washers with some exceptions. A standard cut washer can be used where all of the following are met: anchor bolts are designed to resist shear only; the full-height segment design method is used; hold downs are designed without considering any dead load to resist overturning; the shear wall aspect ratio does not exceed 2:1; and the nominal shear capacity of the shear wall does not exceed 980 plf.

Section 2308 *Conventional Light-Frame Construction*

The prescriptive conventional construction provisions originated with the repetitive light-frame wood construction provisions of the UBC. Early editions of the UBC had a section entitled "wood-joisted dwelling construction," which later became "light-frame construction," and later changed to "conventional construction provisions" in the 1970 UBC. The conventional construction provisions of the UBC have always been entirely prescriptive and were intended to apply to buildings constructed of repetitive light wood-framing members consisting of studs, joists, and rafters.

The provisions of Section 2308 are based on experience gained over the last 60 years or more. In general, the provisions are reasonably easy to modify where experience shows they are inadequate or in need of modification. An example of such modification is the change in lateral bracing requirements that resulted from experience gained in the 1971 San Fernando, 1987 Whittier, and 1994 Northridge, California, earthquakes.

The term *prescription* means, "the action of laying down authoritative rules or directions." The term *prescriptive* means, "acquired by, founded on, or determined by prescription or by long-standing custom." Together these two definitions clearly describe the nature of the conventional wood-frame construction provisions in the IBC: they are a set of rules based on long-standing custom that have evolved over time.

The underlying philosophy of prescriptive conventional construction was clearly defined in Section 2518(a) of the 1970 UBC as follows: "The requirements contained in this section are intended for light-frame construction. *Other methods may be used provided a satisfactory design is submitted showing compliance with other provisions of this code.*" Although the conventional construction provisions in the UBC were modified and expanded over the years, the introductory language of this section remained unchanged through all editions of the UBC up to and including the 1997 edition. The first sentence of IBC Section 2308 is essentially the same.

A key feature of the conventional construction provisions is in this statement: "Other methods may be used provided a satisfactory design is submitted showing compliance with other provisions of this code." In other words, one need not conform to the restrictions, limitations, and requirements of the provisions if a design is submitted to the jurisdiction that conforms to the engineering requirements of the code. In addition to Section 2518(a), the 1970 UBC also included an exception in the engineering chapter, which stated, "Buildings or portions thereof that are constructed in accordance with the conventional light-framing requirements specified in Chapter 25 of this code shall be deemed to meet the requirements of this section." The 1997 UBC has the same exception, with the added phrase, "*Unless otherwise required by the building official* buildings or portions thereof that are constructed in accordance with the conventional light-framing requirements . . . " What these two sections mean is simple: a wood-frame building must either conform to all of the restrictions, limitations, and requirements (rules) prescribed in the conventional construction provisions, or engineering must be provided that demonstrates compliance with the engineering requirements of the code. Where engineering is provided, the designer is required to determine all applicable gravity and lateral loads that act on the structure, and design and detail the various structural systems to resist these loads. This requirement is embodied in Section 1604.4, which states, "Load effects on structural members and their connections shall be determined by methods of structural analysis that take into account equilibrium, general stability, geometric compatibility and both short- and long-term material properties. Any system or method of construction to be used shall be based on a rational analysis in accordance with well-established principles of mechanics. Such analysis shall result in a system that provides a complete load path capable of transferring loads from their point of origin to the load-resisting elements."

Most of the seismic-related provisions in Section 2308 originated with the 1997 UBC and the 1997 NEHRP *Provisions* (FEMA 302). New provisions have been added through the ICC code development process from the initial 2000 IBC to the current 2012 IBC. A few significant changes were made in the 2009 and 2012 IBC, which will be discussed in the applicable subsections.

23 Wood

2308.1 General. The provisions of Section 2308 provide prescriptive construction details and methods for light-frame wood construction. Typical construction details as required by the provisions are shown in Figures 2308-1 through 2308-5. Light-frame wood construction consists of 2x construction with walls of 2-inch nominal thickness studs spaced at 16 or 24 inches on center, and roofs and floors framed of 2-inch nominal thickness rafters or joists spaced at 12, 16, or 24 inches on center. Studs are typically 2 by 4 or 2 by 6. Construction that meets the prescriptive provisions of Section 2308 is deemed to comply with the intent of the code without requiring engineering. Perhaps most important is Section 2308.2, which contains the specific restrictions and limitations for the conventional construction provisions.

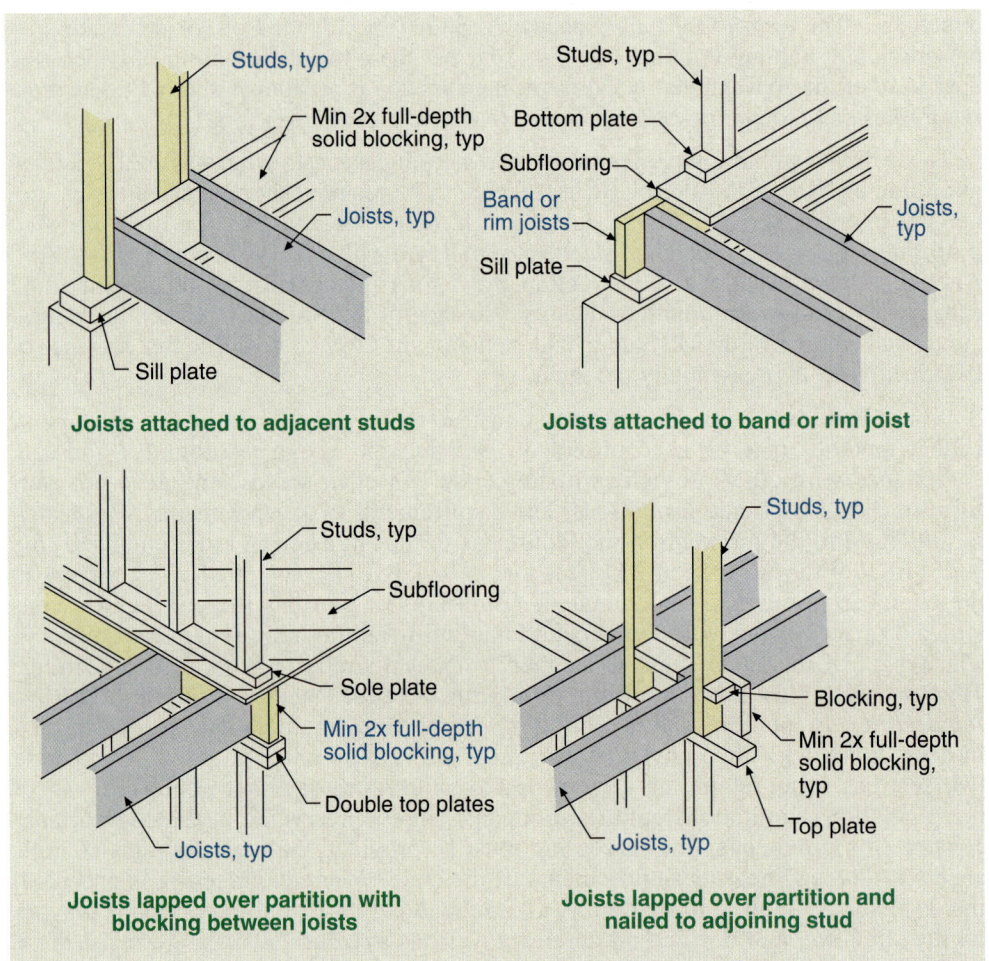

Figure 2308-1 Typical details—joist framing.

In the 2006 IBC, the language was changed so that detached one- and two-family dwellings and townhouses not more than three stories in height with a separate means of egress are specifically required to comply with the *International Residential Code* (IRC). In fact, that language in the exception to Section 101.2 is identical to the last sentence of Section 2308.1. In addition, the definition of the term *townhouse* was added to Section 202 of the 2006 IBC because the term is used in this section. The language in the 2006 IBC differs from previous editions of the IBC in that dwellings and townhouses are specifically required to conform to the IRC. The purpose of the change in language was to make clear that Section 2308 is not an alternative for detached one- and two-family dwellings and townhouses, because Section 101.2 requires the use of the IRC for those buildings. Only buildings that are within the scope of the IBC and Section 2308.2 are permitted to

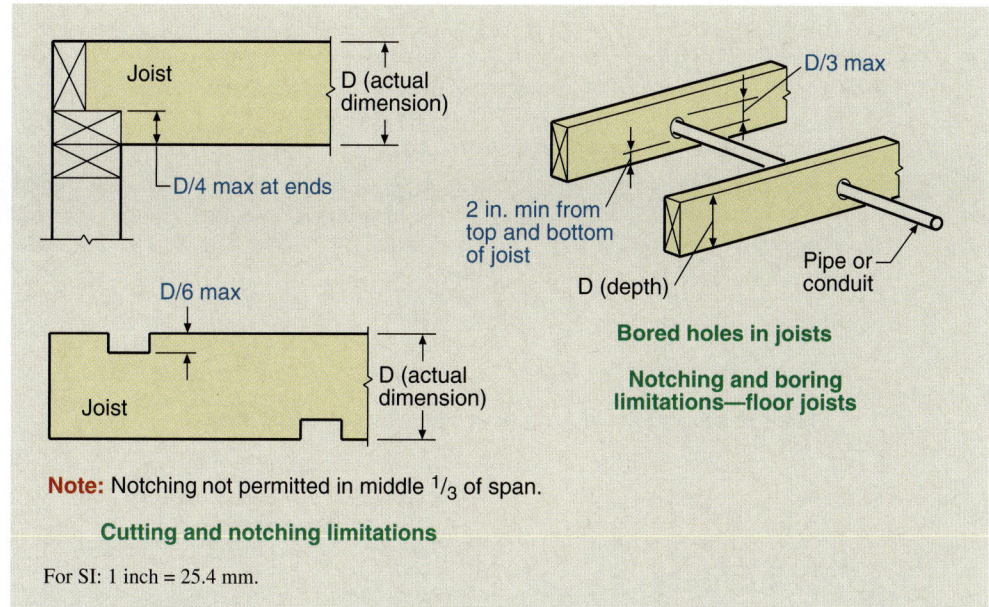

Figure 2308-2
Typical details—cutting, notching, and boring joists.

use the conventional construction provisions of Section 2308. However, Section R301.1.3 of the IRC requires structural elements that do not conform to the limits of the IRC to be designed in accordance with accepted engineering practice under the engineering provisions of the IBC. Because Section 2308.1 also permits compliance with the AF&PA *Wood Frame Construction Manual* (WFCM) as an alternative, the apparent intent is for detached one- and two-family dwellings and townhouses to either conform to the IRC or the AF&PA WFCM or be designed in accordance with accepted engineering practice under the engineering provisions of the IBC.

Another code change removed the story height exception for one- and two-family dwellings in Seismic Design Categories C, D, and E in Sections 2308.11 and 2308.12. According to the code change proponent, the story height exception for detached one- and two-family dwellings was removed because detached one- and two-family dwellings are within the scope of the IRC. Thus, the 2006 IBC created a peculiar situation: One- and two-family wood-frame dwellings are not permitted to be designed by the conventional construction provisions in the IBC but must comply with either the IRC or the WFCM, or be engineered. Yet hotels and multifamily dwellings (apartments) are allowed to use the conventional construction provisions of Section 2308. This is a significant departure from the past because the conventional wood-frame construction provisions of the IBC originated with the UBC, which were primarily used for one- and two-family dwellings, and hotels and multifamily dwellings (apartments) were usually engineered. As a result, the conventional wood-frame construction provisions in the IBC no longer have the broad application they once did. Unfortunately, there are still some remnants of requirements for one- and two-family dwellings such as Table 2308.9.5, which according to Section 2308.9.5 only apply to headers in one- and two-family dwellings.

Another significant change in the 2006 IBC pertains to portions of buildings that do not conform to the limitations for conventional construction. The section essentially expanded and clarified Section 2308.4.1, which requires *portions* of buildings of otherwise conventional construction that exceed the limits of Section 2308.2 to have those *portions* and the supporting load path be designed in accordance with accepted engineering practice and other provisions of the code. The code change clarified the "design of portions"

23 Wood

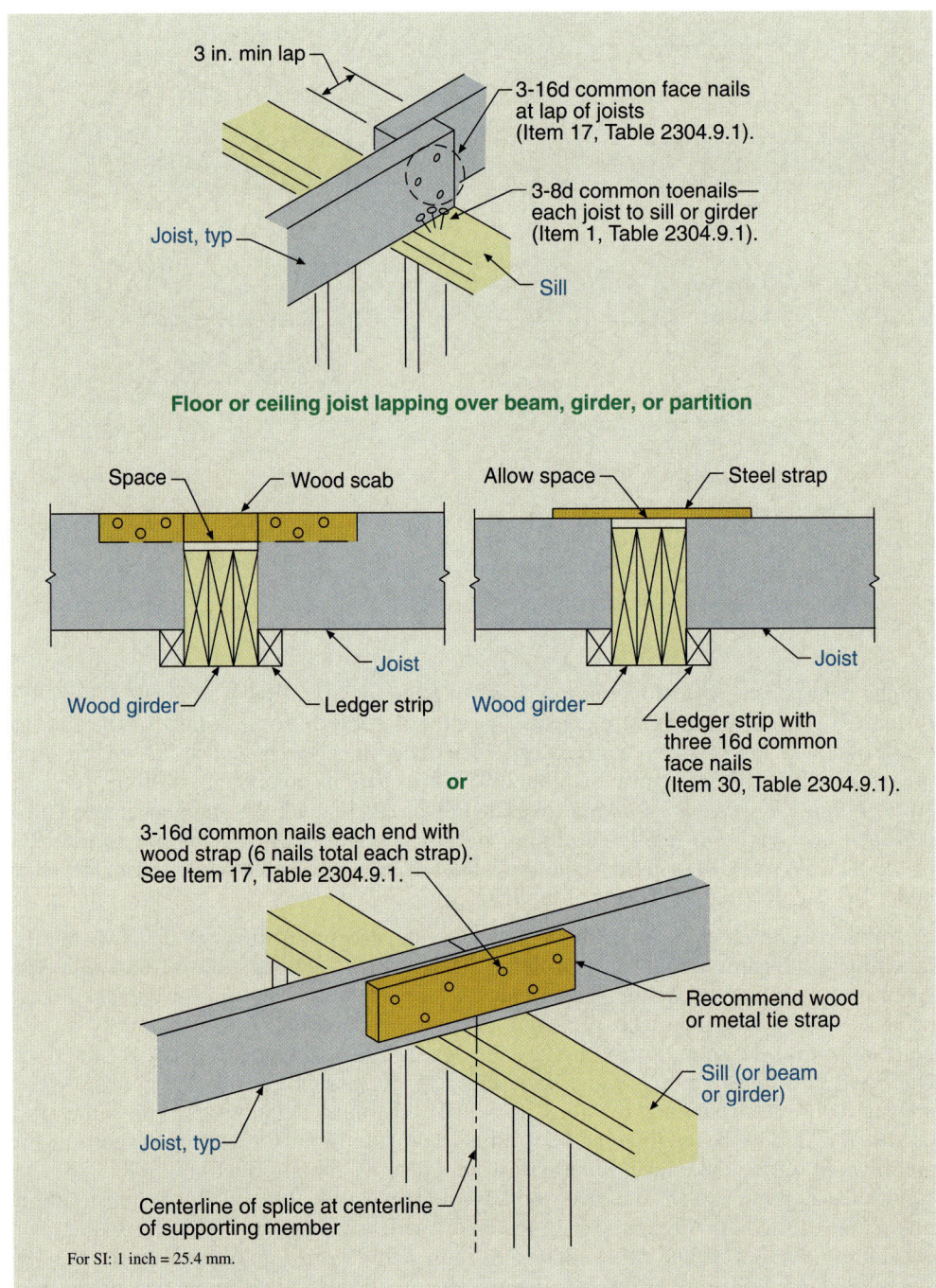

Figure 2308-3
Typical details—floor or ceiling joists.

requirements by distinguishing between *portions* of a structure and *elements* of a structure. The code permits elements and members as well as rooms or a series of rooms to be engineered in an otherwise conventionally constructed building. In addition, the code change added the phrase "and supporting load path" to emphasize the importance of providing a continuous load path for the engineered portions or engineered elements of a structure.

Conventional Light-Frame Construction

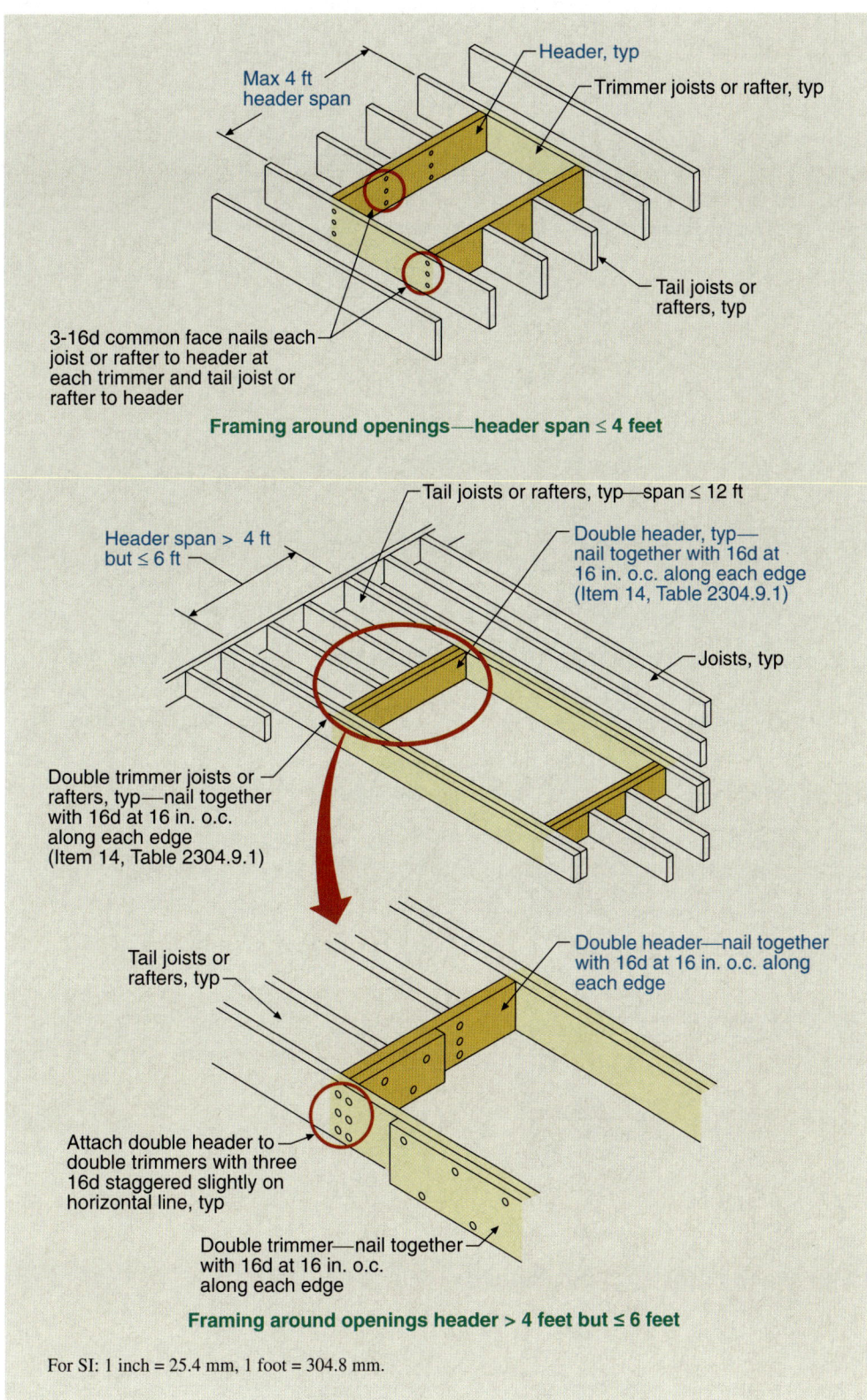

Figure 2308-4 Typical details—framing around openings.

23 Wood

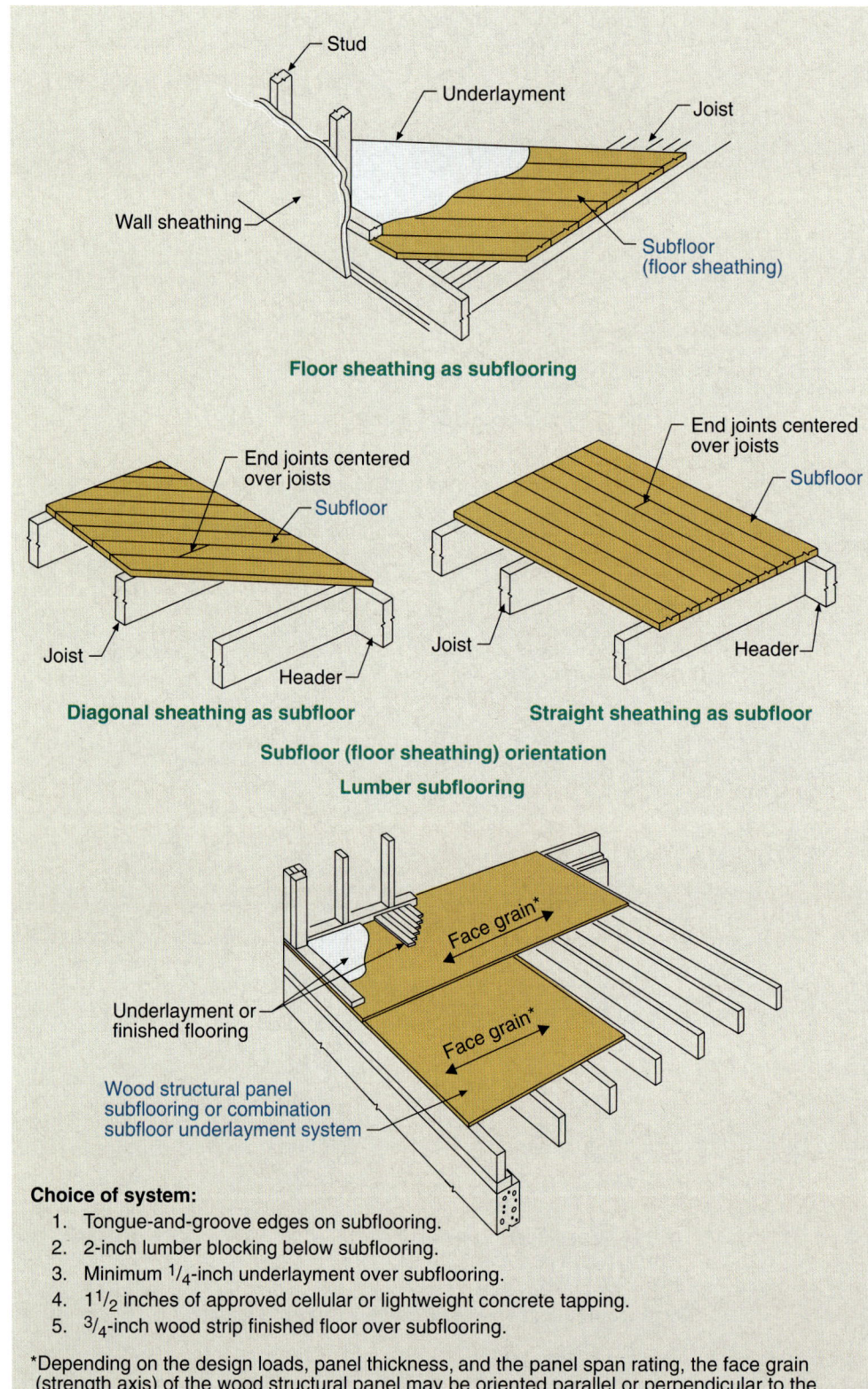

Figure 2308-5 Typical details—plywood subfloor.

Portions exceeding limitations of conventional construction. This is a new section in the 2006 IBC pertaining to conventional buildings that have portions that exceed the limitations imposed by Section 2308.2. In this case, those portions as well as the supporting load path must be designed in accordance with the engineering provisions of the code. In this section, the term *portions* does not refer to structural elements but to parts of the building that contain volume and area such as a room or a series of rooms. See Section 2308.4 regarding design of elements or members that exceed the limitations of conventional construction.

2308.1.1

Limitations. One of the most important aspects of most prescriptive methods is meeting restrictions and limitations required to use the method. The structures for which conventional construction is applicable are described in this section. The limitations are broken into categories as follows: (1) maximum building height, (2) maximum story height, (3) maximum dead, live and snow loads, (4) maximum nominal wind speed, (5) maximum roof span, (6) highest permissible seismic design category, and (7) permissible structural irregularities. Item 6 essentially restricts the use of conventional construction to Seismic Design Categories A through E, and Item 7 refers to Section 2308.12.6 for limitations on structural irregularities in Seismic Design Categories D and E. Additional requirements for structures in Seismic Design Category B or C are covered in Section 2308.11, and additional requirements for structures in Seismic Design Category D or E are covered in Section 2308.12. The conventional construction provisions are not permitted to be used for structures in Risk Category IV, which corresponds to Seismic Design Category F. This means that an engineering design is required for structures in Seismic Design Category F. Risk categories are determined from Table 1604.5 based on the use of the structure and the seismic design category of the structure. The seismic design category of a structure is determined in accordance with Section 1613.3.5 based on the risk category (formerly called occupancy category), soil properties (site class), and potential earthquake ground motion at the site. Note that in order for a structure to be assigned to Seismic Design Category F, it must be in Risk Category IV, which are not permitted to use the conventional construction provisions.

2308.2

Item 1 restricts the building height to three stories above grade plane. This applies to Seismic Design Category A because Section 2308.11 restricts the building height to two stories above grade plane in Seismic Design Categories B and C, and Section 2308.12 restricts the building height to one story above grade plane in Seismic Design Categories D and E. In Seismic Design Category D or E, cripple walls exceeding 14 inches are considered a story for the purposes of determining the number of stories. Note that cripple walls exceeding 14 inches high in structures classified in any seismic design category are considered an additional story for purposes of bracing. See discussion of Sections 2308.9.4.1 and 2308.12.4. Typical conventional raised-floor construction generally uses post and girder construction to support the floor joists within the interior of the structure. Because interior-braced wall lines may not necessarily extend down to the foundation, lateral loads must be transferred through the cripple walls to the foundation. Because cripple walls are at the lowest level of the structure, they are the most heavily loaded, relative to the braced panel walls. Thus, cripple walls are very important to satisfactory performance of the structure when subjected to lateral loads from wind pressure or earthquake ground motion. By classifying the cripple walls as a story, higher bracing strength criteria may be required to resist the higher lateral loads. See, for example, Section 2308.12.4.

The maximum floor-to-floor height was changed in the 2009 IBC. Earlier editions of the IBC had a maximum floor-to-floor stud height of 10 feet plus an allowance of 16 inches maximum for floor framing. The 2009 IBC has two conditions: a maximum floor-to-floor height of 11 foot 7 inches and a maximum bearing wall stud height of 10 feet. See Figure 2308-6.

Section 2308.2 limits the average dead load to 15 psf for roofs and exterior walls, floors, and partitions. The 2006 IBC clarified that the maximum dead load limit is for the combined roof and ceiling load. There are two exceptions: (1) Stone or masonry veneer up to the lesser of

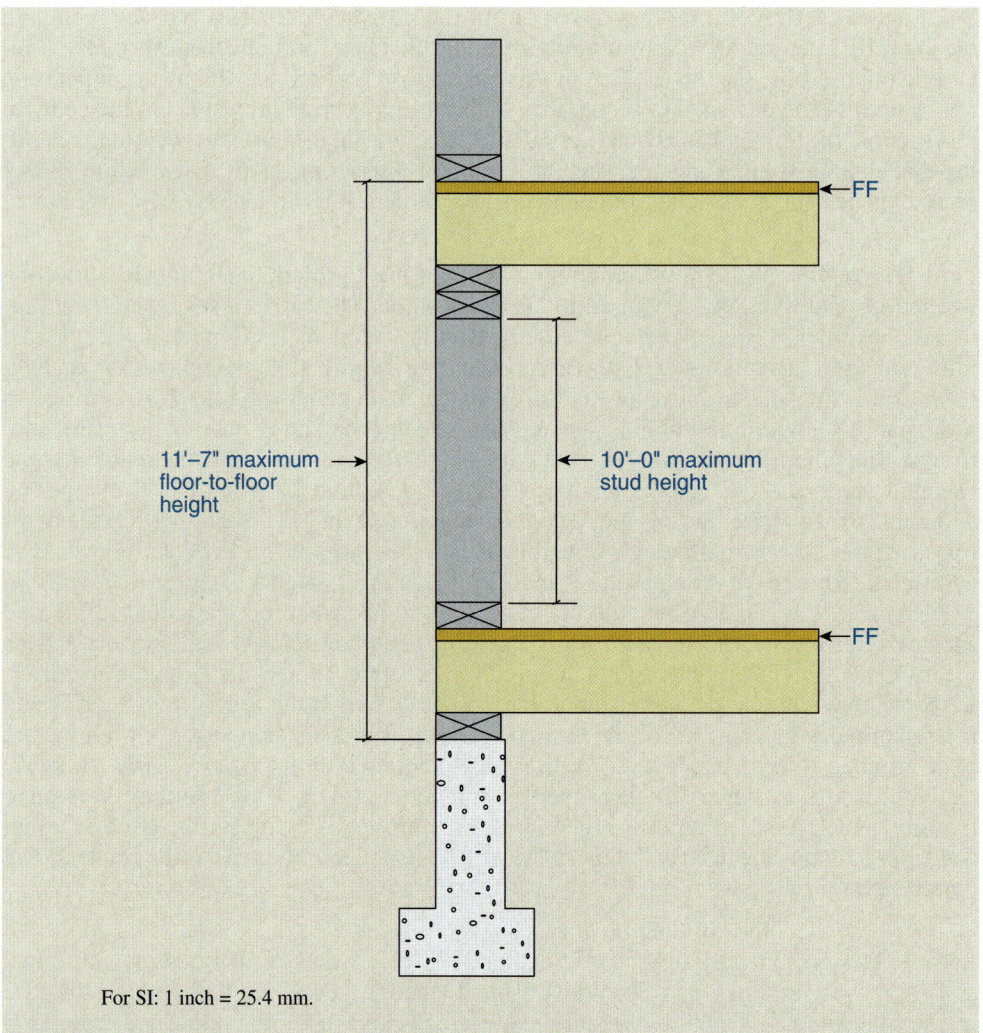

Figure 2308-6
Maximum floor-to-floor stud height.

5 inches thick or 50 psf is permitted to a height of 30 feet above a noncombustible foundation with an additional 8 feet permitted for gable ends, and (2) concrete or masonry fireplaces, heaters, and chimneys are permitted.

A change in the 2009 IBC involves the maximum wind speeds that are permitted with the conventional wood-frame construction provisions. The maximum wind speed is 100 mph with an exception allowing 110 mph if the building is sited in Exposure Category B. In the 2009 IBC, the exception was changed to apply only to buildings that are outside of the hurricane-prone region. In other words, buildings sited in the hurricane-prone region such as along the U.S. Atlantic Ocean and Gulf of Mexico coasts or in Hawaii, Puerto Rico, Guam, Virgin Islands, and American Samoa cannot use the exception. See Section 202 for definition of *hurricane-prone region*. The change to ultimate design wind speeds in the 2012 IBC will be discussed in more detail under Section 2308.2.1.

2308.2.1 Basic wind speed greater than 100 mph (3-second gust). A significant change to the wind design provisions of ASCE 7-10 involves the introduction of ultimate design wind speed, v_{ult}, as described in Section 1609.3. Because of the change to ultimate design wind speed, it was necessary to introduce the nominal design wind speed, v_{asd}, to relate to the basic wind speed in previous editions of the code. In other words, a *basic wind speed* of 100 mph in the 2009 IBC corresponds to a 100-mph *nominal design wind* speed in the 2012 IBC.

See Section 1609.3.1 for specific details on converting between ultimate design wind speed and nominal design wind speed and vice versa. Note that where wind load design is concerned, the ultimate design wind speed is used in conjunction with the wind load provisions in ASCE 7-10.

Where the nominal design wind speed, v_{asd}, exceeds 100 mph, special details are necessary to resist the higher wind pressures and ensure adequate load path continuity. This section allows the AWC WFCM or ICC 600 to be used where the nominal design wind speed exceeds 100 mph. In the 2009 IBC, the reference to SSTD 10 was deleted and replaced with the new ICC 600 standard, *Standard for Residential Construction in High Wind Regions*. The 2012 IBC references the 2012 edition of the AWC WFCM and the 2008 edition of the ICC 600 standard.

Buildings in Seismic Design Category B, C, D, or E. Additional requirements for structures in Seismic Design Categories B and C are covered in Sections 2308.11 and in Section 2308.12 for structures in Seismic Design Category D or E. — **2308.2.2**

Braced wall lines. Braced wall lines are the lateral-force-resisting elements in conventional construction and are analogous to shear wall lines in engineered structures. Braced wall lines are illustrated in Figure 2308.9.3. The structural elements must be adequately connected to ensure a continuous load path to effectively transfer wind or seismic lateral loads through the roof and floor systems to the braced wall panels that make up the braced wall lines. Braced wall panels may be any of the eight types described in Section 2308.9.3 subject to the requirements in Table 2308.9.3(1), as well as the additional seismic-related requirements in Sections 2308.11 and 2308.12. — **2308.3**

Spacing. The maximum 35-foot braced wall line spacing originated with the NEHRP. For buildings in Seismic Design Category D or E, Section 2308.12 reduces the maximum braced wall line spacing to 25 feet. — **2308.3.1**

Braced wall line connections. The general intent of the provisions in this section is to provide prescriptive requirements for achieving a continuous load path for lateral loads from the roof and floor system to the braced wall lines and braced wall panels. In order to clarify the intent, the section was subdivided into two parts in the 2012 IBC: Section 2308.3.2.1 contains requirements for the bottom plate connections, and Section 2308.3.2.2 contains requirements for the top plate connections. Recommended connection details for various framing conditions are shown in Figures 2308-7 through 2308-12. Figures 2308-13 through 2308-16 illustrate methods of transferring lateral loads from the roof system to braced wall panels. — **2308.3.2**

These connection details are designed to transfer shears of approximately 100–200 pounds per foot, which is consistent with the allowable shear resistance of the prescriptive braced wall panels.

A code change in the 2009 IBC clarified the connection requirements by referencing specific items in the prescriptive fastening Table 2304.9.1. The new language improves application of connection requirements at braced wall lines and eliminates redundant requirements. The section states that the purpose of the connections is to transfer wind and seismic lateral forces. The first paragraph clearly states that the connections apply to braced wall lines and not the braced wall panel portion of a braced wall line. For example, the connection between blocking or floor framing at a braced wall line top plate does not occur only at the braced wall panel, nor does the connection of the wall bottom plate to the foundation occur only at the braced wall panel. These connections are required along the entire braced wall line.

Bottom plate connection. The bottom plate must be connected to the supporting elements below such as joists, blocking, or the foundation plate. The section refers to Item 6 of Table 2304.9.1, which requires nails or staples spaced at 16 inches on center. — **2308.3.2.1**

Top plate connection. This section requires braced wall line top plates to be connected over their entire length to the framing above and refers to various fastening options in Table 2304.9.1. — **2308.3.2.2**

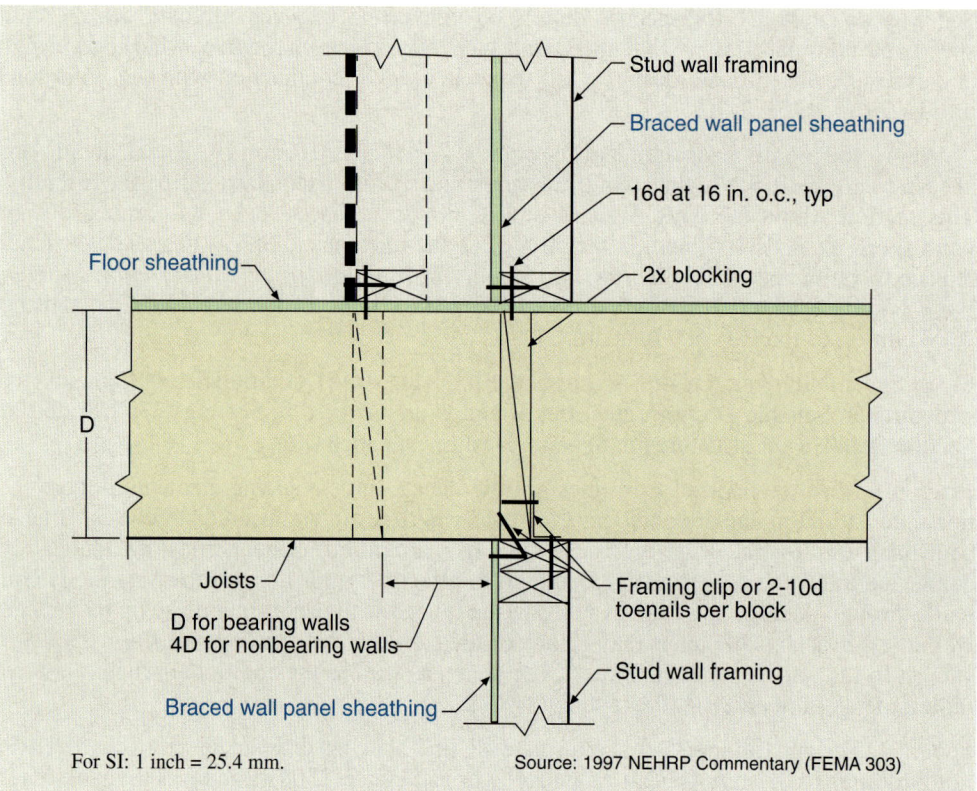

Figure 2308-7 Interior braced wall at perpendicular joist.

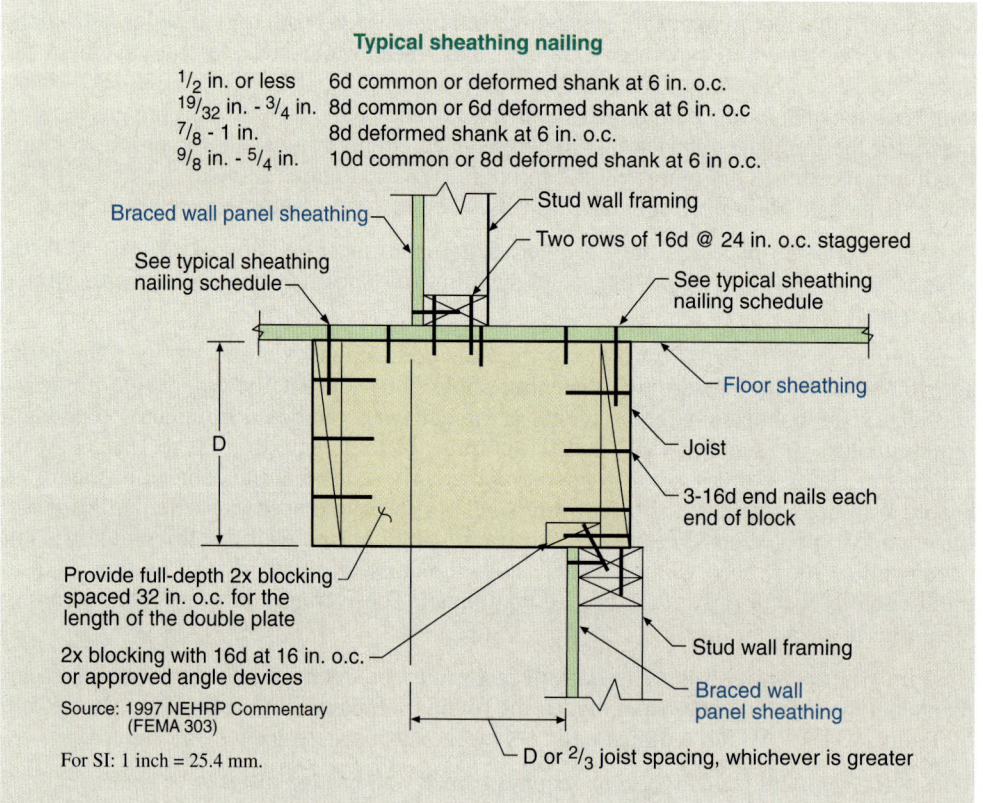

Figure 2308-8 Offset at interior braced wall.

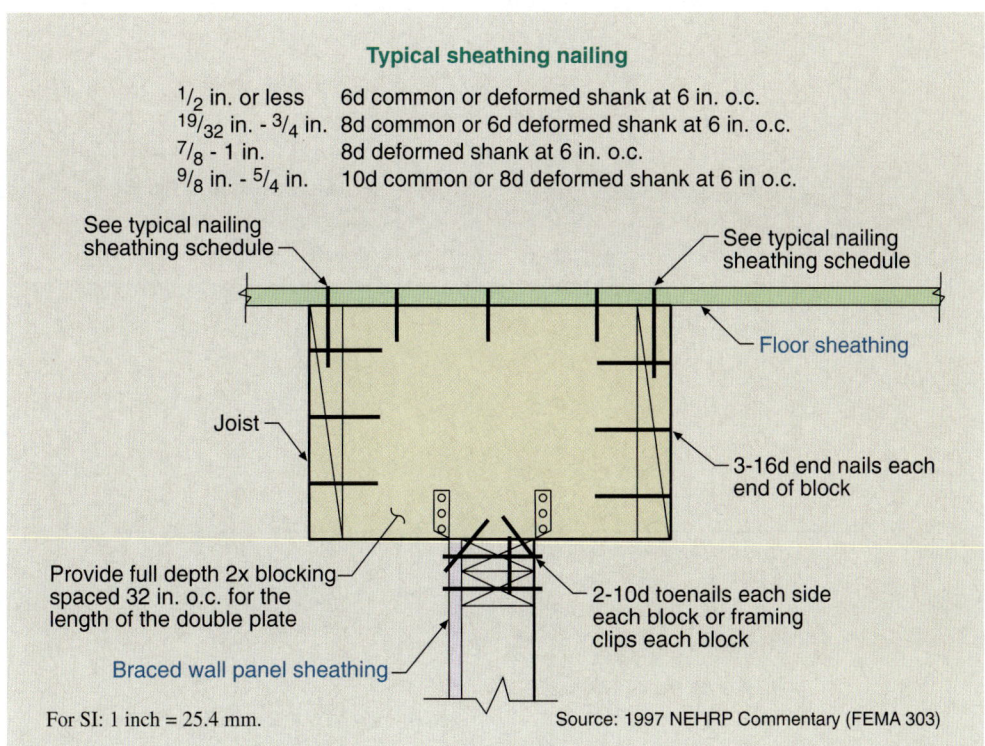

figure 2308-9
Diaphragm connection to braced wall below.

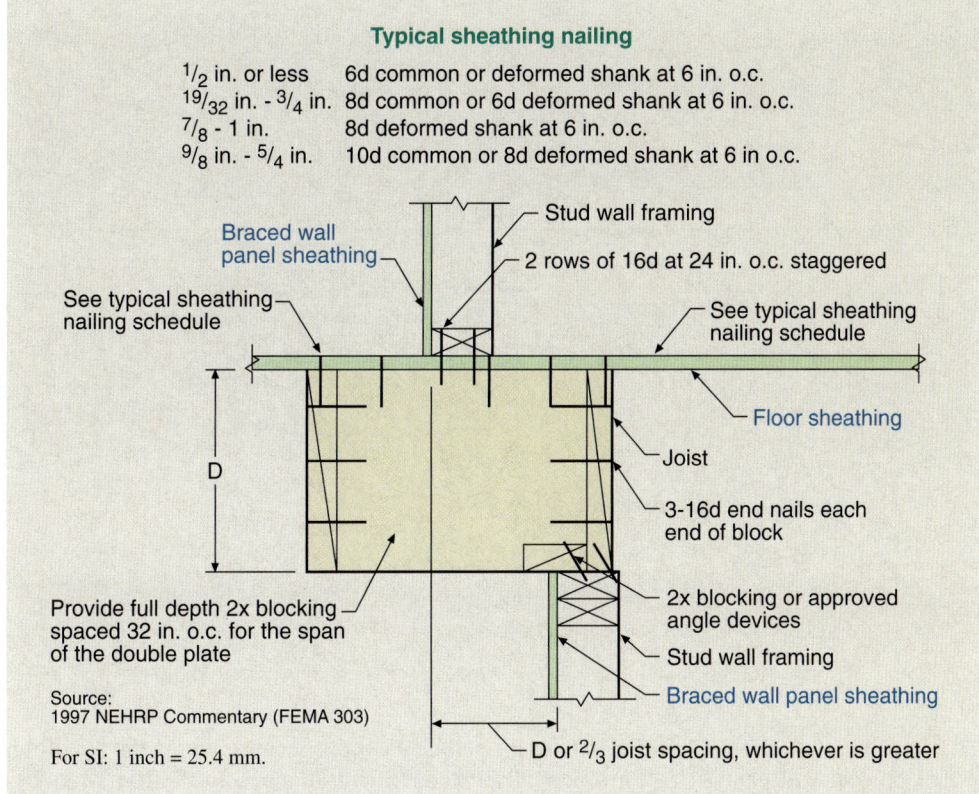

figure 2308-10
Offset at interior braced wall.

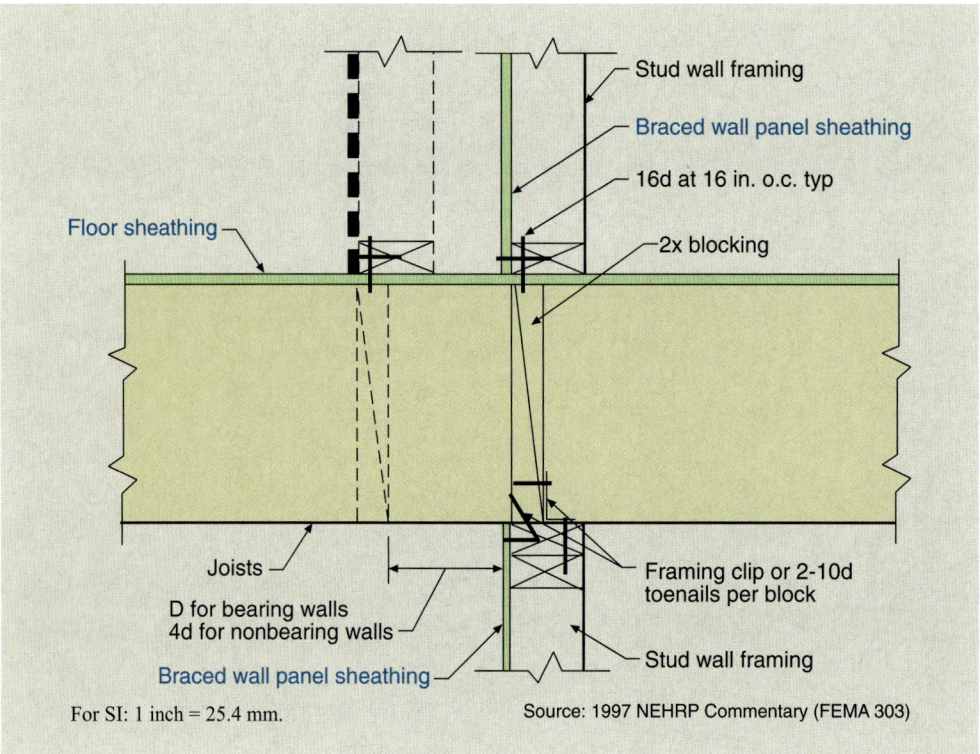

Figure 2308-11
Interior braced wall at perpendicular joist.

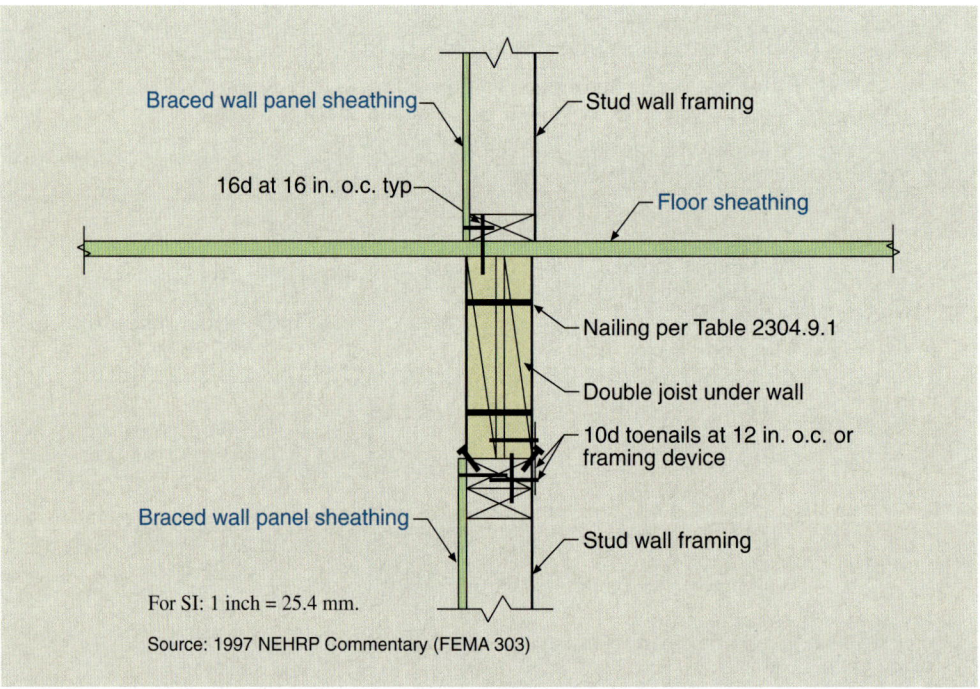

Figure 2308-12
Interior braced wall at parallel joist.

The wording was changed to refer to exterior gable end walls because that is the condition where braced wall panels can be practically extended and attached to the roof framing. The phrase "braced wall panel sheathing in the top story shall be extended" is to provide a clearer description of the requirement. The end of the first sentence of the exception

Conventional Light-Frame Construction

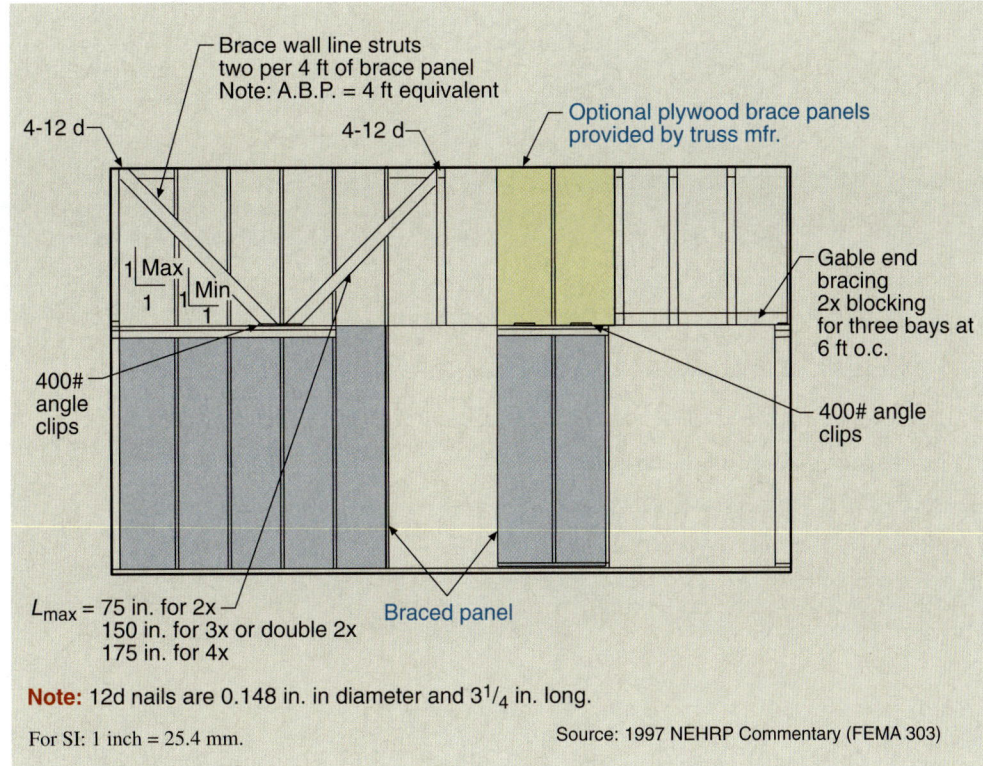

Figure 2308-13
Suggested method for transferring roof diaphragm loads to braced wall panels.

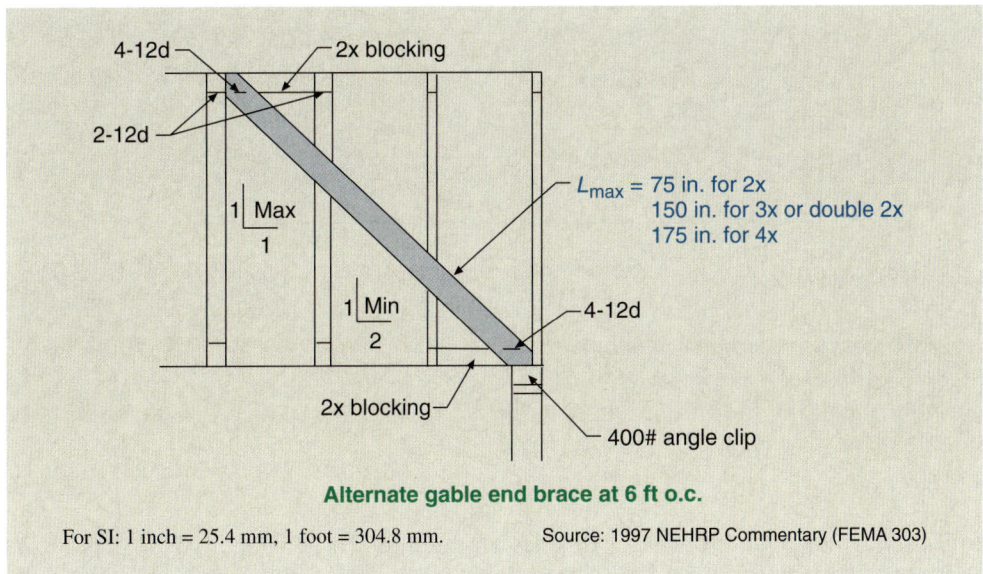

Figure 2308-14
Alternate gable end brace.

was modified to include the words "providing equivalent lateral force transfer" to give the building official a basis for accepting methods other than blocking between ends of trusses. Language was added to specify the required minimum blocking size and the connection of the blocking to the braced wall top plate because the minimum size of blocking was not previously specified. Additional guidance on permissible notching or drilling through the blocking was added in the 2012 IBC by reference to Section 2308.8.2 (floor joists) and Section 2308.10.4.2 (ceiling joists and rafters).

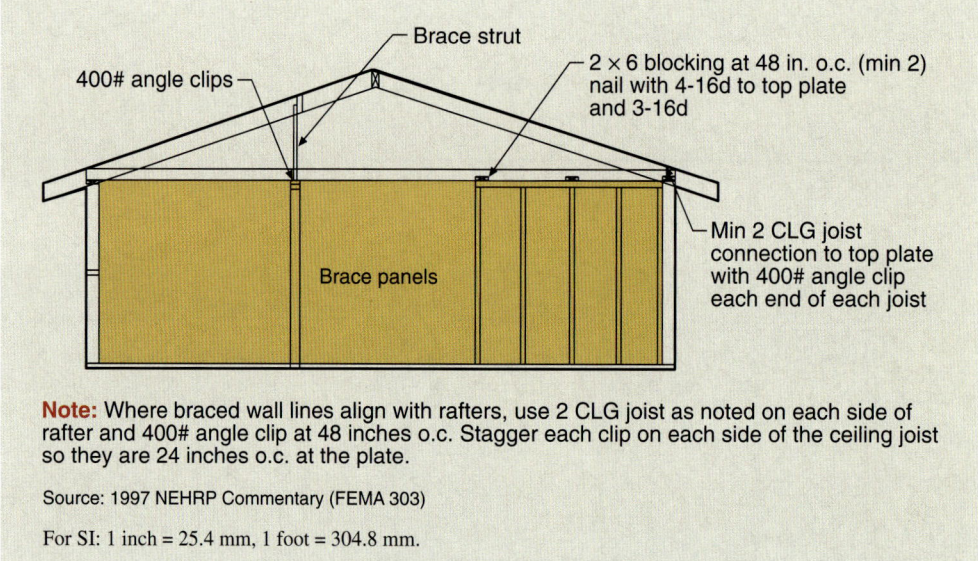

Figure 2308-15 Wall parallel to truss bracing detail.

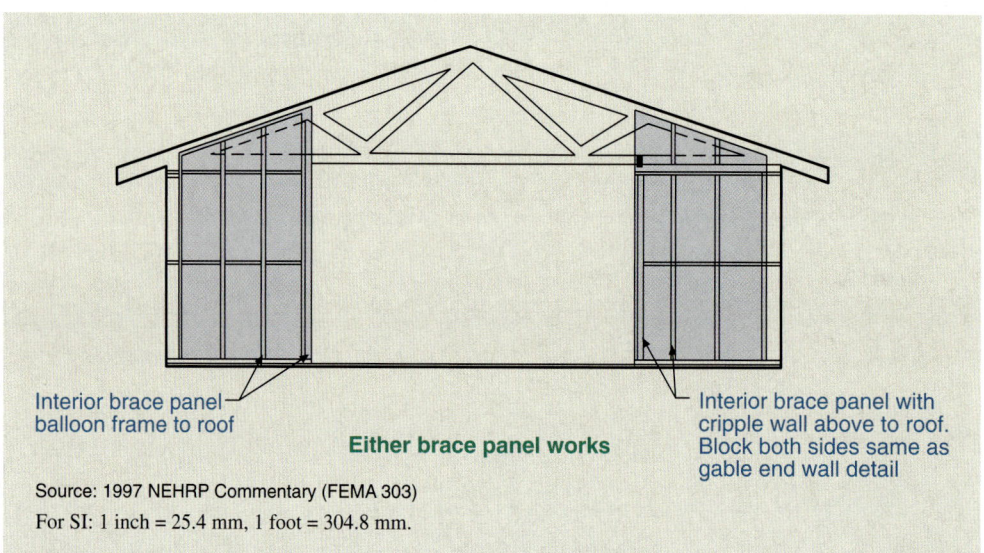

Figure 2308-16 Wall parallel to truss alternate bracing detail.

2308.3.3 Sill anchorage. Where braced wall lines are required to be supported by continuous footings, sills must be anchored to the foundation with $^1/_2$-inch-diameter anchor bolts with a maximum spacing of 6 feet on center, or 4 feet on center for structures over two stories in height. Buildings in Seismic Design Category E require $^5/_8$-inch-diameter anchor bolts in accordance with Section 2308.12.9. See Section 2308.3.4 for braced wall line support requirements, and discussion under Section 2308.6 for anchorage requirements.

2308.3.3.1 Anchorage to all wood foundations. This section requires that anchorage to wood foundations be designed to have at least as much capacity as is required by Section 2308.3.3 for conventional concrete or masonry foundations.

2308.3.4 Braced wall line support. Braced wall lines must be supported by continuous footings, except that interior braced wall lines need not be supported by continuous footings if the maximum plan dimension of the structure does not exceed 50 feet. In this case, the lateral loads on interior-braced wall lines are assumed to be transferred to the exterior foundation by the floor system or slab on grade.

2308.4 Design of elements. Where a structure consists of a combination of engineered and conventional construction, only those engineered elements must be designed to resist the loads and forces specified in Chapter 16. The performance of engineered elements should be compatible with the performance of a conventionally framed system. For example, the stiffness of the engineered elements should be approximately the same as the stiffness of the conventional construction unless a stiffness analysis for proportioning the loads between the conventional and engineered construction is performed. The engineering needs only to demonstrate compliance of the nonconventional elements with applicable provisions of the code. A significant change in the 2006 IBC expanded and clarified Section 2308.4.1, which requires elements of buildings of otherwise conventional construction that exceed the limits of Section 2308.2 to have those elements and that the supporting load path be designed in accordance with accepted engineering practice and other provisions of the code. The code change clarified the "design of portions" (Section 2308.1.1) requirements by distinguishing between *portions* (Section 2308.1.1) and *elements* (Section 2308.4.2). The new language also clarifies that the code permits elements and members as well as rooms or a series of rooms to be engineered in an otherwise conventionally constructed building. Thus, many so-called conventional wood-framed buildings have elements and portions that are engineered.

2308.4.1 Elements exceeding limitations of conventional construction. When a conventionally constructed building contains structural elements that exceed the limits of Section 2308.2, the elements and their supporting load path must be designed in accordance with the engineering provisions of the code.

2308.4.2 Structural elements or systems not described herein. When a conventionally constructed building contains structural elements or systems that are not specifically covered by Section 2308, the elements and their supporting load path must be designed in accordance with the engineering provisions of the code. Some examples of such elements are engineered floor and roof trusses, and glued laminated or LVL beams, which are very common nowadays.

2308.5 Connections and fasteners. The minimum connections and fastener requirements for conventional construction are the same as the minimum required for engineered construction. See Section 2304.9 and Table 2304.9.1 for minimum fastening requirements. See Table 2308.10.4.1 for rafter tie connections as a function of rafter slope and snow load. The number of 16d nails required by Table 2308.10.4.1 for rafter tie connections can be significantly greater than the minimum nailing requirements of Table 2304.9.1.

Improper nail sizes have reportedly been used in wood-frame building construction because the pennyweight system of specifying nail sizes is not universally understood. Code users often focus on pennyweight (8d - 8 penny, 16d - 16 penny, etc.) rather than pay proper attention to the specific style of nail such as common, box, cooler, sinker, finish, and so on. A typical example is substitution of box nails for common nails of the same pennyweight. The specific style of nail is critical because there can be significant differences in strength properties of connections fastened with nails of the same pennyweight but of different nail style. A code change to the 2006 IBC added the nominal dimensions of nails in the fastening tables in an effort to avoid confusion and help reduce the number of misapplications in nailed connections. The fastener industry expected some reluctance on the part of some in the building construction community to completely abandon the pennyweight system of designating nail sizes, so the IBC continues to maintain the pennyweight system designations. Because nominal dimensions are not as subject to misinterpretation, the actual shank length and diameter were added in parentheses for the various styles of nails used in wood connections.

2308.6 Foundation plates or sills. Prescriptive requirements for foundation plates and anchorage are given in this section, although these provisions apply only to conventional construction. The requirements in Section 2304.3.1 apply to both engineered and conventional construction. Foundation (footing) requirements for light-frame construction are contained in Section 1809.7.

A change in the 2009 IBC added the phrase, "spaced to provide equivalent anchorage as the steel bolts" to give a criterion for determining the spacing requirements for other anchorage methods such as anchor straps. See Sections 2308.12.8 and 2308.12.9 for discussion of sill plate anchorage requirements for buildings in Seismic Design Categories D and E.

2308.7 Girders. The design of girders to support floor loads involves so many variables that the tables in the code are only practical for the most simple cases. In many cases, girders should be engineered because of unusual configurations or loading conditions. Figures 2308-17 and 2308-18 show acceptable details deemed to comply with the splice and tie requirements for posts and girders.

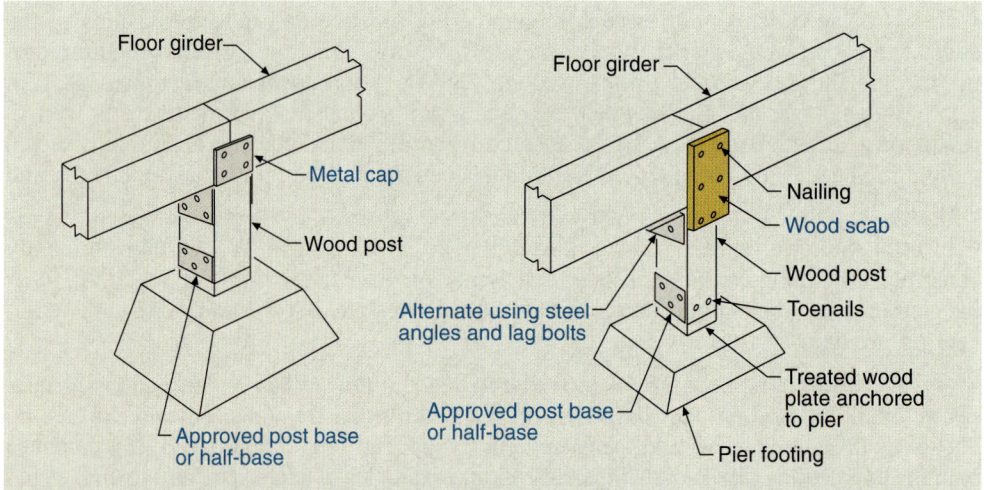

Figure 2308-17 Post-to-girder connection.

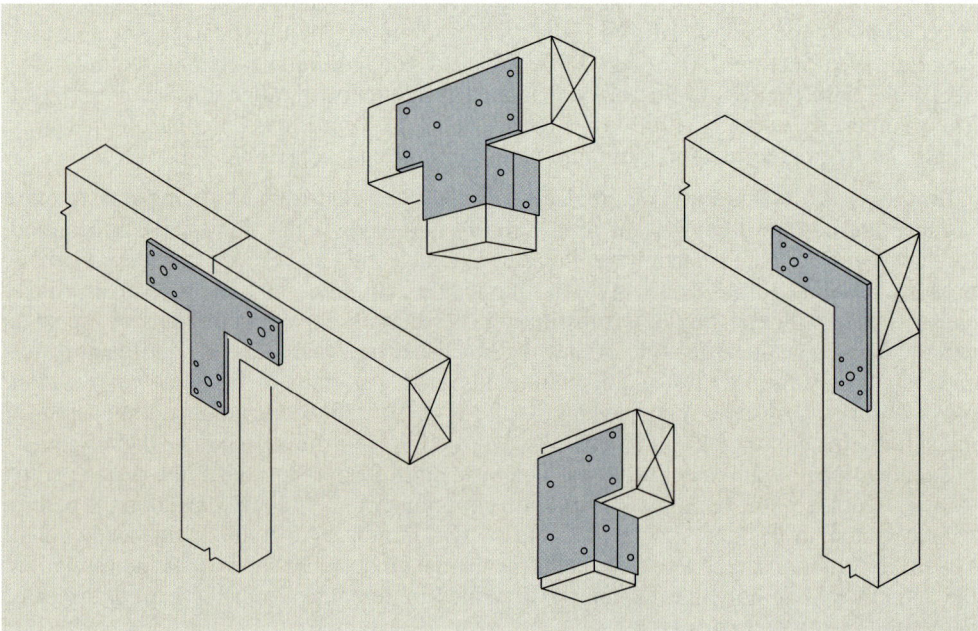

Figure 2308-18 Post-to-girder connection.

Floor joists. Tables 2308.8(1) and 2308.8(2) are excerpted from the AF&PA span tables for joists and rafters. Joists selected from these tables are deemed to comply with code requirements for both strength and deflection. Note that Table 2308.8(1) is for a floor live load of 30 psf because Table 1607.1 allows a 30 psf floor live load in residential one- and two-family sleeping areas and habitable attics, which originated with the *National Building Code*. The other legacy codes typically require a 40 psf floor live load throughout the residence.

2308.8

Bearing. The code requires a minimum of $1\frac{1}{2}$ inches of bearing on wood or metal and 3 inches of bearing on concrete or masonry. The 1×4 ribbon (ledger) nailed with 3–16d common nails to the studs was common in older balloon-framed multistory structures but is uncommon today. Figures 2308-19 and 2308-20 illustrate minimum bearing requirements.

2308.8.1

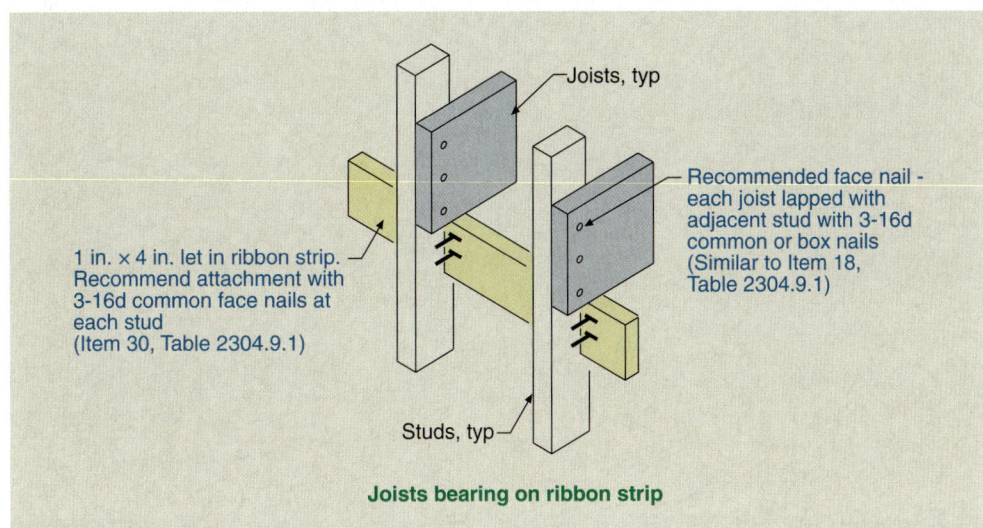

Figure 2308-19
Bearing requirements.

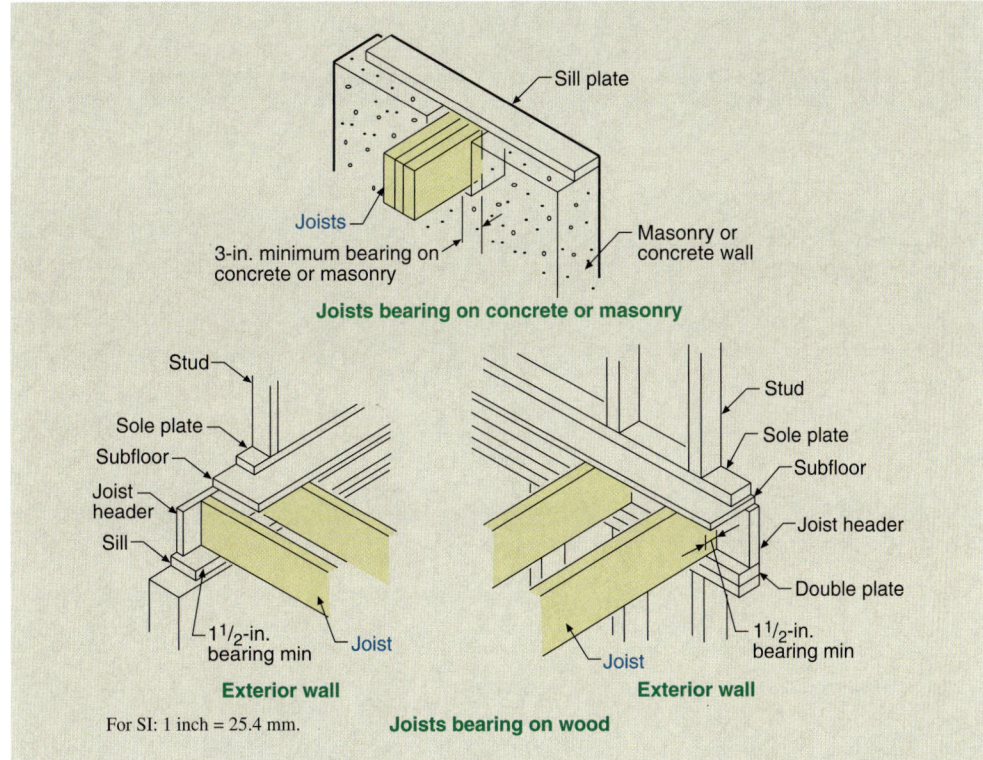

Figure 2308-20
Bearing requirements.

2308.8.2 Framing details. This section contains basic requirements for framing floor joist systems. Figure 2308-21 depicts the provisions for support by solid blocking or other means, and joists framing into a girder. Figure 2308-22 illustrates the allowed notches or bored holes. Figure 2308-23 depicts the requirements for lapping and tying joists over a beam or partition.

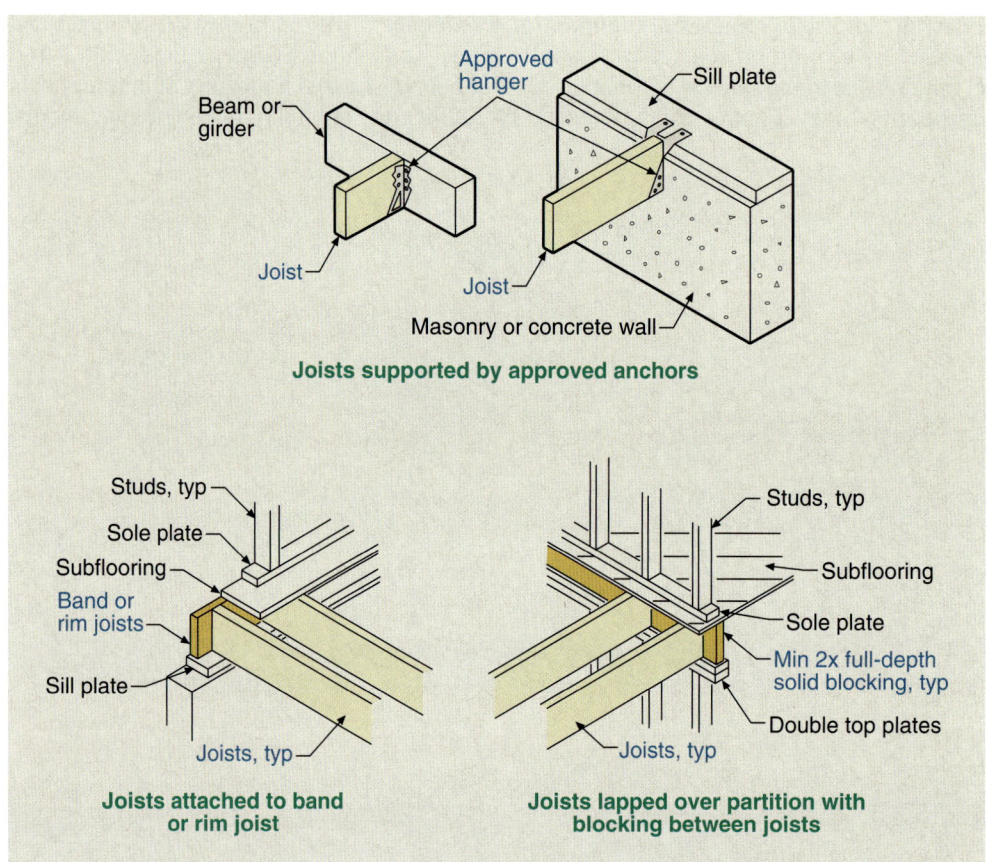

Figure 2308-21 Framing details.

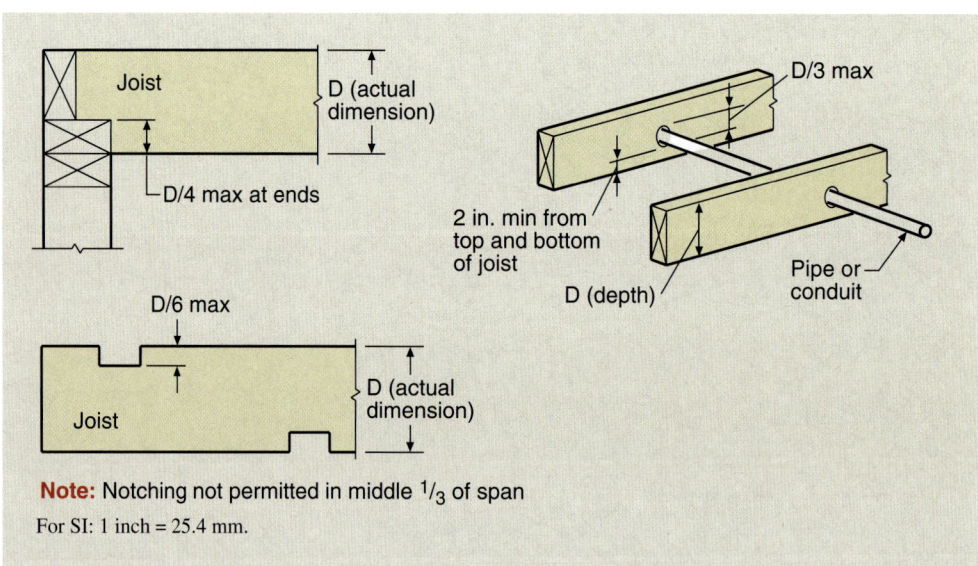

Figure 2308-22 Cut, notched, or bored holes.

Conventional Light-Frame Construction

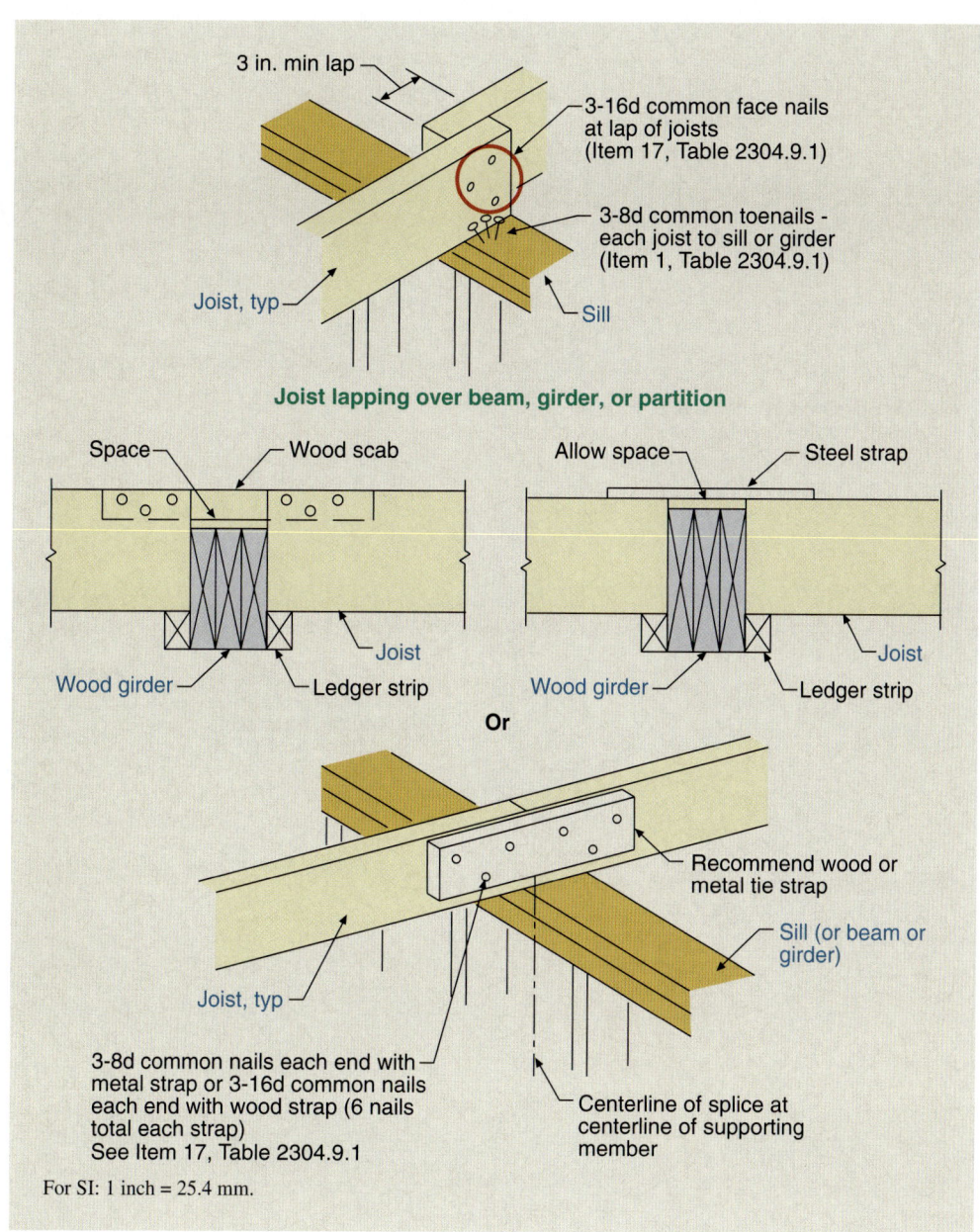

Figure 2308-23
Floor joists tied over wood beam, girder, or partition.

Engineered wood products. This section specifically prohibits cuts, notches, and holes bored in trusses, structural composite lumber, structural glue-laminated members, or I-joists, unless specifically permitted by the manufacturer or where the registered design professional has considered their effects on the stiffness and resistance capacity of the member.

2308.8.2.1

Framing around openings. Figures 2308-24 through 2308-26 illustrate typical framing details for floor openings. Figure 2308-24 depicts acceptable framing details for openings not greater than 4 feet. Figure 2308-25 depicts acceptable framing for openings greater than 4 feet but not greater than 6 feet. Figure 2308-26 shows framing and hangers for openings greater than 6 feet.

2308.8.3

Supporting bearing partitions. Figure 2308-27 depicts the requirement for double joists supporting bearing partitions.

2308.8.4

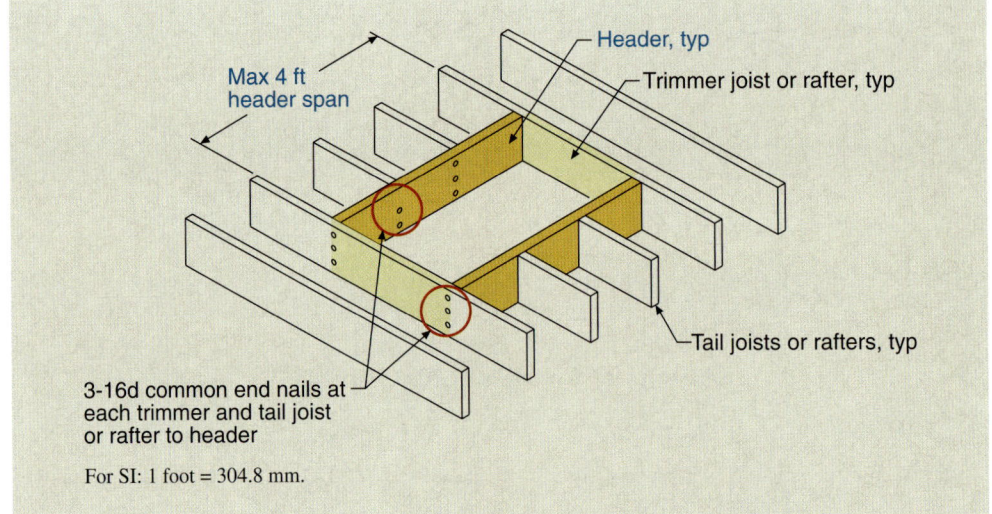

Figure 2308-24 Framing around openings— header span ≤ 4 feet.

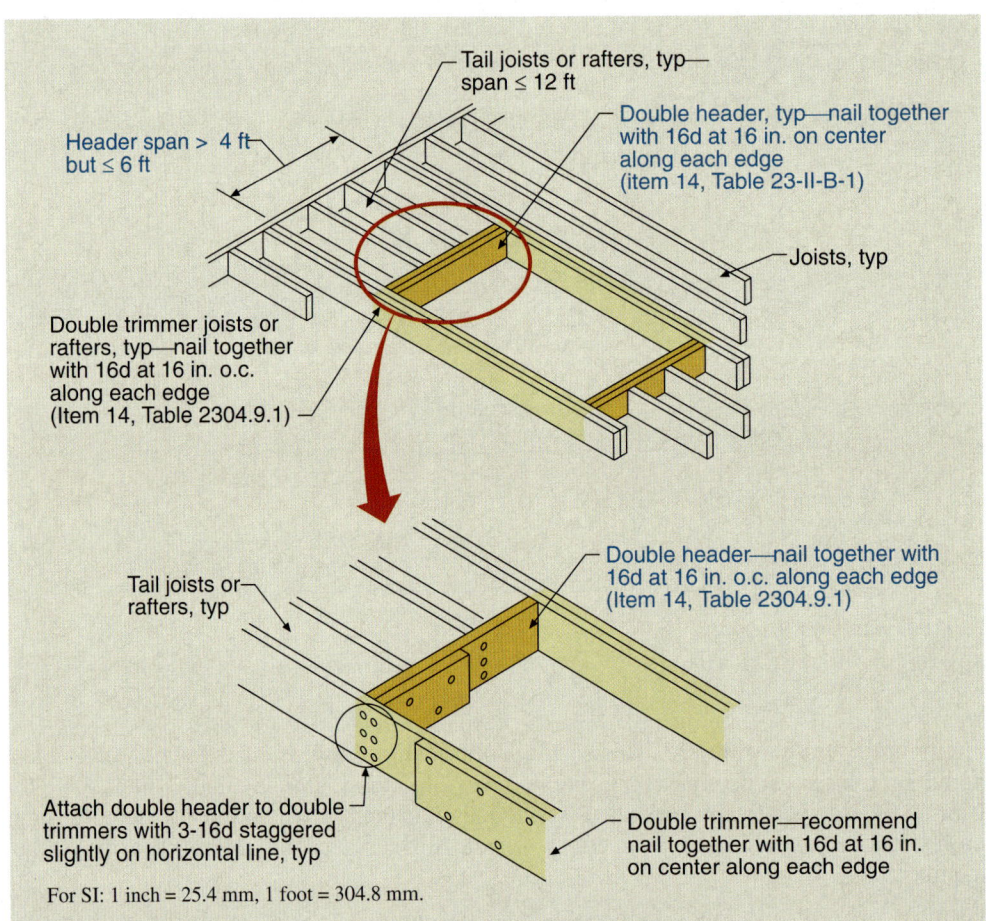

Figure 2308-25 Framing around openings header > 4 feet but ≤ 6 feet.

2308.8.5 Lateral support. Rectangular members require lateral support to avoid buckling in the weak direction. The requirements are based on the *nominal* depth-to-thickness (d/b) rather than the actual dimensions of the member. Thus, a 2 × 12 has a d/b ratio equal to 12/2 = 6. Figure 2308-28 depicts the lateral support requirements.

Conventional Light-Frame Construction

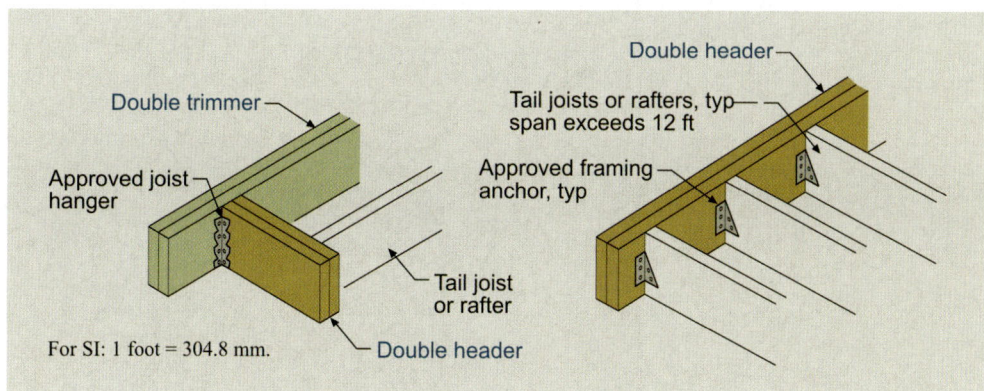

Figure 2308-26
Framing around opening—header span > 6 feet.

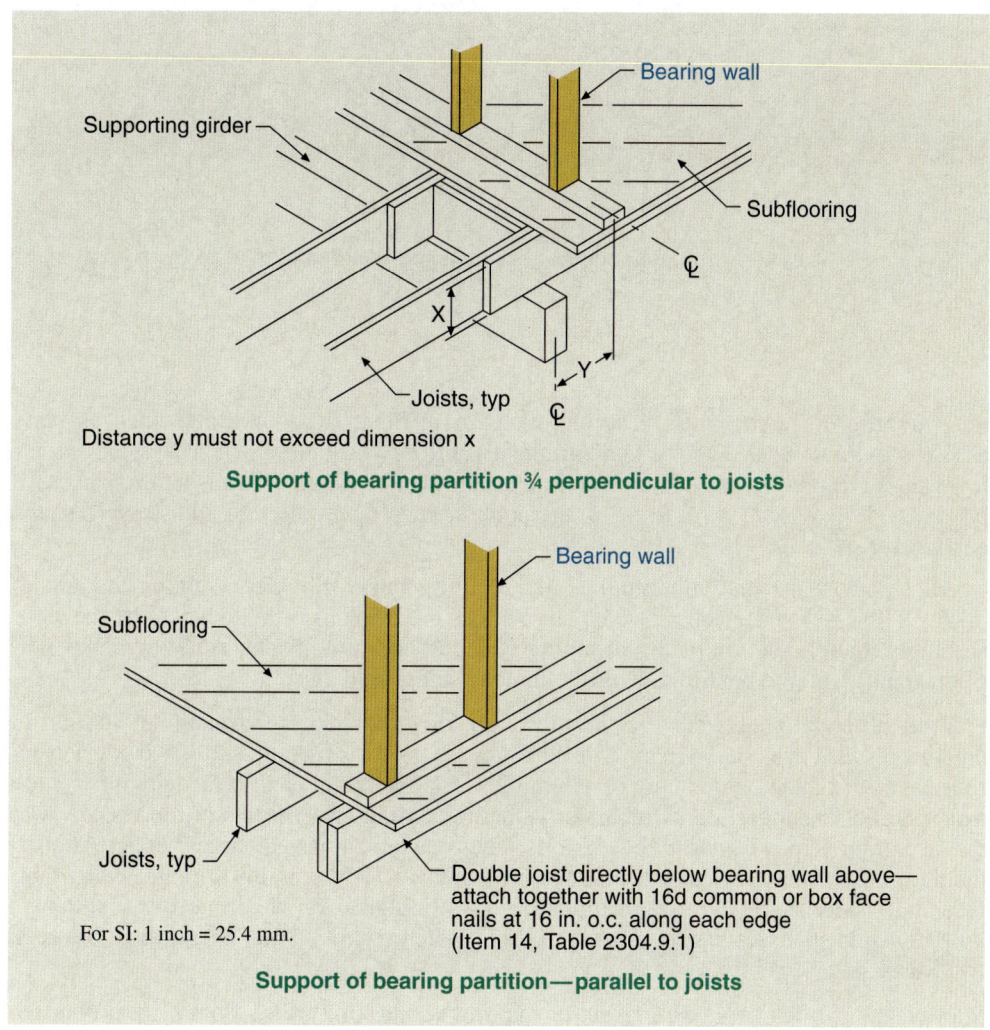

Figure 2308-27
Supporting bearing partitions.

2308.8.6 **Structural floor sheathing.** This section references Section 2304.7.1 and Tables 2304.7(1) through 2304.7(4) for installation requirements for floor sheathing. Flooring systems may consist of a subfloor on which may be placed an underlayment for the floor covering or a combined subfloor and underlayment system, upon which a finished floor-surfacing material is applied. The finished floor-surfacing material may be either wood-strip flooring,

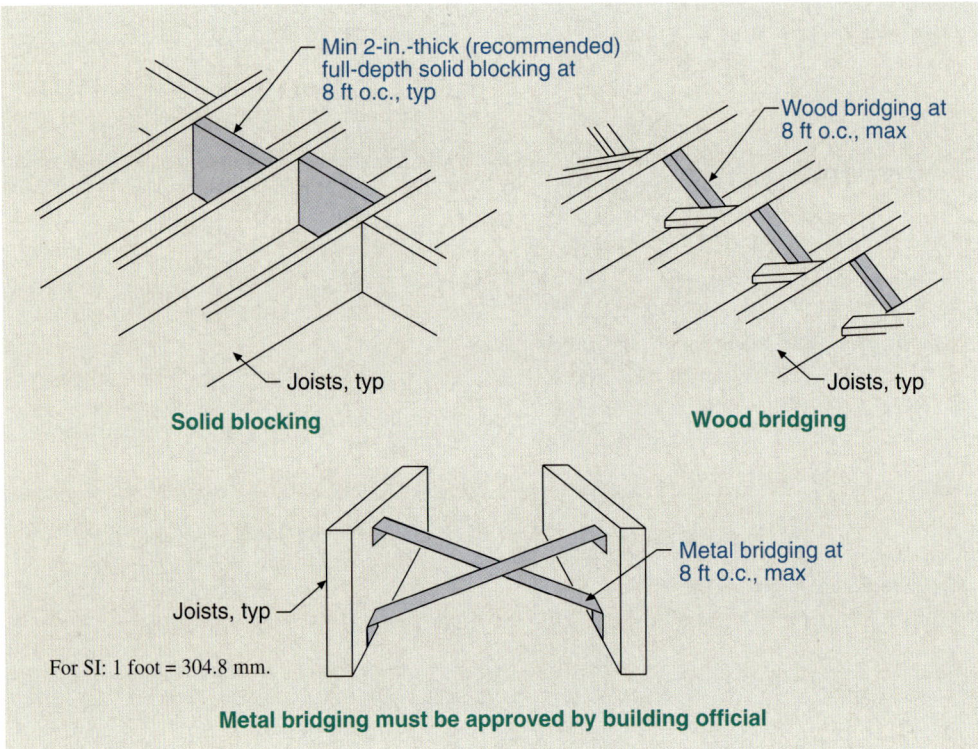

Figure 2308-28 Lateral support requirements.

tongue-and-groove flooring, or various types of resilient floor coverings such as vinyl asbestos tile or carpet. For the noncombined subfloor and underlayment system, underlayment is required to provide a smooth, even surface to which the finished flooring will be attached. However, in some cases, the underlayment is required to add strength to the subflooring.

Allowable spans and minimum grade requirements for lumber subflooring are in Tables 2304.7(1) and 2304.7(2). The span tables are based on the thickness of the floor sheathing, the orientation of the sheathing with respect to the joists (either perpendicular or diagonal), and the board grade of the lumber being used.

Wood structural panels may be manufactured for use as either structural subflooring or combined subfloor-underlayment. The allowable spans for structural subflooring and combined subfloor-underlayment are based on the wood structural panel's face grain (strength axis) parallel to supporting members or its being continuous over two or more spans with the face grain perpendicular to the supports. These qualifications are critical in determining the permissible spans. Most wood structural panels are considerably stronger when their face grain is perpendicular to the supports and continuous over two or more spans. Panels with multiple spans have greater capacity than when they are simply supported between two joists.

To create a stiffer floor and prevent or minimize squeaking of the floor system after the building has been in use, the subfloor may be glued to the joists. This gluing prevents the relative movement between the panel and the joist that takes place when loads are placed on the floor, and provides additional stiffness.

Particleboard can be used as underlayment, structural subflooring, or as combined subfloor-underlayment. Where used as underlayment, the code permits Type PBU particleboard in accordance with ANSI A208.1.

Conventional Light-Frame Construction

Under-floor ventilation. See discussion of Section 1203.3 for under-floor ventilation requirements. — **2308.8.7**

Wall framing. This section contains typical framing requirements for conventionally framed stud walls. — **2308.9**

Size, height, and spacing. Figure 2308-29 depicts typical stud requirements. A code change in the 2009 IBC added the last two sentences in this section, requiring wall studs to be continuous from a support at the sole plate to a support at the top plate to resist out-of-plane wind loads perpendicular to wall. The requirement does not apply to the trimmers (jack studs) and cripple studs at openings in walls. Where scissor trusses are used to create a vaulted ceiling, a scissor truss is often used at the gable end wall, and the top plate is not to be laterally supported because the ceiling follows the profile of the bottom chord of the truss. See Figure 2308-30. The new code language clarifies that the studs are required to be laterally supported at the sole plate and top plate to resist out-of-plane loads perpendicular to wall. Where scissor trusses are used to create a vaulted ceiling profile, the wall studs at the gable end wall should be balloon framed up to the bottom chord of the truss or to the roof sheathing. If the studs are not supported at the top by a ceiling or roof deck in accordance with this requirement, then an engineered design should be provided for the gable end wall studs to resist out-of-plane (component and cladding) wind loads perpendicular to wall. — **2308.9.1**

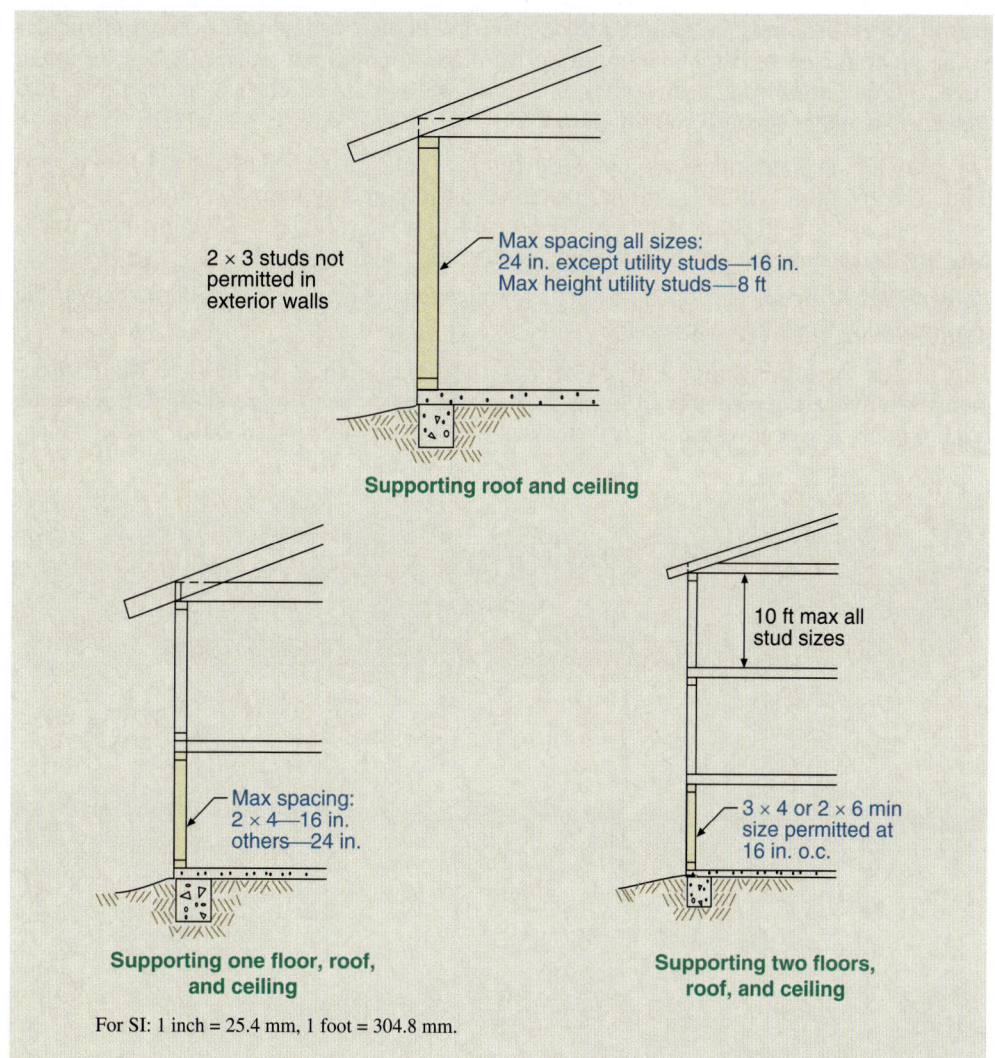

Figure 2308-29
Stud requirements.

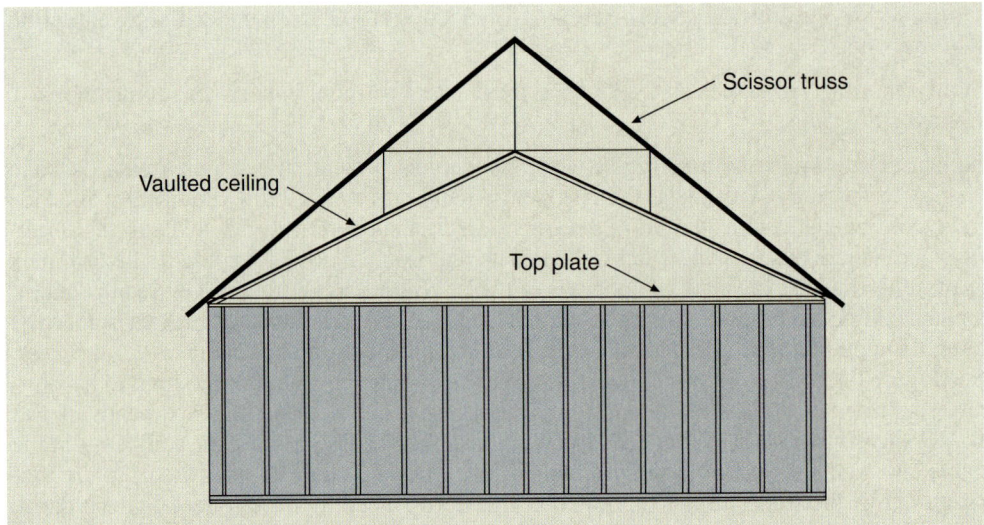

Figure 2308-30
Vaulted ceiling at gable wall.

2308.9.2 Framing details. Studs are required to be placed with their wide dimension perpendicular to the plane of the wall for maximum out-of-plane strength and stiffness. Not less than three studs are installed at each corner of exterior walls, although the exception allows two stud corners under specific conditions.

2308.9.2.1 Top plates. Top plate splices require eight 16d face nails on each side of the 4-foot lapped joint. The 16d common nails or other fasteners are required by Item 10 of Table 2304.9.1. Single plate splices must have at least six 8d nails on each side of the splice with a 3-inch-wide by 6-inch-long by 0.036-inch-thick steel strap.

Figure 2309-31 depicts the double top plate splice requirement; Figure 2308-32 depicts the recommended single top plate splice.

2308.9.2.2 Top plates for studs spaced at 24 inches (610 mm). The code has specific requirements where bearing wall studs are spaced 24 inches on center. Figure 2308-33 depicts the requirements for locating the joists or trusses where studs are spaced 24 inches on center.

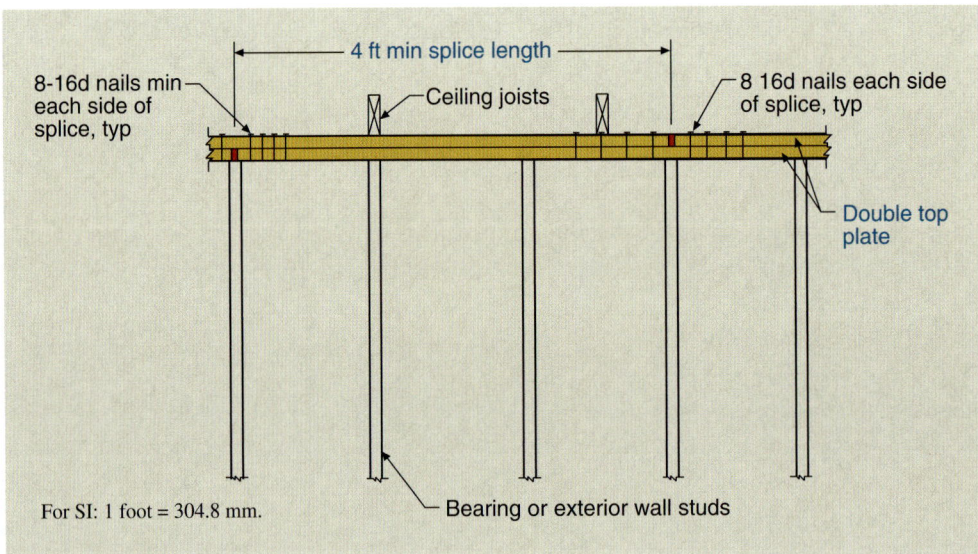

Figure 2308-31
Double top plate splice.

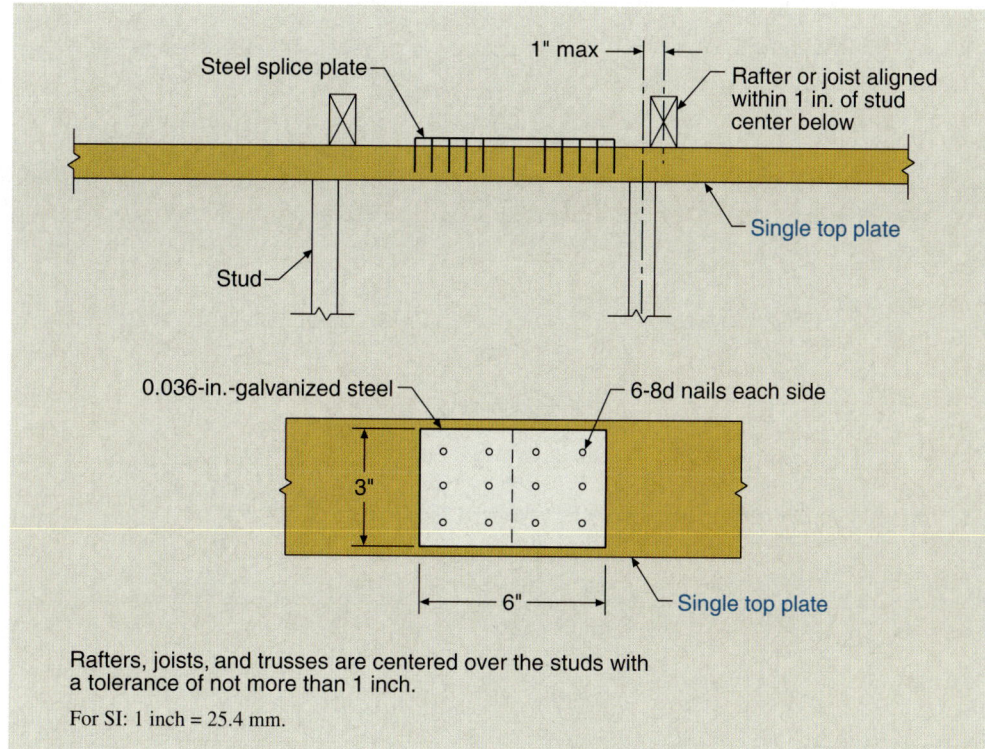

Figure 2308-32
Single top plate splice—bearing and exterior walls.

Nonbearing walls and partitions. This section allows for increased stud spacing on nonbearing walls, and studs may be turned with the long dimension parallel to the wall for construction of plumbing chases. See Section 202 for definition of nonbearing wall. (For wood-frame construction, a nonbearing wall is a stud wall that supports 100 plf or less superimposed vertical load.) Note that sheathing materials must be increased in thickness when wider stud spacing is used. Partitions must be capped with at least a single top plate to provide overlapping at corners and at intersections with other walls and partitions and continuously tied at joints by 16-inch-long solid blocking and equal to the plate size, or by $1/2$-inch by $1\frac{1}{2}$-inch metal ties fastened with two 16d nails on each side of the joint.

2308.9.2.3

Plates or sills. Studs are required to have full bearing on a 2-inch nominal plate or sill at least as wide as the supported stud.

2308.9.2.4

Bracing. The bracing requirements for conventional construction are depicted in Figure 2308-34. As required by Section 2308.3, braced wall lines are required at interior and exterior wall lines at a maximum spacing of 25 or 35 feet, depending on the seismic design category assigned to the building. The maximum spacing of braced wall lines in Seismic Design Categories A, B, and C is 35 feet and in Seismic Design Categories D and E is 25 feet. Prescriptive requirements for eight different braced wall panel types are described in this section. Method 3 is wood structural panel sheathing. The minimum thickness of wood structural panel sheathing was changed in the 2009 IBC from $5/16$ inch to $3/8$ inch to reflect current manufacturing practice because $5/16$-inch wood structural panels are a very small fraction of the panels produced today.

2308.9.3

Early editions of the IBC require braced wall panels to start not more than 8 feet from each end of a braced wall line. A code change to the 2003 IBC revised the maximum distance that braced wall panels are permitted to start from the end of a braced wall line to be 12.5 feet, which is consistent with the provisions in the IRC for SDCs A, B, and C. The requirement that a designed collector be provided if the bracing begins more than

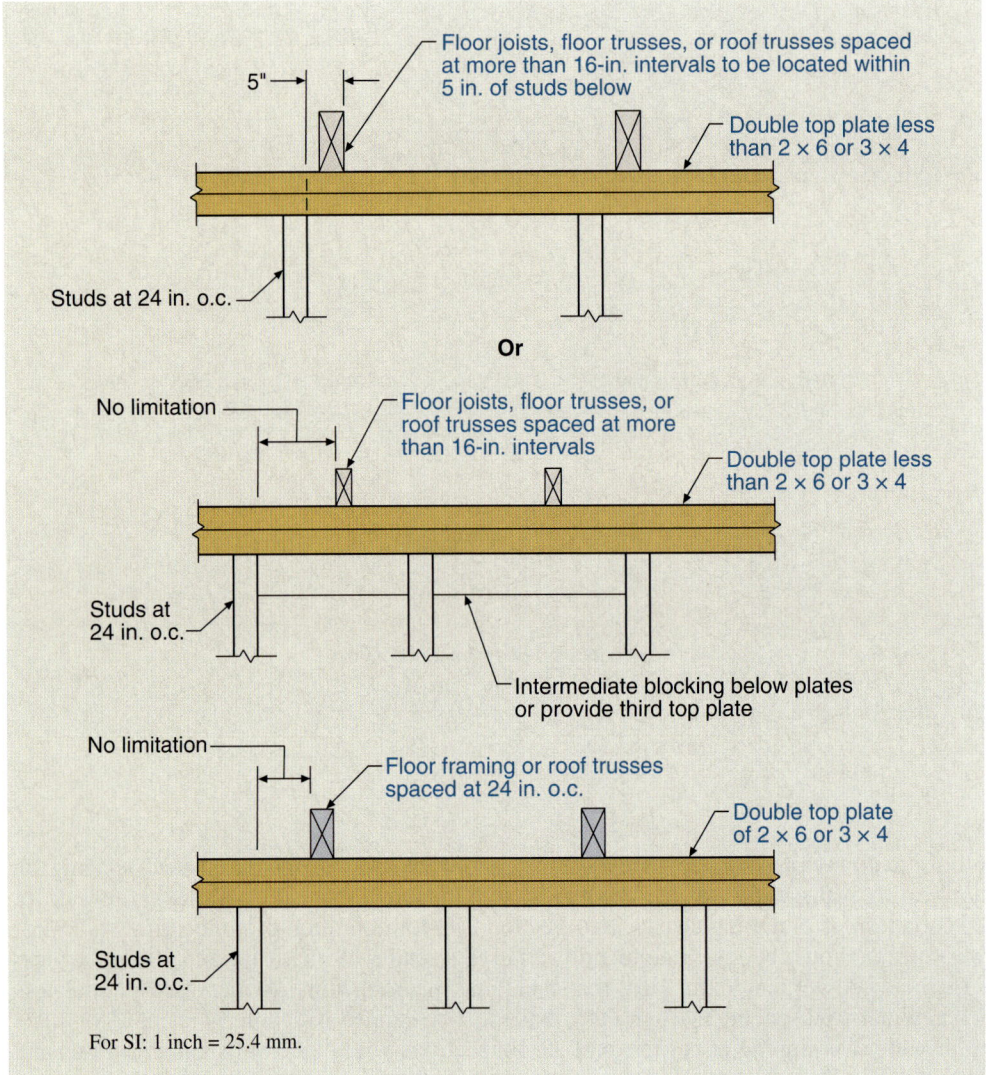

Figure 2308-33
Top plate— limitations bearing.

12.5 feet from the end of a braced wall line was deleted in the 2006 IBC because it made no sense within the context of prescriptive conventional construction. The collector is an engineered component that must be designed to resist the governing lateral load, wind, or seismic, and requires the determination and distribution of the governing lateral load to properly design it. In addition, the result is an engineered component that is designed to transfer engineered lateral loads to prescriptive braced wall panels. The solution was to either provide prescriptive rules for collectors in various configurations based on the capacity of the wall line or delete the provision from the code. A code change resolved the problem in the 2006 IBC by deleting the collector requirement. The code now requires braced wall panels to be located not more than 12.5 feet from the ends of braced wall lines, and if braced wall panels do not meet this requirement, then the bracing system does not conform to prescriptive requirements and engineering is required. In this case, the design of the collector is one essential part of the engineering for the lateral-force-resisting system. Because the code allows partial engineering, the design could consist of the collector and shear wall line. Figure 2308-35 depicts the braced wall panel location requirements.

Conventional Light-Frame Construction

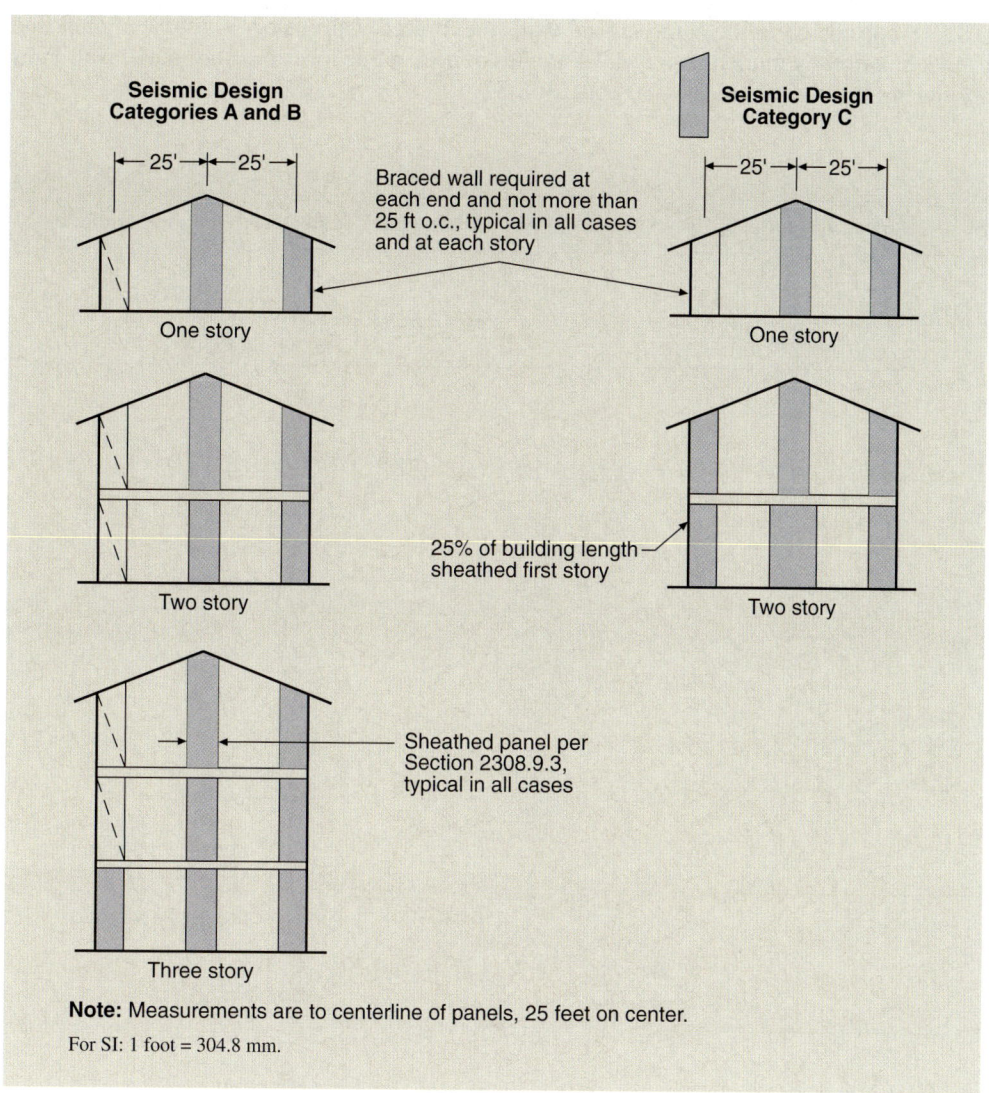

Figure 2308-34
Wall bracing panel.

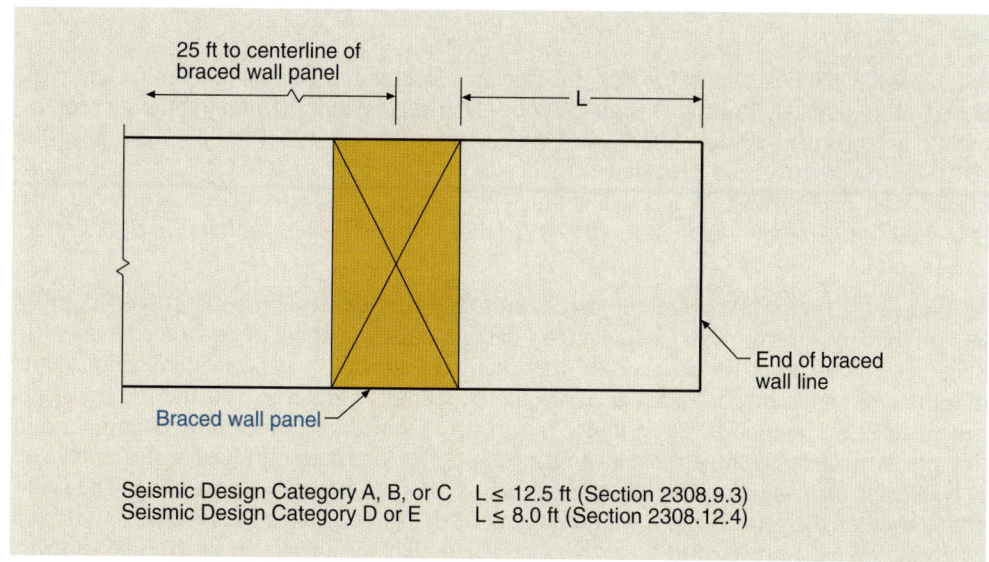

Figure 2308-35
Wall bracing panel location.

To be considered in the same braced wall line, braced wall panels in a braced wall line cannot be offset by more than 4 feet. Examples of acceptable locations for braced wall lines are illustrated in Figures 2308-36 and 2308-37.

Seismic Design Category	Maximum Wall Spacing (feet)	Required Bracing Length, b
A, B, and C	35'-0"	Table 2308.9.3(1) and Section 2308.9.3
D and E	25'-0"	Table 2308.12.4

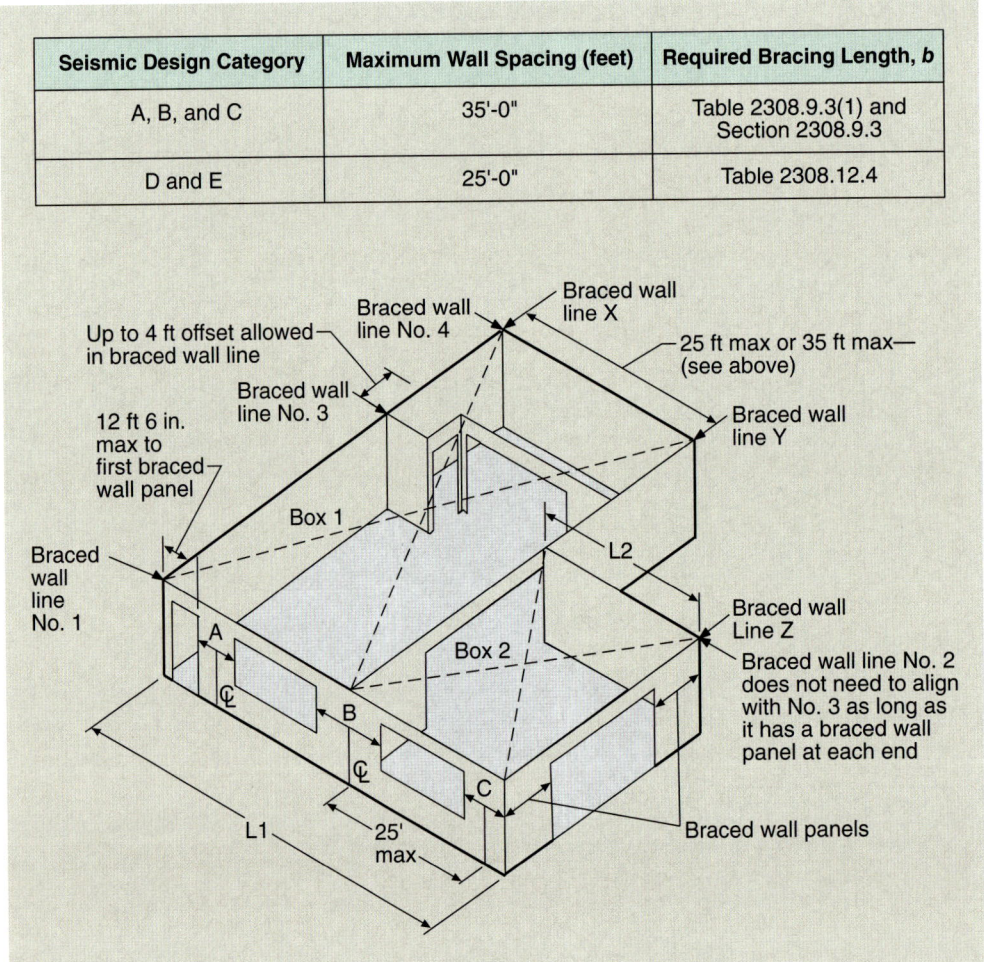

Figure 2308-36
Basic components of the lateral bracing system—one story.

Table 2308.9.3(1) is applicable only to structures assigned to Seismic Design Category A, B, or C. Structures in Seismic Design Category D or E have more stringent bracing requirements, as shown in Table 2308.12.4. The language in Table 2308.9.3(1) of the 2006 IBC was revised to read "located in accordance with Section 2308.9.3," which means braced wall panels must be located within 12.5 feet from the end of a braced wall line. Section 2308.12.4 was changed in the 2012 IBC, which will be discussed in detail under that section.

2308.9.3.1 Alternative bracing. Alternative braced wall panels were developed to solve the problem of having narrow walls adjacent to garage door openings. The braced panels in Section 2308.9.3 may be replaced by the alternative braced panels of this section. Note that the alternative braced panel is a complete assembly; therefore, all the requirements described in the section must be met. The specific requirements for constructing alternative braced panels are illustrated in Figure 2308-38. Alternative braced wall panels can only be used on a one-story building or on the first story of a two-story building. They cannot be used on the second story, because the code does not address the specific details and load path connections required for in-plane shear transfer and uplift anchorage to resist

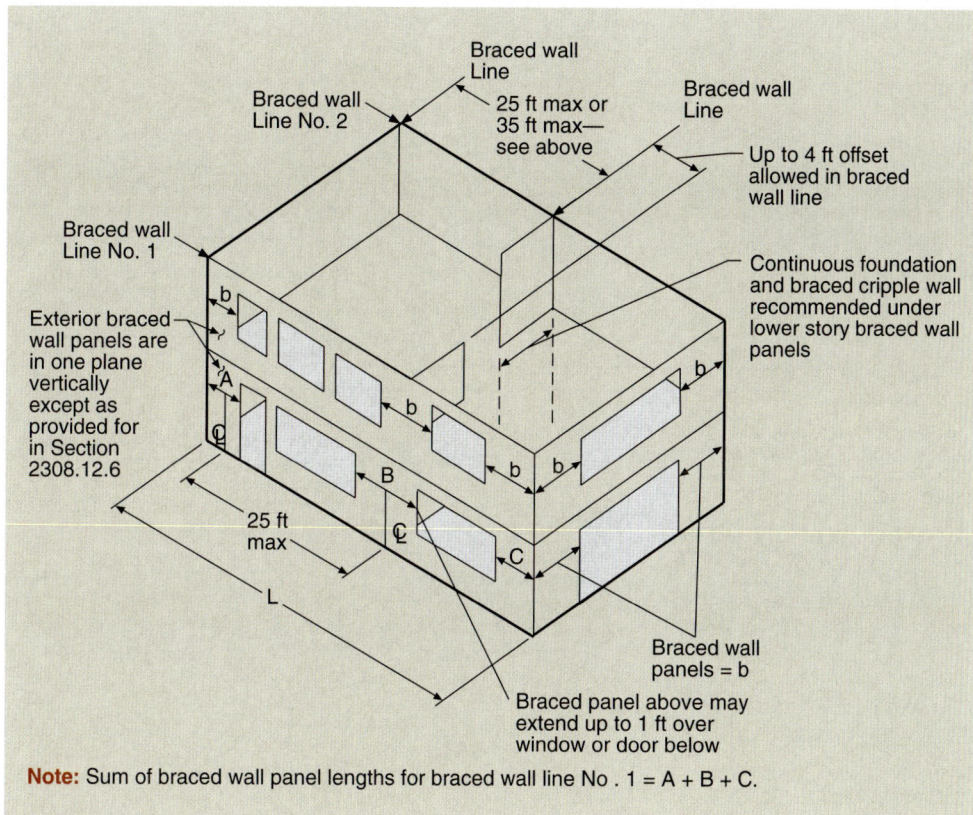

**Figure 2308-37
Basic components of the lateral bracing system—two stories.**

overturning forces at a second-story condition. See discussion under Section 2308.9.3.2 for requirements for alternative bracing wall panel adjacent to a door or window opening that were first introduced in the 2006 IBC.

Alternative bracing wall panel adjacent to a door or window opening. Section 2308.9.3.1 allows alternative braced wall panels to be used in lieu of braced wall panels described in Section 2308.9.3. The alternative braced wall panel in Section 2308.9.3.1 is required to be a minimum of 32 inches in length. Although this allows the alternative braced wall panel to fit in tight conditions in comparison to the 48-inch-long braced wall panel, the 32 inches can be difficult to achieve when located adjacent to openings. The new alternative allows a reduction of the width of the full-height segment of alternative braced wall panels to 16 inches in width for a one-story building, and to 24 inches in width for the first story of a two-story building. Because it is a prescriptive alternative bracing method, no engineering is required if constructed strictly in accordance with all of the requirements of the section, as shown in Figure 2308-39.

2308.9.3.2

This so-called "portal frame" alternative was first developed in the State of Washington, and builders in the Pacific Northwest were constructing these narrow *portal frames* in the field based on engineering. The popularity of the detail grew until it was routinely permitted in a variety of jurisdictions in California and the Pacific Northwest. Although the original portal frame design was based on engineering and monotonic testing, the APA Laboratory performed a series of cyclic tests using the Structural Engineers Association of Southern California (SEAOSC) cyclic testing protocol. During these tests, the single-sided/single-story 32-inch-wide alternative braced wall panel covered by Section 2308.9.3.1 was tested along with the proposed alternative portal frame system. The test results of the double-sided/two-story 32-inch-wide alternative braced wall panel were compared with the 24-inch-wide alternative portal frame. In both cases, the proposed alternative system

23 Wood

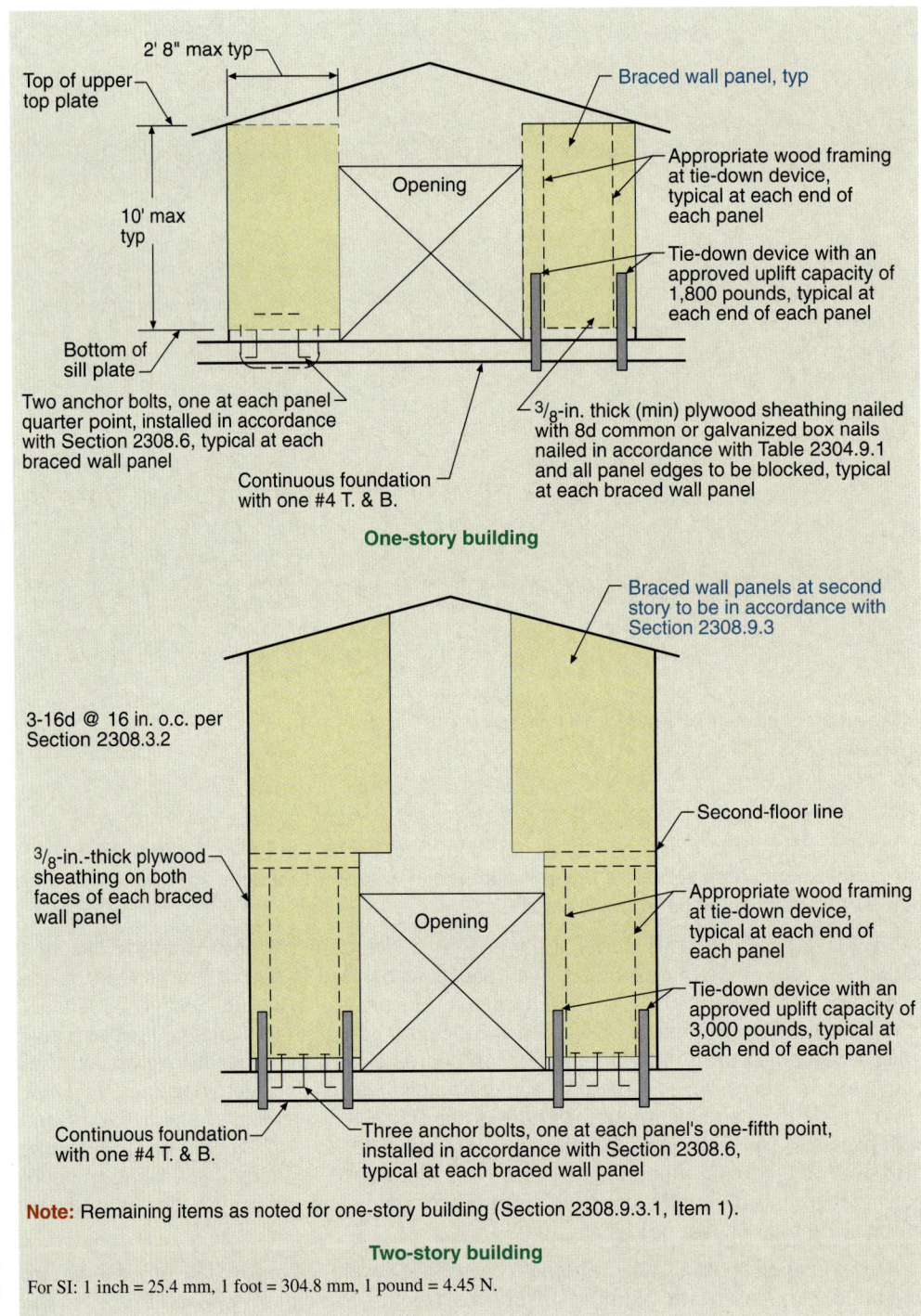

Figure 2308-38
Alternate braced panels.

performed significantly better in terms of both strength and stiffness than did the 32-inch alternative bracing panel that was already permitted by the code.

The new alternative bracing method uses the header over the adjacent opening by running the header the length of the sheathed bracing panel to the first full-length stud. Where the sheathing and header overlap, the header is fastened to the full-height sheathing with a grid nailing pattern that provides a moment-resistant connection at the top.

Conventional Light-Frame Construction

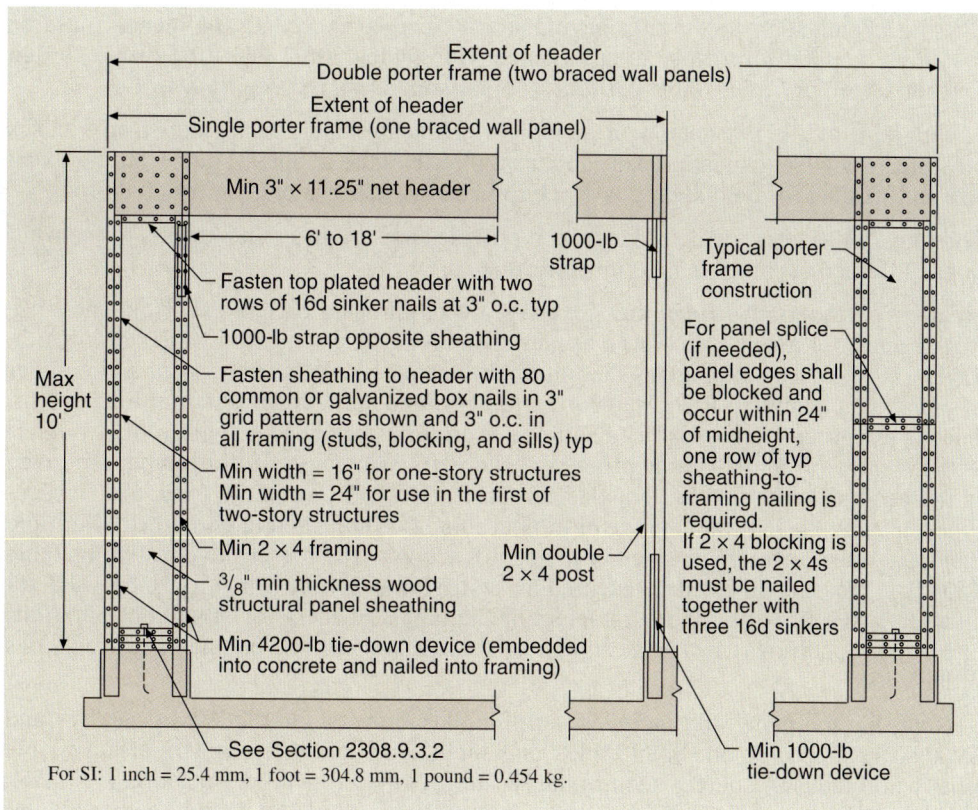

Figure 2308-39
Alternate braced wall panel adjacent to a door or window opening.

At the base of the sheathed section, embedded framing anchors nailed to the edge studs provide a moment-resistant connection at the base. The framing anchors are sized to provide uplift and shear resistance capacity, resulting in a reduction in anchoring requirements at the plate. Additional straps are required as shown in Figure 2308-39.

Cripple walls. See the commentary for Section 2308.2. Cripple walls (sometimes referred to as foundation stud walls or knee walls) are stud walls usually less than 8 feet in height that rest on the foundation plate and support the first immediate floor and/or wall above.

2308.9.4

The code requires a minimum height of 14 inches for cripple wall studs, and this minimum height is based on the length necessary to properly fasten the studs to the foundation wall plate and the double-wall plate above. Where the 14-inch minimum is not possible, the code requires that the cripple wall be framed with solid blocking. In this case, the cripple wall studs, even though shorter than 14 inches in length, should be installed with wall plates as required by Section 2308.9.2 and with the solid blocking tightly fitted between each stud. This solid blocking performs three purposes: it provides a level uniform bearing surface for the support of the floor above, it transmits lateral forces from the floor to the foundation, and it reduces the *racking* effects of the studs during a seismic or high-wind event. Wood structural panel sheathing may also be used to brace these walls, when adequate nailing is provided along the foundation sill and top plates.

Foundation stud walls exceeding 4 feet in height must be framed with studs having the size required for an additional story. Thus, for a building that would be considered to be two stories in height with a crawl space beneath the first story but having foundation stud walls with the studs more than 4 feet in height, the code would require that the studs be framed with either 3 by 4 or 2 by 6 members, as would be required for the first story of a three-story building.

2308.9.4.1 Bracing. Foundation stud walls having a stud height exceeding 14 inches must be braced as required for a story in accordance with Table 2308.9.3(1) for Seismic Design Category A, B, or C, and Table 2308.12.4 for Seismic Design Category D or E.

2308.9.4.2 Nailing of bracing. Edge nailing for cripple wall bracing should be not more than 6 inches on center along the foundation plate and the top plate of the cripple wall. The required nail size and spacing for field nailing is as required for the specific bracing material used.

2308.9.5 Openings in exterior walls. The section provides prescriptive requirements for conventionally framed openings in exterior stud walls.

2308.9.5.1 Headers. Headers over openings in walls are required to carry loads from the wall, floors, and roof above. Allowable header spans for exterior bearing walls are given in Table 2308.9.5. The table gives the minimum header size, allowable span, and number of jack studs (trimmers) required for various ground snow loads and building widths. The 30-psf ground snow load is also to be used for sizing headers with roof live loads without snow of 20 psf (see Footnote e). The permissible species for lumber are given in the heading on the table (Douglas-Fir Larch, Hem-Fir, Southern Pine, and Spruce-Pine-Fir), and the minimum grade of lumber (No. 2) is specified in Footnote b. Multiple members must be fastened together in accordance with the prescriptive fastening Table 2304.9.1. Headers selected in accordance with Table 2308.9.5 may only be used for structures that fit within the parameters of the table. Headers for other configurations or loading conditions must be designed in accordance with the engineering provisions of the code.

Note that the table only includes multiple 2x members. On the West Coast, 4x and 6x headers are common. Span tables that include 4x and 6x members are available from wood industry sources such as the Western Wood Products Association (WWPA). WWPA publishes a Tech Note, *Design Load Tables for Solid-Sawn Lumber Beams and Headers*, which includes load tables for single 4x and 6x plus built-up double and triple 2x wood members with 3- to 14-foot spans. WWPA resources are available from their website at wwpa.org. Figure 2308-40 illustrates the typical header framing requirements.

2308.9.5.2 Header support. Headers are supported by jack studs (also called trimmers) as required by Table 2308.9.5 and have a minimum of $1^1/_2$ inches of bearing at the supports.

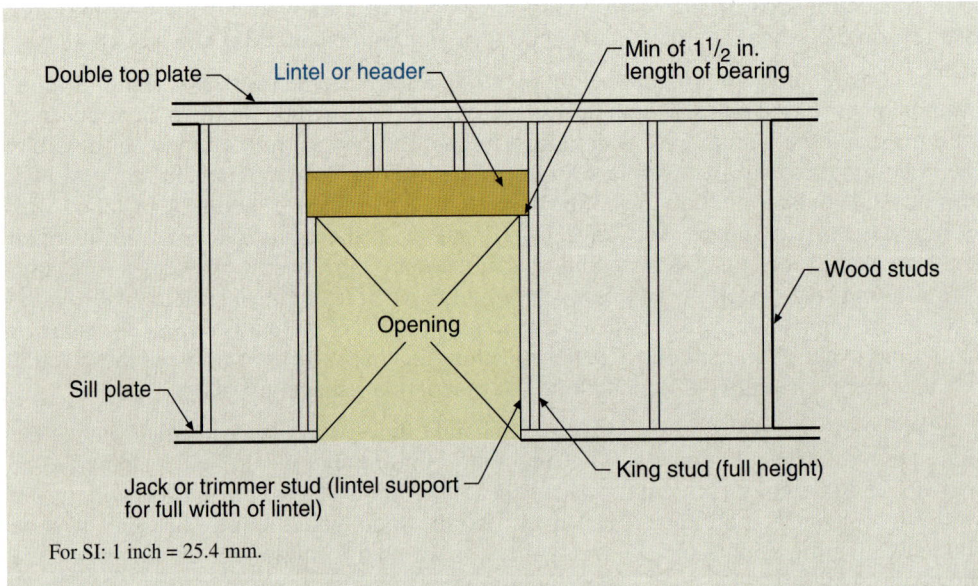

Figure 2308-40 Header over wall opening.

For SI: 1 inch = 25.4 mm.

Openings in interior-bearing partitions. Allowable header spans for interior-bearing walls supporting floor loads are given in Table 2308.9.6. The table gives the minimum header size, allowable span, and number of jack studs required. Headers selected in accordance with Table 2308.9.6 may be used only for structures that fit within the parameters of the table. Headers for other configurations or loading conditions must be designed in accordance with the engineering provisions of the code.

2308.9.6

Openings in interior-nonbearing partitions. Interior-nonbearing walls can have single headers with a minimum of $1^1/_2$ inches of bearing at each support.

2308.9.7

Pipes in walls. This section provides requirements for framing walls with pipes such as plumbing walls. Figure 2308-41 illustrates the requirements of this section for pipes in a wood-framed wall.

2308.9.8

Bridging. Where stud partitions do not have adequate sheathing to brace the studs laterally in their weak (smaller) dimension, and the studs have a height-to-least-thickness ratio exceeding 50, the studs are required to have bridging or solid blocking with a minimum nominal thickness of 2 inches and a width the same as the studs. This blocking should be installed at heights that reduce the height-to-least-thickness ratio below 50.

2308.9.9

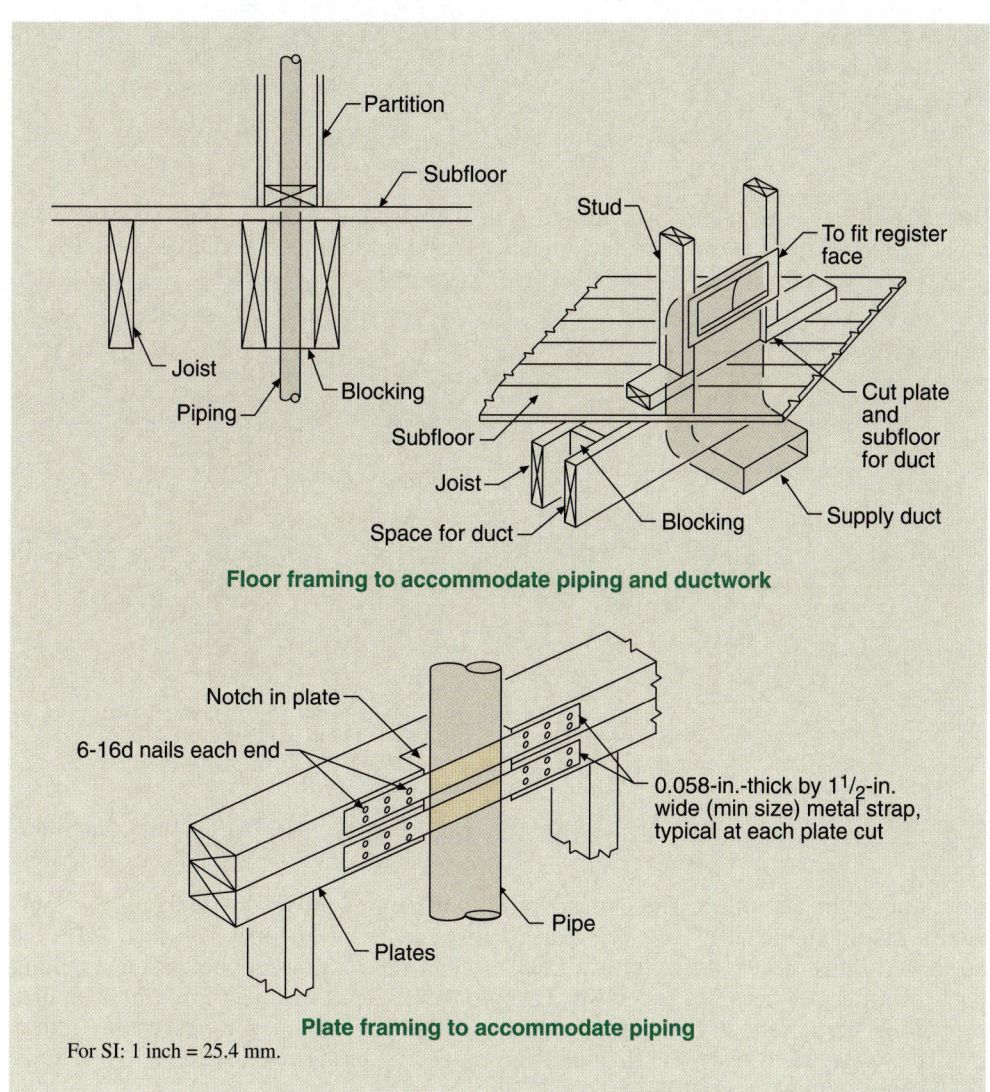

Figure 2308-41
Pipes in walls.

The l/d ≤ 50 originated with the NDS. Use of 2 by 4 studs will require placement of the blocking when the maximum height of the wall is over 6 feet 3 inches (1.5 × 50 = 75 inches = 6 feet 3 inches).

2308.9.10 Cutting and notching. Studs are often notched to accommodate wiring and the like. Figure 2308-42 illustrates acceptable cutting or notching of studs.

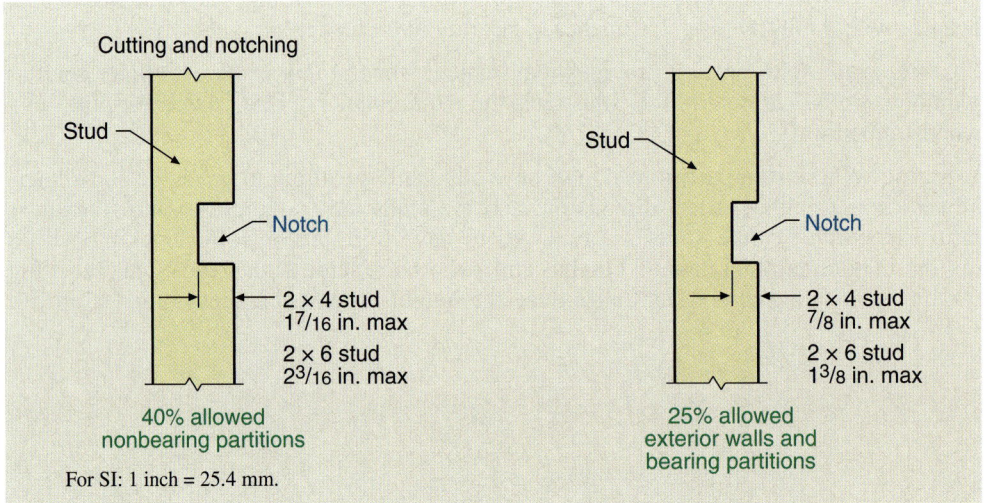

Figure 2308-42
Cutting and notching of studs.

2308.9.11 Bored holes. Studs often have drilled holes to accommodate wiring and the like. Figure 2308-43 illustrates acceptable dimensions and location of bored holes in studs.

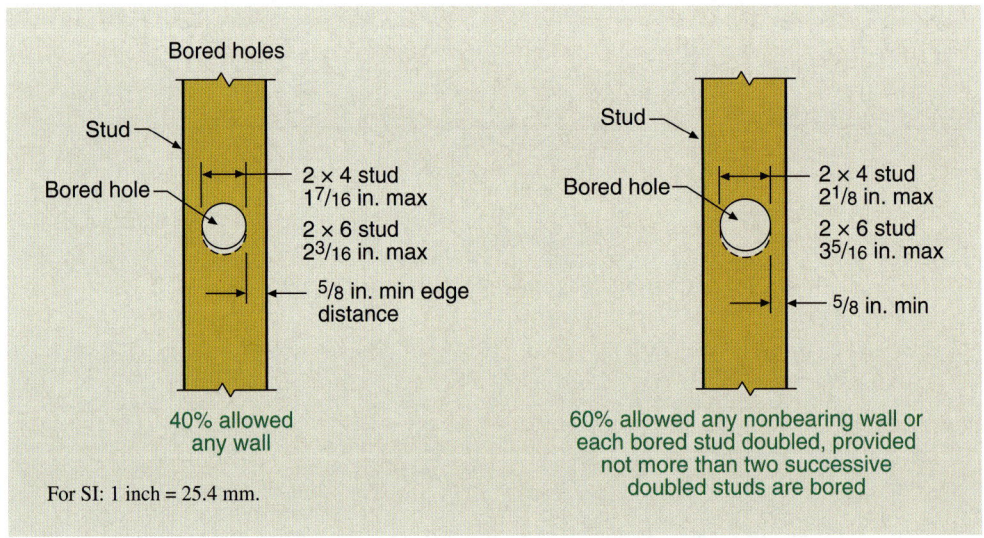

Figure 2308-43
Bored holes in studs.

2308.10 Roof and ceiling framing. These prescriptive roof framing provisions apply only to roofs with a minimum slope of 3:12 or greater. Where roofs have slopes less than 3:12, the horizontal thrust necessary to form a truss mechanism with the ridge board and ceiling joists or rafter ties becomes excessive. Thus, for roof slopes less than 3:12, the members supporting the rafters and ceiling joists such as the ridge beam, hip, and valley rafters must be designed to resist the gravity loads as beams instead of behaving like a truss. See Figures 2308-44 and 2308-45.

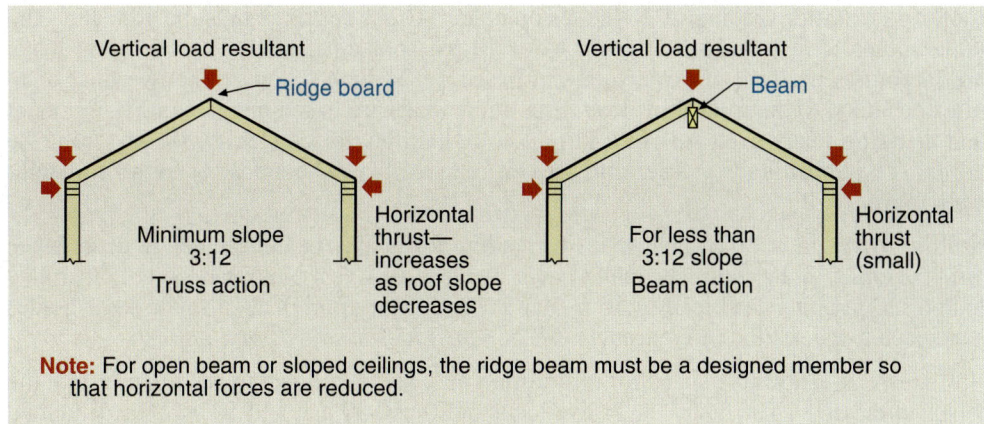

Figure 2308-44
Roof framing.

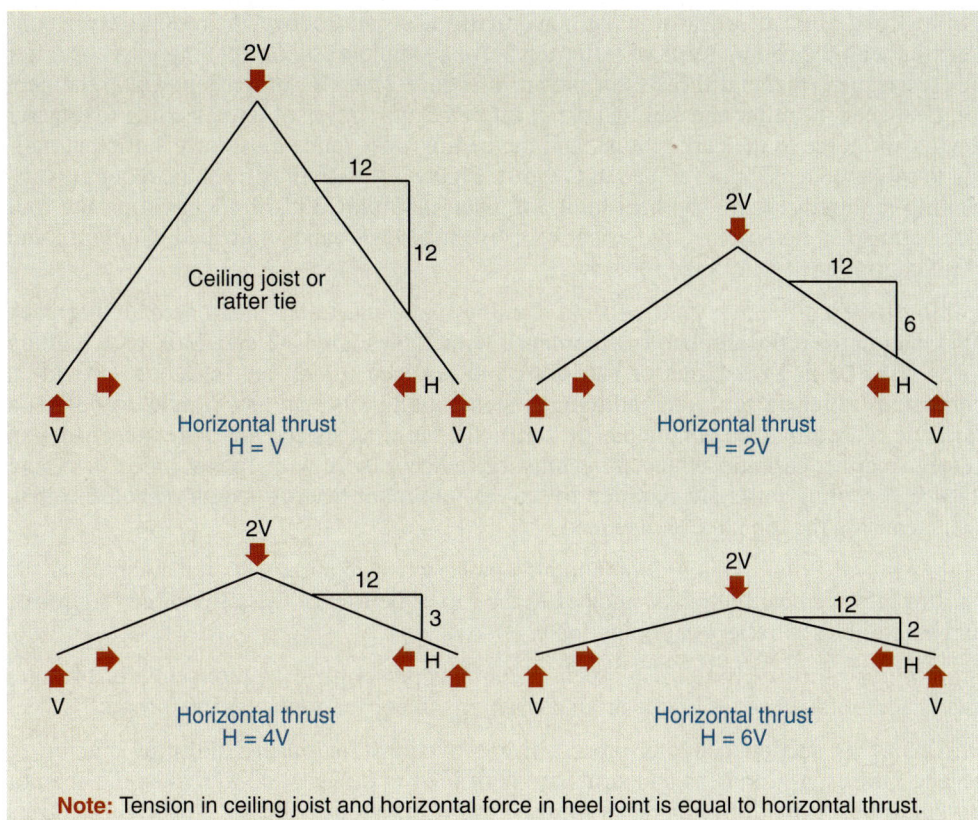

Figure 2308-45
Roof framing thrusts—truss action.

Wind uplift. Wind suction (component and cladding pressures) can cause considerable uplift forces on roof framing. The uplift loads must be positively transferred into the structure below to achieve sufficient gravity loads to resist the uplift. See also Section 2304.9.6 for additional discussion. Note that the uplift forces on low slope roofs such as those less than 3:12 will be larger than the forces for roofs with slopes equal to or greater than 3:12. Although this section only applies to roofs with a slope greater than or equal to 3:12, ties of at least the same strength should be used on these lower slope roofs. The key features of the table are covered in the footnotes. The uplift loads are in pounds and are based on

2308.10.1

roof truss or rafter spacing of 24 inches on center. Other spacings must be adjusted. The table values have a built-in allowance for 10-psf roof dead load. The overhang loads are in pounds per foot of projection and based on 24-inch spacing. The overhang load must be added to the roof uplift load. The uplift loads are based on end zone component and cladding loads from ASCE 7 and allow reductions for connections away from the corner. The uplift loads are permitted to be reduced by 100 pounds for each full-height wall above.

2308.10.2 Ceiling joist spans. Allowable spans for ceiling joists may be determined in accordance with the tables in the code. The tables cover live load of 10 psf plus dead load of 5 psf, live load of 20 psf plus dead load of 10 psf, and a deflection limit of $L/240$. For other grades and species, the section references *AF&PA Span Tables for Joists and Rafters*.

2308.10.3 Rafter spans. Allowable spans for roof rafters may be determined in accordance with the tables in the code. The tables cover live loads of 20 psf, 30 psf, and 50 psf plus dead load of 10 psf and 20 psf. The deflection limit of $L/180$ is used for rafters without a ceiling attached to the rafters. See Table 1604.3 for deflection limits. For other grades and species, the section references *AF&PA Span Tables for Joists and Rafters*.

2308.10.4 Ceiling joist and rafter framing. As noted in the discussion above, the roof rafters in conjunction with the ceiling joists or rafter ties form a simple truss. The ceiling joists or rafter ties resist the horizontal thrusts, as shown in Figure 2308-46, in tension. Thus, the heel joint between the rafter and ceiling joist or rafter tie must have sufficient nailing to transfer the tension force to the ceiling joist, and the ceiling joist splice must have sufficient nailing to resist the tension developed in the joist. Figure 2308-46 illustrates the requirements. Nailing requirements for the heel joint are given in Table 2308.10.4.1 based on the roof slope and span, tie spacing, and either roof live load or ground snow load. Guidance and clarifications are given in the footnotes.

2308.10.4.1 Ceiling joist and rafter connections. Ceiling joists and rafters must be nailed to each other and to the top wall plate in accordance with Tables 2304.9.1 and 2308.10.1. Ceiling joists must be tied over interior partitions and fastened to adjacent rafters to provide a continuous rafter tie across the building. Where ceiling joists are not parallel to rafters, a rafter tie must provide a continuous tie across the building spaced not more than 4 feet on center. In either case, the connections must be in accordance with Tables 2308.10.4.1 and 2304.9.1. Ceiling joists are required to have a minimum bearing length of not less than $1^1/_2$ inches on the top plate at each end.

2308.10.4.2 Notches and holes. This section contains prescriptive requirements for notching at the ends of rafters or ceiling joists, notches in the top or bottom of the rafter or ceiling joists, and bored holes in rafters or ceiling joists.

2308.10.4.3 Framing around openings. This section contains prescriptive requirements for trimmer and header rafters used to frame around openings in conventionally framed roofs.

2308.10.5 Purlins. This section contains prescriptive provisions for purlin and strut bracing to reduce rafter spans. Purlins and struts are permitted to be installed to reduce the span of rafters to be within allowable span limits. Purlins are required to be supported by struts to bearing walls, which means a purlin that is braced to an interior wall creates a bearing wall. The maximum span of 2-inch by 4-inch purlins is 4 feet, and the maximum span of 2-inch by 6-inch purlins is 6 feet. However, in no case can the purlin be of smaller size than the supported rafter. Struts cannot be less than 2-inch by 4-inch members. The maximum unbraced length of struts is 8 feet, and the minimum slope of the struts cannot be less than 45 degrees from the horizontal. Figure 2308-47 illustrates purlin and strut requirements.

2308.10.6 Blocking. Roof rafters and ceiling joists must be supported laterally to prevent rotation and lateral displacement in accordance with Section 2308.8.5, which contains lateral support requirements for floor, roof, and ceiling framing.

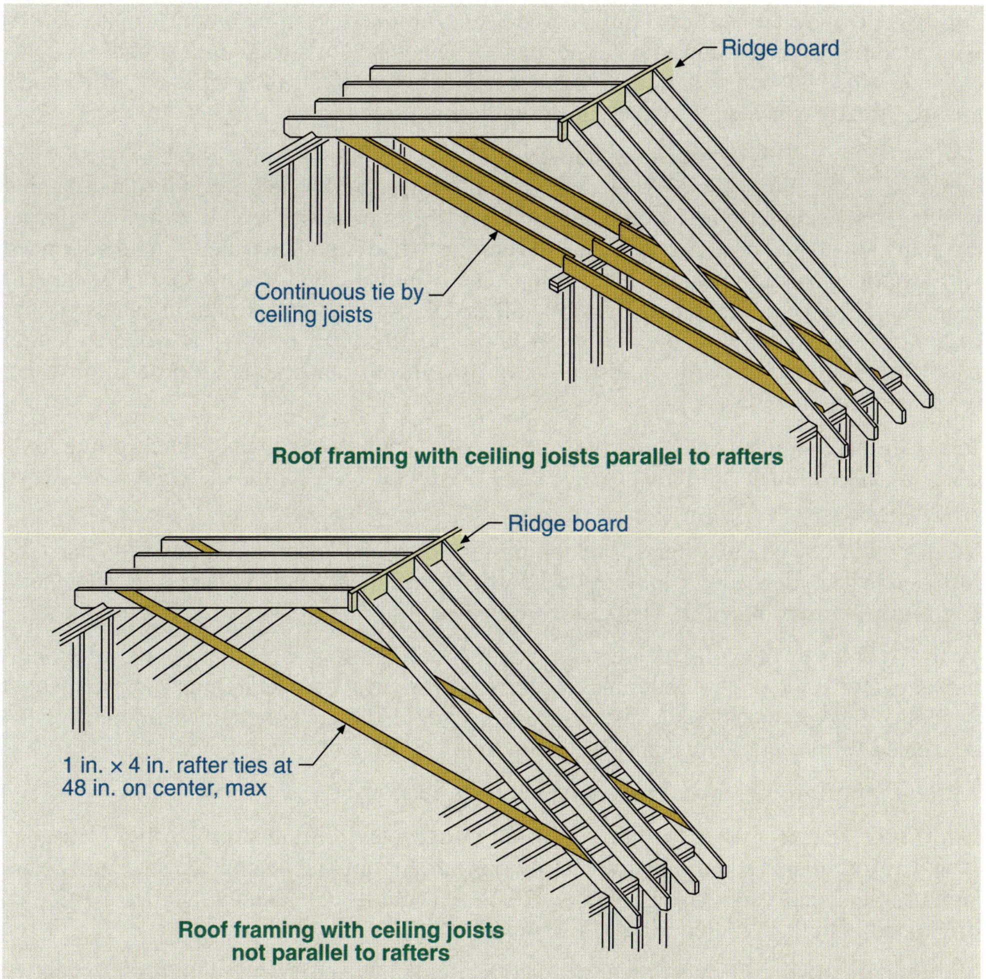

Conventional Light-Frame Construction

Roof framing with ceiling joists parallel to rafters

Labels: Ridge board; Continuous tie by ceiling joists

Roof framing with ceiling joists not parallel to rafters

Labels: Ridge board; 1 in. × 4 in. rafter ties at 48 in. on center, max

Note: Locate ties as near as practical to the top of ceiling joists.

See Tables 2304.9.1 and 2308.10.4.1 for nailing requirements.

For SI: 1 inch = 25.4 mm.

Figure 2308-46 Ceiling and rafter framing.

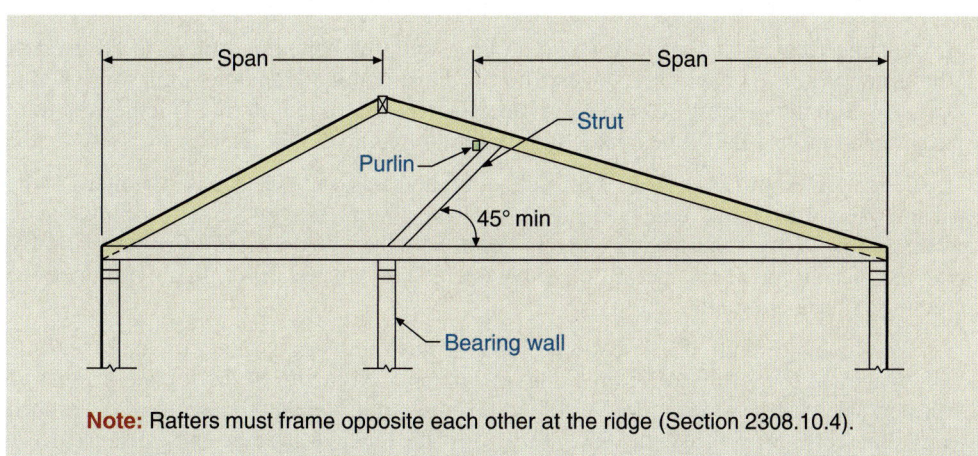

Note: Rafters must frame opposite each other at the ridge (Section 2308.10.4).

Figure 2308-47 Rafter and purlin framing.

2308.10.7 Engineered wood products. Engineered wood products such as wood I-joists, glued-laminated timber, and composite lumber cannot be notched or drilled except where specifically permitted by the product manufacturer or where the effects of notches or drilled holes are specifically considered by the registered design professional.

2308.10.8 Roof sheathing. This section references the appropriate tables for wood structural panel and lumber roof sheathing. The code requires that wood structural panels used for roof sheathing be bonded by exterior glue because moisture often gets beneath the roof covering, causing delamination of plies or strands over time. Therefore, wood structural panels should be bonded with exterior glues to prevent delamination. The wood structural panels should be labeled Exterior or Exposure 1. See Sections 2304.6.1 and 2304.7.2. Wood structural panels permanently exposed to outdoor conditions should be Exterior type, except that wood structural panels used on the exposed underside of roof overhangs are permitted to be Exposure 1 type. See Section 2303.1.4.

2308.10.8.1 Joints. Joints in lumber sheathing must occur over supports unless end-matched lumber is used. Where end matched lumber is used, each individual piece must bear on at least two supports.

2308.10.9 Roof planking. Where 2-inch tongue-and-groove wood planking is used, it must be installed in accordance with Table 2308.10.9 or be designed in accordance with the general engineering provisions of this code.

2308.10.10 Wood trusses. Wood trusses are required to be designed in accordance with the engineering provisions of the code and conform to the detailed requirements prescribed in Section 2303.4. As noted in the discussion of Section 2308.4.2, pre-engineered metal plate–connected wood trusses are considered structural elements that must be designed.

2308.10.11 Attic ventilation. Refer to the discussion under Section 1203.2.

2308.11 Additional requirements for conventional construction in Seismic Design Category B or C. Additional seismic-related restrictions and requirements apply to structures classified as Seismic Design Category B or C. The exceptions contained in Section 2308.11.2 originated with the 1997 NEHRP (FEMA 302).

2308.11.1 Number of stories. Conventional wood-frame buildings are not permitted to exceed two stories in height in Seismic Design Category C. An exception in the 2003 IBC that allowed an additional story for detached one- and two-family dwellings was deleted in the 2006 IBC because detached one- and two-family dwellings are within the scope of the IRC.

2308.11.2 Concrete or masonry. Section 2308.11.2 of the 2000 IBC imposed restrictions on the use of concrete or masonry walls and stone and masonry veneer above the basement in Seismic Design Categories B and C. A code change to the 2003 IBC expanded the provisions of Section 2308.11.2 as follows:

In Seismic Design Category B, stone and masonry veneer is permitted to be used in three stories above grade if the lowest story is constructed of concrete or masonry walls. Other wall bracing must be wood structural panels with a length that is increased to at least one- and one-half times the length required by Table 2308.9.3(1).

In Seismic Design Categories B and C, stone and masonry veneer is permitted to be used in two stories above grade provided the lowest story is constructed of concrete or masonry walls. Where the lowest story is not constructed of concrete or masonry walls, stone and masonry veneer is permitted to be used in two stories above grade, provided the following criteria are met:

(1) The bracing required by Section 2308.9.3 must be wood structural panels with an allowable shear capacity of 350 plf minimum; (2) the bracing of the top story must be located at each end, at least every 25 feet o.c. and not less than 40 percent of the braced wall line. The bracing of the first story must be located at each end, at least every 25 feet o.c. and not less than 35 percent of the braced wall line; (3) 2000-pound capacity

hold-down devices are required at the ends of braced wall panel segments from the second floor to the first floor wall, and 3900-pound capacity hold-down devices are required from the first floor to the foundation; (4) cripple walls are not permitted.

Framing and connection details. The provisions for stepped footings in Section 2308.11.3.2 originated with the 1997 NEHRP (FEMA 302), which recommends that the provisions apply to all seismic design categories. However, in the IBC, these requirements only apply to conventionally constructed buildings in Seismic Design Categories B, C, D, and E and do not apply to Seismic Design Category A. **2308.11.3**

Anchorage. There are no special anchorage requirements in Seismic Design Category B or C. Braced wall lines are required to be anchored to the foundation in accordance with the general requirements prescribed in Section 2308.6. **2308.11.3.1**

Stepped footings. IBC Figure 2308.11.3.2 illustrates the requirements of this section. Where a stepped foundation creates cripple walls that vary in height by more than 4 feet, this section has specific detailing requirements. The sill must be bolted to the foundation as required by Section 2308.3.3. This is somewhat redundant because this is a general requirement that applies to all sills bolted to foundations. If there is a segment at least 8-foot long where the floor framing bears directly on the sill bolted to the foundation, then the braced wall line is considered sufficiently braced, and the double plate of the cripple wall beyond this segment of footing must be spliced to the sill plate with metal ties as shown in the figure. Item 2 describes the requirements for the splice strap and minimum nailing. Item 3 requires cripple walls to meet the bracing requirements for a story, which is essentially the same as is required by Section 2308.9.4.1. **2308.11.3.2**

Openings in horizontal diaphragms. Blocking and strapping are required to transfer shear around openings in diaphragms as shown in IBC Figure 2308.11.3.3. **2308.11.3.3**

Additional requirements for conventional construction in Seismic Design Category D or E. Additional restrictions and requirements are imposed on structures assigned to Seismic Design Category D or E, as prescribed in Sections 2308.12.1 through 2308.12.9. The requirements in this section are in addition to the requirements in Section 2308.11 for buildings in Seismic Design Categories B and C. As noted in the discussion of Section 2308.2, conventional construction is not permitted in structures assigned to Seismic Design Category F. **2308.12**

Number of stories. Conventional wood-frame buildings are not permitted to exceed one story in height in Seismic Design Categories D and E. The 2003 IBC had an exception that allowed an additional story for detached one- and two-family dwellings that was deleted in the 2006 IBC because detached one- and two-family dwellings are within the scope of the IRC. **2308.12.1**

Concrete or masonry. Stone and masonry veneer cannot be used above the foundation, with some exceptions. The exceptions are based on the NEHRP *Provisions*. The exception allows stone or masonry veneer if certain conditions are met. **2308.12.2**

In the 2000 IBC, masonry veneer above the foundation (basement) level was prohibited entirely in Seismic Design Categories D and E. This restriction on the use of masonry veneer above the basement was removed in the 2003 IBC and an exception was added allowing masonry veneer to be used in the first story above grade in Seismic Design Category D, provided the following criteria are met: (1) Wall bracing must be wood structural panels with a minimum allowable shear capacity of 350 plf; (2) the bracing on the first story must be located at each end and at least every 25 feet on center and not less than 45 percent of the braced wall line; (3) 2100-pound capacity hold-down devices are required at the ends of braced walls from the first floor to the foundation; (4) cripple walls are not permitted.

Braced wall line spacing. The maximum spacing between braced wall lines is limited to 25 feet for buildings in Seismic Design Category D or E. **2308.12.3**

2308.12.4 Braced wall line sheathing. The minimum percentage of bracing Seismic Design Categories D and E must be in accordance with Table 2308.12.4, based on the type of sheathing and the value of the short-period spectral response acceleration, S_{DS}. Prior to the 2012 IBC, Table 2308.12.4 prescribed the number of feet of required bracing per 25 feet of braced wall length but did not give a minimum percentage. Because there was no percentage given, it was not clear how to properly apply the requirements of the table. For example, for a building sited where $S_{DS} > 1.00g$, the table required 12 feet of wall bracing for each 25 feet of wall. It is clear that a 25-foot-long wall requires 12 feet of bracing and a 50-foot-long wall requires 24 feet of bracing, but it was not explicitly clear what amount of bracing is required for a 40-foot-long wall. To resolve this, Section 2308.12.4 and Table 2308.12.4 were revised to specify the minimum percentage of wall bracing instead of a minimum length of bracing per 25 feet of wall line. For example, for a building sited where $S_{DS} > 1.00g$, the table now requires 48 percent of wall bracing. Thus, a 40-foot-long braced wall line requires $0.48 \times 40 = 19$ feet of wall bracing.

One interesting aspect of the second paragraph of this section is that cripple walls having a stud height exceeding 14 inches must be considered an additional story, yet conventional wood-frame buildings in Seismic Design Categories D and E are not permitted to exceed one story in height. The section requires cripple wall to be braced with the percentage of bracing required by Table 2308.12.4. Thus, if a one-story building is supported by a cripple wall below, both the cripple wall and the first-story wall would require the same amount of bracing.

The length of the cripple wall braced panels must be increased to $1^1/_2$ times the value required by Table 2308.12.4 if the interior braced wall lines are not supported on a continuous foundation. Note that interior braced wall lines must be on continuous foundations in accordance with Section 2308.3.4, unless the structure is not more than 50 feet in any plan dimension. Where type S-W sheathing is used (wood structural panel or diagonal wood sheathing), and it is not practical to fit $1^1/_2$ times the bracing in the braced wall line, a reduced nail spacing of 4 inches on center is a permissible alternative.

2308.12.4.1 Alternative bracing. This is a new section in the 2012 IBC that is intended to clarify that alternative braced wall panels constructed in accordance with Section 2308.9.3.1 or 2308.9.3.2 are permitted to be substituted for braced wall panels required by Table 2308.12.4. For buildings in Seismic Design Categories A, B, and C, Sections 2308.9.3.1 and 2308.9.3.2 are quite clear that alternative braced wall panels and alternative braced wall panels adjacent to a door or window opening can be substituted for the wall bracing required by Section 2308.9.3. However, for buildings in Seismic Design Categories D and E, the 2009 IBC was not clear that these alternative braced wall panels could be substituted for the wall bracing required by Table 2308.12.4. Although it was the intent of the code, it was difficult to conclude this by simply reading the language in the section. To resolve this issue, the new Section 2308.12.4.1 has been added that clearly states that alternative braced wall panels constructed in accordance with Section 2308.9.3.1 or 2308.9.3.2 are permitted to be substituted for braced wall panels required by Table 2308.12.4.

Footnote a of Table 2308.12.4 states that the height-to-width ratio for braced wall panels cannot exceed 2:1. For a typical 8-foot-high wall, this means the minimum length of the braced wall panel must be 4 feet. The alternative braced wall panel described in Section 2308.9.3.1 is 2 feet 8 inches in length, and the alternative braced wall panels adjacent to an opening in Section 2308.9.3.2 can be only 16 inches in length for a one-story building. In the 2009 IBC, it was not clear whether Footnote a in effect meant that the alternative braced wall panels of Section 2308.9.3.1 and 2308.9.3.2 cannot be used in Seismic Design Categories D and E because of the restriction on height-to-width ratio. The table in the IBC is derived from Section 12.4 and Table 12.4-2 of the 2003 NEHRP *Provisions*, which do not require overturning restraint and do not include alternative braced wall panels. According to the NEHRP Commentary, it appears that the primary concern that led to the aspect ratio requirement is aimed at minimizing overturning demand due to the lack of overturning restraint. The alternative braced wall panels address overturning directly by specifically requiring overturning restraint devices; thus, they should not be subject to the 2:1 h/w limit. To resolve this, Footnote a in the 2012 IBC was modified so that the 2:1 height-to-width ratio limitation does not apply to the alternative braced wall panels.

Attachment of sheathing. Wall sheathing must be fastened as specified in Table 2308.12.4 or the general fastening Table 2304.9.1 without substitution. Adhesives are not permitted to be used to fasten the sheathing in Seismic Design Category D or E. Nails or staples yield when the braced panels are subjected to high seismic demands from earthquake ground motion and the panels undergo ductile behavior. Panels attached by adhesives are not permitted because they can behave in a brittle manner when subjected to cyclic loading imposed by high seismic demand. **2308.12.5**

Irregular structures. Several types of structural irregularities are not permitted in conventionally constructed wood-frame buildings in Seismic Design Category D or E. IBC Figures 2308.12.6(1) through 2308.12.6(8) illustrate the various conditions to aid the code user. The conditions describing irregular structures originated with the 1997 NEHRP *Provisions* (FEMA 302) and are very similar to the UBC conditions, except that the NEHRP *Provisions* apply the restrictions to Seismic Design Category C, D, or E. The figures in the IBC are intended to facilitate understanding the intent of the requirements. **2308.12.6**

Structures with geometric discontinuities in the lateral-force-resisting system sustain more earthquake and wind damage than structures without discontinuities. They have also been observed to suffer concentrated damage at the discontinuity location. For Seismic Design Category D or E, this section translates applicable irregularities from ASCE 7 Tables 12.3-1 and 12.3-2 into limitations on conventional wood-frame construction. When a structure has an irregularity, either the entire structure or the nonconventional (irregular) portions must be engineered in accordance with the engineering provisions of the code. Although conceptually these are equally applicable to all seismic design categories, they are most critical in areas of high seismic risk, where damage caused by irregularities has repeatedly been observed in past earthquakes.

Providing an engineered design of the nonconventional portions in lieu of the entire structure is permitted by the code (see Section 2308.1.1) and is a common practice in some regions. The registered design professional must judge the extent of the portion required to be designed. This often involves design of the nonconforming element, adequate force transfer into the element, and a complete load path from the element to the foundation. In some cases, the nonconforming portion may have enough of an impact on the behavior of a structure to warrant that the entire lateral-force-resisting system be engineered.

Item 1. This limitation applies when braced wall panels are offset out-of-plane from floor to floor. In-plane offsets are discussed in Item 3. Ideally, braced wall panels should always stack above each other from floor to floor with the length stepping down at upper floors where less length of bracing panel is required.

Because cantilevers and setbacks are very often incorporated into residential construction, the exception offers rules by which limited cantilevers and setbacks can be considered conventional. Floor joists are limited to 2 by 10 (actually $1^1/_2$ inches by $9^1/_4$ inches) or larger and doubled at braced wall panel ends to accommodate the vertical overturning reactions at the end of braced wall panels. In addition, the ends of the cantilever are attached to a common rim joist to allow for redistribution of load. For rim joists that cannot run the entire length of the cantilever, the metal tie is intended to transfer vertical shear as well as provide a nominal tension tie. Limitations are placed on gravity loads to be carried by the cantilever or setback floor joists so that the joist strength will not be exceeded. The roof loads discussed are based on the use of solid sawn members where allowable spans limit the possible loads. Where engineered framing members such as trusses are used, gravity load capacity of the cantilevered or setback floor joists should be carefully evaluated.

Item 2. This limitation applies to open-front structures or portions of structures. The conventional construction bracing concept is based on using braced wall lines to divide a structure into a series of boxes of limited dimension, with the seismic force to each box being limited by the size. The intent is that each box be supported by braced wall lines on all four sides, limiting the amount of torsion that can occur. The exception, which permits portions of roofs or floors to extend past the braced wall line, is intended to permit construction such as porch roofs and bay windows, as illustrated in IBC Figure 2308.12.6(4). Wall segments that are not considered braced panels, that is, with minimal lateral resistance, are allowed in areas where braced wall panels are prohibited. See the lower right-hand portion of IBC Figure 2308.12.6(4).

Item 3. This limitation applies when braced wall panels are offset in-plane. Ends of braced wall panels supported on window or door headers below can transfer large vertical reactions to headers that may not be of adequate size to resist these reactions. The exception permits the braced wall panel to extend over an opening with a nominal 4 by 12 or larger header on the basis that the vertical reaction will not result in critical shear or flexure. Other header conditions that do not meet the exception require an engineered design. Wall segments that are not considered braced panels, that is, with minimal lateral resistance, are allowed over openings.

Item 4. This limitation results from observation of damage that is somewhat unique to split-level wood-frame construction. If floors on either side of an offset move in opposite directions because of earthquake ground motion or wind loading, the short bearing wall in the middle becomes unstable and vertical support for the upper joists can be lost, resulting in a collapse. If the vertical offset is limited to a dimension equal to or less than the joist depth, and a simple strap tie directly connecting joists on different levels can be provided, then the instability is eliminated. Where the floor framing on each side is directly supported by a foundation, no tie is required.

Item 5. This limitation applies to nonperpendicular-braced wall lines. When braced wall lines are not perpendicular to each other, engineering is required to determine the distribution of forces and the required lateral-support system. There are no exceptions to Item 5.

Item 6. This limitation attempts to place a practical limit on the size of openings in floors and roofs. See Section 2308.11.3.3 for detailing requirements for permitted openings in diaphragms. There are no exceptions to Item 6.

The NEHRP also considers a structure irregular if there are differences of more than 6 feet in height of the shear panels in a single story. Where the heights of a braced wall panel vary significantly, the stiffness and thus the distribution of lateral forces will also vary. The usual assumption of a uniform shear per foot will not be correct, and the net result is a possible torsional irregularity. For example, if a structure on a hill is supported on 2-foot-high braced cripple wall panels on one side and 8-foot-high panels on the other, torsion and redistribution of forces will occur. Although the code does not currently require it, an engineered design is recommended to evaluate force distribution and provide adequate wall bracing and anchor bolting in this situation.

2308.12.7 Anchorage of exterior means of egress components. Positive anchorage to the main structure is required for exterior egress balconies, exterior exit stairways, and similar means of egress components. The anchorage must be spaced at no more than 8 feet on center. If anchorage to the main structure is not provided, then the egress component itself must be designed to resist lateral seismic forces. Toenails and nails in withdrawal are not permitted because they do not perform well when subjected to cyclic loads from seismic demands.

Steel plate washers. The requirement for plate washers originated in response to the extensive splitting of wood sills during the Northridge earthquake. The splitting was caused by a variety of factors including cross-grain tension and oversized holes for anchor bolts, in addition to splitting from insufficient edge distance in heavily loaded walls. Subsequent cyclic testing of wood shear panel assemblies showed that use of 2-inch by 2-inch by $3/16$-inch, or larger, steel plate washers at the anchor bolts helps prevent the sill from splitting. The *IBC Final Draft*, which was used as the base document to develop the 2000 IBC, required $3 \times 3 \times 1/4$-inch thick steel plate washers, which was based on the 1997 NEHRP *Provisions* (FEMA 302). The size of the plate washers was changed from $3 \times 3 \times 1/4$-inch to $2 \times 2 \times 3/16$-inch thick during the 2000 IBC code development process, to be consistent with the 1997 UBC. The reason given was adequate performance of sill plates tested with $2 \times 2 \times 3/16$-inch steel plate washers. A code change to the 2003 IBC changed the square plate washer size to $3 \times 3 \times 1/4$-inch thick to be consistent with the NEHRP *Provisions*. A subsequent code change modified the minimum thickness from $1/4$ inch to 0.229 inches in order to allow the plate washers to be manufactured from cold-rolled steel instead of $1/4$-inch hot-rolled steel. Cold-rolled steel of this thickness is more readily available, more economical, and can be ordered with a hot-dipped galvanized finish from the steel mill. Hot-rolled steel is not galvanized and must be galvanized after fabrication. Some steel hardware manufacturers recommend that galvanized plate washers be used where in contact with preservative-treated wood. The hole in the plate washer is permitted to be diagonally slotted with a width up to $3/16$-inch wider than the bolt diameter with a length up to $1^3/4$ inches provided a standard cut washer is used between the square plate washer and the nut. The provision permitting slotted holes was taken directly from 2003 NEHRP *Provisions*.

2308.12.8

The phrase, "or approved anchor straps load rated in accordance with Section 1716.1," was added to this section in the 2009 IBC. The addition of anchor straps is based on recent cyclic testing of walls using foundation anchor straps by Simpson Strong-Tie that showed that they perform very well under cyclic loading. Their performance is partly because they wrap around the sill plate at the sheathing nailing location, thereby preventing cross-grain bending of the sill plate. Because prevention of cross-grain bending is the primary reason for using plate washers, anchor straps can be substituted for anchor bolts with square plate washers.

Anchorage in Seismic Design Category E. Typical sill anchorage is $1/2$-inch-diameter spaced not more than 6 feet on center as required by Section 2308.6. For conventional wood-frame buildings in Seismic Design Category E, $5/8$-inch-diameter anchor bolts are required. The larger-diameter bolts are required to transfer the increased in-plane shear load from the anticipated increase in seismic demands. The phrase, "or approved foundation anchor straps load rated in accordance with Section 1716.1 and spaced to provide equivalent anchorage," was added to this section in the 2009 IBC. The addition of anchor straps is based on recent cyclic testing of walls using foundation anchor straps by Simpson Strong-Tie that showed that they perform very well under cyclic loading.

2308.12.9

KEY POINTS

- Chapter 23 contains requirements for the design and construction of wood-frame buildings and other wood structures regulated by the IBC.
- The 2012 IBC references ANSI/AWC NDS—2012 *National Design Specification (NDS) for Wood Construction* with 2012 Supplement for general design of wood structures and ANSI/AF&PA SDPWS—2008 *Special Design Provisions for Wind and Seismic* (SDPWS) for lateral design of wood structures.
- In many cases, the IBC refers to specific sections in the referenced standards for detailed design and construction requirements.

- Wood structures must be designed by allowable stress design (ASD) in accordance with Section 2306, load and resistance factor design (LRFD) in accordance with Section 2307, or the prescriptive conventional construction provisions of Section 2308.

- Regardless of the design method used, all wood structures must conform to the minimum quality standards in Section 2303 and the general construction requirements of Section 2304.

- Buildings designed using the 2012 *Wood Frame Construction Manual for One- and Two-Family Dwellings* (WFCM) are deemed to meet the requirements for conventional construction prescribed in Section 2308.

- The quality of wood elements must conform to the requirements and standards referenced in Section 2303.

- Wood trusses must comply with Section 2303.4, which includes requirements for design, truss design drawings, truss bracing, restraint and anchorage, truss placement diagrams, alterations, and truss quality assurance.

- Wood-frame structures designed by ASD in accordance with Section 2306 or LRFD in accordance with Section 2307 must be designed and constructed in accordance with the ANSI/AWC NDS—2012 and Section 2305, which references AF&PA SDPWS-08.

- Sections 2306 and 2307 both reference the AWC/ANSI NDS—2012, which is a dual-format standard allowing both ASD and LRFD procedures.

- Buildings that meet the scope, restrictions, and limitations of Section 2308 may be designed using the prescriptive conventional construction provisions without engineering.

- Portions and elements in otherwise conventionally constructed buildings may be engineered without requiring the entire building to be engineered.

- Wood-framed buildings of conventional construction in Seismic Design Category B or C must meet the additional seismic requirements in Section 2308.11.

- Wood-framed buildings of conventional construction in Seismic Design Category D or E must meet the additional seismic requirements in Section 2308.12.

REFERENCES

1. NEHRP, *Recommended Provisions for Seismic Regulations for New Buildings and Other Structures*, Building Seismic Safety Council, Washington, DC, 1997, 2000, 2003.

2. AWC, *ANSI/AWC National Design Specification (NDS) for Wood Construction*, American Wood Council, Leesburg, VA, 1997, 2005, 2012.

3. AF&PA, *ANSI/AF&PA Special Design Provisions for Wind and Seismic (SDPWS)*, American Wood Council, Leesburg, VA, 2005, 2008.

4. ASCE-16, *Standard for Load and Resistance Factor Design for Engineered Wood Construction: AF&PA/ASCE 16-95*, American Society of Civil Engineers, New York, NY, 1995.

5. FPL, *Wood Handbook: Wood as an Engineering Material*, reprinted from Forest Productions Laboratory General Technical Report FPL-GTR-113 with the consent of the USDA Forest Service, Forest Products Laboratory, Forest Products Society, 1999.

6. AITC, *Timber Construction Manual, 5th edition*, American Institute of Timber Construction, New York, NY, John Wiley and Sons, Inc., 2004.

7. SEAOC, *Recommended Lateral Force Requirements and Commentary of the SEAOC Seismology Committee* (SEAOC Blue Book), Structural Engineers Association of California, Sacramento, CA, 1999.

BIBLIOGRAPHY

AF&PA, *Manual of Wood Construction: Load and Resistance Factor Design (LRFD)*, American Forest & Paper Association, Washington, DC, 1996.

APA, *Northridge California Earthquake, T94-5*, American Plywood Association, Tacoma, WA, 1994.

APA, *Proposed Cyclic Testing Standard for Shear Walls*, American Plywood Association, Tacoma, WA, 1996.

APA: Rose, J.D., *Preliminary Testing of Wood Structural Panel Shear Walls Under Cyclic (Reversed) Loading. Research Report 158*, American Plywood Association, 1996, Tacoma, WA, 1996.

ASCE, *A Pre-Standard Report American Society of Civil Engineers*, ASCE Committee on Wood Load and Resistance Factor Design for Engineered Wood Construction, New York, NY, 1988.

ATC, *ATC-R-1: Cyclic Testing of Narrow Plywood Shear Walls*, Applied Technology Council, Redwood City, CA, 1995.

Breyer, Donald E. et al., *Design of Wood Structures,* Third Edition, McGraw-Hill Book Company, New York, NY, 1980, 1988, 1993, 1998, 2003, 2007.

CWC, *Wood Design Manual*, Canadian Wood Council, Ottawa, Canada, 1990.

CWC, *Wood Reference Handbook*, Canadian Wood Council, Ottawa, Canada, 1991.

DofANAF, *Seismic Design for Buildings*, TM5-809-10 (Tri-Services Manual), Department of the Army, Navy and Air Force, U.S. Government Printing Office, Washington, DC, 1992.

Dolan, J.D., "Proposed Test Method for Dynamic Properties of Connections Assemblies with Mechanical Fasteners," *ASTM Journal of Testing and Evaluation* 22(6): 542–547, 1994.

Dolan, J.D., and Johnson, A.C., *Cyclic Tests of Long Shear Walls with Openings*, Virginia Polytechnic Institute and State University, Timber Engineering Report No. TE-1996-002, Blacksburg, VA, 1996.

Dolan, J.D., and Johnson, A.C., *Monotonic Tests of Long Shear Walls with Openings*, Virginia Polytechnic Institute and State University, Timber Engineering Report No. TE-1996-001, Blacksburg, VA, 1996.

EERI, *Northridge Earthquake Reconnaissance Report, Chapter 6, Supplement C to Volume 11, pp 125 et seq., Earthquake Spectra*, Earthquake Engineering Research Institute, 1996.

Faherty, Keith F., and Williamson, T.G., *Wood Engineering and Construction Handbook*, McGraw-Hill, New York, NY, 1989, 1995, 1997.

FPL, *Wood: Engineering Design Concepts*, Forest Products Laboratory, Materials Education Council, The Pennsylvania State University, University Park, PA, 1986.

Goetz, Karl-Heinz, Dieter Hoor, Karl Moehler, and Julius Natterer, *Timber Design and Construction Source Book: A Comprehensive Guide to Methods and Practice*, McGraw-Hill, 1989, New York, NY, 1989.

Hoyle and Woeste, *Wood Technology and Design of Structures*, Iowa State University Press, 1989.

HUD, *HUD Minimum Property Standards, Vol. I, II, and III*, U.S. Department of Housing and Urban Development, United States Government Printing Office, Washington, DC, 1984.

ICBO, *Handbook to the Uniform Building Code: An Illustrative Commentary*, International Conference of Building Officials, 1998, Whittier, California, copyright held by International Code Council, 1998.

Keenan, F.J., *Limit States Design of Wood Structures*, Morrison Hershfield Limited, 1986.

NEHRP, *Recommended Provisions for Seismic Regulations for New Buildings*, Building Seismic Safety Council, Washington, DC, 1997, 2000, 2003.

NOAA, *San Fernando, California, Earthquake of February 9, 1971*, U.S. Department of Agriculture, National Oceanic and Atmospheric Administration, 1971, Washington, DC, 1971.

Sherwood and Stroh, "Wood-Frame House Construction" in *Agricultural Handbook 73*. U.S. Government Printing Office, 1973, Washington, DC, 1973.

Somayaji, Shan, *Structural Wood Design*, West Publishing Co., St. Paul, Minnesota, 1992.

Stalnaker, Judith J., and E.C. Harris, *Structural Design in Wood*, Second Edition, McGraw-Hill, New York, NY, 1996.

WWPA, *Western Wood Use Book*, Western Wood Products Association, Portland, Oregon, 1983.

Additional references are listed in John Peterson, "Bibliography on Lumber and Wood Panel Diaphragms," *Journal of Structural Engineering*, Vol. 109 No. 12. American Society of Civil Engineers, New York, NY, 1983.

CHAPTER 24

GLASS AND GLAZING

Section 2403 General Requirements for Glass
Section 2404 Wind, Snow, Seismic, and Dead Loads on Glass
Section 2405 Sloped Glazing and Skylights
Section 2406 Safety Glazing
Section 2407 Glass in Handrails and Guards
Section 2408 Glazing in Athletic Facilities
Key Points

24 Glass and Glazing

> Chapter 24 regulates glass and glazing for essentially two important reasons:
> 1. To protect against breakage that is due to building distortion under lateral loads or that is due to wind loads applied perpendicularly to the surface of the glass.
> 2. To protect against breakage that is due to accidental impact by individuals adjacent to the glazing.
>
> Glass, light-transmitting ceramic panels, and light-transmitting plastic panels are regulated by Chapter 24 for their use in both vertical and sloped applications.

Section 2403 General Requirements for Glass

2403.1 Identification. Because glass is not manufactured at the building site and is usually incorporated into assemblies that are not manufactured at the building site, the code requires that the glass and glazing be identified as to type and thickness so that it is possible to determine compliance with this chapter as far as permitted areas of glass are concerned. The identifying information furnished on each light may be removable or permanent. Where approved by the building official, an affidavit furnished by the glazing contractor is acceptable in lieu of the manufacturer's mark located on each pane of glass or glazing material. The affidavit must certify that each light is glazed in accordance with approved construction documents, as well as the provisions of Chapter 24. Where an identification mark is utilized, it shall designate both the type and thickness of the material.

Unless used in a spandrel application, tempered glass is required to be permanently identified by the manufacturer. The identifying mark is to be acid etched, sandblasted, ceramic fired, embossed, or of any other type that is unable to be removed without being destroyed. A removable paper marking provided by the manufacturer is permitted, but only for tempered spandrel glass.

2403.2 Glass supports. Glass firmly supported on all edges is considerably stronger than a glass light with one or more free edges. As a result, where the glass does not have firm support on all edges, the code requires that a design for the glass be submitted to the building official for approval. In this latter case, the design would be based on the number and location of any free edges.

2403.5 Louvered windows or jalousies. The requirements for louvered windows are based on there being no edge support on the longitudinal edges. Moreover, for safety purposes, the code requires that the exposed edge be smooth. For the same reason, wired glass, when used in jalousie or louvered windows, shall have no exposed wire projecting from the longitudinal edges. Where a louvered window complies with the provisions of this section, such an application is exempt from the requirements for safety glazing for use in hazardous locations.

Section 2404 Wind, Snow, Seismic, and Dead Loads on Glass

Exterior glass and glazing are subject to the same loads as the exterior cladding of the building; therefore, the code requires that glass and glazing in a vertical or near-vertical position should be designed for the same wind loads as specified in Section 1609 for components

24 Sloped Glazing and Skylights

and cladding. For seismic considerations, ASCE 7 is referenced for the design of glass in glazed curtain walls, glazed storefronts, and glazed partitions. This design will also include the increased pressures on local areas at discontinuities. As the slope of the glass increases, it may be necessary to address other loads as well, such as the dead load and any snow load.

Section 2405 Sloped Glazing and Skylights

By application, sloped glazing and skylights consist of glazing installed in roofs or walls that are on a slope of more than 15 degrees (0.26 rad) from the vertical. See Figure 2405-1. The provisions of the *International Building Code®* (IBC®) are intended to protect these glazed openings from flying fire brands and to provide adequate strength to carry the loads normally attributed to roofs. The provisions are also intended to protect the occupants of a building from the possibility of falling glazing materials.

Allowable glazing materials and limitations. Glazing materials and protective measures for sloped glazing and skylights are outlined in this section. The materials and their characteristics and limitations are as follows: 　　2405.2

Laminated glass. Laminated glass is usually furnished with an inner layer of polyvinyl butyral, which has a minimum thickness of 30 mil (0.76 mm). Such glass is highly resistant to impact and, as a result, requires no further protection below. When used within dwelling units of Group R-2, R-3, and R-4 occupancies without a protective screen, laminated glass is permitted to have a 15-mil (0.38-mm) polyvinyl butyral inner layer, provided

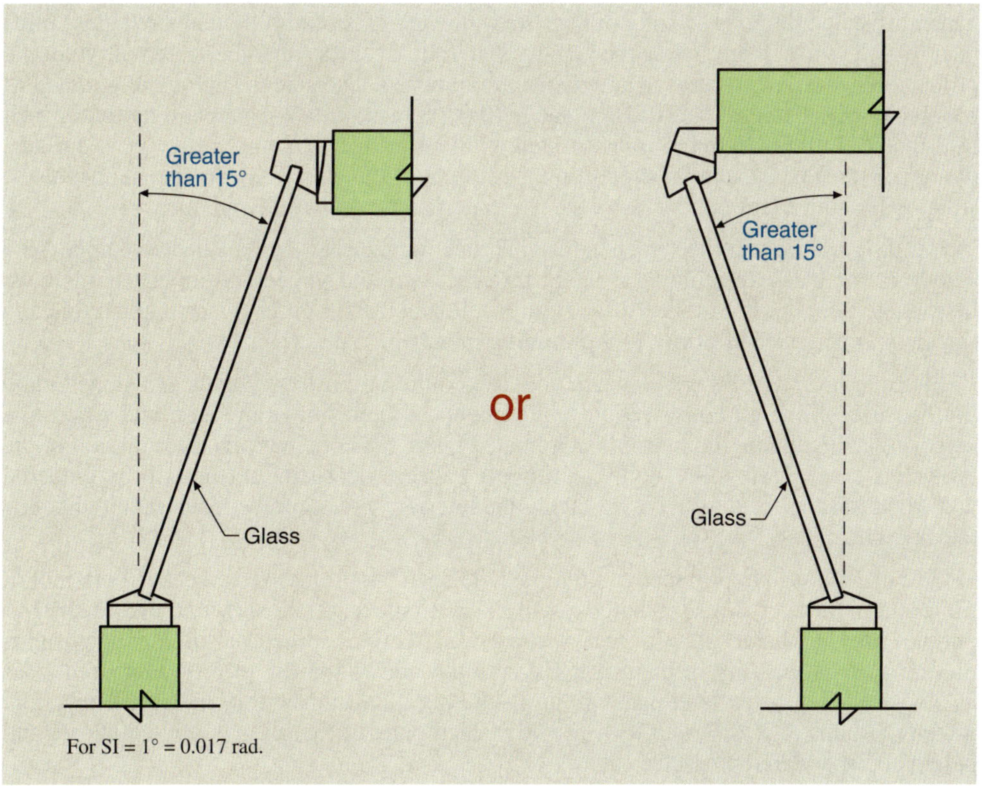

For SI = 1° = 0.017 rad.

Figure 2405-1
Sloped glazing.

each pane of glass is 16 square feet (1.5 m²) or less in area, and the highest point of the glass is no more than 12 feet (3,658 mm) above a walking surface or other accessible area.

Wired glass. Wired glass is resistant to impact and, when used as a single-layer glazing, requires no additional protection below.

Tempered glass. Tempered glass is glass that has been specifically heat-treated or chemically treated to provide high strength. When broken, the entire piece of glass immediately breaks into numerous small granular pieces. Because of its high strength and manner of breakage, tempered glass has been considered in the past to be a desirable glazing material for skylights without any protective screens. However, as a result of studies by the industry that show that tempered glass is subject to spontaneous breakage such that large chunks of glass may fall under this condition, the IBC requires screen protection below tempered glass unless specifically exempted.

Approved light-transmitting plastic materials. See discussion in this handbook under Section 2609.

Heat-strengthened glass. Heat-strengthened glass is glass that has been reheated to just below its melting point and then cooled. This process forms a compression on the outer surface and increases the strength of the glass. However, heat-strengthened glass has the unsatisfactory characteristic of breaking into shards, as does annealed glass. Thus, as a general rule, heat-strengthened glass requires screen protection below the skylight to protect the occupants from falling shards.

Annealed glass. Annealed glass is subject to breakage created by impact, has very low strength, and is unsatisfactory as a glazing material in skylights and sloped glazing. Annealed glass has a further unsatisfactory characteristic for use as a skylight because it breaks up under impact into large sharp shards which, when they fall, are hazardous to occupants of a building. Annealed glass is permitted to a limited degree where used as sloped glazing and skylights, as discussed under Section 2405.3.

2405.3 Screening. The use of protective screens was briefly discussed in Section 2405.2. As a general rule, single-layer glazing of heat-strengthened glass and fully tempered glass must be provided with screens below the glazing material. To ensure that the screen provides the protection necessary, several requirements are imposed. The screen shall be of a noncombustible material at least No. 12 B&S gauge (0.0808 inch) with mesh not larger than 1 inch by 1 inch (25 mm × 25 mm). To be located within 4 inches (102 mm) of the glass surface, the screen must be capable of supporting the weight of the glass. In corrosive atmospheres, structurally equivalent noncorrosive screen materials are to be utilized.

It is also critical that the screen is installed in a manner that will adequately support the weight of the glass. In utilizing a safety factor of two, the screen and its fastenings must be capable of supporting twice the weight of the glazing. In order to accomplish this, the screen is to be fastened firmly to the framing members.

Multiple-layer glazing systems are now used quite often for skylights and sloped glazing because of energy-conservation requirements. Where this is the case and where the layer of glazing facing the interior is laminated glass, the code permits the omission of the protective screen below the skylight. However, where heat-strengthened, fully tempered and wired glass is used as the layer facing the interior, screen protection is required below the skylight.

Five exceptions are provided to eliminate the need for protective screens in both monolithic and multilayer sloped-glazing systems. The first exception, shown in Figure 2405-2, permits fully tempered glass in near vertical wall sections, based on the low height plus the very low probability of breakage. Exception 4 allows the use of fully tempered glass as single glazing or as both panes in an insulating glass unit within individual dwelling units in Groups R-2, R-3, and R-4, provided each pane of the glass is limited in size and height above a walking surface.

Sloped Glazing and Skylights 24

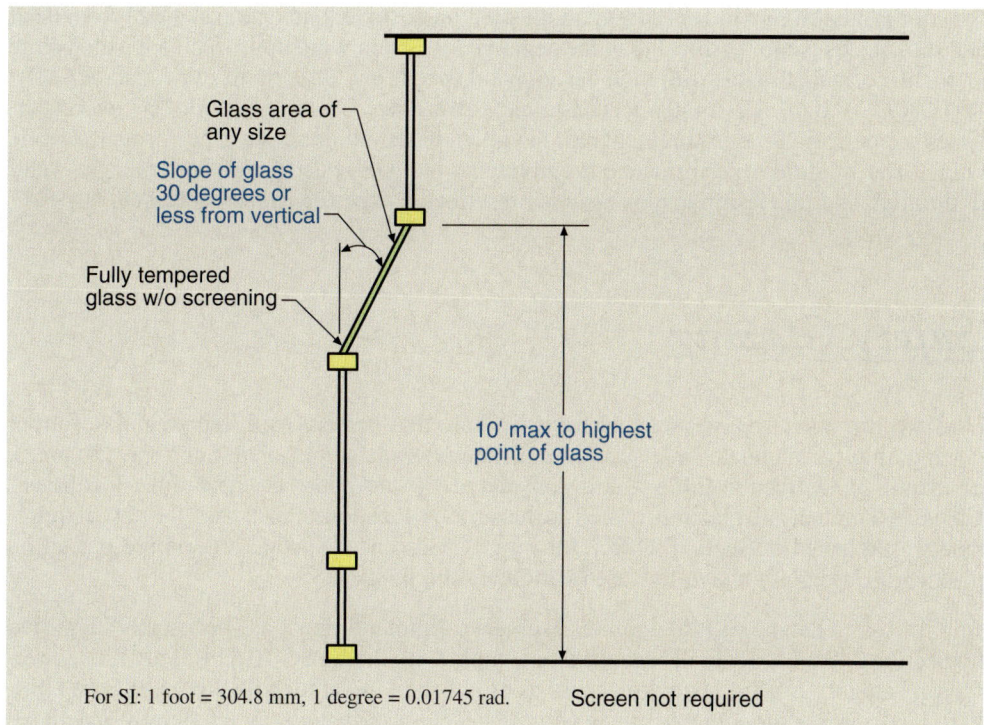

Figure 2405-2
Sloped glazing without screen.

Annealed glass may be used only (and is permitted without screening) under the following two circumstances:

1. Where the accessible area below is permanently protected from falling glass.
2. In greenhouses under the limitations outlined in Exception 3.

As previously discussed, screens are also not required below laminated glass with a 15-mil (0.38-mm) polyvinyl butyral inner layer within dwelling units of specific Group R occupancies. Such glazing is also limited in size and height above the walking surface.

Framing. This is an omnibus section containing requirements related to: **2405.4**

1. Combustibility of materials, and
2. Leakage protection of the skylight juncture with the roof.

This section logically requires that skylight frames be constructed of noncombustible materials where erected on buildings for which the code requires noncombustible roof construction—that is, Type I and II construction. Where combustible roof construction is permitted by the code, combustible skylight frames are also permitted.

The provision requiring a curb at least 4 inches (102 mm) above the plane of the roof for mounting skylights is intended to provide a means for flashing between the skylight and the roofing to prevent leaks around the margin of the skylight. The 4-inch (102-mm) curb then provides a vertical surface up which the flashing can be extended and to which the counterflashing may be attached to cover the flashing.

Unit skylights. A unit skylight is considered a factory-assembled, glazed roof unit containing a single panel of glazing. Its purpose is simply the introduction of natural daylight through the roof while maintaining the roof assembly's weather-resistant barrier. Unlike windows, the most critical uniform load on a skylight (positive or negative) will depend on the climate in which it is installed. In most cold climate areas where heavy snow loads are **2405.5**

common, and design wind speeds are moderate, the positive load on the skylight from dead and snow loads is more critical than the negative load from wind uplift. The opposite is true in warm, coastal climates with high design wind speeds and little or no snow load. AAMA/WDMA/CSA 101/I.S.2/A440 *Specifications for Windows, Doors, and Unit Skylights* establishes the performance requirements for skylights based on the desired performance grade rating. The minimum performance requirements include resistance to air leakage, water infiltration, and the design load pressures prescribed by the *International Codes*.

Section 2406 *Safety Glazing*

2406.1 Human impact loads. In areas where it is likely that persons will impact glass or other glazing, the IBC mandates that specific glazing materials be installed. Such specific areas, as identified in Section 2406.4, are considered by the code as "hazardous locations." Unless exempted, all glazing located in hazardous locations must pass the test requirements established in Section 2406.2 for impact resistance. Special criteria are placed on plastic glazing, glass block, louvered windows, and jalousies.

2406.2 Impact test. Glazing installed in areas subject to human impact as specifically identified by Section 2406.4 is generally required to comply with the CPSC 16 CFR, Part 1201, criteria for Category I or II glazing materials as established in Table 2406.2(1). SPSC 16 CFR, Part 1201, *Safety Standard for Architectural Glazing Materials,* is a federally mandated safety regulation of the U.S. Consumer Products Safety Commission. The exception also allows the installation of glazing materials that have been tested to a different standard, ANSI Z97.1 *Safety Glazing Materials Used in Buildings—Safety Performance Specifications and Methods of Test*, in limited applications. Glazing tested under the ANSI Z97.1 standard must meet the test criteria for Class A or B as set forth in Table 2406.2(2).

For the most part, the differences between the CPSC's 16 CFR Part 1201 standard and the ANSI Z97.1 standard relate to their scope and function. The CPSC standard is not only a test method and a procedure for determining the safety performance of architectural glazing, but also a federal standard that mandates where and when safety glazing materials must be used. It preempts any nonidentical state or local code. In contrast, ANSI Z97.1 is only a voluntary safety performance specification and test method. It does not indicate where and when safety glazing materials must be used.

Glazing in compliance with the appropriate test criteria of CPSC 16 CFR Part 1201 may be used in all hazardous locations. It is also acceptable to utilize safety glazing materials complying with ANSI Z97.1, but only to the extent of those applications other than storm doors, combination doors, entrance-exit doors, sliding patio doors, and closet doors. In addition, such glazing is not permitted in doors and enclosures for hot tubs, whirlpools, saunas, steam rooms, bathtubs, and showers. In all other areas subject to human impact (hazardous locations) as specified in Section 2406.4, required safety glazing is permitted to comply with the applicable requirements of either CPSC 16 CFR 1201 or ANSI Z97.1. See Figure 2406-1.

Where glazing that is tested in accordance with CPSC 16 CFR Part 1201 is utilized, the base requirement is that Category II glazing be used. Only where permitted by Table 2406.2(1) is the use of Category I glazing permitted. A similar approach is provided for those locations where glazing complying with the ANSI Z97.1 criteria can be applied. [See the discussion of Tables 2406.2(1) and 2406.2(2) for information on glazing categories.] As an example, glazing adjacent to stairs as regulated by Section 2406.4.6 is not addressed in Table 2406.2(1). Therefore, glazing tested in accordance with CPSC 16 CFR 1201 must meet the test criteria for Category II. If tested in accordance with ANSI Z97.1, Category A glazing is required. Where Tables 2406.2(1) and 2406.2(2) are applicable, such as for

Safety Glazing 24

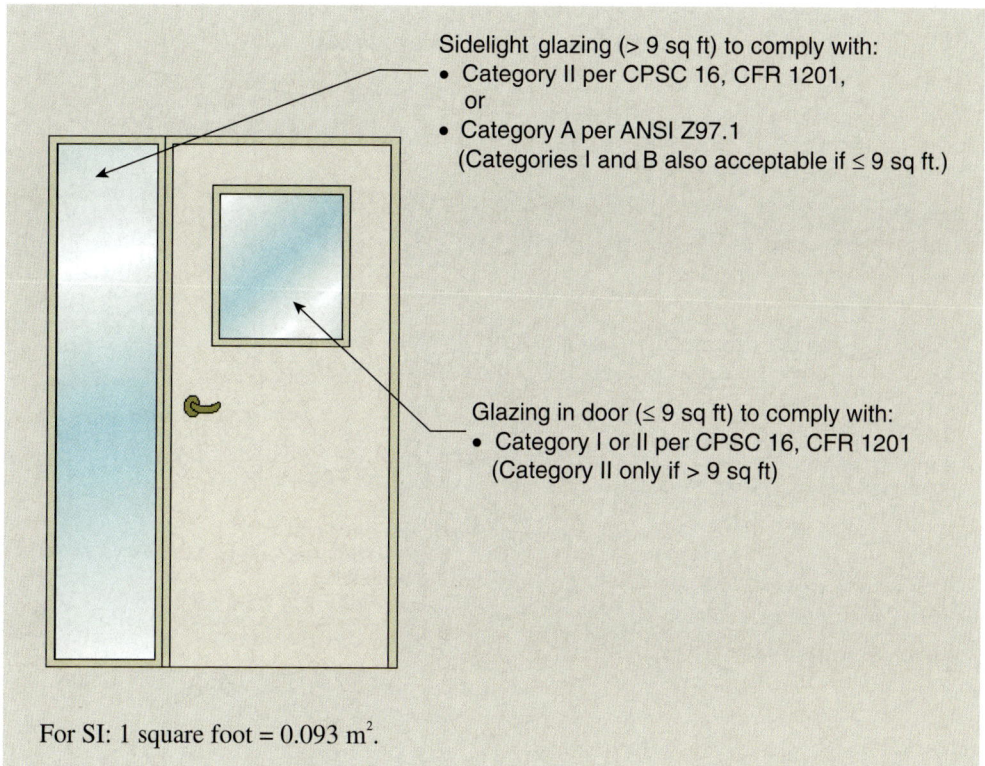

Figure 2406-1
Safety glazing classification.

glazed panels adjacent to doors as regulated by Section 2406.4.2, the minimum required category classification is provided based on the size of the glass lite.

Table 2406.2(1)—Minimum Category Classification of Glazing Using CPSC 16 CFR 1201. Glazing tested in accordance with CPSC 16 CFR 1201 is classified as either Category I or II, based on the specifics of the test. The Category II classification is more difficult to achieve and is thus required in those areas where the highest degree of protection is required. Although Category II material is acceptable in all hazardous locations, Category I glazing may only be installed as relatively small lites in doors and areas adjacent to doors. Otherwise, only Category II glazing is acceptable. See Figure 2406-2. As addressed in the discussion of Section 2406.2, Category II glazing is required in those hazardous locations not addressed in Table 2406.2(1).

Table 2406.2(2)—Minimum Category Classification of Glazing Using ANSI Z97.1. The minimum required category classification of glazing tested to ANSI Z97.1, established in Table 2406.2(2), is also based on the glazing location and size of the lite. This approach is similar to that for determining the minimum required category classification of glazing tested to CPSC 16 CFR 1201 as set forth in Table 2406.2(1). ANSI Z97.1 addresses three separate impact categories or classes, based on impact performance. ANSI Z97.1 Class A glazing materials are comparable to the CSCS Category II glazing materials and ANSI Z97.1 Class B glazing materials are comparable to the CPSC Category I glazing materials. For those hazardous locations not addressed in Table 2406.2(2), the use of Category A glazing is required. Although there is also a Class C category recognized by ANSI Z97.1, applicable only for fire-resistant glazing materials, it is not viewed by the code as an acceptable safety glazing material. Only Class A and B ANSI Z97.1 glazing materials are recognized in the table.

Identification of safety glazing. The code requires the identification of safety glazing for the same reasons as for ordinary annealed glass not subject to human impact. However, in the case of safety glazing, the requirement carries more detail insofar as improperly

2406.3

24 Glass and Glazing

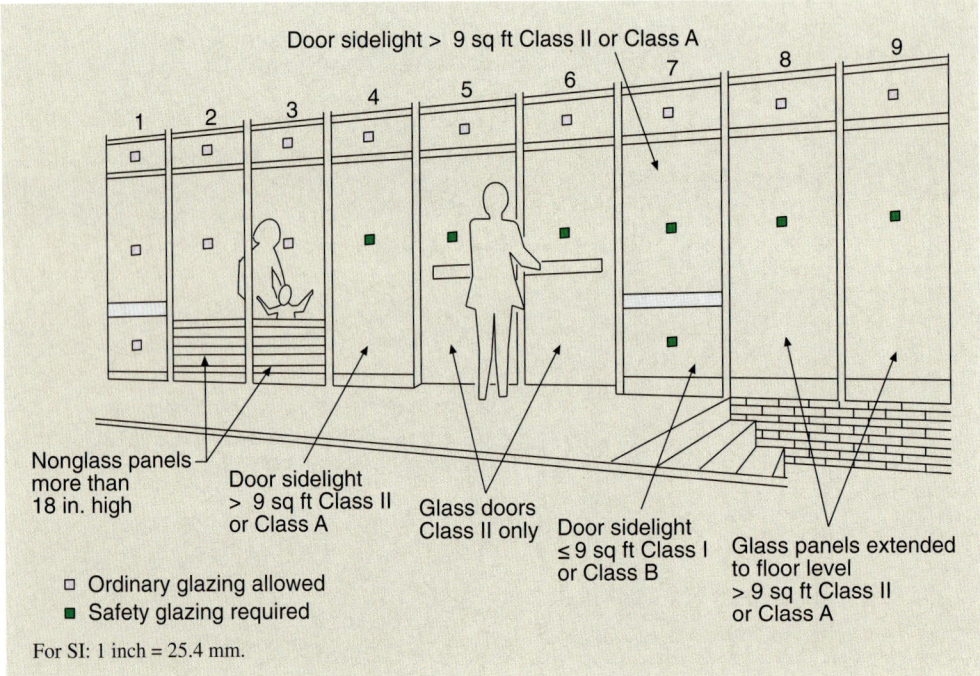

Figure 2406-2
Hazardous locations.

installed annealed glass in areas of human impact can create a serious hazard. Therefore, it is doubly important that the glazing material be further identified to ensure that the proper glazing material is in place. Not only does proper marking assist in the inspection process, it could also help identify a location where safety glazing must be installed should future replacement be required. The code specifically requires the use of identification marks for glazing installed in hazardous locations.

Each pane of safety glazing must be individually identified with a manufacturer's designation. It shall specify who applied the designation, the manufacturer or the installer, and the safety glazing standard with which the glazing complies, as well as the type and thickness of the glass or glazing material. Acceptable methods for the designation include acid etching, sand blasting, ceramic firing, and embossing. Additionally, the designation is permitted to be of a type that cannot be removed without being destroyed. This limitation ensures that a designation cannot be transferred to any other glazing materials. As another option, the building official may be willing to accept a certification or affidavit from the supplier and/or installer indicating that the appropriate glazing was provided and installed. Under such a situation, it is critical that the building official be completely satisfied that the appropriate material is utilized for the installation location. The acceptance of an affidavit or similar document is appropriate for all safety glazing materials other than tempered glass.

The only variation to the general requirement for individual identification is for multi-pane glazed assemblies such as French doors. Where the individual panes do not exceed 1 square foot (0.0929 m^2) in exposed area, only one pane in the assembly is required to be marked as described above. However, all other panes in the assembly must still be identified with "CPSC 16 CFR, Part 1201." See Figure 2406-3. Where the multiple panes occur in other than a door or an enclosure regulated by Item 5 of Section 2406.4, glazing tested and labeled as "ANSI Z97.1" is also acceptable.

2406.4 Hazardous locations. This section lists those specific hazardous locations where safety glazing is required. Some of these locations are shown in Figures 2406-5 through 2406-13. In addition to the hazardous locations shown in the various illustrations, safety glazing is

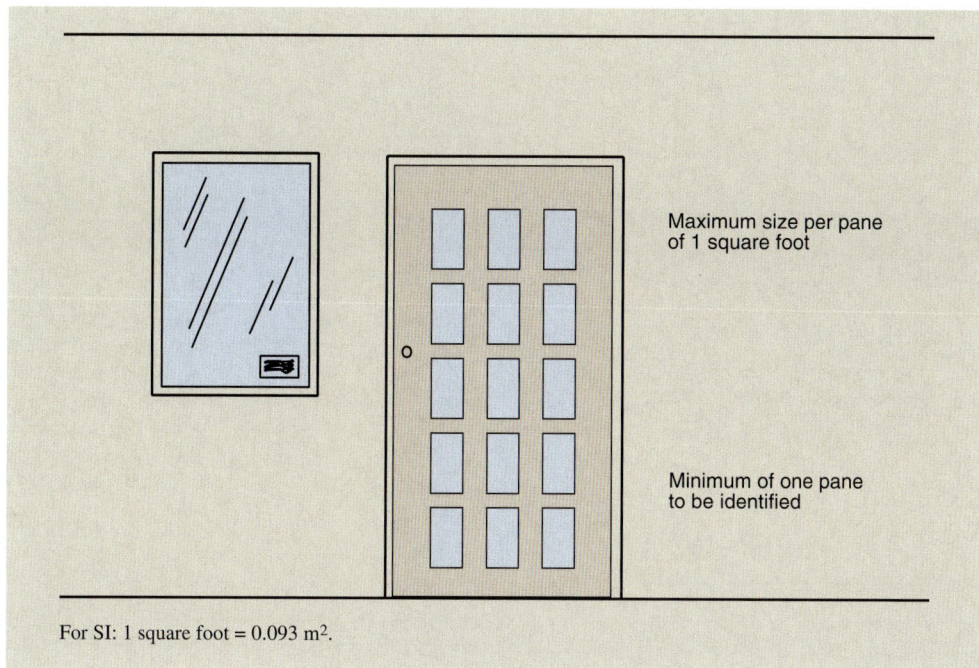

For SI: 1 square foot = 0.093 m².

**Figure 2406-3
Safety glazing identification.**

also required for a number of other conditions, including fixed and sliding panels of sliding door assemblies, storm doors, and glass railings.

Figure 2406-2 also illustrates several locations where safety glazing may or may not be required. To facilitate discussion, each panel has been numbered. Panels 1, 2, 3, 8, and 9 are addressed under Section 2406.4.3. Under this provision, all four stated conditions must occur before safety glazing is required. These conditions are as follows:

1. The area of an individual pane must be greater than 9 square feet (0.84 m²);
2. The bottom edge must be less than 18 inches (457 mm) above the floor;
3. The top edge must be more than 36 inches (914 mm) above the floor; and
4. One or more walking surfaces must be within 36 inches (914 mm), measured horizontally, of the glazed panel.

Panel 1 is not required to have safety glazing, because a protective bar has been installed in compliance with Exception 2 to Section 2406.4.3, the requirements of which are illustrated in Figure 2406-4. Panels 2 and 3 do not require safety glazing, because their bottom edges are not less than 18 inches (457 mm) from the floor. If Panels 8 and 9 have a walking surface within 36 inches (914 mm) of the interior, safety glazing would be required. This would be true even though the bottom of the panel appears to be greater than 18 inches (457 mm) above the exterior walking surface, as the exterior condition would have no bearing on the determination. However, the exterior condition, because it is adjacent to a stairway, would be regulated by Section 2406.4.6 and/or Section 2406.4.7 for glazing adjacent to stairways. Therefore, multiple conditions mandate the need for safety glazing in Panels 8 and 9. Panels 4 and 7 require safety glazing because they are door sidelights. The exception mentioned above does not apply to panels adjacent to a door; therefore, although Panel 7 has been provided with a protective bar, safety glazing is still required.

Figures 2406-5 and 2406-6 illustrate when safety glazing is required for panels adjacent to a door. This requirement applies to both fixed and operable panels. Where there is an intervening wall or permanent barrier as shown in Figure 2406-7, safety glazing would not be required. If the door serves only a shallow storage room or closet, adjacent glazing need not

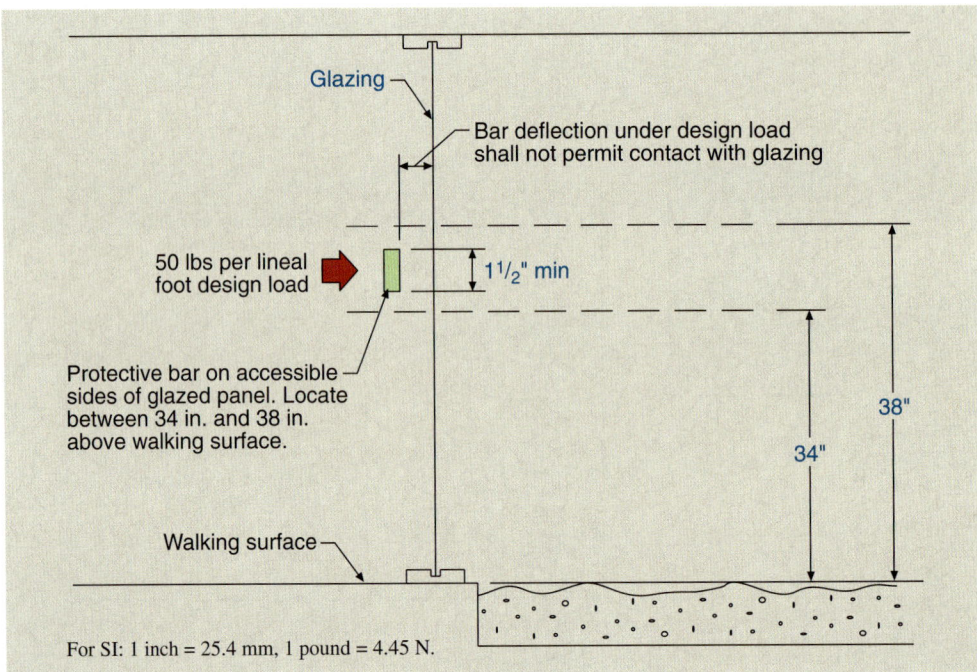

Figure 2406-4
Protective bar alternative.

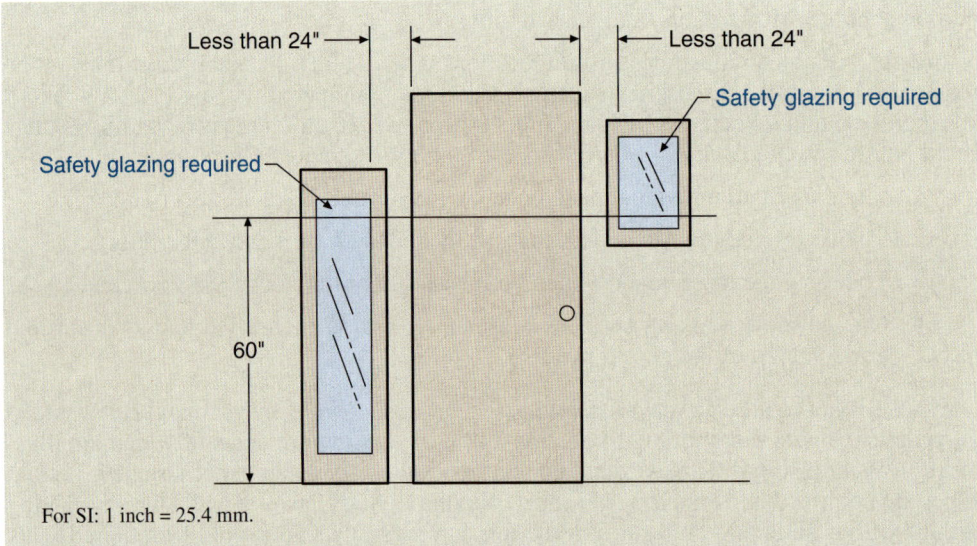

Figure 2406-5
Glass in sidelights—elevation.

be safety glazing, as depicted in Figure 2406-8. Figure 2406-9 illustrates an exception applicable only to one- and two-family dwellings and within dwelling units in Group R-2 uses.

Panels 5 and 6 in Figure 2406-2 are glass doors, which require safety glazing based on the provisions of Section 2406.4.1. All ingress and egress doors, unframed swinging doors, and glazing in storm doors require safety glazing, as well as any other swinging, sliding, and bifold doors with fixed or operable glazed panels. See Figure 2406-10. There are several exceptions. If openings in a door will not allow passage of a 3-inch-diameter (76-mm) sphere, the glazing is exempt, as are assemblies of leaded, faceted, or carved glass used for decorative purposes. The latter exception not only applies to doors, but also to sidelights and other glazed panels covered by Sections 2406.4.2 and 2406.4.3.

Safety Glazing 24

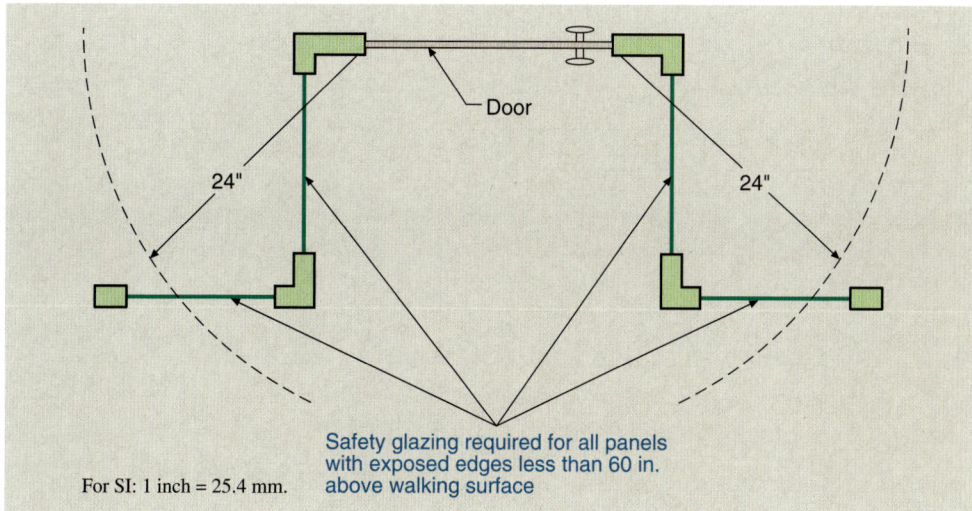

Figure 2406-6
Glass in sidelights—plan.

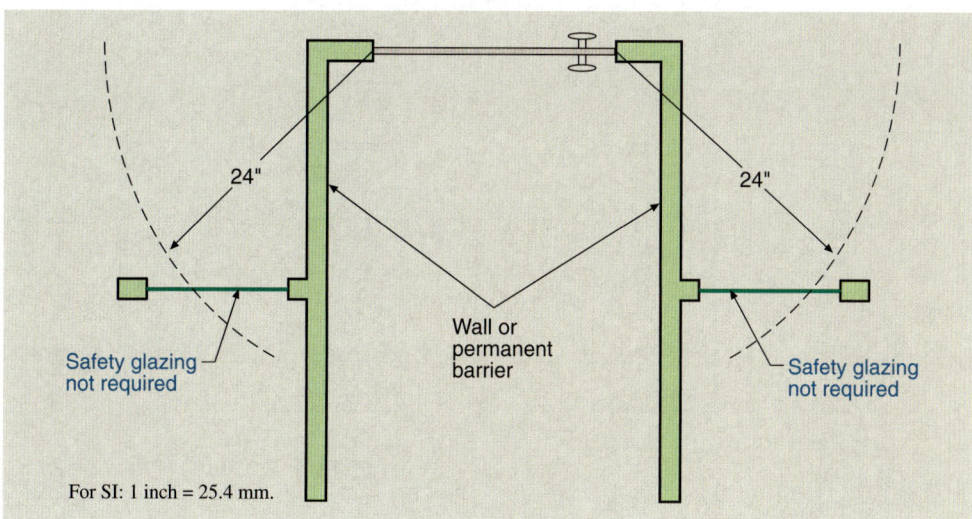

Figure 2406-7
Barrier between glazing and door.

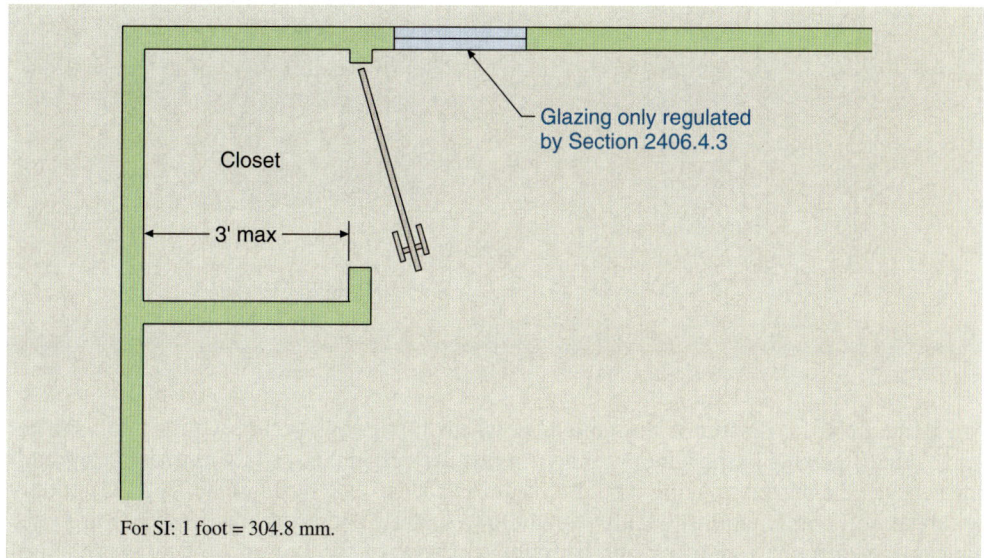

Figure 2406-8
Glazing adjacent to closet door.

24 Glass and Glazing

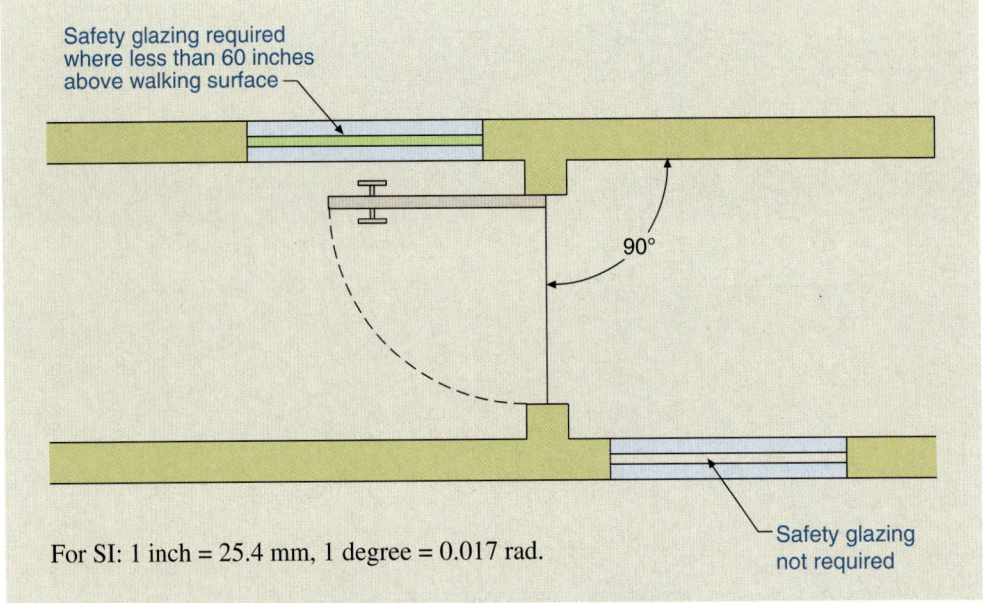

Figure 2406-9 Applicable to dwelling units of Group R-2 and R-3.

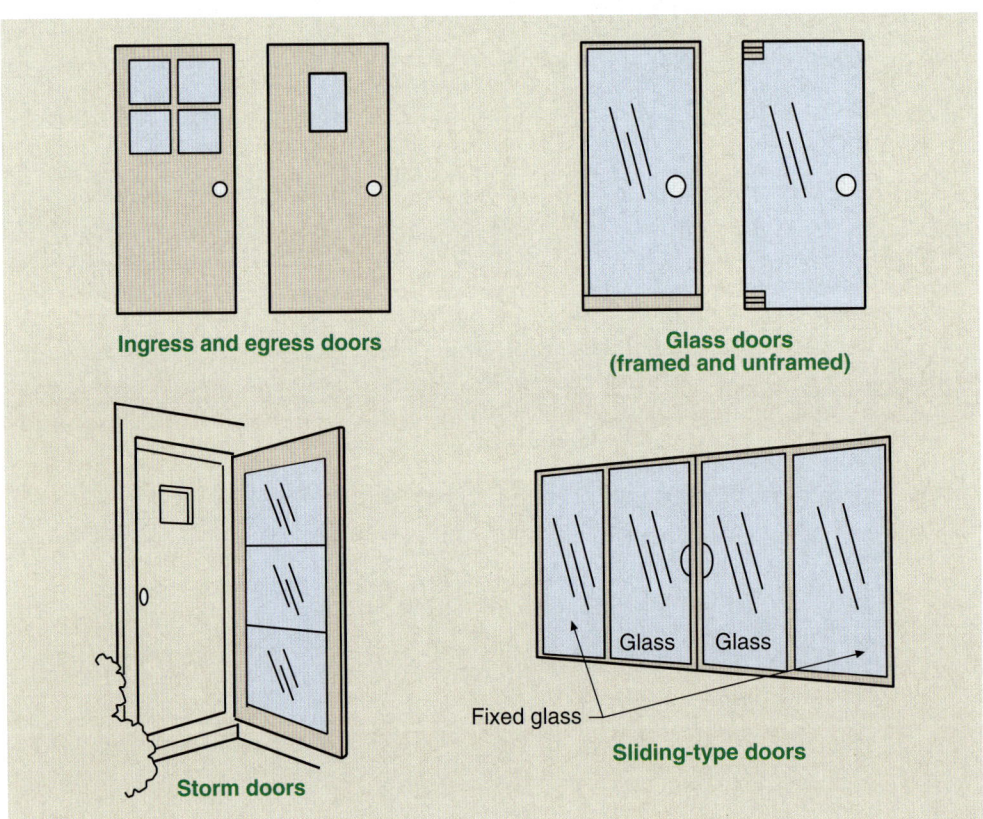

Figure 2406-10 Glazing in doors.

Figure 2406-11 illustrates the condition where a window occurs within a tub/shower enclosure. If glazing in the window shown is less than 60 inches (1,524 mm) above a standing surface, then safety glazing would be required. This same requirement applies not only to tub/shower combinations, but also to windows installed adjacent to hot tubs, whirlpools, saunas, steam rooms, showers, and bathtubs. Because of the presence of moisture, all of

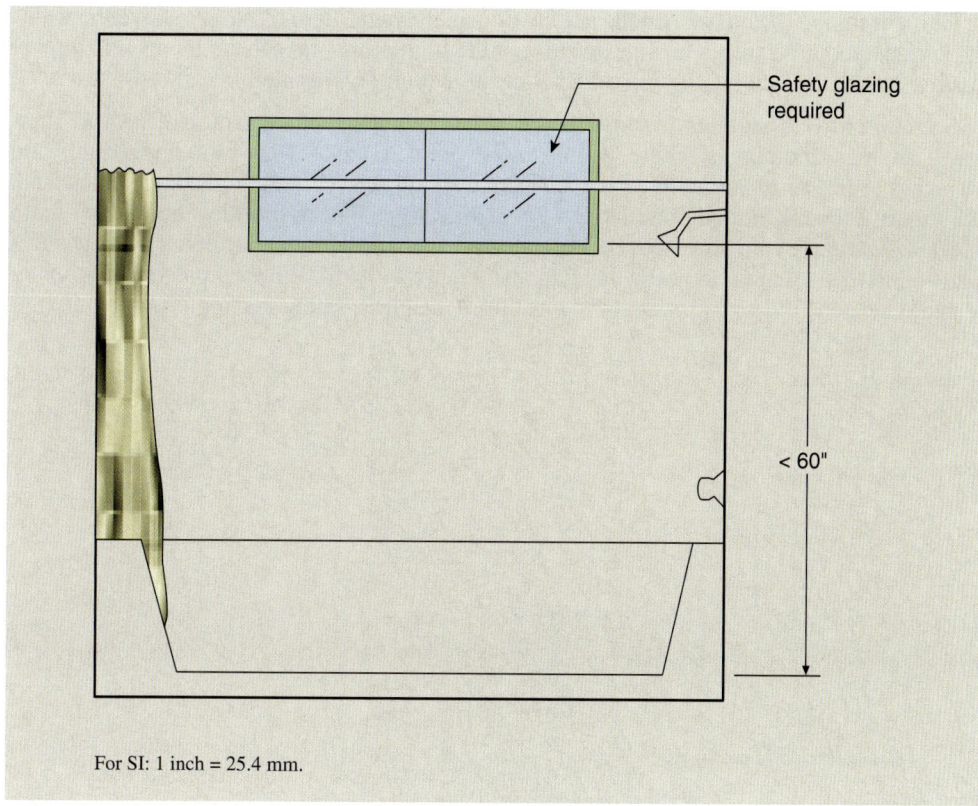

Figure 2406-11
Glazing within a shower enclosure.

these locations represent slip hazards and need safety glazing to prevent injury in case of a fall. The provisions of Section 2406.4.5 not only address glazing adjacent to bathtubs, showers, hot tubs, and whirlpools, but also any glazing adjacent to indoor and outdoor swimming pools, as shown in Figure 2406-12.

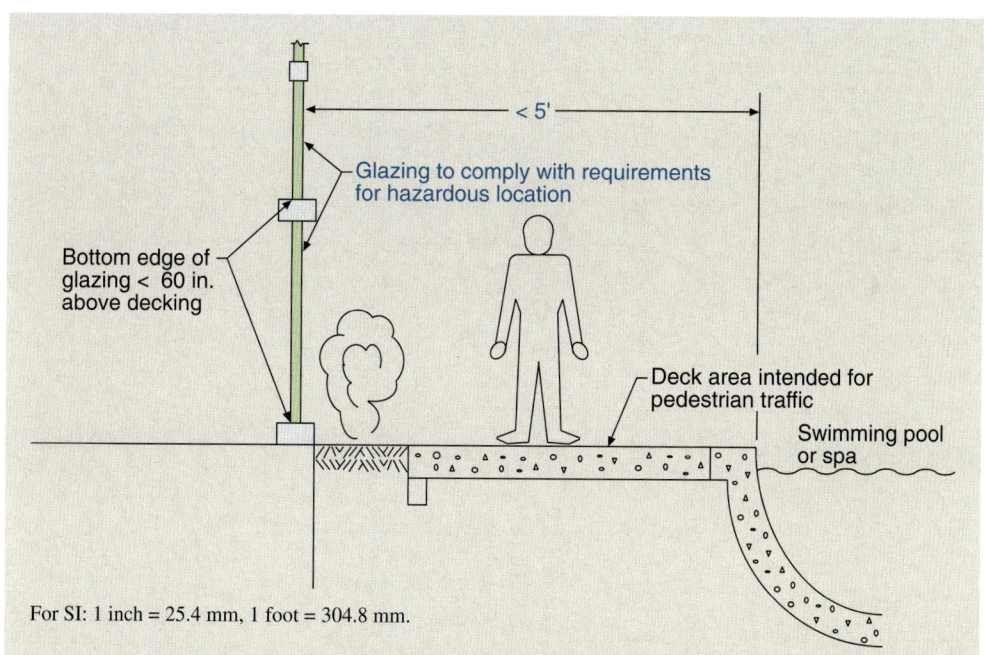

Figure 2406-12
Glazing adjacent to swimming pool or spa.

Glass in railings, baluster panels, and in-fill panels, regardless of height above the walking surface, require safety glazing. Because of the high probability that persons will impact guards, it is critical that an increased level of protection be provided.

Section 2406.4.6 requires the installation of safety glazing for glazing adjacent to stairways, landings, and ramps. Where the glazing is within 36 inches (914 mm) horizontally and 60 inches (1,524 mm) vertically of the adjacent walking surface, safety glazing is mandated. See Figure 2406-13. In addition, Section 2406.7 identifies a hazardous location to be within 5 feet (1,524 mm) of the bottom tread of a stairway when the bottom edge of the glazing is less than 36 inches (914 mm) above the landing. See Figure 2406-14. Where protected by a railing or guard located at least 18 inches (457 mm) from the glass, safety glazing is not required.

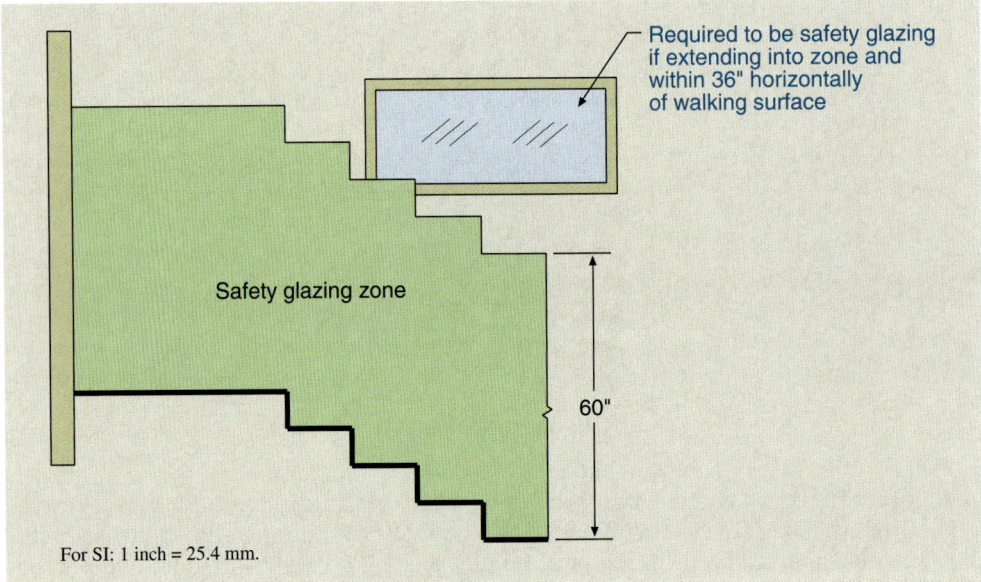

Figure 2406-13
Glazing adjacent to stairways.

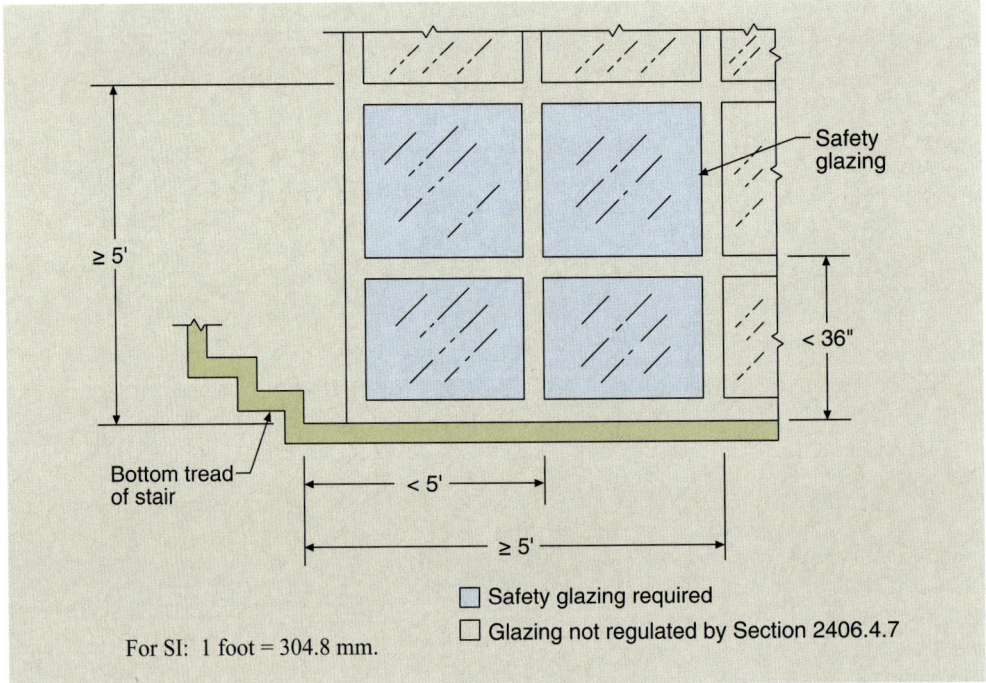

Figure 2406-14
Glazing adjacent to bottom stair landing.

Fire department access panels. There may be conditions under which glass panels will be utilized for fire department access purposes. Certainly, such panels shall be of the type that will provide safe conditions under which fire department personnel can enter a building or a specific area of a building. For this reason, glass access panels must be of tempered glass. Where the glazed unit is an insulating panel, all panes shall be tempered. Because access must be provided completely through the panel, it is necessary that all panes be of this specific safety glazing material.

2406.5

Section 2407 Glass in Handrails and Guards

The increased use of glass (usually tempered) in handrail assemblies and guard sections prompted the inclusion of these provisions in the IBC. These provisions provide uniform regulations identifying the specific types of safety glazing that may be used structurally. Fully tempered, laminated tempered, and laminated heat-strengthened glass are the only types considered by the code to be structurally adequate for this use so critical to life safety.

Only glazing conforming to the provisions of Section 2406.1.1 is permitted. This would limit glazed railing in-fill panels to materials that have passed the applicable test requirements of CPSC 16 CFR 1201 or ANSI Z97.1. Regardless of the type of glazing, the minimum nominal thickness of the structural balustrade panels in railings shall be $1/4$ inch (6.4 mm). The panels and their supporting system must be designed to withstand the loads as specified in Section 1607.8, utilizing a safety factor of four. Not permitted to be installed without an attached handrail or guardrail, at least three glass balusters shall be used to support each handrail or guard section. The purpose for requiring at least three balusters is that, should one fail, the remaining two balusters will continue to support the handrail or guard section. If another method is devised to provide continued support should a single baluster panel fail, such a method is acceptable.

Section 2408 Glazing in Athletic Facilities

Where glazing forms entire or partial wall sections, or is used as a door or as part of a door, in racquetball courts, squash courts, gymnasiums, basketball courts, and similar athletic facilities subject to impact loading, it shall comply with Section 2408. In racquetball and squash courts, glass walls and glass doors must pass specific test criteria above and beyond those typically required of safety glazing materials. Such special test criteria are necessary to address those glazed areas where impact with the glass is not merely accidental, but rather expected because of the nature of the physical activities involved. Special conditions for compliance are also set forth for glazing subject to human impact in gymnasiums, basketball courts, and other high-intensity activity areas where it is expected that contact with the glazing will occur more often, and with more force, than in most other hazardous locations addressed by the code. In such facilities, all glazing in hazardous locations identified by the code must meet the Category II requirement of CPSC 16 CFR 1201 or the Class A requirements of ANSI Z97.1. This would include glazing both in doors (only CPSC 16 CFR 1201 glazing permitted) and adjacent to doors.

24 Glass and Glazing

KEY POINTS

- Glass and glazing must resist lateral loads in a manner consistent with other building components.
- Skylights and other sloped glazing are regulated as to the type of glazing material and the need for protective screening below the skylight.
- Safety glazing materials are to be installed in those areas subject to human impact, referred to as hazardous locations.
- The minimum required classification category of safety glazing materials is based on the size and location of the glazing material.
- Common safety glazing materials include tempered or laminated glass, as well as approved plastic.
- In order to verify compliance, the code specifically requires the use of identification marks for glazing installed in hazardous locations.
- Some of the most common locations identified as hazardous include those glazed areas in and around doors.
- Glazing adjacent to showers and bathtubs, where located in a position where impact is likely, must be safety glazed.
- Glass in handrails and guards is regulated for both structural adequacy and human impact.
- Special requirements are mandated for glazing subject to human impact in athletic facilities.

CHAPTER 25

GYPSUM BOARD AND PLASTER

Section 2501 Scope
Section 2502 Definitions
Section 2504 Vertical and Horizontal Assemblies
Section 2506 Gypsum Board Materials
Section 2508 Gypsum Construction
Section 2509 Gypsum Board in Showers and Water Closets
Section 2510 Lathing and Furring for Cement Plaster (Stucco)
Section 2511 Interior Plaster
Section 2512 Exterior Plaster
Key Points

25 Gypsum Board and Plaster

> This chapter regulates the covering materials for walls and ceilings:
>
> 1. To provide weather protection for the exterior of the building.
> 2. To secure the material to the wall and ceiling framing so that it will remain in place during the expected life of the building.
>
> Where these materials are used or required for fire-resistance-rated construction, the code requires that they also comply with the provisions of Chapter 7.

Section 2501 Scope

This chapter of the *International Building Code*® (IBC®) covers the installation requirements for wall- and ceiling-covering materials, including their method of fastening and, in the case of plaster, the permitted materials for lath, plaster, and aggregate.

Although plaster has many uses, including ornamental and decorative work, its use in the IBC is regulated purely as a wall- and ceiling-covering material.

The IBC regulates the installation of wall- and ceiling-covering materials as well as quality standards for the materials themselves. The primary wall- and ceiling-covering material in use today is gypsum wallboard; however, lath, plaster, and wood paneling are sometimes utilized. As wood paneling is covered in Chapter 23, it follows that Chapter 25 only regulates gypsum wallboard, lath, and plaster. However, in this section the code permits the installation of other wall- and ceiling-covering materials, provided the materials have been approved. On this basis, the manufacturer's recommendations and conditions of approval should be consulted.

Gypsum wallboard is a relatively new material for covering walls and ceilings. On the other hand, plaster is among the oldest of building materials still in use. The use of gypsum plaster dates back to about 4000 B.C., when the Egyptians applied it to the interior and exterior of the pyramids.

Section 2502 Definitions

Definitions specific to the provisions addressing gypsum board and plaster are listed in this section and defined in Chapter 2. In short, the determination of whether or not a surface is considered interior or exterior is based on how it is viewed in relationship to the definition for weather-exposed surfaces. Any surface that can be considered weather-exposed under the definition in Section 202 is considered an exterior surface. Surfaces other than weather-exposed surfaces are viewed as interior surfaces.

Surfaces of walls, ceilings, floors, roofs, soffits, and similar elements, where exposed to the weather, are typically considered weather-exposed surfaces. There are three exceptions to the general criteria that would define such surfaces as interior where applying the code provisions. Those exterior conditions considered other than weather-exposed surfaces are illustrated in Figure 2502-1.

25 Vertical and Horizontal Assemblies

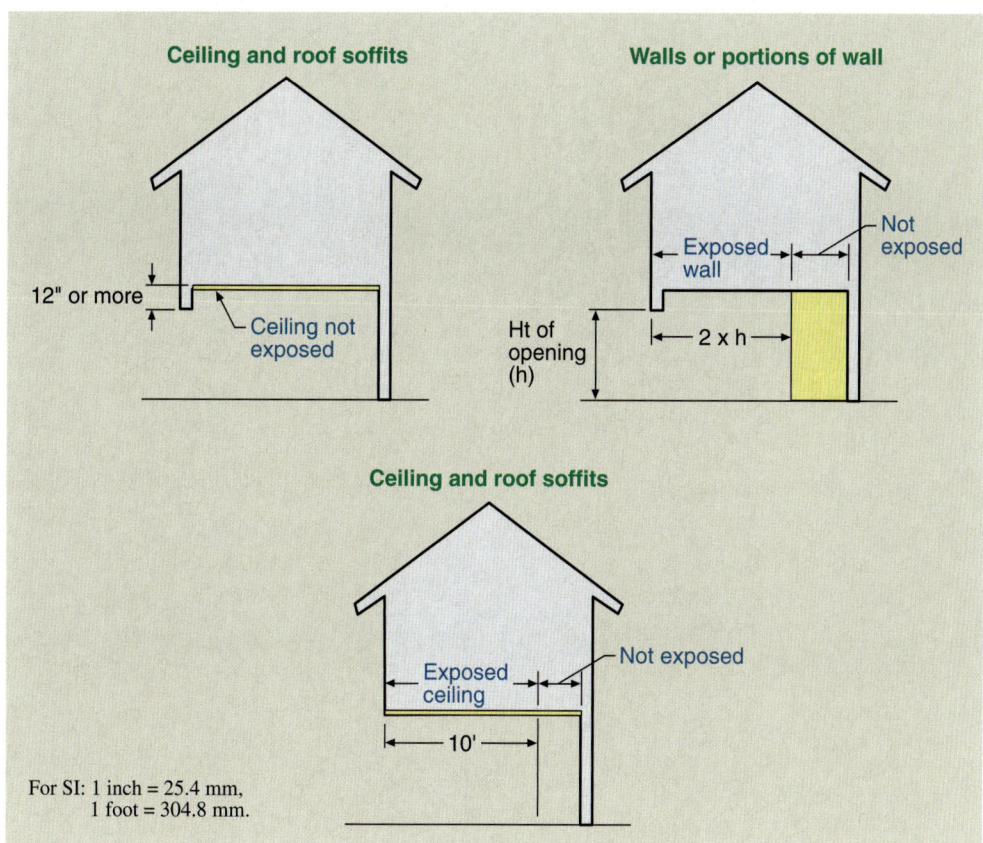

Figure 2502-1 Weather-exposed surfaces.

Section 2504 Vertical and Horizontal Assemblies

The minimum dimensions required by this section for supports of lath or gypsum board are intended to prevent fastener failures and surface distortion that are due to warping of the wood supports. Where supporting lath or gypsum board, wood supports, stripping, or furring is to be at least 2 inches (51 mm) nominal thickness in the least dimension. An exception permits a 1-inch by 2-inch (25-mm by 51-mm) wood furring strip to be installed over a solid backing.

Figure 2504-1 shows a solid plaster studless partition, permitted where constructed in compliance with the conditions set forth in Section 2504.1.2. As its name implies, it is a partition constructed solidly of plaster and lath without framing studs. Thus, the lath and plaster form the structural elements of the partition to resist lateral loads. These partitions are used as nonbearing partitions, but they must meet the horizontal load requirements specified in Section 1607.14.

25 Gypsum Board and Plaster

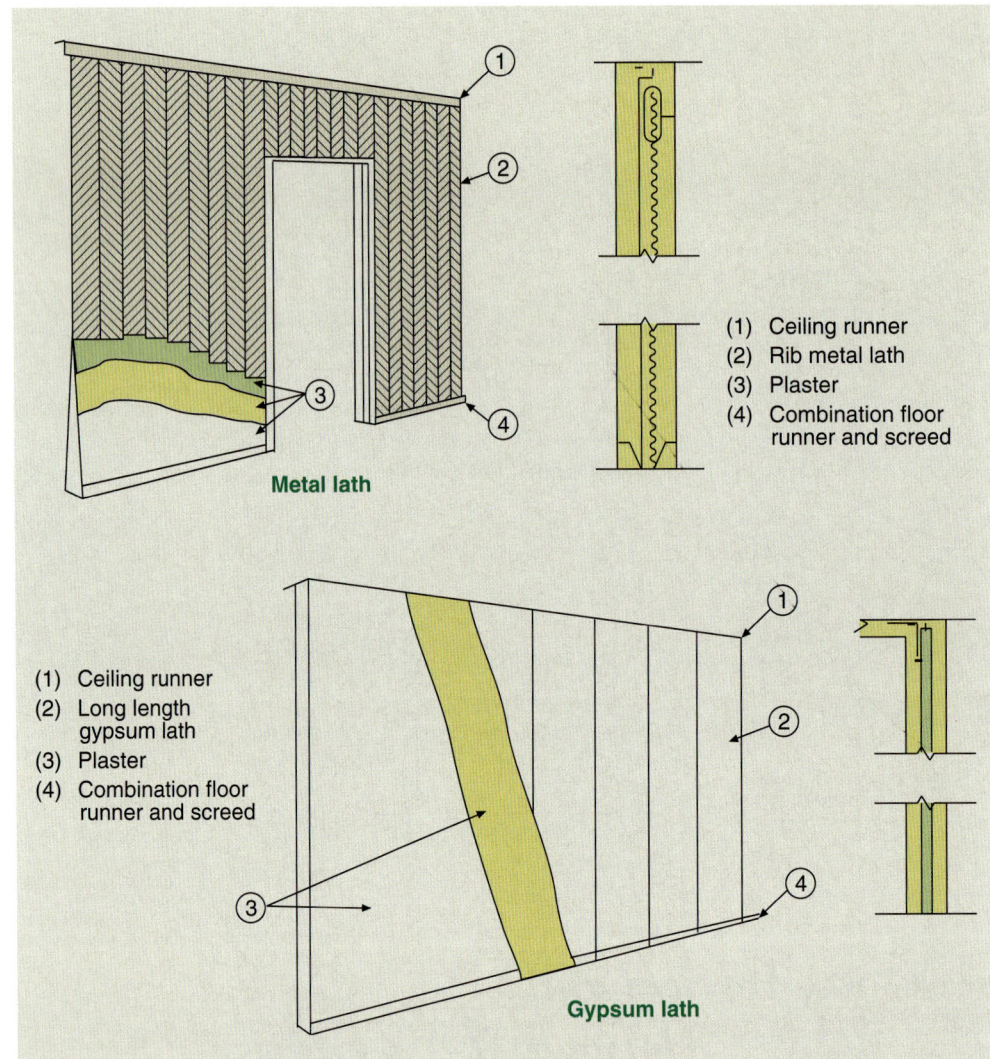

Figure 2504-1
Solid plaster studless partition.

Section 2506 *Gypsum Board Materials*

To indicate compliance with the appropriate materials standards, gypsum board materials and related accessories shall be identified by the manufacturer's designation. Such standards, including those for gypsum sheathing, water-resistant gypsum backing board, nonload-bearing steel studs, steel screws, and nails for gypsum board are referenced in Table 2506.2. It is also noted that any gypsum board materials utilized for fire-protection purposes must conform to the provisions found in Chapter 7.

Where acoustical and lay-in panel ceilings are provided, the metal-suspension systems for such ceilings are to conform with ASTM C 635 and Section 13.5.6 of ASCE 7 for installation in high seismic areas.

Section 2508 *Gypsum Construction*

This section addresses the installation of gypsum materials, primarily that of gypsum wallboard installed as wall and ceiling membranes. As gypsum board is a construction material utilized in almost every construction project, it is important that the application is in compliance with the IBC and the appropriate referenced standards.

2508.1 General. The primary installation criteria for various gypsum materials are found in the referenced standards identified in Table 2508.1. The application of gypsum board varies based on gypsum board thickness, wall or ceiling installation, orientation of gypsum board to the framing members, and maximum spacing of the framing members. Based on the specific conditions encountered, the maximum fastener spacing and size of fasteners are identified.

2508.2 Limitations. Because gypsum plaster and gypsum board are subject to deterioration from moisture, the code restricts their use to interior locations and weather-protected surfaces. The definition for weather-exposed surfaces is found in Section 202 and illustrated in Figure 2502-1. Even where installed in interior locations, it is important that gypsum wallboard not be installed in areas of continuous high humidity or wet locations. The IBC further requires that interior gypsum board, gypsum plaster, and gypsum lath shall not be installed until the installation has been weather protected.

2508.3 Single-ply application. The application of gypsum wallboard is specified in this chapter for locations where fire-resistance-rated construction is not provided or for construction where diaphragm (shear wall) action is not required. Chapter 7 and fire-test reports will establish the means of fastening and supporting the ends and edges of gypsum wallboard for fire-resistance-rated assemblies. Table 2508.1 provides the installation standards for various types of gypsum construction.

The code requires that the fit of gypsum wallboard sheets be such that the edges and ends are in moderate contact. However, wider gaps are permitted in concealed spaces where fire-resistance-rated construction or diaphragm action is not required. This requirement is based primarily on appearance. Therefore, where the wallboard application is concealed, it is not objectionable to have wider gaps than those resulting from moderate contact. However, where the wallboard surface is exposed as it normally is, moderate contact is required so that there will be no objectionable cracking when the joint between the sheets is finished.

Unless the wallboard is considered a shear-resisting element or an element of a fire-resistance-rated assembly, fasteners may be omitted at certain locations. It should be emphasized that where a fire-test report or other installation standard indicates that fasteners are required on supports or edges, the fastening pattern may not be modified. Otherwise, those fasteners located at the top and bottom plates of vertical assemblies are permitted to be omitted. In addition, fasteners need not be provided at the edges and ends of horizontal assemblies perpendicular to supports and at the wall line. Note that fasteners are to be applied in a manner in which the face paper is not fractured by the fastener head. The intent of the requirement is to provide a tight fastening but not damage the gypsum board to the extent its nail-holding power may be affected. Proper construction procedure for the nailing of gypsum wallboard panels is to use a drywall hammer that has a crowned head and use wallboard nails that have concave heads. Figure 2508-1 illustrates the case where a drywall hammer is used and the drywall nail is not overdriven. The intent is to create a dimple in the plaster board with no projection of the nail head above the plaster board.

2508.4 Joint treatment. Although as a general rule the IBC requires joint and fastener treatment for fire-resistance-rated assemblies, the code exempts in Exception 1 those locations where the wallboard is to receive a decorative finish or any other similar application, which is considered to be equivalent to the joint treatment. Also, joint treatment is not required where joints occur over wood framing members, or where square-edge or tongue-and-groove edge gypsum board is used. In addition, Exception 5 indicates that joint treatment is not required for assemblies tested without joint treatment. In general, joint treatment does not materially increase the fire rating, and many partitions have passed the fire test without joint treatment. As indicated earlier in this section, joint treatment is primarily used for aesthetic reasons. One further exception addresses the condition where a multilayer system is constructed. Where two or more layers of gypsum board are utilized in the assembly, joint and fastener treatment is not required where the joints of adjacent layers are offset from each other.

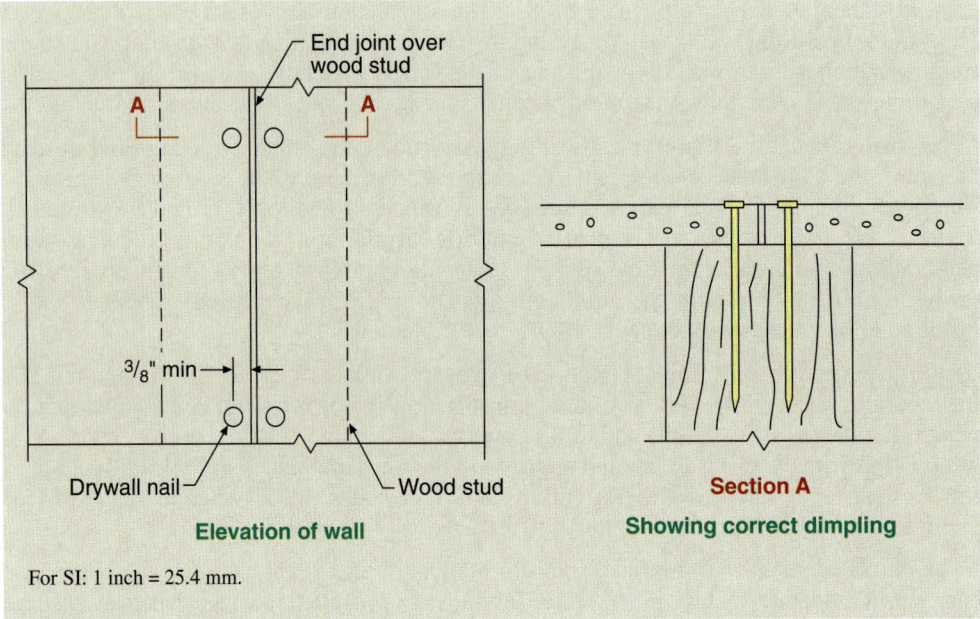

Figure 2508-1 Gypsum wallboard nailing.

Section 2509 *Gypsum Board in Showers and Water Closets*

Special consideration is given to toilet and bathing areas subject to some degree of exposure to water or high humidity. In shower areas, it is mandated that complying glass mat water-resistant gypsum backing panels, discrete nonasbestos fiber-cement interior substrate sheets, or nonasbestos fiber-mat reinforced cementitious backer units be used as a base for wall tile and wall panels, as well as a base for ceiling panels. In tub areas, such backers are required as a base for wall tile. Where wall tile is installed over a gypsum board base on water closet compartment walls, it is mandated that the tile be installed over a base of water-resistant gypsum backing board. The use of glass mat water-resistant gypsum backing panels, discrete nonasbestos fiber-cement interior substrate sheets, or nonasbestos fiber-mat reinforced cementitious backer units is also permitted in such locations.

Two locations are identified in Section 2509.3 as areas where water-resistant gypsum backing board is not ever to be used. In shower or bathtub compartments, water-resistant gypsum backing board shall not be installed over a vapor retarder. In addition, such materials are prohibited in locations subject to direct water exposure or continuous high humidity, such as in saunas, steam rooms, gang showers, or indoor pools. An additional provision permits the use of water-resistant gypsum backers on ceilings, but only under limited conditions. Because of the potential for sagging, the backer boards must be supported at close intervals. One-half-inch thick (12.7-mm) water-resistant gypsum backing board is prohibited on ceilings where the framing members are spaced in excess of 12 inches (305 mm) on center. Such gypsum board, when $^5/_8$ inch (15.9 mm) in thickness, is permitted only where ceiling members are spaced a maximum of 16 inches (406 mm) on center.

Except for tub, shower, and water closet compartment walls, and ceilings of shower areas, regular gypsum board may be used as a backing for tile or wall panels. An example would be a backsplash at the rear of a kitchen counter. Even if this area is to be covered with wall panels or tile, regular gypsum board is permitted as the backing material.

Section 2510 *Lathing and Furring for Cement Plaster (Stucco)*

Cement plaster (stucco) used in both exterior and interior locations must be installed in accordance with this section. In order to ensure reliability and consistency, the materials of construction must comply with the appropriate standards listed in Table 2507.2. Of particular importance is the installation method of the water-resistive barrier that is required over wood-based sheathing where exterior plaster is applied. The installation must have a performance level "at least equivalent to two layers" of Grade D paper.

The greatest benefits of using two layers of water-resistive barrier (WRB) can only be realized if the method and manner of the installation establish a continuous drainage plane, separated from the stucco. In a two-layer system, each layer provides a separate and distinct function. The primary function of the inboard layer is to resist water penetration into the building cavity. This interior layer should be integrated with window and door flashings, the weep screed at the bottom of the wall, and any through-wall flashings or expansion joints. This inner layer becomes the drainage plane for any incidental water that gets through the outer layer or at one of the joints or openings or where the outer layer is damaged. The primary function of the outboard layer (layer that comes in contact with the stucco) is to separate the stucco from the water-resistive barrier. This layer has historically been called a sacrificial layer, intervening layer, or bond break layer. See Figure 2510-1.

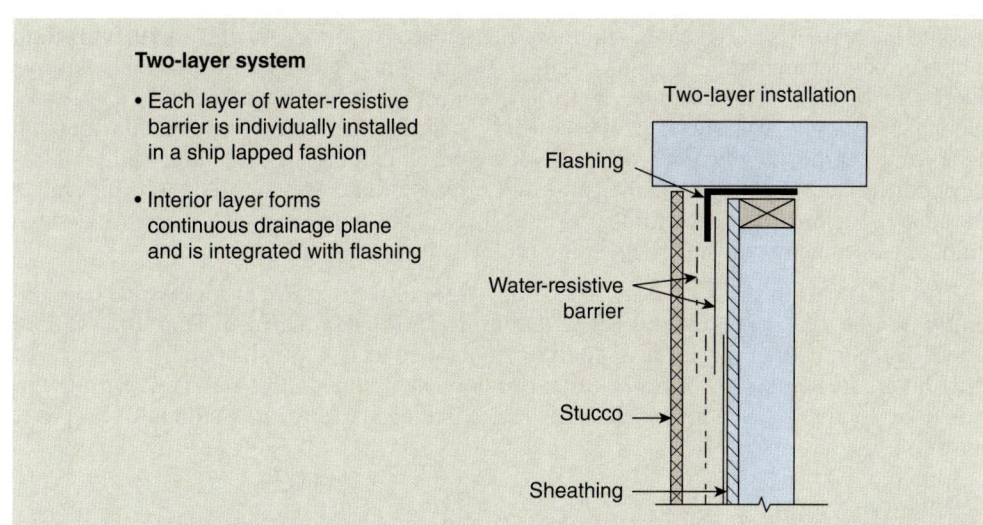

Figure 2510-1
Installation of water-resistive barrier.

25 Gypsum Board and Plaster

Section 2511 *Interior Plaster*

Multicoat plastering has been the standard in the western world for over 100 years. It is generally the consensus of the industry that, particularly where plaster application is by hand, multicoat work is necessary for control of plaster thickness and density. Most of the materials used for plaster densify under hand application because of the pressure applied to the trowels, and it is believed that this change in density is more controllable and will be of a more uniform nature where the plaster is applied in thin, successive layers. For these reasons, the IBC requires three-coat plastering over metal or wire lath and two-coat work applied over other plaster bases approved for use by the code. Reducing the requirement for plaster bases other than metal or wire lath is based on the rigidity of the plaster base itself. More rigid plaster bases are not as susceptible to variations in thickness and flatness of the surface. In fact, it may be considered that the first coat applied in three-coat work on a flexible base, such as wire lath, is used to stiffen that base to provide the rigidity necessary to attain uniform thickness and surface flatness.

Fiber insulation board does not have the qualities necessary for a good performing plaster base. It absorbs excessive moisture from the plaster mix, creating problems of workmanship, and it does not have the stability and rigidity required for a proper functioning plaster base. Also, in the colder and damper climates, the fiberboard insulation retains the moisture absorbed from the plaster for a relatively longer period of time than other bases, causing premature failure of the plaster. For these reasons, the code prohibits its use as a plaster base.

Because portland cement plaster does not bond properly to gypsum plaster bases, the code prohibits its use over gypsum plaster bases. However, the code permits exterior plaster to be applied on horizontal surfaces, soffits, and so on, over gypsum lath and gypsum board when used as a backing for metal lath.

Plaster grounds are utilized to establish the thickness of plaster and usually are wood or metal strips attached to the plaster base. The intent is that plaster grounds are used as a guide for the straightedge in determining the thickness. In many cases, door and window frames are used as plaster grounds.

In plaster work, a base coat is any coat beneath the finish coat. This is true whether the plaster is of two-coat or three-coat application. In three-coat work, the first coat is usually referred to in the trade as the *scratch* coat. It is usually applied over flexible bases, such as metal or wire fabric lath, and is intended to stiffen the base and provide a mechanical bond to the base. Also, as its name implies, the first coat is scratched with a scarifying tool, which provides horizontal ridges or scratches that are intended to provide mechanical keys for the application of the second coat (or brown coat). The brown coat usually constitutes the major bulk of the plaster and, consequently, materially affects the membrane strength. As a result, proportioning and workability are critical, and the mix should have high plasticity for proper application. The term *brown coat* is utilized by the trade to differentiate the relative color of the second coat to the finish coat, which is usually much lighter in color and is sometimes white, depending on the constituents.

The base coats in plaster work provide the strength for the plaster membrane but generally do not provide a proper surface texture for a finished surface. Therefore, a thin, almost veneer, coat of plaster is applied to the base coats as a finish coat. The finish coat may be applied in such a manner as to provide an ornamental or decorative finish, or it may be applied as a smooth surface to act as a flat base over which paint and wallpaper may be applied.

Section 2512 *Exterior Plaster*

General. Portland cement plaster is the only material approved by the code for exterior plaster. Gypsum plaster deteriorates under conditions of weather and moisture, which are prevalent on the exterior surfaces of buildings. For this reason, Section 2512.3 states that gypsum plaster cannot be used on exterior surfaces. Exterior portland cement plaster is required by the code to be applied in not less than three coats when applied over metal lath, wire-fabric lath, or gypsum board backing for the same reasons as discussed for interior plaster. When the portland cement plaster is applied over other approved plaster bases, the code requires only two-coat work. The code permits plaster work that is completely concealed to be of only two coats, provided the total thickness is that required by ASTM C 926, insofar as the finish code of plaster is to provide a surface for exterior finishes (such as paint) and to provide an aesthetic appearance. Thus, where the plaster surface is to be completely concealed, it is not necessary to provide a finish coat.

2512.1

The code requires that the exterior plaster be installed to completely cover, but not extend below, the lath and paper on wood or metal-studded exterior wall construction supported by a nongrade concrete floor slab. This requirement, combined with the requirement in Section 2512.1.2 for a weep screed, is intended to prevent the entrapment of free moisture and the subsequent channeling of the moisture to the interior of the building. This requirement is depicted in Figure 2512-1.

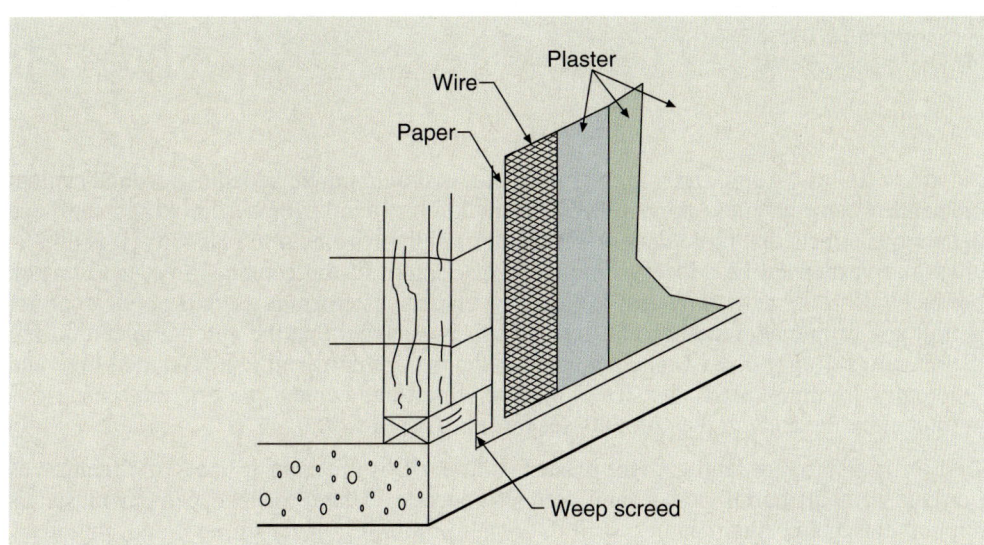

Figure 2512-1
Termination of exterior plaster at on-grade concrete floor for stud walls.

Weep screeds. Water can penetrate exterior plaster walls for a variety of reasons. Once it penetrates the plaster, the water will run down the exterior face of the water-resistive barrier until it reaches the sill plate or mudsill. At this point, the water will seek exit from the wall, and if the exterior plaster is not applied to allow the water to escape, it will exit through the inside of the wall and leak into the building. Thus, the IBC requires a weep screed that, when constructed as shown in Figure 2512-2, will permit the escape of the water to the exterior of the building. In addition, where weep screeds are not provided for plaster exterior walls constructed in cold-climate areas, it is possible that the trapped moisture will freeze and

2512.1.2

cause a premature failure of the exterior plaster. The water-resistive barrier required by the code must lap the weep screed's vertical flange. Although this section does not specify the amount of overlap, at least 2 inches (51 mm) should be adequate in keeping with the typical weather-resistive-barrier lap requirements.

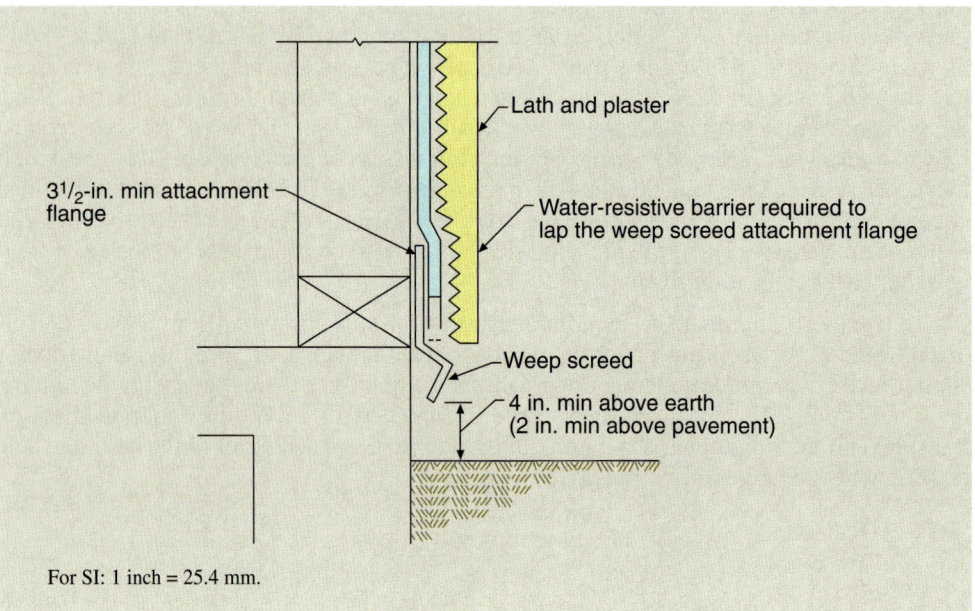

**Figure 2512-2
Weep screed
installation.**

2512.2 Plasticity agents. Admixtures such as plasticizers should not be added to portland cement or blended cement unless approved by the building official. Some admixtures can have deleterious effects that more than offset the desired improvement in plasticity. It is preferable that plasticizers be added during the manufacture of the cement in order to ensure product uniformity and proper proportions. When plastic cement is used, the code does not permit any further additions of plasticizers as it is assumed that the amount added during the manufacturing process is adequate and is the maximum permitted. Hydrated lime and lime putty are time-tested plasticizers used with portland cement plaster, and their use is permitted by the code in the amounts set forth in ASTM C 926.

2512.4 Cement plaster. Portland cement plaster is affected by freezing in the same manner as portland cement mortar or portland cement concrete. When portland cement plaster is applied during freezing weather, it loses a high proportion of its strength and, therefore, does not meet the intent of the code. In addition to protecting the plaster coats from freezing for at least 24 hours after set has occurred, application of the plaster should only be done when the ambient temperature is higher than 40°F (4°C). Plaster may be applied in colder temperatures where provisions are made to keep the cement plaster work above 40°F (4°C) during application and for at least 48 hours thereafter.

It is also important that the plaster not be applied to frozen bases or those covered with frost, which will not only weaken the bond of the plaster to its base but will also freeze the layer of plaster adjacent to the frozen base. In those cases where portland cement plaster is mixed with frozen ingredients or applied to a frozen base, it loses a high percentage of its strength.

KEY POINTS

- Provisions for wall and ceiling coverings are expanded when used on weather-exposed surfaces.

- Because gypsum plaster, gypsum lath, and gypsum board are subject to deterioration from moisture, the code limits their use to interior locations and weather-protected surfaces.

- When used as a base for tile or wall panels, or shower compartment walls, the code intends that glass mat water-resistant gypsum backing panels, discrete nonasbestos fiber-cement interior substrate sheets, or nonasbestos fiber-mat reinforced cement substrate sheets be used.

- For exterior plaster walls, the IBC requires a weep screed that will permit the escape of water to the exterior of the building.

CHAPTER 26

PLASTIC

Section 2603 Foam Plastic Insulation
Section 2604 Interior Finish and Trim
Section 2605 Plastic Veneer
Section 2606 Light-Transmitting Plastics
Section 2607 Light-Transmitting Plastic Wall Panels
Section 2608 Light-Transmitting Plastic Glazing
Section 2609 Light-Transmitting Plastic Roof Panels
Section 2610 Light-Transmitting Plastic Skylight Glazing
Key Points

26 Plastic

> This chapter covers several topics, all related to the use and installation of various types of plastic materials. Included are foam plastic insulation, light-transmitting plastics, plastic veneer, and interior plastic trim.

Section 2603 Foam Plastic Insulation

During the early 1970s, the Federal Trade Commission (FTC) investigated claims made by some manufacturers in the plastics industry of "slow burning" or "nonburning" as related to foam plastic insulation materials. With assistance from the former National Bureau of Standards, now called the National Institute of Standards and Technology, the FTC concluded that these claims were erroneous because of improper testing. Because of the earlier criticisms aimed at the claims, the code changes that were finally adopted into the codes were, of necessity, somewhat conservative.

The provisions were developed by the Society of the Plastics Industry, Inc. (SPI) after numerous meetings, hearings, and seminars relating to the hazardous characteristics of the materials. During this time, SPI funded an extensive program of research that reviewed the then-current test procedures, with a goal of establishing new test procedures where necessary to properly reflect the hazards of the material as it would actually be used in buildings.

The code provisions developed as a result of the extensive research were centered on two basic concepts:

1. An index limitation of the flame spread and smoke developed to 75 and 450, respectively.
2. Separation of the foam plastic insulation from the interior of the building by an approved thermal barrier. The adequacy of the thermal barrier is related to the time during which the thermal barrier is expected to remain in place under fire conditions.

2603.2 Labeling and identification. In addition to the flame-spread and smoke-developed criteria, the code also requires that the containers of foam plastic and foam plastic ingredients be labeled by an approved agency to show that the material is compliant. There are many foam plastic products on the market that do not comply with the code and that were not intended for use in construction. The labeling requirement is intended to prevent the misapplication of products not designed for this use.

2603.3 Surface-burning characteristics. It is important that any foam plastic insulation or foam plastic core material found in manufactured assemblies be limited in flame spread and smoke development. In this section, the code limits such foam plastic materials to a flame-spread index of 75 and a smoke-developed index of 450 where tested at the maximum intended thickness of use. Various exceptions are provided for interior trim, cold-storage buildings and similar facilities, interior signs in covered mall buildings, listed roof assemblies, and special approvals.

2603.4 Thermal barrier. Because of the potential hazards involved, foam plastic must typically be separated from the interior of a building by an approved thermal barrier. Gypsum wallboard at least $1/2$ inch (12.7 mm) in thickness satisfies this requirement, as does any equivalent thermal barrier material complying with the criteria of this section. It must be demonstrated by approved testing that the thermal barrier will remain in place for the required 15-minute time period.

When the following conditions are met, the thermal barrier described above is not required; however, some form of protective membrane is typically mandated:

Masonry or concrete construction. See Figure 2603-1. When foam plastics are encapsulated within concrete or masonry walls, or floor or roof systems, the code does not require a thermal barrier as long as the foam plastic is covered by a minimum 1-inch (25-mm) thickness of the masonry or concrete.

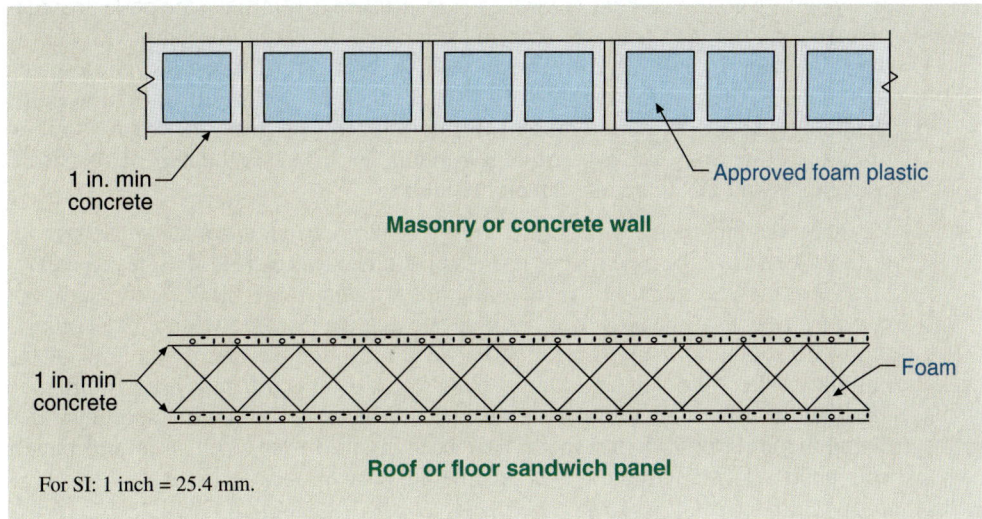

Figure 2603-1
Encapsulated foam plastic.

Cooler and freezer walls. Cold storage uses provide a unique condition for foam plastic insulation in that thicknesses are generally required to be greater than 4 inches (102 mm) for proper thermal insulation, although 4 inches (102 mm) is about the maximum that can be tested. The code, in all other cases, places limits on the thickness of the foam plastic insulation to that which was tested. However, because of the nature of the use in cold-storage facilities, the ignition hazards are not great. Therefore, the code permits foam plastic insulation in greater thicknesses, up to a maximum of 10 inches (25 mm), even though tested in a thickness of 4 inches (102 mm). The intent is that the foam plastic will be provided with a complying protective thermal barrier. In the case of interior rooms within a building, the foam plastic is required to be protected on both sides with a complying thermal barrier.

Provisions are included to permit cooler and freezer walls without a thermal barrier, provided the foam plastic has a flame-spread rating of 25 or less, has minimum allowable flash and self-ignition temperatures of 600°F and 800°F (316°C and 427°C), respectively, and the foam is protected by 0.032-inch thick (0.8-mm) aluminum or 0.0160-inch thick (0.4-mm) steel. The cooler or freezer and the portion of the building where the cooler or freezer is located must be sprinklered in this case. Again, the code presumes that with a low-hazard use, such as a cold storage and freezer box, the metal covering will prevent the actual impingement of any flames on the foam plastic, and the sprinkler system will provide the cooling necessary to maintain proper low temperatures to prevent ignition of the foam plastic.

Walk-in coolers. Where freestanding coolers and freezers have an aggregate floor area not exceeding 400 square feet (37 m²), the code contains an exception that, in effect, provides for no thermal barrier and no sprinkler protection as long as the foam plastics comply with the general provisions of Section 2603.4.1.3. The foam plastic must be covered by an aluminum or steel facing of appropriate thickness. If the foam plastic material is over 4 inches (102 mm) in thickness, a complying thermal barrier must enclose the material.

Exterior walls—one-story buildings. For one-story buildings, metal-clad sandwich panels with foam plastic cores with thicknesses up to 4 inches (102 mm) are permitted to be installed without a thermal barrier, provided the metal cladding complies with the provisions outlined in Section 2603.4.1.4 and, furthermore, the building is protected with automatic fire sprinklers. In this case, the code assumes that the protection and cooling effect provided by automatic sprinklers is a reasonable alternative to the thermal barrier.

Roofing. This item covers two different cases involving roof coverings or roof assemblies:

1. The first case involves nonclassified roof assemblies or roof coverings. As there are generally no test standards for these prescriptive assemblies, the code provides that they may be applied over foam plastic when the foam is separated from the interior of the building by minimum 0.47-inch (12-mm) wood structural-panel sheathing bonded with exterior glue. The edges of the wood structural-panel sheathing must be supported by blocking or be of tongue-and-groove construction or of any other approved type of edge support. In this case, the thermal barrier is waived, as well as the smoke-developed index.

   Based on the fact that a wood structural panel provides an adequate separation for ordinary roof-covering assemblies, it is also considered acceptable for a tested assembly. Thus, it is the intent of the *International Building Code*® (IBC®) that any roof covering assembly installed over foam plastic may be installed with only a complying wood structural-panel separation between the assembly and the interior of the building. Where the wood structural-panel separation is utilized, it is important to recognize that the joints must be protected even though the roofing specimen used during the fire-retardancy test might have been installed over wood structural panels with abutted joints without any supplemental protection.

2. The second case involves the use of Class A, B, or C roof-covering assemblies in which the foam plastic insulation is also considered to be an integral part of the assembly. See Figure 2603-2. Here, a nationally recognized test standard for insulated roof decks is to be utilized. The test standards for insulated roof-deck construction are adequately conservative so that assemblies passing either of the two test standards are considered to meet the intent of the code without any limit on flame spread or smoke development. Furthermore, no thermal barrier is required.

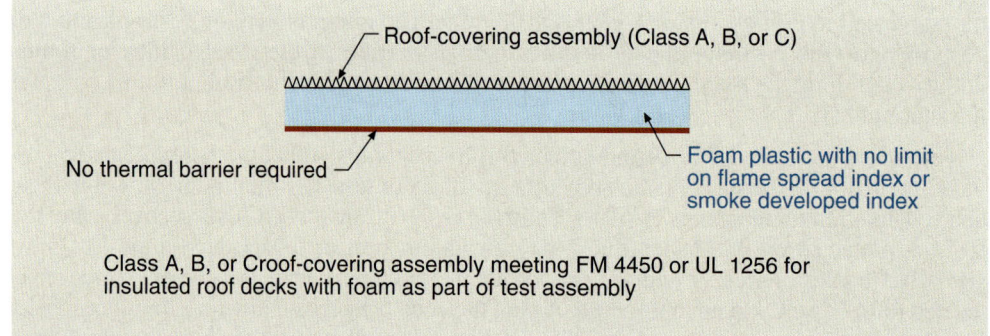

**Figure 2603-2
Foam plastic used with roof covering.**

It should be noted that most insulated metal decks that are listed require that the deck be nonperforated—essentially a nonacoustical deck. Acoustical decks are commonly proposed in gymnasiums and auditoriums for sound control purposes and would therefore require a thermal barrier unless specifically listed under UL 1256 or FM 4450.

Attics and crawl spaces. See Figure 2603-3. This item describes specific methods used to protect foam plastics located within attics and crawl spaces (in lieu of a complying thermal barrier) where entry is provided only for service of utilities. The phrase "where entry is provided only for service of utilities" is intended to restrict these reduced requirements for a thermal barrier to those unused areas where there are no heat-producing appliances.

In addition, drop lights or portable service lights are often utilized when serving equipment in such concealed spaces, and such lighting devices pose an ignition threat to the foam plastic. Thus, the reduced provisions are intended to provide a barrier whose only purpose is to prevent the direct impingement of flame on the foam plastic.

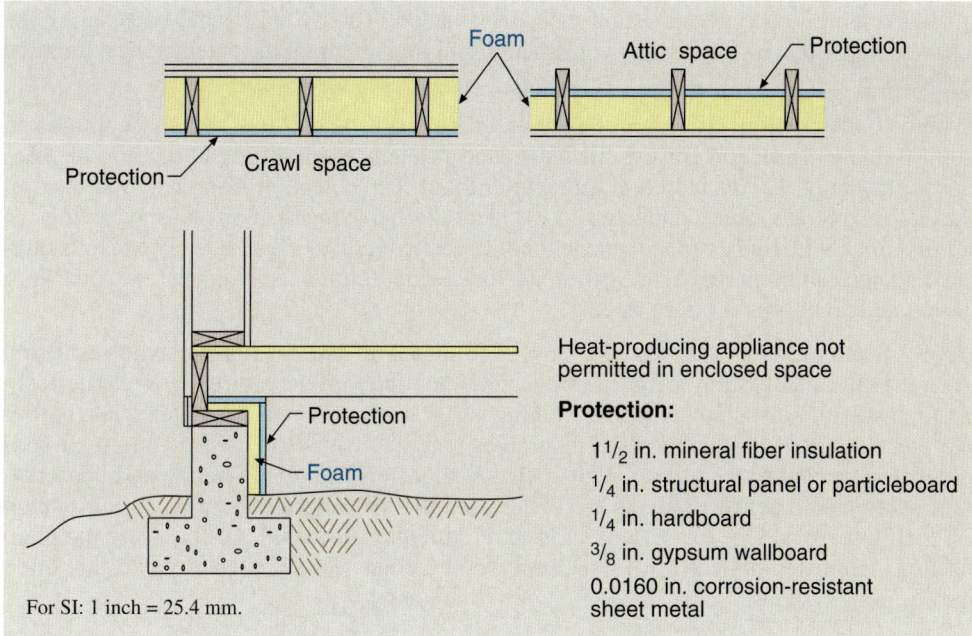

Figure 2603-3
Foam plastic in attic and crawl spaces.

The reduced level of protection is also applicable in those situations where the service of utilities is not an issue. If there are no utilities within the attic space or crawl space that require service, the minimum described degree of separation between the foam plastic and the enclosed space must still be provided. Where the attic or crawl space provides a suitable area that exists for a purpose other than the access to utilities, such as storage, a thermal barrier complying with Section 2603.4 is required.

Doors not required to have a fire-protection rating. Pivoted or side-hinged doors not required to have a fire protection rating are permitted to be installed without the thermal barrier, provided the door facings are of sheet metal of the thicknesses prescribed in this section. The rationale behind the waiver of the thermal barrier is that the foam plastic is completely encapsulated within the sheet-metal facings, and the quantity of foam plastic in protected doors is quite small.

Exterior doors in buildings of Groups R-2 and R-3. In specific residential occupancies, exterior doors to individual dwelling units are permitted to have a foam-plastic core. This allowance applies where the doors do not require a fire-resistance rating and the foam is covered with wood or other approved materials.

Garage doors. Garage doors, other than those in garages accessory to dwellings, that contain foam plastic are allowed, provided the door does not require a fire-resistance rating and is faced with materials prescribed by this section. If the garage door containing the foam plastic does not have an aluminum, steel, or wood facing of the minimum thickness prescribed, the door must be tested in accordance with DASMA 107 *Room Fire Test Standard for Garage Doors Using Foam Plastic Insulation*. This provision is intended to regulate the commercial applications of overhead, sectional, and tilt-up types of doors.

Siding backer board. Where it is desired to insulate exterior walls under exterior siding, the code permits foam plastic to be used as a backer board for the siding, provided the insulation has a potential heat of not more than 2,000 Btu per square foot (22.7 mJ/m^2). The thermal barrier is not required under these circumstances as long as the siding backer board has a minimum thickness of $^1/_2$ inch (12.7 mm) and is separated from the interior of the building by not less than 2 inches (51 mm) of mineral-fiber insulation or the equivalent.

The code also permits the siding backer board without a thermal barrier when the siding is applied as re-siding over existing wall construction. This is reasonable considering the separation provided by the existing construction and limitations on the potential heat imposed by the code.

Type V construction. The use of spray-applied foam plastic has become common in wood-frame construction for the sill plates and headers. Such a limited amount of foam plastic insulation is considered acceptable without the protection afforded by a thermal barrier. Testing has been conducted to evaluate the behavior of foam plastic having the density, thickness, flame-spread, and smoke-developed indices stipulated. The results indicated no substantial performance difference between a foam plastic–insulated wood floor system and an all-wood floor system.

Floors. Section 2603.4.1.14 provides a viable means to protect foam plastic insulation when it is installed within a floor system. The thermal barrier required to separate foam plastic insulation installed beneath a walking surface must not only be an adequate barrier to protect the foam plastic, but also be durable enough to withstand the load and wear-and-tear that is needed for the floor. With society's focus on energy efficiency and conservation, many new types of products are being used that incorporate foam plastic insulation for energy reasons. One example is the use of structural-insulated panels where the foam plastic is laminated between two structural wood facings. These types of panels are often used in floor systems.

Although a $^1/_2$-inch (12.7-mm) wood structural panel (i.e., plywood or oriented strand board) is not by itself considered as a complying "thermal barrier" as required by Section 2603.4, it will fulfill the dual need for structural strength and thermal protection of the foam insulation. In the case of a floor, the use of such panels will provide sufficient protection because, in the event of an interior fire, the floor faces a reduced exposure and is typically the last building element to be significantly exposed by the fire. See Figure 2603-4.

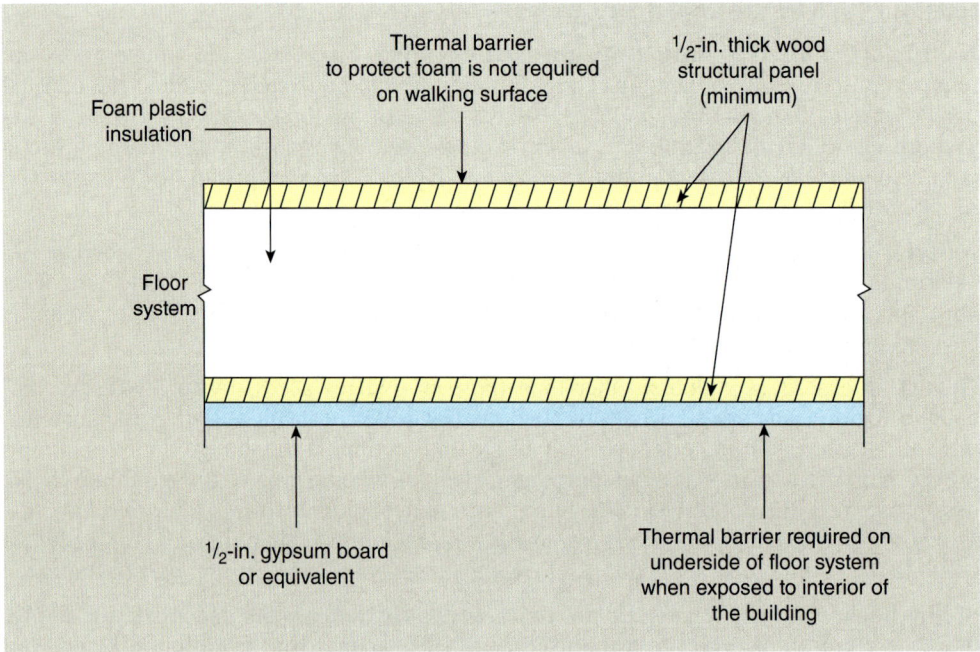

Figure 2603-4
Floors with foam plastic insulation.

It is important to note that the protection provided by wood structural panels is only accepted on the walking surface side of the floor. If the floor is used in multistory construction, then the underside of the floor system (ceiling of the room below) must be covered by the typically required thermal barrier. The thermal barrier protection on the bottom side of the assembly does not get to take advantage of the reduction because it will not be used as a walking surface does, and it will face a more severe exposure to an interior fire.

The exception is intended to address items such as carpet padding, and others, that do not need to be covered by a thermal barrier.

Exterior walls of buildings of any height. The provisions for foam plastic insulation also allow such material in the exterior walls of buildings required to have noncombustible exterior wall construction (Types I, II, III, and IV). Applicable to such buildings of any height, an important provision of this section requires that the wall be tested in accordance with NFPA 285. This test provides a method of evaluating the wall's flammability characteristics that are due to the combustible foam plastic materials within the wall. Wall assemblies need not be tested where they can comply with the provisions of Section 2603.4.1.4; however, this allowance is only applicable to fully sprinklered, one-story buildings. Section 2603.5.1 also requires that test data be provided to show that if a fire-resistance rating is required, the rating of the wall containing the foam maintains the required rating. Moreover, the foam plastic insulation must: **2603.5**

1. Be separated from the interior of the building with a thermal barrier meeting the provisions of Section 2603.4.

2. Not have a potential heat content exceeding that of the foam plastic insulation contained in the wall assembly as tested.

3. Have a maximum flame-spread index of 25 and smoke-developed index of 450. Exterior coatings and facings, tested individually, must also comply with these flame-spread and smoke-development limitations.

4. Be labeled by an approved agency.

5. Comply with the ignition limitations imposed by Section 2603.5.7.

Roofing. As previously addressed, complying foam plastic insulation may be utilized as a portion of a roof-covering assembly, provided the assembly with the foam plastic insulation has been tested in accordance with ASTM E 108 or UL 790, and has been listed as a Class A, B, or C roofing assembly. **2603.6**

Interior finish in plenums. The installation of foam plastic insulation within concealed plenum spaces creates a unique challenge due to the general requirements for protection of the insulation and how the protection is accomplished. This section establishes the three permissible options, as shown in Figure 2603-5, where foam plastic insulation is used as interior wall and ceiling finish within plenums. Note that Section 2603.8 indicates that foam plastic insulation used as trim in plenums is also subject to the provisions of Section 2603.7. **2603.7**

Based on Item #2, foam plastic insulation that is exposed within the plenum either as an interior finish material or as an interior trim must have a flame-spread index of 25 or less and a maximum smoke-developed index of 50. In addition, it must meet the requirements of the full-scale room-corner fire test (NFPA 286) with provisions for flame spread, heat release, smoke release, and inability for flashover.

In lieu of leaving the foam plastic insulation exposed and imposing fairly restrictive flame and smoke ratings, this section provides two alternative means for protection. Item #1 requires the use of the thermal barrier as a means of protection. This will allow the foam plastic to have a maximum flame-spread rating of 75 and a smoke-developed rating of 450, with no mandate to comply with the NFPA 286 test. Item #3 will allow the same increased rating values if the insulation is covered by a layer of corrosion-resistant steel.

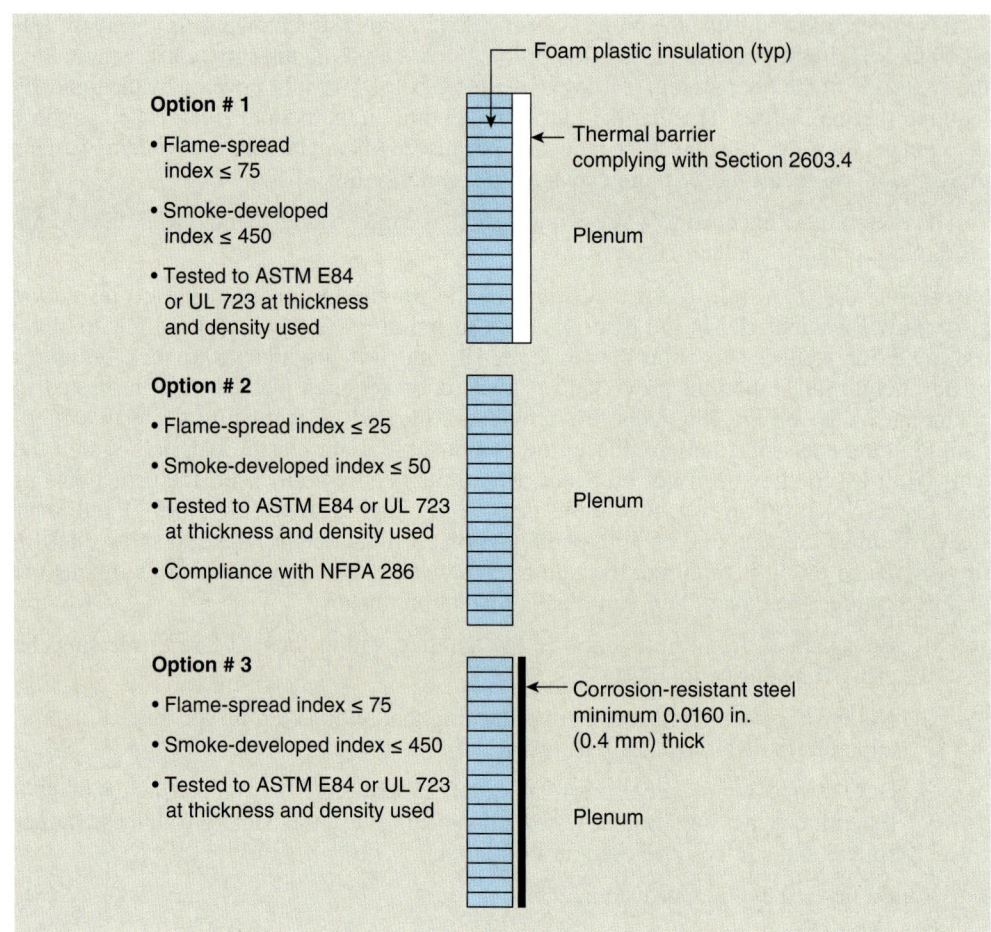

Figure 2603-5
Interior finish in plenums.

As mentioned previously, the use of trim made of foam plastic insulation should be regulated the same as when the insulation is used as a wall or ceiling finish material within the plenum. Section 2603.8 refers to the three protection options that are allowed by Section 2603.7 and will therefore eliminate the possibility that the trim may be allowed by any type of reference to Section 2604.

2603.10 Specific approval. In this section, the code provides for those cases where foam plastic products and protective coverings do not comply with the specific requirements of Sections 2603.4 through 2603.7. The specific approvals are based on testing that is related to the actual end use of the products. The code refers to a number of test standards for determining specific approvals and, in addition, there are others that utilize some variation of the room test and are designed for testing exterior wall applications.

Section 2604 *Interior Finish and Trim*

The provisions of Chapter 8 for wall and ceiling finishes are applicable to those plastic materials installed as trim or interior finishes. In addition, foam plastics used as interior finish and trim must be in compliance with the provisions of this section, as well as the flame-spread index requirements of Chapter 8. By limiting the density, thickness, wall and ceiling area, as well as flame spread of foam plastic materials, the hazard level created by the exposure of foam plastics is low.

Section 2605 *Plastic Veneer*

Because it is a combustible material, plastic veneer used in the interior of a building is required by the code to comply with the interior finish requirements of Chapter 8.

Where plastic veneer is used on the exterior of a building, the code requires that the veneer be of approved plastic materials as defined in Section 202. This places severe restrictions on the combustibility and smoke development of the plastic materials. Because plastic materials are combustible, the code limits their attachment on any exterior wall to a height no greater than 50 feet (15,240 mm) above grade. Furthermore, the IBC limits the area of plastic veneer to 300 square feet (27.9 m^2) in any one section and requires each section to be separated vertically by a minimum of 4 feet (1,219 mm). The 4-foot (1,219-mm) separation helps control the rapid vertical spread of fire. The code anticipates that local fire-fighting forces can effectively fight a fire that involves plastic veneer up to a height of about 50 feet (15,240 mm). Also, if the plastic veneer involves too large an area, it is conceivable that a fire could overtax local fire-fighting forces. The exception applies to Type VB buildings where the walls are not required to have a fire-resistance rating. In this case, the plastic materials do not present a greatly different hazard than the unprotected wood construction.

Plastic siding used on the exterior of a building is regulated separately from plastic veneer. The provisions of Section 2605 are not appropriate for exterior plastic siding, as the requirements for exterior wall coverings established in Sections 1404 and 1405 are to be applied.

Section 2606 *Light-Transmitting Plastics*

It is the intent of this section and Sections 2607 through 2611 to regulate the use of light-transmitting plastics—those plastics used in the building envelope or with interior lighting to transmit light to the interior of the building. Light-transmitting plastics are regulated because they are combustible materials. The unregulated use of combustible materials in the roof structure and for the exterior walls can possibly defeat the intent of the provisions of the code relating to types of construction. Thus, these six sections regulate these materials so that they do not materially affect the other requirements of the code regarding types of construction.

Any use of light-transmitting plastic materials must be approved by the building official and be based on technical data submitted to substantiate their use. As a basis of this approval, the building official should refer to Section 202 for the definition of "Plastic, approved." The definition refers to the criteria of Section 2606.4 for the combustibility classifications of approved plastic materials, determined to be either Class CC1 or CC2 in accordance with ASTM D 635.

Materials of light-transmitting plastic, such as lenses, panels, grids, or baffles, located below independent light sources are thought of as creating a light-diffusing system. Light-diffusing systems are specifically regulated in Section 2606.7. Regulated as to occupancy, location, support, installation, and size, light-diffusing systems pose potential hazards that are due to their combustibility.

Section 2607 Light-Transmitting Plastic Wall Panels

Exterior wall panels are regulated for the same reason as plastic glazing in openings. However, because exterior wall panels are sheet materials, they generally constitute larger unbroken areas than plastic glazing for openings do; as a result, their burn-rate characteristics are more critical than those for plastic glazing in openings addressed in Section 2608.

Section 2608 Light-Transmitting Plastic Glazing

Because plastic glazing materials are combustible, their use is limited to openings not required to be fire protected. In the case of building construction other than Type VB, their use is further restricted. The glazing of openings not required to be fire protected in Type VB construction is essentially unlimited as to the area, height, percentage, and separation requirements applicable to the individual glazed openings.

For plastic-glazed openings in buildings other than Type VB, restrictions are placed on the area, height, percentage, and separation requirements for the individual glazed openings because plastic glazing materials are combustible. In other types of construction, unprotected combustible materials must be limited in accordance with their real extent and separation. Because of the combustibility of plastic glazing, the code requires flame barriers at each floor level for nonsprinklered multistory buildings to prevent the transmission of flame from one story to another by way of combustible openings.

As with other provisions of the code limiting the height of combustible materials above grade, this section also limits the height of plastic materials above grade to 75 feet (22,860 mm) unless the building is sprinklered throughout.

Section 2609 Light-Transmitting Plastic Roof Panels

Plastic panels are regulated on the basis of three conditions, of which only one needs to be met in order to utilize light-transmitting plastic panels in roofs of all occupancies other than Groups H, I-2, and I-3. Light-transmitting plastic roof panels may be installed in buildings equipped throughout with an automatic sprinkler system, in buildings where the roof construction is not required to have a fire-resistance rating, or where the roof panels meet the requirements for roof coverings in accordance with Chapter 15.

Because plastic roof panels constitute unprotected openings in the roof, the code requires that they be separated from each other by 4 feet (1,219 mm) horizontally. The minimum 4-foot (1,219-mm) separation is not mandated for fully sprinklered buildings, nor is it required in buildings housing low-hazard occupancies as limited by Exception 2 or 3 to Section 2609.4. Furthermore, their location on the roof is regulated based on the building's location in respect to lot lines. Roof panels shall be located at least 6 feet (1,829 mm) from exterior walls that are located in a manner to require protected wall openings.

Because Class CC1 plastics have a slower burn rate than Class CC2 plastics, the code limits Class CC2 plastics to smaller areas than allowed for Class CC1 materials. The area limitations may be doubled based on the installation of a sprinkler system, whereas plastic roof panels in low-hazard occupancy buildings, greenhouses, and patio covers are not limited in area under specific conditions. As with exterior wall panels, the actual numbers relating to area, height, and separation requirements must be somewhat arbitrary but are reasonable code limits determined by a consensus of knowledgeable experts.

Section 2610 *Light-Transmitting Plastic Skylight Glazing*

In this section, the requirements for skylights are more detailed than those for roof panels in Section 2609 because there is no limit on the type of construction or fire-protection requirements for the roof assembly. Furthermore, skylights have unique requirements, such as those for flashing and resistance to burning brands. Also, as plastic-glazed skylights provide an unprotected combustible assembly in the roof, limitations must be placed on the area, percentage, and separation of each unit. Each unit's location on the roof relative to lot lines is regulated in a manner consistent with that for plastic roof panels.

Two of the primary concerns of plastic-glazed skylights are related to flashing at the intersection with the roof and their ability to resist the effects of flying, burning brands. Therefore, with one exception, the code requires that they be mounted on a curb at least 4 inches (102 mm) above the plane of the roof so that proper flashing may be accomplished. The exception involves skylights on roofs that have a minimum slope of 3 units vertical in 12 units horizontal (25-percent slope) and applies only to Group R-3 occupancies and on buildings having unclassified roof coverings. This slope should provide adequate roof drainage to accommodate skylights. The slope requirements for flat or corrugated plastic-glazed skylights and the rise requirement for dome-shaped skylights are based on the skylights' ability to shed flying brands. However, when the glazing material in the skylights can pass the Class B burning brands test specified in ASTM E 108 or UL 790, there is no limitation on slope, either of flat or corrugated glazed skylights, or on rise in the case of dome-shaped skylights.

The requirement for the protection of edges of plastic-glazed skylights or domes is to prevent the rapid spread of fire along the roof, as the edges of the plastic glazing material ignite more readily than the interior portions. Under those conditions where unclassified roof coverings are permitted, the metal or noncombustible edge material is not required.

As with roof panels, the various limitations on area, percentage, and separation of skylights are somewhat arbitrary and, as with roof panels, are based on a consensus among knowledgeable experts on what is reasonable.

KEY POINTS

- Foam plastic is regulated for flame spread and smoke development.
- Separation with a thermal barrier must be provided between foam plastic insulation and the interior of the building.
- Containers of foam plastic and foam plastic ingredients must be labeled to prevent the misapplication of products not designed for their use.

- Foam plastics used in several applications, such as masonry or concrete construction, cooler and freezer walls, roofing, attics and crawl spaces, and doors not required to have a fire-protection rating, may be installed without a thermal barrier under specified conditions.
- When properly tested, foam plastic insulation is permitted in exterior walls of buildings required to have noncombustible exterior wall construction.
- Foam plastic used as interior finish and trim is acceptable where the density, thickness, wall and ceiling coverage, and flame spread of the foam plastic materials is limited.
- Light-transmitting plastics are regulated in part because they are combustible materials.

CHAPTER 27

ELECTRICAL

Section 2702 Emergency and Standby Power Systems

27 Electrical

> All electrical components, equipment, and systems are to be designed and constructed in accordance with the provisions of NFPA 70: *National Electrical Code* (NEC). The *International Building Code®* (IBC®) specifically references the NEC for its technical provisions. The only electrical issues addressed within this chapter of the IBC are those emergency and standby power systems that are required by other provisions of the code. It should be noted that these provisions are maintained through the code change process of the *International Fire Code®* (IFC®).

Section 2702 *Emergency and Standby Power Systems*

NFPA Standards 70, 110, and 111 regulate the installation of emergency and standby power systems. This section identifies 20 situations where such systems must be in place. Stationary engine generators, where used to provide emergency and standby power, must comply with the requirements of UL 2200. This UL standard provides a benchmark for the evaluation of the safety and reliability of such generators.

Emergency power shall be provided as follows for:

1. Emergency voice/alarm communication systems in Group A occupancies with an occupant load of 1,000 or more.
2. All required exit signs, other than those that are self-luminous.
3. Means of egress illumination in occupancies where two or more means of egress are required.
4. Semiconductor fabrication facilities per Section 415.10.10.
5. Occupancies with highly toxic or toxic materials per IFC Section 6004.2.2.8.
6. Power-operated sliding doors or power-operated locks for swinging doors in Group I-3 occupancies, unless a remote mechanical operating release is provided.

Standby power is required under the following conditions for:

1. Smoke-control systems.
2. Elevators and platform lifts that are a portion of accessible means of egress.
3. Horizontal sliding doors utilized as a component of a means of egress.
4. Auxiliary-inflation systems in membrane structures exceeding 1,500 square feet (140 m^2) in area.
5. Occupancies where Class I and unclassified detonable organic peroxides are stored.
6. Emergency voice/alarm communication systems in covered and open mall buildings greater than 50,000 square feet (4,645 m^2) in floor area.
7. Pressurization equipment, mechanical equipment, lighting, elevator-operator equipment, fire-alarm systems, and smoke-detection systems in airport traffic-control towers more than 65 feet (19,812 mm) in height.
8. For operation of one or more elevators in a building.
9. Mechanical vestibule and stair shaft ventilation systems, and automatic fire detection systems for smokeproof enclosures.

Both emergency power and standby power shall be provided in the following situations for:

1. High-rise buildings (with exceptions), defined as those structures having occupied floors more than 75 feet (22,860 mm) above the lowest level of fire-department vehicle access.

2. Underground buildings (with exceptions), defined as those building spaces having a floor level used for human occupancy more than 30 feet (9,144 mm) below the lowest level of exit discharge.

Either emergency power or standby power shall be provided, in Group H occupancies where inside storage, dispensing, or use of hazardous materials occurs, where the following systems are required: mechanical ventilation, treatment, temperature control, alarm, detection, or other electrically operated systems as identified in Section 414.5.3.

CHAPTER 28

MECHANICAL

28 Mechanical

This chapter merely references the *International Mechanical Code®* (IMC®), *International Fuel Gas Code®* (IFGC®), and Chapter 21 of the *International Building Code®* (IBC®) for various requirements relating to the construction, installation, and maintenance of mechanical equipment and systems. The appropriate code shall be utilized to address heating; air conditioning; refrigeration; mechanical and natural ventilation; plenums; and factory-built chimneys, fireplaces and barbecues. Reference is made to the IMC and Chapter 21 of the IBC for the regulation of masonry chimneys, fireplaces, and barbecues.

CHAPTER 29

PLUMBING

Section 2902 Minimum Plumbing Facilities
Key Points

29 Plumbing

> The intent of Chapter 29 is to reference the *International Plumbing Code®* (IPC®) for the construction, installation, and maintenance of plumbing systems and equipment, and the *International Private Sewage Disposal Code®* (IPSDC®) for the regulation of private sewage disposal systems. In addition, the provisions of this chapter provide for the determination of plumbing fixture counts based on occupancy classification and occupant loads.
>
> The provisions of Chapter 29 are maintained through the code change process of the IPC.

Section 2902 *Minimum Plumbing Facilities*

2902.1 Minimum number of fixtures. Section 2902.1 establishes the minimum number of plumbing fixtures that must be provided for various occupancies based on Table 2902.1. The table is based on the distinct occupancy classifications identified in Chapter 3 and the corresponding occupant loads calculated for the building. Those fixtures required in most occupancies include water closets, lavatories, drinking fountains, and service sinks. In addition, bathtubs or showers, automatic clothes-washer connections, and kitchen sinks are mandated in some residential occupancies. It is assumed that at least one fixture of each type as required by Table 2902.1 will be provided in a building designed for human occupancy.

When determining the proper occupant load to be utilized in calculating plumbing fixture count, there is no specific methodology referenced. However, because the only provisions in the IBC addressing occupant load calculation are found in Chapter 10, it is typically assumed that the approach established in Section 1004 for the means of egress should also be used for plumbing fixture count. The basis for most occupant load determinations, other than areas with fixed seating, is the density factor established in Table 1004.1.2. An occupant load is determined by dividing the floor area under consideration by the appropriate density factor. It should be noted that the building official also has the authority as established in the exception to Section 1004.1.2 to base the occupant load on the actual number of occupants anticipated, rather than the calculated number. Although the use of this exception is typically inappropriate for egress and fire safety purposes, it is commonly applied for plumbing fixture count. The occupant load utilized for egress and fire protection requirements is purposely conservative by most counts because of the life safety concerns. The occupant load to be used in calculating the minimum plumbing fixture count should be based on more of a convenience concern, recognizing the need to satisfy any sanitation issues. Therefore, the occupant load used in the plumbing fixture count could differ from that used as the basis for the design of the means of egress system. The building official should rely on all available information that will assist in the appropriate determination of occupant load for fixture count purposes.

2902.1.1 Fixture calculations. Once the appropriate occupant load is determined, the minimum required number of fixtures is calculated by using Table 2902.1. The provisions of Section 2902.1.1 address the method in which the fixtures are to be distributed between the sexes.

For the determination of required plumbing fixtures, the total occupant load to be served by the plumbing facilities must first be calculated. Unless modified by special conditions, the total occupant load is then halved, assuming 50 percent of the occupants to be male, 50 percent to be female. The resulting occupant loads are then used when applying the table, with fixtures calculated individually for each of the sexes. Where the required number of fixtures contains a fraction, an additional fixture is required. See Application Examples 2902-1 and 2902-2. Note that the provisions of Section 419.2 of the IPC allow urinals as a substitution for water closets on a 1-to-1 basis, provided that, in assembly and educational occupancies, urinals account for no more than 67 percent of the required fixtures. For example, if nine water closets are required in a men's toilet room in a large nightclub, it is acceptable to provide six urinals and only three water closets. Where the occupancy classification is other than Group A or Group E, only one-half of the required number of water closets may be substituted with urinals.

Application Example 2902-1

GIVEN: An exhibition hall classified as a Group A-3 occupancy. The hall's occupant load is determined to be 8,680.
DETERMINE: The minimum required number of (a) water closets for the male occupants, (b) water closets for the female occupants, (c) lavatories for the male occupants and (d) lavatories for the female occupants.

1. Assume occupants as 50% male, 50% female per Section 2902.1.1:
   4,340 males
   4,340 females

2. (a) $\frac{4,340}{125} = 34.72 = 35$ water closets* for males

   * IPC Section 419.2 would allow 12 or more water closets, with the remainder urinals, to make up the 35 fixtures.

   (b) $\frac{4,340}{65} = 66.77 = 67$ water closets for females

3. $\frac{4,340}{200} = 21.7 =$

   (c) 22 lavatories for males
   (d) 22 lavatories for females

MINIMUM REQUIRED PLUMBING FIXTURES

The determination of the minimum plumbing fixture count becomes a bit more complex where the building contains a number of different occupancies. If toilet room facilities are provided independently for each of the occupancies in the building, the basic method of calculating the number of fixtures should be satisfactory. However, if common toilet facilities are designed to serve the occupants from multiple occupancy groups, a different approach is more appropriately warranted. In such a determination, the number of required fixtures for each sex would be calculated for each occupancy, then without rounding, added together to arrive at the minimum fixtures that must be provided. An example of this methodology is shown in Application Example 2902-3. A similar approach could be utilized when substituting urinals for water closets.

Application Example 2902-2

GIVEN: A manufacturing facility classified as an F-1 occupancy. The occupant load is determined to be 684.
DETERMINE: The minimum required number of water closets and lavatories.
1. Assuming a 50:50 split, assign 342 male and 342 female occupants.
2. $\frac{342}{100} = 3.42 = 4$ water closets minimum for each sex

   4 water closets * for males
   4 water closets for females

Same calculation for lavatories, a minimum of 4 for each sex.
   * If urinals are substituted for water closets, a minimum of 2 water closets must be provided.

MINIMUM REQUIRED PLUMBING FIXTURES

Application Example 2902-3

GIVEN: A mixed-occupancy building containing a Group M with an occupant load of 368, a Group B with an occupant load of 56, and a Group S-1 with an occupant load of 78. All of the plumbing fixtures will be located at a single toilet room location.

DETERMINE: The minimum number of water closets that would be required in each toilet room.

SOLUTION:

Group M: 184/500 @ 1 per 500 occupants
Group B: 28/25 @ 1 per 25 for first 50 occupants
Group S-1: 39/100 @ 1 per 100 occupants

$$184/500 + 28/25 + 59/100 =$$
$$0.37 + 1.12 + 0.39 = 1.88$$

Minimum of two water closets required in each toilet room

MINIMUM REQUIRED PLUMBING FIXTURES

2902.2 Separate facilities. In most buildings where plumbing fixtures are required, separate facilities must be available for each sex. Simply, at a minimum, one women's toilet room and one men's toilet room must be provided. There are conditions, however, where only a single toilet room is mandated. Separate-sex facilities need not be provided within dwelling units and sleeping units. In addition, common facilities are permitted where the number of people (both customers and employees) does not exceed 15. In mercantile occupancies with an occupant load of 100 or less, such as a small retail sales tenant, a single toilet room is also permitted.

2902.2.1 Family or assisted-use toilet facilities serving as separate facilities. Where each separate-sex toilet room is required to only have a single water closet, the required separate restrooms are permitted to be designated as "family or assisted-use" toilet facilities versus requiring the facilities to be designated for each sex. Allowing the toilet facilities to be used by either sex provides greater flexibility and increases overall availability.

Where facilities are designated for a specific sex and the toilet room for one of the sexes is occupied or being cleaned, then a person needing that facility is typically forced to either wait, or sneak in and use the toilet room that is designated for the other sex. This provision requires that two separate facilities be provided, but by eliminating the separate-sex designations, both of the restrooms are available to anyone and the overall availability in small establishments is increased.

The selection of the term "family or assisted-use" toilet facilities is important because it influences how the toilet rooms are constructed. Sections 1109.2.1.1 through 1109.2.1.7 provide several details that affect the construction of these toilet rooms. While all of the requirements of these sections are applicable, the main provisions to review are Sections 1109.2.1.1, 1109.2.1.2, 1109.2.1.6, and 1109.2.1.7.

2902.3 Employee and public toilet facilities. Only specific occupancies and uses are required to have public toilet facilities for use by customers, patrons, and visitors of the building. Restaurants, nightclubs, places of assembly, business uses, mercantile occupancies, and similar buildings and tenant spaces intended for public use must be provided with customer toilet facilities located within one story above or below the area under consideration, and with a path of travel not to exceed 500 feet (152,400 mm). Those uses where public use is not expected, such as warehouses, factories, and similar buildings, only require employee toilet facilities.

KEY POINTS

- The minimum required number of plumbing fixtures is based on the use of the building and the anticipated number of occupants.
- Except for a limited number of situations, separate toilet facilities are required for each sex.
- Family or assisted-use toilet rooms are permitted in lieu of separate-sex facilities under limited conditions.

CHAPTER 30

ELEVATORS AND CONVEYING SYSTEMS

Section 3002 Hoistway Enclosures
Section 3003 Emergency Operations
Section 3004 Hoistway Venting
Section 3006 Machine Rooms
Section 3007 Fire Service Access Elevator
Section 3008 Occupant Evacuation Elevators
Key Points

30 Elevators and Conveying Systems

Elevators and other types of conveying systems are regulated under the provisions of this chapter. For the most part, the American Society of Mechanical Engineers (ASME) standards are utilized to address the specifics of elevator safety. ICC A117.1 must also be referenced for all elevators required to be accessible by Chapter 11 of the *International Building Code®* (IBC®).

Section 3002 *Hoistway Enclosures*

3002.1 Hoistway enclosure protection. This section is essentially a cross reference to Section 713 of the code, which contains the specific requirements for the enclosure of shafts in buildings. Elevator shafts are to be fire-resistance-rated enclosures, unless exempted by one of the alternatives established in Section 712.1. If required to be fire-resistance rated, the shaft enclosure must have a 1-hour or 2-hour rating based on the number of stories connected by the shaft enclosure, as well as the required fire-resistance rating of the floor construction. Opening protectives for hoistway enclosures are also regulated by Chapter 7.

3002.2 Number of elevator cars in a hoistway. These provisions were extracted from the elevator code insofar as they are more appropriate as building code requirements. The basis for limiting the number of cars in a single hoistway is to provide a reasonable level of assurance that a multilevel building served by several elevators would not have all of its elevator cars located in the same hoistway. This could result in a single emergency disabling all elevators within the building. For example, if all elevator cars were allowed to be located in the same hoistway, smoke that entered the enclosure during a fire would render all elevator cars unusable. The code provisions will increase the chance that some elevators within a major building would remain operational during a fire or other emergency. See Figure 3002-1.

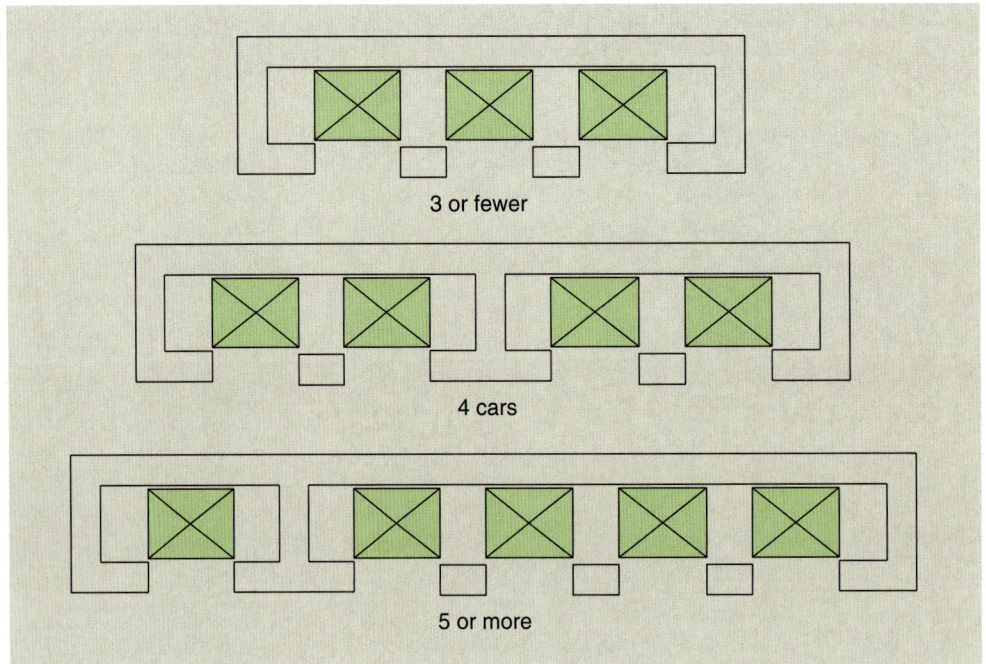

Figure 3002-1
Elevator cars in a hoistway.

Figure 3002-2
Emergency signs.

Emergency signs. In order to alert occupants to the fact that an elevator is not to be used for egress purposes during a fire incident, this section mandates the placement of a sign adjacent to each elevator call station on each floor of the building. The standardized pictorial sign (an example is illustrated in Figure 3002-2) advises occupants to use the exit stairways rather than the elevators in case of a fire. The emergency sign does not need to be installed at those elevators complying as an accessible means of egress per Section 1007.4 or at occupant self-evacuation elevators as described in Section 3008.

3002.3

Elevator car to accommodate ambulance stretcher. Where elevators are provided in buildings of four stories or more in height, this section of the code requires that at least one elevator serving all floors accommodate an ambulance stretcher. The ability to transport an individual on a stretcher in an elevator in a multistory building is a basic life-safety consideration. Immediate identification of elevators that accommodate stretchers is necessary so that emergency-services personnel can quickly respond to emergency conditions. For this reason, an identifying symbol as shown in Figure 3002-3 shall be placed inside on both sides of the hoistway door frame.

3002.4

Minimum elevator car size requirements have been established to ensure that the typical ambulance stretcher can be accommodated. The minimum 24-inch by 84-inch (610-mm by 2,134-mm) size is further described to address stretchers with rounded or chamfered corners with a minimum 5-inch (127-mm) radius. See Figure 3002-4. By specifically identifying the minimum size requirements, flexibility is provided to the elevator industry in its efforts to provide appropriately sized cars. It is also beneficial to stretcher manufacturers by providing direction to aid in the standardization of their products.

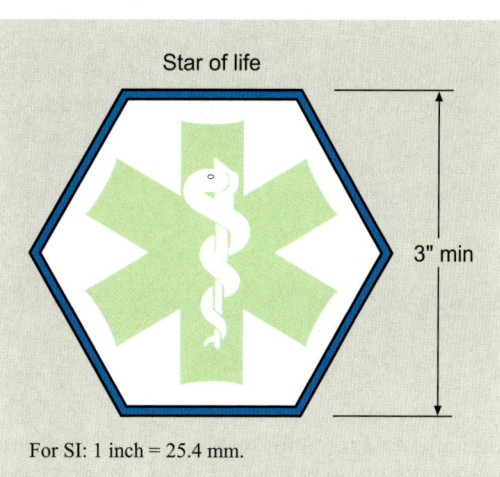

Figure 3002-3
Sign denoting accommodation of ambulance stretcher.

30 Elevators and Conveying Systems

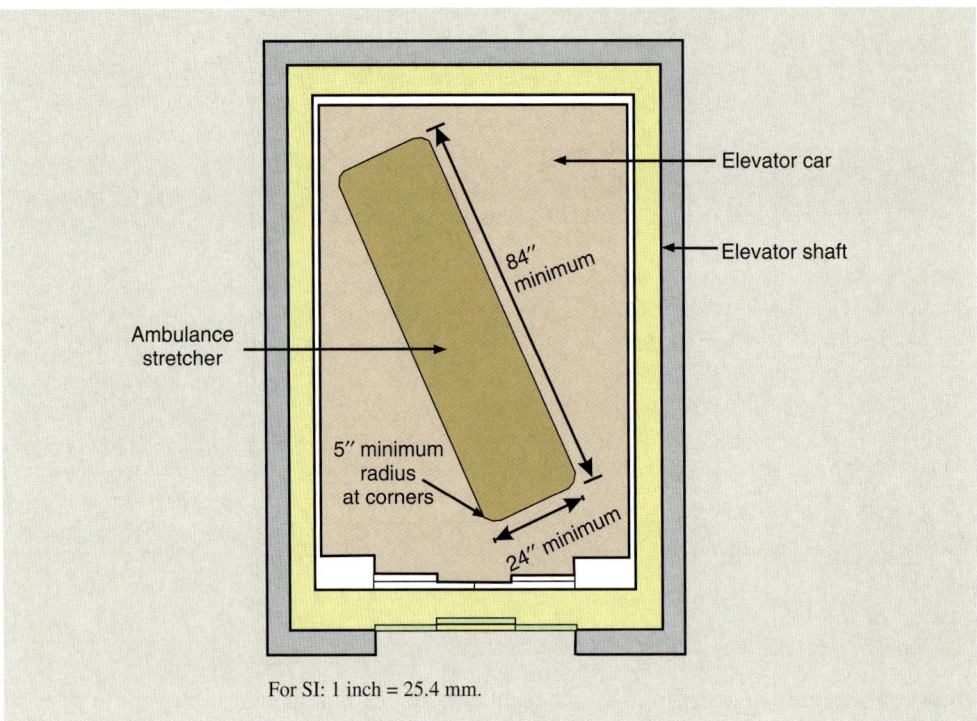

Figure 3002-4
Elevator used for fire department emergency access.

3002.5 Emergency doors. The provisions of ASME A17.1 regulate the installation of an emergency door when required by this section. It is necessary that an emergency door be provided in the blind portion of a hoistway or the blank face of an exterior wall where an elevator is installed in a single blind hoistway or on the outside of a building.

3002.6 Prohibited doors. The prohibition against installing additional doors at the point of access to an elevator car is to minimize the possibility of a person becoming trapped between doors, or the possibility of the elevator car being rendered unusable because of blocked access at a particular floor where such an additional door has been closed and locked. The only condition under which such doors are permitted occurs where the doors are readily openable from the car side without the use of a key, a tool, or any special effort or knowledge.

3002.7 Common enclosure with stairway. To better ensure that a single fire incident will not restrict or eliminate the use of multiple access or egress components within a building, it is important that elevators not be located in the same shaft enclosure as a stairway. Since there is no requirement for a shaft enclosure for an elevator or exit stairway serving an open parking garage, the isolation of the elevator from the stairway is not required in open parking garages.

Section 3003 *Emergency Operations*

Standby power requirements in this section are applicable for those buildings or structures where standby power is furnished to, or required to, operate an elevator. For example, the provisions of Section 1007.4 mandate standby power for any elevator used as an accessible means of egress. The requirement that standby power be manually transferable to all elevators in each bank is necessary to improve their reliability during emergency conditions. As an example, the elevator to which standby power is connected may be in a hoistway that is unusable because of smoke contamination. In this case, the transferability requirement provides for transferring the emergency power from the elevator in the affected hoistway to an elevator in a hoistway in the same bank that is usable.

Section 3004 *Hoistway Venting*

Hoistway ventilation is intended to remove any smoke that may have entered the hoistway, which would render the cars within the hoistway useless. In addition, unvented smoke can eventually spill out of the hoistway into the upper floors of the building. Inadequate top ventilation of the elevator shafts at the MGM Grand Hotel was one cause cited for smoke spreading into upper-story corridors during the 1980 fire event. A portion of the required vent area must always be open so that venting would be immediately available in case of need. However, there is an allowance for maintaining all of the vent openings in a closed position, provided they all open upon actuation of any elevator lobby smoke detector, upon power failure, or upon activation of a manual override control.

Section 3006 *Machine Rooms*

Modern elevator control equipment is solid state and is extremely sensitive to temperature. For this reason, provisions have been developed to keep smoke and heat out of the machine room by requiring that the room be provided with an independent ventilation or air-conditioning system.

Section 3007 *Fire Service Access Elevator*

To facilitate the rapid deployment of fire fighters, Section 403.6.1 contains provisions for fire service access elevators in high-rise buildings that have at least one floor level more than 120 feet (36,576 mm) above the lowest level of fire department vehicle access. Usable by fire fighters and other emergency responders, the specific requirements for the elevators are set forth in Section 3007. See Figure 3007-1.

A fire service access elevator has a number of key features that will allow fire fighters to use the elevator for safely accessing an area of a building that may be involved in a fire or for facilitating rescue of building occupants. The elevator is required to be protected by a shaft enclosure that complies with Section 713. For a building four or more stories in height, Section 713.4 requires that the shaft have a minimum 2-hour fire-resistance rating.

An elevator lobby is mandated as a transitional element between the fire service access elevator and the remainder of the floor. Because the elevator lobby will be the location that fire fighters will use as the point of departure to the floor or area of fire origin, it must be constructed to limit the entrance of fire or smoke. The lobby must be enclosed by a smoke barrier with a minimum fire-resistance rating of 1 hour, and the lobby door is required to

30 Elevators and Conveying Systems

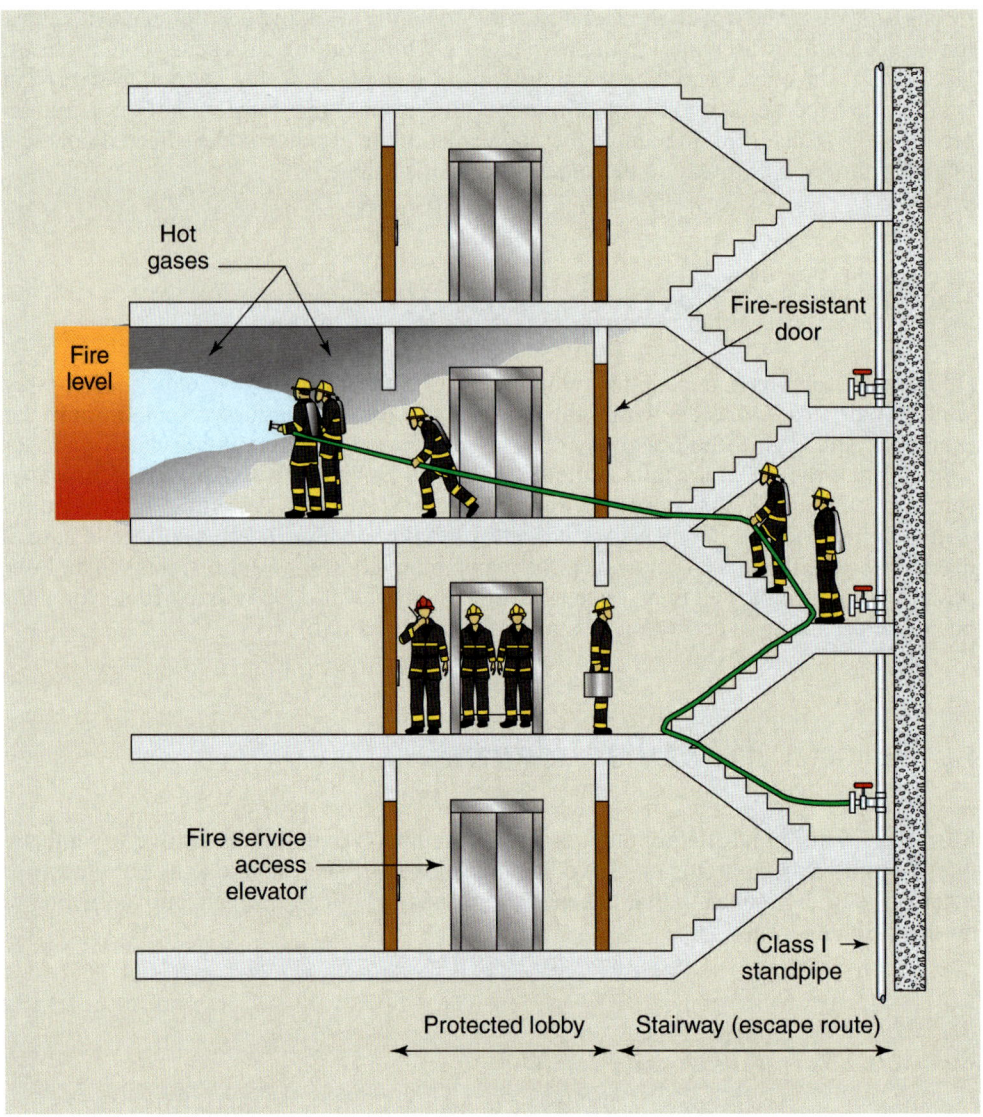

Figure 3007-1
Fire service access elevator.

have a minimum fire-resistance rating of 45 minutes. The lobby for the fire service access elevator must be designed such that it has direct access to an enclosure for an interior exit stairway. By providing a direct connection between the elevator lobby and a stair enclosure, efficient access to and from other floors is increased. In addition, the required standpipe hose valves located within the enclosure for the interior exit stairway will be directly accessible from the fire service access elevator lobby, providing a protected location for fire fighters to prepare for manual fire-fighting operations and to assess interior fire and smoke behavior near the area of fire origin.

A unique requirement for fire service access elevators is that they must be designed so their status can be continuously monitored in the fire command center. The elevator is to be monitored by a standard emergency services interface meeting the requirements of NFPA 72, *National Fire Alarm Code*. The requirements stipulate that such an interface must be designed and arranged in accordance with the requirements of the organization that will use the device. In the case of a fire service access elevator, the fire department or fire service provider will need to be involved in the arrangement of the interface.

A fire service access elevator requires normal utility power and connection to a standby power system. Provisions further stipulate that the transfer switch for the standby power system operate within 60 seconds of utility power failure (Type 60), and that the power source is designed to operate for at least 2 hours under its design load (Class 2) and meet the requirements of NFPA 110 for Level 1 service. Level 1 service, as defined by NFPA 110, indicates that the standby power system is used in a building where the loss of electrical power could result in the death of or serious injury to one or more occupants. Loads that must be connected to the standby power system include the elevator, its machine room ventilation and cooling equipment, and equipment provided to maintain the elevator controller within its temperature limits. An additional requirement for the electrical power system serving the fire service access elevator is the protection of wiring or cables. Electrical conductors that provide normal or standby power to the fire service access elevator must be protected. They should be located by a shaft or similar enclosure having a minimum 2-hour fire-resistance rating; or circuit integrity cable having an equivalent fire resistance must be utilized.

Section 3008 *Occupant Evacuation Elevators*

Under the conditions of Section 3008, public-use passenger elevators are allowed to be utilized for the self-evacuation of occupants in high-rise buildings. Although such elevators may be used by building occupants during building evacuation, they are not intended to replace any means of egress facilities as required by Chapter 10. The only permitted reduction in required egress facilities due to the presence of complying occupant evacuation elevators is the elimination of the extra exit stairway mandated by Section 403.5.2 for high-rise buildings over 420 feet (128 m) in height. Under no conditions is the installation of such elevators required. The allowance for occupant evacuation elevators, although voluntary, provides tools for the architect to consider when designing tall buildings.

Where elevators are to be used for self-evacuation purposes, all passenger elevators in the building must be in compliance with the special provisions of Section 3008. The use of such elevators for occupant self-evacuation is limited to when the elevator is in the normal operating mode prior to Phase I Emergency Recall Operation per ASTM A17.1 and the building's fire-safety and evacuation plan.

The occupant evacuation elevators must open directly into an elevator lobby that conforms to special requirements addressing access, enclosure, size, and signage. Each elevator lobby must be provided with a status indicator arranged to display the following information as applicable:

- Illuminated green light and the message, "Elevators available for occupant evacuation"
- Illuminated red light and the message, "Elevators out of service, use exit stairs"
- No illumination or message when the elevators are in normal service operation

Additional provisions address the design and installation of a two-way communications system and instructions on the use of the system. The elevators must be continuously monitored at the fire command center or an approved central control point. A variety of other conditions are placed on the design and installation of the occupant evacuation elevators to help ensure their reliability under emergency conditions.

30 Elevators and Conveying Systems

> **KEY POINTS**
> - Elevator shafts and elevator machine rooms are typically required to be fire-resistance-rated enclosures, with the rating of either 1 hour or 2 hours based on the number of stories in the building and required fire-resistance rating of the building's floor construction.
> - Additional doors are prohibited at the point of access to an elevator car unless they are readily openable from the car side without the use of a key, a tool, or any special effort or special knowledge.
> - Hoistway ventilation is intended to remove any smoke that may have entered the hoistway, which would render the cars within the hoistway useless.
> - Fire service access elevators help facilitate the rapid deployment of fire fighters in applicable high-rise buildings.
> - Although not intended to replace any required means of egress facilities, public-use passenger elevators are allowed to be utilized for the self-evacuation of occupants in high-rise buildings.

CHAPTER 31

SPECIAL CONSTRUCTION

Section 3102 Membrane Structures
Section 3104 Pedestrian Walkways and Tunnels
Section 3105 Awnings and Canopies
Section 3106 Marquees
Section 3109 Swimming Pool Enclosures and Safety Devices
Key Points

31 Special Construction

> Those special types of elements or structures that are not conveniently addressed in other portions of the *International Building Code®* (IBC®) are found in this chapter. By special construction, the code is referring to membrane structures, pedestrian walkways, tunnels, awnings, canopies, marquees, and similar building features that are unregulated elsewhere.

Section 3102 *Membrane Structures*

3102.1 General. Because membrane structures have several unique characteristics that set them apart from other buildings, they are regulated in Chapter 31 under the provisions for special construction. The regulations cover all such structures, including air-supported, air-inflated, cable-supported, and frame-supported membrane structures. The intent of the provisions is that, except for the unique features of membrane structures, they otherwise comply with the code as far as occupancy requirements, allowable area, and other regulations are concerned. Membrane structures are limited to one story in height insofar as there is insufficient experience to justify multilevel structures enclosed with a membrane.

The membrane structures regulated by the IBC are deemed to be permanent in nature, erected for a period of at least 180 days. Membrane structures in place for shorter periods of time, such as temporary tents, are to be regulated by the *International Fire Code®* (IFC®). Where a membrane structure is erected as a part of a permanent structure, such as a covering for a building, balcony, or deck, it must comply with the provisions of Section 3102 for any time period.

Because of the limited hazards present in structures not used for human occupancy, such as water-storage facilities, water clarifiers, sewage-treatment plants, and greenhouses, only a few provisions are applicable where membrane structures cover these types of facilities. Limitations on the membrane and interior liner material, as well as the structural design, are the only criteria in the IBC that apply to membrane structures covering facilities not typically used for human occupancy.

3102.3 Type of construction. In general, membrane structures are considered to be of Type V construction, except where the membrane structure is shown to be noncombustible. In this case, the membrane structure should be classified as Type IIB construction. Membrane structures supported by heavy-timber framing members are to be considered Type IV construction. The code permits the use of nonflame-resistant plastic material for the membrane of a greenhouse structure that is not available to the general public.

3102.6 Mixed construction. This section permits the use of a noncombustible membrane on a structure that would otherwise comply as Type IA, IB, or IIA where the membrane is used exclusively as a roof or skylight and is located at least 20 feet (6,096 mm) above any floor, balcony, or gallery. This exception is similar to Footnote b of Table 601. This exception will permit nonrated noncombustible membranes to be constructed as roof systems for sports stadiums and similar buildings as well as for atriums. In other types of construction under the same conditions, the membrane need only be flame resistant.

3102.8 Inflation systems. Where membrane structures are air-supported or air-inflated, this section addresses the regulations for equipment, standby power, and support. The primary inflation system shall consist of one or more blowers, designed in such a manner that over pressurization is prevented. Air-supported or air-inflated structures exceeding 1,500 square feet (140 m^2) in floor area must also be provided with an ancillary inflation system. This backup system, connected to an approved standby power-generating system, shall operate automatically to maintain the inflation of the structure if the primary system fails.

Additional support for the membrane must also be provided where covering structures that have occupant loads of more than 50 and where covering swimming pools.

Section 3104 *Pedestrian Walkways and Tunnels*

This section regulates connecting elements between buildings, such as tunnels or pedestrian walkways, that are utilized for occupant circulation. The provisions of this section are only applicable to such tunnels and walkways designed primarily as circulation elements, typically for weather-protection purposes. A covered walkway or bridge connecting two buildings is the most common example of a pedestrian walkway. These elements may be located below, at, or above grade. When in compliance with this section, pedestrian walkways and tunnels are not considered to contribute to the floor area or height of the connected buildings. In addition, those buildings connected by the pedestrian walkway or tunnel are permitted to be considered separate structures. These allowances establish the primary reason for the use of this section. Where multiple structures are attached, they would generally be considered by the code to be a single building and regulated as such. These provisions not only allow each building to be regulated independently, but also limit the requirements of the tunnel or pedestrian walkway to those provisions of this section. A common use of pedestrian walkways is the connection of buildings that are on separate lots, including situations where the buildings are on opposite sides of a public way. Often referenced as *skyways* in northern climates, pedestrian walkways allow for a method to connect buildings across lot lines without the normally mandated fire-resistance-rated exterior walls and opening prohibition associated with a fire separation distance of zero. The pedestrian walkway is treated as a *nonbuilding* and therefore is not regulated where it crosses the lot line. Instead, this approach allows for a level of fire protection at the connection between the buildings and the pedestrian walkway.

It is important to understand that this section is essentially voluntary in application, and is only utilized where the design professional chooses not to consider the walkway and connected buildings a single structure. There is always the option of regulating the entire structure as a single building, in which case this section would have no application. See Figure 3104-1.

This section establishes specific requirements for the protection of walls and openings between the connected buildings and the pedestrian walkway or tunnel, specifies minimum and maximum widths, and limits the length of exit access travel within a pedestrian walkway or tunnel. Section 3104.3 further requires that pedestrian walkways be constructed of noncombustible materials or of fire-retardant-treated wood unless all connected buildings are of combustible construction.

As previously mentioned, where pedestrian walkways and tunnels are designed and constructed in accordance with the provisions of this section, the code intends that they not be considered part of the connected buildings.

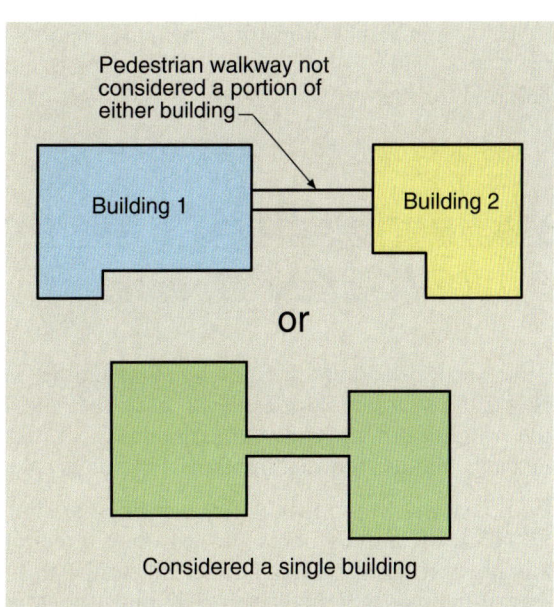

Figure 3104-1
Pedestrian walkways.

31 Special Construction

Furthermore, they need not be considered in the determination of the allowable area or height of either of the connected buildings. In effect, this section is an exception to the general construction requirements, and its use is optional. Otherwise, the designer would consider the connected buildings to be a single building on the same lot, including the pedestrian walkway portion.

Section 3105 Awnings and Canopies

The definition of awning in Section 202 describes it as an architectural projection, comprising a lightweight, rigid skeleton structure over which a covering is attached, that provides weather protection, identity, or decoration, and is completely or partially supported by the building to which it is attached. Retractable awnings are also regulated by this chapter as well as Chapter 16; therefore, a definition is provided. Awnings are regulated for structural and construction features as well as cover materials.

Section 3106 Marquees

A marquee is defined in Section 202 as a canopy with a flat, or relatively flat, roof surface located in close proximity to operable building openings. A canopy is defined as a permanent roof structure or architectural projection that is structurally independent or attached to and supported by a building. A classic example of a marquee is the theater marquee or the entrance marquee at a hotel. The restrictions placed on the construction, projection, and clearances for marquees are intended to prevent interference with:

1. The free movement of pedestrians.
2. Trucks and other tall vehicles using the public street.
3. The fire department in its fire-fighting operations at a building.
4. Utilities.

Thus, the height, construction, and location of the marquee are regulated to meet the intent of the code. Figure 3106-1 depicts the permissible projection and clearances required for marquees.

Section 3109 Swimming Pool Enclosures and Safety Devices

This section addresses barriers surrounding swimming pools, spas, and hot tubs. Regulations are provided for both public and residential swimming pools; however, the primary criteria apply to residential facilities. Because of the supervision provided at public pools, the requirements for public swimming pools are much less extensive.

3109.3 Public swimming pools. A fence, barrier, or similar enclosure method is required around all public swimming pools. No less than 4 feet (1,290 mm) in height, the fence must be equipped with self-closing and self-latching gates. Openings in the fence are also regulated so that the passage of a 4-inch (102-mm) sphere is not possible.

Swimming Pool Enclosures and Safety Devices 31

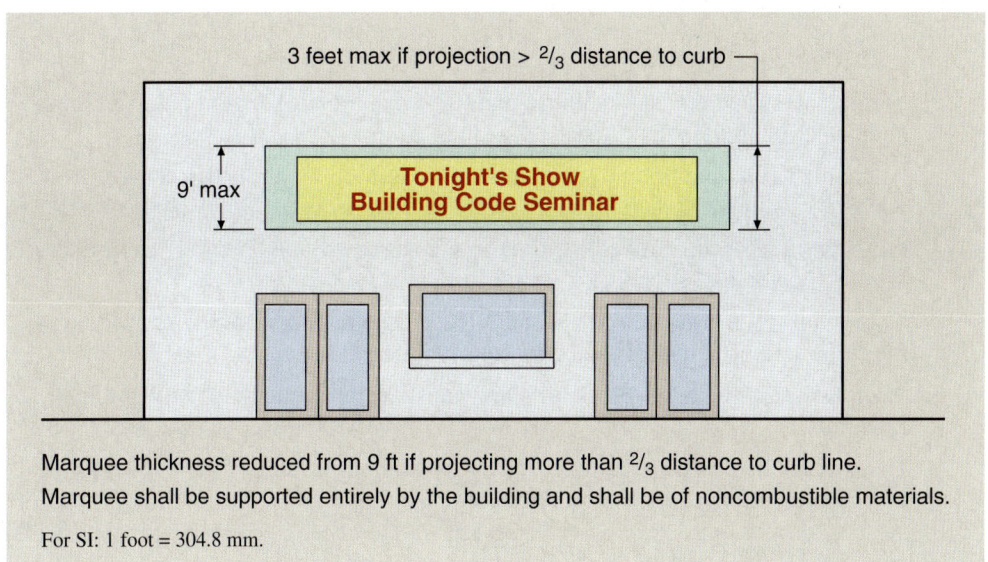

Figure 3106-1
Marquees.

Residential swimming pools. Unless a swimming pool is provided with a power safety cover or a spa is provided with a safety cover complying with the provisions of ASTM F 1346, a residential swimming pool, spa, or hot tub must be enclosed in accordance with this section. The code addresses a number of different barrier designs, each intended to limit access to the pool area. See Figure 3109-1. Where a wall of a dwelling serves as a portion of the barrier, additional requirements are also in place to regulate unsupervised access to the pool or spa.

3109.4

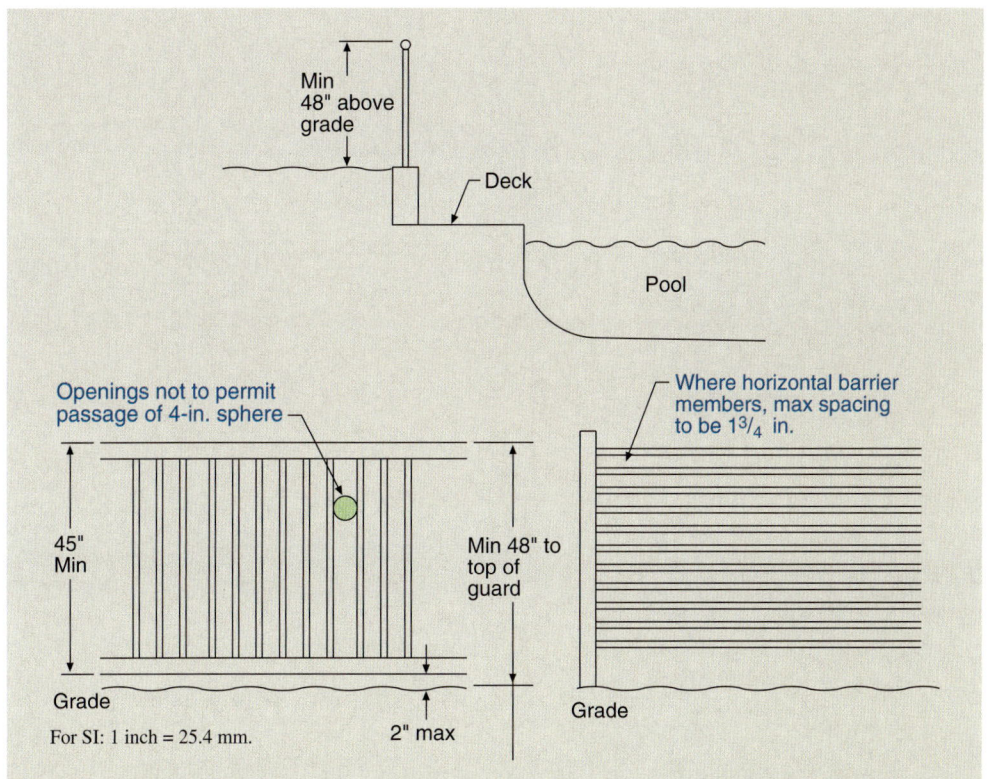

Figure 3109-1
Swimming pool enclosure barriers.

2012 International Building Code Handbook

31 Special Construction

KEY POINTS

- Membrane structures include those buildings that are air-supported, air-inflated, and cable- or frame-supported.
- Noncombustible membrane structures are classified as Type IIB construction, whereas combustible membrane structures are considered Type VB.
- Air-supported or air-inflated membrane structures are regulated for equipment, standby power, and support.
- Complying pedestrian walkways and tunnels are not considered to contribute to the floor area or height of the connected buildings.
- Awnings are regulated for structural and construction features, as well as canopy materials.
- Unless provided with an approved safety cover, a swimming pool or spa must be enclosed by a fence, barrier, or similar enclosure method.

CHAPTER 32

ENCROACHMENTS IN THE PUBLIC RIGHT-OF-WAY

Section 3201 General
Section 3202 Encroachments

32 Encroachments in the Public Right-of-Way

> By using the language "encroachments into the public right-of-way," the code is referring to the projections of appendages and other elements from the building that are permitted to project beyond the lot line of the building site and into the public right-of-way. The intent of the code is that projections from a building into the public right-of-way shall not interfere with its free use by vehicular and pedestrian traffic, and shall not interfere with public services such as fire protection and utilities.

Section 3201 *General*

The *International Building Code®* (IBC®) requires that any construction that projects into or across the public right-of-way be constructed as required by the code for buildings on private property. Furthermore, the intent is that the provisions of this chapter are not considered so as to permit a violation of other laws or ordinances regulating the use and occupancy of public property. Many jurisdictions have ordinances regulating the use and occupancy of public property that go beyond the provisions of this chapter, many of which may be more restrictive and not permit the projections permitted by the IBC. Where there are no other ordinances that prohibit such construction, it is the intent of this chapter to permit the construction of the connecting structure between buildings either over or under the public way.

These types of permanent uses of the public way are often permitted by the jurisdiction under a special license issued by the public works agency. Where these uses create no obstruction to the normal use of the public way, the jurisdiction, by authorizing connecting structures between buildings on either side of the public way, can derive revenue from a licensing agreement, which can be of assistance in maintaining the public right-of-way.

The code prohibits roof drainage water from a building to flow over a public walking surface, as well as any water collected from an awning, canopy, or marquee, and condensate from mechanical equipment. The intent is not necessarily to prevent drainage water from flowing over public property in general, but rather to prevent drainage water from flowing over a sidewalk or pedestrian walkway that is between the building and the public street or thoroughfare.

There are at least two problems that arise when drainage water from a building or structure is allowed to flow over a public sidewalk:

1. Under proper conditions of light and temperature, algae will form where the water flows across the sidewalk and create a hazardous, slippery walking surface.
2. During heavy rain storms, the velocity and force of the water emitting from the drain can create hazardous walking conditions for pedestrians.

Therefore, the usual procedure is to carry the drain lines from the roof or other building elements inside the building through the wall of the building and under the sidewalk through a curb opening into the gutter.

Section 3202 *Encroachments*

The projection of any structure or appendage is generally regulated in regard to its relationship to grade. Those encroachments found below grade, such as structural supports, vaults, and areaways, are governed in one fashion. Those encroachments located above grade, but below 8 feet (2,438 mm) in height, are more severely controlled because of their immediate location adjacent to public areas. Although projections that are located 8 feet (2,438 mm) or more above grade tend to have fewer implications, such encroachments are also regulated.

Encroachments below grade. Footings, foundations, piles, and similar structural elements that support the building and are erected below grade are not permitted to project beyond the lot lines. There is an allowance for footings that support exterior walls, provided the projection does not extend more than 12 inches (305 mm) beyond the street lot line, and that the footings or supports are located at least 8 feet (2,438 mm) below grade. In this exception, the code assumes that there will be no utilities or other public facilities below this depth adjacent to the building and, therefore, permits footings to project a limited amount where they are below the 8 feet (2,438 mm) depth. It should be noted that this encroachment is permitted for street walls only and does not apply to interior lot lines. See Figure 3202-1.

3202.1

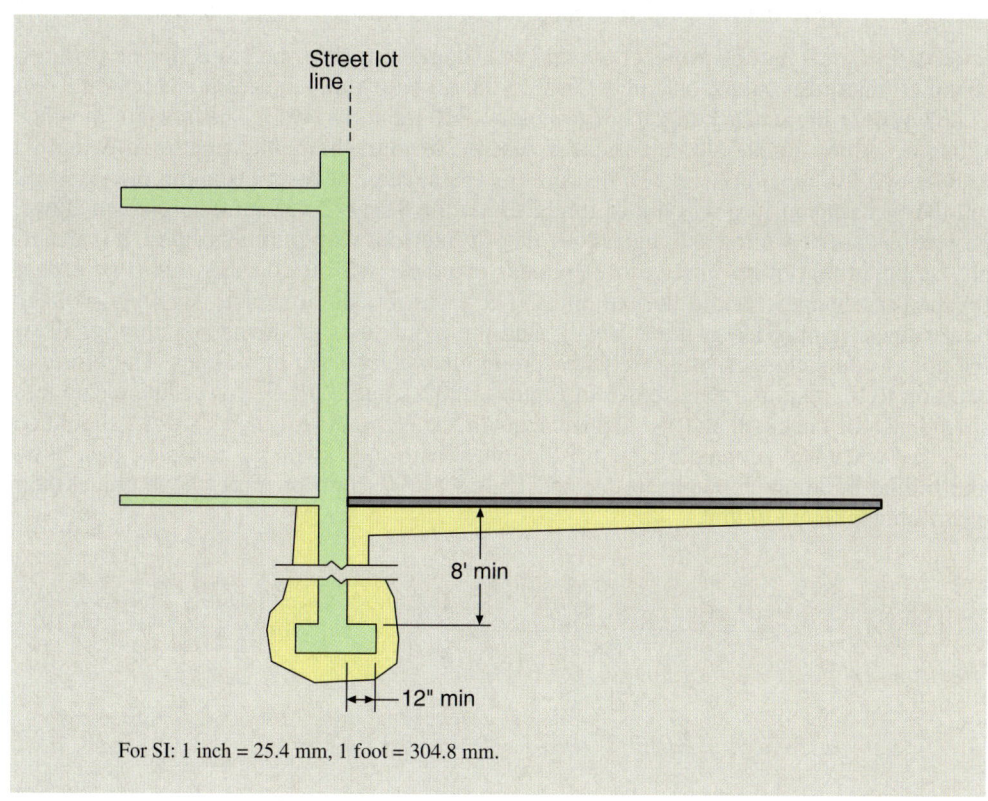

**Figure 3202-1
Encroachments below grade.**

Where vaults, basements, and other enclosed spaces are located below grade in the public right-of-way, they shall be further regulated by the ordinances and conditions of the applicable governing authority. In many cases, the space below the sidewalk or other area adjacent to the building in the public right-of-way is not used for public purposes such as utilities, sewers, and storm drains (except for catch basins); therefore, where local ordinances do not prohibit the use of space beneath the sidewalk or other public area, the code permits its use by adjoining property owners. Usually the basement or other enclosed space of an adjoining building is extended beneath the public right-of-way and is used for any of the uses permitted by the code. This is a revocable permission, as the jurisdiction may find at a later date that there is a public need for the public use of the space beneath grade.

Encroachments above grade and below 8 feet in height. Building components that may arbitrarily project into the public right-of-way pose a considerable hazard to pedestrians or other users of the right-of-way. Therefore, doors and windows are prohibited from

3202.2

opening into such an area. However, certain features of the building that are permanent in nature are permitted to encroach into the public right-of-way to a limited degree. Where steps are permitted by other provisions of the IBC, they may project into the right-of-way up to 12 inches (305 mm), provided they are protected by guards at least 3 feet (914 mm) high or similar protective elements such as columns or pilasters. The intent of the provision is that the encroachment be small enough so as not to obstruct pedestrian or vehicular traffic and yet provide some protection for occupants of the building to look each way before proceeding into the right-of-way.

Encroachments into the public right-of-way for certain architectural features are also permitted, but again, only a limited projection is allowed. For columns or pilasters, the maximum projection is 12 inches (305 mm). Other architectural features such as lintels, sills, and pediments are limited to a 4-inch (102-mm) projection. Awnings, including valances, may only extend into the public right-of-way when the vertical clearance from the right-of-way to the lowest part of the awning is at least 7 feet (2,134 mm).

3202.3 Encroachment 8 feet or more above grade. The code permits the projection of awnings, canopies, marquees, signs, balconies, and similar architectural features at a height of 8 feet (2,438 mm) or more in accordance with Figures 3202-2 and 3202-3. The intent of the code is that no projection should be permitted near the level of the public right-of-way and up to 8 feet (2,438 mm) in height so that the free passage of pedestrians along the sidewalk or other walking surface will not be inhibited. At the 8 feet (2,438 mm) height and above, projections are permitted as long as they do not interfere with public utilities. It is generally assumed that utility lines for telephones or power will not occupy this zone except for the service entrances to the buildings. There are jurisdictions that have high-voltage power lines running along the sidewalk, and the regulations of the agency that regulates the power companies generally require certain clearances from these lines. Therefore, in addition to the requirements shown in Figures 3202-2 and 3202-3, power-line clearances should also be checked, and the requirements of the *National Electrical Code®* should be reviewed when it is adopted by the jurisdiction. Where such awnings, canopies, balconies, and similar building elements are located 15 feet (4,572 mm) or more above grade, their encroachment is unlimited.

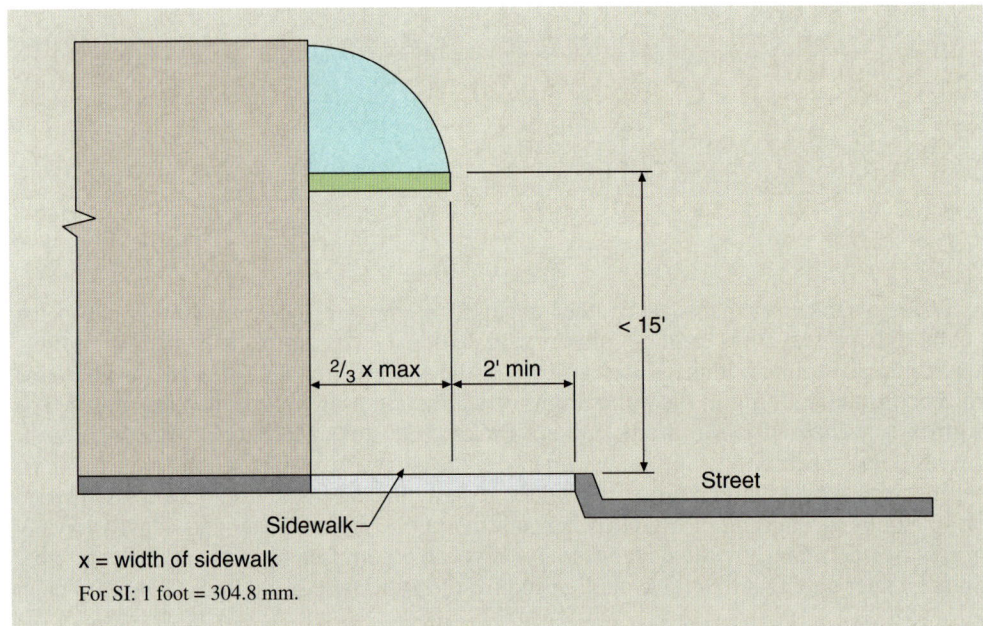

Figure 3202-2 Awing, canopy, marquee, and sign projections.

Encroachments 32

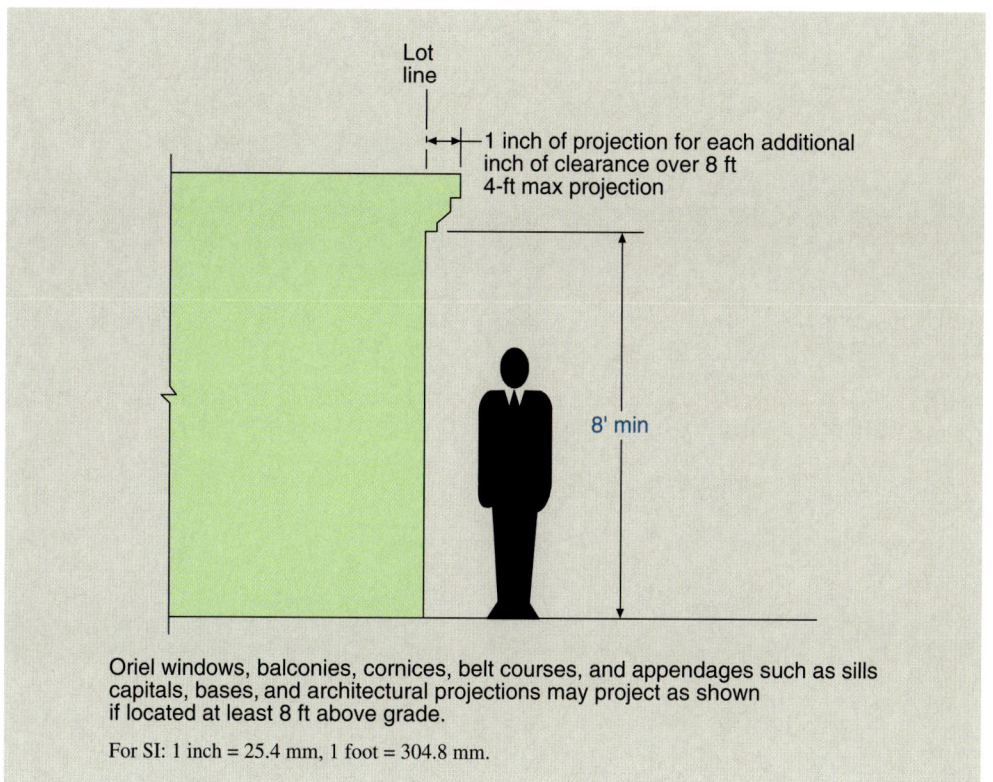

Figure 3202-3
Balcony and appendages projections.

It is not uncommon to find a pedestrian walkway or a similar circulation element to be constructed over a public right-of-way. Such a condition is permitted subject to the approval of the local authority having jurisdiction. In no case, however, should the vertical clearance between the right-of-way and the lowest part of the walkway be less than 15 feet (4,572 mm). A clearance of at least 15 feet (4,572 mm) allows not only for unobstructed pedestrian travel but for vehicular access as well.

CHAPTER 33

SAFEGUARDS DURING CONSTRUCTION

Section 3302 Construction Safeguards
Section 3303 Demolition
Section 3304 Site Work
Section 3306 Protection of Pedestrians
Section 3307 Protection of Adjoining Property
Section 3308 Temporary Use of Streets, Alleys and Public Property
Section 3309 Fire Extinguishers
Section 3310 Means of Egress
Section 3311 Standpipes
Key Points

33 Safeguards During Construction

> This chapter deals with safety practices during the construction process, as well as the protection of property, both public and private, adjacent to the construction site. Included in the provisions are requirements addressing site work; protection of pedestrians; and temporary use of streets, alleys, and public property. Special issues relating to fire extinguishers, exits, standpipes, and automatic sprinkler systems are also addressed.

Section 3302 Construction Safeguards

Where an existing building continues to be occupied during the process of remodeling or constructing an addition, it is very important that the life- and fire-safety features of the building continue to be in place. Required exits shall be maintained, the strength of structural elements shall not be diminished, fire-protection devices are to remain in working condition, and sanitary facilities shall be fully functional. These features must be maintained at all times during any remodeling, alterations, repairs, or additions to the building under consideration. Obviously, there will be times when the specific protective feature is the element or device being altered or repaired. In this case, an equivalent method shall be provided to safeguard the occupants. Chapter 33 of the *International Fire Code®* (IFC®) is referenced for fire safety during construction, as well as for demolition as stated in Section 3303.7.

Section 3303 Demolition

As the work of demolishing a building is subject to so many variations and so many different hazards, the *International Building Code®* (IBC®) authorizes the building official to require the submission of plans and a complete schedule of the demolition. Under certain circumstances, pedestrian protection may be required during the demolition process. Such protection shall be provided in compliance with other provisions of Chapter 33. For some multilevel buildings, and for certain types of demolition operations, it may also be necessary to temporarily close the street. Once the structure has been demolished, any resulting excavation is to be filled consistent with the existing grade, or in any other manner in conformance with the ordinances of the jurisdiction.

Section 3304 Site Work

The provisions in Section 3304.1 for excavations and fills are intended to apply to the specific area where the building or structure will be located. This section is not intended to address massive grading on a site. For those requirements, one would turn to Section 1804, as well as any local grading ordinance that might be in effect. To prevent decay and to eliminate an avenue of entrance for termites and other insects, the code requires that the area of the site occupied by the building be free of all stumps, roots, and any loose or casual lumber. Typically applicable in the construction of wood-framed structures, the requirements also address wood forms that have been used in placing concrete, as well as any loose or casual wood that might be in direct contact with the ground under the building.

The IBC limits the slopes for permanent fills or cut slopes for permanent excavations for structures to two units horizontal in one unit vertical (50-percent slope). Although steeper-cut slopes are permitted where substantiating data are submitted, the limitation on filled slopes of two units horizontal in one unit vertical (50-percent slope) is an absolute limitation, and filled slopes may not be steeper. Although cut slopes into natural soils may be excavated with a slope steeper than 50 percent, depending on the nature of the soil materials and the density, the code reasons that fill slopes must not be steeper than 50 percent to cover unprecedented circumstances such as heavy rains creating overly saturated soils or, in the case of seismic activity, vibration of the soils during an earthquake causing failures of steep-filled slopes.

Understandably, the code requires that fill or surcharge loads should not be placed adjacent to an existing building or structure unless the existing building or structure is structurally capable of resisting the additional loads caused by the fills or surcharge. See Figure 3304-1. Alternatively, the existing building can be strengthened in order to resist the additional loading.

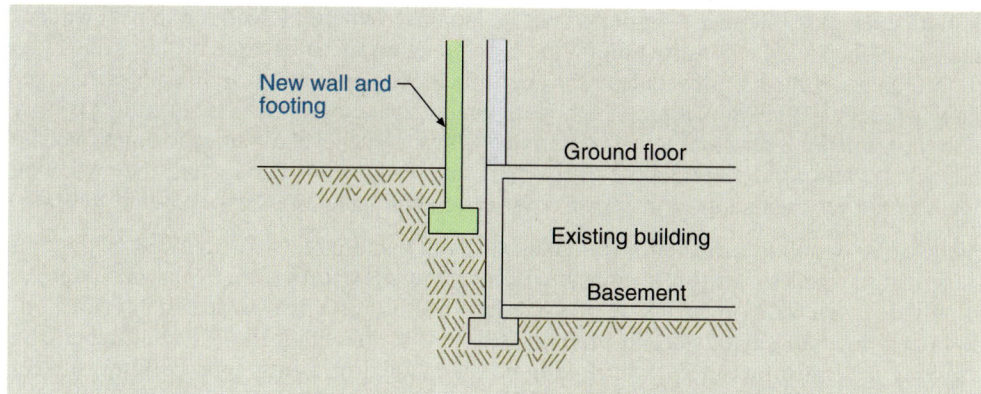

Figure 3304-1
Surcharge load on adjacent building.

The code requires that existing footings or foundations that may be affected by an excavation be adequately underpinned or otherwise protected. This concerns excavations on the same lot and under the control of the same individual who has control of the existing buildings. For obvious reasons, it is the intent of the code that excavations in close proximity to an existing footing shall not be made unless proper protective measures are taken for the existing building. These measures may involve underpinning of the existing foundations, or shoring and bracing of the excavations, so that the existing building foundations will not settle or lose lateral support.

Where the excavation is for a new building foundation and the new footing is at an elevation below, but within reasonably close proximity to, an existing foundation, underpinning is the usual procedure to protect the existing foundation. If the existing footing is for a structure that creates a horizontal thrust, the means of providing lateral support may take the form of a buttressed retaining wall designed to resist the lateral thrust of the existing foundation.

Where buildings are to be supported by fills, the code requires that the fills be placed in accordance with accepted engineering practice. Thus, fills utilized for the support of buildings must be designed by a geotechnical engineer (soils engineer) utilizing the principles of geotechnical engineering so as to provide a proper and adequate foundation for the structure above, and one that will limit settlements to tolerable levels. In order to verify the adequacy of the geotechnical design of the fills, the IBC permits the building official to ask for a soil investigation report outlining the geotechnical design of the fill materials, as well as a report of the satisfactory placement of the fill.

33 Safeguards During Construction

Section 3306 *Protection of Pedestrians*

Both Section 3306 and Chapter 32 regulate the use of public streets and sidewalks. Section 3306 provides those general regulations and criteria for the temporary use of public property, which are generally found to apply to most jurisdictions.

As the nature and philosophy of each jurisdiction varies, so do the regulations each promulgates regarding the use of public property and, in particular, the use of streets and sidewalks. Adjacent property owners have rights of access to their property. The public street also provides access for public services such as fire and police protection, street sweeping and maintenance, street lighting, trash pickup, and other services provided by the jurisdiction or its contractors. Utilities serving adjacent property also use the public street for access.

Pedestrians must be protected from the potential hazards that exist during construction or demolition operations adjacent to the public way. The type of protection depends on the type of operation being conducted. For example, an excavation directly adjacent to a pedestrian path would necessitate a minimum of a construction railing along the side facing the excavation. It is also possible that a barrier would be required. Where the pedestrian path extends into the public street, a construction railing is required on the street side of the walkway to protect the pedestrians from vehicular traffic. Table 3306.1 provides the criteria for determining whether or not additional protection is required, depending on the height and proximity of the construction operations to the pedestrian walkway. The level of protection mandated by the table is shown in Figure 3306-1. Depending on these various parameters, the protection required will vary from none or merely a construction railing, to a barrier or covered walkway.

3306.2 Walkways. A public jurisdiction provides sidewalks and streets for the free passage of vehicular and pedestrian traffic. In the case of pedestrian traffic, the usual procedure is to provide a sidewalk on each side of the street. Therefore, when construction or demolition operations are conducted on property adjacent to the sidewalk, the code requires a walkway at least 4 feet (1,219 mm) wide be maintained in front of the building site for pedestrian use. However, in those cases where pedestrian traffic is unusually light, the jurisdiction may authorize the fencing and closing of the sidewalk to pedestrian use. The walking surface, in addition to being durable, must be in compliance with the accessibility provisions of Chapter 11. The walkway shall also be designed to support all imposed loads, with a minimum design live load of 150 pounds per square foot (7.2 kN/m^2).

3306.3 Directional barricades. Where the temporary walkway used for pedestrian travel around a construction or demolition site extends into the street, it is critical that a sufficient barrier be provided to direct vehicular traffic away from the walkway. The construction of the barrier is to be large enough to ensure that motorists will easily and quickly identify the revised traffic path.

3306.4 Construction railings. The code only requires an open-type guardrail 3 feet, 6 inches (1,067 mm) in height where a construction railing is required. The intent of the railing is to direct pedestrians as they travel adjacent to construction areas, as well as to provide a very limited degree of protection. Where access must be further regulated, such as adjacent to an excavation, additional provisions must be applied. It is incumbent that the jurisdiction adequately protect pedestrians from the hazards associated with construction or demolition work.

3306.5 Barriers. The intent of the code is that a barrier, where required, be solid and sturdy enough to prevent impacts from construction operations from penetrating the barrier and injuring passing pedestrians. Because of this intent, the code also requires that the barrier be at least 8 feet (2,438 mm) in height and extend along the entire length of the building site. Although the IBC requires openings in the barrier to be protected by doors that are normally kept closed, the code does not intend to prevent the use of small viewports at eye level so that passing pedestrians may stop and view the construction operations if they wish.

Protection of Pedestrians   33

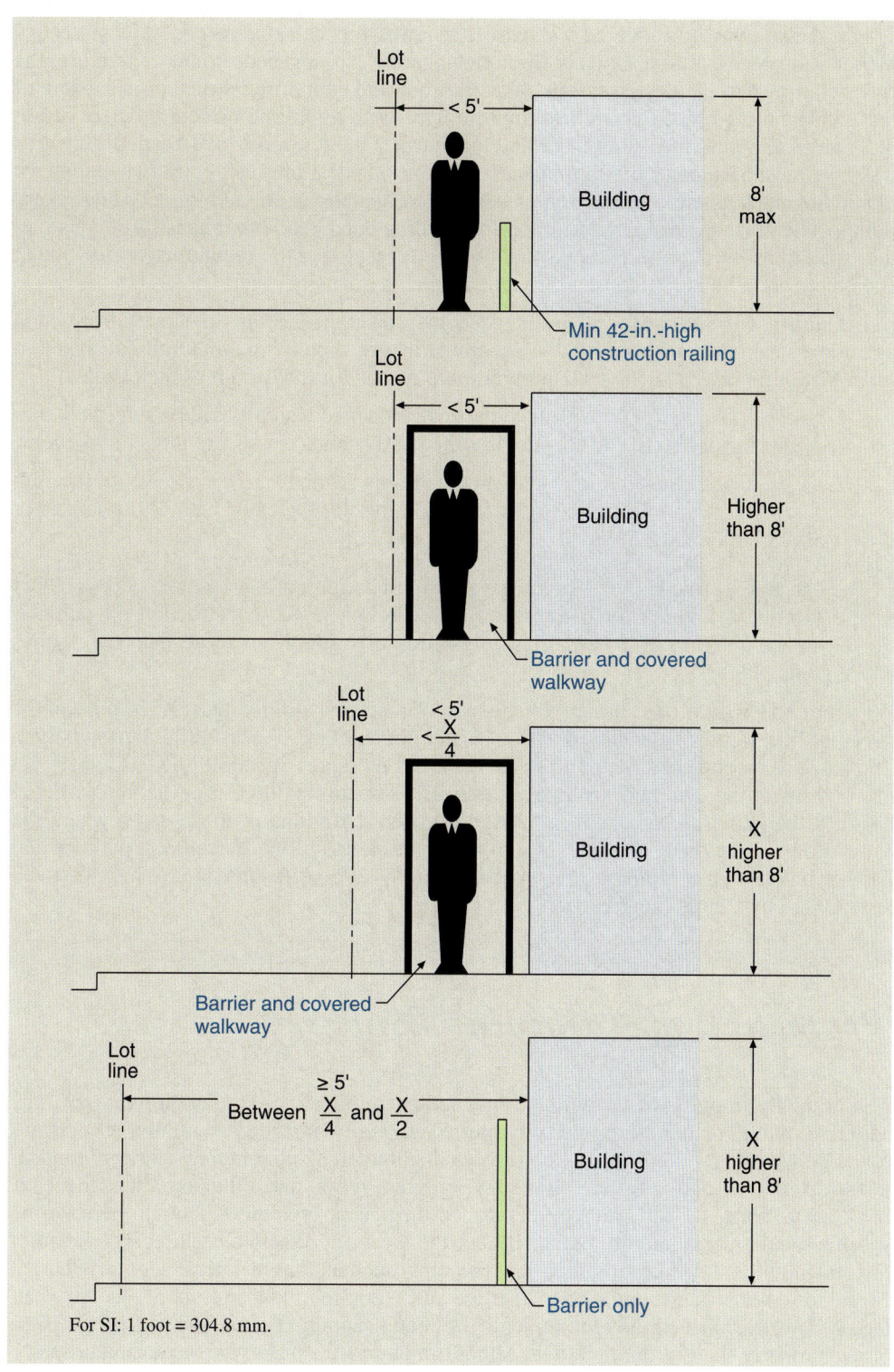

Figure 3306-1
Pedestrian protection.

To provide a reasonable level of protection from the construction operations, barriers are to be designed to resist loads as required in Chapter 16. As an option, specific construction procedures are outlined in Section 3306.6. The procedures address the size and spacing of studs as well as the size and span limitations for wood structural panels used in the barrier.

33 Safeguards During Construction

3306.7 Covered walkways. A covered walkway is required for the same reasons as a protective barrier—to prevent falling objects from endangering passing pedestrians. Therefore, the code requires that the design be such that the roof and supporting structure is capable of preventing falling objects from breaking through. With a minimum required clear height of 8 feet (2,438 mm) measured from the floor surface to the canopy overhead, the covered walkway must be adequately illuminated at all times. Pedestrians using walkways adjacent to building sites where construction or demolition operations are taking place have every right to safe passage and to be protected from falling debris and other hazards of construction operations. Furthermore, they should not be subjected to the indignities of dirt, water, and other foreign material sifting through leaks in the canopy.

Because a covered walkway is intended to protect pedestrians against construction operations, including demolition, the canopy structure should be structurally designed to serve that purpose. Thus, the code provides two means for the design of the canopy:

1. A *structural design* to withstand the actual loads to which it will be subjected, provided the design live load is not less than 150 pounds per square foot (7.18 kN/m^2).

2. A *prescriptive design* in accordance with this section of the code where the covered walkway will not be subjected to a live load greater than 150 pounds per square foot (7.18 kN/m^2).

The exception provides for a permissive design based on a live load of 75 pounds per square foot (3.56 kN/m^2) for the covered walkway where the construction operation is limited to the erection of a new, small, light-frame building not more than two stories above grade plane.

3306.9 Adjacent to excavations. Where a site excavation occurs within 5 feet (1,524 mm) of the street lot line, it must be enclosed with a barrier not less than 6 feet (1,829 mm) in height. The barrier must be constructed to resist the wind pressures specified in Chapter 16, but need not meet the general barrier requirements of Section 3306.6. The building official also has the authority to require a barrier around excavations in those cases where the excavation is more than 5 feet (1,524 mm) from the street lot line. Because of the potential liability involved, it is probable that most excavations of considerable size or hazard would be enclosed by a complying barrier.

Section 3307 *Protection of Adjoining Property*

It is critically important that during construction, remodeling, or demolition work, any adjoining public or private property be protected from damage. Essentially, all portions of the neighboring structure must be protected from damage, including footings, foundations, exterior or party walls, chimneys, skylights, roofs, and other building elements. In addition, water run-off must be controlled to prevent erosion. Although there are no specific requirements laid out in the IBC for the level of protection required, the intent of the protection is based on common-law precepts, such as that of lateral support where it comes to footings and foundations. Over the years, common law and, more recently, statute law have established the requirements for lateral support. That is, the owner of a piece of real property shall be entitled to the lateral support of the property by adjacent property. Responsible practice would also dictate a satisfactory level of protection for other portions of the adjacent site and building during the construction process.

Where excavation work is to be performed, the person responsible for such work shall notify the owners of any adjoining buildings. The written notice, to be delivered no less than 10 days prior to the initiation of the excavation work, shall indicate that excavation work is to take place and that protection of the adjoining buildings should be considered.

Appropriate action should be taken where access to an adjoining site or building is desired in order to provide the proper protective measures.

Section 3308 *Temporary Use of Streets, Alleys and Public Property*

It is important that access to essential facilities be maintained where construction materials and equipment are temporarily stored on public property. Obstructions shall not block access to such fire protection features as standpipes, fire hydrants, or alarm boxes, as well as any catch basins or manholes that might be present. In regard to maintaining safe and effective vehicular traffic in the area where construction materials and equipment are stored, such materials or equipment must be located at least 20 feet (6,096 mm) from the street intersection, or otherwise located where sight lines are not obstructed.

Section 3309 *Fire Extinguishers*

It is quite common for combustible debris and waste materials to accumulate in and around a building under construction. Therefore, at least one approved portable fire extinguisher sized for at least ordinary hazard and complying with Section 906 of the IFC is to be provided at every stairway on all floor levels where combustible material has accumulated. In addition to those extinguishers required within the building, every storage shed and construction shed shall also contain an approved fire extinguisher. Additional extinguishers may be required by the building official where any special or unique hazards exist, such as the presence of flammable or combustible liquids.

Section 3310 *Means of Egress*

Once a building under construction reaches a considerable height, deemed to be 50 feet (15,240 mm) or four stories by the IBC, there is a need to provide at least one stairway that is available and usable for egress purposes. The temporary stairway shall be adequately illuminated during those times where there are occupants who may need to utilize the stairway. A temporary stairway need not be provided where there is at least one permanent stairway that is maintained and usable as the construction progresses. Where an existing building exceeding 50 feet (15,240 mm) in height is undergoing an alteration, at least one lighted stairway must continue to be available.

Section 3311 *Standpipes*

For those buildings required to have a permanent standpipe system per Section 905.3.1, during construction operations a standpipe system must also be installed before the construction height exceeds 40 feet (12,192 mm) above the lowest level of fire department access. As construction continues to proceed upward, the standpipes shall also be extended in a timely manner. A standpipe shall always be available within one floor of the highest point of construction having secured decking or flooring. During construction operations, the amount of combustible materials from concrete forms, scaffolding, plastic and canvas tarpaulins, and other materials are prevalent not only throughout the building, but throughout the construction site itself. Thus, in many cases, the standpipe system provides the only source of water for fire-fighting operations.

A minimum of one standpipe is required, with hose connections to be located in accessible areas adjacent to usable stairs. The standpipes may be either temporary or permanent, but in all cases shall meet the minimum requirements of Section 905 for capacity, outlets, and materials. The water supply may also be either temporary or permanent, as long as it is available at the first sign of combustible material accumulation.

> **KEY POINTS**
>
> - Where an existing building continues to be occupied during the process of remodeling or constructing an addition, it is very important that the life- and fire-safety features of the building continue to be in place.
>
> - Once a structure has been demolished, any resulting excavation is to be filled, consistent with the existing grade.
>
> - The code requires that fills or surcharge loads not be placed adjacent to an existing building unless the existing building is structurally capable of resisting the additional loads caused by the fills or surcharge.
>
> - Pedestrians must be protected from the potential hazards that exist during construction or demolition operations adjacent to the public way.
>
> - Depending on various conditions, required pedestrian protection varies from none, to merely a construction railing, to a barrier, to a covered walkway.
>
> - During construction, remodeling, or demolition work, any adjoining public or private property must be protected from damage.

CHAPTER 34

EXISTING STRUCTURES

Section 3404 Alterations
Section 3408 Change of Occupancy
Section 3411 Accessibility for Existing Buildings
Section 3412 Compliance Alternatives
Key Points

34 Existing Structures

This chapter covers all aspects of existing buildings, including their occupancy and maintenance, as well as additions, alterations, or repairs to existing buildings. In addition, accessibility provisions for existing structures are found in Section 3411.

A building in existence when a new code edition is adopted has the right to have its existing use or occupancy continued, provided all devices or safeguards have been maintained in conformance with the code edition under which they were installed. Should the owner subsequently want to change the use or occupancy, Section 3408 provides the guidelines for doing so. Additional provisions are contained in Section 3412, which are primarily intended to provide alternatives that will maintain or increase the current degree of public safety, and are to be used in those situations where full compliance with the *International Building Code®* (IBC®) is not possible.

In existing buildings, there are times where full and strict compliance with the minimum accessibility requirements for new construction cannot be achieved. There is little chance the alteration can be made to be fully accessible because of existing structural conditions that require removal or alteration of a load-bearing member that is an essential part of the structural frame, or because of the fact that there are other existing physical or site restraints that limit or prohibit a fully accessible structure. The code refers to such a situation as *technically infeasible*. In those situations where full accessibility is deemed to be technically infeasible, exceptions are provided allowing for a reduced level of access.

Unless modified by jurisdictional amendment, the use of this chapter for the regulation of existing buildings is optional. Another code published by the International Code Council, the *International Existing Building Code®* (IEBC®), provides an alternative approach to the regulation of existing buildings that undergo repair, alterations, additions, or a change of occupancy. The IEBC, referenced in Section 3401.6 of the IBC, is deemed to be an acceptable alternative to the provisions of Chapter 34. The primary intent of the IEBC is to encourage the use and reuse of existing buildings by providing flexibility to permit the use of alternative approaches to achieve compliance with the minimum requirements. The value of the IEBC extends beyond its potential as an alternative compliance tool. It can be a valuable resource to assist in the application of IBC Chapter 34, particularly in regard to buildings that undergo a change in use or occupancy.

Section 3404 *Alterations*

The provisions in this section are, for the most part, simple and direct. Any alteration work must be in full compliance with the code requirements for new construction. Therefore, all construction performed in the alteration to an existing building must comply with all of the appropriate provisions of this code. The IBC also intends that the alteration work will not cause the existing building to be in violation of the code. For example, any new alterations should not remove or block existing exits. On the other hand, those portions of an existing building that are not involved or affected in the alteration work are not required to comply with the current code provisions. Simply, the portions of the building under construction are regulated as for a new building; the portions that are not part of the work are exempted.

The basic requirement of Section 3404.3 is that any structural alteration or repair made to an existing building cannot increase the force in any structural element by more than 5 percent, unless the new force levels remain in compliance with the IBC for new structures. The strength of any existing structural elements cannot be lessened to a level below that required by the code for new structures.

Accessibility for Existing Buildings

Section 3408 *Change of Occupancy*

Because each occupancy group contemplates a different level of hazard, the IBC intends that when a use is changed, and particularly when the change in use increases the level of hazard, the building must be brought to conform to the requirements of the code for the proposed occupancy.

The building official is granted broad authority in determining whether or not the new or proposed use is less hazardous than the existing use. Therefore, each change of occupancy must be analyzed based on the hazards consequent to the new use. In addition, the fire- and life-safety and structural features of the building must be evaluated to determine if they are as required by the code for the new use. It is the intent of the code that the change of occupancy shall not be made unless the building is made to comply with the requirements of the code that are related to the proposed use. Thus, the structural requirements of the code must also be met where they are more restrictive for the new use.

Wide authority is also granted to the building official in using judgment to determine which code requirements will be exacted. The building official should develop the rules that are to be followed upon a change of occupancy so that consistent enforcement will result. If it is determined that the proposed or new use is less hazardous than the existing use, the building official may approve a change in occupancy without requiring that the building conform to all the requirements of the code for the new use. Given this wide authority, a building official may require that the building comply with only those requirements for the new use that are important to fire and life safety. Under these circumstances, the IBC permits the building official to choose not to require those changes that are deemed unnecessary as far as fire and life safety are concerned, based on the character of the new use.

A practical application. In closing, let us briefly discuss a change of occupancy determination that building officials face occasionally—the conversion of a single-family dwelling to an office use. Districts that were formerly residential in nature have seen the expansion of neighboring commercial areas; now someone wants to save an attractive old Victorian residence and convert it into offices. To what extent does the building need to be modified to meet the intent of this section? To determine this, the building official must evaluate the fire- and life-safety features and structural features of the building to determine if existing conditions satisfy the code, and to determine to what extent the new use poses a greater or a lesser hazard than the existing use. For example, the adequacy of the floor system would need to be reviewed because a dwelling allows for a lower design live load than does an office. Thus, floor system structural capacity is an important factor to consider. Another item that always arises with such conversions is what to do about exterior wall and opening protection. Many of the dwellings that are converted are historical in nature or, at the very least, are a part of the community's cultural fabric. As a result, there is an understandable desire to maintain their present appearance. In such cases, some jurisdictions permit the use of exterior fire sprinklers to serve as exterior protection. In other jurisdictions, a comparison is made between the fire- and life-safety hazards of the two occupancies. Such things as fire loading, alertness of the building occupants, building size, and similar considerations may be used by the building official in determining if the converted use poses a greater or lesser hazard than the former one.

Section 3411 *Accessibility for Existing Buildings*

This section addresses the level of accessibility required when maintenance or a change of occupancy occurs, or where additions or alterations are made to existing buildings. In general, the provisions contained in this section are less detailed than those for new buildings and place a greater reliance on determinations made by the building official.

Asking a number of questions may be helpful in determining compliance with Section 3411.6 on alterations:

1. What is the current level of accessibility?
2. What is the scope of the alteration?
3. How will the alteration affect the current level of accessibility?
4. Does the alteration increase the accessibility or use of the building?
5. If it is technically infeasible, what is the maximum extent the alteration can be made to meet accessibility requirements?

A number of building elements are addressed in this section because of the importance of their features. To maintain a certain degree of accessibility to these elements, modifications to the standard provisions lessen the degree of alteration required. However, alterations must occur at these minimum levels, regardless of their perceived technical infeasibility. The following elements are addressed in the scoping provisions of Section 3411.8 as they relate to alterations of a building:

1. Entrances
2. Elevators and platform lifts
3. Stairs, ramps, and escalators
4. Performance areas
5. Dwelling or sleeping units
6. Jury boxes and witness stands
7. Toilet rooms
8. Dressing, fitting, and locker rooms
9. Fuel dispensers
10. Thresholds

Historic buildings and facilities that undergo alterations or a change of occupancy are regulated for accessibility by Section 3411.9. The IBC describes a historic building as one that is listed or is eligible for listing in the National Register of Historic Places, or is designated as historic under an appropriate state or local law. The provisions address key building elements such as accessible routes, entrances, and toilet facilities. Where the historic significance of the building will be threatened or destroyed through full accessibility compliance, modifications for these key elements are permitted.

Section 3412 *Compliance Alternatives*

In the conservation and rehabilitation of existing building stock, it is often difficult to fully comply with the provisions of Chapters 2 through 33 of the IBC. This section is designed to provide an alternative means for meeting the intent and purpose of the code, which is safeguarding the public's safety, health, and general welfare. Unless an unsafe condition exists as determined by the building official, compliance with this section shall be considered acceptable. The compliance alternatives addressed in this section apply to all occupancies other than Group H or I.

Under the method presented in this section, the three fundamental categories of fire safety, means of egress, and general safety must be evaluated. These categories are further broken down into more specific issues. Fire safety is considered to be structural fire-resistance, automatic fire detection, fire alarm, and fire suppression system features. Means of egress includes the configuration, characteristics, and support features of exiting systems. The parameters of fire safety and means of egress are included within the general safety category.

In the evaluation process, the following issues are addressed:

1. Building height and area
2. Compartmentation
3. Tenant and dwelling unit separations
4. Corridor walls
5. Vertical openings
6. Heating, ventilating, and air-conditioning systems
7. Automatic fire detection and fire alarm systems
8. Smoke control
9. Means of egress capacity and number
10. Travel distance and dead ends
11. Elevator control
12. Means of egress emergency lighting
13. Mixed occupancies
14. Sprinklers and standpipes
15. Incidental uses

After all of the building safety parameters have been evaluated, the values of the parameters are tabulated based on the three major categories. The score of each category, based on the tabulations, is compared with the mandatory safety score for each category. The result is quite simple—it either passes or fails.

KEY POINTS

- A building in existence when a new code edition is adopted has the right to have its existing use or occupancy continued, provided all devices and safeguards have been maintained in conformance with the code edition under which they were installed.
- The IEBC is deemed to be an acceptable alternative to the provisions of Chapter 34.
- Existing buildings may be altered, repaired, or modified without complying with all provisions of the code as long as the new work complies.
- The IBC intends that when a use is changed, and particularly when the change in use increases the level of hazard, the building must be brought to conform to the requirements of the code for the proposed occupancy.
- The accessibility provisions for existing buildings tend to be less detailed than those for new buildings.
- Through the use of compliance alternatives, an alternate means is available for meeting the intent and purpose of the code.

CHAPTER 35

REFERENCED STANDARDS

35 Referenced Standards

The *International Building Code®* (IBC®) is, for the most part, a performance-based code. Thus, the IBC contains numerous references to standards that are intended to assist in the application of the code. Where standards are referenced in the body of the IBC, they are considered a part of the code. Thus, when a jurisdiction adopts the IBC as its building code, it also adopts the standards identified in this chapter.

Standards can be divided into several different categories—structural engineering standards, materials standards, installation standards, and testing standards. Developed by a consensus process, the standards referenced in the IBC are all identified in this chapter. Organized by the promulgating agency of the standard, the chapter provides the standard number, standard title, and the code section or sections where the standard is referenced in the IBC.

Where one of the listed standards is referenced in the IBC, only the applicable portions of the standard relating to the specific code provision are in force. For example, Chapter 11 references ICC A117.1 for the technical requirements relating to accessibility. The provisions of ICC A117.1 address accessible telephones; however, Chapter 11 does not require telephones to be accessible. Therefore, that portion of the standard is not considered part of the IBC, and accessible telephones are not required. Accessible telephones are regulated in Appendix E, so if the jurisdiction adopts the appendix chapter, the technical provisions of ICC A117.1 for telephones will be in force. As a reminder, Section 102.4 also indicates that where the provisions of the code and a standard are in conflict, the provisions of the code apply.

APPENDIXES

Appendix A Employee Qualifications
Appendix B Board of Appeals
Appendix C Group U Agricultural Buildings
Appendix D Fire Districts
Appendix E Supplementary Accessibility Requirements
Appendix F Rodentproofing
Appendix G Flood-Resistant Construction
Appendix H Signs
Appendix I Patio Covers
Appendix J Grading
Appendix K Administrative Provisions
Appendix L Earthquake Recording Instrumentation
Appendix M Tsunami-Generated Flood Hazard

Appendixes

> The appendix chapters to the *International Building Code®* (IBC®) contain subjects that have been determined to be an optional part of the code rather than mandatory, with each jurisdiction adopting all, parts of, or none of the appendix chapters—depending on its needs for enforcement in any given area. The provisions of IBC Section 101.2.1 indicate that the requirements contained in the appendixes are only applicable where specifically adopted by the jurisdiction. It is important that each jurisdiction review the appendix chapters in detail prior to their adoption to ensure their appropriateness.

Appendix A *Employee Qualifications*

The provisions of this appendix are intended to assist the jurisdiction in qualifying individuals to be employed in key roles in the Department of Building Safety. Employee qualifications are provided for the position of building official as well as those for chief inspectors, inspectors, and plans examiners. An overview of the qualifications is shown in Figure A-1.

In addition to the education and experience criteria, an important consideration is the professional qualification obtained through certification. A comprehensive certification program is available through the International Code Council (ICC) that recognizes individuals for their knowledge relating to code enforcement.

Position	Experience			Certification
	Total Years	Supervisory Years	Type	
Building official	10	5	Architect, engineer, inspector, contractor, superintendent of construction	Certified building official
Chief inspector				Certified inspector for appropriate trade: building, plumbing, mechanical, electrical, combination
Inspector	5	—	Types listed above plus foreman or mechanic in charge of construction	
Plans examiner				Certified plans examiner for appropriate trade

Figure A-1 Employee qualifications.

Appendix B *Board of Appeals*

This appendix expands on the provisions of Section 113 relating to the board of appeals. Issues dealing with the filing of an appeal, the board membership, meeting notices, and board decisions are also addressed.

This appendix specifies that the board consist of five individuals, representing various disciplines or professions. In addition, two alternate members should be appointed to serve in the absence or disqualification of a regular member. All individuals are to be registered design professionals or contractors, qualified by registration or experience to rule on technical matters that may come before the board. The following disciplines are identified in Section B101.2.2:

1. Architecture or building construction
2. Structural engineering

3. Plumbing and mechanical engineering or contracting
4. Electrical engineering or contracting
5. Fire-protection engineering or contracting

It is important that all hearings before the board are considered open meetings and are available to all interested parties. The appellant, the appellant's representative or counsel, the building official, and all other persons who have an affected interest in the decision shall be permitted to address the board. In order to overturn or modify the decision of the building official, a minimum two-thirds vote of the board is required. Unless acceptable to the appellant, all five members must be present for the board to act on an appeal, with at least four concurring votes necessary to modify or reverse the building official's decision.

Appendix C *Group U Agricultural Buildings*

The provisions of Appendix C were developed to address the needs of those jurisdictions (primarily unincorporated county territory) whose primary development is agricultural. In these cases, agricultural property usually consists of large tracts of land on which agricultural buildings are placed, usually with large open spaces and with essentially no congestion. Therefore, the provisions for agricultural buildings classify the structures as Group U occupancies and include barns, shade structures, grain silos, stables, and horticultural structures, as well as buildings used for livestock and poultry shelters, equipment and machinery storage, and milking operations.

Because of the generally large open spaces that usually surround the buildings and the relatively low occupant load, the limitations imposed on construction, height, area, mixed uses, and exiting are generally more liberal than the requirements in the body of the code for Group F or S occupancies that would generally otherwise apply.

Appendix D *Fire Districts*

Maintained by the *International Fire Code®* (IFC®) and its code change committee, this appendix is available for adoption by jurisdictions wishing to establish fire districts. The use of fire districts provides a method to address fire hazards that are created by a variety of conditions, with the primary concerns based on occupancy and structure density. The provisions apply to new buildings built within the fire district, as well as those buildings undergoing alterations or a change of occupancy.

It is necessary to first establish the territory that is to be included within the fire district. The code identifies three basic types of areas that are of importance in the regulation of fire districts. These include adjoining blocks, buffer zones, and developed blocks. The specifics of each of these areas are illustrated in Figure D-1.

Those buildings already existing when a fire district is established may not be increased in height or area unless in compliance with the code for new buildings. Any new construction must also be of a type permitted within the fire district. Alterations may not increase the level of fire hazard in the building, nor may the occupancy be changed to a classification that is not permitted in the fire district. Any buildings located partially in the fire district are to be regulated as if they are in the district, provided at least 50 percent of the

Appendixes

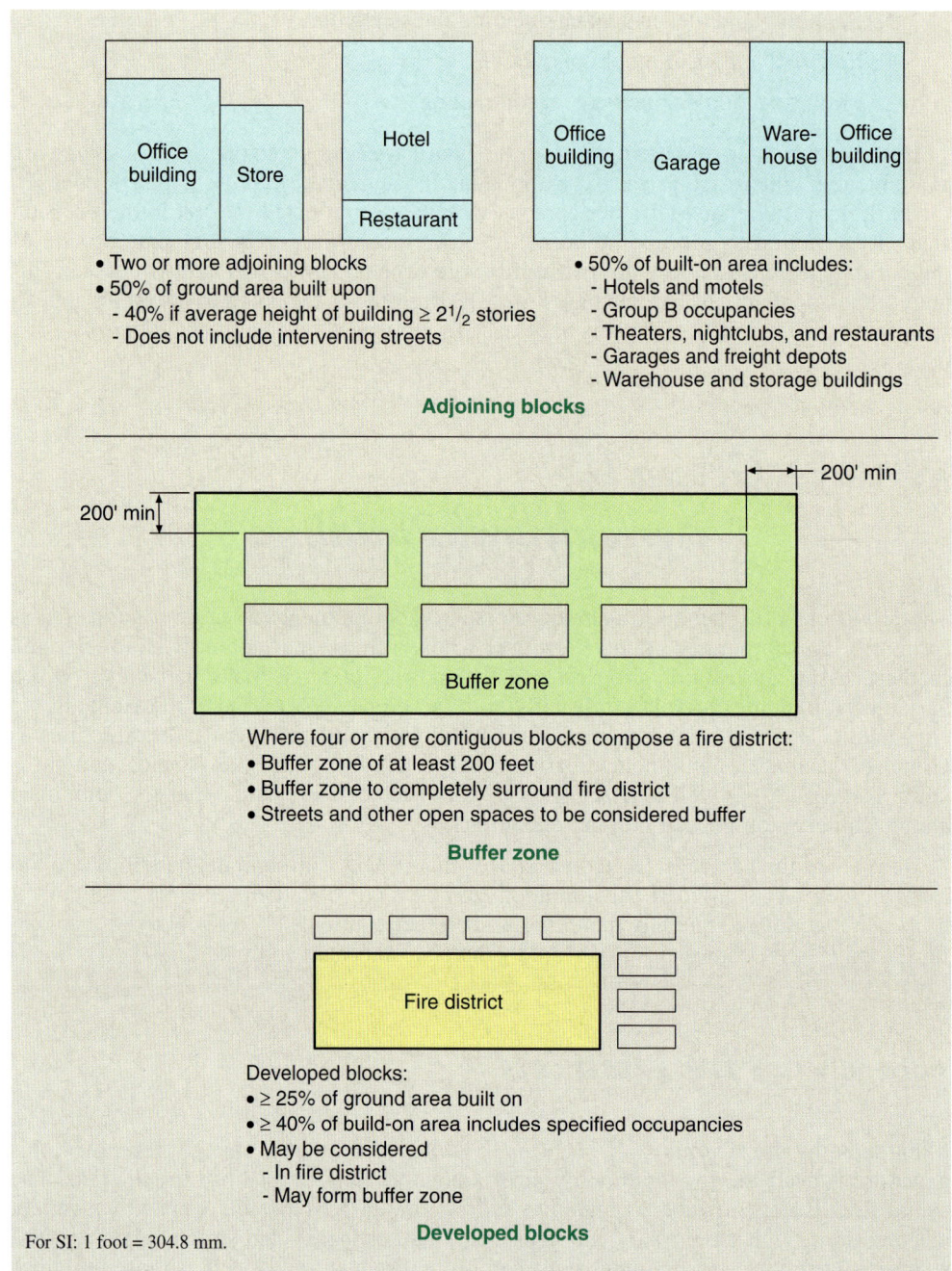

Figure D-1
Fire districts.

structure lies within the district, or the building extends more than 10 feet (3,048 mm) inside the fire district's boundaries.

Section D105 identifies a number of types of uses and structures permitted within a fire district that might otherwise be excluded. The listed uses are all relatively minor in nature and do not pose a significant hazard to the fire-safety level that is mandated. Such uses include small private garages and sheds, fences, tanks and towers, small greenhouses, and wood decks. This section also permits a limited amount of alteration on dwellings of Type V construction.

Appendix E *Supplementary Accessibility Requirements*

This appendix addresses the design and construction of those accessible facilities not typically addressed by a building code. Although it is important that all reasonable efforts are made to make buildings accessible and usable, the body of the code in Chapter 11 contains only those requirements that directly relate to structures. Therefore, additional criteria are provided in this appendix to expand on the other accessibility features of the built environment.

Many of the facilities regulated by this chapter involve furnishings or equipment. In Section E103.1, the code requires that an accessible route be provided to a raised platform used as a head table or speaker's lectern. Section E104.2 regulates transient lodging features such as accessible beds and access to such beds, while communications features in Group I-3 occupancies and transient lodging facilities are addressed in Section E104.3.

Water coolers, portable toilet and bathing rooms, laundry equipment, vending machines, mailboxes, automatic-teller machines, and fare machines are other possible features of a building that, through the regulation of this section, can be made more usable for individuals with physical disabilities. Telephones are fully addressed in Section E106, including provisions for wheelchair access, volume controls, and TTYs. Section E107.2 mandates that where permanent signage designates the use or description of a room or area, tactile identification is also required. Directional and informational signs, as well as other special types of signage, are also addressed.

Three of the remaining sections of this appendix regulate specific types of uses or buildings. Bus stops and bus shelters must be designed and constructed in a manner that makes them accessible. Fixed transportation facilities, such as stations for rapid rail, light rail, commuter rail, high-speed rail, and other fixed-guideway systems, must selectively have station entrances, signage, fare machines, platforms, TTYs, track crossings, public-address systems, and clocks that are accessible or usable. Some of these same features are regulated for airports as well.

Appendix F *Rodentproofing*

In an effort to reduce the possibility of rodents entering a building, this appendix sets forth construction methods to seal those potential entry points. The provisions apply not only to habitable and occupiable rooms but also to any spaces containing feed, food, or foodstuffs. The obvious intent is to prevent unsanitary conditions and the potential spread of disease that may follow.

All openings in the foundation walls are to be covered or sealed in a prescribed manner to prevent the passage of rodents. Doors and windows are regulated when located adjacent to ground level. It is also the intent of this appendix that an apron or similar protective barrier be installed where the foundation wall is not continuous. The intent of this appendix is an attempt to eliminate all potential avenues for rodent entry that occur around the exterior of a building.

Appendixes

Appendix G *Flood-Resistant Construction*

Most jurisdictions in the United States have specific areas that are subject to flood conditions. This appendix is designed to reduce those losses, both public and private, that occur because of flooding. Administrative procedures and land-use limitations are set forth, intending to meet or exceed the regulations of the National Flood Insurance Program (NFIP).

The building-sciences provisions for flood-resistant design and construction are located in Section 1612. In conjunction with the provisions of this appendix, the regulations are consistent with the NFIP regulations.

Appendix H *Signs*

The design and installation of outdoor signs is regulated by this appendix. Signs can come in many shapes and sizes, and are used for many purposes. This appendix classifies signs based on their location such as ground signs, roof signs, wall signs, projecting signs, marquee signs, and portable signs. The types of signs, including internally illuminated signs, combustible signs, and animated devices, are also addressed.

This appendix identifies the areas of concern when signs are placed on structures. It is important that any exit signs, fire escapes, or egress openings remain unobstructed. Required natural ventilation openings must also remain available. Signs must be able to withstand all imposed loads, including any wind or seismic loads that may be encountered. The combustibility of signs is also regulated, with specific provisions for plastic materials.

Appendix I *Patio Covers*

Patio covers regulated by Appendix I are limited to one-story structures not exceeding 12 feet (3,657 mm) in height. Enclosure walls may have any configuration, provided the open area of the longer wall and one additional wall is equal to at least 65 percent of the area below a minimum of 6 feet, 8 inches (2,032 mm) of each wall, measured from the floor. Openings may be enclosed with insect screening, translucent or transparent plastic not more than $^1/_8$ inch (3.2 mm) in thickness, or glass conforming to Chapter 24.

Patio covers may be detached or attached to dwelling units. Patio covers shall be used only for recreational, outdoor living purposes and not as carports, garages, storage rooms, or habitable rooms.

Exterior openings required for light or ventilation may open into a patio structure conforming to this section. Where emergency egress or rescue openings from sleeping rooms lead to a patio structure, the structure shall be unenclosed. Where an exit from the dwelling unit passes through the patio structure, the structure shall be unenclosed or exits shall be provided in conformance with Chapter 10.

Patio covers shall be designed and constructed to sustain the applicable snow loads or all dead loads plus a vertical live load of 10 pounds per square foot (0.48 kN/m^2), whichever is greater. The minimum wind and seismic loads shall also be considered in the design.

A patio cover may be supported on concrete slab on grade without footings, provided the slab is not less than $3^1/_2$-inches (89-mm) thick, and further provided that the columns do not support live and dead loads in excess of 750 pounds (3.34 kN) per column.

Appendix J *Grading*

Not every jurisdiction is located in an area where the topography of the terrain requires extensive grading operations on private property. In those areas where developers need to grade private property, Appendix J provides appropriate administrative and technical regulations to assure the jurisdiction of reasonable safety against slope failure, landslides, and other soil failure hazards.

Appendix K *Administrative Provisions*

Appendix K is provided to allow those communities who adopt NFPA 70, the *National Electrical Code*® (NEC®), to include administrative provisions that will assist in their implementation and enforcement. These provisions assist in the administration of the NEC by providing administrative language that correlates with that of the *International Codes*. In addition, the provisions established in Section K111 address technical issues that are additions or modifications to the requirements of the NEC.

Appendix L *Earthquake Recording Instrumentation*

Earthquake recording instrumentation measurements provide fundamental information needed to cost-effectively improve understanding of the seismic response and performance of buildings subjected to earthquake ground motions. The language of the new provision in the IBC requiring earthquake recording instrumentation originated with the 1997 Uniform Building Code (UBC). The requirement only applies to newly constructed buildings of a specified size and located where the 1-second spectral response acceleration, S_1, is greater than 0.40. Because the provision is in an appendix chapter, it is not mandatory unless specifically adopted by the jurisdiction.

Appendix M *Tsunami-Generated Flood Hazard*

The areas designated on State or National Oceanic and Atmospheric Administration (NOAA) Tsunami Hazard Inundation Maps are most likely to suffer significant damage during a design tsunami event. Given the potentially serious life-safety risk presented to structures within these areas, the intent of the provisions in Appendix M is to limit the presence of high-hazard and high-occupancy structures (Risk Categories III and IV) within the designated Tsunami Hazard Zone. Buildings within the designated hazard zone are only permitted under certain conditions. A vertical evacuation tsunami refuge is permitted when constructed in accordance with FEMA P646 or where critical facilities are located within the hazard zone to fulfill their function, and they incorporate adequate structural and emergency evacuation features. Vertical evacuation is a central part of the National

Tsunami Hazard Mitigation Program, driven by the fact that there are coastal communities along the West Coast of the United States that are vulnerable to tsunamis that could be generated within minutes of an earthquake on the Cascadia Subduction Zone. Vertical evacuation structures provide a means to create areas of refuge for communities in which evacuation out of the inundation zone is not feasible. The referenced FEMA guide includes information to assist in the planning and design of tsunami vertical evacuation structures. Because the provision is in an appendix chapter, it is not mandatory unless specifically adopted by the jurisdiction.

METRIC CONVERSION TABLE

Metric Units, System International (SI)
Soft Metrication
Hard Metrication

Metric Units, System International (SI)

The most widely used system of units and measures around the world is the Systeme International d'Unites (SI), which is the modern form of the metric system. The other system of measurement is the U.S. Customary System Units, also known as "English Units," consisting of the mile, foot, inch, gallon, second, and pound. Although the English system is gradually being replaced by the metric system in some sectors of U.S. industry, the full conversion of the U.S. to the metric system is still incomplete.

Using metric units or providing metric equivalents is important because of the following reasons:

1. Use of metric units facilitates understanding and communication in technical areas such as engineering, architecture, building codes, and other scientific areas at a global level.

2. Use of metric units is simpler because variations from smaller to larger units or vice versa are in multiples of 10, but in English units the multipliers could vary with unit and with subject. For example, smaller units of an inch are $1/2$, $1/4$, $1/8$, or $1/16$ of an inch, each a multiplier of 2 larger than the other. Conversion of inches to feet is at a multiplier of 12 and from feet to yards at a multiplier of 3. For metric, the smaller unit of a centimeter is a millimeter, which is a centimeter divided by 10. Conversion of centimeters to decimeters is by multiplying a centimeter by 10, and conversion of decimeters to meters is by multiplying by 10, and so on. Accordingly, computations and problem solving are prone to less error.

The conversion to metric units can take two forms—soft metrication and hard metrication.

Soft Metrication

Soft metrication is the use of metric units in specifying measurements, sizes, and other dimensions without changing product sizes and without changing the everyday practice of using English units. For example, a wood-stud member commonly used is a 2 × 4, which is actually 1.5 inches × 3.5 inches. To report or to specify the actual size of this member in metric, a soft conversion of 38 mm × 89 mm is used (rounded from actual 38.1 and 88.9). Another example could be the load-bearing pressure of clay soils of 1,500 pounds per square-foot being reported as 72 kPa (kilo pascal).

Hard Metrication

Hard metrication goes beyond soft metrication and converts production and manufacturing based on metric sizes. For example, instead of manufacturing 2 × 4 wood studs of 1.5 inches × 3.5 inches (38.1 mm × 88.9 mm, actual dimensions), wood studs of 40 mm by 90 mm might be manufactured. Another example is $1/2$-inch diameter (12.7-mm) U.S. size automotive bolts versus 13-mm metric bolts, which are manufactured with a diameter of 13 mm. The production of other structural or nonstructural members such as structural steel, plywood, nails, pipes, ducts, insulation panels, and all other such elements would also be done in metric rather than manufacturing in English units and reporting metric equivalents.

Hard Metrication

More information on the SI system in the United States is available from the National Institute of Standards and Technology (NIST), an agency of the U.S. Department of Commerce: http://physics.nist.gov/cuu/Units/index.html or http://physics.nist.gov/Pubs/SP811/contents.html

Unit Conversion Tables
SI Symbols and Prefixes

BASE UNIT		
Quantity	Unit	Symbol
Length	meter	m
Mass	kilogram	kg
Time	second	s
Electric current	ampere	A
Thermodynamic temperature	kelvin	K
Amount of substance	mole	mol
Luminous intensity	candela	cd

SI SUPPLEMENTARY UNITS		
Quantity	Unit	Symbol
Plane angle	radian	rad
Solid angle	steradian	sr

SI PREFIXES		
Multiplication Factor	Prefix	Symbol
1 000 000 000 000 000 000 = 10^{18}	exa	E
1 000 000 000 000 000 = 10^{16}	peta	P
1 000 000 000 000 = 10^{12}	tera	T
1 000 000 000 = 10^{9}	giga	G
1 000 000 = 10^{6}	mega	M
1 000 = 10^{3}	kilo	k
100 = 10^{2}	hecto	h
10 = 10^{1}	deka	da
0.1 = 10^{-1}	deci	d
0.01 = 10^{-2}	centi	c
0.001 = 10^{-3}	milli	m
0.000 001 = 10^{-6}	micro	μ
0.000 000 001 = 10^{-9}	nano	n
0.000 000 000 001 = 10^{-12}	pico	p
0.000 000 000 000 001 = 10^{-15}	femto	f
0.000 000 000 000 000 001 = 10^{-18}	atto	a

Metric Conversion Table

SI DERIVED UNITS WITH SPECIAL NAMES			
Quantity	Unit	Symbol	Formula
Frequency (of a periodic phenomenon)	hertz	Hz	1/s
Force	newton	N	kg•m/s^2
Pressure, stress	pascal	Pa	N/m^2
Energy, work, quantity of heat	joule	J	N-m
Power, radiant flux	watt	W	J/s
Quantity of electricity, electric charge	coulomb	C	A-s
Electric potential, potential difference	volt	V	W/A
Electromotive force	farad	F	C/V
Capacitance	ohm	Ω	V/A
Electric resistance	siemens	S	A/V
Conductance	weber	Wb	V-s
Magnetic flux	tesla	T	Wb/m^2
Magnetic flux density	henry	H	Wb/A
Inductance	lumen	lm	cd•sr
Luminous flux	lux	lx	lm/m^2
Luminance	becquerel	Bq	1/s
Activity (of radionuclides)	gray	Gy	J/kg
Absorbed dose			

Conversion Factors

To convert	to	multiply by
LENGTH		
1 mile (U.S. statute)	km	1.609 344
1 yd	m	0.9144
1 ft	m	0.3048
	mm	304.8
1 in	mm	25.4
AREA		
1 mile^2 (U.S. statute)	km^2	2.589 998
1 acre (U.S. survey)	ha	0.404 6873
	m^2	4046.873
1 yd^2	m^2	0.836 1274
1 ft^2	m^2	0.092 903 04
1 in^2	mm^2	645.16
VOLUME, MODULUS OF SECTION		
1 acre ft	m^3	1233.489
1 yd^3	m^3	0.764549
100 board ft	m^3	0.235 9737
1 ft^3	m^3	0.028316 85
	L(dm^3)	28.3168
1 in^3	mm^3	16 387.06
	mL (cm^3)	16.3871
1 barrel (42 U.S. gallons)	m^3	0.158 9873
(FLUID) CAPACITY		
1 gal (U.S. liquid)*	L**	3.785 412
1 qt. (U.S. liquid)	mL	946.3529
1 pt. (U.S. liquid)	mL	473.1765
1 fl oz (U.S.)	mL	29.5735
1 gal. (U.S. liquid)	m^3	0.003 785 412
*1 gallon (UK) approx. 1.2 gal (U.S.)	**1 liter approx. 0.001 cubic meter	
SECOND MOMENT OF AREA		
1 in^4	mm^4	416 231 4
	m^4	416 231 4 × 10^{-7}
PLANE ANGLE		
1° (degree)	rad	0.017 453 29
	mrad	17.453 29
1′ (minute)	urad	290.8882
1″ (second)	urad	4.8481 37
VELOCITY, SPEED		
1 ft/s	m/s	0.3048
1 mile/h	km/h	1.609 344
	m/s	0.447 04

Metric Conversion Table

To convert	to	multiply by
VOLUME RATE OF FLOW		
1 ft^3/s	m^3/s	0.028 316 85
1 ft^3/min	L/s	0.471 9474
1 gal/min	L/s	0.063 0902
1 gal/min	m^3/min	0.0038
1 gal/h	mL/s	1.051 50
1 million gal/d	L/s	43.8126
1 acre ft/s	m^3/s	1233.49
TEMPERATURE INTERVAL		
1° F	°C or K	0.555 556 5/9 °C = 5/9 K
EQUIVALENT TEMPERATURE (toC = TK − 273.15)		
t_{oF}	t_{oC}	t_{oF} 9/5 t_{o}C + 32
MASS		
1 ton (short***)	metric ton	0.907 185
	kg	907.1847
1 lb	kg	0.453 5924
1 oz	g	28.349 52
*** 1 long ton (2,240 lb)	kg	1016.047
MASS PER UNIT AREA		
1 lb/ft^2	kg/m^2	4.882 428
1 oz/yd^2	g/m^2	33. 905 75
1 oz/ft^2	g/m^2	305. 1517
DENSITY (MASS PER UNIT VOLUME)		
1 lb/ft^3	kg/m^3	16.01846
1 lb/yd^3	kg/m^3	.593 2764
1 ton/yd^3	t/m^3	1.186 553
FORCE		
1 tonf (ton-force)	kN	8.896 46
1 kip (1,000 lbf)	kN	4.448 22
1 lbf (pound-force)	N	4.448 22
MOMENT OF FORCE, TORQUE		
1 lbf•ft	N•m	1.355 808
1 lbf•in	N•m	0.112 9848
1 tonf•ft	kN•m	2.711 64
1 kip•ft	kN•m	1.355 82

To convert	to	multiply by
FORCE PER UNIT LENGTH		
1 lbf/ft	N/m	14.5939
1 lbf/in	N/m	175.1268
1 ton/ft	kN/m	29.1878
PRESSURE, STRESS, MODULUS OF ELASTICITY (FORCE PER UNIT AREA) (1 Pa = 1 N/m^2)		
1 tonf/in^2	Mpa	13.7895
1 tonf/ft^2	kPa	95.7605
1 kip/in^2	Mpa	6.894 757
1 lbf/in^2	kPa	6.894 757
1 lbf/ft^2	Pa	47.8803
Atmosphere	kPa	101.3250
1 inch mercury	kPa	3.376 85
1 foot (water column at 32°F)	kPa	2.988 98
WORK, ENERGY, HEAT (1J = 1N • m = 1W • Ws)		
1 kWh (550 ft•lbf/s)	MJ	3.6
1 Btu (Int. Table)	kJ	1.055 056
	J	1055.056
1 ft•lbf	J	1.355 818
COEFFICIENT OF HEAT TRANSFER		
1 Btu/(ft^2•h•°F)	W/(m•K)	5.678 263
THERMAL CONDUCTIVITY		
1 Btu/(ft•h•°F)	W/(m^2•K)	1.730 735
ILLUMINANCE		
1 lm/ft^2 (footcandle)	lx (lux)	10.763 91
LUMINANCE		
1 cd/ft^2	cd/m^2	10.7639
1 foot lambert	cd/m^2	3.426 259
1 lambert	kcd/m^2	3.183 099

INDEX

Index

A

ACCESS OPENINGS
- Attic 1209.2
- Crawl space 1209.1
- Fire damper 716.4

ACCESSIBILITY . . . . 1007, Chapter 11, 3411, Appendix E
- Assembly 1007.1, 1108.2, 1109.11
- Controls 1109.13
- Detectable warnings 1109.10
- Detention and correctional facilities . . . . 1103.2.14, 1107.5.5, 1108.4.2, 3411.8.7, Appendix E
- Dining areas 1108.2.9, 1109.11
- Dressing rooms 1109.12.1
- Drinking fountains 1109.5
- Dwelling units 1103.2.4, 1105.1.6, 1107, 3411.8.7, 3411.8.8, 3411.8.9
- Egress (see ACCESSIBLE MEANS OF EGRESS) 1007
- Elevators 1007.2.1, 1007.4, 1007.7.3, 1109.7, 3001.3, 3411.8.2
- Employee work areas . . . 907.5.2.3.2, 1103.2.3, 1104.3.1
- Entrances 1105, 3411.8.1, 3411.9.3
- Exceptions 1103.2, 1104.4, 1107.7
- Existing buildings 1007.1, 1103.2.2, 3411, 3412.2.5
- Historic buildings 3411.9
- Judicial facilities 1108.4, 3411.8.7, 3411.8.10
- Kitchens 1109.4
- Lifts 1007.5, 1109.8, 3411.8.3
- Live/work unit 419.7, 1103.2.13
- Parking and passenger loading facilities 1106
- Performance areas 1108.2.8, 3411.8.6
- Press box 1104.3.2
- Ramps 1010, 3411.8.5
- Recreational facilities 1109.15
- Route 1003.3.4, 1104, 1107.4, 3411.7, 3411.9.1, 3411.9.2
- Saunas and steam rooms 1109.6
- Scoping . . . . 108.2, 116.1, 1101, 1103.1, 3411.1, Appendix E
- Seating 1108.2, 1109.11
- Service facility 1109.12, 3411.8.12
- Signage . . . . 1007.8 through 1107.11, 1110, Appendix E
- Sleeping units 1107, 1105.1.6, 3411.8.7, 3411.8.8, 3411.8.9
- Storage 1108.3, 1109.9
- Toilet and bathing facilities . . . 1107.6.1.1, 1109.2, 1109.3, 3411.8.11, 3411.9.4, Appendix E
- Transient lodging . . . . . . 1103.2.11, 1107.6.1, 3411.8.7, 3411.8.9, Appendix E
- Windows 1109.13.1

ACCESSIBLE MEANS OF EGRESS 1007
- Areas of refuge (see AREA OF REFUGE)
- Assembly 1007.1, 1028.8
- Elevators 1007.2.1, 1007.4, 1007.8
- Existing building 1007.1, 3411.6
- Exterior area for assisted rescue (see EXTERIOR AREAS FOR ASSISTED RESCUE)
- Horizontal exit (see HORIZONTAL EXIT)
- Platform lift 1007.5
- Required 1007.1
- Stairs 1007.3
- Signage . . . . . 1007.8 through 1007.11, 3002.3

ACCESSORY OCCUPANCIES 303.1.2, 303.1.4, 305.1.1, 312.1, 419.1, 508.2

ADDITION . . . . 3403, Appendix D
- Accessibility 3411.5
- Means of egress 3302, 3310

ADMINISTRATION . . . . Chapter 1

ADOBE CONSTRUCTION 2109.3

AGRICULTURAL BUILDINGS (see GROUP U) . . . . 312.1, 1103.2.5, Appendix C

AIRCRAFT HANGARS 412.4
- Aircraft paint hangars 412.6, 507.9
- Basements 412.4.2
- Construction 412.4.1, 412.6.2
- Fire area 412.4.6.2
- Fire suppression system 412.4.6, 412.6
- Heliports and helistops 412.7
- Residential 412.5
- Unlimited height and area 504.1, 507.9

AIRCRAFT-RELATED OCCUPANCIES . . . . 412, Appendix E
- Airport traffic control towers 412.3
- Alarms and detection 412.3.4, 907.2.22
- Construction type 412.3.2
- Egress 412.3.3
- Traffic control towers 412.3
- Type of construction 412.3.2

AISLE 1017
- Aisle accessways 1017.4, 1028.10
- Assembly seating 1017.2, 1028.6
- Bleachers 1028.1.1
- Business 1017.3, 1017.4
- Check-out 1109.12.2
- Converging 1028.9.3
- Egress 1017
- Folding and telescopic seating 1028.1.1
- Grandstands 1028.1.1
- Mercantile 1017.4
- Walking surfaces 1028.11
- Width 1028.9

ALARM SYSTEMS, EMERGENCY 908

ALARMS, FIRE (see FIRE ALARM AND SMOKE DETECTION SYSTEMS)

ALARMS, VISIBLE 907.5.2.3
   Common areas 907.5.2.3.1
   Employee work areas 907.2.5.3.2
   Group I-1 . . . . 907.5.2.3.3
   Group R-1 . . . . 907.5.2.3.3
   Group R-2 . . . . 907.5.2.3.4
   Public areas 907.5.2.3.1
ALARMS, VOICE 907.5.2.2
   Amusement buildings, special 411.6, 2702.2.1
   Covered and open mall buildings 402.7.4,
      907.2.20, 2702.2.14
   High-rise buildings 403.4.3, 907.2.13
   Occupant evacuation elevators 3008.10
   Special amusement buildings 411.6
   Underground buildings 405.9.1, 907.2.18,
      907.2.19
ALLOWABLE STRESS DESIGN 1604
   Load combinations 1605.3
   Masonry design 2101.2.1, 2107
   Wood design 2301.2, 2306
ALTERATIONS . . . . 3404, Appendix D
   Accessibility 3411.6, 3411.7, 3411.9
   Compliance alternatives 3412
   Means of egress 3302.1, 3310.2, 3411.6
ALTERNATING TREAD DEVICES 1009.13
   Construction 1009.13.2
   Equipment platform 505.5
   Technical production areas 410.5.3
**ALTERNATIVE MATERIALS,
DESIGN AND METHODS** 104.11
ALUMINUM 1404.5.1, 1604.3.5,
      Chapter 20
AMBULATORY CARE FACILITIES 422
   Alarm and detection 907.2.2
   Smoke compartment 422
AMUSEMENT BUILDING, SPECIAL 411
   Alarm and detection 411
   Classification 411
   Emergency voice/alarm
      communications system 907.5.2.2
   Exit marking 411
   Sprinklers protection 411
**ANCHOR STORE (see COVERED
AND OPEN MALL BUILDINGS)** 402
   Construction type 402.4.1
   Means of egress 402.8
   Occupant load 402.8.2.3
   Separation 402.4.2.2, 402.4.2.3
   Sprinkler protection 402.5
ANCHORAGE 1604.8
   Braced wall line sills 2308.3.3
   Concrete 1911, 1912
   Conventional light-frame
      construction 2308.11.3.1, 2308.12.7,
      2308.12.8, 2308.12.9
   Decks 1604.8.3
   Seismic anchorage for masonry chimneys . . . 2113
   Seismic anchorage for masonry
      fireplaces 2111
   Walls 1604.8.2
   Wood sill plates 2308.6
APARTMENT HOUSES 310.1
APPEALS 113
**ARCHITECT (see definition for
REGISTERED DESIGN PROFESSIONAL)**
ARCHITECTURAL TRIM 603.1
AREA, BUILDING . . . . Chapter 5
   Accessory uses 508.2.1
   Aircraft control towers 412.3.1
   Covered and open mall building 402.4.1
   Enclosed parking garage 406.6.1, 510.3
   Equipment platforms 505.3.1
   Incidental uses 509.3
   Limitations 503, 505
   Membrane structures 3102.4
   Mezzanines 505.2.1
   Mixed construction types 3102.6
   Mixed occupancy 508.2.1, 508.3.2, 508.4.2
   Modifications 506, 510
   Open mall building 402.4.1
   Open parking garage 406.5.4, 406.5.4.1,
      406.5.5, 510.2, 510.3,
      510.4, 510.7, 510.8, 510.9
   Private garages and carports . . . 406.3.1, 406.3.2
   Unlimited area 503.1.1, 503.1.3, 507
**AREA FOR ASSISTED RESCUE,
EXTERIOR (see EXTERIOR AREAS
FOR ASSISTED RESCUE)**
**AREA OF REFUGE (see ACCESSIBLE
MEANS OF EGRESS)**
   Requirements 1007.6, 1007.6.1, 1007.6.2
   Signage 1007.9, 1007.10, 1007.11
   Two-way communication . . . 1007.6.3, 1007.8,
      1007.11
   Where required 1007.2, 1007.3, 1007.4
ASSEMBLY OCCUPANCY (GROUP A) . . . 303, 1028
   Accessibility 1108.2, 1108.4, 1109.2.1,
      1109.15
   Alarms and detection 907.2.1
   Area 503, 506, 507, 508
   Bleachers (see BLEACHERS)
   Folding and telescopic seating (see BLEACHERS)
   Grandstands (see GRANDSTANDS)
   Group specific provisions
      A-1 . . . . 303.2
      A-2 . . . . 303.3
      A-3 . . . . 303.4
      A-4 . . . . 303.5
      A-5 . . . . 303.6
   Motion picture theater 409, 507.11
   Special amusement buildings 411
   Stages and platforms 410

ASSEMBLY OCCUPANCY (GROUP A) *(cont.)*
   Height 503, 504, 505, 506, 508, 510
   Incidental uses 509
   Interior finishes . . . . .Table 803.9, 804
   Means of egress
      Aisles. 1017.2, 1028.9, 1028.10
      Assembly spaces 1028
      Exit signs. 1011.1
      Guards 1013.2, 1028.14
      Main exit 1028.3
      Outdoors 1009.3, 1022.1
      Panic hardware 1008.1.10, 1008.2.1
      Travel distance 1014.3, 1016.2, 1021.1,
                                  1021.2, 1028.7
   Mixed occupancies 508.3, 508.4
      Accessory 508.2
      Education 303.1.3
      Live/work units 419
      Mall buildings. 402
      Other occupancies 303.1.1, 303.1.2
      Parking below/above 510.7, 510.9
      Religious facilities 303.1.4
      Special mixed 510.2
   Motion picture theaters 409, 507.11
   Occupancy exceptions 303.1.1, 303.1.2,
                   303.1.3, 303.1.4, 305.1.1, 305.2.1
   Plumbing fixtures 2902
   Risk category 1604.5
   Seating, fixed (see SEATING, FIXED)
   Seating, smoke-protected 1028.6.2
   Sprinkler protection 410, 507.3, 507.6,
                      507.7, 507.11, 903.2.1
   Stages and platforms 410, 905.3.4
   Standby/Emergency power systems . . . 2702.2.1
   Standpipes 905.3.2, 905.3.4, 905.5.1
   Unlimited area 402, 507.3, 507.3.1, 507.6,
                           507.7, 507.11
ASSISTED LIVING (see GROUP I-1) 308.3,
                                      310.6
   Sixteen or fewer residents
     (see Group R-4) . . . 308.3.1, 308.3.2, 310.5.1
ATMOSPHERIC ICE LOADS 1614
ATRIUM 404
   Alarms and detection 404.4, 907.2.14
   Enclosure 404.6, 707.3.5
   Interior finish 404.8
   Smoke control 404.5, 909
   Sprinkler protection 404.3
   Standby power 404.7
   Travel distance . . . 404.9, 1014.3, 1016.2, 1021.2
   Use 404.2
ATTIC
   Access 1209.2
   Combustible storage 413.2
   Draftstopping 717.4
   Insulation 719.3.1

   Unusable space fire protection 712.3.3
   Ventilation 1203.2
AUDITORIUM 303, 305.1.1
   Accessibility 1108.2
   Foyers and lobbies 1028.4
   Interior balconies 1028.5
   Motion picture projection rooms 409
   Stages and platforms 410
AUTOMOBILE PARKING GARAGE
   (see GARAGE, AUTOMOBILE PARKING) . . . 406
AWNINGS 3105
   Design and construction 3105.3
   Drainage, water 3201.4
   Encroachment, public right-of-way . . . 3202.2.3,
                          3202.3.1, 3202.4
   Materials 3105.4
   Motor vehicle service stations 406.7.2
   Plastic 2606.10

B

BALCONIES
   Accessibility 1108.2.4
   Assembly 1028.5
   Construction, exterior 1406.3
   Guards 1013.2
   Means of egress 1016.2, 1019, 1028.5
   Open mall building 402.4.3, 402.5
   Projection 705.2, 1406.3
   Public right-of-way encroachments . . . 3202.3.2,
                                 3202.3.3
   Travel distance 1016.2.1
BARRIERS
   Fire (see FIRE BARRIERS)
   Pedestrian protection 3306
   Smoke (see SMOKE BARRIERS)
   Vehicle 406.4.3, 1602.1,
                                1607.8.3
BASEMENT
   Aircraft hangars 412.4.2
   Area modification 506.4, 506.5
   Considered a story 202
   Emergency escape 1029.1
   Exits 1021.2.2
   Flood loads 1612
   Height modifications for 510.5
   Prohibited 415.6, 415.7.2, 415.10.5.2,
                                  418.1
   Rodent proofing . . . . Appendix F
   Sprinkler protection 903.2.11.1
   Waterproofing and dampproofing 1805
BASEMENT WALLS
   Concrete 1904.2
   Soil loads 1610
   Waterproofing and dampproofing 1805
BASIC WIND SPEED 1609

BLEACHERS 303.6, 1028.1.1, 3401.1
   Accessibility 1108.2
   Egress 1028.1.1
   Occupant load 1004.7
   Separation 1028.1.1.1
BLOCK (see GLASS UNIT MASONRY)
BOARD OF APPEALS . . . . 113, Appendix B
   Alternate members . . . . Appendix B
   Application for appeal . . . . Appendix B
   Board decision . . . . Appendix B
   Limitations on authority 113.2
   Membership of board . . . . Appendix B
   Notice of meeting . . . . Appendix B
   Qualifications . . . . 113.3, Appendix B
BOLTS 2204.2
   Anchors 1908, 1909, 2204.2.1
BONDING, MASONRY 2103.9
BRACED WALL LINE 2308.3
   Bracing 2308.3
   Seismic requirements 2308.12.2, 2308.12.4,
                                     2308.12.6
   Sill anchorage 2308.3.3
   Spacing 2308.3.1
   Support 2308.3.4
BRACED WALL PANEL 2308.9.3
   Alternative bracing 2308.9.3.1, 2308.9.3.2
   Connections 2308.3.2
BRICK (see MASONRY)
BUILDING
   Area (see AREA, BUILDING) 502.1, 503,
                        505, 506, 507, 508, 510
   Demolition 3303
   Existing . . . . Chapter 34
   Fire walls 706.1
   Height (see HEIGHT, BUILDING) . . . 502.1, 503,
                       504, 505, 506, 508, 510
   Occupancy classification . . . . Chapter 3
   Party walls 706.1.1
BUILDING DEPARTMENT 103
BUILDING OFFICIAL
   Approval 202
   Duties and powers 103
   Qualifications . . . . Appendix A
   Records 104.7
BUILT-UP ROOFS 1507.10
BUSINESS OCCUPANCY (GROUP B) 304
   Alarms and detection 907.2.2
   Ambulatory health care facilities 305.2, 308.6,
                            310.5.1, 422
   Area 503, 505, 506, 507, 508
   Height 503, 504, 505, 506, 508, 510
   Incidental uses 509
   Interior finishes . . . . Table 803.9, 804
   Means of egress
      Aisles 1017.3, 1017.4
      Travel distance 1014.3, 1016.2, 1021.2

   Mixed occupancies 508.3, 508.4
      Accessory 303.1, 508.2
      Ambulatory care facilities 422
      Assembly 303.1.1, 303.1.2
      Educational 303.1, 304.1
      Live/work units 419
      Mall buildings 402
      Parking below/above 509.7, 509.8, 509.9
      Special mixed 510.2
   Occupancy exceptions 303.1.1, 303.1.2
   Plumbing fixtures 2902
   Risk category 1604.5
   Sprinkler protection 903.2.2
   Unlimited area 507.3, 507.4

C

CABLES, STEEL STRUCTURAL 2208
CALCULATED FIRE RESISTANCE
   (see FIRE RESISTANCE, CALCULATED)
CANOPIES 3105
   Design and construction 3105.3
   Drainage, water 3201.4
   Encroachment, public right-of-way 3202.3.1
   Materials 3105.4
   Motor vehicle service stations 406.5.3
   Permanent . . . . Appendix D
   Plastic 2606.10
CARBON MONOXIDE ALARMS
   AND DETECTION 908.7
CARE FACILITIES (see HEALTH CARE)
CARE SUITES 202, 407.4.3
CARPET
   Floor covering 804.2
   Textile ceiling finish 803.6
   Textile wall coverings 803.5
CATWALKS (see TECHNICAL
   PRODUCTION AREAS)
   Construction 410.3.2
   Means of egress 410.6.3
   Sprinkler protection 410.7
CEILING
   Height 406.4.1, 409.2, 909.20.4.3, 1003.2,
                1009.5, 1010.6.2, 1205.2.2, 1208.2
   Interior finish 803
   Penetration of fire-resistant assemblies 713.4,
                                 716.2, 716.6
CELLULOSE NITRATE FILM 409.1, 903.2.5.3
CERAMIC TILE
   Material requirements 2103.6
   Mortar 2103.11
CERTIFICATE OF OCCUPANCY . . 106.2, 111, 3408.2
CHANGE OF OCCUPANCY . . . . . 3408, Appendix D
   Accessibility 3411.4, 3412.2.5
CHILD CARE (see DAY CARE) 305.2,
                            308.6, 310.5.1

CHILDREN'S PLAY STRUCTURES........ 424
    Accessibility.................. 1109.15
    Covered and open mall building..... 402.6.3
CHIMNEYS......... 2101.3, 2111, 2112, 2113
    Factory-built................... 717.2.5
    Flashing........................ 1503.6
    Protection from adjacent construction.... 3307.1
CHURCHES (see RELIGIOUS WORSHIP, PLACES OF)
CIRCULAR STAIRS (see CURVED STAIRWAYS)
CLAY ROOF TILE................... 1507
    Testing........................ 1711.2
CLINIC
    Hospital (see INSTITUTIONAL I-2)..... 308.3
    Outpatient (see AMBULATORY
        HEALTH CARE FACILITIES)....... 202, 304.1, 422
CODES...... 101.2, 101.4, 102.2, 102.4, 102.6, Chapter 35
COLD STORAGE (see FOAM PLASTIC INSULATION)
COLD-FORMED STEEL............... 2210
    Light-frame construction........... 2211
    Special inspection...... 1705.2.2, 1705.2.2.2, 1705.10.2, 1705.11.3
COMBUSTIBLE DUSTS.... 307.4, 414.5.1, 415.8.1, 415.8.2
COMBUSTIBLE LIQUIDS..... 307.1, 307.4, 307.5, 414.2.5, 414.5.3, 415.8.2, 415.9.1, 418.6
COMBUSTIBLE MATERIAL
    Concealed spaces.......... 413.2, 717.5
    Exterior side of exterior wall........ 1406
    High-pile stock or rack storage.... 413.1, 910.2.2
    Type I and Type II............. 603, 805
COMBUSTIBLE PROJECTIONS.... 705.2, 1406.3
COMBUSTIBLE STORAGE....... 413, 910.2.2
COMMON PATH OF EGRESS TRAVEL.... 1014.3
COMPARTMENTATION
    Ambulatory care facilities........... 422.3
    Group I-2................. 407.5, 407.6
    Group I-3....................... 408.6
    Underground buildings....... 405.4, 405.5.2
COMPLIANCE ALTERNATIVES......... 3412
COMPRESSED GAS......... 307.2, 415.8.7.2.2
CONCEALED SPACES........... 413.2, 717
CONCRETE................. Chapter 19
    ACI 318 modifications............. 1905
    Anchorage................. 1908, 1909
    Calculated fire resistance.......... 721.2
    Construction documents........... 1901.3
    Durability....................... 1904
    Exposure conditions............... 1904
    Footings........................ 1809
    Foundation walls......... 1807.1.5, 1808.8
    Materials................ 1705.3.1, 1903

    Pipe columns, concrete-filled......... 1912
    Plain, structural................. 1909
    Reinforced gypsum concrete......... 1911
    Rodent proofing.............. Appendix F
    Roof tile................ 1507.3, 1711.2
    Shotcrete....................... 1910
    Slab, minimum................... 1907
    Special inspections...... 1705.3, Table 1705.3
    Specifications................... 1903
    Strength testing................ 1705.3.1
    Wood support.................. 2304.12
CONCRETE MASONRY
    Calculated fire resistance.......... 721.3
    Construction.................... 2104
    Design............. 2101.2, 2108, 2109
    Materials..................... 2103.1
    Testing............... 2105.2, 2105.3
    Wood support.................. 2304.12
CONCRETE ROOF TILE............. 1507.3
    Wind resistance......... 1609.5.3, 1711.2
CONDOMINIUM (see APARTMENT HOUSES)
CONDUIT, PENETRATION PROTECTION... 713.3, 1022.4
CONFLICTS IN CODE................ 102
CONGREGATE LIVING FACILITIES..... 202, 310
CONSTRUCTION (see SAFEGUARDS DURING CONSTRUCTION)
CONSTRUCTION DOCUMENTS....... 107, 1603
    Alarms and detection............. 907.1.1
    Concrete construction............ 1901.3
    Design load-bearing capacity........ 1803.6
    Fire-resistant joint systems.......... 714
    Masonry..................... 2101.3
    Means of egress................ 107.2.3
    Penetrations..................... 713
    Permit application................ 105.1
    Retention....................... 107.5
    Review......................... 107.3
    Roof assemblies.................. 1503
    Seismic certification............. 1705.12.3
    Site plan...................... 107.2.5
    Soil classification................ 1803.6
    Temporary structures............. 3103.2
CONSTRUCTION JOINTS
    Shotcrete..................... 1910.7
CONSTRUCTION TYPES........... Chapter 6
    Aircraft related................ 412.4.6.2
    Classification.................... 602
    Combustible material in Type I
        and Type II construction......... 603, 805
    Covered and open mall buildings..... 402.4.1
    Fire resistance........ Table 601, Table 602
    High-rise....................... 403.2
    Type I............. Table 601, 602.2, 603
    Type II............ Table 601, 602.2, 603
    Type III................ Table 601, 602.3

Index

Type IV . . . . Table 601, 602.4
Type V . . . . Table 601, 602.5
Underground buildings 405.2
CONTRACTOR'S RESPONSIBILITIES 1704.4
CONTROL AREA 414.2, 707.3.7
    Fire-resistance rating 414.2.4
    Maximum allowed quantities 414.2.2
    Number 414.2.3
CONVENTIONAL LIGHT-FRAME
    CONSTRUCTION 2301.2, 2302.1, 2308
    Additional seismic requirements 2308.11, 2308.12
    Braced wall lines 2308.3
    Connections and fasteners 2308.5
    Design of elements 2308.4
    Floor joists. 2308.8
    Foundation plates or sills 2308.6
    Girders 2308.7
    Limitations. 2308.2
    Roof and ceiling framing 2308.10
    Wall framing. 2308.9
CONVEYING SYSTEMS. 3005
CORNICES
    Draftstopping 717.2.6
    Projection 705.2, 1406.3
    Public right-of-way encroachments. . . 3202.3.2, 3202.3.3
CORRIDOR (see CORRIDOR
    PROTECTION, EXIT ACCESS,
    and FIRE PARTITION). 1018
    Air movement 1018.5
    Continuity 1018.6
    Covered and open mall buildings 402.8.1, 402.8.6
    Dead end 1018.4
    Encroachment. 1018.3
    Elevation change 1003.5
    Group I-2 . . . . 407.2, 407.3, 407.4.1
    Hazardous. . . .415.10.1.2, 415.10.2, 415.10.6.4, 415.10.7.1.4
    Headroom. 1003.2, 1003.3
    Walls 709.1, 1018.1
    Width 1003.3.4, 1003.6, 1005.2, 1005.7, 1018.2, 1018.3
CORRIDOR PROTECTION,
    EXIT ACCESS
    Construction, fire protection. . . . . . . 709.1, Table 1018.1, 1018.6
    Doors 715.4
    Glazing 715.5
    Group I-2 . . . . 407.3
    Interior finish . . . . Table 803.9, 804.4
    Opening protection 715, 716.5.4.1
    Ventilation. 1018.5, 1018.5.1
CORROSIVES . . . . 307.2, 307.6, Table 414.2.5(1), 414.3, 415.9.3, Table 415.10.1.1.1

COURTS (see YARDS OR COURTS). 1206
COVERED AND OPEN MALL BUILDINGS 402
    Alarms and detection 402.7.4, 907.2.20, 2702.2.14
    Children's play structures 402.6.3, 424
    Construction type 402.4
    Fire department 402.3, 402.7.5
    Interior finish 402.6.1
    Kiosk 402.6.2
    Means of egress 402.8
    Occupant load. 402.8.3
    Open mall construction 402.4.3
    Perimeter line 402.1.2
    Separation. 402.4.2
    Signs 402.6.4
    Smoke control. 402.7.2
    Sprinkler protection 402.5
    Standby power 402.7.3, 2702.2.14
    Standpipe system 402.7.1, 905.3.3
    Travel distance 402.8.5, 1014.3, 1016.2, 1021.2
COVERED WALKWAY
    (see PEDESTRIAN WALKWAY) 3306.7
CRAWL SPACE
    Access 1209.1
    Drainage 1805.1.2
    Unusable space fire protection 712.3.3
    Ventilation 1203.3
CRIPPLE WALL 2308.9.4, 2308.12.4
CURVED STAIRWAYS. 1009.11

D

DAMPERS (see FIRE DAMPERS
    and SMOKE DAMPERS) . . . . 716.2 through 716.5
DAMPPROOFING AND WATERPROOFING 1805
    Required 1805.2, 1805.3
    Subsoil drainage system 1805.4
DAY CARE. 305.2, 308.6, 310.5
    Accessibility 1103.2.13
    Adult care 308.6
    Child care 308.6, 310.5
    Egress. . . . . 308.6, Table 1004.1.1, 1015.6
DAY SURGERY CENTER
    (see AMBULATORY CARE FACILITIES)
DEAD END. 1018.4
DEAD LOAD. 1606
    Foundation design load 1808.3
DECK
    Anchorage. 1604.8.3
DEFLECTIONS 1604.3
    Framing supporting glass 2403.3
    Preconstruction load tests. 1710.3.2
    Wood diaphragms. 2305
    Wood shear walls 2305
DEMOLITION 3303

DESIGNATED SEISMIC SYSTEM. 1702.1
    Seismic certification. 1705.12.3
    Special inspection. 1705.11.4
DIAPHRAGMS. 2302
    Special inspection 1705.5, 1705.10, 1705.11
    Wood 2305, 2306.2
DIRECT DESIGN METHOD (masonry). . . . 2101.2.7
DOORS 1008
    Access-controlled 1008.1.9.8
    Atrium enclosures. 404.6
    Dwelling unit separations 406.3.4, 412.5.1
    Emergency escape 1029.1
    Fire (see OPENING PROTECTIVES). . . . 715.4, 1022.4, 1022.5
    Glazing 715.4.7, 715.5, 1405.13
    Hardware (see LOCKS AND
        LATCHES) 1005.7.1, 1008.1.9.8, 1008.1.9, 1008.1.10
    Horizontal sliding 1008.1.4.3
    I-2 occupancies 407.3.1
    I-3 occupancies 408.3, 408.4, 408.7, 408.8.4
    Landings 1008.1.5, 1008.1.6
    Operation 1008.1.3, 1008.1.9, 1008.1.10
    Panic and fire exit hardware 1008.1.10
    Power-operated 1008.1.4.2
    Revolving 1008.1.4.1
    Security grilles. 402.8.8, 1008.1.4.4
    Side swinging 1008.1.2
    Smoke. 710.5, 711.5
    Stairways 1008.1.9.11
    Stairways, high-rise 403.5.3
    Structural testing, exterior. 1710.5
    Thresholds 1003.5, 1008.1.5, 1008.1.7, 3411.8.14
    Underground buildings 405.4.3
    Vestibule 1008.1.8
    Width 1008.1.1, 1008.1.1.1
DRAFTSTOPPING
    Attics 718.4
    Floor-ceiling assemblies 717.3
DRINKING FOUNTAINS. 1109.5, 2902.5
DRY CLEANING PLANTS. 415.8.4
DRYING ROOMS 417
DUCTS AND AIR TRANSFER OPENINGS
    (see MECHANICAL)
DWELLING UNITS. 202
    Accessibility. 1103.2.4, 1103.2.12, 1105.1.6, 1106.2, 1107
    Accessibility, existing 3411.1, 3411.8.7, 3411.8.8, 3411.8.9
    Alarms and detection. . . . 420.5, 907.2.8, 907.2.9
    Area 1208.3, 1208.4
    Group R 310
    Live/work units (see LIVE/WORK UNITS)
    Scoping 101.2

    Separation. 420.2, 420.3
    Sound transmission 1207
    Sprinkler protection 420.4, 903.2.8

E

EARTHQUAKE LOADS (see SEISMIC) 1613
**EARTHQUAKE RECORDING
    EQUIPMENT**. . . . . Appendix L
**EAVES (see COMBUSTIBLE
    PROJECTIONS and CORNICES)**
EDUCATIONAL OCCUPANCY (GROUP E). . . . 305
    Accessibility 1108.2, 1109.5.1, 1109.5.2
    Alarms and detection 907.2.3
    Area 503, 505, 506, 507, 508
    Height 503, 504, 505, 506, 508
    Incidental uses 509
    Interior finishes . . . . .Table 803.9, 804
    Means of egress
        Aisles. 1017.5
        Corridors 1018.1, 1018.2
        Panic hardware 1008.1.10
        Travel distance 1014.3, 1016.2, 1021.2
    Mixed occupancies 508.3, 508.4
        Accessory 303.1, 508.2
        Assembly. 303.1.3
        Day care 305.2, 308.5, 310.1
        Education for students
            above the 12th grade 304
        Gyms (see GYMNASIUMS) 303.1
        Libraries (see LIBRARIES). 303.4
        Religious facilities 303.4
        Stages and platforms 410
    Plumbing fixtures 2902
    Risk category 1604.5
    Sprinkler protection 903.2.3
EGRESS (see MEANS OF EGRESS) . . . .Chapter 10
ELECTRICAL . . . . 105.2, 112, Chapter 27, Appendix K
ELEVATOR. . . . . Chapter 30
    Accessibility 1007.2.1, 1007.4, 1007.8, 1109.6, 3001.3, 3411.8.2
    Car size 3001.3, 3002.4
    Conveying systems . . . . Chapter 30
    Emergency operations 3002.3, 3002.5, 3003, 3007.2, 3008.2
    Fire service access 403.6.1, 3007
    High-rise 403.2.3, 403.4.7, 403.4.8, 403.6
    Hoistway enclosures 403.2.3, 708, 1022.4, 1023.5, 3002, 3007.6, 3008.6
    Hoistway lighting 3007.6.2
    Hoistway pressurization 909.21
    Hoistway venting 3004
    Keys .3003.3
    Lobby 709.1, 1007.4, 1007.8, 3007.7, 3008.7

Machine rooms 3006
Means of egress 403.6, 1003.7, 1007.2.1,
1007.4, 3008
Number of elevator cars
  in hoistway 3002.2
Occupant evacuation elevators . . . 403.6.2, 3008
Roof access. 1009.17
Signs 914, 1007.10, 3002.3, 3007.7.5,
3008.7.5
Stairway to elevator equipment 1009.17
Standards 3001
Standby power 2702.2.5, 2702.2.19,
3007.9, 3008.9
System monitoring 3007.8, 3008.8
Underground 405.4.3
EMERGENCY COMMUNICATIONS
Accessible means of egress 1007.8
Alarms (see FIRE ALARMS)
Elevators, occupant evacuation. 3008.7.7
Fire command center 403.4.6, 911, 3007.8,
3008.7.7, 3008.8
Radio coverage 403.4.4, 915
EMERGENCY EGRESS OPENINGS 1029
Required . . . . Table 1021.2(2), 1029.1
Window wells 1029.5
EMERGENCY LIGHTING 1006.3, 1205.5
EMERGENCY POWER 2702.1, 2702.3
Exit signs 1011.6.3, 2702.2.3, 2702.2.9
Group A. 2702.2.1
Group I-3 . . . . 408.4.2, 2702.2.17
Hazardous. 414.5.4, 415.8.10, 2702.2.8,
2702.2.10, 2702.2.11, 2702.2.13
High-rise. 403.4.8, 2702.2.15
Means of egress illumination 1006.3,
2702.2.4
Semiconductor fabrication 415.8.10, 2702.2.8
Underground buildings 405.9, 2702.2.16
EMERGENCY RESPONDERS
Additional exit stairway 403.5.2
Elevators 403.6, 1007.2.1, 3002.4, 3003,
3007, 3008
Fire command center 403.4.6, 911, 3007.8,
3008.7.7, 3008.8
Fire department access in malls 402.17
Mall access 402.17
Radio coverage 403.4.4, 915
Roof access. 1009.16
Safety features 914
EMPIRICAL DESIGN OF MASONRY . . . 2101.2.4, 2109
Adobe construction 2109.3
General 2109.1
Special inspection. 1705.4
EMPLOYEE
Accessibility for work areas . . . 1103.2.3, 1104.3.1
Deputies to building official 103.3
Liability 104.8
Qualifications . . . . Appendix A

ENCROACHMENTS INTO
  THE PUBLIC RIGHT-OF-WAY. . . . . Chapter 32
END-JOINTED LUMBER 2303.1.1.2
Moved structures . . . . 3410, Appendix D
Rodent proofing . . . . Appendix F
ENERGY EFFICIENCY 101.4.6, 110.3.7,
Chapter 13
ENGINEER (see definition for REGISTERED
  DESIGN PROFESSIONAL)
EQUIPMENT PLATFORM 505.5
EQUIVALENT OPENING FACTOR . . . . Figure 705-7
ESCALATORS. 3005
Accessibility 3411.8.4
Floor opening protection 708.2
Means of egress 1003.7
ESSENTIAL FACILITIES
  (see RISK CATEGORY) 1604.5
EXCAVATION, GRADING AND FILL 1804, 3304
EXISTING BUILDING . . . . 102.6, Chapter 34
Accessibility 1103.2.2, 3411
Additions . . . . 3403, Appendix D
Alteration . . . . 3404, Appendix D
Change of occupancy. . . . . . 3408, Appendix D
Flood-resistant construction. . . . . . Appendix G
Historic 3409, 3411.9
Moved structures . . . . 3410, Appendix D
Repairs 3405
Rodent proofing . . . . Appendix F
EXIT (see MEANS OF EGRESS) 1020
through 1026
Basement 1021.2.2
Configuration 1021.3
Construction. 713.2, 1009.3.1, 1022.2
Dwellings 1021.2.3
Enclosure 707.3, 1009.2.2, 1022.2
Fire resistance 707.3, 1009.3.1, 1022.2
Group H. 415.8.4.4, 415.8.5.5
High-rise buildings 403.5, 403.6, 1024
Horizontal 707.3.5, 1025
Interior finish . . . . Table 803.9, 804
Luminous 403.5.5, 411.7.1, 1024
Mezzanines 505.3, 505.4,
1004.1.1.2
Number, minimum. 403.5, 1015.1, 1021
Occupant load. 1004.1.1
Passageway. 1023
Ramp, interior 1009.2, 1022
Ramps, exterior 1026
Signs 1011
Stairways, exterior 1026
Travel distance 402.8.4, 402.8.5, 402.8.6,
404.9, 407.4, 407.4.2,
407.4.3, 408.6.1, 408.8.1,
410.6.3.2, 411.4, 1014.3,
1016, 1021.2, 1028.7, 1028.8
Underground buildings 405.7

Index

EXIT ACCESS (see MEANS
   OF EGRESS). . . . . 1014 through 1019
    Aisles 1017
    Balconies 1016.2.1, 1019
    Common path. 1014.3
    Corridors 1018
    Doors 1005.7, 1008, 1015, 1020.2
    Intervening space 1014.2
    Path of egress travel, common 1014.3
    Seating at tables 1017.2
    Single exit. 1015.1, 1021.2
    Stairway, interior 1009.3
    Travel distance 402.8.4, 402.8.5,
        402.8.6, 404.9, 407.4, 408.6.1,
        408.8.1, 410.6.3.2, 411.4, 1014.3,
        1015.5, 1016, 1021.2, 1028.7

EXIT DISCHARGE
   (see MEANS OF EGRESS) 1027
    Courts. 1027.4
    Horizontal exit. 1027.1
    Lobbies 1027.1
    Marquees 3106.4
    Public way. 1027.5
    Termination 1022.3
    Vestibules 1027.1

EXIT PASSAGEWAY
   (see MEANS OF EGRESS) 707.3.4, 1023

EXIT SIGNS 1011
    Accessibility 1011.4
    Floor level exit signs 1011.2
    Group R-1. . . . . 1011.2
    Illumination 1011.3, 1011.5, 1011.6
    Required 1011.1
    Special amusement buildings. 411.7

EXPLOSIVES . . . . 202, Table 414.5.1, Table 415.3.2
    Detached building. 415.8
    Explosion control 415.6

EXPOSURE CATEGORY (see WIND LOAD) . . . 1609.4

EXTERIOR AREAS FOR ASSISTED RESCUE
    Requirements 1007.7
    Signage 1007.9, 1007.10, 1007.11
    Where required 1007.2

EXTERIOR INSULATION AND
   FINISH SYSTEMS (EIFS) 1408
    Special inspection. 1705.15

EXTERIOR WALLS (see WALL,
   EXTERIOR) . . . Table 601, 602, 705, Chapter 14

F

FACTORY OCCUPANCY (GROUP F). 306
    Alarm and detection. 907.2.4
    Area 503, 503.1.1, 505, 506, 507, 508
    Equipment platforms 505.2
    Groups
        Low-hazard occupancy 306.3
        Moderate-hazard occupancy. 306.2

    Height 503, 504, 505, 508
    Incidental uses 509
    Interior finishes . . . . Table 803.9, 804
    Means of Egress
        Aisles. 1017.5
        Dead end corridor 1018.4
        Travel distance. . . . 1014.3, 1015.4, 1015.5,
            1015.6, 1016.2, 1021.2
    Mixed occupancies 508.3, 508.4
    Plumbing fixtures 2902
    Sprinkler protection 903.2.4
    Unlimited area. 507.2, 507.3, 507.4

FARM BUILDINGS. . . . . Appendix C
FEES, PERMIT. 109
    Refunds 109.6
    Related fees. 109.5
    Work commencing before issuance. 109.4

FENCES 105.2, 312.1
FIBERBOARD 2303.1.5
    Shear wall 2306.3

FILL MATERIAL 3304
FINGER-JOINTED LUMBER
   (see END-JOINTED LUMBER)

FIRE ALARM AND SMOKE DETECTION SYSTEMS
    Ambulatory care facilities 422.7, 907.2.2.1
    Assembly 907.2.1
    Atriums 404.5, 907.2.14
    Audible alarm 907.5.2.1
    Construction documents 907.1.1
    Covered and open mall building 402.6.2,
        402.7, 907.2.20
    Education 907.2.3
    Emergency system 908
    Factory 907.2.4
    Group H 907.2.5
    Group I 907.2.6, 907.5.2.3.3
    Group M 907.2.7
    Group R 420.5, 907.2.8, 907.2.9, 907.2.10,
        907.2.11, 907.5.2.3.3,
        907.5.2.3.4
    High-rise. 403.4.1, 403.4.2, 907.2.13
    Live/work units 419.5
    Occupancy requirements 907.2
    Special amusement buildings. 411.3, 411.5,
        907.2.12
    Underground buildings. . . . 405.6, 907.2.18, 907.2
    Visible alarm 907.5.2.3

FIRE ALARM BOX, MANUAL. 907.4.2
FIRE AREA 202, 901.7
    Ambulatory care facilities 903.2.2, 907.2.2
    Assembly 903.2.1
    Education 903.2.3
    Factory 903.2.4
    Institutional 903.2.6
    Mercantile 903.2.7
    Residential 903.2.8
    Storage 903.2.9, 903.2.10

Index

FIRE BARRIERS 202, 707
   Continuity 707.5, 713.5
   Exterior walls . . . . . . Table 602, 707.4, 713.6
   Fire-resistance rating of walls 603, 703, 707.3, 713.4
   Glazing, rated 716.6
   Incidental 509.4
   Joints 707.8, 713.9, 715, 2508.4
   Marking 703.7
   Materials 707.2, 713.3
   Opening protection 707.6, 707.10, 713.7, 713.10, 714.3, 716, 717.5.2
   Penetrations 707.7, 713.8
   Shaft enclosure 713.1
   Special provisions
      Aircraft hangars 412.4.4
      Atriums 404.3, 404.6
      Covered and open mall buildings 402.4.2
      Fire pumps 403.3.3, 901.8, 913.2.1
      Flammable finishes 416.2
      Group H-2 . . . . 415.8.1.2, 415.8.2.2
      Group H-3 and H-4 . . . . 415.9
      Group H-5 . . . . . 415.10.1.2, 415.10.1.5, 415.10.5.1, 415.10.6.4
      Group I-3 . . . . 408.5, 408.7
      Hazardous materials 414.2
      High-rise 403.2.1.2, 403.2.3, 403.3
      Stages and platforms . . . 410.5.1, 410.5.2
FIRE COMMAND CENTER . . . 403.4.5, 911, 3007.8, 3008.7.7, 3008.8
FIRE DAMPERS . . . . 717.2 through 717.5
FIRE DEPARTMENT (see EMERGENCY RESPONDERS)
FIRE DETECTION SYSTEM (see FIRE ALARM AND SMOKE DETECTION SYSTEMS)
FIRE DISTRICT . . . . Appendix D
FIRE DOOR (see OPENING PROTECTIVES) . . . 716, 1022.4, 1022.5
FIRE ESCAPE 3406
FIRE EXTINGUISHERS, PORTABLE . . . 906, 3309
FIRE EXTINGUISHING SYSTEMS . . . 416.5, 417.4, 903, 904
FIRE PARTITION 202, 709
   Continuity 708.4
   Exterior walls . . . . Table 602, 709.5
   Fire-resistance rating 603, 703, 708.3
   Glazing, rated 716.6
   Joint treatment gypsum 2508.4
   Joints 715
   Marking 703.6
   Materials 708.2
   Opening protection 709.6, 714.3, 716, 717.5.4
   Penetrations 714, 717
   Special provisions
      Covered and open mall buildings . . . 402.4.2.1
      Group I-3 . . . . 408.7
      Group I-1, R-1, R-2, R-3 . . . . 420.2
FIRE PREVENTION 101.4.5
FIRE PROTECTION
   Explosion control . . . 414.5.1, 415.6, 415.8.1.4, 421.7
   Fire extinguishers, portable 906
   Glazing, rated 716.2
   Smoke and heat vents 910
   Smoke control systems 909
   Sprinkler systems, automatic 903
FIRE PROTECTION SYSTEMS . . . . . . Chapter 9
FIRE PUMPS 403.3.3, 901.8, 913, 914.2
FIRE RESISTANCE
   Calculated 722
   Conditions of restraint 703.2.3
   Ducts and air transfer openings 717
   Exterior walls . . . . . . Table 602, 705.5, 708.5
   Fire district . . . . Appendix D
   High-rise 403.2
   Joint systems 715
   Multiple use fire assemblies 701.2
   Prescriptive 721
   Ratings . . . . . Chapter 6, 703, 705.5, 707.3.10
   Roof assemblies 1505
   Structural members 704
   Tests 703
   Thermal and sound insulating materials 720.1
FIRE RESISTANCE, CALCULATED 722
   Clay brick and tile masonry 722.4
   Concrete assemblies 722.2
   Concrete masonry 722.3
   Steel assemblies 722.5
   Wood assemblies 722.6
FIRE-RETARDANT-TREATED WOOD . . . 2303.2
   Balconies 1406.3
   Concealed spaces 718.5
   Fastening 2304.9.5
   Fire wall vertical continuity 706.6
   Partitions 603
   Platforms 410.4
   Projections 705.2.3
   Roof construction . . . Table 601, 705.11, 706.6, 1505
   Shakes and shingles 1505.6
   Type I and II construction 603
   Type III construction 602.3
   Type IV construction 602.4
   Veneer 1405.5
FIRE SEPARATION DISTANCE . . . . 202, Table 602, 702
   Exterior walls 1406
FIRE SERVICE ACCESS ELEVATORS 403.6.1, 3007
FIRE SHUTTER (see OPENING PROTECTIVES) 716.5, 716.5.10, 716.5.11

FIRE WALLS 706
   Aircraft. 412.6.2
   Combustible framing 706.7
   Continuity 706.5, 706.6
   Exterior walls . . . . Table 602, 706.5.1
   Fire-resistance rating 703, 706.4
   Glazing, rated 716.6
   Inspection 110.3.6
   Joints 706.10, 715
   Marking 703.7
   Materials 706.3
   Opening protection 706.8, 706.11, 714.3,
                            716, 717.5.1
   Penetration 706.9, 714.3
   Special provisions
      Aircraft hangars 412.4.6.2
      Covered and open mall buildings . . . 402.4.2.2
      Group H-5 . . . . 415.10.1.6
      Structural stability 706.2
FIRE WINDOWS (see OPENING PROTECTIVES)
FIREBLOCKING 718.2
   Chimneys 718.2.5.1
   Wood construction . . . 718.2.1, 718.2.7, 1406.2.3
   Wood stairs 718.2.4
FIREPLACES, MASONRY 2111
FLAMESPREAD . . . . . . 802, 803.1.1, Table 803.9
FLAMMABLE FINISHES 307.1, 416
FLAMMABLE LIQUIDS 307.4, 307.5, 406, 412,
                                 414, 415
FLASHING
   Roof 1503.2, 1503.6, 1507.2.9, 1507.3.9,
                1507.5.7, 1507.7.7, 1507.8.8,
                      1507.9.9, 1510.6
FLOOD-RESISTANT CONSTRUCTION
   Flood elevation 107.2.5.1, 1612
   Flood loads 1612, 3001.2, 3102.7
   Flood-resistant construction . . . . . Appendix G
   Site plan 107.2.5
FLOOR/CEILING (see FLOOR CONSTRUCTION)
FLOOR CONSTRUCTION (see FLOOR
   CONSTRUCTION, WOOD)
   Draftstopping 718.3
   Finishes 804, 805, 1003.4, 1210.1
   Fire resistance . . . . Table 601, 711
   Loads (see FLOOR LOADS)
   Materials . . . . Chapter 6
   Penetration of fire-resistant
      assemblies 711, 714.4, 717.2, 717.6
FLOOR CONSTRUCTION, WOOD
   Beams and girders 2304.11.2.1, 2308.7
   Bridging/blocking 2308.8.5, 2308.10.6
   Diaphragms 2305.1
   Fastening schedule 2304.9.1
   Framing 2304.4
   Joists 2308.8
   Sheathing 2304.7

FLOOR LEVEL 1003.5, 1008.1.5
FLOOR LOADS
   Construction documents 107.2
   Live 1607
   Posting 106.1
FLOOR OPENING PROTECTION
   (see VERTICAL OPENING PROTECTION)
FOAM PLASTICS
   Attics 720.1, 2603.4.1.6
   Cold storage 2603.3, 2603.4.1.2, 2603.5
   Concealed 603
   Crawl space 2603.4.1.6
   Doors . . . . 2603.4.1.7 through 2603.4.1.9
   Exterior walls of multistory buildings 2603.5
   Interior finish 801.2.2, 2603.10, 2604
   Label/identification 2603.2
   Roofing 2603.4.1.5
   Siding backer board 2603.4.1.10
   Surface burning characteristics 2603.3
   Thermal barrier requirements 2603.5.2
   Trim 806.3, 2604.2
   Type I and II construction . . . . 603.1(2), 603.1(3)
   Walk-in coolers 2603.4.1.3
FOLDING AND TELESCOPIC SEATING 1028.1.1
                                   3401.1
   Accessibility 1108.2
   Egress 1028.1.1
   Occupant load 1004.7
   Separation 1028.1.1.1
FOOD COURT 202
   Occupant load 402.8.2.4
   Separation 402.4.2
FOOTBOARDS 1028.14.2
FOUNDATION (see FOUNDATION,
   DEEP and FOUNDATION,
   SHALLOW) . . . . Chapter 18
   Basement 1610, 1805.1.1, 1806.3, 1807
   Concrete 1808.8, 1809.8, 1810.3.2.1
   Dampproofing 1805.2
   Encroachment, public right-of-way 3202.1
   Formwork 3304.1
   Geotechnical investigation
      (see SOILS AND FOUNDATIONS) . . . 1803
   Inspection 110.3.1
   Load-bearing value 1806, 1808, 1810
   Masonry 1808.9
   Pedestrian protection 3306.9
   Pier (see FOUNDATION, SHALLOW)
   Pile (see FOUNDATION, DEEP)
   Plates or sills 2308.6
   Protection from adjacent
      construction 3303.5, 3307.1
   Rodent proofing . . . . Appendix F
   Special inspections . . . 1705.3, 1705.4.2, 1705.7,
                            1705.8, 1705.9
   Steel 1809.11, 1810.3.2.2, 1018.3.2.3

Index

Timber	1809.12, 1810.3.2.4
Waterproofing	1805.3

FOUNDATION, DEEP 1802.1, 1810
    Drilled shaft 1802.1
    Geotechnical investigation 1803.5.5
    Grade beams 1810.3
    Helical pile 1810.3.1.5
    Micropile 1810.3.5.2.3
    Piles 1810

FOUNDATION, SHALLOW 1809
    Piers and curtain wall 1809.10
    Slab-on-grade 1808.6.2
    Strip footing 1808.8, 1809

FOYERS
    Assembly occupancy 1028.4, 1028.9.5
    Corridors 1018.6

FRAME INSPECTION 110.3.4
FRATERNITIES 310
FROST PROTECTION 1809.5
FURNACE ROOMS 1015.3

G

GALLERIES (see TECHNICAL PRODUCTION AREAS)
GARAGE, AUTOMOBILE PARKING (see PARKING GARAGES)
GARAGE, REPAIR 406.8
    Floor surface 406.8.3
    Sprinkler protection 406.8.6, 903.2.9.1
    Ventilation 406.8.2

GARAGES, TRUCK AND BUS
    Live load 1607.7
    Sprinkler protection 903.2.10.1

GARAGES AND CARPORTS, PRIVATE
    Area limitations 406.3.1, 406.3.2
    Classification 406.3
    Parking surfaces 406.3.3
    Separation 406.3.4

GATES 1008.2
    Vehicular 3110

GIFT SHOPS 407.2.4

GIRDERS
    Fire resistance . . . . Table 601
    Materials . . . . Chapter 6
    Wood construction 2304.11.2.1, 2308.7

GLASS (see GLAZING)
GLASS BLOCK (see GLASS UNIT MASONRY)
GLASS UNIT MASONRY 2102, 2110
    Atrium enclosure 404.6
    Design method 2101.2
    Fire resistance 2101.2.5
    Hazardous locations 2406.1.3
    Material requirements 2103.7

GLAZING
    Athletic facilities 2408
    Atrium enclosure 404.6
    Doors 705.8, 709.5, 710.5, 716.4.3.2, 1405.13, 1710.5
    Elevator hoistway and car 2409
    Fire doors 716.5.5.1, 716.5.8
    Fire-resistant walls 716.5.3.2
    Fire windows 703.5, 716.5
    Group I-3 . . . . 408.7
    Guards 1013.1.1, 2406.4.4, 2407
    Handrail 1009.15, 2407
    Identification 2403.1, 2406.3
    Impact loads 2406.1, 2407.1.4.2, 2408.2.1, 2408.3
    Impact resistant 1609.1.2
    Jalousies 2403.5
    Label/identification 716.5.7.1, 716.5.8.3, 716.5.8.3.1, 716.6.8
    Louvered windows 2403.5
    Opening protection 716.2
    Replacement 2401.2, 3407
    Safety 716.5.8.4, 716.6.3, 2406
    Security 408.7
    Skylights 2405
    Sloped 2404.2, 2405
    Supports 2403.2
    Swimming pools 2406.4
    Testing 1710.5, 2406.1.1, 2408.2.1
    Veneer 1405.12
    Vertical 2404.1

GRADE, LUMBER (see LUMBER) 2302
GRADE PLANE 202
GRANDSTANDS 303.1, 1028.1.1, 3401.1
    Accessibility 1108.2
    Egress 1028.1.1
    Exit sign 1011.1
    Occupant load 1004.4
    Separation 1028.1.1.1

GREENHOUSES 312.1
    Area 503, 506, 507, 508
    Membrane structure 3102.1
    Plastic 2606.11
    Sloped glazing 2405

GRIDIRON (see TECHNICAL PRODUCTION AREAS)
GRINDING ROOMS 415.6.1.2
GROSS LEASABLE AREA (see COVERED AND OPEN MALL BUILDINGS) 202, 402.3, 402.8.2
GROUT 714.3.1.1, 714.4.1.1, 2103.13
GUARDS 1013
    Assembly seating 1028.1.1, 1028.14
    Exceptions 1013.2
    Glazing 1013.2.1, 1303.1, 2406.4.4, 2407
    Height 1013.3
    Loads 1607.8
    Mechanical equipment 1013.6
    Opening limitations 1013.4

2012 International Building Code Handbook   **949**

GUARDS (cont.)
Parking garage 406.4.2
Ramps. 1010.11
Residential 1013.3
Roof access 1013.7
Stairs 1013.2
Vehicle barrier 406.4.3, 1607.8.3
Windows 1013.8

GYMNASIUMS 303.1
Group E 303.1.3
Occupant load 1004.1

GYPSUM . . . . Chapter 25
Aggregate, exposed 2513
Board . . . . Chapter 25
Concrete, reinforced 1911
Construction 2508
Draftstopping 718.3.1
Exterior soffit . . . . Table 2506.2
2508.1
Fire resistance 719, 722.2.1.4, 722.6.2
Fire-resistant joint treatment 2508.4
Inspection 2503
Lath 2507, 2510
Lathing and furring for cement plaster . . . 719, 2510
Lathing and plastering 2507
Materials 2506
Plaster, exterior 2512
Plaster, interior 2511
Shear wall construction 2306.3, 2505
Sheathing 2304.6
Showers and water closets 2509
Stucco 2510
Veneer base 2507.2
Veneer plaster 2507.2
Vertical and horizontal assemblies 2504
Wallboard . . . . Table 2506.2
Water-resistant backing board . . . 2506.2, 2509.2

H

HANDRAILS 1012
Alternating tread devices 1009.13.1
Assembly aisles 1028.13
Construction 1012.4, 1012.5, 1012.6
Extensions 1012.6
Glazing 2407
Graspability 1012.3
Guards 1013.3
Height 1012.2
Loads 1607.8
Location 1012.1, 1012.7, 1012.8, 1012.9
Ramps 1010.9
Stairs 1009.15

HARDBOARD 1404.3.2, 2303.1.6

HARDWARE (see DOORS and LOCKS AND LATCHES)

HARDWOOD
Veneer 1404.3.2

HAZARDOUS MATERIALS 307, 414, 415
Control areas 414.2
Explosion control . . . . . 414.5.1, Table 414.5.1, 415.8.1.4, 415.10.5.5
Special provisions 415.6, 415.7
Sprinkler protection . . . . . . Table 414.2.5(1), Table 414.2.5(2), 415.4, 415.10.11, 903.2.5 415.10.6.4, 415.10.7, 415.10.10, 1203.5
Weather protection 414.6.1

HAZARDOUS OCCUPANCY (GROUP H), (see HAZARDOUS MATERIALS) 307, 414, 415
Alarm and detection 414.7, 415.3, 415.10.2, 415.10.3.5, 415.10.5.9, 415.10.8, 901.6.3, 907.2.5, 908.1, 908.2
Area 503, 505, 506, 507, 508
Dispensing 414.5, 414.6, 414.7.2, 415.5
Group provisions
H-1 (detonation) 307.3, 403.1, 415.5.1.1, 415.5.2 415.6, 415.6.1
H-2 (deflagration) . . . 307.4, 403.1, 415.7, 415.8
H-3 (physical hazard) 307.5, 403.1, 415.7, 415.9
H-4 (health hazard) 307.6, 415.9
H-5 (semiconductor) 307.7, 415.10
Height 415.6, 415.7.1, 415.8.1.1, 415.8.1.6, 415.8.2.1.1, 503, 504, 505, 506, 508
Incidental uses 509
Interior finishes . . . . 416.2.1, 416.3.1, Table 803.9, 804
Location on property 414.6.1.2, 415.5
Low hazard (See Factory—Group F-2 and Storage—Group S-2)
Means of egress
Aisles 107.5
Corridors 415.10.2
One means of egress . . . . . Table 1015.1, Table 1021.2
Panic hardware 1008.1.10
Travel distance . . . . . Table 1016.2, 1014.3, 1021.2
Mixed occupancies 508.3, 508.4
Accessory 508.2
Moderate hazard [See FACTORY OCCUPANCY (GROUP F) and STORAGE OCCUPANCY (GROUP S)]
Multiple hazards 307.8
Occupancy exceptions 307.1
Plumbing fixtures . . . . Chapter 29
Prohibited locations 419.2
Risk category 1604.5
Smoke and heat vents 910.2
Special provisions—General
Detached buildings 415.5.2, 415.7.1
Fire separation distance 415.5
Separation from other occupancies . . . 415.5.1, 508.2.4, 508.3.3, 508.4

Sprinkler protection 415.2, 415.10.6.4,
415.10.9, 415.10.10.1,
415.10.11, 705.8.1, 903.2.5
Standby, emergency power 2702.2.8,
2702.2.10 through 2702.2.13
Storage 413, 414.1, 414.2.5, 414.5,
414.6, 414.7.1, 415.5, Table 415.5.2,
415.6.1, 415.7, 415.8.1, 415.8.2
Unlimited area. 507.8
HEAD JOINT, MASONRY 2102.1
HEADROOM 406.2.2, 505.1, 1003.2,
1003.3, 1008.1.1, 1008.1.1.1,
1009.5, 1010.6.2, 1208.2
**HEALTH CARE (see INSTITUTIONAL I-1
and INSTITUTIONAL I-2)**
Ambulatory care facilities 202, 422
Clinics, outpatient 304.1
Hospitals 308.4
HEALTH-HAZARD MATERIALS 307.2,
Table 414.2.5(1), 415.2, 415.4,
Table 415.8.2.1.1, 415.8.6.2
HEIGHT, BUILDING 503, 504, 505, 508, 510
Limitations. 503
Mixed construction types 510
Modifications 504
Roof structures 504.3
HELIPAD
Definition 202
Live loads 1607.6
HIGH-PILED COMBUSTIBLE STORAGE. . . . 413,
907.2.15, 910.2.2
HIGH-RISE BUILDINGS. 403
Alarms and detection. . .403.4.1, 403.4.2, 907.2.13
Application 403.1
Construction. 403.2
Elevators 403.6, 1007.2.1, 3007, 3008
Emergency power. 403.4.8, 2702.2.15
Emergency systems. 403.4
Fire command station 403.4.6
Fire department communication. . .403.4.3, 403.4.4
Fire service elevators 403.6.1, 3007
Occupant evacuation elevators . . . 403.6.2, 3008
Smoke removal 403.4.6
Smokeproof enclosure 403.5.4, 1022.10
Sprayed fire-resistant materials (SFRM). . . 403.2.4
Sprinkler protection 403.3, 903.2.11.3
Stairways 403.5
Standby power 403.4.7, 2702.2.5, 2702.2.15
Structural integrity. 403.2.3, 1615
Super high-rise (over 420 feet). . . 403.2.1, 403.2.3,
403.2.4, 403.3.1, 403.5.2
Voice alarm 403.4.3, 907.2.13
Zones 907.6.3.2
HISTORIC BUILDINGS 3409
Accessible. 3411.9
Flood provisions. G105.3

HORIZONTAL ASSEMBLY 711
Continuity . . . 508.2.5.1, 711.4, 713.11, 713.12
Fire-resistance rating . . . . . 603.1(1), 603.1(22),
603.1(23), 703, 707.3.10, 711.3
Glazing, rated 716.6
Group I-1 . . . . 420.3
Group R 420.3
Incidental 509.4
Insulation 720, 807, 808
Joints 715, 2508.4
Opening protection 711.8, 714.4, 716, 717.6
Shaft enclosure 713.1
Special provisions
Aircraft hangars 412.4.4
Atrium 404.3, 404.6
Covered and open mall
buildings 402.4.2.3, 402.8.7
Fire pumps 913.2.1
Flammable finishes 416.2
Group H-2 . . . . .415.8.1.1, 415.8.2.2
Groups H-3 and H-4 . . . . 415.9.2
Group H-5 . . . . . . 415.10.1.2, 415.10.5.1
Group I-2. . . . . 407.4.3
Groups I-1, R-1, R-2 and R-3 . . . . . 420.3
Hazardous materials 414.2
High-rise buildings 403.2.1, 403.3,
403.4.7.1
Stages and platforms 410.4, 410.5.1
HORIZONTAL EXIT 1025
Accessible means of egress . . .1007.2, 1007.2.1,
1007.3, 1007.4, 1007.6, 1007.6.2
Doors 1025.3
Exit discharge 1027.1
Fire resistance 1025.2
Institutional I-2 occupancy 407.4, 1025.1
Institutional I-3 occupancy 408.2, 1025.1
Refuge area (see REFUGE AREAS)
**HORIZONTAL FIRE SEPARATION
(see HORIZONTAL ASSEMBLY)**
**HOSE CONNECTIONS
(see STANDPIPES, REQUIRED)**
HOSPITAL (see INSTITUTIONAL I-2) 308.4, 407
**HURRICANE-PRONE REGIONS
(see WIND LOADS)** 1609.2
HURRICANE SHELTER (see STORM SHELTER)
HYDROGEN CUTOFF ROOMS . . . . . 421, Table 509

I

ICE-SENSITIVE STRUCTURE
Atmospheric ice loads. 1614
Definition 202
IDENTIFICATION, REQUIREMENTS FOR
Fire barriers 703.6
Fire partitions 703.6
Fire wall 703.6
Glazing 2403.1, 2406.3

IDENTIFICATION, REQUIREMENTS FOR (*cont.*)
    Inspection certificate 1702
    Labeling 1703.5
    Preservative-treated wood 2303.1.8.1
    Smoke barrier 703.6
    Smoke partition 703.6
    Steel 2203
IMPACT LOAD 1607.9
INCIDENTAL USES
    Area 509.3
    Occupancy classification 509.2
    Separation and protection 509.4
INCINERATOR ROOMS . . . . Table 509, 1015.3
INDUSTRIAL (see FACTORY OCCUPANCY)
INSPECTIONS 110, 1704, 1705
    Alternative methods and materials 1705.1.1
    Approval required 110.6
    Concrete construction 110.3.1, 110.3.2,
        110.3.9, 1705.3
    Concrete slab 110.3.2
    EIFS 110.3.9, 1705.15
    Energy efficiency 110.3.7
    Fabricators 1704.2.5
    Fees 109
    Final 110.3.10
    Fire-extinguishing systems 904.4
    Fire-resistant materials . . . 110.3.9, 1705.13, 1705.14
    Fire-resistant penetrations . . . 110.3.6, 1705.16
    Footing or foundation . . . 110.3.1, 110.3.9, 1705.3,
        1705.4, 1705.7, 1705.8, 1705.9
    Frame 110.3.4
    Lath or gypsum board 110.3.5, 2503
    Liability 104.8
    Masonry 110.3.9, 1705.4
    Preliminary 110.2
    Required 110.3
    Right of entry 104.6
    Seismic 1705.11
    Smoke control 104.16, 909.18.8, 1705.17
    Soils 110.3.9, 1705.6
    Special (see STRUCTURAL TESTS AND
        SPECIAL INSPECTIONS) 110.3.9, 1704,
        1706, 1707
    Sprayed fire-resistant materials 1705.13
    Sprinkler protection 903.5
    Steel 110.3.4, 110.3.9, 1705.2
    Third party 110.4
    Welding 110.3.9, 1705.2, 2204
    Wind 110.3.9, 1705.10
    Wood 110.3.9, 1705.5
INSTITUTIONAL I-1 [see INSTITUTIONAL
    OCCUPANCY (GROUP I) and
    RESIDENTIAL (GROUP R-4)] 308.3, 420
    Accessibility 1106.7.2, 1107.5.1
    Alarm and detection 420.5, 907.2.6.1,
        907.2.11.2, 907.5.2.3.3
    Combustible decorations 806.1

    Emergency escape and rescue 1029
    Means of egress
        Aisles 1017.5
        Travel distance 1016.2, 1021.2
    Occupancy exceptions 308.3.1, 308.3.2
    Separation, unit 420.2, 420.3
    Sprinkler protection . . . 420.4, 903.2.6, 903.3.2
INSTITUTIONAL I-2 [see INSTITUTIONAL
    OCCUPANCY (GROUP I)] 308.4, 407
    Accessibility . . . 1106.3, 1106.4, 1106.7.2, 1107.5.2,
        1107.5.3, 1107.5.4, E106.4.6
    Alarms and detection . . . 407.7, 407.8, 907.2.6.2
    Care suites 407.4
    Combustible decorations 806.1
    Hyperbaric facilities 408.10
    Means of egress
        Aisles 1017.5
        Corridors 407.2, 407.3, 407.4, 1018.2
        Doors 1008.1.9.6, 1008.1.9.8
        Exterior exit stair 1026.2
        Hardware 1008.1.9.3, 1008.1.9.6
        Travel distance 407.4
    Occupancy exceptions 308.4.1
    Smoke barriers 407.5
    Smoke compartment . . . 407.2.1, 407.2.3, 407.5
    Smoke partitions 407.3
    Sprinkler protection . . . 407.6, 903.2.6, 903.3.2
    Yards 407.9
INSTITUTIONAL I-3 [see INSTITUTIONAL
    OCCUPANCY (GROUP I)] 308.5, 408
    Accessibility 1103.2.14, 1105.4, 1107.5.5,
        1108.4.2, 3411.8.7, E104.3, E104.4, E106.4.8
    Alarm and detection 408.10, 907.2.6.3
    Combustible decorations 806.1
    Means of egress 408.2, 408.3, 408.4
        Aisles 1017.5
        Doors 1008.1.1, 1008.1.2
        Exit sign exemption 1011.1
        Hardware . . . 408.4, 1008.1.9.3, 1008.1.9.7,
        1008.1.9.8, 1009.1.9.10
        Travel distance . . . 408.6.1, 408.8.1, 1016.2,
        1021.2
    Security glazing 408.7
    Separation 408.5, 408.8
    Smoke barrier 408.6
    Smoke compartment . . . 408.4.1, 408.6, 408.9
    Sprinkler protection 408.11, 903.2.6
    Standby/emergency power 2702.2.17
INSTITUTIONAL I-4 [see INSTITUTIONAL
    OCCUPANCY (GROUP I)] 308.6
    Accessibility 1103.2.12
    Alarms and detection 907.2.6
    Corridor rating 1018.1
    Educational 303.1, 304.1
    Means of egress
        Day care 1015.6
        Travel distance 1014.3, 1016.2, 1021.2

Occupancy exceptions	308.6.1, 308.6.2, 308.6.3, 308.6.4
Sprinkler protection	903.2.6

INSTITUTIONAL OCCUPANCY (GROUP I) . . . 308
- Accessory 508.2
- Adult care 308.5.1
- Area 503, 505, 506, 507, 508
- Child care 308.3.1, 308.5.2, 310.1
- Group specific provisions
  - Group I-1 (see INSTITUTIONAL I-1) . . . 308.2
  - Group I-2
    - (see INSTITUTIONAL I-2) . . . 308.3, 407
  - Group I-3
    - (see INSTITUTIONAL I-3) . . . 308.4, 408
  - Group I-4
    - (see INSTITUTIONAL I-4) 308.3.1, 308.5, 310.1
- Height 503, 504, 505, 506, 508
- Incidental uses 509
- Interior finishes . . . . Table 803.9, 804
- Means of egress
  - Corridors 1018.2
  - Travel distance . . . 407, 1014.3, 1016.2, 1021.2
- Mixed occupancies 508.3, 508.4
- Occupancy exceptions . . . . . 303.1.1, 303.1.2, 308.3.1, 308.3.2, 308.4.1, 308.6.1 through 308.6.4, 310.5.1
- Plumbing fixtures 2902
- Risk category . . . . Table 1604.5
- Standby, emergency power 2702.2.1

INSULATION
- Concealed 720.2, 2303.1.5
- Foam plastic (see FOAM PLASTICS) . . . 720.1
- Loose fill 720.4, 720.6
- Roof 720.5, 1508
- Sound 720, 807, 1207
- Thermal 720, 807, 1508

INTERIOR ENVIRONMENT
- Lighting 1205
- Rodent proofing . . . . Appendix F
- Sound transmission 1207
- Space dimensions 1208
- Temperature control 1204
- Ventilation 409.3, 414.3, 415.8.2.6, 1203.4
- Yards or courts 1206.2, 1206.3

INTERIOR FINISHES . . . . Chapter 8
- Acoustical ceiling systems 807, 808
- Application 803.10, 804.4
- Atriums 404.8
- Children's play structures 424
- Covered and open mall buildings 402.6
- Decorative materials 801.1.2, 806
- Floor finish 804, 805
- Foam plastic insulation 2603.3, 2603.4
- Foam plastic trim 806.3, 2604.2
- Insulation 807
- Light-transmitting plastics 2606
- Signs 402.6.4, 2611

Trim	806.5, 806.6
Wall and ceiling finishes	803
Wet location	1210, 2903

INTERPRETATION, CODE 104.1

J

JAILS (see INSTITUTIONAL I-3) 308.3, 408
JOINT
- Gypsum board 2508.4
- Lumber sheathing 2308.10.8.1
- Shotcrete 1910.7
- Waterproofing 1805.3.3

JOINTS, FIRE-RESISTANT SYSTEMS 715
- Special inspection 1705.16

K

KIOSKS 402.11
KITCHENS 303.3, 306.2
- Accessibility 1109.4
- Dimensions 1208
- Means of egress 1014.2
- Occupant load . . . . Table 1004.1.1
- Rooms openings 1210.5

L

LABORATORIES
- Classification of 304.1
- Hazardous materials 414, 415
- Incidental uses . . . . Table 509

LADDERS
- Boiler, incinerator and furnace rooms . . . 1015.3
- Construction 1009.7.2, 1012.2, 1012.6, 1013.3, 1013.4
- Emergency escape window wells 1029.5.2
- Group I-3 . . . . 408.3.5, 1009.14
- Refrigeration machinery room 1015.4
- Stage 410.6.3.4

LAMINATED TIMBER, STRUCTURAL
- **GLUED** . . . 602.4, 2303.1, 2303.1.3, 2304.11.3, 2306.1, 2308.8.2.1, 2308.10.7

LANDINGS
- Doors 1008.1.6
- Ramp 1010.7
- Stair 1009.8

LATH, METAL OR WIRE . . . . Table 2507.2
LAUNDRIES . . . . 304.1, 306.2, Table 509
LAUNDRY CHUTE 713.13, 903.2.11.2
LEGAL
- Federal and state authority 102.2
- Liability 104.8
- Notice of violation 114.2, 116.3
- Registered design professional . . . 107.1, 107.3.4
- Right of entry 104.6
- Unsafe buildings or systems 116
- Violation penalties 114.4

LIBRARIES
    Classification, other than school 303.1
    Classification, school 305.1
LIGHT, REQUIRED 1205
    Artificial 1205.3
    Emergency (see EMERGENCY LIGHTING)
    Means of egress 1006.1, 1006.2
    Natural 1205.2
    Stairways .1205.4
    Yards and courts 1206
LIGHT-FRAME CONSTRUCTION
    Definition 202
    Cold-formed steel 2211
    Conventional (wood) 2308
LIGHTS, PLASTIC CEILING DIFFUSERS . . .2606.7
LINEN CHUTE 713.13, 903.2.11.2
LINTEL
    Fire resistance 704.11
LIQUEFIED PETROLEUM GAS . . . . Table 414.5.1, 415.8.3
LIVE LOADS 1607
    Construction documents 107.2, 1603
    Posting of 106.1
LIVE/WORK UNITS 202, 310.4, 419
    Accessibility 1103.2.13
    Separation 508.1
LOAD AND RESISTANCE FACTOR
DESIGN (LRFD) 1602.1
    Load combinations 1605.2
    Wood design 2301.2, 2307
LOAD COMBINATIONS 1605
    Allowable stress design 1605.3
    Load and resistance factor design 1605.2
    Strength design 1605.2
LOADS 106, 202
    Atmospheric ice 1614
    Combinations 1605
    Dead 1606
    Flood 1612
    Impact 1607.9
    Live 419.6, 1607
    Rain 1611
    Seismic 1613
    Snow 1608
    Soil lateral 1610
    Wind 1609
LOBBIES
    Assembly occupancy 1028.4
    Elevator . . . 713.14.1, 1007.2.1, 1007.4, 3007.7, 3008.7
    Exit discharge 1027.1
    Underground buildings 405.4.3
LOCKS AND LATCHES 1008.1.9, 1008.1.10
    Access-controlled egress 1008.1.9.8
    Delayed egress locks 1008.1.9.7
    Electromagnetically locked 1008.1.9.9

    Group I-2 . . . . 407.4.1.1, 1008.1.9.6
    Group I-3 . . . . 408.4, 1008.1.9.10
    High-rise buildings 403.5.3
    Toilet rooms 2902.3.5
LUMBER
    General provisions . . . . Chapter 23
    Quality standards 2303

M

MAINTENANCE 3401.2
    Accessibility 3411.2
    Means of egress 3310.2
    Property 101.4.4
MALL (see COVERED AND OPEN MALL BUILDINGS)
MANUAL FIRE ALARM BOX 907.4.2
MANUFACTURED HOMES
    Flood resistant G501
MARQUEES 202, 3106, H113
MASONRY
    Adhered veneer 1405.10
    Adobe 2109
    Anchorage 1604.8.2
    Anchored veneer 1405.6
    Architectural cast stone 2103.5
    Autoclaved aerated concrete (AAC) 2103.3
    Calculated fire resistance 722.4
    Chimneys 2113
    Construction 2104
    Construction documents 2101.3
    Design, methods 2101.2
    Fire resistance, calculated . . . 722.3.2, 722.3.4
    Fireplaces 2111
    Floor anchorage 1604.8.2
    Foundation walls 1807.1.5
    Glass unit 2110
    Grouted 202
    Heaters 2112
    Inspection, special 1705.4
    Materials 2103
    Penetrations 714
    Quality assurance 2105
    Rodentproofing . . . . Appendix F
    Roof anchorage 1604.8
    Rubble stone 202
    Seismic provisions 2106
    Serviceability 1604.3
    Support 2304.12
    Test procedures 2105.2, 2105.3
    Veneer . . . 1405.6, 1405.10, 2101.2.6, 2308.11.2
    Wall, composite 202
    Wall, hollow 202
    Wall anchorage 1604.8.2
    Waterproofing 1805.3.2
    Wythe 202

MATERIALS
- Alternates 104.11
- Aluminum . . . . Chapter 20
- Concrete . . . . Chapter 19
- Glass and glazing . . . . Chapter 24
- Gypsum . . . . Chapter 25
- Masonry . . . . Chapter 21
- Noncombustible 703.4
- Plastic . . . . Chapter 26
- Steel . . . . Chapter 22
- Testing (see TESTING) 1711
- Wood . . . . Chapter 23

MEANS OF EGRESS . . . . Chapter 10
- Accessible 1007, 2702.2.5, 2702.2.6, 3411.6, 3411.8.10
- Aircraft related 412.3.2, 412.5.2
- Alternating tread device 412.7.3, 505.3, 1009.3, 1015.3, 1015.4
- Ambulatory care facilities 422.5
- Assembly 1007.1, 1028
- Atrium 404.9, 707.3.6
- Capacity 1005.3
- Ceiling height 1003.2
- Child care facilities (see Day care facilities)
- Construction drawings 107.2.3
- Convergence 1005.6
- Covered and open mall buildings 402.8
- Day care facilities . . . 308.5, 310.1, Table 1004.1.1, 1015.6
- Distribution 1005.5
- Doors . . . 1005.7, 1008, 1015, 1020.2, 2702.2.7
- During construction 3303.3, 3310
- Elevation change 1003.5
- Elevators 403.5.2, 403.6.2, 1003.7, 1007, 3008
- Emergency escape and rescue 1029
- Encroachment 1005.7
- Equipment platform 505.3
- Escalators 1003.7
- Existing buildings 1007.1, 3310, 3406.1, 3411.6, 3412.5, 3412.6.11
- Exit (see EXIT) . . . . 1020 through 1026
- Exit access (see EXIT ACCESS) . . . . 1014 through 1019
- Exit discharge (see EXIT DISCHARGE) 1027
- Exit enclosures 1022.2
- Exit signs 1011, 2702.2.3, 2702.2.9
- Fire escapes 3406
- Floor surface 804, 1003.4
- Gates 1008.2
- Group I-2 . . . . 407.2, 407.3, 407.4
- Group I-3 . . . . 408.2, 408.3, 408.4, 408.6
- Guards 1013
- Handrails 1012
- Hazardous materials 414.6.1.2, 415.10.3.3, 415.10.5.6
- Headroom 1003.2, 1003.3
- High-hazard Group H 415.10.3.3, 415.10.5.6
- High-rise 403.5, 403.6
- Illumination 1006, 2702.2.4, 3412.6.15
- Interior finish 803.9, 804
- Ladders (see LADDERS)
- Live/work units 419.3
- Mezzanines 505.2.2, 505.2.3, 1004.1.1.2, 1007.1
- Occupant load 1004.1, 1004.1.2, 1004.2
- Parking 406.5.7
- Protruding objects 1003.3, 1005.7
- Ramps 1010, 1026
- Scoping 101.3, 105.2.2, 108.2, 1001.1
- Seating, fixed 1007.1, 1028
- Special amusement 411.7
- Stages 410.3.3, 410.6
- Stairways 403.5, 1005.3.1, 1009, 1022.2, 1026
- Temporary structures 3103.4
- Travel distance (see TRAVEL DISTANCE) 1014.3, 1016
- Turnstile 1008.3
- Underground buildings 405.5.1, 405.7
- Width 1005.1, 1005.2, 1005.4, 1009.4, 1010.6.1, 1018.2, 1028.6, 1028.8

MECHANICAL (see REFRIGERATION, AND VENTILATION) 101.4.2
- Access 1009.16, 1009.17, 1209.3
- Air transfer openings . . . . 705.10, 706.11, 707.10, 712.1.8, 713.10, 709.8, 711.8, 711.7, 714.1 .1, 717
- Chimneys (see CHIMNEYS)
- Code . . . . Chapter 28
- Ducts 704.8, 705.10, 706.11, 707.10, 712.1.5, 712.1.16, 713.10, 709.8, 710.8, 711.7, 714.1.1, 717
- Encroachment, public right-of-way 3202.3.2
- Equipment on roof 1509, 1510.2
- Factory-built fireplace 2111
- Fireplaces 2111
- Incidental use room . . . . Table 509
- Motion picture projection room 409.3
- Permit required 105.1, 105.2
- Roof access 1009.16
- Seismic inspection and testing 1705.11.4, 1705.12
- Systems . . . . Chapter 28

MECHANICALLY LAMINATED DECKING . . . 2304.8

MEMBRANE ROOF COVERINGS 1507.11, 1507.12, 1507.13

MEMBRANE STRUCTURES 2702.2.9, 3102

MENTAL HOSPITALS (see INSTITUTIONAL I-2)

MERCANTILE OCCUPANCY (GROUP M) 309
- Accessible 1109.12
- Alarm and detection 907.2.7
- Area 503, 505, 506, 507, 508
- Covered and open mall buildings 402

Index

MERCANTILE OCCUPANCY (GROUP M) *(cont.)*
    Hazardous material display and storage. . . 414.2.5
    Height 503, 504, 505, 506, 508
    Incidental uses 509
    Interior finishes . . . . Table 803.9, 804
    Means of egress
        Aisles 1017.3, 1017.4
        Travel distance 402.8, 1014.3, 1016.2, 1021.2
    Mixed occupancies 508.3, 508.4
        Accessory 508.2
        Live/work units 419
        Mall buildings 402
        Parking below/above. 510.7, 510.8, 510.9
        Special mixed 510.2
    Occupancy exceptions 307.1
    Plumbing fixtures 2902
    Sprinkler protection 903.2.7
    Standby/emergency power 2702.2.14
    Standpipes 905.3.3
    Unlimited area. 507.3, 507.4, 507.12

METAL
    Aluminum . . . . Chapter 20
    Roof coverings 1504.3.2, 1507.5
    Steel. . . . . Chapter 22
    Veneer. 1404.5

MEZZANINES 505
    Accessibility 1104.4, 1108.2.4, 1108.2.9
    Area limitations 505.2.1, 505.3.1
    Egress. 505.2.2, 505.2.3, 1004.6, 1007.1
    Equipment platforms 505.3
    Guards 505.3.3, 1013.1
    Height 505.2, 1003.2
    Occupant load. 1004.1.1.2
    Stairs 712.1.10, 1009.13, 1022.2

MIRRORS 1008.1, 2406.1

MIXED OCCUPANCY
    (see OCCUPANCY SEPARATION)

MODIFICATIONS 104.4, 104.10

MOISTURE PROTECTION 1210, 1403.2, 1503, 2304.11

MONASTERIES 310.4

MORTAR. .2103
    Ceramic tile 2103.11
    Dampproofing 1805.2.2
    Fire resistance 714.3.1, 714.4.1.1
    Glass unit masonry 2110
    Masonry. 2103.9, 2103.10
    Rodent proofing . . . . Appendix F

MOTELS 310.3, 310.4

MOTION PICTURE PROJECTION ROOMS . . . 409
    Construction. 409.2
    Exhaust air 409.3.2, 409.3.3
    Lighting control 409.4
    Projection room 409.3
    Supply air 409.3.1
    Ventilation 409.3

MOTOR FUEL-DISPENSING SYSTEM 406.5
    Accessibility 1109.14, 3411.8.13

MOTOR VEHICLE FACILITIES 304, 311, 406

MOVING, BUILDINGS 3410, D103.3

MOVING WALKS 3005.2
    Means of egress 1003.7

N

NAILING 2303.6, 2304.9

NONCOMBUSTIBLE BUILDING MATERIAL. . . 703.4

NURSING HOMES
    (see INSTITUTIONAL I-2) 308.3, 407

O

OCCUPANCY
    Accessory. 508.2
    Certificates (see CERTIFICATE OF OCCUPANCY)
    Change (see CHANGE OF OCCUPANCY)
    Floor loads 1607

OCCUPANCY CLASSIFICATION 302
    Covered and open mall buildings 402
    Mixed 508, 510
    Mixed occupancy values 3412.6.16

OCCUPANCY SEPARATION
    Accessory. 508.2
    Aircraft related. 412.5.1
    Covered mall and open mall building. . . 402.4.2
    Mixed occupancy 508, 510, 707.3.9
    Parking garages. . . . . . 406.3.4, Table 508.4(d)
    Repair garages 406.6.2
    Required fire resistance. . . . . .Table 508.4, 510

OCCUPANT EVACUATION ELEVATORS. . . . 403.5.2, 403.6.2, 3008

OCCUPANT LOAD
    Actual 1004.1.2
    Certificate of occupancy. 111
    Covered and open mall building 402.8.2
    Cumulative 1004.1.1
    Determination of. 1004.1, 1004.1.1, 1004.6
    Increased 1004.2
    Outdoors 1004.5
    Seating, fixed 1004.4
    Signs 1004.3

OFFICE BUILDINGS
    (See GROUP B OCCUPANCIES)
    Classification 304
    Live loads 1607.5

OPEN MALL BUILDINGS (see COVERED AND OPEN MALL BUILDINGS)

OPENING PROTECTION, EXTERIOR WALLS. . .705.8

OPENING PROTECTION, FLOORS
    (see VERTICAL OPENING PROTECTION)

OPENING PROTECTIVES. 705.8, 706.8, 707.6, 709.5, 711.8, 713.7, 716

Automatic-closing devices 909.5.2
Fire door and shutter assemblies 705.8.2,
711.8, 716.5
Fire windows 716.6
Glazing 716.6
ORGANIC COATINGS. 418

P

PANIC HARDWARE. 1008.1.10
PARAPET, EXTERIOR WALL 705.11, 2109.3.4.3
Construction. 705.11.1
Fire wall 706.6
Height 705.11.1
PARKING, ACCESSIBLE 1106, 1110.1,
3411.4, 3411.7
PARKING GARAGES 406.4
Accessibility 1105.1.1, 1106.1, 1106.7.4,
1110.1
Barriers, vehicle 406.4.3, 1602.1, 1607.8.3
Classification 311, 406.3, 406.4
Construction type . . .406.5.1, Table 503, Table 601
Enclosed (see PARKING GARAGE,
ENCLOSED) 406.6
Gates 3110
Guards 406.4.2, 2407.1.3
Height, clear. 406.4.1
Live loads 1607.10.1.3
Means of egress 1009.3, 1021.1, 1021.4
Occupancy separation 508, 510
Open (see PARKING GARAGE, OPEN) . . . 406.3
Special provisions. 509
Sprinkler protection 903.2.10
Underground 405
PARKING GARAGES, ENCLOSED. 406.6
Area and height [see STORAGE
OCCUPANCY (GROUP S)] 406.6.1
Means of egress 1003.2, 1010.1, 1021.1.2
Ventilation. 406.4.2
PARKING GARAGES, OPEN 202, 406.5
Area and height [see STORAGE
OCCUPANCY (GROUP S)] 406.5,
406.5.1, Table 406.5.4
Construction type 406.3.3
Means of egress 406.5.7, 1003.2, 1007.3,
1007.4, 1010.1, 1016.1, 1018.1,
1022.2, 1024.1, 1027.1
Mixed occupancy 406.5.3
Standpipes 406.5.8
Ventilation. 406.5.10
PARTICLEBOARD. 2302.1
Draftstopping 718.3.1
Quality. 2303.1.7
Wall bracing 2308.9.3
PARTITIONS
Fire (see FIRE PARTITION)
Live loads 1607.5, 1607.14

Materials . . . . . . .602.4.6, 603.1(1), 603.1(11)
Occupancy, specific. 708.1
Smoke (see SMOKE PARTITIONS)
Toilets .1210
PARTY WALLS (see FIRE WALLS). 706.1.1,
Table 716.6
PASSAGEWAY, EXIT (see EXIT) 1023.1
PASSENGER STATIONS 303.4
PATIO COVERS . . . . .2606.10, Appendix I
PEDESTRIAN
Protection at construction site. . . . 3303.2, 3306
Walkways and tunnels 3104, 3202.3.4
PENETRATION-FIRESTOP SYSTEM
Fire-rated horizontal assemblies 714.4.1.2
Fire-rated walls 714.3.2
PENETRATIONS. 714, 717
Fire-resistant assemblies
Exterior wall 705.10
Fire barrier 707.7, 707.10
Fire wall 706.9, 706.11
Horizontal assemblies . . .711.5, 711.7, 714.4
Shaft enclosures 712.1, 713.1,
713.8, 713.10
Smoke barriers. 709.6, 709.8, 714.5
Smoke partitions 710.6, 710.7
Special inspection 1705.16
Walls 714.3
Nonfire-resistant assemblies 714.4.2
PERFORMANCE CATEGORY
Definition 202
Wood structural panels 2302
PERLITE. . . . . .Table 721.1(1), Table 2507.2
PERMITS. 105
Application for 104.2, 105.1, 105.3
Drawings and specifications 107.2.1
Expiration 105.5
Fees. 109
Liability for issuing. 104.8
Placement of permit. 105.7
Plan review 104.2, 107.3
Suspension or revocation 105.6
Time limitations 105.3.2, 105.5
PHOTOVOLTAIC SYSTEMS
Fire classification 1505.8
Modules/shingles 202, 1507.17
Panels/modules 1511
Rooftop mounted 1509.7
**PIER FOUNDATIONS (see FOUNDATION,
SHALLOW)**
PILE FOUNDATIONS (see FOUNDATION, DEEP)
PIPES
Embedded in fire protection. 704.8
Insulation covering 720.1, 720.7
Penetration protection. 714, 1022.4
Under platform 410.4
PLAIN CONCRETE (see CONCRETE) 1906

PLAN REVIEW 107.3
PLASTER
    Fire-resistance requirements 719
    Gypsum 719.1, 719.2
    Inspection 110.3.5
    Portland cement . . . . 719.5, Table 2507.2, Table 2511.1.1
PLASTIC . . . . Chapter 26
    Approval for use 2606.2
    Core insulation, reflective plastic 2613
    Fiber-reinforced polymer 2612
    Fiberglass-reinforced polymer 2612
    Finish and trim, interior 2604
    Light-transmitting panels 2401.1, 2607
    Roof panels 2609
    Signs 402.6.4, 2611, D102.2.10, H107.1.1
    Thermal barrier 2603.4
    Veneer 1404.8, 2605, D102.2.11
    Walls, exterior 2603.4.1.4, 2603.5
PLASTIC, FOAM
    Children's play structures 424.2
    Insulation (see FOAM PLASTICS) 2603
    Interior finish 803.4, 2603.10
    Malls 402.6.2, 402.6.4.5
    Stages and platforms 410.3.6
PLASTIC, LIGHT-TRANSMITTING
    Awnings and patio covers 2606.10
    Bathroom accessories 2606.9
    Exterior wall panels 2607
    Fiber-reinforced polymer 2612.4
    Fiberglass-reinforced polymer 2612.4
    Glazing 2608
    Greenhouses 2606.11
    Light-diffusing systems 2606.7
    Roof panels 2609
    Signs, interior 2611
    Skylight 2610
    Solar collectors 2606.12
    Structural requirements 2606.5
    Unprotected openings 2608.1, 2608.2
    Veneer, exterior . . . . 603.1(15), 603.1(17), 2605
    Wall panels 2607
PLATFORM (see STAGES AND PLATFORMS) . . . 410
    Construction 410.4
    Temporary 410.4.1
PLATFORM, EQUIPMENT
    (see EQUIPMENT PLATFORM)
PLATFORM LIFTS, WHEELCHAIR
    Accessible means of egress . . . 1007.2, 1007.5, 1009.4, 2702.2.6
    Accessibility 1109.8, 3411.8.3
PLUMBING (see TOILET AND TOILET ROOMS) . . . 101.4.3, 105.2, Chapter 29
    Aircraft hangars, residential 412.5.4
    Facilities, minimum 2902, 3305.1
    Fixtures . . . . Table 2902.1
    Room requirements 1210, 2406.2, 2406.4, 2606.9
PLYWOOD (see WOOD STRUCTURAL PANELS) 2302
    Preservative-treated 2303.1.8
PRESCRIPTIVE FIRE RESISTANCE 721
PRESERVATIVE-TREATED WOOD 2303.1.8
    Fastenings 2304.9.5
    Quality 2303.1.8
    Required 1403.6, 2304.11
    Shakes, roof covering 1507.9.6, 1507.9.8
PROJECTION ROOMS
    Motion picture 409
PROJECTIONS, COMBUSTIBLE 705.2.3, 1406.3
PROPERTY LINE
    (see FIRE SEPARATION DISTANCE) 705.3
PROPERTY MAINTENANCE 101.4.4
PROSCENIUM
    Opening protection 410.3.5
    Wall 410.3.4
PSYCHIATRIC HOSPITALS
    (see INSTITUTIONAL I-2) 308.4
PUBLIC ADDRESS SYSTEM
    (see EMERGENCY COMMUNICATIONS)
    Covered and open mall building 402.7, 907.2.20, 2702.2.14
    Special amusement buildings 411.6
PUBLIC PROPERTY . . . . . Chapter 32, Chapter 33
PUBLIC RIGHT-OF-WAY
    Encroachments . . . . Chapter 32
PYROPHORIC MATERIALS . . . . . Table 307.1(1), 307.4

R

RAILING (see GUARDS AND HANDRAILS)
RAMPS 1010
    Assembly occupancy 1028.11
    Construction . . . . 1010.2 through 1010.6.3, 1010.8, 1010.10
    Existing buildings 3411.8.5
    Exterior 1026, 3201.4
    Guards 1010.11, 1013, 1607.8
    Handrails 1010.9, 1012, 1607.8
    Interior 1010.2
    Landings 1010.7
    Parking garage 406.4.4
    Slope 1010.3, 3411.8.5
REFERENCED STANDARDS . . . . Chapter 35
    Applicability 102.3, 102.4
    Fire resistance 703.2
    List . . . . Chapter 35
    Organizations . . . . Chapter 35
REFORMATORIES 308.4
REFRIGERATION
    (see MECHANICAL) 101.4.2
    Machinery room 1015.4

Index

REFUGE AREAS (see HORIZONTAL EXIT, SMOKE COMPARTMENTS, STORM SHELTER) 407.5.1, 408.6.2, 422.4, 423.1.1, 1025.4

REFUSE CHUTE 713.13

REINFORCED CONCRETE (see CONCRETE)
   General 1901.2
   Inspections 1705.3

REINFORCEMENT
   Concrete 1910.4, 1912.4
   Masonry 2103.14

RELIGIOUS WORSHIP, PLACES OF
   Alarms and detection 907.2.1
   Balcony 1028.5, 1108.2.4
   Classification . . . 303.1.4, 303.4, 305.1.1, 305.2.1
   Door operations 1008.1.9.3
   Egress 1028
   Interior finishes . . . . Table 803.9, 804
   Unlimited area 507.6, 507.7

REPAIRS, BUILDING 202, 3405
   Compliance alternatives . . . 3412.1, 3412.2.4, 3412.3
   Flood 1612
   Minor 105.2.2
   Permit required 105.1
   Scope 101.2, 3401.1, 3401.3, 3409.1

RESIDENTIAL OCCUPANCY (GROUP R) 310
   Accessibility . . . 1103.2.4, 1103.2.11, 1003.2.13, 1106.2, 1107.6, 3411.8.7, 3411.8.8, 3411.8.9, E104.2, E104.3
   Alarm and detection . . . 907.5.2.3.3, 907.5.2.3.4, 907.2.8, 907.2.9, 907.2.10, 907.2.11
   Area 503, 505, 506, 508, 510
   Draftstopping 718.3.2, 718.4.2
   Group provisions
      Group R-1 (transient) 310.3
      Group R-2 (apartment) 310.4
      Group R-3
         (two dwellings per building) 310.5
      Group R-4
         (group homes) 310.6, 1008.1.9.5.1
   Height 503, 504, 505, 508, 510
   Incidental uses 509
   Interior finishes . . . . Table 803.9, 804
   Means of egress
      Aisles 1017.5
      Corridors 1018.1, 1018.2
      Doors 1008.1.1, 1008.1.9.5.1
      Emergency escape and rescue 1029.1
      Exit signs 1011.1, 1011.2
      Single exits 1021.2, 1021.2.3
      Travel distance 1014.3, 1016.2, 1021.2
   Mixed occupancies 508.3, 508.4
      Accessory 508.2, G801.1
      Live/work units 419
      Parking, private 406.1
      Parking below/above . . . 510.4, 510.7, 510.9
      Special mixed 510.2

   Plumbing fixtures 2902
   Risk category 1604.5.1
   Special provisions 510.5, 510.6
      Separation 419, 420, 508.2.4
      Swimming pools 3109.4
   Sprinkler protection 903.2.8, 903.3.2

RETAINING WALLS 1807.2, 2304.11.7
   Seismic 1803.5.12

REVIEWING STANDS (see BLEACHERS AND GRANDSTANDS)

RISERS, STAIR (see STAIRWAY CONSTRUCTION)
   Alternating tread device 1009.13.2
   Assembly 1009.3, 1028.6, 1028.7, 1028.9, 1028.11
   Closed 1009.7.5
   General 1009.7
   Spiral 1009.12
   Uniformity 1009.7.4

RISK CATEGORY (Structural Design) 1604.5
   Multiple occupancies 1604.5.1

RODENT PROOFING . . . . Appendix F

ROLL ROOFING 1507.6

ROOF ACCESS 1009.16, 1009.17

ROOF ASSEMBLIES AND ROOFTOP STRUCTURES
   Cooling towers 1509.4
   Drainage 1503.4, 3201.4
   Fire classification 1505
   Height modifications 504.3
   Impact resistance 1504.7
   Materials 1506
   Mechanical equipment screen 1509.6
   Parapet walls 1503.3, 1503.6
   Penthouses 1509.2
   Photovoltaic systems 1509.7
   Tanks 1509.3
   Towers, spires, domes and cupolas . . . 1509.5
   Weather protection 1503
   Wind resistance 1504.1, 1609.5

ROOF CONSTRUCTION
   Construction walkways 3306.7
   Coverings (see ROOF COVERINGS) . . . 1609.5.2
   Deck 1609.5.1
   Draftstopping 718.4
   Fire resistance . . . . Table 601
   Fireblocking 718.2
   Live loads 1607, 1607.12
   Materials . . . . Chapter 6
   Penetration of fire-resistant assemblies . . . 714
   Protection from adjacent construction . . . 3307.1
   Rain loads 1611
   Roof structures 504.3, 1509, D102.2.9
   Signs, roof mounted H110
   Slope, minimum . . . . Chapter 15
   Snow load 1608
   Trusses 2211.3, 2303.4, 2308.10.10
   Wood (see ROOF CONSTRUCTION, WOOD)

ROOF CONSTRUCTION, WOOD 602.4.3, 602.4.5
   Anchorage to walls 1604.8.2
   Attic access 1209.2
   Ceiling joists 2308.10.2
   Diaphragms 2305.1, 2306.2
   Fastening requirements 2304.9
   Fire-retardant-treated . . . . Table 601, 603.1(25)
   Framing 2304.10.3, 2308.10
   Rafters 2306.1.1
   Sheathing 2304.7, 2308.10.8
   Trusses 2303.4, 2308.10.10
   Ventilation, attic 1203.2
   Wind uplift 2308.10.1
ROOF COVERINGS 1507
   Asphalt shingles 1507.2
   Built up 1507.10
   Clay tile 1507.3
   Concrete tile 1507.3
   Fire resistance . . . . 603.1(3), 1505
   Flashing 1503.2, 1503.6, 1507.2.9,
                 1507.3.9, 1507.5.7, 1507.7.7,
                 1507.8.8, 1507.9.9, 1510.6
   Impact resistance 1504.7
   Insulation 1508
   Liquid-applied coating 1507.15
   Membrane 3102
   Metal panels 1507.4
   Metal shingles 1507.5
   Modified bitumen 1507.11
   Photovoltaic modules/shingles 1507.17
   Plastics, light-transmitting panels 2609
   Replacement/recovering 1510.3
   Reroofing 1510
   Roll 1507.6
   Single-ply 1507.12
   Slate shingles 1507.7
   Sprayed polyurethane foam 1507.14
   Thermoplastic single-ply 1507.13
   Wind loads 1504.1, 1609.5
   Wood shakes 1507.9
   Wood shingles 1507.8
ROOF DRAINAGE 1503.4
ROOF REPLACEMENT/RECOVERING 1510.3
ROOF STRUCTURE (see ROOF ASSEMBLIES AND ROOFTOP STRUCTURES)
ROOM DIMENSIONS 1208
ROOMING HOUSE (see BOARDING HOUSE) 310

S

SAFEGUARDS DURING CONSTRUCTION . . . . Chapter 33
   Accessibility 1103.2.6
   Adjoining property protection 3307
   Construction 3302
   Demolition 3303
   Excavations 1804.1
   Fire extinguishers 3309
   Means of egress 3310
   Protection of pedestrians 3306
   Sanitary facilities 3305
   Site work 3304
   Sprinkler protection 3312
   Standpipes 3308.1.1, 3311
   Temporary use of streets, alleys
      and public property 3308
SAFETY GLAZING 716.5.8.4, 2406
SCHOOLS (see EDUCATIONAL OCCUPANCY)
SEATING, FIXED 1028
   Accessibility 1108.2, 1109.11
   Aisles 1028.9, 1028.10
   Bleachers (see BLEACHERS)
   Grandstands (see GRANDSTANDS)
   Occupant load 1004.4
SECURITY GLAZING 408.7
SECURITY GRILLES 402.8.8, 1008.1.4.4
SEISMIC 1613
   Construction documents 107, 1603
   Earthquake recording equipment . . . Appendix L
   Existing building 3404.5, 3405.2, 3408.4
   Geotechnical investigation . . . 1803.5.11, 1803.5.12
   Loads 1613
   Masonry 2106
   Seismic design category 1613.5
   Seismic detailing 1604.10
   Site class 1613.3.2
   Site coefficients 1613.3.3
   Special inspection 1705.11
   Statement of special inspections 1704.3
   Steel 2205.2, 2206.2
   Structural observations 1704.5.1
   Structural testing 1705.12
   Wood 2305, 2308.11, 2308.12
SERVICE SINKS . . . . 1109.3, Table 2902.1
SERVICE STATION (see MOTOR FUEL-DISPENSING FACILITIES)
SHAFT (see SHAFT ENCLOSURE AND VERTICAL OPENING PROTECTION) 202
SHAFT ENCLOSURE (see VERTICAL OPENING PROTECTION) 713
   Continuity 713.5, 713.11, 713.12
   Elevators 713.14
   Exceptions 713.2, 1009.2, 1016.1
   Exterior walls 713.6
   Fire-resistance rating 707.3.1, 713.4
   Group I-3 . . . . 408.5
   High-rise buildings 403.2.1.2, 403.2.3,
                 403.3.1.1, 403.5.1
   Materials 713.3
   Opening protection . . . 713.8, 713.10, 714, 717.5.3
   Penetrations 713.8
   Refuse and laundry chutes 713.13
   Required 713.1

Index

SHEAR WALL
    Gypsum board and plaster 2505
    Masonry. 202
    Wood 202, 2305.1, 2306.3

SHEATHING
    Clearance from earth 2304.11.2.2
    Fastening 2304.9
    Fiberboard. 2306.3
    Floor. 2304.7, 2308.8.6
    Gypsum . . . . Table 2506.2, 2508
    Moisture protection 2304.11.2.2
    Roof. 2304.7
    Roof sheathing 2308.10.8
    Wall 2304.6, 2308.9.3
    Wood structural panels 2303.1.4

SHOPPING CENTERS (see COVERED AND OPEN MALL BUILDINGS)

SHOTCRETE 1910

SHUTTERS, FIRE (see OPENING PROTECTIVES) 716.5

SIDEWALKS. . . . . 105.2(6), G801.4
    Live loads . . . . Table 1607.1

SIGNS . . . . 3107, Appendix H
    Accessibility. . . . 1011.4, 1110, E106.4.9, E107, E109.2.2
    Accessible means of egress . . . . . 1007.8.2, 1007.9 through 1007.11
    Covered and open mall building 402.6.4
    Doors 1008.1.9.3, 1008.1.9.7, 1008.1.9.8
    Elevators 1109.7, 1110.2, 3002.3, 3007.7.5, 3008.7.5
    Encroachment, public right-of-way 3202.3.1
    Exit 1011, 2702.2.3, 2702.2.9
    Floor loads 106.1
    Luminous 403.5.5, 1011.5, 1024
    Obstruction 1003.3.2, 1003.3.3
    Occupant load, assembly 1004.3
    Parking spaces 1110.1
    Plastic. 2611, D102.2.10
    Protruding objects. 1003.3
    Stair identification 1022.8, 1022.9, 1110.2, 1110.3
    Standpipe control valve 905.7.1
    Toilet room 1110.1, 1110.2, 2904, 2904.1
    Variable message 1110.4
    Walls 703.6

SITE DRAWINGS 107.2.5

SITE WORK 3304

SKYLIGHTS 2405, 3106.3
    Light, required. 1205.2
    Loads 2404
    Plastic. 2610
    Protection from adjacent construction 3307.1

SLAB ON GROUND, CONCRETE 1901.2

SLATE SHINGLES. 1507.7

SLEEPING UNITS 202
    Accessibility. 1103.2.11, 1105.1.6, 1106.2, 1106.7.2, 1107
    Accessibility, existing 3411.8.7, 3411.8.8, 3411.8.9
    Group I 308
    Group R 310
    Scoping 101.2
    Separation. 420.2, 420.3

SMOKE ALARMS
    Live/work unit 419.5, 907.2.11.2
    Multiple-station 907.2.11
    Residential occupancies 420.5, 907.2.11.1, 907.2.11.2
    Single-station 907.2.11

SMOKE BARRIERS 202
    Construction. 407.4.3, 709.4, 909.5
    Doors 709.5, 716.5.3, 909.5.2
    Fire-resistance rating 703, 709.3
    Glazing, rated 716.6
    Horizontal assemblies. 711.9
    Inspection 110.3.6
    Joints 709.7, 715
    Marking 703.6
    Materials 709.2
    Opening protection 709.5, 714.3, 714.5, 716, 717.5.5, 909.5.2
    Penetrations. 709.6, 714
    Smoke control. 909.5
    Special provisions
        Ambulatory care facilities 422.2
        Group I-2 . . . . 407.5
        Group I-3 . . . . 408.6, 408.7
        Underground. 405.4.2, 405.4.3

SMOKE COMPARTMENT 407, 408, 422
    Refuge area (see REFUGE AREA)

SMOKE CONTROL 909
    Amusement buildings, special 411.1
    Atrium buildings. 404.5
    Covered and open mall building 402.10
    Group I-3 . . . . 408.9
    High-rise (smoke removal) 403.4.6, 403.5.4, 1022.10
    Special inspections 1705.17
    Stages. 410.3.7.2
    Standby power systems. . . . 909.11, 909.20.6.2, 2702.2.2
    Underground buildings 405.5

SMOKE DAMPERS . . . . 717.2 through 717.5

SMOKE DETECTION SYSTEM (see FIRE ALARM AND SMOKE DETECTION SYSTEMS) 907

SMOKE DETECTORS
    Covered and open mall building 402.8.6.1, 907.2.20
    High-rise buildings 403.4.1, 907.2.13
    Institutional I-2. . . . . 407.8
    Smoke-activated doors 716.5.9.3

Index

SMOKE DETECTORS *(cont.)*
    Special amusement buildings 411.5
    Underground buildings 907.2.18, 907.2.19
SMOKE DEVELOPMENT . . . 802, 803.1.1, Table 803.9
SMOKE EXHAUST SYSTEMS
    Underground buildings 405.5, 907.2.18, 909.2
SMOKE PARTITIONS 202, 710
    Continuity 710.4
    Doors 710.5
    Ducts and air transfer openings 710.8
    Fire-resistance rating 710.3
    Inspection 110.3.6
    Joints 710.7
    Marking 703.6
    Materials 710.2
    Opening protection 710.5, 717.5.7
    Penetrations 710.6
    Special provisions
        Atriums 404.6
        Group I-2 . . . . 407.3
SMOKE REMOVAL (High-rise buildings) . . . 403.4.6
SMOKE VENTS 410.3.7.1, 910
SMOKEPROOF ENCLOSURES . . . 403.5.4, 1022.10
    Design 909.20
SNOW LOAD 1608
    Glazing 2404
SOILS AND FOUNDATIONS
    (see FOUNDATION) . . . . Chapter 18
    Depth of footings 1809.4
    Excavation, grading and fill 1804, 3304, J106, J107
    Expansive 1803.5.3, 1808.6
    Flood hazard 1808.4
    Footings and foundations 1808
    Footings on or adjacent to slopes 1808.7, 3304.1.3
    Foundation walls 1807.1.5, 3304.1.4
    Geotechnical investigation 1803
    Grading . . . . 1804.3, Appendix J
    Load-bearing values 1806
    Soil boring and sampling 1803.4
    Soil lateral load 1610
    Special inspection 1705.6
SORORITIES 310.4
SOUND-INSULATING MATERIALS
    (see INSULATION) 720
SOUND TRANSMISSION 1207
SPECIAL CONSTRUCTION . . . . Chapter 31
    Awnings and canopies
        (see AWNINGS and CANOPIES) 3105
    Marquees (see MARQUEE) 3106
    Membrane structures
        (see MEMBRANE STRUCTURES) . . . 3102
    Pedestrian walkways and tunnels
        (see WALKWAYS and TUNNELED
        WALKWAYS) 3104

    Signs (see SIGNS) 3107
    Swimming pool enclosures and safety
        devices (see SWIMMING POOL) 3109
    Temporary structures
        (see TEMPORARY STRUCTURES) . . . 3103
SPECIAL INSPECTIONS AND TESTS
    (see INSPECTIONS) . . . . . 110.3.9, Chapter 17
    Alternative test procedure 1707
    Approvals 1703
    Contractor responsibilities 1704.4
    Design strengths of materials 1706
    General 1701
    In situ load tests 1709
    Material and test standards 1711
    Preconstruction load tests 1710
    Special inspections 1705
    Statement of special inspections 1704.3
    Structural observations 1704.5
    Test safe load 1708
    Testing seismic resistance 1705.12
SPECIAL INSPECTOR 202
SPIRAL STAIRS 1009.12
    Construction 1009.4, 1009.5, 1009.12
    Exceptions 1009.7.2, 1009.7.3, 1009.7.5, 1009.15
    Group I-3 . . . . 408.3.4
    Live/work 419.3.2
    Stages 410.6.3.4
SPRAY-APPLIED FIRE RESISTANT
    MATERIALS 1705.13
    Inspection 1705.13, 1705.14
    Steel column calculated fire
        resistance 722.5.2.2
SPRINKLER SYSTEM, AUTOMATIC 903, 3312
    Exempt locations 903.3.1.1.1
    Signs 914.2
    Substitute for fire rating . . . . Table 601(4)
    Values 3412.6.17
SPRINKLER SYSTEM, REQUIRED 903
    Aircraft related 412.4.6
    Ambulatory care facilities 422.6, 903.2.2
    Amusement buildings, special 411.4
    Area increase 506.3
    Assembly 903.2.1, 1028.6.2.3
    Atrium 404.3
    Basements 903.2.11.1
    Children's play structures 424.3
    Combustible storage 413
    Commercial kitchen 903.2.11.5
    Construction 903.2.12
    Covered and open mall building 402.5
    Drying rooms 417.4
    Education 903.2.3
    Exempt locations 903.3.1.1.1
    Factory 903.2.4
    Fire areas 707.3.10

Hazardous materials . . . . . . Table 414.2.5(1), Table 414.2.5(2), 903.2.11.4
Hazardous occupancies. 415.4, 415.10.6.4, 415.10.11, 705.8.1, 903.2.5
Height increase 504.2
High-rise buildings 403.3, 903.2.11.3
Incidental uses . . . . Table 509
Institutional 407.6, 408.11, 420.4, 903.2.6, 903.3.2
Laundry chutes, refuse chutes, termination rooms and incinerator rooms 713.13, 903.2.11.2
Live/work units 419.5, 903.2.8
Mercantile. 903.2.7
Mezzanines 505.2.1, 505.2.3, 505.3.2
Multistory buildings 903.2.11.3
Parking garages . . .406.6.3, 903.2.9.1, 903.2.10.1
Residential 420.4, 903.2.8, 903.3.2
Special amusement buildings. 411.4
Spray finishing booth 416.5
Stages. 410.7
Storage 903.2.9, 903.2.10
Supervision (see SPRINKLER SYSTEM, SUPERVISION) 903.4
Underground buildings 405.3, 903.2.11.1
Unlimited area. 507

STAGES AND PLATFORMS 303, 410
Dressing rooms 410.5
Fire barrier wall 410.5.1, 410.5.2
Floor finish and floor covering. . . 410.3, 410.4, 804.4, 805.1
Horizontal assembly. 410.5.1, 410.5.2
Means of egress 410.6
Platform, temporary 410.4.1
Platform construction . . . . . . .410.4, 603.1(12)
Proscenium curtain 410.3.5
Proscenium wall. 410.3.4
Roof vents. 410.3.7.1
Scenery 410.3.6
Smoke control. 410.3.7.2
Sprinkler protection 410.7
Stage construction . . . . . . .410.3, 603.1(12)
Standpipes 410.8, 905.3.4
Technical production areas . . .202, 410.3.2, 410.6.3
Ventilation. 410.3.7

STAIRWAY (see ALTERNATING TREAD DEVICES, SPIRAL STAIRS, STAIRWAY CONSTRUCTION and STAIRWAY ENCLOSURE)

STAIRWAY CONSTRUCTION
Aisle steps. 1028.9
Alterations. 3404.1
Alternating tread. 1009.13
Circular (see Curved)
Construction. 1009.9
Curved 1009.6, 1009.11
Discharge barrier 1022.8
During construction 3310.1
Elevators 1009.17, 1022.4, 3002.7
Enclosure under. 1009.9.3
Existing 3404.1, 3408.3
Exterior exitway 1026.1, 1027.1
Fireblocking. 718.2.4
Guards 1013.2, 1013.3, 1607.7
Handrails 1009.15, 1012, 1607.7
Headroom. 1009.5
Illumination 1006.1, 1205.4, 1205.5
Ladders 408.3.4, 410.6.3.4, 1009.14
Landings 1009.8, 1009.10
Live load. 1607.7
Luminous 403.5.5, 411.7.1, 1024
Roof access. 1009.16, 1009.17
Seismic anchorage 2308.12.7
Spiral (see SPIRAL STAIRS) . . . 408.3.4, 410.5.3, 419.3.2, 1009.12
Treads and risers 1009.6, 1009.7
Width 1005.3.1, 1009.4
Winders 1009.6, 1009.7.2, 1009.7.3, 1009.7.4, 1009.11

STAIRWAY ENCLOSURE 713.1, 1009.2, 1009.3, 1022.1
Accessibility. 1007.3
Construction. 1009.3.1, 1022.2
Discharge 1009.2, 1022.3.1, 1027.1
Doors 716.5.9, 1008.1.9.11, 1009.3.1.4
Elevators within. . . . 1009.3.1.4.1, 1022.4, 3002.7
Exterior walls 705.2, 707.4, 708.5, 713.6, 1009.3.1.8, 1022.2, 1026.6
Fire-resistant construction 1009.3.1.2, 1022.2
Group I-3 . . . . 408.3.8
High-rise buildings 403.5
Penetrations. 1009.3.1.5, 1022.5
Pressurization. 403.5.4, 405.7.2, 909.6, 909.20.5, 1022.10
Smokeproof . . . 403.5.4, 405.7.2, 909.20, 1022.10
Space below, use 1009.9.3
Ventilation. 1009.3.1.7, 1022.6

STANDARDS (see REFERENCED STANDARDS)

STANDBY POWER 2702.1, 2702.3
Aircraft traffic control towers . . . 412.3.4, 2702.2.18
Atriums 404.7, 2702.2.2
Covered and open mall building 402.7.3, 2702.2.14
Elevators 1007.4, 2702.2.5, 2702.2.19, 3003.1, 3007.9, 3008.9
Hazardous occupancy 414.5.4, 421.8, 2702.2.10, 2702.2.12
High-rise buildings 403.4.7, 2702.2.15
Horizontal sliding doors. . . . 1008.1.4.3, 2702.2.7
Membrane structures 2702.2.9, 3102.8.2
Platform lifts. 1007.5, 2702.2.6
Smoke control. 909.11, 2702.2.2
Smokeproof enclosure 909.20.6.2, 2702.2.20
Special inspection. 1705.11.6
Underground buildings 405.8, 2702.2.16

**STANDPIPE AND HOSE SYSTEMS
(see STANDPIPES, REQUIRED)** . . . 905, 3106.4,
3308.1.1, 3311
    Dry 905.8
    Hose connection location 905.1,
905.4 through 905.6, 912
STANDPIPES, REQUIRED
    Assembly 905.3.2, 905.5.1
    Covered and open mall buildings 402.7.1,
905.3.3
    During construction 905.10
    Elevators, fire service access 3007.10
    Parking garages. 406.5.8
    Roof gardens and landscaped roofs . . . 905.3.8
    Stages. 410.8, 905.3.4
    Underground buildings 405.10, 905.3.5
STEEL . . . . Chapter 22
    Bolting. .2204.2
    Cable structures. 2208
    Calculated fire resistance 722.5
    Cold-formed. . . . . 202 , 2210, 2211
    Composite structural steel
       and concrete 2206
    Conditions of restraint. 703.2.3
    Decks 2210.1.1
    Identification and protection. 2203
    Joists 202, 2207
    Open web joist 2207
    Parapet walls 1503.3, 1503.6
    Seismic provisions . . . 2205.2, 2206.2, 2210.2
    Special inspections 1705.3
    Storage racks 2209
    Structural steel 2205
    Welding 2204.1
STONE VENEER 1405.7
    Slab-type .1405.8
STOP WORK ORDERS 115
STORAGE OCCUPANCY (GROUP S) 311
    Accessibility. 1108.3
    Area 406.3.5, 406.3.6, 406.4.1, 503, 505,
506, 507, 508
    Alarm and detection
    Equipment platforms 505.2
    Group provisions
       Hazard storage, low, Group S-2 . . . . 311.3
       Hazard storage, moderate,
          Group S-1 . . . . 311.2
    Hazardous material display
       and storage 414.2.5
    Height 406.3.5, 406.4.1, 503, 504,
505, 506, 508, 510
    Incidental uses 509
    Interior finishes . . . . .Table 803.9, 804
    Live loads .1607
    Means of egress
       Aisles. 1017.5
       Travel distance. . . . 1014.3, 1016.2, 1021.2

Mixed occupancies 508.3, 508.4
    Accessory 508.2
    Parking above/below. . . 510.3, 510.4, 510.7,
510.8, 510.9
    Special mixed 510.2
Plumbing fixtures 2902
Special provisions
    Aircraft related occupancies 412
    High-piled combustible. 413
    Parking garages 406
Sprinkler protection 903.2.10
Unlimited area. 507.2, 507.3, 507.4
STORM SHELTER. 423
    Refuge area (see REFUGE AREA)
STRENGTH
    Design requirements 1604.2
    Masonry. 2102.1
STRENGTH DESIGN 1602.1, 1604.1
    Masonry. 2101.2.2, 2108
STRUCTURAL DESIGN. . . . . Chapter 16
    Aluminum . . . . Chapter 20
    Concrete . . . . Chapter 19
    Foundations . . . . Chapter 18
    Masonry . . . . Chapter 21
    Steel. . . . . Chapter 22
    Wood . . . . Chapter 23
STRUCTURAL OBSERVATION 1702.1,
1704.5
STUCCO. .2512
SUSCEPTIBLE BAY
    Definition 202
    Ponding instability. 161
SWIMMING POOL. 3109
    Accessibility. 1109.15
    Flood provisions. G801.5
    Gates, access 3109.4.1.7
    Glass 2406.4
    Indoor .3109.4.2
    Public 3109.3
    Residential 3109.4

T

TECHNICAL PRODUCTION AREAS 410.3.2,
410.6.3
**TELESCOPIC SEATING (see FOLDING
AND TELESCOPIC SEATING)**
TEMPORARY STRUCTURES. 3103
    Certificate of occupancy. 108.3
    Conformance 108.2
    Construction documents 3103.2
    Encroachment, public
       rights-of-way 3202.3
    Means of egress 3103.4
    Permit 108.1, 3103.1.1
    Power, temporary 108.3
    Termination of approval 108.4

TENANT SEPARATION
Covered and open mall building 402.4.2.1, 708.1
TENTS (see TEMPORARY STRUCTURES)
Standby and emergency power. 2702.2.9
TERMITES, PROTECTION FROM 2304.11
TERRA COTTA 1405.9
TESTING
Automatic fire-extinguishing systems 904.4
Building official required. 104.11.1
Fire-resistant materials 703.2
Glazing 2406, 2408.2.1
Roof tile 1711.2
Seismic 1705.12
Smoke control. 909.3, 909.18, 1705.17
Soils. 1803
Sprinkler protection 903.5
Structural (see SPECIAL INSPECTIONS AND TESTS)
THEATERS [see ASSEMBLY OCCUPANCY (GROUP A), PROJECTION ROOMS and STAGES AND PLATFORMS] 303.2.4.10
THERMAL BARRIER, FOAM PLASTIC INSULATION. 2603.4, 2603.5.2
THERMAL-INSULATING MATERIALS (see INSULATION) 719
TILE 2102.1
Ceramic (see CERAMIC TILE)
Fire resistance, clay or shale 721.1
TOILETS and TOILET ROOMS. . . . Chapter 29, 3305
Accessible. 1109.2, 1607.7.2
Construction/finish materials 1210
Door locking. 1008.1.9.5.1, 1109.2.1.7, 2902.3.5
Family or assisted-use 1109.2.1, 2902.1.2, 2902.2.1
Fixture count . . . . .Table 2902.1
Grab bar live loads 1607.7.2
Location . . . 1210.4, 2902.3.1, 2902.3.2, 2902.3.3
Partitions 1210.3
Privacy 1210.3
Public facilities 2902.3
Signs 1110.1, 1110.2, 2902.4, 2902.4.1
TORNADO SHELTER (see STORM SHELTER)
TOWERS
Airport traffic control. 412.3
TOXIC MATERIALS [see HIGH-HAZARD OCCUPANCY (GROUP H)]
Classification 307.6, 414, 415
Gas detection system. . . . 415.10.7, 421.6, 908.3
TRAVEL DISTANCE
Area of refuge. 1007.6
Assembly seating 1028.7
Atrium 404.9
Balcony, exterior 1016.2.1
Care suites (Group I-2) 407.4.2, 407.4.3
Common path of travel 1014.3
Exit access 1016.2
Mall 402.8.5, 402.8.6
Measurement 1016.3
Smoke compartments (Group I-2 and I-3) 407.5, 408.6, 408.9
Special amusement building 411.4
Stories with one exit. 1021.2
Toilet facilities 2902.3.2, 2902.3.3
TREADS, STAIR (see STAIRWAY CONSTRUCTION)
Concentrated live load 1607
TREATED WOOD 2302
Fire-retardant-treated wood. 2303.2
Pressure-treated wood 2303.1.8
Stress adjustments 2306
TRUSSES
Cold-formed steel 2210.3
Fire resistance 704.5
Materials . . . . Chapter 6
Metal-plate-connected wood 2303.4.6
Wood 2303.4
TSUNAMI-GENERATED FLOOD HAZARD . . . . Appendix M
TUNNELED WALKWAY 3104, 3202.1
TURNSTILES 1008.3

U
UNDERGROUND BUILDINGS 405
Alarms and detection 405.6
Compartmentation 405.4
Construction type 405.2
Elevators 405.4.3
Emergency power loads 405.9, 2702.2.16
Means of egress 405.7
Smoke barrier. 405.4.2, 405.4.3
Smoke exhaust/control 405.5
Smokeproof enclosure 405.7.2, 1022.10
Sprinkler protection 405.3
Standby power 405.8, 2702.2.16
Standpipe system 40510.1, 905.3.5
UNLIMITED AREA BUILDINGS. 507
UNSAFE STRUCTURES AND EQUIPMENT (see STRUCTURES, UNSAFE) 115
Appeals . . . . 113, Appendix B
Restoration 115.5
Revocation of permit 105.6
Stop work orders 115
Utilities disconnection 112.3
UNUSABLE SPACE. 712.3.3
USE AND OCCUPANCY. . . . . Chapter 3
Accessory 508.2
Incidental uses . . . . .509, Table 509
Mixed 508.3, 508.4
Special . . . . Chapter 4

Index

UTILITIES 112
    Service connection 112.1
    Service disconnection. 112.3
    Temporary connection. 112.2
UTILITY AND MISCELLANEOUS
    OCCUPANCY (GROUP U) 312
    Accessibility 1103.2.5, 1104.3.1
    Agricultural buildings . . . . Appendix C
    Area 503, 505, 506, 507, 508
    Flood provisions. G1001
    Height 503, 504, 508
    Incidental uses 509
    Live loads . . . . Table 1607.1
    Means of egress
        Exit signs 1011.1
    Mixed occupancies 508.3, 508.4
    Special provisions
        Private garages and carports 406.1
        Residential aircraft hangers 412.5
    Sprinkler protection 903.2.11
    Travel distance 1014.3, 1016.1, 1021.2

V

VALUATION OR VALUE (see FEES, PERMIT) . . 109.3
VEHICLE BARRIER SYSTEMS . . . 406.4.3, 1607.8.3
VEHICLE SHOW ROOMS 304
VEHICULAR FUELING 406.7
VEHICULAR GATES 3110
VEHICULAR REPAIR 406.8
VENEER
    Cement plaster 1405.15
    Fastening 1405.17
    Fiber-cement siding 1405.16
    Glazing 1405.12
    Masonry, adhered 1405.10
    Masonry, anchored 1405.6
    Metal 1405.11
    Plastic 2605
    Slab-type 1405.8
    Stone 1405.7
    Terra cotta 1405.9
    Vinyl 1405.14
    Wood 1405.5
VENTILATION (see MECHANICAL) 101.4.2
    Attic 1203.2
    Aircraft hangars, residential 412.5.4
    Aircraft paint hangars 412.6.6
    Bathrooms 1203.4.2.1
    Crawl space 1203.3
    Elevator hoistways 3004
    Exit enclosure 1022.6
    Fabrication areas, HPM 415.10.2.7
    Hazardous 414.3, 414.5.3, 415.8.1.4,
            415.8.2.7, 415.10.2.8.1, 415.10.5.8,
            415.10.6.4, 415.10.7, 415.10.9.3
    High-rise stairways 1022.10

    Live/work unit 419.8
    Mechanical 1203.1
    Natural 1203.4
    Parking 406.5.2, 406.5.5, 406.5.10, 406.6.2
    Projection rooms 409.3
    Repair garages 406.8.2
    Roof 1203.2, 1503.5
    Smoke exhaust 910.4
    Smoke removal, high-rise buildings 403.4.7
    Smokeproof enclosures 909.20.3, 909.20.4,
                       909.20.6, 1022.10
    Spray rooms and spaces 416.2.2, 416.3
    Stages 410.3.5, 410.3.7
    Under-floor ventilation 1203.3
VENTS, PENETRATION PROTECTION 714
VERMICULITE, FIRE RESISTANT 721
VERTICAL OPENING PROTECTION
    Atriums 404.6
    Duct penetrations 717.1
    Elevators 713.14, 3007.6.1, 3008.6.1
    Exceptions 1022.1
    Group I-3 . . . . 408.5
    High-rise buildings 403.2.1.2, 403.2.3, 403.5.1
    Live/work units 419.4
    Open parking garages 406.5.9
    Permitted vertical openings 712
    Shaft enclosure . . . 713, 1009.2, 1009.3, 1022.2
VESTIBULES, EXIT DISCHARGE 1027.1
VINYL
    Rigid 1405.14
VIOLATIONS 114
VOICE ALARM (see ALARMS, VOICE)

W

WALKWAY 3104
    During construction 3306
    Encroachment, public right-of-way 3202.3.4
    Fire resistance . . . . Table 601
    Materials per construction type . . . . . Chapter 6
    Opening protection 716, 717
WALL, EXTERIOR 705
    Bearing . . . . Chapter 6
    Coverings 1405
    Exterior Insulation and Finish
        Systems (EIFS) 1408
    Exterior structural members 704.10
    Fire district D102.1, D102.2.6
    Fire-resistance ratings . . . . . Table 602, 703,
                705.5, 706.5.1, 707.4, 1403.4
    Flashing, veneered walls 1405.4
    Foam plastic insulation 2603.4.1.4, 2603.5
    Glazing, rated 715.5
    Joints 705.9, 714
    Light-transmitting plastic panels 2607
    Materials 705.4, 1406
    Nonbearing . . . . Chapter 6

Opening protection 705.8, 705.10, 716.5.6
Parapets. 705.11
Projections 705.2
Structural stability 705.6
Veneer (see VENEER)
Weather resistance 1403.2, 1405.2,
1407.6, 1408.4
Weather-resistant barriers 1405.2
WALL, FIRE (see FIRE WALLS)
WALL, FOUNDATION (see FOUNDATION)
WALL, INTERIOR
Finishes 803, 1210.2
Opening protection 716, 717
WALL, INTERIOR NONBEARING
(see PARTITIONS) WALL, MASONRY 202
Wood contact 2304.11.2.3, 2304.11.2.5
WALL, PARAPET 705.11, 1503, 2109.3.4.3
WALL, PARTY (see FIRE WALLS)
WALL, PENETRATIONS 714.3
WALL, RETAINING (see RETAINING WALL)
WALL, VENEERED (see VENEER) . . . . Chapter 14
WALL, WOOD CONSTRUCTION
Bracing 2308.9.3
Cutting, notching, boring 2308.9.10
Exterior framing 2308.9
Fastening schedule 2304.9.1
Framing 2304.3, 2308.9
Interior bearing partition 2308.9.1
Interior nonbearing partition 2308.9.2.3
Openings 2308.9.5, 2308.9.6, 2308.9.7
Shear walls 2305.1, 2306.3
Sheathing (see SHEATHING)
Studs 2308.9.1
Top plates 2308.9.2.1
WATER-REACTIVE MATERIALS. . . . Table 307.1(1),
415.7
WEATHER, COLD
Masonry construction 2104.3
WEATHER, HOT
Masonry construction 2104.4
WEATHER PROTECTION
Exterior walls 1405.2
Roofs 1503
WELDING 2204.1
Materials, verification of steel
reinforcement 1705.3.1
Special inspections 1705.2.2.1, 1705.11.3
Splices of reinforcement
in masonry 2107.4
Structural testing 1705.12.1
WIND LOAD 1609
Alternate all-heights method 1609.6
Basic wind speed 1609.3
Construction documents 107, 1603
Exposure category 1609.4
Glass block 2110

Glazing 1609, 2404
Hurricane-prone regions 1609.2
Roofs 1504.1, 1609.5, 2308.10.1
Seismic detailing required 1604.10
Special inspection 1705.10
Statement of special inspections 1704.3
Structural observation 1704.5.2
Wind-borne debris region 1609.2
Wind tunnel testing 1609.1.1, 1711.2.2
WINDERS, STAIR
(see STAIRWAY CONSTRUCTION)
WINDOW
Accessibility 1109.13.1
Emergency egress 1029
Exterior, structural testing 1710.5
Fire (see OPENING
PROTECTIVES) 716.5.10, 716.5.11
Glass (see GLAZING) 1405.13
Guards 1013.8
Required light 1205.2
Wells 1029.5
WIRES, PENETRATION PROTECTION 714
WOOD . . . . Chapter 23
Allowable stress design 2306
Bracing, walls 2308.9.3
Calculated fire resistance 722.6
Ceiling framing 2308.10
Connectors and fasteners 2304.9
Contacting concrete, masonry
or earth 2304.11.4
Decay, protection against 2304.11
Diaphragms 2302, 2305.1, 2305.2, 2306.2
Draftstopping 718.3, 718.4
End-jointed lumber 2303.1.1.2
Fiberboard 2303.1.5 2306.3
Fire-retardant treated 2303.2
Fireblocking 718.2
Floor and roof framing (see FLOOR
CONSTRUCTION, WOOD) 2304.4
Floor sheathing 2304.7
Foundation 1807.1.4, 2308.3.3.1
Grade, lumber 2303.1.1
Hardboard 2303.1.6
Heavy timber construction 2304.10
Hurricane shutters 1609.1.2
I-joist 2303.1.2
Inspection, special 1705.5, 1705.10.1,
1705.11.2
Lateral force-resisting systems 2305
Light-frame construction, conventional . . . 2308
Load and resistance factor design 2307
Moisture content 2303.1.8.2, 2303.2.6
Nails and staples 2303.8
Plywood, hardwood 2303.3
Preservative treated . . . 1403.5, 1403.6, 2303.1.8
Roof framing (see ROOF
CONSTRUCTION, WOOD) 2304.4

Index

WOOD (*cont.*)
    Roof sheathing 2304.7
    Seismic provisions 2305, 2306,
                                    2308.11, 2308.12
    Shear walls 2305, 2306.3
    Standards and quality, minimum 2303
    Structural panels 2302, 2303.1.4
    Supporting concrete or masonry 2304.12
    Termite, protection against 2304.11
    Trusses 2303.4
    Veneer. . . . . Chapter 14
    Wall framing (see WALL,
        WOOD CONSTRUCTION) 2304.3
    Wall sheathing 2304.6
WOOD SHINGLES AND SHAKES 1507.8, 1507.9
WOOD STRUCTURAL PANELS
    (see WOOD) 202, 2303.1.4
    Bracing 2308.9.3
    Decorative. 2303.3
    Design requirements 2301
    Diaphragms 2305.2, 2306.2
    Fastening 2304.9
    Fire-retardant-treated 2303.2
    Performance category. 202
    Quality. 2303.1.4
    Roof sheathing 2304.7, 2308.10.8
    Seismic shear panels 2305.1, 2308.12.4
    Shear walls 2306.3
    Sheathing 2304.6.1
    Standards 2306.1
    Subfloors 804.4
    Veneer. 1405.5

Y

YARDS OR COURTS 1206
    Exit discharge 1027.4
    Group I-2 . . . . 407.8
    Group I-3 . . . . 408.3.6, 408.6
    Light, natural 1205
    Motor fuel-dispensing facilities 406.7.2
    Occupant load 1004.5
    Parking garage, open 406.5.5
    Unlimited area building 507.1

 **ICC EVALUATION SERVICE**

HOW DO YOU KNOW WHEN A PRODUCT IS COMPLIANT

with codes, standards or rating systems

Tap into free resources from ICC Evaluation Service® (ICC-ES®)

- Online directory of code-compliant building products and systems

- CEU webinars, presentations and videos

- Unmatched expertise and experience in technical evaluations

ICC-ES is the most widely accepted and trusted brand and industry leader in performing technical and environmental evaluations of building products and systems.

Learn more:

www.icc-es.org
1.800.423.6587 (x42237)
es@icc-es.org

Look for the marks of conformity code officials trust:

ICC INTERNATIONAL CODE COUNCIL

GET AHEAD WITH NEW CODE TOOLS FOR YOUR IBC

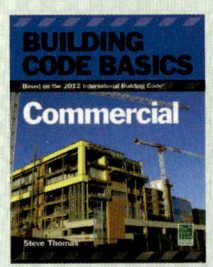

COLOR ILLUSTRATIONS!
BUILDING CODE BASICS: COMMERCIAL, BASED ON THE 2012 IBC

Provides an up-to-date, step-by-step guide to understanding and applying the base requirements of the 2012 IBC from technical jargon to more complex regulations. It simplifies critical concepts so that users can begin to build a foundation for learning and applying the code. Topics are organized to follow the structure of the 2012 IBC, and detailed illustrations and tables enhance comprehension and help students to visualize complex codes. Practical, on-the-job scenarios provide a real-world context for the information. (228 pages)

SOFT COVER #4081S12
PDF DOWNLOAD #8950P255

FREE PREVIEW VIDEO ONLINE!
SIGNIFICANT CHANGES TO THE INTERNATIONAL BUILDING CODE®, 2012 EDITION

This must-have guide provides comprehensive, yet practical, analysis of the critical changes made between the 2009 and 2012 editions of the IBC. Key changes are identified then followed by in-depth discussion of how the change affects real-world application. Coverage reflects provisions with special significance, including new and innovative design ideas and technologies, modern materials and methods of construction, and current approaches to safety and stability. Authored by ICC code experts Doug Thornburg, John Henry and Jay Woodward. (390 pages)

SOFT COVER #7024S12

NEW CODE PLUS COMMENTARY IN ONE!
2012 IBC® CODE AND COMMENTARY

Contains the complete text of the IBC plus expert commentary printed after each code section. The Code and Commentary is an ideal go-to reference for effective design, construction and inspection. Includes:

- ✓ All text, tables and figures from the code.
- ✓ Expert technical commentary printed after each code section.
- ✓ Suggestions for effective application.
- ✓ Historical and technical background.

2012 IBC CODE AND COMMENTARY (2-VOLUME SET)
SOFT COVER #3010S12
PDF CD-ROM #3010CD12
PDF DOWNLOAD #870P12

VOLUMES ALSO SOLD SEPARATELY
2012 IBC CODE AND COMMENTARY, VOLUME 1 (CHAPTERS 1–15)
SOFT COVER #3010S121
PDF DOWNLOAD #870P121

2012 IBC CODE AND COMMENTARY, VOLUME 2 (CHAPTERS 16–35)
SOFT COVER #3010S122
PDF DOWNLOAD #870P122

NEW DOWNLOAD!
INTERACTIVE GUIDE TO THE 2012 IBC: AN ILLUSTRATED CHECKLIST

A new on-screen guide that illustrates key code concepts and organizes code requirements in a way that mirrors the design process. This companion document will help jurisdictions, architects, contractors, and other members of the design team successfully apply the 2012 IBC. Unlike a printed document, this interactive guide takes full advantage of speedy website style interface and is fully searchable to allow a specific concept to be found quickly. Includes:

- 256 explanatory illustrated pages
- 7 checklists
- Accessibility and Building Code requirements
- Index of key terms
- Links to ICC-ES Evaluation Reports

PDF DOWNLOAD #8950P292

Order Your Code Tools Today! 1-800-786-4452 | www.iccsafe.org/books

People Helping People Build a Safer World®

HANDY POCKET GUIDE!
ACCESSIBILITY POCKETBOOK: 2012 IBC AND ICC A117.1-2009

Compiles vital accessibility information you need into one handy source. *Accessibility PocketBook* contains selected provisions from the 2012 IBC and the entire text of *ICC A117.1-2009: Accessible and Usable Buildings and Facilities*. It assists in the design, plan review, construction, and inspection of accessible facilities and features by combining the accessibility provisions of the IBC with the technical requirements of the A117.1 standard into a single easy-to-reference resource. It is an ideal tool for the job site or the office. (396 pages)

SOFT COVER #4028S12
PDF DOWNLOAD #8950P076

100+ WORKED-OUT EXAMPLES!
SEISMIC AND WIND FORCES: STRUCTURAL DESIGN EXAMPLES, 4TH EDITION

This new edition by Alan Williams, Ph.D., S.E., F.I.C.E., C. ENG., has been updated to the 2012 IBC, ASCE/SEI 7-10, ACI 318-11, NDS-2012, AISC 341-10, AISC 358-10, AISC 360-10, and the 2011 MSJC Code. In each chapter, sections of the code are presented and explained in a logical and simple manner and followed by illustrative examples. More than 100 completely worked-out design examples clearly illustrate proper application of the code requirements. Problems are solved in a straight forward step-by-step fashion with extensive use of illustrations and load diagrams. (580 pages)

SOFT COVER #9185S4
PDF DOWNLOAD #8804P

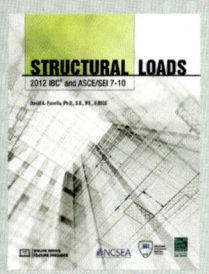

INCLUDES BONUS SOLUTIONS MANUAL!
STRUCTURAL LOADS: 2012 IBC AND ASCE/SEI 7-10

Presents requirements in a straightforward manner with emphasis placed on proper application in everyday practice. The structural load provisions are organized in comprehensive flowcharts that provide a road map to guide the reader through the requirements. A valuable training resource for those who need to understand how to determine structural loads including dead load, occupancy live load, roof live load and environmental loads such as rain, snow, ice, flood, wind and seismic loads. This edition includes a new chapter on load path, a new section on atmospheric ice loads, completely updated discussion of the new wind design procedures in ASCE/SEI 7-10, and problems at the end of each chapter. A complete Solutions Manual is included as an online bonus. (550 pages)

SOFT COVER #4034S12
PDF DOWNLOAD #8950P261

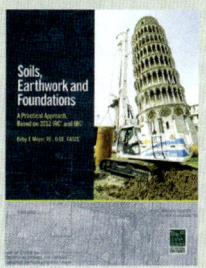

REVIEW QUESTIONS IN EACH CHAPTER!
SOILS, EARTHWORK, AND FOUNDATIONS: A PRACTICAL APPROACH, BASED ON 2012 IRC® AND IBC

Author Kirby T. Meyer, P.E., D.GE, F.ASCE bridges the worlds of geotechnical engineering and foundation design, and construction activities and inspections dealing with soil conditions and building foundations. The book provides illustrations to assist in the understanding of the principles involved in foundations, guidelines for when to call on a geotechnical or foundation design engineer, and tips for effectively communicating with geotechnical and structural professionals. Building department personnel, inspectors, laboratory personnel, and foundation and earthwork contractors will find the book helpful. Architects and engineers will benefit from the information on design and field applications for foundations. Review questions are included at the end of each chapter that will assist those studying for the Soils Special Inspection Certification Exam. References to the 2012 IRC and IBC are provided throughout. (300 pages)

SOFT COVER #4036S12
PDF DOWNLOAD #8950P275

Order Your Code Tools Today! 1-800-786-4452 | www.iccsafe.org/books

People Helping People Build a Safer Wo[rld]

Dedicated to the Support of Building Safety and Sustainability Professionals

An Overview of the International Code Council
The International Code Council (ICC) is a membership association dedicated to developing model codes and standards used in the design, build and compliance process to construct safe, sustainable, affordable and resilient structures. Most U.S. cities, counties, states and U.S. territories, and a growing list of international bodies that adopt building safety codes use ones developed by the International Code Council.

Services of the ICC
The organizations that comprise the International Code Council offer unmatched technical, educational and informational products and services in support of the International Codes, with more than 250 highly qualified staff members at 16 offices throughout the United States, Latin America and the Middle East. Some of the products and services readily available to code users include:

- **CODE APPLICATION ASSISTANCE**
- **EDUCATIONAL PROGRAMS**
- **CERTIFICATION PROGRAMS**
- **TECHNICAL HANDBOOKS AND WORKBOOKS**
- **PLAN REVIEW SERVICES**
- **ELECTRONIC PRODUCTS**
- **MONTHLY ONLINE MAGAZINES AND NEWSLETTERS**
- **PUBLICATION OF PROPOSED CODE CHANGES**
- **TRAINING AND INFORMATIONAL VIDEOS**
- **BUILDING DEPARTMENT ACCREDITATION PROGRAMS**
- **EVALUATION SERVICE FOR CODE COMPLIANCE**
- **EVALUATIONS UNDER GREEN CODES, STANDARDS AND RATING SYSTEMS**

The ICC family of non-profit organizations includes:

ICC EVALUATION SERVICE (ICC-ES)
ICC-ES is the leader in performing technical evaluations of building products and materials for compliance with building and green codes, standards and rating systems.

ICC FOUNDATION (ICCF)
ICCF is dedicated to consumer education initiatives, professional development programs to support code officials and community service projects that result in safer, more sustainable buildings and homes.

INTERNATIONAL ACCREDITATION SERVICE (IAS)
IAS accredits testing and calibration laboratories, inspection agencies, building departments, fabricator inspection programs and IBC special inspection agencies.

NEED MORE INFORMATION? CONTACT ICC TODAY!
1-888-ICC-SAFE (422-7233)
www.iccsafe.org